Vertical and Horizontal Deformations of Foundations and Embankments

Volume 2

Proceedings of Settlement '94

Sponsored by the
Geotechnical Engineering Division of the
American Society of Civil Engineers

in cooperation with the
Federal Highway Administration
National Science Foundation

Hosted by the
Department of Civil Engineering
Texas A&M University

College Station, Texas
June 16-18, 1994

Geotechnical Special Publication No. 40

Edited by Albert T. Yeung and Guy Y. Félio

Published by the
American Society of Civil Engineers
345 East 47th Street
New York, New York 10017-2398

ABSTRACT

This proceedings, *Vertical and Horizontal Deformations of Foundations and Embankments* (Geotechnical Special Publication No. 40) consists of papers presented at the Specialty Conference sponsored by the Geotechnical Engineering Division of the American Society of Civil Engineers held in College Station, Texas, June 1618, 1994. The prime goals of the conference and the resulting proceedings were: 1) to update the progress made since the 1964 Specialty Conference on Design of Foundations for Control of Settlement on prediction methods and mitigation technologies for both vertical and horizontal deformations of different types of foundations and embankments built on different soils; 2) to provide an effective forum for researchers and practitioners to share research results, current practice and recent technological advances; 3) to recognize the limitations in current practice, identify current and future needs of the profession, and assess the potential and feasibility of innovative technologies; and 4) to provide an effective vehicle for technology transfer. The 137 papers included in this publication cover such topics as: 1) embankments, deep fills, and landfills; 2) shallow foundations; 3) deep foundations; 4) soil improvement; 5) numerical and physical modeling; 6) soil dynamics; and 7) expansive soils, residual soils, and rock. This proceedings provides students, educators, researchers, and practitioners of geotechnical engineering an opportunity to assimilate new information and to meet new challenges.

Library of Congress Cataloging-in-Publication Data

Settlement '94 (1994 : College Station, Tex.)
Vertical and horizontal deformations of foundations and embankments : Proceedings of Settlement '94 / sponsored by the Geotechnical Engineering Division of the American Society of Civil Engineers ; in cooperation with the Federal Highway Administration, National Science Foundation ; hosted by the Department of Civil Engineering, Texas A & M University, College Station, Texas, June 16-18, 1994 ; edited by Albert T. Yeung and Guy Y. Felio.
p. cm.
Includes index.
ISBN 0-7844-0027-X
1. Foundation—Congresses. 2. Embankment—Congresses. 3. Settlement of structures—Congresses. 4. Soil mechanics-Congresses I. Yeung, Albert T. II. Felio, Guy Y. III. American Society of Civil Engineers. Geotechnical Engineering Division. IV. United States. Federal Highway Administration. V. National Science Foundation (U.S.) VI. Title.
TA775.S425 1994 94-20011
624.1'5—dc20 CIP

Library of Congress Catalog Card No: 94-20011
ISBN 0-7844-0027-X
Manufactured in the United States of America.

Photo courtesy of Albert T. Yeung

CONTENTS

Opening Lecture

Session 1 Embankments on Clay I

Session 2 Embankments on Clay II

Session 3 Shallow Foundations on Clay

Session 4.1 Shallow Foundations I

Session 4.2 Deep Foundations I

Session 4.3 Embankments, Deep Fills, and Landfills I

Session 4.4 Soil Improvement I

Session 4.5 Numerical and Physical Modeling I

Session 4.6 Soil Dynamics

Session 4.7 Shallow Foundations II

Session 4.8 Deep Foundations II

Session 4.9 Embankments, Deep Fills, and Landfills II

Session 4.10 Soil Improvement II

Session 4.11 Numerical and Physical Modeling II

Session 4.12 Expansive Soils, Residual Soils, and Rock

*Manuscript not available at time of printing

Session 10 Soil Dynamics

Session 11 Expansive Soils, Residual Soils, and Rock

Session 12

VERTICAL DISPLACEMENT INDUCED IN SOIL BY CONICAL SHELL FOUNDATIONS

Mohamed Abdel-Rahman[1] and Adel Hanna,[2] Member, ASCE

ABSTRACT: Shells can hold prospect for adoption in foundation engineering only if they can be projected as economical alternatives to conventional flat foundations. Therefore, the initial enthusiasm for shell foundations, generated by the saving in construction materials, their structural integrity, and the anticipated improvement in their geotechnical behavior, should be tempered by the fact that they need extra expenditure for construction due to the complexity of their geometry. However, with new innovations in construction industry such as robotics and the progress in fabrication process of precast concrete units, the use of shell foundations becomes more competitive and feasible. This paper is part of an extensive experimental and numerical investigations conducted on different types of shell foundations to examine their performance with respect to: settlement, ultimate bearing capacity, contact pressure, stress distribution, and soil displacement. The work presented in this paper was limited to study vertical displacement induced in the soil by conical shell foundations under axial loading condition at working loads level. Parametric study using finite elements technique was performed on four foundation models (one circular flat and three different conical models). The soil behavior was modelled by a perfectly elastic model and soil parameters were defined by modulus of elasticity (E) and Poisson's ratio (ν). The effect of shell configuration, depth of embedment, soil parameters (ν and E) on the distribution of vertical

[1] Research Associate, Dept. of Civil Engrg., Concordia University, 1455 De-Maisonneuve west, Montreal, Quebec, Canada H3G 1M8.

[2] Prof., Dept. of Civil Engrg., Concordia University, 1455 De-Maisonneuve west, Montreal, Quebec, Canada H3G 1M8.

displacement under the models were investigated and the results were presented in graphical format.

Literature Background

It is less than five decades since the Mexican Architect Felix Candela designed and poured the first reinforced concrete shell footing in history on the Mexican soil (Candela 1955). Since then, the use of shells in foundation engineering has drawn considerable interest in Mexico and in different parts around the world. The hollow conical foundation designed by Prof. F. Leonhardt for the Stuttgart TV tower in Germany (Kurian 1982), the hyperbolic paraboloid shell foundations for Sumner High School Stadium in Washington (Anderson 1960), the circular barrel shell foundation used in the Nonalco-Tlaltelolco project in Mexico City (Enriquez and Fierro 1963), and the conical shell substructure for Moscow tower (553.00 m) at Ostankino are few examples of the applications of shell foundations in the field.

Reports on the application of hollow conical shell foundations in engineering practice in China indicated that they have resulted in saving over 50% in construction materials (concrete and steel reinforcement), when compared to conventional flat foundations (He 1984, 1985; Huang 1984). This type of shell foundation, which is composed of a normal conical shell and an inverted one, is specially applicable for chimneys and telecommunication towers.

Design of shell foundations is currently based on the assumption that contact pressure distribution on soil-shell interface is uniform. Experimental investigations were performed to measure the contact pressure under different models of hyperbolic paraboloid footings. The results indicated substantial deviation from the uniform picture and that the actual contact pressure distribution was function of the soil-shell interaction. Contact pressure showed edge concentration in the elastic stage and exhibited a definite tendency for progressive shift of concentration towards central regions at the ultimate stage (Dierks and Kurian 1981, 1988; Kurian and Mohan 1981).

According to several experimental and theoretical investigations conducted on different types of shell models, ultimate bearing capacity as well as settlement were reported to be on the safe side when compared to flat foundations. (Nicholls and Izadi 1968; Kurian and Varghese 1969, 1971; Kurian and Jeyachandran 1972; Jain et al. 1977; Agarwal and Gupta 1983; Hanna and Hadid 1987; Abdel-Rahman 1987; Hanna and Abdel-Rahman 1991). This fact implies the possibility of using higher value for the ultimate bearing capacity in the design of shell foundations, which may lead to avoid the need of using deep foundations.

The results of experimental and theoretical studies conducted on different models of triangular shell strip footings, i.e. plane strain condition, rested on sand indicated that they provided higher bearing capacity and produce lesser settlement when compared with flat strip model of similar width, (Abdel-Rahman 1987; Hanna and Abdel-Rahman 1991). Design charts for modified bearing capacity coefficients were presented taking into consideration the effect of shell configuration on the ultimate bearing capacity.

NUMERICAL INVESTIGATION

General

In the study of determining vertical soil displacement induced by conical shell foundations, the following assumptions have been considered in the analyses:

1. Foundation model is perfectly rigid and loaded by a uniformly distributed load (p) on top of the column as shown in Fig. 1.
2. Soil is an elastic, homogeneous, isotropic, and compressible medium and soil properties are defined by modulus of elasticity (E) and Poisson's ratio (ν).
3. Thickness of the soil layer (H) below the foundation level is limited by a rigid base as shown in Fig. 1.
4. Contact surface between soil and foundation base is rough.

Numerical analyses were conducted on four different foundation models using the Finite Elements package, CRItical State Program "CRISP" developed by the geotechnical group at the department of Engineering, Cambridge University (Britto and Gunn 1987, 1992).

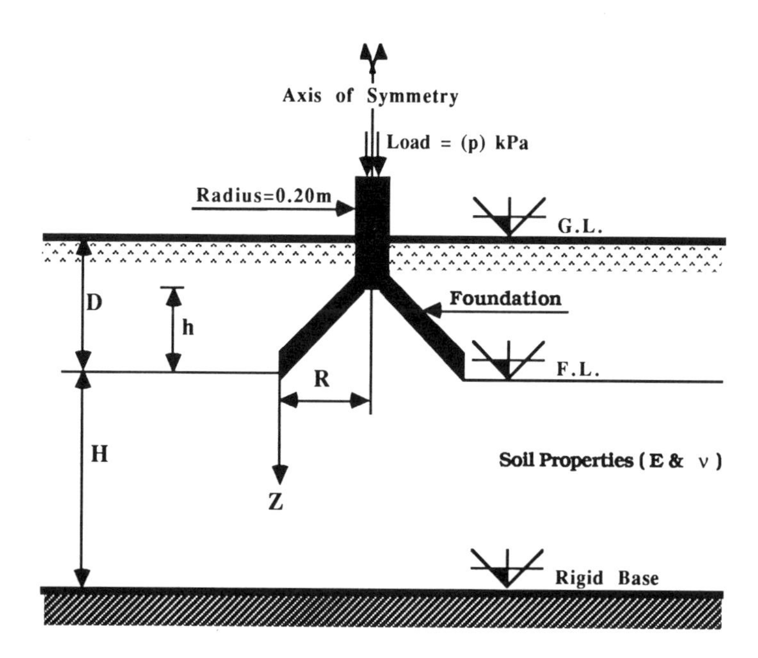

FIG. 1. Geometry of the Problem

Foundation Models

The foundation models used in the present numerical analyses are four models; one circular flat and three conical models. Different (h/R) ratios for the conical models were chosen in order to examine the effect of shell configuration on the vertical soil displacement. Table 1 summarizes the geometrical description of the four foundation models.

TABLE 1. Foundation Models

Model	R (m)	h (m)	h/R
Circular	0.80	0.00	0
Conical1	0.80	0.20	1/4
Conical2	0.80	0.40	1/2
Conical3	0.80	0.80	1

Notes: Refer to Fig. 1 for symbol definitions.

Mesh Design

The choice of number of elements and mesh design has to reflect a compromise between an acceptable degree of accuracy and computing cost. The idealized geometry of the mesh used in the present study is symmetrical about its centre line, therefore the mesh chosen represents only one half of the cross section until the axis of symmetry. The geometry of the problem is considered to be three dimensional case. However, since the foundation models involved in the analyses are circular and conical models loaded centrically with a uniformly distributed load, the analysis is considered to be axisymmetrical (i.e. reduced to two dimensions).

The nodes along the bottom boundary are considered as pinned supports (i.e. no movement is allowed in both vertical and horizontal directions to simulate rigid base behavior), whereas the nodes along the vertical boundary as well as the centre line are free to move in the vertical direction (i.e. rollers support).

The soil and the foundation models are modelled using eight nodded Linear Strain Quadrilateral elements "LSQ" with a quadratic variation for displacement. Smaller size elements for the soil elements located in the vicinity of foundation model were selected, where changes in stresses and strains are expected to be significant.

A depth of 5.00 m for the soil layer until the rigid base (H) is used in the present study, which gives a ratio of H/R = 6.25. Several trials have been performed to determine the mesh width, the optimum number of nodes and elements at which the solution converged and the effect of vertical boundaries on soil displacement becomes negligible. A mesh width of 5.00 m (from axis of symmetry to the vertical boundary) is chosen. The number of nodes and elements is adopted to be 357 nodes and 320 elements for circular model and 375 nodes and 335 elements for conical models.

Soil Behavior Model

The soil behavior in the present investigation is modelled using a perfectly elastic model since the analyses were performed at the working loads level. Table 2 summarizes the soil parameters used in the analyses.

TABLE 2. Soil Parameters for Parametric Study

Parametric Study	E (kPa)	ν
The effect of Poisson's ratio	10,000	0.20
		0.30
		0.40
The effect of modulus of elasticity	10,000	0.30
	30,000	

As part of the global research program conducted on different types of shell foundations, other non-linear soil models such as: elastic perfectly plastic model using Mohr-Coulomb failure criteria and modified Cam-Clay; were used in the analyses to determine the ultimate bearing load of the foundation at failure. It was noticed that the part of the load-settlement plot at the working loads level (at small strains) could be assumed to be the same for different soil models (linear or non-linear). However, non-linear soil models have to be utilized to represent soil behavior when larger deformations at ultimate stage are considered.

Parametric study was conducted to investigate the effect of the soil parameters (E and ν) on vertical soil displacement. To study the effect of depth of embedment on vertical soil displacement, two different depth to width ratios (D/2R) are used; surface foundations; i.e. D/2R = 0.0, and embedded foundations with D/2R = 1.0.

Foundation Behavior Model

The elastic model is used to simulate the behavior of the foundation models. The properties of the Alloy used in manufacturing the foundation models used in the experimental phase of the global research program were adopted in the analysis using modulus of elasticity of 68.95 MPa and Poisson's ratio of 0.33. The foundation is considered to be perfectly rigid with respect to the soil underneath. A uniformly distributed load of 5,000 kPa acting centrically on the foundation models was used.

Results of Parametric Study

Effect of Shell Configuration

For the case of surface foundations, i.e. at Z/R = 0.0, the vertical soil displacement is reduced by 9% from circular to conical2 model. The change from conical1 to conical2 model is about 2%, however, the vertical displacement for conical3 is increased again from conical2 by 3%. At lower soil depths, the effect of shell configuration is started to be eliminated, and the vertical soil displacements are almost the same for all models except for conical3, which has higher displacement value than the other models, see Fig. 2a.

For the case of embedded foundations with D/2R = 1.0, the behavior is similar to the case of surface foundations. However, conical3 model behaves similarly to models conical1, and conical2. The effect of shell configuration is eliminated at lower soil depths, for example at Z/R = 2.0, the soil vertical displacement is the same for all foundation models, see Fig. 2b.

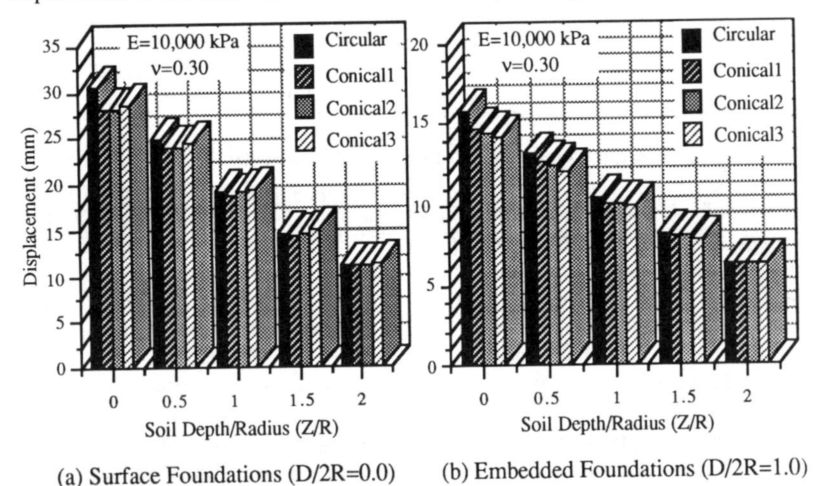

FIG. 2. Effect of Shell Configuration on Soil Displacement

Effect of Depth of Embedment

The depth of embedment has the same effect on all foundation models. At the soil surface, i.e. at Z/R = 0.0, a reduction in the vertical soil displacement of 47-50% when D/2R increased from 0 to 1 is recorded. At lower soil depths, for example at the ratio Z/R = 2.0, the effect of decreasing the vertical soil displacement continues with the same trend, and a reduction in the vertical soil displacement of 41-44% is occurred, see Fig. 3.

Effect of Poisson's Ratio

For the case of surface foundations, the vertical soil displacement decreased when Poisson's ratio increased. This behavior is consistent for all foundation models. The reduction in vertical soil displacement is about 10% when Poisson's ratio ν changed from 0.20 to 0.40, and is decreasing when moving towards lower soil depths. For example, at Z/R = 2.0, a reduction of 5% is recorded, see Fig. 4.

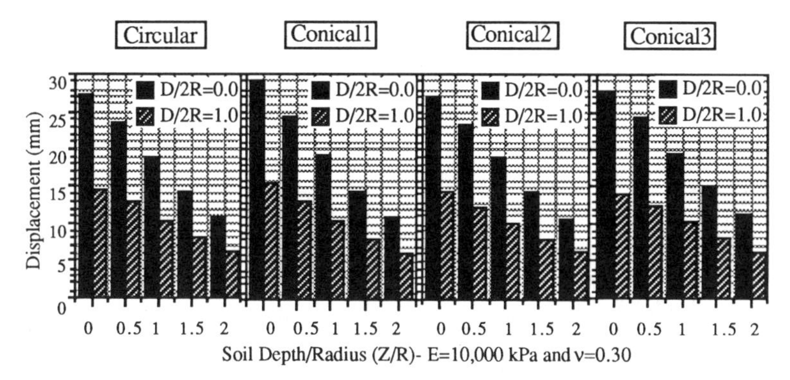

FIG. 3. Effect of Depth of Embedment on Soil Displacement

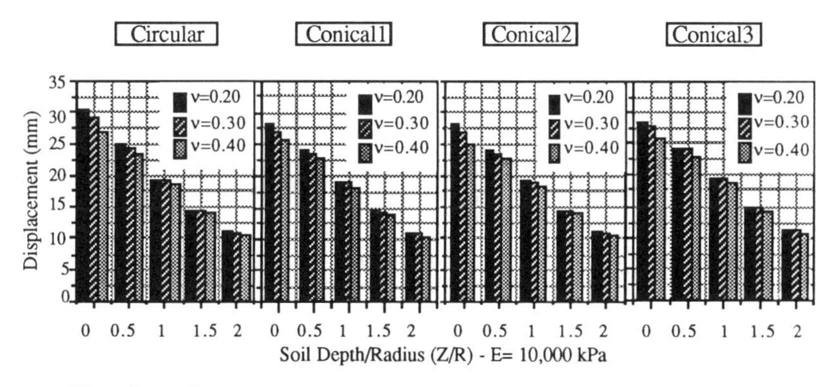

FIG. 4. Effect of Poisson's Ratio-Surface Foundation D/2R = 0.0

For the embedded foundations, the effect of Poisson's ratio is not significant, particularly at lower soil depths. The trend of decreasing vertical soil displacements with increasing Poisson's ratio is not consistent. For example, the vertical soil displacement is larger at Poisson's ratio of 0.30, than at Poisson's ratio of 0.20 for the ratio of Z/R = 0.5 for circular, conical1, and conical3 models, see Fig. 5.

Effect of Modulus of Elasticity

The modulus of elasticity has significant effect on vertical soil displacement. For the case of surface foundation, i.e. D/2R = 0.0, when the modulus of elasticity in-creased from 1,000 kPa to 3,000 kPa, the vertical soil displacement decreased by 67% at the surface, i.e. at Z/R = 0.0, see Fig. 6. The same reduction effect continues at lower soil depths. This effect is exactly the same for the case of embedded foundations. The vertical soil displacement for all models are affected with the same degree with a reduction value of {1 - (E1/E2)}, where E1 is less than E2, see Fig. 7.

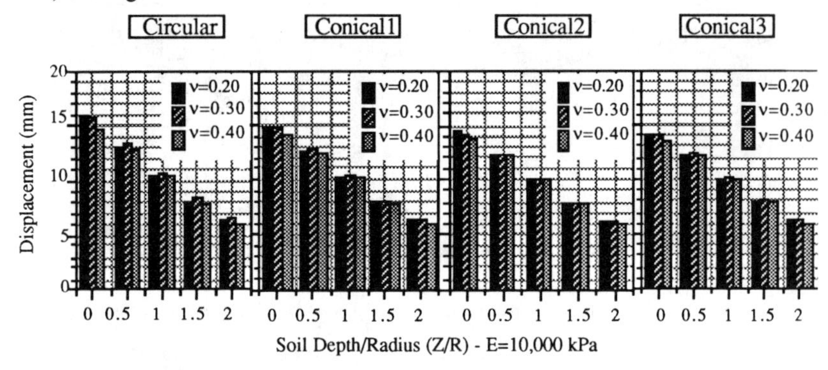

FIG. 5. Effect of Poisson Ratio-Embedded Foundation D/2R = 1.0

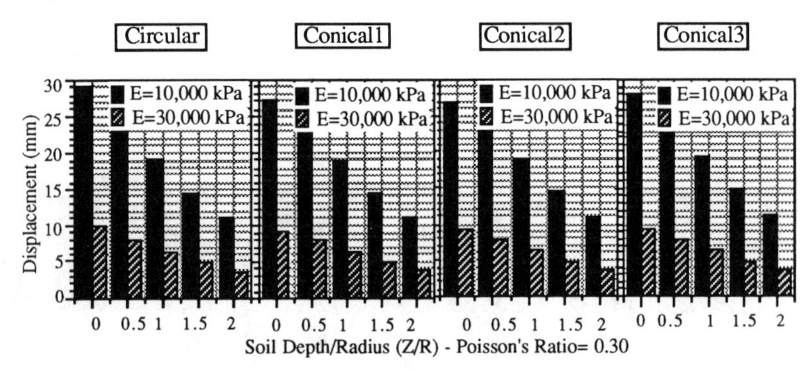

FIG. 6. Effect of Modulus of Elasticity-Surface Footings D/2R = 0.0

CONCLUSIONS

Vertical soil displacement induced in soil by conical shell foundations was investigated by conducting a parametric study using Finite Elements package, CRISP. Shell configuration caused a reduction of 10%. This reduction trend is similar for both surface and embedded foundations. However, there is a limit for the ratio h/R at which the vertical soil displacement increased again. Depth of embedment has a great influence on vertical displacement. A reduction of up to

50% was recorded at Z/R = 0.0. When Poisson's ratio increased, the vertical soil displacement decreased with about 10% for surface foundations. However, this effect is not significant for the case of embedded foundations. Modulus of elasticity has a significant effect on vertical soil displacement, which is reduced by a factor of {1/(E1/E2)}. This effect is the same for surface as well as embedded foundations.

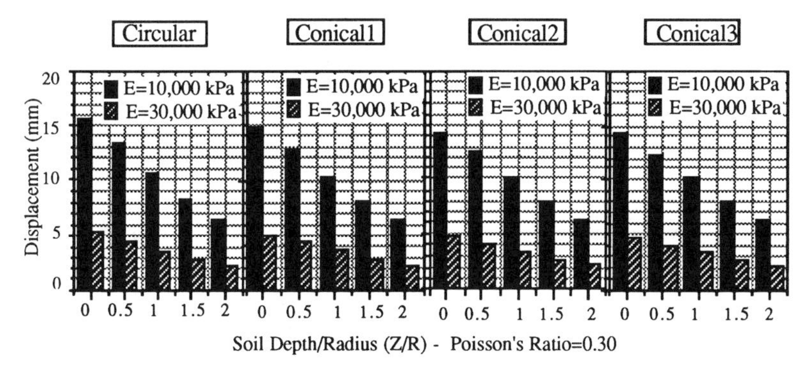

FIG. 7. Effect of Modulus of Elasticity-Embedded Footings D/2R = 1.0

Fig. 8 summarizes the results of the parametric study. The curves in Group (1) and in Group (2) for soil with modulus of elasticity of 10,000 kPa, and Poisson's ratios of 0.20, 0.30, and 0.40 for surface and embedded foundations with D/2R = 1.0, respectively. While the curves in Group (3) and in Group (4) for soil with modulus of elasticity of 30,000 kPa, and Poisson's ratios of 0.20, 0.30, and 0.40 for the surface and embedded foundations with D/2R = 1.0, respectively.

ACKNOWLEDGEMENTS

The financial support of the Natural Sciences and Engineering Research Council of Canada (NSERC) is gratefully acknowledged. The authors would like to thank Prof. Arul Britto, Department of Engineering, Cambridge University, for his cooperation in providing consultation, through electronic mail, on the use of F.E. package "CRISP".

APPENDIX I. REFERENCES

Abdel-Rahman, M.M. (1987). "Ultimate bearing capacity of triangular shell strip footings on sand," M. Eng. thesis, Dept. of Civil Engrg., Concordia University, Montreal, Quebec, Canada.

Agarwal, K. B., and Gupta, R. N. (1983). "Soil structure interaction in shell foundations." *Proc. Int. Workshop on Soil Structure Interaction*, University of Roorkee, Roorkee, India, 1, 110-112.

Anderson, A. R. (1960). "Precast prestressed stadium floats on hyperbolic paraboloids." *Engrg. News Record*, 164(7), 62-63.

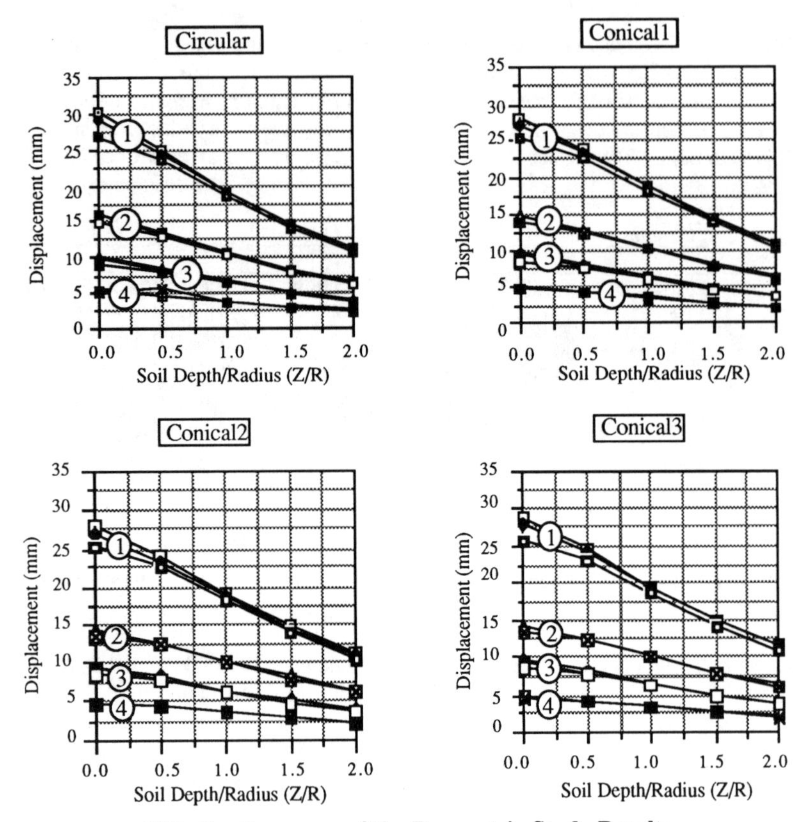

FIG. 8. Summary of the Parametric Study Results

Britto, A. M., and Gunn, M. J. (1987). *Critical State Soil Mechanics via Finite Elements*, Ellis Horwood, Ltd., Chichester, England.

Britto, A. M., and Gunn, M. J. (1992). *CRISP User's and Programmers's Guide*, Dept. of Engrg., Cambridge University, England.

Candela, F. (1955). "Structural applications of hyperbolic paraboloidal shells." *J. ACI*, 26(5), 397-415.

Dierks, K., and Kurian, N. P. (1981). "Zum Verhalten von kegelschalen fundamenten unter zentrischer und exzentrischer Belastung." *Bauingenieur*, 56(2), 61-65 (in German).

Dierks, K., and Kurian, N. P. (1988). "Zum Verhalten von Kugel und Doppelkeg-elschalen fundamenten unter zentrischer und exzentrischer Belastung." *Bauingeneur*, 63(7), 325-333 (in German).

Enriquez, R. R., and Fierro, A. (1963). "A new project for Mexico City." *Civil Engineering*, ASCE, 33(6), 36-38.

Hanna, A. M., and Hadid, W. (1987). "New models of shallow foundations." *Int. J. Math. Modeling*, 9(11), 799-811.

Hanna, A. M. (1988). "Shell foundations: the future alternative." *Int. J. for Housing Science*, 12(2), 289-295.

Hanna, A. M., and Abdel-Rahman, M. M. (1990). "Ultimate bearing capacity of triangular shell strip footings on sand." *J. Geotech. Eng.*, ASCE, 116(12), 1851-1863.

He, C. Z. (1984). "Hollow conic shell foundation and calculation." *Proc. 5th Engrg. Mech. Div.*, Spec. Conf. in Engrg. Mech. in Civil Engrg., Laramie, 1, 535-538.

He, C. Z. (1984). "Rational calculation of conic shells." *Proc. 5th Engrg. Mech. Div.*, Spec. Conf. in Engrg. Mech. in Civil Engrg., Laramie, 1, 547-550.

He, C. Z. (1985). *Hollow Conic Shell Foundation*, Scientific Publisher, Beijing, China (in Chinese).

Huang, Y. (1984). "The theory of conical shell and its applications." *Proc. 5th Engrg. Mech. Div.*, Spec. Conf. in Engrg. Mech. in Civil Engrg., Laramie, 1, 539-542.

Jain, V. K., Nayak, G. C., and Jain, O. P. (1977). "General behavior of conical shell foundation." *Proc. 3rd Int. Symp. on Soil Structure Interaction*, University of Roorkee, Roorkee, 2, 53-61.

Kurian, N. P., and Varghese, P. C. (1969). Discussion on "Design and testing of cone and hypar footings." *J. Soil Mech. Found. Div.*, Proc. ASCE, 95(SM1), 415-416.

Kurian, N. P., and Varghese, P.C. (1971). "Model studies on the elastic behavior of hypar footings in sand." *Bull. Int. Ass. Shell Spatial Structures*, (Madrid), No. 45, 35-40.

Kurian, N. P., and Jeyachandran, S.R. (1972). "Model studies on the behavior of sand under two and three dimensional shell foundations." *Indian Geotechnical J.*, 2(1), 79-90.

Kurian, N. P., and Mohan, C. S. (1981). "Contact pressures under shell foundations." *Proc. 10th ICSMFE*, Stockholm, 2, 165-168.

Kurian, N. P. (1982). *Modern Foundations: Introduction to Advanced Techniques*, Tata McGraw-Hill Co., New Delhi, India.

Nicholls, R. L., and Izadi, M. V. (1968). "Design and testing of cone and hypar footings." *J. Soil Mech. Found. Div.*, Proc. ASCE, 94 (SM1), 47-72.

APPENDIX II. NOTATION

The following symbols are used in this paper:

G.L. = ground level;
F.L. = foundation level;
E = soil modulus of elasticity (kPa);
ν = soil Poisson's ratio;
p = uniformly distributed load (kPa);
R = radius of foundation model (m);
Z = soil depth below foundation level (m);

D	=	depth of embedment (m);
H	=	thickness of soil layer until a rigid base (m);
h	=	height of conical shell model (m).

THE EFFECT OF LIVE LOAD ON DOWNDRAG FORCES

Samuel A. Leifer,[1] Member, ASCE

ABSTRACT: In deep-seated compressible deposits, large downdrag loads can develop. However, when additional load such as transient live load is placed on the pile, downdrag loads can be reduced, eliminated, or reversed as the pile moves down relative to the soil. The results of a parametric study are presented to evaluate the reduction in downdrag as a function of magnitude of the transient load, pile geometry, and pile flexibility. The analyses were conducted using the program APILE2 with special load transfer functions to represent the mobilization of downdrag load and the reduction of downdrag with increasing additional load. The reduction in downdrag is shown to be related to a pile/soil flexibility factor, f , which incorporates the pile geometry, pile elastic properties, and the relationship between mobilized side friction and relative pile/soil displacement. The analyses show that the additional load required to eliminate the downdrag can vary from one to six times the downdrag, depending on the pile flexibility. Design charts for estimation of the reduction in downdrag are presented. Comparisons with reported field measurements are also presented.

INTRODUCTION

When the soil around a pile moves down relative to the pile, negative skin friction or downdrag forces develop and add load to the pile. In deep-seated compressible deposits, large downdrag loads can develop and have occasionally resulted in pile failures. For such soil conditions, consideration of the downdrag load can increase the cost of piled foundations by a considerable amount.

---

[1] Staff Services Engineer, The Port Authority of New York and New Jersey, One World Trade Center, New York, NY 10048.

When additional load is placed on the pile, downdrag loads in the upper portion of the pile are reduced or eliminated. The additional load causes the pile to move down relative to the soil and the frictional forces are reduced or reversed. The results of analyses conducted to investigate the magnitude of reduction in downdrag under application of transient load as a function of pile length and diameter, relative stiffness of the pile-soil system, and the magnitude of the transient load are presented herein.

The reduction of downdrag loads due to the imposition of a temporary live load has been documented in the engineering literature. Bozozuk (1981) and Fellenius (1972) describe long term tests in which downdrag loads of 1.52 MN and 0.46 MN respectively, were measured. Upon imposition of temporary live loads equal in magnitude to the downdrag load, the downdrag load was almost completely released from the pile. York et al. (1974) describe a series of tests conducted at Newark International Airport in the early 1970's. In these tests, several foundation piles supporting elevated roadway structures were instrumented so that the dead load and buildup of downdrag load could be measured. Loaded trucks were then parked temporarily at predetermined locations on the span to add varying amounts of live load to the piles and the foundation response was measured. Although imposition of the live load did result in a reduction in the downdrag load, approximately 40 % of the downdrag load still remained in the pile even when the live load exceeded the downdrag load by 35 %.

The objectives of our analyses were to: (a) evaluate the effect and significance of variations in the parameters that affect the foundation response; (b) establish a relationship between the magnitude of the live load imposed and the percent decrease in the downdrag load for use in preliminary design of pile foundations; and (c) understand the differences between the results of the tests conducted by the Port Authority at Newark International Airport and those reported by Bozozuk (1981) and Fellenius (1972).

Analysis Methodology

The analyses were performed using the computer program APILE2 developed by Prof. L. C. Reese of the University of Texas at Austin. This program determines the load-deflection response and the distribution of load and deflection with depth for a single pile. In addition to the pile geometry and pile elastic properties, the user must input a load-deflection relationship for the soil/rock material supporting the tip of the pile and load transfer curves to describe the development of side friction as a function of relative pile-soil movement for the various strata through which the pile is driven. The deflection of the pile at any point (due to settlement of the tip plus elastic shortening of the shaft up to that point) is assumed to be the relative movement between the pile and the soil.

The various steps in development and release of load in the pile are shown schematically in Fig. 1. In the APILE2 analysis, these steps are modeled by modifying the load transfer curves. Fig. 2 shows three typical load transfer curves. Fig. 2a pertains to the standard case of pile loading in which the soil provides positive side friction to support the pile under an imposed static load. Fig. 2b shows the load transfer curve used for those portions of the pile that are subject to development of downdrag load due to settlement of the ground. The side friction is negative and is independent of the pile deflection since the load transfer curve is intended to model the condition when full downdrag has already been mobilized. Fig. 2c illustrates the load transfer curve used to model the reduction in downdrag due to live load. Initially, the side friction is negative. When the pile deflection exceeds the deflection that would occur under the dead load and downdrag load, the negative skin friction begins to reduce and ultimately becomes positive friction as the pile continues to deflect under larger and larger loads. Upon removal of the live load, the pile would rebound. The upper portion of the pile would move up relative to the soil and the downdrag load would be reimposed.

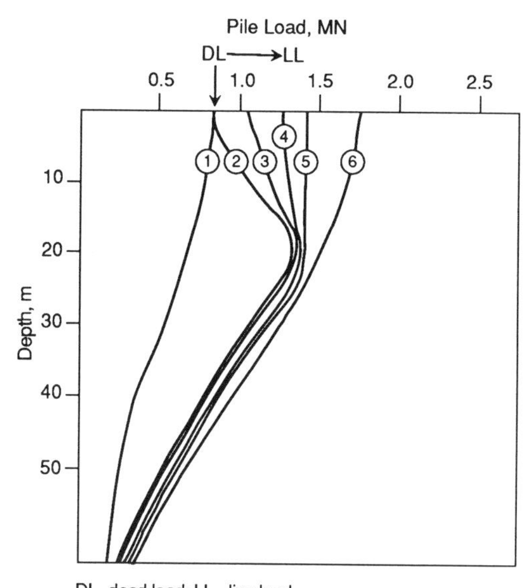

DL=dead load; LL= live load

Curve No.
1 Dead load only at top of pile prior to development of downdrag; load dissipates with depth
2 Dead load at top of pile; downdrag develops within top 20m
3-6 Load at top of pile = dead load + live load

FIG. 1. Pile Load versus Depth at Various Loading Stages

An analysis is first conducted to determine the distribution of load and deflection versus depth, with full downdrag applied to the pile (curve No. 2 in Fig. 1). The load at the top of the pile represents the dead load on the pile. The calculated deflections along the length of the pile are then used to develop the load transfer curves shown in Fig. 2c for the live load analysis from which curve Nos. 3-6 in Fig. 1 are obtained. The downdrag acting on the pile for any given load is

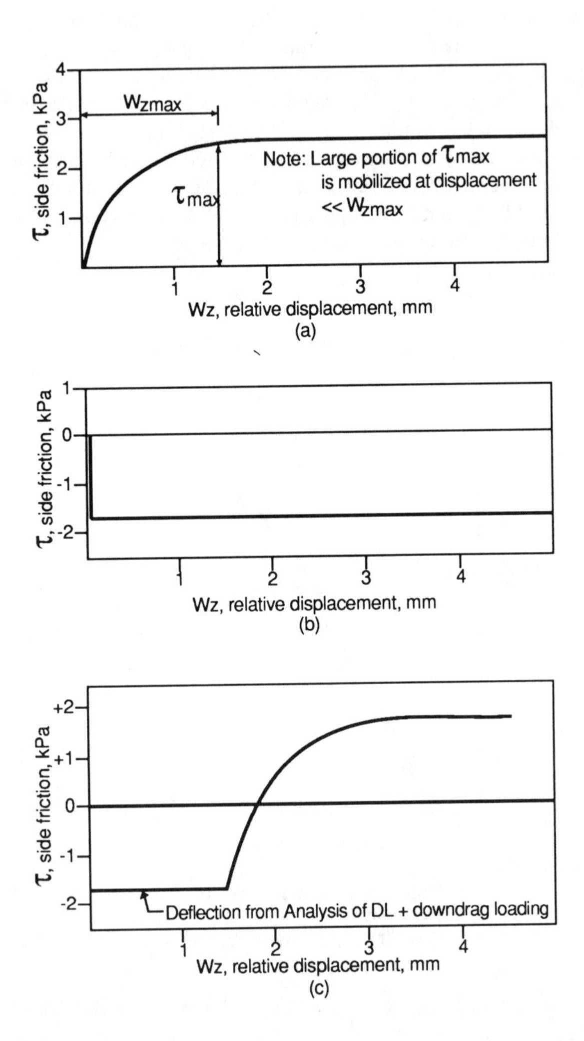

FIG. 2. Load Transfer Curves: (a) Mobilization of Positive Friction; (b) Negative Friction in Upper Portion of Pile; and (c) Release of Negative Friction

the difference between the maximum load in the pile and the load at the top of the pile.

It is important to note that the pile deflection, (i.e. the relative movement between pile and soil assumed to be necessary for mobilization of full friction) can have a significant effect on the results of the analysis, depending on the flexibility of the pile, as described below.

Alonso et al. (1984) present the results of laboratory direct shear tests on low plasticity silty clay that indicate that although the maximum shear stress is mobilized at a relative displacement of about 2.5 mm, about 75% of the maximum shear stress is mobilized at a relative displacement of 1 mm and about 60% at a relative displacement of only 0.2 mm. Tests on sand reported by Clemence and Brumund (1975) also indicate mobilization of most of the shear stress at displacements of less than a millimeter. In our study, analyses were conducted using values of the relative displacement necessary to mobilize the maximum side friction, W_{zmax}, of 1.27 mm (0.05 in.) and 2.54 mm (0.1 in.). However, about 75% of the maximum value was assumed to be mobilized at displacements of less than one millimeter.

PRESENTATION OF RESULTS

A standard generic subsurface profile established for these analyses is shown in Fig. 3. Various parameters that enter into the analysis are also shown. The values for these parameters considered in our analyses are listed in Table 1.

Fig. 4 shows typical results from one set of analyses in which the pile stiffness, AE, varies from 1268 MN to 5625 MN. The reduction in downdrag load, expressed as DD/DD_{max}, is plotted versus the applied live load normalized with respect to the maximum downdrag, LL/DD_{max}, where DD = the downdrag remaining under a given live load; DD_{max} = the maximum downdrag mobilized prior to the application of live load; and LL = live load. When DD/DD_{max} equals zero, the amount of live load that must be added to completely remove the downdrag can be calculated from the corresponding value of LL/DD_{max}. For the cases considered in Fig. 4, a live load equal to between two and six times the downdrag must be applied to the pile to release the downdrag. The data shown in Fig. 4 pertain to a specific combination of pile geometry and soil parameters. It will be shown in subsequent sections that for other conditions, a live load equal to the downdrag is sufficient to release all the downdrag. To provide a design chart that is applicable for a range of conditions, a parameter whose values reflect that range must be developed. It is suggested that soil/pile flexibility is the appropriate parameter.

SOIL/PILE FLEXIBILITY

The reduction of downdrag under the application of live load results from the downward deflection of the pile. The magnitude of the downward deflection of the pile is related to the total pile length, L, the length of pile subject to downdrag, KL, the pile stiffness, AE, the distribution of load in the pile which is determined by the values assumed for τ_{max}, and the magnitude of the load increase. τ_{max} is the maximum side friction mobilized at any given point along the pile.

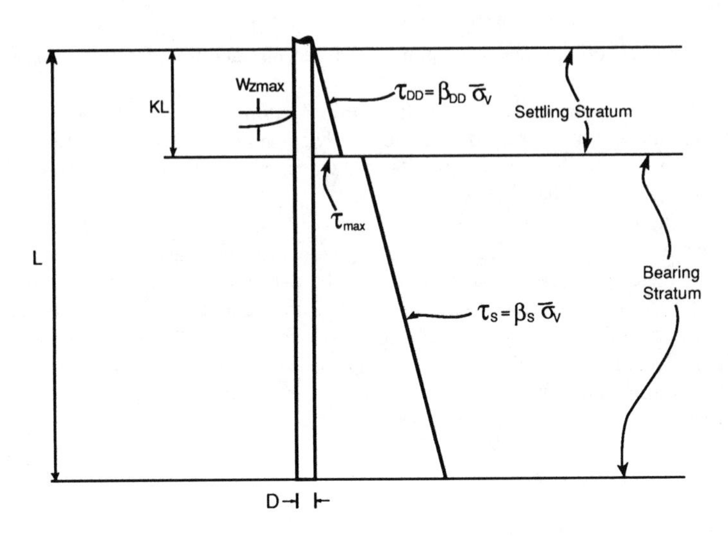

L= Pile Length, m
D= Pile Diameter, mm
τ_{DD}= Load Transfer (kPa) in zone subject to downdrag
τ_s= Load Transfer (kPa) in zone providing support to pile
K= portion of pile length subject to downdrag
β_{DD}= constant relating τ_{DD} to vertical effective stress
β_s= constant relating τ_s to vertical effective stress, $\bar{\sigma}_v$
Wzmax = relative pile-soil movement for mobilizing full τ

FIG. 3. Standard Subsurface Profile for Parametric Study and Analysis Parameters

It would be useful to relate the reduction in downdrag to a pile/soil flexibility factor which considers the elastic compression of the pile that occurs under the application of a load equal to the downdrag load as compared to the amount of pile deflection required to release the downdrag. The deflection of the pile under the downdrag load can be written as:

$$\delta = \left[\frac{\tau_{max}}{2}\right] \left(\pi DKL\right) \left[\frac{L}{AE}\right] \tag{1}$$

The deflection of the pile required to release the downdrag is shown in Fig. 5 to be

$$\left(W_z\right)_{DD} = \frac{\tau_{max}}{E_i} \tag{2}$$

where all of the parameters are as defined in Figs. 3 and 5. The pile/soil flexibility factor is defined as the ratio of these two deflections, or

$$f = \frac{\delta}{(W_z)_{DD}} = \left[\frac{E_i}{AE}\right]\left[\frac{\pi DKL^2}{2}\right] \qquad (3)$$

TABLE 1. Analysis Parameters

Parameter	Description	Values
D	Pile Diameter, mm	324, 457
L	Pile Length, m	15, 46
β_{DD}, β_S	Factor relating load transfer shear stress to vertical effective stress in zone subject to downdrag (DD) or in zone giving positive	β_{DD} = 0.25, 0.5 β_S = 0.25, 0.5
W_{zmax}	Relative displacement between soil and pile	1.27, 2.54
K	Portion of pile length subject to downdrag	0.3, 0.5
AE	Pile Stiffness, MN	1258, 3382, 5625
q_{tip}	Tip bearing capacity, kPa	10.1, 101, 604
Δ_{tip}	Tip settlement at failure	0.06D, 0.10D

Fig. 6 shows the ratio LL/DD_{max} when the downdrag equals zero versus f and includes analyses of piles that vary in length from 15 m to 46 m, that have a stiffness varying from 1268 MN (hollow, thin walled pipe) to 5625 MN (thick-walled, concrete filled pipe), and that are subject to downdrag loads ranging from 3% to 52% of the design load. The upper range shown in the figure is for β_s = 0.50, i.e. stiff or medium dense soil in the portion of the soil profile providing positive support. The lower range is for β_s = 0.25. Fig. 7 shows the range in reduction in downdrag as a function of normalized live load for various f values. The range shown in Fig. 7 encompasses the upper limit of the upper range and the lower limit of the lower range shown in Fig. 6.

Based on Figs. 6 and 7, a preliminary assessment can be made of the percent reduction in downdrag to be anticipated for a given pile size and stiffness and magnitude of live load. A site specific analysis considering the load transfer characteristics of the various strata in which the pile is founded would be required to obtain a more accurate estimate.

Based on examination of Figs. 6 and 7, several general trends can be identified:

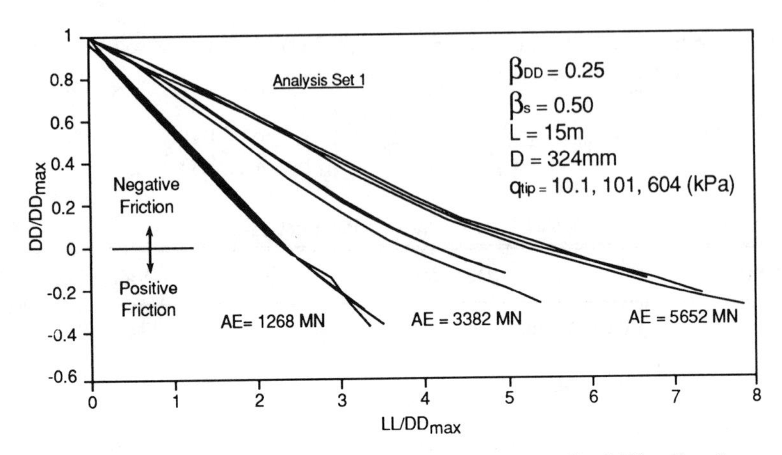

FIG. 4. Reduction in Downdrag versus Normalized Live Load

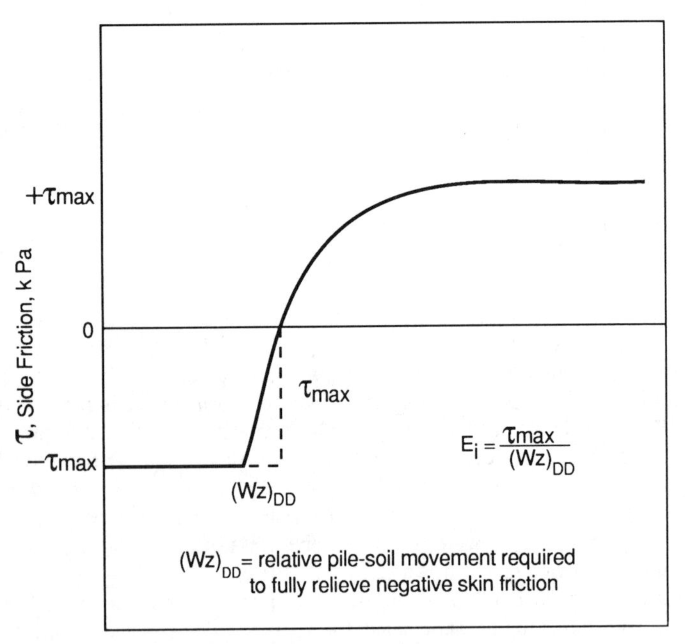

FIG. 5. Deflection Modulus E_i, for Pile/Soil Load Transfer

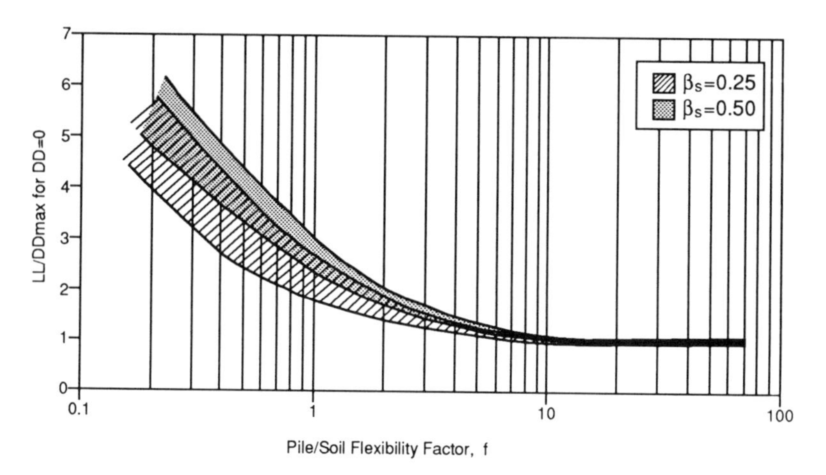

FIG. 6. LL/DD$_{max}$ for DD = 0 versus f

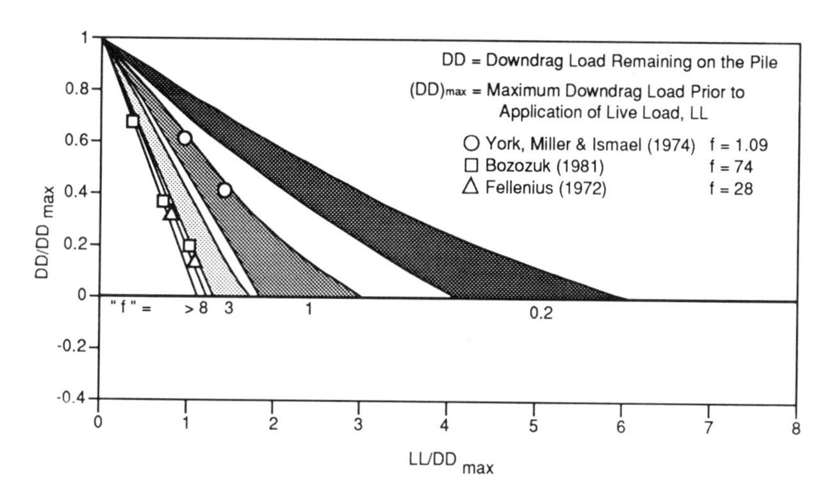

FIG. 7. Reduction in Downdrag versus Normalized Live Load for Various Values of f

1. The more flexible the pile, the greater the reduction in downdrag for a given increase in pile load. Pile flexibility is increased primarily by an increase in pile length or a reduction in AE.
2. As the pile flexibility factor, f , gets smaller, the range in predicted values increases, indicating that the solution becomes more sensitive to the specific parameter values selected. In this case, a site specific analysis may be warranted, depending on the cost implications of the uncertainty represented by the range of values obtained from Figs. 6 and 7.
3. For values of f greater than 2, reasonable variations in the assumed values for the analysis parameters have relatively little impact on the results.

It should be noted that for the cases considered, the tip bearing capacity and load deflection characteristics of the soil/rock material on which the tip bears had virtually no effect on the results. The assumed tip capacities varied from 10 kPa (5 tsf) to 600 kPa (300 tsf). At normal service loads, tip movements are typically quite small and therefore have little impact on the results. It should also be noted that the analyses presented herein pertain to reduction of the downdrag acting at the time the additional load is applied. This could be at some intermediate point when only a portion of the full downdrag has been applied or after the maximum downdrag load has been mobilized.

EVALUATION OF CASE HISTORIES

Significant differences in pile length, and hence, pile flexibility exist between the piles tested in the case studies reported by Bozozuk (1981) and Fellenius (1972) and the piles tested at Newark International Airport. The pertinent facts from these three case histories are summarized on Table 2. Also shown is the calculated pile/soil flexibility factor, f, for each of the three cases and the predicted reduction in downdrag estimated from Figs. 6 and 7 as compared to the actual measurements. The f values for these points were calculated using a load transfer curve based on W_{zmax}=1.27 mm. The difference in magnitude of the reduction in downdrag between the Newark International Airport study and the tests reported by Bozozuk (1981) and Fellenius (1972) is clearly reflected in the difference in f values. As seen from Fig. 7, DD/DD_{max} is essentially zero at LL/DD_{max} = 1 for f values of 28 and 74 whereas for an f value of 1, DD/DD_{max} is about 0.4 to 0.5 for LL/DD_{max} = 1.35. On Fig. 7, data points from the three case histories showing the reduction in downdrag for various magnitudes of live load are shown. The data in Fig. 7 suggest that the calculation of f based on W_{zmax} = 1.27 mm results in a better correlation between predicted and measured reduction in downdrag than would be obtained with W_{zmax} = 2.54 mm. For example, the data points for Newark International Airport fall within the band corresponding to f = 1. If W_{zmax} = 2.54 had been used to calculate f, the resulting value would be 0.5 and one would have predicted a much smaller reduction in downdrag than was actually measured. This conclusion is consistent with the laboratory test data cited above that indicate mobilization of a large portion of the maximum side friction with very small relative displacements.

TABLE 2. Review of Case Histories

References	Length, m	Pile Type & Dia., mm	Pile Tip Bearing in	DD_{max}, MN	Live Load, MN	f	DD/DD_{max}	
							Predicted Figs. 6 & 7	Measured
Bozozuk (1981)	49	conc. filled pipe 324	clay	1.52	1.52	74	0	0
Fellenius (1972)	55	perim. = 1050*	clayey silt	0.49 0.31	0.43 0.35	28	0	0.1
York et al. (1974)	15	conc. filled pipe 324	weathered Shale	0.066	0.089	1.09	0.4-0.5	0.4

* hexagonal, precast concrete

Areas for Further Study

Several factors exist that may affect the magnitude of reduction in downdrag due to application of live load:

1. The rate of application of live load, e.g., an instantaneous spike due to wind loading versus gradual filling and emptying of a storage area over a period of several days.
2. The presence of bitumen coatings to reduce downdrag.
3. The mechanism causing settlement, e.g., placement of new fill over compressible soils, consolidation under self weight, or reconsolidation after pile driving.

The analysis methodology presented herein is applicable to all of these cases. However, the parameters describing the load transfer, β_{DD} and W_{zmax} , as well as the shape of the load transfer curve would have to be defined. It would seem reasonable to assume that the relative displacement required to mobilize or release the side friction would be smaller with a sudden, instantaneous application of load than with a gradual application.

Conclusions

1. The computer program APILE2 can be used to model the reduction of downdrag loads due to transient live load.
2. The amount of live load required to eliminate the downdrag load can vary from one to six times the downdrag load, depending on the pile flexibility. A pile/soil flexibility factor, f, that relates the deflection of the pile under load to the relative pile/soil displacement required to mobilize or release the downdrag load, has been defined.
3. The reduction in downdrag load as a function of live load has been defined and is illustrated in Fig. 7 for several values of f. The greater the pile flexibility (higher f) the smaller the amount of live load required to relieve the downdrag.
4. Figs. 6 and 7 can be used to make a preliminary assessment of the reduction in downdrag load when live load is applied.

Acknowledgments

The writer wishes to thank the Port Authority of New York and New Jersey (PANYNJ) for their support and encouragement in the preparation of this paper. In particular, the writer is indebted to Donald L. York, Chief Geotechnical Engineer for PANYNJ who urged the writer to undertake the analyses described herein and who provided insightful comments on the drafts of this paper. The writer also acknowledges the assistance of K. Kostic, a summer intern at PANYNJ who performed most of the APILE2 analyses.

Appendix - References

Alonso, E. E., Jose, A., and Ledesma, A. (1984). "Negative skin friction on piles: A simplified analysis and prediction procedure." *Géotechnique*, 34(3), 341-357.

Bozozuk, M. (1981). "Bearing capacity of a pile preloaded by downdrag." *Proc.10th ICSMFE*, San Francisco, 2, 631-636.

Clemence, S. P., and Brumund, W. F. (1975). "Large scale model test of drilled pier in sand." *J. Geotech. Eng. Div.*, Proc. ASCE, 101(GT6), 537-550.

Fellenius, B. H. (1972). "Downdrag on piles due to negative skin friction." *Can. Geotech. J.*, 9(4), 323-337.

York, D. L., Miller, V. G., and Ismael, N. F. (1974). "Long term load transfer in end-bearing pipe piles." *Transportation Research Record 517*, Transportation Research Board, Washington, D. C., 48-60.

Settlement of Structures Supported on Marginal or Inadequate Soils Stiffened with Short Aggregate Piers

Evert C. Lawton,[1] Member, ASCE, and
Nathaniel S. Fox,[2] Associate Member, ASCE

Abstract: A short aggregate pier system, which was developed to provide an economical alternative to the overexcavation/replacement technique, has been used since 1988 to control settlement of structures located at sites with near-surface deposits consisting of marginal or inadequate soils. In this system, highly densified aggregate piers are incorporated within the marginal or inadequate soils, which results in a composite bearing material that is substantially stiffer than the unimproved soil, and on which shallow foundations can be supported with tolerable settlements. Three case histories are described in which the viability and effectiveness of the aggregate pier system in reducing settlements of shallow foundations bearing on the composite material are illustrated. Methods for analyzing and predicting settlements of footings supported on aggregate pier-reinforced soils are also discussed.

Introduction

A short aggregate pier system has been used since 1988 to control settlement of structures located at sites with near-surface deposits consisting of marginal or inadequate soils. In this system, highly densified aggregate piers are incorporated within the marginal or inadequate matrix soils, which results in a composite bearing material that is substantially stiffer than the unimproved soil, and on which shallow foundations can be supported with tolerable settlements. Buildings supported by

---

[1] Assistant Professor, Department of Civil Engineering, University of Utah, 3220 Merrill Engineering Building, Salt Lake City, UT 84112.

[2] President, Geopier Foundation Company, Inc., 769 Lake Drive, Lithonia, GA 30058.

soils that were reinforced with aggregate piers have ranged from a large, single story, lightly loaded steel and glass greenhouse on deep, highly organic fills, to a 16 story tower on relatively strong and stiff residual soils. Column loads for the buildings have varied from 89 to 8,900 kN (20 to 2,000 kips), typically ranging from 222 to 4,450 kN (50 to 1,000 kips).

BACKGROUND

This short aggregate pier system was developed to provide an economical alternative to the overexcavation/replacement technique that has been widely used throughout the world for many years to improve bearing soils beneath shallow foundations. The practical limitations of the overexcavation/replacement technique include the following: (1) large volumes of excavated soils and replacement aggregate are required; (2)excavations may collapse, necessitating the use of sloped or braced excavations deeper than 1.5 m (5 ft) in those cases where personnel are needed to compact the loosened bottom surface soils and/or place the aggregate in lifts; (3) support capacity may be limited unless the aggregate is compacted in thin lifts; (4) the bottom of the excavation may become softened by seeping groundwater or infiltrating rainwater; and (5) underpinning is needed if the project is a building addition or if another structure is located nearby.

Aggregate piers are constructed of well-graded aggregate, which typically consists of crushed stone as used for highway base course material. A pier is formed by creating a cavity within the matrix soil by augering or trenching, and densely compacting the aggregate within the cavity in thin lifts. The nominal shape of the pier may vary, but is generally either cylindrical or rectangularly prismatic. The matrix soil at the bottom of the cavity and the aggregate lifts are compacted using a special high energy, relatively high frequency, impact tamper with a 45° beveled foot. During these compaction processes, the soil at the bottom of cavity is prestrained and prestressed, and the aggregate displaces the matrix soil outward, resulting in a buildup of horizontal confining pressures prior to structural loading. The product is a very stiff, very dense, short pier element with an irregular (undulating) perimeter surface.

Although appearing similar in some ways to stone columns formed by vibro-replacement methods, aggregate piers differ in a number of significant ways, including the following:

1. The piers are designed primarily to stiffen the subgrade soil. Although some strengthening of the subgrade soil and increased radial drainage within the subgrade soil may occur, these are secondary considerations.
2. Aggregate piers are short, typically only two to three times as tall as they are wide. The piers are normally not extended to stronger, deeper soil zones.
3. Construction of aggregate piers involves the formation of a cavity by removal of matrix soil, rather than by lateral or vertical soil displacement, thus preserving to a large extent the soil's natural cementation and fabric.
4. Radial drainage is not a primary factor in aggregate pier design, allowing the aggregate comprising the pier to be well-graded crushed stone that includes

a fine and medium sand fraction, and which can be highly densified during compaction.

5. Aggregate piers are constructed using impact densification methods with relatively high impact frequency, rather than vibratory methods.
6. Aggregate piers are densified in thin lifts, prestraining, prestressing, and densifying adjacent matrix soils and producing very dense and very stiff foundation elements, thereby reducing vertical displacements upon application of structural loads.

The unique differences inherent within the aggregate pier system compared to other foundation types or ground improvement methods have resulted in award of a U. S. patent, with international patents pending (Fox and Lawton 1993).

Load Transfer Mechanisms in Aggregate Pier-Reinforced Soils

The transfer of load from the footing to the aggregate pier-reinforced soil can be classified by three general cases: (1) the footing is supported by one pier with the same diameter or width as the footing; (2) the footing is supported by one pier with a smaller diameter or width than the footing; and (3) the footing is supported by two or more piers with smaller diameters or widths than the footing. Cases 2 and 3 are the most common. The percentage of the foundation load carried by the aggregate piers is primarily a function of two factors - the areal coverage of the piers within the footprint of the footing and the relative stiffness of the piers compared to the matrix soil.

In Case 1, the pier carries the entire load and the response of the pier to the applied load can be reliably modeled by conducting static load tests (ASTM D-1194) on the pier using plates with the same diameter or width as the pier. The results from selected static load tests on aggregate piers are shown in Fig. 1. The load-deflection response is a function of the density and type of aggregate within the pier, the lateral stiffness of the matrix soil, and the lateral confining stresses acting along the pier-matrix interface. The lateral pressures along the interface depend on the initial stresses (before construction of the piers), stress relief due to excavation of the cavity, the installation process, and the stiffness and type of matrix soil. The magnitude of lateral stress increase generated during the pier construction process is limited by the maximum passive resistance of the matrix soil. Comparison of results from static load tests on aggregate piers and unreinforced matrix soil using plates with the same diameters as the piers shows that the piers are 5 to 40 times as stiff as the matrix soil (Fig. 2).

In Cases 2 and 3, the areal coverage of the aggregate piers (area of piers divided by the total area of the footing) may vary from about 20 to 40%. Using the range of typical relative stiffness ratios given above (10 to 20) and assuming a rigid footing, the portion of the total load carried by the piers typically ranges from 71 to 93%. Numerical analyses using the finite grid method (Bowles 1988) show that the ratio of bearing stress applied to the piers to the bearing stress applied to the adjacent matrix soil is approximately equal to the relative stiffness ratio. In

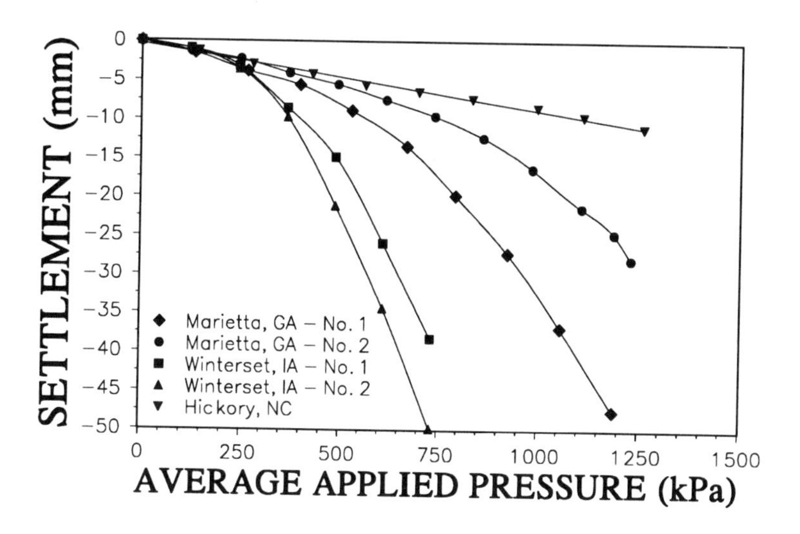

FIG. 1. Results from Selected Static Load Tests on Aggregate Piers

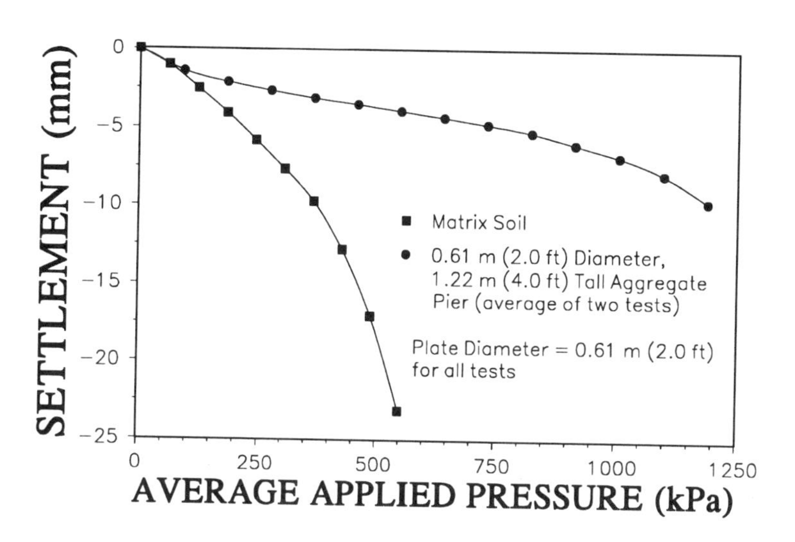

FIG. 2. Results from Static Load Tests on Unreinforced Matrix Soil and Aggregate Piers (Columbia, SC Project)

addition, even for fairly large footings that are relatively flexible, the bearing stress on each pier is roughly the same for a centric, vertical load.

At working loads, the stresses are transmitted through the aggregate piers in a manner similar to friction piles, but with two important differences. Most of the load is transmitted through the pier to the matrix soil by shear stresses along the pier-matrix interface, similar to the skin resistance that develops along the perimeter of friction piles. However, the aggregate piers also develop bearing resistance along the underside of the undulations at the perimeter of the pier, which transfers the potential shear surface out into the matrix soil rather than along the pier-soil interface. The result is greater shear resistance than that which would develop along a regular surface. A second important difference is that as settlement takes place, deformations occur near the top of the pier in the form of bulging, which displaces the matrix soil outward and increases the lateral pressures along the interface. The increased lateral pressure stiffens the pier similar to the stiffening that occurs in a strain-hardening material.

Settlement Analyses in Aggregate Pier-Reinforced Soils

The settlement of an aggregate pier-supported footing or mat is a complex soil-structure interaction problem consisting of interaction between footing and piers, footing and matrix soil, and matrix soil and piers. These complex interactions are not yet completely understood, and the preliminary methods described herein for estimating settlement provide rough estimates of settlement.

The reduction in settlement for an aggregate pier-reinforced soil compared to an unreinforced soil is related primarily to two factors: (1) The stiffness of the composite soil (matrix soil plus aggregate piers) within the reinforced zone is stiffer than the unreinforced soil; and (2) the stresses transmitted to the soil beneath the reinforced zone are lower than those transmitted through the comparable unreinforced zone. Therefore, settlement predictions are made separately for the upper and lower zones, and the two values are summed to yield the total predicted settlement.

Upper Zone Analyses

The upper zone (UZ), also called the aggregate pier influence zone, consists of the composite soil zone plus the soil beneath the composite zone that is densified and prestressed during the construction process. Densification and prestressing of this soil occurs by compacting the soil at the bottom of the excavated cavities with the tamper prior to placement of the aggregate lifts; some additional densification and prestressing also occurs during compaction of the first aggregate lift. During compaction of the aggregate lifts, the beveled foot of the tamper pushes some of the aggregate laterally against the adjacent matrix soil, resulting in densification of the matrix soil immediately surrounding the pier with a concomitant buildup of lateral stresses along the pier-matrix interface. The depth of this upper zone is considered to be equal to the height of the composite zone plus the width of the aggregate piers.

Although several methods have been used to estimate the settlement of the upper zone, the method determined to be the most practicable is the finite grid method (Bowles 1988). Details of the method can be found in the reference. Values of subgrade modulus for the pier and the matrix soil are either (1) obtained from the results of static load tests conducted on aggregate piers and the specific soils, (2) estimated from previous static load tests on similar soils, or (3) estimated from other available information such as recommendations for allowable bearing pressure provided by the project geotechnical consultant. For the pier moduli, results from static load tests using steel plates of the same diameter as the pier are used, which is somewhat conservative because it neglects the beneficial effect of the additional confining pressure produced from the footing bearing stress acting on the matrix soil. Because the aggregate piers are so stiff compared to the matrix soil, the subgrade modulus used for the matrix soil generally has limited influence on the magnitude of predicted settlement.

Stress Distribution at Interface of Upper and Lower Zones

Since the upper zone is substantially stiffer than the lower zone (LZ), the use of Boussinesq-type equations for stress distribution is inappropriate because these equations are based on the assumption of a homogeneous semi-infinite material. The actual stress distribution at the UZ/LZ interface is quite complex because it involves a stiff composite zone of finite horizontal and vertical extent with less stiff soils below and surrounding this stiff zone. The authors are currently conducting some finite element analyses and are planning to conduct field tests to study the stress distribution along the UZ/LZ. Some pertinent information can be obtained from Burmister's (1958) analysis of circular loads founded on a stiff upper layer of infinite horizontal extent underlain by a less stiff lower layer of infinite horizontal extent and semi-infinite vertical extent. Although this work was conducted relative to wheel loads on highway pavement systems, it clearly shows that the presence of a stiff upper layer reduces the magnitude of vertical stress transmitted to the lower layer compared to a homogeneous soil with the stiffness of the lower layer, and that the reduction in vertical stress increases with increasing stiffness of the upper layer.

The general procedure presently used by the authors to estimate the vertical stress intensity at the UZ/LZ interface is a modification of the 2:1 method and involves engineering judgment. In this approach, the area over which the applied load is distributed at the UZ/LZ interface is a function of depth below the bearing level and the stiffness ratio. For example, for a stiffness ratio of 10, a slope of 1.67:1 (vertical to horizontal) is typically used by the authors. The actual values used for any particular foundation depends on the type of aggregate, pier installation equipment and procedures, matrix soil type and condition, and engineering judgment. The magnitude of vertical stress increase and area of stress application are used to compute settlement of the LZ.

Lower Zone Analyses

For prediction of settlement within the lower zone, a number of methods are used depending upon available soil information. For granular soil strata, typically only Standard Penetration Test (SPT) blow counts (N) are available, although sometimes Cone Penetration Test (CPT) point resistance (q_c) data are also available.

Laboratory one-dimensional consolidation tests are usually performed for cohesive soil strata, although more sophisticated testing techniques can be used but are rarely justified economically unless a cohesive stratum is very soft or unusually thick and close to the UZ/LZ interface.

Both Schmertmann's strain influence factor method (Schmertmann 1970; Schmertmann et al. 1978) and Bowles' modified elastic theory method (Bowles 1988) are used to estimate settlements within the LZ if all the relevant strata are granular. For both these methods, the soil must be divided into layers of assumed constant elastic modulus (E_s), and estimates of E_s made for each layer based on regional or published correlations based on SPT N or CPT q_c (e.g. Bowles 1988).

Bowles' method can also be used to predict settlements within subsoils containing cohesive layers, but it is not recommended unless the stiffness of the cohesive layer(s) is of the same order of magnitude as the other layers and reasonable estimates of drained E_s for cohesive layers can be obtained from laboratory or field testing. The settlement of cohesive layers is more commonly predicted from the results of one-dimensional consolidation tests using appropriate modifications for variations from one-dimensional loading conditions [e.g. Skempton and Bjerrum (1957) or Leonards (1976)]. Likewise, Schmertmann's method may be used with cohesive soils with the realization that greater predictive error may occur. For major projects, the settlement of cohesive soils may be better approximated using the stress path method (Lambe 1967).

Case Histories

In the six years since the first short aggregate pier system was installed, thousands of aggregate piers supporting over one thousand footings, mats, and grade beams, have been constructed in six states. Although the initial idea of this system was to improve on the overexcavation/replacement method, the most common use has been to replace the need for deep foundations. Aggregate piers have also been used to improve soils that otherwise would require larger footings, resulting in the use of more economical smaller footings as well as reducing settlements.

General project information, foundation data, and predicted and actual values for settlement are summarized in Table 1 for ten projects in which shallow foundations have been supported on aggregate pier-improved soils. Settlements for the pier-reinforced soils were estimated using the methods described previously. Predicted settlements for the unreinforced soils were conducted using the same methods as for the lower zone analyses of the pier-reinforced soils, and are for the actual average bearing pressure and foundation dimensions listed in the table.

Comparison of the actual settlement values with the predicted values shows that the inclusion of aggregate piers within the matrix soil was successful in substantially reducing settlements for a variety of loading and matrix soil conditions. It also appears that the method used by the authors for predicted settlement of pier-reinforced soils provides a reasonable estimate of settlement in these cases.

TABLE 1. Predicted and Actual Settlements from Ten Aggregate Pier Projects

Project Description	Typical Foundation Description	Load kN (kips)	Bearing Pressure kPa (ksf)	Settlement, mm (in.)		
				Predicted		
				Unreinforced Matrix Soil	Reinforced with Piers	Actual
5 story office bldg. Columbia, SC	3.66 m (12 ft) square footing	3,560 (800)	266 (5.6)	33 to 102 (1.3 to 4.0)	18 (0.7)	< 1.5 (< 0.06)
12 m (40 ft) tall milk silo Atlanta, GA	4.57 m (15 ft) square footing	3,010 (675)	144 (3.0)	48 to 104 (1.9 to 4.1)	13 (0.5)	< 1.8 (< 0.07)
46 x 91 m (150 x 300 ft) greenhouse, Atlanta, GA	0.91 m (3 ft) diameter footing	160 (35)	244 (5.0)	58 to 79 (2.3 to 3.1)	5 (0.2)	< 6 (< 0.25)
Industrial warehouse Winterset, IA	1.52 m (5 ft) square footing	445 (100)	193 (4.0)	150 to 230 (6 to 9)	23 (0.9)	< 19 (< 0.75)
Office addition Orangeburg, SC	1.07 x 2.13 m (3.5 x 7 ft) footing	801 (180)	352 (7.3)	41 to 112 (1.6 to 4.4)	13 (0.5)	< 13 (< 0.50)
Hospital addition Hickory, NC	12.2 m (40 ft) square mat	47,150 (5,300)	317 (3.3)	61 to 109 (2.4 to 4.3)	10 (0.4)	3.3 (0.13)
Hospital addition Hickory, NC	2.74 m (9 ft) square footing	1,824 (410)	242 (5.1)	30 to 104 (1.2 to 4.1)	13 (0.5)	< 6 (< 0.25)
16 story tower Atlanta, GA	15.2 x 30.5 m (50 x 100 ft) mat	66,720 (15,000)	144 (3.0)	20 to 89 (0.8 to 3.5)	10 (0.4)	3 to 8, avg. 6 (0.1 to 0.6, avg. 0.25)
12 story tower Atlanta, GA	3.66 m (12 ft) square footing	4,448 (1,000)	332 (6.9)	61 to 66 (2.4 to 2.6)	10 (0.4)	< 6 (< 0.25)
7 story parking deck Marietta, GA	4.27 m (14 ft) square footing	5,782 (1,300)	318 (6.6)	124 to 188 (4.9 to 7.4)	38 (1.5)	20 to 33 (0.8 to 1.3)

Case I: Five Story Office Building, Columbia, South Carolina

This 91 m by 49 m (300 ft by 160 ft), five story steel-framed structure with full basement, was constructed in Columbia, South Carolina in 1991. The site was within the Piedmont geological province, and all soils below the foundation bearing elevations were virgin residual soils. The site was within a seismic 2 zone, which by building code required horizontal support between pile or pier caps in the form of tie-beams to control lateral displacements if deep foundations were used. Column loads varied from 222 to 3,560 kN (50 to 800 kips), with wall loads of 58 to 102 kN/m (4 to 7 kips/ft). The predominant soil type within the upper 9 to 12 m (30 to 40 ft) was a very loose to firm silty fine to medium sand (SM). SPT blow counts varied from 2 to 20 with an average of 8. The geotechnical consultant determined that the soil could not satisfactorily support shallow isolated spread footings. Underlying this unsuitable stratum was a zone of stiffer clayey fine to coarse sand (SC) with N varying from 12 to 37, averaging 20. No groundwater was encountered to the maximum drilling depth of 15 m (50 ft). As is often the case in the southeastern U. S., no laboratory testing of soils was performed.

The geotechnical consultant recommended using either (1) a mat foundation with compacted subgrade, (2) pressure injected footings, or (3) drilled piers. Driven piles were eliminated from consideration because the structure was adjacent to church and office facilities, and the noise and vibrations accompanying driven piles would have been objectionable. Cast-in-place concrete piles were not mentioned as a foundation solution in the geotechnical consultant's report. The bid documents, however, did ultimately specify auger-cast piles to an estimated depth of 18 to 21 m (60 to 70 ft). The low bid exceeded the owner's budget by about $1.5 million, so the owner chose the lowest two bidders and had them each provide a Value Engineering study to get the project within budget. Aggregate piers were approved during the Value Engineering study, and all piles were eliminated along with 85% of the tie-beams, at a total cost savings estimated at $250,000.

Two static load tests were performed on 0.61 m (2.0 ft) diameter, 1.2 m (4 ft) deep aggregate piers (Fig. 2), and the results from the two tests were nearly identical. The pressure-settlement plot was fairly linear up to a pressure of 1,000 kPa (21 ksf), and the calculated subgrade modulus of 149 MN/m^3 (550 pci) was nearly twice the initial estimate of 76 MN/m^3 (280 pci). The final design bearing pressure of 287 kPa (6.0 ksf) for the aggregate pier system was four times the 72 kPa (1.5 ksf) allowable bearing pressure estimated by the geotechnical consultant for the soil without piers. The estimated settlement for the pier-reinforced soil was 18 mm (0.7 in.), somewhat less than the maximum tolerable settlement of 25 mm (1.0 in.) specified in the contract.

Nominal dimensions of the installed aggregate piers were 0.76 m and 0.91 m (2.5 ft and 3.0 ft) in diameter and 1.5 m and 1.8 m (5 ft and 6 ft) in height, respectively. Compaction of the matrix soil at the bottom of the holes prior to placement of the first aggregate lift increased the depth of the cavity by 100 to 180 mm (4 to 7 in.), averaging about 130 mm (5 in.). The quality of the compacted aggregate in the piers was determined by performing dynamic penetration tests on selected lifts. Comparative tests indicated that 15 blows from the dynamic penetrometer correlated to modified Proctor maximum dry density (100% modified Proctor relative compaction). Except for the top lift on the piers, blow counts for the compacted aggregate ranged from 18 to 46, indicating that densities greater than 100% relative compaction were achieved. Blow counts as low as 8 were measured within the upper 150 mm (6 in.) of the piers, primarily as a result of the aggregate being loosened during subsequent footing excavations to the top of the piers. After light surface compaction with a hand-held mechanical compactor, the top of each pier was densified to the same condition as the rest of the pier.

Six months after the building was completed, the general contractor performed a settlement survey on twelve instrumented columns, with a representative of the authors as an observer. The maximum settlement of any footing was 1.6 mm (0.06 in.), and most registered zero. The survey was repeated three times. The Project Manager stated that he had "never before seen shallow foundations behave this well."

Case II: Forty Foot High Milk Silo, Atlanta, Georgia

This project is the smallest aggregate pier project performed to date, but is also one of the most interesting. No preconstruction load test was performed, but the project itself constituted a full-scale footing load test on aggregate-pier reinforced soil since a single footing supported a known vertical load.

A 12 m (40 ft) high, 3.7 m (12 ft) diameter steel silo was designed to store milk at a dairy products company in Atlanta, Georgia. The foundation was a 4.57 m (15 ft) square, 1.22 m (4.0 ft) thick footing, which fit tightly into an L shaped area with two sides of the footing less than 50 mm (2 in.) from an existing two story brick and masonry block building. The subsoils were identified by two SPT borings performed in close proximity on the tiny site. Each boring showed consistently low blow counts (average N = 3) to a depth of 8 m (25 ft), below which blow counts increased to 8 and 10. The soils classified as soft, fine sandy highly micaceous silts (ML). The site was within the Piedmont geological region, and the bearing strata were virgin residual soils. An indication of their high compressibility and low strength was observed in the 20 year old adjacent structure, in which numerous cracks had occurred from excessive settlements.

Three rectangularly prismatic piers, each 0.61 m (2.0 ft) wide and 1.5 m (5.0 ft) deep, were constructed diagonally across the footprint of the footing to the edges of the existing wall footings, exposing limited sections of the footings. Vibrations were minimized by use of a small installation apparatus, which coupled with the relatively high impact frequency of 500 cycles per minute, resulted in low intensity vibrations. Existing cracks in the adjacent building were monitored and none showed any signs of worsening or widening. Dynamic penetration blow counts for the pier aggregate varied from 20 to 42 blows, indicating that densification of aggregate exceeded 100% modified Proctor relative compaction by a considerable margin.

The bearing pressure producing settlement (144 kPa = 3.0 ksf) was accurately calculated based on the weights of the silo and the milk stored within it when filled. The predicted settlement based on this bearing pressure was 13 mm

(0.5 in.). Elevation readings were taken on the footing prior to construction of the silo. Three months after the silo had been filled, a final settlement survey was performed. The three readings varied from 1.5 to 1.8 mm (0.06 to 0.07 in.). Observations of relative displacements of the footing and adjacent concrete pavement also indicated that less than 2.5 mm (0.1 in.) settlement had occurred.

Case III: Industrial Manufacturing Building, Winterset, Iowa

This project represents one of the poorest soil conditions within which aggregate piers have been installed to date. The structure was a large, one story, steel framed manufacturing building with column loads varying from 180 to 800 kN (40 to 180 kips). The soils were soft aeolian silts (loess) that classified as either CL-ML, CL, or CH, overlying stiffer glacial till (CH), with the groundwater table located between 0.08 m and 0.9 m (6 in. and 3 ft) below the ground surface. The loesses were moderately to highly plastic, with plasticity indices as great as 45. The foundation was designed as shallow continuous footings for walls, and isolated footings for columns, and was based on 96 kPa (2.0 ksf) allowable bearing pressure. In addition, overexcavation of footings and backfilling with compacted aggregate had been recommended by the project geotechnical consultant.

When excavation for the footings began, problems occurred almost immediately as walls of the excavations collapsed. The bottoms of the excavations extended below groundwater in many cases, which combined with low strength soils, caused the collapses. Construction was halted while a solution was sought. The two alternatives considered were piles and short aggregate piers. Aggregate piers were selected because of lower cost, shorter time of construction, and lesser time required to redesign the footings.

Two static load tests were performed on 0.61 m (2.0 ft) diameter aggregate piers. In the first test, well-graded stone with a fine to medium sand fraction was used in the first lift before groundwater began to seep into the cavity. None of the overlying lifts were affected by the groundwater. For the second load test, it was decided to make the cavity and pier construction represent a potential worst case scenario. After drilling the cavity, water was poured into the hole to a depth of 0.5 m (18 in.), and left in place for four hours. Water was then pumped out, with a residual level of 80 mm (3 in.) of water remaining. The first aggregate lift consisted of an open graded stone without fine or medium graded sand, which was not sensitive to water. The remaining lifts were well-graded stone including the fine to medium sand fraction that is sensitive to seepage and water content during compaction. Unintentionally, some excess water saturated the lower portions of the second lift during placement. Somewhat surprising was the relatively close results from the two load tests (Fig. 1). The second test was poorer, but the difference was not as great as anticipated. Actual production pier construction proved to be closer in quality to the first test, and no piers were installed under conditions as severe as the second. This was accomplished by leaving each cavity open for short time periods only, and, where necessary, pumping water out of the cavity before placing the aggregate.

Unpublished data from field tests conducted on soils located near the two static load tests by Handy et al. (1993) using the K_o-stepped blade show a substantial increase in soil lateral stress after installation of the aggregate pier down to and several feet below the depth of the bottom of the pier. The lateral stress decreased radially with distance outward from the surface of the aggregate pier, where an extrapolation of the data indicated about a 48 kPa (7 psi) increase in stress at the top of the pier along the interface with the matrix soil, and a total lateral stress of 90 kPa (13 psi), the additional lateral stress being attributed to the expansive nature of the clay.

Borehole shear tests conducted in undisturbed loess at the site gave a drained friction angle of 37° and a cohesion intercept of 10 kPa (200 psf). These data, combined with a unit weight of 14.3 kN/m^3 (91 pcf), indicate that at a depth of 0.91 m (3 ft, the depth to the top of the aggregate pier), the limiting passive pressure was 90 kPa (13 psi), the same as was determined from the stepped blade tests. That limiting passive pressure was induced in the adjacent matrix soil during installation of the aggregate piers also was suggested by observations of radial tensile cracks in the surficial soil adjacent to the piers.

The final design bearing pressure of 192 kPa (4.0 ksf) with an estimated settlement of 23 mm (0.9 in.) was twice the design bearing pressure without aggregate piers. 0.61 m (2.0 ft) and 0.76 m (2.5 ft) diameter aggregate piers were installed, with the nominal height of the piers equal to twice the diameter. Compaction of the soft soil at the bottoms of the cavities typically increased the height of the cavity by about 0.2 m (8 in.). This was accomplished by placing open-graded stone as a first lift and tamping the stone downward, forming an aggregate bulb at the bottom of the pier. The project was completed in nine days. Data provided by the general contractor indicated that the maximum settlement was 19 mm (0.75 in.) six months after completion.

Summary and Conclusions

Short aggregate piers, with height to width ratios of 2:1 to 3:1, can be incorporated within existing soils to control settlements under various subsoil and structural loading conditions. In some instances, this solution can be used either as an alternative to overexcavation/replacement or in place of deep foundations. Future plans include

(1) conducting a full-scale load test on a footing supported by three aggregate piers to determine how a group of aggregate piers behaves in comparison to a single pier;

(2) undertaking field tests and numerical analyses to determine the stress distribution characteristics along the interface between the upper reinforced zone and the lower unreinforced zone;

(3) performing field tests and numerical analyses to better define the load-transfer characteristics between the aggregate piers and the adjacent matrix soil; and

(4) conducting additional field tests to better understand the magnitude of lateral stress increase along the pier-matrix interface during installation of the piers,

and how these increased lateral stresses improve the stiffness of the pier-reinforced soil.

ACKNOWLEDGMENT

The assistance of Dr. Richard L. Handy in reviewing the manuscript and providing constructive comments is greatly appreciated.

APPENDIX - REFERENCES

Bowles, J. E. (1988). *Foundation Analysis and Design*, 4th Edition, McGraw-Hill, New York, New York.

Burmister, D. M. (1958). "Evaluation of pavement systems of the WASHO road test by layered system methods." *Highway Research Board Bulletin 177*, 26-54.

Fox, N. S., and Lawton, E. C. (1993). "Short aggregate piers and method and apparatus for producing same," U. S. Patent No. 5,249,892.

Handy, R. L. (1993). *Personal Communication*, September 22.

Lambe, T. W. (1967). "Stress path method." *J. Soil Mech. Found. Div.*, Proc. ASCE, 93(SM6), 309-331.

Leonards, G. A. (1976). "Estimating consolidation settlements of shallow foundations on overconsolidated clays." *Special Report 163*, Transportation Research Board, Washington, D. C., 13-16.

Schmertmann, J. H. (1970). "Static cone to compute static settlement over sand." *J. Soil Mech. Found. Div.*, Proc. ASCE, 96(SM3), 1011-1043.

Schmertmann, J. H., Hartman, J. P., and Brown, P. R. (1978). "Improved strain influence factor diagrams." *J. Geotech. Eng. Div.*, ASCE, 104(GT8), 1131-1135.

Skempton, A. W., and Bjerrum, L. (1957). "A contribution to the settlement analysis of foundations on clay." *Géotechnique*, 7(4), 168-178.

DISPLACEMENTS OF BATTER PILES IN LAYERED SOIL

A. S. Yalcin[1] and G. G. Meyerhof,[2] Fellow, ASCE

ABSTRACT: The displacements of flexible vertical and batter model piles under eccentric and inclined loads in layered soil of clay overlying sand is studied. Horizontal and vertical displacements are found to agree fairly well with theoretical estimates using an effective pile depth for elastic conditions. The proposed analysis is also compared with some field cases.

INTRODUCTION

After Poulos and Davis (1980) presented a theoretical analysis of displacements of flexible vertical and batter piles under vertical and horizontal loads in elastic soils, Sastry and Meyerhof (1990) investigated the displacements of an instrumented flexible vertical model pile in sand and clay under inclined loads and compared the observed behavior with theoretical estimates and some field test results. Yalcin and Meyerhof (1988) found from model pile tests that the vertical and horizontal secant moduli of the soil varied considerably with load eccentricity and inclination. More recently, the behavior of flexible batter piles under central inclined load was investigated (Meyerhof and Yalcin 1993). In continuation of this work, the present paper is based on the results of an extensive study of the load-displacement behavior of model vertical and batter piles in layered soil of clay overlying sand and subjected to the more general case of eccentric and inclined loads. The test results are compared with theoretical estimates and some field cases.

---

[1] Research Associate, Civil Engineering Department, Technical University of Nova Scotia, Halifax, NS, Canada B3J 2X4.

[2] Research Professor, Civil Engineering Department, Technical University of Nova Scotia, Halifax, NS, Canada B3J 2X4.

TESTS

Soil Characteristics

The layered soil consisted of a soft clay layer of varying thickness over a loose sand bed. The inorganic clay (liquid limit of 43%, plastic limit of 21%) was prepared at a moisture content of 55% and mixed with 2% lime on dry weight basis to increase the brittleness of the clay. It was compacted into wooden boxes about 425 mm square and 48 to 190 mm or more high and was then cured for 7 days to ensure the reaction of the lime with the clay. The average total unit weight of clay was 17 kN/m^3 and its average undrained shear strength was 22 kN/m^2. The average deformation moduli of the clay from plate load tests were $\bar{E}_h = 1.2$ MPa in the horizontal direction and $\bar{E}_v = 5.5$ MPa in the vertical direction (Yalcin and Meyerhof 1988). The dry quartz sand used was medium to coarse grained with an effective size D_{10} = 1.5 mm and a total unit weight of 14 kN/m^3. Its friction angle, ϕ, determined by triaxial tests was about 30° (plane strain angle of 33°, approximately) when the average porosity was 45% and the relative density, D_r, was about 0.2. The average deformation moduli of the sand from plate load tests were $\bar{E}_h = 0.5\ MPa$ in the horizontal direction and $\bar{E}_v = 1\ MPa$ in the vertical direction (Yalcin and Meyerhof 1988). The box of clay was attached over the sand box giving a total height of 350 mm. The ratio H/D of the thickness of H of the clay layer to the embedment depth, D, of the pile varied from 0 (sand) to ∞ (clay) with an intermediate value of 0.5.

Pile Characteristics

The flexible model piles were made of nylon with insignificant short-term creep and an elastic modulus of 3×10^3 MPa, a diameter B of 12.5 mm and a length D of 190 mm. The pile head was attached to a rigid 100 mm wide square aluminum cap about 27 mm above ground surface. The cap carried an aluminum bar with grooves along its length for the application of the eccentric and inclined loads. The free-head piles were jacked into the soil either vertically or at a batter angle $\beta = 30°$, using an uniform penetration rate if 0.6 mm/sec. The load was applied in increments centrally and eccentrically on the vertical piles and centrally on the batter piles. The ratio of load increments in the vertical and horizontal directions provided an angle of load inclination, α, which varied from 0° (vertical) to 90° (horizontal) with intermediate values of 26°, 45° and 63°. A similar loading method was used for eccentric loading with an eccentricity/depth ratio, e/D, ranging from 0 to 0.82 with intermediate values of 0.14 and 0.41.

Test Results

Typical load-displacement relations for vertical and batter piles appear in Fig. 1. The safe load Q represented generally about 50% of its ultimate value Q_u determined from the load-displacement relation according to Terzaghi and Peck (1967) as the load corresponding to the initial linear part of the curve.

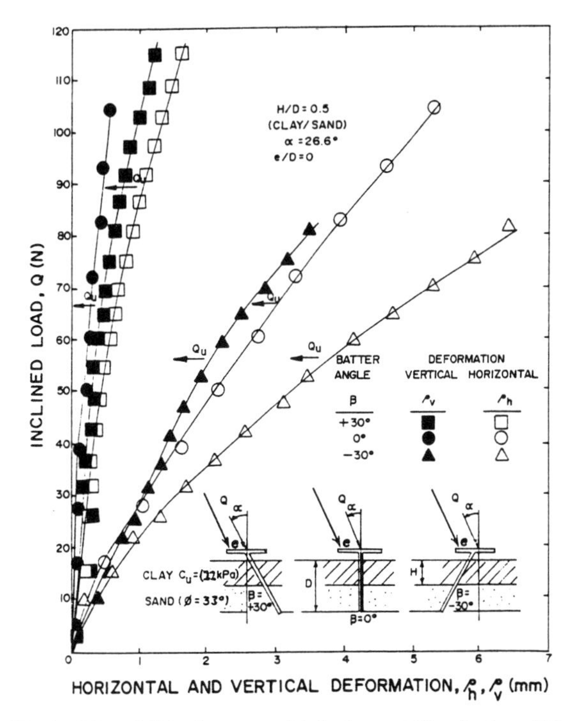

FIG. 1. Typical Load-Displacement Relations of Vertical and Batter Piles in Layered Soil Under Eccentric and Inclined Load

Figs. 2a, 2b and 2c for clay thickness ratios H/D = 0, 0.5 and ∞, respectively, show for vertical and batter piles the horizontal ground-line displacement ρ_h either versus the horizontal component $Q_h = Q\sin\alpha$ of the eccentric and inclined resultant load Q, or versus the moment M = Qe at the pile cap in cases of vertical eccentric loads Q only. The batter angle of a pile is called positive, $+\beta$, if the load acts in the direction of the batter, and negative, $-\beta$, vice versa as shown at the bottom of Fig. 1, which also defines the eccentricity e and inclination α of the load. In the present tests the batter angle was varied from $\beta = +30°$ to $\beta = -30°$, with an intermediate value of $\beta = 0°$ for a vertical pile. Due to the small scale of the model piles there is, unfortunately, a considerable scatter of the test results. The theoretical relationships (dash lines) shown are based on elastic conditions (Poulos and Davis 1980), and considering the effective embedment depth, D_e, of a equivalent rigid pile (Meyerhof and Yalcin 1984); (Meyerhof et al 1988) as discussed in the next section, and using the soil moduli from plate load tests. The values of the average overall horizontal soil moduli $\bar{E}_h$ at mid-embedment depth of the piles are also shown in Figs. 2a, 2b and 2c. It will be observed that generally the back-calculated average horizontal soil modulus $\bar{E}_h$ deduced from the pile load tests (solid lines) is roughly 50% greater than that determined from the plate load tests. This difference may be explained by the compaction of the sand and prestressing of the clay by installation of the piles.

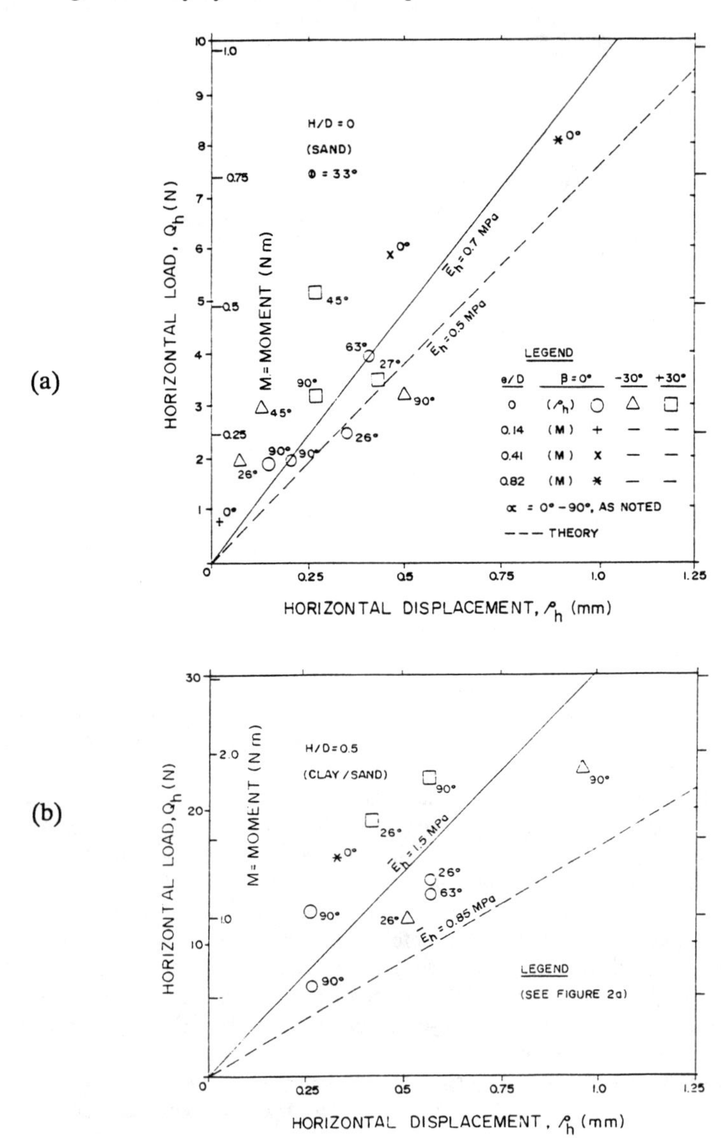

(c)

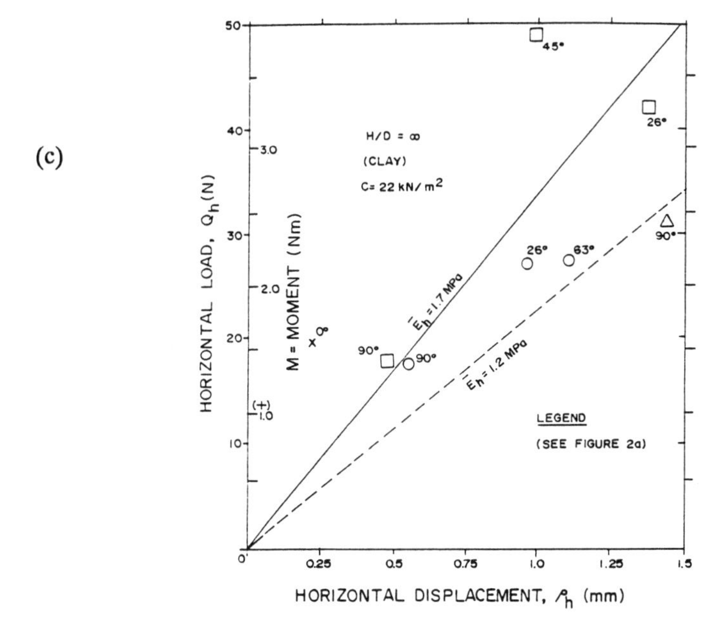

FIG. 2. Horizontal Load-Displacement Relationships of Piles under Eccentric and Inclined Loads: (a) Sand, (b) Layered Soil, and (c) Clay

Figs. 3a, 3b, 3c show the vertical displacement, ρ_v, of the pile cap under the vertical load, or the vertical component, $Q_v = Q\cos\alpha$ of the eccentric and inclined resultant load Q, for H/D = 0, 0.5 and ∞, respectively. The values of the average vertical soil moduli $\bar{E}_v$ at the mid-embedment depth of the piles are also shown. For piles fully or partly embedded in sand, compaction of the soil is found to raise the corresponding back-calculated vertical soil modulus $\bar{E}_v$ above that deduced from plate load tests, as would be expected, while for piles in clay the scatter of the test results does not permit any conclusions about the effect of prestressing of the soil.

Analysis of Test Results

Figs. 4a and 4b show the displacement of a vertical pile under eccentric inclined load and those of a batter pile under central inclined load, respectively. In previous studies (Meyerhof and Yalcin 1984; Meyerhof et al. 1988), it was found that the behavior of a flexible pile of depth D under a lateral load and moment can be related to an effective depth, D_e, of an equivalent rigid pile on which the lateral earth pressures govern most of the lateral displacements. In the theoretical analysis of elastic displacements of piles the relative stiffness K_r of the pile to that of the soils is introduced (Poulos and Davis 1980), such that

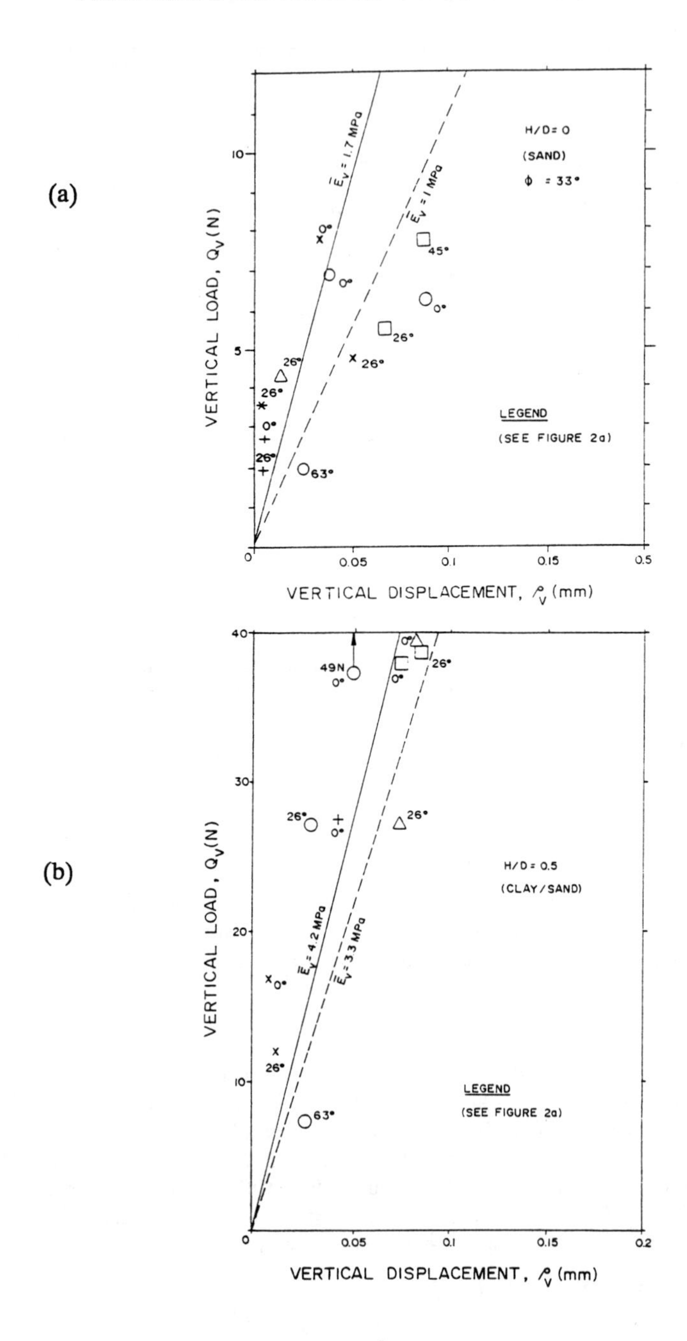

(a)
VERTICAL LOAD, Q_V(N)
VERTICAL DISPLACEMENT, ρ_V (mm)
0
5
10
0.05
0.1
0.15
0.5
$\bar{E}_V$ = 1.7 MPa
$\bar{E}_V$ = 1 MPa
H/D = 0
(SAND)
ϕ = 33°
0°
45°
26°
63°
LEGEND
(SEE FIGURE 2a)
(b)
VERTICAL LOAD, Q_V(N)
VERTICAL DISPLACEMENT, ρ_V (mm)
0
10
20
30
40
0.05
0.1
0.15
0.2
49N
0°
26°
63°
$\bar{E}_V$ = 4.2 MPa
$\bar{E}_V$ = 3.3 MPa
H/D = 0.5
(CLAY/SAND)
LEGEND
(SEE FIGURE 2a)

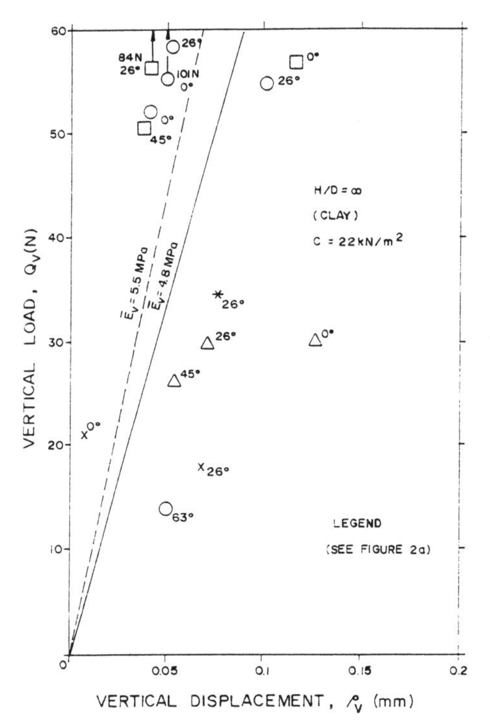

FIG. 3. Vertical Load-Displacement Relationships of Piles under Eccentric and Inclined Loads: (a) Sand, (b) Layered Soil, and (c) Clay

$$K_r = \frac{E_p I_p}{\bar{E}_h D^4} \tag{1}$$

where E_p = modulus of elasticity of pile; I_p = moment of inertia of pile; and $\bar{E}_h$ = average value of the horizontal soil modulus within pile depth D.

In a recent study (Liu and Meyerhof 1988), a relationship of D_e/D with K_r, pile to that of the soils is introduced (Poulos and Davis 1980), such that considering also certain dimensional properties, was established and it was found that

$$\frac{D_e}{D} = 2K_r^{0.2} \leq 1 \quad \text{approximately} \tag{2}$$

Theoretical load-displacement relationships of vertical piles under horizontal and vertical loads, Q_h and Q_v, respectively, have been presented previously (Poulos and Davis 1980). With D_e of an equivalent rigid pile replacing D of a flexible pile the same relationships may be used (Meyerhof and Yalcin 1984). Thus, the horizontal groundline displacement of the pile is, approximately,

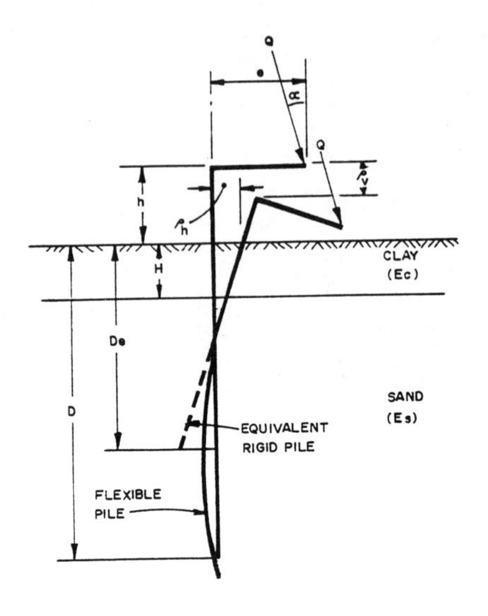

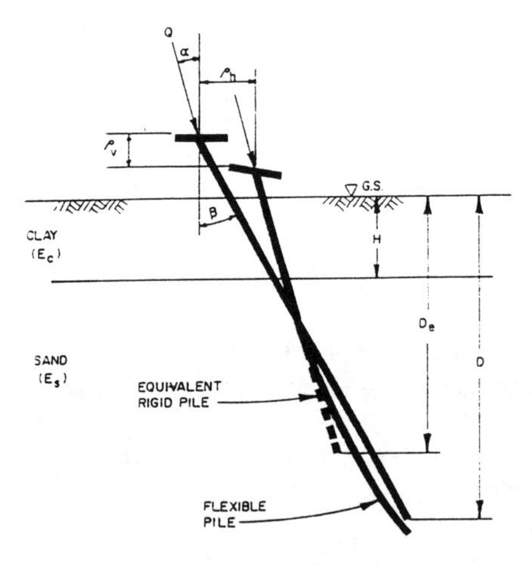

FIG. 4. Displacement of Flexible Pile and Equivalent Rigid Pile: (a) Vertical Pile Under Eccentric Inclined Load, and (b) Batter Pile Under Central Inclined Load

$$\rho_h = I_{\rho h}\left[\frac{Q_h}{\bar{E}_h D_e}\right] + I_{\rho M}\left[\frac{M}{\bar{E}_h D_e^2}\right] \tag{3}$$

where $I_{\rho h}$ and $I_{\rho M}$ are influence factors for a rigid pile under horizontal load and moment, respectively, and M is the moment at the pile cap. The average overall value of the horizontal soil modulus, $\bar{E}_h$, of layered soil may be approximated by the weighted average E_h values of the individual soil layers.

Similarly, the vertical pile displacement is, approximately,

$$\rho_v = I_{\rho v}\left[\frac{Q_v}{\bar{E}_v D}\right] \tag{4}$$

where $I_{\rho v}$ is the influence factor of a vertical load and $\bar{E}_v$ is the average overall vertical soil modulus for layered soil. In the present study, the values of influence factors for horizontal deflection, $I_{\rho h}$ and $I_{\rho M}$ are 7.5 and 9.0, respectively, for sand and, 4 and 5, respectively, for clay (Poulos and Davis 1980). Further, for vertical displacements the value of $I_{\rho v}$ ranges from 1.8 for sand to 1.2 for clay, with intermediate values for layered soil (Poulos and Davis 1980).

It was also shown (Meyerhof and Yalcin 1993) that the theoretical horizontal displacement of a batter pile (angle β) can be estimated from ρ_h of a vertical pile [Eq. (3)] by using a factor of I_β/I_o where I_β and I_o are resultant influence factors for a batter pile and a vertical pile, respectively, under similar loading and soil conditions (Poulos and Davis 1980). Similarly, it was shown that the theoretical vertical displacement of a batter pile can be estimated from ρ_v [Eq. (4)] by using a factor of I_β'/I_o' where I_β' and I_o' are resultant influence factors for a batter pile and vertical pile, respectively.

The above relationships have been used to back-calculate the observed values of $\bar{E}_h$ and $\bar{E}_v$ as shown in Figs. 2 and 3, respectively, for vertical and batter piles in comparison with the theoretical relationships based on plate load tests. In spite of some scatter of the results the average values of the soil moduli, $\bar{E}_h$ and $\bar{E}_v$ deduced from test results on batter piles are found to support the corresponding values for vertical piles.

Field Cases

Although several results of lateral load tests on flexible batter piles have been published, only a few piles were instrumented to determine the bending moments and other information form which the effective embedment depth of equivalent rigid piles can be deduced to estimate their corresponding behavior. At one of these test sites (Alizadeh and Davisson 1970) two steel H batter piles (piles 12 and 13) of 355 mm width and a flexural rigidity E_pI_p = 86 MNm^2 were driven at batter angles of β = +18° and β = -18°, respectively, to a depth of 13 m into the soil. The soil consisted of submerged medium to fine sands and silty sands with an average standard penetration resistance N = 27 blows/0.3 m in the upper 18 m and an average friction angle ϕ = 33°.

The increase of the horizontal soil modulus with depth was about n_h = 30 MN/m^3 at the average design deflection of about 10 mm at the pile head, and the corresponding average horizontal soil modulus along the piles is estimated (Terzaghi 1955) to be about $E_h = 8\,MPa$. For the relative pile stiffness $K_r = 4\times10^{-4}$ [Eq. (1)] the theoretical elastic effective depth was $D_e = 9$ m [Eq. (2)] compared with an average depth of 6 m found from the measured bending moments in the piles. Under a normal load of 200 kN applied at 450 mm above ground level with e/D = 0.05 the lateral deflections of both piles were practically identical at 13 mm which is less than the estimated deflection $\rho_h = 20$ mm [Eq. (3)]. This difference would indicate that the proposed relationship (Terzaghi 1955) between the values of n_h and $\bar{E}_h$ is on the safe side, and the back calculated modulus $\bar{E}_h$ deduced from the pile load tests is about 50% greater than estimated.

CONCLUSIONS

Series of tests on model vertical and batter piles were performed to investigate their displacements for the general case of eccentric and inclined loads in a layered soil of soft clay overlying loose sand. The test results indicate that the horizontal and vertical displacements depend on the type and layered structure of the soil, eccentricity and inclination of the load and the pile batter. Horizontal and vertical displacements of flexible vertical piles under eccentric inclined loads and batter piles under central inclined soils in layered soil are found to agree fairly well with theoretical estimates based on the effective depth concept for elastic soil conditions. The average soil moduli deduced from the pile load tests are larger than those of corresponding plate load tests due to compaction or prestressing of the soil by pile installation. The theoretical analysis is also compared with some lateral load tests on batter piles in the field. Further field observations on instrumented batter piles under eccentric and inclined loads in layered soils are required for analysis of pile displacement behavior under field conditions.

ACKNOWLEDGEMENTS

The research program at the Technical University of Nova Scotia was carried out with the financial support of the Natural Sciences and Engineering Research Council of Canada.

APPENDIX - REFERENCES

Alizadeh, M. and Davisson, M. T. (1970). "Lateral load tests on piles - Arkansas River project." *J. Soil Mech. Found. Div.*, Proc. ASCE, 96(SM5), 1583-1627.

Liu, Q. F. and Meyerhof, G. G. (1988). "New method for non-linear analysis of laterally loaded flexible piles." *Computer and Geotechnics*, 4(3), 151-169.

Meyerhof, G. G. and Yalcin, A. S. (1984). "Pile capacity for eccentric inclined load in clay." *Canadian Geotechnical Journal*, 21(3), 389-396.

Meyerhof, G. G. and Yalcin, A. S. (1993). "Behaviour of flexible batter piles under inclined loads in layered soil." *Can. Geotech. J.*, 30(2), 247-256.

Meyerhof, G. G., Sastry, V. V. R. N., and Yalcin, A. S. (1988). "Lateral resistance and deflection of flexible piles." *Can. Geotech. J.*, 25(8), 511-522.

Poulos, H. G. and Davis, E. H. (1980). *Pile Foundation Analysis and Design*, John Wiley and Sons, New York, New York.

Sastry, V. V. R. N. and Meyerhof, G. G. (1990). "Behaviour of flexible piles under inclined loads." *Canadian Geotechnical Journal*, 27(1), 19-28.

Terzaghi, K. (1955). "Evaluation of coefficients of subgrade reaction". *Géotechnique*, 5(4), 297-326.

Terzaghi, K. and Peck, R. B. (1967). *Soil Mechanics in Engineering Practice*, 2nd Edition, John Wiley & Sons, New York, New York.

Yalcin, A. S. and Meyerhof, G. G. (1988). "Movements of rigid piles under eccentric and inclined loads in sand and clay." *Soils and Foundations*, 28(2), 25-34.

DOWN DRAG ON FRICTION PILES: A CASE HISTORY

Yalcin B. Acar,[1] R. Richard Avent,[1] Members, ASCE
and Mohd R. Taha,[2] Student Member, ASCE

ABSTRACT: The down drag settlements experienced by the foundation of a supermarket constructed in 1972 in Gretna-New Orleans, Louisiana are presented. Two different lengths of piles (12.2 m and 21.3 m) were used to support the foundation. Negative skin friction on the piles due to 1.83 m fill was not considered in the design. The foundation system, specifically the floor slab supported by the shorter piles, experienced excessive settlements and failed. The case history is presented including an account of differential settlements, rotations, structural damages and legal aspects. The case history displays the significance of the care and attention that need to be dedicated to preparing a geotechnical site investigation report.

INTRODUCTION

An increase in vertical effective stress in soft, cohesive deposits subsequent to installation of friction piles may result in *down drag* of the pile system possibly leading to undesirable and excessive pile settlements or pile capacity failures. There are several causes for the increase in the vertical effective stress: a surcharge load from a fill or a nearby foundation, lowering of the groundwater table, progress in consolidation of underconsolidated, recent fills and generation and dissipation of pore pressures subsequent to pile driving.

This paper presents a case history of *down drag* experienced by the foundation of a supermarket supported by piles. The down drag, in this specific case, is found to be due to the placement of about 1.83 m (6 ft) of fill over most of

---

[1] Professor, Dept. of Civil and Environmental Engineering, Louisiana State University, Baton Rouge, LA 70803.

[2] Graduate Research Assistant, Dept. of Civil and Environmental Engineering, Louisiana State University, Baton Rouge, LA 70803.

the site immediately prior to placement of the piles. An evaluation of the progress of foundation settlements over the years, the structural damage experienced by the building due to these settlements, and an assessment of the cause and magnitude of the foundation failure are presented. The progress of the total and differential settlements are evaluated in light of the present state of understanding of *down drag* in friction piles. This case history shows that the state of practice lagged the state of the art in design and analysis of the pile foundation system against a down drag. Some legal aspects and outcomes of the case are also reviewed.

LOCATION AND HISTORY

The supermarket building was constructed in 1972 in Gretna-New Orleans within 2 km (1.24 miles) south of the Mississippi River. The location plan of the area and the site together with the schematic diagram of the dimensions of the building and the location of the borings/cone penetration tests conducted throughout the history of investigations at the site are shown in Fig. 1. The slightly overconsolidated pleistocene era deposits are found to lie within 40 m to 100 m depth on this side of Lake Pontchartrain (Kuecher et al. 1993). Therefore, the recent alluvial deposits of the river directly underlie the city in this area.

In 1971, prior to the construction, a geotechnical investigation report was prepared and presented to the architect by a geotechnical firm. This report was based upon two soil borings (B1 and B2 shown in Fig. 1) which extended to a depth of 21 m. The idealized vertical soil profile obtained by this initial investigation is also provided in Fig. 1. The initial site investigation report recommended that the building be supported by piles "*to avoid excessive settlements under the load of the superstructure*." Variable length piles were recommended possibly for economy; 21.3 m (70 ft) steel/timber composite piles were recommended under the column loads and 12.2 m (40 ft) wood piles were chosen under the slab. The allowable pile bearing capacities were estimated to be 142 kN and 62 kN for the longer and shorter piles respectively, implying that both were intended to be of friction type. This report recognized that a fill will be required for grade and drainage purposes at this site and provided cautionary remarks for the consolidation settlements "*specifically in areas not supported by piles*." The report did not show any anticipation of *down drag* or negative skin friction in the piles supporting the superstructure. The construction reports indicated that a 1.83 m (6 ft) of granular fill was placed over the entire site prior to the construction. The architect approved and certified all construction operations including the structural plan of the building. Construction was started soon after the completion of the filling operation without allowing adequate time for preloading.

A few years subsequent to the completion of the construction and opening of the store for business in 1972, frequent alignments were required in the doors, gutters and the water lines. As settlements further progressed, the differential settlements became very evident to the naked eye across the aisles in the supermarket. A lawsuit was filed by the owner of the building in 1984. Under the supervision of the first author, supplementary geotechnical investigations and surveys were conducted from 1985 to 1987 to establish and document the cause and

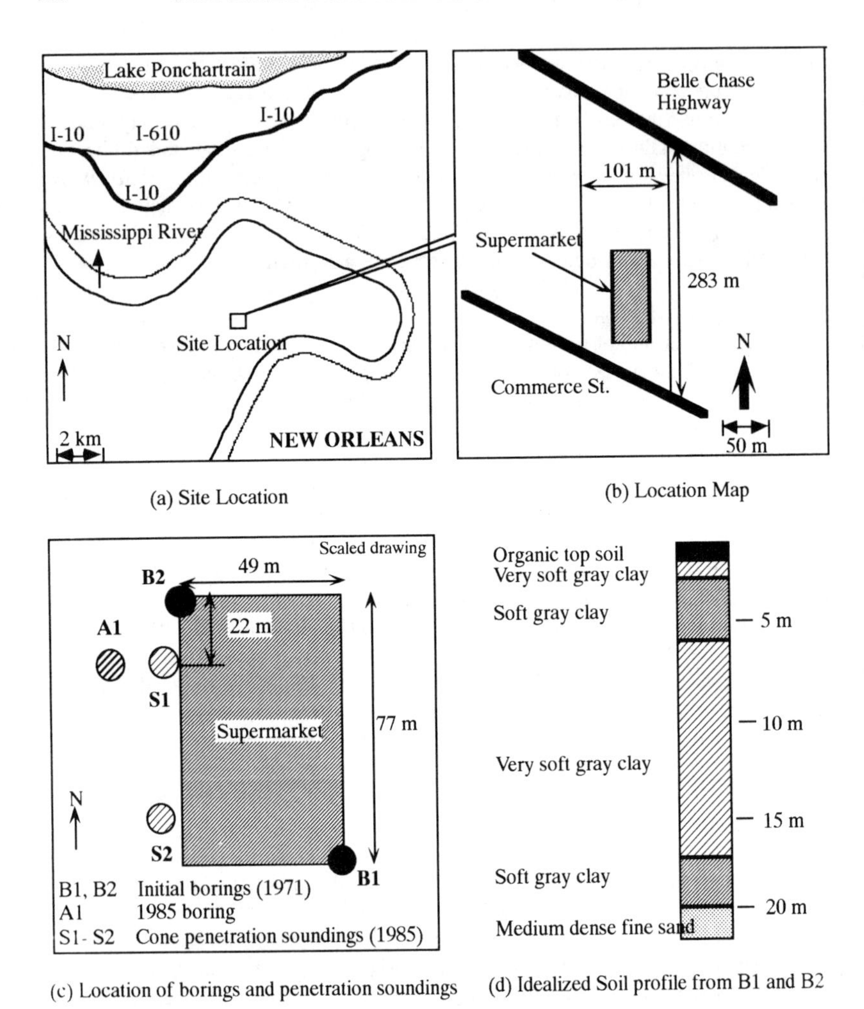

(a) Site Location

(b) Location Map

(c) Location of borings and penetration soundings

(d) Idealized Soil profile from B1 and B2

FIG. 1. Site Location, Location of Borings and Penetration Soundings and an Idealized Soil Profile

progress of the settlements and report the extent of the structural damages. In 1985, the entire floor slab was cracked, five step cracks were observed in the exterior walls, the interior walls were separated at five different locations, the overhead crane was derailed, ceilings were stained, most doors were misaligned and continuous alignments were needed and gutters had to be shortened several times. Service was hindered due to further progress of these damages after 1985, and the building was ultimately condemned and closed to the public in 1987. The judgement given in a lower court in 1991 was in favor of the plaintiff. The court of appeals stayed the judgement indicating that both the engineer and the architect were at fault in failing to communicate possible outcomes of placing a 1.83 m (6 ft) thick fill. The owner was compensated for the cost of the building and the accumulated interest in 1993.

GEOTECHNICAL CHARACTERISTICS OF THE UNDERLYING DEPOSITS

The geotechnical investigations of 1985 included two piezocone penetration tests and one boring with continuous sampling. The locations of these soundings are also depicted in Fig. 1. Table 1 presents the idealized soil profile obtained from this soil boring and the two soundings. Comparisons with the 1979 records indicated very good agreement for the first 21 m. The idealized average tip resistance (q_c) and friction ratio (f_s/q_c) values for the soundings are also provided. It is noted that high tip resistance together with low friction ratios (f_s/q_c) display sandy and silty deposits. The thickness of the fill is clearly determined to be 1.83 m (6 ft) from the cone penetration soundings. This fill is underlain by a thick very soft to soft compressible clay deposit with occasional silt seams appearing towards the bottom. At around 17 m, a silty clay deposit appears and transcends into a dense sand layer of about 2 m in thickness. This sand layer is underlain by a silty clay deposit. Specimens taken from this silty clay deposit indicated an overconsolidation ratio of less that one and lower compression indices and coefficients of consolidation than the upper soft clay layer.

Undrained shear strength and internal friction angles of the underlying deposits are also depicted in Table 1. Undrained shear strength values are estimated from the tip resistance values with,

$$s_u = \frac{(q_u - \sigma_v')}{N} \tag{1}$$

where s_u = undrained shear strength, σ_v' = effective overburden pressure; and N is the cone factor. The cone factor is taken as 18 based upon previous studies correlating the undrained shear strength of soft cohesive soils in Louisiana to cone penetration resistance (Acar 1981). s_u values are not corrected for any increase due to consolidation of the fill as the inaccuracy due to the variability in N is more significant than any inaccuracy due to neglecting the realized increase in undrained shear strength. The drained friction angles are estimated by correlations proposed by Schmertmann (1976) and Masood (1990) for the sand and clay layers, respectively.

The upper alluvial deposits display a slightly lower shear strength than the lower deposits. The upper soft clay layer is the compressible layer and constitutes the lead cause of major settlements under the surcharge of the fill.

PILE FOUNDATION SYSTEM

The base slab of 15 cm (6 in.) thickness is supported on 10.7 m (35 ft) to 12.2 m (40 ft) ASA Class 5 timber piles arranged in a group of one to two and placed with a pile cap. The columns in the building are supported by a group of two to three composite (steel can and timber) piles (ASTM D25) ranging in length from 18.3 m (60 ft) to 21.4 m (70 ft).

TABLE 1. Soil Profile and Geotechnical Characteristics of the Underlying Deposits

Depth (m)	Soil type	Unit Weight (kN/m^3)	Consistency Limits				Penetration test results		Strength Parameters		Consolidation Characteristics		
			PL	w	LL	LI	q_c (MPa)	f_s/q_c (%)	S_u (kPa)	ϕ' (°)	e_o	C_c	C_v (m^2/y)
1.83	Fill	17.3					2.6	2		44			
	Very soft clay with silt seams towards the bottom CH	15.1	20	44	45	.96	0.6	3.5	22-29	27	2.1	.95	.24
16.75			31	73	84	.79							
20.42	Silty clay CL	15.7	30	64	83	.64	1.0	1.75	40-50	26			
22.25	Dense sand	18.9					6.0	0.75		41			
	Medium soft clay with sillt/sand seams CL/CH	15.7	29	70	82	.77	1.0	2.25	25-38	18	1.9	.60	.026

PL Plastic Limit
w Moisture content
LL Liquid Limit
LI Liquidity index

q_c tip resistance
f_c/q_c friction ratio

S_u undrained shear strength
ϕ' peak drained friction angle
e_o initial void ratio
C_c compression index
C_v coefficient of consolidation

Fig. 2 presents pile load tests conducted in July 1972 during the construction of the supermarket which demonstrate the ultimate bearing capacity of an 12.3 m timber pile and a 21.4 m composite pile to be 125 kN and 320 kN, respectively. The capacities predicted using average soil properties presented in Table 1 and simplified techniques cited by Poulos and Davis (1980) agree quite well with the experimental values. Allowable working loads of 62 kN and 142 kN are recommended by the geotechnical engineer for a 12.2 m and 21.4 m.pile, respectively. These measured capacities and the recommended working loads indicate that the piles installed at two different depths are both designed to function in friction. It is noted that when the 1.83 m (6 ft) fill is accounted, the longer piles (21.4 m) would rest very close to the 2 m thick dense sand layer.

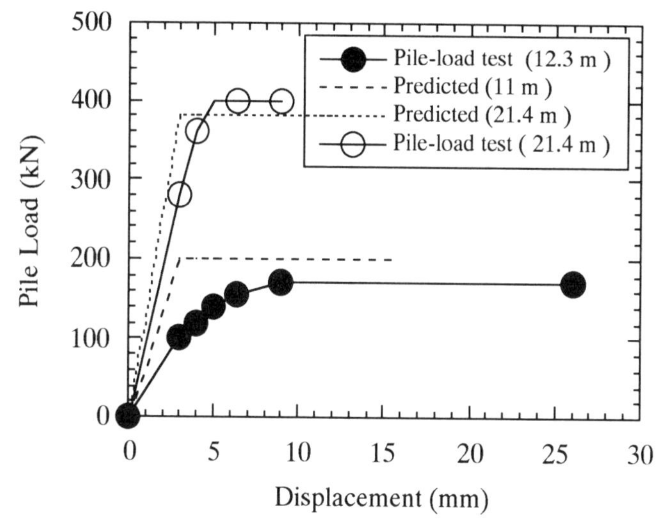

FIG. 2. Comparison of Pile Load Tests with Simplified Predictions Using Average Soil Properties

However, since the pile load tests do not display any significant tip resistance contribution and since these piles are not embedded into the sand deposit, they are expected to perform mostly as friction piles. An analysis of the existing loads based upon the floor plan indicated that the *apparent* factors of safety for the shorter piles ranged between 1.5 to 4.7 while the longer piles displayed factors of safety ranging between 1.7 to 2.1. In estimating these factors of safety, the slab dead and live loads are taken as 3.6 kPa (75 psf), the roof dead load is 0.96 kPa (20 psf) and the exterior block wall load is 3.1 kPa (65 psf). The *apparent* safety factors calculated are lower than those recommended by the parish code (Acar 1985).

SETTLEMENT

Differential settlements across the floor were recorded on January of 1983 and 1985 and November 1991. An assessment of these differential settlements requires an estimate of total settlements expected across the site under the load of the fill.

Estimated Total Settlements

Consolidation settlements due the 1.83 m fill are estimated using the idealized soil profile presented in Table 1 together with the depicted average soil properties. Conventional one-dimensional Terzaghi consolidation theory is used. Settlement of only the very soft clay layer within the first 18 m is considered.

Immediate elastic settlements are neglected since the building was constructed subsequent to placement of the fill. The estimated progress of settlements over the years is presented in Fig. 3 in comparison with the maximum differential settlements recorded across the building. The total settlement is expected to be 60 cm. It is estimated that 58 cm of this settlement will be realized during a design period of 60 years. A total of 45 cm of settlement would be expected during the period between 1972 to 1991. The agreement between the rate of increase in estimated total and recorded differential settlements displays the accurate estimate of the time rate of consolidation. The maximum differential settlements are about 68 % to 78 % of the estimated total settlements.

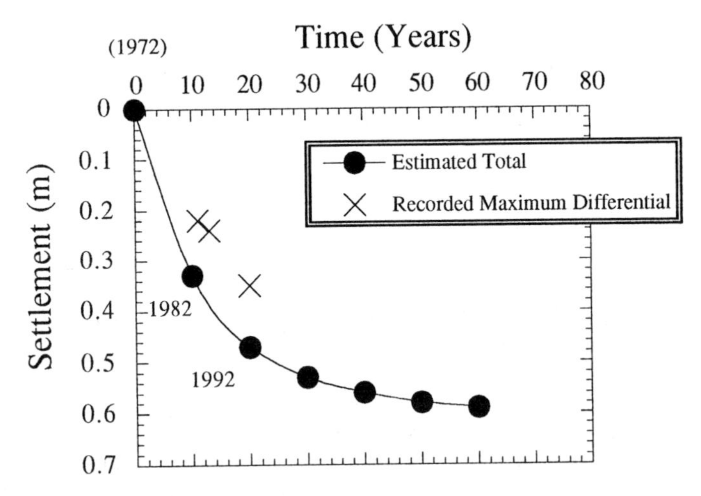

FIG. 3. Comparison of Estimated Total Settlements with Recorded Maximum Differential Settlements under a 1.83 m Fill

Differential Settlements

It was not possible to find a benchmark that would allow measurement of total settlements experienced. Therefore, only the relative displacement of one point with respect to a benchmark was recorded. The readings were taken with respect to the top right corner of the building plan and the highest point shot across the floor in 1991 is taken as the updated benchmark. The coordinates of this point are 37 m horizontal and 59.5 m vertical. This benchmark point has a column supported by the longer piles. All differential settlements are calculated with respect to this benchmark. Therefore, the term "*differential settlement*" is used to describe the displacement of a point across the slab with respect to the benchmark. An isocontour of the differential settlements recorded in November 1991 across the floor is shown in Fig. 4.

It is interesting to note that the area between the columns is supported by the shorter piles and this area has experienced differential settlements of up to 350 mm, almost as much as the 1991 settlements predicted in Fig. 3. The immediate vicinity of the columns, along the center of the building and the sides, however, is supported by the longer piles. Substantially lower differential settlements are recorded within these areas. One other point of interest is the increase in differential settlements towards the lower left (southwest) corner of the building, possibly due to thicker fill or a thicker consolidating deposit underlying the structure. A comparison of differential settlements recorded among the areas supported by the longer piles, the shorter piles and a comparison of differential settlements recorded among these two areas are essential for a better assessment of the performance of the foundation system.

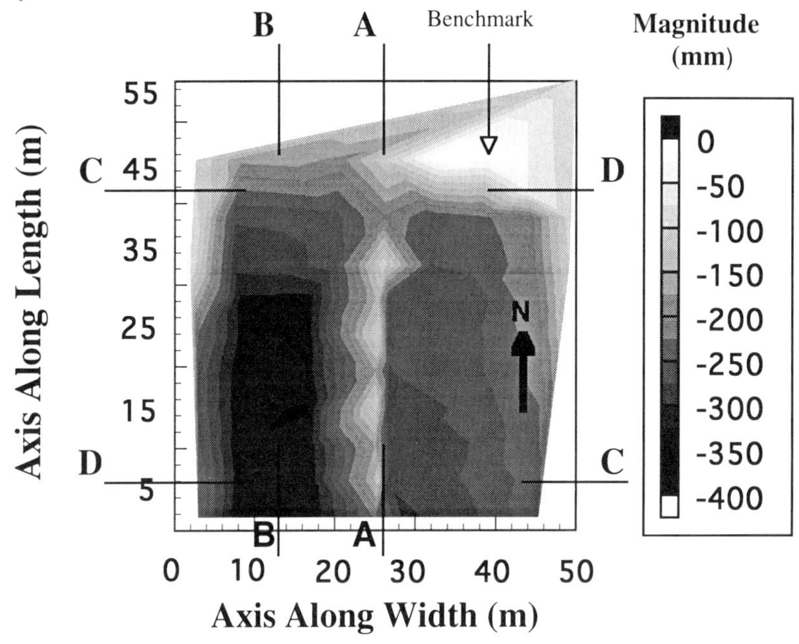

FIG. 4. Isocontours of Differential Settlements (November 1991)

The differential settlements among the areas supported by the shorter and the longer piles can best be evaluated along Section A-A and B-B, respectively. Section A-A represents the points under the columns. Section B-B runs parallel to A-A and represents the locus of points along the floor slab supported by the shorter piles. A comparison of differential settlements along these lines is presented in Fig. 5. The following observations are made:

Floor slab supported by the shorter piles

This section displays differential settlements of up to 35 cm. More than half of the differential settlements have been realized from 1972 to 1983. The rear of the building settled more than the front, possibly due to larger thickness of the fill or a thicker consolidating layer. Although the differential settlements in the rear of the building have increased from about 21 cm to 35 cm during the period extending from 1983 to 1991, the relative displacement of the points along this line have only increased from 11 cm to 18 cm. If it is assumed that the benchmark had no settlements, the differential settlements of 23 cm and 35 cm for 1983 and 1991 corresponds to 68% to 78% of the total estimated settlements (Fig. 3).

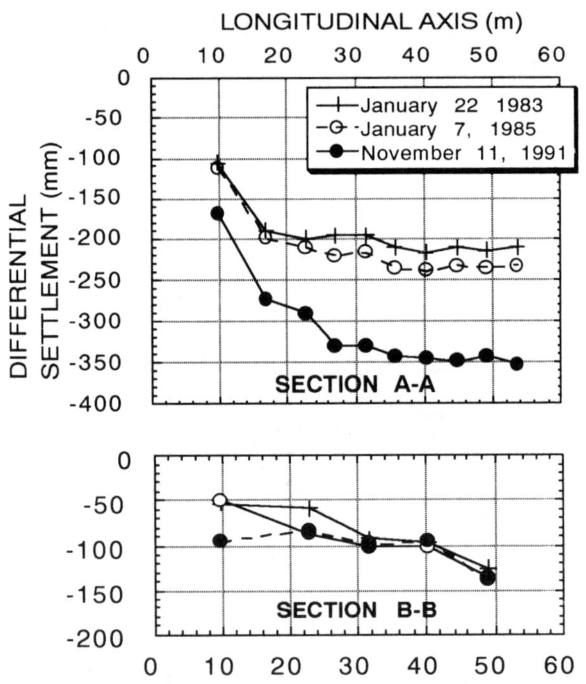

FIG. 5. Progress of Settlements with Respect to the Benchmark from 1983 to 1991: Section A-A: along the Floor Slab Supported by the Shorter Piles; Section B-B: along the Middle Columns Supported by the Longer Piles

Floor in the vicinity of the columns supported by the longer piles

This section displays differential settlements of up to 14 cm. Similar to the findings along the section supported by the shorter piles, the rear of the building displays more settlements than the front. The relative displacement of the points along this section has remained around 4 cm to 9 cm. The maximum differential settlement of 14 cm corresponds to 40% of the total estimated settlement in 1983. Almost all the settlements had been realized by 1983. Most piles along this section did not display any further displacement since that date. In an attempt to evaluate whether this was the case for all sections supported by the longer piles, the differential settlements along Section D-D are presented in Fig. 6. In this section, differential settlements are recorded after 1983 although the magnitude is in the order of only 5 cm. The maximum differential settlement is about 19 cm, or the same order of magnitude encountered along Section B-B.

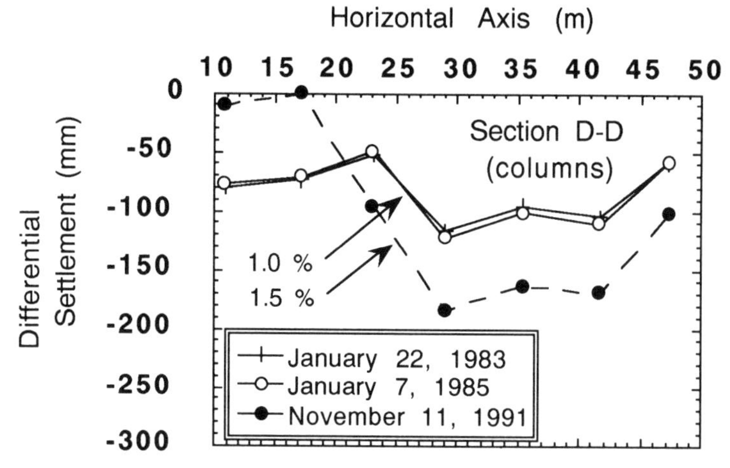

FIG. 6. Differential Settlements Along Section D-D Depicting Relatively Lower Differential Settlements of Sections Supported by the Longer Piles

The fact that limited differential settlements are recorded after 1983 may possibly be due to both the increased length of these piles together with their tip resting close to the dense sand deposit. The settlements realized until 1983 seem to have resulted in sufficient resistance close to the tip of these piles to support the increase in downdrag force without any further yielding. Differential settlements that run up to 20 cm however, have resulted in damage to the roofing, the cranes, the separations, etc. while significant structural damage has not been observed in load carrying members.

ROTATIONS

Section C2-C2 passing along the interior columns supported by the longer piles and Section C1-C1 along the slab between these columns are selected to assess the magnitude of the rotations. A comparison of the differential settlements and rotations along the two is presented in Fig. 7. The rotations of the order of 2% to 11 % across the slab would be sufficient to transfer both some of the slab and the fill weight on the section carrying the longer piles (Terzaghi and Peck 1948) decreasing the "apparent" factor of safety of these piles. However, the fact that these longer piles display lower differential settlements may also be indicative of their resting in the lower silty deposit and/or the sand deposit which in both cases would increase the bearing capacity beyond which is anticipated as friction piles resting in the upper soft clay layer.

The rotations along the sections carrying the column loads have been substantially lower (1-1.5%). However, even these rotations are beyond the tolerable limits of 0.1 to 0.6% encountered in the profession (Das 1990).

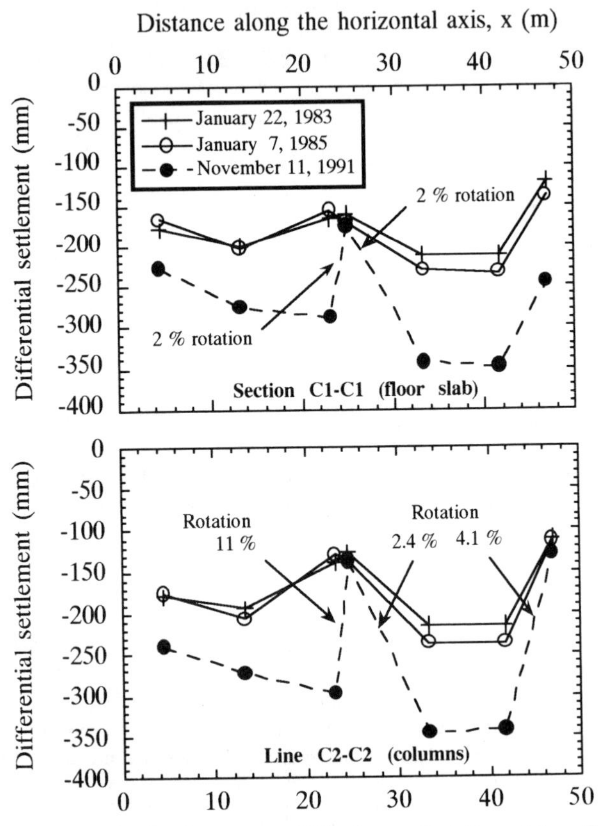

FIG. 7. Differential Settlements and Maximum Rotations Along Section C-C

Proposal for Repair

As most of the settlements are confined to the floor slab and the settlements of the piles underneath the columns carrying the structural loads may be considered relatively *more tolerable*, an engineer retained by the defendants recommended that the fill underneath the supermarket be removed to cease the ongoing down drag in the shorter piles and allow an effective rehabilitation and repair program. The existing floor would then be jacked up to its original level and covered with a new, level slab. The argument was made that the 12.2 m shorter piles would not be subjected to any further downdrag as the surcharge imposed by the fill would be removed. It was further argued and assumed that the load carrying capacity of the jacked up piles would not be different than their design values. However, the fill at this site is not confined only to the region underneath the building. The construction records indicate that it extends to an area far in excess of the dimensions of the building possibly covering the parking lot and the lot immediately beside the building. One other point of concern would be the downdrag force realized on the longer piles and any additional weight that may have superimposed on these piles. An appropriate assessment of the technical feasibility of this redesign would necessitate pile load tests on these longer piles carrying the column loads. The proposal was not found technically feasible and cost-effective.

LEGAL ASPECTS

The soil investigation report was deficient and the geotechnical engineer was found liable for preparing a defective subsoil report and piling design. The court noted that the soil report should have addressed the effect of a fill that would be placed at the site as it was essential to place a fill in most construction in this area. The court also concluded that recommending different pile lengths was inappropriate under the circumstances. The inconsistencies in the geotechnical report which in one section assumed a minimum tip embedment of 15 m and in another recommended that the floor slab be pile-supported with 12.2 m piles were also noted. The geotechnical engineer's decision to extend the borings only as deep as the recommended pile lengths was found contradictory to the practice by the court.

The engineering design consultant was also found liable for negligence in that they failed to consult and seek additional advice or information regarding the site at the design stage. Better communication between the engineers was expected to avoid this failure. The design engineer was expected to notice that down drag could be a problem under the present conditions at the site. The state of the practice should have been far beyond that which was practiced in construction of this building. Also found responsible was the building architect. The architect was not involved in almost any engineering design decisions. However, the design documents signed and approved by the architect clearly indicated the assumption of responsibility for all the documents in the packet, including the foundation plans, the electrical plans, and the mechanical and plumbing plans.

The court also heard about options for repair or for the demolition of the building. A repair method, as mentioned in the preceding section was suggested by the defendants. Expert witnesses were concerned about the feasibility and cost-effectiveness of the scheme. The parish requires that the entire building, not just the repaired portion, conform to the current code requirements if the cost of repairs run more than 50% of the current value of the building. If the repair option were taken, the building elevation had to be increased, electrical wiring and electrical lines, among others, had to be replaced and pile load capacities had to be reestablished to meet the requirements of the code. The court finally decided that the supermarket could not be adequately repaired necessitating its reconstruction. The case was concluded with a final judgement from the Court of Appeals in early

1992 compensating the plaintiff (Nicholson & Loup, Inc. vs Carl E. Woodward, Inc., et al. 1992).

SUMMARY AND CONCLUSIONS

History of extensive differential settlements leading to an unserviceable building is documented. The cause of the failure was unaccounted and/or underestimated negative skin friction. The case history demonstrates that there is the need to bridge the gap between the state of practice and the state of the art in understanding and analysis of negative skin friction and down drag.

A soil investigation report is a very important document. It should be carefully written by considering all possible aspects of the objectives of the construction and it should be used with great care by the design engineer in an attempt to eliminate the possibility of a major failure. It is essential to disclose the possible effects of all major environmental factors that may influence the performance of the civil engineering structure. Communication and collaboration between the geotechnical and the design engineer are essential. The case history also displays the continued need for a comprehensive and broad undergraduate education in civil engineering.

ACKNOWLEDGEMENTS

The authors would like to thank Mr. Scott Slaughter and Mr. Harouch Ignacio for the 1991 survey. Mr. John Nicholson and Mr. David Hebert, attorneys for the plaintiff are acknowledged for their cooperation and collaboration during the course of the investigation.

APPENDIX - REFERENCES

Acar, Y. B. (1981). "Piezo-cone penetration testing in soft cohesive soils," *Fugro Postdoctoral Fellowship Activity Report No. 4*, Dept. of Civil Engrg., Louisiana State University, Baton Rouge.

Acar, Y. B. (1985). "Geotechnical investigations and analysis for Nicholson and Loup Supermarket: Main and Addendum Report," Baton Rouge, Louisiana.

Das, B. M. (1990). *Principles of Foundation Engineering*, 2nd Edition, PWS-Kent, Boston, Massachusetts.

Keucher, G. J., Chandra, N., Roberts, H. H., Suhayda, J. H., Williams, S. J., Penland, S. P., and Autin, W. J. (1993). "Consolidation settlement potential in South Louisiana." *Proc. Coastal Zone 93,* ASCE, 1197-1214.

Masood, T. (1990). "Comparisons of in situ methods of determining lateral earth pressure at rest in soils," Ph.D. dissertation, University of California, Berkeley.

Nicholson & Loup, Inc. versus Carl E. Woodward, Inc., Eustis Engineering Co., Employers Commercial Union Insurance Co. and Larry H. Case (1992). *4th District Court of Appeal, Case No. 91-CA-1525*, State of Louisiana.

Poulos, H. G., and Davis, E. H. (1980). *Pile Foundation Analysis and Design*, John Wiley & Sons, New York, New York.

Schmertmann, J.H. (1976). "An updated correlation between relative density and Fugro type electric cone bearing," *Contract Report No. DACW 9-76-M, 6646*, Waterways Experimentation Station, Vicksburg, Mississippi.

Terzaghi, K., and Peck, R. B. (1948). *Soil Mechanics in Engineering Practice*, John Wiley & Sons, New York, New York.

DESIGN OF UNDERPINNING PILES TO REDUCE SETTLEMENTS

H. G. Poulos,[1] Fellow, ASCE

ABSTRACT: Piles are frequently used to underpin structures which have suffered, or are expected to suffer, damage due to excessive settlement or differential settlement. However, existing methods of designing such piles are rather simplified and do not generally give proper consideration to the process of load transfer from the existing foundation to the underpinning elements. This paper describes an approach to analysing such load transfer for the case where the loads remain constant but it is desired to limit future settlements and differential settlements. The approach involves the use of a soil-structure interaction analysis in which an existing shallow foundation is modelled as a strip or raft foundation, while the underpinning piles are treated as interacting spring elements of appropriate stiffness.

The paper discusses the application of this approach to a large raft slab supporting a racecourse grandstand in Brisbane, Australia. It is demonstrated that, by a process of trial and error, a system of underpinning piles can be designed which reduces future settlements and minimises future differential settlements.

INTRODUCTION

Circumstances may arise where the settlements and/or differential settlements of an existing foundation may become excessive and additional settlements may be expected to occur in the future e.g. Ergun and Uygurer (1991). Under such circumstances, it is usually necessary to undertake some form of foundation treatment in order to minimise the likely future settlements and/or differential settlements.

---

[1] Senior Principal, Coffey Partners International Pty Ltd, 12 Waterloo Road, North Ryde, Australia 2113; and Professor of Civil Engineering, University of Sydney.

Traditional methods of underpinning (e.g. Thorburn and Hutchinson 1985) usually aim to almost completely arrest the foundation settlements. Despite the frequent use of such techniques, procedures for the design of underpinning piles are generally relatively simplified, and little attention is paid to the complex interaction process which develops when piles are installed beneath a foundation already undergoing settlement. A numerical study of this interaction has been presented recently by Makarchian and Poulos (1994).

This paper considers an alternative option for foundation treatment, in which piles are introduced to reduce the future differential settlement, without necessarily arresting completely the overall settlements. The general philosophy of this approach is described, and a simplified method is presented for analysing the interaction between a foundation and underpinning piles. The application of this method is illustrated with reference to a large raft slab supporting a racecourse grandstand in Brisbane, Australia.

THE CONCEPT OF ENGINEERED REMEDIAL PILES TO REDUCE DIFFERENTIAL SETTLEMENTS

The foundation treatment system proposed is illustrated schematically in Fig. 1 which shows a section of a raft slab which will potentially suffer differential settlements. The system involves the installation of a series of piles across the slab section, such that the additional settlement of the slab-pile system is uniform. It is important to emphasise that the piles are not meant to completely arrest the settlement, but to even out the distribution of settlement. Consequently, it is generally necessary to vary the length and/or stiffness of the piles across the slab section, with the stiffest piles being located near the centre, where the potential settlements are greatest, and the least stiff piles (or no piles at all) near the edges of the slab, where the potential settlements are least.

The design of such a foundation system requires the assessment of the required length, diameter and spacing of the additional piles to achieve a uniform additional settlement across the slab. This assessment requires an analysis which can properly consider interaction between the slab, piles and soil, under the effects of the applied loading and the potential settlements of the slab without piles. One means of carrying out such an analysis is described briefly below.

ANALYSIS OF SLAB-PILE-SOIL INTERACTION

The analysis employed considers a strip section of the entire slab-pile foundation system, as illustrated in Fig. 2, and has been described in detail by Poulos (1991). A simplified form of boundary element analysis is used in which the soil can be modelled as a layered elastic continuum. Limiting compressive and uplift contact stresses can be specified below the strip elements, so that simple nonlinear behaviour of the strip foundation may be considered. The strip section is modelled as a beam, composed of a series of rectangular elements, whose stiffness can vary along its length. The piles are modelled as springs whose head stiffness can either be input (e.g. from load test data or independent calculations) or else computed from elastic theory. The force developed in each pile is "smeared" across

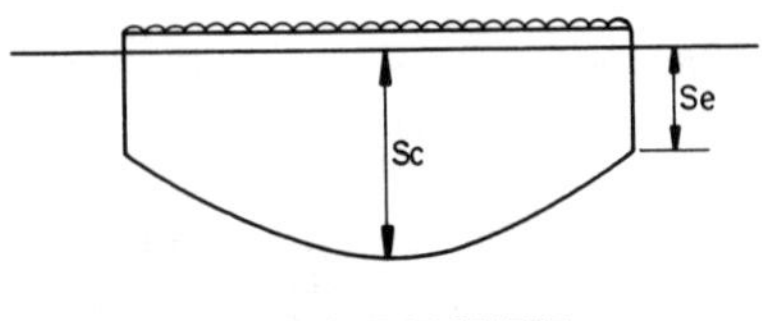

Sc= FINAL SETTLEMENT AT CENTRE
Se= FINAL SETTLEMENT AT EDGE

a) ORIGINAL SLAB & PROFILE OF EXPECTED SETTLEMENT

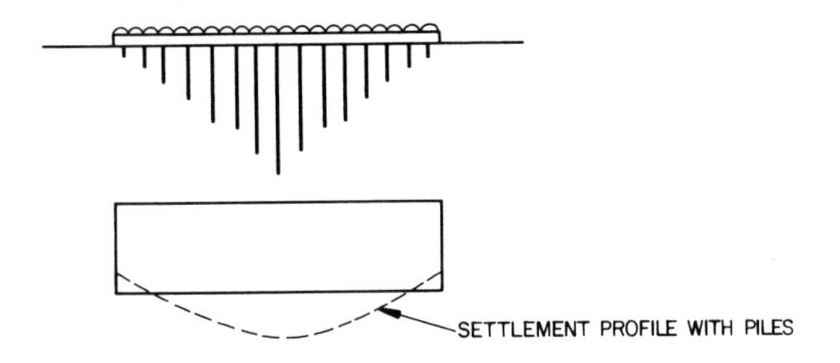

b) SLAB WITH ENGINEERED REMEDIAL PILES & PROFILE OF SETTLEMENT

FIG. 1. Concept of Engineered Remedial Piles

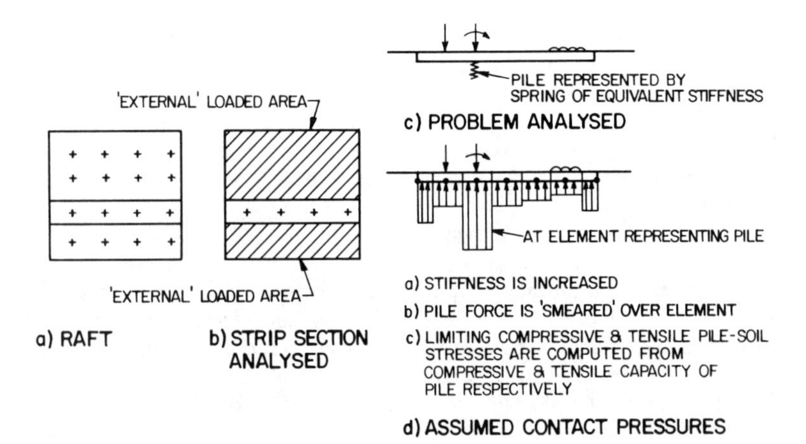

FIG. 2. Simplified Model for Slab-Pile Interaction Analysis

the strip element within which the pile is located, and is treated as an equivalent contact pressure (Fig. 2c). The ultimate pile capacity in both compression and tension is specified, and converted into equivalent limiting contact stresses below the strip in compression and tension. In the analysis, the computed contact stresses are not permitted to exceed these limiting values.

The analysis involves computing the vertical displacement of the foundation at each element, equating this to the displacement of the soil beneath each element, and solving the subsequent equations for the contact pressures and displacements. The contact pressures and pile forces are not allowed to exceed the limiting values specified. An iterative process is necessary once one or more elements reach their limiting state.

The displacement of the strip itself is described by the simple equation of bending, which can be written in finite difference form (Poulos 1984). The vertical displacement of the soil can be computed from elastic theory, with consideration being given to interaction among elements of the strip, interaction among piles, the effect of piles on strip elements, and the effect of strip elements on piles.

The analysis has the capacity to handle a strip or piled strip subjected to concentrated and/or uniformly distributed loading, to moment loading, and to externally imposed vertical *free field* soil movements. It can also consider a strip or piled strip located adjacent to one or more loaded areas or concentrated loadings. This can be used to consider the effect of a new structure on an existing foundation, and for the present purposes, a strip section within a raft or slab foundation. As indicated in Fig. 2b, the portions of the foundation outside the strip section are treated as external loaded areas which apply additional "free-field" soil movements to the strip. These free-field soil movements are computed from elastic theory, and to allow approximately for the effect of piles, the applied loads on piled areas are assumed to be applied at a depth below the soil surface of two-thirds of the pile length.

In the application of the analysis to an engineered pile-slab system, the settlements which would occur without the presence of piles are imposed as free-field soil movements on the piled-slab system. Differing pile lengths and diameters along the slab section can be considered, and the analysis evaluates the distributions of settlement, rotation, bending moment, shear and contact pressure beneath the raft, and the loads developed in the piles. The pile characteristics and spacing can be adjusted by trial and error until a uniform (or nearly uniform) distribution of additional settlement is obtained.

The analysis described above has been implemented in a computer program GASP (Geotechnical Analysis of Strip with Piles). The accuracy of the analysis has been examined by Poulos (1991) and found to be reasonable, particularly for displacement prediction.

Application to Raceway Grandstand

Introduction

A practical application of the concept and analysis outlined above has been made for a grandstand at the Albion Park Raceway in Brisbane Australia. Settlement measurements of the grandstand extension indicated continuing settlements of the order of 1.5 mm per month near the centre of the grandstand slab and 0.5 mm near the edges. The rate of settlement had not decreased with time to the extent which had been anticipated, and problems had been encountered with drainage of surface water after rainfall.

Because of concerns about possible distress of the grandstand structure supported on the foundation slab, an assessment was made of the settlement problem and a solution was developed for arresting future additional differential settlements. The approach adopted is described below.

Geotechnical Information

Fig. 3 shows a general plan of the site and the location of the three boreholes in the above investigations. A summary of the geotechnical data obtained from these investigations is shown in Fig. 4. All three boreholes showed the presence of soft to very soft, highly plastic marine clay with a high organic content. The deepest borehole was terminated at about 27.5 m and showed no evidence of stiffer underlying strata. Laboratory triaxial tests and field vane shear tests indicated undrained shear strengths in the range of 5 to 22 kPa. The extremely low shear strengths encountered at depth in all three holes suggested that the clay may have been underconsolidated at the time of investigation, i.e. still consolidating under its own weight, or else very sensitive to disturbance. There was clear evidence from the investigations that significant settlements of the structures built on the site should be anticipated.

Time-Settlement Data for Grandstand Extensions

Settlement measurement points were established at several locations in December 1982 and January 1983. Time-settlement relationships for various points are plotted in Fig. 5 on a log time scale. The locations of the points are indicated in Fig. 3. These relationships show settlement between the beginning of 1983 and the end of 1990 of nearly 320 mm near the centre of the grandstand extensions, and between about 90 and 180 mm near the corners. The shape of the time-settlement relationship suggested that soil consolidation was continuing and that settlements were likely to continue to develop for several years to come.

Approximate contours of the total settlement measured between January 1983 and December 1990 are shown in Fig. 6. These contours demonstrate the anticipated *dishing* behaviour of relatively flexible foundations subjected to relatively uniform loading, with maximum settlements near the centre and minimum settlements near the corners of the slab.

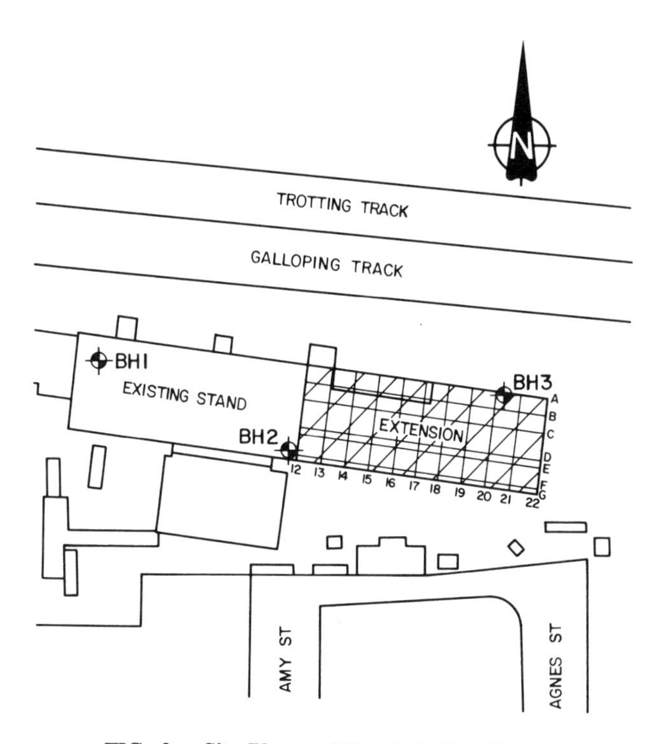

FIG. 3. Site Plan and Borehole Location

Estimated Final Settlements of Slab

The available time-settlement data were used to estimate the final settlements of the foundation (excluding long-term creep), using the method developed by Asaoka (1978). For typical points, D18 (near the centre) and F22 (near the corner), the following estimates of final settlement were obtained:

D18: 450 mm
F22: 200 mm

The additional settlements at these points (subsequent to 12/12/90) were therefore estimated to be about 143 mm and 99 mm respectively. Consequently, it was expected that, over the next few years, both total and differential settlements would increase.

In view of the drainage problems and some structural cracking which had already developed in the grandstand extension, and which were believed to arise from differential settlements, it appeared highly desirable to try and avoid significant increases in differential settlements in the future.

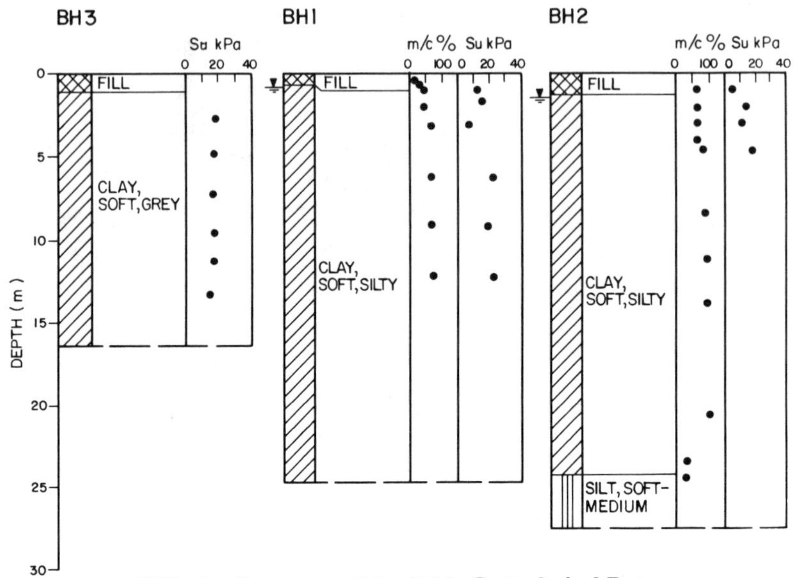

FIG. 4. Summary of Available Geotechnical Data

Among the solutions considered was the engineered remedial pile concept, and the procedure used to investigate this option is described below.

Analysis of Feasibility of Engineered Remedial Pile Option

Because of the considerable depth of the soft clay layer responsible for the majority of the settlement of the grandstand extension, it was not considered feasible to adopt a conventional underpinning solution which would attempt to completely stop all future settlements. Instead, an approach was adopted in which the use of engineered remedial piles was investigated, the objective being to minimise the future **differential** settlements while allowing some additional total settlements to occur.

In making a feasibility assessment of this option a central longitudinal section of the foundation slab, 72 m long by 4.56 m wide was analysed. For the soil. values of Poisson's ratio of 0.35 and Young's modulus of 1.4 MPa were adopted. The latter value was derived by back-calculating the modulus to give a final settlement of 450 mm (as estimated from Asaoka's method) at the centre of the slab.

The ultimate bearing pressure below the slab was taken as 110 kPa, as estimated from bearing capacity theory. Bored piles of 250 mm diameter, at centre-to-centre spacings of 4 m, were introduced, with lengths varying between 5 m near the edges to 15m near the centre. The ultimate load capacity of the piles was assessed on the basis of an ultimate skin friction of 12 kPa in the upper 5 m and 20 kPa below 5 m, and an ultimate end-bearing pressure of 110 kPa. The pile-slab

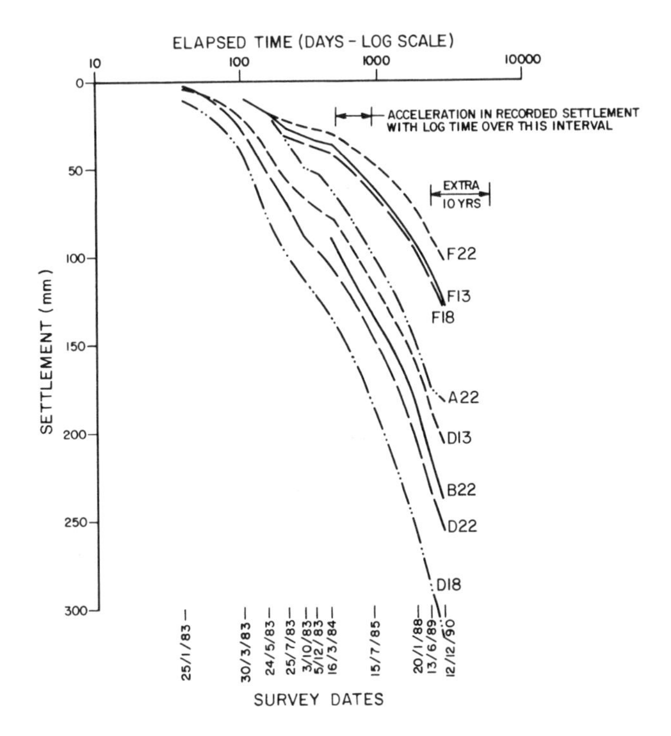

FIG. 5. Time-Settlement Data

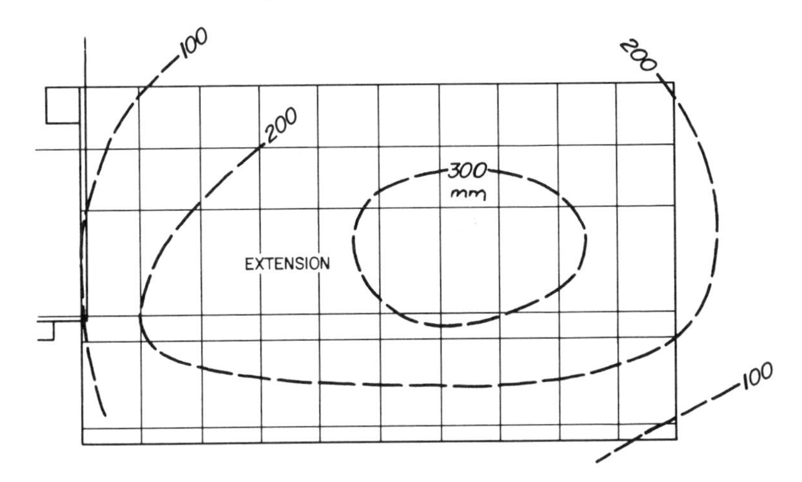

FIG. 6. Contours of Total Settlement Between 1/83 and 12/90

system was subjected to the estimated **additional** settlements expected to occur beneath the slab. For simplicity, this additional settlement profile was assumed to be parabolic, varying from 150 mm at the centre to 100 mm at the edges. These settlements were applied as *free-field* vertical soil movements to the strip section in the analysis, using the program GASP.

Figs. 7, 8 and 9 show the results of the analysis. Fig. 7 shows the system analysed and the corresponding loads developed in each pile. The pile load, as a proportion of ultimate pile capacity, is also shown. The largest loads were developed in the central piles, and in this case the ultimate load capacity of the seven central piles was reached. The pile load progressively decreased as the edge was approached. It should be emphasised that, since the piles were meant to act as settlement-reducing foundation elements, there was no difficulty in having some of the piles loaded to their ultimate capacity.

The computed profile of additional deflection of the piled-slab system is shown in Fig. 8, together with the additional settlement profile which would occur if no piles are added. The effect of the piles was clearly to reduce both the maximum settlement and, in particular, the differential settlement. For the pile configuration analysed, the maximum additional settlement was reduced from 150 mm to 107 mm, while the maximum additional differential settlement was reduced from 50 mm to 12 mm. By progressive adjustment of the pile length, it would be possible to reduce the additional differential settlements to a value approaching zero.

The computed distribution of additional bending moment in the slab is shown in Fig. 9. Significant additional positive moments were developed in the outer third of the slab, while small negative moments were developed near the centre. These additional moments would be additional to the bending moments existing in the slab prior to installation of the piles.

No attempt was made to refine the preliminary calculations by exploring different combinations of pile length, diameter and spacings, nor were other strip sections of the slab analysed. It was felt that the analysis confirmed the feasibility of the proposed foundation treatment involving the use of engineered remedial piles to arrest the development of future differential settlements, while not attempting to completely inhibit the overall settlement of the grandstand.

CONCLUSIONS

The paper has described an approach to underpinning a foundation in which the primary objective is to minimise future differential settlements, rather than completely stop future overall settlements. This approach may involve the use of piles of varying length and/or diameter in order to achieve the desired *evening-out* of future settlement.

A simplified analysis has been outlined which is capable of analysing the interaction between a strip section of a slab, the supporting soil, and piles installed below the slab. This analysis has been applied to the case of a slab supporting a grandstand on a deep soft clay deposit. It is demonstrated that a potential future additional differential settlement of about 50 mm can be reduced to a few millimetres by the use of relatively widely-spaced 250 mm diameter bored piles

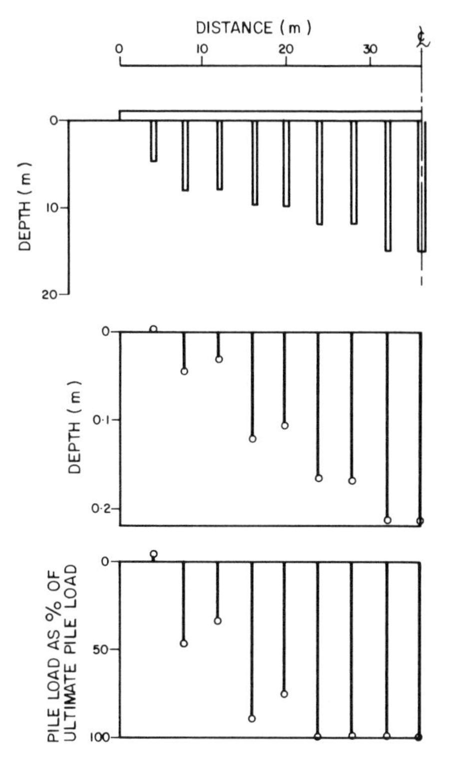

FIG. 7. Pile Arrangement and Computed Pile Loads

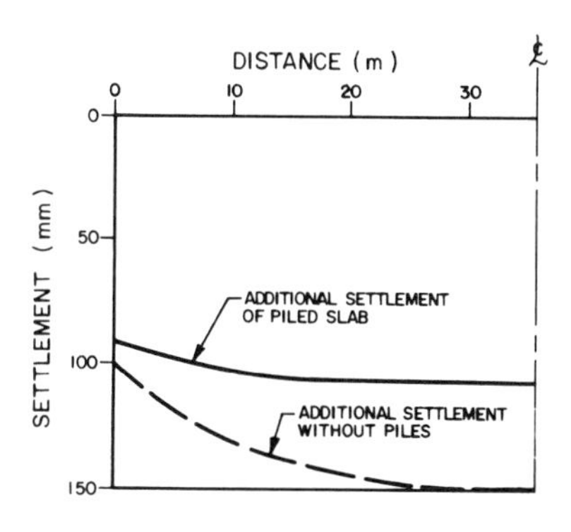

FIG. 8. Computed Profiles of Additional Settlement

whose length varies from 5 m near the edges to 15 m near the centre of the slab. At the same time, the maximum future settlement is reduced from about 150 mm to about 107 mm. In the proposed scheme, several of the piles will reach their ultimate load capacity, but nevertheless serve their primary purpose of settlement reduction.

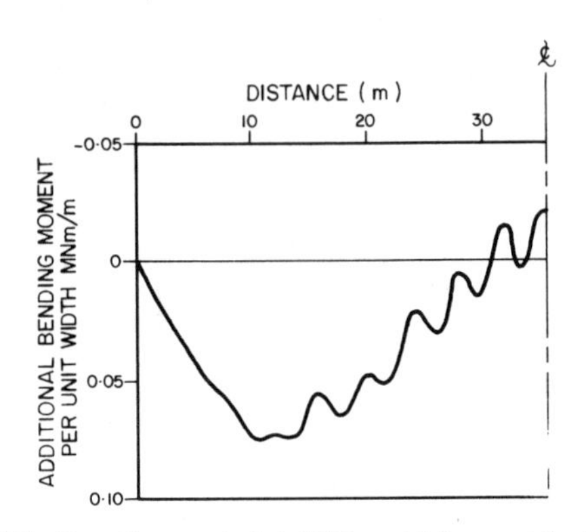

FIG. 9. Computed Additional Moment in Slab

ACKNOWLEDGMENTS

The author is grateful to the Albion Park Raceway Trust for permission to publish this paper, to Mr. D. Lewis of Lewis WCBM Pty Ltd for his advice in relation to several aspects of the projects, and to Dr. P. Shaw and Mr. P. J. N. Pells of Coffey Partners International Pty Ltd Brisbane for their constructive comments.

APPENDIX - REFERENCES

Asaoka, A. (1978). "Observational procedure of settlement prediction." *Soils and Foundations*, 18(4), 87-101.

Ergun, M. V., and Uygurer, C. (1991). "Excessive settlement and underpinning of a raft." *Proc. 10th Eur. Conf. Soil Mechanics and Foundation Engineering,* Florence, 1, 389-392.

Makarchian, M., and Poulos, H. G. (1994) "Underpinning by piles: A numerical study." *Proc. 13th ICSMFE*, New Delhi, 4, 1467-1470.

Poulos, H. G. (1991). "Analysis of piled strip foundations." *Comp. Methods and Advances in Geomechanics*; Balkema, Rotterdam, The Netherlands, 1, 183-191.

Poulos, H.G. (1984). "Parametric solutions for strip footings on swelling and shrinking soils." *Proc. 5th Int. Conf. on Exp. Soils*, Adelaide, 149-153.

Thorburn S., and Hutchison, J. F. (1985). *Underpinning*, Surrey Univ. Press, London, England.

MEASURED DOWNDRAG ON SEVEN COATED AND UNCOATED PILES IN NEW ORLEANS

Randall K. Bush,[1] Associate Member, ASCE,
and Jean-Louis Briaud,[2] Fellow, ASCE

ABSTRACT: A site outside of New Orleans, Louisiana was selected as part of a research effort to study the phenomenon of downdrag on bitumen-coated piles. Two steel pipe piles, two timber piles, and three prestressed, precast concrete piles were instrumented, installed and monitored. Some of the piles were coated with bitumens and some were not. The soil conditions were determined prior to pile installation by means of soil borings and cone penetrometer soundings. A surcharge layer of sand was placed on the site to facilitate settlement of the soil and the subsequent downdrag on the piles. All piles were load tested in both compression and tension, and monitored periodically throughout the test period of two years. The soil response was monitored by means of piezometers and a series of settlement points. The results of the settlement study and its effect on downdrag are presented.

INTRODUCTION

A field testing program was performed as part of a study involving downdrag on bitumen-coated piles. One site for this field testing program was located in New Orleans, Louisiana. The Louisiana Department of Transportation and Development (LDOTD) was planning a bridge project which would require constructing an approach embankment over an area of soft clay. The surcharge from the approach embankment would cause a large amount of settlement in the soft

---

[1] Project Engineer, Law Engineering, Inc., 5500 Guhn Road, Houston, TX 77040.
[2] Buchanan Professor of Civil Engineering, Texas A&M University, College Station, TX 77843-3136.

clays, possibly resulting in significant downdrag loads on the piles. As part of the downdrag study, an instrumentation program was developed to monitor downdrag on the piles and the soil behavior, including settlement. The data presented in this paper was collected after conducting load tests. Some of the test piles were coated with bitumen and some were not. The purpose of this paper is to present the data collected.

Site and Soil Conditions

The site was located east of the downtown area of the city of New Orleans near the intersection of State Highway 47 (Paris Road) and U.S. Highway 90 (Chef Menteur) as shown in Fig. 1. A section of elevated roadway, to become part of Interstate 510, was to be constructed parallel to State Highway 47. The specific location of the pile test area is shown in Fig. 2 with the detailed locations of the soil borings, piles, and soil instrumentation presented in Fig. 3.

A special test site was constructed that would not interfere with the construction progress, since it was to be a long-term study. A layer of sand approximately 2.1 m thick was placed as a surcharge on the test area to model an approach embankment. The site was located in a reclaimed marsh area with the elevation of the natural ground at approximately -1.3 m. The soil at the site is a normally consolidated soft clay interbedded with thin layers of sands and silts. The clays become medium stiff at a depth of about 16 m below the ground surface and stiff below about 24 m.

Various laboratory and field tests were performed. The laboratory tests included various classification tests, undrained shear strength and consolidation. Additionally, a Cone Penetration Test (CPT) sounding was performed. Figs. 4 and 5 present the results of the tests pertinent to this study.

Piles

A total of eight test piles and four reaction piles were included at the test site. The test piles consisted of two closed-end steel pipe, three Precast, Prestressed Concrete (PPC), and three treated timber piles. The piles were all 21 m long. Some of the piles were coated with a bitumen and some were not. The pile dimensions are presented in Table 1. Some pertinent physical properties of the bitumens used on the piles are presented in Table 2. The physical properties of Gulfseal Mastic, as supplied by the manufacturer, were of the base asphalt and not the composite mixture, and are not included in Table 2.

Instrumentation

The piles were instrumented with seven levels of strain gages, for measurement of axial strain, and a tell-tale near the point. At each level, two gages were mounted 180 degrees apart around the circumference to compensate for bending and for redundancy.

The soil instrumentation was designed specifically for the soft soil conditions at the site. The locations of soil instrumentation were presented in Fig. 3. Three pneumatic transducers (piezometers) were installed at depths of 6.1, 15.2, and

24.4 m to monitor the in situ pore water pressures in the vicinity of the test piles. The piezometer readings were obtained periodically after their installation, after the pile driving operations, during load testing, and subsequently as part of the long-term monitoring program.

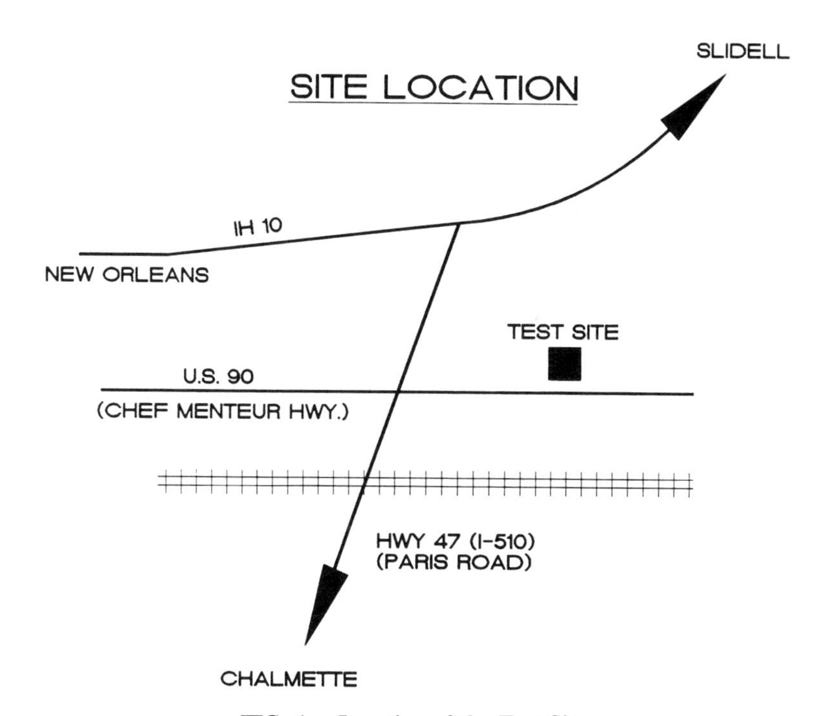

FIG. 1. Location of the Test Site

A new system for monitoring large magnitudes of settlement in soft soils was utilized at this site. A Magnetic Extensometer System developed by Geo Group, Inc., Gaithersburg, Maryland, was used. Settlement was assessed by monitoring the location of magnetic targets which were positioned over a near-vertical access tube. The depths of the magnetic targets and typical installation profile are presented in Fig. 6. The location of the magnetic targets was accomplished by passing a reed switch probe through the access tube. When the probe enters the magnetic field generated by a magnetic target, an audible signal is emitted at ground level. Measurements made with an integral steel tape were then related to a convenient survey datum.

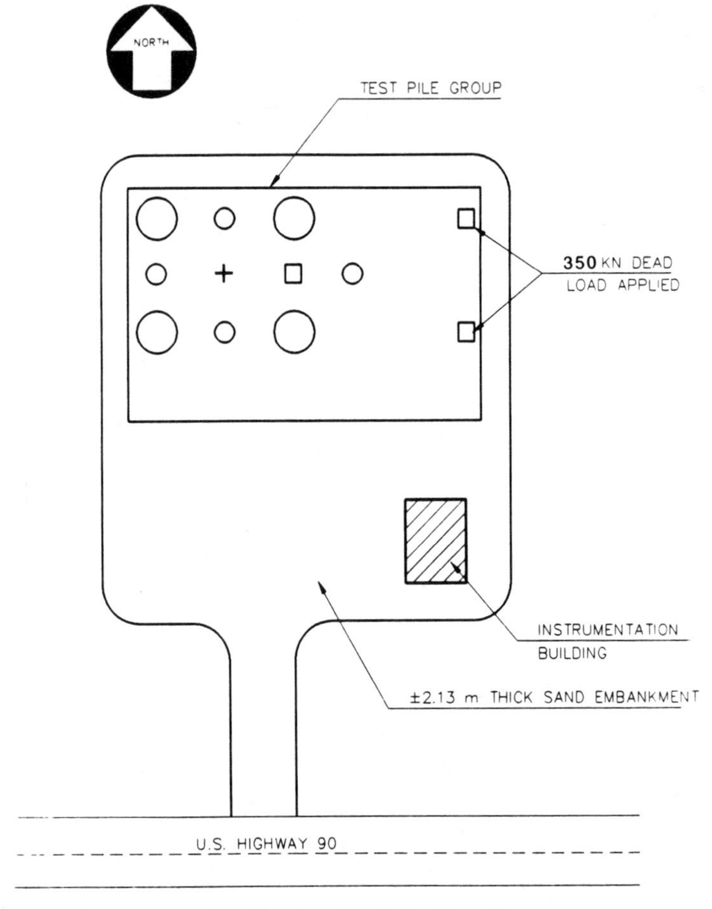

FIG. 2. Details of the Test Site

FIELD MONITORING

Pile top and tip movements, pile strain gages, piezometers and soil settlement instrumentation were all monitored over a period of approximately two years. A survey elevation was obtained on the pile top and tell-tale immediately after the load testing program. The difference between these initial survey elevations and subsequent survey elevations allowed the pile movements to be determined. The top movement of the piles over time is presented in Fig. 7. This figure indicates that the piles are settling, but less than the soil. Piles CPM and CPI were placed under a constant dead load of approximately 350 kN to model piles in service.

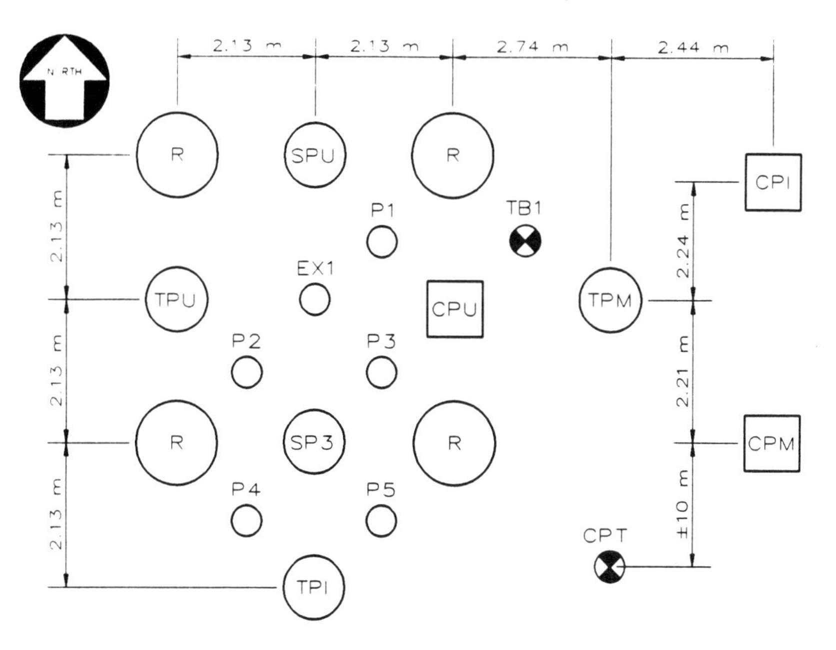

LEGEND

P1 – PIEZOMETER 1 (20 FT. DEPTH)
P2 – PIEZOMETER 2 (50 FT. DEPTH) STOPPED WORKING 7/31/90
P3 – PIEZOMETER 3 (80 FT. DEPTH) STOPPED WORKING 7/27/90
P4 – PIEZOMETER 4 (45 FT. DEPTH) INSTALLED 9/5/90
P5 – PIEZOMETER 5 (70 FT. DEPTH) INSTALLED 9/6/90
EX1 – SETTLEMENT EXTENSOMETER
CPT – CONE PENETROMETER TEST
TB1 – TEST BORING 1
SPU – STEEL PIPE, UNCOATED
TPU – TIMBER PILE, UNCOATED
CPU – PPC PILE, UNCOATED
SP3 – STEEL PIPE, TYPE 3
TP1– TIMBER PILE, TYPE 1, ROOFING COMPOUND
TPM – TIMBER PILE, WATERPROOFING MASTIC
CPI – PPC PILE, U.S. INTEC BLUE COMPOUND
CPM – PPC PILE, U.S. INTEC BLUE MEMBRANE
R – REACTION PILES

FIG. 3. Detailed Layout of the Test Pile Group

All pile strain gages were also monitored periodically. These readings indicate the amount of axial load (downdrag) that has been induced on the pile due to soil settlement. Plots of downdrag load versus depth for some of the individual piles are presented in Figs. 8 through 12. The instrumentation on the timber piles did not function properly throughout the long-term monitoring. Consequently, the downdrag loads for the timber piles are not presented. The strain gages on the lower portion of piles CPM and CPI (Figs. 9 and 10) were erratic or non-functional after a period of time. The dotted lines on these figures indicate an interpreted pile tip load.

TABLE 1. New Orleans Pile Dimensions and Coatings

Pile Type	Pile Symbol	Diameter or Width, mm	Wall Thickness	Bitumen Coating
Closed-end Steel Pipe	SPU	324 (O.D.)	10 mm	None
	SP3			Trumbull Type 3
Square Precast, Prestressed Concrete	CPU	356	N/A	None
	CPI			Intec Blue Compound
	CPM			Intec Blue Membrane
Timber	TPU	432 (Top) 229 (Tip)		None
	TPM			Gulfseal Mastic
	TP1			Trumbull Type 1

TABLE 2. Bitumen Properties from the Manufacturer and from TAMU

Bitumen	Viscosity, Poise	Penetration at 25°C	Softening Point, °C
Intec Blue	6.52×10^5	27-35	153-156
Trumbull Type 1	1.50×10^4	18-60	57-66
Trumbull Type 3	2.00×10^4	15-35	85-96

A plot of soil settlement versus depth at various time intervals is presented in Fig. 13. The surface settlement as a function of time is presented in Fig. 14. The settlement shown in these plots begins at the same time that the long-term monitoring of the piles began, i.e., immediately after load testing.

As can be seen from the concrete piles, the downdrag loads were actually higher on the coated piles than for the uncoated pile. The U.S. Intec products were stiff bitumens and the higher loads could be attributed to the increased surface area from the coating. The Type 3 Roofing Compound on pile SP3 appeared to provide the most reduction in downdrag; about 50 percent. It is the researcher's opinion that the softer bitumens that were applied to the timber piles would have provided the best reduction in downdrag. However, due to the instrumentation not functioning properly throughout the monitoring period, this can not be verified.

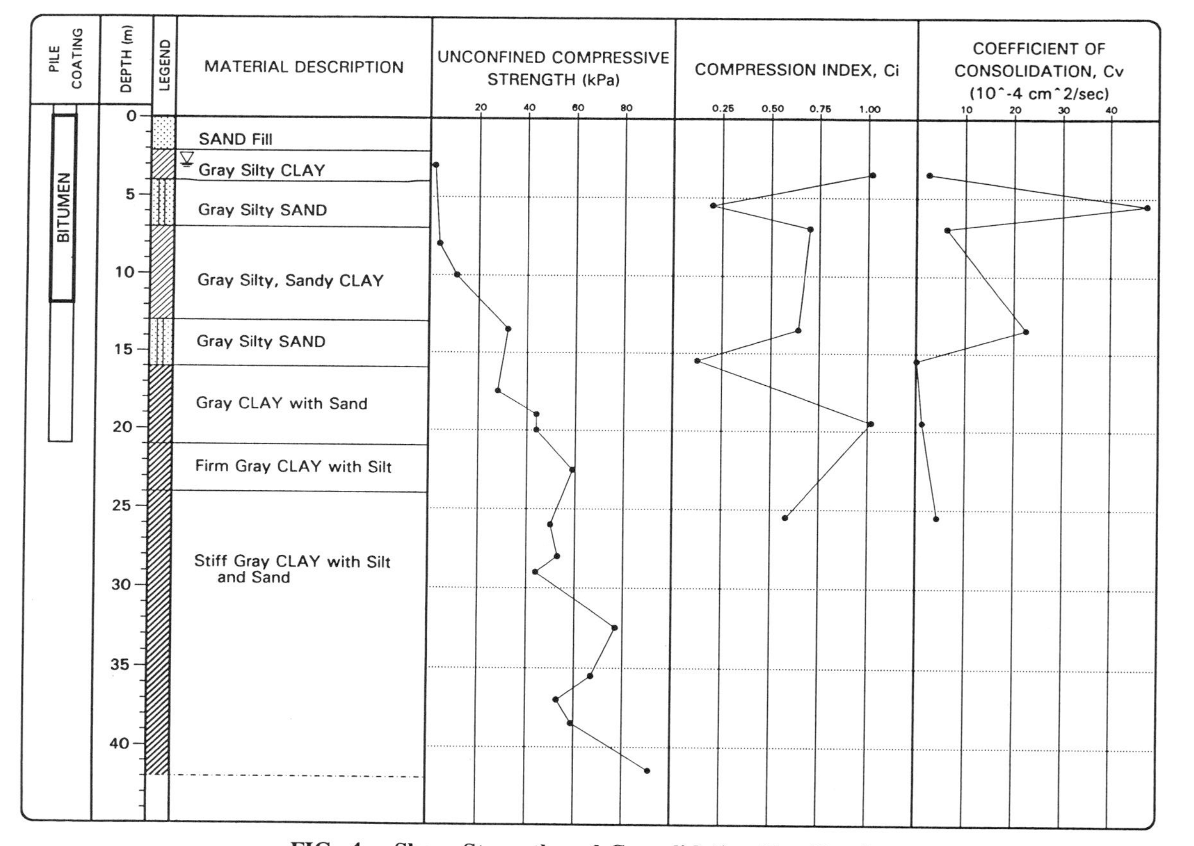

FIG. 4. Shear Strength and Consolidation Test Results

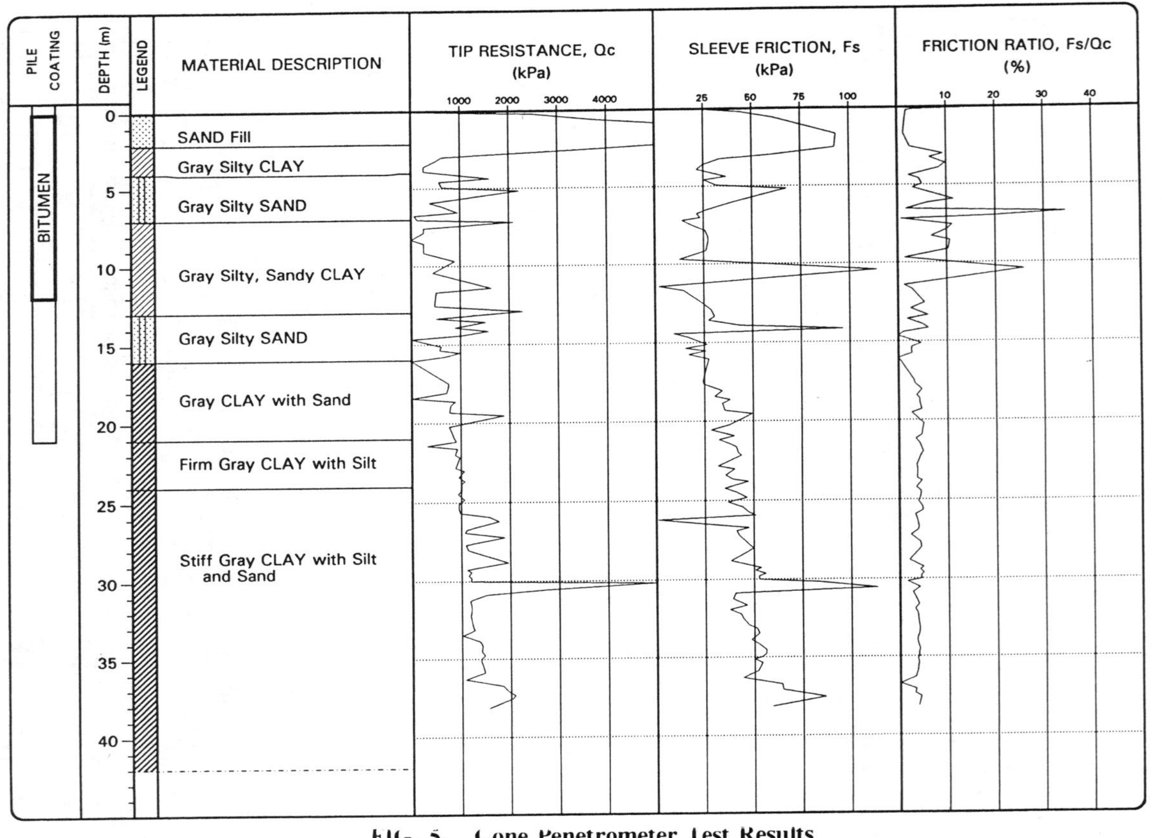

FIG. 5. Cone Penetrometer Test Results

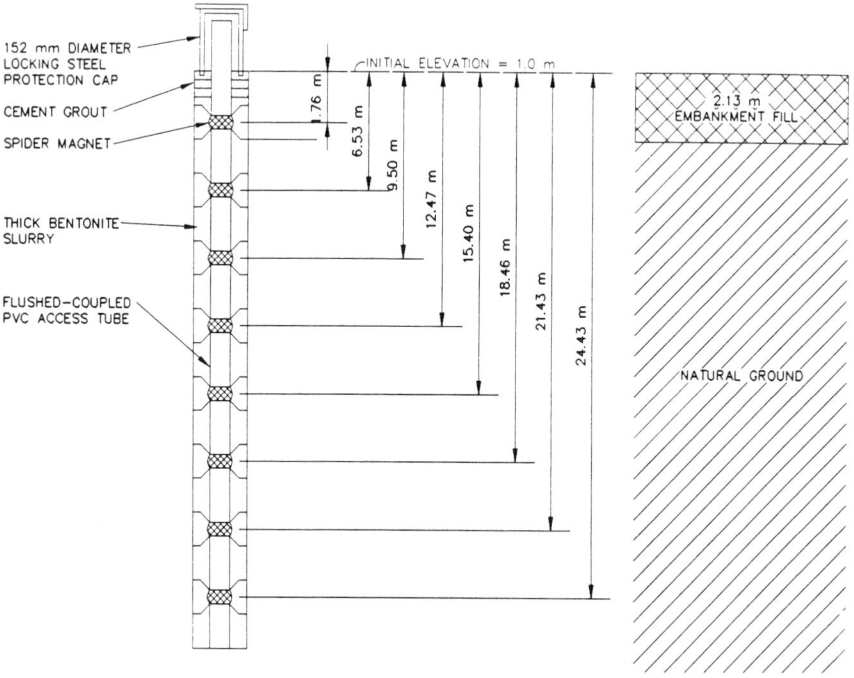

FIG. 6. Magnetic Extensometer Settlement System

At the end of the monitoring period, all of the piles were extracted and bitumen coatings examined. All of the coatings were found to be intact on the piles.

CONCLUSIONS

Settlement of the soft soils at this site induced downdrag forces on the piles. As can be seen from the downdrag plots, the bitumens used were moderately to not effective in reducing the downdrag forces at this site. The downdrag forces varied from about 200 to 400 kN on the piles. For more details on the results the reader is referred to the research report (Bush et al. 1991).

ACKNOWLEDGEMENTS

This work was sponsored by the American Association of State Highway and Transportation Officials (AASHTO), in cooperation with the Federal Highway Administration (FHWA), and was conducted in the National Cooperative Highway Research Program (NCHRP) which is administered by the Transportation Research Board (TRB) of the National Research Council (NRC). The NCHRP was lead by George Machan.

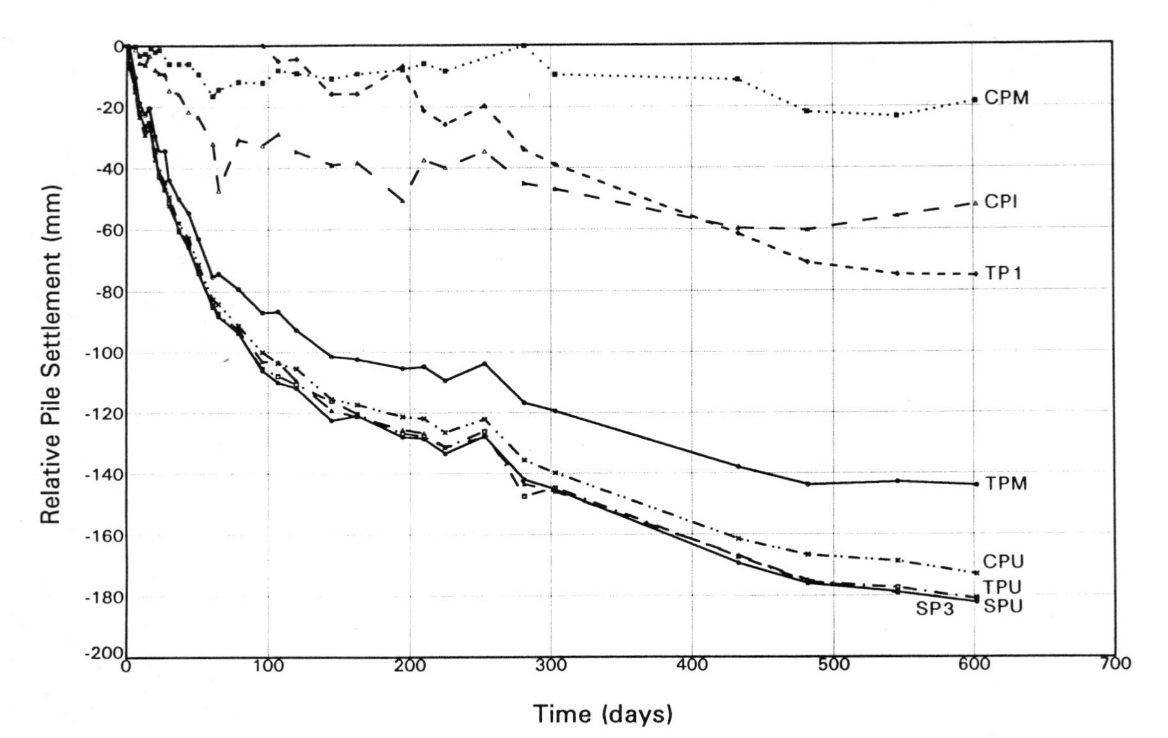

FIG. 7. Pile Settlement Relative to the Soil

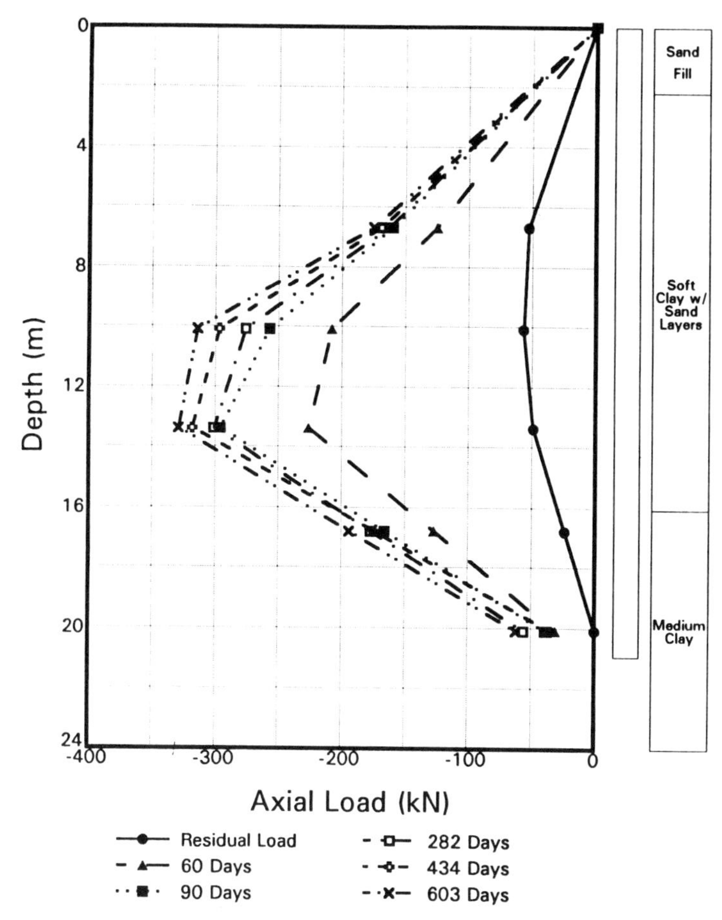

FIG. 8. Pile CPU Downdrag Loads

Many individuals and companies have contributed time, money, materials, and services to make this project possible. They include Louisiana DOT (Mark Morvant, J. B. Esnard), FHWA (Carl Ealy, Al Dimillio), GeoResource Consultants (Eric Ng, Steve Tsang), Pittman Corporation (Albert Pittman), National Timber Pile Council (Jim Graham), and S. K. Whitty & Co., Inc. (Henry and Steve Whitty). At Texas A&M University many individuals have participated. They include Larry Tucker, Rajan Viswanathan, Sangseom Jeong, Zaid Al Gurgjia, and Mohamed Quraishi.

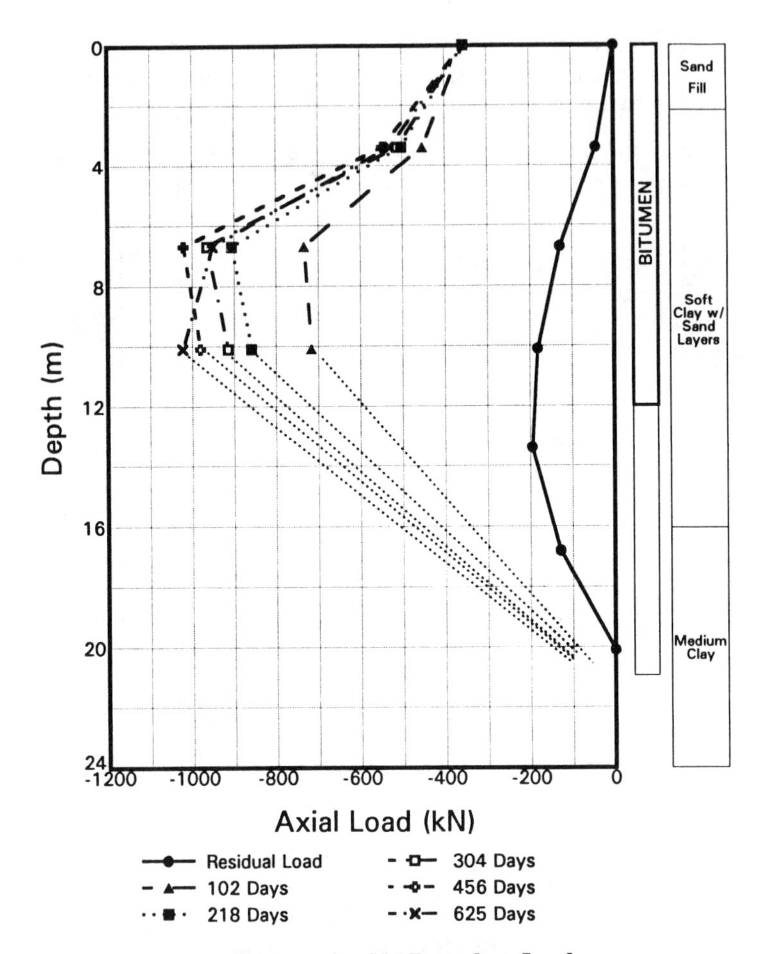

FIG. 9. Pile CPI Downdrag Loads

APPENDIX - REFERENCES

Bush, R. K., Viswanathan, R., Jeong, S., and Briaud, J.-L. (1991). "Downdrag on bitumen-coated piles," *Preliminary Report*, NCHRP, Washington, D. C.

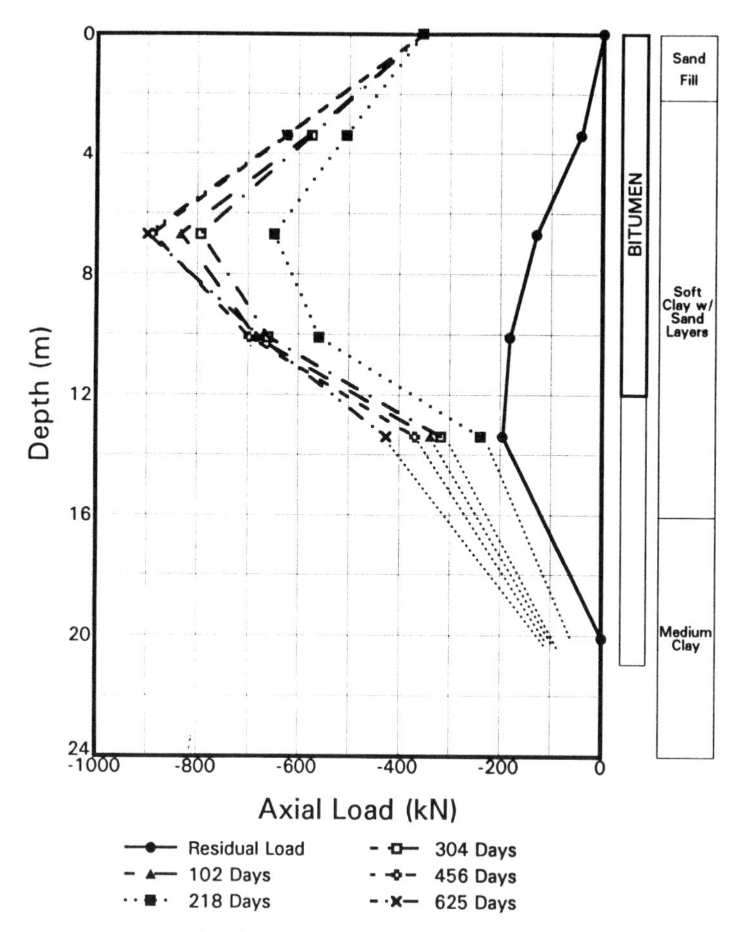

FIG. 10. Pile CPM Downdrag Loads

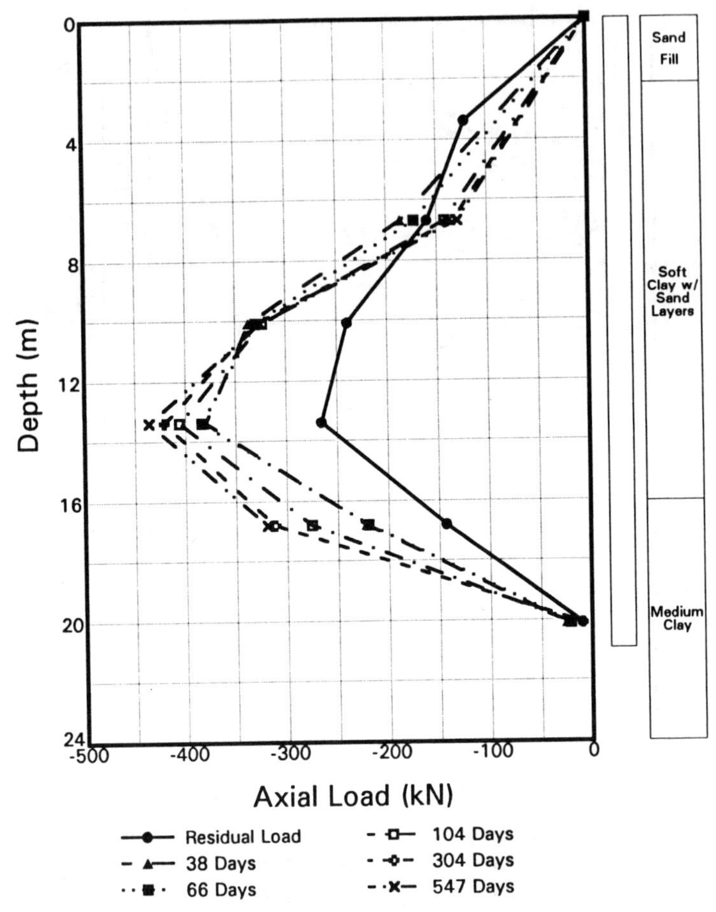

FIG. 11. Pile SPU Downdrag Loads

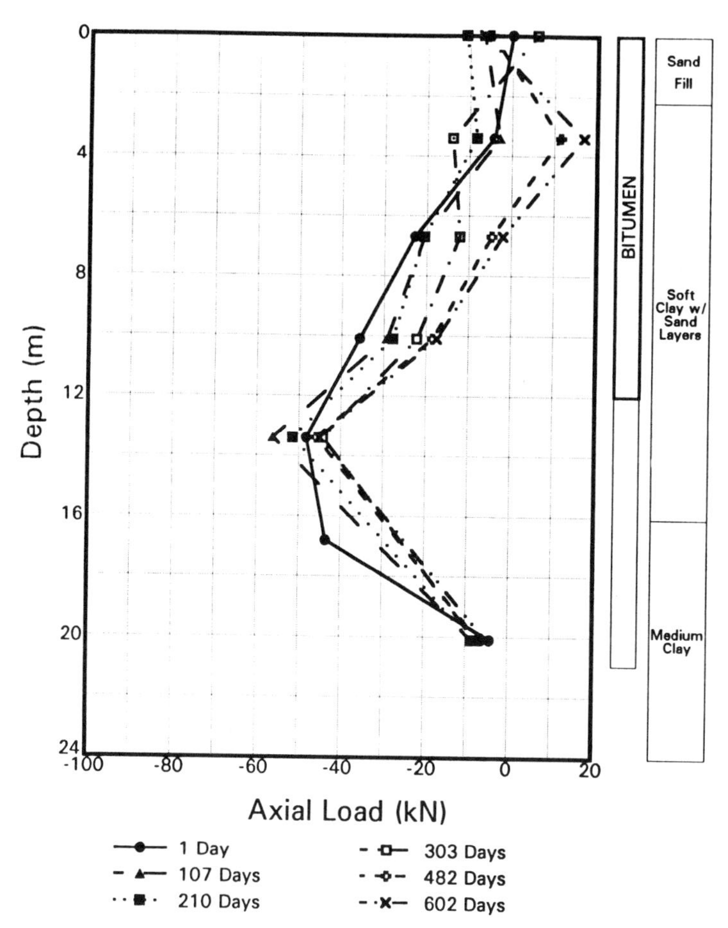

FIG. 12. Pile SP3 Downdrag Loads

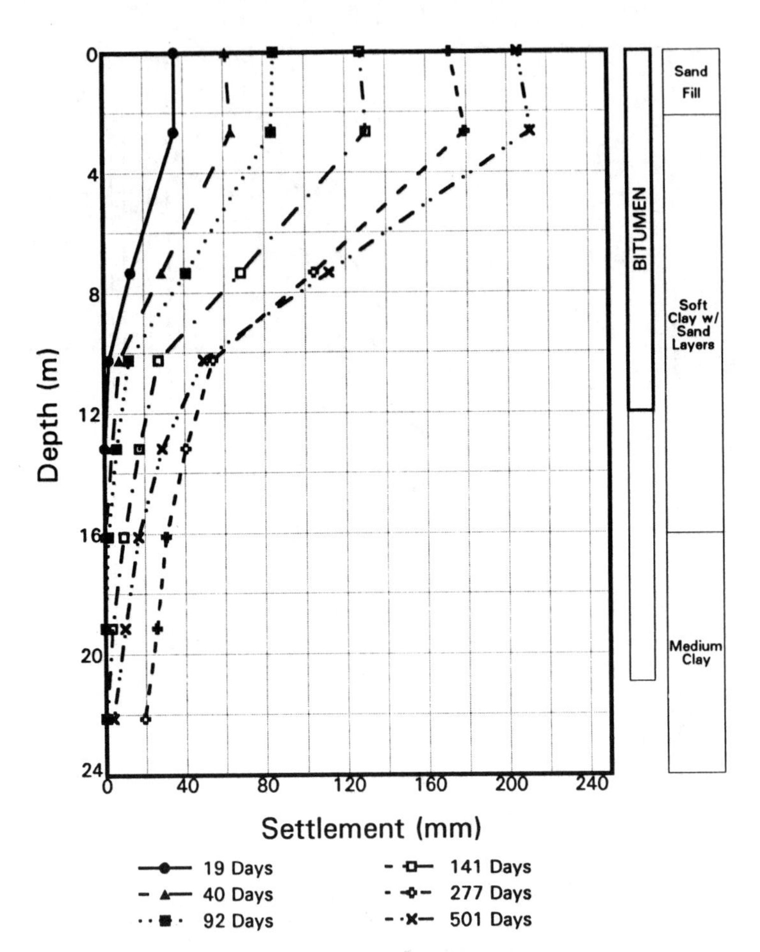

FIG. 13. Soil Settlement versus Depth

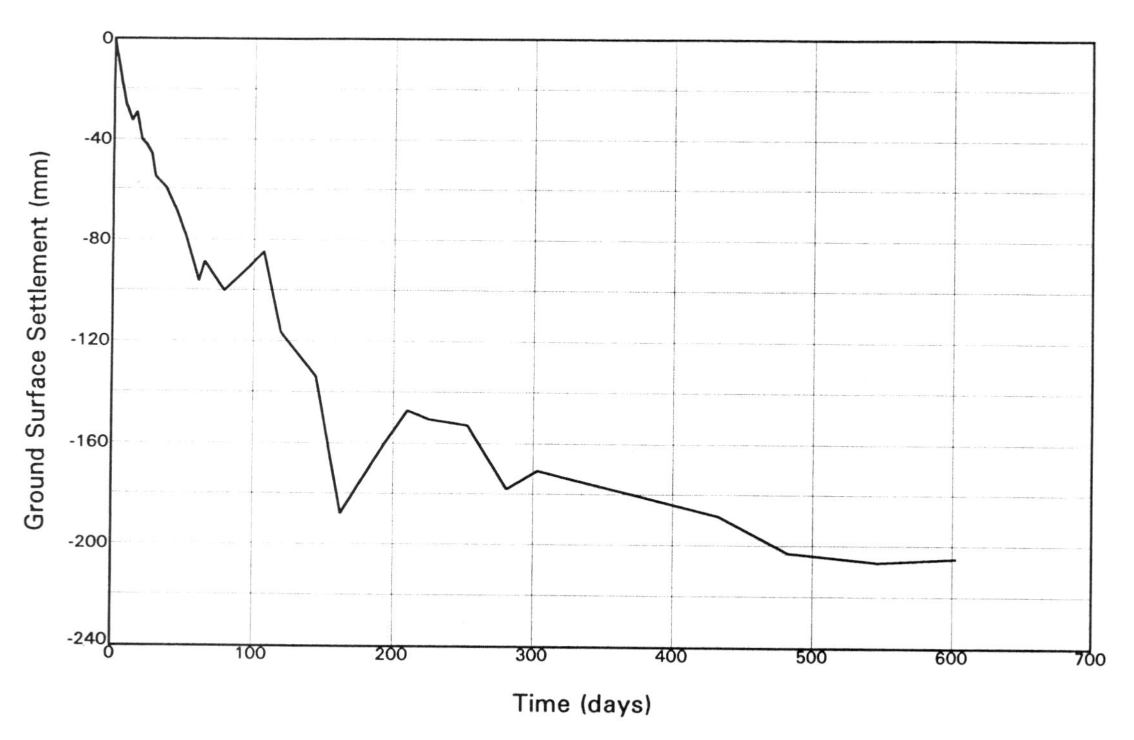

FIG. 14. Ground Surface Settlement over Time

Comparison of Settlement Predictions for Single Piles in Sand Based on Penetration Test Results

James B. Nevels, Jr.,[1] Member, ASCE,
and Donald R. Snethen,[2] Fellow, ASCE

ABSTRACT: The analysis of settlement for driven piles in cohesionless soils is a complex soil structure interaction problem. The total settlement for such a pile is generally considered to be the sum of the following three components (Vesic 1977):

1. Settlement due to axial deformation of the pile shaft
2. Settlement of the pile point caused by load transmitted at the pile tip
3. Settlement of the pile point caused by load transmitted along the pile shaft.

A recent pile load test program was conducted to assist in the design of the foundation for a new bridge over the Cimarron River in northwest Oklahoma. The load tests were conducted on a 0.66 m closed-end steel pipe pile and a 0.61 m octagonal prestressed concrete pile. Prior to and following the pile load tests, a number of in situ penetration tests were conducted at the site. They included: Standard Penetration Test (SPT), mechanical cone penetration test (MCPT) and electric cone penetration tests (ECPT). This study compares the total settlement and components of settlement measured during the pile load test on the two types of pile with that estimated by the various penetration test results. The Coyle method of analysis was used for the SPT data, and the Verbrugee method was used for the mechanical and electric cone data. This study reports the details

[1] Engineer, Materials Division, Oklahoma Dept. of Transportation, Oklahoma City, OK 73105-3204.

[2] Professor, School of Civil and Environmental Engineering, Oklahoma State Univ., Stillwater, OK 74078-0327.

of the site investigation and in situ penetration tests, the comparisons between predicted settlements and pile load test results, and documentation of a case study of the use of pile load tests to design bridge foundations.

INTRODUCTION

Purpose and Scope

This paper uses the results of a pile load test program conducted in October 1988 to evaluate settlement predictions for single piles in sand using the results of different penetration tests (i.e., electric cone penetration test (ECPT), mechanical cone penetration test (MCPT) and Standard Penetration Test (STP)). Various methods for predicting pile settlement were applied to two test piles driven into a well-defined subsurface profile. The results of the settlement predictions were compared to the load-settlement responses for the test piles. Two piles were tested at the specific site: a 0.66 m closed-end steel pipe pile and a 0.61 m octagonal prestressed concrete pile. The purpose of the comparisons was to evaluate the accuracy of the selected pile prediction methods.

Project Description

The pile load tests were conducted in conjunction with the relocation of the U.S. Highway 64 bridge over the Cimarron River in far northwest Oklahoma. The bridge relocation is approximately 30 miles north of Woodward, Oklahoma, and 14 miles east of Buffalo, Oklahoma. The pile load tests were conducted in the alluvial valley of the Cimarron River adjacent to the proposed bridge relocation. The pile load tests were conducted under the direction of the Oklahoma Department of Transportation (ODOT) as part of a Federal Highway Administration (FHWA) Demonstration Project, Demo 66—"Design and Construction of Driven Pile Foundation" (FHWA 1989). A total of four instrumented piles were driven to complete the demonstration project: two on the west bank and two on the east bank of the Cimarron River. The focus of this paper is the results of the pile load tests conducted on the west bank at Station 999+46.

Pile Load Tests

The pile load tests included dynamic pile monitoring using a pile driving analyzer and static load testing using the FHWA mobile pile load testing frame. The static load tests were conducted according to ASTM D1143. All four test piles were instrumented with strain gages, toe load cells, earth pressure cells, pneumatic piezometers, and telltales. The type and location of the various instruments for the 0.66 m pipe pile and 0.61 m concrete pile are described elsewhere (Geo/Resources 1988). Both piles were driven their full design length with 3-foot stickup. After initial driving, each pile was restuck.

Site Geology and Soil Profile Conditions

The project site is located, stratigraphically, within the Flowerpot shale formation of the Permian age. Quaternary terrace and alluvium deposits overlay the Flowerpot shale over the site. The Flowerpot shale is approximately 122 m thick and is a reddish-brown gypsiferous silty shale in this area. The Cimarron River deposits are approximately 9 m to 15 m thick and are comprised of mixtures and layers of sand and gravel with silt or clay fines.

Subsurface conditions for the project were determined from soil and rock borings and penetration tests. Soil samples for identification and classification were retrieved using a split spoon sampler, and rock (shale) cores were recovered using an NX size core barrel with rotary wash drilling. Specific subsurface conditions at the pile load test site were defined using one continuous split spoon sampled boring (No. 202) using ASTM Method D1586, two mechanical cone penetration soundings (MCPT 1 and 2), and two electric cone penetration soundings (ECPT 1 and 2) using ASTM Method D3441. Locations of the five borings/soundings are shown in Fig. 1. Because of the extensive field exploration conducted at the site, boring 202, MCPT 1, and ECPT 2 were taken as typical and used to provide data for the various settlement prediction methods discussed later in the paper. The subsurface profile, soil descriptions, SPT N-value, soil properties, and test pile depths are shown in Fig. 2. The results of the mechanical (MCPT-1) and electrical (ECPT-2) cone penetration soundings are shown in Figs. 3 and 4, respectively. In general, the soils at the test site are alluvial river deposits consisting of intermixed beds of sands, gravels, clays, and silts. At the location of the pile load tests, the soils are classified as poorly graded and well graded sands with silt (SP-SM, SW-SM) interbedded throughout the major portion of the profile. The alluvial deposits overlay the Flowerpot shale.

Analysis of Settlement Behavior

The analysis of settlement for driven piles in cohesionless soils is a complex soil structure interaction problem. Settlement predictions for pile foundations on sand lead to the many uncertainties as noted in the literature (Vesic 1977; Poulos and Davis 1980; Prakash and Sharma 1990). The most reliable method for estimating pile settlement is the pile load test which, of course, is expensive, involved, and time consuming. The static cone penetrometer is considered a model pile when pushed into homogeneous cohesionless soils (Meyerhoff 1976). This study looks at the applicability of using the static cone penetrometer for predicting the load-settlement relationship of a driven pile. For comparison, SPT data and associated calculations are included in the analysis.

For the cone penetrometer test—mechanical or electric cone—the Laboratoire des Ponts et Chaussles (LPC) method, described in Briaud and Miran (1992), can be used to predict vertical pile capacity. An interpretation of the LPC method by Verbrugge involved a method for analyzing the load transfer characteristics of the soil based on cone penetrometer test results (Riaund and Miran 1992). The entire top-load settlement curve can be estimated knowing the load transfer curves, ultimate bearing capacity, and pile dimensions. The computer program (PILECPT)

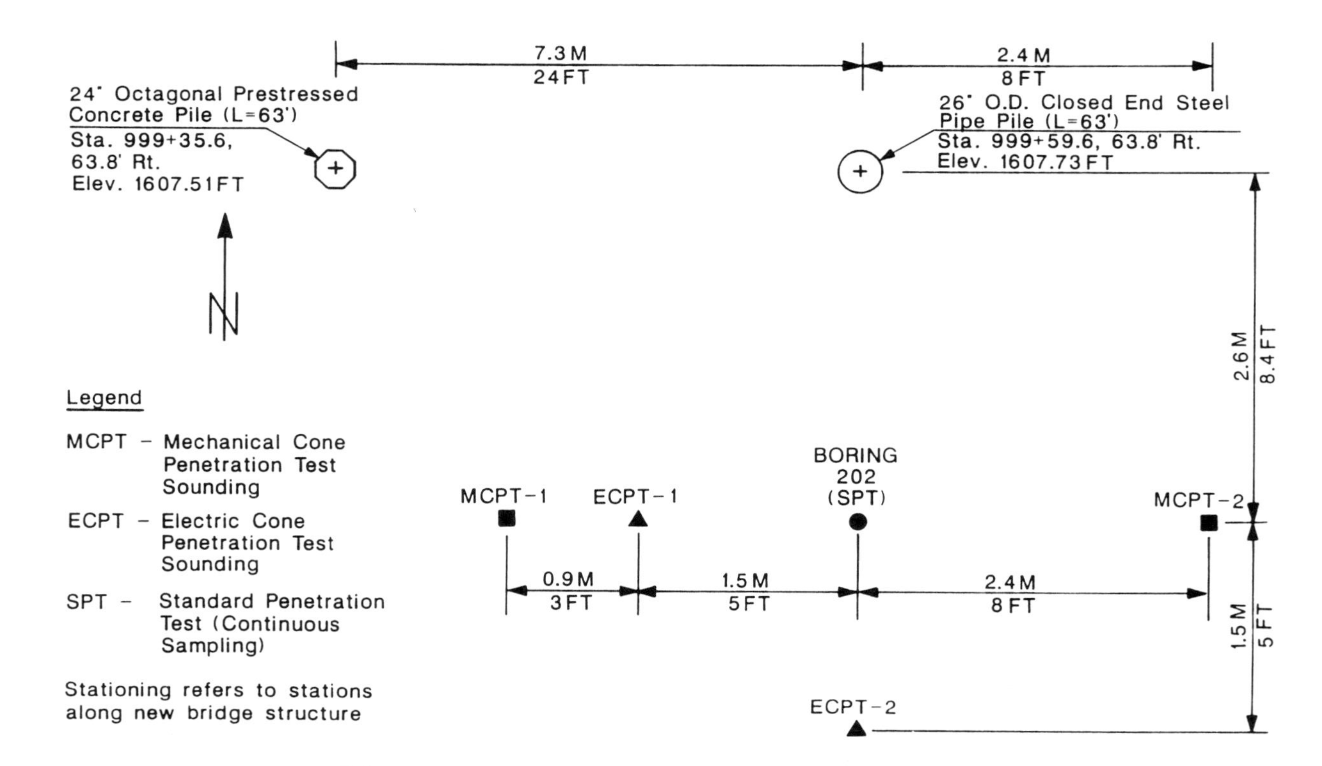

FIG. 1. Layout of Test Piles and Exploration Soundings

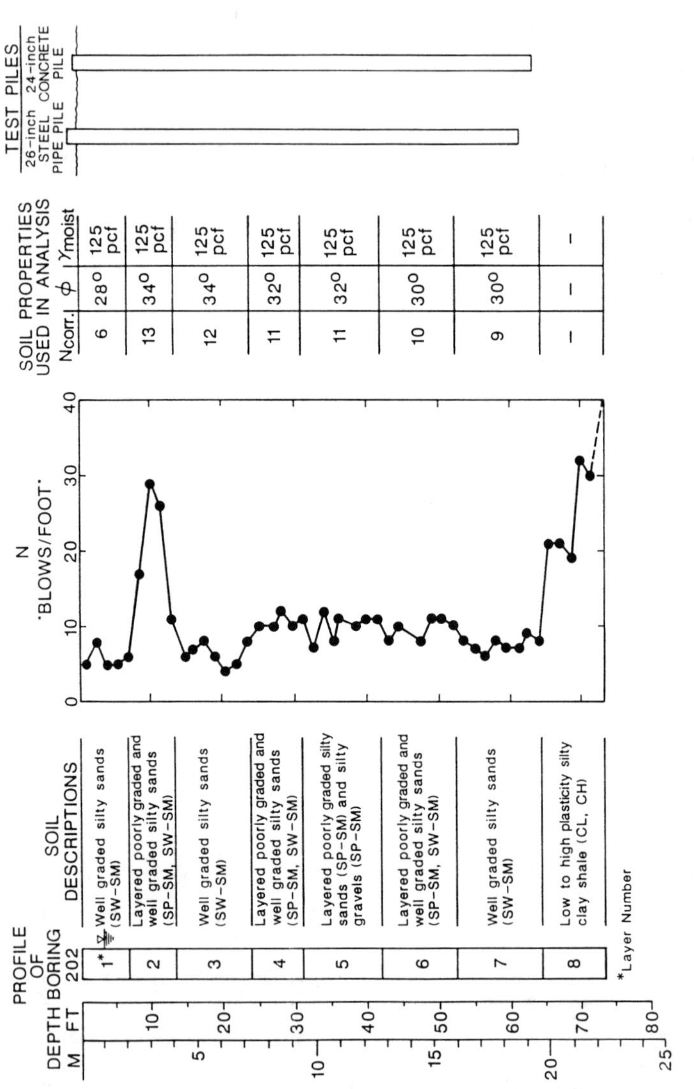

FIG. 2. Soil Profile, Standard Penetration Test Data, and Soil Properties for Boring 202

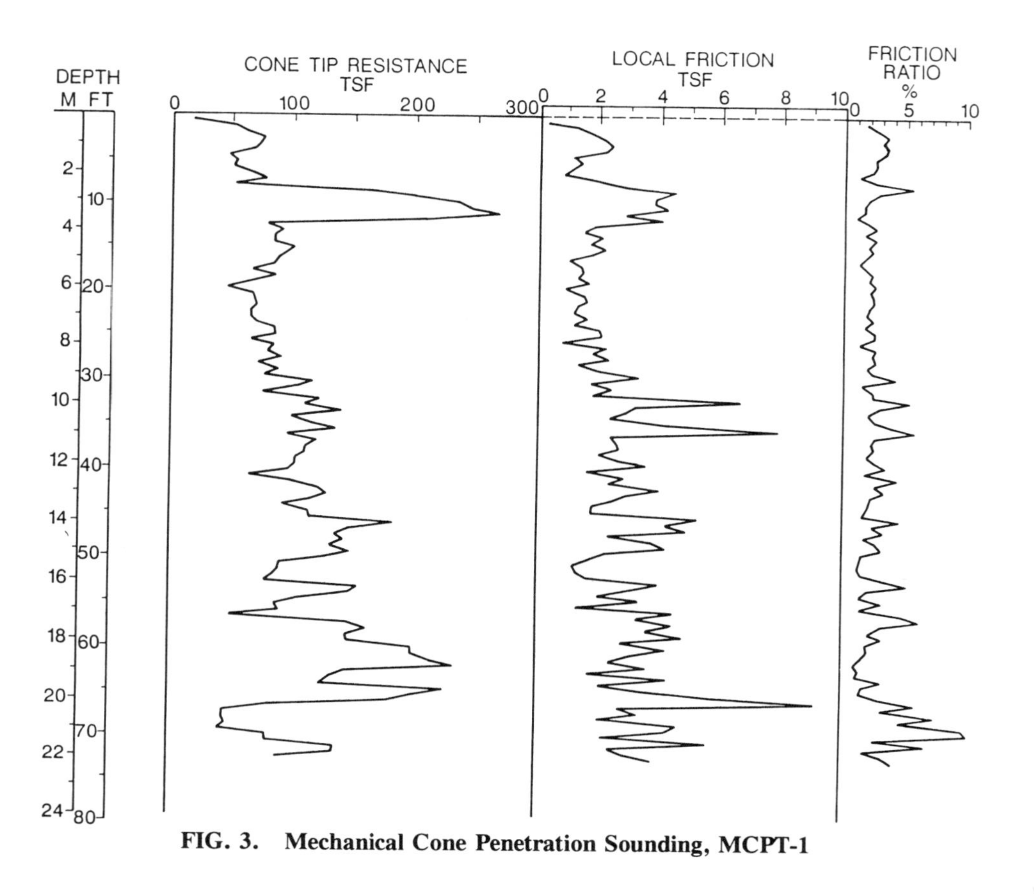

FIG. 3. Mechanical Cone Penetration Sounding, MCPT-1

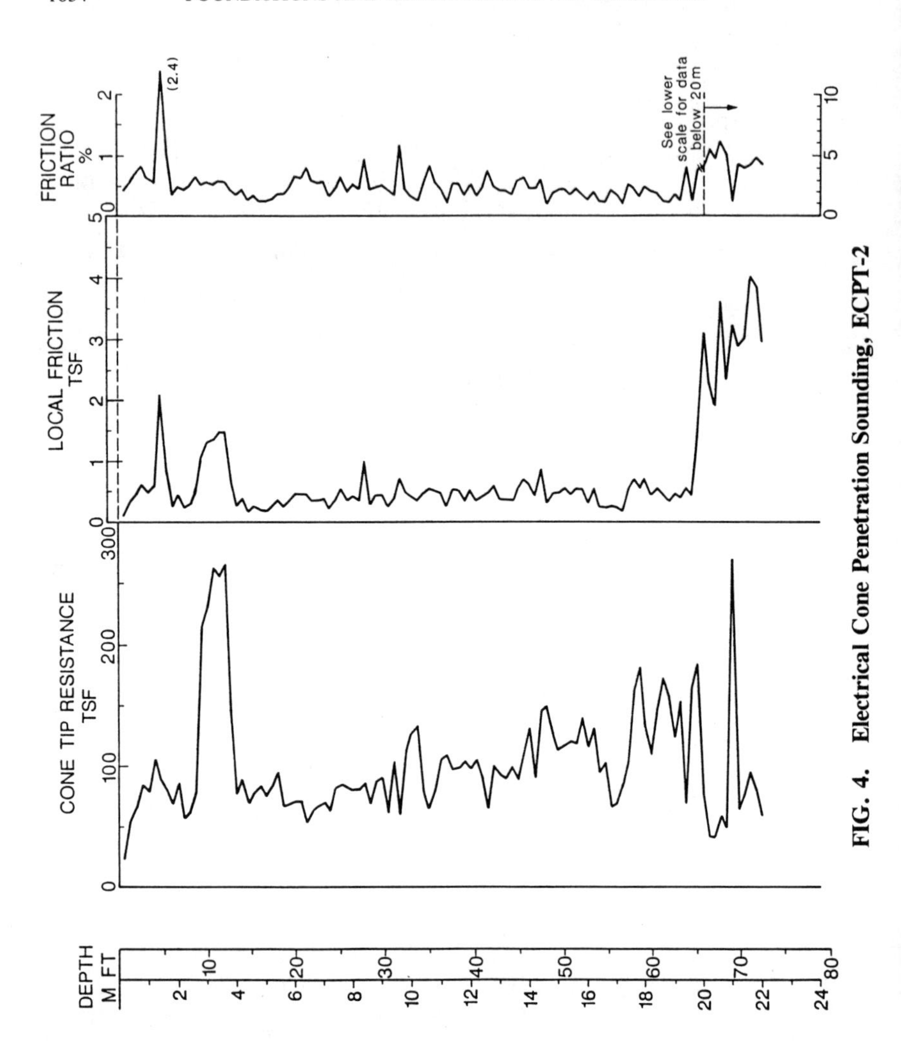

FIG. 4. Electrical Cone Penetration Sounding, ECPT-2

developed by Briaud and Tucker (1985, 1986) combines the LPC and Verbrugge methods to produce the top-load settlement curve.

Applying the PILECPT program to the mechanical cone and electrical cone soundings shown in Figs. 3 and 4, respectively, produces the top-load settlement curves given in Figs. 5 and 6. These two figures show that both mechanical and electric cone soundings approximate the shape of the respective pile load test load-deformation curves. Also, the relative amount of deformation occurring at plunging failure in the cone soundings approximates that of the respective pile load tests. Table 1 presents the numerical results for the PILECPT program.

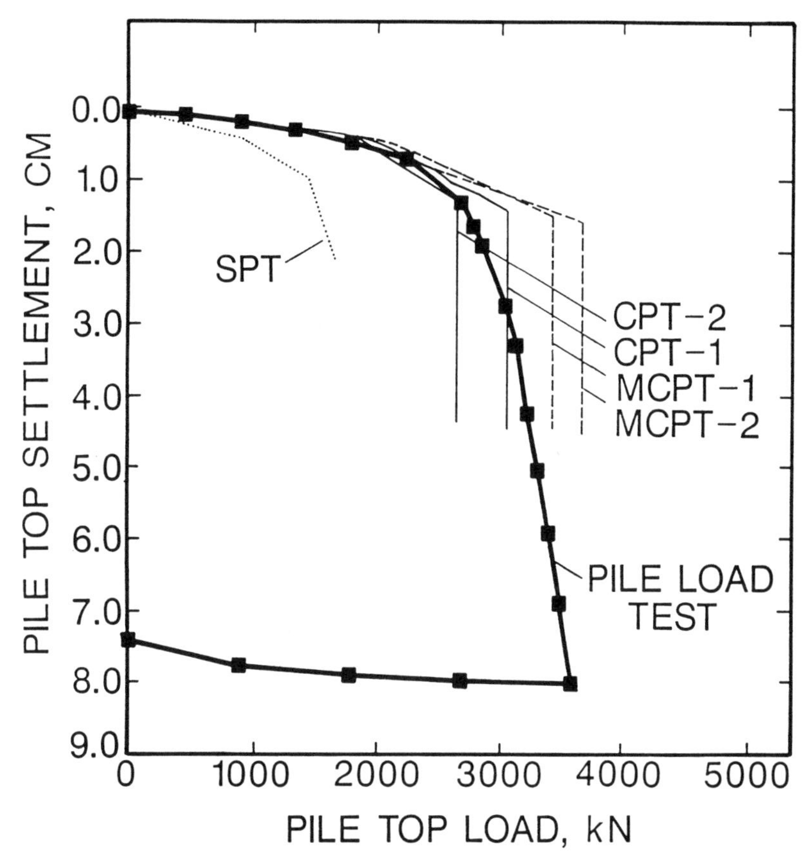

FIG. 5. Pile Load Test Results and Predicted Pile Settlement for 0.66 m Closed-End Pipe Pile

Since the SPT has been historically involved with settlement, the study looked at a prediction using the SPT N-value. The Meyerhoff approach of estimating settlements for pile groups was used to make an approximation for total settlement (Meyerhoff 1976). Results indicated top settlement of 1.47 and 1.42 cm for the octagonal and pipe piles, respectively. A further review of the literature yielded another method for predicting the ultimate load for piles in sand and the load transfer characteristics (Coyle and Castello 1981; Coyle 1985) using the SPT N-value. A computer program (COYLE2) combining these methods to calculate a top-load settlement curve has been developed by Briaud and Tucker (1985). Applying the COYLE2 to SPT data shown in Fig. 2 produces the top-load settlement curve also shown as Figs. 5 and 6 and summarized in Table 1.

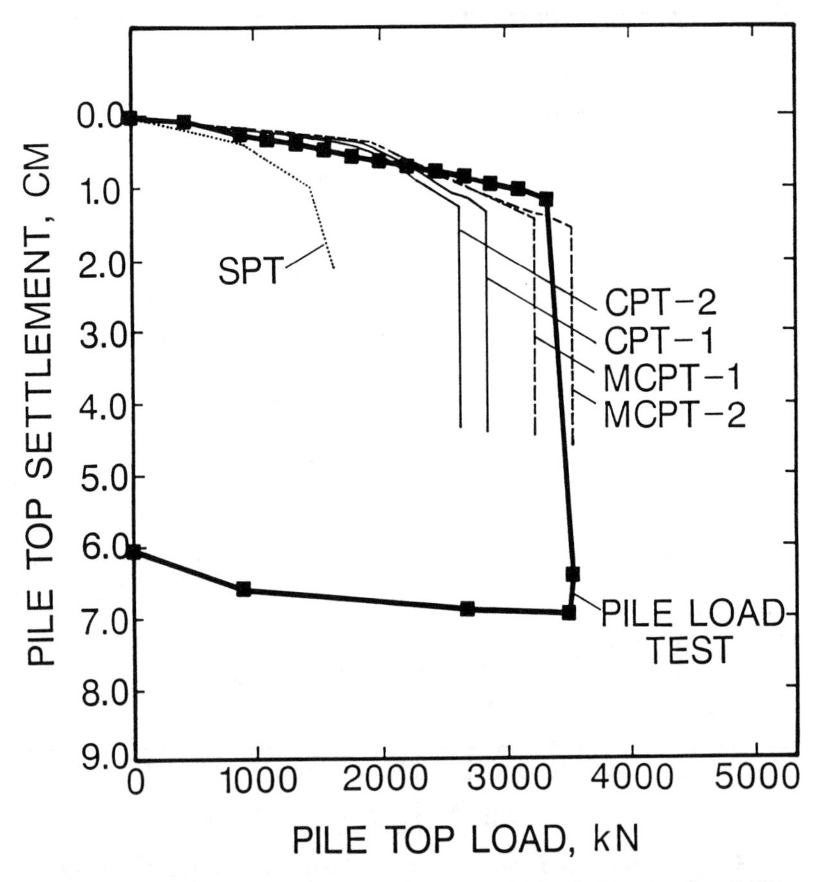

FIG. 6. Pile Load Test Results and Predicted Pile Settlement for 0.61 m Octagonal Prestressed Concrete Pile

TABLE 1. Deformation Summary

Type of Test	0.61 m Octagonal Concrete Pile		0.66 m Steel Pipe Pile	
	S_P[1]	S_T[2]	S_P	S_T
Pile Load Test	1.18	6.37	1.25	7.92
CPT-1	1.24	4.29	1.24	4.29
CPT-2	1.32	4.37	1.35	4.39
MCPT-1	1.42	4.47	1.42	4.47
MCPT-2	1.55	4.60	1.50	4.55
SPT	1.50	2.16	0.91	2.11

[1] Settlement at plunging in centimeters.
[2] Total settlement in centimeters.

Summary

The data presented in Figs. 5 and 6 and Table 1 exhibit a predicative trend in both the shape of the load-settlement curve and in the amount of deformation prior to plunging failure based on using the in situ cone soundings to estimate load-settlement curve behavior, i.e., assuming the cone as a model pile. The bracketing of the pile load test curve with the mechanical and electric cones is attributable to some variations in the soil profile, and in the selection of soil layers and corresponding cone tip resistances values required in the computer programs.

Appendix - References

Briaud, J. L., and Tucker, L. (1985). "User's manual for COYLE," *Geotechnical Engineering Research Report,* Dept. of Civil Engineering, Texas A&M University, College Station, Texas.

Briaud, J. L., and Tucker, L. (1986). "User's manual for PILECPT," *Geotechnical Engineering Research Report,* Dept. of Civil Engineering, Texas A&M University, College Station, Texas.

Coyle, H. M. (1985). *Unpublished Class Notes From Graduate Course CE 687, Marine Foundation Engineering*, Dept. of Civil Engineering, Texas A&M University, College Station, Texas.

Coyle, H. M., and Castello, R. (1981). "New design correlations for piles in sand." *J. Geotech. Eng. Div.*, Proc. ASCE, 107(GT7), 965-986.

Federal Highway Administration (1989). *Dynamic Pile Monitoring and Pile Load Test Report—Proposed Cimarron River Bridge Over U.S. 64*, Demonstration Projects Division, Washington, D. C.

Geo/Resource Consultants, Inc. (1988). "Pile instrumentation and data collection for pile load tests — U. S. Highway No. 64, Harper and Woods Counties, Oklahoma," *Report to Federal Highway Administration*, Washington, D. C.

Meyerhoff, G. G. (1976). "Bearing capacity and settlement of pile foundation." *J. Geotech. Eng. Div.*, Proc. ASCE, 102(GT3), 197-228.

Poulos, H. G., and Davis, E. H. (1980). *Pile Foundation Analysis and Design*, John Wiley & Sons, New York, New York.

Prakash, S., and Sharma, H. (1990). *Pile Foundations in Engineering Practice*, John Wiley & Sons, New York, New York.

Briaud, J. L., and Miran, J. (1992). "The cone penetrometer test," *Report No. FHWA-SA-91-043,* Federal Highway Administration, Washington, D. C.

Vesic, A. S. (1977). "Design of pile foundations," *NCHRP Synthesis of Highway Practice No. 42*, Transportation Research Board, Washington, D. C.

THE TIP DISPLACEMENT OF DRILLED SHAFTS IN SANDS

V. N. Ghionna,[1] M. Jamiolkowski,[2] S. Pedroni,[3] and R. Salgado[4]

ABSTRACT: The design load of large diameter drilled shafts is determined so that the settlements do not exceed the values associated with the ultimate and serviceability limit states. Based on the results of deep plate loading tests, performed in a calibration chamber on two different sands, a procedure is proposed for the evaluation of the relationship between the mobilized base capacity and the corresponding settlement of the pile tip of the drilled shaft. The procedure requires the knowledge of either the static cone penetration resistance or of the seismic shear wave velocity.

INTRODUCTION

A comprehensive description of the drilled shafts also referred to as large diameter bored piles can be found in the work by Reese and O'Neill (1988). The techniques used for their installation cause soil unloading and some degree of disturbance at the base of the borehole. As a result, when the drilled shaft is loaded, large relative displacements are required for the mobilization of the base resistance in cohesionless soils. In this respect, drilled shafts show a different behavior than driven piles for which the mobilization of the ultimate base resistance is attained at much smaller relative displacement of the tip.

---

[1] Associate Professor, Dipartimento di Ingegneria Strutturale, Politecnico di Torino, Corso Duca degli Abruzzi 24, 10129 Torino, Italy.

[2] Professor, Dipartimento di Ingegneria Strutturale, Politecnico di Torino, Corso Duca degli Abruzzi 24, 10129 Torino, Italy.

[3] Research Engineer, ENEL-CRIS.

[4] Assistant Professor, School of Civil Engineering, Purdue University, West Lafayette, IN 47907-1284; formerly Visiting Fellow, Dipartimento di Ingegneria Strutturale, Politecnico di Torino, Corso Duca degli Abruzzi 24, 10129 Torino, Italy.

The large number of well documented load tests performed on instrumented drilled shafts have permitted the development of an adequate understanding of load transfer mechanisms (Reese and O'Neill 1988). The *load* Q applied at the head of the drilled shaft is absorbed by a combination of the mobilized shaft and base capacities, Q^s and Q^b respectively.

Presently it is generally accepted that the *ultimate* value Q_u^s for large bored piles is mobilized after a small (10 to 20 mm) displacements of the shaft with respect to the surrounding soil. On the other hand, the *ultimate* value Q_u^b is only reached at very large relative displacements of the pile tip (De Beer 1986; Reese and O'Neill 1988; Ghionna et al. 1993; Fioravante et al. 1994).

In this circumstance the assessment of the overall ultimate bearing capacity $Q_u = Q_u^s + Q_u^b$ of a drilled shaft is of minor relevance for design purpose. In fact, long before Q_u is attained a settlement is reached that may cause the collapse of the supported structure. Therefore, the design of a drilled shaft should be made with reference to the two limit states (Wright 1979; Reese and O'Neill 1988; Franke 1989):

The load Q_{uls} corresponding to the *ultimate limit state* (ULS), also named the *critical load* Q_{cr}, causes the critical settlement s_{cr}. Such settlement at which the *critical base capacity* Q_{cr}^b is mobilized would lead to the collapse of the superstructure.

The load Q_{SLS} corresponding to the *serviceability limit state* (SLS) also defined as the impairment load Q_{imp} induces the impairment settlement s_{imp} such settlement endangers functionality of the constructed facility.

The qualitative significance of these two limits states ($Q_{cr} = Q_{cr}^b + Q_u^s = Q_{ULS}$ and $Q_{imp} = Q_{imp}^b + Q_u^s = Q_{SLS}$) as defined by many modern design codes is shown in Fig. 1 referring to the *relative settlement* $s_R = s/D$, being s and D the settlement and the diameter of the pile respectively. The above expressions for Q_{cr} and Q_{imp} imply that the settlements s_{cr} and s_{imp} are always larger than the settlement necessary to mobilize the ultimate shaft capacity.

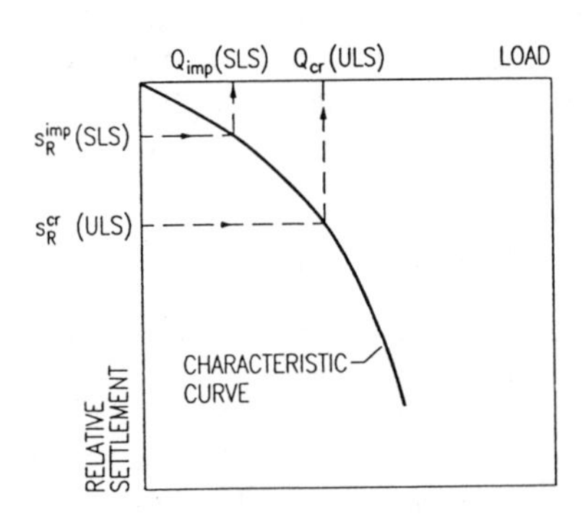

FIG. 1. Displacement Analysis of Piled Foundation

As structures are supported by many piles, accomplishment of the ULS and of the SLS is a function more of differential settlements of the distortional type rather than of the rigid body translation and/or rotation. So the evaluation of s_{cr} and s_{imp} requires complex soil-structure interaction analyses. Due to their inherent complexity, these analyses are rarely carried out. Although there are indications about the levels of potentially damaging differential settlements and angular distortions for different types of buildings (Wahls 1981), it became common in design to assume the critical values of s_R^{cr} as being either 5% (Reese and O'Neill 1988) or 10% (Franke 1989). Therefore, the criteria used to design drilled shaft foundations are based on S_R being not larger than a *critical relative displacement* s_R^{cr} equal to either 5 or 10% (in principle this value should refer to the tip of the piles, for large bored piles, due to the inherent rigidity of the pile shaft the difference between the settlement of the head and the base can be neglected) depending on the criterion selected.

Conventionally, the design or working load Q_d of a bored pile is then determined by dividing the Q_u by an appropriate *global safety factor* F_s, usually ranging between 2 and 3.

In this paper, based on the results of the *plate loading tests* PLT's on dry silica Ticino (TS) and carbonatic sands (QS) performed in the calibration chamber (CC), the values of the *operational secant Young's modulus* E_s for a range of relative densities and stress conditions are derived. The obtained values of E_s can then be used when estimating the mobilized base capacity versus settlement relationship of drilled shafts in sands.

BACKGROUND

Except for an initial loading up to an elastic threshold shear strain approximately equal to 10^{-5}, the stress-strain relationship for soils is strongly non-linear. For shear strains below the threshold strain the soil behaves as a linear elastic material with a shear modulus G_o referred to as the *small-strain shear modulus*. A variety of experimental techniques exist for determining in the laboratory and in the field the value of G_o. Laboratory techniques, which include the resonant column, the torsional shear, and bender element tests, have made it possible to establish the following relationship between G_o the stress state, and the void ratio e or relative density D_R for granular soils (Hardin and Black 1966; Hardin and Drnevich 1972):

$$G_o = C_g \frac{(e_g - e)^2}{1 + e} (\sigma_m')^n \qquad (1)$$

where C_g, e_g = material constants;
e = void ratio;
σ_m' = mean effective stress

Because of the difficulties in obtaining undisturbed samples of granular soils, in-situ seismic tests, such as the seismic CPT (Robertson et al. 1984; Baldi et al. 1988), the down-hole and the cross-hole shear wave measurement techniques, are used to obtain in-situ values of G_o. These tests actually determine the shear wave velocity V_s to which the small-strain shear modulus is related through the following expression:

$$G_o = \rho \cdot V_s^2 \tag{2}$$

where ρ = saturated mass density of the soil.

The small-strain *Young's modulus* E_o can be estimated from G_o using the following expression valid for an isotropic elastic medium:

$$E_o = 2(1 + \nu_o)G_o \tag{3}$$

where ν_o is the *small strain Poisson's ratio*, which in sands ranges from 0.1 to 0.2 (Jamiolkowski et al. 1994; Fioravante et al. 1994), it has been assumed here equal to 0.15.

Apart from the already mentioned dependence of G_o and E_o on e and σ_m', the initial stiffness is also controlled by the intrinsic characteristics of a given sand (grain-size distribution, grain shape, angularity, surface roughness and mineralogical composition) and its structure (fabric, inter-particle forces and bonds). Unless significant grain crushing occurs, the intrinsic characteristics of soil do not change with changes of e and σ_m', while its structure is subject to modifications. Despite this, in many uncemented sands the initial stiffness can be regarded as a state parameter appropriate for the normalization of deformation moduli measured at strain levels higher than the elastic threshold. For larger strains a pronounced modulus degradation occurs as a result of soil non-linearity. This means that if a problem is to be solved using the formulae of the theory of elasticity, *a secant deformation modulus* E_s smaller than E_0 must be used in the expressions.

Although it is not strictly correct to use the formulae of the elasticity theory beyond the elastic threshold strain, because the non-linearity is associated with irreversible plastic deformations, boundary value problems, of practical interest, can still be solved with an acceptable degree of reliability provided that an *appropriate operational value* of E_s is introduced into computation.

Considering that, as previously indicated, at the design load Q_d or even at the critical load Q_{cr} only a fraction of the Q_u^b of a drilled shaft is mobilized, it is suggested that the following formula of the theory of elasticity be used to predict the mobilized tip resistance q vs corresponding settlement:

$$s = \frac{\pi}{4}\frac{qD}{E_s}(1 - \nu_s^2)f\left(\frac{z}{D}\right) \tag{4}$$

where ν_s = operational value of the secant Poisson's ratio, here assumed equal to 0.25

$f(z/D)$ = non-dimensional coefficient taking into account the embedment z of the pile tip in the bearing layer; for the PLT's performed in the CC, $z/D \geq 6$ and f (z/D) results equal to 0.65 (see Donald et al. 1980)

The value of E_s in Eq. (4) that yields the desired value of the tip displacement s is referred to as the equivalent or *operational* Young's modulus.

So the concept of E_s reflects an "average" stiffness of the soil for the given boundary value problem (Jardine et al. 1986), function of the strain level induced in the surrounding soil mass. In the next sections, the results of the PLT's performed in the CC, see Fig. 2, are examined with the aim of determining the values of E_s useful for computing the tip settlement of drilled shafts and, after an appropriate normalization, also for evaluating the settlements of shallow foundations in sands.

Calibration Chamber Plate Loading Tests

A series of PLT's were performed in the ENEL-CRIS calibration chamber, housing a sample 1.2 m diameter and 1.5 m in height. The samples were formed by stationary pluviation, which was stopped at an elevation where a rigid plate having a diameter of 104 mm was installed. After the plate was put in place, the inner rod and the outer casing with the same diameter as the plate were installed, completing thereafter the sand deposition, see Fig. 2.

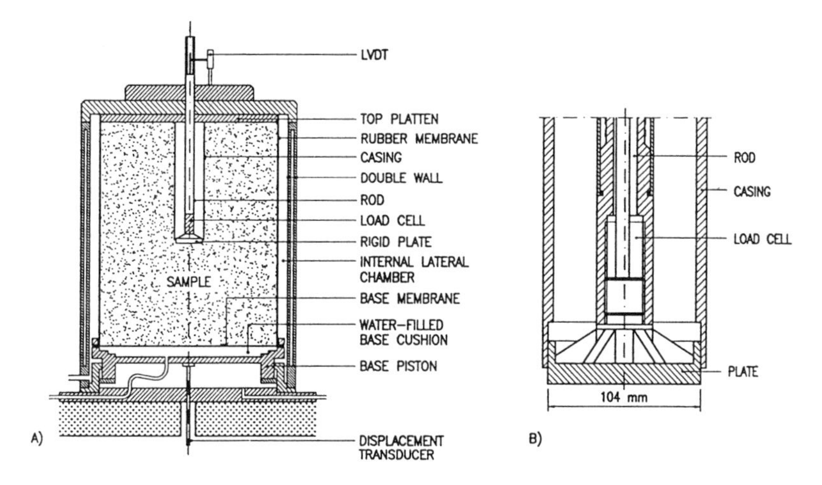

FIG. 2. Plate Load Tests in Calibration Chamber: (A) Plate Installation; (B) Plate Assembly

Thereafter, the sample with the plate inside, was consolidated under K_o - conditions, obtaining normally consolidated (NC) or overconsolidated (OC) sand samples at a desired relative density D_R.

After the consolidation the plate was loaded via the pushing rod so that the relationship between the stress acting on the plate and its settlement may be obtained. The PLT's were conducted controlling either the stresses and/or the strains at the sample boundaries, see Table 2 and Bellotti et al. (1982).

The PLT's examined in the following sections of the paper were made in silica TS and carbonatic QS whose properties are summarized in Table 1.

TABLE 1. Properties of Ticino and Quiou Sands (see also Golightly 1988)

	$\gamma_{d,min}$ kN/m³	$\gamma_{d,max}$ kN/m³	G_s	e_{min}	e_{max}	U	D_{50} mm	D_{10} mm	ϕ'_{cv} °	n	C_g	e_g
TS	13.65	16.68	26.23	0.922	0.573	1.5	0.54	0.36	34.8	0.44	647	2.27
QS	12.45	14.76	26.68	1.123	0.809	2.9	1.04	0.42	40.0	0.62	197	4.16

where $\gamma_{d,min}$ = minimum dry density;
$\gamma_{d,max}$ = maximum dry density;
G_s = specific gravity;
e_{min} = minimum void ratio;
e_{max} = maximum void ratio;
D_{50} = grains diameter at which 50% of the soil weight is finer;
D_{10} = grains diameter at which 10% of the soil weight is finer;
U = uniformity coefficient;
σ'_{cv} = friction angle at critical state;
n = material constants in Eq. (1); and
C_g, e_g = material constants in Eq. (1).

ULTIMATE BEARING CAPACITY

The plate *ultimate bearing capacity* q_u, i.e. the pressure under which the plate would plunge through the soil, can be computed using the penetration resistance analysis proposed by Salgado (1993), According to this theory a deep foundation element assimilated by the plate embedded in the CC sample will reach the stage of penetration at a constant rate by expanding a cylindrical cavity from an initial radius equal to zero. So the stress and strain fields in a zone P sufficiently far from the base of the plate (Fig. 3) will be equivalent to those generated by the expansion of a cylindrical cavity in the soil mass, with a horizontal major principal stress related to the limit cavity pressure p_c^u. In the conic region Q underneath the plate the major principal stress is vertical and related to q_u. The apex angle of zone Q for a plate or drilled shaft tip is that of a passive Rankine zone (90° - ϕ'), where ϕ' is the friction angle representative for zone Q. In the immediate vicinity of the plate or shaft tip the principal stresses experience a rotation. A stress rotation analysis between zones P and Q allow the computation of q_u based on the value of p_u. Both the cavity expansion and the stress rotation analyses consider material non-linearities. The computations of p_u and q_u were carried out using the program CONPOINT (Salgado 1993).

The difference between q_u and q_c obtained from a cone penetration test is small (Salgado and Mitchell 1994); Ghionna et al. 1993), and for design purposes it is possible to substitute one for the other without incurring significant errors.

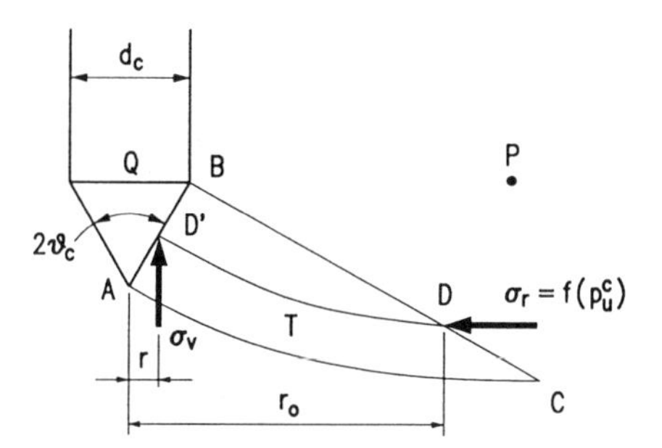

FIG. 3. The Failure Pattern underneath the Tip of a Pile or a Plate after Salgado (1993)

Plate Load Tests Results

Table 2 summarizes the key information gathered from a total of 34 chamber PLT's performed on TS and QS. It reports the following information:

- initial samples and tests conditions,
- values of q at s_R of 5 and 10% respectively and the ultimate bearing pressure q_u, computed according Salgado (1993),
- operational stiffness E_s for the two above mentioned values of s_R,
- small strain Young's modulus E_o obtained from Eqs. (1) and (3),
- normalized plate pressure $q_N = q/q_u$,
- normalized deformation modulus $E_N = E_s/E_o$ and normalized modulus numbers $K[E_o]$, $K[E_s]$ obtained from Eqs. (8) and (6) respectively,
- ratio of $K[E_s]$ to $K[E_o]$ referred to as K_N.

Fig. 4 shows for TS how the ratio of the operational stiffness E_s to the small-strain elasticity modulus E_o varies with relative displacement s_R. The same plot in double logarithmic scale is shown in Fig. 5. The referenced figures contain results of all tests on TS, except tests No. 315 and 316, see Table 2, which have been disregarded because of experimental irregularities.

Although the above mentioned figures display the results of 28 PLT's run on both NC and OC samples of TS, the scatter in the experimental data is quite small. Moreover, this scatter can be further reduced normalizing both E_s and E_o with respect to the current level of the effective stress.

A rational normalization stress should be the mean effective stress σ_m'. However, in case of E_s being the evaluation of σ_m' under plate during loading too complex to be followed, it is a matter of practicality to normalize it with respect to the current average vertical effective stress v'_{va} which, in first approximation, (Berardi and Lancellotta 1991) can be assumed equal to:

$$\sigma'_{va} = \sigma'_{vo} + I_\sigma \left[\frac{q}{2}\right] \tag{5}$$

where σ'_{vo} = effective overburden stress at the depth 0.5 D below the plate;

TABLE 2. Summary of Plate Loading Tests in Calibration Chamber

Test No.	Soil	Cond.	BC	DR %	σ'_v kPa	σ'_h kPa	OCR	q_u MPa	E_o MPa	$K[E_o]$
300	TS	dry	1	51.0	115.0	51.0	1.00	14.1	173.1	1645
301	TS	dry	1	92.0	115.0	40.0	1.00	34.6	215.4	2048
302	TS	dry	1	50.4	113.0	51.0	2.73	13.9	171.6	1644
303	TS	dry	1	55.2	313.0	140.0	1.00	25.4	276.9	1694
304	TS	dry	1	55.2	214.0	92.0	1.00	20.9	232.3	1980
305	TS	dry	1	58.4	512.0	223.0	1.00	33.9	349.1	1720
306	TS	dry	1	49.4	116.0	76.0	2.70	16.1	187.9	1779
307	TS	dry	1	92.8	314.0	123.0	1.00	49.0	344.5	2105
308	TS	dry	1	92.5	216.0	85.0	1.00	43.5	292.0	2103
309	TS	dry	1	92.5	115.0	77.0	2.73	40.9	249.1	2367
310	TS	dry	1	90.9	66.0	25.0	1.00	29.3	170.2	2066
311	TS	dry	1	91.7	63.0	25.0	1.00	29.7	169.2	2095
312	TS	dry	1	55.5	513.0	235.0	1.00	32.8	346.5	1706
313	TS	dry	1	51.7	62.0	26.0	1.00	10.5	130.9	1633
314	TS	dry	1	49.1	54.0	52.0	7.61	12.9	148.6	1970
315	TS	dry	1	43.5	410.0	189.0	1.00	23.1	290.3	1577
316	TS	dry	1	56.8	512.0	184.0	1.00	30.4	332.8	1640
317	TS	dry	1	55.2	62.0	24.4	1.00	11.4	132.3	1651
318	TS	dry	1	59.0	216.0	92.0	1.00	22.8	238.7	1719
319	TS	dry	1	58.0	65.0	53.0	6.34	16.6	163.2	1994
320	TS	dry	1	59.3	63.0	50.0	6.54	16.8	161.1	1996
321	TS	dry	1	60.2	412.0	177.0	1.00	31.5	320.3	1736
322	TS	dry	1	92.2	314.0	129.0	1.00	49.2	346.5	2117
323	TS	dry	1	92.0	66.0	64.0	6.27	37.6	215.0	2609
324	TS	dry	1	91.2	66.0	27.0	1.00	30.0	173.1	2101
325	TS	dry	3	91.4	116.0	47.0	1.00	35.6	221.8	2100
326	TS	dry	3	91.2	65.0	25.0	1.00	29.4	169.9	2076
327	TS	dry	2	90.9	65.0	26.0	1.00	29.5	170.9	2088
328	TS	dry	4	90.9	65.0	26.0	1.00	29.5	170.9	2088
329	TS	dry	1	90.6	65.0	26.0	1.00	29.3	170.6	2084
401	QS	dry	1	72.6	121.1	44.7	1.00	10.9	190.0	2441
402	QS	sat	1	73.6	101.2	42.6	1.00	11.8	176.6	2447
403	QS	sat	3	47.5	96.5	39.6	1.00	10.5	149.3	2146
404	QS	sat	3	37.5	97.9	46.7	1.00	11.3	149.5	2038

BC = Boundary conditions during PLT: (1) constant σ'_v and σ'_h; (2) $\epsilon_v = \epsilon_h = 0$; (3) σ'_v = constant, $\epsilon_h = 0$; and (4) σ'_r = constant, $\epsilon_v = 0$.

QS = Carbonatic Quiou Sand; TS = Silica Ticino Sand; ϵ_v = vertical strain; ϵ_r = horizontal strain.

TABLE 2. Summary of Plate Loading Tests in Calibration Chamber (Cont'd)

Test No.	s/D = 0.05					
	q (MPa)	E_s (MPa)	q_N	E_N	$K[E_s]$	K_N
300	1.43	13.7	0.101	0.079	57.3	0.035
301	2.42	23.2	0.070	0.108	77.7	0.038
302	1.18	11.2	0.085	0.066	50.9	0.031
303	3.08	29.5	0.121	0.107	81.9	0.048
304	2.07	19.9	0.099	0.085	67.1	0.040
305	2.21	21.1	0.065	0.060	60.6	0.035
306	1.43	13.7	0.089	0.073	57.1	0.032
307	3.99	38.2	0.081	0.111	95.8	0.046
308	3.60	34.4	0.083	0.118	93.1	0.044
309	3.49	33.4	0.085	0.134	95.2	0.040
310	2.00	19.1	0.068	0.112	72.0	0.035
311	2.02	19.4	0.068	0.115	72.6	0.035
312	3.36	32.2	0.102	0.093	80.8	0.047
313	1.20	11.4	0.114	0.087	54.3	0.033
314	1.11	10.6	0.086	0.071	52.4	0.027
315	1.00	9.6	0.043	0.033	35.5	0.023
316	4.41	42.2	0.145	0.127	96.3	0.059
317	1.00	9.6	0.087	0.072	48.9	0.030
318	1.70	16.3	0.075	0.068	59.1	0.034
319	1.27	12.1	0.076	0.074	55.9	0.028
320	1.62	15.5	0.097	0.096	64.3	0.032
321	2.73	26.2	0.087	0.082	73.0	0.042
322	4.67	44.7	0.095	0.129	105.2	0.050
323	2.96	28.3	0.079	0.132	89.0	0.034
324	1.99	19.1	0.066	0.110	71.9	0.034
325	2.75	26.3	0.077	0.118	83.3	0.040
326	1.92	18.4	0.065	0.108	70.5	0.034
327	2.07	19.8	0.070	0.116	73.4	0.035
328	1.98	19.0	0.067	0.111	71.7	0.034
329	1.91	18.3	0.065	0.107	70.2	0.034
401	0.81	7.8	0.074	0.041	39.8	0.016
402	0.74	7.1	0.063	0.040	38.6	0.016
403	0.65	6.2	0.062	0.042	35.7	0.024
404	0.44	4.2	0.039	0.028	27.3	0.018

TABLE 2. Summary of Plate Loading Tests in Calibration Chamber (Cont'd)

Test No.	s/D = 0.10					
	q (MPa)	E_s (MPa)	q_N	E_N	$K[E_s]$	K_N
300	–	–	–	–	–	–
301	3.59	17.2	0.104	0.080	48.3	0.024
302	1.58	7.5	0.114	0.044	30.4	0.037
303	4.00	19.1	0.157	0.069	48.0	0.029
304	2.97	14.2	0.142	0.061	41.6	0.025
305	3.17	15.6	0.093	0.043	38.8	0.022
306	1.95	9.3	0.121	0.050	34.3	0.020
307	6.25	29.9	0.127	0.087	62.1	0.030
308	5.35	25.6	0.123	0.087	58.3	0.028
309	5.01	24.0	0.123	0.096	58.8	0.025
310	2.77	13.2	0.095	0.087	42.9	0.021
311	2.83	13.5	0.095	0.080	43.5	0.021
312	4.42	21.6	0.135	0.061	48.2	0.028
313	1.53	7.3	0.146	0.056	31.2	0.019
314	1.40	6.7	0.108	0.045	29.9	0.015
315	1.30	6.2	0.056	0.021	21.7	0.014
316	5.83	27.9	0.191	0.084	57.2	0.035
317	1.36	6.5	0.119	0.049	29.2	0.017
318	2.32	11.1	0.102	0.046	35.8	0.021
319	1.69	8.1	0.101	0.050	37.8	0.017
320	2.06	9.9	0.123	0.061	36.7	0.018
321	3.73	17.8	0.118	0.056	44.6	0.026
322	6.70	32.0	0.136	0.092	64.6	0.030
323	4.05	19.9	0.108	0.090	52.4	0.020
324	2.74	13.1	0.091	0.076	42.7	0.021
325	4.18	20.0	0.117	0.090	52.4	0.025
326	2.90	13.9	0.099	0.082	44.1	0.021
327	3.25	15.6	0.110	0.091	46.8	0.022
328	2.71	13.0	0.092	0.076	42.4	0.022
329	2.68	12.8	0.091	0.075	42.1	0.020
401	1.26	6.1	0.116	0.032	26.8	0.011
402	0.97	4.7	0.082	0.027	22.9	0.009
403	0.95	4.5	0.091	0.019	22.7	0.011
404	0.57	2.8	0.050	0.027	16.8	0.008

q = current vertical stress imposed on the plate; and
I_σ = Boussinesq's influence factor for the vertical stress at a depth equal to half the diameter of the plate.

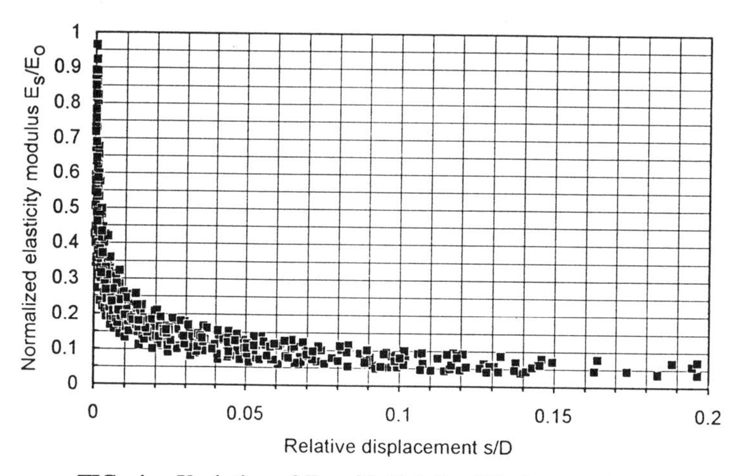

FIG. 4. Variation of E_N with Relative Displacement s_R

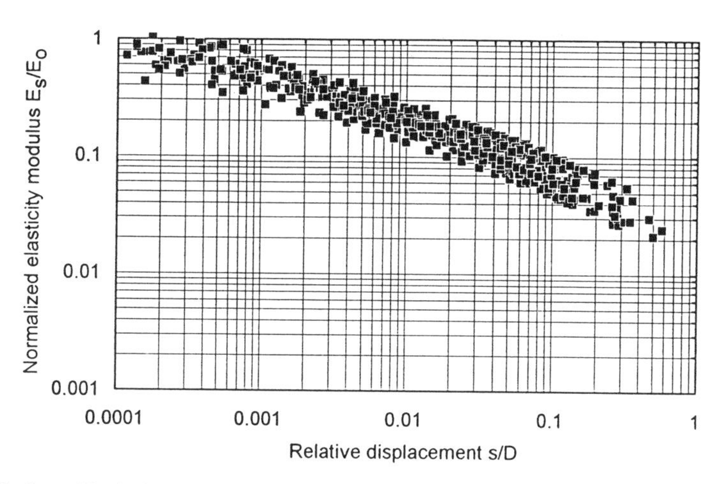

FIG. 5. Variation of E_N with Relative Displacement s_R in Double Logarithmic Scale

The normalized operational stiffness E_s becomes therefore:

$$K[E_s] = \frac{E_s}{(p_a)^{1-n}(\sigma'_{va})^n} \tag{6}$$

where P_a = reference stress = 100 kPa.

Assuming, in a first approximation, that the stress exponent n is equal to 0.5, Eq. (6) becomes:

$$K[E_s] = \frac{E_s}{\sqrt{p_a \sigma'_{va}}} \tag{7}$$

where $K[E_s]$ = non-dimensional modulus number.

As to the normalization of E_o it can be as usual normalized with respect to the *initial* mean effective stress σ'_{mo} which existed at the depth of 0.5D below the plate before starting the PLT:

$$K[E_o] = \frac{E_o}{(p_a)^{0.56}(\sigma'_{mo})^{0.44}} \tag{8}$$

assuming for TS the value of stress exponent n = 0.44 resulting from triaxial and resonant column tests.

Otherwise, for sake of consistency with Eq. (6), E_o can be normalized with respect to the *initial* vertical stress at the depth of 0.5D as above, preserving, in first approximation, the same value of n = 0.44,

$$K[E_o] = \frac{E_o}{(p_a)^{0.56}(\sigma'_{vo})^{0.44}} \tag{9}$$

Figs. 6 and 7 show the ratio $K[E_s]/K[E_o] = K_N$ as function of s_R, computed using Eq. (7) combined with Eqs. (8) and (9) respectively. Both Figs. 6 and 7 contain the results of 28 PLT's on TS, six of which performed on overconsolidated samples having two levels of OCR, 2.7 and 7 corresponding to the values of the earth pressure coefficient at rest K_o close to 0.65 and 0.95 respectively, as opposed to $0.35 \leq K_o \leq 0.45$ of the NC samples. Although not shown in Figs. 6 and 7, the points corresponding to OC samples plot are higher than those of the NC samples. This difference is more pronounced at values of $s_R \leq 0.01\%$ becoming negligible when the relative displacement exceeds 0.5%.

Figs. 8 and 9 show for TS the values of q_N as a function of relative displacement s_R in arithmetic and double logarithmic scales, respectively. In the plot shown in Fig. 9 a straight line drawn through the bulk of the experimental points in the range from s_R = 0.1 to 10%, for which the chamber size effect is expected to be small, yields on average the values of q_N of 0.09 and 0.13 at s^{cr}_R equal to 5 and 10% respectively. These values are slightly smaller than those suggested by Jamiolkowski and Lancellotta (1988) and Ghionna et al. (1993) for the ratio of the critical base resistance q to cone penetration resistance q_c. This is mainly due to the fact that the theoretical value of q_u is 10 to 15 % larger than the values of q_c from CC tests after their correction for the chamber size effect.

The evidence of the relatively small chamber size effect on the results of the PLT's run in CC for s_R values below 10 to 15% is provided by the finite element analyses carried out by Bocchio (1993), and by the comparable q vs s curves obtained by performing chamber PLT's on four samples at the same relative density (about 90%) and consolidation stress under each of the possible boundary conditions, see Table 2 and Bellotti et al. (1982).

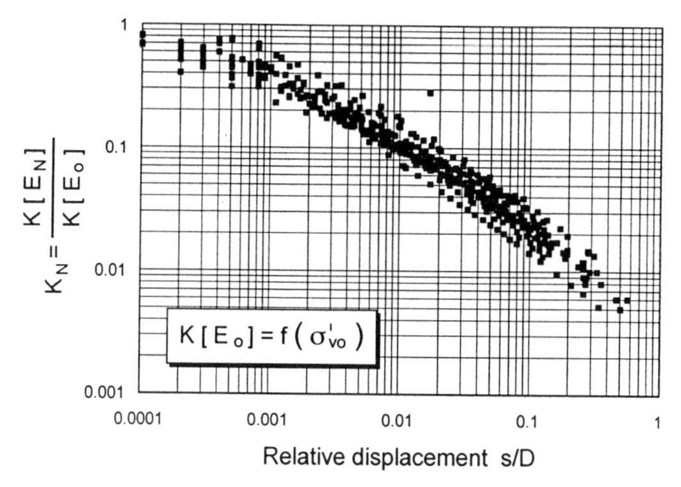

FIG. 6. Variation of K_N with Relative Displacement s_R

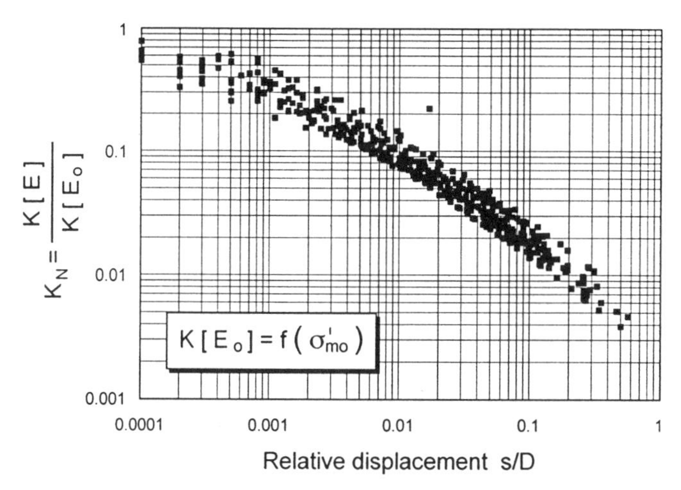

FIG. 7. Variation of K_N with relative displacement S_R

During the PLT's run in the CC, unload-reload loops imposing to the plate the amplitudes of the relative cyclic displacements $\Delta s/D$ ranging from 0.05 to 0.09% have been carried out. The value of the deformation modulus E_{ur}, secant to the above mentioned unload-reload loops, inferred by means of the Eq. (4) approximate 0.9 E_o.

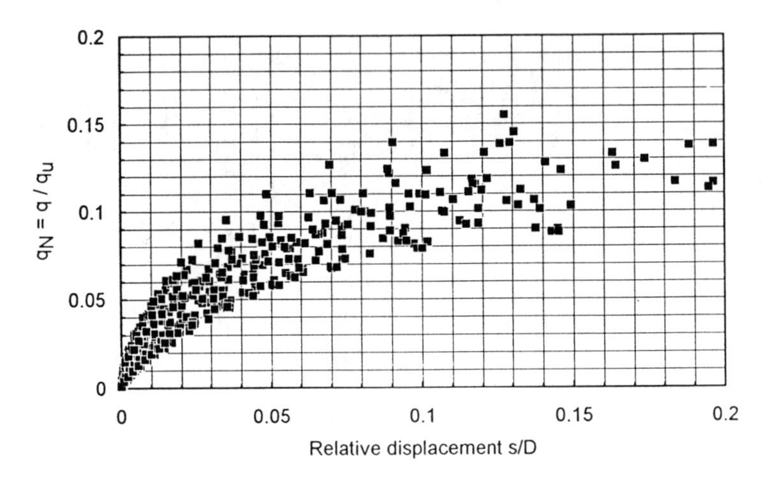

FIG. 8. Variation of q_N with Relative Displacement s_R

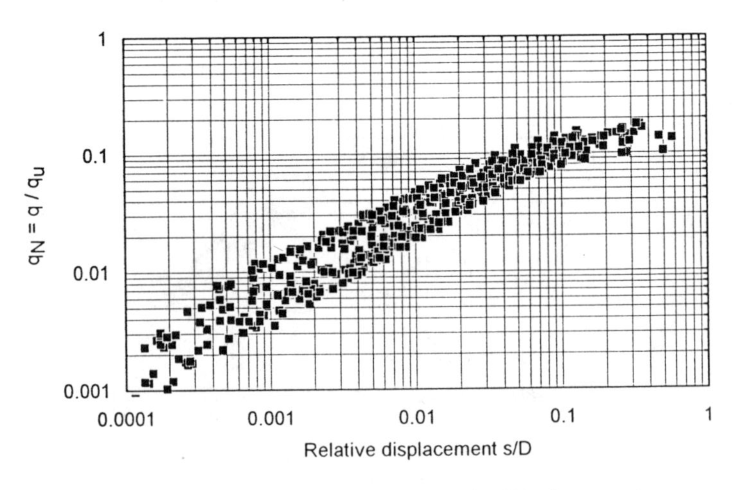

FIG. 9. Variation of q_N with Relative Displacement s_R

As far as carbonatic QS is concerned, the limited numbers of the PLT's so far available do not allow any conclusions. The experimental results given in Table 2 suggest that in the uncemented carbonatic sands the values of q_N and E_N at s_R equal to 5 and 10% might be appreciably lower than those encountered for silica sands. This is linked to the more pronounced compressibility of the QS if compared with that of the typical silica sands like TS as indicated for the two PLT's performed at the comparable values of D_R shown in Fig. 10.

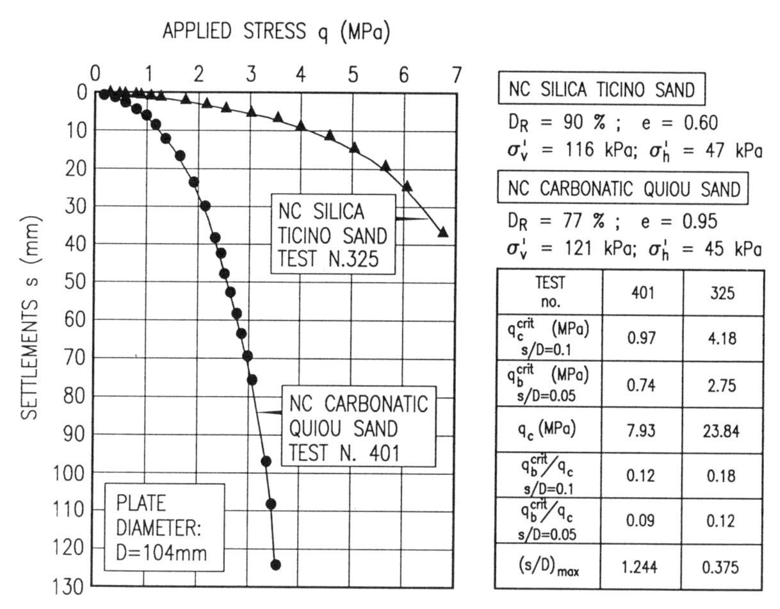

TEST no.	401	325
q_c^{crit} (MPa) s/D=0.1	0.97	4.18
q_b^{crit} (MPa) s/D=0.05	0.74	2.75
q_c (MPa)	7.93	23.84
q_b^{crit}/q_c s/D=0.1	0.12	0.18
q_b^{crit}/q_c s/D=0.05	0.09	0.12
$(s/D)_{max}$	1.244	0.375

FIG. 10. Deep Plate Loading Tests in the Calibration Chamber for Two Dry Sands

BASE RESISTANCE OF A DRILLED SHAFT

The results of the chamber PLT' s previously examined offer the possibility to evaluate the mobilized base capacity $Q^b = A_b\, q_b$ (A_b = area of the pile tip) of the drilled shafts following the suggestions given by Jamiolkowski and Lancellotta (1988) and by Ghionna et al. (1993).

Admitting that only CPT results are available, it is best to define the critical base resistance q^b_{cr} as a fraction of q_c. Based on the experimental results discussed in the previous sections the recommended values of q_N (assuming implicitly $q_N \approx q^b_{cr}/q_c$ which is slightly on the safe side) for use in design are:

$$q_N = \begin{cases} 0.090 \pm 0.02, & \text{for } s/D = 0.05 \\ 0.13 \pm 0.02, & \text{for } s/D = 0.10 \end{cases}$$

If in-situ shear wave velocity measurements at the depth near the tip of the drilled shaft are available, then E_o can be computed and the design values of E_N or K_N are estimated. On average the experimental data available lead to the following results:

$$E_N = \begin{cases} 0.10 \pm 0.018, \text{ for } s/D = 0.05 \\ 0.07 \pm 0.018, \text{ for } s/D = 0.10 \end{cases}$$

$$K_N = \begin{cases} 0.037 \pm 0.007, \text{ for } s/D = 0.05 \\ 0.023 \pm 0.006, \text{ for } s/D = 0.10 \end{cases}$$

Despite the limited data presently available for QS, an estimate of q_N, E_N and K_N allow to figure out the following preliminary values:

$$q_N = \begin{cases} 0.07 \text{ for } s/D = 0.05 \\ 0.10 \text{ for } s/D = 0.10 \end{cases}$$

and

$$E_N = \begin{cases} 0.038 \text{ for } s/D = 0.05 \\ 0.027 \text{ for } s/D = 0.10 \end{cases}$$

$$K_N = \begin{cases} 0.009 \text{ for } s/D = 0.05 \\ 0.006 \text{ for } s/D = 0.10 \end{cases}$$

CONCLUSIONS

Based on the results of the of plate loading tests performed in CC on the pluvially deposited sand the following conclusions can be drawn:

a. For the examined boundary value problem i.e. PLT, the operational stiffness E_s inferred from Eq. (4) exhibits an extremely pronounced non-linearity for the values of SR < 1%.

b. Both E_N and K_N values appear to be relatively insensitive to the magnitudes of the D_R and the σ' within the range of s_R from 0.1 to 10% of practical interest.

c. The deformation modulus E_{ur} secant to the unload-reload loops performed during PLT's run in the CC results on average equal to 0.9 E_o.

d. The mechanical overconsolidation has only a moderate influence on E_N, K_N and q_N for s^{cr}_R values of 5 and 10% respectively. Table 3 reports the ratio of $K_N(OC)$ to $K_N(NC)$ at different levels of s_R.

TABLE 3. Influence of Overconsolidation on K_N

s_R (%)	$\frac{K_N(OC)}{K_N(NC)}$
0.01	1.87
0.1	1.67
1	1.36
2	1.29
5	1.21
10	1.15

e. The critical base resistance of a drilled shaft at the desired level of s_R can be evaluated by means of the following formula:

$$q_{cr}^{b} = E_s \frac{\frac{s_{cr}}{D}}{\frac{\pi}{4}(1-\nu^2)f(\frac{z}{D})} \tag{10}$$

following the iteration procedure outlined in Appendix II.

f. Hence the experimental relationships $E_N = f(s_R)$ and $K_N = f(s_R)$ have been obtained for freshly deposited sands their application to natural aged sand deposits should be done with caution and subject to field verifications.

g. Considering the feature of the normalization adopted to infer the values of K_N, the formulae (A-1) and (A-2) could also at least in principle, be used to obtain E_s for evaluating the settlements of shallow foundations in sands.

APPENDIX I. REFERENCES

Baldi, G., Bruzzi, D., Superbo, S., Battaglio, M., and Jamiolkowski, M. (1988). "Seismic cone in Po River sand." *Penetration Testing*, Proc. ISOPT-I, Orlando, 2, 643-650.

Bellotti, R., Bizzi, G., and Ghionna, V. (1982). "Design, construction and use of a calibration chamber." *Penetration Testing*, Proc. ESOPT II, Amsterdam, 2, 439-446.

Berardi, R. and Lancellotta, R. (1991). "Stiffness of granular soil from field performance." *Géotechnique*, 41(1), 149-157.

Bocchio, G. L. (1993). "Numerical simulation of the plate loading tests performed in the calibration chamber," Ph.D. thesis, Technical University of Torino, Torino, Italy, under preparation.

De Beer, E. (1984) "Different behavior of bored and driven piles." *Proc. 6th Danubian Conf. on Soil Mech. Found. Eng.*, Budapest, 307-318.

Donald, I. B., Chiu, H. K., and Sloan. S. W. (1980). "Theoretical analyses of rock socketed piles." *Proc. Int. Conf. on Structural Foundation on Rock*, Sidney, Balkema.

Fioravante, V., Jamiolkowski, M., and Lo Presti, D. C. F. (1994). "Stiffness of Carbonatic Quiou Sand." *Proc. 13th ICSMFE*, New Delhi.

Fioravante, V., Jamiolkowski, M., and Pedroni, S. (1994a). "Modelling the behavior of piles in sand subjected to axial load." *Centrifuge '94*, Singapore

Franke, E. (1989). "Prediction of the bearing behavior of piles, especially large bored piles." *Proc. 12th ICSMFE*, Rio de Janeiro, 2, 915-918.

Golightly, C. R. (1988). "Engineering properties of carbonate sands," Ph.D. thesis, University of Bradford, United Kingdom.

Ghionna, V. N., et al. (1993). "Base capacity of bored piles in sands from in-situ tests." *Proc. 2nd Int'l. Geotechnical Seminar on Deep Foundation on Bored and Auger Piles*, Ghent.

Hardin, B. O. and Black, W. L. (1966). "Sand stiffness under various triaxial stresses." *J. Soil Mech. Found. Eng. Div.*, Proc. ASCE, 92(SM2), 27-42.

Hardin, B. O. and Drnevich, V. P. (1972). "Shear modulus and damping in soils: Design equations curves." *J. Soil Mech. Found. Eng. Div.*, Proc. ASCE, 98(SM7), 667-692.

Jamiolkowski, M., et al. (1985). "New developments in field and laboratory testing of soils." *Proc. 11th ICSMFE*, San Francisco, 1, 57-153.

Jamiolkowski, M. and Lancellotta, R. (1988). "Relevance of in-situ tests results for evaluation of allowable base resistance of bored piles in sands." *Proc. 1st Int'l. Geotechnical Seminar on Deep Foundation on Bored and Auger Piles*, Ghent, 107-120.

Jamiolkowski, M., et al. (1994). "Stiffness of Toyoura sand at small and intermediate strain." *Proc. 13th ICSMFE*, New Delhi.

Jardine, R. J., Potts, D. M., Fourie, A. B., and Burland, J. B. (1986). "Studies of the influence of non-linear stress-strain characteristics in soil-structure interaction." *Géotechnique*, 36(3), 377-396.

Reese, L. C. and O'Neill, M. W. (1988). "Drilled shafts: Design procedures and design methods," *Report No. FHWA-HI-88-42*, U. S. Dept. of Transportation, Washington, D. C.

Robertson, P. K., et al. (1984). "Seismic CPT to measure in-situ shear wave velocity," *Soil Mechanics Series No. 85*, Dept. of Civil Engineering, Univ. of British Columbia, Vancouver, Canada.

Salgado, R. (1993). "Analysis of penetration resistance in sands," Ph.D. thesis, University of California at Berkeley.

Salgado, R. and Mitchell, J. K. (1994). "Extra-terrestrial soil property determination by CPT." *Proc. 8th Int'l. Conf. of the Association for Computer Methods and Advances in Geomechanics*, Morgantown, in press.

Wahls, H. E. (1981). "Tolerable settlements of buildings." *J. Geotech. Eng. Div.*, Proc. ASCE, 107(GT11), 1489-1504.

APPENDIX II. OUTLINE OF THE CALCULATION PROCEDURE

Eq. (10) establishes a relationship between q^b_{cr} and the adopted level of the relative critical tip displacement s^{cr}_R. Using this formula it is possible to outline the following procedure for evaluating the critical base resistance of a drilled shaft:

1. With the value of V_s obtained from in situ seismic tests the value of E_o representative for the soil conditions in the vicinity of the pile tip is assessed from Eqs. (2) and (3).
2. The design value of $s^{cr}{}_R$ is selected.
3. The values of K_N for the desired values of $s^{cr}{}_R$ is obtained from Figs. 6 or 7.
4. Knowing the value of E_o, the value of E_s is computed by means of one of the following expressions, introducing a preliminary value of $q^b{}_{cr} = q$

$$E_s = K_N \left[\frac{E_0}{(\sigma'_{mo})^{0.44} \cdot (p_a)^{0.56}} \right] \cdot \left[\left(\sigma'_{vo} + I_\sigma \cdot \frac{q}{2} \right)^{0.5} (p_a)^{0.5} \right] \qquad \textbf{(A-1)}$$

$$E_s = K_N \left[\frac{E_0}{(\sigma'_{vo})^{0.44} \cdot (p_a)^{0.56}} \right] \cdot \left[\left(\sigma'_{vo} + I_\sigma \cdot \frac{q}{2} \right)^{0.5} (p_a)^{0.5} \right] \qquad \textbf{(A-2)}$$

5. With the value of E_s the value Of $q^b{}_{cr}$ for the preselected value of $s_{cr/D}$ is computed from Eq. (10).
6. If the obtained value of $q^b{}_{cr}$ is not sufficiently close to that of q or used when assessing E_s, the iteration through steps 4 through 5 is repeated until the convergency on the value of q is reached.

FOOTING MODULUS NEAR SLOPES

Mark C. Gemperline,[1] Member, ASCE

ABSTRACT: A method is described for estimating the footing modulus of shallow spread foundations located near slopes of cohesionless material. The ratio of footing modulus for a shallow foundation located near a slope to the modulus of a similar footing on level ground is graphically presented. Footing modulus for a footing on level ground (the reference condition) may be estimated either from local experience or using a method presented herein. Equations are presented which relate footing modulus to the footing length, width, depth and proximity to a slope. This method compliments a similar technique recently proposed to account for the effect of slopes on footing bearing capacity. Equations and charts are empirically based on the results of 215 centrifuge model tests. Uncertainties in footing modulus estimates are discussed.

INTRODUCTION

In 1988, the author summarized the results of tests on 215 centrifuge model footings located on, and beneath, horizontal ground surfaces near the crest of slopes of cohesionless sand (Gemperline 1988). An equation was proposed to describe the footing modulus, E_m, for footings of different size and shape located on a horizontal surface near slopes. A similar equation was proposed for bearing capacity. These empirical equations represent observed test results.

Shields et al. (1990) showed that the equation for bearing capacity cannot be extended to include footings located beneath the slope. Unreasonable values are calculated for this condition. It can be shown that the equation for footing modulus is invalid in the region beneath the slope for similar reasons.

---

[1] Civil Engineer, Bureau of Reclamation, Mail Code D-3734, P. O. Box 25007, Denver, CO 80225-0007.

The footing modulus ratio, M_R, is introduced as the ratio of the footing modulus of a shallow footing located near a slope to the modulus of a similar footing on a level ground surface. An equation is presented which fits calculated model footing modulus ratios and can be reasonably extended beneath the slope thereby eliminating the problem identified for similar bearing capacity equations by Shields et al. (1990) . Design charts are presented which show contours of M_R in the vicinity of the slope crest. The design charts compliment similar charts presented by Shields et al. (1990) to account for the effect of slopes on footing bearing capacity factor.

This paper presents a footing modulus equation which is simpler than the original equation presented by the author in 1988 (Gemperline 1988). The equation, derived by regression analysis of model test results, considers only the main effects of footing depth, width, and length on footing modulus. The previous equation considered both main effects and first order interactions between these variables adding to its complexity.

MODEL TESTS

The centrifugal model testing technique was used to experimentally determine the response of shallow footings to a number of conditions (Gemperline 1988). The conditions were created by varying the ground surface geometry and the footing geometry and using four different sands representing cohesionless foundation soils. The geometric variables include footing width (B), footing depth to width ratio (D/B), footing length to width ratio (L/B), slope angle (α), and the ratio of the distance between the slope crest and the footing edge to the width of the footing (b/B). The terms are defined in Fig. 1. Four different sands were placed at different dry unit weights so that there were a total of nine different foundation conditions with properties shown in Table 1.

Footing load (q) and displacement (d) were the measured parameters. Relative displacement (d/B) is introduced as the ratio of the vertical footing displacement to footing width. Data were collected from 215 model tests representing 194 test variable combinations. Variable combinations were selected that would provide information necessary to evaluate main and coupled effects of variables on the footing load displacement response for each of the nine foundation conditions. Table 2 lists the test variable and range for modeled prototype dimensions.

Aluminum model containers were constructed that had inside dimensions of 29.8 cm deep, 43.2 cm wide, and 40.6 cm long. A hydraulic cylinder was rigidly attached to the container and to aluminum blocks representing the model footings. Fig. 2 shows an assembly drawing of the model container and loading system. Two LVDT's (linear variable differential transformers, not shown in Fig. 2) and two load cells were used simultaneously to measure the load-displacement response of model footings. Model footing displacement rate was controlled by regulating the rate of oil flow into the hydraulic cylinder. Twenty to thirty minutes were generally required for each test. Models were constructed of air-dried sand that was poured into the model container in shallow lifts. The unit weight was adjusted by changing

Table 1. General Characteristics of the Nine Foundation Conditions

1/	2/	3/	Mean particle size (mm)	5% smaller particle size (mm)	Maximum particle size (mm)	Specific gravity	Maximum index dry unit weight (kN/m^3)	Minimum index dry unit weight (kN/m^3)
4H	Sa/A	G	0.35	0.20	1.19	2.69	15.9	13.0
4M	Sa/A	G	0.35	0.20	1.19	2.69	15.9	13.0
4L	Sa/A	G	0.35	0.20	1.19	2.69	15.9	13.0
7H	Sa/A	G	0.60	0.11	2.38	2.68	19.3	14.9
7L	Sa/A	G	0.60	0.11	2.38	2.68	19.3	14.9
8H	R/Sr	S	0.20	0.08	0.59	2.66	17.0	14.1
8M	R/Sr	S	0.20	0.08	0.59	2.66	17.0	14.1
8L	R/Sr	S	0.20	0.08	0.59	2.66	17.0	14.1
9H	Sr	S	0.64	0.30	1.19	2.63	16.2	14.3

1/ Foundation condition. - All sands were placed by pluviation in an air-dried condition: H = high unit weight, relative density greater than 90 percent. M = medium unit weight, relative density between 55 and 65 percent. L = low unit weight, relative density less than 5 percent.

2/ Angularity. - Sa/A = subangular to angular. R/Sr = rounded to subrounded. Sr = subrounded.

3/ Predominant mineralogy. - G = granitic rock fragments, quartz, and feldspar; some mica. S = silica.

Note: 1 lbf/ft^3 = 0.16 kN/m^3; 1 lbf/in^2 = 6.9 kPa.

Table 1. General Characteristics of the Nine Foundation Conditions (continued)

1/	Average placement unit weight (kN/m^3)	Average placement relative density (%)	Placement void ratio	ϕ Triaxial shear test friction angle $\sigma_3 = 2$ atm (degrees)	E_t Triaxial shear test secant modulus at $\sigma_3 = 2$ atm $\epsilon_a = 2$ % (kPa)
4H	15.7	95	0.680	43.2	46970
4M	14.1	63	0.795	39.9	35830
4L	13.2	5	1.00	34.8	18770
7H	17.1	55	0.540	42.2	40700
7L	15.1	2	0.753	36.3	17310
8H	16.7	90	0.569	38.7	35270
8M	15.6	57	0.670	36.3	28340
8L	14.3	3	0.834	33.0	20760
9H	16.2	100	0.587	41.3	40950

1/ Foundation condition. - All sands were placed by pluviation in an air-dried condition: H = high unit weight, relative density greater than 90 percent. M = medium unit weight, relative density between 55 and 65 percent. L = low unit weight, relative density less than 5 percent.

Note: 1 $lbf/ft^3 = 0.16$ kN/m^3; 1 $lbf/in^2 = 6.9$ kPa.

the height from which the sand was poured. Sand deposits near 0 percent relative density were made by: (a) placing a 4.75 mm sieve screen [a 9.5 mm screen was used for the 7L foundation condition] in the bottom of the model container; (b) pluviating the sand into the container using a sand pouring device; and (c) slowly pulling the screen vertically through the sand.

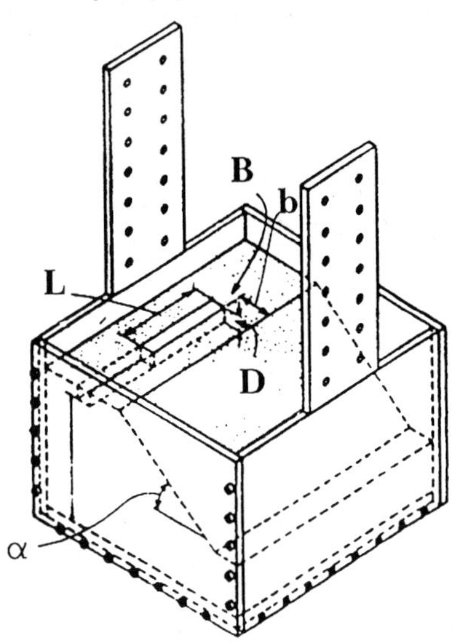

FIG. 1. Definition of Terms

Table 2. Test Variables and Prototype Dimensions

Test variable	Symbol	Prototype dimensions
Footing width	B	0.6, 1.2, and 1.8 meters
Footing length to width ratio	L/B	1, 3, 6
Footing depth to width ratio	D/B	0, 0.5, 1.0
Footing location to width ratio	b/B	0, 1.5
Slope angle	α	0°, 26.6°, 33.7°

Excavation necessary to shape slopes and bury footings was accomplished using a vacuum cleaner to remove most of the sand followed by careful hand trimming.

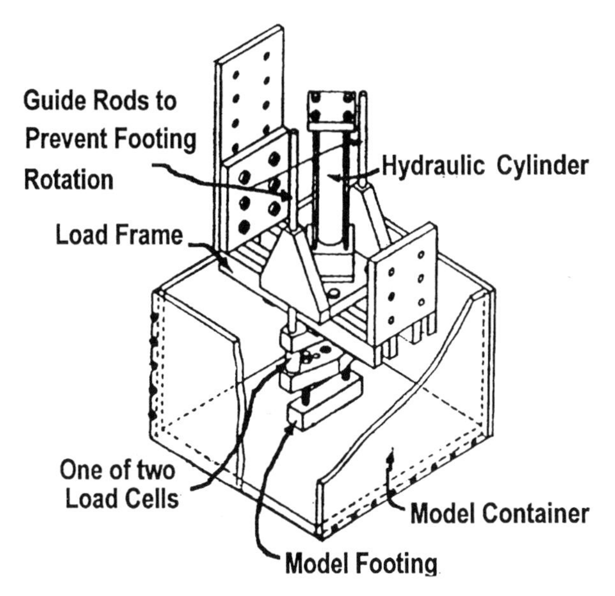

FIG. 2. Assembly Drawing of the Model Container and Loading System

The experimental method was validated using the concept of modeling-of-models (Gemperline 1988). Modeling-of-models involves testing models of the same prototype at different scales, in different testing environments, or with different procedures in order to evaluate the effect of these variables on the applicability of centrifugal modeling. Similar test results for the same prototype footing, tested at several scales, validated the testing technique. Modeling-of-models was also used to demonstrate grain size and model boundary effects (Gemperline 1988).

Triaxial Shear Tests

Consolidated-drained triaxial shear tests were performed on compacted specimens of dry sand. Sixty-eight conventional 5.1 cm diameter by 12.7 cm high triaxial shear specimens were tested at effective minor principle stresses equal to 172, 345, 689, and 1310 kPa. These effective minor principle stresses were selected to include minor principle stresses expected in model testing. Dry unit weights were varied by changing compaction effort, pluviating the sand from various heights, vibrating the specimen in the mold, and/or pulling a screen through the specimen to force dilation. Table 2 lists the friction angle (ϕ) and secant modulus (E_t) interpreted from the triaxial shear tests. These parameters are used to represent the foundation conditions in the empirical analysis of the experimental footing test data. Values of ϕ and E_t are interpreted from test results to represent triaxial test conditions at two atmospheres (atm) confining pressure. Secant modulus (E_t) was interpreted from the test results to represent 2 percent axial strain. A table in a previous paper by the author (Gemperline 1988) presents the same E_t values in units of lbf/in^2, however, that table incorrectly indicates the triaxial shear test condition of one atm confining pressure.

FOOTINGS NEAR SLOPES

Footing modulus, E_m, is defined as the ratio of the bearing pressure, Q_1, at 2.5 cm prototype footing displacement to the corresponding relative displacement (d/B). The modulus ratio, M_R, is the ratio of the footing modulus near a slope to the footing modulus representing the same footing on an infinite horizontal ground surface.

A representative value of the modulus ratio is calculated for all footing locations for model footings having a length to width ratio of 6 and a prototype width of 1.22 meters. Footing modulus is calculated by approximating the load displacement curve to failure as a second degree polynomial, using this equation to calculate the bearing pressure, Q_1, and then dividing Q_1 by the appropriate relative displacement. The average modulus ratio, corresponding standard deviation, and number of values used in developing these statistics, are presented in Table 3 for various D/B and b/B ratios and slopes.

Contour lines representing locations of equal modulus ratios within the slopes are expected to become parallel to the horizontal ground surface as the distance from the slope crest increases. Likewise, it is expected that the contour lines will become parallel to the slope surface, beneath the slope, as the distance from the slope crest increases. A properly rotated and translated hyperbolic function demonstrates this character.

The following equation was fit to the experimental data using multivariable non-linear regression.

$$1 = \left[\frac{\left[\frac{b}{B}+N(0.27M_R-3.8)\right]K-\left[3(M_R-1)-\frac{D}{B}\right]P}{N(4M_R-1.8)}\right]^2 - \left[M\,\frac{\left[\frac{b}{B}+N(0.27M_R-3.8)\right]P+\left[3(M_R-1)-\frac{D}{B}\right]K}{N(4M_R-1.8)}\right]^2 \tag{1}$$

where

$$M = \tan\left(\frac{\alpha}{2}\right)$$

$$K = \cos\left(\frac{\pi-\alpha}{2}\right)$$

$$P = \sin\left(\frac{\pi-\alpha}{2}\right)$$

$$N = \sin(\alpha)$$

A comparison of the experimental M_R with the M_R corresponding to Eq. (1) is shown in Table 3. The greatest difference between these pairs of values is 0.06. This small difference indicates that Eq. (1) reasonably explains the data trend. Design charts based on Eq. (1) are presented as Figs. 3a and 3b. Positive values of b/B indicate the footing is on or beneath the horizontal surface. Negative values indicate the footing is on or beneath the slope. These charts model similar charts developed by Shields et al. (1990) to describe the influence of slopes on bearing capacity.

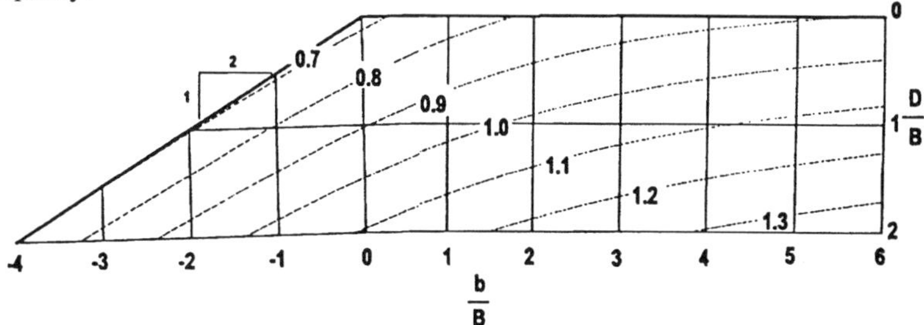

FIG. 3a. Design Chart for 26.6 Degree Slope – Contours of Equal Modulus Ratio

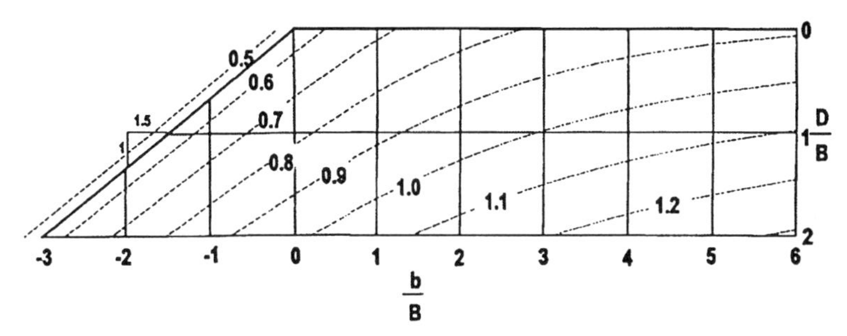

FIG. 3b. Design Chart for 26.6 Degree Slope – Contours of Equal Modulus Ratio

Eq. (1) was derived by regression analysis using the experimental data subject to the following constraints. It was necessary to assure that the modulus ratio contour representing the value 1 approaches the horizontal ground surface at a reasonable rate. A reasonable rate for contours to approach lines parallel to the slope was also required. Experimental data is not adequate to develop these constraints. Imposed constraints are based on the following logic.

If the presence of a nearby slope does not noticeably affect the footings ultimate bearing capacity then it is not expected to noticeably affect the footing modulus. For clarity, the term bearing capacity ratio is introduced to describe the ratio of the ultimate bearing capacity of a footing near a slope to the ultimate bearing capacity of a similar footing on an infinite horizontal surface. It is expected that the modulus ratio approaches unity at the same rate or faster than the bearing capacity ratio.

Shields et al. (1990) reported a 90% reduction in ultimate bearing capacity for a large scale model footing test on dense sand, α = 26.6 degrees, D/B = 0, B = 0.3 m, L/B = ∞, and b/B = 5 (Shields et al., 1990). Consequently the first constraint is to make the modulus ratio approximately 0.9 for the same condition. The second constraint results from the desire not to allow the modulus ratio on the slope surface to exceed the modulus ratio for a similar footing at the crest of the slope.

Contours of M_R greater than 1 in Fig. 3 are not supported by test data in the proximity of the slope. Numerical or physical modeling is required to validate these contours and constitutes a need for future work. The field behavior of foundations should be compared with Eq.(1) predictions to validate design chart usefulness in engineering practice.

FOOTINGS ON HORIZONTAL GROUND SURFACES

A regression analysis was performed using the results of all model footing tests representing infinite horizontal ground surface conditions to develop the following equation relating footing modulus to test variables.

$$E_m = f(E_t)\, f(B)\, f(\frac{D}{B})\, f(\frac{B}{L})\, f(\frac{D}{L}) \qquad (2)$$

where

$$f(E_t) = 0.5\, E_t - 8445$$

$$f(B) = \frac{2B}{B+1}$$

$$f\left[\frac{D}{B}\right] = 1 + 0.3\,\frac{D}{B}$$

$$f\left[\frac{B}{L}\right] = 1 + 0.4\,\frac{B}{L}$$

$$f\left[\frac{D}{L}\right] = 1 + 0.1\,\frac{D}{L}$$

E_t is in kPa and B , D, and L are in meters.

The equation is different in several aspects from a previous equation presented by author to represent the same data (Gemperline 1988). The standard condition chosen for analysis is a 1 m wide footing having a length to width ratio of 6 located on the surface of a horizontal soil layer. The coefficients in the functions representing the main variable effects are determined by regression analysis independently for each variable. This is accomplished using data differing only in the variable of concern from the standard condition. Previous analysis (Gemperline 1988) considered the main effect and first order interaction of several variables. Eliminating variable interactions results in a simpler equation.

Note that the drained triaxial shear test secant modulus (E_t), representing the condition of minor principle stress equal to two atmosphere and two percent axial strain, is selected as an index value to represent sand compressibility. Obviously the function $f(E_t)$ must be greater than zero to yield reasonable results. It is recommended that $f(E_t)$ not be used for estimates of footing modulus less than 700 kPa.

Table 3. Comparison of Calculated and Experimental Footing Modulus Ratio

Slope Angle	$\frac{D}{B}$	$\frac{b}{B}$	Number of Samples near Slope*	Calculated M_R**	Average Experimental M_R***	Standard Deviation Experimental M_R
1½H:1V	0	0	9	0.54	0.54	0.08
1½H:1V	0.5	0	4	0.66	0.71	0.19
1½H:1V	1	0	4	0.78	0.84	0.22
1½H:1V	0	1.5	6	0.72	0.67	0.05
1½H:1V	0.5	1.5	3	0.82	0.81	0.11
1½H:1V	1	1.5	3	0.92	0.89	0.14
2H:1V	0	0	11	0.67	0.66	0.07
2H:1V	0.5	0	5	0.79	0.80	0.10
2H:1V	1	0	3	0.90	0.94	0.13
2H:1V	0	1.5	8	0.78	0.82	0.13
2H:1V	0.5	1.5	3	0.89	0.85	0.13
2H:1V	1	1.5	3	0.99	0.97	0.14

* 18 Tests were performed on horizontal ground surface footings representing footings having prototype dimension B = 1.22 meters and L/B = 6.

** Values calculated by Eq. (1).

*** Values of experimental M_R were calculated by dividing the footing modulus of tests conducted using similar sand type and then averaging. When duplicate tests were available, the average modulus was used in the calculation.

This equation can be used to estimate the footing modulus of shallow footings on and beneath horizontal ground surfaces of cohesionless soil. Care should be taken not to include the depth term when calculating the reference surface footing modulus for application with the design charts and Eq. (1). The ranges of test variables in table 1 indicate the range which may be used with Eq. (2). Future work should include a comparison of the field behavior of foundations with Eq. (2) predictions to validate its usefulness in engineering practice.

Engineers may prefer to use local experience or other methods as a guide to predicting the modulus of a footing on a horizontal ground surface . Still, Eq. (1) and the design curves on Fig. 3 could serve to identify M_R and, thereby, account for the proximity of the slope.

UNCERTAINTIES

Intuitively, it is expected that the error associated with predicting the footing modulus for field conditions is different than the corresponding error associated with the equation fit to experimental data. Fig. 4 graphically displays the latter error by comparing calculated modulus with experimental results. The uncertainty can by described by placing 99% confidence limits on the expected value of the logarithm of the ratio of calculated footing modulus, E_{MC}, to the measured footing modulus, E_{MF}. These limits, based on results of 215 model tests, are given by

$$-0.43 \quad < \quad Log_{10}\frac{E_{MC}}{E_{MF}} \quad < \quad 0.43 \tag{3}$$

Dashed lines on Fig. 4 represent these limits.

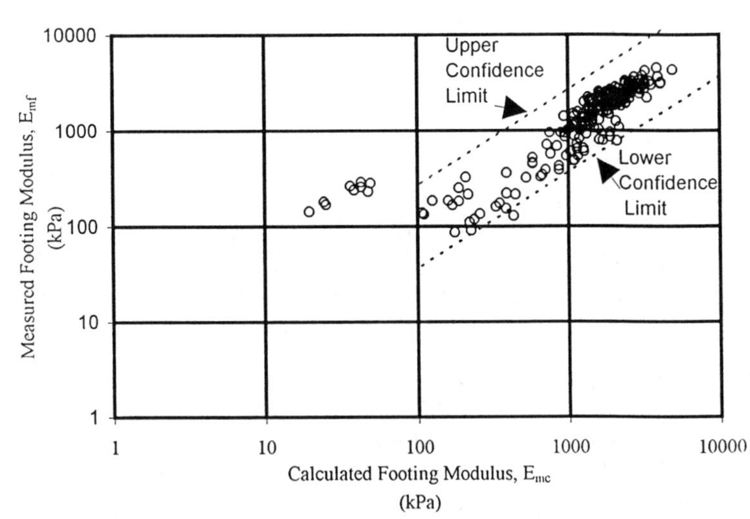

FIG. 4. Measured Footing Modulus Compared with Calculated Footing Modulus

CONCLUSION

Equations and design charts are presented to assist engineers with evaluating the effect of slopes on the expected value of footing modulus. These are empirically based on centrifugal model tests conducted using cohesionless soils and shallow footing conditions. Reasonable boundary conditions pertaining to expected limits of the footing modulus on the slope and at a distance from the slope were implemented in the analysis. These boundary conditions were based, in part, on the work of Shields et al. (1990). Consequently, although model tests were carried out in a narrow region near the crest of the slope, the design charts and equations presented can be used to represent a larger region.

The cluster of data points on Fig. 4 between measured modulus 100 and 1000 kPa and calculated modulus 10 to 100 kPa characterize tests with foundation condition 7L. Foundation condition characteristics are indicated in Table 2. The large deviation between measured and calculated values exemplifies the uncertainty associated with the calculation at low modulus values.

Design charts for footing modulus ratio compliment similar charts for bearing capacity presented by Shields et al (1990). Together, these charts are a useful design tools. They also provide a basis for evaluating the appropriateness of proposed analytical methods.

Numerical and physical modeling are required to validate the presented equations. The field behavior of foundations should also be compared with equation and design chart predications to confirm their usefulness in engineering practice.

ACKNOWLEDGEMENTS

The author would like to thank Mr. Richard Young for his helpful comments and suggestions during the preparation of this paper.

APPENDIX I. REFERENCES

Gemperline, M. C. (1988). "Centrifugal modeling of shallow foundations." *Soil Properties Evaluation from Centrifugal Models and Field Performance*, Geotechnical Special Publication No. 17, ASCE, New York, New York, 45-70

Shields, D., Chandler, N., and Garnier, J. (1990). "Bearing capacity of foundations in slopes." *J. Geotech. Eng.*, ASCE, 116(3), 528-537.

APPENDIX II. NOTATION

The following notation is used in this paper:

B = footing width;
b = distance from the slope crest to the footing edge;
D = footing depth;
d = footing vertical displacement;
E_m = footing modulus
E_{mf} = experimental footing modulus;

E_{mc} = footing modulus calculated by equation;
E_t = triaxial shear secant modulus;
$f(i)$ = function describing the effect of the ith variable on footing modulus;
L = footing length
K,M,N,P = functions of slope angle used in the modulus ratio equation;
M_R = footing modulus ratio;
q = footing load;
Q_1 = bearing pressure at 2.5 cm prototype footing displacement;
α = slope angle;
ϕ = soil friction angle.

Settlement and Deformation Response of a Bridge Footing on a Sloped Fill

Guy Y. Félio,[1] Member, ASCE and Gunther E. Bauer[2]

Abstract: The problem of settlement of spread footings located within or near granular slopes has been investigated in the laboratory using a large testing facility. Based on these results, footings for a prototype bridge abutment were dimensioned. The settlement behaviour of the abutment was monitored over 18 months. The results are given and discussed.

Introduction

Based on test results of large-scale footing tests in sloped granular backfills, it was found that settlements will control the design of spread footings for bridge abutments. These results of the footing tests have been reported previously (Bauer et al. 1979; Bauer et al. 1981; Bauer 1982). But due to scale factors and the lack of methods of analysis, settlement predictions for prototype footings are difficult to make however. Field measurements are generally necessary in many cases to verify predictions.

A prototype bridge abutment, founded on spread footings, was instrumented and its settlement performance was monitored during and after construction. The instrumentation program for movement consisted of monitoring the following quantities: (a) surface settlement of the abutment footing; (b) compression of the underlying soil layers, and (c) horizontal and rotational movements of the abutment. The results of these measurements are discussed with regard to the performance of the bridge.

---

[1] Head, Infrastructure Laboratory, Institute for Research in Construction, National Research Council of Canada, Ottawa, Ontario, Canada, KlA OR6.

[2] Professor, Department of Civil and Environmental Engineering, Carleton University, Ottawa, Ontario, Canada, KlS SB6.

BRIDGE SITE

The bridge was built as an overpass for a four lane highway and is located north of Toronto, Ontario, Canada. The two abutments are founded on spread footings in compacted granular fill. Only the east abutment was instrumented. The length of the span was 82.6 m and was supported at the centre by two piers founded on a square footing. Figs. 1 and 2 show a longitudinal and a cross-section of the abutment face, respectively. These figures also give the overall dimensions of the east abutment. The overall height of the wall was 5.8 m, including the 0.9 m thick footing which was 3.8 m wide. The total length of the abutment, perpendicular to the alignment of the bridge, was 12.8 m. The granular pad on which the abutment was founded was 2.6 m thick and was compacted to 95% of modified Proctor density.

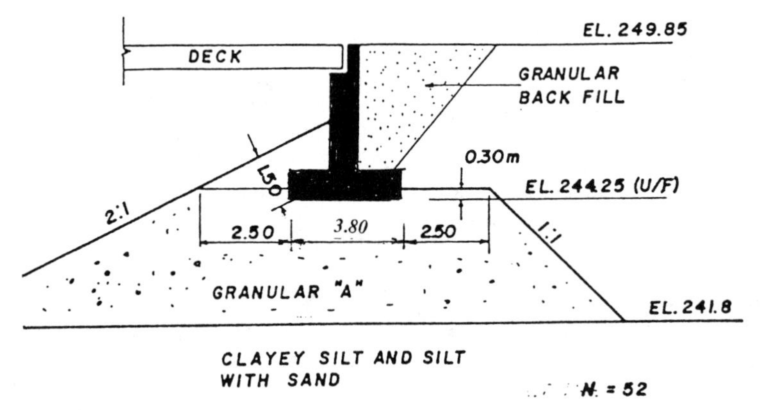

FIG. 1. Longitudinal Section through Abutment

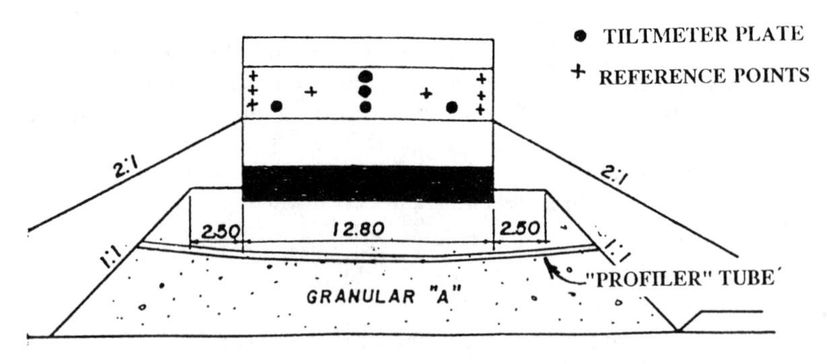

NOTE: ALL DIMENSIONS AND ELEVATIONS IN METRES.

FIG. 2. Cross-Section at Abutment Face

SOIL CONDITIONS

The surface topography in this area is generally flat with an approximate elevation above sea level of 257 m. A dense till deposit of an average thickness of 6 m consisting of clayey silt with varying percentages of sand and gravel. This was underlain by compact to very dense gravelly sand. The blow count from Standard Penetration testing averaged 70 blows per 0.3 m in the till and was in many instances greater than 100. The allowable bearing capacity of the granular fill was 380 kPa based on the model footing tests and no stability problem was anticipated for the approach fills. In the sand the blow count averaged 50 at the till/sand interface and increased rapidly with depth to over 100 blows per 0.3 m. The groundwater level recorded in the boreholes varied from 0.3 m to 0.9 m below the original ground surface.

The compacted granular fill on which the abutment was founded (Figs. 1 and 2),consisted of a well graded crushed stone material. Its grain size distribution is shown in Fig. 3. The maximum dry density (modified Proctor) was 22.8 kN/m^3 with a corresponding moisture content of 7.8%. The specified field density was 95% dry of optimum. The measured field densities varied between 21.7 and 22.4 kN/m^3 with moisture contents in the order of 5 to 6%. The angle of internal friction was determined from large diameter (0.25 m) drained triaxial compression tests on specimens compacted to a dry density of 22.0$\pm$0.4 kN/m^3 with a corresponding moisture of 5%. The internal friction angles varied from 53.5° to 57.5° for confining pressures of 552 and 34 kPa, respectively.

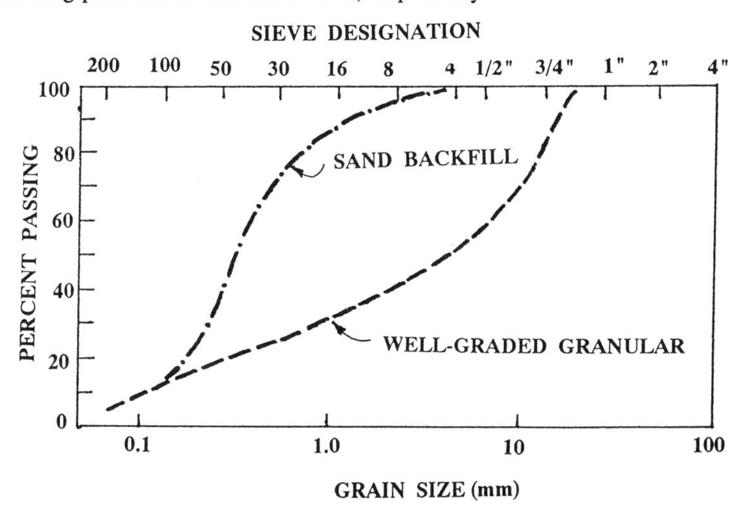

FIG. 3. Grain Size Distribution

The backfill material against the abutment wall consisted of a coarse sand having a grain size distribution which is also shown in Fig. 3. Measured field densities, corresponding to 100% Standard Proctor, varied between 17.5 and 18.8

kN/m^3 with moisture contents between 4 and 5%. Triaxial compression specimens (0.25 m in diameter) compacted to 18.0 kN/m^3 and 4.5% moisture yielded friction angles of 42.0° and 47.5° corresponding to confining pressures of 552 and 40 kPa.

CONSTRUCTION SEQUENCE

The layer of top soil was removed before the granular material for the footing pad was compacted. Compaction was done by a vibratory roller using 0.3 m lifts. The material was compacted to 95% modified Proctor density. When the granular pad reached a height of 2.6 m, the abutment footing was poured directly on the compacted granular fill as indicated in Fig. 4. After the abutment wall was poured the backfilling and compaction operation began behind the wall.

The common backfill against the wall consisted of a coarse sand to an approximate height of 6 m above the footing base (Fig. 4). Compaction close to the wall (< 1.2 m) was done with jumping jacks in order to prevent overstressing of the abutment wall. The wooden formwork for the bridge deck close to the abutment wall was supported on top of the footing by vertical studs.

INSTRUMENTATION FOR MOVEMENT

Settlement Gauges

The settlement of the concrete abutment was measured by means of settlement rods extending from the top of the footing at the toe to the surface. Readings were taken with a precise level during and after completion of construction. Settlement plates were placed at different elevations below the footing within the granular pad. These gauges consisted of a 250 × 250 mm steel base plate to which a 25 mm steel pipe was welded (Fig. 4). This pipe extended through the footings to the surface. A protective 38 mm outer pipe sleeve was used over the inner pipe to prevent the soil from touching the inner measuring rod. The annular space between the two pipes was filled with heavy motor oil for lubrication. The differential settlement between the inner pipe and the outer sleeve, embedded in the concrete footing, was measured with a portable dial gauge. A calibration cup was made for the dial gauge in order to calibrate the setting before and after each measurement.

Settlement Profiler

This system was described by Bozozuk (1969) in detail and consisted of a sensitive pressure transducer, referred to as "torpedo" being pulled through a liquid (ethylene glycol) filled plastic tubing. The pressure transducer measures the differential pressure between any location within the tube and a reference point, usually the open end(s) as the extremity of the tube. The pressure difference was then correlated to vertical movement. The tubing was installed under the centre of the abutment wall perpendicular to the alignment of the bridge as indicated in Fig. 2. The settlement readings recorded from the device were found to be influenced considerably by atmospheric pressure variations. The readings were found to be erratic and were considered not reliable.

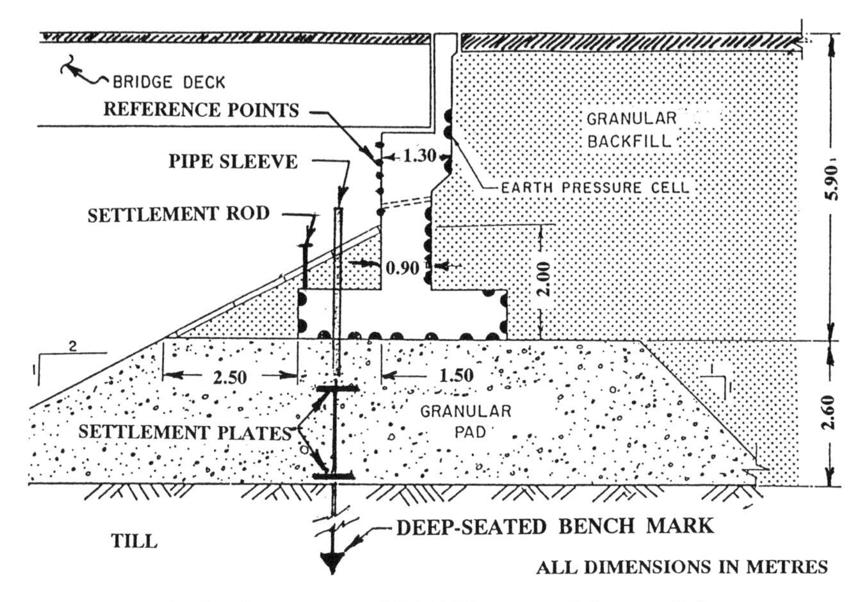

FIG. 4. Settlement and Wall Movement Reference Points

Tiltmeter

A commercially available tiltmeter (SINCO Digitilt) was employed to monitor any change in vertical alignment of the face of the abutment wall. Five ceramic tilt plates were installed on the open-face of the wall as shown in Fig. 2. The change in vertical alignment was measured by holding the Digitilt instrument against the prongs of the tiltplate.

Horizontal Wall Movement

The total horizontal movement of the wall was measured by offsets from an optical reference line using a theodolite. A series of measuring points were marked on the vertical face of the wall. Of course, these points could only be located on the wall above ground surface. The locations of these points are also shown in Fig. 2.

Presentation and Discussion of Results

Vertical Settlement

As mentioned in the previous section, the settlement of the structure was measured by means of three independent systems, settlement plates, settlement rods and a settlement profiler. The latter system was inconsistent for most of the time and no reliable measurements could be obtained. The results of the settlement measurements on the structure and the compressibility of the underlying soil is shown in Fig. 5. The settlement of Layer A refers to the compression of the soil between the base of the footing and the level of the first settlement plate located at 1.25 m below the base. The pouring of the concrete bridge deck started on day 35 and is reflected by a rapid increase in the rate of settlement. After completion of the bridge construction and backfill operation on day 168, very little additional settlement took place. The total settlement measured after 700 days is in the order of 8 mm, well within the limits of acceptability of 25 mm for such a structure (Bozozuk 1978).

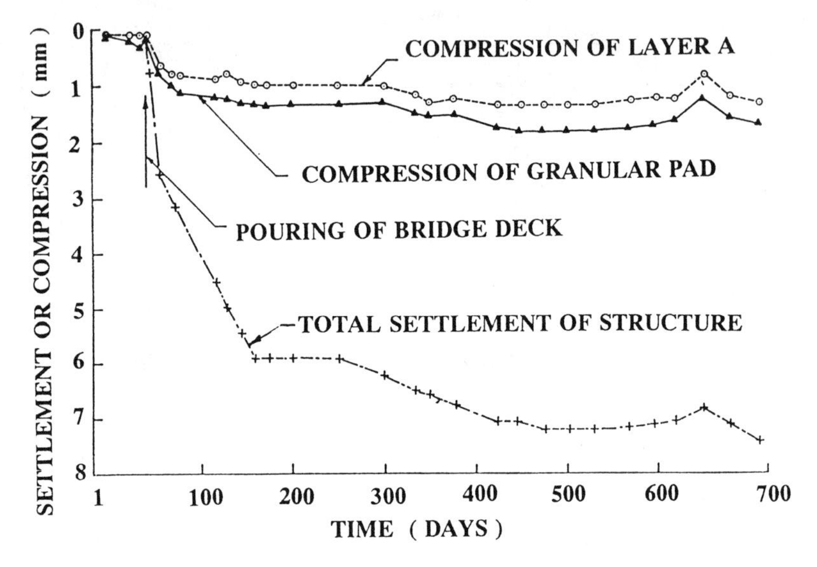

FIG. 5. Observed Settlements

The results of the settlement rods embedded at the toe of the footing are shown in Fig. 6. The variation of settlements with time, as given in this figure, shows clearly that a rapid increase in settlement occurred during the pouring of the bridge deck. The maximum settlement of the footing was recorded as 22 mm compared to 8 mm for the settlement plates. This difference is quite logical since the formwork close to the abutment carrying the wet concrete of the bridge deck was supported by props located at the top and towards the toe of the footing. It necessarily settled more than the centre portion of the footing where the settlement plates were located. As soon as the load came off the formwork, due to the hardening of the wet concrete, the footing rebounded as shown in Fig. 6. The final settlement was in the order of 10 to 12 mm which agrees reasonably well with those values recorded by the settlement plates shown in Fig. 5.

A settlement analysis using Schmertmann's method (1970) was carried out using the following relationship:

$$S = C_1 C_2 \Delta p \sum_{0}^{2B} \frac{I_z \Delta z}{E} \tag{1}$$

where C_1 = embedment factor (C_1 = 1.0); $C_2 = 1+0.2 \log_{10} (t_{years}/0.1) = 1.26$ for case on hand; Δp = applied stress; I_z = strain factor; and E = soil modulus. The geometry of the problem is shown in Fig. 7 and the information used in the calculation is given in Table 1. The modulus of deformation of the granular pad was obtained from the triaxial test results. A secant modulus at one-half of the failure stress was used. The moduli of the underlying subsoil were obtained from the blow count of the Standard penetration test profile. The resulting modulus values are listed in column 4 of Table 1. The average applied contact stress from dead load considerations was 158 kN/m^2. Using this value one would obtain a total settlement of 13.8 mm. In order to compare the observed settlements (Figs. 5 and 6) with the calculated settlement one must keep in mind that settlement readings were started to be recorded several weeks after the granular pad and the 1 m thick footing had been constructed. Taking this into consideration the precompression settlement caused by the combined weight of the footing and the pad on the underlying soil layers, would amount to 2.6 mm. This will result in a net settlement of 11.2 mm. This value should be used when comparing the settlement behaviour of the measured values in Figs. 5 and 6. One, therefore, can state that the calculated and observed values are in good agreement. One should also keep in mind the relatively small vertical deformation the structure underwent.

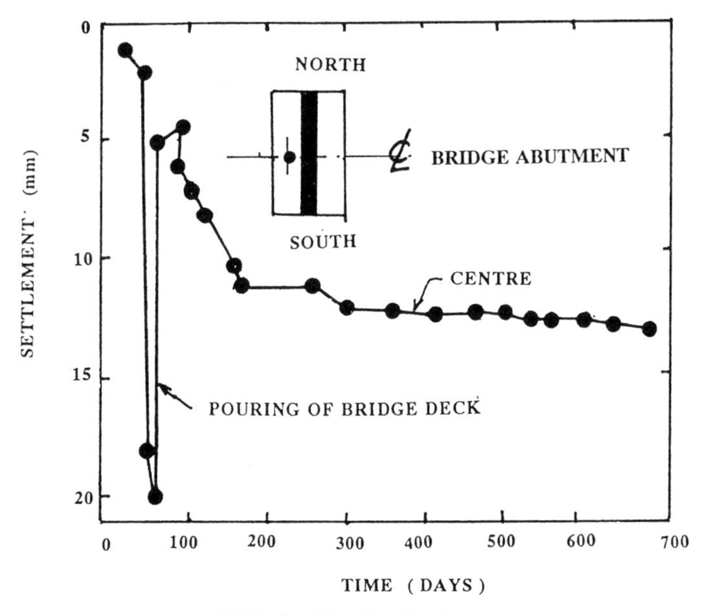

FIG. 6. Footing Settlement

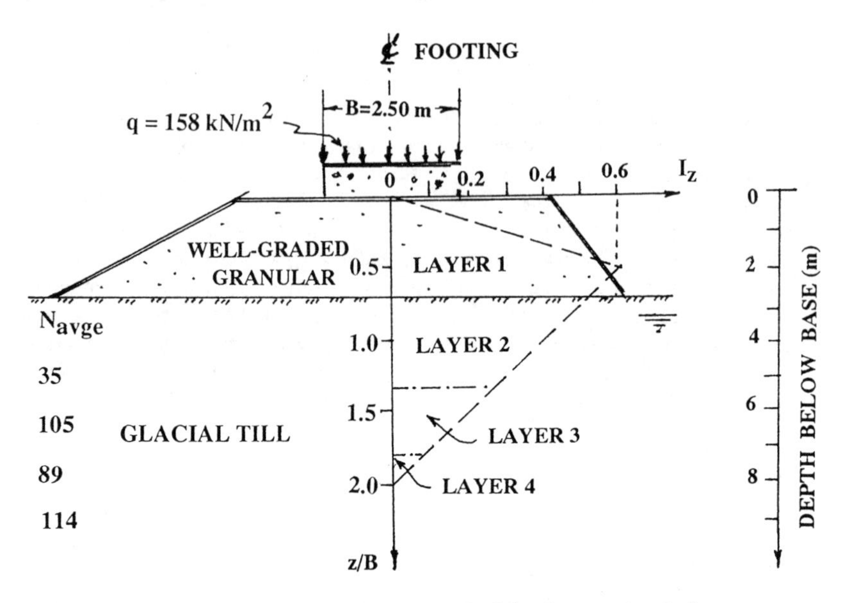

FIG. 7. Schmertmann's Method of Settlement Analysis

TABLE. 1. Values for Settlement Calculation

Layer No. (1)	Δz (m) (2)	N (blows) (3)	E (kPa) (4)	Z_c (m) (5)	I_z (6)	$(I_z/E)\Delta z$ (m/kPa) (7)
1	2.6	–	4.2×10^4	1.3	0.415	2.57×10^{-5}
2	2.4	35	2.4×10^4	3.6	0.39	3.9×10^{-5}
3	1.8	9.7	6.7×10^4	5.9	0.175	0.47×10^{-5}
4	0.8	114	7.8×10^4	7.2	0.04	0.04×10^{-5}
					Σ =	6.98×10^{-5}

Wall Movement

The translational movement of the vertical face of the abutment wall was obtained by measuring optically the movements of reference points located at the face of the wall using a precision theodolite. These movements are given in Fig. 8. The wall was moving away from the backfill and at the same rotating about the toe. The maximum observed horizontal Movement occurred at the top of the wall and was approximately 23 mm. Generally, horizontal movements of less than 25 mm are acceptable for bridge abutment walls and should not cause any structural distress (Bozozuk 1978). The average horizontal movement at midheight was about

11 mm. The variation of the horizontal movement of the abutment wall at midheight perpendicular to the bridge alignment (skewing effect) is shown in Fig. 9. The north end of the wall was moving slightly more than the south end (by approximately 3 mm) away from the backfill.

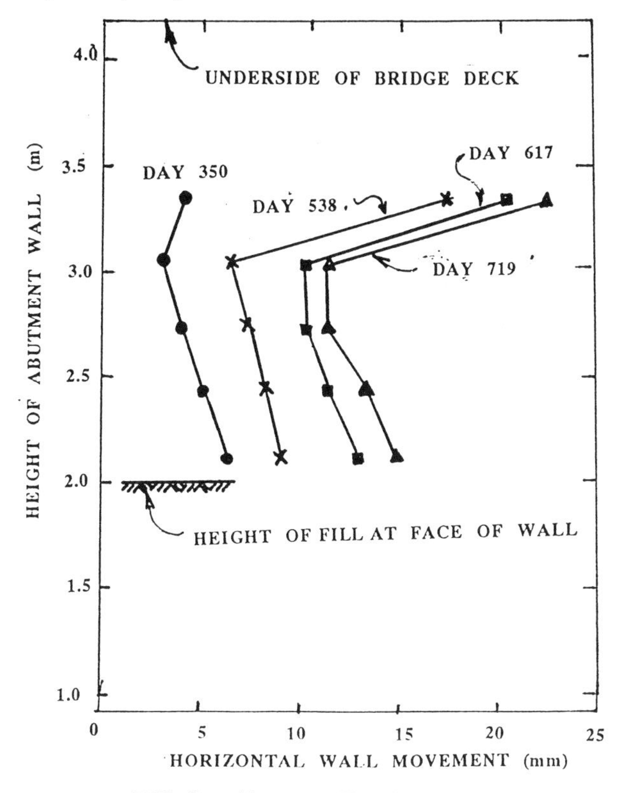

FIG. 8. Abutment Face Movement

The data from the five tiltmeter plates were inconclusive. Only a few readings could be obtained due to damage to the reference pegs of the ceramic plates (vandalism). The readings which were obtained confirmed the measurements from the optical instrument (theodolite) indicating that the wall rotated about its toe and away from the backfill.

CONCLUSIONS

The performance of the bridge abutment must be considered satisfactory. Tolerable vertical and horizontal displacements were recorded over the 18 month period following the construction of the bridge. According to the classification given by Bozozuk (1978), the movements of the bridge abutment under study was considered acceptable and no major distress occurred due to differential or rotational movements. The most reliable data were obtained from the settlement plates installed at different depths below the base of the footing.

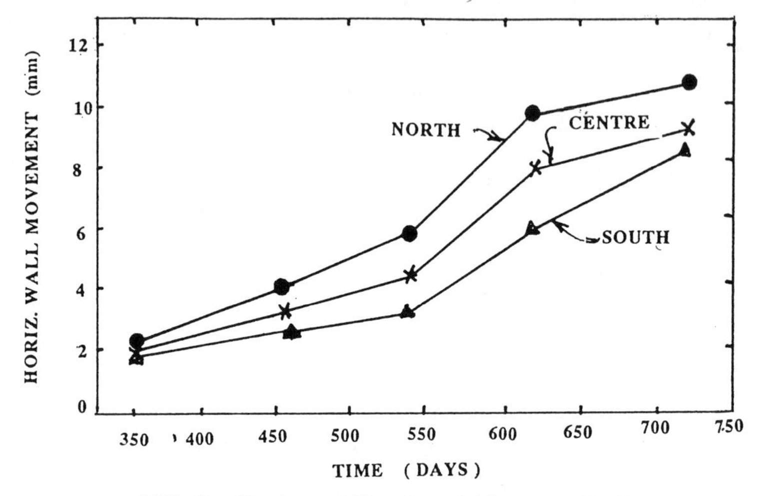

FIG. 9. Horizontal Skewing of Abutment Wall

The settlement rods yielded also reliable reading using precise levelling equipment. Horizontal movement measurements proved less reliable due to disturbance of the baseline benchmarks by construction equipment and due to vandals destroying the reference pegs of the tiltmeter plates.

APPENDIX - REFERENCES

Félio, G. Y. and Bauer, G. E. (1984). "Monitoring and performance of a bridge abutment." *Proc. Int'l. Conf. on Case Histories in Geotechnical Engineering*, St. Louis, 1, 235-240.

Bauer, G. E. (1982). "Design guide for abutment footings located in compacted granular fills," *Final Report*, Ministry of Transportation of Ontario, Downsview, Ontario, Canada

Bauer, G. E., Shields, D. H., Scott, J. D., and Gruspier,J. E. (1981). "The bearing capacity of footings in granular slopes." *Proc. 10th ICSMFE*, Stockholm, 5, 431-438.

Bauer, G. E., Scott, J. D., and Shields, D. H. (1979). "Testing for soil parameters bearing capacity of foundations in granular soils." *Proc. 7th European Conf. on Soil Mech. and Found. Eng.*, Brighton, 1, 138-143.

Bozuzuk, M. (1969). "A fluid settlement gauge." *Can. Geotech. J.*, 6(3), 362-364.

Bozuzuk, M. (1978). "Bridge foundations move." *Tolerable Movements of Bridge Foundations, Sand Drains and Culverts*, Transportation Research Record 678, Transportation Research Board, Washington, D. C., 17-21.

Félio, G. Y. (1980). "Monitoring of a bridge abutment founded on granular compacted fill," M.Eng. thesis, Department of Civil Engineering, Carleton University, Ottawa, Canada.

Schmertmann, J. H. (1970). "Static cone to compute static settlement over sand." *J. Soil Mech. found. Div.*, Proc. ASCE, 96(SM3), 1011-1023.

SETTLEMENT OF TEST FILLS FOR CHEK LAP KOK AIRPORT

Dominic O. Kwan Lo[1], A.M., ASCE, and Gholamreza Mesri[2], M. ASCE

ABSTRACT: Embankments were constructed about 200 m offshore from Chek Lap Kok to examine the feasibility of using vertical drains to allow successful airport reclamation over the very soft marine clay forming the seabed. Settlement and pore water pressure observations are available over a period of about 600 days. ILLICON methodology was used to evaluate oedometer data, to obtain information on compressibility and permeability of seabed, and to predict and interpret pore water pressure and settlement profiles under the embankments. At Chek Lap Kok, the mobilized discharge capacity, q_w(mob) of Alidrains was less than the initial minimum discharge capacity, q_w(min) required for negligible well resistance; however, as horizontal permeability of marine clay decreased during consolidation, the required q_w(min) decreased and Alidrains functioned quite adequately. The sand drains which were installed after significant undrained loading of the soft seabed, were intruded by the soft clay. The mobilized permeability of contaminated sand drain, k_w(mob) was significantly less than the permeability of clean sand. However, as k_w(min) required for negligible well resistance decreased substantially during consolidation, the sand drains did enhance consolidation rate of the soft marine clay.

INTRODUCTION

In 1970s, a replacement airport was proposed to be built at Chek Lap Kok, Hong Kong. This operation involved reclamation of about 600 ha of land from the

---

[1] Project Engineer, Golder Associates, Inc., 305 Fellowship Rd., Mt. Laurel, New Jersey 08054.

[2] Professor of Civil Engineering, University of Illinois at Urbana-Champaign, 205 North Mathews Avenue, Urbana, Illinois 61801.

sea, with fill thickness reaching as much as 15 m. A preliminary site investigation revealed the presence of a very soft marine clay forming the seabed. A geotechnical investigation was commissioned by the government of Hong Kong to examine the feasibility of using vertical drains to allow successful construction over the marine clay (Koutsoftas et al. 1987). A test embankment, 100 m^2 in plan, was constructed between 1981 and 1982 about 200 m offshore from Chek Lap Kok (Foott et al. 1987). The southeast (SE) quadrant of the test fill was the control area with no vertical drains. Displacement type sand drains of 50 cm diameter were installed at 3 m spacing in the SW quadrant. Alidrains were installed in the NW and NE quadrants at 1.5 and 3 m spacing, respectively. A haul road, 35 m wide, was also constructed adjacent to the test embankment. Alidrains were installed at 1, 2 and 3 m spacing under three different sections of the haul road embankment. The test embankment and haul road which were constructed during a period of 350 and 210 days, respectively, were instrumented to monitor behavior of seabed during and after construction (Handfelt et al. 1987). The instrumentation included piezometers at 12 different elevations below seabed, from 1.5 m down to 25.0 m depth, settlement plates, and subsurface settlement anchors at 8 elevations from 1.1 to 17.5 m depth below seabed. Settlement and pore water pressure observations are available over a period of about 600 days.

A typical subsurface profile is shown in Fig. 1. The maximum sea water depth in the reclamation area is 10 m. The seabed consists of 4 - 7 m of very soft upper marine clay with occasional organics and shells. This post-glacial clay which is about 8000 years old, has values of σ'_p/σ'_{vo} in the range of 1.5 to 2. In some areas, a distinct bed of marine sand with shells separates the upper marine clay from the upper alluvial crust. The upper alluvium with thickness of 2 - 8 m and σ'_p/σ'_{vo} values in the range of 4 - 8, consists of interbedded layers of stiff clay and dense sand. The sand layers which are typically less than 2 m thick are believed to be of limited lateral extent. However, pore water pressure measurements suggest that the sand layers acted as an impeded drainage boundary to the upper marine clay.

The lower marine stiff clay, which is 5 - 15 m thick, is about 30,000 years old. It was preconsolidated by desiccation, and has values of σ'_p/σ'_{vo} in the range of 2 - 3. The lower alluvium consists of very dense coarse to fine sand with occasional pockets of clay. Based on field observations during drilling, Dames and Moore (1982a) concluded that the lower alluvium has low permeability. The lower alluvium grades into the decomposed granite through a layer of gravel and cobble.

ILLICON methodology (Mesri and Choi 1985; Mesri and Lo 1987) was used to evaluate oedometer data (Dames and Moore 1982a) to obtain information on compressibility and permeability of seabed and to predict and interpret pore water pressure and settlement profiles at seven locations which are identified in Table 1 (Lo 1991). Information and assumptions on subsurface conditions, loading, and drainage boundary conditions, required for ILLICON analysis, are described in the following sections.

TABLE 1. Compressible Stratum and Vertical Drains

	Test Embankment				Haul Road		
Drain Type	No Drains	Alidrains	Alidrains	Sand Drains	Alidrains	Alidrains	Alidrains
Drain Spacing (m)	–	1.5	3.0	3.0	1.0	2.0	3.0
Depth of Drain Tip (m)	–	17.8	7.8	7.6	9.5	9.5	9.2
Thickness of Upper Marine Clay (m)	6.9	6.9	6.6	7.1	6.5	6.4	3.9
Thickness of Stratum (m)	25.0	26.8	26.8	26.4	24.7	20.0	14.4
Number of Layers (ILLICON)	10	11	11	11	10	9	8

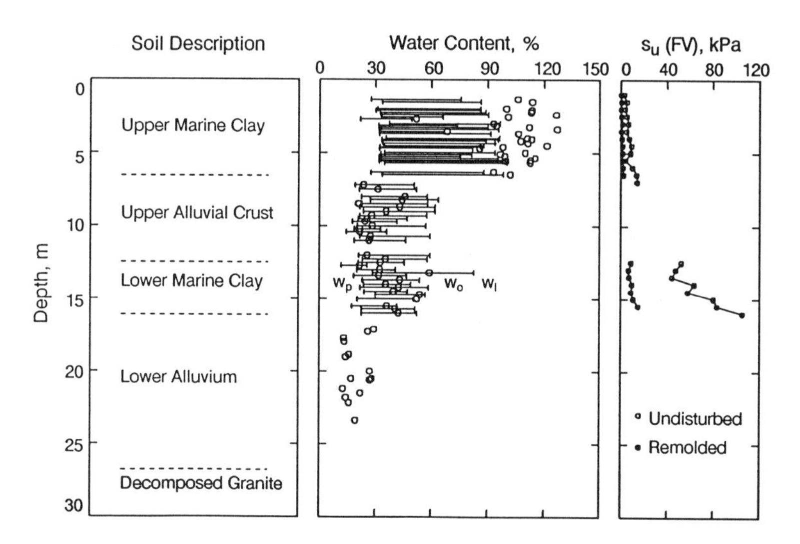

FIG. 1. Typical Soil Profile Under Test Embankment and Haul Road

Initial Void Ratio

Values of initial void ratio, e_o, for all samples used in laboratory tests were reported by Dames and Moore (1982a). Samples from different boreholes were used to prepare e_o profiles for the seven embankment areas in Table 1. The data for test embankment area with 1.5 m Alidrain spacing (NW quadrant) are shown in Fig. 2. Based on profiles of initial void ratio, preconsolidation pressure and coefficients of permeability, for ILLICON analysis the soil profile was divided into a number of layers as indicated in Table 1. The e_o of each layer was taken as the arithmetic average of the initial void ratio data within that layer.

Initial Effective Vertical Stress

The σ'_{vo} profile at each embankment area was computed using the average unit weights of soil layers, reported by Dames and Moore (1982a). The average unit weights of the upper marine clay, upper alluvium, lower marine clay and lower alluvium are 15.0, 19.0, 18.0, and 20.0 kN/m^3, respectively. The mean sea level was at elevation +1.5 m P.D. (Hong Kong Principal Datum) which was about 4.5 m above the sea bed in the vicinity of test areas. The σ'_{vo} profile for the NW quadrant is shown in Fig. 2b.

Preconsolidation Pressure

According to Dames and Moore (1982a) and Koutsoftas et al. (1987), 222 oedometer tests were performed on undisturbed samples of upper marine clay (97 tests), lower marine clay (78 tests) and upper alluvium (47 tests). For ILLICON analysis, the quality of the oedometer specimens was evaluated using the vertical strain that occurred when the soil specimens were reloaded to σ'_{vo}, and was categorized according to the scheme proposed by Andresen and Kolstad (1977). Fig. 3 shows that on the whole, the quality of the oedometer specimens was very good. However, the A to B quality assigned for the specimens of the upper alluvial crust may be misleading, since the specimen quality scheme by Andresen and Kolstad (1977) was intended for soft clays, and is expected to overestimate the quality of the less compressible stiff clays. The end-of-primary (EOP) e - log σ'_v curves were used together with the Casagrande construction (Casagrande 1936) to define preconsolidation pressure, σ'_p. Fig. 2b shows the significant scatter in the preconsolidation pressure data, as well as in the values used in ILLICON analysis at the NW quadrant.

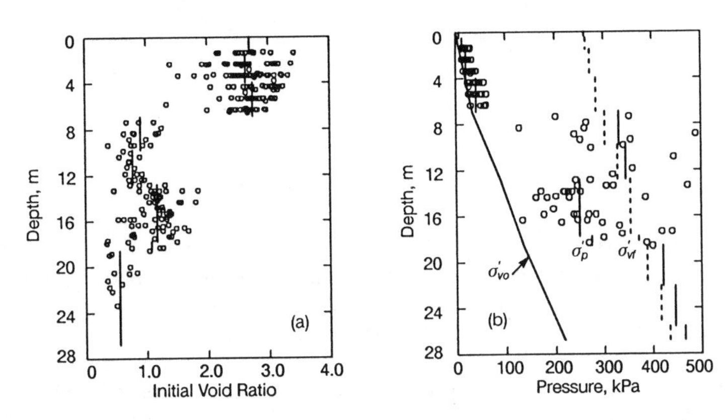

FIG. 2. (a) Initial Void Ratio Profile; (b) Profiles of σ'_{vo}, σ'_p and σ'_{vf}, at NW Quadrant

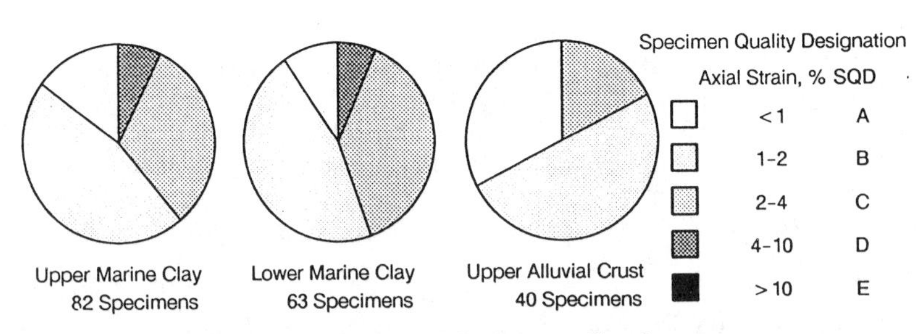

FIG. 3. Quality of Oedometer Specimens

Final Effective Vertical Stress

The embankment construction schedules reported by Dames and Moore (1982b) and Foott et al. (1987) were used to compute $\Delta\sigma_v$. The value of σ'_{vf} at mid-depth of each layer was computed from $\sigma'_{vf} = \sigma'_{vo} + \Delta\sigma_v$. For each area and construction schedule, $\Delta\sigma_v$ profile was computed using the Boussinesq elastic stress distribution for homogeneous half space. The hydraulic sand fill was assumed over a rectangular area. The decomposed granite fill at the haul road was represented by a strip load and that at the test embankment by a rectangular load. The unit weight of the hydraulic sand fill and of decomposed granite fill was assumed to be 19 kN/m^3 (Dames and Moore 1982b). The values of σ'_{vf} for the NW quadrant are shown in Fig. 2b.

EOP e - log σ'_v

The EOP e - log σ'_v curves of the 222 oedometer tests were reported by Dames and Moore (1982a). These were used to prepare C'_c - log σ'_v/σ'_p data corresponding to different layers (Mesri and Choi 1985a). RMP ENCON (1983) reported that an unusually high accumulation of shells and shell fragments under the haul road contributed to a lower compressibility of the upper marine clay. Therefore, the average C'_c - log σ'_v/σ'_p curve was used to construct EOP e - log σ'_v for the upper marine clay under the test embankment, and a lower C'_c - log σ'_v/σ'_p was used for the upper marine clay under the haul road. Based on oedometer test results, recompression index, C_r, values for the upper marine clay, the upper alluvium, the lower marine clay and lower alluvium, of 0.09, 0.05, 0.08, and 0.01, respectively, were used. The EOP e - log σ'_v for ILLICON analysis was constructed for each layer starting from (e_o, σ'_{vo}) with a recompression slope C_r, which extended to σ'_p, beyond which the compression curve was constructed using the corresponding C'_c data (Mesri and Choi 1985). The EOP e - log σ'_v curves for settlement analysis of the NW quadrant are shown in Fig. 4a.

Coefficient of Permeability

Dames and Moore (1982a) reported values of k_v which were calculated from the laboratory c_v data of incremental loading oedometer tests. For the present analysis, for each specimen an e - log k_v was plotted and extrapolated to zero compression to obtain k_{vo}. The values of C_k (= $\Delta e/\Delta \log k_v$) were also determined from the slope of these e - log k_v plots. These values of k_{vo} and C_k for the NW quadrant are plotted as hollow symbols in Fig. 4b. An average value of c_v = 27 m^2/yr from Dames and Moore (1982a) was used together with compressibility data to estimate a $k_v = 2\times10^{-7}$ cm/s for the upper alluvium. Dames and Moore (1982b) assumed a $k_v = 10^{-6}$ cm/s for the lower alluvium, which was adopted in the present analysis.

Dames and Moore (1982a) reported in situ variable head permeability tests at 6 and 7 m depths in the upper marine clay. Rising head tests were carried out in the upper marine clay at five locations, and in the lower marine clay at two locations. In some locations, these tests were followed by falling head tests. RMP ENCON (1983) reported additional rising head and constant head in situ permeability tests in the upper marine clay at depths of 0.5 to 5 m, using hydraulic piezometers installed under the test embankment. For the present analysis, e - log k_h plots were constructed with the help of EOP e - log σ'_v at the particular depth, and were extrapolated to e_o to estimate k_{ho}. These values of k_{ho} together with in situ data reported by Dames and Moore (1982a) are plotted as solid symbols in Fig. 4b. It can be seen that the k_{ho} data from in situ tests are within the scatter of k_{vo} data from laboratory c_v values. Dames and Moore (1982a) reported values c_h computed from piezocone pore water pressure dissipation tests. The value of c_h in the compression range was estimated by multiplying the piezocone c_h by C_r/C_c. These values of c_h together with c_v in compression range from laboratory oedometer tests lead to c_h/c_v values of about 3 for both the upper and lower marine clays. However, Dames and Moore (1982a) cautioned that the piezocone c_h is likely to overestimate the actual value, as an initial sudden drop in the measured pore water pressure was caused by stopping the penetration of the piezocone rather than by dissipation of the pore water pressure.

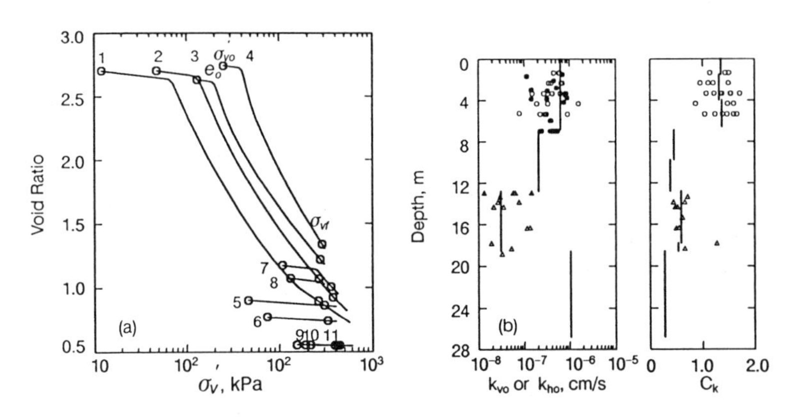

FIG. 4. (a) EOP e - log σ'_v Used in ILLICON; (b) k_{vo}, k_{ho} and C_k Profiles, for NW Quadrant

The values of k_{ho}/k_{vo} for the upper and lower marine clay from permeability tests and piezocone tests, respectively, are about 1 and 3. However, the field permeability tests probably underestimated k_{ho} as most data are from rising head tests which cause a consolidation of clay adjacent to the piezometer, whereas the piezocone dissipation tests probably overestimated c_h. Hence, for the present analysis a value of k_{ho}/k_{vo} equal to 2 was used for both the upper and lower marine clays. For the alluviums, no data on k_{ho}/k_{vo} was available and a value of 1 was used. The values of k_{vo} which were used in the ILLICON analysis of the NW quadrant are shown in Fig. 4b. These were used together with k_{ho}/k_{vo} to compute k_{ho} profiles.

The C_k data in Fig. 5a confirm $C_k = 0.5\ e_o$ for the upper and lower marine clays (Tavenas et al. 1983; Mesri et al. 1994). The same relationship between C_k and e_o was assumed for the alluviums for which no C_k data were available. This relationship, together with the e_o profiles, was used to determine C_k profile for each area such as the one for the NW quadrant shown in Fig. 4b.

In the present analysis, the hydraulic fill above the upper marine clay and the decomposed granite underlying the lower alluvium were considered to be freely draining. Pore water pressure measurements suggest that the sand layer between upper marine clay and upper alluvium acted as an impeded drainage boundary through horizontal water flow. The effect of this drainage boundary is significant in the control SE quadrant with no vertical drains, and is less important after vertical drains were installed. In the present analysis, in the control area and other areas before the vertical drains were installed, pore water pressure at this impeded drainage boundary was set equal to the measured pore water pressures. After the vertical drains were in place, this boundary condition was ignored in areas with vertical drains.

COMPRESSIBILITY WITH RESPECT TO TIME

In the ILLICON procedure, the parameter describing the contribution of compressibility with time during the primary consolidation stage and for computing secondary compression is C_α/C_c, where secondary compression index $C_\alpha = \Delta e/\Delta \log t$ and compression index in the compression and recompression range $C_c = \Delta e/\Delta \log \sigma'_v$. Dames and Moore (1982a) reported secondary compression data at various loads from the incremental loading oedometer tests on undisturbed specimens of the upper and lower marine clays and the upper alluvium. For the present analysis, the values of C_α were computed and the values of the corresponding C_c were determined from the EOP e - log σ'_v curves. The data for the upper marine clay are shown in Fig. 5b. Similar plots for the upper alluvial crust and the lower marine clay (Lo 1991), respectively, resulted in values of C_α/C_c of 0.03 and 0.05. Since there was no C_α/C_c data for the lower alluvium, it was assumed to be the same as that of the upper alluvium.

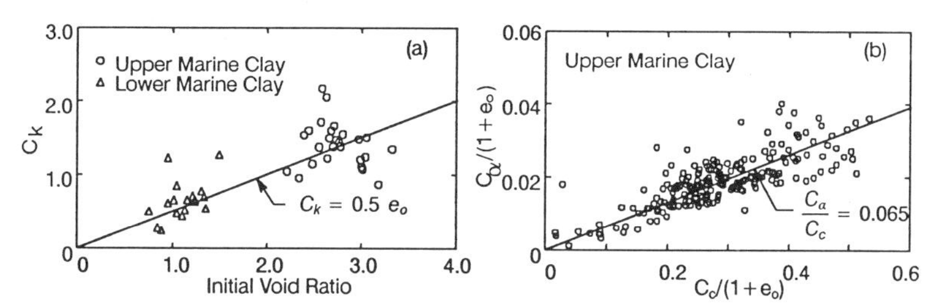

FIG. 5. (a) Relationship Between C_k and e_o; (b) Relationship Between C_α and C_c

DISCHARGE CAPACITY OF VERTICAL DRAINS

The function of a vertical drain is to accept radial flow of water from the consolidating ground, transport it in the vertical direction, and discharge it into top and/or bottom drainage layers, with as little hydraulic resistance as possible. Well resistance refers to the finite permeability of vertical drain with respect to that of the soil. Well resistance depends on the amount of water the consolidating soil discharges into the vertical drain and on the discharge capacity, q_w, and maximum drainage length of the vertical drain, ℓ_m. Mesri and Lo (1991) defined a discharge factor, $D = q_w/k_{ho} \ell^2_m$, where k_{ho} is the initial horizontal permeability of the clay. For a wide range of subsoil conditions and vertical drain geometries, well resistance is negligible when D is greater than 5. Thus, the minimum discharge capacity required for negligible well resistance is $q_w(\text{min}) = 5\ k_{ho}\ \ell^2_m$. The values of q_w(min) for the practical range of values of k_{ho} and ℓ_m are shown in Fig. 6a.

The threshold value of D is a function of the magnitude of decrease in horizontal permeability during the primary consolidation stage, expressed in terms of k_{ho}/k_{hf}, where k_{hf} is horizontal permeability at the end-of-primary consolidation. In other words, the threshold value of D and the corresponding required q_w(min) decrease during the process of consolidation. As k_h decreases during consolidation, less water enters the vertical drain at a given time, and therefore, a smaller q_w(min), is required to discharge water with negligible hydraulic resistance. Thus D = 5 and Fig. 6a specify the initial value of q_w(min) which may be reduced with the help of the D against k_{ho}/k_{hf} relationship in Fig. 6b.

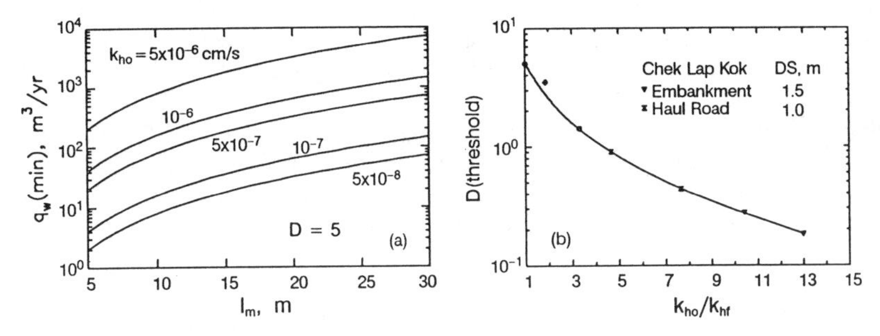

FIG. 6. (a) q_w(min) As a Function of ℓ_m and k_{ho}; (b) Threshold D as a Function k_{ho}/k_{hf}

A number of factors related to installation procedure, filter action, creep and chemical deterioration of filter fabric, clogging of filter and/or core channels, and bending of the vertical drains, may operate in field situations and result in values of mobilized discharge capacity, q_w(mob), significantly less than those measured by laboratory longitudinal flow tests. For the present analysis of Chek Lap Kok embankments, values of q_w(mob) were computed by comparing the predictions and measurements of settlement as well as pore water pressure profiles. A constant value of q_w(mob) was required at each of the four Alidrain sections to obtain agreement between predictions and observations. However, at the NW quadrant of test fill, where the Alidrains were almost twice as long as in other areas and the embankment loads were the highest, q_w(mob) decreased to 1/2 of its initial value during about 600 days of primary consolidation. The values of q_w(mob) as compared to the required values of q_w(min), as well as the k_w(mob) of the contaminated sand drains, are presented in a later section.

VERTICAL DRAIN INSTALLATION AND SMEAR ZONE

Installation of vertical drains creates a disturbed soil of external radius r_s around the drain, which is called the smear zone. A displacement type of installation procedure using a mandrel (of radius r_m) produces the most severe disturbance. The outward displacement of soil distorts the adjacent ground and a zone of soil close to the drain is remolded and dragged downward and then upward as the mandrel is pushed into and then pulled out of the ground. The overall effect is to produce a disturbed soil zone of reduced permeability and preconsolidation pressure, and increased compressibility. Field measurements suggest values of r_s/r_m in the range of 2 to 4 (Mesri and Lo 1991). Note that whereas for sand drains r_w nearly the same as r_m, the equivalent radius of a wick drain ($r_w = (a+b)/\pi$, where a and b are the thickness and width of the drain, respectively, Hansbo 1979) is significantly smaller than r_m. Therefore, $s = r_s/r_w$ for wick drains may have magnitudes significantly greater than r_s/r_m values of 2 to 4.

A diamond shaped mandrel with external dimensions of 75 and 166 mm was used to install the partially penetrating 7.6 to 17.8 m long Alidrains (triangular pattern). The values of a = 4 mm and b = 100 mm lead to r_w = 0.034 m. The equivalent radius of the mandrel is r_m = 0.063 m, which together with an assumed value of r_s/r_m = 2 results in radius of smear zone of r_s = 0.126 m. In the case of the sand drain r_s = 0.50 m which is twice the radius of the drains. In ILLICON analysis the vertical and horizontal permeability of the smear zone was assumed to be equal to the vertical permeability of the undisturbed soil and to decrease during consolidation according to C_k of the undisturbed soil. The compressibility of the smear zone was defined by a linear e -log σ'_v relationship from (σ'_{vo}, e_o) to (σ'_{vf}, e_p), where e_p is the EOP void ratio at final effective vertical stress.

OBSERVED AND COMPUTED SETTLEMENTS AND PORE WATER PRESSURES

ILLICON procedure which is used in the present analysis has been described in detail by Mesri and Choi (1985) and Mesri and Lo (1987) and further illustrated by Mesri and Lo (1986) and Mesri et al. (1988). The present computer program is based on the formulation of vertical and radial flow described in Mesri and Lo (1987), except that a smear zone of radius r_s and compressibility and permeability different from the undisturbed soil, surrounds the fully or partially penetrating vertical drain with or without well resistance.

Detailed settlement and pore water observations at the test embankment and the haul road were reported by Dames and Moore (1982b) and RMP ENCON (1983). ILLICON procedure was used to compute settlement and excess pore water pressure profiles under the center of seven areas identified in Table 1 (Lo 1991). In most areas, vertical drains were installed after placing part of the fill, and this was included in the ILLICON analysis. At the haul road, settlement measurements started 70 days after embankment construction began. Therefore, the 70-day settlement computed by ILLICON was added to the measurements. Observed and computed settlements and pore water pressures of the SE quadrant with no drains and the NW quadrant with 1.5 m spaced Alidrains are shown in Figs. 7, 8, and 9. Settlement and pore water pressure predictions for the NW quadrant correspond to q_w(mob) that was 1 m^3/yr up to 350 days after construction began,and decreased to 1/2 m^3/yr after an additional 5.7 m of fill was placed. The agreement between the computed and observed surface and subsurface settlements at all seven locations on test embankment and haul road is good. ILLICON slightly underestimated the settlements at early stages of loading possibly because it does not include settlements resulting from undrained lateral deformation. Total undrained settlements of 13 to 22 cm were estimated by Foott et al. (1987) from lateral deformations recorded by inclinometers on the periphery of the test embankment area. Some surface settlement probably originated from the fill material as several settlement markers were embedded within the hydraulic fill.

In ILLICON procedure, the excess pore water pressure increase in response to an increment of loading is assumed to be equal to the increase in vertical stress, $\Delta\sigma_v$, computed from elastic stress distribution. Comprehensive records of pore water pressure measurements under the test embankment and haul road were reported by Dames and Moore (1982b) and RMP ENCON (1983). The pore water pressure increase during each loading stage, measured at different depths at all areas, are plotted against the corresponding vertical stress increase in Fig. 8a. The measured pore water pressure response is in general smaller than the computed $\Delta\sigma_v$ because some dissipation of pore water pressure took place during the application of the pressure increment. Fig. 8b takes into account pore water pressure dissipation and compares the computed and observed excess pore water pressure increment during each stage of loading. It is apparent that the procedure for computing pore pressure response according to $\Delta u = \Delta\sigma_v$ is quite reasonable. This approach has been confirmed for two other major embankment construction projects (Lo 1991). There is in general good agreement between computed and observed pore water pressure dissipation with time. According to ILLICON analysis primary settlement with 1.5 m Alidrain spacing is completed in less than 3 years, whereas without vertical drains it will require about 9 years.

Mobilized Discharge Capacity of Vertical Drains

A mobilized discharge capacity, q_w(mob), was determined at each of the seven areas to obtain agreement between observed and computed settlements as well as between observed and computed pore water pressures, at all depths and time. These are plotted in Fig. 10a against the required initial q_w(min) computed using D = 5 at $k_h = k_{ho}$ of upper marine clay. Also shown in Fig. 10a are the values of q_w(min) computed using the D from Fig. 6b, corresponding to k_{ho}/k_{hf} of upper marine clay at the end-of-primary consolidation. Fig. 11 compares the predictions of settlement as well as of pore water pressure using a discharge capacity of vertical drain equal to either q_w(mob) = 0.5-1 m^3/yr or to the initial q_w(min) = 90 m^3/yr. At Chek Lap Kok, in general q_w(mob) was much less than the initial q_w(min). However, as q_w(min) decreased significantly during consolidation and approached q_w(mob), the Alidrains functioned quite adequately.

At Chek Lap Kok different values of q_w(mob) were computed for different

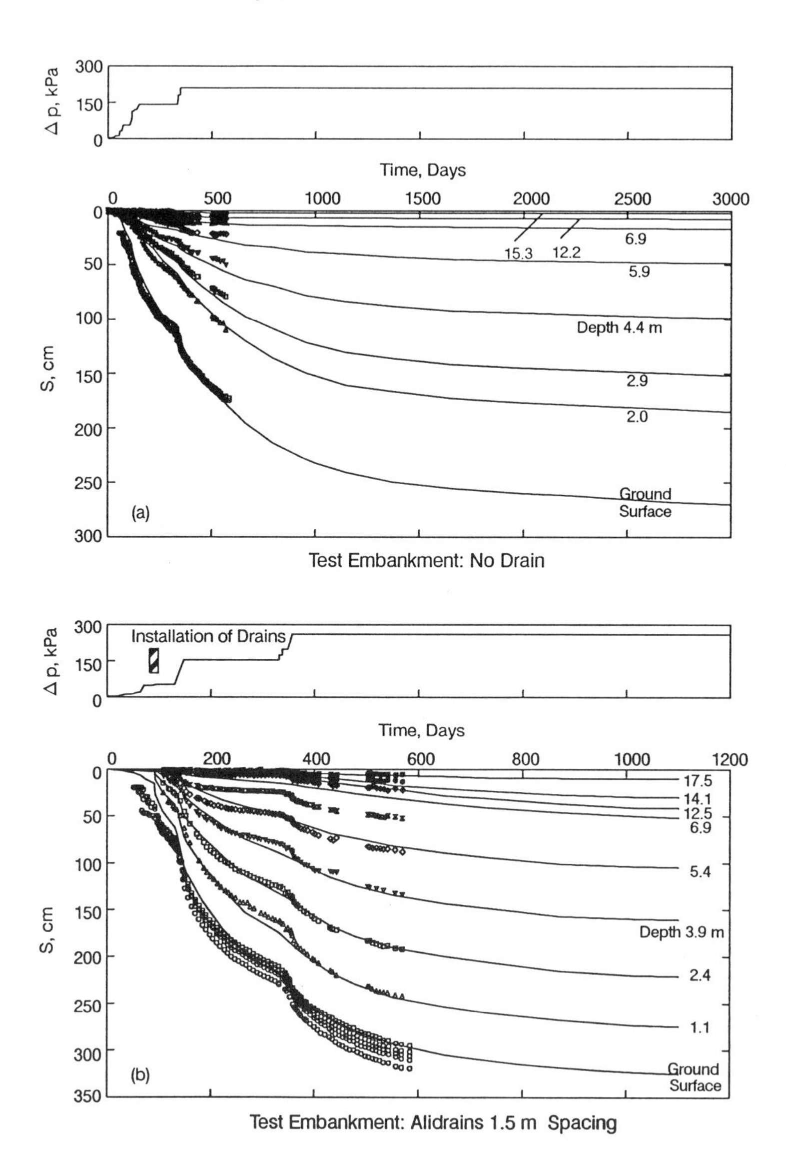

FIG. 7. Observed and Computed Settlements (Data Points: Observed, Curves: ILLICON)

embankment loads. These are plotted against EOP vertical compressive strain of upper marine clay in Fig. 10b. There is an indication that q_w(mob) decreases with the magnitude of vertical compressive strain of soil, possibly because of bending of the vertical drains. Mobilized discharge capacity at a location may also decrease with time, as dispersed clay particles from the smear zone enter and clog the filter and/or the core.

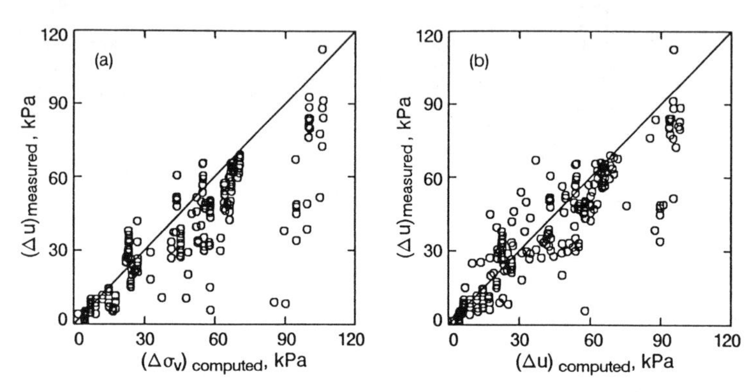

FIG. 8. Computed and Observed Excess Pore Water Pressure Increase under Test Embankment and Haul Road.

The 50 cm diameter sand drains at the SW quadrant of test embankment, which were installed after significant undrained loading of the very soft seabed, were intruded by the soft clay. The permeability of the contaminated sand drain was back-calculated, using ILLICON together with settlement and pore water pressure observations, to be k_w(mob) = 1.5×10^{-5} cm/s. This is much less than $k_w = 8\times10^{-2}$ cm/s estimated from the Hazen equation for uniform clean sand. However, since k_w(min) started at 1.5×10^{-3} cm/s and decreased to 7.6×10^{-5} cm/s during consolidation, the sand drains did enhance consolidation rate of the upper marine clay. In fact, Fig. 12 shows that by the end-of-primary consolidation, sand drains with 3 m spacing are almost as efficient as the 3 m spaced Alidrains.

CONCLUSIONS

Geotechnical investigations at Chek Lap Kok were sufficiently detailed and comprehensive to provide subsurface information for successful settlement and pore water pressure analysis using the ILLICON procedure at seven locations at the test embankment and haul road. The embankment locations were extensively instrumented for the measurement of surface and subsurface settlements as well as pore water pressures at many depths, to allow, together with ILLICON, an evaluation of the performance of vertical drains. Although the mobilized discharge capacity of Alidrains, q_w(mob), was less than the initial minimum discharge capacity, q_w(min), required for negligible well resistance, as q_w(min) decreased during consolidation toward q_w(mob), overall the Alidrains functioned quite

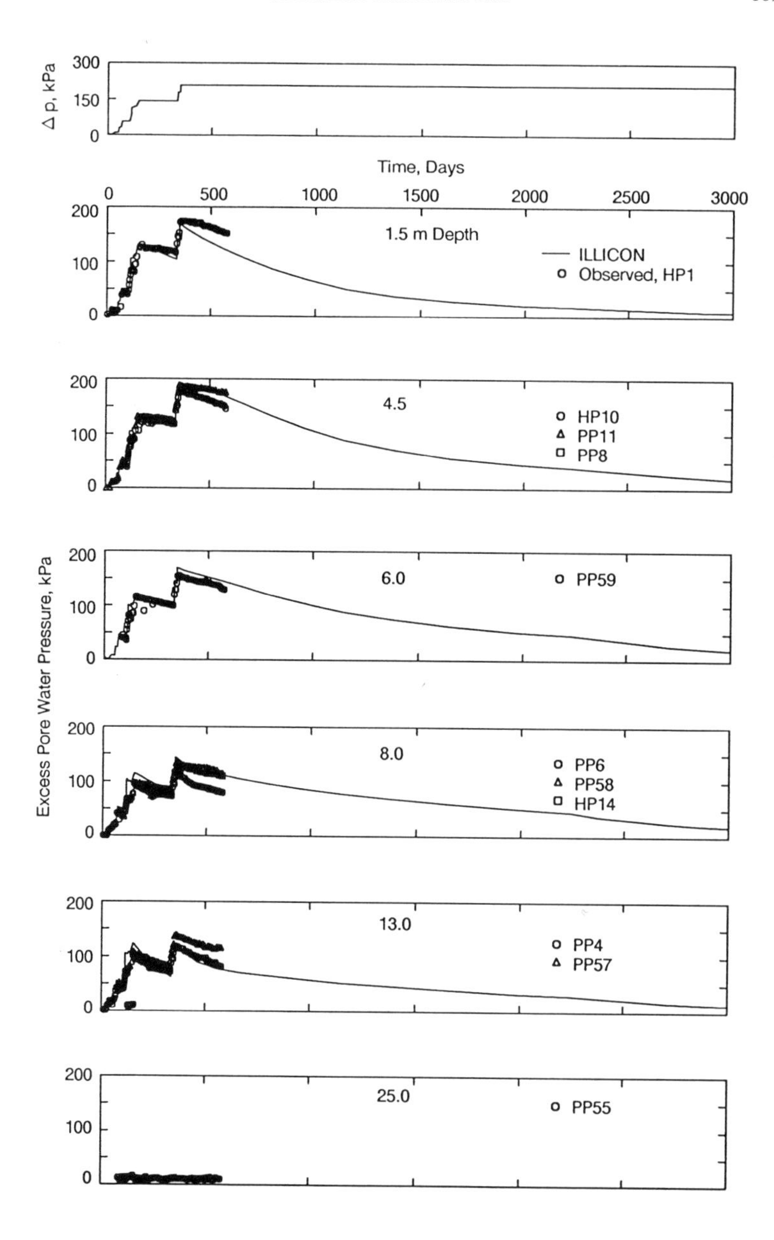

FIG. 9a. Observed and Computed Excess Pore Water Pressures Under Test Embankment Without Vertical Drains

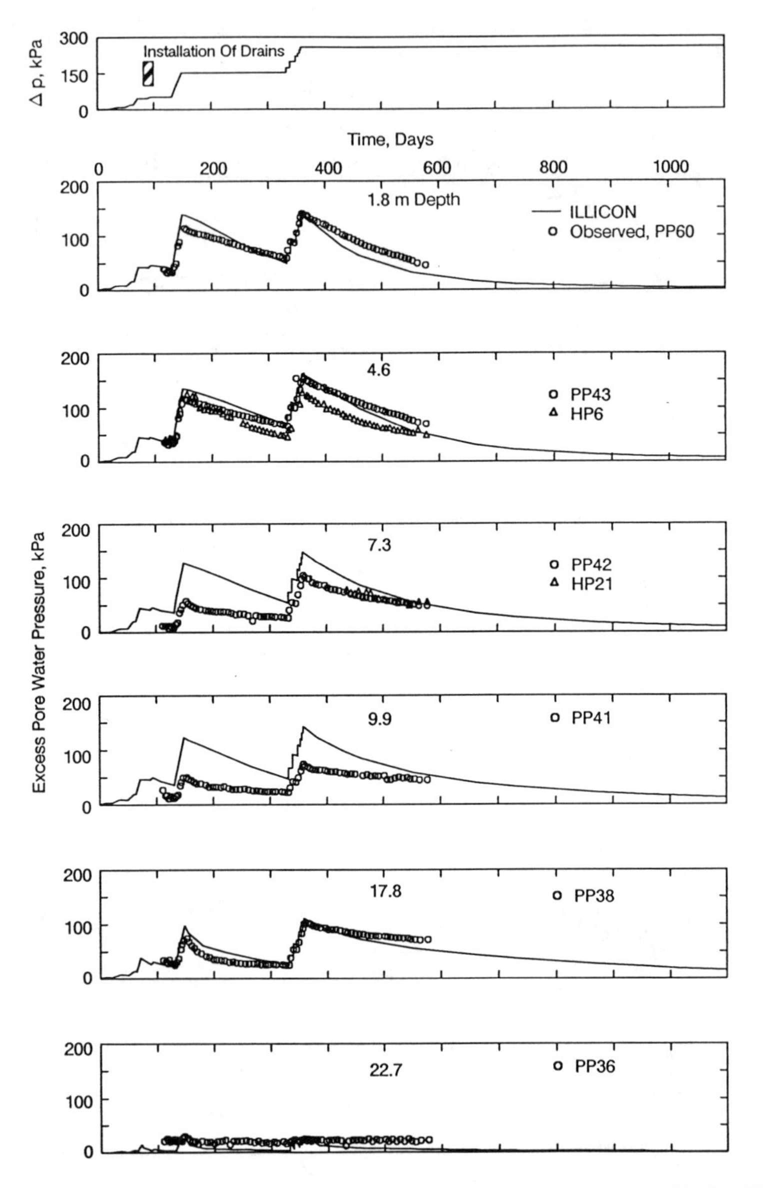

FIG. 9b Observed and Computed Excess Pore Water Pressures Under Test Embankment With Alidrains at 1.5 m Spacing

adequately. The less successful experience with sand drain installation procedure suggests that sand drains should not be installed after undrained embankment loading of soft clays.

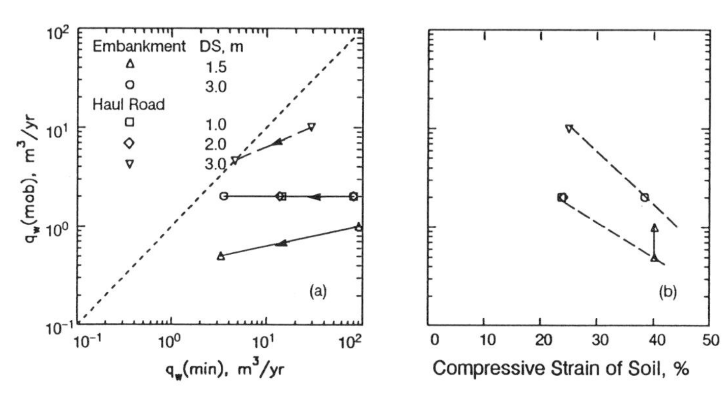

FIG. 10. (a) Values of q_w(mob) and Decrease q_w(min) During Consolidation; (b) Decrease in q_w(mob) With Vertical compressive strain in upper Marine clay

ACKNOWLEDGMENTS

This research program was sponsored by the U.S. National Science Foundation, Grant MSS-8911899. Roger Foott of Roger Foott Associates is gratefully acknowledged for providing the Dames and Moore (1982a & b) and RMP ENCON (1983) reports to the authors.

APPENDIX - REFERENCES

Andresen, A. and Kolstad, P. (1977). "The NGI 54-mm sampler for undisturbed sampling of clays and representative sampling of coarser materials." *Proc. International Symposium of Soil Sampling*, Singapore, 13-21.

Casagrande, A. (1936). "The determination of the preconsolidation load and its practical significance." *Proc. 1st ICSMFE*, Cambridge, 1, 60-64.

Dames and Moore (1982a). "Replacement Airport at Chek Lap Kok - Site Investigation," *Report No. 1*, San Francisco, California.

Dames and Moore (1982b). "Replacement Airport at Chek Lap Kok - Test Embankment," *Report No. 2*, San Francisco, California.

Foott, R., Koutsoftas, D. C., and Handfelt, L. D. (1987). "Test fills at Chek Lap Kok, Hong Kong." *J. Geotech. Eng.*, ASCE, 113(2), 106-126.

Hansbo, S. (1979). "Consolidation of clay by band-shaped prefabricated drains." *Ground Engineering*, 12(5), 16-25.

Handfelt, L. D., Koutsoftas, D. C., and Foott, R.(1987). "Instrumentation for test fill in Hong Kong." *J. Geotech. Eng.*, ASCE, 113(2), 127-146.

Koutsoftas, D. C., Foott, R., and Handfelt, L. D. (1987). "Geotechnical investigations offshore Hong Kong." *J. Geotech. Eng.*, ASCE, 113(2), 87-105.

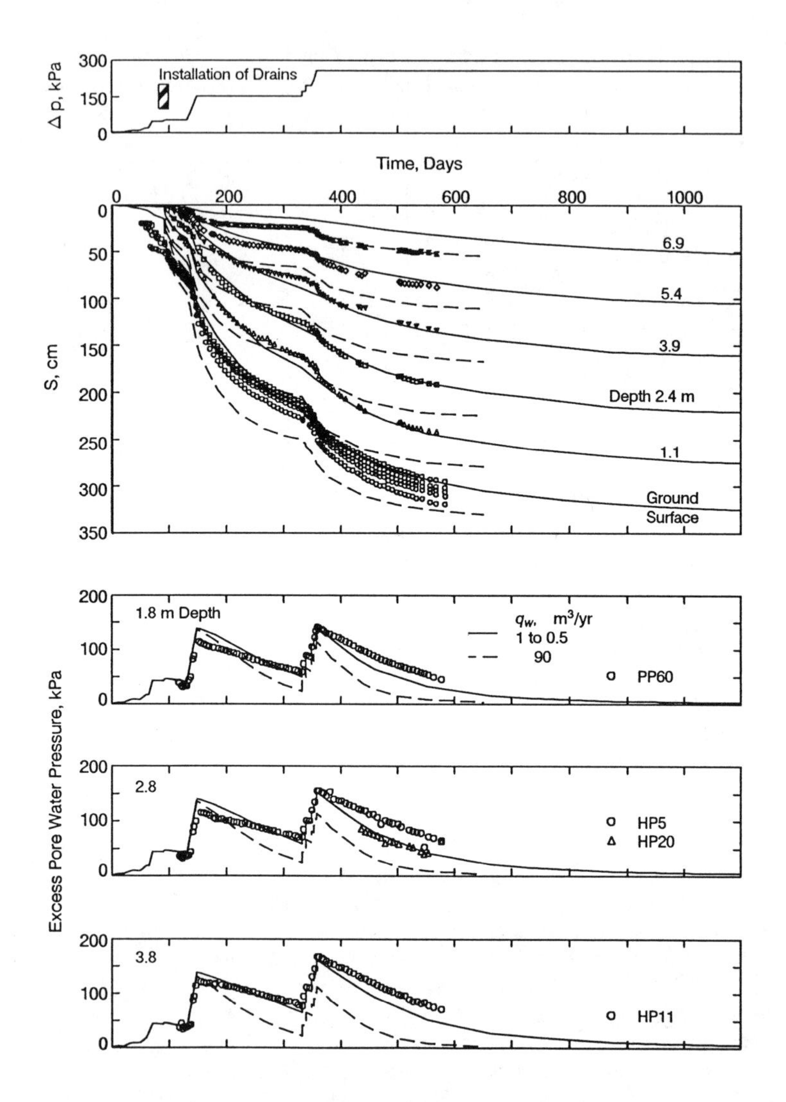

FIG. 11. Computed Settlements and Excess Pore Water Pressures Under Test Embankment With Alidrains at 1.5 m Spacing, Assuming Different Values of Discharge Capacity

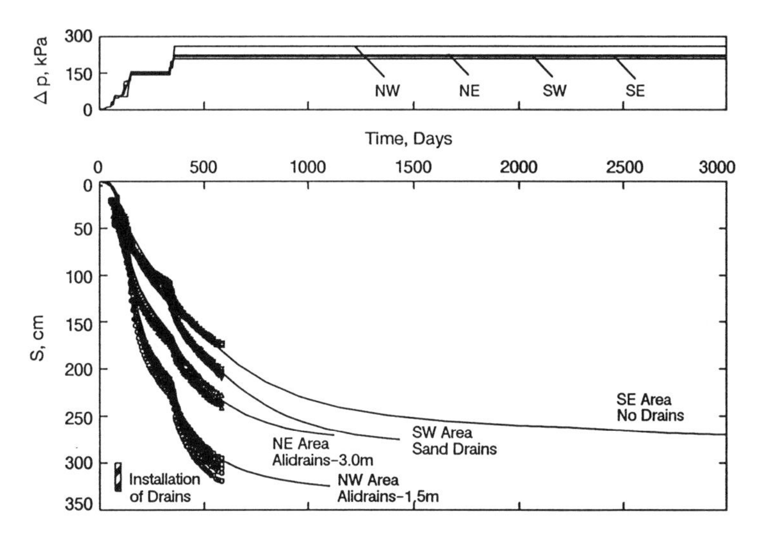

FIG. 12. Rate of Settlement Enhanced by Vertical Drains

Lo, D. O. K. (1991). "Soil improvement by vertical drains," Ph.D. thesis, University of Illinois at Urbana-Champaign, Illinois.

Mesri, G., and Choi, Y. K. (1985). "Settlement analysis of embankments on soft clays." *J. Geotech. Eng.*, ASCE, 111(4), 441-464.

Mesri, G., and Lo, D. O. K. (1986). "An analysis of the Gloucester test fill using ILLICON." *Proc. Gloucester Test Fill Symposium*, NRC, Ottawa, Canada, 1-15.

Mesri, G., and Lo, D. O. K. (1987). "Subsoil investigation: the weakest link in the analysis of test fills." *The Art and Science of Geotechnical Engineering*, Prentice-Hall, Englewood Cliffs, New Jersey, 309-335.

Mesri, G., and Lo, D.O.K. (1991). "Field performance of prefabricated vertical drains." *Proc. GEO-COAST '91*, Yokohama, 231-236.

Mesri, G., Lo, D. O. K., and Karlsrud, K. (1988). "ILLICON settlement analysis of Ellingsrud test fill." *Proc. 2nd Int'l. Conf. on Case Histories in Geot. Engrg.*, Rolla, 3, 1765-1770.

Mesri, G., Feng, T. W., Ali, S., and Hayat, T.M. (1994). "Permeability characteristics of soft clays." *Proc. 13th ICSMFE*, New Delhi, India, 1, 187-192.

RMP ENCON, Ltd. (1983). "Replacement Airport at Chek Lap Kok - Test Embankment." *Study Report No. 2A*, Hong Kong.

Tavenas, F., Jean, P., Leblond, P., and Leroueil, S. (1983). "The permeability of natural soft clays - permeability characteristics." *Can. Geotech. J.*, 20(4), 645-660.

Consolidation Settlements and Pore Pressure Dissipation

Demetrious C. Koutsoftas,[1] M. ASCE and Raymond K.H. Cheung[2]

Abstract: A test fill was constructed off the coast of one of the outer islands in Hong Kong to evaluate the consolidation characteristics of a soft plastic clay deposit known as Hong Kong mud. Extensive instrumentation was installed to monitor surface and subsurface settlements of the soft mud, as well as pore pressures at various depths in the mud. The results of the measurements were analyzed, correlating the observed pore pressures with corresponding settlements and their distribution with depth. The test fill involved a main test area with four quadrants, each 50 m × 50 m in plan. One of the quadrants served as control to evaluate the consolidation of the mud without drains. Plastic band chains were used in two of the other quadrants and large diameter (500 mm) sand drains were used in the fourth quadrant, to accelerate the consolidation of the mud. At various times after filling, piezoprofile tests were carried out to supplement the measurements made with permanent piezometers, to better define the distribution of pore pressures with depth. From the pore pressure profiles, settlements were calculated using 1-dimensional compression theory. The distribution of calculated settlements with depth was compared with measured settlements. There was good agreement between the measurements and the calculated settlements. A value of compression ratio, CR, of 0.40 was backcalculated from these analyses. Settlements and pore pressures were also measured at four areas outside the main test area. They included an untreated area (where no drains were installed) and three areas where band drains had been installed at 1 m, 2 m and 3 m triangular spacings, respectively. Measured excess pore pressures and consolidation settlements are discussed. In three areas of the test fill where the primary consolidation settlements had been completed, the measurements showed substantial residual excess pore pressures. The causes for and significance of the residual pore pressures are discussed.

Introduction

The proposed replacement airport at Chek Lap Kok, Hong Kong, is planned to be

[1] Principal, Dames & Moore, San Francisco, CA 94105-1917

[2] Manager, Standardization/Certification Department, Singapore Institute of Industrial Standards and Research, 1 Science Park Drive, Singapore, 0511

constructed by leveling two small islands and reclaiming approximately 600 ha (1,500 acres) of land from the surrounding sea. The site is located in the South China Sea, approximately 25 km west of Hong Kong Island. Water depths range up to 10 m with a tidal range of 2 m. The reclamation would involve placement of 80,000,000 m^3 of fill up to 20 m thick.

The subsurface conditions consist of a very soft and compressible mud layer at the seabed underlain by a sequence of stiff alluvium over an older marine clay deposit, over an older alluvium over decomposed and weathered granite overlying granitic bedrock.

The placement of heavy fills directly on the soft mud layer would result in settlements of up to 4 m in the mud layer with additional settlements of up to 1 m in the lower marine clay layer. Filling could also result in mud waves due to disturbance and remodeling of the seabed mud layer that would cause large and erratic long-term settlements that would impact the runways and affect airfield facilities.

As part of the design studies, a test fill was constructed to evaluate methods of placing the fill without causing mud waves and to accelerate the consolidation of the seabed mud layer so that post-construction settlements could be controlled within airport design tolerances. The performance of the test fill was monitored over a period of 2$^1/_2$ years and included measurements of pore pressures, settlements, lateral deformations, measurements of in-situ strength gain with consolidation and other in-situ and laboratory tests.

The objective of this paper is to present measurements of pore pressures and settlements of the soft mud layer and to correlate the dissipation of the excess pore pressures with the measured consolidation settlements, with particular emphasis on the distribution of mud settlements with depth.

Subsurface Conditions

The subsurface conditions at the airport site in general, and at the test fill site in particular, were presented and discussed in detail by Koutsoftas et al. (1987) and Koutsoftas (1994). The subsurface stratigraphy is briefly described below.

The seabed is covered by a layer of very soft and compressible plastic clay of Holocene age. On the average the thickness of this layer ranges from 6 m to 8 m. At the shoreline of the surrounding islands the mud is absent but its thickness increases rapidly as one moves offshore. Mud thicknesses of up to 15 m were encountered at some locations. Key components of the test fill are located 200 m offshore, where the mud is about 7 m thick.

Characteristic geotechnical properties of the mud are illustrated in Figure 1. Moisture contents range from 90% to 120% and total densities range from 14 to 15 kN/m^3. Moisture contents are typically higher than the liquid limit, indicating a sensitive material. Sensitivity values measured from vane shear tests range from 7 to 15. The results of consolidation tests indicate that the mud is lightly overconsolidated with overconsolidation ratio (OCR) values ranging from 1.5 to 2.0, which is typical for Holocene marine clays, and is attributed to aging (see Bjerrum 1972). The maximum past pressure profile selected for settlement calculations is indicated in Figure 1. The mud is highly compressible when stressed beyond the maximum past pressure ($\overline{\sigma}_{vm}$) with Compression Ratios (CR = $C_c/1 + e_o$) ranging from 0.3 to 0.5. Field vane tests indicate shear strengths increasing linearly with depth from 3 kPa at a depth of 1 m to 18 kPa at a depth of 9 m.

The mud layer is underlain by a stiff alluvial crust (probably of Pleistocene age), a desiccated clay with undrained shear strengths of 70 kPa or higher and preconsolidation

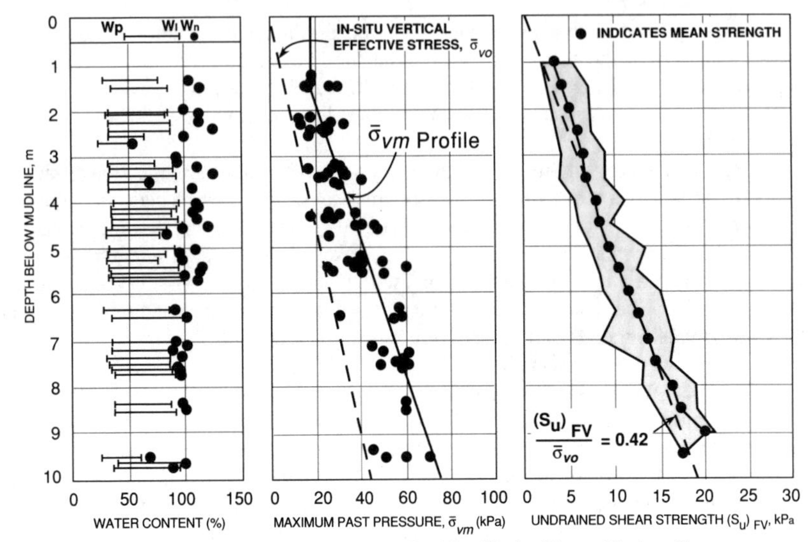

FIG. 1. Characteristic Geotechnical Profile for Upper Marine Clay

stresses of 300 kPa or higher. It ranges in thickness from 2 m to 8 m. The alluvial crust is underlain by a layer of older marine clay, a lean clay layer with moisture contents in the range of 30% to 50%. Undrained shear strengths range from 40 kPa at a depth of 14 m below seabed to 100 kPa at a depth of 22 m. Preconsolidation stresses range from 200 kPa to 450 kPa with corresponding OCRs of 2 to 3. Below the marine clay layer is an older alluvium consisting primarily of very dense coarse to fine sand with some silt and clay, grading into a layer of cobbles and gravels towards the base of the layer. The alluvium ranges in thickness from 0 to 10 m. A layer of completely decomposed granite (CDG) underlies the lower alluvium. It is a very dense, brown-reddish, silty sand to sandy silt with occasional lenses and pockets of clay. The CDG grades into weathered granite of variable thickness over granitic bedrock.

Test Fill Construction and Instrumentation

A detailed description of the construction of the test fill was presented by Foott et al. (1987). A plan and section of the test fill in the north-south direction are shown on Figure 2. It consists of the Main Test Area, 100 m × 100 m in plan, a protective seawall along the north and west sides and the Haul Road and south slope along the south side.

The Main Test Area (MTA) was divided into four quadrants, each 50 m × 50 m in plan. Vertical band drains were installed in two of the quadrants at 1.5 m and 3 m triangular spacings, respectively. Displacement type sand drains, 500 mm in diameter and 13 m long, were installed in the third quadrant. A steel mandrel driven with a vibrating hammer was used to install the sand drains. The fourth quadrant was left untreated for control purposes. The area along the Haul Road was also treated with band drains installed at triangular spacings of 1 m, 2 m, and 3 m as shown in Figure 2. The area between the MTA and the Haul Road was left untreated.

The test fill was constructed in two phases. During the first phase, the seawall and south slope were completed and the MTA was filled to elevation +6.5 m P.D. (principal

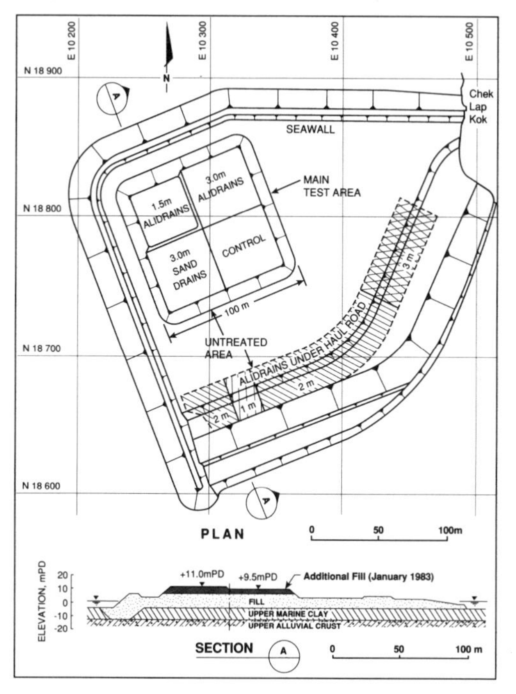

FIG. 2. Plan and Cross Section of the Test Fill

datum) approximately 2.5 above the fill surrounding the MTA. After approximately 6 months of monitoring, the Phase II fill was placed, raising the MTA to elevation +9.5 m P.D. in three of the quadrants and to +11.0 m P.D. in the quadrant with 1.5 m Alidrains. Monitoring continued for approximately 24 months after placing the Phase II fill.

Extensive instrumentation had been installed to monitor the performance of the test fill. Details of the monitoring program were presented by Handfelt et al. (1987). A large number of piezometers was installed to monitor pore pressures in the subsurface soils, with particular emphasis in the distribution of the pore pressures with depth in the mud. Extreme care was taken to make certain that the piezometers were installed in the center of the triangle defined by the drains. At various times during the monitoring period, piezoprofile tests were performed to supplement the results of the measurements from piezometers to better define the distribution of excess pore pressures with depth. The piezoprofile test consists of pushing a hydraulic piezometer to the desired depth and allowing the excess pore pressures generated by pushing to dissipate until the equilibrium pore pressure value was reached. Sondex anchors placed around inclinometer casing at various depths provided measurements of settlements below the mudline.

Analysis of Pore Pressure and Settlement Measurements

The availability of detailed measurements of pore pressure and settlement profiles with depth provided an opportunity to correlate the observed settlements at various stages of monitoring with the observed dissipation of pore pressures. Consolidation settlements were calculated from measured excess pore pressures using 1-dimensional compression theory as follows:

1. First the effective stress profile was calculated from measured excess pore pressures using the expression:

$$\bar{\sigma}_v = \bar{\sigma}_{vo} + (\Delta\sigma_v - \Delta_u) \qquad (1)$$

where $\bar{\sigma}_v$ is the vertical effective stress, $\bar{\sigma}_{vo}$ is the in-situ vertical effective stress, $\Delta\sigma_v$ is the incremental stress caused by the weight of the fill; and Δ_u is the excess pore pressure, equal to the measured total pore pressure minus the hydrostatic pressure.

2. The mud layer was then subdivided into a number of sublayers (0.5 m to 1.0 thick) and the consolidation settlements of each sublayer were calculated from the following equation:

$$\rho = H\left(CR \log(\bar{\sigma}_v/\bar{\sigma}_{vm})\right) + H\left(RR \log(\bar{\sigma}_{vm}/\bar{\sigma}_{vo})\right) \qquad (2)$$

where ρ is the consolidation settlement, H is the thickness of each sublayer, RR is the recompression ratio, $\bar{\sigma}_{vm}$ is the maximum past pressure; and CR is the Compression Ratio.

3. The settlement profile is obtained by summing the settlement of each sublayer.

Typical results of the analyses and comparisons with measured settlements are presented below.

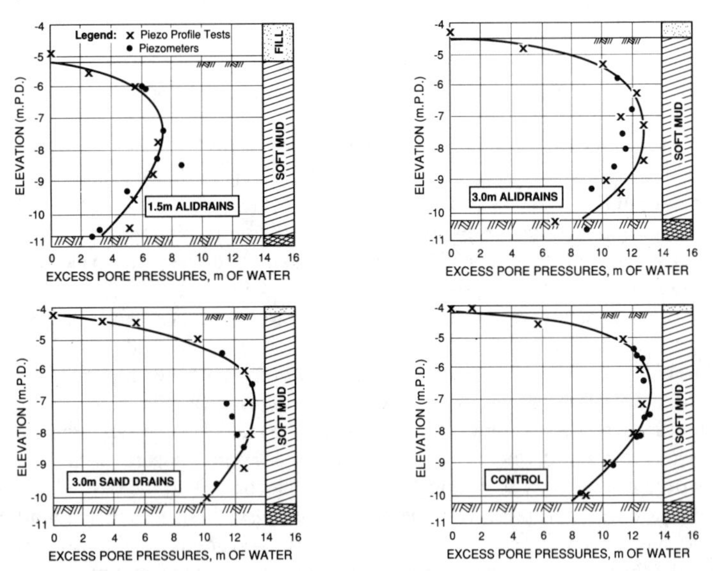

FIG. 3. Excess Pore Pressures in the Mud Layer: Phase I Monitoring (Dec. 1982)

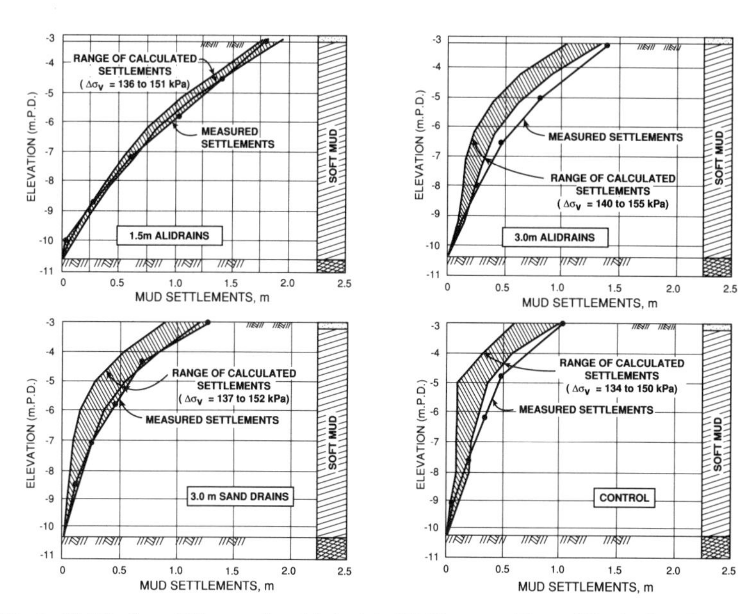

FIG. 4. Distribution of Measured and Calculated Settlements: Phase I Monitoring (Dec. 1982)

Figure 3 presents excess pore pressures measured in the MTA in December 1982, approximately 6 months after the Phase I fill was placed. Measurements from permanent piezometers and from piezoprofile tests are shown. Figure 4 shows the measured and calculated settlements and their distribution with depth in the mud layer. The calculations are based on assumed values of RR = 0.025 and CR = 0.40. The value of incremental stress, $\Delta\sigma_v$, was calculated from the known thicknesses of fill placed in each quadrant and from measured densities of the fill. A range of $\Delta\sigma_v$ values consistent with the measured variations in fill densities was used. In the 1.5 m Alidrain quadrant, the calculated settlements are in excellent agreement with the measured values; however, in the other three quadrants the analyses based on the lower values of incremental stresses underestimate the settlements by 30% to 50%. The higher values of incremental stress give settlements in reasonably good agreement with the measurements.

Figure 5 shows the excess pore pressures measured during the Phase II monitoring period in June 1983 and November 1984. The effects of the different methods of treatment on the dissipation of excess pore pressures are evident. Figure 6 presents the measured and calculated settlements and their variation with depth, corresponding to the pore pressures shown on Figure 5. Excellent agreement was obtained in the 1.5 m Alidrain area and the 3.0 m sand drain area. Reasonable agreement was obtained in the other two areas for the June 1983 measurements. There was good agreement in all four quandrants for the November 1984 measurements. For clarity only the calculated values corresponding to the highest fill densities are shown for the November 1984 measurements.

Figure 7 presents the distribution of pore pressures measured in the four areas south

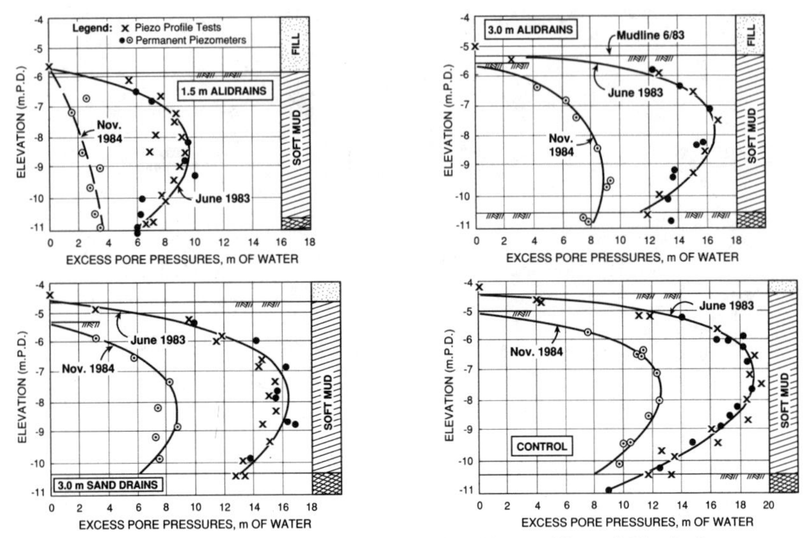

FIG. 5. Excess Pore Pressures in the Mud Layer: Phase II Monitoring

of the MTA and Figure 8 presents the consolidation settlement versus time. It is clear that by June 1983, primary consolidation was complete in the 3 m Alidrain area, and by November 1984, primary consolidation was also completed in the 1 m Alidrain area. Yet

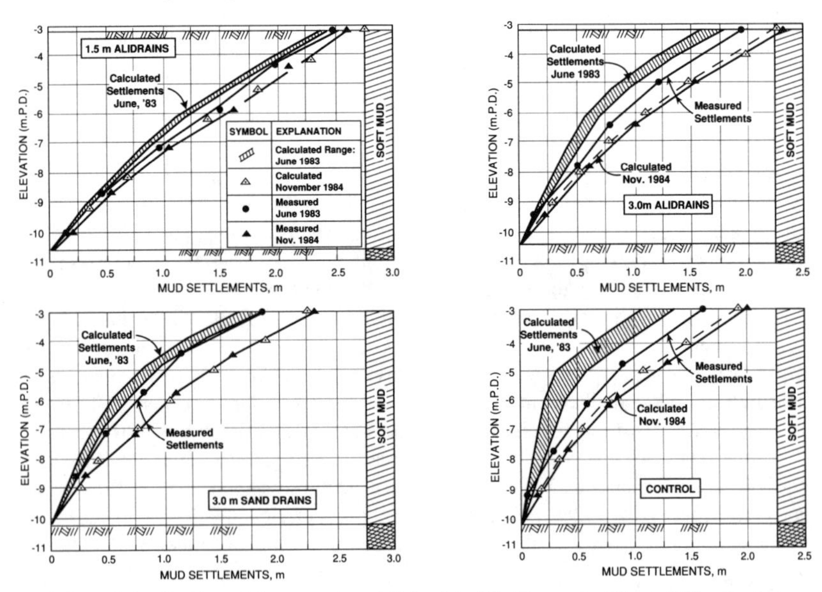

FIG. 6. Distribution of Measured and Calculated Settlements: Phase II Monitoring

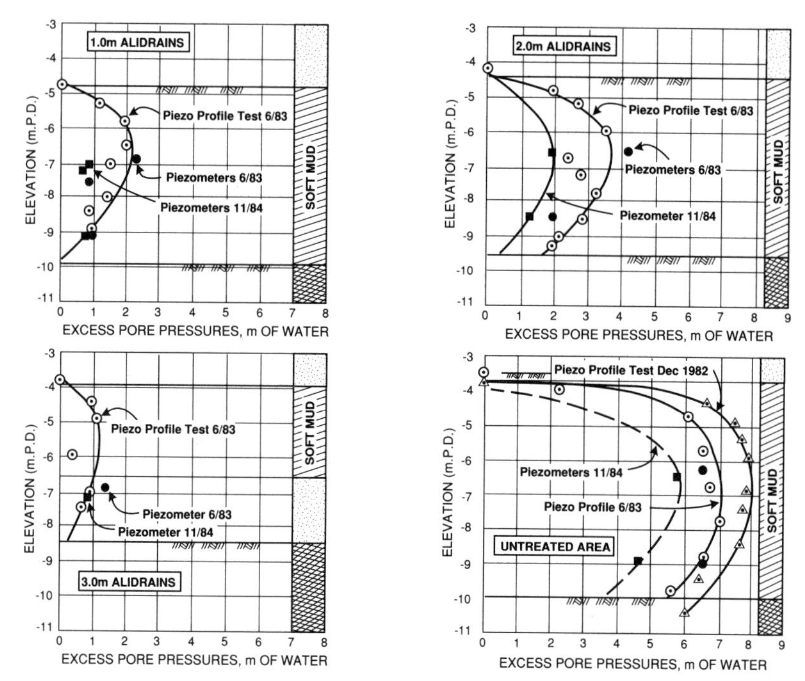

FIG. 7. Excess Pore Pressures Measured in the Haul Road Area

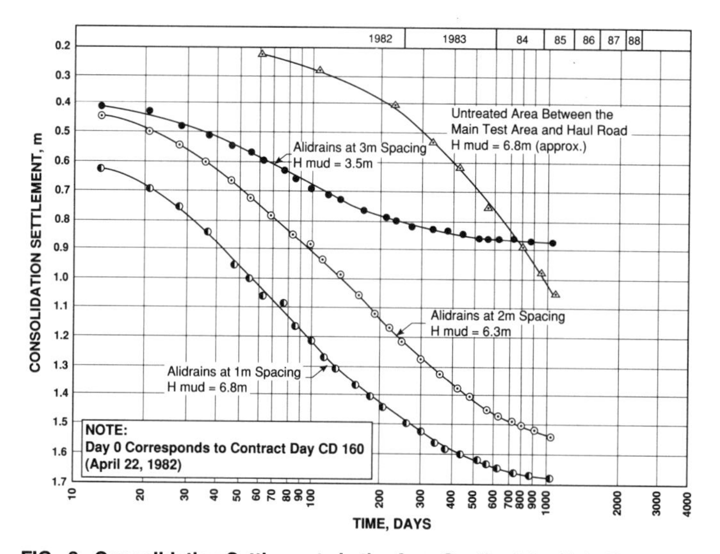

FIG. 8. Consolidation Settlements in the Area South of the Main Test Area

the measurements from piezoprofile tests and from piezometers show residual pore pressures of over 1 m of water. Although the measurements were initially dismissed as probably erroneous, careful examination of the measurements in the 1.5 m Alidrain quadrant of the MTA suggests otherwise.

Figure 9 shows the full profile of excess pore pressures under the 1.5 m Alidrain quadrant and Figure 10 shows the corresponding settlement of the mud versus time. Clearly, by November 1984, consolidation of the mud was well into secondary compression while the pore pressure measurements indicate excess pore pressures of up to 4 m of water. It is evident from Figure 9 that whereas during most of the consolidation period the underlying upper Alluvium had consolidated rapidly enough to act as a partial vertical drainage boundary for the mud, towards the end of the monitoring period the rate of consolidation of the deeper strata controlled the dissipation of the excess pore pressure in the mud. It appears that the excess pore pressures in the mud represent a condition of steady state seepage with flow from the bottom to the top of the mud layer and in effect counteract the tendency of the mud to consolidate further. The mud began to settle due to secondary compression, as can be seen from Figure 10, because no further dissipation of pore pressures could occur without substantial dissipation of the pore pressures in the

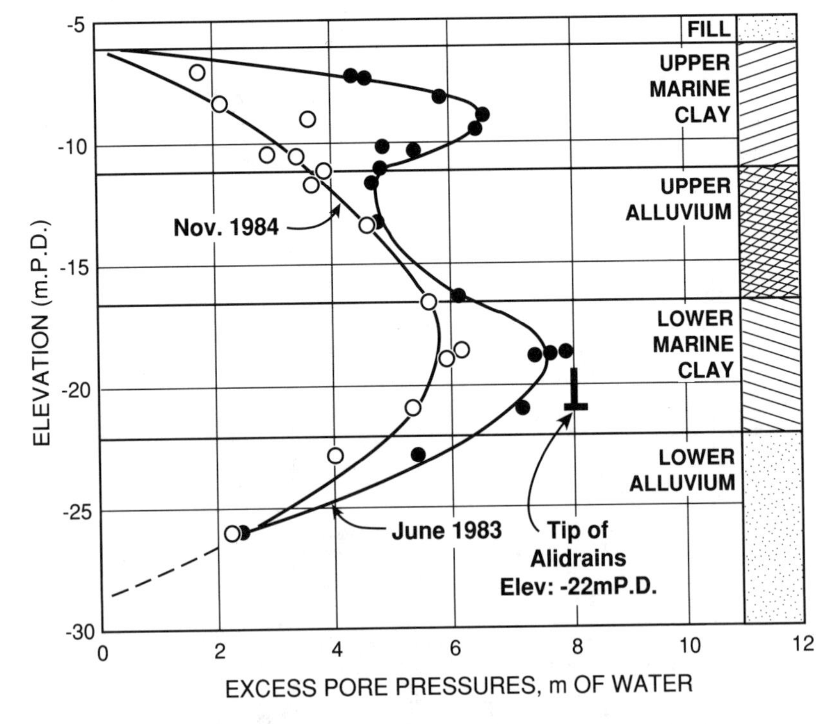

FIG. 9. Distribution of Pore Pressures under the 1.5m Alidrain Quadrant: Phase II Monitoring (Jan. 1983-Nov. 1984)

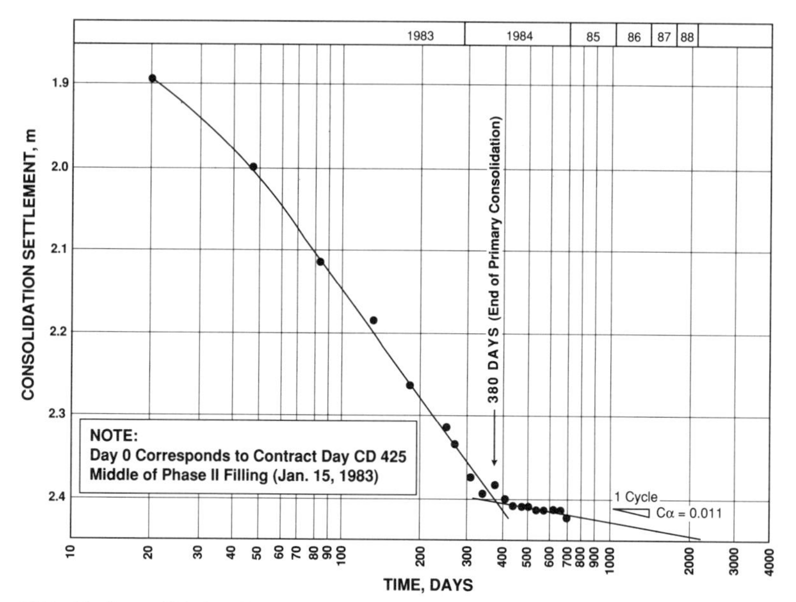

FIG. 10. Consolidation Settlements in the 1.5m Alidrain Quadrant during Phase II Monitoring (Jan. 83-Dec. 84)

underlying layers. Presumably, when the excess pore pressures in the deeper strata dissipate, the excess pore pressures in the mud will also dissipate, causing additional settlement due to primary consolidation.

A similar phenomenon is probably responsible for the observed residual pore pressures in the Haul Road area.

Conclusions

A comprehensive set of measurements of pore pressures and settlements shows a close correlation between pore pressure dissipation and consolidation settlements that allows back calculation of the compressibility characteristics of the mud. A compression ratio of 0.40 gives good agreement between measured and calculated settlements. The analyses also show the importance of having reliable information on the density of the fills for calculating the incremental stresses caused by filling.

Residual pore pressures were observed in some areas even though the settlement measurements show clearly that primary consolidation was complete and secondary compression was well under way. The slower dissipation of pore pressures in the underlying strata creates a back pressure effect retarding the primary consolidation process in the mud layer.

Acknowledgments

Many Dames & Moore employees contributed to the success of the project. Former colleagues Roger Foott, Roy Bell, Leo Handfelt and E. M. Thomas made significant contributions to the success of the project. Mr. John Dunnicliff was consultant to Dames & Moore on instrumentation. At least 15 other Dames & Moore employees contributed significantly to the success of the project by their dedication and hard work.

Appendix I. References

Bjerrum, L. (1972). "Embankments on Soft Ground." State-of the-Art Report, *Proc. ASCE Specialty Conference on Performance of Earth and Earth Supported Structures*, Lafayette, Vol. II, 1-54.

Foott, R., Koutsoftas, D. C., and Handfelt, L. C. (1987). "Test Fill at Chek Lap Kok, Hong Kong." *J. Geotech. Eng.*, ASCE 113(2), 106-126.

Handfelt, L. D., Koutsoftas, D. C., and Foott, R. (1987). "Instrumentation for Test Fill in Hong Kong." *J. Geotech. Eng.*, ASCE 113(2), 127-146.

Koutsoftas, D. C., Foott, R., and Handfelt, L. D. (1987). "Geotechnical Investigations Offshore Hong Kong." *J. Geotech. Eng.*, ASCE 113(2), 87-105.

Koutsoftas, D. C. (1994). "Lateral Deformations Under and Around a Test Fill." *Vertical and Horizontal Deformations of Foundations and Embankments*, Geotechnical Special Publications, ASCE, New York, New York.

Appendix II. Notations

The following symbols are used in this paper:

C_c = compression index

CR = compression ratio

$C_{\propto}$ = coefficent of secondary consolidation

e_o = initial void ratio

H = thickness of each sublayer

OCR = overconsolidation ratio

RR = recompression ratio

$(Su)_{FV}$ = undrained shear strength from field vane test

W_l = liquid limit

W_n = natural water content

W_p = plastic limit

Δ_u = excess pore pressure

$\Delta\sigma_v$ = incremental stress caused by the weight of the fill

ρ = consolidation settlement

$\bar{\sigma}_v$ = vertical effective stress

$\bar{\sigma}_{vm}$ = maximum past pressure

$\bar{\sigma}_{vo}$ = in-situ vertical effective stress

Lateral Foundation Deformations for a Marine Test Fill

Demetrious C. Koutsoftas,[1] Member, ASCE

Abstract: A test fill was constructed off the coast of one of the outer islands in Hong Kong, involving filling in stages over a layer of soft marine clay, referred to as "Hong Kong mud." Twenty-one inclinometers were installed to monitor lateral deformations of the subsurface soils during and after filling. The test fill, which was constructed along the west coast of Chek Lap Kok island, consisted of a Main Test Area (MTA), a protective seawall along the north and west sides of the test fill and a flat slope along the south side. The test fill had two objectives; first, to develop a method for filling over the soft mud without causing mudwaves, and second, to evaluate the effectiveness of vertical drains to accelerate consolidation of the mud. Lateral deformations measured during construction of the south slope illustrate the response of the soft mud as it was stressed to incipient failure, as well as the slowdown in deformations resulting from the construction of a stabilizing berm built to arrest the slope movements. Lateral deformations measured during and after filling of the Main Test Area are presented and discussed to illustrate the mobility of the mud even under a large test fill intended to simulate 1-dimensional consolidation conditions. The development of lateral deformations with time is also discussed to draw attention to the long duration of lateral movements after filling.

Introduction

The proposed replacement airport at Chek Lap Kok, Hong Kong, is planned to be constructed by leveling two small islands and reclaiming approximately 600 ha (1,500 acres) of land from the surrounding sea. The site is located in the South China Sea, approximately 25 km west of Hong Kong Island. Water depths range up to 10 m with a tidal range of 2 m. The reclamation would involve placement of 80,000,000 m^3 of fill up to 20 m thick.

The subsurface conditions consist of a very soft and compressible mud layer at the seabed underlain by a sequence of stiff alluvium over an older marine clay deposit, over an older alluvium over decomposed and weathered granite overlying granitic bedrock.

The placement of heavy fills directly on the soft mud layer would result in predicted

---

[1] Principal, Dames & Moore, 221 Main Street, San Francisco, California 94105.

settlements of up to 4 m in the mud layer with additional settlements of up to 1 m in the lower marine clay layer. Filling could also result in mud waves due to disturbance and remolding of the seabed mud layer that would cause large and erratic long-term settlements that would impact the runways and affect airfield facilities.

As part of the design studies, a test fill was constructed to evaluate methods of placing the fill without causing mud waves and to accelerate the consolidation of the seabed mud layer so that post-construction settlements could be controlled within airport design tolerances. The performance of the test fill was monitored over a period of 2½ years, and included measurements of pore pressures, settlements, lateral deformations, measurements of in-situ strength gain with consolidation and other in-situ and laboratory tests.

The objective of this paper is to present the lateral deformations measured during filling and subsequent consolidation and discuss their significance in terms of the behavior of the test fill.

SUBSURFACE CONDITIONS

The subsurface conditions at the test fill site are illustrated on Fig. 1. The various strata found at this site are briefly described below. Detailed geotechnical data have been presented elsewhere by Koutsoftas et al. (1987).

The seabed is covered by a layer of very soft and compressible plastic clay of Holocene age. On the average it ranges in thickness from 6 m to 8 m, but in places it is up to 15 m thick. It has moisture contents ranging from 90% to 120%, that are typically above the liquid limit, indicating a very sensitive material. Sensitivity values measured from field vane tests ranged from 7 to 15. The results of consolidation tests indicate that the material is lightly overconsolidated with overconsolidation ratio (OCR) values ranging from 1.5 to 2.0. It is highly compressible when stressed in the virgin compression zone with Compression Ratios, $C_c/(1 + e_o)$, ranging from 0.3 to 0.5. Field vane shear strengths indicate strengths increasing linearly with depth from 3 kPa at a depth of 1 m to 18 kPa at a depth of 9 meters. The mud layer is underlain by a stiff alluvial crust, a desiccated clay with undrained shear strengths of 70 kPa or higher and preconsolidation

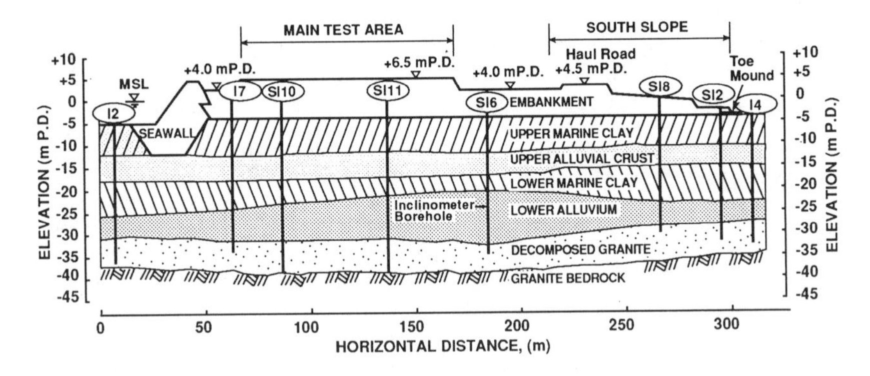

FIG. 1. Subsurface Section Through the Test Fill

stresses of 300 kPa or higher. It ranges in thickness from 2 m to 8 m. The alluvial crust layer is underlain by a layer of older marine clay, a lean clay with moisture contents in the range of 30% to 50%. Undrained strengths range from 40 kPa at a depth of 10 m to 100 kPa at a depth of 22 m. Preconsolidation stresses ranged from 200 kPa to 450 kPa with corresponding OCRs ranging from 2 to 3. Below the marine clay is a layer of older alluvium consisting primarily of very dense coarse to fine sands, silty sands and clayey sands, grading into a layer of cobbles and gravels towards the base of the layer. The lower alluvium is up to 10 m thick. A layer of completely decomposed granite (CDG) underlies the lower alluvium. It is a very dense, brown-reddish, silty sand to sandy silt with occasional lenses and pockets of clay. The CDG grades into weathered granite of variable thickness over granitic bedrock.

Test Fill Construction Sequence and Instrumentation

The plan of the test fill is shown on Fig. 2. It consists of a Main Test Area (MTA) 100 m $\times$ 100 m in plan, a protective seawall to the north and west of the MTA, and the haul road and south slope areas.

The MTA was divided into four quadrants, as shown on Fig. 2, each 50 m $\times$ 50 m in plan. Vertical drains at various spacings were installed in three of the four quadrants to evaluate the feasibility of vertical drains to accelerate the consolidation of the mud. The fourth quadrant was left untreated as control. The primary function of the seawall was to limit the size of the test fill (by eliminating the volume of fill that would be required to build the very flat slopes around the MTA) and to provide protection from wave action,

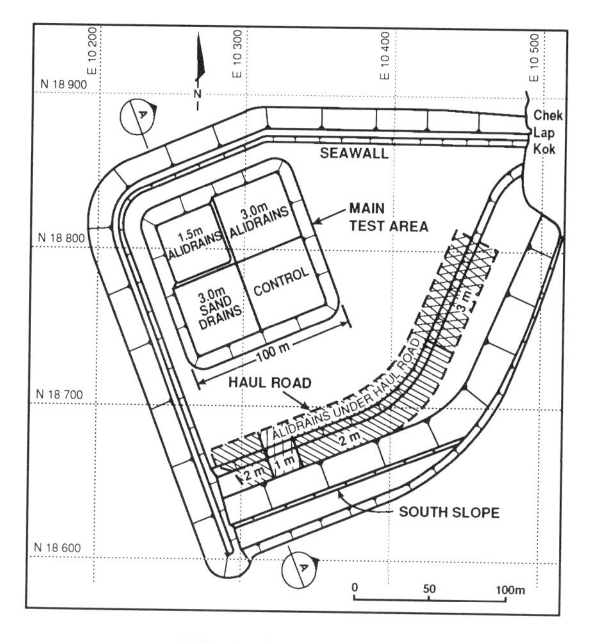

FIG. 2. Plan of Test Fill

particularly during the typhoon seasons. The south slope was constructed in layers to evaluate methods of fill placement without causing mud waves or slope failures. The slope was designed to be loaded to incipient failure, with provisions made in the contract for construction of a stabilizing berm to arrest the failure, once the lateral deformations indicated the onset of incipient failure.

The test fill was constructed in two phases. During the first phase, the seawall and south slope were completed, and the MTA was filled to elevation +6.5 m P.D. (principal datum), approximately 2.5 m above the level of the fill surrounding the MTA. During the second phase of filling, the MTA was raised to elevation +9.5 m P.D. in three of the four quadrants and to elevation +11.0 m P.D. in the fourth quadrant. After Phase I construction was completed, the test fill was monitored for a period of about 200 days before placing additional fill in the MTA. After the Phase II fill was placed, the test fill was monitored for two more years.

Fig. 3 summarizes the sequence of constructing the test fill. After constructing the seawall to mean sea level, and after building a rock fill toe mound along the south side of the test fill, wick drains were installed in the haul road area. A blanket layer of hydraulic sand fill approximately 2 m thick was then placed on the seabed to cover the entire area enclosed by the seawall and the toe mound. Then the haul road was constructed in thin lifts, while the lateral deformations of the south slope were closely monitored. Construction continued until the lateral deformations reached 300 mm and showed a rapid acceleration as additional fill was being placed. At the same time, cracks were observed on the haul road parallel to the south slope. At that point the construction of the haul road was interrupted and a stabilizing berm was constructed south of the toe mound to arrest the movement of the slope. With the south slope secure against

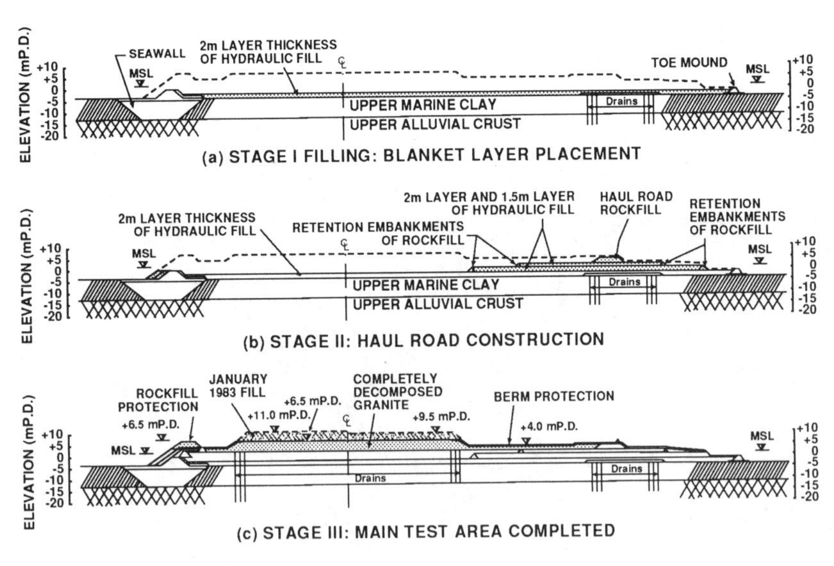

FIG. 3. Sequence of Test Fill Construction

instability, additional hydraulic sand fill was placed between the haul road and the seawall to raise the fill above sea level to provide a platform for the installation of vertical drains in the MTA and the additional instrumentation. At the same time, the seawall was raised above sea level to protect the work area from wave action. In two of the three quadrants, Alidrain type wick drains were installed at 1.5-m and 3-m triangular spacings, respectively. In the third quadrant, displacement-type sand drains 500 mm in diameter were installed at 3-m triangular spacings. The fourth quadrant was left untreated for control purposes.

After the drains had been installed and the instrumentation completed, the fill in the MTA was raised to elevation +6.5 m P.D., and the seawall was completed. Approximately 6 months after the Phase I construction was completed, additional fill was placed in the MTA to raise the fill levels to the final planned grades.

The test fill was heavily instrumented, with 21 inclinometers, numerous piezometers, settlement points, deep multipoint settlement anchors, settlement tubes placed

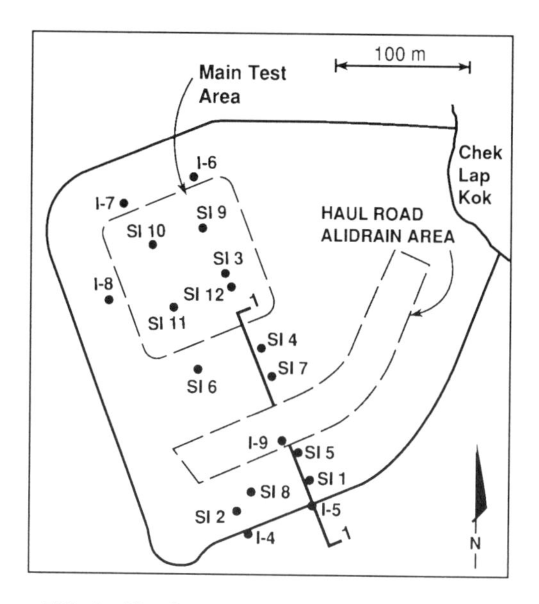

FIG. 4. Plan Showing Locations of Inclinometers

under the test fill, and survey markers installed at the surface. Details of the instrumentation and monitoring program were presented by Handfelt et al. (1987). Fig. 4 shows the locations of the inclinometers under and around the test fill. The measured lateral deformations are described below.

Lateral Deformations in the Haul Road Area

Fig. 5 shows the approximate sequence of construction and development of lateral deformations as construction of the haul road approached completion. After placement of the initial 2-m-thick hydraulic sand layer, a second layer of hydraulic fill 2 m to 3 m

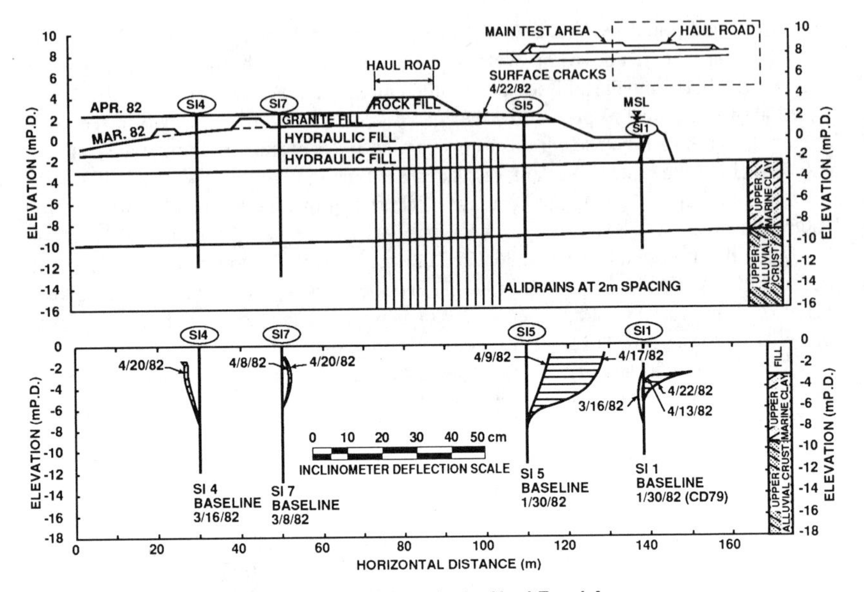

FIG. 5. Deformations in the Haul Road Area

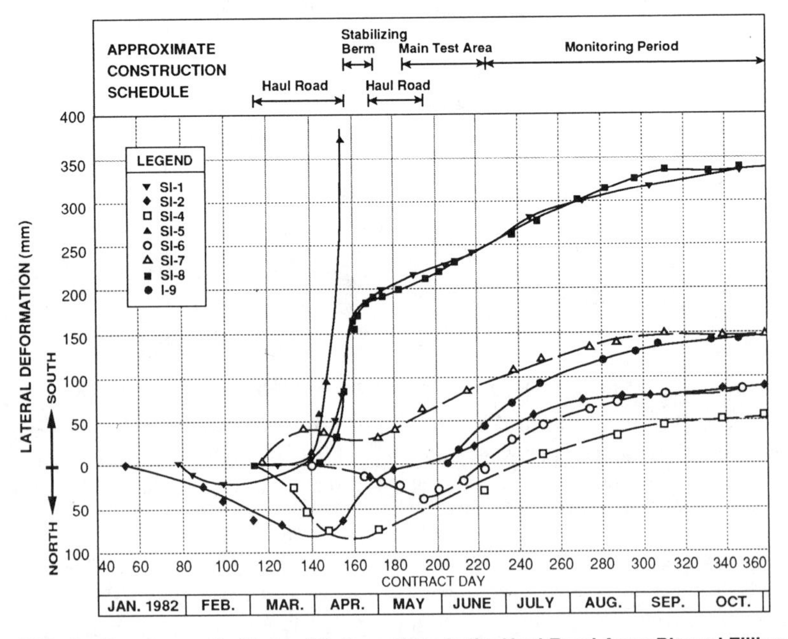

FIG. 6. Development of Lateral Deformations in the Haul Road Area: Phase I Filling

thick was placed in March 1982, followed by construction of two rockfill retention dikes towards the north end of the south slope. The deflections of SI4 and SI7 are in opposite direction and of similar magnitude, reflecting the placement of those rockfill dikes. In early April 1982, a lift of completely decomposed granite was placed to raise the area to about elev. +3 m. P.D. This was followed by construction of the last lift of rockfill to complete the haul road. Up to that point, lateral movements were fairly small, typically less than 50 mm. Construction of the last lift of the haul road proceeded from east to west, starting near the shore of Chek Lap Kok Island and moving west along the alignment of the haul road. By the time the placement of the last lift of fill reached the area of Section 1 (see Fig. 4), large lateral deformations developed and cracks appeared at the surface of the decomposed granite fill just south of the haul road. At that point construction of the haul road was interrupted and a hydraulic fill berm was constructed to stabilize the south slope.

During construction of the remainder of the haul road and the MTA, additional lateral movements developed but they were substantially smaller than the movements that developed prior to construction of the stabilizing berm. Fig. 6 shows the development of movements with time in the area of the haul road. Key construction events are also noted in the figure. The rapid increase in lateral movements between contract day (CD) 140 and CD 160 is evident as the south slope approached incipient instability, as well as the immediate impact of the placement of the berm in slowing down and finally stabilizing the slope movements. It should be noted that lateral deformation continued to increase due to placement of fill in the MTA. The lateral movements also continued

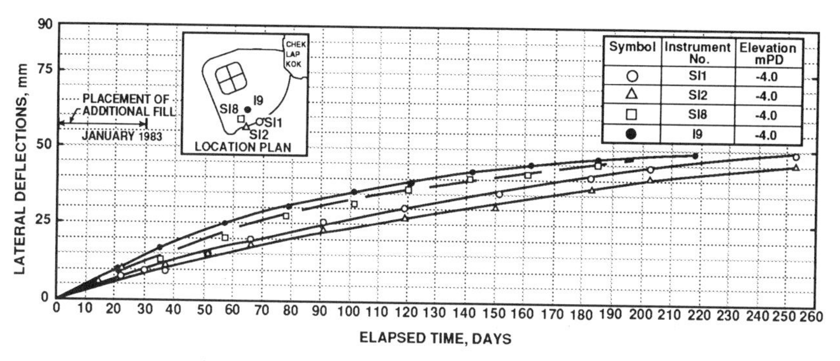

FIG. 7. Lateral Deformations of the South Slope Resulting from Phase II Filling

for several months after completion of the MTA fill.

Fig. 7 shows the effects of raising the fill in the MTA during Phase II. Lateral movements of up to 50 mm developed over a period of 250 days after beginning the filling for the MTA.

Lateral Deformations in the MTA Area — Phase I

Fig. 8 shows a cross section of the MTA area through the NW and SW quadrants (north to south section) and the lateral movements under and around the MTA in the vicinity of the section. The sequence of filling and the distribution of deformations with depths at various stages of filling are shown in the figure. Similar measurements under

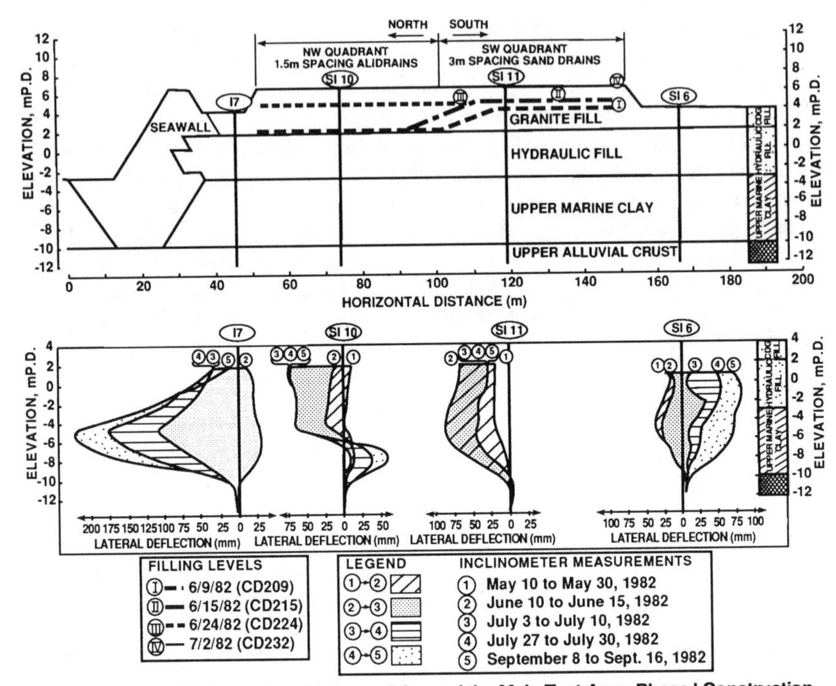

FIG. 8. Lateral Deformations Under and Around the Main Test Area: Phase I Construction

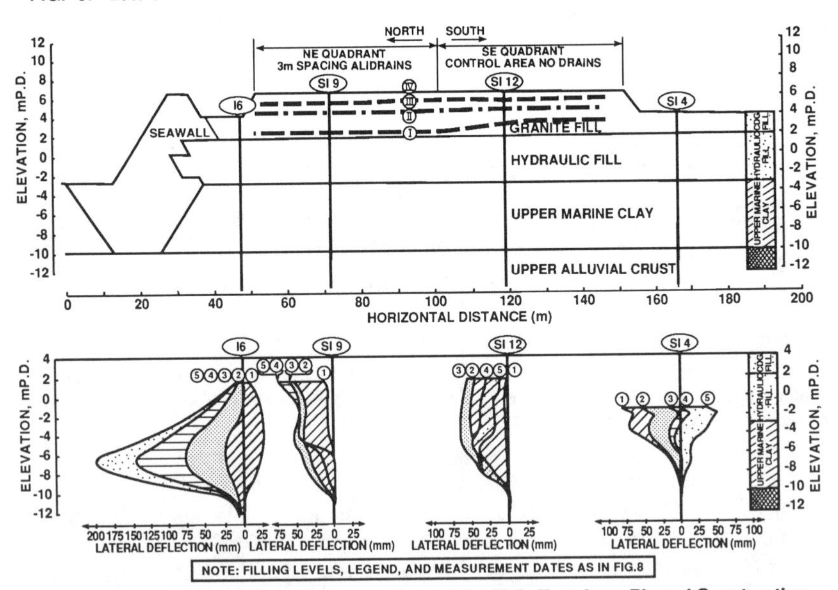

FIG. 9. Lateral Deformations Under and Around the Main Test Area: Phase I Construction

and around the NE and SE quadrants, in a north to south direction, are shown on Fig. 9.

The initial deformations under the MTA that developed towards the north, reflect the effects of the sloping fill front as it advanced from south to north. The direction of movement reversed when the MTA area fill was raised above sea level. In the area north of the MTA, the movements were initially towards the south, reflecting the movement of the inboard slope of the seawall. However, after the MTA was filled, the movements reversed direction. They increased quite rapidly during filling and continued to increase for a considerable time while filling. Apparently these very large movements reflect the squeezing of the mud layer into the rock fill seawall.

South of the MTA the movements were initially towards the north, consistent with the advancement of the filling operations from south to north. After filling of the MTA

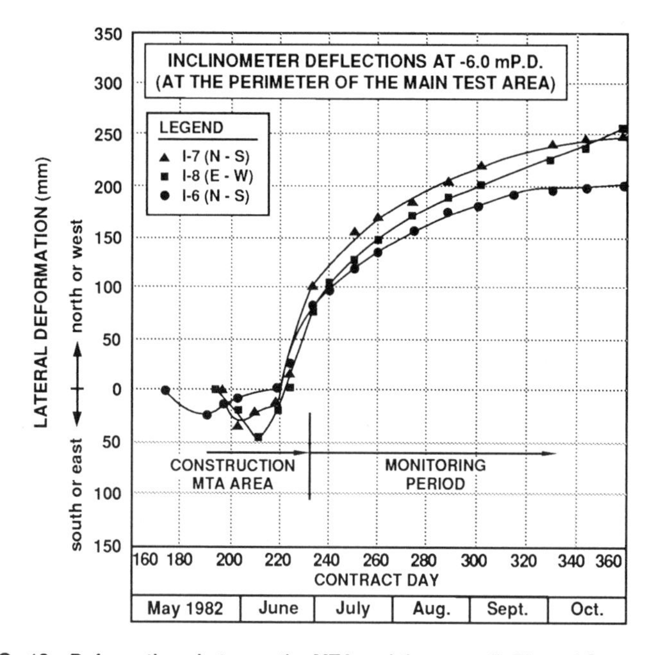

FIG. 10. Deformations between the MTA and the seawall: Phase I Construction

the direction of movement reversed towards the south. Initial maximum movements of 25 mm to 75 mm towards the north reversed to maximum movements of 50 mm to 75 mm towards the south.

The development of lateral deformations with time as measured by the three inclinometers installed between the MTA and the seawall is shown in Fig. 10. The movements continued to increase during the entire period of monitoring after the Phase I filling. Fig. 11 shows lateral deformations versus time as measured by the inclinometers installed under the MTA. The measurements show insignificant changes after the Phase I fill was placed in the MTA. This is an indication that in the central area of each quadrant,

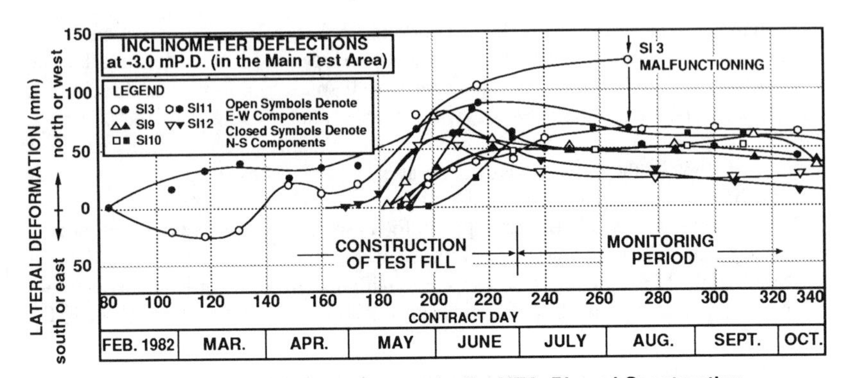

FIG. 11. Lateral Deformations under the MTA: Phase I Construction

where clusters of instruments were installed to evaluate the consolidation of the mud, 1-dimensional consolidation conditions prevailed after filling.

Lateral Deformation in the MTA in Phase II

Incremental lateral deformations in the north-south direction that developed during and after Phase II filling in the MTA are shown in Fig. 12.

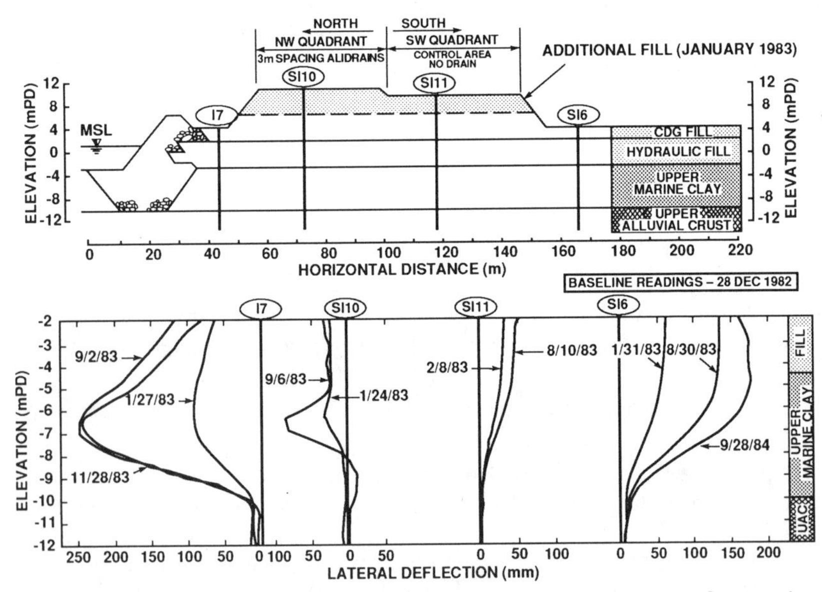

FIG. 12. Lateral Deformations Under and Around the Main Test Area: Phase II Construction

The lateral deformations in the central portion of each quadrant were relatively small (25 mm to 50 mm) while the deformations beyond the toe of the slopes of the MTA were much larger and continued to increase for a considerable time after filling.

In the area between the MTA and the seawall, incremental deformations 250 mm to 300 mm were recorded. These values are of the same order of magnitude as the

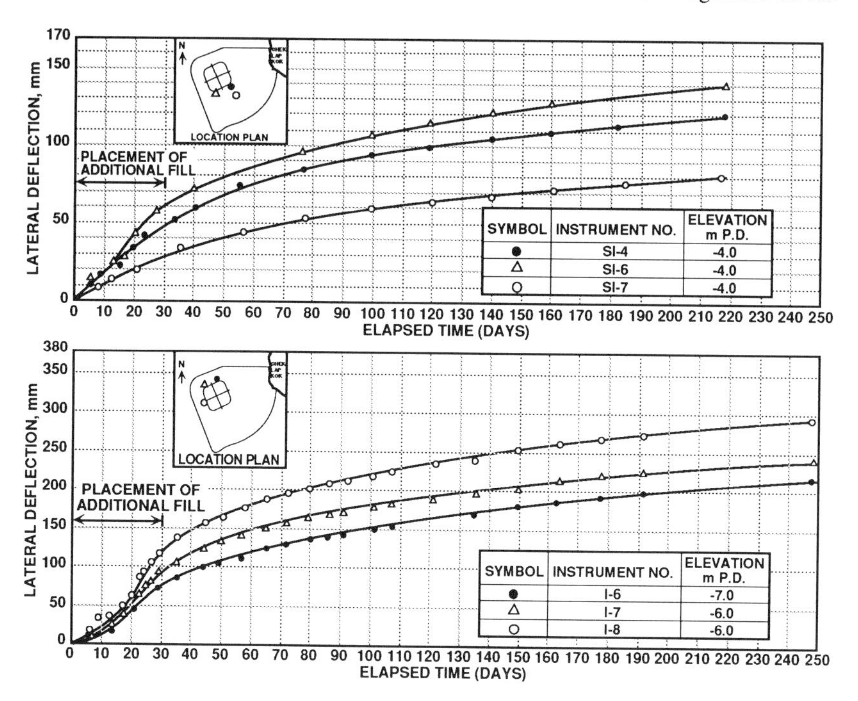

FIG. 13. Lateral Deformations Under and Around the Main Test Area: Phase II Construction

deformations that developed during Phase I filling.

The development of lateral deformations with time is shown in Fig. 13. The upper portion of the figure shows the deflections of the inclinometers along the south toe of the MTA. Although filling was completed in 30 days, the lateral deformations continued to increase for 200 days and at the end of the measurements there was no indication that the movements were about to stop. Maximum values ranged from 100 mm to 150 mm. The lower portion of Fig. 13 shows the lateral deformations measured by the three inclinometers installed between the MTA and the seawall. The movements continued to increase over a period of 250 days. The undrained movements that developed during filling (over a period of 30 days) are less than one-half of the total recorded movements.

Settlements measured in the MTA area are shown on Fig. 14. It is evident from the measurements that there were substantial edge effects over a distance of 20 m to 30 m from the toe of MTA slopes, while in the central portion of each quadrant the settlements were quite uniform. This again reinforces that 1-dimensional consolidation conditions prevailed in the central area of each quadrant. Edge effects are also evident along the boundary between quadrants, reflecting the effects of different rates of settlements and the resulting interaction between adjacent quadrants.

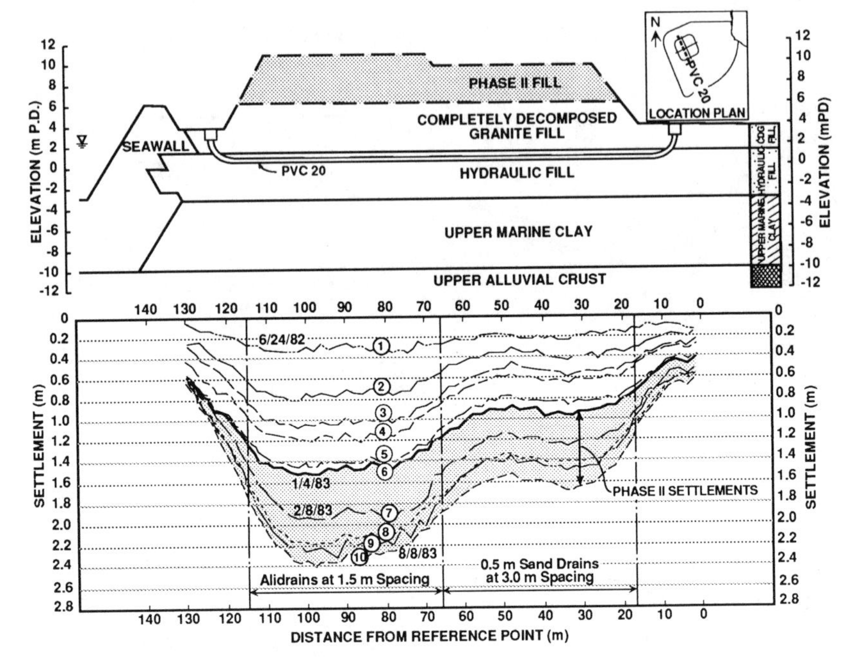

FIG. 14. Settlements under the Test Fill

Summary and Conclusions

The placement of heavy fill over a very soft and compressible mud deposit created very large lateral deflections under the south slope and around the MTA. Because of its very low strength and stiffness, the mud exhibited considerable mobility in response to fill placement. The direction of movements changed rather readily in response to the loading from the fill. In the area of the south slope, lateral movements in excess of 300 mm developed in response to the final 2 m increment of fill, as the slope approached incipient failure. The placement of a stabilizing berm controlled the lateral deflections and stabilized the slope, but it did not prevent the increase of lateral movements, which continued for many months after completion of filling. During and after the Phase I filling in the MTA, approximately 100 m inboard from the south slope, lateral movements in the haul road area continued to increase. During the Phase II filling in the MTA, the south slope began to move again and continued more than 250 days. Lateral movements on the order of 50 mm developed in response to a 3 m increase in fill thickness in the MTA.

In the area between the MTA and the seawall, total lateral movements up to 500 mm were recorded as the weight of the test fill squeezed mud into the rockfill seawall.

Lateral movements in the untreated area south of the MTA were small and stabilized shortly after the Phase I filling in the MTA. However, during and after the Phase II filling in the MTA, the three inclinometers installed between the MTA and the haul road showed relatively large movements, which continued for more than 250 days. The movements that developed after filling were somewhat larger than the "undrained" movements that developed during filling. The mechanism that caused the lateral movements to continue for such a long time, while the mud under the MTA was consolidating and therefore gaining significant strength, is not readily evident.

Appendix I.—References

Foott, R., Koutsoftas, D. C., and Handfelt, L. C. (1987). "Test fill at Chek Lap Kok, Hong Kong." *J. Geotech. Eng.*, ASCE, 113(2), 106-126.

Handfelt, L. D., Koutsoftas, D. C., and Foott, R. (1987). "Instrumentation for test fill in Hong Kong." *J. Geotech. Eng.*, ASCE, 113(2), 127-146.

Koutsoftas, D. C., Foott, R., and Handfelt, L. D. (1987). "Geotechnical investigations offshore Hong Kong." *J. Geotech. Eng.*, ASCE, 113(2), 87-105.

Appendix II.—Notations

The following symbols are used in this paper:

C_c = compression index

e_o = initial void ratio

OCR = overconsolidation ratio

SETTLEMENT OF DEEP COMPACTED FILLS IN CALIFORNIA

Ernesto E. Vicente,[1] Gerald M. Diaz,[2]
and Allen M. Yourman, Jr.,[3] Members, ASCE

ABSTRACT: Measured settlement of deep compacted engineered fills from a single site were summarized and compared to data from other sites. The data indicate that the rate and magnitude of movement of compacted (unsaturated) fills are dependent on thickness of fill, moisture added during the life of the fill and the type of soil. These ground deformation data should assist other engineers in planning construction on deep compacted fills. The data presented include total settlements and settlement rates for compacted engineered fills up to 36 meters thick.

INTRODUCTION AND BACKGROUND

Settlement of deep compacted engineered fills has become a major concern in southern California over the last several years. Land development during the 1970's and 1980's has resulted in compacted engineered fills over 30 meters deep becoming commonplace. Troublesome differential settlement has occurred where the fill thickness varied significantly within the plan area of a single structure or lot.

The subject site is located in the Los Angeles area and was graded in the mid 70's (exact location is confidential). Distress due to settlement and differential settlement was not noticed until approximately 9 years after rough site grading (5 years after building construction was completed). The ensuing litigation regarding the settlement continued for several more years. Forensic investigations were conducted by eight geotechnical consulting firms. These investigations were

---

[1] Geotech. Design Engineer, Fluor Daniel, Inc., 333 Michelson Dr., Irvine, CA 92730.
[2] Pres., Diaz•Yourman & Associates, 17461 Irvine Blvd., Suite E, Tustin, CA 92680.
[3] Vice Pres., Diaz•Yourman & Asso., 17461 Irvine Blvd., Suite E, Tustin, CA 92680.

performed between 10 and 14 years after grading (6 to 10 years after building occupancy). The litigation has now been completed and the data made available.

The site covers approximately 30,000 square meters. A canyon was filled with materials taken from the adjacent ridges to develop relatively level building pads for several large (1,000- to 2,500-square-meter in plan area) condominium buildings. The site grading resulted in fill slopes with inclinations between 1.5:1 and 2:1 (horizontal:vertical). Maximum cut and fill depths were approximately 17 and 36 meters, respectively. Subdrains were placed at the bottom of the canyon. The fill soils are underlain by less than 2 meters of natural fine-grained soils, which are, in turn, underlain by predominantly siltstone bedrock formation. Fill soils, consisting mostly of silts and clays, were placed in lifts approximately 200 millimeters (mm) thick and compacted. A minimum relative compaction of 90 percent based on ASTM D1557 was required. Approximately 1100 field density tests were performed during rough site grading using the sand cone (ASTM D1556) and drive cylinder (ASTM D2937) methods. The average relative compaction was approximately 91 percent. The site grading was completed in general accordance with local practice at that time. The landscaping featured extensive sprinkler irrigation systems to support lawns and lush vegetation. Some of the building footprints spanned the old canyon location, extending from shallow to deep fill or from cut to deep fill. Approximately 9 years after grading, (5 years after occupancy), distress due to settlement and differential settlement was observed by the condominium occupants. Much of the most noticeable distress was in the exterior features such as planter boxes which had separated from building walls.

SITE CHARACTERIZATION

Site characterization was based on field, laboratory, and monitoring data collected during the initial geotechnical investigation, site grading and supplemental forensic investigations. Forensic investigations included approximately 30 borings, cone penetration tests (CPTs), and test pits. In addition, instrumentation (consisting of inclinometers, surface settlement points, and at-depth settlement points) was installed and monitored for several years.

Samples obtained were tested to evaluate soil classification and index properties (moisture content/dry density, grain size distribution, plasticity, and specific gravity), compression characteristics, and relative compaction. Confined compression tests to measure hydrocompression (collapse) or swell upon sample inundation were conducted on two series of reconstituted soil samples simulating insitu compacted engineered fill soil conditions. In addition, intact fill soil samples obtained from the borings 14 years after grading were tested in a similar manner. A summary of general fill soil characteristics is presented in Table 1. A summary of primary, secondary, and hydrocompression (upon sample inundation) characteristics of reconstituted fill soil samples is presented in Table 2 and on Fig. 1.

TABLE 1. Fill Soil Property Characterization

	Range	Number of tests	Average	Std. Dev.
Classification (ASTM D422 and D4318)				
Percent passing 0.075 mm - %	66 to 92	21	83.4	8.1
Percent passing 0.002 mm - %	9 to 34	21	23.6	8.0
Plastic limit	30 to 39	22	34.5	2.9
Liquid limit	45 to 67	22	60.3	5.2
Plasticity index	14 to 37	22	25.8	5.3
Elastic silt (MH) and fat clay (CH) with sand				
Insitu Density and Penetration Resistance				
Dry density (above 12 meters)-kN/m^3	12 to 15	46	13.1	0.93
Dry density (below 12 meters)-kN/m^3	13 to 15	13	13.8	0.78
SPT - N (ASTM D1586)	10 to 25	20	21	10
CPT - q_c (ASTM D3431)	2 to 5	---	---	---
CPT - R_f (ASTM D3431)	4 to 6	---	---	---
Insitu Moisture and Saturation Conditions				
Fill Placement (ASTM D1556)				
Moisture content - %	18 to 27	1100	23	4
Saturation - %	67 to 80	---	74	5
Ten years after fill placement (ASTM D2216)				
Moisture content - %	27 to 36	76	32	4
Saturation - %	81 to 97	---	75	5
Fourteen years after fill placement (ASTM D2216)				
Moisture content - %	27 to 36	71	31	4
Saturation - %	82 to 99	---	87	5
Volummetric indices (fourteen years after fill placement)				
Porosity	0.44 to 0.58	45	0.48	0.03
Void ratio	0.7 to 1.0	45	0.92	0.09
Specific gravity (ASTM D854)	2.57 to 2.71	27	2.65	0.04
Moisture-Density Relationship (ASTM D1557; 11 tests)				
Maximum dry density - kN/m^3	14.5 to 16.5	10	15.5	0.6
Optimum moisture content - %	19 to 28	10	23.1	2.6
Volume Change Characteristics				
Clay activity	0.72 to 1.6	20	1.2	0.52
Expansion index (ASTM D4829)	54 to 71	15	58	8.4
Confined compression (ASTM D2435) (fourteen years after fill placement)				
Compressibility index, $C_c/(1+e_0)$	0.10 to 0.15			
Rebound Index, $C_r/(1+e_0)$	0.02			
Secondary compression, C_α	0.0001 to 0.0003			

TABLE 2. Reconstituted Fill Soil Sample Characteristics

	Location 1	Location 2
Classification (ASTM D422 and D4318; 21 tests)		
Passing 0.075 mm - %	89	87
Passing 0.002 mm - %	17	27
Plastic limit	36	35
Liquid limit	63	60
Elastic silt with sand (MH)		
Moisture-Density Relationship (ASTM D1557)		
Maximum dry density - kN/m^3	15.5	15.0
Optimum moisture content - %	23	25
Saturation - %	91	91
Sample Reconstitution Conditions		
Series 1		
Dry density - kN/m^3	13.8 to 14.3	14.1 to 14.4
Relative compaction - %	88 to 91	90 to 92
Moisture content - %	19 to 23	21 to 24
Saturation - %	60 to 69	64 to 69
(Reconstituted - Opt.) Moisture - %	-4 to 0	-4 to 0
Series 2		
Dry density - kN/m^3	13.2 to 13.5	12.9
Relative compaction - %	84 to 86	85
Moisture content - %	28 to 31	27 to 30
Saturation - %	79 to 83	64 to 69
(Reconstituted - Opt.) Moisture - %	5 to 8	3 to 5
Volume Change Characteristics (ASTM D4829; 15 tests)		
Series 1 - Confined compression		
Compressibility index, $C_c/(1+e_0)$	0.06	
Secondary compression, C_α	0.0005 to 0.001	
Rebound index, $C_r/(1+e_0)$	0.004	
Hydrocompression	See Fig. 1	
Series 2 - Confined compression		
Secondary compression, C_α	0.001 to 0.003	
Hydrocompression	See Fig. 1	

The fill soils consisted primarily of elastic (clayey) silt (MH) and, to a lesser extent, fat clay (CH) with sand and occasional fragments of siltstone and sandstone generally less than 25 mm in maximum size. Fill conditions were found to be fairly uniform; occasional layers of sandy soils, less than 50 to 100 mm thick, were found generally several meters apart. Occasional perched groundwater was encountered in the otherwise unsaturated fill; standpipe piezometers were either dry or indicated water levels near the bottom of the original canyon.

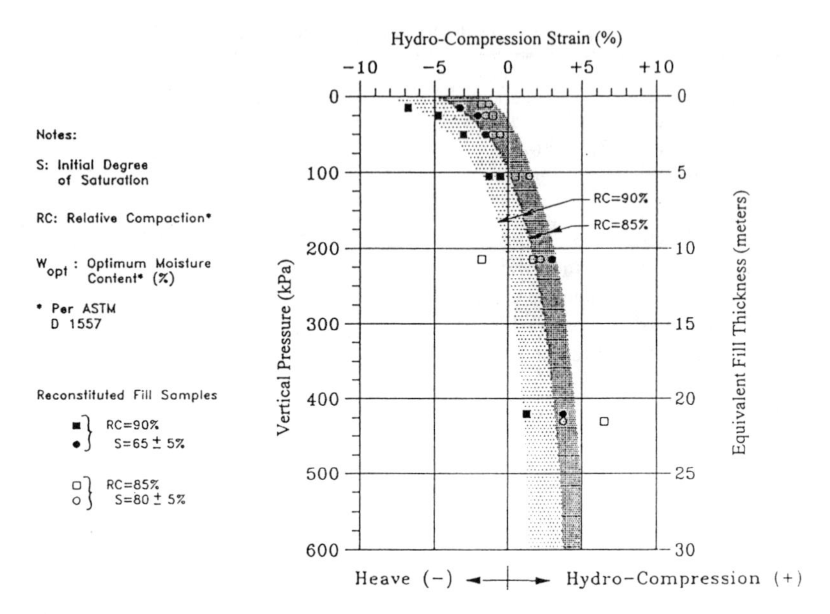

FIG. 1. Vertical Strain upon Wetting (Laboratory Data)

Unsaturated Ground Deformation Mechanisms

Reviews of mechanisms of vertical ground movement of unsaturated soils were recently published by Brandon et al. (1990), Lawton (1986), Lawton et al. (1989, 1991), and Fredlund and Rahardjo (1993). Earlier studies on these mechanisms were reported by Cox (1978). Effect of slopes and stress ratio on compacted fill deformation were investigated by Noorany (1991) and Lawton et al. (1991), respectively.

Because of the unsaturated nature of compacted fill soils, the relatively insignificant changes of effective overburden pressure after completion of grading, and the timing of the compression, it was not reasonable to try to justify the magnitude and rate of measured settlement by primary and secondary consolidation considerations. Hydrocompression was thought to be the primary cause of observed settlements.

Primary and secondary compression of fill materials could not account for more than a small fraction of the observed settlements. Similarly, siltstone bedrock materials exhibited compressibilities one-fifth to one-third of those of the overlying compacted fill materials.

Surface Vertical Movements

The vertical surface measurements plotted versus compacted fill thickness are presented on Fig. 2. The accuracy of vertical measurements is estimated to be on the order of 5 to 10 mm. These vertical settlements were measured between approximately 10 and 14 years after rough grading (6 to 10 years after the buildings were occupied). Therefore, measured settlements are believed to be less than total vertical settlement. No significant ground movements were noticed during the first few years of building occupancy, which is consistent with the findings of other researchers (Brandon et al. 1990).

Measured vertical settlement increased with increasing fill thickness. Some possible swell (heave) was measured where the fill thickness was less than 5 to 10 meters.

Surface vertical movement data from other deep fill sites in southern California are presented on Fig. 3. Predominant fill soils at other these sites were elastic silt (MH) to fat clay (CH) with sand and clayey sand (SC) to sandy lean clay (CL).

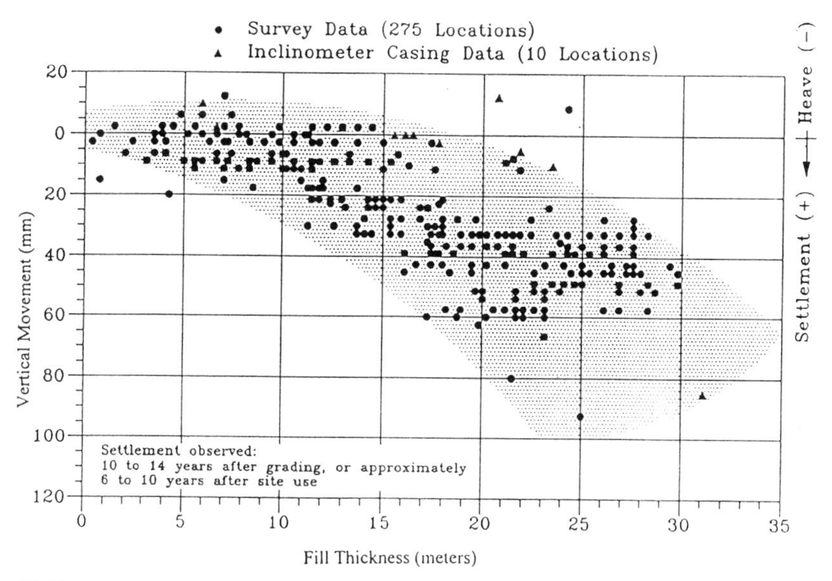

FIG. 2. Accumulated Vertical Movement - Site 1 (MH, Ch Soils) (4-Year Monitoring Period

Total surface vertical movement was calculated based on the confined compression tests performed on reconstituted samples of the fill soils. The comparison of the calculated to the measured movements is presented in Fig. 3.

RATE OF SURFACE VERTICAL MOVEMENT

The ground surface's average rate of vertical movement versus thickness of fill is presented on Fig. 4a for our primary study site (Site 1) for a period of 10 to 14 years after rough grading. Rates of vertical movement were for Site 1 were collected from the same 285 instrumentation and surveying locations, and averaged within the same four-year time period. Fig. 4b shows the maximum rate of ground surface settlement for other sites (Sites 2 and 3).

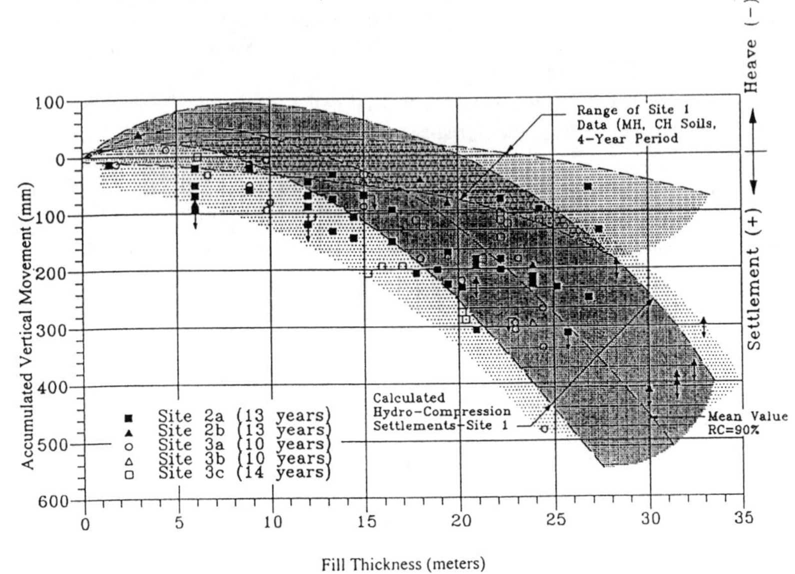

FIG. 3. Accumulated Vertical Movement - Sites 2 and 3 (10- to 14-Year Monitoring Period

Similar to results for the magnitude of settlement, average rates of measured vertical ground movement were found to increase with increasing fill thickness. Settlement rates in the range of 10 to 30 mm per year were measured for fill thickness greater than 25 meters. Therefore, the average movement was approximately 0.4 to 1.2 mm per year per meter thickness of compacted fill. The lower and higher values of average settlement rate roughly correspond to compacted fill thicknesses on the order of 10 meters and 25 meters or greater, respectively. Swell (heave) rates of less than approximately 3 to 6 mm per year were measured at stations with fill thickness less than 5 to 10 meters.

MAGNITUDE AND RATE OF GROUND SURFACE HORIZONTAL MOVEMENT

A limited amount of horizontal movement data was collected from slope inclinometers. The amount of data appears to be insufficient to draw accurate quantitative conclusions on the effect of slopes on hydrocompression-induced horizontal ground surface movement, but general qualitative conclusions are as follows.

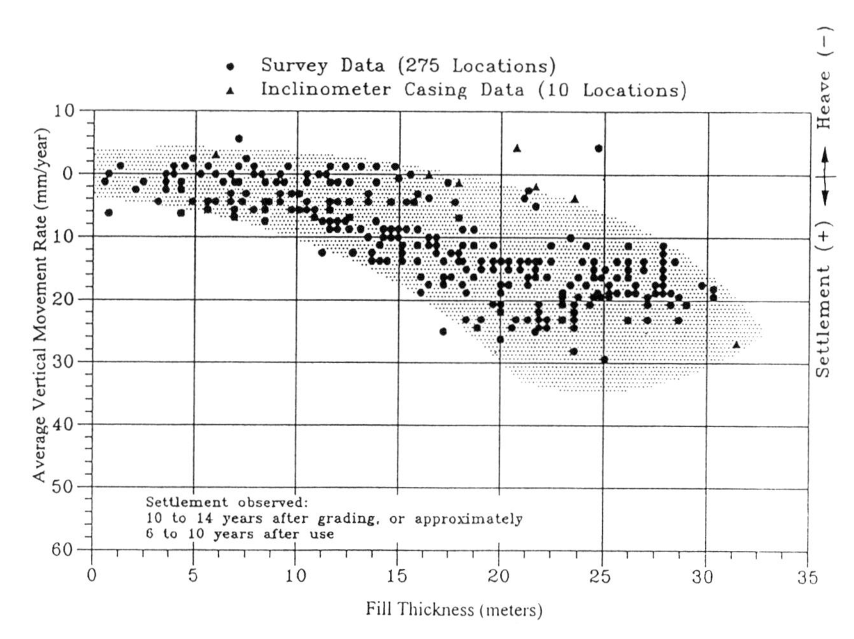

FIG. 4a. Average Vertical Movement Rate - Site 1 (MH, CH Soils) (4-Year Period)

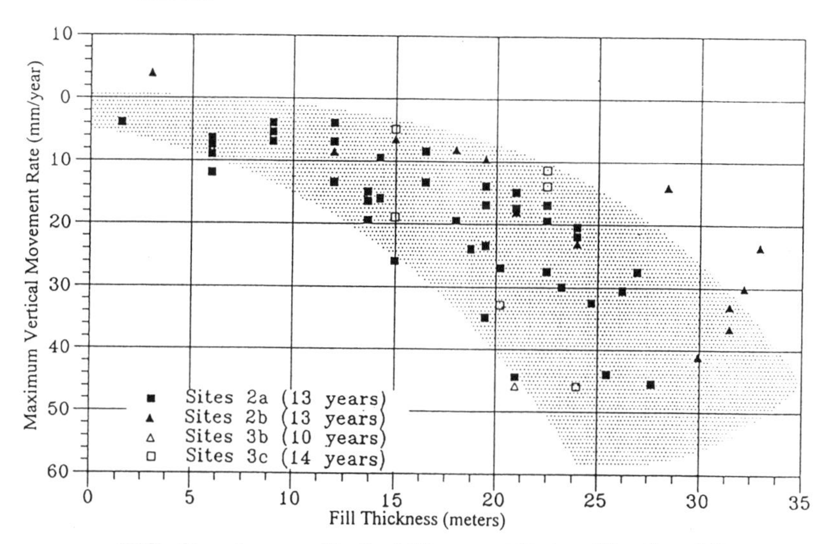

FIG. 4b. Average Vertical Movement Rate - Sites 2 and 3

Present and pregrading topography (ground surface and canyon bottom original slopes) affect measured rate and direction of ground surface deformation. Measurements suggested that horizontal deformations were roughly in the direction of present ground surface maximum gradient, but in some cases also followed the canyon bottom (bedrock) maximum gradient. Horizontal movements up to 25 mm were measured on 1.75:1 slopes. Measured horizontal movement rates were approximately 3 to 6 mm per year for compacted fill thicker than approximately 25 meters.

Ground Surface Vertical Movement Stabilization Time

A limited amount of data on vertical ground movement stabilization time, defined as the end of the full rate and beginning of the final slowdown ranges, was collected and is presented in Fig. 5. At many locations, it was not clear that the final slowdown range had been reached (sometimes because additional data were not available). These are indicated by the open data points. Minimum stabilization times on the order of 7 to 8 years are required for compacted fills on the order of 15 to 25 meters thick; longer stabilization times, probably more than 12 years, are interpreted (extrapolated) to be necessary for the stabilization of compacted fills thicker than approximately 30 meters. Continued monitoring, however, is required to further define data presented in Fig. 5, which should be considered preliminary.

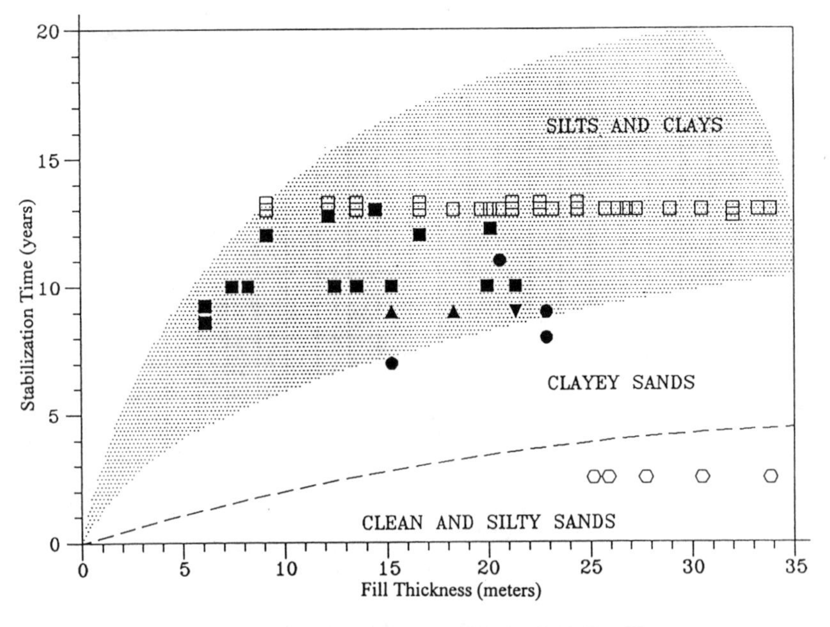

FIG. 5. Stabilization Time (Years) after Grading

CONCLUSIONS

The scatter of ground deformations measured in the field may appear discouraging, particularly if it is compared to tests or models under laboratory-controlled conditions. This scatter, however, reflects the variability of field conditions over a relatively large site, and an "uncontrolled" water infiltration and soil saturation process (mostly as a result of landscape irrigation). Consequently, providing data backfitting functions for estimating ground deformation is not an easy task. Figs. 2 through 4 illustrate the magnitude of ground deformation for various thickness of fill; however, for specific applications, local field monitoring and laboratory testing data should be gathered.

Compacted fill behavior data from field monitoring and laboratory testing were collected and summarized to aid in assessing magnitude and average rate of accumulated ground surface settlements as a result of hydrocompression of compacted fills up to approximately 36 meters thick consisting of elastic silts (MH) and fat clays (CH). Field data were compared with those collected from other areas for clayey sand (SC) to sandy clay (CL) fills, and with laboratory-testing-based procedures. It is believed that these ground deformation data will assist other engineers in planning construction on compacted deep fills to anticipate approximate amount and rate of movement as a function of fill thickness.

ACKNOWLEDGEMENTS

Data presented in this paper, gathered mostly from one well documented site, is the result of compiling work of nine geotechnical consulting firms, including our own research, as well as eight civil, structural, architectural, and surveying companies, and two government agencies in southern California, resulting in over fifty technical reports and other related documents. Most of the compilation work was conducted under contract between the Community Redevelopment Agency of the City of Los Angeles and Harding Lawson Associates. The contribution of data from all those organizations and the permission to publish the data, are gratefully acknowledged. The conclusions noted in this paper, however, are solely those of the authors.

APPENDIX - REFERENCES

Brandon, T. L., Duncan, J. M., and Gardner, W. S. (1990). "Hydrocompression settlements of deep fills." *J. Geotech. Eng.*, ASCE, 116(10) 1536-1548.

Cox, D. W. (1978). "Volume change of compacted clay fill." *Proc. Conference on Clay Fills*, London, Institution of Civil Engineers, 79-87.

Fredlund, D. G., and Rahardjo, H. (1993). *Soil Mechanics for Unsaturated Soils*, John Wiley & Sons, New York, New York.

Lawton, E. C. (1986). "Wetting-induced collapse in compacted soil," Ph.D. dissertation, Department of Civil Engineering, Washington State University, Pullman, Washington.

Lawton, E. C., Fragazzy, R. J., and Hardcastle, J. H. (1989). "Collapse of compacted clayey sand." *J. Geotech. Eng.*, ASCE, 115(1), 1252-1267.

Lawton, E. C., Fragazzy, R. J., and Hardcastle, J. H. (1991). "Stress ratio effects on collapse of compacted clayey sand." *J. Geotech. Eng.*, ASCE, 117(5), 714-730.

Noorany, I. (1991) "Deformation of compacted fill slopes caused by wetting," *Research Report* (dated September 11, 1991), Department of Civil Engineering, San Diego State University, San Diego, California.

MODELING SETTLEMENTS OF AN EXISTING MUNICIPAL SOLID WASTE LANDFILL SIDESLOPE USING AN EARTHEN SURCHARGE PILE

William L. Deutsch, Jr.,[1] Member, ASCE, Owen R. Esterly,[2] Member, ASCE, and John Vitale[3]

ABSTRACT: This paper discusses the results of a large-scale field compressibility test completed on municipal solid waste (MSW). The test consisted of surcharging the waste mass with an earthen surcharge pile. A survey control system permitted the total settlement of the waste mass induced by the surcharge loading to be determined. Mathematical models for predicting MSW settlement under an applied surcharge loading were subsequently developed using normally consolidated soil settlement and multiple linear regression theories.

INTRODUCTION

The large-scale field study discussed in this paper was completed for the purpose of determining design parameters for potential overfilling of existing landfill sideslopes at the Lanchester Landfill facility in Honey Brook, Pennsylvania. In particular, to maximize future waste disposal quantities at this facility, the Chester County Solid Waste Authority (CCSWA) is proposing to overfill additional MSW within the available air space between the eastern sideslope of their closed Municipal Site Landfill and the adjacent western sideslope of Cell No. 1 of their currently active Area B landfill. In accordance with current Pennsylvania Department of Environmental Resources (DER) MSW landfill lining system regulations, CCSWA

---

[1]Technical Director, Geotechnical Engineering, Roy F. Weston, Inc., West Chester, PA 19380-1499.

[2]Facility Engineer, Chester County Solid Waste Authority, Lanchester Landfill, Honey Brook, PA.

[3]Assistant Engineer, Roy F. Weston, Inc., West Chester, PA 19380-1499.

must initially install a geosynthetic lining/leachate collection system over the existing landfill sideslope prior to placement of additional waste at this location. The intent of this project was to simulate the total settlements and lateral movements of the existing waste mass when subjected to the additional applied stress of overfilled waste materials. These vertical and lateral movements would allow the maximum strains generated within the geosynthetics of an overfill lining system to be calculated. Based on these calculated strains, appropriate geosynthetic reinforcement criteria for the overfill lining system could be developed. The reinforcement would protect the geosynthetic components of this lining system, in particular, the geomembrane, from excessive straining and possible tear resulting from underlying waste settlements.

To complete this study, an earthen surcharge pile, whose maximum height modeled the stress of the proposed overfilled waste, was constructed on the eastern sideslope of the closed Municipal Site Landfill following an initial survey of the topography of this slope. This pile was removed at a later date and the area was resurveyed. This database allowed total settlements and lateral movements of the underlying waste mass to be determined, as discussed in the remainder of this paper.

EXISTING SITE CONDITIONS

The site of the field study consisted of the eastern sideslope of a closed portion of the landfill. The length of the referenced sideslope is approximately 250 feet (ft) (76.2 m). The inclination of this sideslope is approximately 3H:1V. At the time of this study, the landfill sideslope was covered with an intermediate soil layer and a vegetative cover.

FIELDWORK

The following activities were completed as part of the construction of the earthen surcharge pile on the eastern sideslope of the Municipal Site Landfill and are discussed in detail below.

I. Test Pit Investigation

Two test pits were excavated along the eastern sideslope of the closed landfill to determine the typical composition and degree of degradation of the buried waste. It was observed that the encountered waste was relatively "fresh" (i.e., very little biodegradation of the waste had occurred), and that newspapers dated from the mid-1970s were intact and readable. The generally "fresh" condition of this waste was an indication that settlement of this mass under the surcharge loading would occur within a reasonably short period of time consistent with the project schedule.

II. Settlement Plates Installation

Six settlement plates were installed on the landfill sideslope in order to monitor the settlement of the MSW as a function of time, both during construction and upon completion of the earthen surcharge pile. The steel settlement plates of 3-ft (0.9-m)-square plan dimensions were set on level earthen benches at approximately 25-ft (7.6-m) intervals along the center line of the surcharge pile in the direction from the toe to the top of the landfill sideslope. Vertical extension rods consisting of 1-inch (2.54-cm)-diameter galvanized pipe were fastened to the plates and then surrounded by a 3-inch (7.62-cm)-diameter pipe casing that directly contacted the adjacent fill soils. This construction allowed the inner pipe, which was used to obtain the survey control measurements, to move freely (i.e., settle) in response to underlying waste settlements.

III. Construction of Survey Control System

Subsequent to installation of the settlement plates, a nonwoven needlepunched (NWNP) geotextile was placed directly on the landfill sideslope beneath the entire footprint of the surcharge pile. The geotextile provided a clean working surface upon which the survey control system discussed below could be established. The geotextile fabric was extended approximately 5 ft (1.5 m) beyond the footprint of the surcharge pile on all sides and secured to the sideslope to avoid potential uplift from wind.

Sixty-four (64) metal pins that served as survey nodal points were subsequently nailed through the geotextile and into the landfill sideslope. The pins were installed in a grid-like pattern at a 25-ft (7.6-m)spacing interval over a 175-ft (53.3-m)-square footprint. Each nodal point was identified with a specific alphanumeric designation (e.g., A1, F6), as shown in Fig. 1. The geotextile surrounding each nodal point was subsequently painted so that these survey points could be easily located during the "post-settlement" survey. A second geotextile was installed atop this geotextile as a protective cover for the survey control system.

Prior to constructing the earthen surcharge pile, the project surveyors determined the northern and eastern coordinates (i.e., plan location) and the elevation of each nodal point and the top of riser pipe for each of the six settlement plates. The "pre-settlement" plan coordinates and elevation of each nodal point and settlement plate were tabulated and later compared to the "post-settlement" survey measurements (i.e., once the surcharge pile was removed from the sideslope). This comparison permitted computation of the total settlement and lateral movements of the underlying waste mass that had occurred under the loading, as discussed later in this paper.

IV. Construction of Surcharge Pile

The geometry of the soil surcharge pile was selected to model the applied stresses resulting from the proposed future overfilled MSW loadings. Recognizing that the unit weight of the soils used to construct the surcharge pile would be approximately twice that of typical MSW, the design height of the surcharge pile was selected to be approximately 50% of the proposed maximum height to which future overfilled refuse would be landfilled. A cross section showing the approximate design dimensions of the earthen surcharge pile in relation to the eastern sideslope of the closed landfill is presented in Fig. 2.

Construction of the surcharge soil pile was completed using two different colored and textured soils that were obtained from on-site borrow areas. The initial soil lift (i.e., approximately 1 ft (0.3 m) of material) placed directly upon the upper

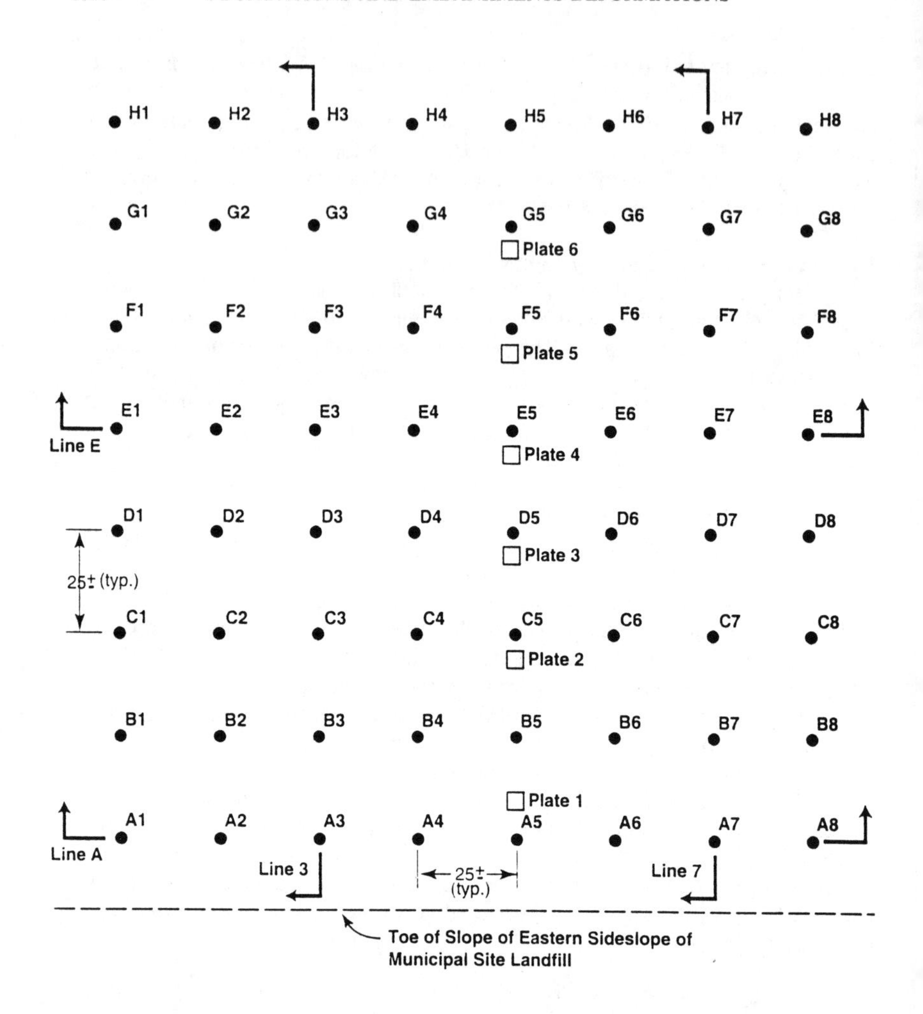

FIG. 1. Survey Control System, CCSWA, Lanchester Landfill

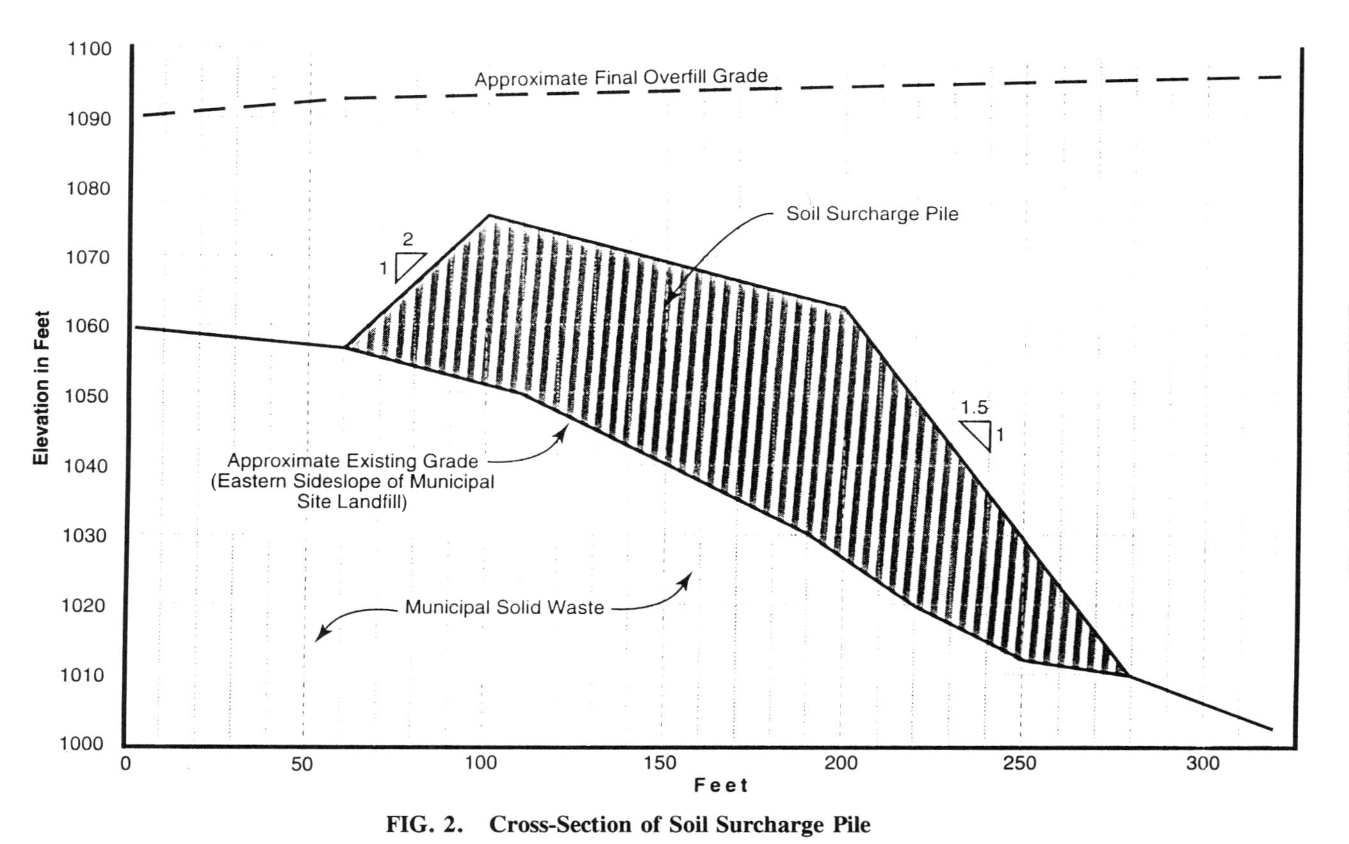

FIG. 2. Cross-Section of Soil Surcharge Pile

exposed geotextile was a dark reddish-brown, predominantly fine-grained soil. The second soil material, a tan predominantly nonplastic granular soil, was placed in near horizontal lifts of approximately 12 inches (30.5 cm) in loose thickness beginning at the landfill toe of slope and compacted until dense and stable. In this manner, the earthen surcharge pile was constructed to the approximate design dimensions presented in Fig. 2. Following completion of the pile, survey spot elevations were determined on the "flattop" of the surcharge pile to permit its thickness, and therefore its loading, to be calculated as discussed subsequently.

Throughout the placement and compaction of the surcharge pile fill materials, the in-place compacted wet and dry density and the moisture content of the fill soils were periodically measured using a Troxler nuclear moisture-density gauge. From these data, it was determined that the average unit weight of the soil surcharge pile was 124.2 pcf (1,990 kg/m^3). This value is slightly less than twice that of typical wet, biodegraded MSW (i.e., approximately 70 pcf (1,121 kg/m^3)).

V. Settlement Plate Monitoring

During construction of the surcharge pile, the settlement plate extension rods were periodically lengthened with additional pipe sections such that they would extend several feet above the top of installed fill soils. Immediately following completion of the pile, the top elevations of the rods were also determined by survey. Periodic measurements of the top elevations of these rods were also obtained at regular time intervals subsequent to completion of the pile. This allowed monitoring of the time distribution of the settlement of the underlying waste mass. A plot showing the total settlement of each settlement plate as a function of time is presented in Fig. 3.

Once the settlement plate data indicated that settlement of the waste mass under the surcharge loading had stabilized (i.e., total settlements had become essentially constant with increasing time, indicating that the waste had fully consolidated under the applied loading), the surcharge pile was removed from the landfill sideslope. As shown in Fig. 3, full consolidation of the waste mass under the surcharge loading occurred approximately 120 days (i.e., 4 months) after construction of the pile began.

VI. Removal of Surcharge Pile

Following completion of waste settlement under the surcharge loading as determined by the settlement plate data, the surcharge pile fill soils were excavated with heavy-duty construction equipment. Once the dark reddish-brown initial lift of cover soil was encountered, careful hand excavation of this material at each of the 64 nodal point locations was completed to uncover these survey control points.

VII. Post-Load Survey

Following exposure of the 64 nodal point locations, the northern and eastern coordinates and elevations of these control points were resurveyed and tabulated for analysis. Based on the data from the pre- and post-load surveys, as well as the topographic map of the surcharge pile, cross sections of the surcharge pile/landfill

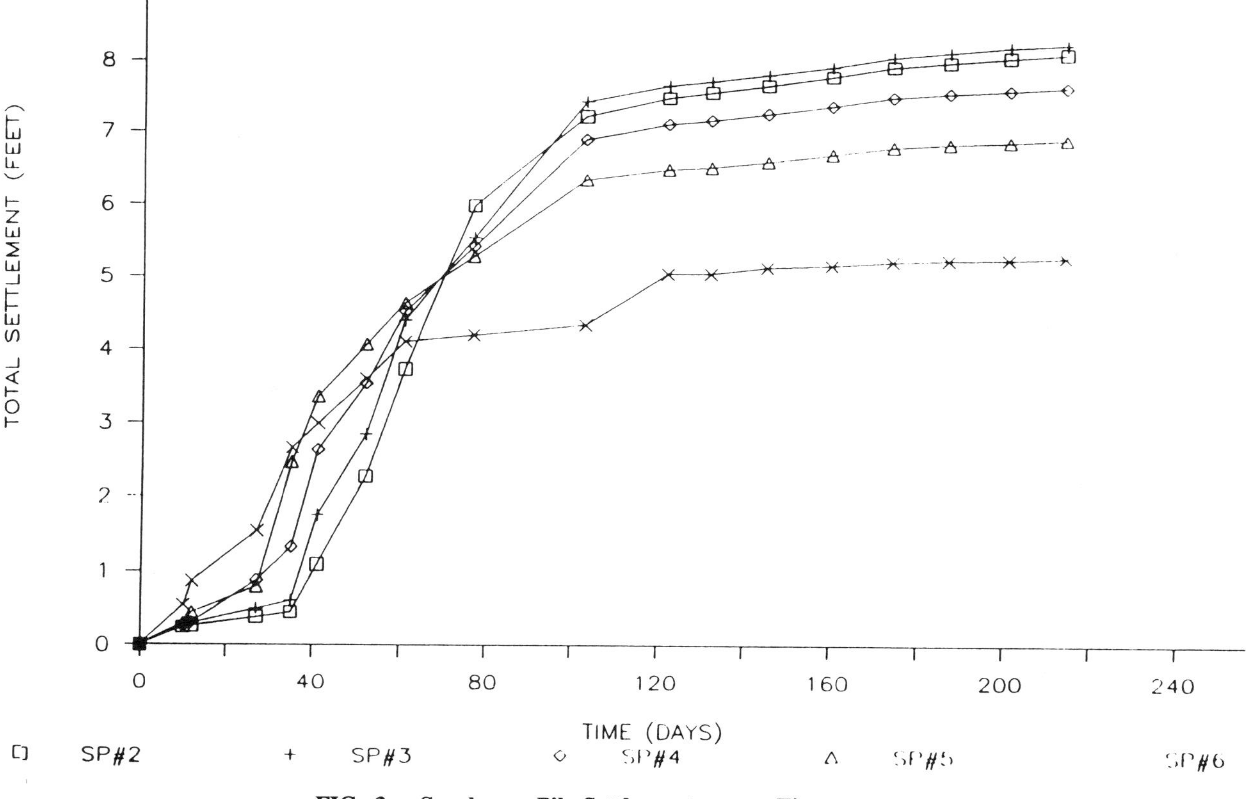

FIG. 3. Surcharge Pile Settlement versus Time

sideslope geometry were developed. Two of these are presented in Figs. 4 and 5. The locations of these cross sections are shown in Fig. 1. Fig. 4 illustrates the geometry of the surcharge pile in a direction parallel to the toe of slope of the closed landfill's eastern sideslope, while Fig. 5 illustrates the geometry of the surcharge pile in a direction perpendicular (i.e., upslope) to the toe of slope. Note that both the pre- and post-settlement profiles of the eastern sideslope of the landfill are shown in Figs. 4 and 5. The elevation difference between these two profiles at any location represents the waste settlement that occurred at this location as a result of the surcharge pile loading.

Data Reduction

The total settlement of the waste mass that resulted from the applied stress of the earthen surcharge pile was determined by subtracting the elevation of a given nodal point after the pile was removed from the sideslope (i.e., the "post-settlement" elevation) from the elevation of this same nodal point prior to construction of the surcharge pile (i.e., the "pre-settlement" elevation). Based on these data, it was determined that total settlement of the waste mass due to the surcharge loading was significant, ranging from approximately 1.5 ft (0.46 m) to approximately 6.5 ft (2.0 m) in magnitude. As anticipated, these settlements were generally lowest for nodal point locations along the perimeter of the survey control system and increased in magnitude for nodal point locations near the center of the survey control system. It is obvious that this is a result of the sloping geometry of the surcharge pile around its perimeter, which resulted in less height (and therefore loading) of this mass at these locations than near the center of the pile, where the height of this mass was fairly constant and of greatest magnitude. A comparison of the total settlements of the waste mass at adjacent nodal points also indicates that differential settlements between these survey control points were reasonably uniform in both the downslope and cross-slope direction. This is graphically illustrated in Figs. 4 and 5, which show the geometry of the settled landfill sideslope. It is noted from these figures that the geometry of the settled sideslope is reasonably parallel to the original sideslope, indicating the generally uniform nature of the settlement.

It was also determined from a comparison of the pre-settlement and post-settlement survey data (i.e., the northern and eastern coordinates of the 64 nodal points) that the lateral movements of the sloping waste mass, both parallel and perpendicular to the toe of the slope of the landfill, were minimal and generally less than 6 inches (15.2 cm) in magnitude. These data, therefore, indicate that movements of the waste mass upon loading were primarily in the vertical direction (i.e., settlement) rather than in the downslope or cross-slope direction.

Data Analysis

Two quantitative analyses of the 64-point survey database were completed as part of this study. These are discussed in the following subsections of this paper.

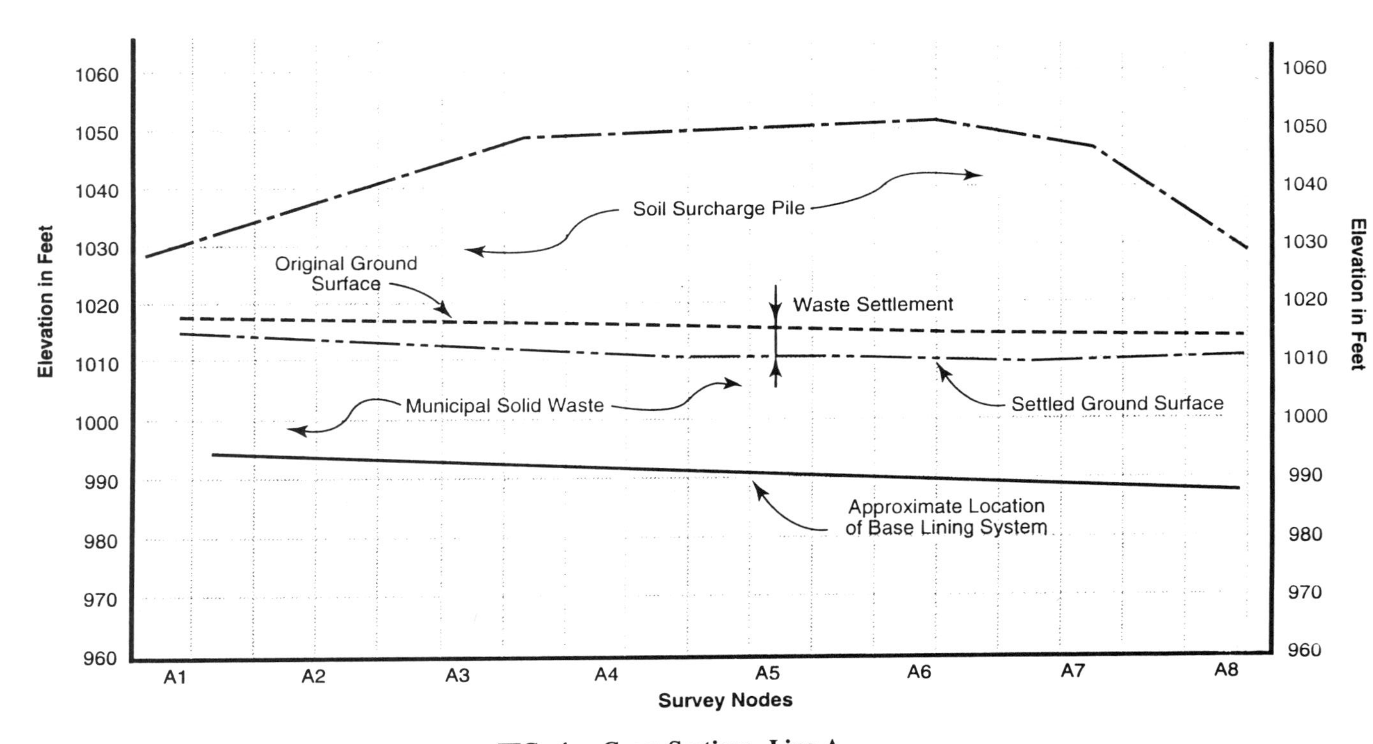

FIG. 4. Cross-Section: Line A

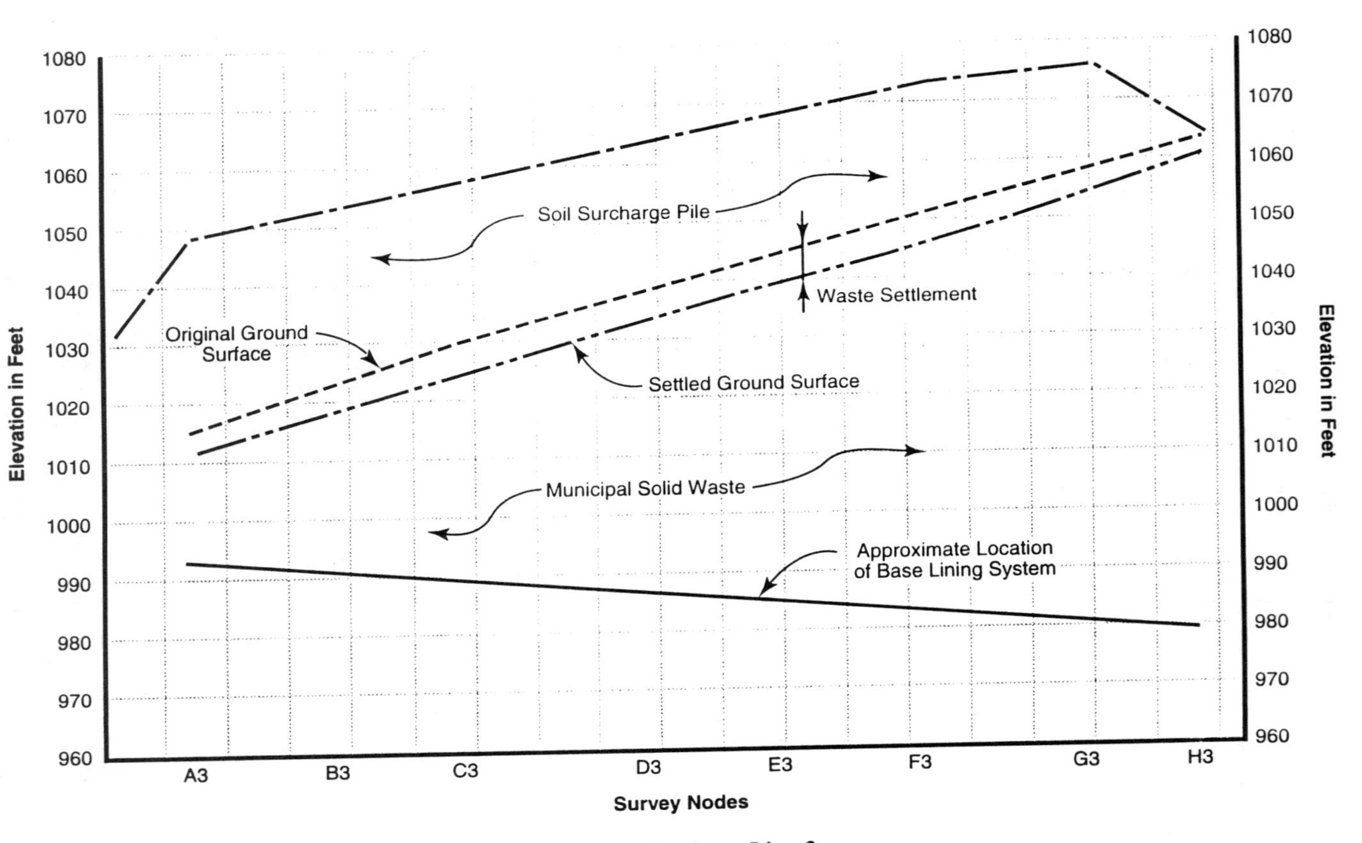

FIG. 5. Cross-Section: Line 3

I. Analysis Using Normally Consolidated Soil Settlement Equation

The first analysis assumed that the settlement behavior of the waste was similar to the settlement of a normally consolidated soil stratum, and could therefore be predicted by the following equation (Holtz and Kovacs 1981):

$$\Delta H_w = \frac{H_w C_c}{(1+e_o)} \log \left[\frac{\sigma_{o_w} + \Delta\sigma_s}{\sigma_{o_w}} \right] \quad (1)$$

where:
- ΔH_w = waste settlement, feet (meters);
- H_w = waste thickness, feet (meters);
- C_c = compressibility index of the waste;
- e_o = in situ void ratio of the waste before loading;
- σ_{o_w} = in situ effective vertical overburden pressure at the midheight of the waste stratum, psf (kg/m^2);
- $\Delta\sigma_s$ = applied surcharge loading at the midheight of the waste stratum, psf (kg/m^2).

It is noted that the concept of normally consolidated settlement behavior assumes that the existing overburden stress condition at any point within the consolidating stratum is the maximum stress condition that these materials have ever experienced. In addition, normally consolidated behavior also implies that the compressible materials have fully consolidated under this overburden condition. Both of these conditions are reasonable assumptions for a MSW landfill that has been sequentially raised in height over many years, and that has existed in its present geometrical configuration for more than a decade.

As noted in Eq. (1), the parameters C_c and e_o directly relate to the compressibility properties of the waste. These parameters represent unknowns in this study. The parameters ΔH_w, H_w, σ_{o_w}, and $\Delta\sigma_s$ represent parameters that can be directly measured from field data obtained in this study (i.e., ΔH_w) or calculated (H_w, σ_{o_w}, and $\Delta\sigma_s$) from existing landfill records data. Therefore, a "modified compressibility index" (C') for the waste materials, defined as shown below, can be calculated from the known parameters of this study using the following equation:

$$C' = \frac{C_c}{1+e_o} = \frac{\Delta H_w}{H_w \log \left[\frac{\sigma_{o_w} + \Delta\sigma_s}{\sigma_{o_w}} \right]} \quad (2)$$

The following paragraphs describe the manner in which the four known parameters discussed above were determined.

The final equilibrium values of waste settlement (i.e., ΔH_w) were determined as discussed previously. The thickness of the waste (H_w) at each nodal point was determined by subtracting the known elevation of the top of landfill lining system at the respective nodal point location from the pre-settlement elevation of the nodal point located on the landfill sideslope. The former value was estimated from

CCSWA drawings of the base lining system for this area of the landfill. The calculated H_w values were not modified to account for daily and final cover soil thickness. It is noted that the waste thickness is approximately equal along each alpha nodal point line and increases in thickness from nodal point line A to H.

The in situ effective vertical overburden pressure at the midheight of the waste stratum was calculated as follows:

$$\sigma_{o_w} = \frac{H_w \gamma_{MSW}}{2} \tag{3}$$

where: γ_{MSW} = unit weight (density) of the waste, which was assumed to be 70 pcf (1,121 kg/m³) for this calculation based on the authors' extensive experience in landfill design work. This value is representative of wetted, "fresh" MSW combined in approximately 9 to 1 proportions with daily/final cover soil.

It is also noted that the calculation of vertical effective stress at the midheight of the compressible waste layer thickness as quantified by Eq. (3) assumes that no liquid (i.e., leachate) level, which would generate hydrostatic pressures and buoyancy effects, exists within the waste mass. This is a reasonable assumption since it is reported that a leachate collection and removal system located directly above the cell's base lining system was installed as part of the construction of the landfill.

The applied surcharge loading at the midheight of the waste stratum (i.e., $\Delta\sigma_s$) was calculated as follows:

$$\Delta\sigma_s = T_{sp} \gamma_s \tag{4}$$

where: T_{sp} = height of the surcharge pile at the nodal point location, feet (meters); and

γ_s = average unit weight of the surcharge pile soil = 124.2 pcf (1,990 kg/m³), as discussed previously.

The T_{sp} parameter was calculated as the difference between the elevation of the top of the surcharge pile at a respective nodal point location and the pre-settlement ground surface elevation of this same nodal point. It is noted that this calculation neglects edge effects for nodal points located adjacent to the sloping faces of the surcharge pile. However, it is believed that these effects are not significant when compared to the degree of accuracy of the other measured and calculated independent variables.

Values of the independent variables for each of the 64 nodal point locations were subsequently input into Eq. (2) in order to calculate the "modified compressibility index" (C′) of the waste. The calculated values of C′ define a fairly limited range of values, with the exception of nodal points H1 through H8. In particular, the C′ values for the 56-nodal point database consisting of lines A through G varied from 0.15 to 0.39, averaging 0.22, while the C′ database consisting of H line values varied from 0.66 to 0.98, averaging 0.80. It is clear that the H line values are anomalous. This is believed to be due to the minimal and highly variable surcharge pile thicknesses that existed in the vicinity of the nodal points along the H line resulting from the sloping geometry of the back face of the pile at these locations. As a result, the surcharge loading varies considerably in the vicinity of each H line nodal point location. Therefore, there is significant error in assuming that the load at these nodal point locations can be accurately calculated using Eq. (4). Based on the above discussion, it was believed appropriate to disregard the H line C′ values in generating pertinent statistics for this database as discussed in the following paragraph.

The 56-point database was analyzed statistically. A quantitative procedure was used to confirm that the database is normally distributed. The mean ($\bar{x}$), standard deviation (σ), variance (σ^2), and coefficient of variation (C_v) of the database were also calculated. These values are as follows:

$$\bar{x} = 0.22; \quad \sigma = 0.0402; \quad \sigma^2 = 0.0016; \quad \text{and } C_v = 10.27\%$$

Based on the properties of a normal distribution, it is known that 97.5% of the database lies below the random variable value of $\bar{x}+2\sigma$. For this database, this value is equal to 0.22 + 2(0.402) = 0.30. Therefore, it can be concluded that calculation of surcharge load induced MSW settlement using Eq. (1) with known values of H_w, σ_{o_w}, and $\Delta\sigma_s$ will yield a conservative estimate of this settlement 97.5% of the time if a value of $C' = C_c/(1+e_o) = 0.30$ is used in the equation.

II. Analysis Using a Multiple Linear Regression Model

The 56-point database was also analyzed using a multiple linear regression model, in which the dependent variable, waste settlement (ΔH_w), was assumed to be a linear function of the independent variable's waste thickness (H_w) and applied surcharge loading ($\Delta\sigma_s$). The form of this equation is as follows (Walpole and Myers 1972):

$$\Delta H_w = b_0 + b_1 H_w + b_2 \Delta\sigma_s$$

The regression coefficients (b_0, b_1, and b_2) were determined using acceptable "least squares estimation" regression procedures. The resulting regression coefficients are as follows:

$$b_0 = -0.35366 \qquad b_1 = +0.04325 \qquad b_2 = +0.000836$$

Therefore, the "best fit" multiple linear regression equation that most accurately models the measured waste settlements is:

$$\Delta H_w = 0.04325 H_w + 0.000836 \Delta\sigma_s - 0.35366 \tag{5}$$

In this equation, the terms ΔH_w and H_w are in units of feet, and the term $\Delta\sigma_s$ is in units of pounds per square foot (psf). It is noted that different regression coefficients would result from using metric units for the input variables ΔH_w, H_w, and $\Delta\sigma_s$. Regression analysis using metric units was not completed as part of this study.

Statistics pertinent to this multiple linear regression model include the standard error (S.E.), the Coefficient of Determination (R^2), and the Correlation Coefficient (R). These values are as follows:

S.E. = 0.65 $\qquad$ $R^2 = 0.72$ $\qquad$ R = 0.85

CONCLUSIONS

The results of this large-scale field study have allowed predictor equations to be developed for estimating the settlement of MSW induced by an applied surcharge loading. A predictor equation was developed under the assumption that the waste mass behaved as a normally consolidated soil stratum. It was shown that a "modified compressibility index" (C′) of 0.30, when used in the normally consolidated soil settlement equation, would provide a conservative estimate of waste settlement 97.5% of the time. A second predictor equation was also developed that utilized multiple linear regression theory to correlate waste settlement (i.e., the dependent variable) to waste thickness and applied surcharge loading (i.e., the independent variables).

APPENDIX - REFERENCES

Holtz, R. D., and Kovacs, W. D. (1981). *An Introduction To Geotechnical Engineering*, Prentice-Hall, Englewood Cliffs, New Jersey.

Walpole, R. E., and Myers, R. H. (1972). *Probability and Statistics for Engineers and Scientists*, Macmillan Publishing Inc., New York, New York.

SETTLEMENT MEASUREMENTS OF 50-FT HIGH EMBANKMENTS AT THE CHESAPEAKE & DELAWARE CANAL BRIDGE

Ed Brylawski,[1] Poh C. Chua,[2] and Edward S. O'Malley,[3] Members, ASCE

ABSTRACT: This is a case history of instrumented settlement measurements on 49 to 56′ (15-17 m) high bridge approach embankments of Delaware State Route 1 crossing of the Chesapeake & Delaware Canal. Details of instrumentation and its installation are stressed. The project's instrumentation measured settlement by three independent methods: (1) liquid-filled tubes connecting reservoirs outside the fills to pressure sensors buried beneath, (2) horizontal inclinometer casing extending across the fill, and (3) settlement plates at the base of the fill with survey rods extending to the surface. The settlement measurements had the dual purposes of indicating when settlement rates had slowed enough to begin construction of pile-groups for bridge abutments that terminate within the embankment fill, and to provide comparisons between settlement results, ease of use, and costs and labor required to get data from the three systems. The results from the three measurements agreed closely. The south side where soil was sandy and settled approximately 4 in (10 cm) whereas the north side with clayier more compressible soil settled approx. 8 in. (20 cm). Measured settlements were significantly less than calculated settlements.

[1] Geotech. Engr., President, Geonor, Inc., P. O. Box 903, Milford, PA 18337-0903.

[2] Geotechnical Engineer, Law Engineering, 4465 Brookfield Corporate Drive, Chantilly, VA 22021.

[3] Principal Engineer, Law Engineering, 4465 Brookfield Corporate Drive, Chantilly, VA 22021.

INTRODUCTION

This paper presents a case history of instrumented settlement measurements of the highway embankment approaches to a major bridge. Three independent methods of measuring settlement were used. Details are presented of (a) installation, (b) logistics, (c) costs, (d) data collection, and (e) measurements. Measured and calculated settlements are compared. The emphasis is on the types of instrumentation used and practical details of obtaining field measurements.

It was critical to have measurements that provided information about the rate of settlement to decide when certain construction operations could proceed; i.e., settlement of approach ramps had to decrease to an acceptable rate before pile driving could begin, since bridge abutment pile groups were founded within the fill embankment. Secondly, the owner and designer wanted to compare two remote settlement measurement techniques, new to them, with the more familiar method of surveying buried settlement plates and riser pipes.

PROJECT DESCRIPTION, SITE AND SOIL CONDITIONS

Located south of Wilmington Delaware, the project consists of a $60-million six-lane concrete segmented cable-stayed bridge crossing of the Chesapeake & Delaware Canal and its approach embankments and roadways. The 4650′ long bridge has a main span of 750′. The south and north embankments are 49 and 56′ high, respectively with 3% grades for the traffic, and a toe-to-toe width of 300′ at their highest points. Side slopes are 2H:1V. The embankment construction sites were level to gently rolling agricultural land.

Soil Investigations

Prior to design, Delaware DOT contracted for 54 test borings up to 130 ft deep and 12 cone penetration tests (CPT) soundings along, or west, of centerline. The borings included standard penetration testing (SPT), split spoon and "undisturbed" sampling. Laboratory tests included: classification tests, triaxial, and consolidation testing.

Fig. 1 shows a profile of the soil looking west along the centerline. Investigations indicated a layer of fill and 6 underlying soil strata, excluding topsoil. More clay and more variable layering underlie the north embankment, whereas south side soils are sandier and the layering more uniform. Ground water depths ranged from surface to 44.0 ft. Tidal variation occurred near the canal but not under the embankments. The soil profile is described briefly as follows:

Fill dredged from the canal, consists of fine to coarse sands, silt, clay, organic matter and debris. Up to 65-ft (20 m) thick, it extends from under the high end of the north embankment to the north bank of the canal. Typically loose, soft, and moist, its SPT values ranged from 0 to 25 blows per foot (bpf) with the majority less than 10 bpf.

Peat (stratum I) is firm to very hard. Its varies in thickness up to 25 ft (8 m), with SPT's between 11 and 52 bpf.

Sand (stratum II) is variable, very loose to dense, and fine to coarse with gravel, silt, clay, and shells. SPT's range from 2 to 40 bpf. CPT, SPT and classification test data were used to estimate friction angles, ϕ, from 25° to 35°.

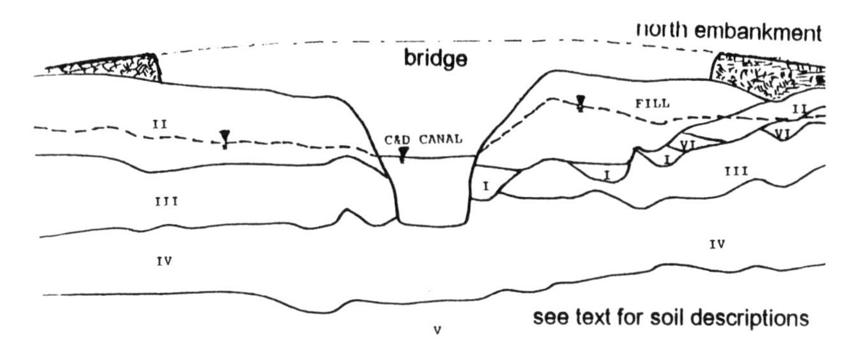

FIG. 1. Representative Soil Conditions

Sand (stratum III) is fine, and medium to very dense, with gravel, silt, and clay. SPT's ranged from 13 to over 100 bpf. CPT's could only penetrate to the top of this stratum. $35° < \phi < 45°$ was inferred from SPT's.

Clay and Silt (stratum IV) is very stiff to very hard with clay and sand. Several types of laboratory undrained shear strength (c_u) tests indicated 1,000 psf (50 kPa) $< c_u <$ 5,000 psf (250 kPa). Piles penetrating this stratum were designed using 1,400 psf (70 kPa) $< c_u <$ 2,600 psf (130 kPa). SPT's ranged from 15 to over 100 bpf.

Silt and Clay (stratum V) is hard to very hard with sand, gravel, and clay. c_u ranged from 1,000 psf (50 kPa) to 9,000 psf (450 kPa). SPT's ranged from 40 to over 100 bpf.

Clay (stratum VI) is soft to very stiff with gravel, sand, and silt, and appears to have caused most of the settlement of the north embankment. Ranging from 0 to 16-ft thick, it was generally interbedded between Strata II and III. SPT's ranged from 4 to 20 bpf. No c_u data was obtained from the few soil samples retrieved.

INSTRUMENTATION USED

Comprehensive discussions of geotechnical instrumentation include DiBiagio (1992), Dunnicliff (1988), and Hanna (1985). Three types of geotechnical instrumentation were used to measure settlement of the embankments and underlying natural strata at the C&D Canal site:

1. conventional surveying of vertical riser pipes connected to *settlement plates* buried in or at the base of the fill;
2. a cable pulled *inclinometer* sensor traveling through horizontal casing buried at the base of the fill (Wilson 1967);

3. buried pressure sensors that settle with time, while measuring the head of *liquid in tubes* that extend out of the fill to a liquid reservoir with a known elevation (Burn 1959; Bergdahl and Broms 1967; Bozozuk 1969).

Table 1 summarizes characteristics of each system from the users viewpoint. Fig. 2 shows instrument locations and the location of survey points on the surface of the embankment. Measurements were made at the ends, closest to the bridge where the fill had its greatest height. The function of each of the three systems is described below.

TABLE 1. Characteristics of the Settlement Monitoring Systems

	Settlement Plates and Vertical Riser Pipes	**Horizontal Inclinometer Casing**	**Liquid Settlement Tubes w/ Pneumatic Pressure Sensors**
Installation Procedure	Simple	Complex	Complex
Maintenance & Installation Requirements	Throughout construction	Initial lifts only	Initial lifts only
Labor required for data collection	A survey crew	Laborious, needs 2-person crew	Easy, 1-person
Time required for data collection	Time consuming	Time consuming	Minimal
Data reduction	Simple	Complex and time consuming	Simple
Data interpretation	Easy	Easy	Requires experience
Inconvenience to construction	Yes	No*	No*

* No significant inconvenience to the contractor once the devices have been installed and several feet of fill have been placed.

Fig. 3 shows a schematic drawing of the settlement plate and riser pipe systems. Plates or anchors can be placed at the base of the fill, or at intermediate depths. Riser pipes must be repeatedly extended upward, typically in 5′ (1.5-m) sections as fill is placed. Changes in the top elevation of the riser pipes were surveyed. The system is simple, but requires added work of hand tool compaction around the risers, and a lot of caution on the part of equipment operators to avoid damage.

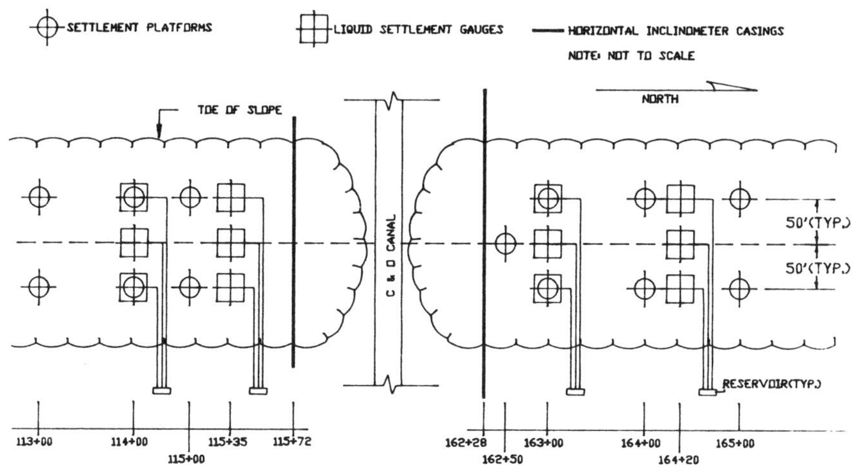

FIG. 2. Location of Measurement

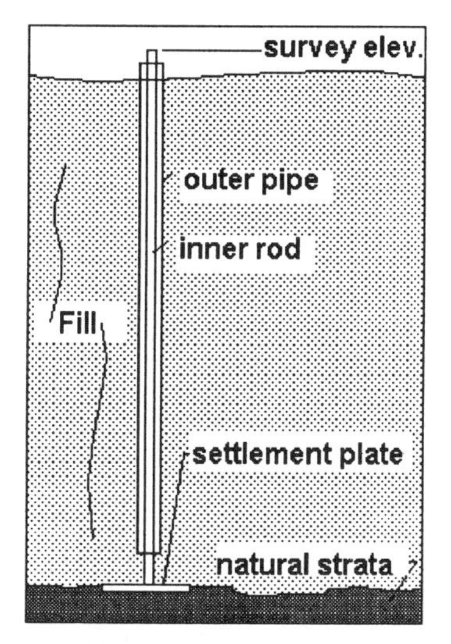

FIG. 3. Settlement Plate

The settlement plates were ½″ (12-mm) thick steel, 2′ (0.6 m) square, with 5′ (1.5 m) lengths of 1″ (25 mm) schedule 40 riser pipe attached. A 2″ (50 mm) protective outer steel pipe was placed over the inner riser and set with a 1′ (0.3 m) gap above the settlement plate such that it was free to settle with the fill, and protect the inner riser from down-drag force.

Fourteen settlement plates were used on the project, seven on each embankment. Two plates were placed at the base of each embankment. The remaining plates were placed higher up in the fill 13′ (3.9 m) below the final embankment grade.

Horizontal Inclinometer System

Fig. 4 shows a schematic of the horizontal inclinometer system. This system consists of plastic casings extending across the base of the embankments. An inclinometer sensor is pulled through the casings stopping at short intervals to measure casing inclination. A trigonometric solution of the inclination data yields a profile of the casing. Comparing sets of subsequent measurements to the initial set yields a settlement profile along casing. Although the tilt measurement interval can be as short as 2′ (0.6 m), on this project, measurements were made at 5′ (1.5-m) intervals, to save labor. It was demonstrated that the profile computed with the larger interval provided nearly the same profile.

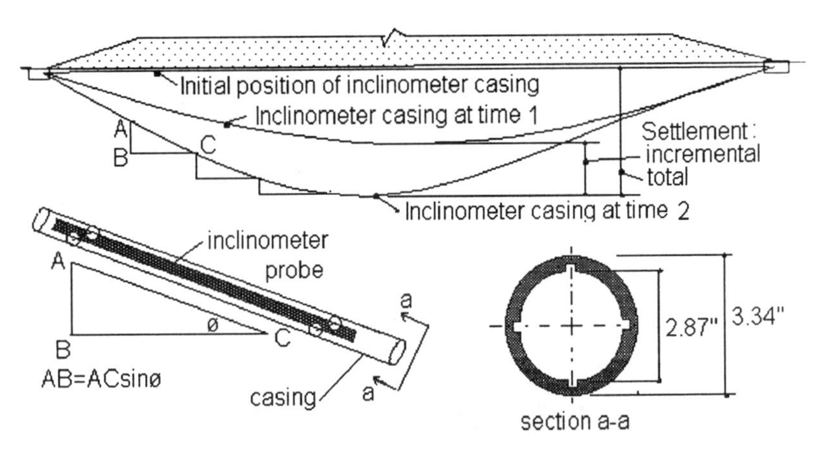

FIG. 4. Schematic of Horizontal Inclinometer Instrumentation

Two 320′ (98-m) lengths of ABS plastic casing were laid in trenches from toe to toe at the base of the embankments. The casing was installed on a slight slope to drain water that might enter through a loose joint, or from flooding at the ends. The north and south embankments each had one casing, located where the fill was highest at the bridge abutments. The casing was placed in 1½′ to 2′ (0.45 to 0.6 m) deep trenches on a 6″ (150 mm) bed of sand with a similar sand cover placed above.

The casing has precision cut grooves 90° apart. The inclinometer sensor has a wheel set that tracks in one pair of grooves as it is pulled through the casing. The casing must be laid with the grooves vertically to allow the instrument to track in a vertical plane. Supplied in 10′ (3 -m) sections, the casing has a 3.34″ (85-mm) OD. The casing sections were joined with telescopic couplings having a travel of 6″ (150 mm). Telescopic couplings are recommended to take displacement when axial strain may exceed 1%.

Liquid Tube Connected to Buried Pressure Sensors

Fig. 5 shows a schematic of the liquid settlement monitoring system. A pressure sensor is buried where settlement measurements are desired. The sensor measures the head in a liquid-filled tube connected to a small cylindrical clear plastic reservoir of liquid at a higher level outside the embankment. Under the influence of the embankment, the pressure sensor settles and the change in liquid head measured equals the settlement. Several buried sensors can be placed along a cross-section.

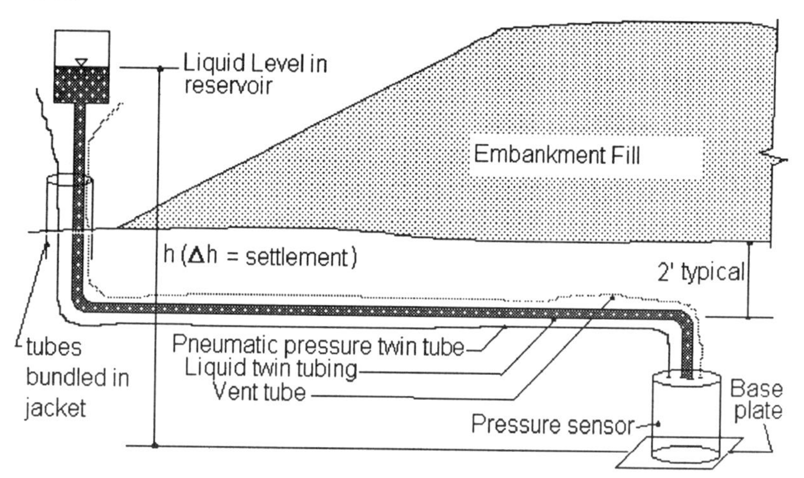

FIG. 5. Liquid Settlement Instrument

For liquid tube systems, steps must be taken to keep the liquid as free of gas bubbles as possible. Bubbles in the liquid column, being less dense than the liquid, would result in an error in the direction of less settlement than actual. Bubbles can be minimized using deaired liquid, placing the tube with a continually upward slope from sensor to reservoir and avoiding intermediate high points so that bubbles migrate to the reservoir and out of the tube. Also, bubble problems are minimized using ¼″ (6-mm) tubes. Smaller tubes cause surface tension problems that restrict bubbles from moving. Larger tubes allow water to pass by under trapped bubbles making it difficult to purge them. Nold (1992) reports success applying a back pressure to the liquid to absorb gas bubbles. Using tubing with a low porosity to air keeps air from permeating through tube walls.

For these installations pneumatic pressure sensors measured the head of liquid. Sensors were installed on square steel plates in shallow excavations at the base of the embankments. Care was exerted to compact the backfill beneath the plates to minimize the effect of local settlement under the loads of the embankments. The liquid tubes extended along the cross-section in a shallow trench with a slight upward gradient out past the edge of the embankment to a reservoir 8 to 12′ (2.4 to 3.7 m) above the sensor. The reservoirs were mounted in wooden terminal stations. Two parallel 1/4″ (6-mm) liquid-filled tubes were used to facilitate flushing in case air bubbles enter the system from an accidental break in the liquid tube or by permeating through the tube walls.

To avoid freezing, the liquid used was a 50%-50% deaired mix of water and pure ethylene glycol. The pure ethylene glycol (antifreeze) was used to get a uniform specific gravity (SG). Liquid deairing was accomplished with a Nold® deaerator. The deaerator is able to deair the liquid an order of magnitude more than boiling under a vacuum. (Nold, 1992).

The liquid settlement instrumentation system included 12 systems, on 4 cross-sections where the fill was highest. There were 6 systems on each side of the canal. Each cross-section had 3 sensors, on centerline and offset ± 50′ (15 m).

INSTRUMENTATION INSTALLATION

Successful measurements from field instrumentation depend, not only on quality instruments, and proper design; but largely on attention to numerous details, (Brylawski 1986) among the most critical of which is establishing good communications with the operators and trades-people that work around the instrumentation. Most failures result from accidental damage to cabling and tubing. Damage is minimized if people on site are aware of the instrument systems. Repairs may be costly. Sometimes replacement of damaged instruments is the only remedy. This section addresses some of these details as they relate to this case history.

On-Site Storage

Instrumentation was ordered during January 1992, and delivered to the site in March. The instrumentation was stored in a locked container except for the 700 ft of plastic inclinometer pipe which was stored outdoors protected from the sun to prevent warping. Readouts and sensors were stored in the field office with surveying equipment.

Field Checks and Calibration

The liquid settlement equipment field calibration procedure consisted of laying out the liquid tubing of each instrument on sloping ground with the sensor at the downhill end some 10 ft (3 m) below the reservoirs. After temperatures stabilized overnight, the instruments were calibrated at dawn while temperatures remained stable. The heads on the columns of liquid were read with the pneumatic pressure readout unit while each reservoir was lifted and held at several elevations 2′ (0.6 m) apart using a survey rod. Results were within the acceptance criteria, ± 0.02 ft = ±¼″ (±6 mm).

The horizontal inclinometer was checked for repeatability by placing it in a near level piece of inclinometer casing periodically. A quick check of the inclinometer was made by tilting it to known angles marked on the wall of the filed office. Calibration was not required during the project, but would have had to be performed off site on a precision jig with an error of $\pm 1''$ of arc.

Planning and Communications

An instrumentation kickoff briefing helped orient project management, superintendents and foremen. To minimize the risk of damaging the instrumentation it is best if operators and laborers that actually work around the instrumentation are also briefed, but this suggestion was not followed up and accidental damage occurred as explained below.

Installation Schedule

The installation schedule was planned around the earthworks subcontractor's embankment construction schedule. The instrumentation was installed at different times between May and June 1992 and required about 8 days total during this period.

The liquid settlement and horizontal inclinometer systems were each installed in about 3 days, and the settlement plates required about 2 days. The average crew size including engineers, technicians, operators and laborers varied from about 4 to 8 persons.

Installation Equipment

The equipment used included a backhoe/front end loader, sled compactors, hand tools, surveying equipment, deaerators, vacuum pumps and portable generators.

Protection and Damage

Installed instrumentation was damaged 3 times, but was repairable at about 7-10% of the instrumentation costs.

Flooding

The horizontal inclinometer casings were installed as specified, at the bottom of the embankment. The end stations daylighted in excavated depressions. Wooden retaining structures and work platforms were constructed. Drainage was not adequate, and during a heavy downpour flooding floated and displaced the south platforms displacing several feet of the horizontal inclinometer casing, and required that it be reset and re-surveyed.

Breaking the inclinometer pipe due to heavy traffic and too thin a cover

Operating heavy construction equipment across a horizontal inclinometer casing trench where the cover was a scant 18″ (0.45 m) caused the casing to break. Repairs required excavating, removing, replacing and re-surveying 30′ (9 m) of casing, and backfilling and recompacting the trench. The work was complicated by the need to retain a 1/8″ stainless steel pull cable inside the inclinometer casing. The repair took about 2 days. Additional fill, or plates or timbers, to bridge the traffic loads over the trench had been requested, but because of a communication gap the subcontractor responsible did not comply.

Ripping out liquid settlement tubing while grading slope with dozer blade

Failure to clearly flag an area at the toe of the slope where 3 liquid tubes daylighted, resulted in damage some 3 months after installation. Unaware of a shallow covering over the liquid tubes, a dozer operator, assigned to grade the slope, ripped out 2 of the 3 liquid tubes at that cross-section. The repair required 4 days. The tubes had to be excavated, spliced and refilled with deaired water, a laborious process requiring a vacuum pump to draw air out of the tube on one end while new deaired liquid was fed in the other end. One of the 2 damaged systems functioned again. The other may have been pulled and broken farther in under the embankment.

Costs

Table 2 summarizes engineering time, and cost of the field instrumentation program. The costs include hardware, installation, monitoring and reporting. Additional costs incurred by the contractor during installation for excavation and backfilling equipment, and materials were not recorded, but if subcontracted out might have cost an additional $10,000.

TABLE 2. Instrumentation Time and Costs

Description	Hours	Cost ($)	% of Total
Installation	305	12,500	26%
Monitoring	350	22,500	49%
Reporting	240	12,000	25%
Total	895	47,000	100%

DATA COLLECTION

Each of the three instrumentation systems used a different type of data collection system. The settlement plates were surveyed, whereas portable readout units were used to collect data from the liquid tubes and horizontal inclinometer. Because elevations changed under the influence of the settlement of the embankment, a survey had to be performed periodically to record elevation changes of the reservoirs at the ends of the liquid settlement tubes and of the ends of the horizontal inclinometer casings.

The pneumatic pressure sensors were read by circulating dry (to absorb condensation) nitrogen through a pair of tubes to the buried pneumatic sensor. Introduced into the system initially at a pressure somewhat higher than the head of liquid, the input valve was closed and the gas pressure bled down, and stabilized when it balanced the head of liquid. The stable gas pressure is measured with an electrical pressure transducer in the readout box to a resolution of 0.01 ft (3 mm) of liquid column. The readout unit is portable and contains a small high pressure nitrogen supply tank. The tank gets periodically recharged from a standard tank of commercial dry nitrogen.

The pressure sensor in the readout was programmable for the density of the liquid used. To avoid daytime sun changing the density of the liquid, readings were made early before the sun heated the liquid column. Although gas bubbles coming out of solution in the liquid are a potential error source, several months of successful measurements were obtained without having to flush and refill the liquid tubes. In practice, although single readings of water column head varied from reading to reading by $\pm 0.2' = \pm 2''$ (± 0.05 m), after taking a few readings one could plot a trend and determine the settlement within $\pm$ 0.04′ = $\pm 1/2''$ (± 0.01 m), as shown in Fig. 6.

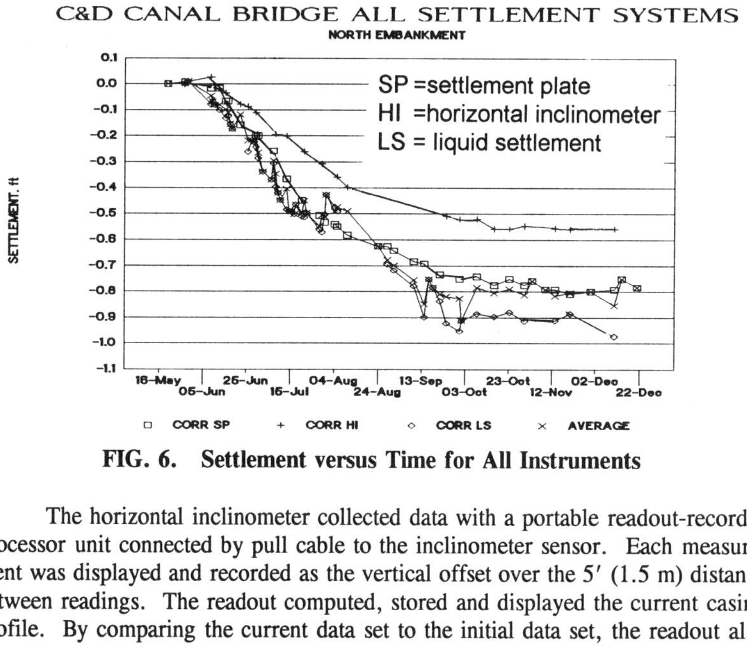

FIG. 6. Settlement versus Time for All Instruments

The horizontal inclinometer collected data with a portable readout-recorder processor unit connected by pull cable to the inclinometer sensor. Each measurement was displayed and recorded as the vertical offset over the 5′ (1.5 m) distance between readings. The readout computed, stored and displayed the current casing profile. By comparing the current data set to the initial data set, the readout also displayed a settlement profile. The data was downloaded to a PC for graphing, and further evaluation. The resolution of the horizontal inclinometer readout is 0.0001′=0.001″ (0.03 m). Typical errors are $\pm$ 0.02′ = $\pm 1/4''$ (± 6 mm) per 100′ (30 m) of casing.

Settlement plate data was recorded by the project survey crews to a resolution of $\pm$ 0.01′ (± 3 mm) and an error of $\pm$ 0.02′ (6 mm).

Spreadsheets were used for data processing, graphing, evaluation and reporting of the data from all three instrument systems.

Effort required to collect data varied. The liquid system was read easily by a single person. The horizontal inclinometer required a minimum two-person crew.

Using a reel and 1/8″ stainless steel pull cable, one person pulled the sensor to the far end through the 300+ ft (90+ m) of casing. Then, while the far end maintained tension on the cable, the other person took up on the rubber-jacketed signal cable with a slip-ring winch stopping at intervals to take readings. Winching is tedious and hard requiring 50 to 60 lb (200 to 300 N) of force to overcome friction between the pull cable and inclinometer casing .

MEASUREMENT RESULTS

Magnitude of Measured Settlement

Table 3 shows a comparison of settlements measured by the three systems. These readings, taken in mid November 1992, are the horizontal inclinometer readings, and the averages for settlement plate and liquid tube readings on each embankment. Space limits the presentation of the complete data set (see Chua and O'Malley 1993). At the end of the monitoring program the south embankment had settled an average of 3.8″ (97 mm), not quite half the 8.3″ (211 mm) average settlement measured for the north embankment.

TABLE 3. Summary of Measured Settlement

Measurement System	Settlement, inches (mm)			
	South Embankment		North Embankment	
	Raw Data	Corrected	Raw Data	Corrected
Liquid Tubes	12.3 (312)*	4.5 (114)*	16.5 (419)*	8.7 (221)*
Horizontal Inclinometer	3.3 (84)	3.3 (84)	6.7 (170)	6.7 (170)
Settlement Plates & Risers	3.7 (94)	3.7 (94)	9.5 (241)	9.5 (241)
Average of all systems		3.8 (97)		8.3 (211)
Minimum of all systems		3.3 (84)		6.7 (170)
Maximum of all systems		4.5 (114)		9.5 (241)
Error: ½(Max. - Min.)		±0.6 (±15)		±1.4 (±36)
Error as a pct. of average		13%		15%
Standard deviation		*16%*		*17%*

* The pneumatic readout unit contains a standard electrical pressure sensor with a digital readout programmed to read ±0.01′ of water column. The factory offset was set at 0.00′ of water column at standard temperature and pressure and indicated 0.00′ water column initially. The readout display is reset to zero each time it is read or the offset is noted for each reading and subtracted to correct the reading. The offset drifted linearly with time and was +0.65′ = +7.8″ (+198 mm) by November 17th, 1992.

Time Rate of Measured Settlement

The embankment construction was completed in September 1992. Fig. 6, for the north embankment, shows a plot of settlement versus time and fill height versus time, for all settlement systems. Settlement slowed down substantially at the end of September when the fills were completed. Fig. 7 shows settlement profiles obtained with the horizontal inclinometer across the north embankment. Fig 6 also shows the average settlement at the centerline from the horizontal inclinometer data, and the average data from the settlement plates. Settlement data for the south embankment are similar with smaller magnitudes. Additional settlement became negligible after mid November 1992, and the measurement program was terminated in December of 1992 except for a survey of the settlement rods in January 1993, which showed no additional settlement (Chua 1993).

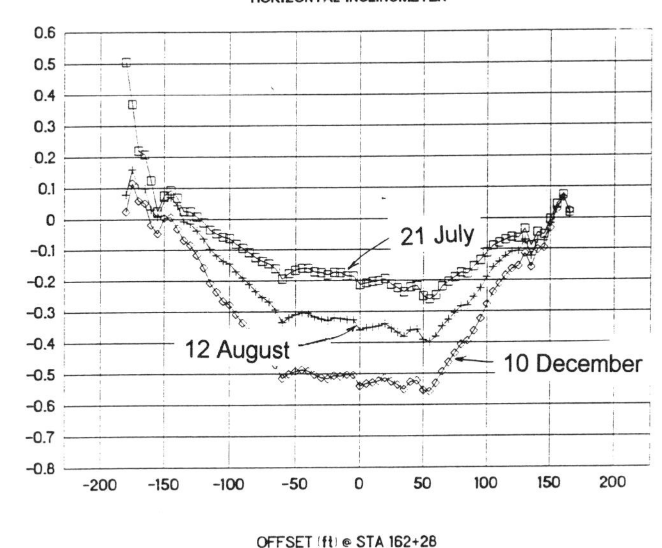

FIG. 7. Settlement Profile from Horizontal Inclinometer

Comparisons of Results Obtained by the 3 Systems

The error, taken as 1/2 the difference between the maximum and minimum values of settlement, is ±0.6″ (±15 mm) or 13% of the average measured settlement for the South embankment on sandier soils and ±1.4″ (±36 mm) or 15% on the north embankment on clayier soils. The error and standard deviation have about the same order of magnitude.

Calculated Settlement Estimates

Table 4 shows calculated immediate, consolidation, and total settlements for the north and south embankments. Both immediate and post-construction settlements for the approach embankments had been estimated (Giese and O'Malley 1990) using available subsurface information. For the 49′ (15-m) high south embankment, calculated immediate settlements were 8½″ (220 mm) due to compression of the underlying sands and silty sands. Calculated consolidation settlements were about 8″ (200 mm) due to compression of the lower clay strata (stratum VI). For the 56′ (17-m) north embankment, calculations showed immediate settlements of about 8″ (200 mm), and long-term consolidation settlements of about 13″ (330 mm).

TABLE 4. Calculated Settlement

Embankment	Settlement, inches (mm)		
	Immediate	Consolidation	Total
South	8½ (220)	8 (200)	16½ (420)
North	8 (200)	13 (330)	21 (530)

For the south embankment, on sandier soils, measured settlement of approximately 4″ (100 mm) were about half of the calculated immediate settlements, and about 25% of the 16½″ (420 cm) calculated total settlement. Settlement on the north embankment, where clayier soils were evident, measured 8″ (200 mm) or about equal to the calculated immediate settlement, and about 38% of the 21″ (530 mm) calculated total settlement.

Table 5 shows time rates of settlement for the highest sections of the embankments. It appears that the clays underlying the site did not consolidate to the magnitude calculated. This could be due to a variety of factors including sampling methods used, and sample disturbance, handling, preparation and testing techniques. The same factors probably led to predicting consolidation times longer than actual. It also appears that the FHWA method (FHWA 1982) overestimated the compressibility of the sands and silts.

TABLE 5. Calculated Time Rates of Settlement

Embankment	Consolidation, days	
	50%	90%
South	360	1440
North	360	1800

Conclusions

1. The instrumentation systems worked and provided the information needed.
2. The different systems agreed well in their measurement of the magnitude of settlement.
3. A significant part of success in field instrumentation measurements depends on operators, trades-people and others working in the vicinity of the instrumentation being aware of its presence so as to avoid accidental damage.

4. Measurements at a few selected positions with either the liquid settlement system or settlement plates would have provided the answers needed with less time for monitoring.
5. The "horizontal" inclinometer or other "horizontal" pipe system would be useful if settlement distribution across the cross-section were needed due to, i.e., variable soil conditions, or for an embankment with embedded structures whose performance were susceptible to differential settlements.

ACKNOWLEDGMENTS

The project is owned by Delaware DOT. The bridge was designed by Figg Engineering Group. The authors acknowledge the contributions of Law Engineering and Geonor, Inc., and appreciate the site support of prime contractor Recchi America's Emilio Rosiello, Dick Lauer and Brian West; Mr. Al Donofrio, Century Engineering; Paul Connor, Delaware DOT; Tom DeHaven, Figg Construction Services; and Ron Beckman, Julian Construction.

APPENDIX - REFERENCES

Bergdahl, U. and Broms, B. (1967). "New method of measuring In-situ measurements." *J. Soil Mech. Found. Div.*, Proc. ASCE, 93(SM5), 51-57.

Bozozuk, M. (1969). "A fluid settlement gauge." *Can. Geotech. J.*, 8(3), 362-364.

Brylawski, E. (1986). "Going afield with instrumentation." *Civil Engineering*, ASCE 56(8), pp. 49-51

Burn, K. N. (1959). "Instrumentation for consolidation study of a clayey deposit beneath an embankment." *Géotechnique*, 9(3), 136-142.

Chua, P. and O'Malley, E. (1993). *Final Report of Settlement Monitoring - SR- 1 Bridge over C&D Canal, New Castle Co. Delaware*, Contract 88-110-08 Report to Figg Construction Services, February 25.

DiBiagio, E. (1992). "Field Instrumentation and performance observations to reduce cost and increase reliability." *Publication No. 187*, Norwegian Geotechnical Institute, Oslo.

Dunnicliff, J. (1988). *Geotechnical Instrumentation for Monitoring Field Performance*, Wiley, New York, New York.

Giese, J. and O'Malley, E. (1990). *Report of Geotechnical Exploration, C&D Canal Bridge*, Law Engineering, Proj. 482-6451-01, Report to Figg Engr. Group, January 1.

Federal Highway Administration (1982). *Soils and Foundation Workshop Manual*, Publication No. FHWA-HI-88-009, McLean, Virginia.

Hanna, T. H. (1985). *Field Instrumentation in Geotechnical Engineering*, Clausthal, Germany.

Nold, Walter (1992). *The Nold DeAERATOR*, Walter Nold Co. Inc., Rev. R., Natick Massachusetts.

Wilson, S. D. (1967). "Investigation of embankment performance." *J. Soil Mech. Found. Div.*, Proc. ASCE, 93(SM4), 135-156.

COMPARISON OF PREDICTED AND MEASURED SETTLEMENT OF A TEST EMBANKMENT OVER SOFT SOIL

Kuo-Hsia Chang,[1] William D. Kovacs,[2] and Ming-Jiun Wu,[3] Members, ASCE

ABSTRACT: A test embankment was instrumented with six (6) settlement plates, three (3) vertical extensometers, two (2) inclinometers, and two (2) horizontal profilers in order to verify settlement predictions for a highway project in Tacoma, WA. The subsoil consists of about 13 meters of soft lacustrine clayey silt and loose silty sand underlain by over 43 meters of alternate layers of medium and dense alluvial silty sand resting on Till. The data obtained offered an unique opportunity to compare different models for their capabilities to predict both the maximum and differential settlements in very heterogeneous soils.

Under such conditions, the cone-penetration test seemed to be the most valuable basis for predicting maximum settlements when there was sufficient local experience to establish empirical relationships between the tip resistance and modulus (Schmertmann 1970), regardless of which model was used. After comparing the three most commonly used models in predicting differential settlements for soil-structure interaction analyses, it was found that in deep alluvial deposits, a modified semi-infinite elastic model assuming that Boussinesq stresses was still valid in a layered soil with variable Young's moduli and limited depth seemed to be better than either the Winkler or the Terzaghi-Taylor model.

[1] Staff Consultant, Shannon & Wilson, Inc., 400 N. 34th St., Seattle, WA 98103.
[2] Prof., Dept. of Civ. & Environ. Engrg., Univ. of Rhode Island, Kingston, RI 02881.
[3] Vice. Pres., Shannon & Wilson, Inc., 400 N 34th Street, Seattle, WA 98103.

INTRODUCTION

Long before Terzaghi started his development of modern soil mechanics and made the important contribution to the theory of consolidation, settlements had been calculated by either the Winkler model using independent spring for soil-structure interaction analyses or the semi-infinite elastic model using a constant Young's modulus and Poisson's ratio. Due to the complex nature of soil and the fact that the assessment of settlement still remains at present an estimation based largely upon local experience, the Winkler model is still the most popular model used in soil-structure interaction analyses for its simplicity, despite the fact that quite a number of much more sophisticated models are in existence.

The aims of this paper are: (1) to present a set of instrumented data from a test embankment that can be used for comparison of different models and for back-calculation of empirical coefficients for assessing the maximum settlement; and (2) to compare the three most commonly used models in soil-structure interaction analyses, namely the Winkler model, the semi-infinite elastic model, and the Terzaghi-Taylor model for their capabilities in predicting differential settlements.

TEST FILL

The test fill is located in area of the proposed SR-509 in Tacoma, WA and is approximately 4.27 meters high. The top of the fill is approximately 18.3 m^2 with side slopes of approximately 2 horizontal 1 vertical (2H:lV). A plan view of the test fill is shown in Fig. 1.

Several different instruments were installed within the test fill to monitor pore water pressure, vertical settlement and lateral movement of the embankment as follows:

- 3-Pneumatic Piezometers, P-1, P-2, & P-3.
- 6-Settlement Plates, S-1 through S-6.
- 2-Horizontal Profilers, H-1 & H-2.
- 3-Probe Extensometers, two of which included
- 2-Inclinometers, X-1, IX-1, & IX-2.

The instrument locations are shown in Plan View in Fig. 1 and on the Generalized Profile in Fig. 2. The instruments were monitored before, during, and after the completion of the test fill, over a period of approximately 390 days.

SUBSURFACE CONDITIONS

The test fill is underlain by 12.2 meters of lacustrine sediments consisting of alternate layers of soft, slightly clayey to clayey silt and loose to medium sand. At a depth of approximately 13.7 to 15.2 m below the ground surface, alluvial soils consist of medium dense to very dense, clean to slightly silty fine sand with occasional thin layers of silt, up to a depth of 60 m (Fig. 2). Fig. 3 shows the log of boring B-29 with water contents and the results of SPT, CPT (Q_c only) and PMT.

METHOD OF SETTLEMENT ANALYSIS

Settlements were estimated by a modified Terzaghi-Taylor method as the sum of immediate, primary and long-term settlements described as follows.

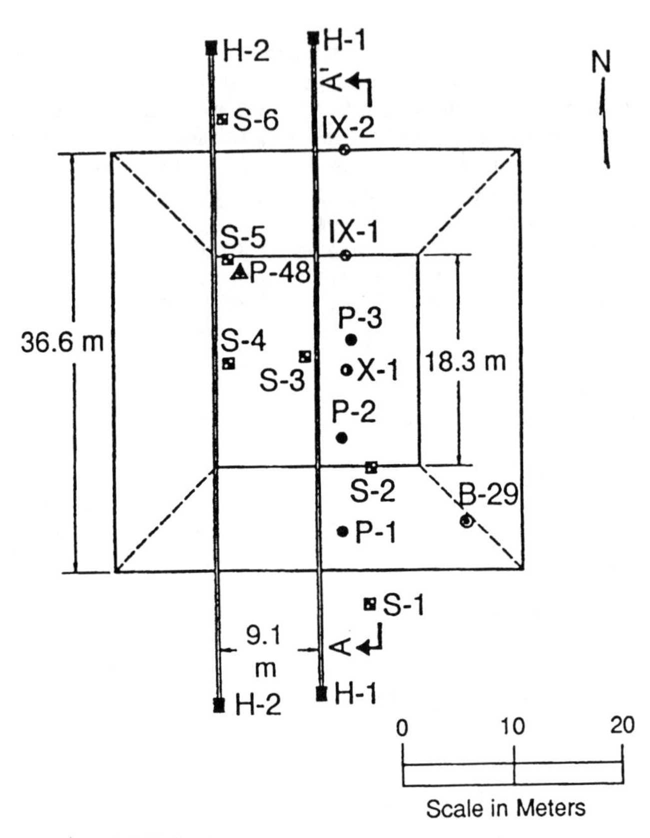

LEGEND

- B-29 Boring Designation and Approximate Location
- P-48 Piezocone Probe Designation and Approximate Location
- S-1 Settlement Plate Designation and Approximate Location
- P-1 Pneumatic Piezometer Designation and Approximate Location
- X-1 Probe Extensometer Designation and Approximate Location
- IX-1 Inclinometer/Probe Extensometer Designation and Approximate Location
- H-1 Settlement Profiler Designation and Approximate Location (Vaults at both ends of Profiler)
- A Generalized Subsurface Profile

FIG. 1. Plan of Test Fill and Location of Instrumentation

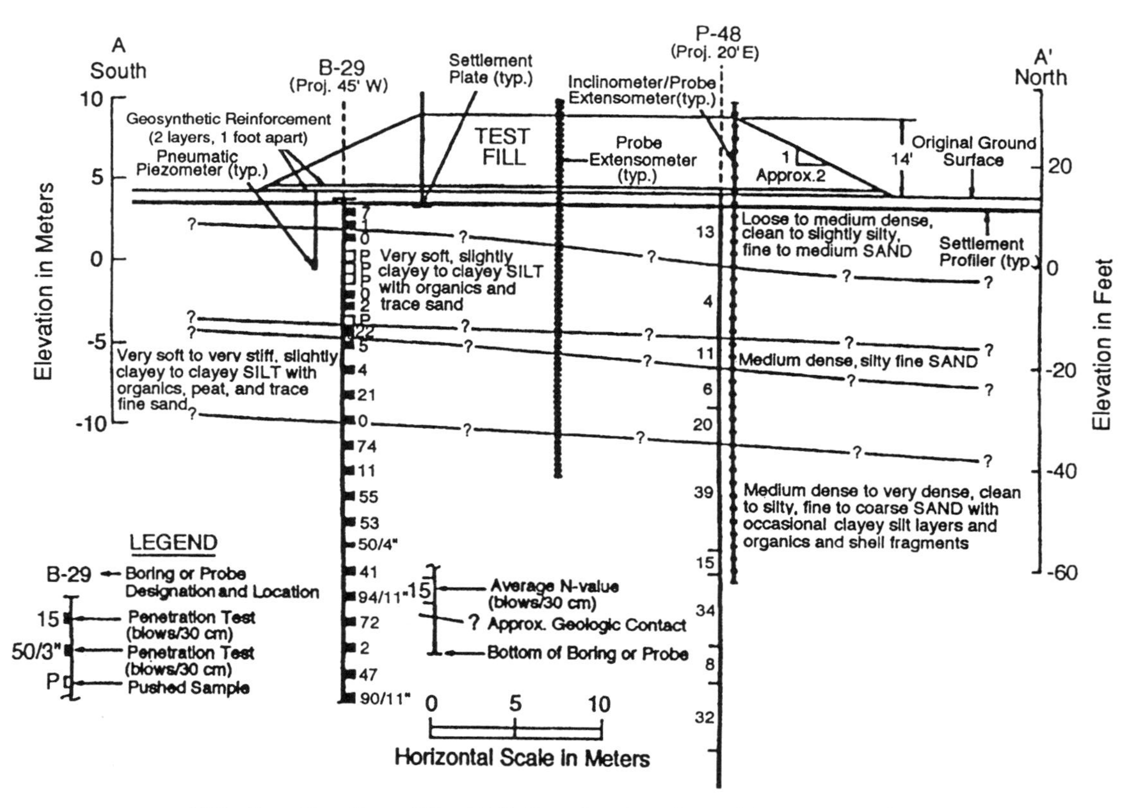

FIG. 2. Elevation of Test Fill Showing Location of Instrumentation

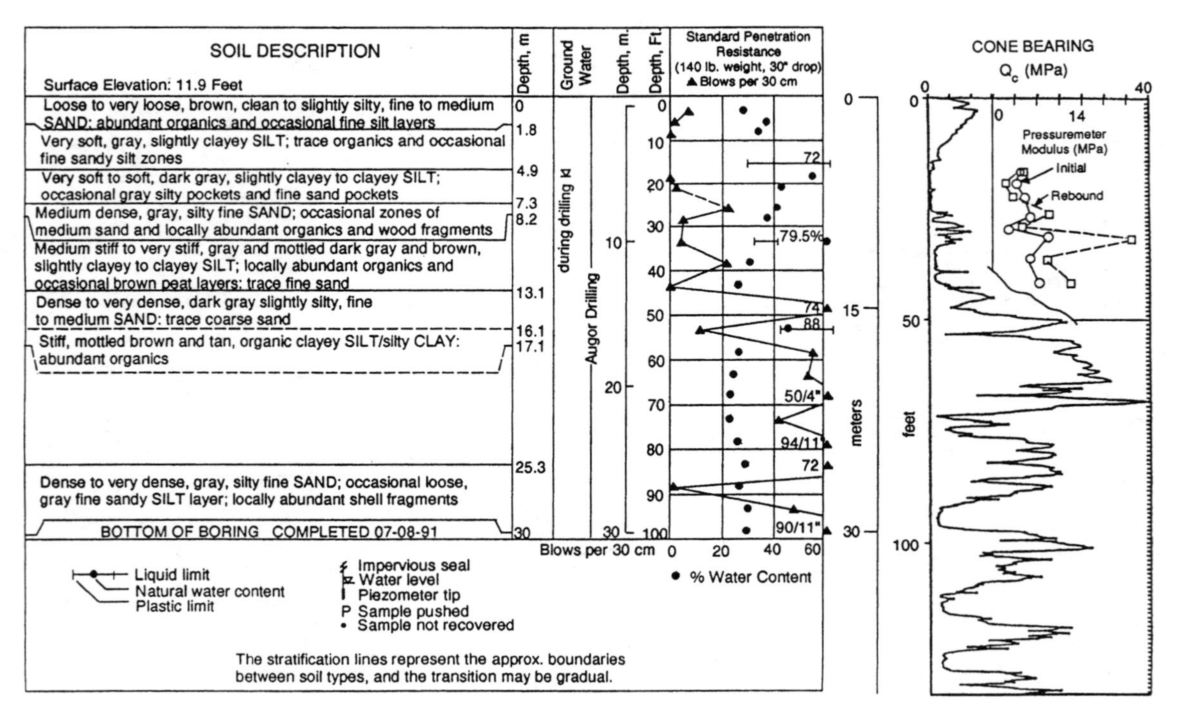

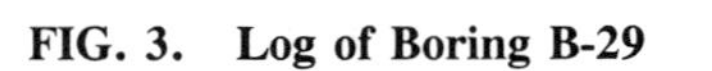
FIG. 3. Log of Boring B-29

- The immediate settlement was calculated by the elastic theory assuming that the stresses determined from the Boussinesq equations were still valid in a layered soil with varying moduli and limited depth (modified semi-infinite elastic model). The Young's moduli of soil layers were determined by multiplying the tip resistances of the cone-penetration test by an empirical coefficient (Schmertmann 1970).
- The primary settlement was calculated by the conventional Terzaghi-Taylor method using an empirical relationship between the constrained compressive modulus and the tip resistance of the cone-penetration test (Yuan 1988).
- The long-term settlement was calculated by assuming a straight line relationship between settlement and the logarithm of time (Gurtowski and Kirkland 1979). The following equation was used in this analysis:

$$S_{t_1 - t_2} = A S_p^B \log \frac{t_2}{t_1} \quad (1)$$

where $S_{t_1 - t_2}$ is the settlement between time t_1 and t_2; S_p is the settlement at the end of primary consolidation; A and B are empirical coefficients determined from local experience, taken to be A=0.5 and B=0.7 in this case.

The basic soil data used in the analysis are listed in Table 1 as follows:

TABLE 1. Basic Soil Data Used in the Analysis

Depth Layer	Interval (m)	Soil type	Average Q_c (MPa)[1]	E_0 (MPa)[2]	E_s (MPa)[3]
(1)	(2)	(3)	(4)	(5)	(6)
1	0 – 3.35	Loose silty sand	5.0	12.5	13.0
2	3.35 – 4.57	Loose silty sand	1.9	4.8	6.3
3	4.57 – 8.53	Soft clayey silt	0.6	1.4	3.2
4	8.53 – 10.36	Loose silty sand	4.0	10.0	10.8
5	10.36 – 12.80	Soft clayey silt	1.1	2.8	4.6
6	12.80 – 16.46	Medium silty sand	7.0	17.5	17.2
7	16.46 – 60.00	Dense sand	20.0	50.0	45.0

[1] Electric Piezocone Tip Resistance
[2] Young's Modulus of Elasticity
[3] Constrained Modulus of Compression

Comparison of Predicted, Back-Calculated and Observed Settlements

A comparison of predicted settlements using the methods mentioned above with results obtained by the horizontal profiler in a one year period is shown in Fig. 4. As shown in Fig. 4, the predicted settlements are 30% less than those observed

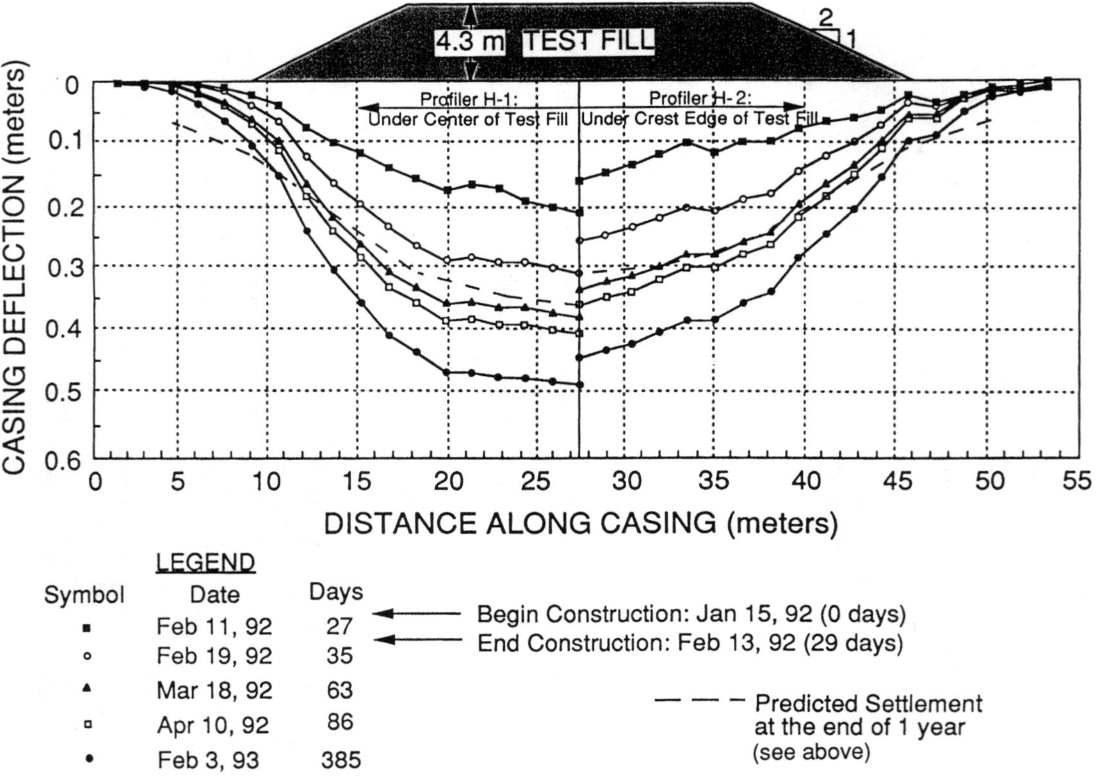

FIG. 4. Comparison of Predicted and Observed Settlements from Horizontal Profilers

at the middle, nearly equal to those observed at the toe of the embankment, but almost 2 times larger than those observed outside of the toe. The results of settlement plate readings and extensometer readings, are presented in Figs. 5 and 6. Due to some malfunction in the pneumatic piezometers, their readings are not presented. The Inclinometer readings are not presented due to lack of space.

In order to compare the capabilities of the three most commonly used models in soil-structure interaction analyses, namely the Winkler model, the semi-infinite elastic model, and the Terzaghi-Taylor model in predicting differential settlements, settlements were back-calculated using the following assumptions:

- All models were adjusted to produce the same maximum settlement as observed.
- For the Winkler model, settlement of a point was assumed to be directly proportional to the load acting on it.
- For the semi-infinite elastic model, stresses determined by the Boussinesq equations were assumed to be still valid for a layered soil with varying Young's moduli and limited depth, and settlements were calculated by the following equation from the theory of elasticity:

$$S = \sum_{i=1}^{n} \left[\frac{S_z}{E} - \nu \frac{S_x}{E} - \nu \frac{S_y}{E} \right]_i H_i \tag{2}$$

  where S is the settlement of a point; S_x, S_y, S_z are the stresses in the x, y, and z direction determined by the Boussinesq equations; E is the Young's modulus of layer i, determined by multiplying the tip resistance of the cone penetration by an empirical constant, ν is the Poisson's ratio of soil, taken to be 0.4, H_i is the thickness of layer i and n is the number of layers under a point.
- For the Terzaghi-Taylor model, settlements were determined by the sum of the settlement of each layer under a point obtained by multiplying the vertical component of the Boussinesq stress at the mid-point of each layer by the thickness of the layer divided by its constrained modulus.

Comparison of settlements back-calculated by these models with those observed settlements are presented in Figs. 6 and 7.

PRELIMINARY CONCLUSIONS

Based upon the comparison of predicted and back-calculated settlements with those observed, the following preliminary conclusions may be drawn:

1. An instrumented test embankment provided a valuable set of data for evaluation of different methods of settlement analysis and for back-calculation of empirical coefficients for assessing the maximum settlement.
2. The cone penetration test appeared to be the most valuable basis for assessing the moduli of soft and heterogeneous soil deposits in providing a continuous profile, regardless of the model used.

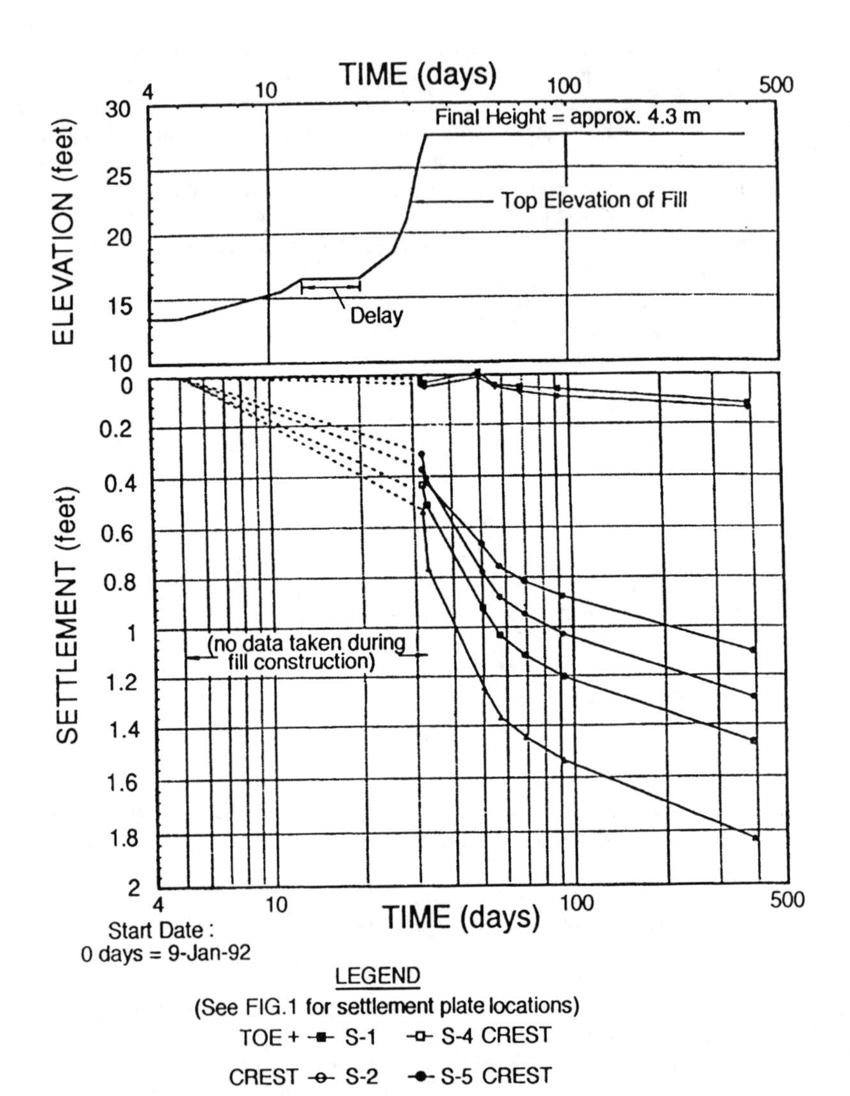

FIG. 5. Observed Settlement Plate Data

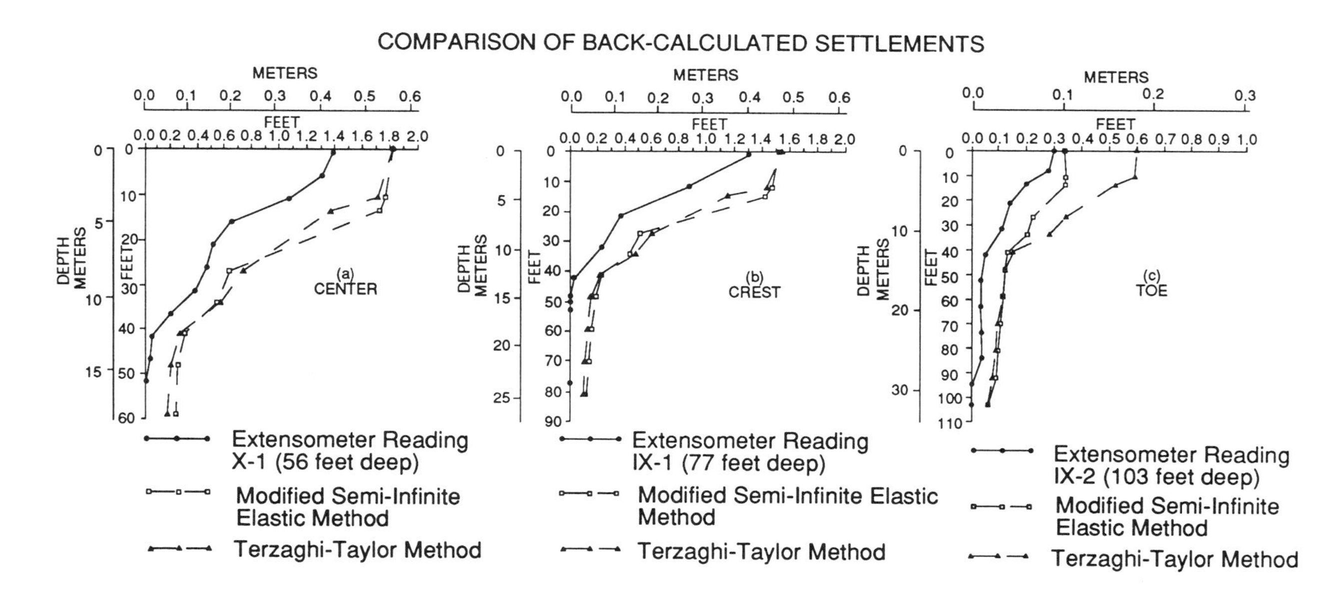

FIG. 6. Comparison of Back Calculated Settlement with Extensometer Readings for: (a) Center; (b) Crest; and (c) Toe of Embankment

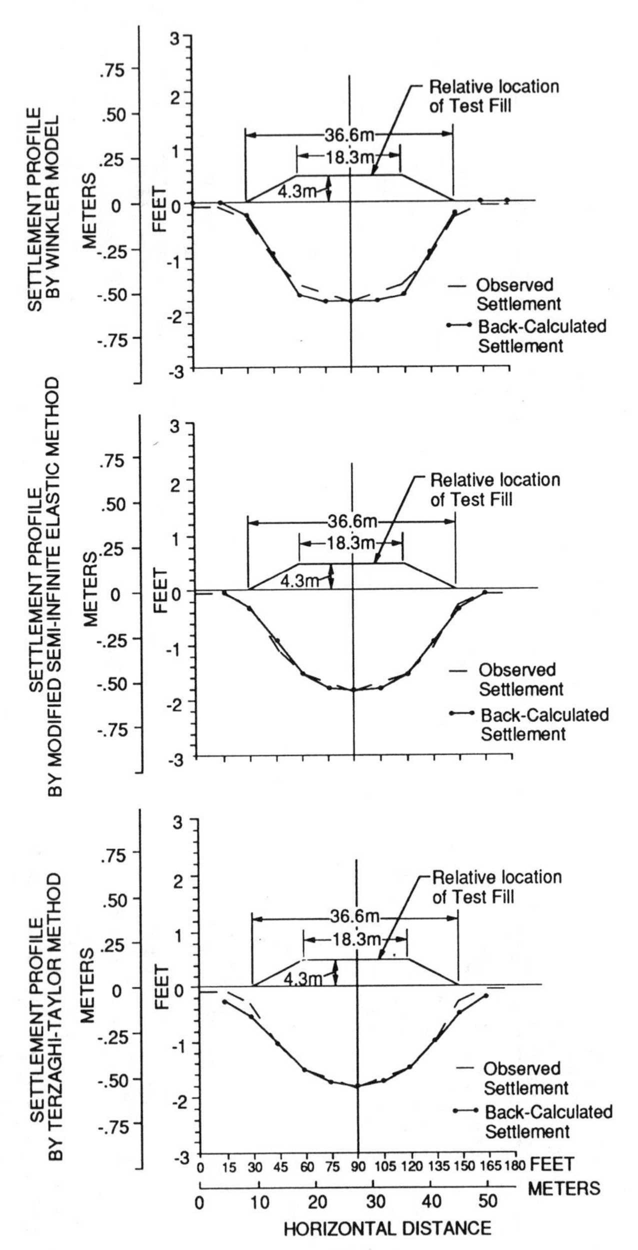

FIG. 7. Comparison of Observed Horizontal Profilers and Back Calculated Settlements

3. After comparing the three most commonly used models in soil-structure interaction analyses, a modified semi-infinite elastic model appears to be better than the Winkler or the Terzaghi-Taylor model in predicting differential settlements for soft and deep alluvial deposits.

Acknowledgements

The authors are grateful to Scott Shimel and Monique Nykamp of Shannon & Wilson, Inc., for monitoring and obtaining the test embankment data. The permission from Sverdrup Corporation and the Washington State Department of Transportation to publish the data is also deeply appreciated. Debbie Gendreau provided the CADs. Artie Martino completed the final typing layout. The Authors thank the Reviewers for their comments and suggestions.

Appendix - References

Gurtowski, T. M. and Kirkland, T. E. (1979). "A simplified preload design." *Proc. Annual Engineering Geology and Soils Engineering Symposium*, University of Idaho, Moscow, 249-273.

Schmertmann, J. H. (1979). "Static cone to compute static settlement over sand." *J. Soil Mech. Found. Div.*, Proc. ASCE, 96(SM3), 1011-1043.

Yuan, J.-X. (1988). "Penetration testing application in China." *Penetration Testing* (Proc. ISPOT-1), Orlando, 1, 389-395.

Zhang, G. X., Zhang, N. R., and Zhang, F. L. (1981). "Nonlinear differential settlement analysis." *Proc. 10th ICSMFE*, Stockholm, 2, 291-294.

COMPACTION GROUTING STOPS SETTLEMENT OF AN OPERATING WATER TREATMENT PLANT

Percy M. Wimberly, III,[1] F. Barry Newman,[2] Kenneth B. Andromalos,[3] and Chris R. Ryan,[4] Members, ASCE

ABSTRACT: In early 1978, about a month into full-time operation, the new filter plant building for the water treatment plant for the City of Glenwood Springs, Colorado, experienced distress due to settlement. Various subsurface investigations and remedial repairs occurred until early 1985, when the intermittent settlement significantly increased. The City sought proposals to investigate and design remedial repairs to stabilize the building. In early 1987, Geo-Con, Inc., a specialty geotechnical contractor, was awarded the contract with GAI Consultants, Inc., being the part of the Geo-Con team charged with investigating and designing the remedial repairs which Geo-Con would install. Earlier subsurface investigations had differed on the depths of settling soil, with estimates from 6 to 27.4 m being given. A 1987 subsurface investigation indicated about 4 to 15.5 m of soil were involved. A compaction grouting program was designed and carried out in 1987-88 involving 65 grout holes outside the building and 48 holes inside, of which about 70 percent were vertical, with the remainder being angled due to physical constraints. Grout takes were relatively high for much of the hole lengths in soil, resulting in significant soil densification and construction of rather large grout columns bearing on rock. After completion, the City

---

[1] Consulting Engineer, 436 Waubonsee Trail, Batavia, IL 60510; formerly Staff Consultant, GAI Consultants, Inc.

[2] Geotech. Grp. Mgr., GAI Consultants, 570 Beatty Road, Monroeville, PA 15146.

[3] Manager of Construction and Remediation, McLaren/Hart Environmental Engineering Corp., Penn Center West III, Suite 106, Pittsburgh, PA 15276.

[4] President, Geo-Con, Inc., P.O. Box 17380, Pittsburgh, PA 15235.

installed a sensitive settlement monitoring system, which has confirmed that the settlement has ceased. This paper discusses the building's settlement background, the 1987-88 subsurface investigation and compaction grouting program, and the results of the program.

INTRODUCTION

The Red Mountain Water Treatment Facility's filter plant building is a rectangular, 2-story, precast and cast-in-place concrete structure, about 11.9 m wide by 33.1 m long. The building houses 4 filter beds, a pipe gallery, an office area, and a chemical storage area. The filter beds and pipe gallery are located in the northern (plant north) two-thirds of the building with the pipe gallery separating the 2 pairs of filters. The filter bed and pipe gallery are cast-in-place reinforced concrete. The southern end of the building contains the chemical storage areas and the second floor office. The building walls are comprised of vertical precast concrete panels, while the roof is comprised of horizontal precast concrete panels spanning east-west, which supports a built-up roof. The general plan of the building is shown on Fig.1.

A short start-up test was followed by the plant starting full 24-hour operations on January 3, 1978. On February 10, 1978, settlement was noted when cracks were discovered in the filter plant building walls. Plant operations were stopped on February 13, 1978, and a subsurface investigation was performed using auger borings inside and outside the building. Leakage estimated to be as much as 8.3 million liters occurred during this initial operating period. In April or May 1978, remedial work to increase soil bearing capacities under much of the east and south wall footings was performed by intrusion gravel packing and grouting. Voids were reportedly detected and grouted below Filter Beds 1 and 2. Grout was injected into the upper 2$\pm$ m of the foundation soils, with relatively high takes occurring in some areas. After the remedial work, no apparent drop of water level in the filter beds was noted overnight. On June 15, 1978, full 24-hour operation resumed.

The plant operation was again stopped on August 28, 1978, when significant leakage in Filter Bed 3 was noted. A geotechnical investigation was carried out by a new consultant and a range of repair options was suggested in the fall of 1978. This investigation indicated that about 6 to 12 m of soil might be settling. In February 1979, leak tests on the filter beds indicated that as much as 7.2 million liters of water may have leaked during the second operational period of the plant. No action was taken on installing the repair options from the fall 1978 investigation, and the building continued to settle.

In early 1983, another geotechnical investigation and discussion of repair options by still a different consultant indicated that as much as 27+ m of soil might be involved in the settlement. A period of relatively little settlement occurred in 1983-84. However, in the spring of 1985, significant settlement occurred at the southeast corner of the building. Due to the resumption of settlement, the City of Glenwood Springs, Colorado, solicited proposals from geotechnical specialty

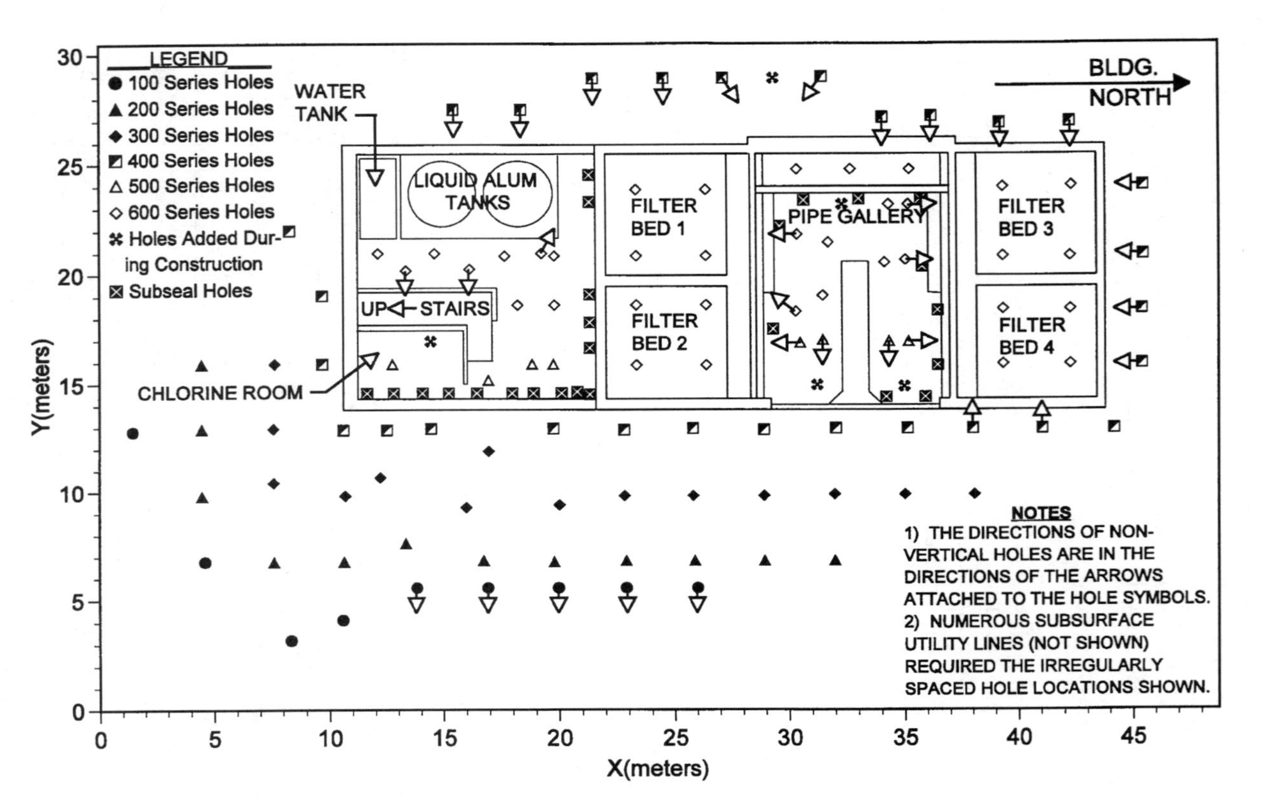

FIG. 1. Plan of Compaction Grouting Holes and Subseal Holes

contractors to evaluate the building settlement problem and provide repair recommendations, with the goal ultimately to have the selected contractor design and install a system to stop the settlement of the building. This paper outlines the design and execution of the selected solution.

Site Characteristics from Available Information and March 1987 Field Reconnaissance

The filter plant building was constructed on the northwest slope of a steep-sided, northeast-trending narrow valley about 1.2 km west of the central business area of Glenwood Springs. The ground floor of the building is constructed at about U.S.G.S. Datum elevation 1862.3 m on a flat area located above what was the lower extension of the narrow valley. The terrain west of the building rises sharply to elevations of 2400+ m.

It is believed that some soil, and perhaps rock, were cut from the mountain slope to the west, north, and south of the building to prepare the flat area. Actual disposition of the cut material is not known. However, much of it was used to construct dikes for Backwash Ponds 1 and 2, which are east of the southern half of the building and northeast of the building, respectively. Some cut material may have been utilized for fill in creating the flat area. Based upon the general terrain, it is believed that little, if any, fill was placed in the area below the building. Shallow fill depths, less than 3 m, were reported in test pits excavated near the building during the 1983 subsurface investigation. In addition, cross-sections based upon the fall 1978 subsurface investigation show fill only outside the building area and above the foundation area. Therefore, most of the soil beneath the building is believed to be residual soil, colluvial soil, slope wash soil, and/or debris fan soil. Soil visible near the building consists of coarse gravel, cobble and boulder regions with finer sandy, silty, and clayey regions, which agrees with the above assessment of the soil's sources. No visible signs of a landslide were noted in the area around the filter plant building.

Bedrock beneath the site is believed to consist mainly of siltstones, sandstones, limestones, shales, and claystones of the Paradox Formation. Rock outcrops occurred in nearby access road cuts north and west of the filter plant building. The bedrock exposed upslope (west) of the building dips generally east at about 39 degrees. In the roadcut upslope and slightly southwest, the bedrock bedding planes were highly contorted with irregular dip angles to the north and south. There is an apparent downdrag character to the bedding, which possibly represents a local fault. Bedrock in the road cut north of the building appeared to exhibit an undetermined dip to the east with a strike of about 9 degrees to the southwest.

Settlement profiles of the east and west walls of the building as of mid-1986, prior to remedial work, are shown in Fig. 2. The differential settlements were observable in the form of displacements and cracks in the building.

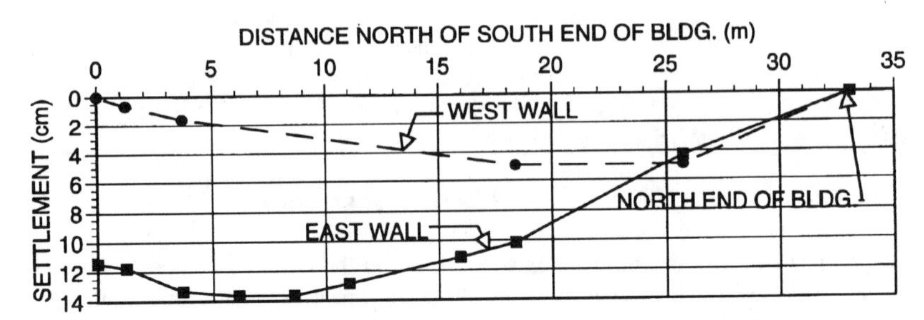

FIG. 2. Settlement of East and West Walls in Mid-1986

Geotechnical Investigation and Subsurface Conditions

Due to differences in soil depths presumed to be involved in the settling presented by earlier consultants, an additional geotechnical investigation was performed by GAI to better define the problem. Five borings were drilled in early March 1987 at the locations shown on Fig. 3. Borings B-1 through B-5 were drilled to depths of 32.0, 36.6, 36.6, 14.6, and 13.7 m, respectively. Continuous standard penetration testing was done in soil, followed by cleaning out of the holes between samples using 15.9 cm O.D. hollow-stem augers, except where boulders were encountered and the augers were used to penetrate the boulders before resuming sampling. Two 6.35 cm O.D. Shelby tube samples were attempted in an auger hole near Boring B-2 between depths of 3.35 to 3.66 m and 7.32 to 7.68 m. Both tubes suffered rock induced deformation, poor recovery, and high sample disturbance. Pocket penetrometer readings of undrained shear strength were obtained for some of the cohesive soil samples. In general, continuous "NQ" rock core samples were obtained when soil sampling met refusal. Drilling mud was used during coring to maintain an open hole and remove cuttings. Special techniques, such as the use of short core runs, a split inner core barrel to aid in core removal, etc., were used to obtain and preserve the rather delicate rock types encountered during coring. Coring began in Borings B-1 through B-5 at depths of 5.6, 14.0, 13.7, 5.3, and 4.2 m, respectively. Soil depths in Borings B-1 through B-5 were estimated to be 4.9, 13.7, 15.6, 4.6, and 3.9 m, respectively. These depths to rock agree roughly with those of the earlier fall 1978 study. The top of rock contours estimated in the area of the filter plant building and shown on Fig. 3 are generally based on the 5 borings and the visible rock outcrops near the building of the 1987 investigation. Top of rock information from earlier borings near the building was given less consideration in developing the top of rock contours shown on Fig. 3, since in cases the earlier data conflicted with that of the 1987 investigation. Core recovery, RQD, fracture angles, clay seams, etc. were recorded for the recovered core. Perforated PVC pipe of 5.1 cm I.D. was inserted to or below the top of rock in Borings B-1, B-3, B-4, and B-5 to permit ground-water level monitoring following drilling. In general, some flushing of the holes using a garden hose and water was required to remove caved plugs and permit the insertion of the PVC pipe to the desired depths. Boring B-2 was reamed to about 15.2 cm I.D., and unperforated 8.6 cm O.D. aluminum slope indicator tubing was installed to the maximum feasible depth of 24.1 m to permit measurements of both settlement and lateral movement. The 24.1 m depth was limited by problems with the hole caving in both the soil and rock zones. Initial settlement data was recorded at the time of installation. Filter plant personnel measured subsequent settlements of the tubing. Inclinometer readings were taken in the slope indicator tubing on April 3 and 24, 1987. Falling head permeability tests were made as the drilling mud level dropped in the open cored holes in rock in Borings B-1 and B-4, yielding an estimated overall rock permeability value of 10^{-6} cm/sec (a value judged to be perhaps an order of magnitude or so low due to the effects of the drilling mud).

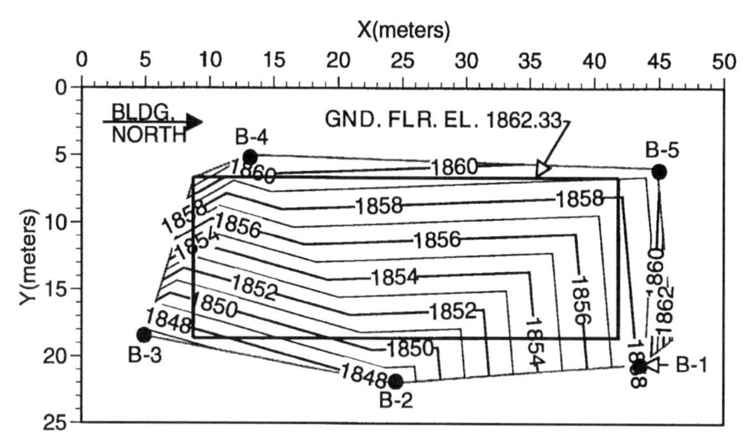

FIG. 3. Plan of Borings and Estimated Top of Rock Elevations in Meters

The soils could be generalized as clayey silt with sandy silt and silty sand zones, with few to many rock fragments. Compactness of granular soils ranged from very loose to very dense. Consistency of cohesive soils ranged from very soft to hard. In general, granular soils were mostly loose to medium dense. Uncorrected standard penetration resistance in the deeper soils of Borings B-2 and B-3 averaged 15, neglecting values greater than 50, which were considered indicative of boulders in the soil. Cohesive soils were mostly soft to stiff. Pocket penetrometer readings in the deeper cohesive soils of Borings B-2 and B-3 ranged from 35.9 to 430.9 kPa, but averaged about 191.5 kPa. Rock varied considerably in type, degree of weathering, existence of and the degree of decomposition of soil seams in the rock, angularity and cementation of fractures, etc. Three prominent rock types encountered in the borings were: (1) Conglomerate comprised of light gray to gray, medium hard to hard siltstone fragments in a poorly cemented matrix of brown to black, very soft to soft (rock classification) or soft to hard (soil classification) sandy or silty clay seams containing a few to extensive calcite stringers along old high and low angle fractures; (2) Gray to brown to black, soft to medium hard, shaly, calcareous siltstone to calcareous, silty shale with weathered brown to black, very soft to soft (rock classification) or soft to hard (soil classification) poorly cemented sandy or silty clay seams containing a few to extensive old calcite cemented high and low angle fractures; and (3) Very soft, fine grained sandstone, which had a white exterior surface when looking at the rock core, but which had a bright orange interior when the core was freshly broken. Except in Boring B-4, this sandstone was so poorly cemented that it crumbled to a highly decomposed, orange sand upon even moderate handling of the core. Ground-water levels in the borings were well below the top of rock.

Once the character of the soil and rock was observed, it was understandable that unless unusual care was taken in soil sampling and rock coring, zones of one could have been mistaken for the other, as was apparently the case in the earlier investigations.

It was noted by plotting top of rock profiles along the east and west walls of the filter plant building that the shapes of the plotted profiles were similar to those of the building settlement profiles along the east and west walls shown on Fig. 2. This suggested that the full soil depth above the top of rock was settling.

Settlement data for the slope indicator tubing in Boring B-2 indicated that no settlement of the soil and rock above the base of the slope indicator tubing occurred during the short monitoring period from March 9 to April 29, 1987. The inclinometer data for the short period from April 3 to 24, 1987, showed no significant lateral movement of the soil and rock above the base of the slope indicator tubing. This lateral movement result supported the earlier reconnaissance judgement that a landslide was not affecting the building.

Laboratory and Field Testing and Estimated Soil Properties

Laboratory and field tests on soils were performed during the various investigations. Testing was for pH, moisture content, unit weight, plasticity indices, grain size, specific gravity, chlorides, sulfate, sulfate sulfur, organic sulfur, pyritic sulfur, and total sulfur. The chemical testing for pH, chlorides, sulfate, sulfate sulfur, organic sulfur, pyritic sulfur, and total sulfur showed no results considered significant in relation to the building's settlement. Table 1 summarizes the estimated physical properties for the site soils.

Settlement Analyses

Since no relatively undisturbed Shelby tube samples could be obtained for testing, it was necessary to utilize literature correlations with standard penetration resistance in estimating parameters for use in settlement analyses. Using the estimated soil properties, settlements were estimated for 7 locations around the building and compared to measured total settlements at 6 locations around the building through mid-1986. In estimating settlements, the following approach was used for each point analyzed. First, assuming soil below the point was deposited in 1.52 m thick layers, elastic settlement of the soil and one-dimensional consolidation settlement of the partially saturated soil were estimated. It was assumed that all of the elastic settlement and one-half to all of the consolidation settlement

TABLE 1. Summary of Estimated Soil Physical Properties

A. From Testing of 1987, 1983 and Fall 1978 Investigations:

	Natural Water Content (%)	Natural Dry Density (kN/m^3)	Specific Gravity	Liquid Limit (%)	Plastic Limit (%)	Grain Size Analysis Percent Passing U.S. Standard Sieve No. 4.76 mm	425μm	75μm
Range	3-15.3	12.95-19.64	2.52-2.56	22.3-25	19-22.8	23-82	19-56	15-45
No. of Tests	38	32	2	5	5	16	16	16
Average	9.1	17.44	2.54	23.7	20.4	~55[(1)]	~36[(1)]	~27[(1)]

B. Estimated Average Values from Calculations with Above Average Data:

Moist Density (kN/m^3)	In-Situ Void Ratio	Degree of Saturation (%)	Water Content for 100% Saturation (%)	Plasticity Index (%)	Liquidity Index
18.85	0.43	53.8	16.9	3.3	-3.4

C. Estimated Average Values from Literature Correlations with Standard Penetration Resistance, Moisture Content, Etc.:

Compression Index	Effective Shear Strength: Friction Angle (degrees)	Effective Shear Strength: Cohesion (kPa)	Undrained Shear Strength: Friction Angle (degrees)	Undrained Shear Strength: Cohesion (kPa)	Elastic Modulus (MPa)	Poisson's Ratio	Overall Permeability (cm/sec)
0.050[(2)]	35	0	0	95.8	9.576	0.3	10^{-4} to 10^{-6}

Notes: (1) On average, the results indicate that overall site soils are about 45 percent gravel and larger size particles, about 28 percent sand size particles, and about 27 percent silt and clay size particles, with about 8 percent clay size particles based on the 4 hydrometer analyses of the 1987 investigation.

(2) Average of 8 different correlations.

occurred prior to building construction. Second, additional post-construction settlement was estimated for the gravelly pockets believed to exist in the soils below the building, assuming that finer particles migrate downward into open voids between gravel particles and that the particles forming the pockets degrade over time due to wetting, etc. Settlements due to this cause are not included in conventional elastic and one-dimensional consolidation settlement estimates, in which the parameters are related mostly to fine material characteristics. Since no well accepted approach exists to assess the settlement of rocky soils susceptible to particle degradation upon wetting, the admittedly crude approach outlined hereafter was applied. The estimated 25 percent thickness of gravelly pockets visible in soil cut heights near the building was applied to the soil depth at the point being analyzed, and the settlement of the gravelly pockets was estimated to be 2 percent of the gravelly pocket thickness at the point. The assumption was made that 30 percent of the gravelly pocket settlement should have occurred over the 9+ years since building construction. These percentages were estimated by extrapolating information in Sherard et al. (1963) on the post-construction settlement of rockfills, tempered by observations from another project discussed in Wimberly et al. (1993), where a building built on an initially rocky 15 m deep fill was settling. It was realized that it was possible that much more than 30 percent of the potential gravelly pocket settlement had occurred, since there had been significant leakage from the plant facilities into the soils below. Third, elastic and one-dimensional consolidation settlements of the partially saturated soil were estimated due to the estimated loads on strip footings alone and on Filter Beds 1 and 2 alone. The consolidation settlement in each case was assumed to include the elastic settlement plus the later time dependent settlement. The post-construction portion of these settlements was taken to be the time dependent settlement and 50 percent of the elastic settlement. Fourth, estimated post-construction total settlements were estimated for the strip footing area alone, for the Filter Beds 1 and 2 areas alone, and for areas where the loaded strip footing and Filter Beds 1 and 2 areas interact. Table 2 summarizes the final computed settlements and measured settlements for the 6 locations where comparisons were made. In general, settlements from the computations exceeded those observed through mid-1986. This supported the idea that significant future settlements were possible, even though the highly approximate nature of the computed values was understood by all involved.

REMEDIAL REPAIR OPTIONS

Brief reviews of most of the available methods to stabilize settling sites were provided to the City, along with a brief discussion of the feasibility of each method for use in stabilizing the settling filter plant building. It was recommended that any system implemented to reduce potential future settlements provide relatively uniformly stiffness conditions below the entire building, in order to not aggravate the existing effects of differential settlements on the structure.

After considering available options, it was recommended that compaction grouting of all soil above the top of rock be carried out in the area below the building and below the planes extending at one horizontal to two vertical downward from the outside edges of the ground floor. In addition, the following 3 drainage options were recommended: (1) Construction of diversion ditches upslope of the filter plant building to divert surface runoff around the building area; (2) Grading of the area around the building to facilitate surface runoff and covering of the area around the building to reduce infiltration by a suitably engineered low permeability soil cap or paving; and (3) Periodically leak testing the filter beds, piping, etc. at the site and performing needed maintenance to minimize leakage. It was also recommended that if voids were encountered below the ground floor in the filter plant building they should be filled with flowable grout.

The City reviewed the recommendations, elected to have the compaction grouting done, and decided to perform the 3 drainage recommendations themselves or with local contractors once the compaction grouting was completed.

TABLE 2. Estimated Post-Construction Total Settlements versus Measured Total Settlements as of Mid-1986

Location	Estimated Settlement With Interaction of Strip Footing and Filter Beds 1 and 2 Loaded Areas (cm)	Measured Settlement (cm)
Southwest Corner of Bldg.	0	0
West Side of Filter Bed 1	10.2[(1)] to 11.5[(2)]	4.3
East Side of Filter Bed 1	18.4[(1)] to 38.5[(2)]	11.9
Southeast Corner of Bldg.	11.2[(1)] to 31.3[(2)]	11.4
Southern 15± m of East Wall	11.2[(1)] to 31.3[(2)] at South End 18.4[(1)] to 38.5[(2)] at North End	12.7
Center of Filter Beds 1 and 2	14.4[(1)] to 22.3[(2)]	8.1

Notes: (1) With no post-building consolidation settlement of soil under its own weight.
(2) With maximum estimated post-building consolidation settlement of soil under its own weight.

Compaction Grouting, A Short Perspective

Briefly, compaction grouting, unlike permeation grouting, involves the controlled injection of stiff, mortar-like grout into previously drilled holes in the soil. The basic concepts were discussed by Graf (1969). Mitchell (1970) compared the approach to other grouting methods. Brown and Warner (1973) discussed how to perform the process and its applicability, as well as the then current technology. Criteria for planning and performing compaction grouting were given by Warner and Brown (1974). Warner (1982) discussed the first 30 years of compaction grouting, and contains excellent summary discussions regarding the planning and execution of compaction grouting programs. Details on the compaction grouting approach used to stabilize the settling water treatment plant building follow, along with an evaluation of its success.

COMPACTION GROUTING PROGRAM

The program was carried out between early July 1987 and mid-March 1988. It consisted of injecting cement grout, having a slump of 2.54 cm or less (usually between 1.25 and 2.54 cm), into the soil to be stabilized via holes drilled to below the top of rock outside and inside the building. The grout was a mixture of cement, sand, fly ash, water, and a minor amount of bentonite and had a minimum unconfined compressive strength of 8.27 MPa after 7 days. The grout injection was limited by: (1) A maximum take of 0.28 m^3 per 0.30 m increment of hole length being grouted; (2) A maximum grouting pressure of 6.90 MPa measured at the top of the hole; and (3) Surface movement as detected by laser levels, or by audible or visible signs. The grout mixer was capable of complete mixing of the stiff grout and had a metered supply system for the ingredients to enable close control of grout consistency. The grout pump was capable of nearly uniform flow rates from 0.003 to 0.15 m^3 per minute. The pump hopper had a force feed mechanism to reduce cavitation of the very stiff grout during pumping. Grout take was measured as the volume displaced by the force feeding piston times the number of piston strokes. Slight cavitation did occur in pumping the grout such that the actual take was less than (within 10$\pm$ percent of) the measured take on the above basis. As such, takes discussed herein are uncorrected for this effect. The grout was injected through minimum 5.1 cm I.D. pressure hose having non-restrictive flow couplings from the pump to the hole and through 5.1 cm I.D. flush joint steel casing placed in the holes to the top of each 0.30 m increment of hole to be grouted.

Holes were drilled using the ODEX over-ream bit method with air as the drilling fluid to remove cuttings. A specially fabricated drill was utilized in the tight access, low overhead clearance locations inside the building. The holes were spaced on a nominal 3.05 m square grid over the area to be stabilized, a spacing in the mid-range of the 2.44 to 3.66 m square grids generally used for initial hole spacings in compacting grouting programs. Hole spacings were adjusted where necessary due to lack of access or buried utility lines. Angled holes were used for relocated holes to try to grout the desired soil depths in locations which would achieve as close to uniform support over the grouted area as feasible. Maximum angles from the vertical were 20 degrees for outside holes and 10 degrees for inside holes. The final hole locations are shown on Fig. 1. The outside holes in Fig. 1 were initially drilled and grouted in the following order to provide support and lateral confinement to the soil below the building: 100 Series (9 holes, 4 vertical and 5 angled to avoid drilling on the steep Backwash Pond 1 slope), 200 Series (12 holes vertical), 300 Series (14 holes vertical), 400 Series (13 holes vertical and 16 angled to avoid buried utilities). The inside holes were then drilled and grouted in the following general order: 500 Series (4 holes vertical and 4 angled) and 600 Series (30 holes vertical and 6 angled). Due to hole relocations and wide spacings in certain areas, 4 vertical inside holes were added during construction. One outside vertical hole was added on the west side of the building, where the 3 nearby holes had relatively high grout takes to the bottom of the holes.

Due to the nature of the soil and rock at the site, it was difficult to detect when the top of rock was reached in drilling the grout holes. Therefore, many of the holes were overdrilled to be sure they were below the top of rock. In rock, as expected, takes were typically low and grouting was usually controlled by high pressures. Some rock-like zones in soil also occurred. In evaluating the grout takes, it was necessary to initially estimate a top of rock from the grouting record of take and pressure per each 0.30 m increment of hole grouted. The evaluation was then carried out only for the depth grouted above the estimated top of rock (nominally that in soil). The 100 Series and 200 Series holes were compaction grouted using the "bottom-up" approach in which the casing was initially located at the drilled hole bottom. It was raised in 0.30 m increments as grout was injected. In only the 100 Series holes, the upper take limit was set as 0.14 m^3 per 0.30 m increment of hole length, as these holes were the first done and were done while the final specifications were still in preparation. The fact that 94 percent of the soil depth grouted in these holes was controlled by the upper take limit resulted in the take limit being doubled for subsequent holes. The 300 through 600 Series and the 5 added holes were grouted using a partial "top-down" and partial "bottom-up" approach. In this approach, an oversized hole was drilled and cased to 5 feet below the surface. Outside the building the casing was held in place by granular backfill around it, while inside the building it was usually cemented in place. Three successively deeper 1.52 m stages of "top-down" compaction grouting were generally carried out between 1.52 m and 6.1 m below the surface to achieve near-surface densification and vertical confinement to permit the normally higher pressures to be used below. In each stage, drilling occurred to the stage base, where the above "bottom-up" method was used in grouting that stage. The stage was then drilled through, after a delay of at least 16 hours, before subsequent work below the stage. After these 3 upper stages were completed, the hole was drilled to its full depth and grouted using the above "bottom-up" approach.

Table 3 summarizes the grout take data in soil. The data in Table 3 indicate that the soil was generally looser or softer in the region outside of the building than it was below the building. This may be partly due to one or more of: (1) The settlement which had occurred under the building; (2) Better initial near-surface compaction in the area of the building than elsewhere; and (3) The earlier grouting and remedial work below the building. Fig. 4 shows a plot of the percent of the soil length grouted for outside, inside, and total holes that were controlled by the upper grout take limit versus the pressure interval occurring during injection when the upper take limit was reached. In general, most of the upper limit takes occurred under relative low grouting pressures, below about 30 percent of the upper pressure limit.

ACCEPTANCE CRITERIA

Ideally, if soil conditions at the site were relatively homogeneous and such that representative undisturbed Shelby tube samples of the soil could have been obtained, it would have been possible to estimate the in-place average unit weight of the settling soil. Modified Proctor tests could then have been performed on soil samples to estimate the maximum dry unit weight for the settling soil. If one assumes that the soil might have settled an acceptable amount if the soil was

TABLE 3. Summary of Grout Takes Per 0.30 m (1-Foot) Increment of Hole Grouted in Soil

For 329 m Length Grouted in Outside Holes (1)	
Percent of Soil Length Grouted	Grout Take
61.4	≥ U.L. (2)
10.8	≥ 0.5 U.L. but < U.L.
8.9	≥ 0.2 U.L. but < 0.5 U.L.
18.9	< 0.2 U.L.
100.0 Total	
For 311 m Length Grouted in Inside Holes	
Percent of Soil Length Grouted	Grout Take
27.6	≥ U.L.
9.5	≥ 0.5 U.L. but < U.L.
9.6	≥ 0.2 U.L. but < 0.5 U.L.
53.3	< 0.2 U.L.
100.0 Total	

Notes: (1) Length of outside holes includes 100 Series Holes for which upper limit on grout take was 0.5 U.L.
(2) U.L. = Upper Limit on grout take = 0.28 m^3 per 0.30 m increment of hole length (equivalent to 0.93 m^3/m of hole and 10 ft^3/ft of hole).

compacted to 90 to 95 percent of the modified Proctor maximum dry unit weight, one could compare the estimated average in-place soil unit weight to the unit weight corresponding to 90 to 95 percent of the modified Proctor maximum dry unit weight. Then, an acceptable increase in the average soil unit weight may be selected, from which the needed volume compression of the soil during injection could be estimated. The injection program could then be planned, monitored, and adjusted, as judged necessary, to try to obtain the desired level of grout take to yield the desired level of soil compression. Unfortunately, the highly variable and rocky nature of the soils at this site prevented the obtaining of representative Shelby tube samples of the settling soil and rendered the above approach impractical.

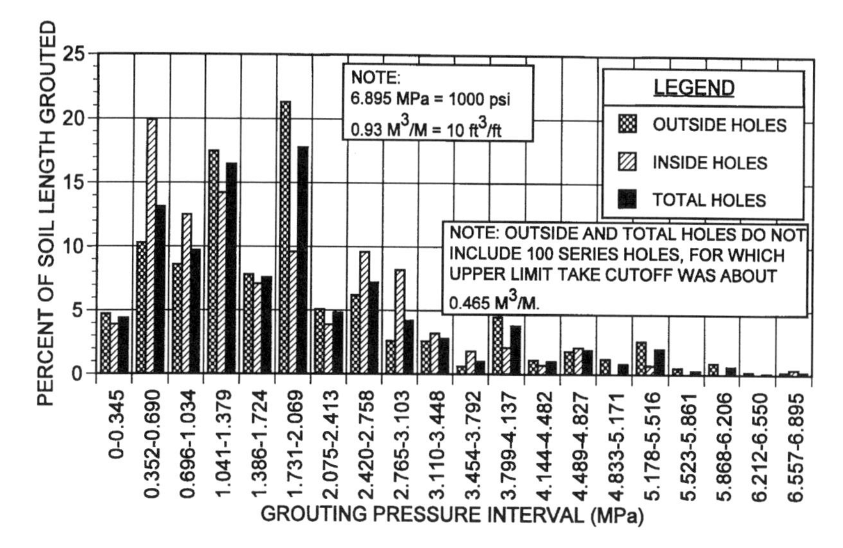

FIG. 4. Percent of Soil Length Grouted When Take ≥ 0.93 m^3/m versus Grouting Pressure Interval

Some judgement of the success of the program may have been achievable by carrying out after grouting standard penetration tests to compare to the data from the earlier field investigations. Such data from below the building were not available from the earlier investigations. Collecting such data before or after the compaction grouting was carried out would have required difficult and expensive inside drilling. In addition, due to the highly variable and rocky nature of much of the soil, it was felt that even outside drilling and comparisons of before and after standard penetration test data, while possible, would be costly and may yield inconclusive results. Therefore, since conditions at this site were far from ideal, an acceptable level of grout take was based largely on judgement, followed by further monitoring to assess if the settlements ceased or slowed sufficiently. If they did, the initial program achieved its goal. If not, a subsequent program, using split-spaced holes, might be required.

A grout take of 0.28 m^3 per 0.30 m increment of hole length for 4 holes at the corners of a 3.05 m square grid implies average theoretical densification of the soil within the grid of about 10 percent; i.e., an increase in the soil's unit weight of about 10 percent, due to compression of the original soil's volume by the injected grout. This level of densification (which also implies soil stiffening) was judged to be a reasonable level of improvement in the site soils and was selected to be the program's upper limit on grout takes, as given above. Such a take also theoretically implies the construction of grout columns at the hole locations having average diameters of about 1.1 m. Due to injection variations from interval to interval, the columns typically have substantial outer surface irregularity. It should be noted, that grout takes and the resulting grout column diameters tend to be larger in more compressible (poorer) soils than in less compressible (better) soils; i.e., the grout tends to go if and where it is needed to affect an improvement in the subsurface conditions. Since the injection of grout can add considerable weight to the settling soil, it is necessary to treat the zone of potentially settling soil down to an essentially incompressible material, such as rock. This is generally the case unless only a relatively thin compressible zone requires treatment, such that the injected grout does not add significant weight to the soil and cause settlement of the soils below the treated zone.

Construction of rather substantial grout columns, such as those theoretically constructed at this site (or even substantially smaller ones) from the top of rock to near the surface over the area treated results in stiff grout columns that help resist soil settlement between the columns by shear stresses along the column-soil interface. Both soil densification and the constructed columns improve the soil's resistance to settling. Which of these is most important in improving a site's stability to acceptable levels often can not be accurately assessed. The rather large takes at this site, implying both substantial soil densification and construction of relatively large grout columns, indicated that major improvement was made in the soil's ability to resist further settlement. However, as explained above, once a site is treated only future monitoring (discussed below) tells the story of success, or failure requiring further treatment.

Subseal Holes

In performing the compaction grouting work, voids were detected below the concrete in the building area. Soundings were done and 25 holes were cored through the concrete to permit high slump (30.48± cm) grout to be injected to fill (seal) the subsurface voids. These hole locations are shown on Fig. 1 as subseal holes. Little to no take occurred in the Pipe and Valve Gallery area. Along the south side of Filter Beds 1 and 2, voids about 2.5 cm deep were grouted. Along the east wall south of Filter Bed 2, a 7.6 to 10.2 cm deep void was grouted for about the northern half. The void depth tapered to about 2.5 cm near the southeast corner. About 2 m^3 of subseal grout were injected to fill the voids.

Settlements During and After the Compaction Grouting Work

During the compaction grouting work, the building continued to be monitored for settlement, along with the slope indicator tubing in Boring B-2. The southern two-thirds of the east side of the building experienced about 1.3 cm of settlement, as did roughly the upper 9.1 m of soil at Boring B-2. At Boring B-2, the soil in the depth range of 9.1 to 12.2 m experienced about 2.5 cm of upward movement. The upward movement was likely associated with rather large grout takes occurring in the area of Boring B-2 (one hole was only about 3 feet to the southeast).

Settlements observed along the southern two-thirds of the east side of the building may be related to compression of the soil mass under the weight of injected grout, but are more likely related to the 1.2 to 2.1 m deep trench excavated parallel to and well below the footings along the east wall by the City in mid-July 1987 to assist in locating utilities so that the grout holes could avoid them. This trench was left open about a week and was backfilled with excavated soil tamped by the backhoe bucket. Also, after several rains, 2 drain pipes from the roof discharged into the trench area until the pipes were extended to the east to drain to Backwash Pond 1.

The City performed or had performed the recommendations regarding drainage control and had a settlement monitoring system installed after the compaction grouting work was completed. The actual data from this system were not provided to the authors. However, from telephone discussions with City personnel, the system is reportedly sensitive enough to detect very small movements, including the thermal movements of the structure. This system, which has gathered data for about 5 years, has reportedly confirmed that no significant additional settlement of the filter plant building has occurred; i.e., no movements not believed to be associated with thermal changes in the structure have occurred over this 5-year period.

Conclusions

Compaction grouting was effective in stabilizing the settling filter plant building. Close cooperation and communication between the City and the Geo-Con/GAI team facilitated the work at this difficult site being completed within City budgetary constraints and without interruption of the City's water supply. The investigation and implementation of the compaction grouting and subseal hole grouting program cost about $525,000 when it was done in 1987-88. The original estimate was about $470,000, which did not fully account for the high grout takes and required subseal holes. This grouting program cost about one-tenth of the cost the City would have incurred if settlement had continued such that construction of a new filter plant building was necessary. The City's cost in implementing the other recommendations is not available to the authors.

Acknowledgements and Comments

The authors thank the City of Glenwood Springs, Colorado, and in particular Mr. Michael Kopp, City Manager, for permission to publish this paper and Mrs. Arlene Wimberly (formerly of GAI) for the preparation of this manuscript. The authors acknowledge the value of the information provided for use by the City from the 1983 and fall 1978 subsurface investigations, as well as from other reports and drawings, and choose not to identify the firms involved in view of differences noted between the various investigations. Overall, the subsurface conditions at this site were highly complex; and the soil and rock were difficult to drill and sample well, and to test effectively. In view of this, the differences are not surprising or particularly unusual in the authors' opinion.

Appendix - References

Brown, D. R., and Warner, J. (1973). "Compaction grouting." *J. Soil Mech. Found. Div.*, Proc. ASCE, 99(SM8), 589-601.

Graf, E. D. (1969). "Compaction grouting technique." *J. Soil Mech. Found. Div.*, Proc. ASCE, 95(SM5), 1151-1158.

Mitchell, J. K. (1970). "In-place treatment of foundation soils." *J. Soil Mech. Found. Div.*, ASCE, 96(SM1), 73-110.

Sherard, J. L., Woodward, R. J., Gizienski, S. F., and Clevenger, W. A. (1963). *Earth and Earth-Rock Dams*, John Wiley and Sons, New York, New York.

Warner, J. (1982). "Compaction grouting - the first 30 years." *Proc. ASCE Specialty Conf. on Grouting in Geotechnical Engineering*, ASCE, New York, New York, 694-706.

Warner, J., and Brown, D. R. (1974). "Planning and performing compaction grouting." *J. Geotech. Eng. Div.*, Proc. ASCE, 100(GT6), 653-666.

Wimberly, P. M., III, Mazzella, S. G., and Newman, F. B. (1994). "Settlement of a 15-meter deep fill below a building." *Vertical and Horizontal Deformations of Foundations and Embankments*, Geotechnical Special Publication No. 40, ASCE, New York, New York.

Performance of Wick Drains Installed by Vibration

Richard P. Long,[1] Fellow ASCE, Leo F. Fontaine,[2] and Bruce Olmstead[3]

Abstract: Settlement and piezometer data from construction of fills on Route I-291 in South Windsor, CT is analyzed for total settlement and coefficient of consolidation. The results are compared with values analyzed from projects at other locations along the lower Connecticut River Valley at which fills were built over varved clay. The present analysis shows the wick drains to be at least as effective as sand drains and there are no detectable adverse affects from vibrating the drains into place. The amount of settlement indicates that the varved clay is overconsolidated.

Introduction

Traffic in the Greater Hartford, CT area requires alternate East-West routes. In 1991, the Connecticut Department of Transportation began constructing the I-291 expressway in the town of South Windsor. This new expressway required filling greater than 12.2 m in height to allow proper clearance between the structures carrying I-291 and CT Routes 5 and 30 beneath them. At the location of this structure, the soil profile contains a clay layer greater than 36.6 m thick covered with 15.2 m of sand and silt. It was important that the construction be kept on schedule which required the use of vertical wick drains to accelerate the consolidation. The standard method of pushing the mandrel through the deposit had been found adequate at other locations. At this location in South Windsor the contractor had to induce vibration in the mandrel to overcome the resistance of the sand layer and penetrate the full depth of the clay. The effect of vibrations on the clay around the mandrel was unknown and the design predictions were further complicated by

---

[1] Professor of Civil Engineering, University of Connecticut, Storrs, CT 06269-3037.

[2] Supervising Engineer, Soils Division, Connecticut Department of Transportation, Berlin, CT.

[3] Engineer III, Soils Division, Connecticut Department of Transportation, Berlin, CT.

uncertainties resulting from sample disturbance indicated by the laboratory compression test results. To develop the anticipated settlements as quickly as possible, the sites were surcharged with 3 m of fill. The fills were instrumented so that an analysis could determine the effect of the vibration on the performance of the wick drains and to verify the stress history of the deposit in this area.

Soil Properties and Profiles

Fig. 1 shows the typical soil profile for the project area. The upper 9.1 m of soil consists of a sand having a permeability great enough to function as the sand blanket for the vertical drains. Below this was a layer up to 6.1 m of gray silt that rests directly on the varved clay. As can be seen in the soil profile, the section west of Route 5 had some existing fill placed over twenty years ago as part of earlier construction activities. The clay layer has had enough time to dissipate excess pore pressures from this applied load.

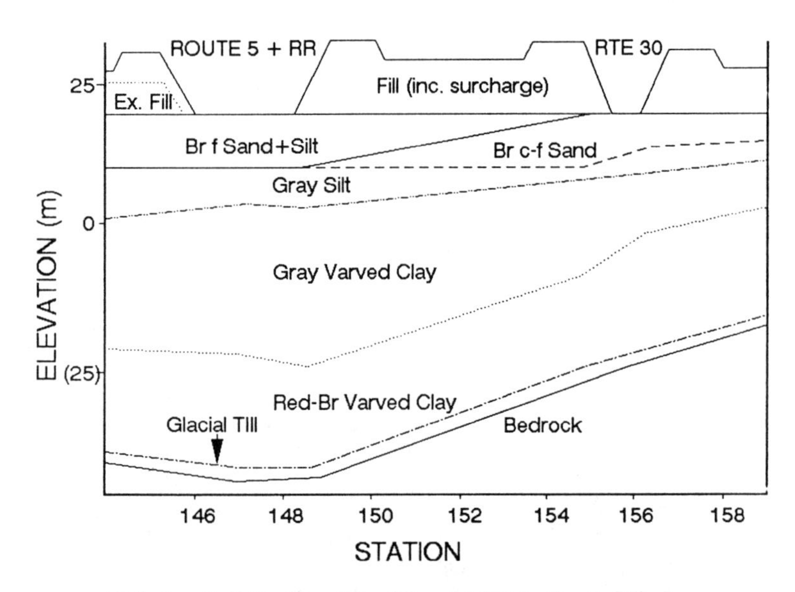

FIG. 1. Soil Profile of the Site of I-291 in South Windsor

Most properties of the clay measured in the laboratory testing program are within the same range as the previously tested samples of varved clay from other areas along the Connecticut River Valley. For example the strength from the unconfined compression tests averaged about 38.1 kPa and the coefficient of consolidation, C_v, averaged about 0.21 mm^2/s. This agrees with overconsolidated results from previous investigations that showed the varved clay in the lower Connecticut River Valley to be relatively homogenous (Ladd and Foott 1977). The compression plots, however, show a difference from previous results in that they indicated that the clay in this area is normally consolidated.

Previous research and other subsurface investigations involving the varved clays of the Connecticut River Valley indicated that the deposit is overconsolidated by 95.8 kPa over most of its depth (Healy and Ramanjaneya 1970; Ladd and Foott 1977; Long et al. 1978). It was suspected that some disturbance occurred to the samples recovered for this project and it was decided to analyze the field settlement data to determine the maximum precompression pressure. An extensive study by Long et al.(1978) analyzed the settlement behavior of fills built over varved clays in the vicinity of Hartford, CT. The field data showed that the average recompression factor $RR = C_r/(1+e_o) = 0.017$ and the average virgin compression factor $CR = C_c/(1+e_o) = 0.26$. It was decided, therefore, to use these values in the analysis of the field settlement data to estimate the maximum past pressure experienced by the clay.

Installation of Drains

The contractor was required to install the vertical drains through the entire depth of the clay layer. The clay layer was approximately 36.6 to 42.7 m thick with a considerable cap of granular soil over the clay, making the installation depth as long as 51.8 m in some sectors. The sand layer and varved clay showed considerable resistance to penetration by a mandrel large enough to contain a wick drain, therefore, it was proposed to assist the installation by attaching the mandrel to an ICE 612 vibratory hammer and a set of 0.9 m fixed leads. The tapered-point mandrel, 102 mm by 150 mm by 10 mm thick, fit inside the leads. The hollow mandrel allowed for continuous internal feed of the wick drain through its center. This is not the first installation of wick drains in this state, however it was the first time the drains had been installed by inducing specific vibrations with a hammer usually used for driving piles.

Triangular spacings of 3.7 m and 6.1 m were used on this project. The drains under the surcharged areas were spaced at 3.7 m and spaced at 6.1 m throughout the remaining area in which they were needed. The location of each drain was determined by survey and numbered. The location of the drained areas is shown in Fig. 2.

Instrumentation and Data Collection

Inclinometers, piezometers, and settlement platforms were installed at these sites to control the construction process and monitor the consolidation of the underlying clay by the applied loads. Fig. 2 shows the location of the instrumentation. The success of the pneumatic piezometers was somewhat limited at this site; of the 13 piezometers installed only 5 yielded consistent data which could be used in this study.

Nine settlement platforms were placed below the fills. The platforms consisted of metal pipe risers attached by a flange to a wooden base about one meter square. The platforms were placed at the ground surface prior to the placement of any fill. Observations on the platforms measured the settlement of the original ground surface. These settlements consisted of compression of the granular layers, as well as the combined settlements of the clay. The observations on the elevations of the platforms continued on a weekly basis for over one year.

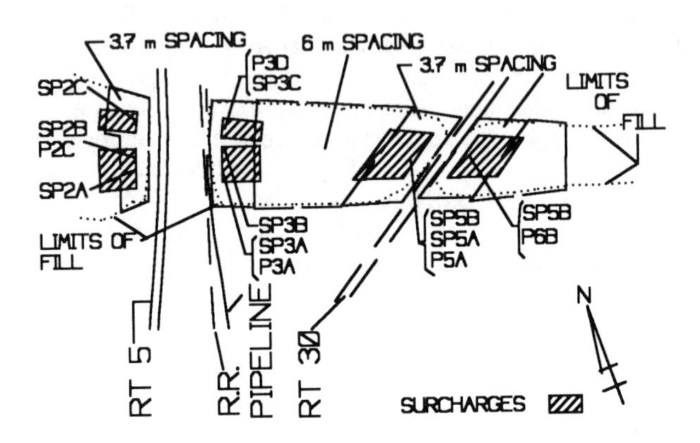

FIG. 2. Location of Instrumentation I-291 South Windsor

The section to the west of Route 5 having an existing fill required less additional fill for this project. Design calculations indicated that the fills in this area would be stable under the additional fill. The section just to the east of route 5 was of concern due to the height of the fill with surcharge, 12.2 m, and the proximity of a major jet fuel pipeline. To insure stability, a filling rate of only one foot per day was used. To monitor the stability, the piezometers and inclinometers were observed daily during the filling process. Observations on the instruments were less frequent after filling was complete and changes occurred at a slower rate. Two inclinometers were installed at the surcharge in this eastern section. The inclinometers showed about 51 mm of movement between the depths of 7.6 and 12.2 m. Most of the movement occurred between the clay and the fine sand layer.

The data obtained from this instrumentation is used to determine the consolidation parameters for comparison with values determined in the laboratory. In addition, the parameters determined from this project are compared with values analyzed from other projects built on the varved clay in Connecticut. These analyses address the possible variation of the soil properties as well and disturbance effects caused by drain installation. The projects included in the comparison have vertical drains using both conventional sand drains as well as the more modern wick drains.

DATA ANALYSIS

After the Filling Period

The field data included piezometer readings, settlement platform observations, and slope indicator measurements. Data from each instrument was analyzed separately. The analyses used piezometer and settlement data after filling was complete and the applied load reasonably constant, to determine the coefficient of consolidation. During filling each increment of applied load caused some amount of initial settlement, an increase of pore pressure, and some consolidation. After filling the load remains reasonably constant and the increase in settlements and the decrease in pore pressures are due to the progress of consolidation. Analysis of the piezometer readings yields information about the apparent coefficient of consolidation in the radial direction and the analysis of the settlement data provides an independent evaluation of the coefficient of consolidation and a value for the total settlement. The influence of the vertical dissipation of pore pressure was considered negligible, because of the thickness of the clay layer. The validity of the methods of analysis have been previously demonstrated (Long and Hover 1984; Long 1990; Long and Fontaine 1992). The data collected at this site indicated that the coefficients of consolidation were essentially constant during the period of observation.

Piezometer Data

The dissipation of pore pressures in the radial direction for a soil having constant properties can be described by the equation (Barron 1948) :

$$\ln(u) = \ln(B) - (Lt) \tag{1}$$

where u = excess pore pressure at time t; B = a constant;

$$L = \frac{2C_R}{F(n)R_e^2} \tag{2}$$

$$F(n) = \frac{n^2}{n^2-1}\ln(n) - \frac{3n^2-1}{4n^2} \tag{3}$$

where C_R= apparent coefficient of radial consolidation; n = R_e/R_w; R_e= effective radius of the area serviced by the vertical drain; R_w= the equivalent radius of the vertical drain. For this analysis the equivalent radius of the vertical drains was computed from the equation (Long and Fontaine 1992):

$$R_w = \frac{w+b}{4} \tag{4}$$

where w = the width of the wick drain; and b = its thickness.

Eq. (1) indicates that a plot of the natural logarithm of the excess pore pressure versus time results in a straight line. The slope of the line can be analyzed for C_R. There is no explicit term in Eq. (1) for the radial distance from the vertical drain at which the piezometer is positioned. The rate of dissipation of the pore pressure is related to C_R through the slope of the straight line.

Settlement

Assuming that the coefficient of consolidation is constant, the progress of settlement after filling is complete can be described by the equation:

$$S = S_t - S_c \exp(-Lt) \tag{5}$$

where S = observed settlement; S_t= total settlement; and S_c= consolidation settlement.

Eq. (5) can be used to analyze settlement data in several ways (Long 1990). The method selected to analyze the data reported here involves the derivative of the Eq. (5), which takes the form of a straight line thus:

$$\ln\left[\frac{\delta S}{\delta t}\right] = C - (Lt) \tag{6}$$

where C = constant and other variables as defined before.

The similarities between Eqs. (1) and (6) result from the fact that the rate of settlement at any time is proportional to the excess pore pressure as shown by Barron's (1948) formulation of the problem.

The observations continued past primary consolidation and the total settlements could be estimated directly from a plot of the settlement data against the logarithm of time.

Secondary Compression

The settlement observations continued beyond the time at which the dissipation of excess pore pressures was complete. Information was therefore obtained on the rates of secondary compression which were calculated according to the equation:

$$\frac{\Delta S}{h} = C_s \log\left[\frac{t_2}{t_1}\right] \tag{7}$$

where h = thickness of the clay layer; C_s= coefficient of secondary consolidation and other symbols as before.

During the Filling Period

Analyses during the filling period are limited because of the uncertainty with respect to initial settlements that occur as the load is placed. When the initial settlements are known or can be estimated with confidence, an average value for the coefficient of consolidation can be calculated based on the observed settlement at the end of filling. This observation contains all of the initial settlement and a portion of the consolidation settlement. Subtracting the initial settlement from the observed settlement at the end of the filling period, yields the consolidation settlement that has occurred during filling. Dividing this consolidation settlement by the maximum consolidation settlement produced by the load, yields a percent average consolidation taking place during the filling period. The filling time must be modified to account for the time rate of loading. The method used to adjust the filling time was that of Taylor (1948).

TABLE 1. Results of Analysis for Route I-291

Instrument Type & ID	Coefficient of Consolidation mm^2/s		Total Settlement m	Coefficient of Secondary Compression
	During Filling	After Filling		
Set Pl 2A	1.31	0.78	0.37	–
Set Pl 2B	1.11	0.78	0.27	–
Set Pl 2C	1.15	0.88	0.20	–
Set Pl 3A	1.32	0.58	0.61	0.007
Set Pl 3B	1.08	1.43	0.48	0.007
Set Pl 3C	1.29	1.56	0.55	0.003
Set Pl 5A	–	1.06	0.87	0.013
Set Pl 5B	–	1.06	0.67	0.009
Piez 2C		0.69		
Piez 3A		0.69		
Piez 3D		1.08		
Piez 5A		0.81		
Piez 5B		0.74		

Although initial settlements could not be separated from the total settlements in this analysis, previous analyses have shown that the initial settlements in the varved clay are, on average, about 25% of the total settlement (Long et al. 1978). To estimate the coefficient of consolidation during filling, the initial settlement at each settlement platform was computed according to this average and subtracted from the observed settlement at the end of filling. The filling time was modified by multiplying by 0.5 after the method of Taylor (1948). The time factor appropriate for the percent consolidation was used with the modified filling time to compute the coefficient of consolidation.

Precompression Pressure

It was decided to backfigure the preconsolidation pressures at this site based on a careful estimate of the increased vertical stress and the amount of consolidation settlement. The stress was calculated using a series of rectangularly loaded areas to simulate the fills, and the equations of elasticity. The initial stress conditions were determined from the field samples. The value of RR, $C_r/(1+e_o)$, was taken as 0.017, and the value of CR, $C_c/(1+e_o)$, used in the calculations was 0.26. Both of these values came from the study by Long et al. (1978).

RESULTS

Values for the data collected at this site are shown in Table 1. The "n" value for this installation equals 71. The analyzed values of C_R during and after filling are approximately equal. The C_R values from the settlement platforms and piezometer data compare. Analysis of the settlements indicated that the precompression pressure was between 48 and 72 kPa. This is slightly smaller than indicated from previous compression tests and may indicate slight disturbance of the varved clay during installation of the wick drains. The coefficients of secondary compression as shown in Table 1 are small, also indicating some overconsolidation of the clay layer. The values shown in Table 1 compare favorably with the results of analysis of previously installed vertical drains in the varved clay.

This is the first use of vibration to place the wick drains. Therefore comparison of the results from this area with results analyzed from field data at other locations is warranted. Such a comparison is shown in Fig. 3. Included in Fig. 3 are averaged results of C_R for times after filling is complete and includes results from sand drains as well as wick drains installed with and without vibration. The coefficient of consolidation is plotted as a function of the F(n) parameter. The coefficients of consolidation shown in Fig. 3 are analyzed from data after the total fill is in place and can be considered as overconsolidated values although stresses in the upper portion of the layer are probably in the normally consolidated region. As can be seen from Fig. 3, the efficiency of the wick drains is not adversely affected by the use of vibration in installing them. It is also apparent that the wick drains are at least as effective in the varved clay as sand drains.

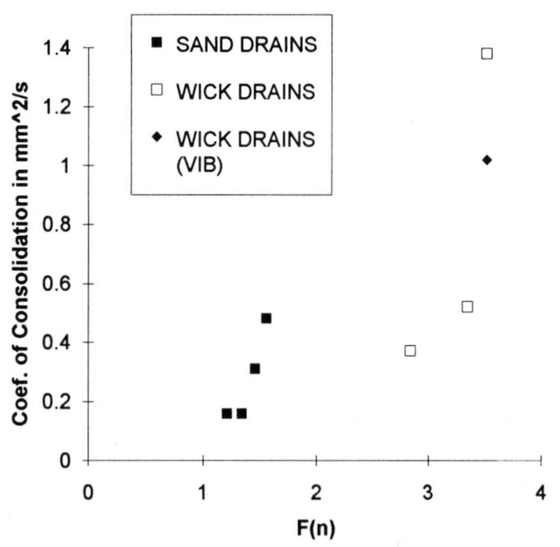

FIG. 3. Plot of Average Values of C_R versus F(n)

CONCLUSIONS

1. Vibrations do not seriously affect the performance of wick drains placed at 3.7 m on center in a triangular pattern.
2. Wick drains are at least as effective as sand drains in the varved clay.
3. There appears to be some disturbance as witnessed by slightly increased settlements but the rate of consolidation remains high.

ACKNOWLEDGEMENT

The Connecticut Department of Transportation made the data for this paper available, as well as the opportunity to analyze it. The authors would also like to thank the field crews for their interest and care in collecting the data.

APPENDIX - REFERENCES

Barron, R. A.(1948). "Consolidation of fine-grained soils by drain wells." *ASCE Trans.*, 113, 718-742.

Healy, K. A. and Ramanjaneya, G. S. (1970)."Consolidation characteristics of a varved clay," *Final Report, Research Report JHR 70-30,* Department of Civil Engineering, University of Connecticut, Storrs.

Ladd, C. C. and Foott, R. (1977). "Foundation design of embankments constructed on varved clays," *Report No. FHWATS-77-214,* United States Department of Transportation, Washington, D. C.

Long, R. P. and Carey, P. J. (1978). "Analysis of settlement data from sand-drained areas." *Tolerable Movement of Bridge Foundations, Sand Drains, K-Test, Slopes, and Culverts*, Transportation Research Record 678, TRB, National Research Council, Washington D. C., 36-40.

Long, R. P. and Hover, W. H. (1984). "Performance of sand drains in a tidal marsh." *Proc. 1st Int'l. Conf. on Case Histories in Geotechnical Engineering*, St. Louis, III, 1235-1244.

Long, R. P. (1990). "Techniques of backfiguring consolidation parameters from field data." *Modern Geotechnical Methods: Instrumentation and Vibratory Hammers 1990,* Transportation Research Record 1277, TRB, National Research Council, Washington, D. C., 71-79.

Long, R. P. and Fontaine, L. L. (1992). "Performance of wick drains at Windsor, Connecticut." *Advances in Geotechnical Engineering*, Transportation Research Record 1369, TRB, National Research Council, Washington, D.C. 1-7.

Long, R. P., Healy, K. A., and Carey, P. J. (1978). "Field consolidation of varved clays," *Final Research Report JHR 78-113*, Department of Civil Engineering, University of Connecticut, Storrs.

Taylor, D. W. (1948). *Fundamentals of Soil Mechanics*, John Wiley and Sons, New York, New York.

BUILDING PERFORMANCE FOUNDED ON AN IMPROVED SAND IN RECIFE, BRAZIL

J. F. T. Jucá,[1] Member, ASCE, A. O. C. Fonte[1], and I. D. S. Pontes Filho[1]

ABSTRACT: Studies performed on a fourteen-storey building founded on an improved sand are presented. These studies included soil profile characterization, the choice of foundation type, the ground improvement, the numerical method used in settlement predictions, and the field instrumentation to monitoring vertical deformations. The effects of the soil-structure interaction and construction sequence are evaluated in the settlement predictions. The settlements measured were closely related to superstructure loading. A limited agreement is obtained between measured and calculated settlements.

INTRODUCTION

The city of Recife is situated on the North-eastern coast of Brazil (Fig. 1). The building under consideration is located in a high population density area near the beach in South Recife. The subsoil of this area consists of loose sands overlying soft to medium clays. This subsoil has caused serious problems in the behavior of tall buildings when the alternative design is shallow foundation. In general, it is necessary to either use deep foundations or some form of ground improvement for support of the structures.

FIG. 1. Site Location

The purposes of this paper are to describe the soil conditions and the structure of the fourteen-

[1] Professor of Civil Engineering Department, Federal University of Pernambuco, Centro de Tecnologia, Cidade Universitária, 50740-530 - Recife - PE, Brazil.

storey residential building in order to evaluate the performance of the alternative foundation design adopted. The suitability of compaction piles as one of the alternatives to ground improvement is presented. Settlements were calculated using the finite element method. The results of this were compared with the settlements measured during all steps of building construction.

LOCAL CONDITIONS

The subsoil of Recife is composed of young unconsolidated alluvial sediments, resulting from a regression of the sea in which the deltas of the two river systems merged and formed one large deltaic fan. Climate is classified as warm and humid with an average temperature that varies from 23°C to 27°C, while the relative humidity is 78% (yearly average). The average annual rainfall is around 1750 mm.

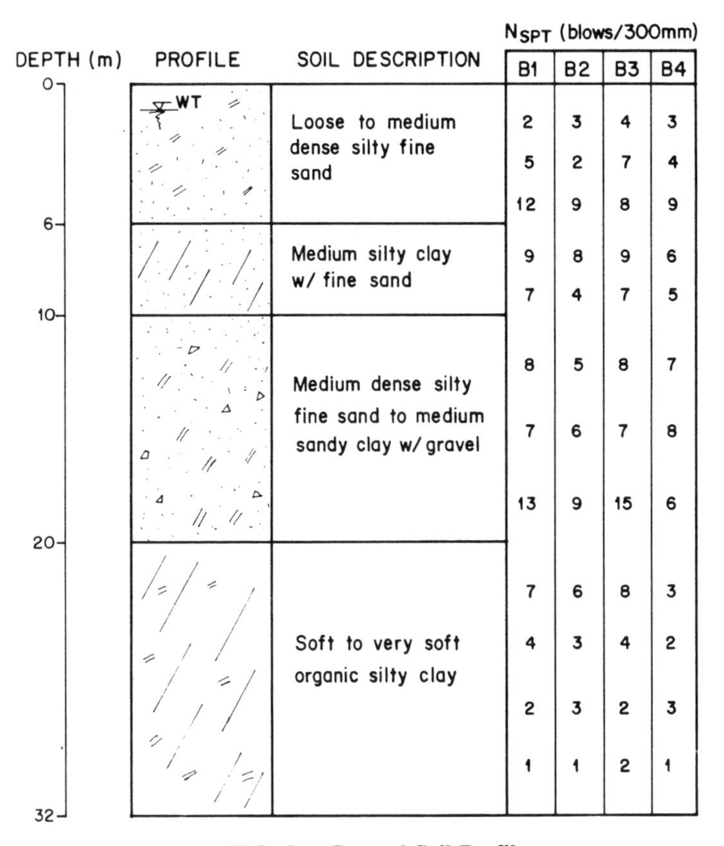

FIG. 2. General Soil Profile

The soil profile at the building site is indicated in Fig. 2. The first layer is a loose to medium dense sand 6 m thick, overlying 4 m of medium silty clay. Below the silty clay is a medium silty sand to medium dense sandy clay with gravel to a depth of 20 m. Below the sand layer is a organic soft clay which exists until a depth of at least 32 m. The groundwater table before the compaction is around 1.2 m depth.

The previous geotechnical investigation was performed by four boreholes using the SPT test. The results of these tests indicated low values of blow counts on the upper sandy layer. The in situ relative density (Dr) was estimated from SPT blow counts as being from 15% to 40%.

The foundation alternatives were analyzed in order to obtain a good cost/benefit ratio for these subsoil conditions. In this area pile foundations can be found only at greater depth (over 45 m). The cost of this was three times more than shallow foundation associated with ground improvement. Footings with treatment of the loose sand was adopted as the final foundation design.

The building comprises a fourteen-storey reinforced concrete structure 38 m long, 22 m wide and 48 m height. Footings with relatively rigid strap-beams were adopted in order to increase the stiffness of structure. A typical lateral view and the foundation layout of the structure are shown in Fig. 3, where the columns, footings and loads can be seen. All footings were founded at a depth of 1.5 m in the sand. After ground treatment, the foundation design pressure was up to 380 kPa.

Construction activities were initiated with the ground improvement during the first two months of 1992. The footing system was completed in March. Concrete to ground floor was completed at the end April. From April to October, 1992, the building superstructure was concluded. All construction activities were completed within one year.

Settlement data presented in this paper were taken during a building construction period of six months. The building was not occupied during this period and accordingly, only dead loads were considered on the foundations.

Ground Improvement

The method applied to densify the sand before construction was compaction piles (Murayama 1958; Nakayama et al. 1973; Ichimoto and Suematsu 1982; Wallays 1982). The execution procedure consisted of driving a casing into the soil, closed at the lower end by a temporary gravel plug. When the depth of treatment was achieved, the casing was filled with a sand and gravel mixture which was introduced into the ground by hammering, and then progressively lifting up the casing. The improvement was originated by addition of the material and the pile driving effects in the adjacent granular soil. In other words, the ground compaction was achieved by soil displacement and vibration.

The piles were driven in a square arrangement spacing 0.85 m. About 1200 piles were compacted with a rammer of 25 kN in weight falling 3 m height. The depth of treatment achieved was 4 m. The casing used was 320 mm diameter and the compose of the sand:gravel mixture was 2:1 in volume. The ratio between the

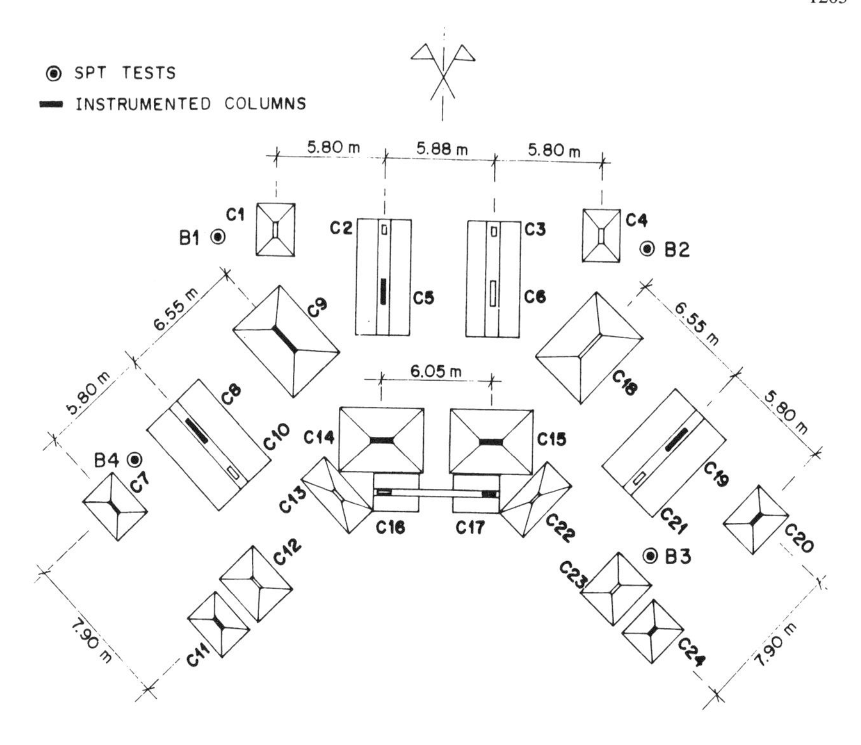

a) Foundation layout

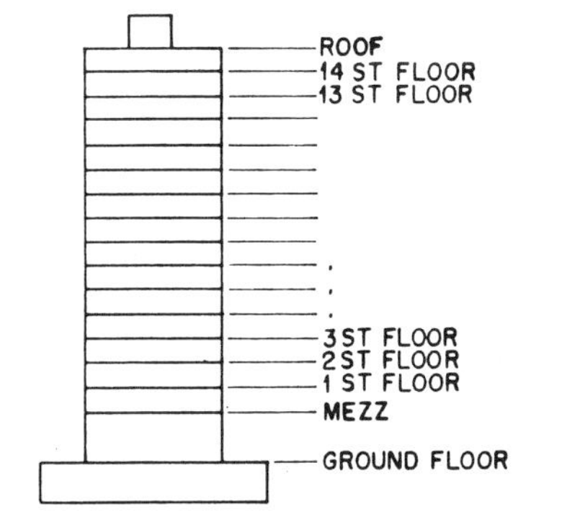

b) Typical lateral view.

COLUMN	DEAD LOAD
C1=C4	1930 kN
C2=C3	1200 kN
C5=C6	3870 kN
C7=C20	2000 kN
C8=C19	4040 kN
C9=C18	4700 kN
C10=C21	1870 kN
C11=C24	1750 kN
C12=C23	2160 kN
C13=C22	2190 kN
C14=C15	4080 kN
C16=C17	1870 kN

c) Superstructure dead loads

FIG. 3. Foundation Layout, Typical View and Loads

sand pile area (Ac) and the reinforced soil surface (Ac+As) given a substitution coefficient of 0.11.

In order to check the densification results, SPT tests were also performed immediately after the treatment. These tests were carried out at the centers of the square grid of compaction points.

The results presented in Fig. 4 show that the upper sandy soils were effectively densify, showing an increase in blow count of about 5 times, particularly from 2 m to 4 m depth. At depths greater than 6 m, where the soil changes to a silty clay, the penetration resistance was not appreciably increased, as would be expected.

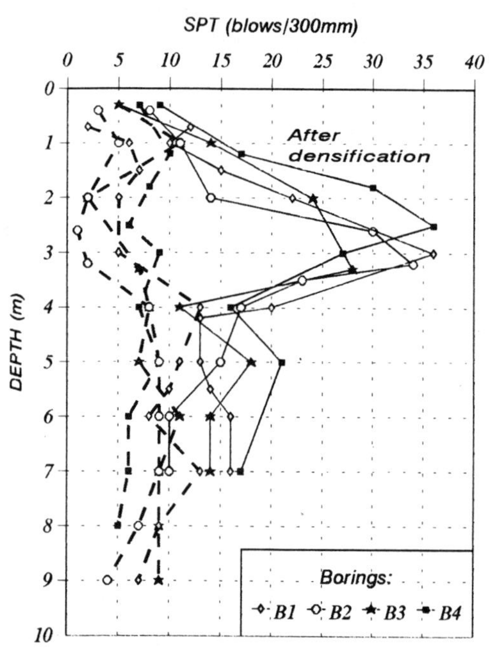

FIG. 4. SPT Tests before and after Sand Densification

It is generally difficult to make an evaluation of design parameters required for settlement predictions on improved soils. This fact can be explained by the complex stress state in the soil, which is radically modified by the construction operation. Such operations may completely destroy the natural soil structure. In this case, field test data, including the blow count N_{SPT}, were used to derive representative parameters of the soil.

SETTLEMENT PREDICTIONS

Settlement predictions were made by the finite element method using a *Building System* computer program for the structural design of tall buildings. This program was developed by the Civil Engineering Department of Federal University of Pernambuco, Brazil (Fonte 1992). A complete three-dimensional frame model was adopted for the structure considering the hypothesis of the rigid diaphragm for the slabs.

Appropriate elements for the simulation of the soil-structure interaction were incorporated in the program based on the Poulos formulation (Poulos 1975). The technique of sequential loading application was used to account for the effects of construction on the mechanical behavior of the structure-foundation-soil system (Soriano and Fonte 1989). The simulation of construction sequence and the gradual application of loads allowed for a more realistic model, leading to a more appropriate analysis of the effect of the structural stiffness on the settlement distribution.

In this way, the elastic settlement caused by dead loads were predicted by four models. The first three models used the finite element method: in Model 1, the adopted construction sequence was one story at a time; while in Model 2, this sequence was two stories at a time. In Model 3, the structure was admitted with instantaneous loading. In the last, Model 4, the current empirical methods were used to estimate settlements from N_{SPT} values (Parry 1971; Shultze and Sherif 1973; Parry 1978; Burland and Burbidge 1984).

The results of the elastic settlement predictions are summarized in Table 1 for different models. It may be seen that the variation among values predicted by the finite element models is small, while the average results from the empirical methods are significantly different. These results are presented in Fig. 5 along the C13, C12, C11, C7, C8, C9, C1, C2, C5, C14, C16 rectified line columns.

MEASURED PERFORMANCE

A program of settlement measurement was elaborated in order to monitoring the allowable movements and to check the settlement predictions. Settlements were measured at eleven column locations (Fig. 3). Stainless steel pins were incorporated into the structure, as the ground-floor building was constructed. Two benchmarks were used to support the levelling system. These were located outside the perimeter of the building influence. Details of the instrumentation design and techniques follows previous buildings controls (Jucá 1987). Measurements were made using a precise levels with parallel-plate optical micrometers. The initial readings were made after concrete in the ground-floor was set and access to column points could be secured. The frequency of reading was dictated by the construction sequence. In general, for each two additional floor levels constructed, one levelling was made.

Results of the settlement readings are presented in Fig. 6. Its values were related to the increase of estimated load on the foundation as construction proceeded. The small settlements measured were in accordance with the good performance observed on the structure during the period of control. The rate of settlements measured was closely related to superstructure load increases during the construction period. The results of this are presented in Fig. 7, where a good agreement between settlement velocities and construction sequence can be seen. Differential settlements and the frame relative rotation were analyzed in order to check limiting movements and damage possibility.

TABLE 1. Summary of Settlement Predictions

Columns	Settlement Predictions (mm)			
	Finite Element Method			Empirical Method
	Model 1	Model 2	Model 3	Model 4
C1	10.7	10.8	11.9	8.6
C2	20.3	20.3	20.0	21.0
C5	21.1	21.1	21.0	20.9
C7	10.6	10.7	11.4	8.8
C8	26.0	25.9	24.8	28.4
C9	21.4	21.4	21.1	22.2
C10	26.0	25.8	24.5	29.9
C11	10.5	10.6	10.9	9.2
C12	12.9	12.8	12.8	13.4
C13	14.2	14.3	14.4	13.6
C14	25.7	25.7	25.1	28.6
C16	17.4	17.5	18.3	13.9

Analysis of Results

The settlements measured were compared with the predictions from the methods previously mentioned. Fig. 8 shows the settlements predicted using Models 1 to 4, and the measured along the C7, C8, C9, C1 rectified line columns, when the building superstructure was concluded. The interaction models show smaller differential settlements. This indicates a transference of load from the intermediate columns (C8 and C9) to the exterior columns (C7 and C1). The average values were higher than estimated. However, relatively good conformance with the predicted curve is apparent. This small difference can be attributed, in part, to the consolidation of the clayey soil underlying the sandy layer.

Fig. 9 shows the differential settlement between the columns C9, C8, C7, and C1, for the four models mentioned above and the settlements measured when the building superstructure was concluded. Model 4, which does not consider the interaction effects, represents the upper settlement values. Model 3 (with interaction and instantaneous loading) shows the smallest values, while Models 1 and 2 show intermediate results. In situ, the maximum differential settlement was about 12 mm.

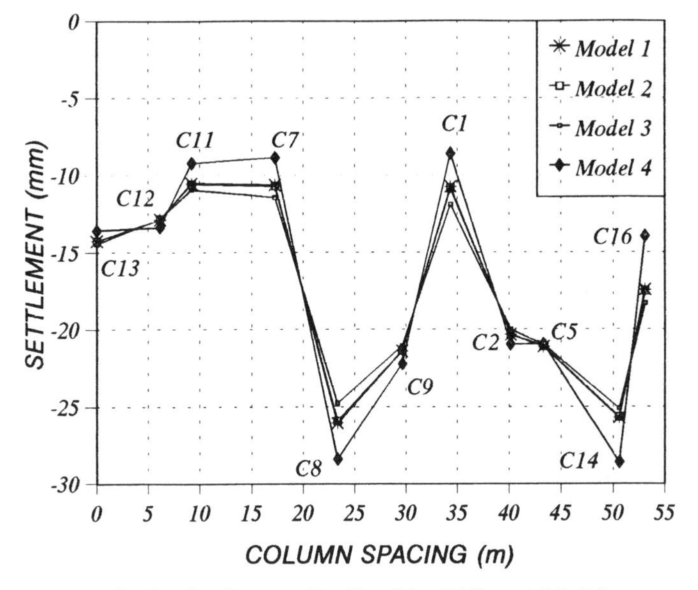

FIG. 5. Settlements Predicted by Different Models

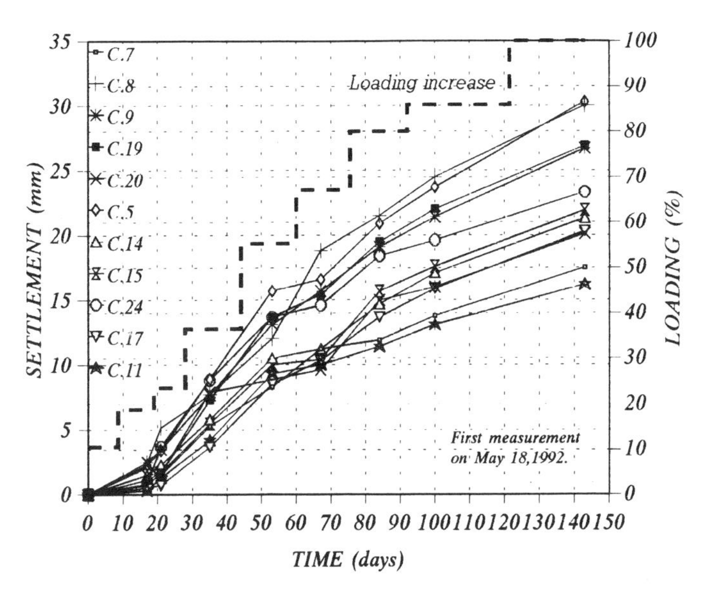

FIG. 6. Measured Settlements

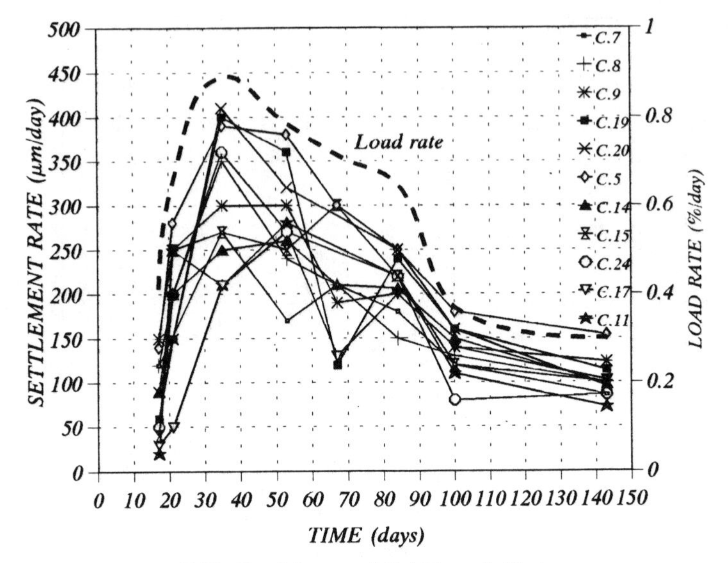

FIG. 7. Measured Settlements Rate

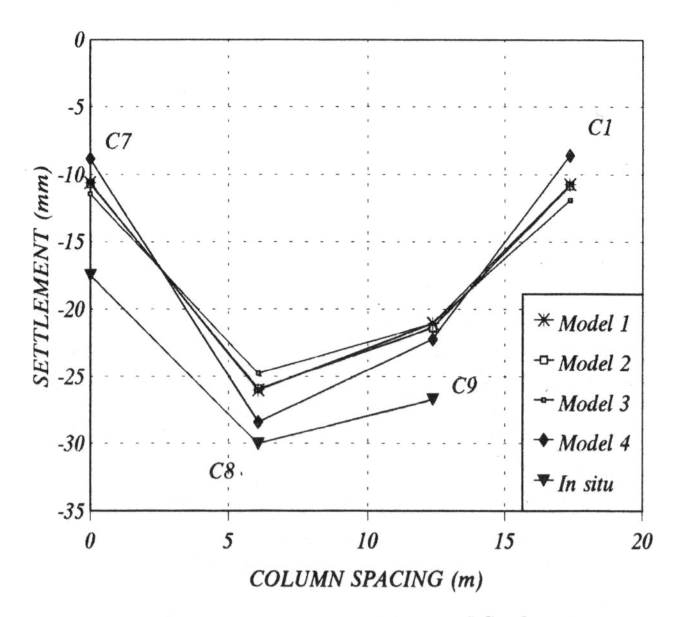

FIG. 8. Predicted and Measured Settlements

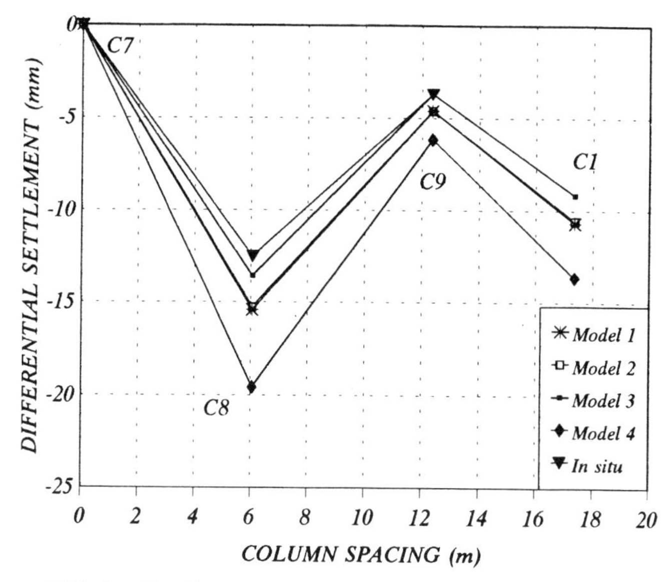

FIG. 9. Predicted and Measured Differential Settlements

The maximum difference between the settlement analyzed with and without interaction models was 38%, which indicates the importance of considering the soil-structure interaction. On the other hand, the influence of the sequence of the loading application reached a maximum of 10.8%. This influence is observed by the difference between Model 3 and Models 1 and 2 of the estimated settlements.

Some angular distortions calculated by the models without interaction presented values over that which is allowable. However, for the models with interaction, all the values were smaller than those allowable for this type of structure. These values were close to the angular distortions calculated from measured settlement data.

In order to determine an indication of the elastic parameters of the improved sand, the settlements measured are plotted against the estimated loading increases for all footings of the structure (Fig. 10). As may be expected in this type of correlation, there is considerable scatter of results. The reasonable and useful curves, however, were obtained for these site conditions. The relationship between settlement and increased load was practically linear for each column. The estimated soil modulus varies from 35 MPa to 50 MPa. These values are considerably lower than the field tests would indicate for the upper layers. This fact could be due to compressibility of the clay underlie the improved sand, which reduces the stiffness of the subsoil. Despite of this, the settlements measured presented a relatively low values in accordance with improved sandy soils.

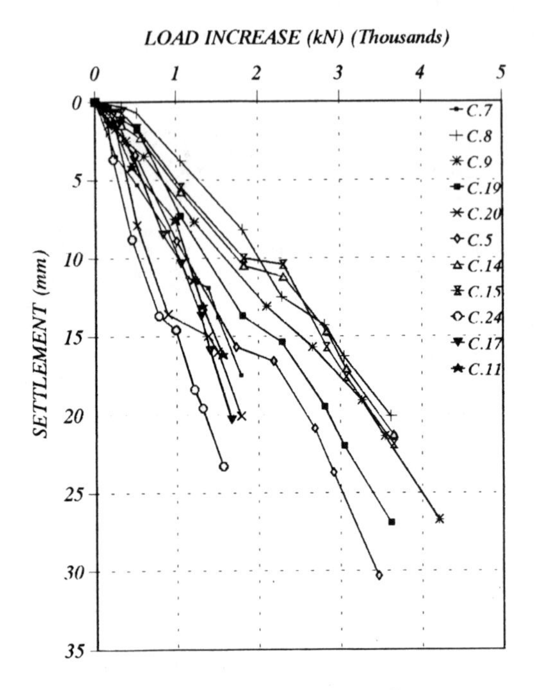

FIG. 10. Load-Settlement Curves

CONCLUSIONS

Measured performance shows that the treatment of the loose sand was effective in order to reduce the settlements of the building. Results of settlement observations could be used to obtain full scale soil parameters.

In relation to the settlements predictions, it can be affirmed that, theoretically the model for instantaneous loading without soil-structure interaction (Model 4) overestimates the differential settlements. The instantaneous loading model, with soil-structure interaction, underestimates those settlements due to theoretical considerations increasing the stiffness of structure more than its real value. Accurate results were given by Models 1 and 2, which consider both the interaction effect and the gradual application of loads simulating the construction sequence.

In the results obtained in this paper, it has been demonstrated that the alternative design, technique of settlement prediction, and the construction quality control were very appropriate in construction of tall buildings.

ACKNOWLEDGMENT

A majority of the data described in this paper was given by the Construction and Quality Control Department of the Oliveira Maciel Group and Engedata Structural Engineering.

APPENDIX - REFERENCES

Burland, J. B., and Burbidge, M. C. (1984). "Settlement of foundations on sand and gravel." *Papers Invited Lecturers Centenary Celebrations*, Glasgow and West Scotland Assoc. of ICE, 5-66.

Fonte, A. O. C. (1992). "Geometric nonlinear analysis of tall buildings," Doctor of Science Thesis, COPPE/UFRJ, Rio de Janeiro, Brazil (in Portuguese).

Ichimoto, E., and Suematsu, N. (1982). "Sand compaction pile (the Compozer method)." *Proc. of Symp. on Recent Developments in Ground Improvement Techniques*, Bangkok, 71-78.

Jucá, J. F. T. (1987). "Effects of braced excavation on nearby building settlements." *Ingeniería Civil*, 64, 19-37, Madrid (in Spanish).

Murayama, S. (1958). "Compaction of soil and compaction machinery." *Proc. of Meeting of Kansai Branch of Japanese Society of Civil Engineering*, 30-65.

Nakayama, J., Ichimoto, E., Kamada, H., and Taguchi, S. (1973). "On the stabilization characteristic of sand compaction piles." *Soils and Foundation*, 13(3), 61-68.

Parry, R. H. G. (1971). "A Direct method of estimating settlements in sand from SPT values." *Proc. Symp. on Interaction of Structures and Foundation*, Birmingham, 29-32.

Parry, R. H. G. (1978). "Estimating foundation settlement in sand from plate bearing tests." *Géotechnique*, 28(1), 107-118.

Poulos, H. G. (1975). "Settlement analysis of structural foundation systems." *Proc. IV Southeast Asian Conf. on Soil Engineering*, Kuala Lumpur, IV, 52-62.

Shultze, E., and Sherif, G. (1973). "Prediction of settlements from evaluated settlement observations for sand." *Proc. 8th ICSMFE*, Moscow, 1, 225-230.

Soriano, H. L., and Fonte, A. O. C. (1989). "Incremental construction effects in design of tall buildings." *Proc. X Ibero-latino American Congress on Engineering Computer Methods (MECON)*, Lisbon, 3, A999 (in Portuguese).

Wallays, M. (1982). "Deep compaction by casing driving." *Proc. of Symp. on Recent Developments in Ground Improvement Techniques*, Bangkok, Thailand, 39-51.

SOIL IMPROVEMENT TO MITIGATE SETTLEMENTS UNDER EXISTING STRUCTURES

H. R. Al-Alusi,[1] Member, ASCE

ABSTRACT: Settlements of sites with existing structures are more difficult to mitigate than those of sites without structures. Equipment access, work area, noise, dust, vibrations, and cost are amplified and become more critical.

Following are three case histories involving mitigation of settlements under three different types of structures.

The first case, an office building in the San Francisco Bay Area, involved soil densification under piles to mitigate further settlements. The compaction grout densification process was extended beyond the bottoms of the piles to treat fill materials under the footprint of the building. Additional lense grout reinforcement was required to reinforce the hillside soils to reduce downward movements. Five years after completion of remediation work, the site showed no detectable movement.

The second case concerned a maintenance facility at the June Lakes Ski Resort in the Sierra Mountains, where a structure had been built on top of a fill that was underlain by a layer of gravel and cobbles. Within a year of construction, signs of structural distress were evident. Geotechnical investigations revealed that settlements were caused by at least two factors; the downward migration of the upper fill layer into the large pores of the lower layer, and the possible densification of the upper fill under its own weight. The remedial work consisted of providing a barrier between the two layers to allow for an effective compaction grout densification effort of the upper fill layer and to prevent further migrations into the

[1] President, Pressure Grout Co., 1975 National Avenue, Hayward, CA 94545-1709.

gravel and cobbles layer. No structural distress or any movement has been detected since the remedial work was completed six years ago.

The third case presents the treatment of the old and new footings of the Rose Bowl Stadium in Southern California. A permeation grouting system was selected, designed, and implemented to solidify zones of the sand-gravel-cobbles mixture of the foundation soils to act as pedestals for underpinning the old footings and supporting the new ones.

INTRODUCTION

Settlement of structures can be caused by a number of factors. These factors include the settlement of the soil caused by its own weight, loads applied by and through the structure, vibrations, change in groundwater levels or other less known factors such as plant root moisture extraction, erosion of a soil layer into a coarser particle layer, chemical reactions, thermal exchange, mineral dissolution, underground erosion due to migration of smaller soil particles caused by groundwater gradients, and many others.

Loads applied by and through a structure may include its own dead load, live loads, wind, seismic, impact and other functional loads. A frequently encountered settlement problem is the inadequacy of soil density/strength resulting in soils consolidating or compacting under their own weight. A soil improvement can be affected by simply densifying the soil mass in-situ without removing the soil or affecting the structure.

Mitigation of soil settlement under existing structures by in-situ pseudo-static densification has been used for more than forty years in the U.S.A. These solutions are achieved by compaction grouting (further detailed in case history No. 1). Other lesser known methods include soil solidification, soil reinforcement, soil sealing, and other methods of soil treatment. Each one of these approaches has several critical details that demand the engineer's and contractor's full attention to achieve successful completion. The three cases presented in this paper represent soil improvements to mitigate settlements caused by several factors. Each case involves an existing structure where on-going settlements needed to be halted.

CASE NO. 1

A two-story office building, measuring 24.4 × 76.3 meters, exhibited continuous settlements within five years of construction completion. When the differential settlement reached 100 mm it became evident that a remedial work program was necessary.

The site was resting on a two-stage graded fill (Fig. 1). Fill thicknesses (wedges) of less than 1.5 meters and up to 6.1 meters underlaid the footprint of the subject structure. Upon completion of construction the longest side of the building was parallel to a heavily vegetated slope of about 1:1, with a height of 4.8 to 6.1 meters. The building was resting on drilled piles of varying depths from 2.8 to 4.9

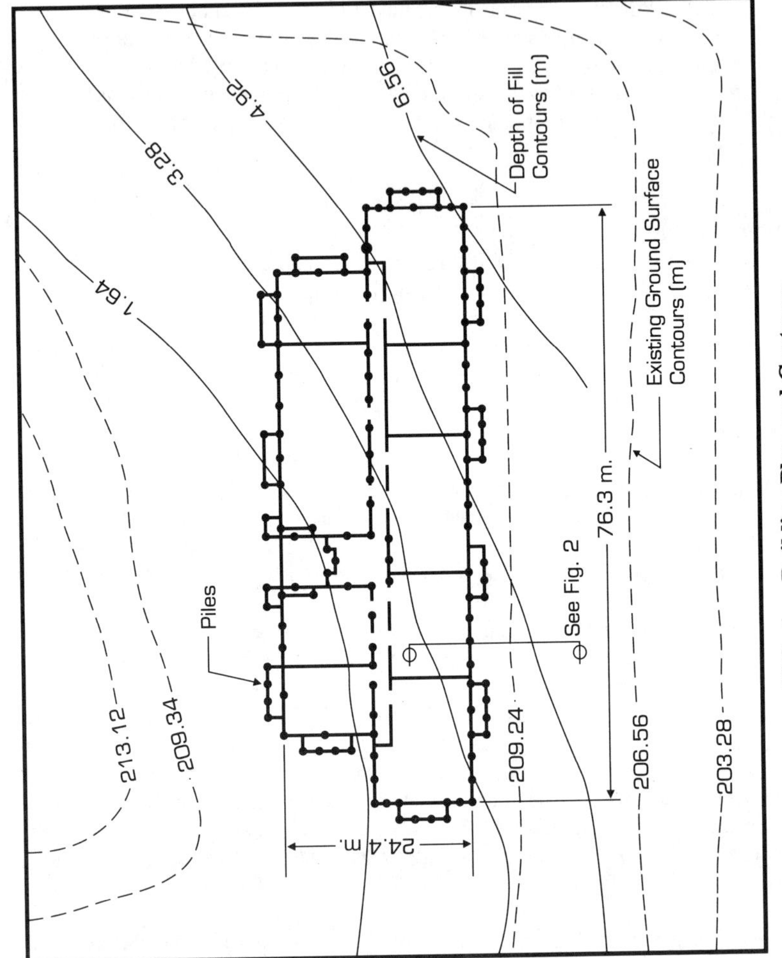

FIG. 1. Building Plan and Contours

meters and with diameters of 0.46 to 0.61 meters. The piles were connected by grade beams, with the floor slabs doweled to the beams.

Twenty years before completion of the building, rough grading had been completed; the final grading and building construction were completed about twelve years before this remedial work started.

Settlement monuments on the grade beams showed a maximum differential settlement of 100 mm across the building, (Fig. 1). Observation of the soil surface in comparison to the grade beam showed a difference of an additional 200 mm of soil movement downward relative to the grade beams. Slabs were exhibiting sagging of up to 70 mm between the grade beams.

Continuous monitoring showed that there were two types of movement. The first was downward, which was attributed to the compaction and consolidation of the fill and native soils. This movement was detected by settlement monuments and the generation of voids below slabs. The second was a hillside creep movement caused by the seasonal drying and wetting of the near surface soils of the slope. A typical soil profile of this site is shown in Fig. 2. The pile settlements were attributed to the additional loading imposed by the negative skin friction generated by the downward movement of the fill materials.

Approach Concept

A number of solutions were considered, among them were:

a. Removal and preservation of building, excavating and re-compacting soils, and resetting building back on same location.
b. Re-supporting building on additional and deeper piles.
c. Improvement of soils by in-situ densification and mitigation of downhill soil movement, by soil reinforcement.

Solution (a) was quickly discarded because of its prohibitive cost and time requirements. Solution (b) was estimated to be many times higher in cost and time requirements than Solution (c).

The remedial work consisted of two major items, namely:

1. Densification of soils below the bottom of the piles and under the rest of the building using compaction grouting, and
2. In-situ soil reinforcement using deep lense grouting under the hillside area.

Fig. 2 represents a conceptual sketch of the remedial work undertaken for this building.

Compaction grouting is the injection of a highly viscous sand-cement mixture designed to volumetrically displace and densify the soils around the point of injection. Compaction grout by definition (Committee on Grouting 1980) is a grout with 50 mm or less slump per ASTM C143-78. Grout materials, pressures and rate of injection were designed to prevent the permeation of the grout into the soil mass and to prevent the fracturing of the soil itself. The strength of the grout material is irrelevant in the compaction grouting process. The amount of densification and the extent of the densification process are the crucial elements in this operation.

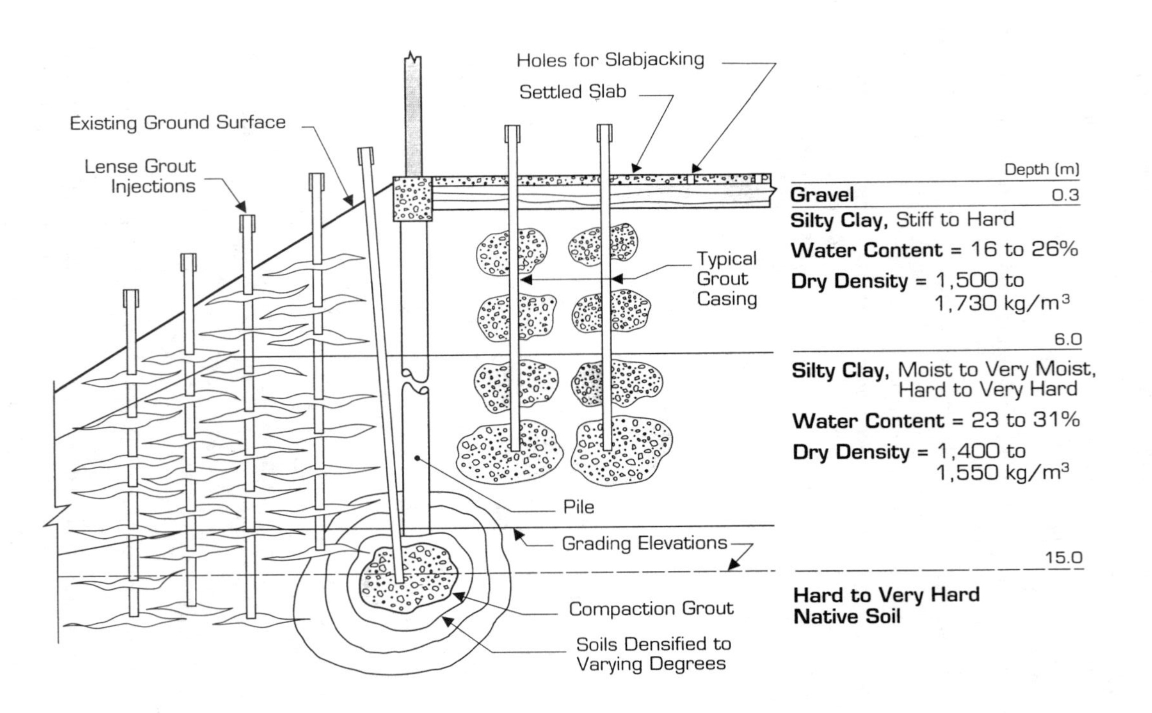

FIG. 2. Cross-Section through Building Foundation and Slope Area (for Location See Fig. 1)

Soils closest to the grout bulb will exhibit highest densification with a diminishing effect away from the point of injection.

Compaction Grout Materials

Materials used in compaction grouting have a wide range of properties. Theoretically, any material that will not permeate, spread, or fracture the soil when injected is an acceptable material. For cost considerations, local materials for a given project site are usually given first priority. Additives can be used to improve the grout material pumpability. As an additive, portland cement is widely used with sufficient water to effect a workable mix. The use of cement is strictly for the workability and pumpability of the material and does not affect the required degree of densification. Compaction grout materials with no cement content or other additives have been reported (Stoker and Wardwell 1987). A set of particle size distributions of materials used in compaction grouting, compiled by the author, is given in Fig. 3. A cement content of five to fifteen percent has been used with these materials.

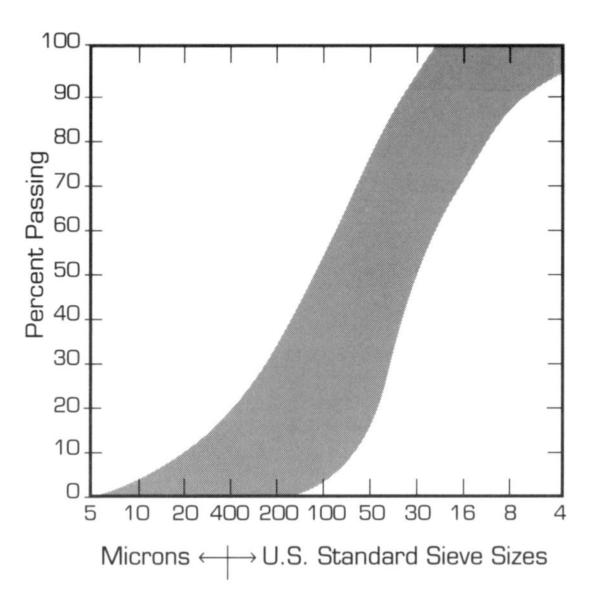

FIG. 3. Gradation for Sand Materials for Compaction Grouting

Soil Densification

The remedial work included only a portion of the building as seen in Fig. 2. This portion represents approximately two-thirds of the total area of the building.

Based on the available settlement records, 63 piles were found to be in need of re-support. A single injection point was used for each pile. Injection points were designed so that the tip of each grout casing was between the center of the pile bottom and the top of the competent soil layer. At each injection point grouting continued until pile upheave or a maximum grout pressure of 4100 kPa at the point of injection was detected.

The aim of this portion of the treatment was to create a grout bulb (footing) as large as possible under the pile until refusal criteria, as given above, were met. The grout take was largely dependent upon the consistency of the material below the bottom of the pile and the distance between it and the competent soil layer below it. Grout takes ranged between 0.17 to 4.73 cubic meters with an effective spherical bulb diameter of 0.67 to 2.11 meters.

For the remaining soil mass (Fig. 2), a grid pattern with a spacing of 1.83 × 1.83 meters was established. The sequence of injections was designed to first create a confinement of the soil mass to be densified, then to proceed with the remaining densification process. Each injection point was driven to the target depth. Grout extrusion started in stages of 0.61 meters in the vertical direction without stopping until a maximum pressure of 4100 kPa was reached or a ground upheave was detected. A total grout take of five to seven percent of the volume of the treated soil was accomplished resulting in four to eight percent increase of the soil dry density.

Soil Reinforcement

The deep soil reinforcement included injections of grout lenses to a maximum depth of 11 meters. Lenses were installed at 0.31 meters intervals vertically. Each lense was designed to fracture the soil and install grout to create a lense of 3 meters in diameter with a thickness of 3 to 6 mm. Injections were installed in a grid of 1.83 × 1.83 meters, Fig. 2. The over-lapping of these lenses provided a continuation of the reinforcement to resist the small but on-going creep movement. A slurry grout mix of a water/cement of 2 was used together with additives, as needed, to provide for the pumpability of grout and to facilitate fracture initiation.

The mechanism of soil reinforcement is based on the friction/bond between the hardened grout lense and the soil, much the same as metal strips in reinforced earth applications (Tabbal 1983).

Performance

No detectable settlements have been observed in the four years since completion of the work.

The hillside showed minor movement for a few months after work completion, but even those movements were greatly reduced to hardly detectable amounts since then.

CASE NO. 2

This case presents remediation of a condition of downward migration of a finer grained fill soil into a layer of gravel and cobbles. A concrete-block building of 18.3 × 39.7 meters with a slab-on-grade floor was constructed in 1986. A cut-fill approach was used to create the original level pad. The gravel and cobble layer was covered with additional fill of silty sand 4.58 meters thick, Fig. 4. Within a year after construction of this building, cracks in the walls and the concrete slab appeared. A maximum differential settlement of more than 76 mm across the building was measured before the initiation of the remedial work. A trench excavated adjacent to one of the footings, just before undertaking the remedial work, revealed a substantial void between the bottom of the footing and the soils below it. Four borings drilled around the building showed evidence of extensive intrusion of the silty sand layer into the gravel and cobble layer.

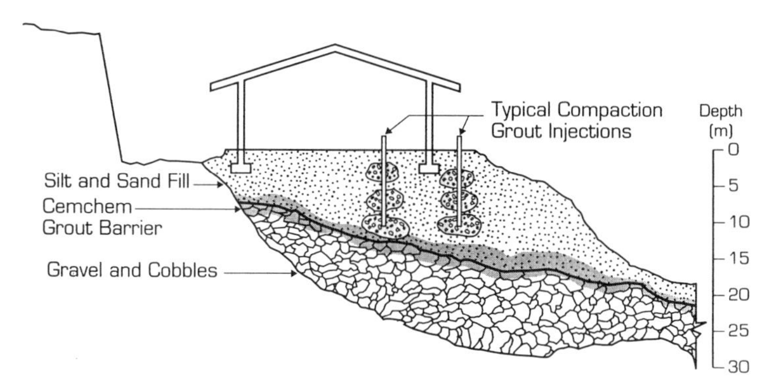

FIG. 4. Section through Foundation Soils Showing Remedial Work

Approach Concept

The lateral and vertical extension of the gravel and cobbles zone, coupled with its very high permeability, necessitated the installation of a grout blanket (barrier) just under the silty sand fill to block the silt and sand intrusion into the gravel and cobbles layer. Cemchem, a controlled fast-gel grout, was selected for this situation. This proprietary system can be controlled to set between twenty seconds and one hour after mixing. With proper mix design, length of grout pipe, depth of soils to be injected and equipment arrangement, the grout can be designed to set within a few seconds after it leaves the tip of the grout pipe.

The remedial work consisted of installing this barrier, then densifying the soils above it using compaction grouting. This was followed by void filling, and structural lifting using compaction grouting techniques to lift and level the structure and slabs.

Remedial Work

Using CemChem grout a blanket with a nominal thickness of 0.3 to 0.6 meters was established in the areas that needed it. By probing in a grid of 0.92 × 1.22 meters, it was first determined whether or not the sand intrusion had reached a point where it had already established a barrier. If the grout permeated the soil, it was assumed that a barrier had not yet been created. If, on the other hand, it did not permeate the soil, the assumption was that the sand had already created a barrier within the gravel and cobble layer, thus no further work would be required in the vicinity of that injection point. The establishment of the blanket required more than 30,300 liters of CemChem grout. This procedure was then followed by compaction grouting to densify the loose soils above the barrier. A grid of 3.66 × 4.27 meters for casing injections was first driven and pumped, followed by splitting this spacing into 1.83 × 2.14 meters. Total compaction grout injected was about 140 cubic meters, which resulted in an improvement of the bulk density of the treated soils of between 15 and 21 percent. The compaction grout was injected in stages of 0.3 to 0.6 meters starting at the top of the blanket and moving upwards to the bottom of the floor slabs or footings. Grout injections were terminated when the grout pressure reached 3445 kPa or the surface lifted to an unacceptable level.

Grout Materials

For the CemChem system, portland cement is the base material. Portland cement Types I, II or V have been successfully used together with bentonite and additives to produce the required gel time.

For the compaction grouting, locally available silty sand was used. This material was found to satisfy the criteria given in Fig. 3. Portland cement was added at the rate of ten percent by weight of grout.

Performance

More than six years after undertaking this remedial action, no distress or movement has been reported.

CASE NO. 3

Loose to medium granular soils undergo volumetric changes (settlements) under additionally imposed loads, vibrations, and seismic activities. If such soils contain larger particles of gravel and boulders in a heterogeneous formation, settlement predictions become highly complicated.

This case history involves the 1915 Rose Bowl Stadium in Southern California (listed as a historical monument). The stadium was undergoing an expansion project involving the press box and new executive suites which resulted in additional loading on the footings. Portions of the new expansion would be supported by some of the old stadium footings and others by the new footings. It was determined that the old footings themselves rested on loose uncompacted fill, making it impossible to underpin the old stadium footings without damage to the structure unless and until the soils below these footings were given additional support.

Approach Concept, Remedial Work and Materials

The solution to this condition was through a permanent solidification system using permeation grouting with chemicals. Injections designed to create "solidified pedestals" of about 1.22 meters in diameter were used. For permanency, strength, and environmental considerations, an ultrafine cement grout was selected. Before finalizing the designs, a pilot test program was undertaken at the subject site. The results revealed satisfactory grout permeation into the soils with an unconfined compressive strength exceeding 1380 kPa. The geotechnical design proceeded with 1,000 pedestals (injections) under the old and new footings.

The site soils immediately below the footings had a gradation that ranged between silt particles and cobbles. Less than five percent of the particles passed the 200 U.S. Standard Sieve (0.075 mm) while the largest particles were up to 100 mm. In ultrafine cement more than 80 percent of its particles are smaller than 6 microns. A water/cement ratio of 4:1 was used. Each injection required 170.5 liters. Nominal pressure used for these injections was 345 kPa.

Performance

The program proved to be successful in terms of being able to affect the required solidification. More than a year has passed since the completion of this work. Full loads have been imposed with no signs of any settlement. It is fair to assume that the designs and remedial work will perform successfully based on the excellent grout take that was recorded at the site and the strength of the obtained samples.

CONCLUSIONS

Mitigation of settlements of existing structures involves stringent requirements to satisfy the site, soil, and structural specifics. The three case histories presented in this paper show how such specialized methods can be used to halt the settlement of structures in a cost-effective way. Replacing structures, re-excavating or re-supporting existing structures on piles are not the only solutions available to the geotechnical engineer today.

APPENDIX - REFERENCES

Committee on Grouting of the Geotechnical Engineering Division (1980). "Preliminary glossary of terms relating to grouting." *J. Geotech. Eng. Div.*, Proc. ASCE, 106(GT7), 803-815.

Stoker, G. G. and Wardwell, S. R. (1987). "Compaction grouting of the Phoenix drain tunnels." *Proc. 1987 Rapid Excavation and Tunneling Conf.*, New Orleans, 1, 575-582.

Tabbal, M. A. (1983). "The study of cement grout reinforcement in slopes of soft clay," Engineer thesis, Department of Civil Engineering, Stanford University, Stanford, California.

A COMPARISON OF THREE SOIL IMPROVEMENT TECHNIQUES TO DENSIFY LIQUEFACTION SUSCEPTIBLE SANDS

Alec D. Smith,[1] and Edward B. Kinner,[2] Members, ASCE

ABSTRACT: Deep soil densification programs using vibratory probe compaction, vibro-replacement and terraprobe methods have been completed on a naval submarine base to limit potential settlements associated with densification of loose fine sands, as a result of seismically-induced vibrations and/or heavy vehicular traffic. The surficial 6.1 to 9.1 m (20 to 30 ft) of saturated, very loose to medium dense fine sands (relative density, D_r = 45 to 70%) were densified to a minimum criterion (D_r = 65%) to limit settlements due to traffic vibrations and the potential for liquefaction. Using a performance specification, the field program involved constructing test areas using each improvement technique with variations of critical parameters (probe spacing, vibration duration, etc.) to demonstrate the relative effectiveness of the proposed methods. Standard Penetration tests (SPT) and cone penetration tests (CPT) were performed before and after densification in both the test areas and the final production areas to assess compliance with the design criteria. The final production areas were completed using vibratory probe compaction and vibro-replacement methods. The relative performance of the improvement techniques is compared.

INTRODUCTION

The project site is located in coastal Georgia on an existing submarine base (see Fig. 1). The project consisted of the design and construction of two new wharf

---

[1] Vice President, Haley & Aldrich, Inc., 58 Charles Street, Cambridge, MA 02141.
[2] Executive Vice President, Haley & Aldrich, Inc., 58 Charles Street, Cambridge, MA 02141.

structures with associated access roads, completed in 1987 and 1993, for the purpose of loading submarines. The missile transporters that will use the roads have 10 axles with axle loads of 30,844 kg (68 kips). This paper addresses the soil improvement programs for the road construction which involved a total of about 1340 m (4400 ft) of roadway 13.4 m (44 ft) wide.

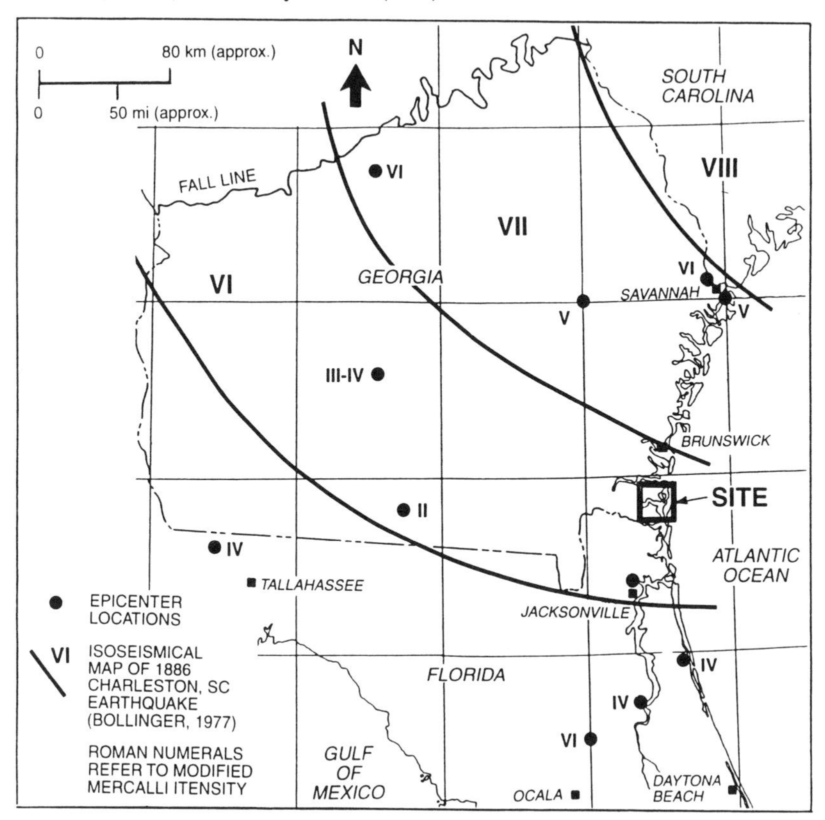

FIG. 1. Site Location and Seismic Activity Summary (after Gee and Jenson, Inc. 1980) (after Bollinger et al.)

SUBSURFACE CONDITIONS

The proposed access roads were to be built for the most part near existing grade (approximately Elev. +3.7 m) with approach embankments in the wharf areas. A subsurface exploration program consisting of 21 SPT borings, spaced about 46 m (150 ft) apart along the road centerline and extending to depths of about 15 m (48 ft), was performed.

The observed surficial stratigraphy consisted of surficial fill/topsoil underlain by sand. The thickness of the sand varied from about 6 m (20 ft) to an observed maximum of greater than 12.3 m (40.5 ft). The density of the sand, based on the SPT "N" values, ranged from very loose to medium dense. Although rather erratic, the sand stratum typically had an upper and lower medium dense sub-stratum. The upper medium dense sand was slightly cemented. Very loose to loose sands were encountered in many of the test borings between Elev. -1.5 m and Elev. -4.6 m, and erratically at other depths. The observed groundwater level varied between Elev. 1.5 m and Elev. 2.4 m (about 2.1 to 1.5 m (7 to 5 ft) below the proposed roadway profile). The general stratigraphy and observed SPT "N" values are summarized in Fig. 2. Although considered to be preferable to SPTs, CPTs were not performed during the design period since rather extensive SPT/CPT work had been completed elsewhere on the base.

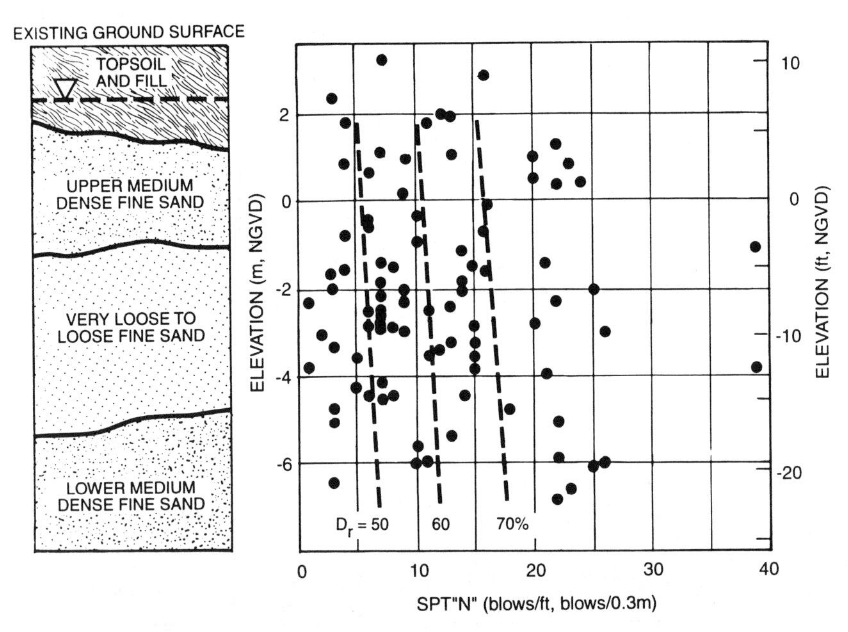

FIG. 2. Summary of Subsurface Conditions

Laboratory tests (Atterberg limits, gradation tests and maximum/minimum density (ASTM D4253 and D4254)) were performed. Based on these analyses, the sands are classified as SP or SP-SM using USCS. The gradation results, summarized in Fig. 3, indicated that the sands had gradations similar to those that liquefied during earthquakes in Japan (Seed and Idriss 1967; Kishida 1970). The densities observed were very uniform ranging from maximums of 1581 to 1651 kg/m^3 (98.7 to 103.1 pcf) to minimums of 1288 to 1342 kg/m^3 (80.4 to 83.8 pcf).

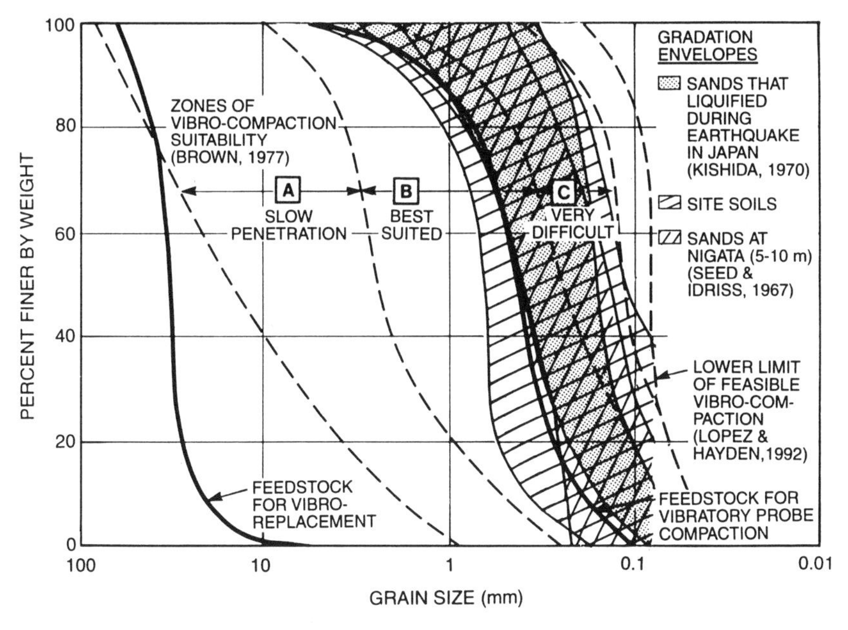

FIG. 3. Summary of Gradation Test Results

Project Design Requirements

In order to maintain missile transporter access to the wharf at all times one design requirement was to limit potential total and differential settlements of the roadway to those tolerable for the transporters in the event of an earthquake.

Gee and Jenson, Inc. (1980) included a review of the earthquake history of the project area. The study concluded that the maximum intensity of earthquake shaking in the recorded history of the area resulted from the 1886 Charleston, SC earthquake. The maximum Modified Mercalli Intensity that the site experienced as a result of that earthquake is VI to VII, corresponding to ground accelerations of approximately 0.12 g to 0.16 g (see Fig. 1).

The layer of loose to medium dense fine sand encountered in the wharf access road area was indicated to have a D_r of 45 to 55%, using correlations presented by Marcuson and Bieganousky (1977) (see Fig. 2). The Marcuson and Bieganousky methods were used, rather than others (Gibbs and Holtz 1957; Bazaraa 1967), because they were developed using sands with similar gradation to the site soils.

The layer was expected to undergo considerable compression due to repeated loading from the missile transporter and from seismically-induced settlements or liquefaction. The loose sands were considered to be susceptible to liquefaction at levels of ground shaking somewhat less than the site is thought to have experienced during the 1886 Charleston, SC earthquake. It was judged that roads and embankments could experience total and differential settlements of several inches to more than a half foot if this layer were to undergo such liquefaction.

Accumulated compression of the deep, loose fine sands under multiple passes of the missile transporters was estimated to amount to approximately 0.05 m (2 in.) of total settlement across the roadway site, with up to 0.025 m (1 in.) differential settlement over the length of the transporter. The amount of settlement would vary across the site depending on the thickness of the loose layer and the degree of looseness. Settlement estimates were based, in part, on methods proposed by Schmertmann (1977).

It was therefore recommended that the entire layer of loose fine sands be densified to $D_r > 65\%$ prior to construction of the wharf access roads. This degree of densification would minimize the risk of liquefaction of the loose sands due to earthquakes producing Modified Mercalli intensities of VI to VII site ground shaking. It was predicted to also limit post construction total settlement due to repeated static transporter loading to less than 0.025 m (1 in.).

Previously completed densification test programs in the area (Hussin and Ali 1987) demonstrated that the required densification could be accomplished by any of several available methods including: deep dynamic compaction, vibratory probe compaction (including vibro-replacement), or compaction grouting. However, because deep dynamic compaction may result in significant vibrations to existing structures, the Navy directed that it not be used.

In vibratory probe compaction (VPC), a cylindrical probe with rotating eccentric weights penetrates into the ground and densifies the surrounding soils. The resulting hole is filled with sand and gravel or stone which is similarly densified as the probe is withdrawn. If stone is added, the method is commonly referred to as vibro-replacement (VR). With compaction grouting, a cement grout is injected into the soil under pressure. The resulting soil displacement by the grout causes densification.

All three methods have been used at the base (Hussin and Ali 1987). Reportedly the success of each method has depended in part on the density and degree of cementation of the upper layer of sand that overlies the loose zone and on the silt content and depth of the loose sand. Dense and cemented layers of sand above the loose sands, such as occur erratically along the wharf access roads, reportedly have hampered densification in a depth interval immediately below the cemented zone, when the VPC procedure was used. However, compaction grouting of this interval has apparently been successful in achieving required densification. In other instances, only marginal densification has been accomplished by VPC where the silt contents have exceeded about 15%. The wharf access road borings encountered sands with generally less than 10% finer than a No. 200 sieve.

CONTRACT DOCUMENTS

Contract documents were developed using a performance specification approach to the soil densification, and the Navy process of Contractor Quality Control (CQC). With the CQC approach, the Contractor provides specified data/reports demonstrating compliance with the specifications, and the designers have a limited field presence. As a result, the level of detail in the field data varied between the two separate construction contracts.

Both wharf projects were bid as lump sums. Therefore, costs for the soil improvement programs were not identified.

The completed contract documents are summarized as follows:

1. Minimum density - overall average $D_r \geq 65\%$ full depth for each CPT plan location, and the average $D_r \geq 60\%$ for any 3 ft depth interval of each CPT location.
2. Limits of densification - from ground surface to Elev. -4.6 m to Elev. -7.3 m (depending on location) within the roadway limits and continuing outward 3.7 m (12 ft) beyond the edge of pavement.
3. Methods of Densification - either VPC (including VR) and/or compaction grouting with the means and methods left to the Contractor.
4. Test Areas - two test areas 279 m^2 (3,000 ft^2) each at least 244 m (800 ft) apart for each proposed method prior to commencement of production work to demonstrate the ability of the method; required to perform CPT tests (a min. of 3 locations prior and 3 after for each test area) to demonstrate the method is producing the required minimum criteria. Specific criteria relating cone tip resistance to relative density, based on Schmertmann (1977), were provided.
5. Vibration Monitoring - densification procedures to be modified as needed to limit ground vibrations to no greater than 48 mm/s (1.9 in./s) on or adjacent to any utility or structure, as measured with a 3-direction seismograph.
6. Protection of Existing Utilities - prior to compaction within 15.2 m (50 ft) of any operating utility, excavate to expose the utility lines and temporarily support the utility to avoid settlement or lateral displacement; monitor utility lines for movement during deep soil densification.
7. Additional general concerns were identified during the design and addressed in the geotechnical design reports. The deep soil densification, depending on the method, was estimated to cause ground settlement as a direct result of the densification of the underlying soils. Based on volumetric strains due to increases in relative density, settlement caused by VPC was expected to average about 0.15 m (6 in.) with localized settlements of up to 0.3 m (1 ft). The settlements were anticipated to be non-uniform and would depend on local variations in soil gradation and initial density, as well as specific compaction procedures.
8. VPC and VR could result in considerable amounts of surface water, primarily from operations of the probe. The Contractor was required to control surface water. Densification which is conducted by VPC/VR methods within a 30 m (100 ft) distance from the new wharf access abutments had to be conducted prior to dredging of the side slopes, which was required as part of the wharf construction.

DENSIFICATION METHODS

The densification methods proposed by the Contractors are summarized as follows:

Vibratory Probe Compaction (VPC)

A 100 HP eccentric mass vibratory probe consisting of a conical tip, provided with water jet nozzles to facilitate penetration into the soil mass, and a 0.4 m (16 in.) O.D. steel casing was used. Outlets were provided at the connection between the probe and the extension shaft to supply circulation water in the hole and to create a passageway for the feed material falling from grade towards the vibrator. The feed material filled the cavity between the probe and the soil and transferred the vibratory energy into the sands being densified. During insertion and densification, the probe was suspended from the end of the 0.3 m (12 in.) O.D. extension shaft graduated at 0.3 m (1 ft) intervals to indicate the probe penetration depth and guide the crane operator.

On the basis of compatibility with the existing soil and convenient suitability number (Brown 1977), the Contractor selected Georgia DOT aggregate no. 10 for feed material. The gradation curve (see Fig. 3), obtained from random samples taken from the stockpiles on site, indicated suitability numbers in the neighborhood of 10. This index rates this sand as having good to excellent feedability.

The vibratory probe was inserted at each probe location, from grade down to 0.3 m (1 ft) below the prescribed densification elevation. As soon as full depth penetration had been reached, feed material was unloaded around the probe and watered down the hole continuously. In the test areas, approximately 6.9 to 9.2 m^3 (9 to 12 cu. yd.) of sand was added at each probe location. The probe was vibrated between 30 and 60 seconds at the bottom of the penetration. Through the densification process, the probe was lifted 0.3 m (1 ft) at a time, and maintained vibrating at each foot for a period of approximately 30 seconds, or until the amperage gauge indicated a significant reading increase. The average range was typically 95 to 140 amps with some values as high as 190 amps. The total time required was about 25 min. per probe location.

Test areas were performed with probes spaced at 2.4, 3.0, 3.7 and 4.3 m (8, 10, 12 and 14 ft) in an equilateral triangle pattern. The production work was completed with a 3.7 m (12 ft) equilateral triangular spacing. Based on limited data provided by the Contractor, the diameter of the densified column was estimated to be 1 m (3 ft).

Vibro-Replacement (VR)

The densification was performed utilizing a 165 horsepower vibratory probe and No. 4 granite ballast stone (see Fig. 3) to create a 1.07 m (3.5 ft) dia. stone column. At each probe location, the probe penetrated the ground to Elev. -4.6 m by means of its weight, vibrations and water jets in its tip. The backfill stone was then added to the hole formed by the probe and the stone and surrounding soil were compacted in lifts from the bottom up. The time to perform each probe was approximately 15 to 20 min.

The probe penetrations were located on an 3.35 m (11 ft) square grid over the test areas and for the production work. The Contractor chose not to vary the spacing in the test sections, based on prior site experience.

Terraprobe (TP)

The TP method was proposed by the Contractor as an alternate to VPC or VR methods. The TP method of compaction has been discussed by others (Anderson 1974; Janes 1973; Brown and Glenn 1976; and Mitchell 1978).

On this project a 0.76 m (30 in.) O. D., open-ended pipe pile with a 9.5 cm (3/8 in.) wall thickness was suspended from a vibratory pile driving hammer (ICE 812, operating at 25 Hz., amplitude of 1.2 to 2.5 cm (0.5 to 1.0 in.), eccentric moment of 46 m-kg (4000 in-lbs.). The pipe pile was penetrated into the soil, without jetting or adding replacement material, at various penetration rates (about 0.3 to 1.2 m (1 to 4 ft) per min.), maximum depths (0 to 3 m (0 to 10 ft) below the lower limit of the required densification zone) and hold times (30 to 300 seconds at the lowest depth). The pipe was withdrawn over a period of 1 to 3 minutes.

Test areas were completed with square and equilateral triangle spacings of 1.2 to 1.5 m (4 and 5 ft), respectively.

DENSIFICATION RESULTS

The three methods were implemented using test areas as specified. The results of the test areas and some of the production work are summarized in Figs. 4 and 5, and discussed below:

General

1. All VPC and VR methods were capable of consistently satisfying both the overall and interval density requirements, and were used for production work. The TP method was quite erratic with respect to both criteria, especially the interval criterion, and subsequently was not used for production.
2. The results from all three test programs indicated that the amount of densification was a function of the soil type. Each of the methods demonstrated limited ability to densify/strengthen the thin silt and clay layers that were encountered in some of the test borings and CPTs.
3. Vibration monitoring was performed during the application of all three methods (see Fig. 4); however, specifics relative to the VPC method are not available. The TP and VR methods demonstrated similar vibration levels within the range for vibro-compaction given by Dobson and Slocombe (1982). None of the observed vibrations at utility locations exceeded the specifications. No utility damage was observed.

Vibratory Probe Compaction (VPC)

1. The results of the test areas demonstrated a clear relationship between spacing and achieved density as indicated in Fig. 5.

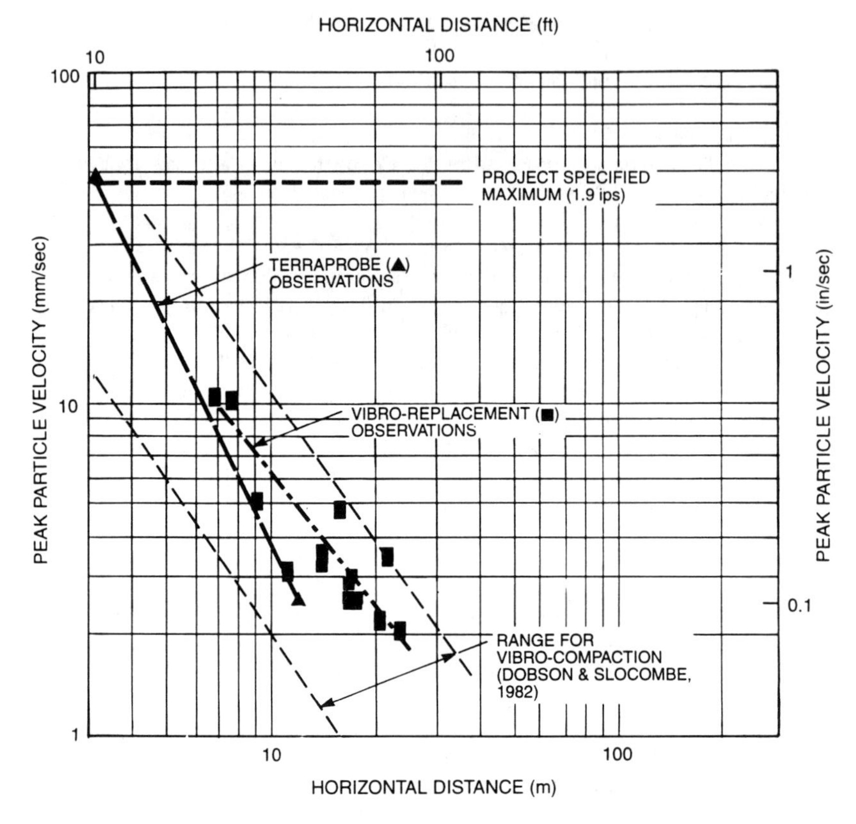

FIG. 4. Observed Ground Vibrations (TP and VR Methods)

2. The Contractor selected the 3.7 m (12 ft) equilateral triangle pattern for production work to address concerns with the specified interval density, and to reduce his potential for re-densification work if the initial densification results were unsatisfactory.

Vibro-Replacement (VR)

1. The test area and production spacing of 3.4 m (11 ft) square pattern was selected by the Contractor based on his previous experience on the base (Hussin and Ali 1987).
2. The actual demonstrated performance both in the test areas and production work was above the overall (full depth) specified minimum, but much closer to the 0.9 m (3 ft) interval criterion.
3. The volume of stone used exceeded that calculated for the 1.07 m (3.5 ft) dia. stone columns.

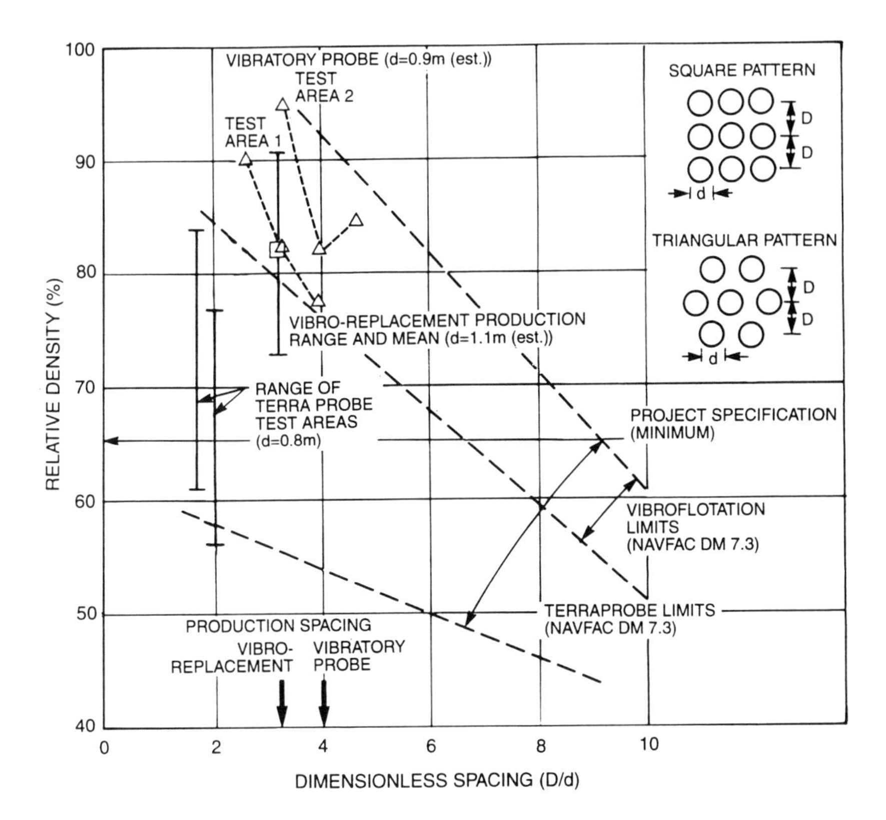

FIG. 5. Summary of Results for Various Methods and Spacings

Terraprobe (TP)

1. The TP results were quite erratic (see Fig. 5). Compared to the TP limits given in NAVFAC DM-7.3 (see Fig. 5), the performance of the TP method on this site was on the low side of the range even though relatively close spacings were used. CPT probes inside the pipe pile footprint indicated the specified minimum density was achieved.
2. After the TP application, the ground surface was dimpled with observable settlement of about 0.3 to 0.45 m (12 to 18 in.) inside the pipe pile footprint only. Survey elevations determined before and after the TP application indicated that the average settlement over the treated area was about 0.08 m (3 in.) (roughly 1% volume reduction). These measurements confirmed the CPT results indicating that the desired increase in in-situ density was not being achieved.

3. CPT values were obtained at time intervals of 3 to 10 days following TP densification to assess general "aging", and possible benefits. Although the testing was limited, there did not appear to be any significant change within that time interval.

SUMMARY AND CONCLUSIONS

Both the vibratory probe compaction and vibro-replacement methods achieved the specified minimum densification, and were ultimately used for production densification. The terraprobe method was erratic with respect to achieving the specified density, and therefore, was not used for production densification. The test area and production results demonstrate the relative performance of the three methods over a relatively large site.

The soil densification work was completed successfully. Portions of the wharf access roads have been in place and exposed to the missile transporter traffic for nearly six years without any reported settlement.

ACKNOWLEDGEMENTS

The authors wish to thank the U. S. Navy for granting permission to publish this data. The views expressed in this paper are the opinions of the authors and do not reflect the official policy of the Department of Defense or the U.S. Navy.

The authors also wish to thank Fay, Spofford and Thorndike, Inc., the wharf designer; and the contractors: Sayler-Tyger, a Joint Venture; Wood-Hopkins Contracting Company; and Hayward-Baker, Inc. for their cooperation and assistance while performing the densification. Additional credit is extended to Mr. Ken Recker of Haley & Aldrich, Inc., who served as the project manager for the first portion of this work.

APPENDIX I. REFERENCES

Anderson, R. D. (1974). "New method for deep sand vibratory compaction." *Journal of Construction Division*, Proc. ASCE, 100(CO1), 79-95.

Baez, J. I., and Martin, G. R. (1992). "Liquefaction observations during installation of stone columns using the vibro-replacement technique." *Geotechnical News*, 10(3), 41-44.

Baez, J. I., and Martin, G. R. (1992). "Quantitative evaluation of stone column techniques for earthquake liquefaction mitigation." *Proc. 10th World Conference on Earthquake Engineering,* Madrid, 1477-1483.

Barksdale, R. D., and Bachus, R. C. (1983). "Design and construction of stone columns, Volume I," *Report No. FHWA/RD-83/026*, Federal Highway Administration, Washington, D.C.

Bazaraa, A. R. S. S. (1967). "Use of standard penetration test for estimating settlements of shallow foundations on sand," Ph.D thesis, the University of Illinois, Urbana-Champaign.

Bollinger, G. A., Nutting, O. W., and Herman, R. B. (1986). "The 1886 Charleston South Carolina, earthquake - a 1986 perspective." *U.S. Geological Survey Circular 985*.

Brown, R. E. (1977). "Vibroflotation compaction of cohesionless soils." *J. Geotech. Eng. Div.*, Proc. ASCE, 103(GT12), 1437-1451.

Brown, R. E., and Glenn, A. J. (1976). "Vibroflotation and terra-probe comparison." *J. Geotech. Eng. Div.*, Proc. ASCE, 102(GT10), 1059-1072.

Department of the Navy, Naval Facilities Engineering Command (1983). "Soil dynamics, deep stabilization, and special geotechnical construction." *Design Manual 7.3 (NAVFAC DM-7.3)*, Alexandria, Virginia.

Dobson, T. (1987). "Case histories of the vibro systems to minimize the risk of liquefaction." *Soil Improvement - A Ten Year Update*, Geotechnical Special Publication, No. 12, ASCE, New York, New York, 167-183.

Dobson, T., and Slocombe, B. (1982). "Deep densification of granular fills." *Proc. 2nd Geotechnical Conference & Exhibit on Design & Construction*, Las Vegas.

Gee and Jenson, Inc. (1980). "Site investigation, waterfront configuration study." Naval Submarine Support Base, Kings Bay, Georgia," *Consulting Report*, 105 pp.

Gibbs, H. J., and Holtz, W. G. (1957). "Research on determining the density of sands by spoon penetration testing." *Proc. 4th ICSMFE*, London, 1, 35-39.

Hussin, J. D., and Ali, P. E. (1987). "Soil improvement at the trident submarine facility." *Soil Improvement - A Ten Year Update*, Geotechnical Special Publication, No. 12, ASCE, New York, New York, 215-231.

Janes, H. W. (1973). "Densification of sand for drydock by terra-probe." *J. Soil Mech. Found. Div.*, Proc. ASCE, 99(SM6), 451-470.

Kishida, H. (1970). "Characteristics of liquefaction of level sandy ground during the Tokachioki earthquake." *Soils and Foundations*, 10(2).

Lopez, R. A., and Hayden, R. F. (1992). "The use of vibro systems in seismic design." *Grouting, Soil Improvement and Geosynthetics*, Geotechnical Special Publication No. 30, ASCE, New York, New York, 2, 1433-1445.

Marcuson, W. F., and Bieganousky, A. M. (1977). "Laboratory standard penetration tests on fine sands." *J. Geotech. Eng. Div.*, Proc. ASCE, 103(GT6), 565-588.

Mitchell, J. K. (1978). "Improving soil conditions by surface and subsurface treatment methods - overview." ASCE Metropolitan Section, Foundation and Soil Mechanics Group Seminar, New York, Oct. 25-28.

Mitchell, J. K. (1986). "Material behavior and ground improvement." *Proc. Int'l. Conf. on Deep Foundations*, Beijing.

Schmertmann, J. H. (1977). "Guidelines for cone penetration test performance and design," *Report No. FHWA-TS-78.209*, Federal Highway Administration, U. S. Department of Transportation, Washington, D. C.

Seed, H. B., and Idriss I. M. (1967). "Analysis of soil liquefaction: Niigata earthquake." *J. Soil Mech. Found. Div.*, Proc. ASCE, 93(SM3), 83-108.

APPENDIX II. NOTATION

CPT = cone penetration test

D_r = relative density

SPT	=	Standard Penetration test
TP	=	terraprobe compaction
VPC	=	vibratory probe compaction
VR	=	vibro-replacement compaction

Vacuum Consolidation Technology - Principles and Field Experience

J. M. Cognon,[1] I. Juran,[2] Member, ASCE, and
S. Thevanayagam,[3] Assoc. Member, ASCE

Abstract: Vacuum consolidation is an effective means for improvement of saturated soft soils. The soil site is covered with an airtight membrane and a vacuum is created underneath it by using a dual venturi and vacuum pump. Experience gained from several site experiments and full scale field implementation indicate that this technology can provide an equivalent preloading of about 4.5 m high conventional surcharge fill. The effectiveness can be increased when applied with combination of a surcharge fill. Field experience indicate a substantial cost and time savings by this technology compared to conventional surcharging. This paper presents the principle behind vacuum consolidation, the technological know-how and several case histories from France where this technology has been successfully implemented. The performance of the vacuum consolidation system is evaluated based on settlement as well as pore pressure measurements at these sites.

Introduction

Vacuum consolidation was proposed in the early 1950s by Kjellman (1952), the developer of the prefabricated vertical "wick" drain. In the 1960s, the Corps of Engineers investigated the feasibility of vacuum consolidation of hydraulic fill. Isolated studies of vacuum induced or assisted consolidation continued for the next

---

[1] President, Menard Sol Traitement, P. O. Box 530, 91946, Les Ulis Cedex, France.

[2] Prof., Dept. of Civil and Env. Engrg., Polytechnic Univ., Brooklyn, NY 11201; and President, GIT Consultants Inc., 707 Westchester Ave., White Plains, NY 10604.

[3] Asst. Prof., Dept. of Civil and Env. Engrg., Polytechnic Univ., Brooklyn, NY 11201.

two decades (Halton et al. 1965; Holtz 1975). However, except for specialized applications like landslide stabilization, vacuum consolidation was not seriously investigated as an alternative to conventional surcharging until recently due to the low cost of placing and removing surcharge fills and the difficulties involved in applying and maintaining the vacuum. The steadily increasing direct and indirect costs of placing and removing surcharge fill and the advent of technology for sealing landfills with impervious membranes for landfill gas extraction systems have now made vacuum-assisted consolidation an economically viable method as a replacement for or supplement to surcharge fill.

Recent field trials in China (Choa 1989; Liu, undated) and France (Cognon 1991), and USA (Jacob et al. 1994, TETC 1990) have verified the effectiveness of vacuum-assisted consolidation in conjunction with vertical drains for site improvement. Cost estimates based on these recent projects indicate potential for cost savings over conventional surcharge fill for an equivalent surcharge height of 4.5 m for the vacuum system.

This paper presents the principle behind vacuum consolidation, the technological know-how and several case histories from France where this technology has been successfully implemented. The performance of vacuum consolidation system is evaluated based on settlement as well as pore pressure measurements at these sites.

CONCEPTUAL DESIGN

Vacuum-assisted consolidation provides an effective alternative to surcharging for preloading soils. Instead of increasing the effective stress in the soil mass by increasing the total stress by means of a conventional mechanical surcharging, vacuum-assisted consolidation preloads the soil by reducing the pore pressure while maintaining a constant total stress. Fig. 1 presents a typical layout of a vacuum-assisted prefabricated vertical drain consolidation scheme.

Fig. 2 graphically portrays the initial total stress in the ground and pore pressures induced in due to: (a) conventional surcharge, and (b) vacuum loading applied at the ground surface. Fig. 3 shows a typical pore pressure response at the end of vacuum consolidation measured at a site in China (Choa 1989) during vacuum consolidation. It consists of two profiles of initial hydrostatic pressures and final pore pressures measured during vacuum consolidation after 110 days and 180 days of vacuum application at the site. The straight lines indicate theoretical pore pressures under various vacuum pressures. An atmospheric pressure corresponds to about 100 kPa. For a site where the water level is at the ground surface, cavitation of water at negative 1 atmosphere (gage pressure) theoretically limits on-land vacuum-consolidation to an effective surcharge pressure of about 100 kPa, equivalent to approximately 6 m of surcharge. Practical problems in maintaining the efficiency of a vacuum system may reduce its effectiveness in the field. A system with an efficiency of 75 percent results in only 4.5 m of equivalent surcharge height. For the case shown in Fig. 3 about 70 to 80% efficiency is evident. Vacuum consolidation in underwater site conditions (off-shore land reclamation) can yield much higher equivalent preloads (Thevanayagam 1993).

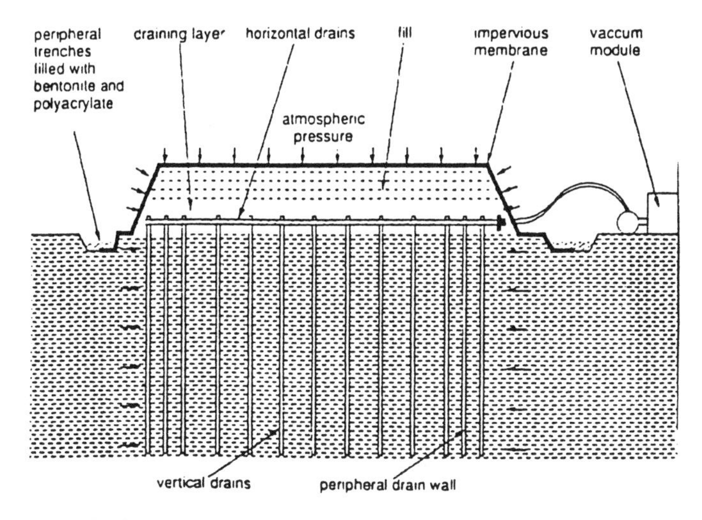

FIG. 1. Schematic Layout - Vacuum Consolidation

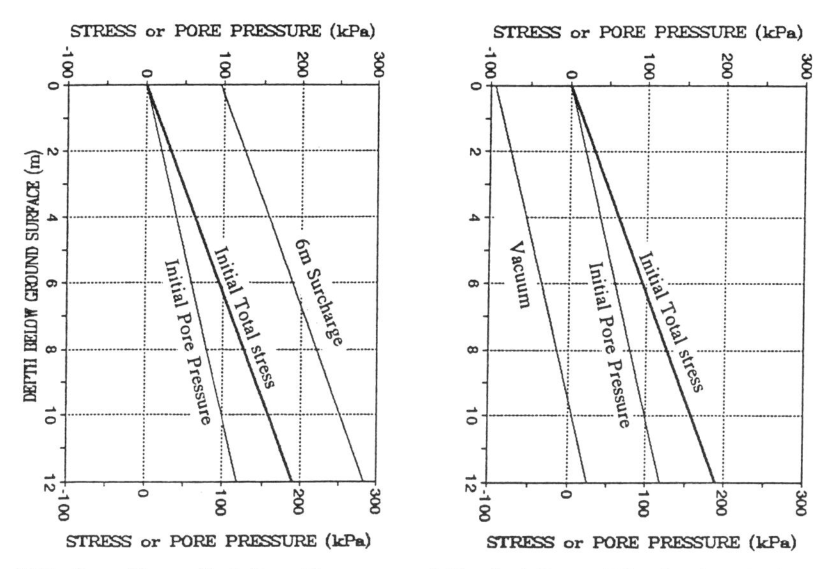

FIG. 2. Theoretical Pore Pressure and Vertical Stress Distribution during Surcharge and Vacuum Consolidation (Assuming 100% efficiency)

In essence, geotechnical design analyses used to evaluate wick drain spacing, and strength gain for preload fills are equally applicable to the engineering design

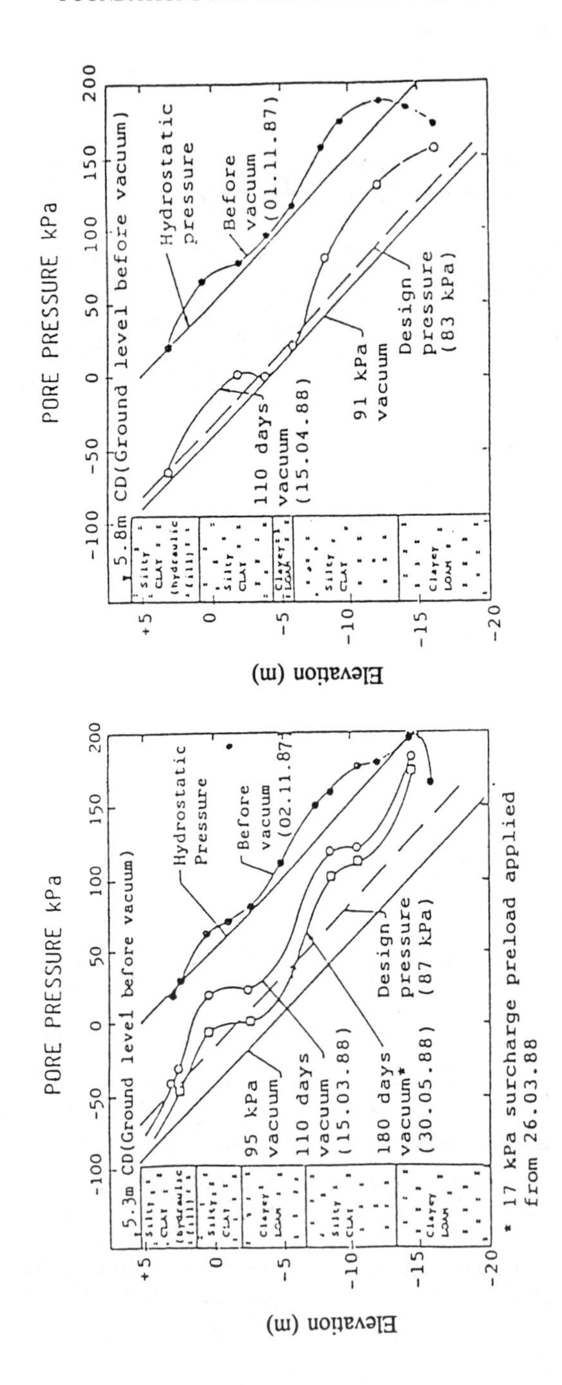

FIG. 3. Initial and Final Pore Pressure Data During Vacuum-Preloading: Tian Jin Harbor (after Choa 1989)

of vacuum consolidation system. There are many technical and operational factors which play important roles in vacuum consolidation. Primary considerations governing the effectiveness and economy of scheme include: (1) integrity of the membrane at the ground surface, (2) seal between the edges of the membrane and the ground, (3) soil stratification including permeable sand seams within the clay deposit, and (4) depth to groundwater.

Breaks in the membrane, a poor seal between the membrane and the ground, and wick drains extending into layers of high hydraulic conductivity all tend to reduce vacuum efficiency, reducing equivalent surcharge height and increasing pumping yields and pumping costs. The success of a vacuum consolidation system depends upon a combination of technological know-how and careful implementation of design details.

French Experience with Vacuum Consolidation

Fig. 4 illustrates a typical drainage scheme used in a vacuum consolidation project. The different stages of a vacuum consolidation project involve:

(1) Placing a free drainage sand blanket (60-80 cm thickness) above the saturated ground in order to provide for a working platform.

(2) Installation of vertical drains, generally of 5 cm in equivalent diameter, as well as relief wells from the sand blanket.

(3) Installation of closely spaced horizontal drains at the base of the sand blanket using a special laser technique to maintain them horizontal as shown in Fig. 4.

(4) The horizontal drains in the longitudinal and transverse directions are linked through connections as shown in Fig. 4.

(5) Excavation of trenches around the perimeter of the preload area to a depth of about 50 cm below the groundwater level and filled with an impervious Bentonite Polyacrolyte slurry for subsequent sealing of the impermeable membrane along the perimeter.

(6) The transverse connectors are linked to the edge of the peripheral trench. They are then connected to a prefabricated module designed to withstand future pressure due to the vacuum.

(7) Installation of the impermeable membrane on the ground surface and sealing it along the peripheral trenches. The membrane is delivered to the site folded and rolled in elements of approximately 1000 m^2. The membrane elements are welded together and laid in the peripheral trench where they are sealed with the Bentonite Polyacrolyte slurry. The trenches are backfilled and filled with water to improve the tight sealing between the membrane and the Bentonite Aquakeep slurry as shown in Fig. 5.

(8) Vacuum pumps are connected to the prefabricated discharge module extending from the trenches. The vacuum station as shown in Fig. 5 consists of specifically designed high-efficiency vacuum pumps acting solely on the gas phase in conjunction with conventional vacuum pumps allowing liquid and gas suction.

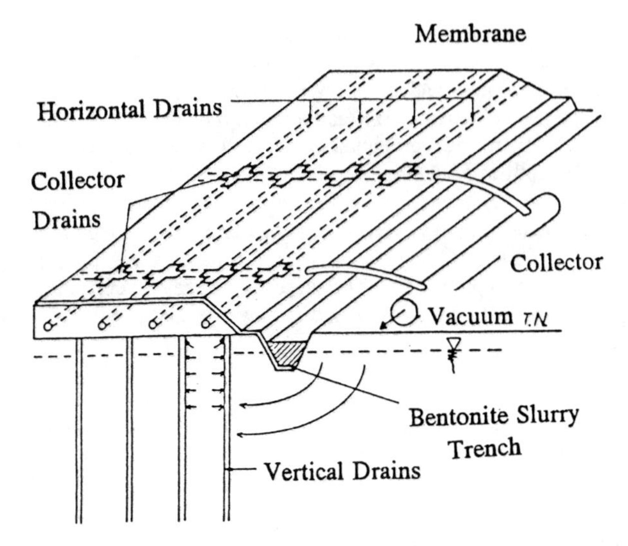

FIG. 4. Drainage System Layout for Vacuum Consolidation

FIG. 5. Vacuum Pump Station and Discharge System

The process combines dewatering and vacuum action to maintain the water table at the base of the granular platform during the entire application of the consolidation process. Eventually an additional drainage system is installed at a required depth to allow for a conventional de-watering under the membrane.

Indeed, the fill will maintain a non-submerged action even when it has settled below the original ground water level. Therefore, with this technology, unlike the case of a surcharge preloading, the load intensity will not decrease during the vacuum application. The discharge drains are manufactured by extrusion of cylindrical and perforated PVC. Use of a suitable filter cloth with proper filtering properties to cover the perforated PVC avoids infiltration of sand and fines during vacuum application. The discharge drains are brought to the surface at every 150 meters spacing where they are connected by transverse drains to the vacuum station as shown in Fig. 5.

The vacuum consolidation technique is often combined with surcharge preloading either by placing an additional surcharge by backfilling or using water placed on top of the impervious membrane. The major practical advantage of the vacuum consolidation is that it generates in the granular layer an apparent cohesion due to the increase of the effective stress and the granular layer provides a useful working platform to accelerate the surcharge backfilling process. Experience indicates that within days after vacuum pump is turned on, construction vehicles can maneuver on the top of the membrane. Table 1 summarizes field experiences from several sites regarding the efficiency of vacuum consolidation.

TABLE 1. Typical Design Parameters and Performance Data at Selected Sites, France

Site No.	Vacuum Consolidation Duration (months)	Spacing of Vertical Drain (m)	$\frac{C_c}{1+e_o}$ (average)	Average Settlement Ratio* (%)
1	4.0	1.5	0.21	12.5
2	4.5	1.8	0.17	8.5
3	3.0	1.5	0.22	9.5
4	2.5	1.3	0.27	13.5
5	3.5	1.5	0.18	7.5
6	3.5	1.5	0.2	10.5
7	3.0	1.4	0.56	18.0
8**	2.5	1.3	0.1	4.0

Notes: * Settlement Ratio = Ratio of Ground surface settlement to the length of the vertical drain within the saturated clay layer being improved.

** Soil at this site is probably overconsolidated.

The efficiency of this technology has been demonstrated under different site conditions where it has successfully provided cost effective solutions to substantially accelerate the consolidation process while leading to significant savings in project costs. Furthermore it does not pose any stability concerns, unlike in the case of a conventional surcharge, while resolving the environmental problems associated with the conventional method of surcharge preloading, specifically the post consolidation need to remove and dispose the fill materials. Several monitored case studies are described below to illustrate the efficiency of this technique.

CASE HISTORIES

Highway Construction Site

The first case study concerns a pilot testing conducted in Ambes, France, for the construction of a highway embankment on a very compressible saturated clayey soil. The soil profile at this site as shown in Fig. 6a indicates the presence of about 1.7 m thick peat layer with a moisture content ranging from 400% to 900 % underlain by about 2.0 m thick highly organic, compressible clay layer with moisture content ranging from 140% to 210%.

The project plan calls for the construction of a highway embankment, 2.15 m high, across the site in order to protect the highway from floods. Alternative solutions, including soil replacement, raft foundation supported by piles and conventional surcharge preloading were considered and rejected for economic reasons. For example, due to stability concerns, the conventional preloading system required very gently sloping embankment with a basewidth of 65 m, which was extremely expensive. As vacuum consolidation has not been previously applied under such soil and moisture conditions, a pilot field testing program was implemented. The vacuum consolidation process at this site involved:

(i) placing 80 cm of free draining sand blanket at the surface;
(ii) installation of the vertical drains with a grid spacing of 1.4 m × 1.4 m (total length of 1640 m);
(iii) installation of horizontal drains (total length of 220 m) and connections to the vertical drains;
(iv) placing an additional granular surcharge fill of 0.5 m in thickness;
(v) installation of the PVC membrane over a surface area of 390 m^2, sealed along the peripheral trench with the Bentonite Aquakeep slurry;
(vi) vacuum consolidation during a period of 3 months using a 25 kW pump; and
(vii) instrumentation and site monitoring of settlements and pore water pressure.

A reference test embankment was considered in order to assess, through comparisons of the recorded settlements, the efficiency of the vacuum consolidation technique.

During the first day of operations, 6 cm of surface settlement were already recorded which corresponds to a water flow of 600 m^3 per day per hectare. The water air pump, backed by the Venturi pumps induced vacuum under the impermeable membrane, and consequently caused settlement at the site. The combined application of surcharge (1.3 m) and vacuum provided an equivalent reload of about

150 kPa after only two months of vacuum application, whereas under conventional surcharge, slope failure would have occurred at this site under a preloading stress of 35 kPa.

Fig. 6b depicts the monitored settlement record during the consolidation process. It indicates that the recorded settlement with vacuum consolidation is approximately equivalent to that induced by a 4.5 m high surcharge embankment. Analysis of the settlement records indicated that at the end of the consolidation process the settlement of the peat layer reached about 80% of the reference settlement that would have been induced by a 4.5 m high surcharge embankment, while the settlement of the underlying highly organic clay reached about 50% of that reference settlement. At the end of the vacuum application the ground surface rebounded by about 3 cm during 48-hour period and then stabilized. During the vacuum application the groundwater level rose by about 40 cm up to the level of the horizontal drains while the upper part of the granular fill remained dry.

The measured pore pressure data are shown in Fig. 7a. Fig. 7b shows initial effective stress, maximum effective stress attained after 40 days of vacuum consolidation, the anticipated effective stress due to design (2.15 m) highway embankment, and due to reference embankment (4.5 m). The results appear to be quite consistent with the percentage of consolidation established for each layer from the settlement records as described earlier. In particular, the effective stresses attained after 40 days of vacuum applications correspond to the final effective stresses anticipated under the highway embankment. Following the pilot testing the vacuum consolidation was selected as the best available solution for the consolidation of about 17,550 m^2 along this highway construction site.

Oil Tank Farm Site

Fig. 8 shows the settlement records obtained at an oil tank farm construction site in Ambes, France where a vacuum consolidation was applied in combination with conventional surcharging. The site, used to erect oil tanks with 27 to 34 m diameter with an effective usable height of 12 to 15 m, consists of about 12 m thick fine compressible layer of saturated alluvium (silt, peat and highly organic clay). For this case the vacuum consolidation replaced the originally foreseen solution of piles and resulted in substantial cost and time savings.

Lemantin Airport Site

Fig. 9 shows the settlement records obtained at the Lemantin Airport freight terminal construction site during vacuum consolidation combined with conventional surcharge preloading. The site consisted of about 10 m deep compressible alluvium over a total area of about 17,700 m^2. Curve A in Fig. 9 corresponds to settlement records obtained at the site using 1 m and 2 m high conventional surcharge embankments placed successively without vacuum application. Curve B corresponds to settlements at an adjacent location where vacuum was applied with combination of the same staged surcharge loading. The settlement records (curve B) indicate that the vacuum consolidation had a major impact on the settlement rate as compared to the successive stage loading (curve A).

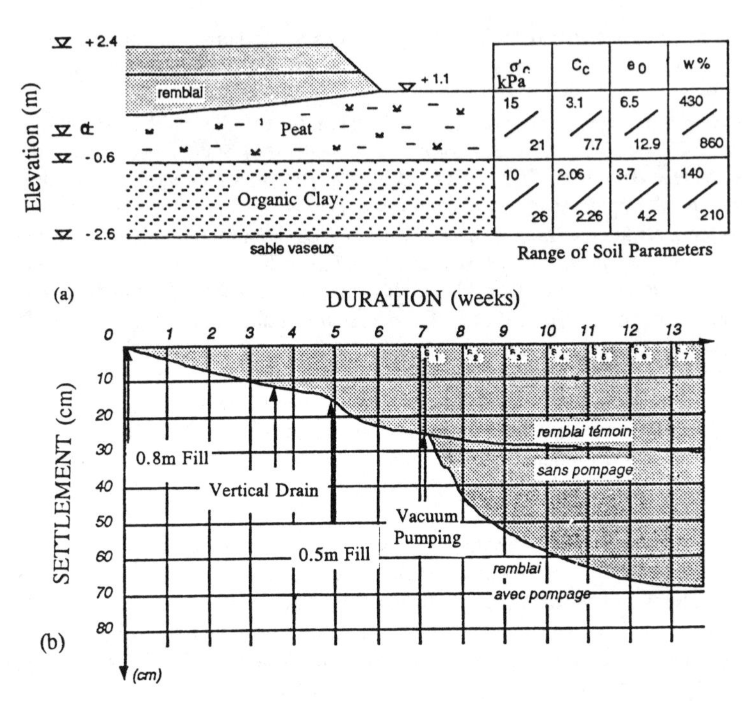

FIG. 6. (a) Typical Soil Profiles at Ambes Site, France, and (b) Measured Settlement versus Time Record during Vacuum Consolidation

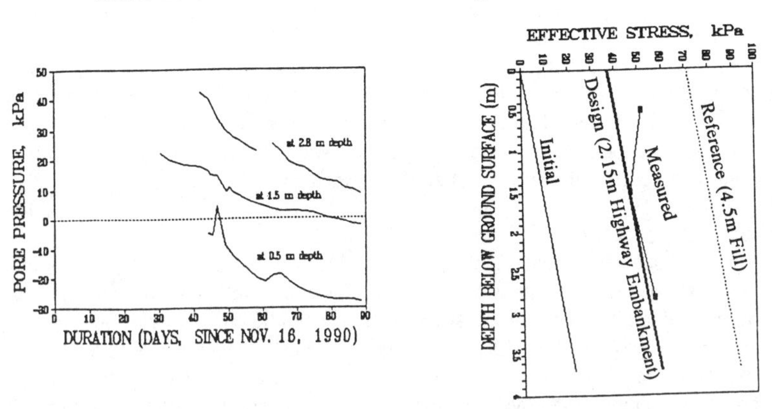

FIG. 7. Measured Pore Pressures and Effective Stresses at Ambes Site, France

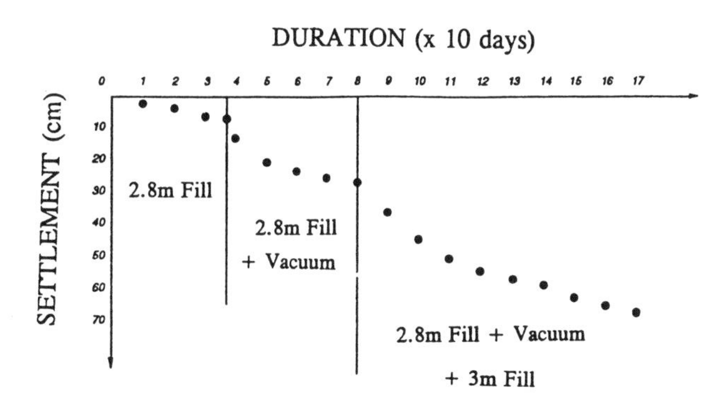

FIG. 8. Settlement vs. Time Record, Oil Tank Farm Site, Ambes, France

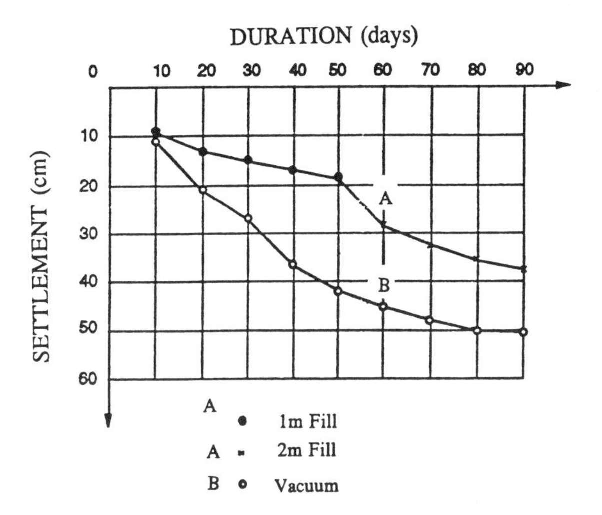

FIG. 9. Comparison of Settlement versus Time Record: Vacuum versus Conventional Surcharge, Lemantin Airport Site

CONCLUSIONS

Vacuum consolidation is an effective means for improvement of highly compressible soft soils. In essence, vacuum consolidation can yield an effective equivalent preload of about 4 to 5 m of conventional surcharge fill. A combination of conventional surcharge with vacuum application can yield much higher equivalent preload. Recent experience from US and China, and the case histories from France described herein indicate that this technology can be applied cost effectively under various challenging site conditions. In certain difficult site conditions where stability under the conventional surcharge is of concern, it allows to cost-effectively accelerate the consolidation process as compared to conventional stage-loading.

In Europe, the engineering use of vacuum consolidation is currently rapidly expanding and it is of interest to note that this technology has been used to cost effectively replace conventional surcharge preloading for the development of about 57,000 m^2 of industrial area at the Channel Tunnel Terminal.

APPENDIX - REFERENCES

Choa, V. (1989). "Drains and vacuum preloading pilot test." *Proc. 12th ICSMFE*, Rio de Janeiro, 1347-1350.

Cognon, J. M. (1991). "Vacuum consolidation." *Rev. French Geotechnique*, 57 (October), 37-47.

Halton, G. R., Loughney, R. W., and Winter, E. (1965). "Vacuum stabilization of subsoil beneath runway extension at Philadelphia International Airport." *Proc. 6th ICSMFE*, Montreal, 62-65.

Holtz, R. D. (1975). "Preloading by vacuum: current prospects." *Transportation Research Record No. 548*, Transportation Research Board, National Research Council, Washington, D. C., 26-29.

Jacob, A., Thevanayagam, S., and Kavazanjian, E. (1994) "Vacuum-assisted consolidation of a hydraulic landfill." *Vertical and Horizontal Deformations of Foundations and Embankments*, Geotechnical Special Publication No. 40, ASCE, New York, New York.

Kjellman, W. (1952). "Consolidation of clay soil by means of atmospheric pressure." *Proc. Conf. on Soil Stabilization*, Massachusetts Institute of Technology, Cambridge, 258-263.

Liu, Yi-Xong. (Undated). "History and present situation of soft soil foundation treatment in Xingang, Port of Tianjin," *Research Report*, First Navigation Engineering Bureau, Ministry of Communications, Tianjin, Peoples' Republic of China.

TETC (1990). "Geotechnical investigation for the Pier 300 42-acre landfill ground modification project, Vol. I and II," Report prepared by Earth Technology Corporation, for the Port of Los Angeles, Los Angeles, California.

Thevanayagam, S. (1993) "Prospects of vacuum-assisted geo-systems for ground improvement at Pier-400, 2020 Plan, Recommendations for field trials and technology transfer," Report prepared for the Port of Los Angeles, Los Angeles, California.

VACUUM-ASSISTED CONSOLIDATION OF A HYDRAULIC LANDFILL

A. Jacob,[1] Member, ASCE, S. Thevanayagam,[2] Assoc. Member, ASCE
and E. Kavazanjian, Jr.,[3], Member, ASCE

ABSTRACT: As part of the Pier 300 42-acre site ground modification project, a field test of vacuum-assisted preconsolidation of hydraulic fill was studied. The objective of the test was to assess the feasibility of this technology as an alternative or supplement to conventional surcharge techniques for ground improvement of future hydraulic fill sites at the Port of Los Angeles. The test section was about 30×30 m^2 consisting of prefabricated wick drains capped with a geomembrane, a slotted pipe collector system embedded in a sand blanket beneath the geomembrane, an anchor trench and protective cover medium for the membrane, and a vacuum pump. Instrumentation included high air-entrainment piezometers to measure negative pore pressures, monitoring of the ground surface settlement and multi-point borehole extensometers for deep settlement measurements. Results indicate that, due to less than optimum site conditions, the efficiency of vacuum consolidation was about 50 percent, or equivalent to about 3 meter of surcharge fill at the test site. Under better conditions or with a more effective membrane seal, higher efficiency and greater equivalent surcharge heights can be achieved, making vacuum-assisted consolidation an economically viable alternative or supplement to mechanical surcharges.

---

[1] Civil Engineering Associate, Port of Los Angeles, San Pedro, CA 90733.

[2] Asst. Prof., Dept. of Civil Engrg., Polytechnic Univ., Brooklyn, NY 11201.

[3] Associate, Geosyntec Consultants Inc., 16541 Gothard St., Huntington Beach, CA 92647.

INTRODUCTION

The Port of Los Angeles is presently formulating plans to expand its facilities by reclaiming hundreds of additional acres, within the outer harbor, using hydraulic landfill (POLA 1987). Traditionally, POLA has relied on conventional mechanical surcharge methods to improve fine-grained hydraulic landfills before building on them. Past experience indicates that a waiting period of two to three years or more of desiccation and consolidation due to self-weight, is required before construction equipment can access the site for installation of vertical drains and surcharging. Once the site is 'walkable', even with vertical drain installation, stage construction of the preload is generally required due to the low initial shear strength of the hydraulic fill. It is not unusual for a 200-acre site to require about a decade or more of post-landfilling development before it can be used. The steadily increasing direct and indirect costs of ground improvement by surcharge fill, particularly in urban areas, including environmental constraints on emissions from earth- moving equipment, dust control, the delays associated with improvement of the site in several stages, and the need to remove and dispose of surcharge fill materials following ground improvement have created a demand for new technologies to expedite and supplement the traditional technique of stabilization by surcharge.

BACKGROUND

Vacuum-assisted consolidation can be an effective alternative to surcharging for preloading soils. Instead of increasing the effective stress in the soil mass by increasing the total stress as in conventional mechanical surcharging, vacuum-assisted consolidation relies on increasing effective stresses by decreasing the pore pressure while maintaining a constant total stress. Fig. 1 compares the total stress, pore pressure, and effective stress profiles under (a) conventional mechanical surcharge, (b) vacuum loading applied at the ground surface, and (c) a combination of surcharge and vacuum loading. Cavitation of water at negative 1 atmosphere (gage pressure) limits on-land vacuum-consolidation to an effective surcharge pressure of about 100 kPa, equivalent to approximately 6 meter of surcharge, at the vacuum-water interface. The location of this interface thus can significantly influence the effectiveness of the vacuum consolidation system. Practical problems in maintaining the efficiency of a vacuum system may reduce its effectiveness in the field. A system with an efficiency of 75 percent results in only 4.5 m (15 ft) of equivalent surcharge height. However, by applying the vacuum below the water table so that it is used in combination with dewatering, the equivalent preload can be increased significantly.

Vacuum consolidation was first proposed by Kjellman (1952) in the early 1950s. In the 1960s, the Corps of Engineers investigated the feasibility of vacuum consolidation of hydraulic fill. Isolated studies of vacuum induced or assisted consolidation continued for the next two decades (Halton et al. 1965; Holtz 1975). However, except for specialized applications like landslide stabilization, vacuum consolidation was not seriously investigated as an alternative to conventional surcharging until recently due to the low cost of conventional surcharge technology, the lack of membrane technology, and associated difficulties involved in applying and maintaining the vacuum. The advent of technology for sealing landfills with impervious membranes for landfill gas extraction systems have now made vacuum-assisted consolidation an economically viable method as a replacement for or supplement to surcharge fill.

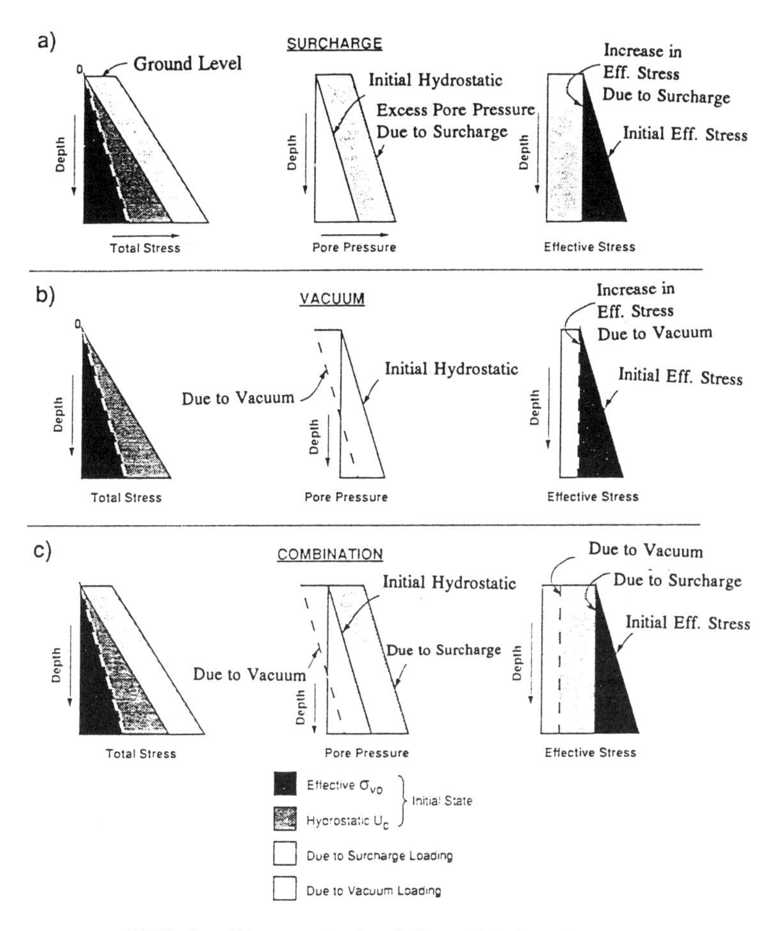

FIG. 1. Vacuum-Assisted Consolidation Concept

Recent field trials in China (Choa 1989; Liu, undated), Japan (Shinsha et al. 1991), and France (Cognon 1991) have verified the effectiveness of vacuum-assisted consolidation in conjunction with prefabricated drains for site improvement. Preliminary cost calculations indicate the potential for cost and time savings over conventional surcharge fill, provided a 75% vacuum efficiency can be achieved, for an equivalent surcharge height of 4.5 m.

DESIGN ASPECTS

The mechanism of consolidation and the rate of settlement in vacuum-assisted consolidation are the same as for conventional surcharging. Hence, geotechnical consolidation design analyses used to evaluate wick drain spacing, settlement rate, and strength gain for preload fills are equally applicable to vacuum consolidation. Furthermore, unlike surcharging, which may cause lateral spreading of the soft soil and pose stability concerns, vacuum technology does not impose any external load and therefore does not pose any stability problems. However, there are many other technical and operational factors which play important roles in vacuum consolidation. The success of a vacuum consolidation system depends upon implementation of a combination of technological know-how and implementation of design details. The potential for successful application of this technology is very much dictated by site-specific soil types, soil stratigraphy, and groundwater conditions. Therefore implementation strategies are likely to differ significantly from place to place.

Fig. 2 presents a schematic diagram of a vacuum-assisted prefabricated vertical drain consolidation scheme. Primary considerations governing the effectiveness and economy of scheme include: (a) integrity of the membrane itself and the seal between the edges of the membrane and the ground, (b) interaction between wick drains and subsurface layers of high hydraulic conductivity which extend beyond the bounds of the membrane, and (c) depth to groundwater. Leakage through defects in the membrane or the seal between the membrane and the ground, and wick drains extending through layers of high hydraulic conductivity within or below the cohesive soil deposit being consolidated can inhibit efficiency, reducing equivalent surcharge height.

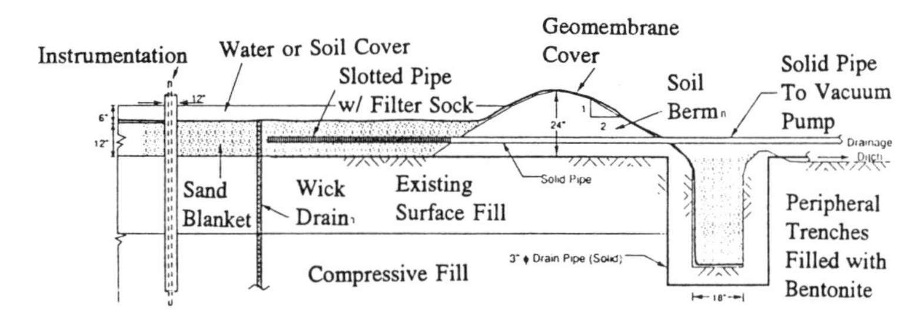

FIG. 2. Conceptual Design - Vacuum Assisted Consolidation

Membrane Integrity and Seal

Experience with landfill gas extraction systems indicates that membrane integrity and sealing between the membrane and soil are not major problems if the proper design details and construction procedures are used. A vacuum-assisted consolidation system should be, to some extent, self-sealing around the edges of the membrane as the system will tend to suck the membrane inward, closing any gaps between soil and membrane, rather than pushing the membrane away from the soil and opening or widening gaps. The greatest threat to membrane integrity are improperly joined field seams and construction activities which could puncture the membrane. Good quality control on field seaming activities and a protective cover placed on top of the membrane as soon as possible should minimize these problems.

In vacuum-consolidation field trials in China, water extracted by the vacuum system was used as a covering medium for the membrane (Liu, undated). This eliminates the need to dispose of the extracted water and simultaneously increases the surcharge load. Other advantages of water as a cover are the low cost and the ability to see small leaks when they develop. The disadvantage of a water cover is that it restricts access to instrumentation and mobility across the membrane and prohibits application of a supplemental mechanical surcharge fill. If a large area is to be surcharged, lack of mobility could hinder site operations. Because vacuum-assisted system is not meant to be a permanent system, ultraviolet degradation of the membrane is not typically an issue.

Horizontal Permeable Layers

Horizontal lenses of pervious cohesionless soils have the potential for either increasing or decreasing the effectiveness of a vacuum-assisted consolidation system, depending upon their lateral extent and location within the subsurface. Layers of high permeability that extend beyond the edge of the membrane can potentially provide drainage paths for water or air and reduce vacuum efficiency and/or increase required pump capacity and power consumption.

If the pervious layers are discontinuous, and/or do not extend beyond the edge of the membrane, horizontal layers of pervious soil can be beneficial, increasing the rate of consolidation by inducing vertical as well as radial consolidation and thus increasing the allowable vertical drain spacing and shortening the duration of vacuum system operation. This, in turn, will reduce the cost of operation by reducing pumping costs and/or drain installation costs. In fact, an engineered fill with continuous horizontal pervious layers could conceivably be vacuum-consolidated with only a limited number of large diameter vertical-drains. The balance between the beneficial effects of pervious layers and loss of efficiency due to leakage depends on site conditions.

Depth to Groundwater Level

A significant depth to groundwater can substantially reduce the cost-effectiveness of vacuum consolidation by increasing the cost of obtaining an adequate seal and reducing vacuum efficiency. Even when a proper seal is attained, a large depth to groundwater will reduce the effectiveness of a system where the vacuum is applied at the ground surface. Figs. 3a and 3b illustrates the effect of depth to groundwater on the resulting equivalent surcharge, Fig. 3a shows the initial pore pressures and the pore pressure at the end of vacuum consolidation. Fig. 3b shows initial effective stress in the soil for four different groundwater levels (0, 3, 6, and 10 m depth) and the final effective stress attained at the end of vacuum consolidation. In each case, 100% vacuum is assumed at the ground surface. As the depth to groundwater level increases the equivalent preload decreases. For example, in the case where the groundwater level is at 10 m, there is no change in effective stress at depths below 10 m (Fig. 3b). However, by modifying the vacuum extraction pipe layout, the efficiency of the system can be increased.

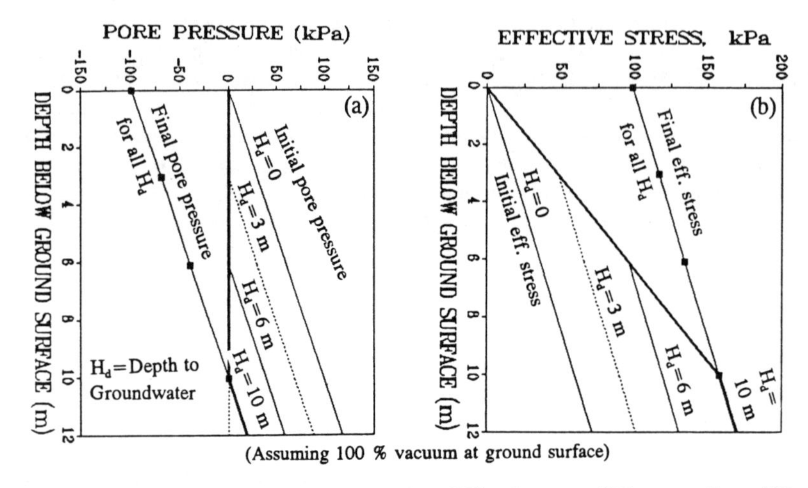

FIG. 3. Effect of Groundwater Level on Effectiveness of Vacuum Consolidation: (a) Pore Pressure Response; and (b) Effective Stress Response

PILOT TEST SITE CONDITIONS

The field test for vacuum-assisted consolidation was performed at the Pier 300, 42-acre hydraulic landfill site located on Terminal Island. Due to the unengineered nature of hydraulic fill construction and large variation in the source material during dredging operations, the stratigraphy over the site varies considerably and consists of intermittent layers of clayey and sandy soils. CPT logs, x-ray images of "undisturbed" tube samples, and examination of extruded soil specimens indicate intermittent thin seams of sandy silt or clayey silt within even the relatively homogeneous clay layers (TETC 1990). Below about 10.5 m (35 ft) to 12 m (40 ft), the soil profile consists of natural deposits of granular marine sand.

Due to presence of a thick granular layer at the surface of the landfill, placed as a working cap during previous site improvement activities, the vacuum test pad area was lowered by about 2 m (6 ft) to EL. 4.5 m (15 ft) MLLW prior to start of the field test. For comparison purposes, a 4 m (13-ft) high instrumented test embankment was constructed next to the test pad. Fig. 4 shows the typical soil profile across the test pad area after grading. Pre-construction CPT soundings and borings indicated the surficial granular layer extended about 1.5 m (5 ft) to 2.5 m (8 ft) below the test pad surface and was underlain by about 4.5 m (15 ft) to 6 m (20 ft) of cohesive hydraulic fill at this area. The groundwater elevation was at about EL. +1.2 m (4 ft) MLLW or about 3.4 m (11 ft) below the test pad surface.

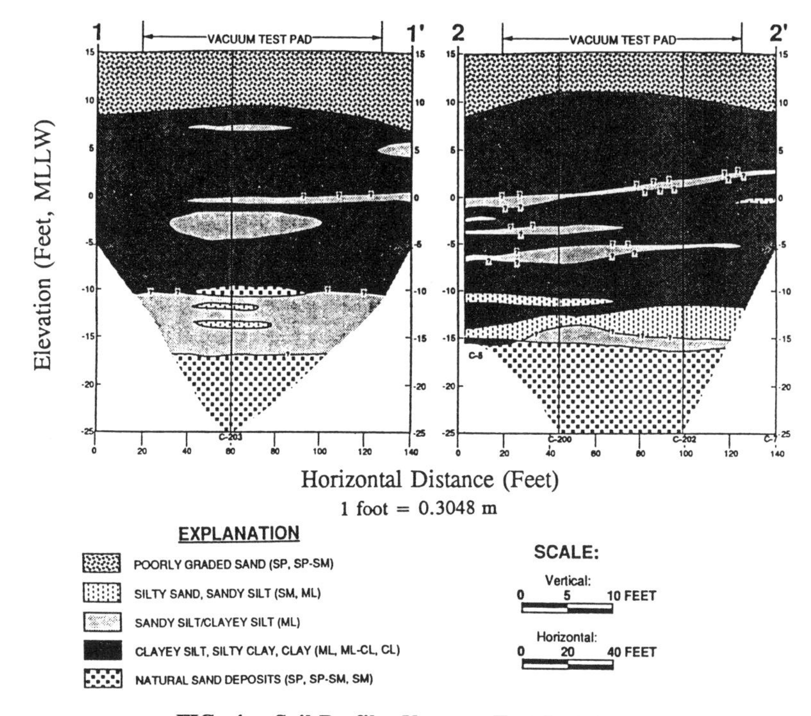

FIG. 4. Soil Profile, Vacuum Test Pad Area

Vacuum System Design Details

The vacuum-assisted consolidation test section can be divided into six essential components: (i) Wick drains and drainage blanket, (ii) Impermeable membrane, (ii) Membrane-ground anchor, (iv) Membrane cover medium, and (v) Vacuum pump and extraction system.

Wick Drains and Drainage Blanket

A wick drain spacing of 1.5 m was used to limit the duration of the test to less than 2 months. As penetration of the wick drains into the bottom sand would provide a direct path for water to flow into the extraction system and potentially overwhelm the capacity of the vacuum pump, the wick drains were terminated at about 0.6 m above the bottom sand. Due to the variability of the depth to this sand deposit, termination depth for the drains was determined on a location-by-location basis based on the penetration resistance felt by the operator of the wick drain installation equipment.

Vacuum Pump and Extraction System

The vacuum extraction system consisted of a vacuum pump, a trap, or "knockout pot", to collect extracted water, and slotted PVC collector pipes to extract water from the sand blanket beneath the membrane. Fig. 5a shows the layout of the vacuum-pump system. Fig. 5b shows the details of the collector system. Ideally, the vacuum pump system for this purpose should have a fluid-vapor extraction system built-in to avoid fouling the pump. However, such a pump is very costly and was not deemed necessary for this experimental field test of limited size and duration.

Design of the membrane seal and selection of the pump size were both determined by the contractor. It was desirable to have a vacuum extraction system capable of continuously maintaining a full (100 kPa) vacuum over the test pad area while accommodating the fluid extraction rates accompanying consolidation. Previous trials on similar size test pads (Choa 1989) indicated that at least a 7.5 horse power pump would be required. However, the contractor chose to conduct initial trials with a 1/2 horse power pump. Based on initial field trials using the small pump, a 10 horse power pump was used to initiate the vacuum underneath the membrane. Once the vacuum was established, it was only necessary to install a 1/2 horse power pump to sustain the vacuum.

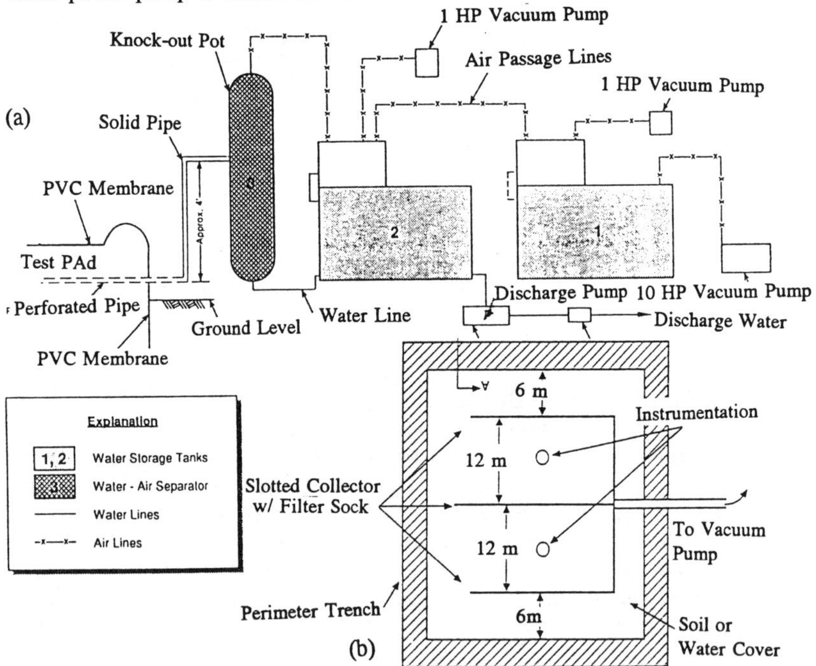

FIG. 5. (a) Schematic Vacuum Pumping System, and (b) Slotted Collector Pipe Layout

Impermeable Membrane

A variety of geosynthetic materials, including polyvinyl chloride (PVC), chlorosulfonated polyethylene synthetic rubber (CPE) or high density polyethylene (HDPE), can be used for the membrane. HDPE membranes are somewhat more resistant to ultraviolet degradation, while PVC and CPE membranes are slightly more puncture resistant. HDPE seams and patches must be thermally welded, while PVC and CPE seams and patches can be glued. Since a vacuum preload system is temporary, ultraviolet degradation was not a consideration in membrane selection.

Based primarily on cost, a 20-mil (0.05 cm thick) PVC membrane was chosen for the project. Key factors in maintaining the integrity of the impermeable membrane are the boots, or cut-outs, where collector pipes and instrumentation pass through the membrane, and the anchor trench details.

Membrane - Ground Anchor

For optimum efficiency, the peripheral trench and/or membrane should form a seal with an impervious layer below the water table. A seal above the water table or into a pervious layer below the water table could result in air or water leakage beneath the edge of the membrane. For the initial trials, the contractor chose to use a shallow, 1 m anchor trench and rely on vacuum pump capacity to establish a satisfactory vacuum level beneath the membrane. However, initial trials indicated an excessive degree of leakage around the membrane. After supplemental CPT soundings indicated the presence of pockets of the surficial cap material to depths of about 3 m below the test pad surface, a cut-and-cover anchor trench system was designed and constructed up to a depth of about 3.3 m. The membrane was covered with water retained by a small (0.3 m high) perimeter berm.

Instrumentation

To provide redundancy and enhanced reliability for the monitoring program, two instrumentation stations, each consisting of 3 pore pressure transducers and 3 deep settlement points (multi-point extensometers, MPBX), were installed at the vacuum test pad. The outer casing of the instrumentation was sealed to the membrane and embedded at a depth sufficient to prevent short-circuiting the vacuum.

OPERATION

After several initial trials in which the pump size was increased and the perimeter anchor trench was extended to a depth of 3.3 m, a good seal was obtained and a vacuum pressure of about -80 kPa (80% efficiency) was attained beneath the membrane. However, this initial pressure lasted only for about a day, and with time, the vacuum pressure rose to about -65 kPa (65% efficiency) and was sustained thereafter to the end of the test.

Measured Pore Pressures and Effective Stresses

Figs. 6a and 6b show the profile of measured pore pressures and interpreted effective stresses at the end of the test. For comparison, Fig. 6b also shows effective stress profiles for the following cases: (a) prior to grading , original ground EL. +6 m, (profile A), (b) after grading the site to EL. +4.5 m, (profile B), (c) at the end of the test (profile C), and (d) theoretical maximum effective stress corresponding to 100 percent efficiency in vacuum consolidation (profile D).

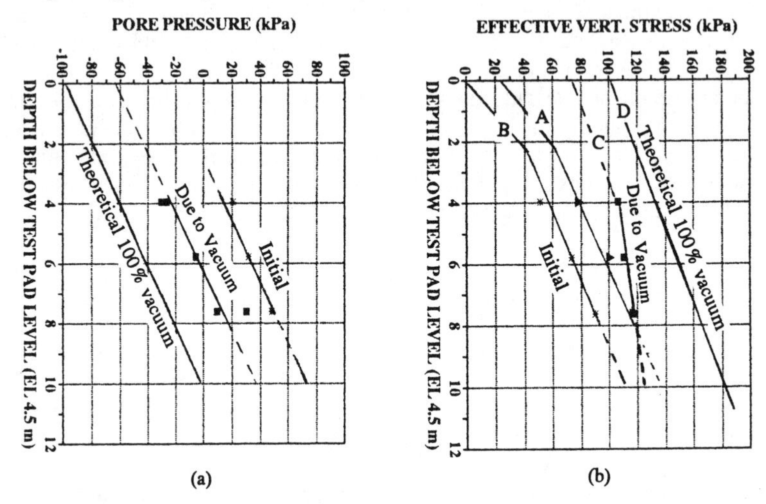

FIG. 6. (a) Measured Pore Pressures, and (b) Effective Stresses

Vacuum Consolidation Efficiency

Based on the pore pressure and effective stress profiles, the overall efficiency (average increase in effective stress divided by atmospheric pressure of 100 kPa), was estimated as about 40 to 50 percent. The efficiency varies with depth between 65 percent at the top of the cohesive hydraulic fill to less than 30 percent at the bottom of the cohesive layer. It is important to note that, based upon a depth to groundwater beneath the test pad at 3 to 3.3 m, the maximum vacuum efficiency in the steady state at the top of the cohesive soil at a depth of 3.3 m would be about 65 percent. Examination of the effective stress profiles (C in Fig. 6b) at the end of the test and the subsurface stratigraphy (Fig. 4) indicates that the losses in efficiency at large depths depth could be due to one of or a combinations of the following two factors: (i) seepage through the pervious lenses within the cohesive fill, and (ii) seepage from the bottom marine sand layer into the wick drains.

Seepage from pervious lenses within the cohesive fill would reduce efficiency if the lenses extend laterally beyond the test pad area. Fig. 4 strongly indicates that this is highly likely to be a contributing factor. Loss of efficiency due to seepage from the marine sand is possible if the wick drains punched into the bottom sand. Precautions were taken to prevent such a possibility during the installation of the wick drains in the test pad areas. However, one cannot completely rule out this possibility due to natural variations in the depth to the bottom sand and operator errors in identifying wick drain termination depths.

Size effects of the test area also impact the efficiency. Efficiency loss due to pervious lenses is only a factor at the site boundaries. The ratio of relative displacement capacity of the pump to the volume of leakage should control the decrease in efficiency of the system from these losses. In a small test site like the one used in this study, this boundary leakage would have a relatively large influence. As the site area increases, the losses across the boundary become a proportionally smaller percentage of the total volume of fluid extracted during consolidation and the relative efficiency of the system will increase. Alternatively, a deeper perimeter cut-off system would also reduce the loss due to perimeter leakage through pervious lenses. In such a case, the presence of pervious layers would become beneficial in accelerating the consolidation process.

Settlements

Settlement measurements indicate that the total ground surface settlement was about 7.5 to 9 cm at the two MPBX's. Due to the excavation of 2 m of soil in the test area, subsurface soils were subjected to stress relief prior to application of the vacuum consolidation load. Subsequent vacuum-consolidation induced recompression settlement by increasing effective stress from the initial excavated state of ground at the test pad (at EL. 4.5 m) (profile B, Fig. 6b) to the state of stress (A, Fig. 6b) prior to excavation, and then virgin compression settlement from the effective stress prior to excavation (A) to the final effective stress condition (C) at the end of the test. During the recompression portion of consolidation, only small settlements occurred. Based on the representative recompression ratio of 0.02 (TETC 1990), the estimated recompression settlement is about 2 cm. Subsequent virgin consolidation settlement was estimated to be about 7.5 cm (virgin compression ratio of 0.15 in the test pad area; TETC 1990) leading to a total settlement of about 3.75 inches, similar to the observed total settlement of 9 cm inches. The total rebound measured after releasing the vacuum pressure was about 1.5 cm. This compares favorably with the estimated rebound value of about 2.5 cm.

For comparison purposes, the reference embankment settled about 15 cm, whereas the vacuum test pad settled only 9 cm. However, it should be noted that the test embankment was sized for a vacuum efficiency of 70 percent, compared to the average efficiency of only about 40 to 50 percent attained by vacuum.

CONCLUSIONS

The test pad achieved an average vacuum efficiency of between 40 and 50 percent compared to the target value of 70 percent. This field efficiency is equivalent to approximately 3 m of surcharge. The membrane seal, location of the vacuum-water interface, depth to the water table, and depth to an impervious stratum below the water table appeared to be contributing factors in the loss of efficiency of the vacuum-assisted consolidation system at the test site.

The potential for pre-installation of the vacuum (or under-pressure) collection system, either within an underlying drainage layer or using prefabricated geosynthetic drainage elements within the hydraulic fill, appears to offer considerable advantages in increasing system efficiency and the maximum achievable equivalent surcharge for future projects. Prefabricated elements within the fill during placement would offer the added potential benefit of accelerating the rate of self weight consolidation of the dredge slurry and, if spaced closely enough, eliminating the need to install vertical wick drains after fill placement to accelerate consolidation. Thus, such a system offers the benefit of further cost reduction due to elimination of the need for vertical drains, increased improvement by mobilizing hydrostatic water pressure for improvement, and significantly reduced time for fill stabilization due to acceleration in the rate of self weight consolidation. Prospects of such a system is summarized elsewhere (Thevanayagam 1993).

ACKNOWLEDGMENTS

The authors wish to acknowledge all individuals who generously participated in the field testing program. Special thanks are due to Messrs. A. Birkenbach, Asst. Chief harbor Engineer, and G. Alberio, Contract Administrator, of Port of Los Angeles for their support and contributions. The engineering services described herein were provided by The Earth Tech. Corp., Long Beach, California, and the construction services were provided by The Nilex Corp. as part of the Pier 300 42-acre Site Ground Modification Project for the Port of Los Angeles.

APPENDIX - REFERENCES

Choa, V., (1989). "Drains and vacuum preloading pilot test." *Proc. 12th ICSMFE*, Rio de Janeiro, 1347-1350.

Cognon, J. M. (1991). "Vacuum consolidation." *French Geotechnique*, 57(3), 37-47.

Halton, G. R., Loughney, R. W., and Winter, E. (1965). "Vacuum stabilization of subsoil beneath runway extension at Philadelphia International Airport." *Proc. 6th ICSMFE*, Montreal, 62-65.

Holtz, R. D. (1975). "Preloading by vacuum: current prospects." *Transportation Research Record 548*, TRB, Washington, D. C., 26-29.

Kjellman, W. (1952). "Consolidation of clay soil by means of atmospheric pressure." *Proc. Conf. on Soil Stabilization*, MIT, Cambridge, 258-263.

Liu, Yi-Xong. (Undated). "History and present situation of soft soil foundation treatment in Xingang, Port of Tianjin," *Research Report*, First Navigation Engineering Bureau, Ministry of Communications, Tianjin, China.

Port of Los Angeles (1987). "2020 OFI Study: cargo handling operations, facilities, and infrastructure requirements," Prepared by Vickery-Zackerman-Miller for the Port of Los Angeles, Los Angeles, California.

Shinsha, H., Watari, Y., and Kurumada, Y. (1991). "Improvement of very soft ground by vacuum consolidation using horizontal drains." *Proc. Int'l. Conf. on Geotech. Eng. for Coastal Development (Geo-Coast 91)*, Yokahama, Japan, I, 387-392.

The Earth Technology Corporation (TETC), (1990). "Geotechnical investigation for the Pier 300 42-acre landfill ground modification project, Volumes I and II," Report submitted to the Port of Los Angeles, Los Angeles, California.

Thevanayagam, S. (1991). Personal Communications with V. Choa, Nanyang Institute of Technology, Singapore, April.

Thevanayagam, S. (1993). "Prospects of vacuum assisted geosystems for ground improvement at Pier 400," *Internal Report*, prepared for Port of Los Angeles, Los Angeles, California.

CONSOLIDATION OF GROUND RECLAIMED BY INHOMOGENEOUS CLAY

Hiroyuki Tanaka[1]

ABSTRACT: A large amount of prefabricated band drain was installed to improve the ground reclaimed by soft soils at the enlargement project of the Tokyo International Airport. By analyzing data obtained at this site, the paper presents how to determine the design parameters for the ground with a large variation in consolidation properties. It is found that the coefficient of consolidation, c_h, which was back-calculated from the observed settlement curve, depends on drain spacing. Characteristics of the secondary consolidation is also studied by using the gradient of the settlement curve after the primary consolidation.

INTRODUCTION

The coastal areas in Japan are often reclaimed by soft dredged soils because it becomes difficult to obtain sand due to environmental restrictions. For the utilization of the land reclaimed by such soils, a vertical drain method is used to reduce residual settlement. However, contrary to natural deposit soil, a large variation in properties of the reclaimed soils harasses geotechnical engineer how to determine design parameters.

Prefabricated band drains are extensively used in Japan due to light weight of construction machines and shortage of good quality sand as drain material. Although there are design codes for the sand drain method and its applicability has been verified by many observations, there still remain some uncertainty; i.e., an effect of smeared zone and the secondary consolidation. In addition to such classical

---

[1] Chief of Geotechnical Survey Laboratory, Port and Harbour Research Institute, Ministry of Transport, Nagase 3-1-1, Yokosuka, Japan.

problems, the shape of the band drain gives rise to other problems; i.e., the equivalent diameter and the well resistance.

A large amount of band drain was installed and settlement was observed at many points at the offshore expansion project of the Tokyo International airport, which is being undertaken by the Ministry of Transport. This paper presents the performance of the band drain at the reclaimed land by dredged soils.

GROUND CONDITIONS

The Tokyo International Airport is under construction to transfer its facilities to the reclaimed land on the offshore of the present airport (Katayama 1991). After the completion of the project as shown in Fig. 1, the airport will have three runways to deal with 1.5 times the present capacity of airplanes. The project has been carried out in three phases. A runway was constructed as the first phase, where the band drain method was employed. The construction was started in January, 1984 and was completed in July, 1988.

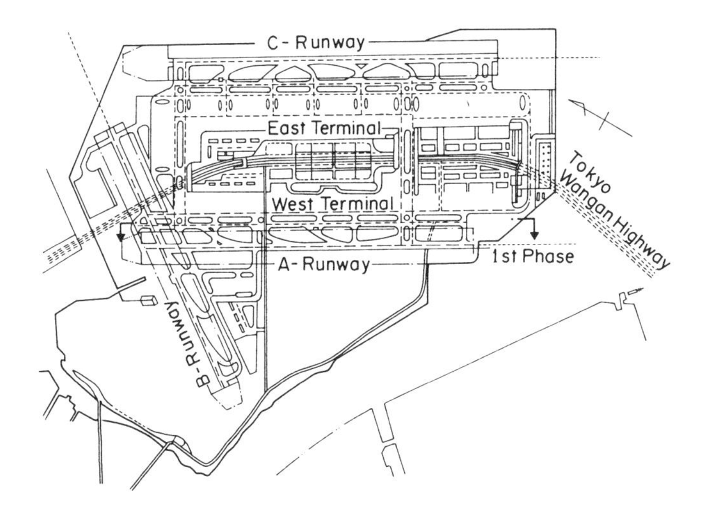

FIG. 1. Tokyo International Airport Expansion Project

The process of the reclamation at this site is illustrated in Fig. 2. Before 1965, this area was covered with a thick sand layer (AS layer) to the sea surface. Between 1966 and 1970, this sand layer was excavated as fill material for the present airport. Afterward the excavated area was reclaimed by clayey soils (AC_1 layer) which was dredged at Tokyo Bay. The surface of the site was then covered with sandy soils (BS layer) which was transferred from the excavation work at the urban area of Tokyo.

Fig. 3 shows soil properties before the BS layer was filled. The AS layer consists of relatively uniform grained sand so that it can be considered as a permeable layer for the band drain. Under the AS layer a alluvium clay layer (AC_2) is found which is slightly overconsolidated against the original ground level due to the secondary consolidation or the cementation effect. After the BS layer is filled, however, the AC_2 layer becomes normally consolidated, and the final settlement in this layer is predicted to be about 60 cm.

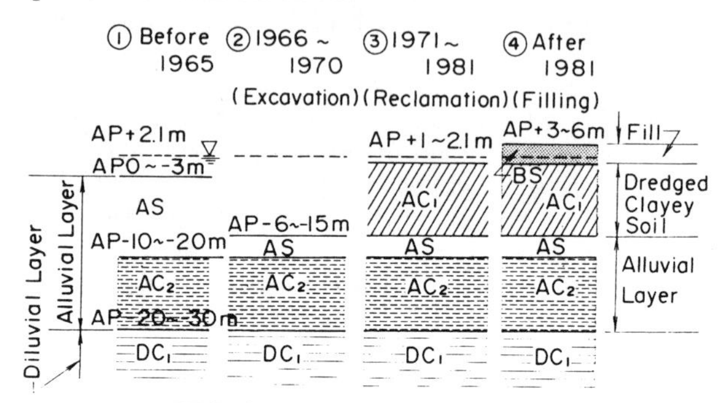

FIG. 2. Reclamation Process

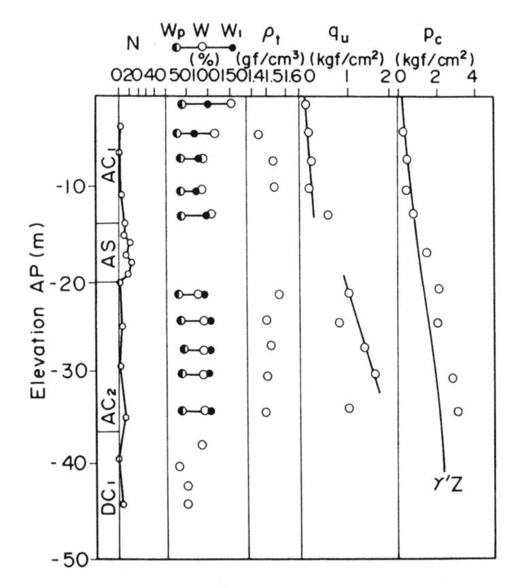

FIG. 3. Typical Soil Profile

Since the AC_1 layer was reclaimed by dredged soils, its properties varies greatly horizontally and vertically. The values of unconfined compression strength, q_u are less than 30 kPa and the water content exceeds its liquid limit. Fig. 4 shows coefficient of consolidation in the normally consolidated state, c_v obtained by laboratory oedometer tests for the AC_1 layer.

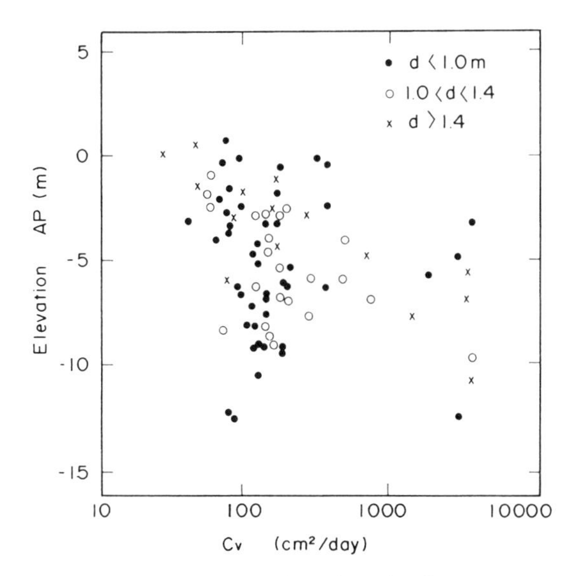

FIG. 4. Distribution of c_v of AC_1 Layer

According to the code of the Civil Aviation Bureau of the Ministry of Transport, the maximum gradient of a runway should be less than 0.5% for safe operation. The consolidation of the AC_2 layer was predicted to cause small differential settlement on the runway since the thickness and consolidation properties of the AC_2 layer, which is a natural deposit, do not significantly change in places. Therefore, it was decided to apply the vertical drain method only to the AC_1 layer.

INSTALLATION OF VERTICAL DRAIN

Considering workability and cost of drain installation, prefabricated band drains were adopted as a vertical drain. While many types of the band drain are available in Japan, four types were used for this project. Three of the drains (B, C, D types in Fig. 8) have a synthetic resin core wrapped with non-woven synthetic filter. Type A consists of non-woven synthetic material with no core. The procedure of the installation of drain is illustrated in Fig. 5 and the brief explanation is as follows:

(1) The surface of the AC_1 layer was covered with the BS layer to the plan elevation, AP (Arakawa Pale) +5.5 m.

(2) The part of the ground surface was excavated to make a subgrade for the pavement of the runway.

(3) The subgrade, which played a role of sand mat for vertical drain, was spread.

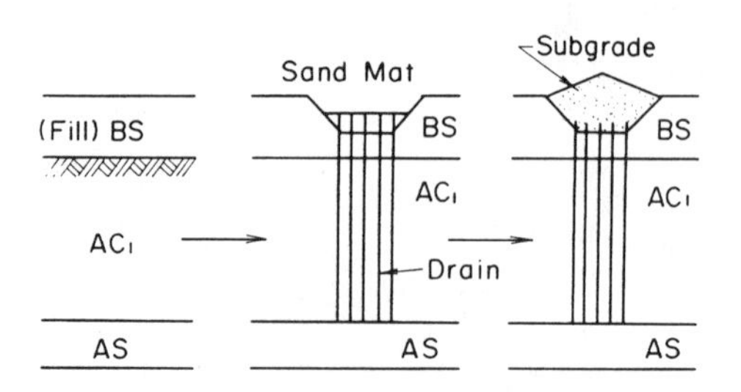

FIG. 5. Drain Installation

(4) Band drain was statically installed to the AC_1 layer until the end of drain reached the AS layer, by the displacement method with a mandrel whose size and shape were slightly different with types of band drain.

(5) Immediately after the installation, a surcharge equivalent to the weight of the pavement was applied.

Settlement was observed at many points to assure the end of the primary consolidation and to evaluate the design parameters, especially c_h, for the next construction phases. The settlement was measured independently for each soil layer. This report analyzes the settlement yielded in the AC_1 layer.

Observed Settlement

Until now, settlement has been observed at many construction sites, and many papers have been reported on consolidation of the ground treated by vertical drains. Compared with the previous studies, the data analyzed in this paper have the following important features from a geotechnical view point:

(1) The layer treated by vertical drain is young; i.e., it was placed 10 or 20 years ago.

(2) Due to filling with dredged soils, the properties are extremely variable.

(3) The consolidation in the vertical direction may be considered to be one-dimensional because the width of treated area was wide enough.

(4) As shown in Fig. 5, vertical drain was installed after spreading BS layer. This means that the consolidation proceeded under a constant loading condition.

(5) At most points, the term of observation of settlement was long enough to confirm the end of the primary consolidation.

Fig. 6 is a typical example of the observed settlement curve plotted against the elapsed time in logarithm scale. In the figure the elapsed time is plotted as the starting time is one when the vertical drain was installed. The following values are defined to analyze the consolidation behavior. Referring to Casagrande's method, which is usually used in the oedometer test, the curve is divided into two parts by the intersecting point of their tangents. The first portion may be considered to be the primary consolidation and the second one is the secondary consolidation. An elapsed time at the intersecting point is defined as t_{100} and the settlement at this point is as ρ_{100}. Furthermore, a half of ρ_{100} is as ρ_{50}, and its corresponding time is as t_{50}. The rate of the secondary consolidation, s is defined as the gradient of the settlement curve at the secondary consolidation.

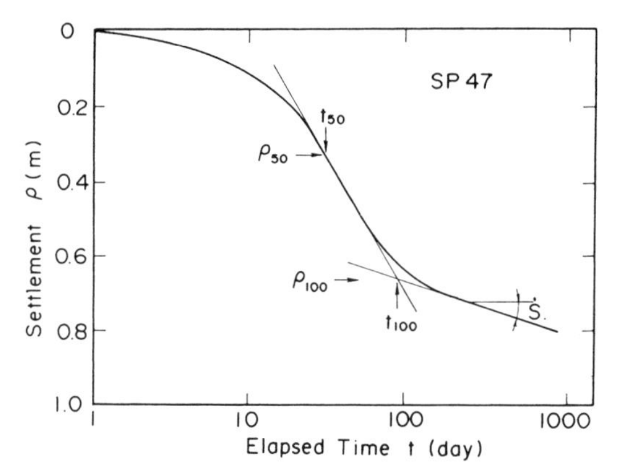

FIG. 6. Observed Settlement Curve

Effect of Drain Length

The well resistance has been one of the most important problems in the design of the band drain, because the cross sectional area of the drain is considerably smaller than that of the sand drain. Effect of the drain length on the rate of consolidation is examined by Fig. 7. A coefficient of consolidation in the horizontal direction, c_h was calculated using Barron's equation, assuming that an equivalent diameter of the band drain is 5 cm and t_{50} (Fig. 6) is the elapsed time at 50 % of degree of consolidation, U_{50}.

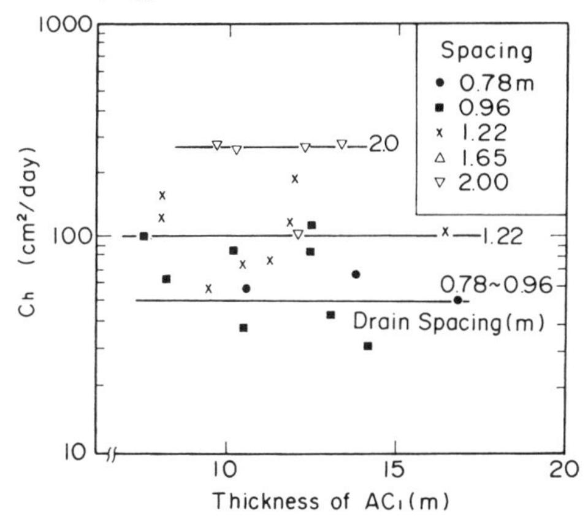

FIG. 7. Effect of Drain Length on c_v

It can be seen in Fig. 7 that the calculated c_h is constant for the thicknesses of the AC_1 layer, which are the same as the length of band drain. It may be concluded that the well resistance is negligible provided that the drain length is less than 17 m and double drainage is expected.

EFFECT OF DRAIN SPACING

The values of c_h seem to be considerably dependent on drain spacing, d, as shown in Fig. 7. To examine in more detail an influence of drain spacing on the rate of consolidation, the values of c_h are plotted against d in Fig. 8. When d is wider than 1.5 m, the values of c_h seem to be constant, and its values is around 250 cm^2/day. On the other hand, when d becomes less than 1.5 m, the values of c_h decreases with decreasing d. Aboshi et al. (1984) also reported the similar dependency on the drain spacing in the case of sand drain. As in Figs. 4 the values of c_v are plotted using different symbols according to d, it cannot be recognized that c_v with some drain spacing is differently distributed. In addition, although four types of drain were used at this site, it does not seem that c_h values are affected by drain types.

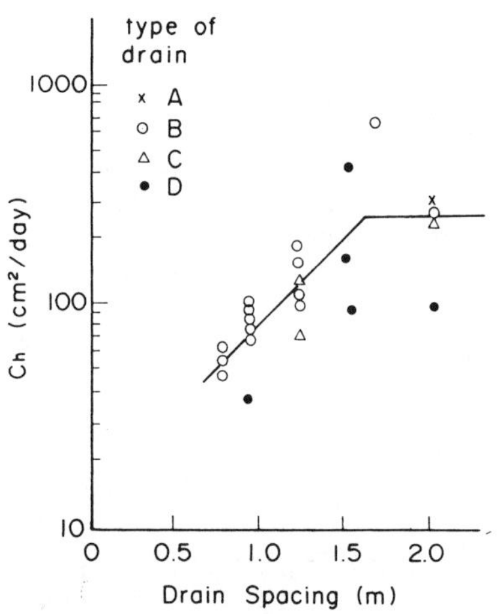

FIG. 8. Effect of Drain Spacing on c_v

It has been said that the installation of vertical drain disturbs the soil structure and reduces permeability of the soil adjacent to the drain. This disturbed zone is called a smear. Since the size of the smeared zone may be independent of drain spacing, the effect of smear becomes relatively great in the narrow drain spacing.

Hansbo (1981) proposed a consolidation formula which can take account of the effect of smear.

$$U\,(T_h) = \exp\left[\frac{-8T_h}{m_s}\right] \tag{1}$$

where $m_s = \ln(n/s) + (k_c/k_c')\ln(s) - 3/4$
$T_h = c_h t/d_e^2, \; n = d_e/d_w, \; s = d_s/d_w$

T_h is the time factor, t is elapsed time, d_e is the diameter of an equivalent cylinder of drained soil (=1.13d), d_w, is the equivalent diameter of band drain (5 cm), d_s is the diameter of smeared zone, k_c and k_c' are coefficients of permeability of undisturbed soil and of smeared zone, respectively.

Fig. 9 shows the calculation result where t'_{50} is a time to reach U_{50} without consideration of the smear effect, and t_{50} is a time for U_{50} calculated by Eq. (1). When the ratio of k_c/k_c' or values of s are larger, the ratio of t_{50}/t'_{50} decreases with decreasing drain spacing. Nevertheless, the reduction rate of c_h is not so large as the observed value in Fig. 8. It may be impossible to explain the reduction of c_h at narrow drain spacing only by the smear.

Fig. 10 shows a result of the site investigation using the piezocone which can measure pore water pressures, u, as well as penetration resistance, q_t and friction, f_s. It can be seen that the values of u drop at some depths of the AC_1 layer, which means that there are some permeable layers in the AC_1 layer. It is considered that these layers were formed in the process of the reclamation of the AC_1 layer. The layer may be not extensive enough to have a function of drainage for the consolidation of the AC_1 layer in the vertical direction. However, once vertical drain is installed, in addition to the horizontal flow to the drain, the vertical water flow occurred as shown in Fig. 11. The influence of permeable layers is studied using the Carrillo's formula, assuming that c_h is equal to that in the vertical direction. The calculation result is shown in Fig. 12, where t_{50} and t'_{50} so are the same definitions as in Fig. 9, and $2H_d$ is a distance of permeable layers in the vertical direction. If H_d is assumed to be 1 or 2 m, the change in the ratio of t_{50}/t'_{50} so are very similar to observed one. It may be concluded that the dependency of c_h on drain spacing is not attributed to the smear, but the existence of permeable layers in the AC_1 layer.

Comparison with Laboratory Tests

Yoshikuni (1979) presented that for the naturally deposited layer improved by vertical drain, the values of c_h calculated by the observed settlement are in the range of 0.5 to 2.0 times those of c_v from laboratory tests. On the other hand, for the AC_1 layer which was artificially reclaimed, the values of c_v vary horizontally and vertically. Fig. 13 shows the histogram of c_h from observed settlement in comparison with c_v obtained from laboratory oedometer tests. The data from the field observation are restricted to those whose spacing is less than 1.4 m, because c_h with drain spacing larger than 1.4 m seems to be affected by the permeable layers, as mentioned above.

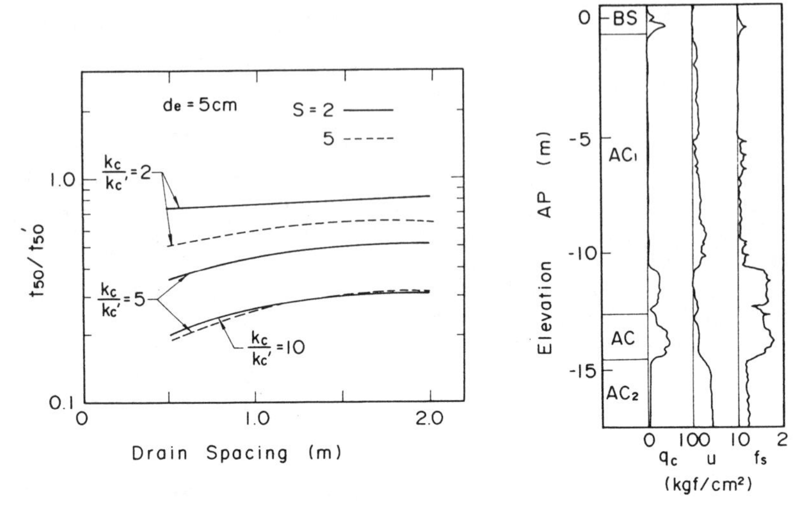

FIG. 9. Examination of Smear Effect by Hansbo Method

FIG. 10. Site Investigation from Piezocone

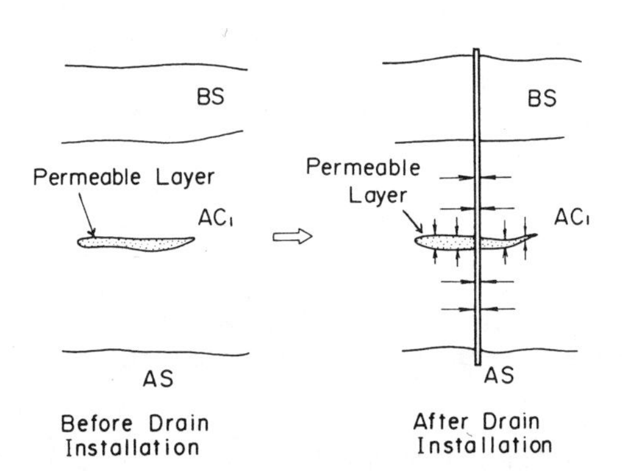

FIG. 11. Effect of Permeable Layer

The values of c_h obtained from the settlement are distributed on the left side of c_v from laboratory tests. That is, the consolidation does not proceed as the ground with the average value of c_v, but as with the minimum values of c_v. This may be because since the soil with a large value of c_v has a tendency to show low compressibility, settlement is apparently governed by the layers whose c_v is small.

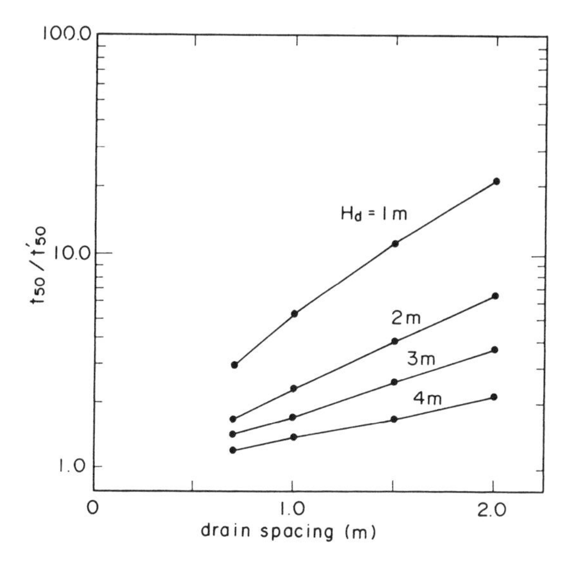

FIG. 12. Examination of Permeable Layer

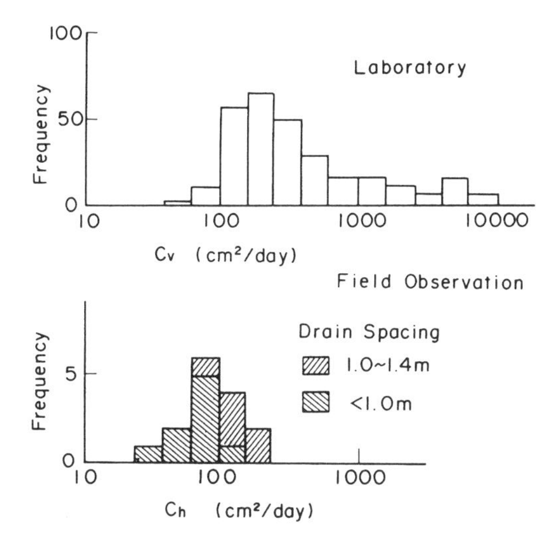

FIG. 13. Comparison of c_v Obtained from laboratory and Field

SECONDARY CONSOLIDATION

It is widely recognized that the settlement due to the secondary consolidation is proportional to the logarithm elapsed time. There do not exist any serious problems when it takes long term to complete the primary consolidation, for example, the vertical drain is not installed. However, if the vertical drain is adopted to accelerate primary consolidation, then the settlement due to the secondary consolidation cannot be ignored.

Characteristics of the secondary consolidation is commonly expressed by a term of the coefficient of secondary consolidation, C_α. C_α can be related to other parameters, as indicated in Eq. (2).

$$\frac{s}{H} = \epsilon = \frac{C_\alpha}{(1 + e)} \tag{2}$$

where s is the gradient of the settlement curve at the secondary consolidation, as defined in Fig. 6 and ϵ is a strain rate due to the secondary consolidation. H is a thickness of the AC_1 layer and e is a void ratio.

Fig. 14 shows the relationship between ϵ and the settlement during the primary consolidation, ρ_{100}. In the figure a special relation of $\epsilon = 0.2\epsilon_c$ can be seen, where ϵ_c is the strain during the primary consolidation. From Terzaghi's equation, ϵ_c can be obtained by Eq. (3).

$$\epsilon_c = \frac{C_c}{(1+e)} \log\left[\frac{\Delta p + p_o}{p_o}\right] \tag{3}$$

where p_o is effective overburden pressure before consolidation pressure is applied, and Δp is consolidation pressure. When Eqs. (2) and (3) are combined, Eq. (4) can be obtained.

$$\frac{s}{\rho_{100}} = \frac{\epsilon}{\epsilon_c} = \frac{\dfrac{C_\alpha}{C_c}}{\log\left[\dfrac{\Delta p + p_o}{p_o}\right]} \tag{4}$$

Taking into consideration that the values of $(\Delta p + p_o)/p_o$ are in the range of 1.5 and 2.0 at this site, and using the relation of $\epsilon = 0.2\,\epsilon_c$ from Fig. 14, the ratio of C_α/C_c becomes 0.036 to 0.060 from Eq. (4). The range of C_α/C_c ratio agrees fairly well with that from the laboratory test reported by Mesri and Choi (1985). It can be concluded that ρ_{100} is a very useful value to predict the settlement after the primary consolidation.

CONCLUSIONS

From the observed settlement at the ground improved by prefabricated band drain, the following conclusions can be drawn:

(1) The well resistance can be ignored. However, it must be noted that at the site of the case study, the drain length is less than 17 m and double drainage is expected.

(2) The value of c_h back-calculated from the observed settlement depends on drain spacing. The main reason may be due to the existence of permeable layers in the AC_1 layer.

(3) The values of c_h from observed settlement correspond to the minimum values of c_v from the laboratory test.

(4) The rate of settlement due to the secondary consolidation is proportional to the settlement during the primary consolidation.

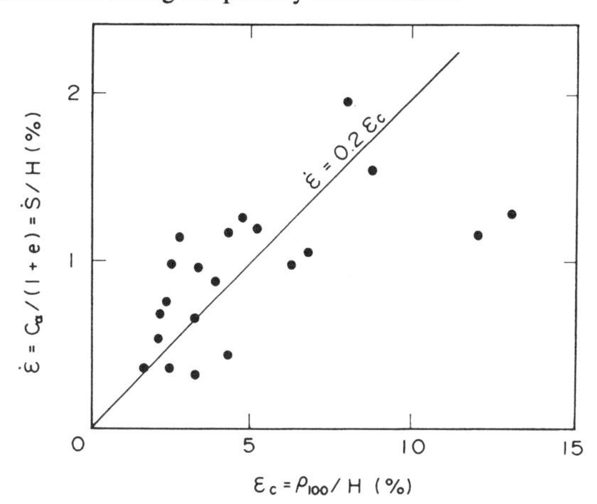

FIG. 14. Relationship between ϵ_c and ϵ

Appendix - References

Aboshi, H., et al. (1984). "The change of consolidation coefficient in driving sand drains." *Proc. 19th Annual Convention of the Japanese Society of Soil Mechanics and Foundation Engineering*, 1573-1574, (in Japanese).

Hansbo, S. (1981). "Consolidation of fine-grained soils by prefabricated drains." *Proc. 10th ICSMFE*, Stockholm, 1, 677-862.

Katayama, T. (1991). "Meeting the challenge to the very soft ground - the Tokyo International Airport offshore expansion project." *Proc. of Geo-Coast '91*, Yokohama, 954-967.

Kobayashi, M., et al. (1990). "Determination of the horizontal coefficient of consolidation c_h.' *Report of Port and Harbour Research Institute*, 29(2), 63-84, (in Japanese).

Mesri, G. and Choi, Y. K. (1985). "Settlement analysis of embankments of soft clay." *J. Geotech. Eng.*, ASCE, 111(4), 441-464.

Yoshikuni, H. (1979). "The role of consolidation theory (part 1)." *Tsuchi-to-kiso JSSMFE*, 27(2), 105-111, (in Japanese).

FIELD TEST OF THERMAL PRECOMPRESSION

Tuncer B. Edil,[1] Member, ASCE, and
Patrick J. Fox,[2] Associate Member, ASCE

ABSTRACT: A new concept for improvement of soft ground using moderate heating to control post-construction settlement is presented. The long-term creep behavior of a peat is presented based on laboratory oedometer tests involving step-stress and step-temperature changes. Field evidence with regard to the effect of temperature on one-dimensional creep of peat deposits is given based on the performance of two test fills, one of which is equipped with a ground heating system. The feasibility of thermal precompression as a method of accelerating in situ settlement has been discussed.

INTRODUCTION

With the recent advances in soil reinforcement, construction of embankments over soft ground has become primarily a problem of controlling excessive settlements. Peats and organic soils are well known for their high compressibility and long term settlement. In many cases, the majority of settlement results from creep at constant vertical effective stress. Although extensive studies of this creep behavior have been performed, work has been limited to isothermal conditions. It is known that temperature effects are important for compression of inorganic soils, however, the effects of temperature on settlement behavior of peats have not been reported.

For field applications, peat soils must generally be improved for construction. Preloading techniques, in which a surcharge load is placed on a site prior to construction, have been used successfully in this regard. The main disadvantage of

---

[1] Professor, Department of Civil and Environmental Engineering, University of Wisconsin-Madison, 1415 Johnson Drive, Madison, WI 53706.

[2] Assistant Professor, School of Civil Engineering, Purdue University, West Lafayette, IN 47907.

the technique is the time required for settlement to occur prior to construction. A new technique, called the "thermal precompression method", is described herein to accelerate the preloading process for organic soils. Long-term settlement has been shown to be practically eliminated in the laboratory using this procedure (Fox 1992).

The objective of this paper is to present laboratory and field evidence with regard to the effect of temperature on one-dimensional settlement of peat deposits from Middleton, Wisconsin. The thermal precompression method is introduced as a means to accelerate settlement of organic soils and control post-construction creep.

BACKGROUND

Most observers have found that, for clays, an increase in temperature during an oedometer test causes a small reduction in void ratio but a negligible change in compressibility (Gray 1936; Laguros 1969; Mitchell 1969; Plum and Esrig 1969). Using data from drained triaxial tests on illite, Campanella and Mitchell (1968) showed that a significant permanent volume decrease may result from increasing temperature. This behavior is explained in terms of irreversible physico-chemical structural adjustments that are necessary for the soil to carry the effective stress at the higher temperature.

Disagreement exists in the literature concerning the effect of temperature on the rate of secondary compression for inorganic soils. Gray (1936) reported that a change of temperature during secondary compression causes a corresponding change in settlement rate. Lo (1961) states that a temperature increase of only 3°C alters the shape of the secondary compression settlement curve. Mesri (1973) reported that the coefficient of secondary compression for both normally and overconsolidated clays and silts is independent of testing temperature. Houston et al. (1985) found that, when heated under triaxial conditions, smectite-rich ocean floor sediments undergo a thermal secondary compression in which volumetric strains are typically linear with the logarithm of time. They also report a significant reduction in the rate of thermal secondary compression when a sample is cooled.

LABORATORY EXPERIMENTS

The site from which peat samples were obtained and test embankments were later constructed is currently farmland adjacent to U. S. Highway 14 in Middleton, Wisconsin. The site is underlain by two peat deposits; an upper fibrous peat (1.5 to 2.3 m) and a lower sedimentary peat (2.3 m to 4.3 m). There is an 0.6 m thick organic silt layer under the peats. Other soils are inorganic (see Fig. 6). The fibrous peat is a dark brown, poorly-humified peat which has the following average index properties: fiber content = 50%, water content = 550%, organic content = 93%, and initial void ratio = 10.5. It is lightly overconsolidated and has a preconsolidation stress of approximately 25 to 50 kPa. The sedimentary peat is an amorphous peat which has the following average properties: fiber content = 20%, water content = 436%, organic content = 60%, and initial void ratio = 6.3.

Variable-temperature oedometer tests were conducted on standard-size specimens (diameter = 63 mm, height = 25 mm) cut from hand-excavated block samples in identical brass oedometers. One-way drainage was permitted with pore pressure measurements taken at the base of each specimen. Loads were applied using dead weights and temperatures were controlled by immersing each cell in a heated water bath. Using this system, accurate temperatures were maintained ($\pm 1°C$) and specimen temperature could be changed rapidly, usually within 30 minutes. The impact of oedometer ring expansion due to temperature changes was investigated and found to account for less than 5% of the measured settlement for the thermal tests conducted (Fox 1992).

The variable-temperature testing program was designed using a procedure called the "step-temperature" test. A step-temperature test is performed by rapidly increasing soil temperature during creep and evaluating the resulting step change of void ratio rate, $\dot{e}$ (i.e. time rate of void ratio change, de/dt) at a given value of void ratio and vertical effective stress, σ'_v. In this way, two void ratio rates are obtained for two corresponding temperatures under conditions of essentially the same soil structure (Mitchell et al. 1968). For each peat specimen, step-temperature tests were performed at increasing vertical stress levels. To avoid complications arising from overconsolidation, normally consolidated stress conditions were maintained for each specimen.

Fig. 1 illustrates the typical response of Middleton peat to the step-temperature procedure. This schematic diagram is a plot of void ratio rate versus void ratio for a single step-temperature increment. The progression of the test on Fig. 1 is from right to left, in the direction of decreasing void ratio. EOP indicates end of primary consolidation and the beginning of creep settlement and it is defined as the time when the excess pore pressure is reduced to 0.7 kPa (the limit of accuracy of the measuring device). Temperature is increased at void ratio e_1 and the values $\dot{e}_1$, T_1 and T_2 are known. The void ratio rate $\dot{e}_2$ is found by extrapolating the creep curve under the higher temperature backwards to the void ratio at which the temperature change occurred (Mitchell et al. 1968; Andersland and Douglas 1970). Primary consolidation does accompany such a temperature increase at constant stress. However, for peat, permeability is relatively large (typically 1×10^{-4} cm/s at low stress levels) and these effects do not prevent extrapolation to find the strain rate at the point of temperature increase. Similar to the step-temperature tests, "step-stress" tests were performed to investigate the stress effects on creep rate. In the latter tests, stress is increased while maintaining constant temperature. Using a procedure similar to the one described for the step-temperature test in Fig. 1, two void ratio rates are determined at a single void ratio corresponding to two effective stress levels (measurements are made in the creep range where pore pressure is zero). The analysis of these step-stress and step-temperature tests is described elsewhere in more detail (Fox 1992; Edil et al. 1994; Fox and Edil 1994).

TEMPERATURE AND STRESS EFFECTS ON CREEP RATE

Fig. 2a shows a time-settlement curve for the 100-104-108.2 kPa load increment of a test on the fibrous peat. At the beginning of this increment, vertical stress is 100 kPa and specimen temperature is 24°C. At 600 minutes of elapsed time, a step-stress test was conducted and vertical stress was increased to 104 kPa.

At 1500 minutes of elapsed time, a step-temperature test was conducted and specimen temperature was rapidly raised from 24 to 35°C. The resulting break in slope indicates an increase in creep rate. Pore pressure also increased in response to the temperature change, as shown in Fig. 2b, and the end of primary consolidation (EOP) is marked. Another step-stress test, from 104 to 108.2 kPa, was later performed at the higher temperature. Fig. 2b shows the complete pore pressure and temperature history for the same load increments. Fig. 2c shows the log $\dot{e}$ vs. e plot for the step-temperature portion only. The effect of increased temperature on settlement rate is clearly seen. For the 11°C temperature rise, void ratio rate increased by a factor of 16.

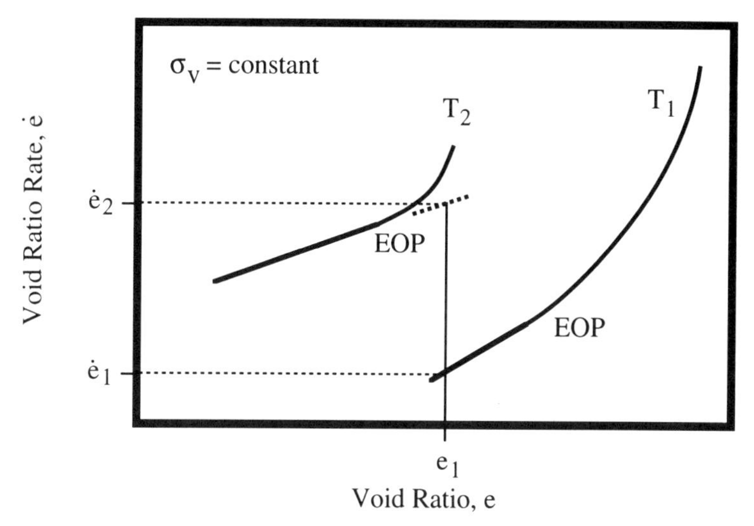

FIG. 1. Step-Temperature Analysis Procedure

Two void ratio rates ($\dot{e}_1$ and $\dot{e}_2$), corresponding to two temperatures (T_1 and T_2) for each step-temperature test and to two effective stresses (σ'_{v1} and σ'_{v2}) for each step-stress test, can be obtained at a given void ratio. To correlate the temperature and stress changes to void ratio rate increases, a "thermal coefficient of creep", C_T, and a "stress coefficient of creep", C_σ are defined as,

$$C_T = \frac{\ln\dot{e}_2 - \ln\dot{e}_1}{T_2 - T_1} \quad (1)$$

and

$$C_\sigma = \frac{\ln\dot{e}_2 - \ln\dot{e}_1}{\sigma'_{v2} - \sigma'_{v1}} \quad (2)$$

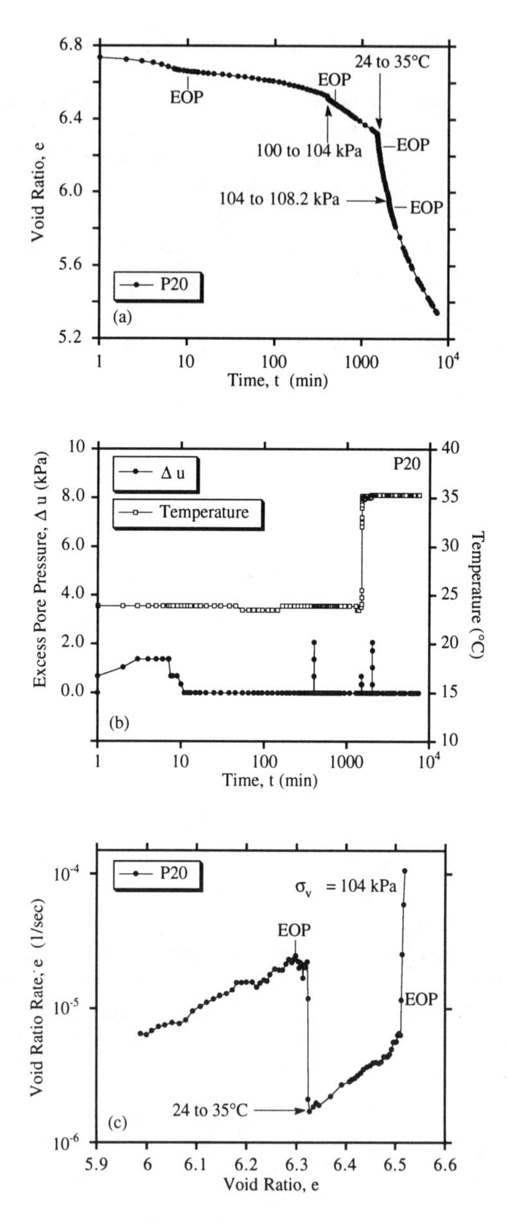

FIG. 2. (a) Time-Settlement Curve; (b) Temperature and Excess Pore Pressure; (c) Void Ratio Rate versus Void Ratio

The step-temperature test shown in Fig. 2 gives $C_T = 0.252$. Fig. 3 shows that C_T is independent of void ratio for all step-temperature tests performed covering a temperature range of 15 to 65°C and including temperature steps of 10°C to 45°C. Similar plots indicate that C_T is also independent of vertical stress level and magnitude of temperature change. In all cases, $C_T = 0.26 \pm 0.02$ for fibrous Middleton peat.

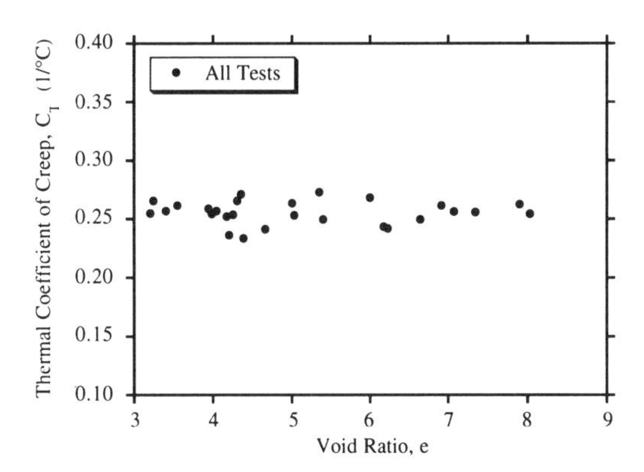

FIG. 3. Thermal Coefficient of Creep versus Void Ratio

Numerous step-stress tests indicated that C_σ is independent of temperature and stress history under monotonic, normally consolidated loading conditions but it depends on void ratio in accordance with the following relationship for fibrous Middleton peat (Fox 1992):

$$C_\sigma = \frac{0.0128}{\text{kPa}} \exp(0.367\,e) \tag{3}$$

Based on extensive laboratory tests performed on fibrous Middleton peat, the following equation is proposed to describe creep rate under constant vertical effective stress (Fox 1992),

$$\dot{e} = -C_f \exp(C_T T) \sinh(C_\sigma \sigma'_v) \tag{4}$$

where C_f is the one-dimensional fabric coefficient of creep and is a function of void ratio. This relationship has not yet been verified in the field because the field test and the laboratory characterization of the sedimentary Middleton peat are still underway.

THERMAL PRECOMPRESSION

It has been shown that raising the temperature of a peat sample under constant vertical stress causes an increase in settlement rate. It has also been found that when the sample is cooled again, the settlement rate reduces to a value lower than before it was heated. The effect is herein termed "thermal precompression" and has not been previously reported for peat soils.

On the basis of laboratory experiments, the thermal precompression method is proposed for reduction of long-term creep settlement of peats in situ. A peat site is first heated and then later cooled under moderate surcharge loading. During the heating phase, settlement is accelerated over that obtained from conventional preloading techniques alone. After the soil is cooled, further long-term settlement is greatly reduced. Creep rate does not completely cease upon cooling and limited data suggest that the C_T value for cooling is 0.18 (Fox 1992). Fig. 4 illustrates the typical laboratory response of fibrous Middleton peat to the thermal precompression method. Vertical stress was approximately 250 kPa throughout the course of heating and cooling. One day after the load was applied, the specimen was rapidly heated from 14 to 60°C and the settlement rate increased drastically. After allowing one week of settlement at this temperature, the soil was cooled to room temperature. Over the next nine weeks, virtually no additional settlement or rebound occurred. Using this method, the long term compression tendency of Middleton peat is virtually arrested.

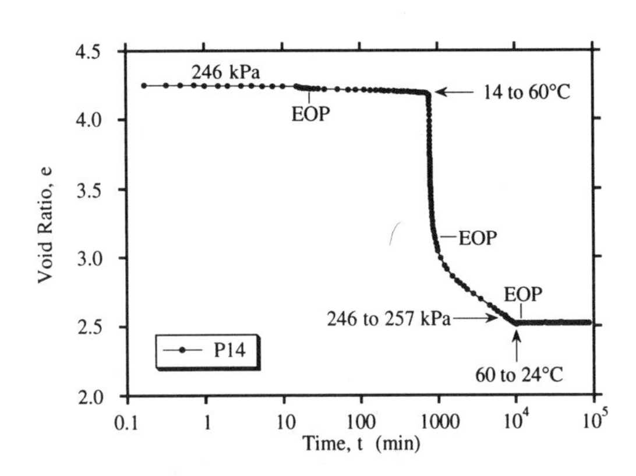

FIG. 4. Thermal Precompression Effect

For field applications, it is of interest to know the effect of additional loads on thermally precompressed peats. Fig. 5 shows a plot of e vs. log σ_v for a test on the fibrous peat. Temperature and stress were sequentially increased as shown. Temperature was then rapidly decreased from 60 to 26°C at 294 kPa. Thereafter, vertical stress was increased in increments to 490 kPa at 26°C. As seen from Fig. 5, little additional settlement occurred. It appears that soils subjected to thermal precompression develop a significant quasi-preconsolidation effect (Leonards and Ramiah 1959) and have very low compressibility with respect to further loading. This illustrates the potential of the thermal precompression method for improvement of peat soils.

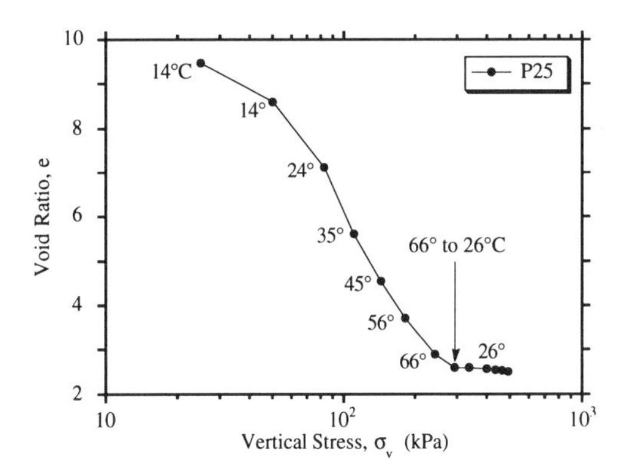

FIG. 5. Thermal Quasi-Preconsolidation Effect

FIELD TEST

In order to investigate the effectiveness and feasibility of the thermal precompression method of improving soft ground, two test fills were constructed at the Middleton site. Test Fill A, 13×13 m in plan, was constructed in 1990 to a height of 1.25 m after removing 0.3 m-top soil (Edil et al. 1991). Test Fill B, which constructed in 1992 adjacent to Test Fill A, had the same dimensions but was equipped with 12 U-shaped hot water circulation wells (each 6 m deep) in a pattern as shown in Fig. 6. Two additional heating wells were also installed outside the fill in an unloaded area. Both test fills were heavily instrumented by survey markers (on the surface of the fill), settlement plates (at the base of the fill), settlement forks (at the stratigraphic interfaces), piezometers (in various strata under the fills as well as in unloaded areas outside the fills), inclinometers (at the edge of the fills), and thermocouples (at various depths in numerous locations under Test Fill B as well as outside the heated area). The instrumentation is shown in Figs. 6 and 7.

Test Fill B was allowed to settle for three months before the heating of the ground was initiated on September 1, 1992. The heating system consisted of a propane gas water heater (12,500 W), a hot water circulation pump, heat control units, gas and water storage tanks, and the necessary piping network. The pipe network consisted of several branches serving groups of heating wells and it was equipped with sufficient number of critically located vents to de-air the system. Heat-resistant 13-mm CPVC pipes were used for the surface pipes and 25-mm steel pipe was used in the heating wells. The heating wells were installed by drilling through the upper silt layer (1.5 m thick) and then jacking the U-shaped pipes down to the desired depth The surface pipes and the fill surface were covered with insulation to minimize heat loss. Anti-freeze was added during winter months to prevent freezing of the circulating water in the closed system. The heating system was set to deliver hot water at a temperature of 65°C at the outlet of the heater and it was operated continuously insofar as possible. There were some periods it could not be operated due to maintenance problems as described below.

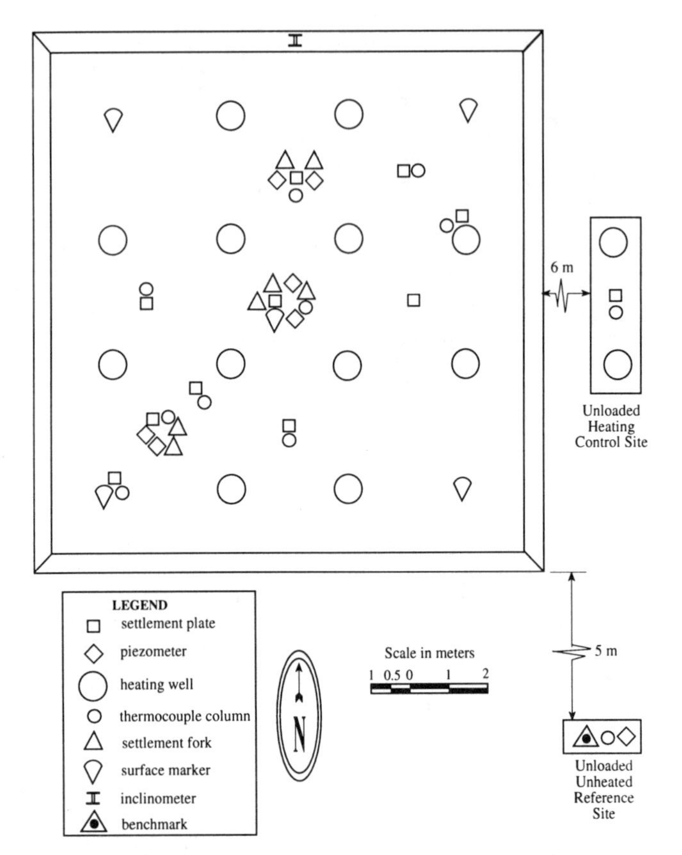

FIG. 6. Test Fill B Site Plan and Instrumentation Layout

Field Results

Fig. 8 shows the settlement at the center of Test Fills A and B due to equal fill loads. Also shown in Fig. 8 is the average peat temperature at the center of Test Fill B. Once heating commenced, the ground temperature rose steadily to a peak value of 21°C at 200 days. Over the next 150 days, the temperature dropped about 4°C due to several problems with the heater. Heating then resumed and the temperature rose steadily to about 30°C. Ground heating was then discontinued at about 500 days of surcharging. Based on the piezometer measurements, the excess pore pressures dissipated rapidly in the fibrous peat layer and more slowly in the sedimentary peat layer; however, EOP was reached within 75 days of surcharging in the upper 5-m compressible zone.

The area between the heating temperature and the ambient ground temperature curves is termed "heating index" and was calculated to be about 5,000

Celsius degree-days at the center of Test Fill B. Fig. 8 shows that the increase in rate of settlement coincides with the rise in ground temperature. The settlement-time curve for Test Fill B also reflects the relative effect of the heating and cooling periods. The rate of settlement as obtained from the settlement-time curves of the two test fills is plotted in Fig. 9. After 80 days, the rise in the rate of settlement of Test Fill B relative to the unheated Test Fill A indicates the effectiveness of heating in accelerating compression. Settlement continues during the cooling periods (between 200 to 350 days and after 550 days of surcharging) because temperatures remain elevated in the peat zone. Eventually, however, the rate of settlement does decline upon cooling.

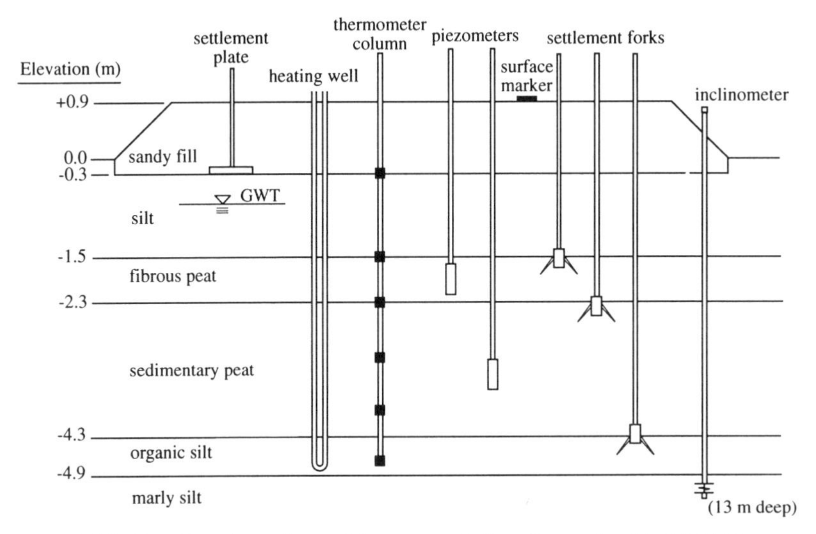

FIG. 7. Stratigraphy and Instrumentation Types for Test Fill B

To provide another measure of the effectiveness of moderate heating on acceleration of compression, the net settlement profiles of the unheated Test Fill A and the heated Test Fill B during the heating period (80 to 550 days) are given in Fig. 10. The center settlement of Test Fill B during this period was 118 mm more than that of Test Fill A. This corresponds to 100% additional settlement when considered over the entire area of the fills. This difference may seem less dramatic than the differences observed in the laboratory step-temperature tests. This is probably a reflection of the small temperature differential achieved in the field and the non-uniform and time dependent field heating compared to laboratory heating.

The field experiment is still underway. Laboratory tests are being performed to determine the thermal and mechanical characteristics of all soil units. An analytical model is being developed for the purpose of quantitative evaluation and optimization of this promising method of ground improvement. Development of efficient and cost effective ground heating technology is expected to follow the demonstration of the effectiveness of the thermal precompression method.

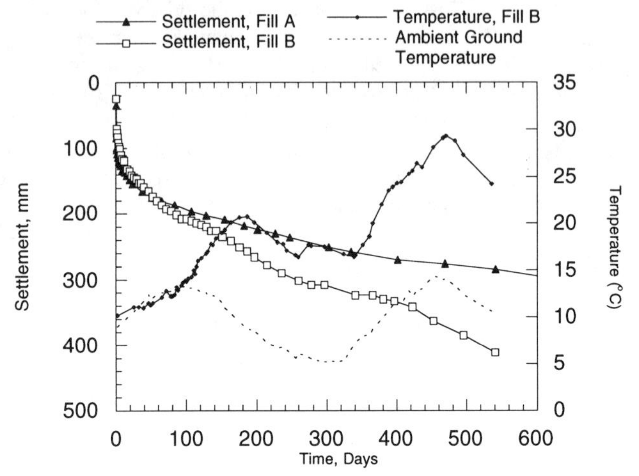

FIG. 8. Settlement-Time Curves for the Test Fills

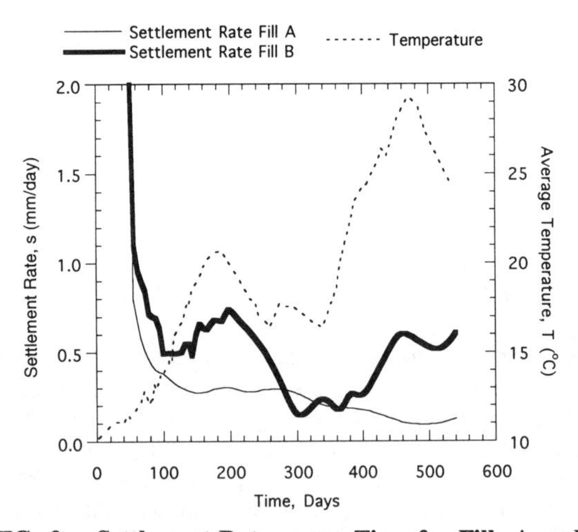

FIG. 9. Settlement Rate versus Time for Fills A and B

CONCLUSIONS

Laboratory and field evidence has been presented indicating that one-dimensional creep rate of Middleton peat is strongly affected by soil temperature. The magnitude of void ratio rate increase is exponentially related to the magnitude of temperature increase, suggesting that the creep behavior of peat is thermally activated. The thermal coefficient of creep, C_T is constant and is independent of void ratio, vertical effective stress and magnitude of temperature change for monotonic loading and heating conditions.

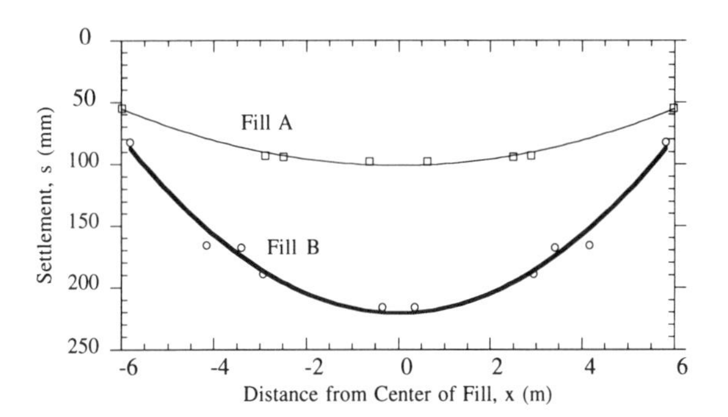

FIG. 10. Net Settlement Profiles from Day 80 to 550

A thermal precompression method, in which a peat sample is heated and then cooled at a later time, significantly reduces the deformation of Middleton peat. In essence, heating causes a drastic increase in strain rate and a corresponding rapid decrease in void ratio. Once the soil is cooled, it has a substantially reduced void ratio and creeps very slowly. In addition, the soil develops a significant quasi-preconsolidation effect and has a very low compressibility with respect to further loading. The data collected in a field test of thermal precompression indicates that moderate ground heating accelerates compression significantly. Efforts are underway to refine the procedure for cost-effective applications and to assess the post-thermal compression behavior upon cooling.

ACKNOWLEDGMENTS

This investigation is based upon work supported by the U. S. National Science Foundation Grants No. MSM-8617238 and MSS-9115315. The authors acknowledge Mr. Norman H. Severson, Mr. James Hanson and Dr. Li-Tus Lan for their assistance with the project and Mr. Sylvester Ziegler for allowing the use of his property for this project.

APPENDIX - REFERENCES

Andersland, O. B., and Douglas, A. G. (1970). "Soil deformation rates and activation energies." *Géotechnique*, 20(1), 1-16.

Campanella, R. G., and Mitchell, J. K. (1968). "Influence of temperature variations on soil behavior." *J. Soil Mech. Found. Div.*, Proc. ASCE, 94(SM3), 709-734.

Edil, T. B., Fox, P. J., and Lan, L. T. (1991). "Observational procedure for settlement of peat." *Proc. Geo-Coast Conf.*, Yokohama, Japan, 2, 165-170.

Edil, T. B., Fox, P. J., and Lan, L. T. (1994). "Stress-induced one-dimensional creep of peat." *Advances in Understanding and Modeling the Mech. Behavior of Peat*, Balkema, Rotterdam, The Netherlands, 3-18.

Fox, P. J. (1992). "An analysis of one-dimensional creep behavior of peat," Ph.D. thesis, University of Wisconsin-Madison.

Fox, P. J., and Edil, T. B. (1994). "Temperature-induced one-dimensional creep of peat." *Advances in Understanding and Modeling the Mech. Behavior of Peat*, Balkema, Rotterdam, The Netherlands, 27-34.

Gray, H. (1936). "Progress report on the consolidation of fine-grained soils." *Proc. 1st ICSMFE*, Cambridge, 2, 138-141.

Houston, S. L., Houston, N. W., and Williams, N. D. (1985). "Thermo-mechanical behavior of seafloor sediments." *J. Geotech. Eng.*, ASCE, 111(11), 1249-1263.

Laguros, J. G. (1969). "Effect of temperature on some properties of clay soils." *HRB Special Report 103*, Washington, D. C., 186-193.

Leonards, G. A., and Ramiah, B. K. (1959). "Time effects in the consolidation of clays." *Symposium on Time Rate of Loading in Testing of Soils*, ASTM STP 254, ASTM, Philadelphia, Pennsylvania, 116-130.

Lo, K. Y. (1961). "Secondary compression of clays." *J. Soil Mech. Found. Eng. Div.*, Proc. ASCE, 87(SM4), 61-87.

Mesri, G. (1973). "Coefficient of secondary compression." *J. Soil Mech. Found. Eng. Div.*, Proc. ASCE, 99(SM1), 123-137.

Mitchell, J. K. (1969). "Temperature effects on engineering properties and behavior of soils." *HRB Special Report 103*, Washington, D. C., 9-27.

Mitchell, J. K., Campanella, R. G., and Singh, A. (1968). "Soil creep as a rate process." *J. Soil Mech. Found. Eng. Div.*, Proc. ASCE, 94(SM1), 231-253.

Plum, R. L., and Esrig, M. I. (1969). "Some temperature effects on soil compressibility and pore water pressure." *HRB Special Report 103*, Washington, D.C., 231-242.

Settlement of Footings on Granular Materials: Low and Large Strain Parameters

F. Ahtchi-Ali,[1] and J. C. Santamarina,[2] Members, ASCE

Abstract: Many methods have been proposed to predict the settlement of shallow foundations on granular soils. The analysis of more than 200 case histories shows that there is no correlation between penetration testing and back-calculated soil deformation moduli, and that settlements are most often overpredicted for large footings (more than 3 m in width), and underpredicted in the case of small footings. Experimental data found in the literature are used to tabulate the effect of soil and in situ parameters on low and large strain soil moduli and in situ tests. The combined use of penetration testing and wave propagation is concluded from this analysis, and experimentally tested. Buried plates were included for the evaluation of mid-strain deformation modulus. It is shown that the correlation of deformation modulus with cone resistance is dependent on relative settlement, and that the best correlation is obtained by normalizing measurements with respect to G-max, determined with shear waves polarized in the vertical plane. The experimental study includes the testing of model footings.

Introduction

During the last six decades, many methods have been developed in an attempt to predict the settlement of shallow foundations on sands. Empirical formulations, settlement performance from case histories, and numerical models, have been used to assess soil deformation moduli. In the case of footings on sands,

---

[1] Formerly Graduate Student, Dept. of Civil Engrg., Polytechnic University, Brooklyn, NY 11201.

[2] Associate Prof., Dept. of Civil Engrg., Univ. of Waterloo, Ontario N2L 2V8 Canada.

settlement usually poses little difficulty; not only for its limited magnitude, but also because it takes place immediately as the load is applied. In some cases, however, the magnitude and timing are critical, e.g., tanks, silos, and settlement affecting neighboring structures. Regardless of the criticality of settlement, the poor correlation between measured and predicted values is intriguing.

Some of the inherent difficulties in predicting the settlement of footings on free-draining granular materials are the immediate change in effective stresses as the load is applied and the consequent change in the stiffness of the material, the non-linear non-elastic behavior of soils, and the extreme difficulty in obtaining undisturbed samples for the laboratory determination of soil deformation modulus under appropriate stress conditions and stress history. Hence, efforts have been centered around the interpretation of in situ test results. Most settlement methods rely on penetration testing, i.e. SPT and CPT. Other in situ tests such as plate load (PLT), screw plate (SP), pressuremeter (PMT), dilatometer (DMT), and wave propagation tests (WPT) have also been used. This paper begins with an evaluation of case histories and in situ tests. Then, the combined use of penetration and wave propagation data is investigated.

Analysis of Case Histories

Over two hundred case histories were collected from the literature (Gifford et al. 1987; Jeyapalan and Boehm 1986; Burland and Burbidge 1985; Schmertmann 1970; and others). For each case, footing width and shape, depth of embedment and water table, blow count (N) from standard penetration test (SPT), applied pressure, and measured settlement were given. Case histories include footings for tanks, bridge abutments, chimneys, multi-story buildings, nuclear reactors, and large-scale plate load tests. An average deformation modulus D was computed for each footing from:

$$D = \frac{q \cdot I_f}{S_m / B} \tag{1}$$

where q is the applied pressure; S_m is the measured settlement; B is the footing width; (S_m/B) is the relative settlement, and I_f is a correction factor for embedment and footing shape (Poulos and Davis 1974). Computed moduli are plotted in Fig. 1 vs. SPT blow count corrected for overburden N_c. Virtually no correlation is observed.

The effect of footing size and strain level on settlement prediction was further studied using the database of case histories. From Eq. (1), settlements were calculated for all cases in the data base, assuming $S_p = qBI_f/D$, where D was estimated using standard D-vs-N correlations (data shown are for Schmertmann's 1970, correlation - conclusions are not affected by this choice). Results of predicted settlement S_p versus footing width B are plotted in Fig. 2a. While the scatter in this log-log plot cannot be overlooked, the high correlation that is observed is surprising because the plotted data include different geological formations, loads, factors of safety, relative settlement S/B, and variations in other relevant factors. Indeed, penetration parameters present relatively narrow variations as compared to other parameters involved in settlement calculations. The trend appears to indicate that designs are meant to predict settlements in the order of 1%B. In contrast, Fig. 2b plots measured settlement versus footing width for the same cases, and shows a very poor correlation (the scatter shown is equivalent to that in Fig. 1, but plotted in log-log scale).

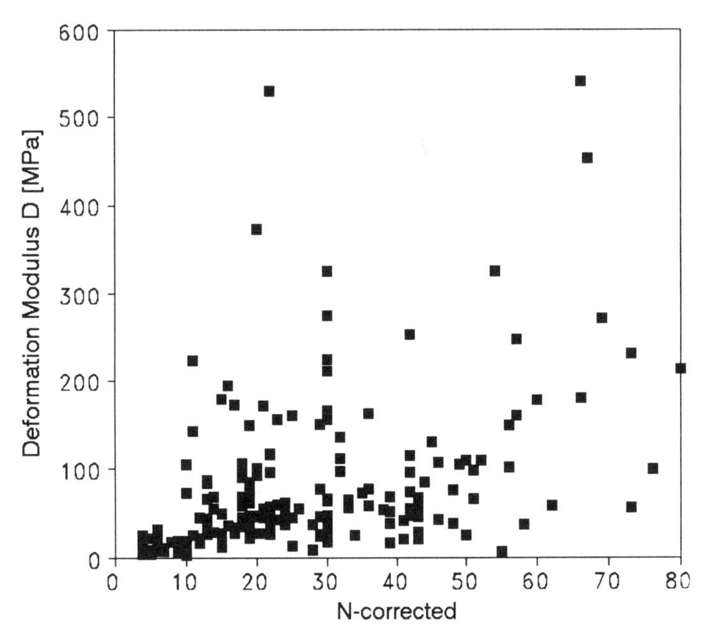

FIG. 1. Global Deformation Modulus versus Penetration Resistance Analysis of Case Histories - Back-calculated

Finally, the same data are plotted in terms of predicted settlements S_m/S_p and the observed relative settlement S_m/B. It can be concluded that for the assumptions made in most penetration-based methods, the estimated settlement S_p is overpredicted for low relative settlements $S_m/B < 1\%$, and underpredicted for high relative settlements, $S_m/B > 1\%$. Adequate estimation is obtained at a relative settlement $S_m/B \approx 1\%$. In practice, the average footing size ranges between 2 m to 4 m; hence, this observation indicates that settlements are well predicted for a design value between 2 cm to 4 cm, which is in the allowable range. In conclusion, typical correlations and current design practice appear to be adjusted to common conditions. However, the scatter is significant even in this case, and major deviations should be expected in all other cases. Unfortunately, the plot presented in Fig. 2c is not bijective; thus, it cannot be inverted to determine a correction factor.

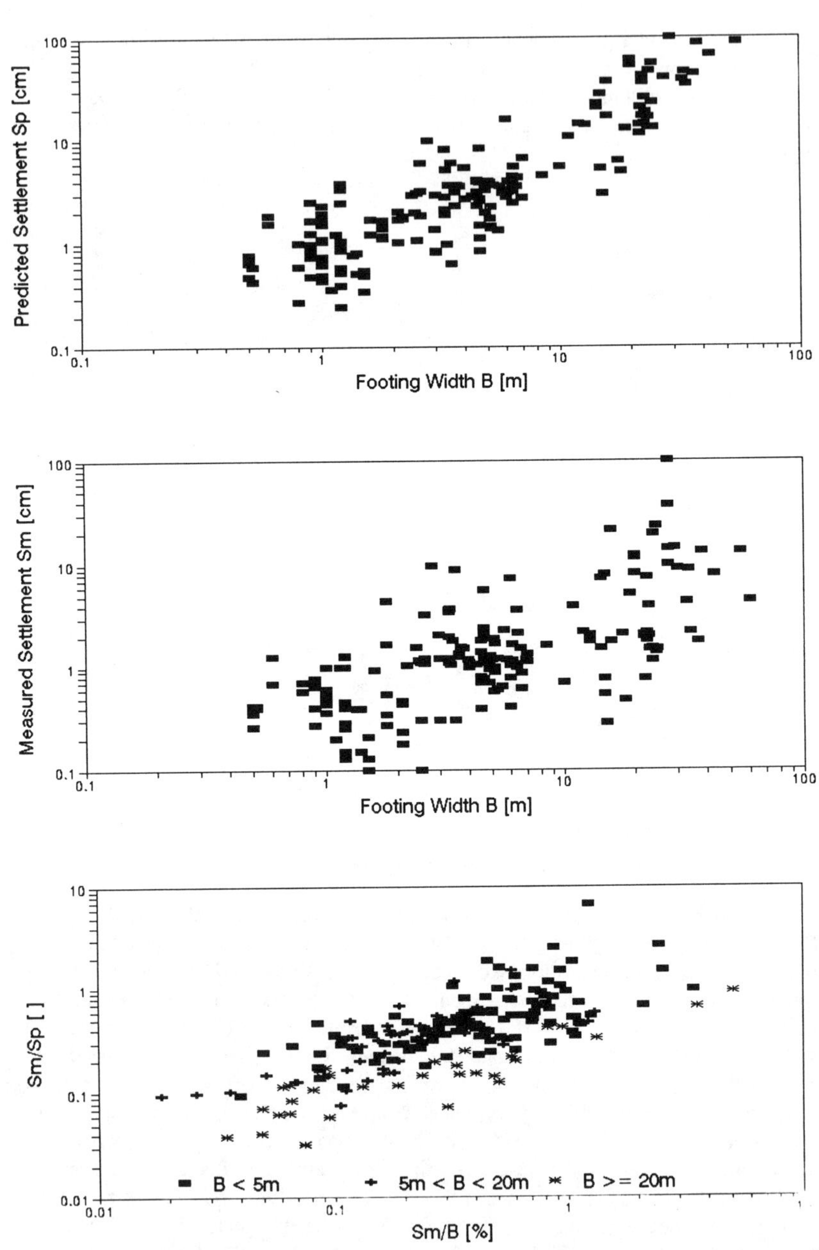

FIG. 2. Analysis of Case Histories: (a) Predicted Settlement; (b) Measured Settlement; and (c) Relative Settlement and Footing Size Effect

Factors Affecting Soil Behavior and In Situ Tests

Experimental data from calibration chambers and other experimental settings were collected and analyzed to assess the sensitivity of constitutive soil parameters and in situ tests to soil characteristics and other formation parameters (e.g., Lunne 1991; Peterson 1991; Schmertmann 1991; Alarcon et al. 1986; Leonards et al. 1986; Clayton et al. 1985; Swiger 1974; and Koerner 1970). Table 1 summarizes, in qualitative terms, the results of this study, including some entries where judgment was needed to overcome lack of data. It is observed that penetration testing is affected by most of the same parameters that affect the large-strain deformation modulus and soil strength, whereas plate load and screw plate testing are similarly affected by those parameters that affect the low-strain modulus. The pressuremeter and dilatometer tests are considered mid-strain tests and have some sensitivity to the effect of prestress. However, installation-induced disturbance and dissimilar stress paths limit the use of these and other tests. While simulation and inversion mathematics may be used to determine constitutive parameters relevant to the settlement of footings, the non-uniqueness of inversion solutions is severe and must be carefully assessed.

Wave propagation testing is a low-strain test which produces virtually no disturbance in the soil mass. It is sensitive to the effect of prestress, at least to residual stresses, and probably to the effect of aging: if the main effect of aging is to relax stresses, then velocity should decrease; however, if aging enhances contacts the main effect will be a reduction in attenuation). Swiger (1974) used wave propagation data to estimate the deformation modulus at very low strains in the order of 0.001%. Finally, recognizing that site investigation is conducted before construction and load application, all tests will, at best, reflect in situ conditions under the in situ state of stress. Yet, stresses induced upon the application of the load will significantly affect the properties of the soil mass.

A combination of low and large-strain tests such as wave propagation and penetration tests, respectively, appears to take into consideration most of the factors that affect the low and high-strain deformation modulus. These two tests offer the additional advantage of being part of the current state of the art. Furthermore, they can be simultaneously run, e.g. seismic cone. This approach is examined next.

Low and Large-Strain Parameters

An extensive test program was designed to study the transition between low strain to large-strain parameters, and the combined use of wave propagation and penetration testing. It included penetration testing, wave propagation, and buried plates to bridge the strain range in the mid-strain region. All tests were conducted in a 2 m^3 cubical soil box, filled with dry, uniform, medium to fine silica sand. Three relative densities were used, in different tests. The soil placement procedure was by pluviation through sieving for the 48% and 70% relative densities. The highest relative density of 100% was achieved by pluviation followed by rodding. Four 6.4 cm diameter plates were embedded at 20 cm, 40 cm, 60 cm, and 80 cm respectively.

TABLE 1. Factors Affecting Soil Behavior and In Situ Tests

Factors	ϕ	D-L	D-H	SPT	CPT	PLT	SP	PMT	DMT	WPT
Soil properties										
Void ratio (-)	+	+	+	+	+	+	+	+	+	se
Angularity (+)	+	-	?	+	+	fs	fs	fs	se	-
Particle size (+)	+	+	+	+	+	+	+	+	+	+
Well graded (+)	+	+	+	+	+	+	+	+	+	+
Roughness (+)	+	-	ne	+se	+se	fs	fs	fs	se	-
Hardness	+	+	ne	+se	+se	+	+	+	+	+
In situ stress and applied load										
Mean Stress (+)	-	+	+	+	+	+	+	+	+	+
σ_v/σ_h (+)	n/a	ne	-	-	-	fs	fs	-	se	*
Others										
Prestress	ne	+	se	ne	ne	+	se	se	se	+
Aging	ne	+	se	ne?	ne?	se	se?	se	ne	-**
Cementation	ne	++	ne	se	se	+	+	+	se	++
Soil fabric	ne	se	ne	ne	ne	se	se	se	ne	se

Legend:

+	increases	-	decreases
fs	function of strain level	n/a	non applicable
se	some to significant effect	ne	little or no effect

* depends on mode of propagation and polarization plane.
** it is assumed that long-term aging leads to stress relaxation

Tests:

D-L deformation modulus - low strain
D-H deformation modulus - high strain
SPT standard penetration test
CPT cone penetration test
SP screw plate
PLT plate load test
PMT pressuremeter test
DMT dilatometer
WPT wave propagation - velocity

Low Strain: Wave Propagation Test

Shear and compression wave velocities were obtained from crosshole and downhole methods, respectively. These measurements involved four 1D-geophones as receivers, placed in the vertical direction at the center of the box. The geophones were installed at four different depths, coinciding with the depth of the buried plates. The plates themselves were used as sources for the measurements of horizontally propagating shear waves, polarized in the vertical plane. Wave measurements were conducted after loading the buried plates. The accuracy in time measurements was about 1%, and compatible with the accuracy in the location of the instrumentation. The same procedure was applied for the three relative densities of 48%, 70%, and 100%.

Mid Strain: Buried Plate Load Test

The loading of the four buried plates is equivalent to the screw plate load test, but it has the advantage of no soil disturbance because plates are buried during the pluviation of the soil bed. The buried plates were loaded separately, until a failure load was reached corresponding to a settlement one-fifth the diameter of the plate. Note that preliminary tests were conducted to select the size of these plates so that no scale effects would be avoided.

Large Strain: Cone Penetration Test

A portable cone penetrometer was adapted to the box/reaction frame system. It consisted of a 10 cm^2 use-and-loose tip with 60° attack angle. The diameter of the cylindrical sleeve (3 cm) was smaller than the diameter of the cone, leading to minimum sleeve friction. The cone penetration test was performed at two locations for three relative densities. The location of the test was selected to minimize boundary effects and residual disturbance from the loading of buried plates.

Interpretation of Test Results

Test results obtained from embedded plate load tests, wave propagation tests, and cone penetration tests were processed in conjunction. The horizontal shear and the vertical compression wave velocities (V_s and V_p) were used to compute the shear modulus G on the xz plane, and the constrained modulus M along the vertical direction.

The "global" secant deformation modulus (D) was calculated from the buried plate load tests at three relative settlements (S/B of 0.001, 0.01, and 0.1) using Eq. (1). Test results for all three relative densities were analyzed in an attempt to determine a general methodology to estimate the mid-strain deformation modulus D for the 6.4 cm diameter buried plates, with the large-strain cone resistance and the low strain wave data. Fig. 3 shows the plot of the deformation modulus normalized with respect to $E_{max}(s)$ obtained from horizontal shear waves polarized in the vertical plane. The use of shear wave velocity provides better correlation than V_p, probably because the horizontal shear wave velocity polarized in the vertical plane is affected by horizontal and vertical stresses, whereas the compressional wave velocity is only dependent on the vertical stresses. Regression equations based on the data in Fig. 3 are also shown (penetration resistance q_c is in psi - [1 psi = 6.895 kPa]).

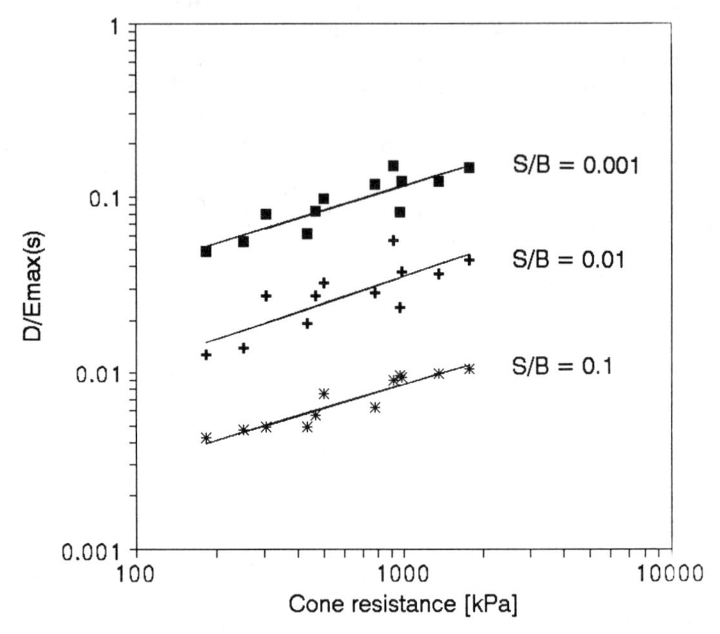

FIG. 3. Low and High-Strain Parameters (Dr= 48%, 70%, and 100%) Wave Propagation, Buried Plates, and Penetration Testing

$$
\begin{aligned}
&\text{For } S/B < 10^{-5} && D_{25} = E_{\max(S)} \\
&\text{For } S/B = 10^{-3} && D_{25} = E_{\max(S)} \times 10^{-1.956+0.469\log(q_c)} \\
&\text{For } S/B = 10^{-2} && D_{25} = E_{\max(S)} \times 10^{-2.544+0.504\log(q_c)} \\
&\text{For } S/B = 10^{-1} && D_{25} = E_{\max(S)} \times 10^{-3.047+0.451\log(q_c)}
\end{aligned}
\qquad (2)
$$

SETTLEMENT OF FOOTINGS

Following previous results, the deformation modulus for a buried 6.4 cm plate at any given relative settlement S/B is readily obtained: (1) measure the cone resistance q_c at B/2 beneath the plate, (2) estimate Emax(s) from shear wave data at the same depth, and (3) compute D for the 6.4 cm plate at the desired relative strain. These correlations permitted the accurate calculation of the load deformation curve for all four buried plates, in the three different density beds.

For large, shallow footings, this procedure must include adequate correction factors for embedment and size. Embedment factors have limited variability and can be estimated from theory of elasticity in a first order approximation. On the other hand, a limited test program was conducted to assess the effect of size. Foundation models, 15 cm diameter, were tested at two relative densities, 70% and 100%. For each relative density two embedment depths were used, 7.5 cm and 15 cm. Vertical and horizontal wave measurements were taken using down hole and cross hole methods, respectively. Two pairs of geophones were installed during sample preparation at depths of 0.5B and 1.5B below the base of the plates. The cone penetration test was performed after the plate load test was completed, at locations that minimized the effect of boundaries and plate load testing. The correction factor, defined in terms of loads, is plotted at pre-selected factors of safety, where the ultimate load was selected at S/B = 20%. Results are shown in Fig. 4. The fact that settlement increases proportionally to the footing size but for large areas it becomes correspondingly smaller, has been recognized for more than 60 years (Kogler 1933; Terzaghi and Peck 1948; Bjerrum and Eggestad 1963; Oweis 1979). Several coexisting phenomena have been identified and analyzed (Ahtchi-Ali 1992): depth of influence and modulus depth variation, relative settlement and strain level, stiffening effect of the loaded footing, factor of safety dependency of the strain influence diagram when used in depth averaging of parameters, dilative tendency and deformation-failure mode.

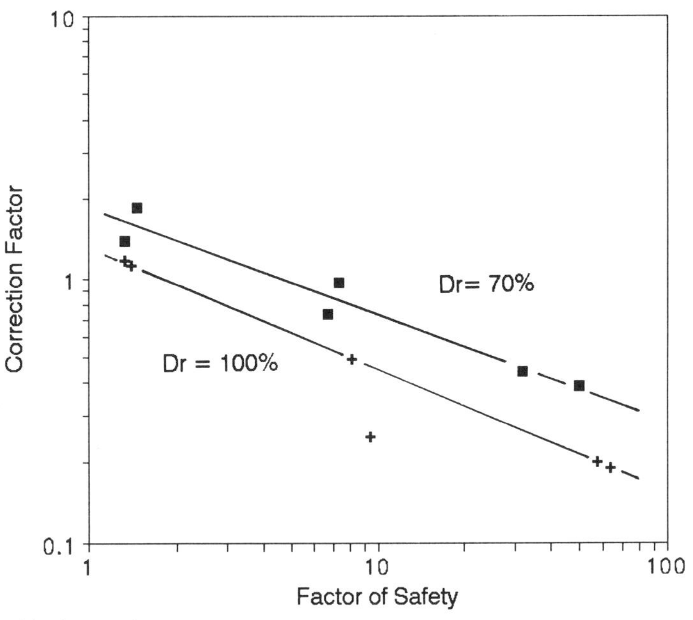

FIG. 4. Correction Factor For Size Effect (15 cm Footing - Dr = 70 & 100%)

CONCLUSIONS

There are large discrepancies between measured and predicted settlements. Over two hundred well-documented case histories were analyzed, and results confirmed that any direct correlation of the soil deformation modulus with penetration testing is very weak, at best.

Penetration testing is affected by most of the same parameters that affect the large strain modulus and soil strength, whereas wave propagation is sensitive to similar parameters that affect the low-strain modulus.

The correlation of the back-calculated secant deformation modulus of the 6.4 cm diameter plate with cone resistance is relative-settlement dependent. Wave propagation data allow adequate normalization of the modulus for varying soil conditions. The horizontal S-wave velocity polarized in the vertical plane is preferred because it is sensitive to both horizontal and vertical stresses.

A methodology proposed based on wave propagation and cone penetration tests successfully predicted the behavior of the 6.4 cm diameter plate at three relative densities in the full range of relative settlement S/B. The correction for size is experimentally obtained for larger size plates.

ACKNOWLEDGMENTS

This study was conducted by the authors at Polytechnic University as part of a study on wave geomedia interaction and applications, with support from the National Science Foundation.

APPENDIX - REFERENCES

Ahtchi-Ali, F. (1992). "An experimental study of settlement of footings on sands (penetration and geophysical testing)," PhD dissertation, Polytechnic University, Brooklyn, New York.

Alarcon, A., Leonards, G. A., and Chameau, J. L. (1986). "Shear modulus and cyclic undrained behavior of sands." *Soils and Foundations*, 29(4), 105-119.

Bjerrum, L., and Eggestad, A. (1963). "Interpretation of loading test on sand." *Proc. European Conf. Soil Mechanics and Found. Eng.*, 2, 135.

Burland, J. B., and Burbridge, M. C. (1985). "Settlement of foundations on sand and gravel." *Proc. Inst. Civil Engineers*, 78, 1325-1381.

Clayton, C. R. I., Hapapa, M. B., and Simons, N. E. (1985). "Dynamic penetration resistance and the prediction of the compressibility of a fine-grained sand -A laboratory study." *Géotechnique*, 35(1), 19-31.

Gifford, D. G., Wheeler, J. R., Kraemer, S. R., and McKown, A. F. (1987). "Spread footings for highway bridges." *Publication No. FHWA/RD-86/185*, U. S. Department of Transportation, McLean, Virginia.

Jeyapalan, J. K., and Boehm, R. (1986). "Procedures for predicting settlements in sands." *Settlement of Shallow Foundations on Cohesionless Soils: Design and Performance*, Geotechnical Special Publication No. 5, ASCE, New York, New York, 1-22.

Kogler, F. (1933). "Research in soil engineering (Foundations)." *Transactions of ASCE*, 98, 299-301.

Koerner, R. M. (1970). "Effect of particle characteristics on soil strength." *J. Soil Mech. Found. Div.*, Proc. ASCE, 96(SM4), 1221-1234.

Leonards, G. A., Alarcon, A., Frost, D., Mohamedzein, Y. E., Santamarina, J. C., Thevanayagam, S., Tomaz, J. E., Tyree, J. L. (1986). Discussion on "Dynamic penetration resistance and the prediction of the compressibility of a fine-grained sand - A laboratory study." *Géotechnique*, 36(3), 275-281.

Lunne, T., (1991). "Practical use of CPT correlations in sand based on calibration chamber tests." *Proc. 1st International Symposium on Calibration Chamber Testing*, Potsdam.

Oweis, I. S. (1979). "Equivalent linear model for predicting settlements of sand bases." *J. Geotech. Eng. Div.*, Proc. ASCE, 105(GT12), 1525-1544.

Peterson, R. W., (1991). "Penetration resistance of fine cohesionless materials," U. S. Army Engineer Waterways Experiment Station, Vicksburg, Mississippi.

Poulos, H. G., and Davis, E. H. (1974). *Elastic Solutions for Soil and Rock Mechanics*, John Wiley & Sons, New York, New York.

Schmertmann, J. H. (1991). "The mechanical aging of soils." *J. Geotech. Eng.*, ASCE, 117(9), 1288-1330.

Schmertmann, J. H. (1970). "Static cone to compute static settlement over sand." *J. Soil Mech. Found. Div.*, Proc. ASCE, 96(SM3), 1011-1043.

Swiger, W. F. (1974). "Evaluation of soil moduli." *Proc. Conference on Analysis and Design in Geotechnical Engineering*, Austin, 2, 79-92.

Terzaghi, K., and Peck, R. B. (1948). *Soil Mechanics in Engineering Practice*, John Wiley & Sons, New York, New York.

COMPRESSIBILITY FOR SAND UNDER ONE-DIMENSIONAL LOADING CONDITION

Ching S. Chang,[1] Member, ASCE

ABSTRACT: Empirical equations for one dimensional compression of sand do not include parameters characterizing properties of sand grains (e.g. moduli, size and shape of sand grains, and friction properties between grains). In this paper, close-form expressions of compressibility and lateral earth pressure are derived for spherical particles in a body centered arrangement. The predicted results are compared with experimental measurements for sands under both loading and unloading conditions. The effects of micro-scale properties and macro-scale responses are discussed to provide insights for future improvements in abilities of predicting settlement under complicated loading conditions.

INTRODUCTION

Estimation of settlement of foundation in sand is an intricate problem due to the complex behavior of sand, namely, the pressure dependent moduli and strength, the inelastic behavior, and the volume dilation during shear. To limit the scope, this paper is focused on the compressibility of sand under a simple one-dimensional loading condition. Numerous experimental work on the compressibility of sand in a consolidometer can be found in the literature. Based on experimental results, useful empirical equations have been established for compressibility and for earth pressure coefficient K_o (e.g. Hendron 1967; Mayne and Kulhaway 1982). However, these empirical equations do not include parameters characterizing properties of san grains (e.g. moduli, size and shape of sand grains, and friction properties between grains). It is thus desirable to analyze soil behavior by explicitly treating soil as a collection of particles.

---

[1] Professor of Civil Engineering, University of Massachusetts, Amherst, MA 01003.

Along this line, some earlier attempts were made by Duffy and Mindlin (1957), Duffy (1959) and Hendron (1963). Recently, more attention has been given to this type of problem. Approaches can be classified into two categories, namely, direct computer simulation (Cundall and Strack 1979; Kishino 1987; Ting et al. 1989; Ng 1989; Chang and Misra 1989; etc.) and microstructural mechanics (Walton 1987; Jenkins 1988; Chang 1988; Chang et al. 1992). The direct computer simulation approach utilizes discrete element method and solves for the movement of every discrete particle. The discrete analytical method can be used as a tool for numerical experiment. The microstructural mechanics approach bridges discrete and continuum analysis. It aims for a constitutive model explicitly in terms of parameters of discrete particles.

In this paper, we attempt to examine the effect of grain properties (e.g., grain size, moduli and inter-grain friction) on compressibility and earth pressure based on the microstructural mechanics approach (Chang et. al. 1992). A body centered cubic packing is used to illustrate the role of grain properties and inter-particle friction. Discussion is also given on the comparison of experimental and theoretical results.

INTER-PARTICLE BEHAVIOR

In a dry granular aggregate of particles, the resultant deformation are derived from two sources: (1) elastic and inelastic deformation of each particle; and (2) relative movement of particles due to sliding and rolling. Strain of particle distortion is usually concentrated at the inter-particle contacts. Resistance to the compression of two particles is represented by a contact stiffness k_n in the normal direction. Resistance to the shearing of two particles is represented by a contact stiffness k_s in the tangential direction. The relative compression δ_n and relative shearing δ_s of two particles in contact generate the contact forces through contact stiffness, defined in the following:

$$\begin{aligned} F_n &= k_n \delta_n \\ F_s &= k_s \delta_s \end{aligned} \tag{1}$$

where F_n and F_s are respectively the forces in the directions normal and tangential to the contact plane. Based on Hertz theory of contact, the contact stiffness k_n of two elastic spheres in compression can be described in a normalized form as follows:

$$\left(\frac{k_n}{P_a r} \right) = c_n \left(\frac{F_n}{P_a r^2} \right)^{\beta} \tag{2}$$

where r is the radius of particles; P_a is the atmospheric pressure; and β and c_n are dimensionless parameters depending on the elastic properties of mineral type. Based on Mindlin-Hertz's theory, the shear contact stiffness can be approximated by

$$k_s = c_r k_n \tag{3}$$

For frictional inter-particle contacts without cementation bonds, particles can not take tension. Thus $k_n = 0$ when the particles start to lose contact and the inter-particle force becomes zero. Denote inter-particle sliding friction angle as ϕ_μ. When contact forces exceeds the frictional strength, particles slide and $\phi_\mu = 0$.

To summarize, the inter-particle contacts can be represented by four constants: c_n, c_r, β, ϕ_μ. Values of c_n, c_r, and β have been back calculated from experimental results for different types of sand, in resonant column device and in triaxial cell under isotropic loading (Chang 1994). A survey of the published values of ϕ_μ for two sand particles is given by Kabir (1992). Range of these parameters are listed in the following table.

TABLE 1. Range of Inter-particle Parameters

c_n	1300 - 2300
c_r	0.5 - 1.0
β	0.3 - 0.6
ϕ_μ	15° - 30°

SIMPLE PACKING MODEL

A simple regular packing of spheres is used here for illustration. Each particle in the packing has eight neighbors. The centers of these neighboring spheres form a unit cell as shown in Fig. 1. For a regular packing, the behavior of one unit cell represents the behavior of the whole packing. It is noted that such a packing is a highly idealized microstructure for real soil. However, analysis of this packing provides a useful analogy for understanding soil behavior. Furthermore, this packing is a simple tractable system for which the close-form results can be derived.

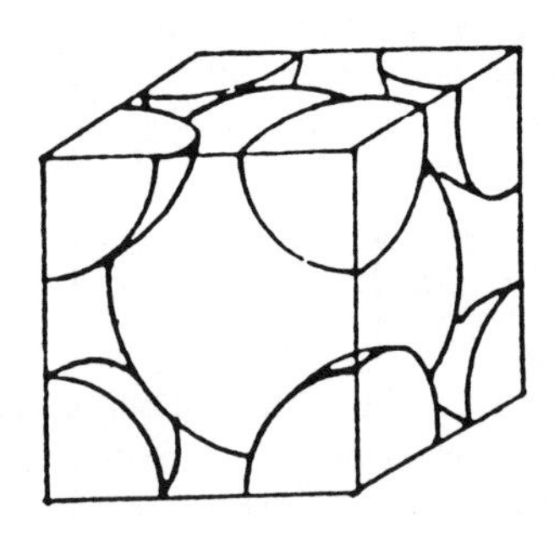

FIG. 1. Unit Cell of a Body-Centered Packing

Let α be the angle between the vertical axis and the vector joining the centers of any two neighboring particles in the unit cell. The volume of the unit cell is

$$\begin{aligned} V &= (2\sqrt{2}\,r\sin\alpha)(2\sqrt{2}\,r\sin\alpha)(4r\cos\alpha) \\ &= 32\,r^3\sin^2\alpha\cos\alpha \end{aligned} \tag{4}$$

When the packing angle α is 54.7 degree, the configuration of the packing represents a body centered cubic packing with a void ratio 0.47.

LATERAL EARTH PRESSURE

Under one-dimensional compression loading, all particles move downwards with zero horizontal movement. Let the downward vertical movement be δ_z of particle A relative to particle B as shown in Fig. 2, the relative compression and shearing displacements of the two particles are given by

$$\begin{aligned} \delta_n &= \delta_z\cos\alpha \\ \delta_s &= \delta_z\sin\alpha \end{aligned} \tag{5}$$

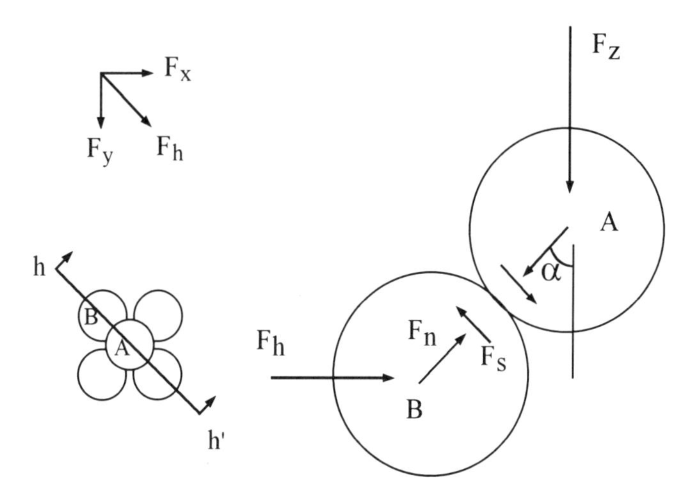

FIG. 2. Forces in the Particulate System

where δ_n is the relative compression of particles normal to the contact area, δ_s is the relative shearing of particles tangential to the contact area. The inter-particle forces can be computed from the stiffness at the contact, given by

$$F_n = k_n \delta_z \cos\alpha$$
$$F_s = k_s \delta_z \sin\alpha \tag{6}$$

From Eq. (6), the force ratio F_s/F_n is a function of packing angle α and contact stiffness. It is convenient to define an angle κ, given by

$$\frac{F_s}{F_n} = \tan\kappa = \frac{k_s}{k_n}\tan\alpha = c_r \tan\alpha \tag{7}$$

where c_r is the constant given in Eq. (3). Note that Eq.(7) is valid only when particles do not slide. If particle sliding has occurred, the ratio of contact forces is governed by the inter-particle sliding angle ϕ_μ. We then define a mobilized angle ϕ_{mob} which can be classified into two conditions, given by

$$\frac{F_s}{F_n} = \tan\phi_{mob} = \begin{cases} \tan\kappa\,; & \text{if } \kappa \leq \phi_\mu \\ \tan\phi_\mu; & \text{if } \kappa > \phi_\mu \end{cases} \tag{8}$$

From the principle of force equilibrium, the vertical force and horizontal force in Fig. 2 are related to the inter-particle forces, given by

$$F_z = 4F_n(\cos\alpha + \tan\phi_{mob}\sin\alpha)$$
$$F_x = 2\sqrt{2}\,F_n(\sin\alpha - \tan\phi_{mob}\cos\alpha) \tag{9}$$

It is interesting to note that the applied vertical force is transferred by arching into horizontal forces through the inclined inter-particle contact. The magnitude of horizontal force depends on the packing structure and the shear forces mobilized on the contact.

Taking into account the representative area on which the forces act, the stresses are as follows:

$$\sigma_z = \frac{F_z}{8r^2\sin^2\alpha}$$
$$\sigma_x = \frac{F_x}{8\sqrt{2}\,r^2\sin\alpha\cos\alpha} \tag{10}$$

The ratio of stresses, thereby the coefficient of earth pressure at rest, is given by

$$K_o = \frac{\sigma_x}{\sigma_z} = \frac{1}{2}\tan(\alpha - \phi_{mob})\tan\alpha \tag{11}$$

where ϕ_{mob} is the mobilized angle for shear force, given in Eq. (8). The value of K_o is a function of packing structure and the mobilized shear force at contacts. The value of K_o is plotted in Fig. 3 for three different packing structures with various values of inter-particle friction angle. The range of measured values of K_o for sand, reported by Mayne and Kulhawy (1982), is from 0.3 to 0.55 with a mean value of

0.405. For the range of inter-particle sliding friction angle ϕ_μ between 18 to 28 degrees, the values of K_o computed from Eq. (11) agree with that experimentally observed from sand. It is noted that the inter-particle sliding friction angle ϕ_μ in Fig. 3 is not same as the friction angle ϕ in Jaky's equation.

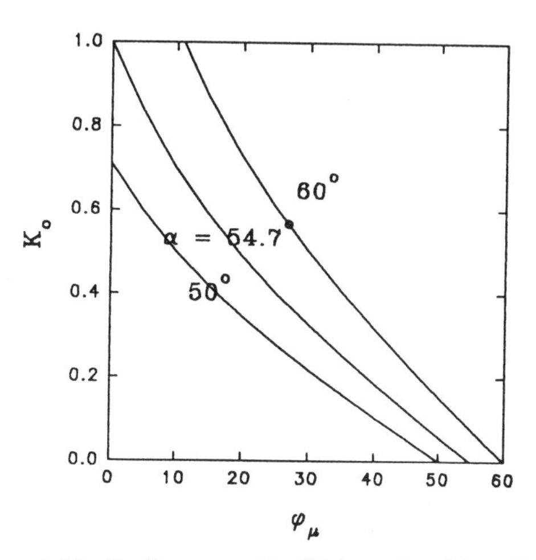

FIG. 3. Lateral Earth Pressure Coefficient for Materials with Different Inter-Particle Friction Angle.

COMPRESSIBILITY

The relative movement of the two particles, under one-dimensional compression loading, can be related to strain by

$$\Delta\epsilon_z = \frac{\Delta\delta_z}{2r\cos\alpha} \tag{12}$$

Then, using Eqs. (6) through (10), we derive the stress-strain relationship for the one-dimensional loading condition,

$$\Delta\sigma_z = D\Delta\epsilon_z \tag{13}$$

where the constrained modulus D is given by

$$D = \frac{k_n}{r}(\cos\alpha + \tan\phi_{mob}\sin\alpha)\cot^2\alpha \tag{14}$$

Using the non-linear contact law of Eq. (2), the stiffness k_n is a function of the contact normal force F_n. Note that, from Eq. (10), the contact normal force is a function of stress expressed by

$$F_n = \frac{\sigma_z 8r^2 \sin^2\alpha}{4(\cos\alpha + \tan\phi_{mob}\sin\alpha)} \tag{15}$$

Substituting Eq. (15) into Eq. (14), the constrained modulus becomes

$$D = 2^\beta c_n \cos^2\alpha \left(\frac{\cos\alpha + \tan\phi_{mob}\sin\alpha}{\sin^2\alpha}\right)^{1-\beta} \left(\frac{\sigma_z}{P_a}\right)^\beta \tag{16}$$

The constrained modulus is a function of stress, packing structure, inter-particle friction angle, and the properties of grains. The stress-strain curve can be obtained from integration of Eq. (13), given by

$$\epsilon_z = \frac{1}{2^\beta c_n (1-\beta)\cos^2\alpha} \left(\frac{\sin^2\alpha}{\cos\alpha + \tan\phi_{mob}\sin\alpha} \frac{\sigma_z}{P_a}\right)^{1-\beta} \tag{17}$$

The computed constrained moduli from Eq. (16) are compared with the experimental data for various types of sand as shown in Fig. 4. The computed stress-strain curve from Eq. (17) are compared with the experimental data for various types of sand as shown in Fig. 5. The derived moduli and stress-strain curves are in reasonable agreement with the range of measured response.

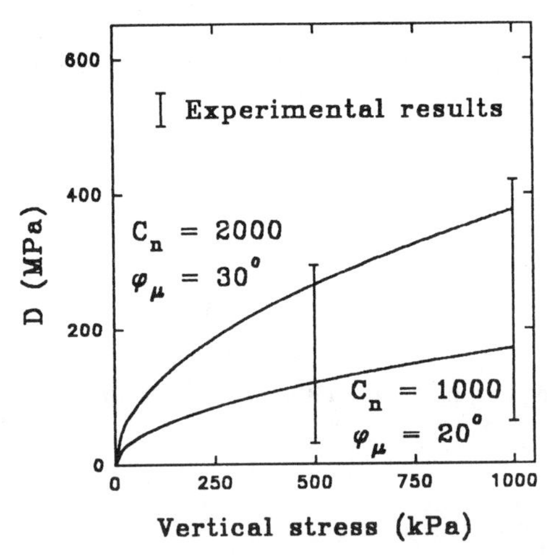

FIG. 4. Computed and Measured Constrained Modulus

Under loading process, particles sliding contributes to plastic deformation and develop horizontal earth pressure. Therefore, the compressibility and the coefficient of earth pressure at rest K_o are greatly influenced by the frictional properties of the inter-particle contact.

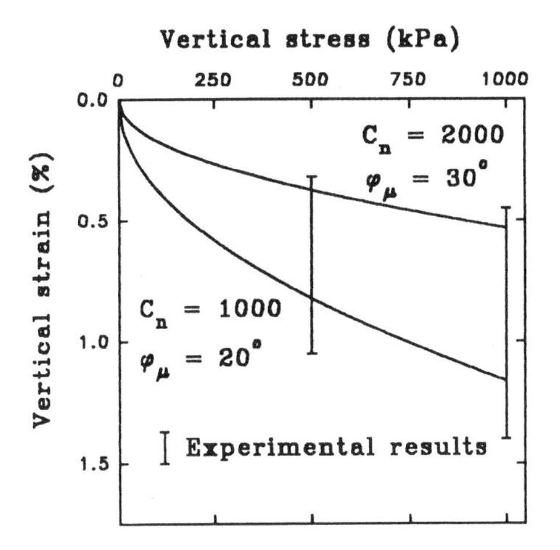

FIG. 5. Computed and Measured Stress-strain Curves

UNLOADING BEHAVIOR

Subsequent to the shear sliding during loading process, two stages of mechanism may occur during unloading: (1) the elastic unloading, and (2) the passive sliding. In the beginning of unloading, the inter-particle tangential forces are reduced and particles begin to shear in the opposite direction in an elastic manner. As the unloading continues, particles overcome the locked-in shear forces and start to slide in the direction opposite from that of loading condition. During unloading, the sliding of particles is locked-in, thus produce a permanent settlement and a higher value of earth pressure coefficient K_o.

The stress state at the onset of unloading is denoted by σ_z^o and σ_x^o. During unloading process, σ_z^o decreases to σ_z; and σ_x^o to σ_x. The degree of unloading is defined by the over consolidation ratio, OCR, given by

$$OCR = \frac{\sigma_z^o}{\sigma_z} \tag{18}$$

To distinguish the differences, we denote k_o^{NC} for the lateral earth pressure coefficient during loading process and k_o^{OC} for the lateral earth pressure coefficient during unloading process, i.e.,

$$K_o^{NC} = \frac{\sigma_x^o}{\sigma_z^o}\ ; \qquad K_o^{oc} = \frac{\sigma_x}{\sigma_z} \tag{19}$$

We denote F_x^o, F_z^o to be the horizontal and vertical forces at the onset of unloading. During unloading, particles move upwards. Let $\Delta\delta_z$ be the upward movement of particle A relative to particle B. The change of forces are discussed for the following two stages:

(1) Elastic Unloading:

The horizontal and vertical forces shown in Fig. 6 are decreased due to the upward movement of particles, given by

$$F_x = F_x^o - \sqrt{2}\,\Delta\delta_z(k_n - k_s)\sin 2\alpha$$
$$F_z = F_z^o - 4\Delta\delta_z(k_n\cos^2\alpha + k_s\sin^2\alpha) \tag{20}$$

(2) Passive Sliding:

When passive sliding occur, the vertical force and horizontal force, as shown in Fig. 6, can be related to the inter-particle forces similar to Eq. (9), given by

$$F_z = 4F_n(\cos\alpha - \tan\phi_{mob}\sin\alpha)$$
$$F_x = 2\sqrt{2}\,F_n(\sin\alpha + \tan\phi_{mob}\cos\alpha) \tag{21}$$

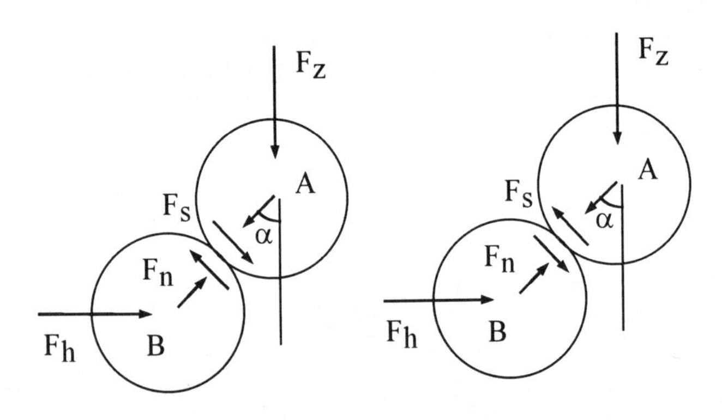

FIG. 6. Forces for Loading and Unloading Process

Forces in Eqs. (20) and (21) can be expressed in terms of stress using Eq. (10). The relationship between vertical and horizontal stresses are discussed for the following two stages:

(1) Elastic unloading:

$$\sigma_x = K_o^{NC}(1-b)\sigma_z^o + b\sigma_z$$
$$b = \frac{\tan(\alpha-\kappa)}{\tan(\alpha-\phi_\mu)}; \qquad \tan\kappa = c_r \tan\alpha \tag{22}$$

In terms of lateral earth pressure coefficient, the value of k_o^{OC} is no longer a constant. The lateral earth pressure coefficient becomes stress dependent, given by

$$\sigma_x = K_o^{OC}\sigma_z$$
$$K_o^{OC} = K_o^{NC}\left(OCR\,(1-b) + b\right) \tag{23}$$

(2) Passive sliding:

$$\sigma_x = K_o^{OC}\sigma_z$$
$$K_o^{OC} = \frac{1}{2}\tan(\alpha+\phi_\mu)\tan\alpha \tag{24}$$

Eqs. (22) and (24) represent respectively the unloading stress path for elastic unloading stage and for passive sliding stage. A typical loading and unloading stress path computed from Eqs. (22) and (24) are plotted in Fig. 7. The slope of the stress path for the elastic unloading stage is influenced by the value of b in Eq. (22) which is governed by the inter-particle stiffness ratio c_r. Slope of the stress path for the passive sliding stage is governed by the value of ϕ_μ. For the specific packing structure discussed in this paper, θ_2 is not equal to θ_1 in Fig. 7.

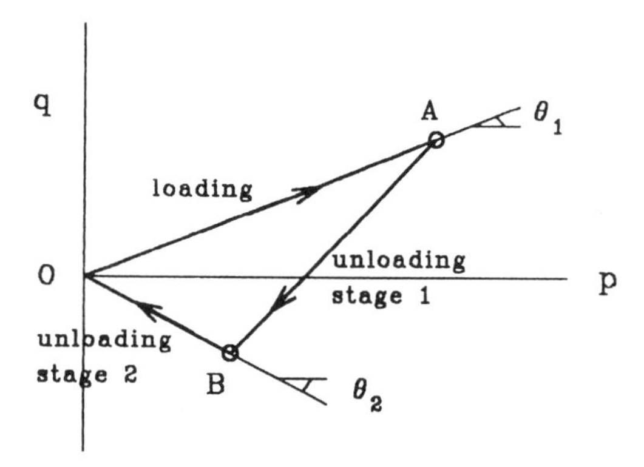

FIG. 7. Stress-path of Loading-unloading Condition

According to Eq. (24) for the case of passive sliding, when the vertical stress is unloaded to zero, the horizontal stress becomes zero. It is noted that, for certain combination, (e.g. ϕ_μ $-$ α), the passive sliding will never occur during the unloading process. In this case, when the vertical stress is unloaded to zero, the horizontal stress remains to be $K_o^{NC}\sigma_z^o(1-b)$, according to Eq. (22). This results to an infinite stress ratio, a stress state beyond the Mohr-Coulomb envelop while the sample is stable. This indicates that the notion of an unique Mohr-Coulomb envelop for the failure of a soil element may be questionable.

Unfortunately, very few measurements of horizontal stress in an oedometer test are reported in the literature, especially for samples unloaded to zero vertical stress. This may be due to difficulties of data interpretation which arise from complex factors such as the compliance of measuring device and the boundary condition between soil and consolidometer. Although it is expected that passive sliding occurs in most cases, it is not unreasonable to expect that, for heavily loaded samples, there exist horizontal stress when vertical stress is unloaded to zero.

A widely used empirical expression for the effect of OCR on the lateral earth pressure coefficient is given as follows (Schmidt 1966; Alpan 1967):

$$K_o^{OC} = K_o^{NC} OCR^m \tag{25}$$

The value of m varies from 0.4 to 0.68 (Alpan 1967; Schmertmann 1975; Sherif et al. 1974). Using these values the empirical equation is calculated and plotted in dash lines in Fig. 8. The calculated results of Eq. (23), using inter-particle friction angle equal to 22° and 30°, give approximately the same range.

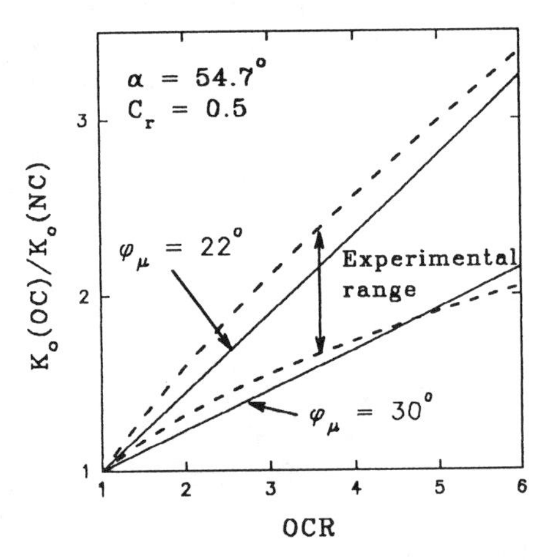

FIG. 8. Effect of OCR on the Lateral Earth Pressure Coefficient

The corresponding strain due to unloading can be straightforwardly obtained using the same analysis. However, the expression is tedious, thus is not given here. A typical stress-strain curve is plotted in Fig. 9.

SUMMARY

Behavior of a body-centered cubic packing under one dimensional loading is analyzed. Note that different behavior are expected for other types of packing such as the face-centered cubic (FCC), or Hexagonal close-packed (HCP). The derived equations for the behavior of BCC are by no means representative for real sand since the microstructure is not realistic for soil. However, the behavior of this packing the microstructure is not realistic for soil. However, the behavior of this packing show reasonable agreement with that observed from experiments in both loading and unloading conditions. The simple analogy provides useful insights to the role of inter-particle properties on macro-scale response. A realistic packing structure for soil should be of a random type. Stress-strain modeling for randomly packed granules with frictional contacts was attempted by Chang et al (1992).

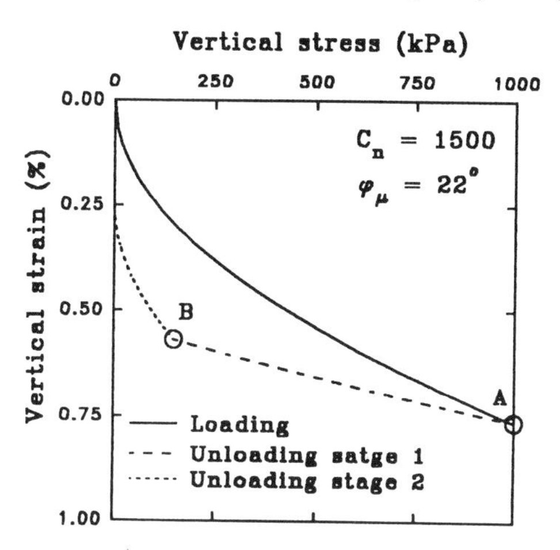

FIG. 9. Loading-unloading One-dimensional Stress-strain Curve

Compressibility of the BCC packing of elastic spheres under one-dimensional loading is stress dependent, influenced by the inter-particle stiffness and sliding friction angle. It is interesting to note that the nature of deformation under the one-dimensional loading condition is not elastic. Significant amount of inter-particle sliding has occurred during the loading process. Evidence of sliding can be observed from the OCR effect and the permanent strain in a loading-unloading stress-strain curve. Therefore application of elastic theory to the analysis of lateral earth pressure and settlement requires a careful evaluation.

The particle sliding generates locked-in forces upon unloading. This mechanism has significant effects on deformation and lateral earth pressure behavior. For the BCC packing, it is possible to have large horizontal stress when the sample in consolidometer is completely unloaded to zero vertical stress. This phenomenon, although unable to be verified in soil experiments, brings a theoretical question on the uniqueness of Mohr-Coulomb failure envelop. In view of the important role of particle sliding in the deformation behavior, it is essential to include this factor in the analysis of settlement and lateral earth pressure of granular media.

APPENDIX - REFERENCES

Alpan, I. (1967). "The empirical evaluation of the coefficient K_o and K_{or}." *Soils and Found.*, 7(1), 31-40.

Chang, C. S. (1988). "Micromechanical modelling of constitutive equation for granular material." *Micromechanics of Granular Materials*, Elsevier Science Publishers, Amsterdam, The Netherlands, 271-278.

Chang, C. S. (1994). "Inter-particle properties and elastic moduli for sand." *Mechanics of Granular Material,* International Society of Soil Mechanics and Foundation Engineering Technical Committee Report.

Chang, C. S., and Misra, A. (1989). "Computer simulation and modelling of mechanical properties of particulates." *Journal of Computer and Geotechniques,* 7(4), 269-287.

Chang, C. S., Chang, Y., and Kabir, M. (1992). "Micromechanics modelling for the stress-strain behavior of granular soil - I: theory." *J. Geotech. Eng.,* ASCE, 118(12), 1959-1974.

Cundall, P. A., and Strack, O. D. L. (1979). "A discrete numerical model for granular assemblies." *Géotechnique,* 29(1), 47-65.

Duffy, J. (1959). "A differential stress-strain relation for the hexagonal close packed array." *Journal of Applied Mechanics,* Trans. ASME, 26(1), 88-94.

Duffy, J., and Mindlin, R. D. (1957). "Stress-strain relations and vibrations of granular media." *Journal of Applied Mechanics,* ASME, 24(4), 585-593.

Hendron, A. J. (1963). "The behavior of sand under one-dimensional compression," Ph.D. thesis, Department of Civil Engineering, University of Illinois, Urbana.

Kabir, M. G. (1992). "Micromechanics modeling for the stress-strain-strength behavior of granular materials," Ph.D. Dissertation, University of Massachusetts, Amherst.

Kishino, Y. (1988). "Disc model analysis of granular media." *Micromechanics of Granular Materials,* Elsevier Science Publishers, Amsterdam, The Netherlands, 143-152.

Mayne, P. W., and Kulhawy, F. H. (1982). "K_o - OCR relationship in soil." *J. Geotech. Eng.,* ASCE, 108(6), 851-872.

Ng, T. T. (1989). "Numerical simulation of granular soil under monotonic and cyclic loading: A particulate mechanics approach," Ph.D. Dissertation, Renesselaer Polytechnic Institute, Troy, New York.

Sherif, M. A., Ishibashi, I., and Ryden, D. E. (1974). "Coefficient of lateral earth pressure at rest in cohesionless soils." *Soil Engrg. Research Report No. 10,* University of Washington, Seattle, Washington.

Schmid, B. (1966). Discussion on "Earth pressure at rest related to stress history." by E. W. Brook and H. O. Ireland, *Can. Geotech. J.,* 3(4), 239-242.

Schmertmann, J. H. (1975). "Measurement of in situ shear strength." *Proc. Specialty Conf. In Situ Measurement of Soil Properties,* Raleigh, 2, 57-138.

Ting, J. M., Corkum, B. T., Kauffman, C. R. and Greco, C. (1989). "Discrete numerical model for soil mechanics." *J. Geotech. Eng.*, ASCE, 115(3), 379-398.

Walton, K. (1987). "The effective elastic moduli of a random packing of spheres." *Journal of Mechanics and Physics of Solids,* 35(3), 213-226.

Application of the "Diflupress L.D." Field Test to Settlement Calculation

C. Leidwanger,[1] E. Flavigny,[2], R. Chambon,[3]
J. L. Giafferi,[4] P. Catel,[4] and G. Bufi[5]

ABSTRACT: Electricité de France developed a new in situ measurement apparatus (Diflupress L.D.) in order to predict the delayed settlements that can appear under heavy constructions. From the experimental curves we can obtain a characteristic of the test which embodies the viscous behavior of soil. We use this characteristic to calculate the delayed settlements using an adaptation of a classical settlement calculation. This involves simplifications but gives a good approach of delayed settlement. We also perform a complete calculation with an elasto-viscoplastic model implemented in a finite element code. The results are interesting too. It appears that the results of the test with the Diflupress L.D. allow rather good predictions of the delayed settlements.

Introduction

Some buildings exhibit non negligible secondary consolidation. Usually laboratory tests data (oedometer, uniaxial compression ...) are used to define constants of secondary settlement models, but in most of the cases these tests are not reliable and require very long term experiments. Faced with this situation, Electricité de France was prompted to attempt a new approach of evaluation of viscous characteristic of soil based on the pressuremeter: the Diflupress L.D. First, a characteristic similar to a classical creep characteristic is obtained from the

---

[1] Student, [2] Asst. Prof., and [3] Prof., respectively, Soils, Solids, Structures laboratory, B.P. 53 X Grenoble Cedex, France.

[4] Engr., Electricité de France, T.E.G.G, 905 av. du Camp de Menthe, B.P. 605 13093 Aix-en-Provence Cedex 2, France.

[5] Engr., Electricité de France, C.I.G., Marseille, France.

experimental data. This characteristic has been related to the delayed settlements really observed on nuclear plant sites. This relation allows for a simple method of determination of delayed settlements. Second, a pressiometric modulus is evaluated from the curves and settlements are calculated with a classical formulation used to calculate settlements, namely the oedometric calculation. To complete this study, a calculation of secondary settlement from a finite element code with an elasto-viscoplastic model is performed. A comparison of the results with the measured settlements is done.

The tests performed with the Diflupress L.D. and the validation of this new field test were first presented in the 13th ICSMFE, New Delhi 1994 (Leidwanger et al. 1994). Let us recall here the principle of the apparatus. Interpretation of the test proposed in this paper is different as here the main objective is the use of the experimental data to calculate settlements. This new interpretation is made in terms of deformation in order to allow for the identification of a soil modulus to predict settlements with an adaptation of the classical oedometric method.

THE DIFLUPRESS L.D.

The basic principle adopted in these trials consists of performing in-situ creep tests (constant internal pressure), by means of an inflatable probe, as used in a pressuremeter test.

Principle

The pressuremeter test consists of studying the relationship between the volume of the liquid injected into a constant level borehole pressurized probe whose dimensions are one meter long and 60 mm of diameter, and the duration of the application of the pressure.

The test allows for the possibility to carry out a fully autonomous and automatic test during a very long period (several weeks or months if necessary). The apparatus keeps the pressure of the injection constant and records data by computer acquisition. The pressure system consists of a gravity loading system made up of lead plates which apply a constant force on a piston by means of pulley block. The piston injects the liquid contained in a pressure chamber into the probe. A set of electronic sensors, which include a rectilinear potentiometer, thermometer and pressure sensors, is driven by a data acquisition and storage facility equipped with programmable memory (see Fig. 1).

Method of Testing

The first test applies to the borehole is a relaxation test. This rebuilds the stress in the borehole after the probe is put in it. In order to have a good contact between the probe and the hole, a certain volume of fluid is injected. Then the probe is disconnected from the fluid charge apparatus. The evolution of the pressure into the probe is then observed and when the pressure stabilizes, we consider the pressure obtained as the initial soil pressure at rest. For the second part of the test, the injection tap is opened and the constant gravity-induced pressure is instantly transmitted to the probe, which begins to inflate. The data acquisition and storage facility scrutinizes the sensors with a predetermined frequency, and records the data. The displacement of the piston is recorded in function of time. The probe is loaded with new plates and the displacements are recorded.

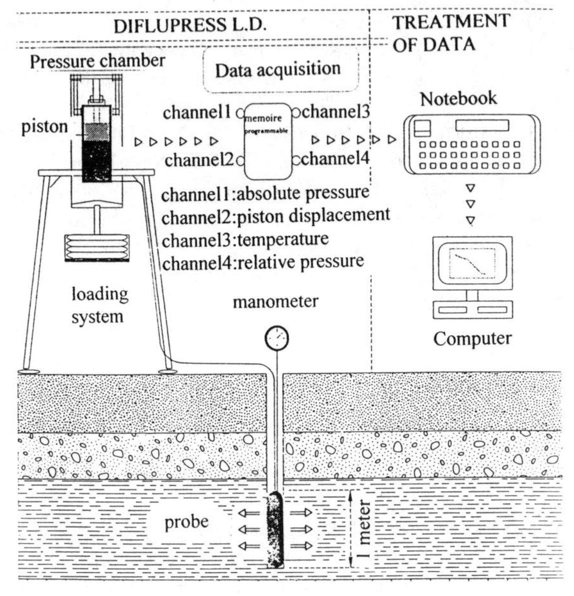

Fig. 1. The Diflupress L.D.

The displacement measured is used to calculate the strain of the probe. On Figs. 2 and 3, the curves of strain in function of logarithm of time for the nuclear plant sites of Nogent and St-Laurent are presented.

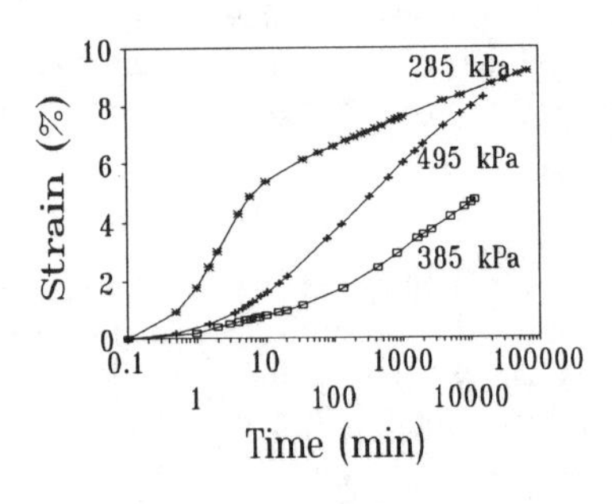

FIG. 2. Strain in Function of Time: St-Laurent

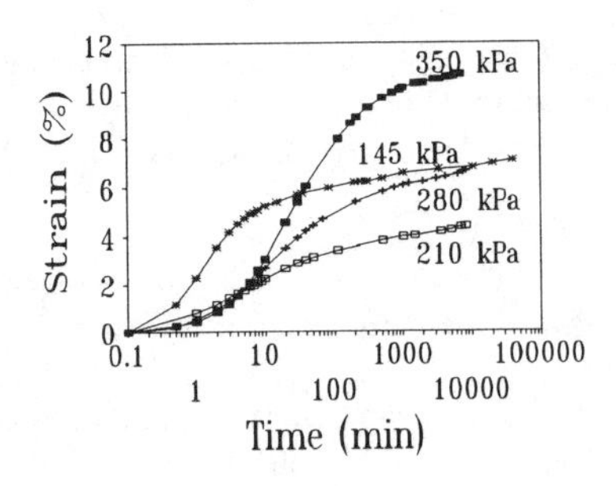

FIG. 3. Strain in Function of Time: Nogent

For each stress level, the strain is calculated using large deformations assumption after the start time is initialized at the beginning of each level of loading. Towards the end of each constant stress loading test, all the curves show the same straight line shape. A linear strain rate as a function of the logarithm of time can be defined. We are interested in this linear part that looks like a characteristic of creep. We will call "creep slope", β, the slope of this linear section. Furthermore, we can see that this slope depends on the pressure applied to the probe.

Data shown in Figs. 2 and 3 indicates data up to 100000 minutes. The test is measuring volumetric strain over long periods of time but no volumetric corrections due to temperature fluctuations with time are made for both sites presented here. A fine study (Leidwanger 1993) made on a third site showed that the temperature fluctuations induce fluctuations of the volumetric strain over short periods of time, so that the global volumetric strain curves are not modified. The linear part of the curves is not influenced by temperature fluctuations.

Modelling the Diflupress L.D. Test

In order to determine the relations between the "creep slope" β and the pressure, we need to evaluate the actual stress applied to the soil. As for the pressuremeter measurements, corrections are made to the pressure read on the manometer to obtain the stress exerted on the soil (Baguelin et al. 1978).

The inertia of probe is determined from tests made on new probes before the campaign of creep tests. Several types of testing are performed, and it appears that the inertia of probe does not depend on the time factor but highly depends on the injected volume of fluid (Leidwanger 1993; Leidwanger et al. 1994). It is interesting noticing that the inertia pressure does not vary more than 5 kPa at the same level of pressure so we can conclude that the stress applied to the soil is really constant.

We chose to relate the "creep characteristic", β, to the incremental pressure (ΔP) that induces the time dependent deformation of soil. Before applying any load, the soil is only subjected to the soil pressure at rest (P_o), evaluated by the relaxation test. The pressure read on the manometer is the soil pressure at rest plus the incremental pressure, ΔP. The incremental stress that results in deformation is the corrected pressure (P) minus the soil pressure at rest (P_o):

$$\Delta P = P - P_o \tag{1}$$

Despite the fact the stress state around the pressuremeter probe is not homogeneous, we adapted relations usually used to express the creep observed on homogeneous samples tested in the laboratory. This strong assumption is justified by the fact satisfying results are obtained. The mathematical expression of the end of the curves of creep drawn in a semi-logarithmic diagram is the following:

$$\epsilon_1(t) = \epsilon_{10} + b \log \frac{t}{t_o} \tag{2}$$

where ϵ_1 is the axial strain of the sample; ϵ_{10} is the strain at the beginning of the creep and b is the slope of the linear part of the curve.

Relation derived from Singh-Mitchell formula

From creep tests performed in a triaxial apparatus, Singh and Mitchell (1968) proposed to write the axial strain rate as:

$$\dot{\epsilon}_1 = A^{\alpha q}\left[\frac{t_1}{t}\right]^m \tag{3}$$

where q is the applied deviator, t is the time, A, t_1, m and α are constants.

For the Diflupress, the linearity of strain with the logarithm of time leads to m=1, so

$$\epsilon(t) = \epsilon_o(t_o) + \beta \log\frac{t}{t_o} \tag{4}$$

where $\epsilon_o(t_o)$ is the strain at time t_o.

We choose the following empirical expression for the "creep slope" β

$$\beta = \exp\frac{(P - P_o) - a'}{b'} \tag{5}$$

where a′ and b′ are constants to be fitted.

Relation derived from Schultze formula

Vuaillat (1980) takes a formulation close to the formulation of Schultze (1961), for triaxial creep tests:

$$\epsilon_1 = a + \beta \log(t - R) \tag{6}$$

where a and R are constants, and ϵ_1 is the axial strain of the triaxial sample. The analytical formulation of β proposed by Schultze is

$$\beta = m\left[\frac{\psi}{1 - \psi}\right]^n \tag{7}$$

where ψ is the ratio between the creep deviator and the failure deviator. Vuaillat proposed a small change by inducing a limit for β when ψ approaches 1. He makes the assumption that the variation law of the coefficient β is good as far as the sample remains homogeneous and slightly deformed. In the Diflupress case, values of both deviators are meaningless because the stress state of the soil is not homogeneous. By analogy with the classical formula, we define ψ as the ratio between the current stress value and the stress value at failure for the considered soil layer. Finally the formulation for the "creep characteristic" β is assumed to be

$$\beta = m\left[\frac{\psi}{1.02 - \psi}\right]^n \quad \text{with} \quad \psi = \frac{P - P_o}{P_{lim} - P_o} \tag{8}$$

where P_{lim} is the limit pressure of the tested soil layer, m and n are constant we need to determine. Around the pressuremeter probe, the failure is reached for a stress state close to the plastic failure. This explains our choice of P_{lim} as a value of the pressure at failure.

The results obtained for the two relations are presented on Figs. 4 and 5 for the nuclear plant site of Nogent. The "creep characteristic" β obtained from tests with the Diflupress L.D. can be related to the stress applied to the soil using relationships that are identical to ones developed for laboratory creep tests.

Now we show the relation between the delayed settlement per log cycle observed on the two nuclear plant sites and the "creep slope" measured with the Diflupress.

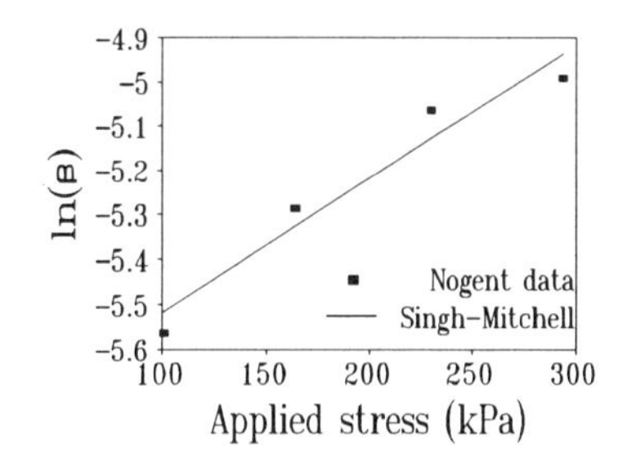

FIG. 4. **Relation of Singh-Mitchell: Nogent Site**

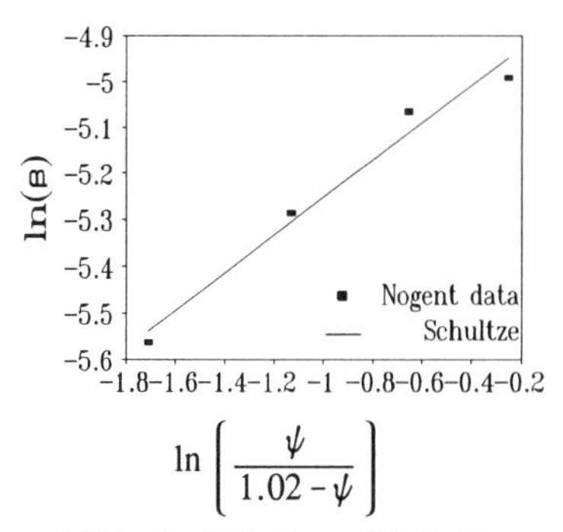

FIG. 5. **Relation of Schultze: Nogent Site**

SIMPLE METHOD OF DETERMINATION OF DELAYED SETTLEMENT

Let us recall here conclusions made in a previous paper (Leidwanger et al. 1994). We found a relation between the vertical delayed deformation per log cycle of time observed for each building of both nuclear plant sites of St-Laurent-des-Eaux and Nogent-sur-Seine and the "creep slope" β measured with the Diflupress L.D. test performed on both sites. This linear relation is presented on Fig. 6. This diagram is obtained first by calculating the vertical delayed deformation ($\Delta W/H_c$) per log cycle for each building and by evaluating the "creep slope" β for a level of stress which characterizes the load applied in surface, for each building.

In order to predict delayed settlements, we propose the following method. If the load applied at the surface by the building which has to be constructed is known, the stress induced in the compressible layer can be evaluated by an elastic calculation of transmission of stress (Poulos and Davis 1974). When this stress value is known, a Diflupress L.D. test is performed on the site by applying a pressure equal to the stress value previously calculated. The "creep slope" β is deduced from the experimental data. This value, referred to the diagram on Fig. 6, allows for the estimation of the vertical delayed deformation of the compressible layer. The delayed settlement per log cycle of time is then obtained since the thickness of the compressible layer is known from usual construction project studies.

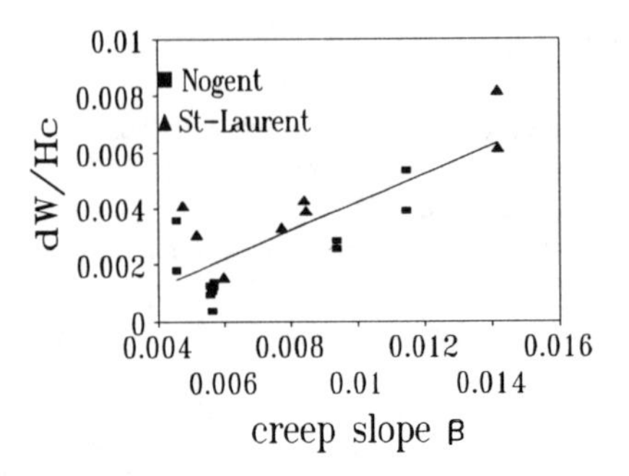

FIG. 6. Relation Deformation - "Creep Slope": Nogent and St-Laurent

This simple method assumes a one-dimensional analysis which is represented by the foundation conditions at Nogent and St-Laurent sites (foundation of about forty meters width, thickness of compressible layers of about few meters, at about ten meters depth).

Adaptation of the Oedometric Formulation

In this section, we present the results of delayed settlements estimations by log cycle obtained from an adaptation of a classical formulation used to calculate settlements, the oedometric calculation.

We adjusted the oedometric method in order to use it to calculate delayed settlements. We used a pressiometric modulus, function of the time, obtained from the results of the tests made with the Diflupress L.D.

Modulus of the Soil from the Tests

For the classical pressiometric test, a shear modulus G is obtained from the curve of the variation of injected volume as a function of the applied pressure (Briaud 1992). In the same way, Baguelin et al. (1978) proposed a shear modulus obtained from the curve of strain in function of the pressure:

$$G = \frac{dP}{2\alpha_o} \tag{9}$$

where α_o is the strain, expressed in large deformations, at the wall of the borehole, $\alpha_o = dV/(2V)$. In the case of Diflupress, we saw above that the strain can be written as a function of time according to Eq. (2). The shear modulus becomes

$$G = \frac{dP}{2\alpha_o(t_o,P)\left[1 + \frac{\beta(P)}{\alpha_o(t_o,P)}\log\frac{t}{t_o}\right]} \tag{10}$$

The shear modulus depends on the time but its value at time t_o is rather difficult to evaluate because we have only three different applied pressures. We make the assumption that the three points are in the pseudo-elastic phases, so the quantity $dP/[2\alpha_o(t_o,P)]$ is the slope of the curve of strain in function of applied pressure at time t_o. Finally, the shear modulus is

$$G = \frac{G_o}{1 + \frac{\beta(P)}{\alpha_o(t_o,P)} \log\frac{t}{t_o}} \tag{11}$$

where G_o is the shear modulus at time t_o. The pressiometric modulus is obtained from the shear modulus by

$$E_p = 2(1 + \nu)G \tag{12}$$

Oedometric Formulation

The oedometric formulation of the settlement takes into account the thickness of the compressible layer. For a foundation that applies a stress σ_{zz} in the compressible layer of thickness H_c, the settlement is

$$W = H_c \frac{\sigma_{zz}}{E_{oed}} \tag{13}$$

where E_{oed} is the oedometric modulus. For elastic media, we can replace E_{oed} as a function of the Young modulus and the Poisson's ratio as suggested in the following.

$$E_{oed} = E_y \frac{1 - \nu}{(1 + \nu)(1 - 2\nu)} \tag{14}$$

In theory the pressiometric modulus is equal to the Young modulus. But studies with self-boring pressuremeter (Powell 1990) showed that the Young modulus is between two and four times higher than the pressiometric modulus. We took a ratio of two between E_y and E_p to evaluate the settlement from the oedometric formulation.

$$W = H_c \frac{\sigma_{zz}}{2E_p} \frac{(1 + \nu)(1 - 2\nu)}{1 - \nu} \tag{15}$$

Delayed Settlement by Log Cycle

By replacing the pressiometric modulus by its value as a function of time, we obtain a settlement function of time and pressure. In order to evaluate the settlement of the heaviest building of the nuclear plant site of St-Laurent-des-Eaux, we evaluated the vertical stress induced by the building into the compressible layer of soil by the elastic Boussinesq relation.

From the relation established between the "creep slope" β and the stress, we can evaluate the "creep slope" β corresponding to σ_{zz}. Finally the total settlement can be expressed by

$$W = W_o + \Delta W \log \frac{t}{t_o} \qquad (16)$$

where W_o depends on the modulus at time t_o and W depends on the "creep slope" β. We are above all interested in the evaluation of the delayed settlement ΔW per log cycle. In the next table, we present the numerical values of initial and delayed settlements obtained by adjustment of the oedometric formulation.

TABLE 1. Results from the Oedometric Formulation

Oedometric formulation
W_o = 0,217 meter
ΔW = 0,077 meter

On the nuclear plant site, the two heaviest buildings (they apply the same surface load of 450 kN/m²) present delayed settlements per log cycle of 0,0494 and 0,0653 meter. The differences between these two identical buildings may be the structure of the subsoil: the thickness of the compressible layer may not be identical under the two buildings. The initial calculated settlement is rather close to the observed settlement at the end of construction. However, we have to keep in mind that the initial settlement calculated with the oedometric formulation represents only the contribution of the compressible layer. The total construction settlement represents contribution from other layers. So the oedometric formulation gives a good approximation of the delayed settlement of the compressible layer but slightly over predicts the settlement at the end of the construction.

To proceed to a simple settlement calculation with the oedometric formulation, we made several assumptions: we considered that the stress path followed under the foundation is an oedometric path, we neglected the influence of the presence of water. In order to evaluate these assumptions and to measure the difference between a complete calculation and a simplified one, we performed settlements calculations with a elasto-viscoplastic model implemented into a finite element code.

Finite Element Calculation

The Model

The chosen elasto-viscoplastic model is the one written by Kaliakin et al. (1987, 1990) and implemented by Li (1992) in the finite element code developed in the University of Liege (Belgium). This model is based on the Cam-Clay model but allows the coupling of plasticity and viscoplasticity theories and incorporates a bounding surface. Inelastic deformations can occur within the bounding surface. The modelling of the settlement of foundation is made for the nuclear plant site of St-Laurent. Usually, the parameters of the model are identified from results of laboratory tests on homogeneous samples. We did not have such tests so we used the tests made with the Diflupress L.D. to identify most of the viscoplastic parameters. The finite element code used allows for the simulation with or without pore pressure. In the first case, this code models a full coupling between the skeleton behavior and the water seepage. We made the two simulations and we present the results obtained with the best set of parameters for each simulation. We point out that difficulties to identify the model parameters were encountered (Leidwanger 1993). Some of them were identified from laboratory tests since they are parameters similar to the one encountered in the elastoplastic model Cam-Clay. Viscous parameters were fitted with the Diflupress L.D. data with a trial-and-error procedure.

Assumptions of Calculation

The Diflupress L.D.

The modelling of the Diflupress L.D. test is made in order to fit the viscous parameters of the constitutive equations. It is not certain that the set of parameters which is able to correctly model the Diflupress L.D. tests is unique.

The finite element model is written to evaluate effective stresses, so when neglecting the pore water pressure, the total stresses are set equal to the effective stresses. When the pore water pressure is not neglected, the total stresses are equal to the effective stresses plus the pore water pressure. Initially, the total vertical stress is determined by the depth of the test and the total weight density of the soil. The total horizontal stress is the soil pressure at rest obtained from the relaxation test made with the Diflupress. The pore water pressure corresponds to a height of water about 9 meters. The field modelled has a radius of 3 meters. We make the assumption of an axisymmetrical stress state. The loading of probe is simulated by a linear loading during 8 minutes. The limit conditions induce that all the nodes at the outer boundary of the mesh are blocked in displacement. In the case of coupling between the mechanical law and the Darcy law, the permeability of soil is taken equal to 1×10^{-8} m/s. This value is obtained from laboratory tests data. A study (Leidwanger 1993) showed that the part of the curves that concerns the secondary compression is not influenced by a variation of the permeability, as long as the value is bigger than 1×10^{-10} m/s.

The foundation

To model the foundation, we use the assumption of axisymmetrical problem. The field concerned by the modelling has a radius of 90 meters and is 45 meters depth. The rock mass is assumed at this depth. The load is applied following the real history of loading of construction. The initial stress field is the same in total stresses in both simulations. The water level is on the surface of the soil for the coupling case. The boundary conditions are applied in stresses: at the outer boundary of the mesh, the stresses are imposed equal to the initial horizontal total stresses. The stratigraphy of the subsoil is divided in six layers and only the behavior of the middle layer is simulated by the elasto-viscoplastic constitutive equations. The behavior of the other layers is assumed to be linear elastic.

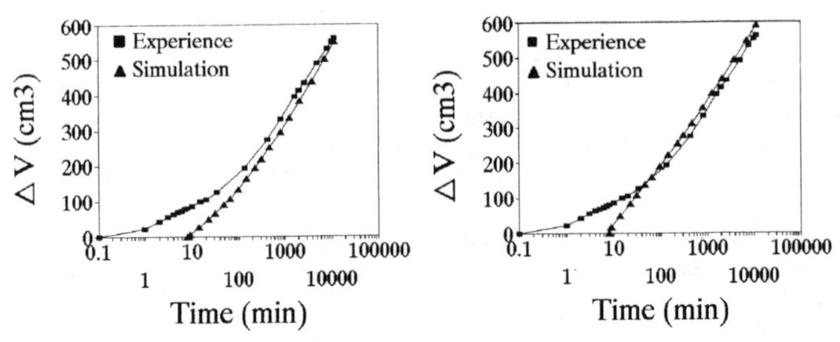

FIG. 7. Second Level of Loading, No Coupling Simulation, Diflupress Test

FIG. 8. Second Level of Loading, Coupling Simulation, Diflupress Test

Results

We studied the influence of loading on the second part of the curves, the part that concerns secondary compression. This study showed (Leidwanger 1993) the loading does not influence this second part of the curves. We present on Figs. 7 and 8, the curves for the second level of loading for the two Diflupress simulations. In these simulations, the optimum sets of parameters are used. Then each set of parameters is used to model the foundation.

The curves obtained for the simulation of the settlement of the heaviest building are presented for the two sets of parameters on Figs. 9 and 10. Regarding the figures, it appears the total settlement obtained from the modelling is bigger than the observed settlement. The secondary settlement per log cycle of time is 37,4 mm for the no coupling simulation whereas in the reality it is of about 49,4 mm per log cycle. For the coupling modelling, the total settlement is slightly lower but the delayed settlement per log cycle is also reduced. It is about 23 mm per log cycle so it is twice smaller than the real delayed settlement per log cycle. The difference between the two calculations is not surprising because the stresses are lower in the coupling case so the viscous part of the response of the model is less important. In the next table are summarized the values obtained for the delayed settlement per log cycle.

TABLE 2. Delayed Settlement per Log Cycle of Time

	No coupled simulation	Coupled simulation	In-situ measurement
W per log cycle	0.0374 m	0.023 m	0.0494 to 0.0653 m

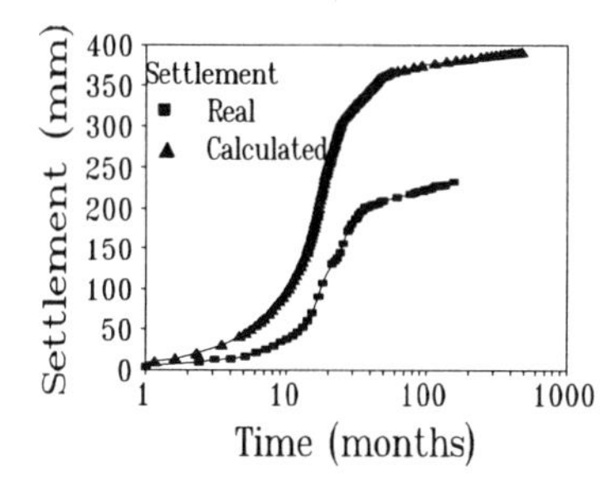

FIG. 9. No Coupling Settlement of Foundation

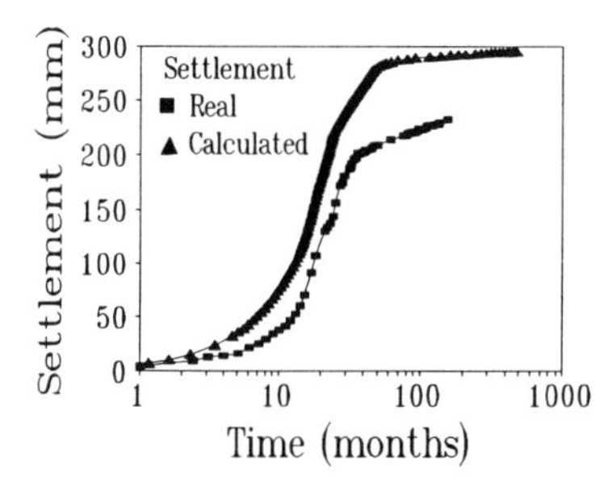

FIG. 10. Coupling Settlement of Foundation

Conclusion

It appears that we can obtain a "creep characteristic" of soil from the data obtained by tests performed with the Diflupress L.D. This characteristic depends on the pressure applied to the soil. By analogy, we adapted relations initially developed for laboratory tests to the case of the field test with the Diflupress L.D. The Diflupress L.D. test is used in three different approaches to calculate delayed settlements. The first approach is the use of the correlation between the vertical delayed deformation and the "creep slope" β observed on both nuclear plant sites of St-Laurent-des-Eaux and Nogent-sur-Seine to define a simple direct method of determination of delayed settlements. The second one consists in performing calculations using an adaptation of the classical oedometric method. The delayed settlements obtained are quite good and this type of calculation could be use in the future to have an indication of the delayed settlement. The third way is the foundation modelling with an elasto-viscoplastic model implemented in a finite element code. With this approach, no assumption on the followed stress path is made. Further studies are necessary to improve our estimates of delayed settlements for heavy structures with such a model.

Appendix - References

Baguelin F., Jezequel J. F., and Shields, D. H. (1978). *The Pressuremeter and Foundation Engineering*. Trans Tech Publications, Clausthal-Zellerfeld, Germany.

Briaud J. L. (1992). *The Pressuremeter*, A. A. Balkema, Rotterdam, The Netherlands.

Bufi, G. (1990a). "New long-term pressuremeter creep test." *Proc. 3rd Int'l. Symposium on Pressuremeters,* Oxford, 95-104.

Bufi, G. (1990b). "Craie pâteuse - Cas de la centrale de NOGENT-SUR-SEINE. Comparaisons d'essais de fluage au pressiomètre, aux essais de laboratoire et au comportement réel des ouvrages." *Mém. Soc. Géol. France, N.S., n° 157.*

Kaliakin, V. N., Dafalias Y. F., and Herrmann L. R. (1987). *Time Dependant Bounding Surface Model for Isotropic Cohesive Soils*, Short Course Notes.

Kaliakin, V. N., and Dafalias, Y. F. (1990). "Theoritical aspects of the elastoplastic-viscoplastic bounding surface model for cohesive soils." *Soils and Foundations,* 30(3), 11-24.

Leidwanger, C., Flavigny E., Giafferi J.L., Catel P., and Bufi, G. (1994). "Delayed settlements and "Diflupress L.D."" *Proc. 13th ICSMFE,* New-Delhi, I, 233-236.

Leidwanger, C. (1993) "Etude des tassements différés à partir de résultats d'essais au Diflupress L.D," Thèse de Doctorat, laboratoire 3S, Université J. Fourier, Grenoble, France.

Li, X. L.(1992). "Contribution à la modélisation du comportement mécanique des argiles avec prise en compte de la consolidation secondaire." *Mémoire de Maîtrise en Sciences Appliquées, Université de Liège,* Liège, Belgium.

Powell, J. J. M. (1990). "A comparison of four pressuremeters and their methods of interpretation in a stiff heavily overconsolidated clay." *Proc. 3rd Int'l. Symposium on Pressuremeters,* Oxford, 287-297.

Poulos, H. G., Davis, E. M. (1974). *Elastic Solutions for Soil and Rock Mechanics*, John Wiley & Sons, New York, New York.

Schultze E. (1971). "Essais de fluage sur des sols normalement compactés." *Compte-rendu des journées françaises de Mécanique des Sols. "Comportement des sols avant rupture".*

Singh, A., and Mitchell, J. K. (1968). "General stress strain time function for soils." *J. Soils Mech. Found. Div.,* Proc. ASCE, 94(SM1), 21-46.

Vuailliat, P. (1980). "Propriétés visqueuses d'une argile: Expériences et fomulation incrémentale," Thèse de Docteur-Ingénieur. IMG, Grenoble, France.

Measurements and Numerical Modelling of High Rise Building Foundations on Frankfurt Clay

Eberhard Franke,[1] Bernd Lutz,[2] and Yasser El-Mossallamy[3]

ABSTRACT: Many high rise buildings were founded on Frankfurt clay on rafts applying average contact stresses in the order of 0.4 MPa. Beside rafts now piled-rafts are often used to reduce total and differential settlements. An extensive research program was conducted to measure the foundation behaviour. Numerical models were developed to simulate the raft and piled raft behaviour. Comparisons between measurements and numerical results are presented.

INTRODUCTION

The piled raft is an interesting example of the soil-structure interaction problem and of the application of the serviceability limit state design as defined by new standards. The piled raft foundations are now widely used in many parts of the world. The main advantages of such a foundation type are:

(a) Reduction of total and differential settlements (s respectively Δs).
(b) Reduction of tilting either due to load eccentricity or due to unpredictable scattering of soil properties.
(c) Decrease of the internal raft stresses and moments.
(d) Prevention of settlement joints between differently loaded parts of the building.
(e) Increase of the global safety factor against bearing failure.

---

[1] Prof. of Geotechnics at Technical Univ. Darmstadt, Troyesstr. 34 A, 64297 Darmstadt, Germany.

[2] Trischler and Partners, Berliner Allee 6, 64295 Darmstadt, Germany.

[3] Institute of Geotechnics, Technical Univ. Darmstadt, Petersenstr. 13, 64287 Darmstadt, Germany.

Until now there is no generally accepted design method for piled raft foundations including all of the above mentioned aspects. Even for high rise buildings with a total height of about 260.0 m above ground level, the foundation design in the past was based more or less on sound engineering judgement.

An extensive research program was performed including measurements of the behavior of several piled raft foundations. Based on the results of these measurements an enhanced numerical model was developed. Further a simplified method is given to determine the settlement reduction and the load shares taken by the contact pressure beneath the raft and by the piles.

ENHANCED NUMERICAL MODEL

A mixed technique using boundary element and finite element methods is used to represent the three dimensional nature of the problem (Fig. 1).

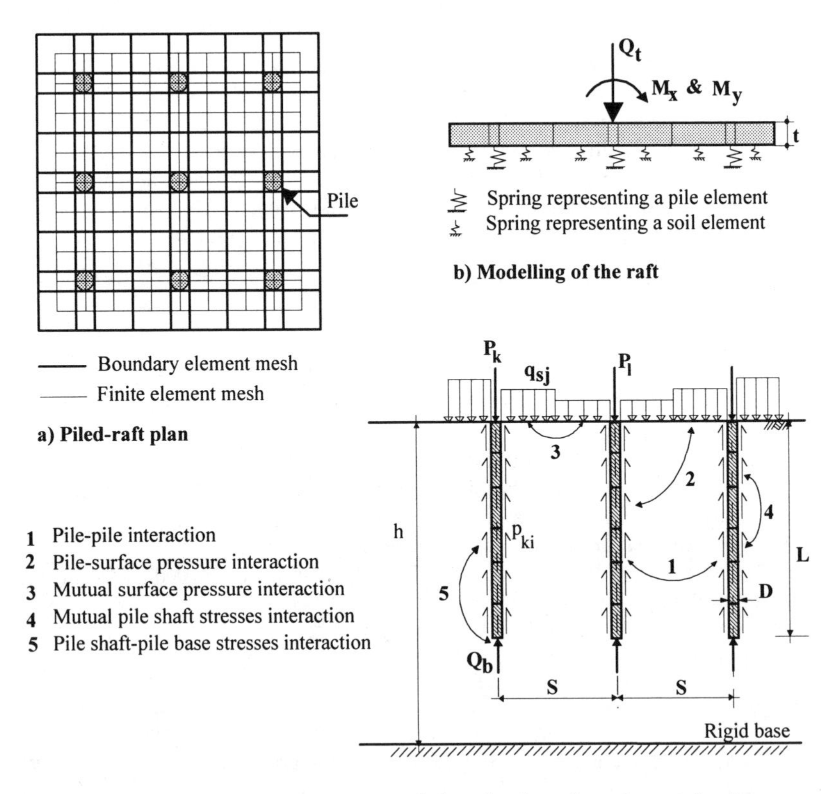

FIG. 1. Modelling of a Piled Raft Foundation

The raft is modelled by the finite element method (FEM) as plate in bending acted on by superstructure loads and supported by nonlinear elastic supports at each node. These nodes (Fig. 1b) represent either a pile or a surface soil element. The spring stiffness depends on the stress and deformation at each node.

The pile group embedded in a finite half space (Fig. 1c) is simulated by a complete boundary element (BEM) structure (Butterfield and Banerjee 1971; Poulos and Davis 1980; El-Mossallamy 1989). The system consists of a pile group acted on by the pile head loads and the raft/soil contact pressure. Nonlinear pile load-settlement behavior is modelled at the pile/soil interface.

An incremental type of nonlinear calculation with iterative adaptation of FEM and BEM in each increment is applied. This mixed technique is used to reduce the computer memory required to simulate this three dimensional problem taking into account the nonlinear behavior.

Nonlinear Pile Response

The nonlinear pile response is caused due to the following influences:

Nonlinear Soil Behavior adjacent to the Pile Shaft

The variation of the shear strain level in the surrounding of the pile shaft causes a nonlinear soil behavior in radial direction (Fig. 2a). A hyperbolic shear stress-shear strain relationship has been suggested to approximate this nonlinear behavior (Randolph 1977; Kraft et al. 1981; Chow 1986).

Another reason causing nonlinear soil behavior is due to pile installation effects. Based on back-calculations this installation effects may be considered to be less important compared with the above mentioned non-linearity (O'Neill 1977; Poulos 1993) especially for large bored piles in overconsolidated Frankfurt clay.

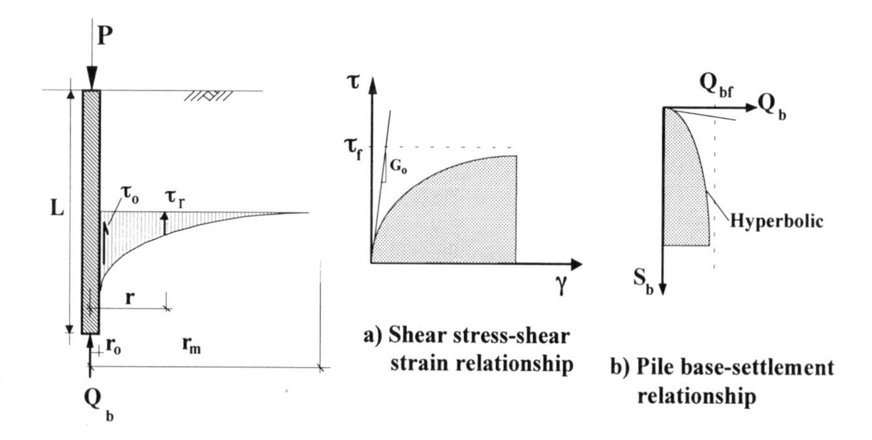

FIG. 2. Nonlinear Pile Response

Slip Effect

The skin friction distribution along each pile is obtained from the complete boundary element solution. If the skin friction at any element exceeds the failure value, the skin friction at that element is set equal to the failure value and the boundary element solution is repeated until stability of the calculated results for all elements is achieved.

To determine the failure value of skin friction at any element undrained and drained conditions are applied alternatively. The effect of unloading by excavation and of loading by raft/soil contact pressure is considered.

Nonlinear Soil Behavior near the Pile Base

The effect of soil non-linearity at the pile base (Fig. 2b) is considered by using a hyperbolic pile base load-settlement relationship (Chow 1986).

Locked Pile Stresses

Due to the installation of multistory underground garages the foundation depth of most high rise buildings in Frankfurt ranges between 14.0 and 20.0 m. The excavation pit is often carried out in two stages. The first stage is the excavation down to the ground water table, then the piles are installed and the excavation is continued down to the final foundation depth.

The stress relief due to pit excavation causes locked stresses in the piles (Sommer 1990; Franke 1991). These locked stresses are taken into account as initial stresses when nonlinear pile response and pile/soil interface slip are considered in the complete boundary element solution.

Loading and Reloading

The unloading stresses due to pit excavation range between 170 and 250 kPa. This large stress relief has to be taken into consideration because the soil stiffness for reloading is much higher than for first loading as given in Table 1.

To take this effect into account approximately, the following procedure (Fig. 3) is applied: The stresses due to excavation and due to locked pile stresses (line I Fig. 3) are calculated along vertical lines beneath every finite element node. The stresses due to the raft/soil contact pressure and due to the pile stresses (skin friction and base stresses) are calculated along the same vertical lines (line II Fig. 3). The vertical stress of line I is compared with that of line II to distinguish between reloading and first loading stresses. According to this comparison different soil stiffness moduli are used to calculate the soil surface settlement at the considered node. A spring stiffness at each finite element node is determined according to settlement and raft/soil contact pressure.

Soil stiffness moduli based on in-situ tests (e.g., pressuremeter) and back-calculated values from observed settlements of structures founded on the same soil (Amann 1975) are used. In addition nonlinear soil model such as the hyperbolic model (Duncan and Chang 1971) is also applied optionally.

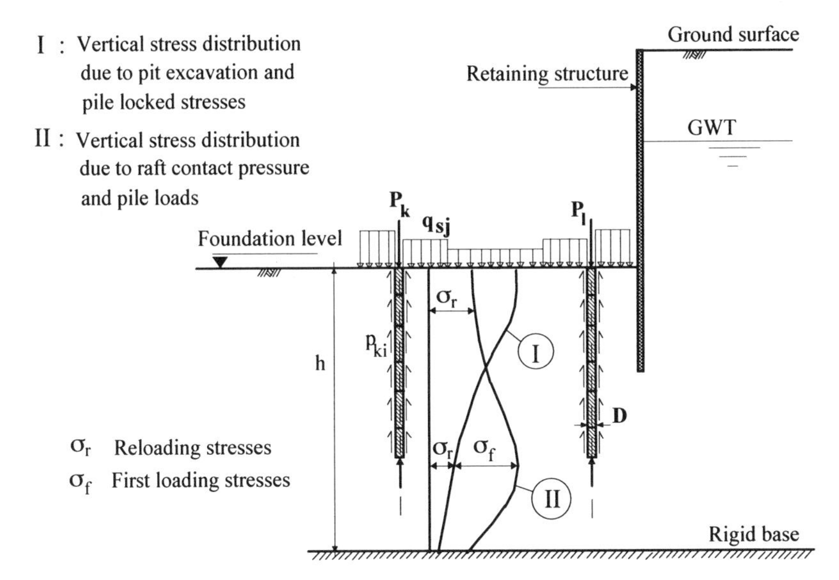

FIG. 3. Effect of Reloading and First Loading

SIMPLIFIED METHOD

For every day practice of piled raft foundation design a simplified method was developed. This method yields the required number of piles, the pile length, the pile diameter and the pile group configuration related to the intended settlement reduction. The pile/raft load shares are also determined.

For such a handy calculation procedure the raft flexibility, the pile stiffness and the interaction between pile forces (pile shaft and pile base) are neglected. Also linear elastic soil behavior is assumed. The application of this model to the observed behavior of high rise buildings approves its precision.

The foundation is divided into a raft and a pile group as shown in Fig. 4.

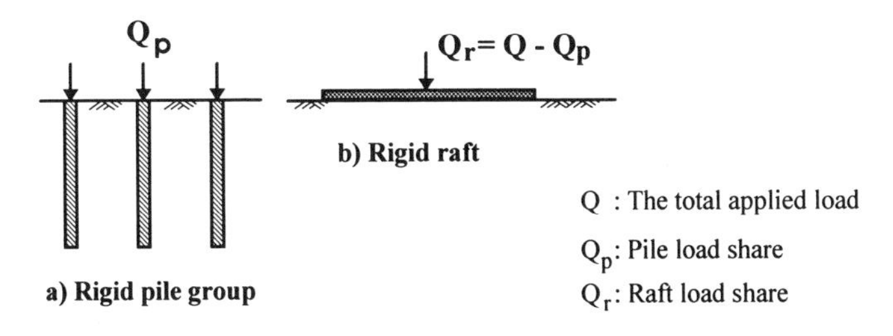

FIG. 4. Dividing the Foundation in Two Systems

For a rigid piled raft the settlement can be determined (similar to Randolph 1983, 1993) as follows:

$$\begin{bmatrix} K_{pp} & K_{pr} \\ K_{rp} & K_{rr} \end{bmatrix} \begin{bmatrix} Q_p \\ Q_r \end{bmatrix} = \begin{bmatrix} S \\ S \end{bmatrix} \tag{1}$$

The K_{pp} value represents the settlement of a rigid free standing pile group connected by a rigid raft without contact to the soil surface due to unit load. The K_{pp} value is calculated using the superposition concept (Randolph and Wroth 1979). Connected research work has revealed that the group influence radius as suggested by Randolph and Wroth (1979) leads to unrealistic results for wide spread pile groups. Applying the single pile influence radius yields better results. Detailed information about this influence radius cannot be given in this brief paper. The raft settlement factor (K_{rr}) is given by Steinbrenner's equation (Steinbrenner 1934) for a surface load on a finite half space with a rigid base. To link the pile group and the raft the interaction between the two systems ($K_{rp} = K_{pr}$) has to be taken into account. A corresponding interaction factor is calculated using the neutral plane concept (Baumgartl 1986; Fellenius 1985) as shown in Fig. 5.

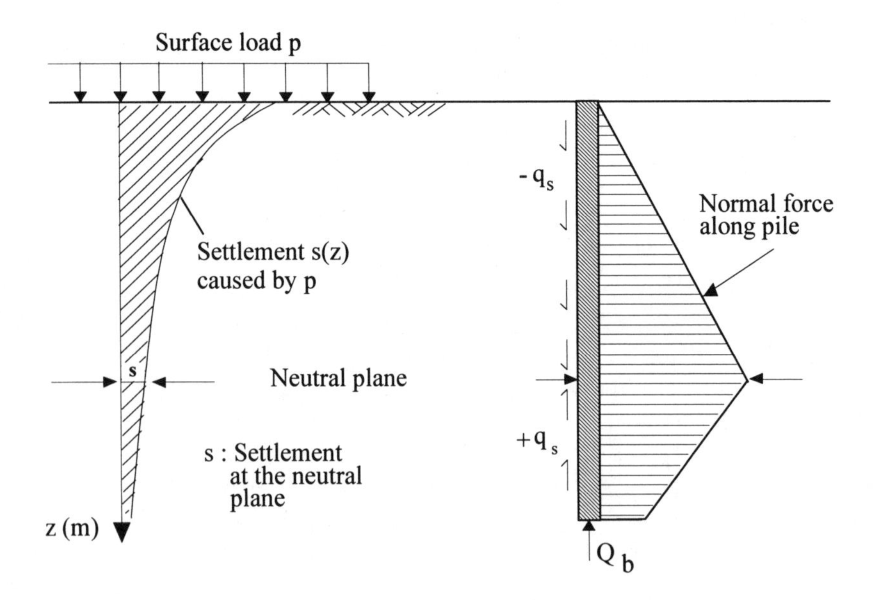

FIG. 5. Settlement and Negative Pile Skin Friction due to Surface Load

COMPARISON BETWEEN MEASUREMENTS AND ANALYSES

Measurements at four high rise buildings which are completed now were conducted in order to extend the experience on piled raft foundations. The main Frankfurt Clay properties are summarized in Table 1. An overview about four existing instrumented foundations is given in Table 2.

TABLE 1. Subsoil Properties of Frankfurt Clay

Clay content	$d < 0.002$ mm = 35-60 %
Atterberg limits	$W_p = 25$ %, $W_l = 70$ %
Natural water content	$W_c = 32$ %
Consistency index	$I_c = 0.85$
Wet unit weight	17.5 - 20.0 kN/m^3
Effective shear parameters	$\phi' = 20.0°$, $c' = 20.0$ kN/m^2
Soil deformation moduli according to pressuremeter:	
Reloading	62.40 MN/m^2
First loading	$8.3 + 1.35 \times Z$ (MN/m^2) Z from clay surface (mean value 8.0 m below ground surface)
Duncan Chang's parameters	$R_f = 0.9$, $n = 0.6$, $K = 250.0$

TABLE 2. Examples of Instrumented Buildings in Frankfurt

	Torhaus	Messe Turm	Westend 1	Th 112
Height	130	256	208	74
Foundation area (m^2)	860	3600	3000	4000
Superstructure load (MN)	400	1750	1420	1000
Number of piles	84	64	40	35
No. of instrumented piles	6	12	6	6
Contact pressure cells	11	11	13	9
Pore water pressure cells	1	2	5	2
Extensometer	3	3	3	–
Inclinometer	–	3	3	6

Beside measuring the subsoil deformations using extensometers and inclinometers, raft contact stresses and pore water pressures were measured. The pile instrumentation includes pile head load cells and concrete strain measurements (Franke 1991; Schwab et al. 1991). The surface and foundation settlements at several points were surveyed additionally.

The 208.0 m high building (Westend 1) will be considered as a representative example in this paper. A cross section and foundation plan are shown in Fig. 6.

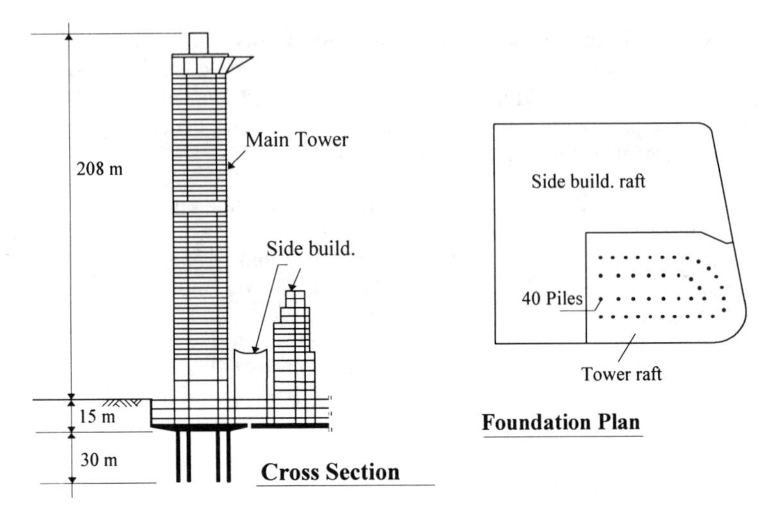

FIG. 6. Cross Section and Foundation Plan of the Building Westend 1

At first the piles are loaded in tension by the pit excavation (subsoil heave) as shown in Fig. 7. Calculated and measured strains along the piles are compared in Fig. 8.

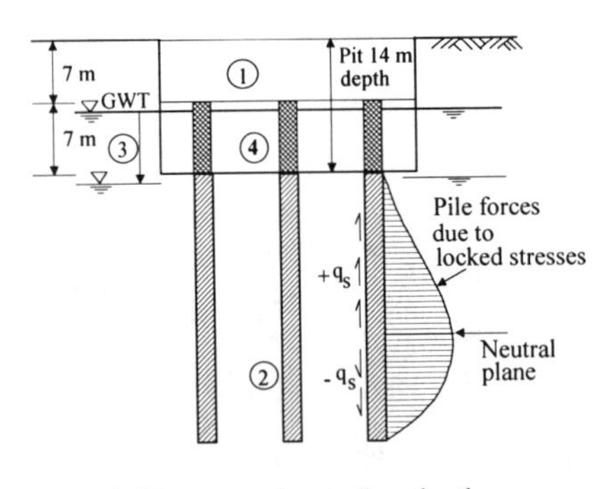

1 Pit excavation to 7 m depth
2 Pile construction
3 Ground water table lowered by 7 m
4 Pit excavation to final depth of 14 m

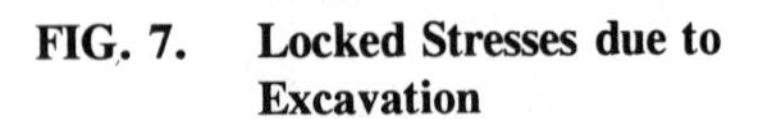

FIG. 7. Locked Stresses due to Excavation

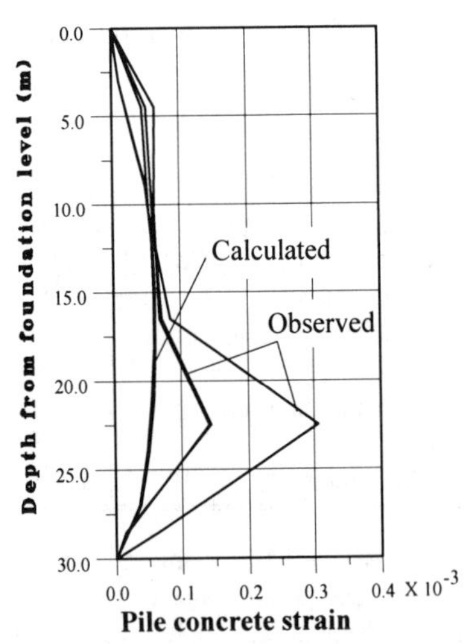

FIG. 8. Pile Concrete Strain due to Pit Excavation

The tension forces of about 4 MN in different piles of the group were rather equal and independent of their position. The concrete tension strains were in the order of up to 0.1 %. The observed larger values are considered to be caused by concrete cracks which were not taken into account in the calculations (Fig. 8).

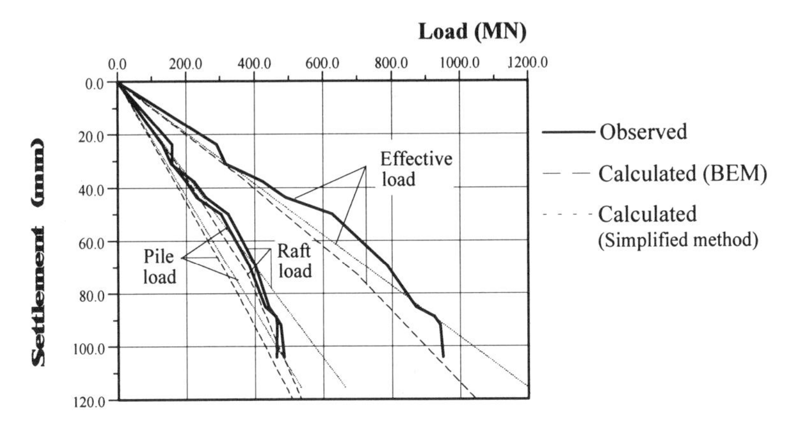

FIG. 9. Load Settlement Relationship for Piled-Raft Elements

In Fig. 9 the load settlement relationship for the piled raft foundation is shown. The observed values are compared with the calculated results by both the more accurate and the simplified method. In the boundary element analysis (enhanced method) the difference effect of first and reloading soil deformation moduli as determined by pressuremeter are considered. The nonlinear pile response is also taken into account. In the simplified method only the reloading soil deformation modulus is considered. Fig. 9 shows that up to the actual applied load the settlement behavior is nearly linear.

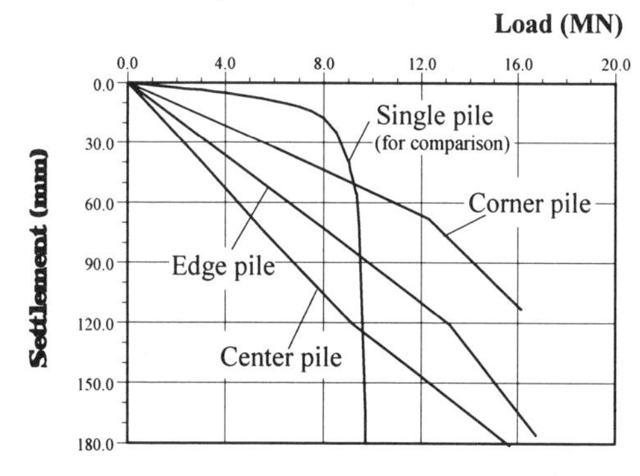

FIG. 10. Load Settlement Relationship for a Single Pile and Individual Piles beneath a Piled Raft

The measured pile head loads range from 10.0 up to 15.0 MN. This is in good agreement with the calculated results. The modulus of pile reaction (load-settlement ratio) of each individual pile in the group depends on its location as shown in Fig. 10.

The average behavior of the piles of a free standing pile group with equally loaded piles and of the piles of a piled raft are compared with the behavior of a single pile in Fig. 11. Due to the interaction of the raft/piles the modulus of foundation reaction is decreased compared with the single pile behavior whilst the ultimate pile load is increased. Similar effects are well known in the case of shallow foundations, where neighboring footings increase the settlement as well as the bearing capacity.

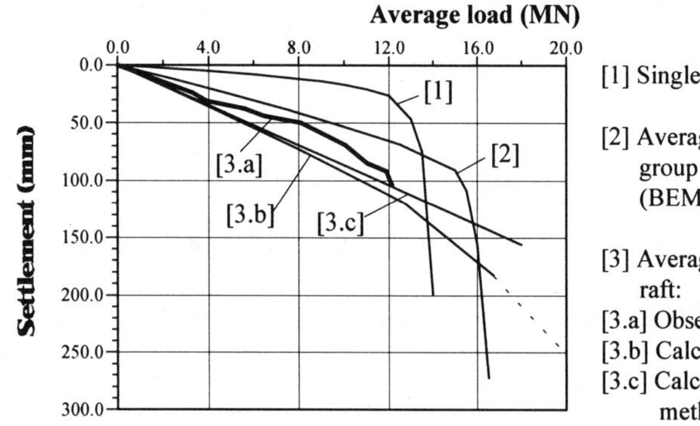

FIG. 11. Load Settlement Relationship for a Single Pile Compared with the Corresponding Average Behavior of the Piles of a Pile Group and the Piles of a Piled Raft

For the back-calculation of the load/settlement behavior good agreement with the observed values has been found (Fig. 9). This is also valid for the load shares of piles and raft and the range of pile head loads (Table 3).

TABLE 3. Comparison of Measurements and Calculation in Case of Westend 1

	Observed	Calculated
Settlement at raft center (mm)	104	109
Raft load (%)	50	51.25
Pilles load (%)	50	48.75
Average pile load (MN)	12.1	11.56
Max. pile load (MN)	15	14.66
Min. pile load (MN)	9.3	8.34

For the load distribution among the different piles no good agreement of measurements and calculations was found. This seems to be caused by pile installation effects and in this special case by a bored pile wall at only one side of the excavation pit very close to the raft.

CONCLUSIONS

The basic aspects of piled raft behavior can be summarized as follows:

(a) Piles are loaded extensively by subsoil deformations in tension due to pit excavation

(b) From experience with different projects the load settlement relationship is found to be nearly linear up to the actual applied loads.

(c) Load settlement behavior and ultimate bearing capacity of the piled raft piles are totally different from those of a corresponding single pile. These differences are dependent on the configuration of the piles forming the group especially on the pile spacings, on the pile length and diameter and on the position of the piles within the group.

ACKNOWLEDGEMENT

The authors gratefully acknowledge the team work at the institute of Geotechnics TH Darmstadt as well as the cooperation with Mr. Wittmann of Trischler and Partners, Darmstadt.

APPENDIX - REFERENCES

Amann (1975) "Über den Einfluß des Verformungsverhaltens des Frankfurter Tons auf die Tiefenwirkung eines Hochhauses und die Form der Setzungsmulde." *Mitteilungen der Versuchsanstalt für Bodenmechanik und Grundbau der TH Darmstadt*, Heft 15, Darmstadt, Germany.

Butterfield, R., and Banerjee, P. K. (1971). "The elastic analysis of compressible piles and pile groups." *Géotechnique*, 21(1), 43-60.

Baumgartl (1986) "Ein einfaches Rechenmodell für Negative Mantelreibung." *Beiträge zum Symposium Pfahlgründungen*, Institut für Grundbau, Boden- und Felsmechanik der Technischen Hochschule Darmstadt, Germany ,71-76.

Chow, Y. K. (1986). "Analysis of vertically loaded pile groups." *Int. J. Num. Anal. methods Geomechs.*, 10 (1), 59-72.

El-Mossallamy, Y. (1989) "Analysis of pile-raft-soil interaction," M.Sc. thesis, Ain Shams University, Cairo, Egypt.

Franke, E. (1985) "Discussion contribution: group action of piles." *Proc. 11th ICSMFE*, San Francisco, 5, 2727.

Franke, E. (1990) "Measurements beneath piled rafts." *Keynote lecture to the ENPC Conf. on Deep Foundations*, Paris, 1-28.

Katzenbach, R. (1993) "Zur technisch-wirtschaftlichen Bedeutung der Kombinierten Pfahl-Plattengründung, dargestellt am Beispiel schwerer Hochhäuser." *Bautechnik*, 70(3), 161-170.

O'Neill, M. W., and Ghazzaly, O. I. (1977) "Analysis of three dimensional pile groups with nonlinear soil response and pile-soil-pile interaction." *Proc. 9th Annual OTC*, Houston, 245-256.

Poulos, H. G., and Davis, E. H. (1980). *Pile Foundation Analysis and Design*, Wiley, New York, New York.

Poulos, H. G. (1993). "Settlement prediction for bored pile groups." *Proc. 2nd Int'l. Seminar Deep foundation*, Ghent, 103-117.

Randolph, M. F. (1977). "A theoretical study of the performance of the piles," Ph.D. thesis, Cambridge University, England.

Randolph, M. F., and Wroth, C. P. (1979). "An analysis of the vertical deformation of pile group." *Géotechnique*, 29(4), 423-439.

Randolph, M. F. (1983) "Design of piled raft foundations." *CUED/D-Soils TR. 163.*

Randolph, M. F. (1993) "Efficient design of piled rafts." *Proc. 2nd Int. Seminar, Deep Foundation*, Ghent, 119-130.

Schwab, H., Gündling, N., and Lutz, B. (1991) "Monitoring pile raft soil interaction." *Proc. 3rd Int'l. Symposium on Field Measurements in Geomechanics*, Oslo, 117-127.

Sommer, H., Katzenbach, R., and DeBenedittis, C. (1990). "Last-Verformungsverhalten des Messeturms Frankfurt/Main." *Vorträge der Baugrundtagung*, Karlsruhe, Germany, 371-380.

Steinbrenner, W. (1934) "Tafeln zur Setzungsberechnung." *Die Straße*, Nr 1, 121-124.

Wittmann, P., and Ripper, P. (1990) "Unterschiedliche Konzepte für die Gründung und Baugrube von zwei Hochhäusern in der Frankfurter Innenstadt." *Vorträge der Baugrundtagung*, Karlsruhe, Germany, 381-397.

GROUND MOVEMENTS FROM MODEL TIEBACK WALL CONSTRUCTION

Christopher G. Mueller,[1] Associate Member, ASCE, James H. Long,[2] Edward J. Cording,[3] and David E. Weatherby,[4] Members, ASCE

ABSTRACT: Four model-scale tieback walls were constructed in a large-scale test facility at the University of Illinois. Ground movement patterns which developed during and after construction are described. The magnitude and distribution of ground movements are compared with observations from two prototype structures. The relationship between the volume of lateral movement at the walls, and the volume of ground surface settlement behind the walls, is examined. The significance of pile bearing and tieback loads on the magnitude of measured ground movements is demonstrated.

INTRODUCTION

Tiebacks can be used to provide support for earth retaining structures where suitable soils are available to develop required anchor capacities. An important consideration in design of tieback walls is the magnitude and distribution of ground movements generated by their construction. As emphasized by Boscardin and Cording (1989), horizontal ground strain and angular distortion control damage potential to adjacent structures. Current procedures for predicting ground movement patterns are based primarily on field observations of both tieback and internally supported walls. Clough and O'Rourke (1990) recently summarized ground movement data. They reported maximum lateral and vertical ground movements for

---

[1] Grad. Res. Asst., Dept. of Civ. Engrg., Univ. of Illinois, Urbana, IL 61801.
[2] Assoc. Prof. of Civ. Engrg., Univ. of Illinois, Urbana, IL 61801.
[3] Prof. of Civ. Engrg., Univ. of Illinois, Urbana, IL 61801.
[4] Vice-President, Schnabel Foundation Company, Sterling, VA 22170.

stiff clays, residual soils, and sands in the range of 0.2% to 0.3% of the wall height for both types of structures.

The magnitude and distribution of ground movements generated by construction of tieback or internally supported walls depend on wall stiffness, support spacing and stiffness, support preloads, and approach to excavation and installation of wall elements. Performance of tieback walls is also influenced by the length and position of the anchors. Liu and Dugan (1972) observed large lateral movements, up to one-half of the maximum lateral wall movement, in the ground behind relatively short tiebacks. Tiebacks can also introduce a large vertical component of load to the structure. Wall settlement can reduce the effectiveness of the tiebacks and produce comparatively large ground movements, (Shannon and Strazer 1970).

Ground movement patterns which developed during construction of four model-scale tieback walls are described. The magnitude and distribution of movements are compared with observations from two prototype structures. The relationship between the volume of lateral movement at the wall (commonly measured in the field) and the volume of surface settlement behind the wall (required to assess the potential for damage to adjacent structures) is discussed. Finally, the significance of pile bearing, and the relationship between tieback loads and the magnitude of ground movements, are evaluated.

The model tieback walls were constructed inside a large model testing facility at the University of Illinois. Model-scale testing can provide useful information on the magnitude and distribution of ground movements which develop during construction of tieback walls. The stiffnesses of wall elements, spacing of supports, and density of the sand deposit can be controlled and varied. Lateral and vertical ground movements can be measured without interference from adjacent structures or pavements, or concerns for seasonal changes in moisture content and temperature. In addition, failure of the walls can be induced without the serious consequences which accompany field-scale walls.

MODEL-SCALE TESTING PROGRAM

The model test facility at the University of Illinois consists of a test chamber measuring 14 ft (4.3 m) by 16 ft (4.9 m) in plan, and 10 ft (3.0 m) in depth, and is equipped with a sand storage bin and conveyors for transporting material. The large size of the test chamber facilitates use of realistic construction procedures and development of significant body forces. The blocks forming the test chamber are post-tensioned laterally and vertically to form "rigid" sidewalls. Interior walls are lined with friction-reducing high density polypropylene.

The model-scale walls constructed in this study consisted of soldier piles and lagging supported by either one or two levels of tiebacks. Each wall was constructed with an exposed height at design grade of 6.25 ft (1.9 m) and pile toe penetration of 1.25 ft (0.39 m). A total of nine piles were used in each construction, and these were spaced on 2 ft (0.61 m) centers. Wall dimensions and the position of the tiebacks along the wall were selected to provide 1/4 scale models of two prototype structures constructed at Texas A & M University and described by

Chung and Briaud (1992). A view of a completed model-scale wall is provided in Fig. 1.

FIG. 1. Photograph of Completed Model-Scale Tieback Wall

In addition to the number of tieback levels, other variables of the testing program included pile stiffness and pile bearing capacity. After reaching design grade, each wall was brought to failure, usually by a combination of unloading ties and extending the excavation. For Model Test No. 1, the cross-sectional area of the pile tip was selected so that the piles would plunge as the excavation was deepened. For the remaining tests, the soldier piles were fitted with a steel plate of sufficient size to support the vertical component of tieback load and reduce wall settlements. Failure of Model Test No. 2 was induced by first decreasing tieback loads to about 30 % of the design load, and then deepening the excavation. In Model Tests No. 3 and 4, the second level of tiebacks was unloaded to observe load transfer and deformation response of the walls. In Model Test No. 3, the second level tiebacks were then reloaded and the excavation extended to the pile tip elevation. In Model Test No. 4, the loads in the upper level of tiebacks were reduced after unloading the second level of tiebacks. A summary of the model testing program is provided in Fig. 2.

Selection of Wall Elements

In addition to scaling model wall geometry, stiffnesses of the elements (soldier piles, lagging, and tiebacks) were selected to ensure wall and soil deformation response consistent with field experience. Beams used in Model Tests No. 1 and 3 were sized to model the prototype piles supported by one and two levels of tiebacks, respectively. The relative stiffness of soil and structure can be defined using beam on elastic foundation theory. If soil stiffness, E_S, is assumed to vary with depth, z,

$$E_s(z) = n\sqrt{z} \tag{1}$$

in which n is a constant, required model pile stiffness can be determined using the following relationship,

$$\left[\frac{E_s L^3}{EI}\right]_{model} = \left[\frac{E_s L^3}{EI}\right]_{prototype} \tag{2}$$

in which E is the modulus of elasticity of the pile material, I is the moment of inertia of the pile divided by pile spacing, and L is the span of the beam. Once the required stiffnesses of the model piles were determined, pile dimensions were selected so their scaled axial and lateral capacities were consistent with the prototype piles. The required stiffness of model lagging was determined using a similar relationship, but constant soil stiffness was assumed.

Test	H (ft)	D (ft)	h_1 (ft)	h_2 (ft)	T_1 (lb)	T_2 (lb)	I_p (in^4)	A_p (in^2)
1	6.25	1.25	2.25	---	3300	---	0.958	1.38
2	6.25	1.25	2.25	---	3500	---	0.337	24
3	6.25	1.25	1.5	2.5	1750	1750	0.337	24
4	6.25	1.25	1.5	2.5	1750	1750	0.096	24

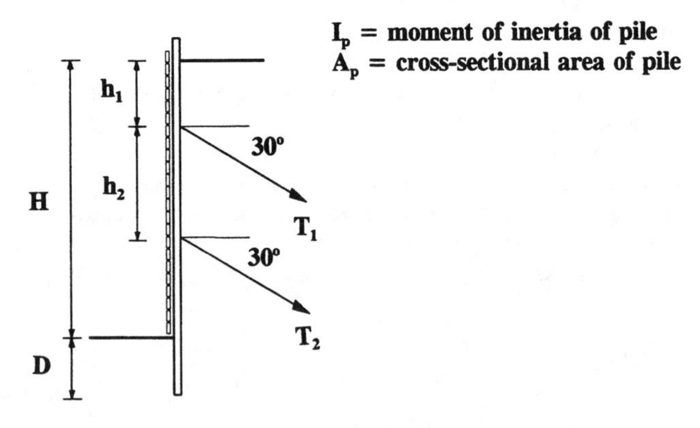

FIG. 2. Summary of Model-Scale Testing Program

Model tieback tendons were selected with axial stiffnesses designed to produce normalized displacements (δ/H), accompanying a scaled change in load, that were compatible with those observed for the full-scale walls. This required consideration of the post-tensioned stiffness of the prototype anchors. The displacement at the head of a post-tensioned tieback, due to change in load, can be expressed as the sum of axial stretch in both the unbonded and anchorage zones,

$$\delta_p = \frac{\Delta P L_{ub}}{A_{ub} E_{ub}} + \frac{1}{A_a E_a} \int_0^{L_a} \Delta P(x)\, dx \tag{3}$$

in which δ_p is the displacement at the head of the tieback for a change in load ΔP, L_{ub} is the unbonded length of the tieback, L_a is the anchorage length, A_{ub} and A_a are the cross-sectional areas of the unbonded and anchorage lengths, and E_{ub} and E_a are the Young's modulii of the unbonded and anchorage lengths. Assuming a triangular distribution of load over one-half of the anchorage length, and a constant cross-sectional area and modulus of the unbonded and anchorage lengths, the displacement at the head of the tieback can be approximated as,

$$\delta_p \approx \frac{\Delta P (L_{ub} + \frac{L_a}{4})}{A_{ub} E_{ub}} \tag{4}$$

The stiffness of the model tieback (for which L_a is zero) can then be related to that of the prototype anchor using the following relationship,

$$\left[\frac{\Delta P L}{AEH}\right]_{model} = \left[\frac{\Delta P}{AEH}\right]_{prototype} \times \left[L_{ub} + \frac{L_a}{4}\right]_{prototype} \tag{5}$$

in which the displacements of the model and prototype anchors are normalized with respect to wall height, H.

In Model Tests No. 2 and 4, the piles were designed by estimating bending moments according to the recommendations of Cheney (1988). A rectangular distribution of earth pressure with an intensity equal to $0.65 \times K_a \gamma H$ was assumed in design. Pile stiffnesses selected to satisfy a strength requirement were more flexible than those used to model the prototype piles.

Measurement of Ground and Wall Movements

Settlement of the ground surface was measured using 36 dial gages. After forming the sand sample, the surface was leveled by hand using a trowel. One and one-half inch square, plexiglass plates were then placed on the surface of the sand and a dial gage, attached to a fixed point of reference, placed on the top of each plate. Measurement of lateral and vertical ground movements inside the sand mass was accomplished electronically using Trans-Tek Model No. 244-0001 DC-DC LVDT's. These gages have a travel of $\pm$ 1 in. (25 mm) and were buried in the sand during deposition. The DCDT's were connected in series to form multiple position extensometers oriented both laterally and vertically. The DC-DC LVDT's were also used to record the displacement and rotation at the top of two of the soldier piles.

Two soldier piles used in each construction were equipped with bonded resistance type strain gages (Measurements Group Inc., Model No. EA-06-250PD) along their full length. Gages were spaced at 6-in. (152 mm) intervals, excluding the toe where 3-in. (76 mm) spacing was used. Each location consisted of two active gages on the front and back flanges of the beams, and two dummy gages to complete the bridge. The dummy gages were used to compensate for temperature variation and drift. The beams were calibrated for both bending moments and axial load. Bending moment distributions were twice integrated to estimate the deflected shape of the piles.

Wall Construction

A soil sample was formed for each model wall by air pluviation of sand from 2 yd^3 (1.5 m^3) buckets hoisted into position with a crane. The sand used in the tests was a dry, poorly graded, river sand. The relative density of the deposit was controlled by fixing the height of fall and mass rate of flow. Based on cone penetration soundings, the relative density of the deposit was estimated in the range of 50 to 60 percent. A medium dense condition was selected to minimize volume change during shear. The angle of internal friction was estimated to be 45 degrees, for the low stress levels in the test chamber.

Some wall elements required installation during sand deposition. Sand was first deposited to the pile tip elevation, and then soldier beams were placed across the test chamber at 2 ft (0.61 m) spacing. The piles were temporarily secured to a wide flange beam. Piles were placed during deposition to avoid damage to instrumentation that might accompany driving. To facilitate installation of lagging during excavation, a thin, heat-bonded, non-woven, geotextile fabric was placed between the piles. The fabric provided the temporary support required to attach the lagging without excessive ground loss, yet permitted a soil deformation condition to develop. Tiebacks rods were also positioned during sand deposition. Each tieback rod was placed inside an extensible tube to isolate the member from the sand mass. The tiebacks were then connected to a reaction frame along the back wall of the test chamber. The tieback rods and soldier piles remained unconnected until later exposed by excavation.

After forming the sand sample, excavation proceeded in small lifts about equal to the width of a lagging board. Model lagging was attached to machine screws installed on the piles and secured with coupler nuts to simulate placement of contact lagging in the field. Any void between the lagging and geotextile fabric was backfilled with a soil-cement prior to deepening the excavation. Measurements of all electronic instrumentation were recorded after excavation and placement of each lagging board. Dial gage readings were recorded following installation of a complete row of lagging, corresponding to an excavation increment of 3 in. (76 mm).

Excavation proceeded in the manner described above until the tiebacks were exposed. A wale, used to span adjacent piles, was then connected to the tieback. Tiebacks were loaded to 1.2 times the design load to replicate the proof or performance tests conducted for prototype anchors, then reduced and locked-off at

75 percent of the design load. Tieback loading was accomplished with a mechanical device similar to a gear puller. After loading all the tiebacks at a given level, the excavation was continued to the next level of tiebacks or completed to design grade by attaching the remaining lagging. After reaching the design grade, each wall was "failed", usually by a combination of reducing tieback loads and over-excavation.

GENERAL PATTERNS OF WALL AND GROUND MOVEMENT

General patterns of wall and ground movement for significant stages of construction are shown in Fig. 3. This wall was constructed with two levels of tiebacks. Excavation to a depth of 1.5 ft (0.46 m), as required for installation of the first level of tiebacks, resulted in outward rotation of the piles. The maximum lateral pile movement of 0.144 in. (3.7 mm) occurred at the top of the wall, with a corresponding maximum surface settlement of 0.073 in. (1.9 mm) immediately behind the wall.

The effect of tieback prestressing was to pull the piles back into the soil. The deflected shape of the piles was concave to a depth of about 3 ft (0.93 m), with a maximum negative pile displacement of 0.014 in. (0.36 mm) just below the tieback level. Small pile settlements of about 0.005 in. (0.13 mm) developed during tieback stressing. The general distribution of ground surface settlement remained unchanged from that which developed during excavation to the first tieback level. Heave of the ground surface was observed, and extended for a distance behind the wall, about equal to the depth of the excavation.

During excavation from the first level of supports to a depth of 4 ft (1.2 m), the tops of the piles rotated toward the soil about the tieback position. Lateral bulging of about 0.039 in. (0.99 mm) developed between the tiebacks and pile toes. The maximum ground surface settlement of 0.089 in. (2.3 mm) was observed immediately behind the wall. As support for the vertical component of tieback force was removed by excavation, axial load was transferred to the pile tips as reflected by the increase in observed pile settlements to 0.015 in. (0.38 mm).

Loading the second level of tiebacks also had the effect of pulling the piles back into the soil. Lateral compression of the soil developed immediately adjacent to the wall and extended behind the wall for a distance of about 18 in. (457 mm). This observation suggests that tieback prestressing may have a very local effect on ground movements behind the wall. Heave of the ground surface was not observed. The maximum ground surface settlement increased from 0.089 in. (2.3 mm) to 0.106 in. (2.7 mm) during stressing of the second level of tiebacks. Pile settlements increased to about 0.037 in. (0.94 mm).

During excavation between the second level of tiebacks and design grade, the piles rotated outward about the toe and lateral bulging developed between the second level of tiebacks and pile tips. The magnitude of lateral bulging between the second level of supports and design grade was about 0.052 in. (1.3 mm). The maximum lateral movement of the piles was 0.143 in. (3.6 mm) and occurred at the top of the wall. Ground surface settlement extended from a maximum of 0.217 in. (5.5 mm) to 0.007 in. (0.18 mm) about 9 ft (2.7 m) behind the wall. The distribution of surface settlement was similar to that which developed during the initial stages of excavation, and was apparently established early in the construction sequence. Pile settlements at the end of construction were on the order of 0.066 in. (1.7 mm).

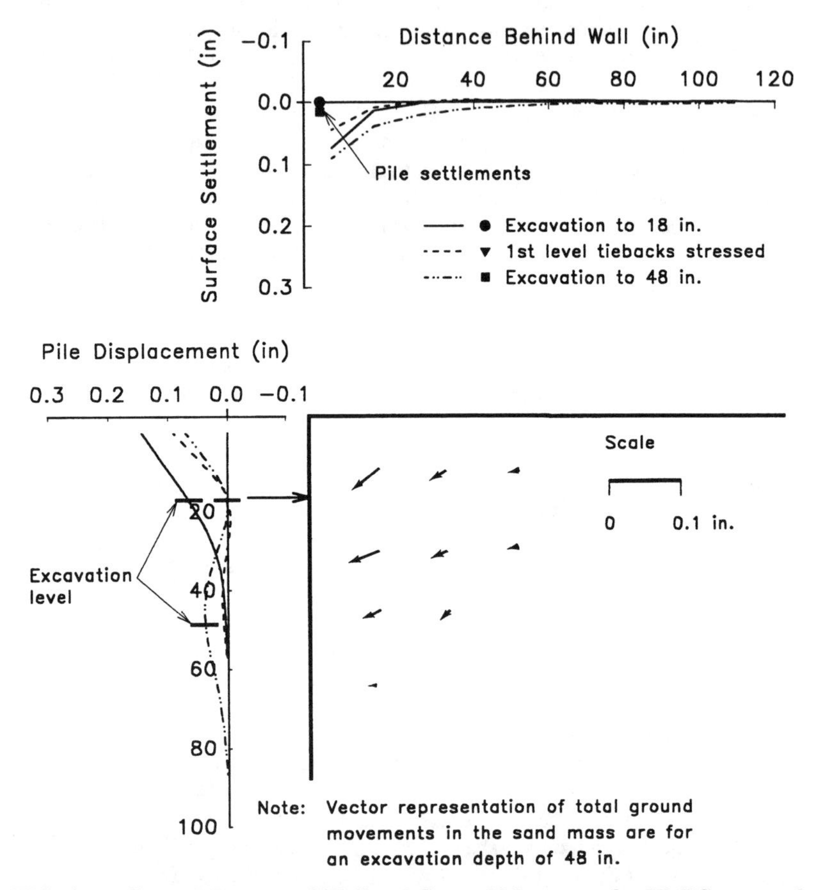

FIG. 3a. General Patterns of Wall and Ground Movement for Wall Supported by 2 Levels of Tiebacks During Excavation to - 4 Feet, Test No. 4

VOLUME RELATIONSHIPS

The relationship between lateral movements at the wall and ground surface settlements behind the wall are shown in Fig. 4. During excavation to the first level of tiebacks, the volume of lateral movement at the wall exceeded the volume of ground surface settlement by a factor ranging from 1.5 to 4. In general, this ratio decreased as the depth of excavation increased. Similar observations were made 18 in. (457 mm) behind the wall suggesting a tendency for the soil to dilate during shear at the low stress levels associated with the initial stages of excavation. Tieback stressing produces large changes in the volume of lateral movements at the wall, with only small changes observed in the volume of surface settlement and lateral movement 18 in. (457 mm) behind the wall. This observation suggests that tieback prestressing may have a very local effect on ground response.

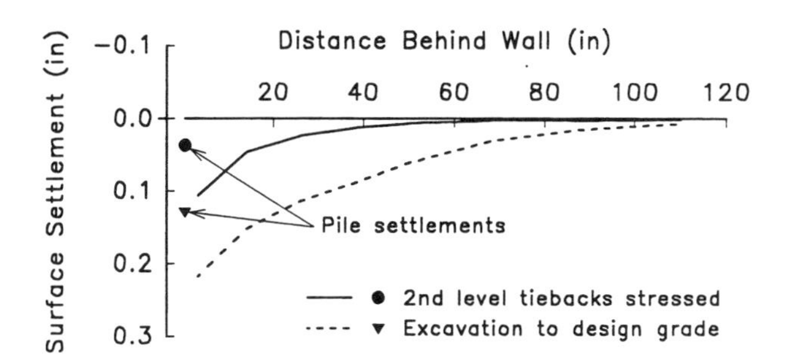

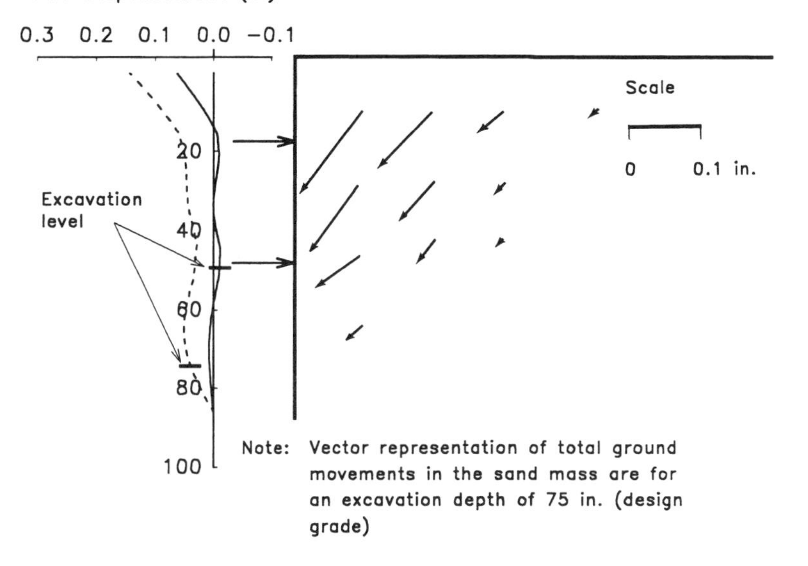

FIG. 3b. General Patterns of Wall and Ground Movement for Wall Supported by 2 Levels of Tiebacks During Excavation from - 4 Feet to Design Grade, Test No. 4

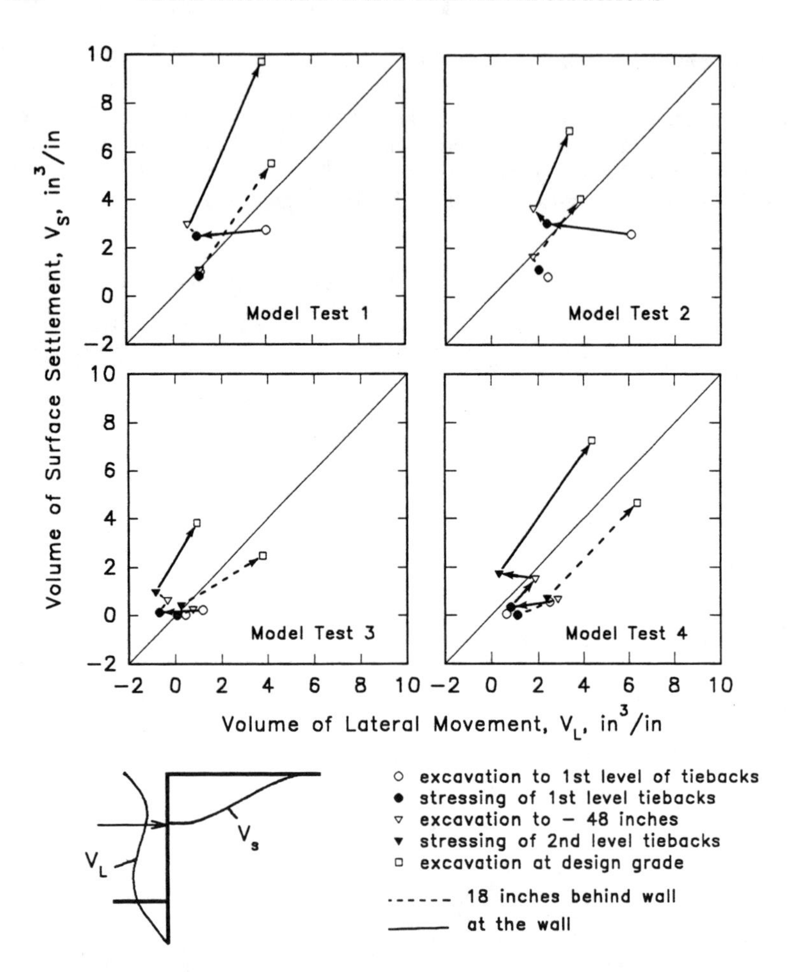

FIG. 4. Relationship Between Volume of Lateral Ground Movement and Volume of Ground Surface Settlement

At design grade, the volume of ground surface settlement behind the wall was between 2 and 4 times the volume of lateral movement measured at the wall. The volume of ground surface settlement observed 18 in. (457 mm) behind the wall, however, was between 0.5 and 1.5 times the volume of lateral movement 18 in. (457 mm) behind the wall. Part of the discrepancy can be explained by considering the change in volume of lateral pile movement associated with tieback stressing. The remaining difference may be associated with procedures used to attach the lagging between the piles. Small voids developed between the geotextile fabric installed between the piles and the lagging. While efforts were made to backfill the voids using a soil cement, some of these may have unintentionally been left unfilled, and as the excavation was deepened, closure of the voids between the fabric and the lagging may have resulted in additional ground surface settlement.

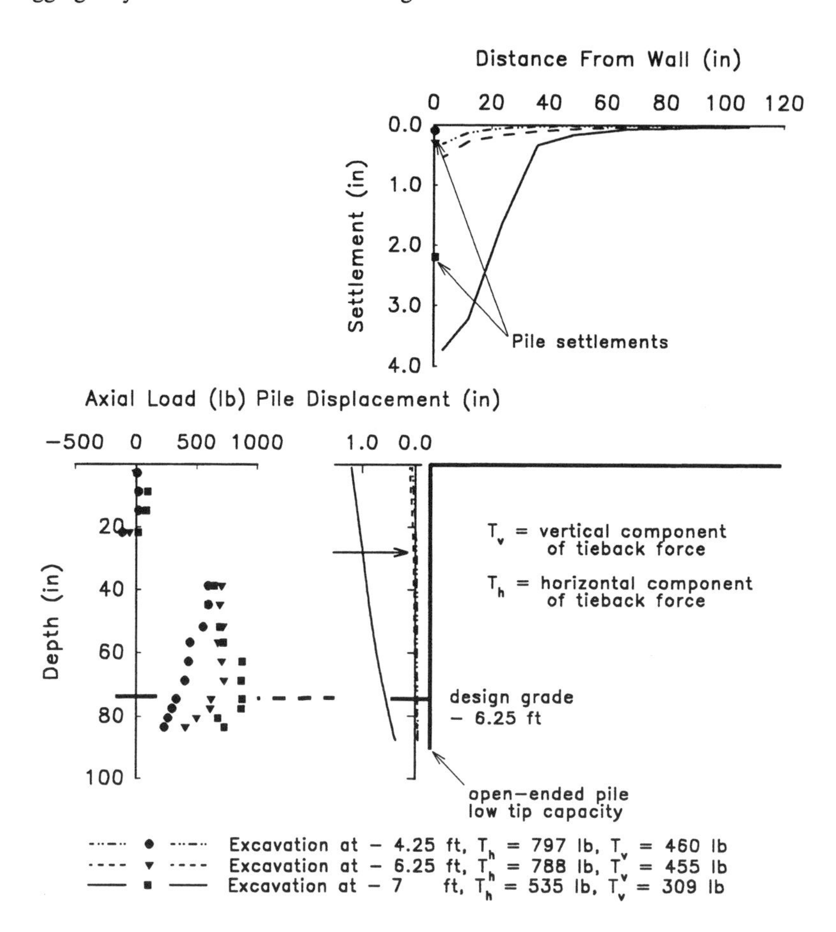

FIG. 5. Influence of Pile Bearing on Wall and Ground Movements, Pile with Low Axial Capacity

IMPORTANCE OF PILE BEARING

The cross-sectional area of the pile tips used in Model Test No. 1 was such that as the excavation was deepened, and support for the vertical component of tieback force was removed, the piles plunged. As shown in Fig. 5, comparatively large lateral and vertical ground movements developed during excavation from 4.25 ft (1.3 m) to 7 ft (2.1 m). Development of the large movements is explained by considering the distribution of axial pile load. As the excavation was deepened, axial load progressively transferred to the pile tip, and pile settlements increased from 0.044 in. (1.1 mm) to 2.27 in. (58 mm). The large pile settlements caused a reduction in tieback loads and resulted in large lateral and vertical ground movements.

Relationship Between Wall/Ground Movements and Tieback Loads

Following excavation to design grade, Model Test No. 2 was failed by unloading the tiebacks. This wall was constructed with a single level of ties. At lock-off, the horizontal component of tieback force was about 567 lb/ft (8.3 kN/m) and the factor of safety with respect to a limit condition was 1.85. In this paper, factor of safety is the ratio of the sum of the horizontal component of tieback force and passive resistance along the pile toe, to the limiting horizontal component of lateral earth pressure.

Wall and ground response to tieback unloading is shown in Fig. 6. At design grade, the estimated factor of safety with respect to a limit condition was about 1.8. During the initial stages of unloading (FS = 1.8 → 1.25), wall movement consisted principally of translation as passive resistance was mobilized along the embedded portion of the piles. Relatively small increases in lateral pile movement and ground surface settlement were observed at this stage. Further reduction in the tieback loads resulted in increased translation and rotation of the wall about the toe. Lateral pile movements and ground surface settlements accelerated as the factor of safety decreased below 1.25.

Comparison of Model and Prototype Walls

The magnitude and distribution of lateral and vertical movements for model and prototype walls were influenced by pile bearing conditions and construction details. In addition, prototype performance was probably affected by development of mass movements behind the tiebacks. A comparison of the movements generated by model and prototype construction is provided in Fig. 7a and 7b.

The distribution of lateral movement at the walls is similar for both model and prototype. Maximum lateral movements occur at the top of the walls, with lateral bulging between levels of support; however, the prototype walls display greater rotation about the toe. Part of the discrepancy may relate to observations made during tieback prestressing. The walls were pulled back into the soil during stressing of tiebacks at all levels in the model studies; however, in prototype construction, only the wall supported by two levels of tiebacks was pulled back into the soil, and only during stressing of the upper level of tiebacks. Additional rotation in the prototype probably resulted from mass movements behind the tiebacks which could not develop in the model test chamber. Although lateral ground movements were not recorded behind the tiebacks, lateral movements at the ground surface of 0.067% H and 0.23% H were observed 15 ft (4.6 m) behind the walls supported by one and two levels of tiebacks, respectively. Large pile settlements and corresponding decreases in tieback load impacted performance of the model-scale wall supported by a single level of tiebacks (Model Test No. 1) and the prototype wall supported by two levels of tiebacks.

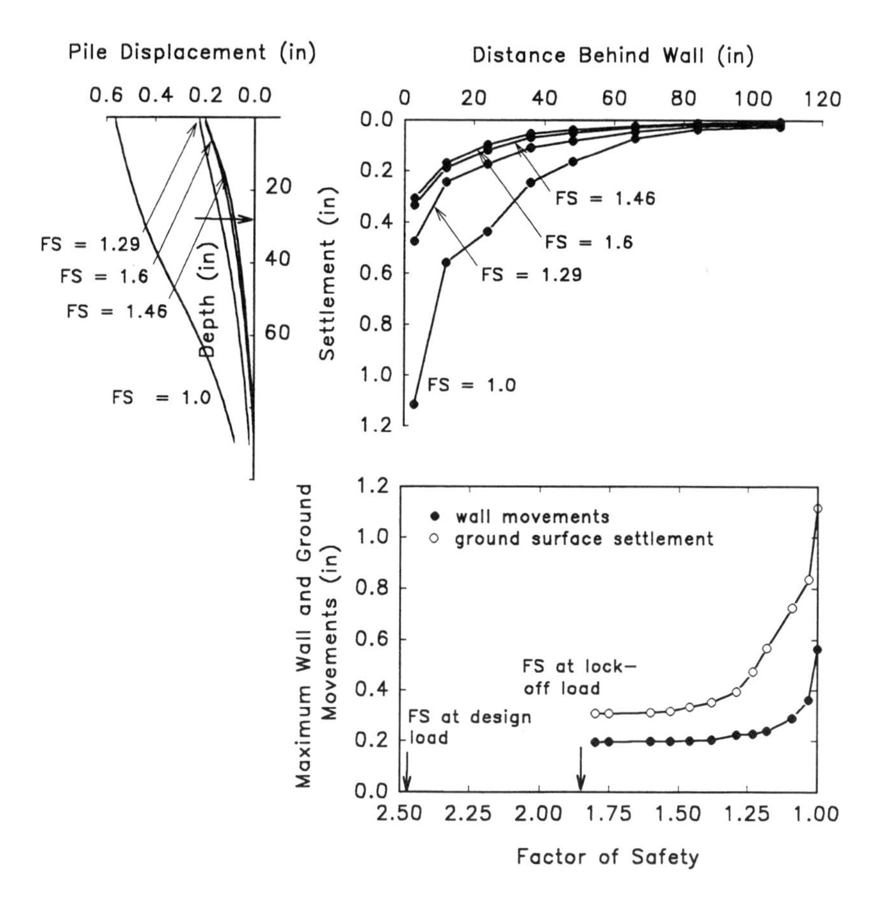

FIG. 6. Relationship Between Wall/Ground Movements and Tieback Loads

The magnitude and distribution of ground surface settlements for the model and prototype walls are shown in Fig. 7a and 7b. In prototype construction, maximum ground surface settlement occurs between the wall and a distance of 0.1% H behind the wall. The maximum slope of the surface settlement trough is observed at a distance between 0.2 % and 0.5% H. In the case of model-scale construction, both the maximum surface settlement and slope of the surface settlement trough occur immediately behind the wall. The magnitude and distribution of surface settlement in model-scale construction was probably influenced by loss of ground associated with connection of lagging to the piles.

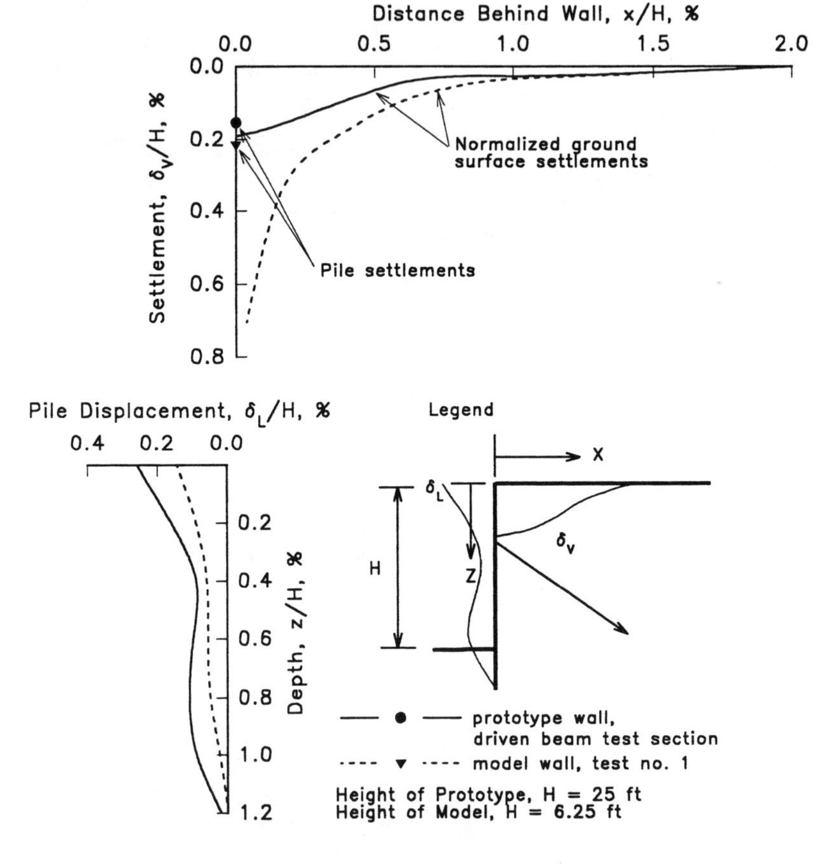

FIG. 7a. Lateral and Vertical Ground Movements for Model and Prototype Walls Supported by a Single Level of Tiebacks

SUMMARY

Ground movement patterns which developed during construction of four, large-scale, model tieback walls have been described in this paper. Lateral and vertical ground movements in the model studies were influenced by tieback stressing, pile bearing conditions, and construction details. Tieback stressing pulled the walls back into the soil without significant change in the magnitude and distribution of lateral and vertical ground movements at distances greater than 18 in.

(457 mm) behind the walls. Pile settlements, which developed during construction of Model Test No. 1, reduced tieback loads and caused development of comparatively large lateral and vertical ground movements. Vertical movements behind the model-scale walls were affected by ground loss associated with connection of lagging to the piles.

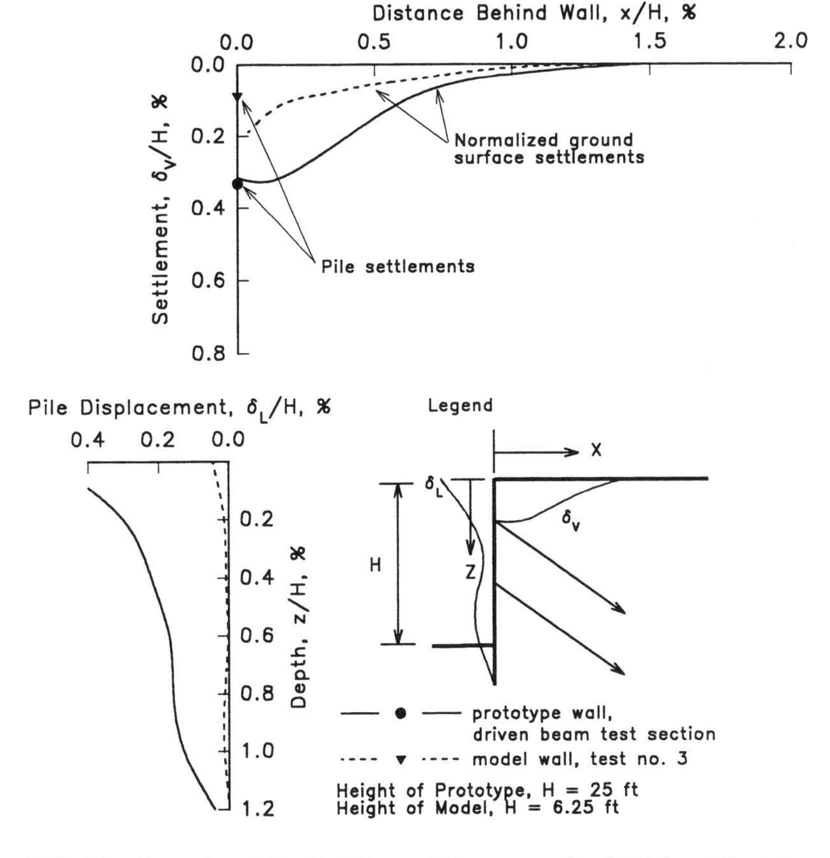

FIG. 7b. Lateral and Vertical Ground Movements for Model and Prototype Walls Supported by 2 Levels of Tiebacks

General patterns of lateral movement at the walls compared favorably with observations made for two prototype structures. The magnitude of lateral movements in both the model and prototype were influenced by pile bearing conditions. In addition, lateral movements in prototype construction were probably influenced by ground movements behind the tiebacks. The distributions of ground surface settlement in model and prototype walls differed in that the maximum slope of the surface settlement trough in the model studies occurred at the wall. The magnitude and distribution of ground surface settlement in the model studies were influenced by ground loss at the walls associated with connection of the lagging to the piles. Maximum lateral movements of the model-scale walls ranged from 0.05% to 0.25% H, compared to 0.25% to 0.5% H for the prototype. Maximum vertical ground movements ranged from 0.2% to 0.7% H, and 0.2% to 0.32% H, in model and prototype construction, respectively.

ACKNOWLEDGEMENTS

The work reported in this paper was conducted as part of a research project entitled "Permanent Tieback Walls" and sponsored by the Federal Highway Administration and Schnabel Foundation Company, FHWA Contract No. DTFH 61-89-C-00038. The large, model testing facility was constructed with support from U.S. Army Research Office through grants DAAl 03-87-k-0006, DAAl 03-85-G-0186, and DAAl 03-86-G-0188 to the University of Illinois Advanced Construction Technology Center. Their support is gratefully acknowledged.

APPENDIX - REFERENCES

Boscardin, M. D. and Cording, E. J. (1989). "Building response to excavation-induced settlement." *J. Geotech. Eng.*, ASCE, 115(1), 1-21.

Clough, G. W. and O'Rourke, T. D. (1990). "Construction induced movement of insitu walls." *Design and Performance of Earth Retaining Structures*, Geotechnical Special Publication No. 25, ASCE, New York, New York, 439-470.

Cheney, R. S. (1988). "Permanent ground anchors," *Report No. FHWA-DP-68-1R*, Federal Highway Administration, U. S. Department of Transportation, Washington, D. C.

Chung, M. and Briaud, J. (1992). "Behavior of a full scale tieback wall in sand," *Research Report*, Department of Civil Engineering, Texas A & M University, College Station, Texas.

Liu, T. K. and Dugan, J. P. (1972). "An instrumented tied-back deep excavation." *Proc. Performance of Earth and Earth-Supported Structures*, West Lafayette, 1323-1339.

Shannon, W. L. and Strazer, R. J. (1970). "Tied-back excavation wall for Seattle First National Bank." *Civil Engineering*, ASCE, 40(3) 62-64.

Management of Dredged Material Placement Operations

Timothy D. Stark,[1] Iván Contreras,[2] and Jack Fowler,[3] Members, ASCE

Abstract: Non-linear finite strain consolidation theory is used to predict the settlement of fine-grained dredged material due to self-weight and surcharge-induced consolidation. Finite strain consolidation theory accounts for the effects of (a) self-weight consolidation, (b) changes in permeability with consolidation, (c) a non-linear void ratio-effective stress relationship, and (d) large strains. An empirical desiccation model is used to describe the removal of water from confined dredged material by surface drying. The combined model has been coded into the microcomputer program PCDDF89 (Primary Consolidation, and Desiccation of Dredged Fill), which is also described. The Craney Island Dredged Material Management Area near Norfolk, Virginia is used to illustrate the use and accuracy of the model. The use of vertical strip drains to increase storage capacity of disposal facilities is also described.

Introduction

U. S. Army Corps of Engineer District Offices are continually addressing the placement of fine-grained dredged material dredged from navigable waterways throughout this country. Increasing environmental concerns together with a general decrease in the number of available placement areas have created the need for maximum utilization of existing and planned dredged material containment areas. In order to maximize the service life of dredged material placement areas, the design and operation of the areas must accurately account for the increase in storage capacity resulting from settlement of confined dredged material. The height of the

---

[1] Asst. Prof. of Civ. Engrg., Univ. of Illinois, MC-250, Urbana, IL 61801.

[2] Research Asst., Dept. of Civ. Engrg., Univ. of Illinois, Urbana, IL 61801.

[3] Research Civil Engineer, U. S. Army Engineer Waterways Experiment Station, Vicksburg, MS 39180.

dredged fill is reduced by sedimentation, consolidation, secondary compression, and desiccation.

Two important natural processes affecting the long-term height of confined dredged material are consolidation and desiccation. Fine-grained dredged materials may undergo axial strains greater than 50 percent during self-weight consolidation. Greater strains are possible, if the disposal site is managed so that surface water is removed and desiccation can occur. The resulting problem is to determine the time rate of settlement for dredged material subjected to the effects of: (a) self-weight consolidation, (b) crust formation induced by desiccation, and (c) additional consolidation due to the surcharge created by the crust and additional dredged material.

This paper provides a brief description of the mathematical model used in the microcomputer program PCDDF89 (Primary Consolidation and Desiccation of Dredged Fill). A field case history is used to illustrate the use and accuracy of the model in estimating the long-term storage capacity of confined dredged material management facilities.

PCDDF89

PCDDF89 simulates the consolidation and desiccation processes in fine-grained soils using the one-dimensional finite strain theory of consolidation (Gibson et al. 1967) and an empirical desiccation model. Secondary compression effects are currently being incorporated into the model. PCDDF89 calculates the total settlement of a dredged fill layer based on consolidation characteristics of the soils below the layer, the consolidation characteristics of the dredged fill layer, local climatological data, and the effectiveness of surface water management within the containment area. This settlement is then accumulated for each dredged fill and compressible foundation layer within the one-dimensional profile to estimate a cumulative settlement.

The major inputs required by PCDDF89 are the void ratio-effective stress and void ratio-permeability relationships obtained from laboratory consolidation tests on dredged fill and compressible foundation materials. In addition, specific gravity, placement void ratio, and desiccation characteristics of the dredged fill material are required. Climatological data, anticipated dredging schedules and quantities, external water surface elevation for under water disposal operations, and drainage characteristics of the containment site are also required.

The original version of PCDDF was developed by Cargill (1985). This version could analyze one dredged fill and one compressible foundation material type. PCDDF was modified by Stark (1991) to account for twenty-five different dredged fill and foundation material types, accelerate interpolation of the void ratio relationships, allow the restart of a previous simulation, and facilitate usage of PCDDF89 by developing a computerized database of climatological, void ratio relationships, and desiccation parameters for various parts of the country. An interactive interface for PCDDF89 was also developed and the program was incorporated into the Corps of Engineers Automated Dredging and Disposal

Alternatives Management System (ADDAMS) software package (Schroeder and Palermo 1990).

Consolidation and Desiccation Models

In PCDDF89 the consolidation and desiccation processes are solved separately to a certain point in time and then the solutions are combined to determine the net settlement of the dredged material. This reconciliation occurs monthly to conform with the availability of average evaporation and rainfall data.

The governing equation of finite strain consolidation theory is:

$$\left(\frac{\gamma_s}{\gamma_w} - 1\right)\frac{d}{de}\left[\frac{k(e)}{1+e}\right]\frac{\partial e}{\partial z} + \frac{\partial}{\partial z}\left[\frac{k(e)}{\gamma_w(1+e)}\frac{d\sigma'}{de}\frac{de}{dz}\right] + \frac{\partial e}{\partial t} = 0 \quad (1)$$

where γ_s = unit weight of solids;
γ_w = unit weight of water;
e = void ratio;
k(e) = permeability as a function of void ratio;
z = vertical material coordinate measured against gravity;
σ' = effective stress; and
t = time.

This equation is suited for prediction of consolidation in thick deposits of soft fine-grained soils, such as dredged material, because it provides for the effects of: (a) self-weight consolidation; (b) permeability varying with void ratio; (c) a non-linear void ratio-effective stress relationship; and (d) large strains. An implicit finite difference scheme is used to solve Eq. (1). After the initial and boundary conditions are defined and appropriate relationships between void ratio and effective stress, and void ratio and permeability are specified, the void ratio in the consolidating layer is calculated using an implicit finite difference technique for any future time. The void ratio distribution in the saturated dredged fill layer is used to calculate the corresponding stresses and pore-water pressures.

At each monthly interval during times when the desiccation process is active, the thickness of the consolidating dredged material will decrease due to evaporation. After desiccation, the layer acts as a surcharge on the consolidating layer and is assumed to be free draining. The empirical desiccation process is divided into two major stages. During the first stage, sufficient free water is available at the surface of the material so that evaporation takes place at the full potential rate. The water lost from a dredged material layer during first-stage drying can be expressed by the following equation and is assumed equal to the vertical settlement:

$$\Delta W' = CS - (C_E')EP + (1 - C_D)RF \quad (2)$$

where $\Delta W'$ = water lost during first-stage drying,
CS = water supplied from lower consolidating soil,
C_E' = maximum evaporation efficiency for soil type,
EP = class A pan evaporation,
C_D = drainage efficiency of containment area, and

RF = rainfall.

Water lost during second-stage drying can be defined as:

$$\Delta W'' = CS - C_E' \left[1 - \frac{h_{wt}}{h_{2nd}} \right] EP + (1 - C_D) \tag{3}$$

where $\Delta W''$ = water lost during second-stage drying,
h_{wt} = depth of water table below surface, and
h_{2nd} = maximum depth of second-stage drying.

The appearance of extensive cracking and the probable loss of saturation within the desiccated material prevents a direct correspondence between water loss and settlement during second stage drying. Therefore, the desiccation settlement during second-stage drying is estimated using the following equation:

$$\delta_D'' = -\Delta W'' - \left[1 - \frac{PS}{100} \right] h_{wt} \tag{4}$$

where δ_D'' = settlement due to second-stage drying, and
PS = gross degree of saturation of desiccated crust that includes cracks.

Determining the second-stage drying settlement involves an iterative process. Additional details of the solution technique for the consolidation and desiccation processes are described by Stark (1991).

Craney Island Dredged Material Management Area

Since 1956 the Craney Island Dredged Material Management Area (CIDMMA) has been used for containment of dredged material from navigable channels and harbors near Hampton Roads, Virginia. The 8.9 km^2 (2200 acre) area annually receives 3.1 to 3.8×10^6 m^3 (4 to 5 million cubic yards) of fine-grained maintenance and new work dredged material (Fig. 1). As a result, a 10.6 to 12.2 m (35 to 40 foot) thick layer of dredged material currently overlies a 30.5 to 36.6 m (100 to 120 foot) thick soft marine clay. The marine clay foundation is underlain by a freely draining dense sand.

With current practices, the CIDMMA is expected to reach ultimate capacity around the year 2000 (Palermo and Schaefer 1990). However, if suitable material is placed in the ocean and unsuitable material is placed in the CIDMMA, the useful life may be extended. An analysis of the service life of the CIDMMA under the proposed Restricted Use Program (RUP) was conducted to evaluate the benefits of the RUP. A simulation of the filling history from 1956 to 1984 was compared to field settlement data to calibrate the input parameters for conditions existing prior to subdivision of the area and implementation of dewatering operations. Simulations of filling history from 1984 (the time of cross dike closure) to 1992 were conducted for each of the three compartments (Fig. 1). Finally, simulations of projected filling rates from 1992 were used to determine the service life of the CIDMMA under the proposed RUP. Two dredging scenarios, Baseline Maintenance Dredging and Worst Case Dredging were considered for the proposed RUP.

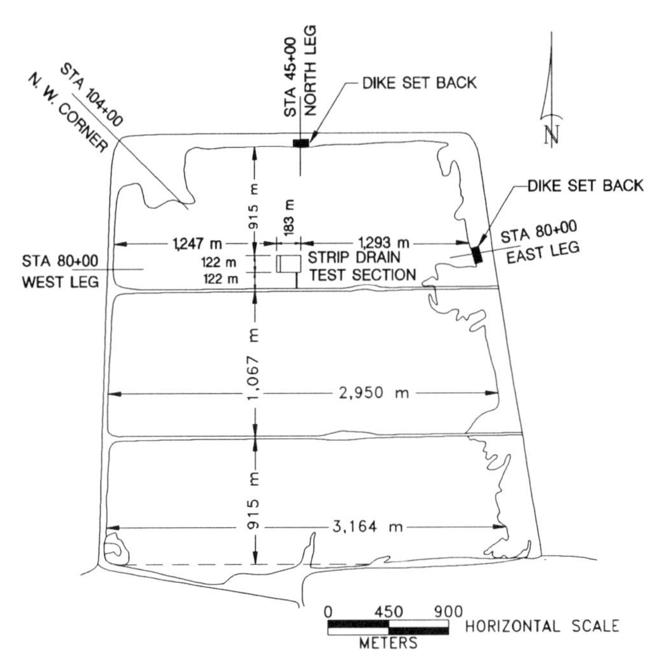

FIG. 1. Plan View of Craney Island and Location of Vertical Strip Drain Test Section

SELECTION OF MODEL PARAMETERS

The consolidation parameters shown in Table 1 were used to evaluate the service life of the CIDMMA under the proposed RUP. The void ratio-effective stress and void ratio-permeability relationships were obtained from the results of self-weight and large strain, controlled rate of strain (LSCRS) consolidation tests (Cargill 1986). The self-weight test yields void ratio relationships from an effective stress of approximately 0.001 kPa to 0.96 kPa (0.02 psf to 20 psf) and the LSCRS test provides data for the effective stress range of 0.96 kPa to 958 kPa (20 psf to 20,000 psf). Conventional oedometer tests were also conducted on samples of dredged material (Cargill 1986) to verify the self-weight and LSCRS test results. The results of the self-weight and LSCRS tests are combined to define the void ratio relationships over the range of effective stresses encountered in a management area.

The desiccation parameters used in PCDDF89 include rate of precipitation, a pan evaporation efficiency, a maximum crust thickness, and a surface drainage efficiency. The desiccation parameters used in the simulations are shown in Table 2 and represent an active dewatering condition. The precipitation and evaporation rates used for the simulations are shown in Table 3 and were obtained from National Climatic Center (1980) and Brown and Thompson (1977), respectively.

TABLE 1. Consolidation Parameters of the Marine Clay Foundation and Dredged Material at the CIDMMA

Marine Clay Foundation			Dredged Material		
Void Ratio	Effective Stress (kPa)	Permeability (cm/sec)	Void Ratio	Effective Stress (kPa)	Permeability (cm/sec)
3.00	0.00	4.27E-07	10.50	0.00	3.30E-04
2.90	0.42	3.64E-07	10.40	0.004	2.90E-04
2.80	0.94	3.12E-07	10.20	0.007	2.34E-04
2.70	1.53	2.69E-07	10.00	0.011	1.86E-04
2.60	2.30	2.26E-07	9.80	0.014	1.48E-04
2.50	3.35	1.84E-07	9.60	0.019	1.17E-04
2.40	4.98	1.49E-07	9.40	0.024	9.14E-05
2.30	7.37	1.22E-07	9.20	0.03	7.38E-05
2.20	11.10	9.64E-08	9.00	0.04	5.86E-05
2.10	16.46	7.62E-08	8.80	0.04	4.59E-05
2.00	24.40	4.94E-08	8.60	0.05	3.71E-05
1.90	37.32	4.66E-08	8.40	0.06	2.95E-05
1.80	55.51	3.64E-08	8.20	0.07	2.29E-05
1.70	81.35	2.72E-08	8.00	0.09	1.83E-05
1.60	121.54	2.05E-08	7.80	0.10	1.45E-05
1.50	179.44	1.52E-08	7.60	0.12	1.14E-05
1.40	265.09	1.09E-08	7.40	0.13	9.14E-06
1.30	406.73	9.53E-09	7.20	0.15	7.13E-06
1.25	497.64	6.71E-09	7.00	0.18	5.68E-06
0.87	2,392.50	3.53E-09	6.80	0.22	4.52E-06
0.80	2,871.00	1.77E-09	6.60	0.28	3.57E-06
			6.40	0.37	2.82E-06
			6.20	0.51	2.23E-06
			6.00	0.70	1.78E-06
			5.80	0.96	1.40E-06
			5.60	1.34	1.11E-06
			5.40	1.87	8.68E-07
			5.00	3.62	5.51E-07
			4.80	5.02	4.34E-07
			4.60	6.65	3.43E-07
			4.40	8.76	2.69E-07
			4.00	15.12	1.68E-07
			3.80	29.57	8.68E-08
			3.00	59.33	3.92E-08
			2.50	115.80	1.34E-08
			2.00	226.81	3.53E-09
			1.00	813.45	1.77E-09

TABLE 2. Consolidation and Desiccation Parameters for the CIDMMA

Parameter	Active Dewatering
Surface drainage efficiency, percent	100
Maximum evaporation efficiency, percent	100
Saturation at end of desiccation, percent	80
Maximum crust thickness, m	0.31
Time to desiccation after filling, days	30
Elevation of fixed water table, m MLW	0.46
Void ratio at saturation limit	6.5
Void ratio at desiccation limit	3.2
In-channel void ratio	5.93
Placement void ratio	10.50
Bulking Factor	1.66
Void ratio of incompressible foundation sand	0.65
Permeability of incompressible foundation sand, cm/sec	1E-07

TABLE 3. Precipitation and Evaporation Rates at the CIDMMA

			Excess Evaporation, (mm)	
Month	Precipitation (mm)	Pan Evaporation (mm)	100-Percent Infiltration	75-Percent Infiltration
January	86.36	0.00	----	----
February	83.82	15.24	----	----
March	86.36	25.4	----	----
April	68.58	114.3	45.72	60.96
May	83.82	177.8	93.98	114.30
June	91.44	195.58	104.14	127.00
July	144.78	195.58	50.80	86.36
August	149.86	167.64	17.78	55.88
September	106.68	124.46	17.78	55.88
October	78.74	91.44	12.70	33.02
November	73.66	30.48	----	----
December	78.74	0.0	----	----
Total	1132.84	1137.92	342.9	533.4

SIMULATION OF DREDGED MATERIAL DISPOSAL

Thicknesses of dredged material for each disposal operation were determined from the actual dredging volumes and the surface areas available for placement in the management area. Since PCDDF89 applies an entire lift instantaneously, the disposal history had to be subdivided and the dredged material applied at the mid-point time of each subdivision.

The height of each lift was obtained by dividing the bulked disposal volume by the surface area of the entire site prior to subdivision. The bulked disposal volume equals the in-channel disposal volume multiplied by the bulking factor (Table 2). The bulking factor is defined as:

$$BF = \left[\frac{1 + e_{pl}}{1 + e_{ic}} \right] \tag{5}$$

where BF = Bulking Factor,
e_{pl} = placement void ratio, and
e_{ic} = in-channel void ratio.

Dredged material was placed using two different filling criteria. In the Baseline Maintenance Dredging Case, dredged material was placed in a compartment until a thickness of approximately 0.3 m (one foot) was obtained. After reaching approximately 0.3 m (one foot), placement in that compartment was discontinued and dredged material was placed in the next compartment. A 0.3 m (one foot) lift was used to investigate the consolidation and desiccation characteristics of thin lifts. In previous years, the filling schedule involved an annual rotation of the compartments. As a result, a large amount of dredged material was usually placed in a compartment causing lift thicknesses of 0.9 m to 1.8 m (3 to 6 ft), which may have slowed the rate of consolidation and desiccation. For comparison purposes, an annual rotation of the compartments was used for the Worst Case Dredging Scenario. This was due to the large quantity of material that was to be placed under the Worst Case Scenario.

CRANEY ISLAND FILLING SIMULATION, 1956 TO 1984

Simulations of the filling history from 1956 to 1984 are shown in Fig. 2. The main objective of this simulation was to calculate the void ratio and effective stress profiles in the dredged fill and compressible foundation in October 1983 (the time of cross dike closure). For discussion purposes, the time of cross dike closure is referred to as 1984 even though the analysis used October 1983 (Fig. 2). The simulation incorporated the affects of desiccation and the results are in excellent agreement with field surface elevations. The calculated void ratio and effective stress profiles reflect the consolidation and desiccation that occurred between 1956 and 1984 and were used as a starting point for subsequent simulations using the restart option in PCDDF89. The excellent agreement with field surface elevations indicates that the input parameters are representative of field conditions and can be used to estimate the service life of the CIDMMA under the proposed RUP.

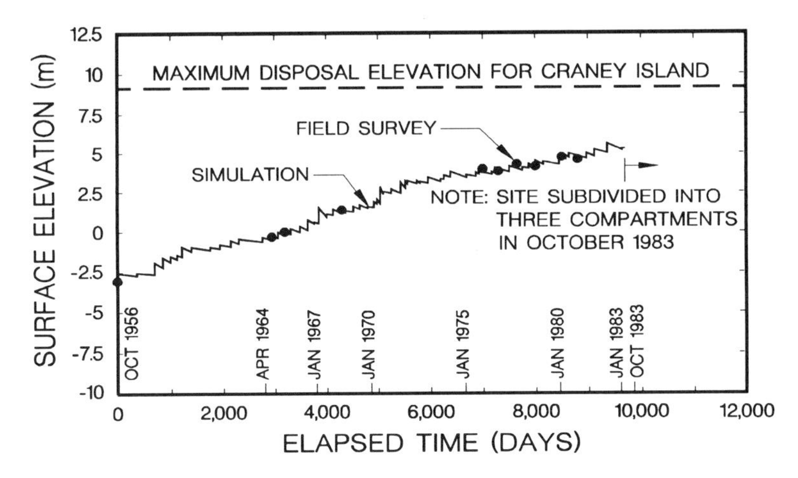

FIG. 2. Fill Rates for Craney Island from 1956 to 1984

Craney Island Filling Simulations, 1984 to 1992

Simulation of the filling history from 1984 to 1992 in the north compartment is shown in Fig. 3. The void ratio and effective stress profiles calculated in the previous simulation were input using the restart option and the surface elevation shown in Fig. 2 at October 1983 was the starting elevation in each compartment. It can be seen from Fig. 3 that the calculated surface elevations are in good agreement with field data.

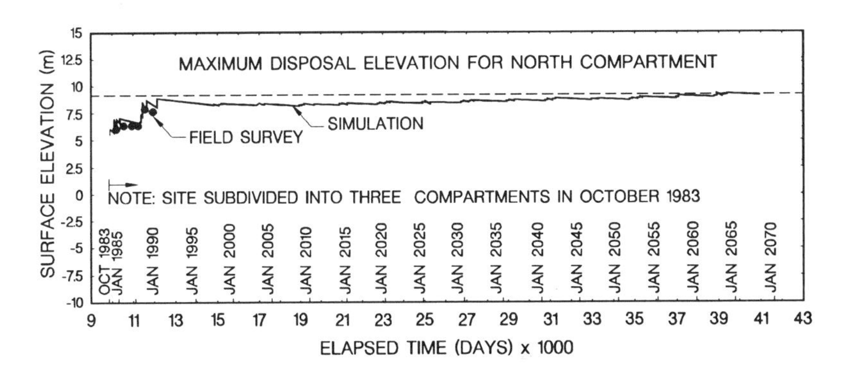

FIG. 3. Simulation of Fill Rates for Baseline Maintenance Dredging Case in North Compartment from 1984 to 2069

BASELINE MAINTENANCE DREDGING SIMULATIONS, 1992 TO 2135

The Baseline Maintenance Dredging simulation from 1992 to 2135 under the proposed RUP is also shown in Fig. 3 for the north compartment. Dredged material was initially placed in the center compartment. After approximately 0.3 m (one foot) of material was placed in the center compartment, dredged material was placed in the south compartment. Placement was moved to the north compartment after a 0.3 m (one foot) thick lift was placed in the south cell and this cycle was repeated until an elevation of +9.2 m (+30 ft) MLW was obtained in all three compartments. It can be seen from Fig. 3 that the north compartment reached capacity in January 2069. After January 2069, all of the dredged material was alternately placed in the center and south compartments using 0.3 m (one foot) thick lifts. The simulation showed that the center compartment reached capacity in May 2131. After May 2131, all of the Baseline Maintenance dredged material was placed in the south compartment. As a result, this compartment reached capacity in May 2132. In summary, the service life of the CIDMMA would be extended to approximately the year 2130 under the proposed Baseline Maintenance Dredging Case of the proposed RUP.

This analysis predicts that the CIDMMA has a service life of approximately 140 years under the Baseline Maintenance Dredging Case of the RUP. Clearly, this prediction is a planning level estimate and is only being used to determine if the RUP deserves further consideration. This prediction involves many assumptions that may not pertain to the CIDMMA around the year 2130. For example, the precipitation and evaporation rates may be different, which would lead to a change in the quantity and character of the dredged material and/or desiccation rate of the confined material. In summary, the results of the Baseline Maintenance Dredging Case clearly show that reducing the amount of dredged material placed in Craney Island under the RUP will significantly extend the service life of this facility.

WORST CASE DREDGING SIMULATIONS, 1992 TO 2085

The Worst Case Dredging simulation from 1992 to 2085 under the proposed RUP is shown in Fig. 4 for the north compartment. Dredged material was placed using an annual rotation starting with the center compartment and ending with the north compartment. The north compartment reached capacity by September 2031. After September 2031 dredged material was placed only in the center and south compartments using an annual rotation schedule. The analysis showed that the south compartment reached capacity in May 2079. After May 2079 all dredged material was placed in the center compartment, causing this compartment to reach capacity in September 2080. Therefore, even under the worst case dredging scenario, the service life of the CIDMMA will be extended to approximately the year 2080 under the proposed RUP.

Comparison of the Baseline and Worst Case Dredging simulation revealed that the use of a one foot lift exhibited better consolidation and desiccation characteristics than thicker lifts. It is anticipated that 0.3 m (one foot) thick lifts would also enhance trenching and crust farming operations in a management area.

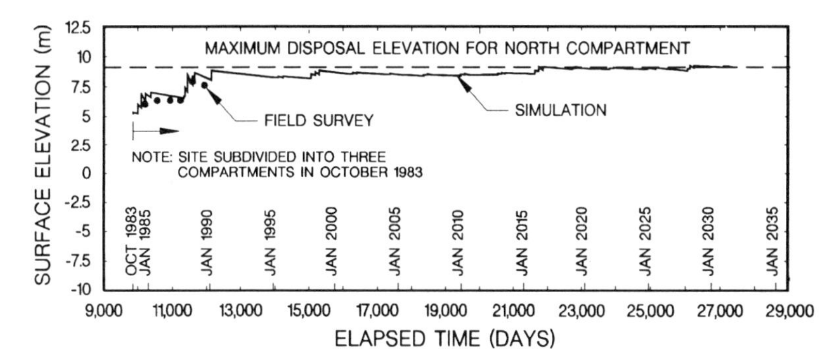

FIG. 4. Simulation of Fill Rates for Worst Case Dredging Scenario in North Compartment from 1984 to 2031

VERTICAL STRIP DRAINS

Review of the calculated void ratio and effective stress profiles in 1992 revealed that the majority of the calculated consolidation occurred in the dredged fill. As a result, large excess pore-water pressures were calculated in the compressible clay foundation. This suggests that the compressible foundation is under-consolidated due to the large drainage path. This is in good agreement with large excess pore-water pressures that were measured in the clay foundation using piezometers in the perimeter dikes. These piezometers showed excess pore-water pressure levels in February 1991 that exceeded the ground surface elevation by 7.6 m (25 feet) in some locations. The installation of vertical strip drains would significantly reduce the drainage path and thus the time required to consolidate the dredged fill and foundation clay. Consolidation of the dredged fill and foundation clay would cause a significant increase in storage capacity and undrained shear strength of these materials. This strength gain would allow the perimeter dikes to be constructed to higher elevations without setbacks or stability berms, which will also provide an increase in storage capacity.

A 183 m by 122 m (600 ft by 400 ft) strip drain test section (Fig. 1) was constructed in the north compartment of the CIDMMA to evaluate the effectiveness of prefabricated strip drains in increasing storage capacity. Strip drain installation commenced on 18 December 1992 and was completed on 26 February 1993. Preliminary results show that the dredged fill and foundation clay underlying the test section are undergoing substantial settlement (1.5 to 1.8 m in thirteen months) as shown in Fig. 5. It is anticipated that consolidation, and thus settlement, will continue through June 1994. Based on these results, the feasibility of installing vertical strip drains throughout the entire management area and/or perimeter dikes is being considered.

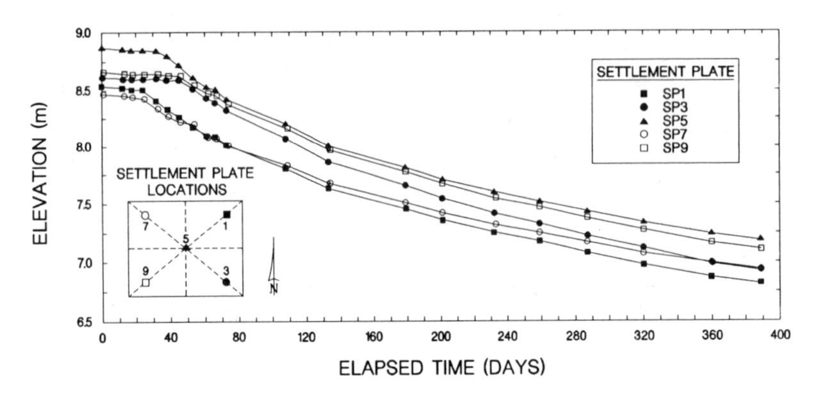

FIG. 5. Settlement Data for CIDMMA Strip Drain Test Section

Conclusions

The service life of the Craney Island Dredged Material Management Area (CIDMMA) under the proposed Restricted Use Program (RUP) was evaluated using the microcomputer model PCDDF89. Field settlement data from 1956 to 1992 was used to verify PCDDF89 input parameters. Based on projections of fill rates under the RUP, the service life of the CIDMMA will be extended significantly by reducing the quantity of dredged material placed. In particular, the CIDMMA will reach capacity around the year 2130 under the Baseline Maintenance Dredging Case and near the year 2080 under the Worst Case Dredging Scenario.

It is also anticipated that the service life of the CIDMMA can be extended by installing vertical strip drains to consolidate the dredged fill and foundation clay. Early results of a strip drain test section show that the dredged fill and foundation clay are undergoing substantial consolidation settlement 1.5 m to 1.8 m (4.9 to 5.9 feet in thirteen months). This consolidation will result in increased storage capacity and an increase in undrained shear strength of the dredged fill and foundation clay. An increase in undrained shear strength should allow the perimeter dikes to be constructed to higher elevations without setbacks or stability berms, which will also provide an increase in storage capacity.

Acknowledgments

The authors wish to acknowledge Ronn Vann, Sam McGee, Tom Friberg, Dave Pezza, and Matt Byrne of the U. S. Army Engineer Norfolk District for their assistance in preparing this paper. Permission was granted by the Chief of Engineers to publish this information.

Appendix I. References

Brown, K. W., and Thompson, L. J. (1977). "General crust management as a technique for increasing capacity of dredged material containment areas," *Technical Report D-77-17*, U. S. Army Engineer Waterways Experiment Station, Vicksburg, Mississippi.

Cargill, K. W. (1985). "Mathematical model of the consolidation desiccation processes in dredged material," *Technical Report D-85-4*, U. S. Army Engineer Waterways Experiment Station, Vicksburg, Mississippi.

Cargill, K. W. (1986). "The large strain, controlled rate of strain (lscrs) device for consolidation testing of soft fine-grained soils," *Technical Report No. GL-86-13*, U. S. Army Engineer Waterways Experiment Station, Vicksburg, Mississippi.

Gibson, R. E., England, G. L., and Hussey, M. J. L. (1967). "The theory of one-dimensional consolidation of saturated clays: I. Finite non-linear consolidation of thin homogeneous layers." *Géotechnique*, 17(3), 261-273.

National Climatic Center (1980). *Climatological Data, National Summary*, Environmental Data and Information Service, NOAA, Denver, Colorado, Volume 31, Number 1-12.

Palermo, M. R., and Schaefer, T. E. (1990). "Craney Island disposal area: Site operations and monitoring report, 1980-1987." *Miscellaneous Paper EL-90-10*, U. S. Army Engineer Waterways Experiment Station, Vicksburg, Mississippi.

Schroeder, P. R., and Palermo M. R. (1990). "Automated dredging and disposal alternatives management system, (ADDAMS)." *Environmental Effects of Dredging Technical Note No. EEDP-06-12*, U. S. Army Engineer Waterways Experiment Station, Vicksburg, Mississippi.

Stark, T. D. (1991). "Program documentation and user's guide: PCDDF89, Primary Consolidation and Desiccation of Dredged Fill." *Instruction Report No. D-91-1*, Environmental Laboratory, U. S. Army Engineer Waterways Experiment Station, Vicksburg, Mississippi.

APPENDIX II. SI UNIT CONVERSIONS

1 psf	= 47.85 Pa
1 foot	= 0.305 m
1 acre	= 4047 m^2
1 cubic yard	= 0.7646 m^3

Nonlinear Three Dimensional Analysis of Downdrag on Pile Groups

Sangseom Jeong,[1] Member, ASCE, and Jean-Louis Briaud,[2] Fellow, ASCE

Abstract: The downdrag on pile groups was investigated by using a numerical analysis and an analytical study. The emphasis was on quantifying the reduction of downdrag on piles in a group due to the group effect. The case of a single pile and subsequently the response of groups is analyzed by using a three dimensional non-linear finite element approach. An approximate closed form solution is also developed for the case of the single pile. It is shown that the downdrag on piles in a group is much less than the downdrag on a single pile. Based on the results obtained, a simple method is proposed to design groups of 9 to 25 piles with spacing-to-diameter ratios varying from 2.5 to 5.0 for downdrag loads.

Introduction

Downdrag develops when a pile is driven through a soil layer which will settle more than the pile. A typical example is a bridge abutment founded on piles driven through a thick soft clay into a hard layer; the approach fill loads the soft clay which settles significantly; the pile resting on the hard layer settles much less and the soft clay drags the pile down.

The principal effect of downdrag is to increase the axial load in the pile. As a result the settlement of the pile increases. Pile settlement due to downdrag has been reported to have caused excessive damage to superstructures (Garlanger 1974). Unanticipated downdrag loads can lead to a structural failure of the pile, a bearing capacity failure in the bearing soil stratum, or an excessive settlement of the pile.

---

[1] Senior Researcher of Civil Engineering, Yonsei University, Seoul, 120, Korea.

[2] Buchanan Professor of Civil Engineering, Texas A&M University, College Station, TX 77843-3136.

The behavior of a pile in a group is influenced by the presence of and loadings on neighboring piles when piles are closely spaced. This is referred to as group effect. There is some evidence that the downdrag force on a group of n closely spaced piles is less than n times the downdrag force on an isolated pile (Okabe 1977). The reason for this reduction is the pile-soil-pile interaction which depends on the load-carrying capacity, the length, the number of piles and the relative position of the piles within the group, and the pile group spacing.

The overall objective of this study (Jeong and Briaud 1992) is to investigate the reduction of downdrag on pile groups based on a numerical simulation and on a close form solution. As a result, a simple method will be proposed to design pile groups for downdrag loads.

Downdrag on a Single Pile

Many methods have been proposed to calculate the downdrag force on single piles. A description of those methods can be found in Lambe and Baligh (1978), Combarieu (1985) or Briaud et al. (1989). A study of existing methods and experimental data lead to the following recommendation. The unit friction, f_{max}, at the pile soil interface should be calculated by an effective stress approach (Zeevaert 1959)

$$f_{max} = \beta \sigma_v' \qquad (1)$$

where β is the friction parameter; and σ_v' is the vertical effective overburden stress. The pile length over which this downdrag friction is to be considered should be determined by properly accounting for the pile compressibility, the soil settlement, the stiffness and strength of the bearing layer and the top load. This is done by using the microcomputer program NEWNEG (Briaud and Porwoll 1989).

Existing Methods for Downdrag on Pile Groups

Several methods have been proposed (Terzaghi and Peck 1948; Broms 1976; Zeevaert 1959; Combarieu 1985). The first method consists of calculating the downdrag on the group F_{ng} as the downdrag on a single pile F_{ns} multiplied by the number of piles in the group, n:

$$F_{ng} = F_{ns} \qquad (2)$$

Terzaghi and Peck method (1948) consists of assuming that the pile group acts as a block and that the downdrag force is developed on the group perimeter.

$$F_n = sLP \qquad (3)$$

where s is the soil shear strength; L the pile length; and P the enclosing perimeter of the group. This method considers that the soil within the group of piles is an integral part of the pile foundation.

Broms (1976) stated that the load increase from the shear stress along the circumscribed area of the group will be carried mainly by the piles located along the perimeter of the group. The interior piles will carry the weight of the soil between the piles plus the superimposed fill, or plus the increase in effective weight of the soil in the event that the ground water table is lowered.

Zeevaert's method (1959) is based on the principle that if the pile carries some of the soil weight, the vertical effective stress σ_{vo}' in the soil is decreased to σ_{vf}' in the vicinity of the pile. Therefore the negative friction is:

$$f = \beta\sigma_{vf}' < \beta\sigma_{vo}' \tag{4}$$

Zeevaert then gives some relatively complex rules to take into consideration the pile spacing and locations, including the case of corner and perimeter piles.

Combarieu (1985) extended the theoretical work of Zeevaert by introducing a hanging coefficient λ and proposed some practical rules for pile spacings between 3 and 5 pile diameters (center to center).

Proposed Method for A Single Pile

It is assumed that during the settlement of the compressible soil layer the shear stresses close to the pile shaft are fully mobilized because the relative movement between the soil and the pile is generally large. Therefore, some of the weight of the surrounding soil will be carried by the pile and as a result, the vertical effective stress will be decreased from $\sigma'(z)$ to $\sigma'(z,R)$ in the vicinity of the pile (Fig. 1).

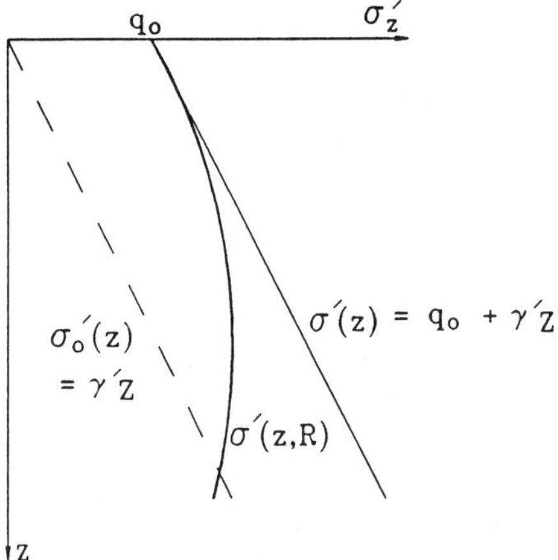

FIG. 1. Variation of Vertical Effective Stress near a Pile

It is also assumed that the shear stress τ on the vertical boundary of the block of the soil as shown on Fig. 2 is zero because of the symmetry within the group and between adjacent piles. Note then that the boundary radius b is equal to one half of the center to center spacing. Alternatively, b is the radius of the zone of influence for a single pile and may be taken as infinity.

To calculate the decrease in vertical effective stress at the pile-soil interface, the vertical force equilibrium that takes into account the negative skin friction at a depth z is derived (Fig. 2):

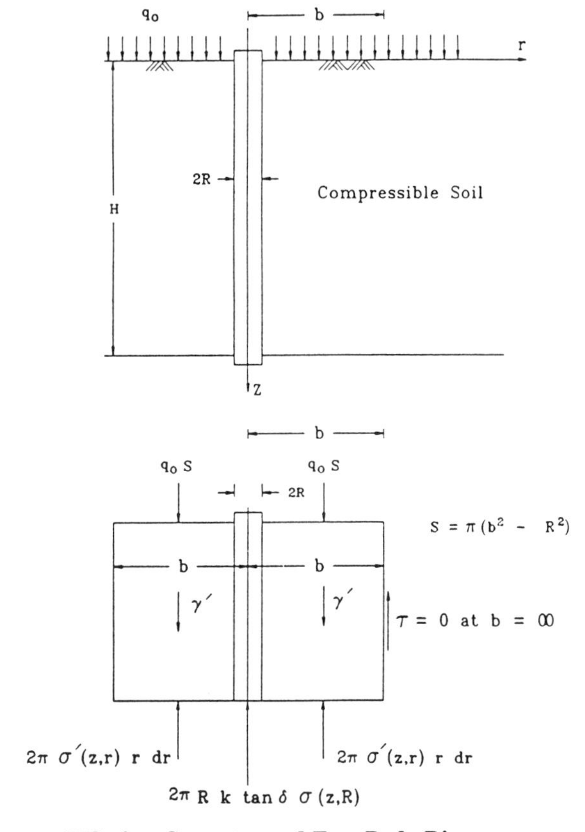

FIG. 2. Geometry and Free Body Diagram

$$2\pi RK\tan\delta\int_0^z \sigma'(z,R)dz + 2\pi\int_R^b \sigma'(z,r)rdr = S(q_0 + \gamma' z) \tag{5}$$

where σ_0' = vertical effective stress (γ'z) at the initial state before pile driving; σ'(z,r) = vertical effective stress that takes into account the effect of the pile evaluated at a distance r from the axis of pile; σ'(z,R) = vertical effective stress at the pile shaft; σ'(z) = vertical effective stress ($q_0+\gamma'z$) at the final stage computed without taking any piles into account and independent of r; K = coefficient of earth pressure, tan δ = friction coefficient at the pile soil interface; b = the radius of influence for a pile; z = the depth where σ'(z,R) is calculated; R = the radius of the pile; γ' = the effective unit weight of the soil; and q_0 = the surcharge applied to the ground surface.

Next, Eq. (5) is differentiated with respect to z:

$$2\pi RK \tan\delta\, \sigma'(z,R) + 2\pi \int_R^b \frac{\partial \sigma'(z,r)}{\partial z} r dr = S\gamma' \qquad (6)$$

In order to determine the decrease in the vertical stress due to downdrag at the pile-soil interface, the radial variation of the effective stress, $\sigma'(z,r)$ is introduced as follows:

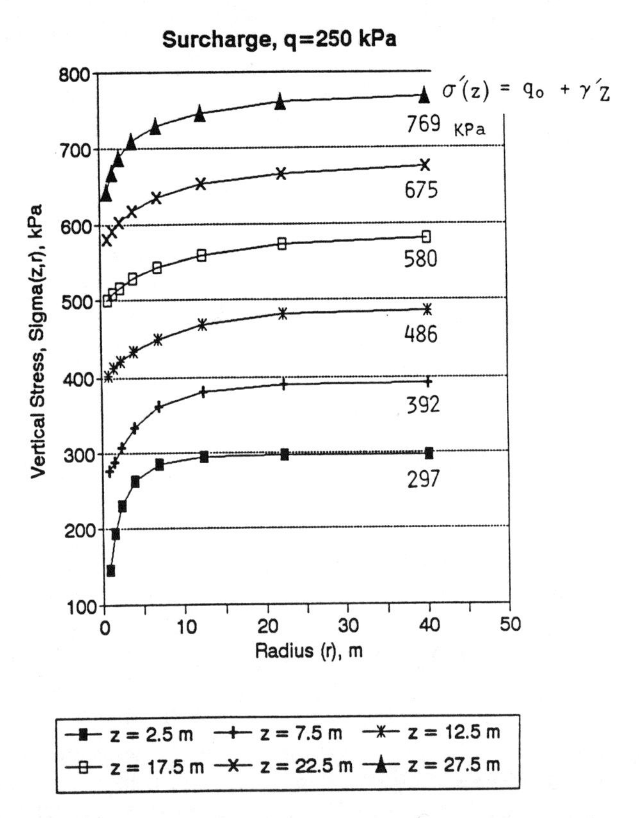

FIG. 3. Vertical Stress Variation at Depth z (3-Dimensional Finite Element Analysis)

The three-dimensional finite element analysis (Hibbitt, Karlsson and Sorensen, Inc. 1991) was done for a single pile which has a diameter of 1.2 m. The pile was driven through 30 m of soft clay with an effective unit weight of γ' = 19 kN/m^3 and the embankment was simulated by applying a surcharge loading, q_0 = 250 kPa, to the ground surface. Fig. 3 shows the variation of the effective stress, $\sigma'(z,r)$ at several depth z. Based on the analysis, the effective stress, $\sigma'(z,r)$ approaches its maximum value $\sigma'(z) = q_0+\gamma'$ at a distance which corresponds to the end of the zone of influence. Close to the pile it has a smaller value which varies as shown on Fig. 3.

The variation of the effective stress $\sigma'(z,r)$ at a distance r from the pile axis and on a horizontal plane at a depth z shown on Fig. 3 can be approximated by the following relation:

$$\sigma'(z,r) = \sigma'(z) - [\sigma'(z) - \sigma'(z,R)] \cdot \exp\frac{\psi(R-r)}{\xi} \tag{7}$$

where ψ = a friction coefficient; ξ = a depth coefficient; L = pile length; $\psi = \alpha(1 - \beta)$; and ξ = L/5 (Table 1).

TABLE 1. Values of the Parameters

β	α	β	α
0.15	2.0	0.3	1.0
0.25	1.5	0.4	0.5

Next, differentiation of Eq. (7) with respect to z is

$$\frac{\partial\sigma'(z,r)}{\partial z} = \gamma' - \left[\gamma' - \frac{\partial\sigma'(z,R)}{\partial z}\right] \cdot \exp\frac{\psi(R-r)}{\xi} \tag{8}$$

Substituting Eq. (8) into Eq. (6) leads to a first order ordinary differential equation with respect to $\sigma'(z,R)$:

$$\frac{\partial\sigma'(z,R)}{\partial z} + f(\psi,b) \cdot \sigma'(z,R) = \gamma' \tag{9}$$

where

$$f(\psi,b) = \frac{RK\tan\delta \cdot \psi}{\xi}\left[R + \frac{\xi}{\psi} - \left[b + \frac{\xi}{\psi}\right] \cdot \exp\frac{\psi(b-R)}{\xi}\right]^{-1} \tag{10}$$

The case of a single pile corresponds to b = ∞ which leads to:

$$f(\psi,b) = f(\psi,\infty) = \frac{RK\tan\delta \cdot \psi}{\xi} \cdot \left[R + \frac{\xi}{\psi}\right]^{-1} \tag{11}$$

The corresponding solution of Eq. (9) for the single pile is:

$$\sigma'(z,R) = \frac{\gamma'}{f(\psi,b)} + \exp^{-f(\psi,b)z}\left[q_o - \frac{\gamma'}{f(\psi,b)}\right] \tag{12}$$

Parametric Analysis for Pile Groups

A series of three dimensional finite element analyses on pile groups were performed for different end bearing conditions, cap rigidity, soil model, and spacing between piles (Table 2). The cases of a single pile and of pile groups are analyzed.

To understand the true behavior, yielding of the soil at the pile-soil interface was considered by taking into account the effective strength parameters of a clay: the effective cohesion, C′ and the effective friction angle, ϕ'. The pile element was assumed to remain elastic at all times, while the surrounding soil was idealized as a linear elastic material first and then as an elastoplastic material. The elastoplastic analyses were run to take into account the local yielding at the pile-soil interface and used an iterative and incremental analysis. The soil model was the extended Drucker-Prager model which was chosen from the soil models in the library of ABAQUS (Hibbitt, Karlsson and Sorensen, Inc. 1991). All the results shown here are the ones obtained by using the nonlinear three-dimensional analysis.

Fig. 4 shows a typical problem for a pile group. It is assumed that the external load is only applied to the ground surface of the soil as a distributed load (i.e., the sum of the pile head loads is zero). To reduce the number of elements the lower rigid boundary for the pile group has been placed in all cases at a depth equal to 1.6 times the length of the piles. The piles are square for simplicity.

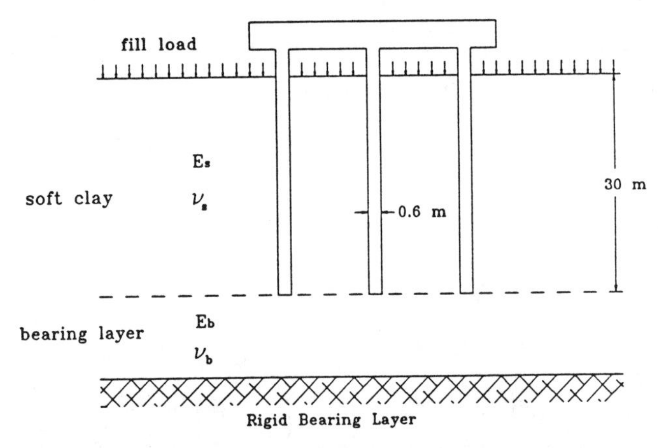

FIG. 4. Typical Geometry of the Pile Group Analyzed

The ultimate downdrag forces result from the long term behavior of the soil and are calculated on the basis of effective stress parameters. The soil properties to represent a soft clay used in this study were selected accordingly. Table 3 shows the material properties and geometries used in this study. Factors which are relevant to the downdrag will be discussed in detail.

Difference between End Bearing and Friction Piles

Fig. 5 shows the influence of the foundation type on the non-dimensionalized

TABLE 2. Numerical Analysis for Pile Groups

Group	End Bearing Condition	Cap Rigidity	Soil Model	Spacing (s/d)
1×1 (single)	E & F	F	Elastic	2.5 & 5.0
	E & F	F	Elastoplastic	2.5 & 5.0
1×2 (G)	E & F	F	Elastic	2.5 & 5.0
	E	F	Elastoplastic	2.5 & 5.0
1×4 (G)	E	F	Elastic	2.5 & 5.0
1×6 (G)	E	F	Elastic	2.5 & 5.0
1×8 (G)	E	F	Elastic	2.5 & 5.0
1×10 (G)	E & F	F	Elastic	2.5 & 5.0
2×2 (G)	E & F	F	Elastic	2.5 & 5.0
	E & F	F	Elastoplastic	2.5 & 5.0
3×3 (G)	E & F	F	Elastic	2.5 & 5.0
	E & F	F	Elastoplastic	2.5 & 5.0
3×10 (G)	E & F	F & R	Elastic	2.5 & 5.0
4×4 (G)	E & F	F	Elastic	2.5 & 5.0
5×5 (G)	E & F	F & R	Elastic	2.5 & 5.0
	E & F	F & R	Elastoplastic	2.5 & 5.0
5×10 (G)	E & F	F & R	Elastic	2.5 & 5.0
6×6 (G)	E & F	F	Elastic	2.5 & 5.0
	E & F	F	Elastoplastic	2.5 & 5.0
8×8 (G)	E & F	F	Elastic	2.5 & 5.0
10×10 (G)	E & F	F & R	Elastic	2.5 & 5.0

Notes: 5×10 (G) = 5 rows and 10 columns of pile groups
s = center-to-center spacing between piles
d = diameter of pile, G = pile group
E = end bearing, F = friction
F = flexible pile (without pile cap)
R = rigid pile (with pile cap)

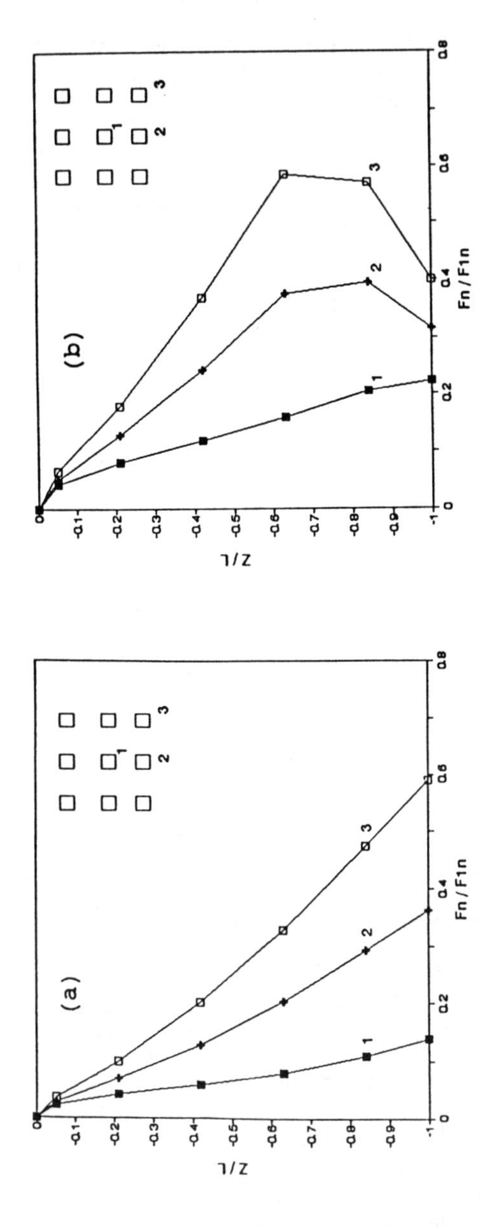

FIG. 5. Effect of Foundation Type: (a) End Bearing Pile (s/d = 2.5); (b) Friction Pile (s/d = 2.5)

final downdrag force profile. Here, the downdrag force, F_n, is normalized by the maximum downdrag force in a single isolated pile, F_{n1}. The downdrag force on the end bearing piles (Fig. 5a) increases continuously with depth. This increase shows a slightly concave curvature because the local yield at the pile-soil interface is controlled by the limiting value of the shear stress. For the friction piles (Fig. 5b), a negative and positive friction is apparent and the neutral point is located at a depth equal to three quarters of the embedded pile length. The effect of downdrag forces is more significant for end bearing piles than friction piles: the stiffer the bearing layer, the smaller the pile settlement, the greater the relative settlement between the piles and the surrounding soil, hence the greater the downdrag force induced.

TABLE 3. Material Properties for Pile Groups

Pile	Area Length E_p ν	0.6 m × 0.6 m 30 m embedding depth 20×10^6 kN/m^2 (concrete) 0.3
Elastic Properties of Soil	E_s E_b ν ν'	20×10^2 kN/m^2 20×10^4 kN/m^2 0.4 9.0 kN/m^3
Plastic Properties of Soil	ρ' β Ψ C'	25° 36.5° 0° 3.0 kN/m^2
Surcharge	q_o	250 kN/m^2

Effect of Number of Piles

Fig. 6 shows the normalized maximum downdrag force as a function of the total number of piles and the relative position of the piles within the group. The figure shows that there is a significant reduction in downdrag forces for the 2.5 d and 5.0 d spacing between piles and for a group size up to 5×5. The reason for this reduction is that as the number of piles increases, the average shear stress along the pile shaft tends to decrease. The settlement of the pile group also decreases. However for groups larger than 5×5, the downdrag force does not reduce any further. In addition, this figure shows that the inner piles of a group have considerably smaller downdrag loads than the outer piles. This is because outside the pile group the soil settles significantly while inside the group the soil settlement is drastically reduced (Fig. 7).

Effect of Spacing (s/d)

Fig. 8 shows the effect of pile spacing and pile location within the group on the maximum downdrag force. For a center to center spacing-to-diameter ratio of

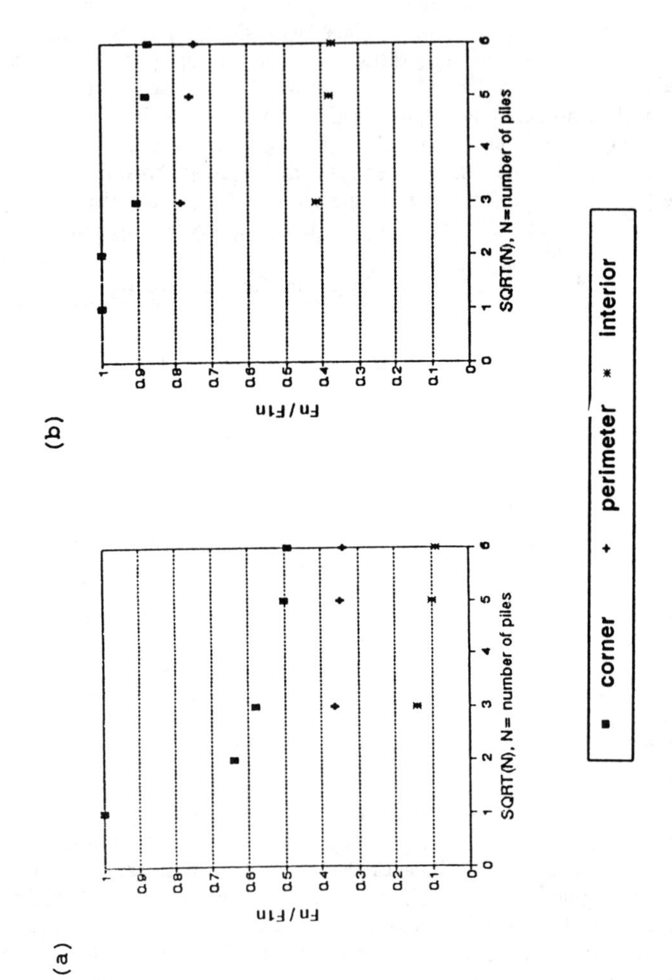

FIG. 6. Effect of Number of Piles in End Bearing Piles: (a) s/d = 2.5; (b) s/d = 5.0

2.5 there is a definite group effect. This is explained by the fact that the downward soil movement between the piles is prevented more and more by the piles as the spacing becomes smaller (Fig. 9). For the ratio of 5.0, there is little difference in downdrag force between the corner pile and the perimeter piles; this shows that there is only a small group effect. For groups larger than 5×5, the group effect is almost constant.

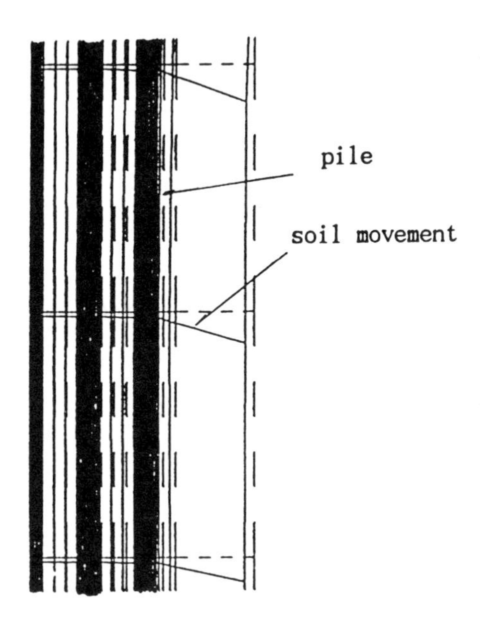

FIG. 7. Soil Movement between Piles

Effect of Cap Rigidity

Fig. 10 shows the load distribution for a rigid pile cap (Fig. 10a) and for a flexible pile cap (Fig. 10b). Here, the cap is not in direct contact with the soil. In the case of a rigid pile cap, a tensile load occurs at the top of the corner piles. This is because the downdrag pulls the corner piles downward while the center piles and the pile cap resist that downward movement. As a result appropriate tensile connections should be designed for the outer piles.

Comparison with Observed Performance

Only a few studies have been reported with measured downdrag forces on pile groups. Downdrag on the pile group reported by Okabe (1977) is compared with the predicted values obtained by the three dimensional finite element analysis.

Fig. 11 shows the pile group configuration. The group is composed of 38 point bearing driven steel pipe piles with a center to center spacing of approximately 2.1 diameters. The piles are 0.7 m in diameter and embedded 40 m in the soil.

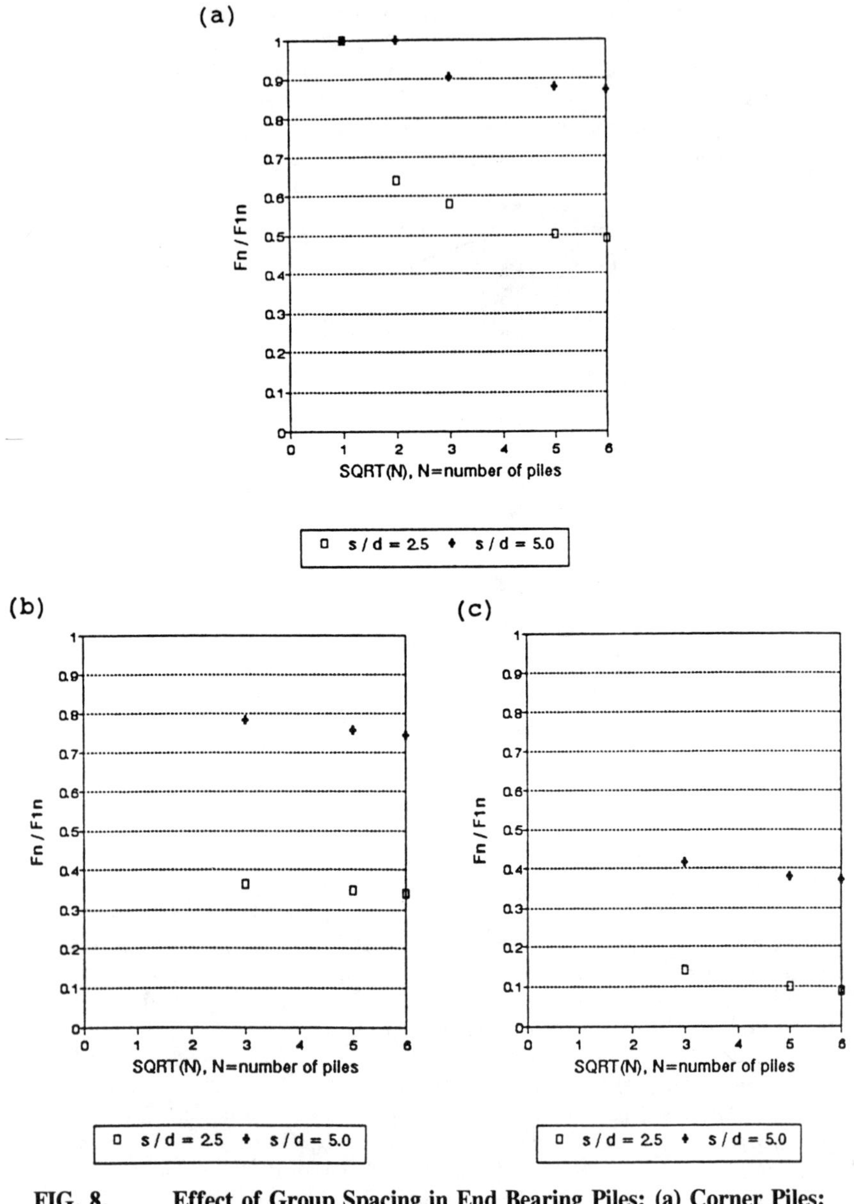

FIG. 8. Effect of Group Spacing in End Bearing Piles: (a) Corner Piles; (b) Perimeter Piles; and (c) Interior Piles

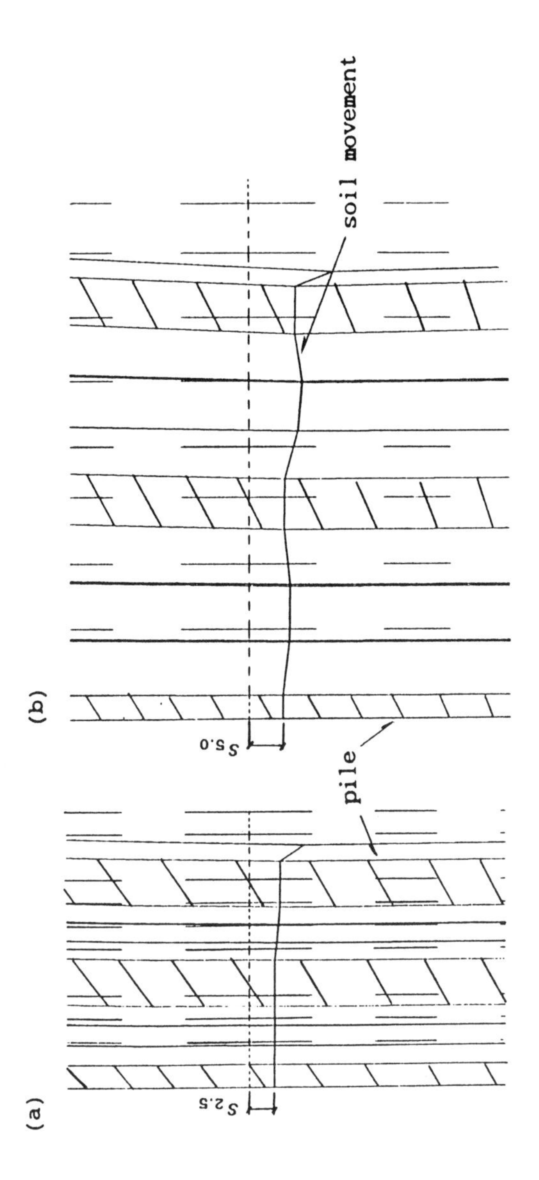

FIG. 9. Soil Movement between Piles

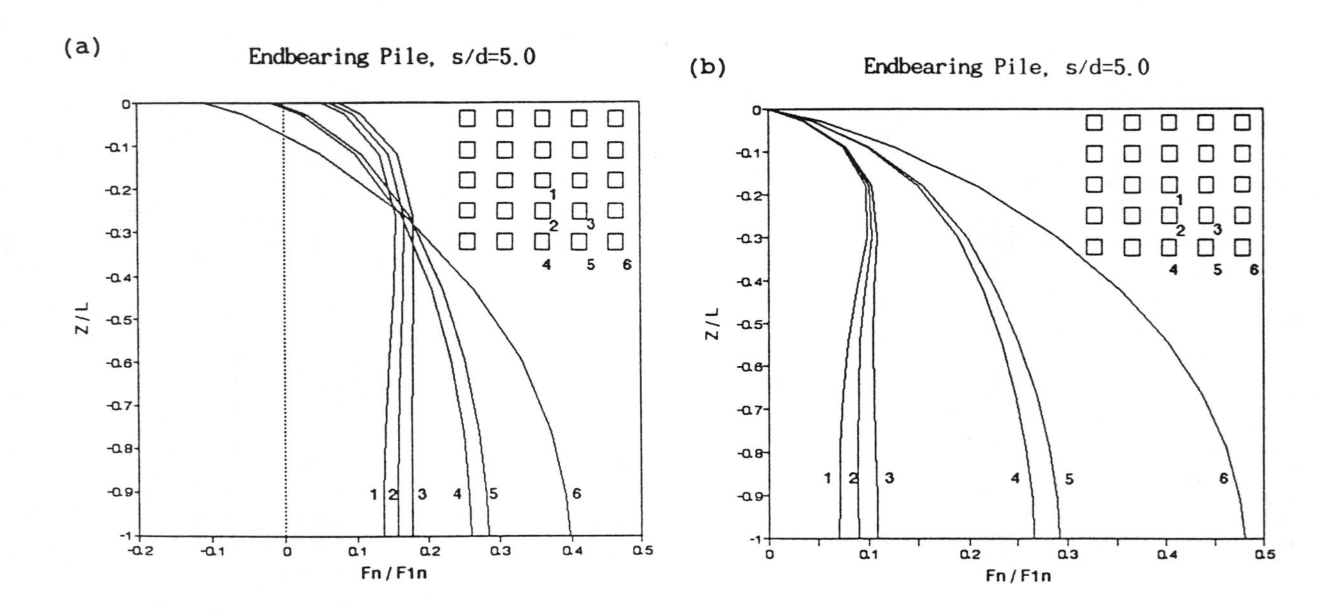

FIG. 10. Effect of Cap Rigidity (End Bearing): (a) Rigid Pile Cap; and (b) Flexible Pile Cap

This group was used for a bridge pier foundation. After installation of the piles, the embankment was placed on a very soft clay layer with 0-10 SPT N-values from 0-40 m depth. Below the soft clay was a medium sandy gravel layer with N-values of about 50. The material properties used for prediction purposes were: the soft clay Young's modulus, E_s = 20 MN/m^2 which corresponds to 200 S_u, poisson's ratio of the soil, ν = 0.3, the effective cohesion, C' = 5.0 kN/m^2 and effective friction angle, ϕ' = 25 degree. A medium sandy gravel layer was assumed as a fictitious rigid foundation. The properties of the soft clay were chosen as reasonable estimates for such a soil.

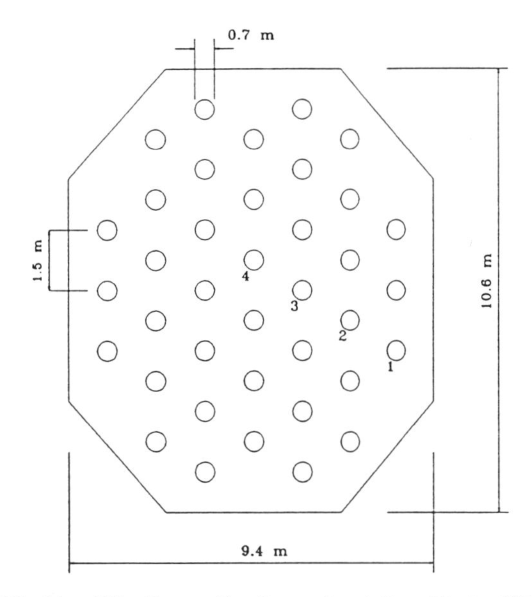

FIG. 11. Pile Group Configuration (after Okabe 1977)

Fig. 12 shows the predicted and observed axial load distribution in the single pile (Fig. 12a) and in the pile group (Fig. 12b). The prediction by the three-dimensional analysis simulates well the general trend observed for the single pile and for the pile group.

SUGGESTED METHOD FOR PILE GROUPS

Based on the results accumulated one might propose the following simple method for groups of 9 to 25 piles. Pile groups that have more than 25 piles show almost the same group effect as those with 25 piles.

At spacings much larger than 5 diameters there is very little group effect. The group should be designed as a cluster of single piles.

At spacings approximately equal to 5 diameters, design square pile groups for the following downdrag loads:

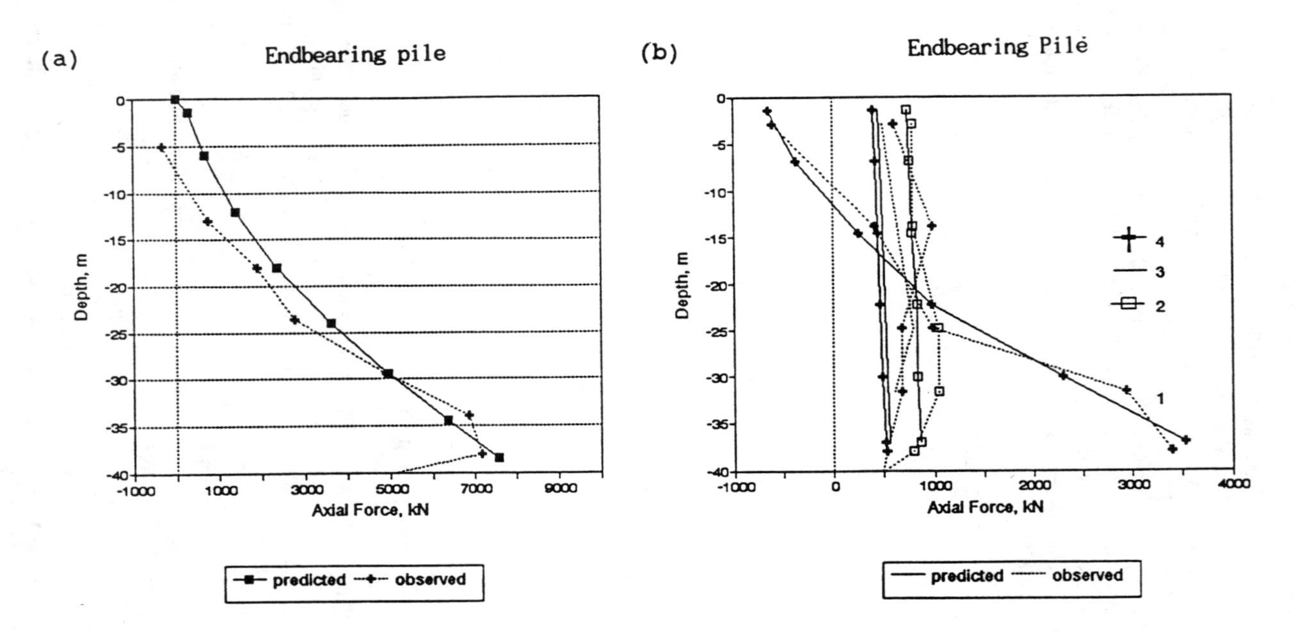

FIG. 12. Comparison of Axial Forces with Observed Performance (End Bearing): (a) Single Pile; and (b) Pile Groups

$$F_{n(corner)} = 0.9 \times F_{n(single)} \tag{13}$$

$$F_{n(side)} = 0.8 \times F_{n(single)} \tag{14}$$

$$F_{n(interior)} = 0.5 \times F_{n(single)} \tag{15}$$

At spacings equal to or smaller than 2.5 diameters there is a definite group effect. Design square pile groups for the following downdrag loads:

$$F_{n(corner)} = 0.5 \times F_{n(single)} \tag{16}$$

$$F_{n(side)} = 0.4 \times F_{n(single)} \tag{17}$$

$$F_{n(interior)} = 0.15 \times F_{n(single)} \tag{18}$$

The neutral point for the group can be taken at the pile point for groups bearing on a perfectly rigid base and at a depth equal to 75 % of the pile group length for friction pile groups with no point bearing resistance.

CONCLUSIONS

The downdrag on pile groups was investigated based on a numerical analysis and on an analytical study. Most instrumented pile groups may confirm the results of this work and lead to significant savings. On the basis of the findings of this study, the following main conclusions are drawn:

(1) The downdrag on piles in a group is much less than the downdrag on a single pile.

(2) The reduction of downdrag forces for center to center spacings of 2.5 and 5 diameter increases as the number of piles in the group increases. This is particularly significant for groups of 9 to 25 piles. Beyond 25 piles, further reduction becomes small.

(3) For center to center spacing-to-diameter ratios of less than or equal to 5.0, there is a definite reduction in the downdrag on pile groups. Thus, designing a group as if it was composed of single piles is unattractive economically. The interior piles in the group develop less downdrag force than the outer piles.

(4) The neutral point for the group can be taken at the pile point for piles bearing on a perfectly rigid base. For friction piles with no point bearing resistance, the neutral point is located at a depth approximately equal to 75% of the pile group length for the corner and perimeter piles but at the pile point for the interior piles.

(5) In the case of a rigid pile cap, the downdrag can induce significant tensile forces at the top of the outer piles.

ACKNOWLEDGMENTS

This study was supported, in part, by the National Cooperative Highway Research Program.

Appendix - References

Hibbitt, Karlsson and Sorensen, Inc. (1991). *ABAQUS User's and Theory Manuals*, Version 4.9, Pawtucket, Rhode Island.

Briaud, J. L., Algurjia, Z. A. K., Quraishi, M. Z. Z., Bush, R. K., and Jeong, S. S. (1989). *Downdrag on Bitumen Coated Piles*, Draft Report to the National Cooperative Highway Research Program, Washington, D. C.

Briaud J. L., and Porwoll, H. (1989). "NEWNEG: Microcomputer program for downdrag on piles." *Foundation Engineering, Current Principles and Practices*, ASCE, New York, New York, 1, 706-718.

Broms, B. B. (1976). "Pile foundation - Pile groups." *Proc. 6th European Conf. Soil Mech. Found. Eng.*, Vienna, 1, 103-132.

Combarieu, O. (1985). "Frottement negatif sur les pieux." *Rapport de recherche LPCN° 136*, Laboratoire Centrale des ponts et Chaussees, Paris, France.

Garlanger, J. E. (1974). "Measurements of pile downdrag beneath a bridge abutment." Transportation Research Record 517, 61-69.

Jeong, S. S., and Briaud, J. L. (1992). "Downdrag on pile groups," *Research Report RF 7112*, Department of Civil Engineering, Texas A&M University.

Lambe, T. W., and Baligh, M. M. (1978). "Negative skin friction downdrag on a pile," *Report No. FHWA/TS - 73/210*, Massachusetts Institute of Technology, Cambridge, Massachusetts.

Okabe, T. (1977). "Large negative friction and friction-free pile methods." *Proc. 9th ICSMFE*, Tokyo, 1, 672-679.

Terzaghi, K., and Peck, R. B. (1948). *Soil Mechanics in Engineering Practice*, John Wiley and Sons, New York, New York.

Zeevaert, L. (1959). "Reduction of point bearing capacity of pile because of negative skin friction." *Proc. 1st Pan Am. Conf. Soil Mech. Found. Eng.*, Mexico City, 1145-1152.

Load-Settlement Prediction of Footings on Steep Slopes

Pedro Arduino,[1] Student Member, ASCE,
Emir J. Macari,[2] Associate Member, ASCE,
and Mark Gemperline,[3] Member, ASCE

Abstract: The use of spread footings on slopes is a common practice in the design of bridge foundations. Footings are generally constructed in the slope or at its crest to minimize the bridge length, thereby minimizing costs. However, the analytical methods that are frequently used to predict the ultimate bearing capacity of shallow spread footings on slopes yield a wide range of solutions. As a result, allowable foundation bearing capacities are selected conservatively leading to the design of spread footings which are larger and more costly than necessary.

There has been an increased effort in recent years to evaluate and improve techniques for predicting the deformation and ultimate bearing capacity of shallow spread footings located on or near slopes. Research efforts currently underway include detailed centrifuge tests on scaled physical models and advanced numerical techniques used in conjunction with sophisticated elasto-plastic constitutive models.

For the work reported here, scaled centrifuge models are used to obtain maximum bearing pressures of prototype shallow spread footings located near sand slopes. These results are compared with classical bearing capacity solutions and numerical predictions obtained with finite element discretizations and a proper constitutive model.

---

[1] Ph.D. Student, School of Civil Engineering, Georgia Institute of Technology, Atlanta, GA 30332-0355.

[2] Associate Professor, School of Civil Engineering, Georgia Institute of Technology, Atlanta, GA 30332-0355.

[3] Civil Engineer, U. S. Bureau of Reclamation, Materials Engineering Branch, Denver, CO 80225.

Deformed meshes, incremental nodal displacements, and stress contours are also presented. The concept of modeling of models is applied to test the validity of the scaling relations in both analyses.

INTRODUCTION

The use of spread footings on slopes is a common practice in the design of bridges. Bridges commonly cross a variety of civil structures, such as canals or highways, which have soil slopes on either side as steep as 1.5H:1.0V. The bridge footings are generally constructed on the slope or at its crest to minimize the bridge length, thereby reducing construction costs.

At the present time, the methods most frequently used to predict the ultimate bearing capacity for design of shallow spread footings on slopes are:

1. Classical bearing capacity equations;
2. Finite element methods; and
3. Centrifuge analysis using scaled models.

Most of the classical bearing capacity equations are based on the assumption that the soil behaves as a rigid- or an elasto-perfectly plastic material, meaning that when the stress intensity reaches a certain critical value, the soil enters the plastic flow range, which indicates continuous deformation at a constant state of stress. This leads to methods of analysis that are relatively simple and amenable to hand calculation. In many cases the solutions obtained are approximations, but since the uncertainties are often greater they are generally sufficient for practical engineering purposes.

These classical methods of determining bearing capacity have proven, in most geotechnical problems over several decades, to be very useful. However, they do not apply when considering contained plastic deformations or complex boundary conditions. In these cases more powerful numerical tools, such as the finite element method, should be used. The main limitation of these methods in the design of slopes is the difficulty encountered when attempting to represent the elasto-plastic behavior of soils. At the present time, sufficient experience exists with the constitutive modeling of soils to obtain accurate predictions of foundation limit loads as well as deformation predictions. In this sense a number of recently developed plasticity-based numerical models along with powerful integration techniques of the incremental constitutive relations have enhanced the versatility of the finite element method.

Among the new developments and advances in the analysis, design, and performance of soil slopes, centrifuge modeling has attained a special place. Centrifuge modeling can be viewed as an alternative or a complement to widely used numerical techniques. Like other physical modeling techniques it has many advantages, the most important of which is its ability to study phenomena for which well defined theories do not exist.

In the work presented here, scaled centrifuge models are used to obtain maximum bearing pressures of prototype shallow spread footings located near sand slopes (Gemperline 1983). These results are compared with classical bearing capacity solutions and numerical predictions obtained with proper finite element discretizations and an adequate constitutive model.

The slope geometry, footing size, and footing location of the prototype were selected to represent conditions common to bridge foundations near canal side slopes. The geometry of the slope prototype used in this analysis is presented in Fig. 1.

A medium sized dense, dry, cleaned poorly graded sand with particle sizes ranging from 0.041 to 0.058 cm., was selected to represent typical situations encountered in many bridge sites (Gemperline 1983). The specific gravity of the sand was 2.66. A relative density test showed the minimum and maximum dry densities to be 13.04 and 15.87 kPa respectively. Direct shear test results indicated a soil friction angle as low as 42.8°, and from drained triaxial compression tests performed at 172.38, 344.76, 689.53, and 1379.06 kPa effective confining pressures, the friction angle computed ranged between 43.7° and 45.3°. For details of the centrifuge testing the reader is referred to Gemperline (1983).

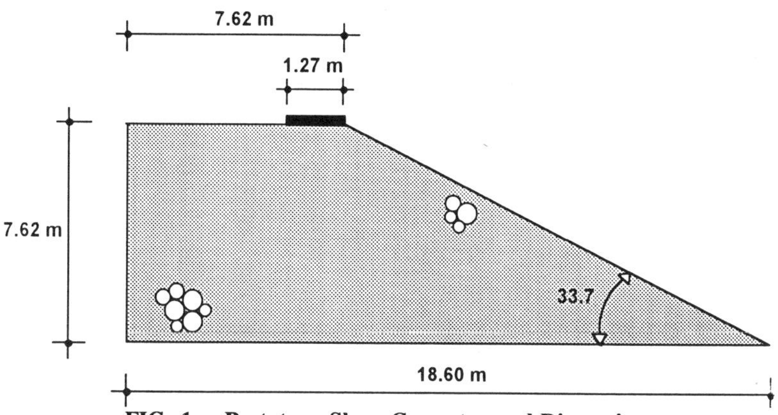

FIG. 1. Prototype Slope Geometry and Dimensions

Classical Bearing Capacity Solutions

In view of the fact that a complete analysis, which includes the true non-linear constitutive response of soils, is far too complicated and impractical for most applications; and that failure by plastic collapse is the governing condition in so many problems in soil mechanics, the development of efficient methods for computing the collapse load in a more direct manner has been of intense practical interest to engineers (Chen and McCarron 1991).

In this sense, the most common theoretical approaches to solving bearing capacity problems have been the slip-line method, the limit equilibrium method, and the limit analysis method. These methods were first applied to the simple case of an infinitely long shallow footing on a horizontal ground surface and later adapted for use with footings of various shapes on ground surfaces having various geometries (Chen 1975).

From this starting point, many authors have derived theoretical and empirical solutions to problems pertaining to the bearing capacity of shallow foundations. Among them are equations presented by Terzaghi, Spencer, and Myslivec using the limit equilibrium method, by Giroud and Hansen using the Slip-Line method, and by Kusakabe and Chen using the limit analysis method (Bureau of Reclamation 1984; Chen 1975).

Most of these analytical methods use the Mohr-Coulomb failure criteria to describe the soil shear strength. Therefore, the selection of the appropriate friction angle is critical. In practice, this friction angle is obtained by triaxial shear tests. However, in the case of shallow spread footings located near sand slopes the predominant plane strain condition makes it necessary to correlate the laboratory friction angles with those encountered in the actual situation. Tests on a very dense sand have shown ϕ to be approximately 7° greater for plane strain shear than for triaxial shear. Some authors have suggested using a correction to the friction angle determined from the triaxial tests, $\phi_{pl} = 1.10\,\phi_{triax}$ where ϕ_{pl} is an approximate plane strain shear friction angle (Bureau of Reclamation 1984).

For the selected sand, direct shear and triaxial shear test results indicate that the soil friction angle, (ϕ_{triax}), may be rationally selected between 42.8° and 45.3°. If the previously mentioned correction for plane strain condition is applied, then ϕ_{pl} ranges approximately between 48.0° and 50.0°. Thus, it is concluded that an appropriate friction angle might range anywhere between 43.0° and 50.0°.

In this work, the ultimate bearing capacities representing solutions to the previously classical methods were computed for the prototype test condition using soil friction angles between 42.0° and 50.0° in increments of 2.0°. The computed ultimate bearing capacities are presented in Fig. 2.

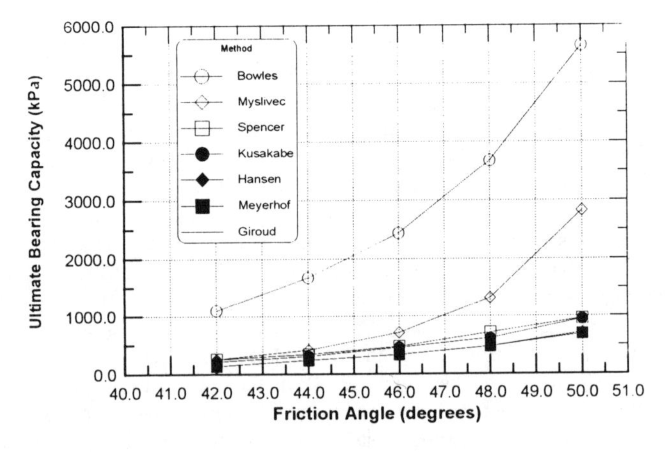

FIG. 2. Ultimate Bearing Capacities using Classical Analytical Methods

NON-LINEAR FINITE ELEMENT ANALYSIS

The solution of boundary-value problems involving the prediction of footing bearing capacity on slopes requires that accurate numerical procedures be adopted in discretizing the slope, solving the equilibrium equations, and implementing an adequate constitutive model (Zienkiewicz and Taylor 1991).

One of the most commonly employed numerical procedures is the finite element method, which consists in the subdivision of the continuum into regions (finite elements) for which the behavior is described by a separate set of assumed functions representing stresses or displacements. These sets of functions are often chosen in a form that ensures continuity of the described behavior throughout the complete continuum. From this starting point and following the conventional elasto-plastic formulation, the resulting equilibrium equations may be written as a residual force relation

$$\mathbf{R}(\mathbf{u},\mathrm{eq}) = \mathbf{F}(\mathbf{u}) - \mathbf{F}(\mathrm{eq}) \approx 0 \tag{1}$$

where $\mathbf{F}(\mathbf{u})$ is the internal force vector

$$\mathbf{F}(\mathbf{u}) = \int_V \mathbf{B}^T \tilde{\sigma}\, dv \tag{2}$$

and $\mathbf{F}(\mathrm{eq})$ are the equivalent applied nodal loads

$$\mathbf{F}(eq) = \int_V \mathbf{N}^T \tilde{f}^b dv + \int_s \mathbf{N}^T \tilde{f}^s ds + \sum_i \mathbf{N}^{i^T} \tilde{f}^i \tag{3}$$

Adopting a constitutive relation

$$\tilde{\sigma} = \mathbf{D}\tilde{\epsilon} \tag{4}$$

the internal force vector takes the form

$$\mathbf{F}(u) = \int_b \mathbf{B}^T \mathbf{D} \mathbf{B}\, dv\ \mathbf{u} = \mathbf{K}\mathbf{u} \tag{5}$$

In Eq. (5) the stiffness is introduced in a rather generic manner. In general, $\mathbf{D}$ is a nonlinear function of strains and the tangent stiffness $\mathbf{K}^t$, or some approximation has to be used to solve Eq. (1). The particular choice of $\mathbf{K}$ gives rise to a variety of iterative solution schemes, such as the Modified Newton Method or the Initial Load Method.

For the Initial Load Method, the iterative analysis requires only the evaluation of the residual force and then back substituting, saving the time-consuming computation and assembly of $\mathbf{K}^t$ in each iteration. Although this is achieved at the expense of the convergence rate, the initial load method has been shown to trace solution paths quite well, even for the case when softening properties are present. Besides, the algorithm needs no knowledge of the material model being used, or the constitutive strategy at the constitutive level (Arduino 1992). This makes the incorporation of additional constitutive models into the finite element formulation quite easy.

Solutions obtained from this method can be precise. However the degree of precision looses its significance if the constitutive assumptions, coupled with the boundary conditions, are inappropriate idealizations of the physical problem. In this context, and in an attempt to simulate the highly nonlinear characteristics of granular materials, a series of constitutive models have been proposed. One of these models is the MRS-Lade model. The MRS-Lade model can be described as a pressure sensitive, three-stress invariant dependent, cone-cap elasto-plastic model. It consists of a piece-wise smooth yield surface formed by a movable curved-cone work-hardening yield surface, which also serves as the ultimate strength or failure surface, and a movable work-hardening cap which is defined as an elliptical surface that extends between the cone and the hydrostatic stress axis. This cap is commonly not activated under shear for dense materials subjected to relatively low effective normal stress states. A non-associated flow rule is defined for the cone and an associated flow rule is adopted for the cap. For more details refer to Macari-Pasqualino et al. (1993) and Arduino (1992).

In this work, the initial load method and the MRS-Lade model have been implemented in a two-dimensional finite element algorithm where plane stress, plane strain and axisymmetric problems can be considered. The selected prototype slope geometry and load conditions defined in the previous section were simulated using a finite element discretization. Fig. 3 shows the mesh (9-node isoparametric elements) of the slope and its boundary conditions.

The material parameters were obtained by physical calibration using three conventional triaxial compression test results corresponding to the selected sand. The comparison of the numerical results and the laboratory test data are shown in Fig. 4.

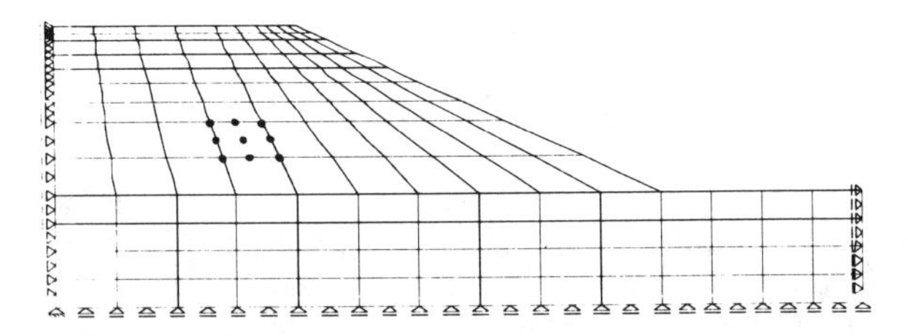

FIG. 3. Undeformed Mesh showing Slope and Boundary Conditions

The loading program consisted of applying the self-weight in several steps to assure convergence. This was in turn followed by a series of small displacement increments applied in the vertical direction to the nodes at the top of the spread footing.

Figs. 5 and 6 show, respectively, the resulting deformed mesh, and the direction and magnitude of the nodal displacements for the prototype finite element model at the post-peak stress state. From this analysis, the ultimate bearing capacity obtained was 1044.46 kPa.

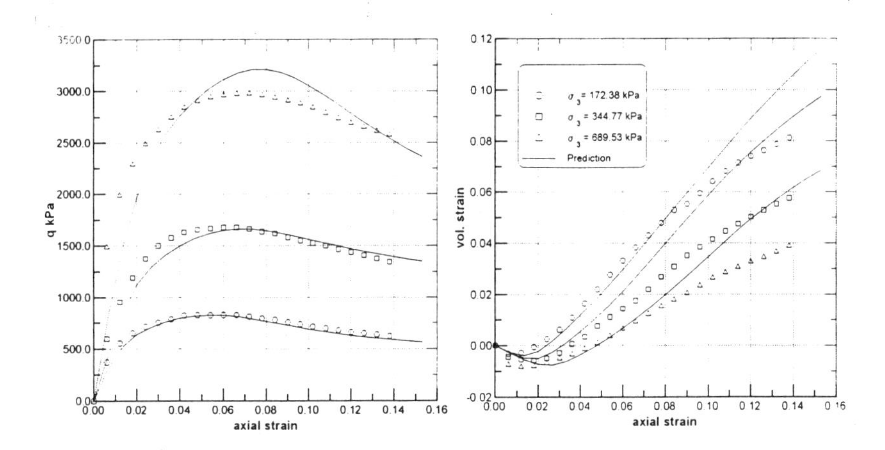

FIG. 4. Material Parameter Calibration

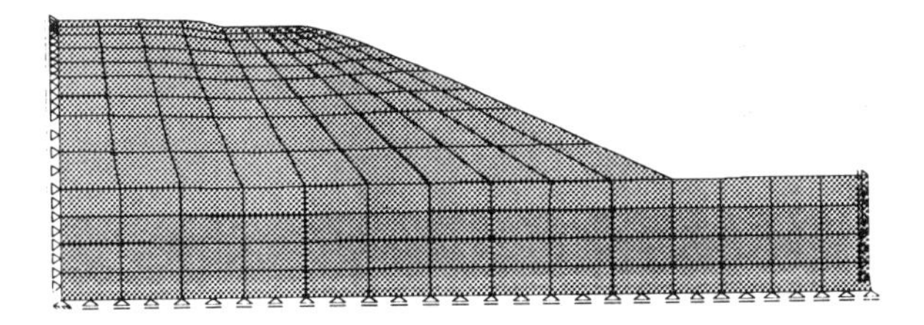

FIG. 5. Original and Deformed Meshes of Slope Specimen at Post-Peak Stress State

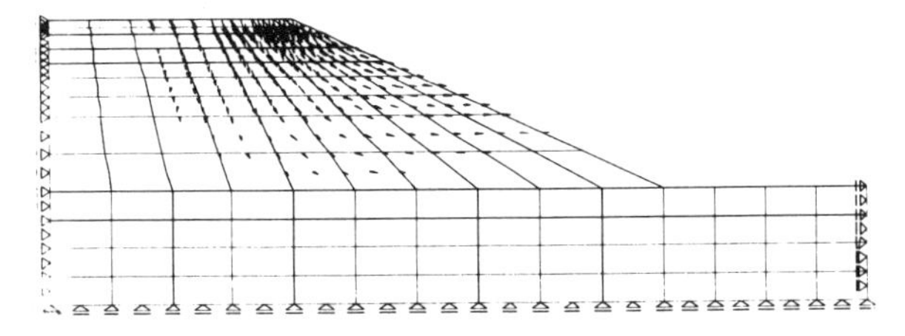

FIG. 6. Original Mesh and Magnitudes of Incremental Nodal Displacements for the Slope Specimen at Post-Peak Stress State

MODEL TESTS

To be useful in practice, these analytical techniques require validation by comparison of computational results with observed soil performance. In this sense, there are few documented cases of measured load bearing capacities of spread footings on steep slopes. As an alternative to field-case histories, laboratory centrifuge tests on prototype scaled models can be used in the validation of analytical procedures (Znidarcic 1992; Bureau of Reclamation 1986).

The idea of using small-scale models to study a physical phenomenon is common in many fields of engineering, including aero-space and structural engineering. However, geotechnical scaled models have a serious lack of similitude because stress levels in the soil model do not match those in the full-scale prototype. Modeling soil stress conditions is important because soil stress-strain and strength characteristics are functions of the stress state. While small physical models of a prototype can duplicate its geometry, the requirement that the stress level be the same at the homologous points between the model and the prototype can be satisfied only if the unit weight of the soil is increased by the same factor by which the dimensions in the model are reduced. The centrifuge is the best tool to produce an artificially high gravity, making model material seem heavier.

The concept of centrifugal model testing as it pertains to footings on sand slopes is to expose a model of a full-scale prototype, built to a length scale of 1/n and keeping the same soil density by using the prototype soil, to an artificial gravity field. This is easily accomplished by spinning the model in a centrifuge to a gravity level that is n times higher than the Earth's gravity.

The centrifugal model deviates from the prototype in two ways: first, the similarity requirement for sand grain size is not fulfilled because the prototype sand is used in the model; and second, the boundary conditions imposed by confinement in a model container are different from the boundary conditions of the prototype. The concept of "modeling of models" is, therefore, applied to test the validity of the scaling relations and the significance of changing boundary conditions. If the model test results from tests conducted in the centrifuge at different scales representing a specific prototype are identical, then it can be concluded that the scaling relations are valid and that differing boundary conditions do not significantly influence the test results.

Many geotechnical problems related to slopes can be modeled in geotechnical centrifuges. In this work, the bearing capacity of spread footings on steep slopes was obtained using models scaled at 1/50, 1/66.7, and 1/100 of the prototype size.

To execute the model testing program, the centrifuge facility at the University of Colorado at Boulder and the previously selected sand were used, to perform 59 model tests representing 18 prototype footing and slope conditions, and scales (Gemperline 1983). For this study, only the results corresponding to plane strain conditions are analyzed and compared with numerical and classical predictions. Table 1 is a summary of the selected test conditions and soil characteristics. A minimum of two tests were conducted for each test condition.

From these tests the average bearing capacities recorded at each model scale are presented in Table 2, and the result of the "modeling of models" process is

TABLE 1. Model Test Summary

Test Condition No.	Prototype Characteristics		Scale	Model Characteristics		Soil Characteristics	
	Footing Width (m)	Footing Length (m)	$\left(\frac{1}{g}\right)$	Footing Width (cm)	Footing Length (cm)	ϕ_{triax} (Degrees)	Density (kN/m^3)
1	1.27	Inf.	1/50	2.54	15.00	43-45	15.87
2	1.27	Inf.	1/50	2.54	15.00	43-45	15.87
3	1.27	Inf.	1/66.7	1.91	15.00	43-45	15.87
4	1.27	Inf.	1/66.7	1.91	15.00	43-45	15.87
5	1.27	Inf.	1/100	1.27	15.00	43-45	15.87
6	1.27	Inf.	1/100	1.27	15.00	43-45	15.87

presented in Fig. 7. This figure, which presents the maximum, minimum and average centrifuge results, shows very good agreement between the 1/66.7 and 1/100 scales. However, test conditions corresponding to the 1/50 scale reveal scale effects which may be caused by the friction developed between the Plexiglass container and the sand, (Bureau of Reclamation 1986).

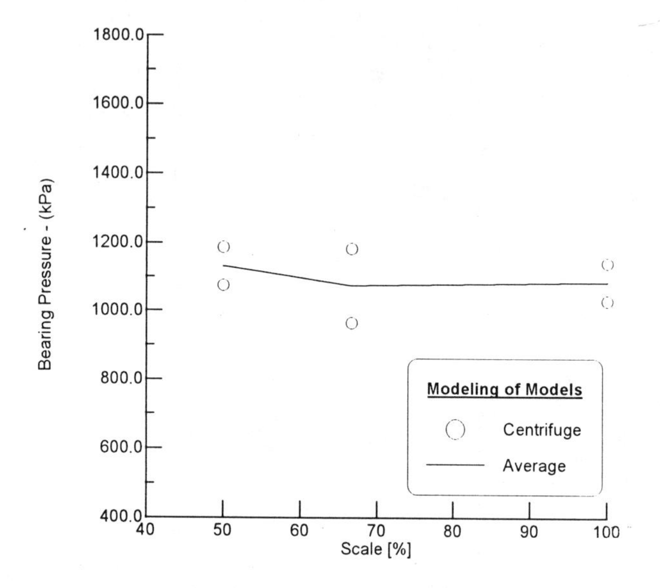

FIG. 7. Modeling of Models

TABLE 2. Average Centrifuge Results

Scale	Q (kPa)
1/50	1101.17
1/66.7	1072.92
1/100	1083.93

Comparison of Centrifuge and Analytical Results

Two main comparisons are made to check the validity of the classical bearing capacity equations and the numerical procedures. First, emphasis is placed on the approximate friction angle that allows each analytical method to predict the model test results with the least error within the range of $\phi = 42°$ to $\phi = 50°$. This is done because, as mentioned earlier, every method can be used to predict the maximum bearing pressure from the model tests by selecting an appropriate friction angle. The analytical method that predicts the observed maximum bearing pressure with the least error when using a friction angle in the plane strain range ($\phi = 48°$

to 50°) will undoubtedly be the best method for the conditions presented. For this purpose, the best-fit friction angle (ϕ_B) is defined as the friction angle which, when used with a given analytical method, yields the ultimate bearing capacity that predicts the average of the model test results with the lowest percentage error which is defined as

$$E = 100 \times \left[\frac{Q_u(\phi) - \bar{Q}}{\bar{Q}} \right] \qquad (6)$$

where $Q_u(\phi)$ is the ultimate bearing capacity computed for a given friction angle (ϕ) using an analytical method, and $\bar{Q}$ is the average of the maximum bearing pressures recorded in the centrifuge tests (Table 3).

TABLE 3. Best-fit Friction Angle and Lowest Average Percent Error

Analytical Method	ϕ_B (degrees)	E (%)	$\bar{Q}$ From centrifuge (kPa)	Q_u $\phi = \phi_B$ (kPa)
Bowles	42.0	4.4	1091.59	1139.47
Myslivec	47.2	1.3	1091.59	1077.23
Spencer	50.0	12.3	1091.59	957.54
Hansen	50.0	31.5	1091.59	746.88
Giroud	50.0	16.2	1091.59	914.45
Kusakabe	50.0	35.9	1091.59	699.00

From this table, the best agreement with model test results, in the plane strain range of friction angles, was achieved by the method recommended by Myslivec and Kysela. In this sense, we should also consider as good bearing capacity predictions the values obtained using the methods proposed by Giroud and Spencer. Although Bowles' method shows to be very sensitive to the friction angle, he explicitly recommends to use the friction angle determined from triaxial shear tests. In this sense, the bearing capacity value obtained using $\phi_{triax} = 42°$ is in very good agreement with the centrifuge results and should also be considered.

In second place, the load-displacement histories and peak stress values of the scaled centrifuge analyses were simulated using the finite element method. The spin effect was simulated by increasing the gravity load value, and the load history was again accomplished by subjecting the footing to a series of small displacements. In Figs. 8, 9, and 10 the load-displacement histories and maximum bearing pressures found by the finite element discretization are compared with the ultimate bearing capacities obtained by the centrifuge tests for the three selected scales. The predictions appear to be initially slightly stiffer than the centrifuge results. However, the peak stress values and in general the overall performance of the predictions are quite good.

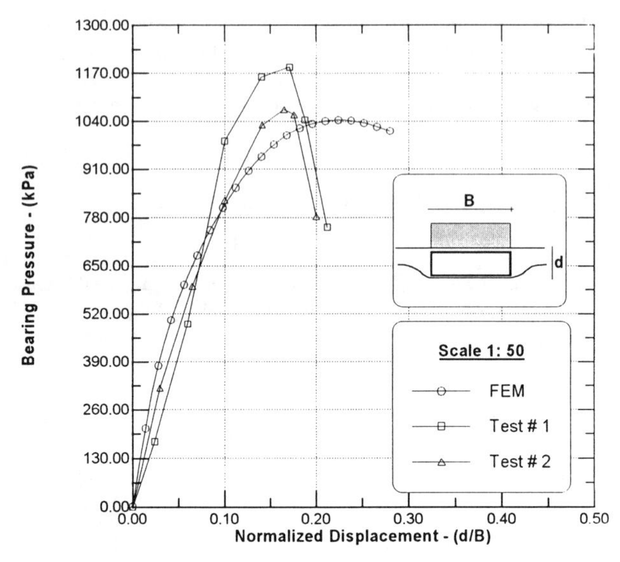

FIG. 8. Comparison between Centrifuge Results and FEM Prediction at a Scale of 1:50

CONCLUSIONS

In this work the methods most frequently used to predict bearing capacities of footings on steep slopes were studied and its advantages and limitations presented. Bearing capacities of footings representing conditions common to bridge foundations near canal side slopes were obtained using classical bearing capacity equations and advanced numerical procedures. The results were successfully compared with centrifuge analysis tests using scaled models. These results show the actual applicability of the classical bearing capacity equations, demonstrate the capabilities of the finite element method to predict the behavior of spread footings on steep slopes, and show the applicability of laboratory centrifuge tests in the validation of analytical procedures. Classical methods commonly predict the ultimate bearing capacity values while the other two techniques model the pre- and post-peak stress-strain behavior of the foundation. Evidently, a design that considers complete load-deformation response of a structure is desirable because large deformations can render a bridge unusable and in some cases cause localized damage that may threaten the integrity of the entire structure.

ACKNOWLEDGEMENTS

Support for this work was received from the Geomechanical, Geotechnical, and Geo-environmental Systems Program of the National Science Foundation under the Presidential Faculty Fellow Grant No. MSS-9253700. This support is gratefully acknowledged.

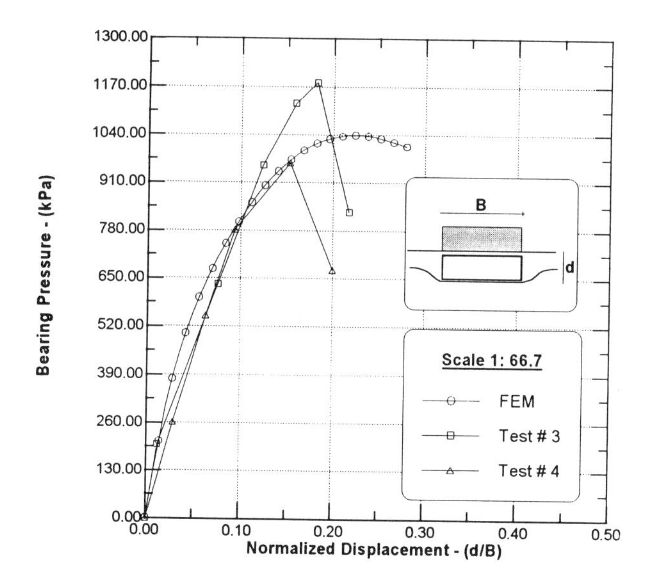

FIG. 9. Comparison between Centrifuge Results and FEM Prediction at a Scale of 1:66.7

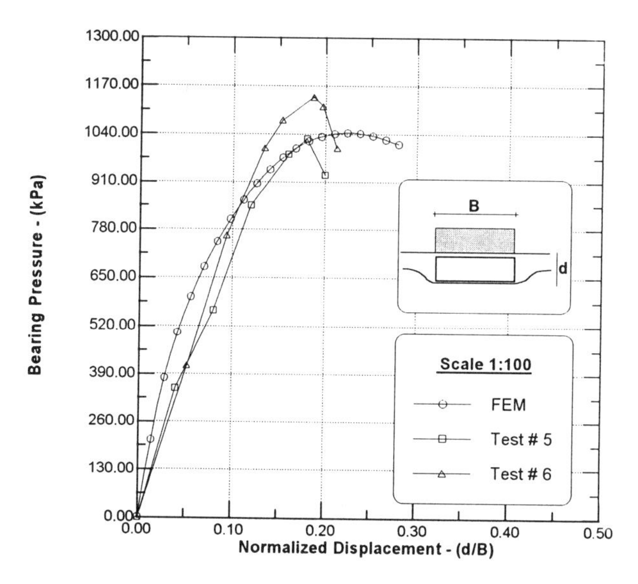

FIG. 10. Comparison between Centrifuge Results and FEM Prediction at a Scale of 1:100

APPENDIX I. REFERENCES

Arduino, P. (1992). "Elasto-plastic characterization of granular materials," M.S. thesis, University of Puerto Rico at Mayagüez, Puerto Rico.

Chen, W. F. (1975). *Limit Analysis and Soil Plasticity*, Developments in Geotechnical Engineering Vol. 7, Elsevier Scientific Publishing Company, Amsterdam, The Netherlands.

Chen, W. F. and McCarron, W. O. (1991). "Bearing capacity of shallow foundations." *Foundation Engineering Handbook*, 2nd Edition, Van Nostrand Reinhold, New York, New York, 144-166.

Bureau of Reclamation (1984). "Centrifugal model test for ultimate bearing capacity of footings on steep slopes in cohesionless soil," Engineering and Research Center, U. S. Department of the Interior.

Gemperline, M. C. (1983). "Centrifugal model tests for ultimate bearing capacity of footings on steep slopes in cohesionless soils," M.S. thesis, University of Colorado at Boulder.

Macari-Pasqualino, E. J., Runesson, K., and Sture, S. (1994). "Response prediction of granular materials at low effective stresses." *J. Geotech. Eng.*, ASCE, in press.

Zienkiewicz, O. C. and Taylor, R. L. (1991). *The Finite Element Method*, 4th Edition, Vol. 1 and Vol. 2, McGraw Hill, London, United Kingdom.

Znidarcic, D. (1992). "Application of centrifuge modeling technique to slopes and embankments." *Stability and Performance of Slopes and Embankments II*, Geotechnical Special Publication No. 31, ASCE, New York, New York, I, 521-537.

APPENDIX II. NOTATIONS

The following symbols are used in the paper:

$\mathbf{R}(\mathbf{u},eq)$	:	Residual force vector
$\mathbf{F}(\mathbf{u})$	:	Internal force vector
$\mathbf{F}(eq)$	:	Equivalent applied nodal loads vector
$\mathbf{B}$	:	Derivative operator matrix
$\mathbf{N}$	:	Shape functions matrix
$\tilde{f}^b$	:	Applied body forces vector
$\tilde{f}^s$	:	Applied surface tractions vector
$\tilde{f}^i$	:	Applied punctual forces vector
$\mathbf{D}$	:	Non-linear constitutive relation for soils
$\tilde{\sigma}$	:	Cauchy stress tensor
$\tilde{\epsilon}$	:	Infinitesimal strain tensor
$\mathbf{u}$	:	Displacement vector
$\mathbf{K}$	:	Stiffness tensor
$\mathbf{K}^t$	:	Tangent stiffness tensor
E	:	Bearing pressure error
$Q_u(\phi)$	:	Ultimate bearing capacity computed using analytical methods

$\bar{Q}$	:	Average maximum bearing pressure recorded in centrifuge tests
ϕ_B	:	Best-fit friction angle
ϕ_{triax}	:	Triaxial friction angle
ϕ_{pl}	:	Plane strain friction angle

NUMERICAL MODELING OF BRIDGE APPROACH SETTLEMENT WITH FOUNDATION-EMBANKMENT INTERACTION

Musharraf Zaman,[1] Member, ASCE, Dinesh Bhat,[2] Student Mem., ASCE and Joakim Laguros,[3] Fellow, ASCE

ABSTRACT: Bridge approach settlement is one of the major problems facing the transportation agencies responsible for road construction and maintenance in the United States and in many other countries of the world. Consolidation settlement of clayey foundation soils caused by embankment construction and the settlement of embankment itself due to its own weight and vehicular traffic load are few major contributors to resulting differential settlement between the bridge approach and the bridge abutment. This paper presents a *public domain interactive software* developed by the authors for the Oklahoma Department of Transportation (ODOT) for the analysis of bridge approach settlements. Nonlinear finite element method (FEM), including infinite element concept, is employed. Modified Cam-clay model is employed to idealize the nonlinear soil behavior. The software is found to be an efficient tool for the analysis of settlement potentials of new as well as existing bridge approaches. An attempt is made to investigate the interaction between the embankment and foundation with respect to a specific approach site in Oklahoma.

---

[1] Professor, School of Civil Engineering and Environmental Science, The University of Oklahoma, Norman, OK 73019.

[2] Graduate Research Assistant, School of Civil Engineering and Environmental Science, The University of Oklahoma, Norman, OK 73019.

[3] David Ross Boyd Professor, School of Civil Engineering and Environmental Science, The University of Oklahoma, Norman, OK 73019.

INTRODUCTION

Bridge approach settlement is one of the major problems facing the transportation agencies in the United States and in many other countries of the world. This problem is created from the differential settlement between a rigid bridge deck and a relatively flexible bridge approach. The excessive settlement, referred as ' bump' at the ends of the bridge, not only leads to an uncomfortable and unsafe riding surface, but also causes damage to the approach slab and creates excessive impact on the bridge structure. Consolidation settlement of the foundation soils due to the construction of the embankment and the embankment settlement due to its self weight and the vehicular traffic loads are the important factors that affect the approach settlement. An analysis procedure is employed in this study based on the nonlinear finite element method (FEM) to analyze the bridge approach settlement problem including the effects of foundation-embankment interaction.

The finite element (FE) model for foundation analysis is based on the Biot's theory of 3-D consolidation (Biot 1941). Plane strain idealization is used for simplicity and computational efficiency. Nonlinear behavior of the soil skeleton is represented by the modified Cam-clay constitutive model (Laguros et al. 1991; Zaman et al. 1991, 1993). The semi-infinite soil domain is divided into a near-field and a far-field. The near-field is discretized by using eight-noded isoparametric quadrilateral elements. The concept of infinite elements is incorporated (Rajapakse et al. 1985) into the FE model to idealize the semi-infinite soil domain of the far-field. The far-field is discretized by using the three-noded horizontal infinite elements. The boundary of the near-field is placed at a distance where displacements and stresses are small enough to be neglected. The FE formulation assumes that the soil medium is fully saturated below the water table and considers the behavior of soil skeleton deformation coupled with the pore pressure effects. The soil above the water table is considered to be unsaturated. Embankment (unsaturated) is discretized by using eight noded isoparametric quadrilateral elements and plane strain idealization. Linear elastic constitutive model is used in the FE model for embankment analysis. The model is presently being extended to account for non-linearity due to the embankment material.

A computer program was developed based on the aforementioned FE formulation. This public domain interactive software, developed by the authors, is available to the public through the Research Division of the Oklahoma Department of Transportation (Study #2188, ORA 125-6074).

The software is divided into four different modules: (i) the preprocessor for facilitating input of data in an efficient manner and for generation of FE mesh and boundary conditions data; (ii) the data convertor for converting the preprocessor data in compatible format for FE analysis; (iii) the FE main program for the analysis of foundation and embankment; and (iv) the post-processor for graphical interpretation of the results from foundation and embankment analysis.

A bridge site in Oklahoma is selected and the approach settlement histories are predicted using the developed software. Field and laboratory tests are conducted at this site to obtain the material parameters used in the analysis. The major field tests conducted are the Standard Penetration Test and the Cone Penetration Test.

The major laboratory tests include consolidation, triaxial compression, permeability, gradation, Proctor density, and unit weight. A section close to the bridge abutment and perpendicular to the direction of the traffic is idealized for the analysis. In the vertical direction, the FE mesh is extended down to the bed rock, where no vertical deformation and no water flow boundary conditions are imposed. In order to investigate the interaction between the embankment and foundation soil layers, a combined finite element analysis of foundation and embankment is carried out first and compared with the results obtained from separate foundation and embankment analyses. The following nomenclature is used in the FE formulation.

NOMENCLATURE

c	= Cohesion of soil
C^{ep}_{ijkl}	= Elasto-plastic constitutive relation tensor;
δ_{ij}	= Kronecker's delta (=1 for i=j and =0 for i=j);
ε_{kl}	= Components of strain tensor;
$\overline{E}_c$	= Displacement function after Laplace transformation of infinite element governing differential equations;
E	= Young's modulus of soil;
e_o	= Initial void ratio of soil;
F_i	= Components of body forces;
ϕ	= Angle of internal friction of soil;
G	= Shear modulus of soil;
H_e	= Height of embankment completed in time = T_c days;
i&j	= Indices in tensor notation;
i,j	= Differentiation with respect to j;
[J]	= Jacobian transformation matrix;
k	= Coefficient of permeability;
k_{ij}	= Permeability tensor;
$[K_1]$	= Stiffness matrix corresponding to soil skeleton;
$[K_2]$	= Matrix corresponding to the pore pressure;
$[K_3]$	= Coupled stiffness matrix between soil skeleton and pore pressure;
κ	= Slope of swelling curve (C_s) of soil;
λ	= Slope of consolidation curve (C_c) of soil;
ν	= Poisson's ratio of soil;
$L_j(\eta)$	= Lagrangian polynomial for node j in an infinite element;
M	= Slope of the failure line in q-p plane;
N_i	= Shape functions;
$\{p_m\}$	= Nodal pore pressure vector;
q_i	= Relative velocity of fluid;
$\{q_m\}$	= Nodal displacement vector;
$\overline{Q}$	= Prescribed fluid flow normal to vertical surface;
$\{\overline{Q}_m\}$	= Prescribed nodal fluid flux vector;
{r}	= Vectors of displacement and pore pressure;
$\{r_p\}$	= Global pore pressure vector;

$\{r_q\}$ = Global displacement vector;
r_* = Real number greater than highest real part of singularities of the Laplace transform of infinite element governing differential equation;
$\{R\}$ = Global load vector ;
S_{ij} = Deviatoric stress tensor;
S^* = Displacement function;
S_c = Displacement function after Fourier transformation of infinite element governing differential equation;
σ_{ij} = Total stress tensor;
$\overline{S}_c$ = Displacement function after Laplace transformation of infinite element governing differential equation;
$\{T_i\}$ = Prescribed traction on horizontal surface;
$\{\overline{T}_m\}$ = Prescribed nodal traction vector;
u_i = Displacement vector;
x&y = Global coordinates;
ξ, η = Local coordinates;
ϵ_{ij} = Strain tensor;
ρ = Mass density of soil;
ρ_w = Mass density of pore fluid;
$*$ = Represents convolution.

FE FORMULATION

The following assumptions were made in the development of the algorithm:

(1) The approach foundation consolidation occurs only due to construction of embankment.

(2) The embankment loading on the foundation is treated as a step loading as depicted in Fig. 1.

(3) The modified Cam-clay model (Roscoe et al. 1963; Schofield et al. 1968) is used to model the nonlinear and inelastic behavior of approach foundation soil matrix.

(4) Stress-strain response of the embankment materials is considered elastic. Study is currently in progress to account for non-linearity of the embankment materials.

(5) The small deformation theory is used in the development of the FE formulation. Geometric non-linearity is not taken into consideration.

(6) The present version of the FE formulation considers only the foundation soil (flow) anisotropy.

(7) Soil below the ground water table is completely saturated and that above unsaturated.

FE Formulation

The Biot's theory of 3-D consolidation as proposed by Sandhu and Wilson (1969) is used in the finite element formulation. Plain strain idealization is adopted for convenience and computational efficiency. Behavior of soil skeleton or matrix deformation coupled with pore pressure effects is expressed in terms of the force equilibrium as:

$$\sigma_{ij,j} + p_{,j}\delta_{ij} + \rho F_i = 0 \tag{1}$$

The flow equilibrium is written in the form

$$\left[K_{ij}(p_{,j} + \rho_w F_i) + \dot{u}_{i,i}\right] = 0 \tag{2}$$

The generalized Darcy's law for the flow behavior is given by

$$q_i = k_{ij}(p_{,j} + \rho_w F_j) \tag{3}$$

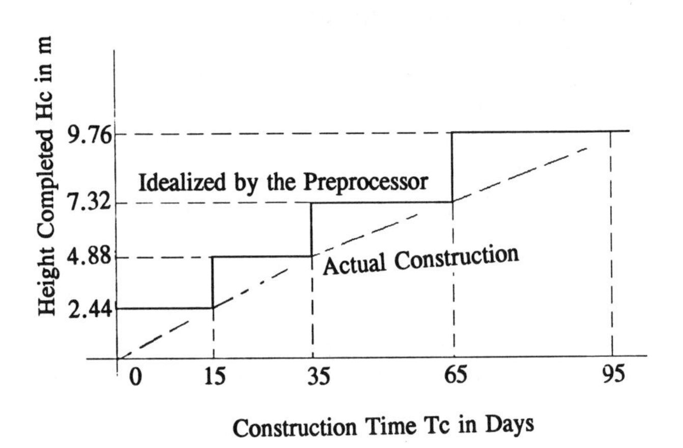

FIG. 1. Foundation Loading due to Embankment Construction

Modified Cam-Clay Model

In the FE algorithm developed, the modified Cam-Clay model is used to account for the elasto-plastic behavior of soil matrix. This model, which belongs to the family of the critical state models, provides a distinction between yielding and ultimate collapse by using the concept of the critical state line in conjunction with a strain dependent yield surface. According to the critical state model, a soil undergoing shear deformation can pass through a yield point without collapse and continue to deform until the critical state line is reached. At this point an ideal plasticity condition exists. Starting from two alternative assumptions regarding the dissipation of energy during plastic yielding, Roscoe, Schofield, Thurairajah and Worth (Roscoe et al. 1963; Schofield et al. 1968) proposed the "Cam-clay" model. Roscoe et al. (1963) extended this model and called it the "Modified cam-clay model". The modified cam-clay model generally fits the experimental data for clayey soils satisfactorily and is used in this study. The constitutive law for the soil skeleton can be written in an incremental form as (Desai and Abel 1972),

$$\Delta\sigma_{ij} = C^{ep}_{ijkl}\,\Delta\varepsilon_{kl} \tag{4}$$

The yield surface f_e is given by

$$f_c = p^2 - p_0 p + \frac{q^2}{M^2} = 0 \tag{5}$$

where

$$q = \sqrt{3J_{2D}} \tag{6a}$$

$$p = \frac{(\sigma'_{11} + \sigma'_{22} + \sigma'_{33})}{3} \tag{6b}$$

$$J_{2D} = \frac{1}{2} S_{ij} S_{ij} \tag{6c}$$

In the above equations, J_{2D} is the second invariant of the deviatoric stress tensor. The following functional is used to derive the FE equations for the analysis of consolidation of a linearly elastic soil skeleton and incompressible pore fluid:

$$\Omega_{(u,p)} = \int_v \left[\frac{1}{2}\overline{\sigma_{ij}}^{*}\varepsilon_{ij} - \rho F_i^{*} u_i + p^{*} u_{i,i} - \frac{1}{2} g^{*} q_i^{*} (p_{,i} + \rho_w F_i)\right] dV - \int_{s1} (\overline{T_i^{*}} u_{i,i}) dS + \int_{s2} (g^{*} \overline{Q^{*}} p) dS \tag{7}$$

Using the standard steps (e.g., the Galerkin method of weighted residuals) of FE formulation, the final expression can be written in matrix form as:

$$\begin{bmatrix} [K_1] & [K_3] \\ [K_3]^T & -\alpha\Delta t[K_2] \end{bmatrix} \begin{Bmatrix} r_q(t_n) \\ r_p(t_n) \end{Bmatrix} = \begin{Bmatrix} R_Q(t_n) \\ R_P(t_n) \end{Bmatrix} \tag{8}$$

Further details on derivation of the constitutive relation and the FE formulation are given by Gopalasingam (1989) and Zaman et al. (1993).

Type of Elements and Displacement Field

Fig. 2 represents the node and element numbering system adopted in this study for finite and infinite elements including the degrees of freedom at different nodes. In order to ensure the same degree of contraction and expansion for effective stresses as well as pore pressures, the nodal displacements are considered as primary unknowns at all the eight nodes while pore pressures are considered to be unknowns at the corner nodes only in near field elements. The displacement functions (u,v) for the eight noded quadrilateral elements are given by

$$u(\xi,\eta) = C_1 + C_2\xi + C_3\eta + C_4\xi\eta + C_5\xi^2 + C_6\eta^2 + C_7\xi^2\eta + C_8\xi\eta^2 \tag{9a}$$

$$v(\xi,\eta) \quad = \quad C_1 + C_2\xi + C_3\eta + C_4\xi\eta + C_5\xi^2 + C_6\eta^2 + C_7\xi^2\eta + C_8\xi\eta^2 \qquad (9b)$$

where

$$u \quad = \quad \sum_{i=1}^{8} N_i(\xi,\eta)u_i \quad = \quad [N_1, N_2,\ldots,N_8]\,[u_1\ u_2\ldots u_8]^T \qquad (10a)$$

$$v \quad = \quad \sum_{i=1}^{8} N_i(\xi,\eta)v_i \quad = \quad [N_1, N_2, \ldots, N_8][v_1\ v_2\ldots v_8]^T \qquad (10b)$$

Substituting the expressions for nodal displacements from Eqs. (10a) and (10b) into Eqs. (9a) and (9b), the values of C_i can be obtained. Pore pressure at a particular node is given by the expression

$$p(\xi,\eta) \quad = \quad C_1 + C_2\xi + C_3\eta + C_4\xi\eta \qquad (11)$$

and Eq. (11) can be written in a convenient form as

$$p \quad = \quad \sum_{i=1}^{i} N_i(\xi,\eta)p_i \quad = \quad [N_1\ N_2\ \ldots\ N_4]\,[p_1\ p_2\ \ldots\ p_4]^T \qquad (12)$$

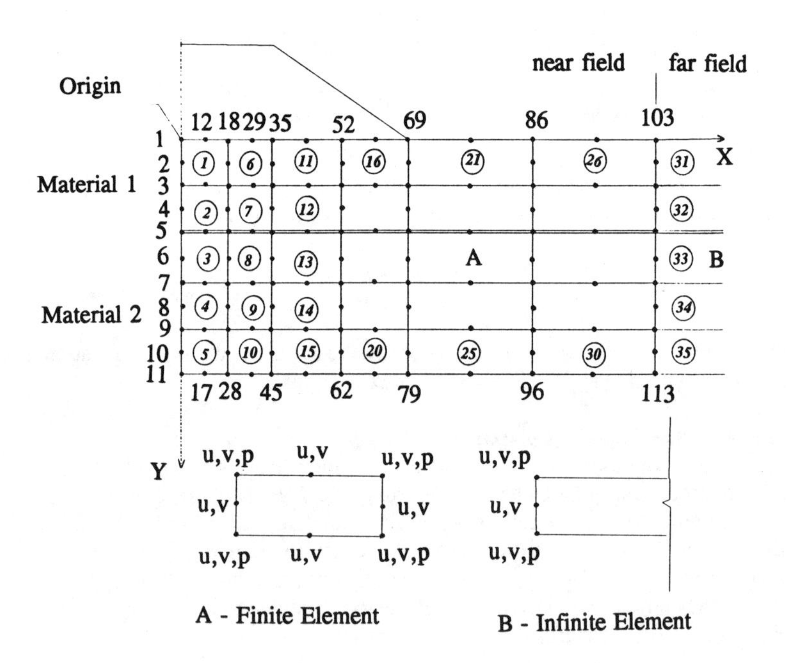

FIG. 2. Node Numbering System and Types of Elements Used

Infinite Elements

The soil domain is considered to be a semi-infinite mass. Finite elements used in discretizing the near field do not usually adequately simulate the semi infinite nature of soil mass. Therefore, the concept of infinite elements are used to idealize the far field behavior as shown in Fig. 2. The interpolation functionals selected for the infinite element must decay monotonically with distance and must approach zero at infinity. The infinite element algorithm presented by Rajapakse et al. (1985) is used in this study. This algorithm, called 'Finite Elements by Singular Contraction', is employed to model the elasto-static far-field behavior of an infinite soil domain. Fig. 3 represents the singular contraction of an infinite element. The following equations are used in the singular contraction algorithm

$$x = \sum_{j=1}^{3} \frac{2}{(1-\xi)} L_j(\eta) x_j \,, \qquad y = \sum_{j=1}^{3} L_j(\eta) y_j \tag{13}$$

Nonlinear Behavior

The stress-strain relation can be expressed as

$$[\sigma] = [C]\,[\varepsilon^e] = [C]\,([\varepsilon] - [\varepsilon_p])$$

Since only the strain energy portion is affected by the stress-strain, the strain energy part in the functional of Eq. (7) is changed to account for the non-linearity (Gopalasingam 1989). This can be rewritten as

$$\begin{aligned} U = {} & \frac{1}{2}\int\!\!\int\!\!\int_v [\varepsilon]^T[C][\varepsilon]dV - \int\!\!\int\!\!\int_v [\varepsilon]^T[C][\varepsilon^p]dV \\ & + \frac{1}{2}\int\!\!\int\!\!\int_v [\varepsilon^p]^T[C][\varepsilon^p]dV \end{aligned} \tag{15}$$

According to the initial strain method used here, the plastic strain can be taken into consideration by adding a load term $[Q_0]$, to the stiffness equation. The residual load terms are calculated for the contribution of plastic strains at different time steps after the application of external load increment. The modified governing equation including material non-linearity becomes

$$\begin{bmatrix} [K_1] & [K_3] \\ [K_3]^T & -\alpha\Delta t[K_2] \end{bmatrix} \begin{Bmatrix} r_q(t_n) \\ r_p(t_n) \end{Bmatrix} = \begin{Bmatrix} R_Q(t_n) + Q^p_{0n} \\ R_p(t_n) \end{Bmatrix} \tag{16}$$

The same algorithm is used for embankment analysis with a slight modification. The pore water pressure terms are not considered. Therefore, the effect of pore fluid and coupling effects between soil skeleton and pore fluid terms are considered as zero. Since the embankment is above the ground, the embankment soil is treated as unsaturated.

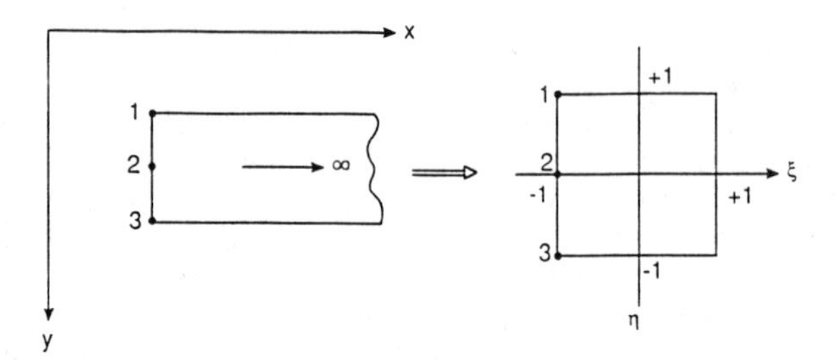

FIG. 3. Singular Contraction of Infinite Elements

THE SOFTWARE

Fig. 4 represents the flow-chart showing the sequential operation of different modules of the software for the analysis of a given bridge approach site. The preprocessor is developed basically for the preparation of data for the finite element analysis which otherwise is very time consuming and prone to manual errors. This user-friendly software was developed using a WINDOW-based software ACTOR IV for the on-screen display of the cross-section and FE mesh. The information required for the preprocessor include the geometric details of the cross-section, the material properties of all soil layers, and the construction history of embankment as well as vehicle loading on the embankment itself.

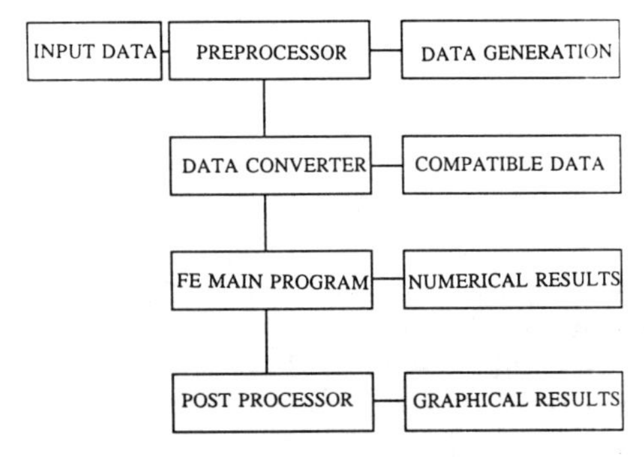

FIG. 4. Flowchart Showing Software Operations

The Data converter was developed in FORTRAN 77 in order to separate the data files for foundation and embankment analysis in the FE Main program in a compatible format. The FE Main program was also developed in FORTRAN 77 which runs both in WINDOW as well as DOS environments. Fig. 5 indicates the organization of subroutines called by the FE Main program in a sequential order. The FE Main program also prepares data files for graphical interpretation of results by the Post-processor. The Post-processor was developed using a software package called "Graphic 6.0". For different plots, customized programs are written in "C", which call several in-built subroutines from the "Graphic 6.0" software library. The results presented in this paper are obtained from the Post-processor.

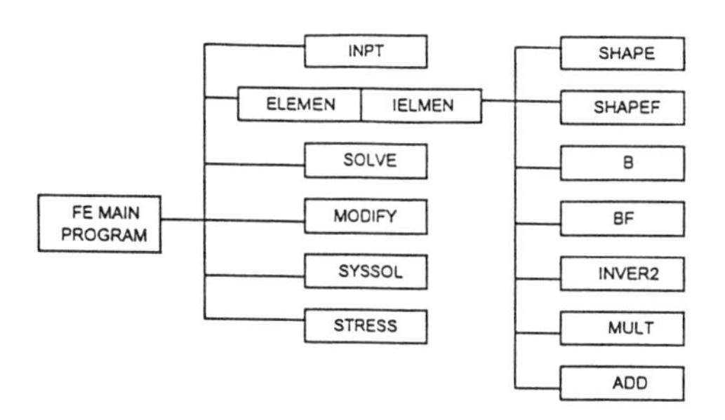

FIG. 5. Subroutine Organization of the FE Main Program

RESULTS AND DISCUSSION

To demonstrate an application of the software, a bridge approach site in Oklahoma was analyzed and the results are presented in this section. The bridge # 10-35 × 0736 on highway US142 located in Carter County, Oklahoma, which was built about 33 years ago, has experienced a total settlement of 25 cm. (10 in.). Fig. 6 shows the cross-section of the approach with soil properties of different layers. In order to study the effect of mesh size on the results, two different foundation meshes were selected: a coarser mesh with 163 nodes and 48 elements; and a finer mesh with 33 nodes and 102 elements. Based on the field and laboratory test results, the foundation soil domain was discretized by one clayey layer of 3.66 m. (12 ft.) thick, while, the 9.16 m. (32 ft.) thick embankment was discretized by three layers, the top 0.61 m. (2 ft.) being pavement and the remaining two clayey soil layers. The embankment mesh was idealized by 199 nodes and 56 elements. In order to study the interaction between the foundation and the embankment soils, an analysis was carried out where the embankment and the foundation were idealized as one mesh. In the earlier embankment analysis, fixed boundary conditions were enforced at the bottom most nodes although it is not the case in reality. In the latter analysis the relative movement of these nodes with the top most nodes of the foundation were given due consideration thus representing the actual field conditions better.

The Foundation Analysis

Figs. 7a and 7b represent the settlement histories during various time periods for the coarse and the fine mesh, respectively. Both meshes yielded same maximum settlements. However, the finer mesh gave a smooth curve due to a lower aspect ratio. A maximum settlement of 4.5 cm. (0.15 ft.) was observed compared to a maximum total settlement of 25 cm. (0.83 ft.) constituting 18% of the total settlement. This demonstrates that for all practical purposes the coarser mesh, which is economic in terms of computer time, yields satisfactory results. Figs. 8a and 8b, respectively, represent excess pore pressure distribution for the coarser and the finer meshes. Both the meshes yielded similar results except that in case of the finer mesh, primary consolidation was completed at the end of 180 days while in case of coarser mesh it took only about 120 days, and the curves were smooth.

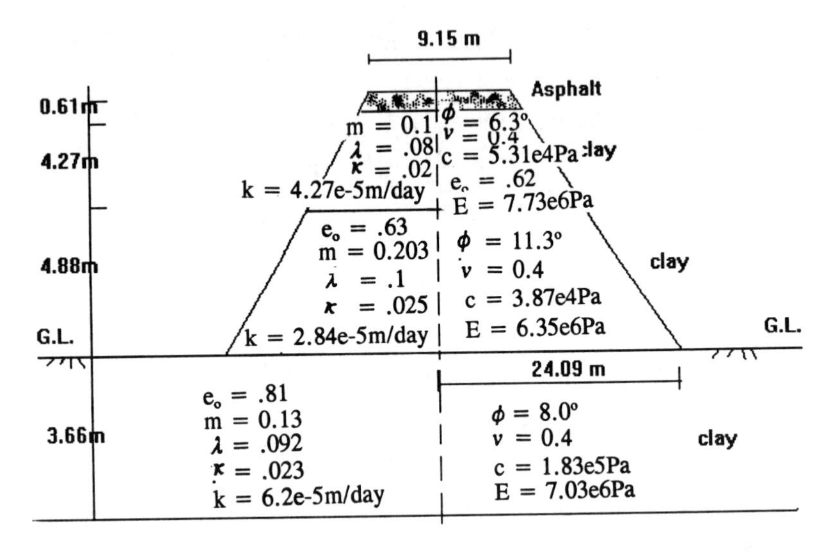

FIG. 6. Cross-Section of the Approach Showing Material Properties

The Embankment Analysis

Figs. 9a and 9b represent the surface settlement profile and settlement along the depth of the embankment, respectively. A maximum of 12.3 cm. (0.41 ft.), constituting 49.2% of the total settlement. It is interesting to note that due to higher Young's modulus value of the top most layer (pavement), the variation of settlement within this layer is negligibly small as compared to the variation in other two layers.

Foundation-Embankment Interaction Results

In order to compare the results from separate foundation and embankment analysis with that of the combined analysis, the results are plotted in same plot given in Fig. 10. As expected, the combined analysis leads to a higher settlement due to the embankment base movement. The difference between the two analyses was not very significant in this particular case as the foundation settlements were not very high. Where the consolidation settlement of the foundation is higher due to thicker soil layers, the combined analysis may give significantly higher settlements as compared to that of individual analyses of the foundation and the embankment. However, more study needs to be done to draw proper conclusions.

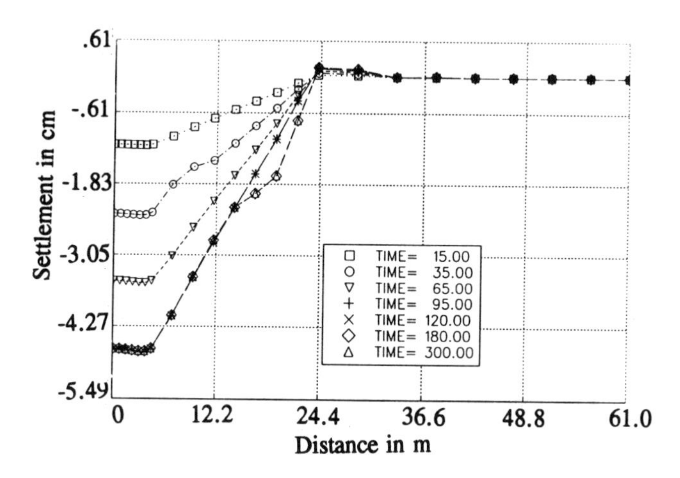

a. Coarser Mesh

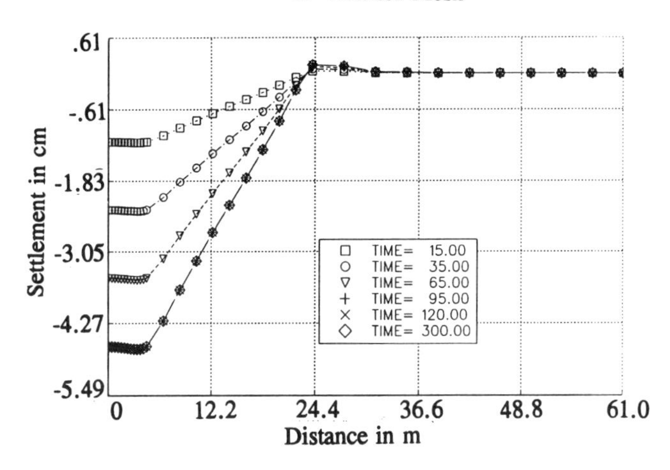

b. Finer Mesh

FIG. 7. Surface Settlement Histories

CONCLUSIONS

Based on the results of the above research, following conclusions were drawn:

1. The software is a useful tool for evaluating the settlement potential of existing as well as new bridge approaches.

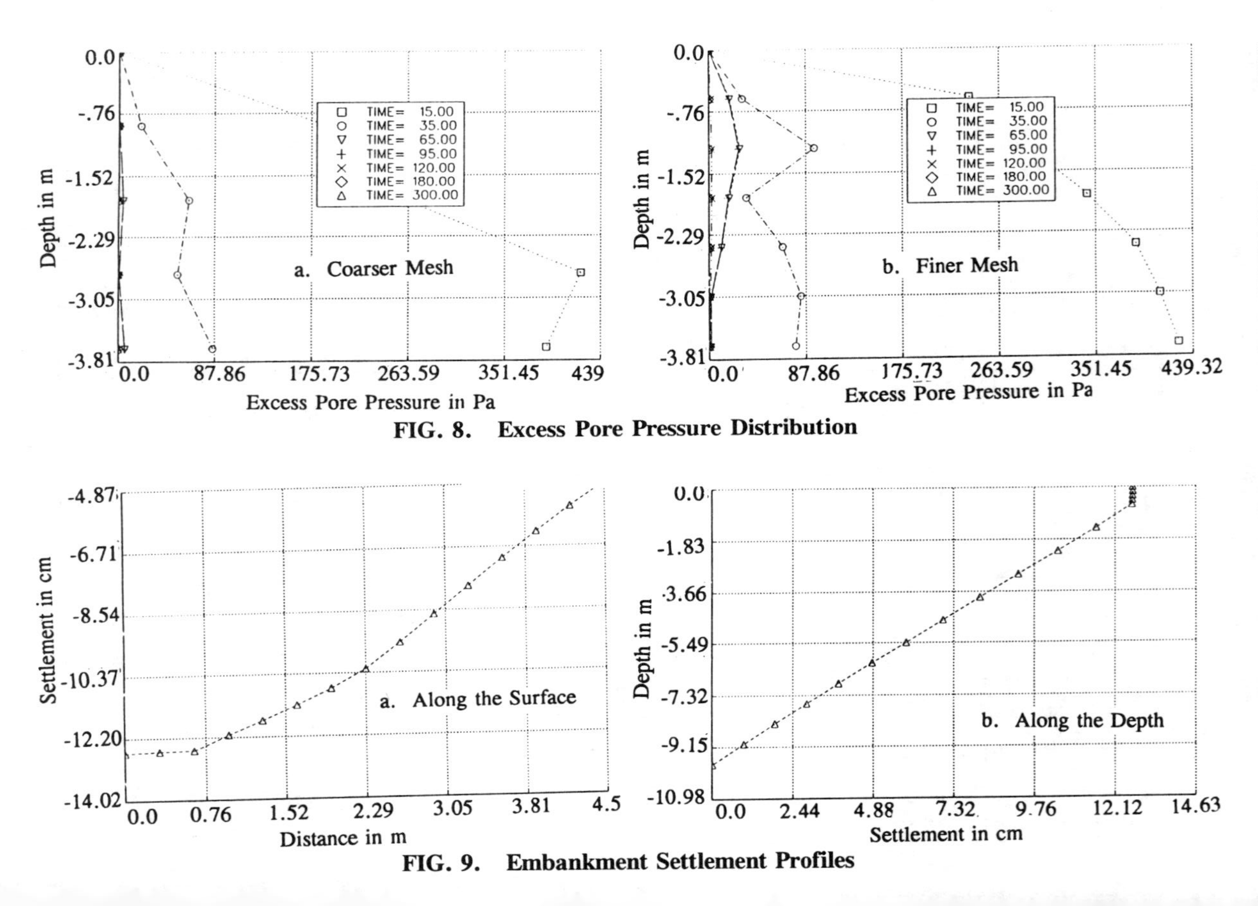

FIG. 8. Excess Pore Pressure Distribution

FIG. 9. Embankment Settlement Profiles

2. The User-friendly nature of the software helps in analyzing a given bridge approach settlement problem by an engineer/designer without an in-depth knowledge of FE formulation.
3. Incorporation of infinite elements helps in restricting the horizontal extent of the foundation soil domain, so that the software could be run on a microcomputer which otherwise would be difficult.

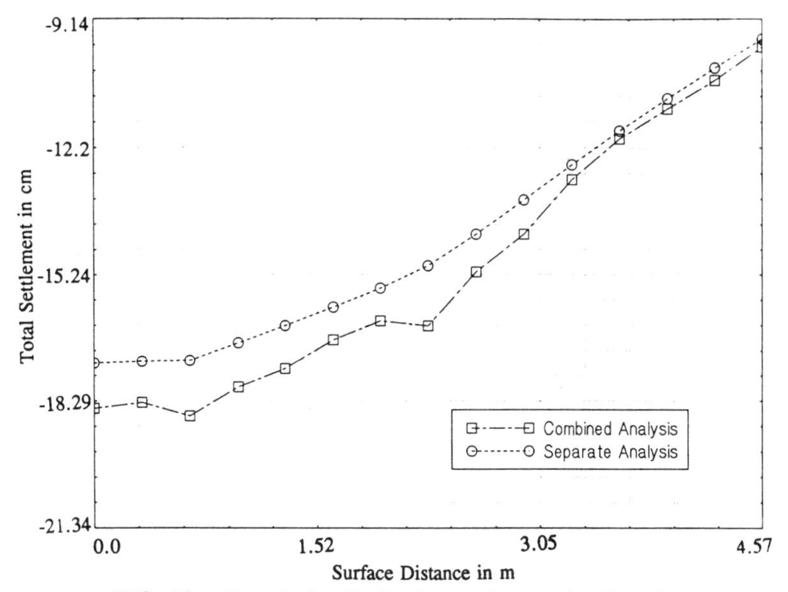

FIG. 10. Foundation Embankment Interaction Results

ACKNOWLEDGEMENTS

Financial support for this study was provided in part by the Oklahoma Department of Transportation (ODOT). Prof. C.S. Desai from University of Arizona provided the original FE code. Technical assistance of Dr. A. Alvappillai and Mr. A. Gopalasingam is appreciated. Special thanks are due to Dr. Jim Nevels, Materials Division, ODOT, Mr. Curt Hayes and Ms. Jennifer Martin, Research Division, ODOT.

APPENDIX - REFERENCES

Bhat, D., Zaman, M., and Laguros, J. G. (1993). "Computer aided analysis of nonlinear consolidation." *Proc. Computers in Engineering - 1993, ASME*, 515-523.

Biot, M. A. (1941). "General theory of three dimensional consolidation." *J. Applied Physics*, 12, 155-164.

Booker, J. R., and Small, J. C. (1975). "An investigation of stability of numerical solutions of Biot's equations of consolidation." *Int. J. Solids and Structures*, 151-172.

Desai, C. S., and Abel, J. F. (1972). *Introduction to Finite Element Method*, Van Nostrand Reinhold Company, New York, New York.

Hopkins, T. C. (1985). "Long term movements of highway bridge approach embankments and pavements," *Report UKTRP-85-12*, Kentucky Transportation Research Program, Lexington, Kentucky.

Gopalasingam, A., (1989), "Analysis of consolidation settlement of bridge approach foundation by nonlinear finite element method including infinite elements," M.S. Thesis, The University of Oklahoma, Norman.

Laguros, J. G., Zaman, M. M., Alvappillai, A., and Kyriakos, E. V. (1991). "Evaluation of causes of excessive settlements of pavements behind bridge abutments and their remedies - Phase III," *Progress Report*, The University of Oklahoma, Norman.

McNamee, J., and Gibson, R. E. (1960). "Displacement functions and linear transforms applied to diffusion through porous elastic media." *Q. J. Mech. Appl. Math.*, 13, 98-111.

Rajapakse, R. K. N. D., and Karasudhi, P. (1985). "Elastostatic infinite elements for layered half spaces." *J. Eng. Mech.*, ASCE, 111, 1144-1158.

Roscoe, K. H., Schofield, A. N., and Thurairajah, A. (1963). "Yielding of clays in states wetter than critical." *Géotechnique,* 13(3), 211-240.

Schofield, A. N., and Worth, C. P. (1968). *Critical State Soil Mechanics*, McGraw Hill, London, England.

Sandhu, R. S., and Wilson, E. L. (1969). "Finite element analysis of flow of saturated porous elastic media." *J. Eng. Mech. Div.*, Proc. ASCE, 95(EM3), 641-652.

Zaman, M. M., Gopalasingam, A., and Laguros, J. G. (1991). "Consolidation settlement of bridge approach foundation." *J. Geotech. Eng.*, ASCE, 117(2), 219-240.

Zaman, M. M., Laguros, J. G., Bhat, D., and Hua, C. (1993). "Userfriendly software for prediction of bridge approach settlements." *Progress Report, No. ORA 125-6074, Study# 2188,* University of Oklahoma, Norman.

Zaman, M. M., Bhat, D., and Laguros, J. G. (1993). "Numerical evaluation of bridge approach settlement." *Proc. Impact of Computational Mechanics to Engineering Problems*, Sydney, 175-182.

CONSOLIDATION CHARACTERISTICS OF AN OFFSHORE CLAY DEPOSIT

J. A. R. Ortigao[1] and A. S. F. J. Sayao[2]

ABSTRACT: This paper summarizes the results from a comprehensive laboratory and in situ investigation of an offshore clay deposit in the northeastern coast of Brazil. Laboratory permeability data from oedometer and triaxial cell are presented and compared with in situ test data from piezocone and piezometers. The coefficient of consolidation obtained from various methods is compared with settlement analysis based on Asaoka's (1978) method.

INTRODUCTION

A comprehensive site investigation program was carried out at an offshore clay deposit in northeastern Brazil (Fig. 1), 2.7 km from the shoreline, where the water depth is about 10 m. An offshore breakwater was proposed to be constructed in stages to allow for partial drainage and consolidation.

A preliminary site investigation was carried out from a diving bell (Ortigao 1988) and included SPT's, field vane tests (FVT) and soil samples. Poor ground conditions were detected. A detailed offshore site investigation program was then carried out on board the drilling vessel Mariner (Ortigao et al. 1985) and included piezocone (CPTU), FVT's and undisturbed sampling. Triaxial and oedometer tests were carried out at the onshore laboratory.

The first loading stage was applied at the mudline for a period of 6 months. At the end of this period, a thorough reassessment of the foundation properties was carried out. The purpose of this paper is to summarize the results of this investigation.

---

[1] Associate Professor, Federal University of Rio de Janeiro, Rio de Janeiro, Brazil.

[2] Assistant Professor, Pontificial Catholic University, Rio de Janeiro, Brazil.

The behavior of the additional construction stages are out of the scope of this paper and will be described in forthcoming publications.

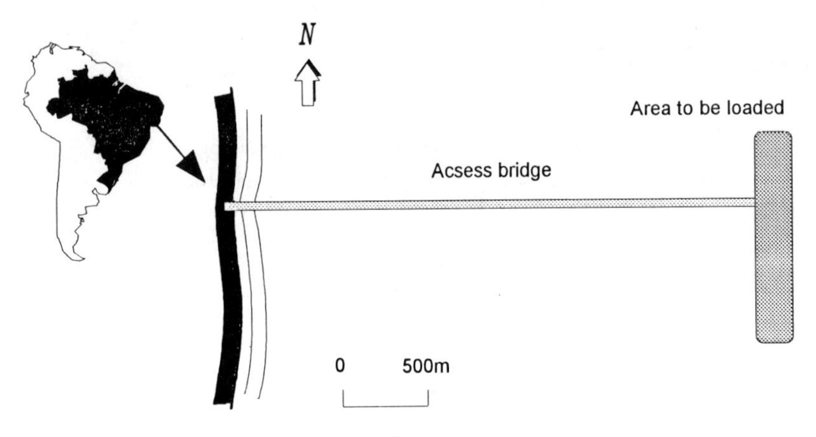

FIG. 1. Site Location

Soil Conditions

Soil profile at the offshore loading area is presented in Fig. 2. A first loading stage of 50 kPa was applied at the elevation -10 m (mudline). It was designed as a berm to stabilize further loading stages. Soil conditions consisted of a 4 m thick upper fine sand layer followed by a 8 m thick fairly homogeneous soft clay overlying dense clayey sand. The upper sand layer can be described by a friction angle of 30-32° and a total unit weight of 19 kN/m³.

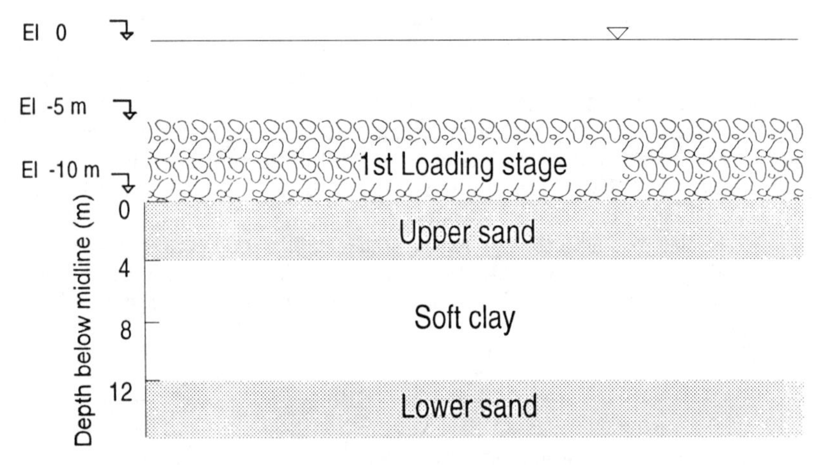

FIG. 2. Soil Profile

Clay properties are summarized in Fig. 3 and a CPTU profile is presented in Fig. 4. The clay is fairly homogeneous and presents a liquid limit LL of 60%, a plastic limit PL of 22%, resulting in a plastic index PI of 38%. The mean water content is 64%. The stress history of the deposit was evaluated through oedometer tests. It was concluded that the clay may be considered normally consolidated. Only at the top, the overconsolidation stress σ'_{vm} exceeds slightly the in situ overburden stress σ'_{vo}, as shown in Fig. 3. The unit weight of the soft clay is 17 kN/m^3 and its undrained strength c_u ranges from 15 to 20 kPa.

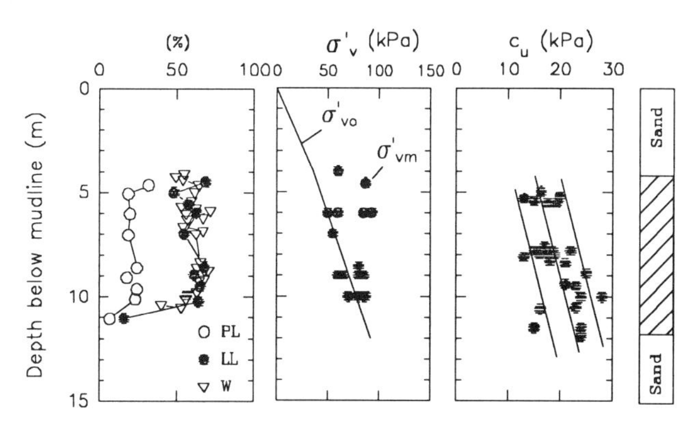

FIG. 3. Soft Clay Properties

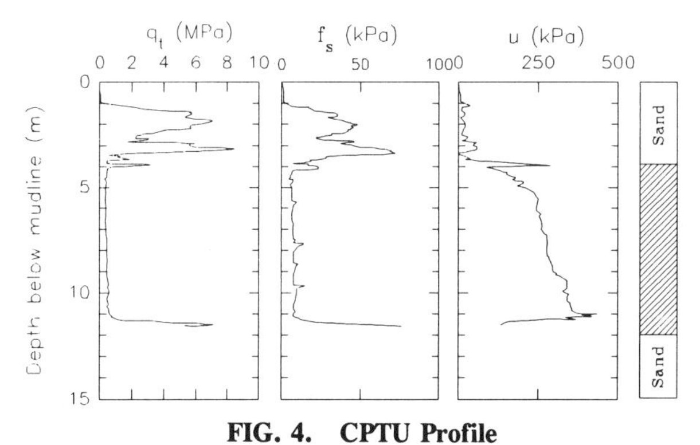

FIG. 4. CPTU Profile

INSTRUMENTATION

An access bridge was constructed prior to the placement of loading. Two open-ended 1.5 m diameter steel casings, named *instrument towers* (Fig. 5), were tipped into the mudline and enabled the instrumentation to be installed. They also provided protection to the instruments during the placement of loading. The towers were free to move vertically, therefore provided a practical way to monitor settlements.

Each instrumentation tower included an inclinometer casing, 4 Casagrande piezometers, one magnetic extensometer comprising 6 settlement detection devices (*spiders*) similar to the instrument reported by Campanella et al. (1994), and a pneumatic piezometer column. The piezometer column consisted of a 100 mm diameter steel pipe pile with piezometers attached flush to the outside. This instrument configuration has been used with success for porepressure monitoring around piles (e.g., Dias and Soares 1992) but, in this particular case, the results were very poor due to instrument malfunctioning.

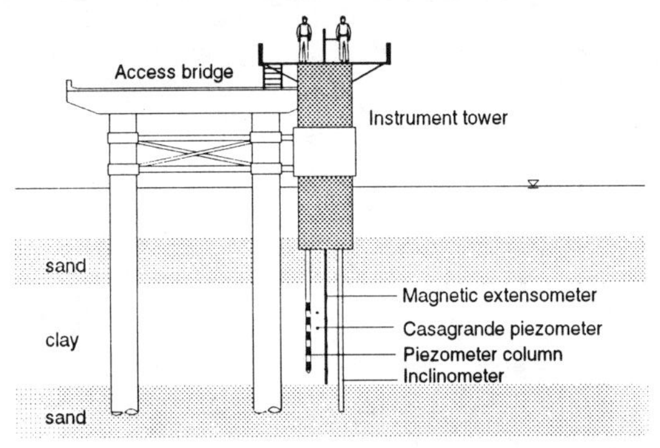

FIG. 5. Instrumentation

The Casagrande piezometers comprised a 32 mm diameter 600 mm long tips made of porous bronze. They were connected to EX drilling rods that enabled the unit to be pushed into clay. The access tubing consisted of a 12.5 mm ID plastic tubing protected inside the drilling rods.

Results of the instrumentation during the first loading stage are presented in Fig. 6. Surface settlements were about 180 mm over a period of 180 days.

Consolidation Properties of the Clay

Consolidation properties of the soft clay layer were assessed from several sources, namely: standard oedometer tests, a combination of in situ permeability with laboratory compressibility data, CPTU dissipation tests and analysis of settlement data based on Asaoka's method. Results are as follows:

• **Oedometer tests**: Oedometer tests on 75 mm undisturbed samples yielded the results shown in Fig. 7. The coefficient of compressibility C_c presents a mean value of 0.91 and the swelling index C_s is 0.09. The in situ void ratio ranges from 1.5 to 2. A linear regression was fitted trough all data points and gave:

$$e_0 = 2.14 - 0.06z \pm 0.28 \qquad (z = \text{depth in meters}) \qquad (1)$$

The coefficient of consolidation calculated from these tests ranged from 0.6 to 2 m²/year.

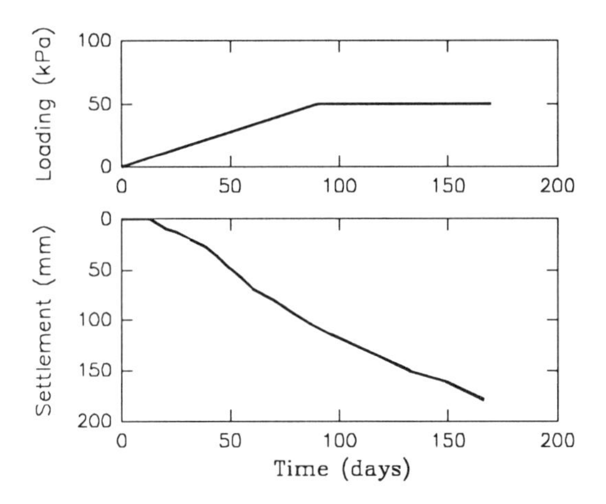

FIG. 6. Settlement Records

• **Combined method**: Values of the coefficient of consolidation were obtained by combining in situ permeability and laboratory compressibility through Eq. (2):

$$c_v = \frac{kM}{\gamma_w} \tag{2}$$

where k is the coefficient of permeability; γ_w is the unit weight of water; and M is the constrained modulus from oedometer data.

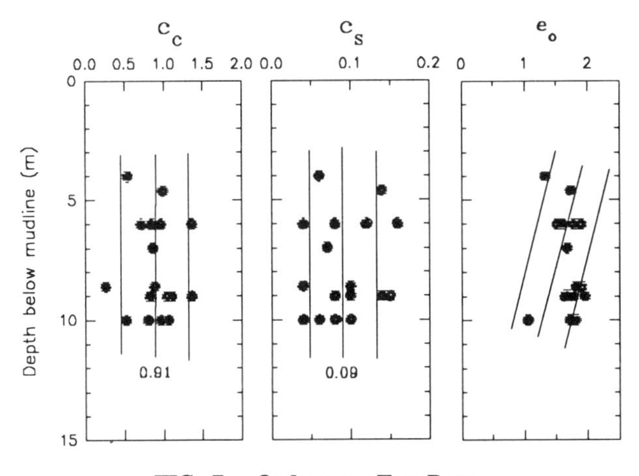

FIG. 7. Oedometer Test Data

Permeability was evaluated from in situ tests and compared to laboratory data. Falling head tests were carried out in all Casagrande piezometers after installation and a resting period of one week.

Direct measurement of permeability in the laboratory by falling head tests at the end of consolidation stages in the oedometer tests gave poor data. Typical results are shown in Fig. 8a. A considerable amount of scatter, probably due to leakage, may be seen. Selection of values for design was a difficult task, contrary to Tavenas et al. (1983) conclusions. Alternatively, more consistent results (Fig. 8b) were obtained in isotropic consolidation in the triaxial cell of 100 mm diameter by 100 mm high specimens backpressured to 200 kPa. Volume changes were measured both at the entrance and at the exit of the cell, providing a cross-check against leakage.

Eq. (3) was fitted to the data in Fig. 8b leading to:

$$\log k = \log k_0 + C_k(e - e_0) \tag{3}$$

where: $k_0 = 9\times10^{-10}$ m/s is the permeability calculated from the in situ void ratio e_0 and $C_k = 0.52$ is the slope of the fitted straight line, as described by Eq. (3).

Combining Eqs. (1) and (3), a relationship between initial permeability and depth can be obtained as a straight line (Fig. 9). This figure also includes data from in situ permeability tests for comparison. The agreement is remarkable, despite the fact that in situ measurements correspond predominantly to radial flow, opposite to the vertical flow in laboratory tests. This is an indication that the permeability ratio in vertical and horizontal directions is close to one.

The results of coefficient of consolidation from Eq. (2) led to values from 0.7 to 1.2 m²/year.

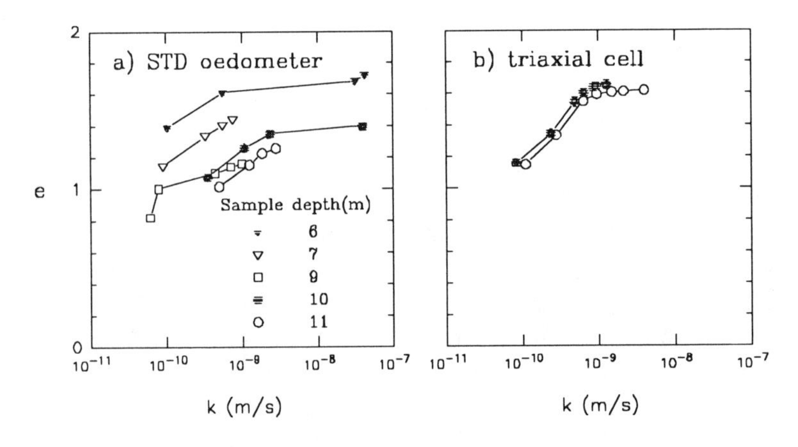

FIG. 8. Permeability Data

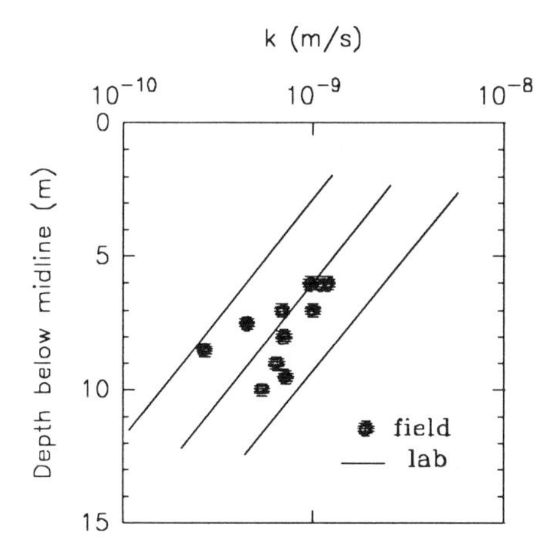

FIG. 9. Comparison between In Situ and Laboratory Permeability

- **CPTU**: dissipation test data are shown in Fig. 10. The coefficient of consolidation was estimated by Houlsby and Teh's (1988) theory, resulting in values that ranged from 3 to 10 m^2/year.
- **Field settlement data**: Asaoka's (1978) method was applied to field settlement data for the evaluation of final settlement and the coefficient of consolidation. This resulted in 0.6 m and 0.8 m^2/year, respectively.

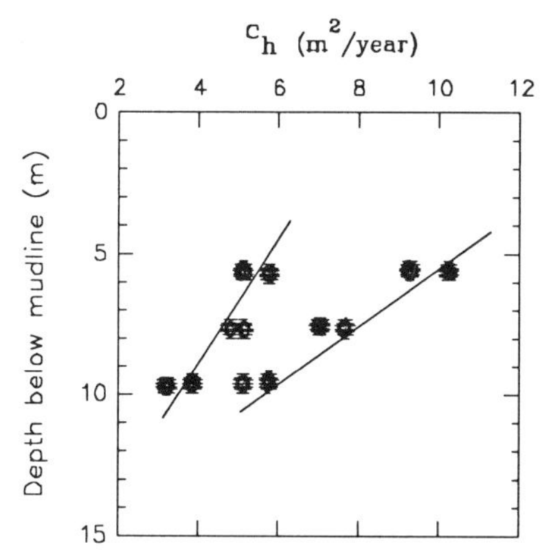

FIG. 10. Results from CPTU Dissipation Tests

Comparison of the Results

A comparison of coefficient of consolidation results is presented in Fig. 11. There is a fair agreement among three of the methods used, but the CPTU data seems to overpredict the coefficient of consolidation by a factor of 4 to 12.

Correction of CPTU data for the analysis of vertical settlement is frequently necessary to account for (a) *direction of flow*: radial consolidation taking place in CPTU dissipation; and (b) *stress level*: dissipation occurs in the recompression range, rather than in the normally consolidated range. Correction for direction of flow seems to be less necessary for this clay, as suggested by isotropy in the coefficient of permeability. For the stress level correction, Baligh and Levadoux (1986) proposed the following equation to transform c_h to normally consolidated conditions:

$$c_h(NC) = c_h(CPTU)\frac{C_s}{C_c} \tag{4}$$

where C_c is 0.91 and C_s is 0.09. A factor of 10 is obtained for the ratio of c_h(CPTU) to c_h(NC). This brings the CPTU data to an agreement with the results from the other three methods (Fig. 11). It should be noted that Eq. (4) gave a reasonable indication of the correction for this particular clay deposit. In many other cases, it may be rather difficult to evaluate the necessary magnitude of correction (Almeida and Ferreira 1992; Ortigao 1994).

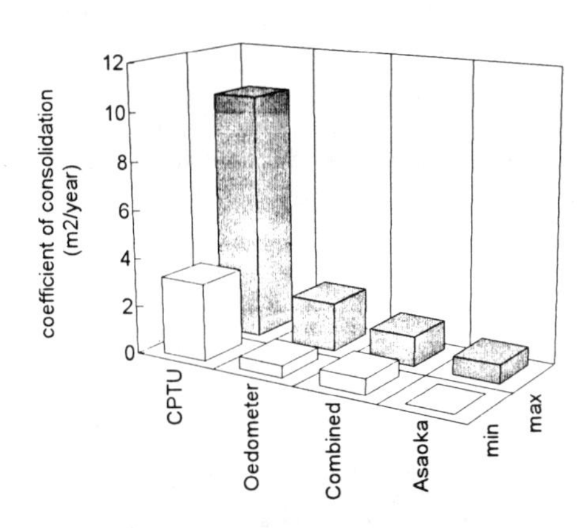

FIG. 11. Comparison between Values of Coefficient of Consolidation

CONCLUSIONS

Field observations during loading on an offshore soft clay deposit were carried out. The instrumentation comprised inclinometers, magnetic settlement gauges and Casagrande and pneumatic piezometers. This paper focused on the analyses of settlements and consolidation properties of the soft clay.

Settlement properties of the clay deposit were evaluated by a comprehensive site investigation program comprising in situ and laboratory tests. Direct measurement of permeability in the standard oedometer yielded poor data, as opposed to high quality data obtained from isotropic consolidation tests in the triaxial cell.

Analyses of the coefficient of consolidation obtained from various methods were presented and the results compared. A reasonable agreement was observed among the results from field data, laboratory tests and the combined method.

On the other hand, CPTU dissipation tests yielded very high values of the coefficient of consolidation and their application in settlement analyses requires correction. In this particular case, Baligh and Levadoux's (1986) equation seems to give a correct indication of the magnitude of the correction.

ACKNOWLEDGMENTS

The work reported herein was sponsored by Petrobrás SA and Construtora Norberto Odebrecht (Brazil). Geomecânica SA (Brazil) and Fugro Ltd (Holland) carried out laboratory and field investigations. Piezocone dissipation tests were conducted by the Federal University of Rio de Janeiro. The authors are indebted to Prof. Raymundo Costa, consultant to this project, and to Dr. B. Danziger for their support throughout the work. Also, the assistance by Lucia Alves and Alejandro Far during the preparation of this paper is appreciated.

APPENDIX - REFERENCES

Almeida, M. S. S. and Ferreira, C. A. M. (1992). "Consolidation parameters of a very soft clay from field, in situ tests and laboratory tests." *Proc. Wroth Memorial Symposium*, St Catherine's College, Oxford, England, 1-20.

Asaoka, A. (1978). "Observational procedure of settlement prediction." *Soils and Foundations*, 18(4), 87-101.

Baligh, M. M. and Levadoux, J. N. (1986). "Consolidation after undrained piezocone penetration II: interpretation." *J. Geotech. Eng.*, ASCE, 112, 727-745

Campanella, R. G., Ortigao, J. A. R., Crawford, C., and Jackson, S. (1994). "Design and installation of a new settlement-inclinometer device." *Vertical and Horizontal Deformations of Foundations and Embankments*, Geotechnical Special Publication No. 40, ASCE, New York, New York.

Dias, C. R. R. and Soares, M. M. (1992). "Instrumentation of a metallic test pile in soft clay." *Proc. ABMS Simp. on Field Geotechnical Instrumentation*, Rio de Janeiro, 1, 71-84 (in Portuguese).

Houlsby, G. and The, C. I. (1988). "Analysis of the piezocone test." *Proc. 1st Int'l. Symp. on Penetration Testing*, Orlando, 2, 777-783

Ortigao, J. A. R., Capellao, S. L. F., and Delamonica, L. (1985). "Marine site investigation and assessment of calcareous sand behavior at Campos basin, Brazil." *Proc. Int'l. Symp. on Offshore Engineering, Brazil Offshore '85*, Rio de Janeiro, Pentech Press, London, England, 238-255.

Ortigao, J. A. R. (1988). "Experience with field vane tests onshore and offshore." *Proc. ABMS Simp. on New Concepts in Laboratory and Field Tests*, Rio de Janeiro, 3, 157-180 (in Portuguese).

Ortigao, J. A. R. (1994). *Soil mechanics in the Light of Critical State Theories*, A. A. Balkema, Rotterdam, The Netherlands, in press.

Tavenas, F., Leblond, P., Jean, P., and Leroueil, S. (1983). "The permeability of natural soft clays. Part I: Methods of laboratory measurements." *Canadian Geotechnical Journal*, 20(4), 629-644.

THE ROLE OF SHALE PORES IN SETTLEMENT

Luis E. Vallejo,[1] Member, ASCE, Michael K. Robinson,[2] and Ann C. Stewart-Murphy[3]

ABSTRACT: Waste fills resulting from the mining of coal in Appalachia should ideally consist of large, free-draining sedimentary rock fragments. The successful performance of these embankments is directly related to the strength and durability of the individual rock fragments. When shale is the predominant rock in the fills, some shale fragments will degrade into soil size particles. This degradation is the result of slaking and point load crushing. Slaking takes place when the shales absorb water. Point load crushing is the result of gravity-induced loads that the individual fragments exert on each other at their points of contact. The slaked and crushed material fill the void spaces between the intact shale fragments. This material rearrangement causes settlement. A laboratory testing program, with point load and slake durability tests, as well as thin section examinations on sixty-eight shale samples from the Appalachian region, revealed that pore microgeometry of shales has a major influence on degradation. The smaller the pores, the more the shales slaked in water. Although not generally true for all the samples, the testing program indicated that the larger the pores, the lower the crushing strength under point loads. Thus, the size of the pores in shales appears to have a direct relationship to degradation and settlement.

---

[1] Asso. Prof., Dept. of Civil Engrg., Univ. of Pittsburgh, Pittsburgh, PA 15261; and Physical Scientist, Office of Surface Mining, Ten Parkway Center, Pittsburgh, PA 15220.

[2] Supervisory Physical Scientist, U. S. Office of Surface Mining, Ten Parkway Center, Pittsburgh, PA 15220.

[3] Mining Engineer, Office of Surface Mining, Ten Parkway Center, Pittsburgh, PA 15220.

INTRODUCTION

The successful performance of any embankment is related to the strength and durability of the individual rock fragments within the fill. The breakdown of shale into soil-size particles occurs because of: (1) compressive loads acting at the points of contact between fragments; and (2) slaking when the shale fragments absorb water. Shale degradation causes settlement of waste fills as illustrated in Fig. 1 (Sowers et al. 1965; Strohm 1978). Embankments for the storage of waste rock from the mining of coal (excess spoil fills) are composed of durable and non-durable shale fragments with void space between them (Fig. 1a). With time, and through the effect of pressures exerted by the overlying fill mass, slaking and crushing of the non-durable shale fragments fills these voids. This process is the main cause for settlement of shale embankments (Fig. 1b).

Prediction of settlement amounts should be based upon laboratory tests that simulate field conditions. For shale embankments, tests should simulate the crushing by point loads and slaking of rock fragments. The soaked compression test has been used to model the crushing and slaking processes. The soaked compression test was developed by Strohm (1968) for the design of shale highway embankments. Material passing a 1.91 cm (3/4 inch) screen is placed at a desired density in a 15.24 cm (6 inch) diameter mold. The material is surcharged by a load which represents the overburden load at the mid-height of the planned embankment (Fig. 1). When the sample height stabilizes under the surcharge, it is flooded with water and vertical compression is measured against time. When equilibrium is again reached, the final vertical compression of the sample is determined. Using this vertical compression, a percent compression of the specimen is calculated. This percent compression is then multiplied by the total height of the embankment in order to calculate its predicted settlement (Fig. 1b).

In order to understand the shale disintegration observed in the field and in soaked compression tests, a laboratory testing protocol was performed on sixty-eight shale samples from Kentucky, Tennessee, Virginia and West Virginia. The point load and the jar slake tests were chosen since they simulate the crushing and the slaking experienced by shale fragments. Thin section examination provided information on shale pore geometry and influence of this geometry on degradation mechanisms.

POINT LOAD TESTING

Point load testing involves the loading of a rock lump of average diameter, D, between two cone-shaped platens. The load, P, necessary for breaking the rock is measured. This load value, divided by the square of the diameter, D , provides the point load index, I_s (a measure of the tensile strength of the rock). I_s is standardized by conversion to a value which is the equivalent of testing a 50 mm diameter rock core (Broch and Franklin 1972). For this testing program, the I_s for each of the sixty-eight shale samples from the Appalachian region was calculated as the mean of twenty values of I_s determined from testing twenty lumps of the same rock sample. According to Oakland and Lovell (1972), twenty lumps is the minimum number needed to obtain a statistically representative value of I_s for one rock sample.

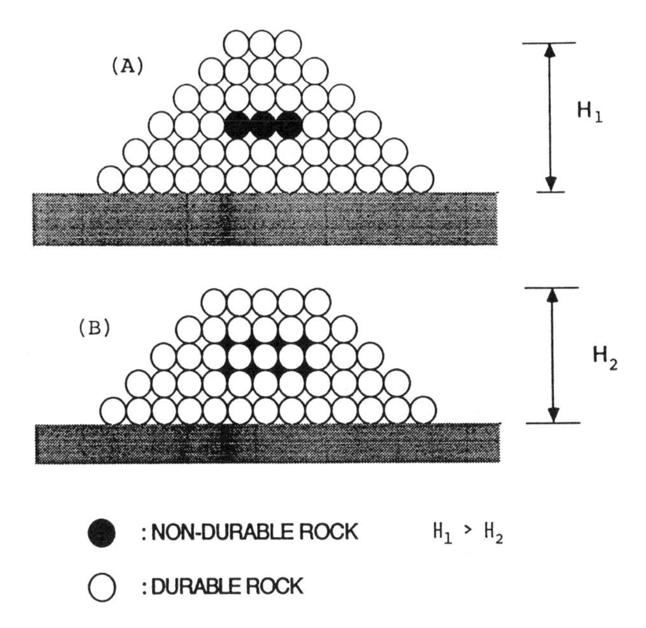

FIG. 1. Settlement in Waste Fills from Crushing and Slaking of Non-durable Rock Fragments

The shales were tested in the point load apparatus at two different water contents. The first water content was the as-received in the laboratory moisture content. The second water content corresponded to near saturated conditions. Near saturation was obtained by immersing the samples in water for a period of 24 hours. As expected, point load values decreased following the immersion of the samples. Due to space constraints, the recorded point load values and the measured moisture contents are not presented in this study. These values have been reported elsewhere (Office of Surface Mining 1992).

DURABILITY TESTS

The slake durability of the sixty-eight shale samples was calculated using the jar slake test. This test provides a qualitative measure of rock behavior after immersion in water for a period of 24 hours. This test has been recommended by Strohm (1978) as the basic screening test for non-durable shales, since it provides quick and inexpensive results. The test is very useful for a continuous characterization of shale durability. A jar slake index, I_j, developed by Lutton (1977), is determined from the appearance of samples after the soak test (Table 1).

TABLE 1. Jar Slake Ranking (Lutton 1977)

Jar Slake Index I_j	Behavior
1	Degrades to a pile of flakes or mud
2	Breaks rapidly and/or forms many chips
3	Breaks slowly and/or forms chips
4	Breaks rapidly and/or develop several fractures
5	Breaks slowly and/or develops few fractures
6	No change

The Lutton ranking system (Lutton 1977) tabulated in Table 1 was applied to the sixty-eight shale samples for this study. The full range of slaking behavior, from $I_j = 1$ to $I_j = 6$, was observed in the tested rock. Of the sixty-eight samples tested by the jar slake method, fourteen samples were rated at an I_j of either 1 or 2, four samples were rated at an I_j equal to 3, and fifty samples were rated at an I_j of either 5 or 6 (Office of Surface Mining 1992).

PORE GEOMETRY

To study the pore geometry of the sixty-eight shale samples, photographs were taken of the thin sections (30 μm in thickness) through a polarizing microscope at two different magnifications (25X and 63X). The cross-sectional areas and the shape of the macropore perimeters in the shales were determined from the photographs of the thin sections using a digitizer (Stanton 1987). Fig. 2 shows typical results of the process for 2 shale samples.

The "equivalent diameter" was computed for each macropore by determining the diameter of an equivalent circular area. From each photograph, forty representative macropores were chosen to calculate the overall average "equivalent diameter" for each shale sample. For example, the average "equivalent diameter" for the two shale samples in Fig. 2 are 0.18 mm and 0.1232 mm respectively.

RELATIONSHIP BETWEEN PORE GEOMETRY AND POINT LOAD STRENGTH

This study determined that the shale pore diameter appears to influence crushing. Point-load strength values were compared with macropore diameters for each shale sample. Since the water content has a major influence on the point load strength of rocks (Broch and Franklin 1972), samples with very similar water contents were chosen for correlation. Fig. 3a shows the relationship between point load index, I_s, and the overall average pore diameter, d, for five samples that have as-received water contents, w , ranging from 0.29 to 0.32%. Fig. 3b shows the relationship between I_s and d for samples that have saturated water contents ranging from 1.4% to 1.6%. Fig. 3a indicates that the larger the pore diameter, the smaller the corresponding point load strength. A linear regression analysis of the results shown in Fig. 3a indicates a high correlation (correlation coefficient R = 0.92) between I_s and d. However, the plot of Fig. 3b indicates that no correlation exists (R = 0.007). Even though only a limited number of samples indicate that pore diameter directly influences crushing, one would intuitively expect the fabric of the

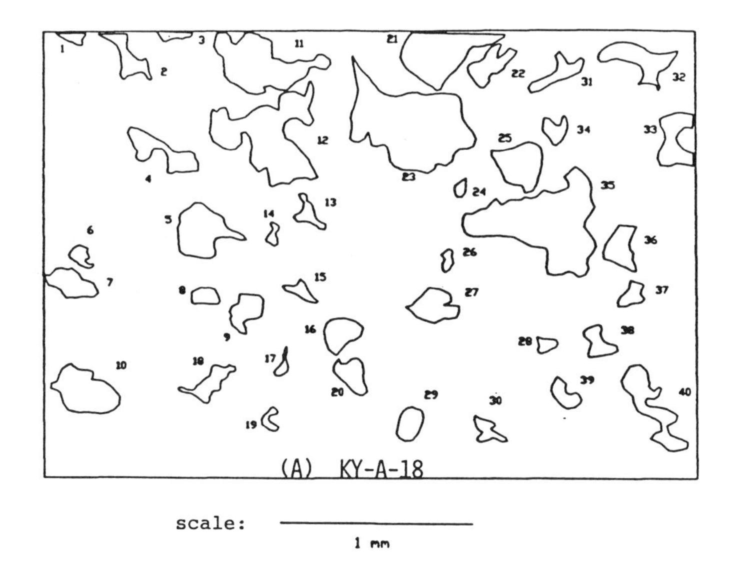

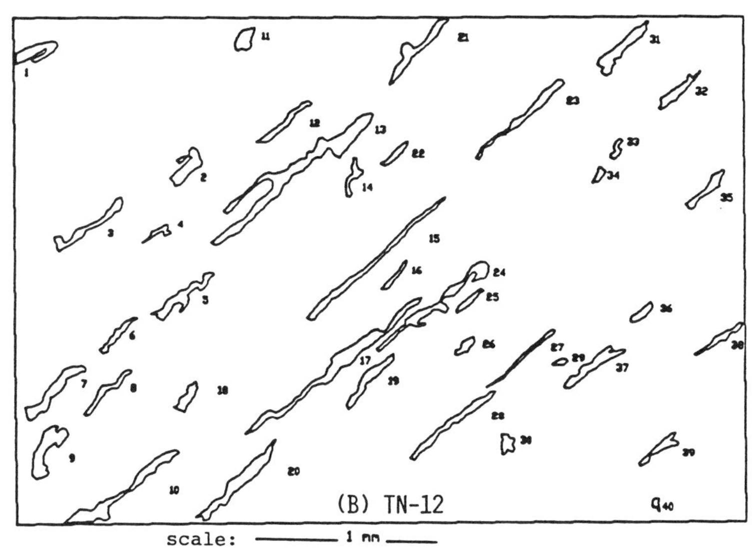

FIG. 2. Pore Geometry for Shale Samples KY-A-18 and TN-12

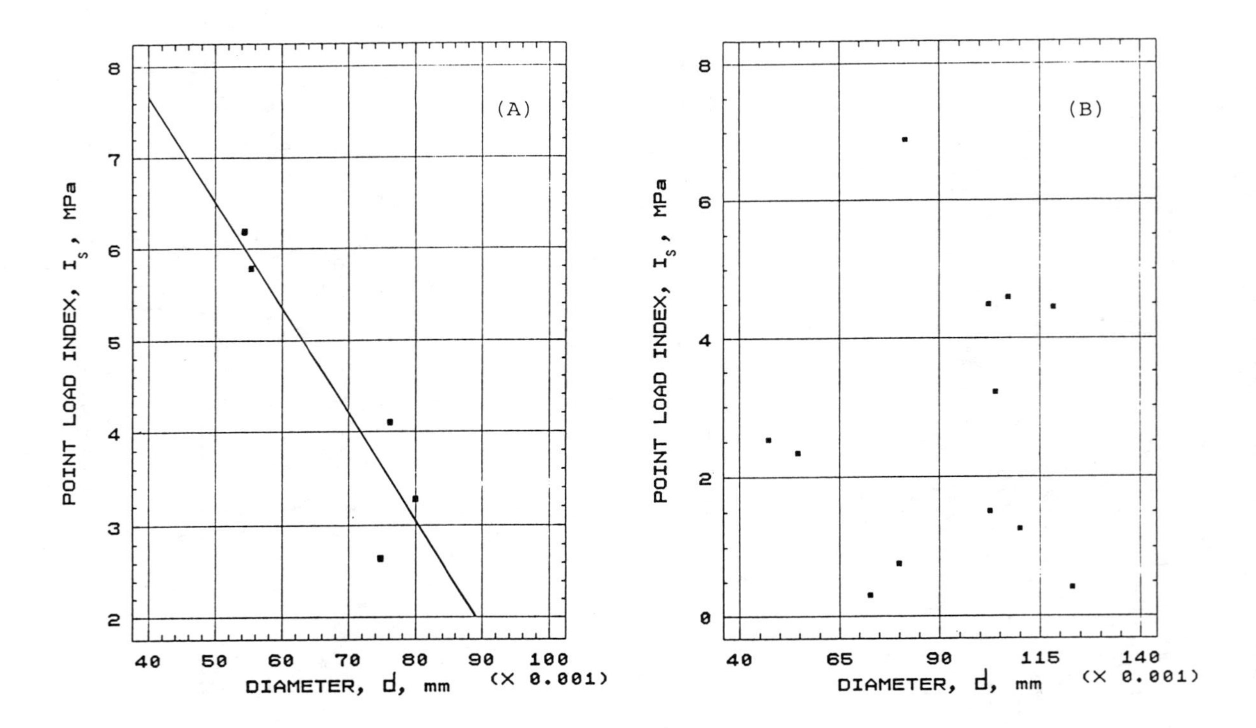

FIG. 3. Relationship between Point Load Index and Pore Diameter: (a) Kentucky Shales (w=0.29 to 0.32%); (b) Kentucky, Virginia and West Virginia Shales (w=1.4 to 1.6%)

shales as reflected by the pore structure to play a role. Reasons for the general lack of correlation can be attributed to differences of fabric and composition among the samples. X-ray diffraction analysis conducted by Vallejo et al. (1993) on the sixty-eight samples revealed that the mineral composition varied among the samples. The shape of the voids would also have an influence on point load strength. Eighty percent of the samples had voids with a vesicular shape (Fig. 2a). Twenty percent of the shales had voids resembling elongated cracks (Fig. 2b). Thus, shale composition, void shape, and average void diameter appear to influence the point load strength of shales.

Relationship between Pore Geometry and Degree of Slaking

It was also found in this study that a direct correlation exists between pore geometry and the slaking of the shales (Table 2). The slake durability of the sixty-eight shale samples (measured by the jar slake index I_j) was correlated with the average diameter of the shale macropores. Three natural groupings of shales were established with respect to durability from this comparison. The jar slake tests showed that fourteen samples completely disintegrated into soil-size particles (I_j = 1 or 2). Four samples developed little change other than small fractures and had an I_j of 3. Fifty shale samples experienced no degradation at all and had an I_j of 5 or 6.

An analysis of Table 2 indicates that disintegrated samples had a system of pores with average diameter equal to 0.06 mm. The samples that developed small fractures had pores with an average diameter equal to 0.07 mm. The fifty samples that experienced no disintegration at all had pores with an average diameter of 0.092 mm. Thus, the size of the pores in the shales apparently has a marked influence on shale durability (resistance to slaking) in the soaked compression test as well as within shale embankments of any type.

TABLE 2. Slaking Behavior and Pore Diameter

Number of Samples	Slake Index I_j	Maximum Diameter (mm)	Minimum Diameter (mm)	Standard Deviation	Mean Diameter (mm)
14	1 or 2	0.081	0.043	0.0125	0.060
4	3	0.082	0.050	0.0136	0.070
50	5 or 6	0.180	0.050	0.0300	0.092

Pressures Causing the Breaking of the Shales

From x-ray diffraction analysis of the sixty-eight shale samples used for this study, Vallejo et al. (1993) determined that the predominant clay mineral in the samples was kaolinite, a non-expansive clay. Shales with non-expansive clays slake because of air compression (Moriwaki 1974). When a shale is below the phreatic surface within a fill, it will absorb water as a result of capillary forces. This suction process is illustrated in Fig. 4, which represents a shale sample with a system of continuous pores. These pores can be assumed to resemble small, non-interacting cylindrical tubes. When under water, the water is pulled into the individual macropore as a result of capillary forces, and the air within the macropore will be subjected to compression (Fig. 4b).

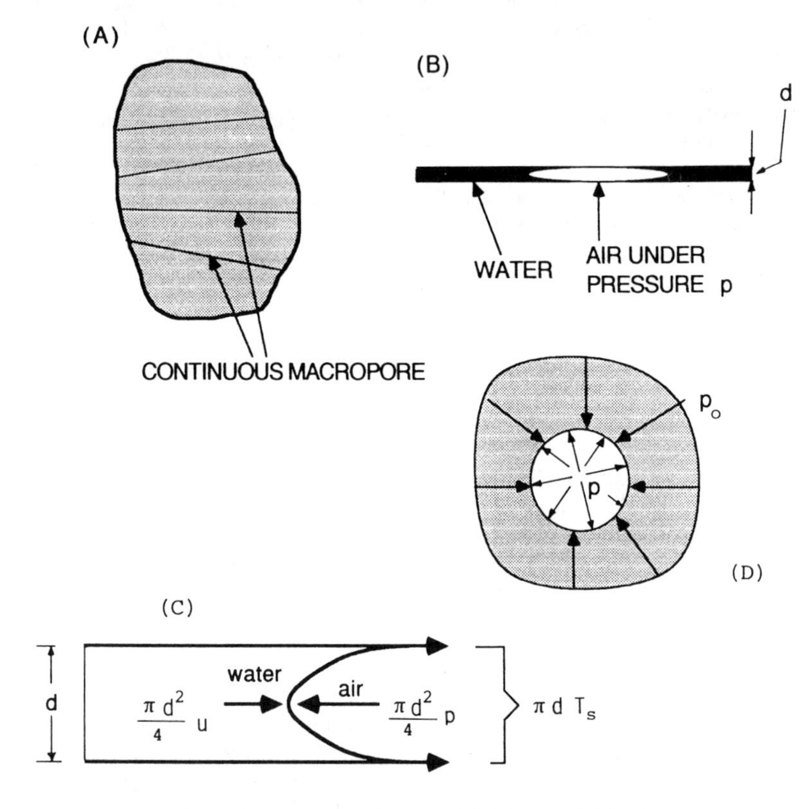

FIG. 4. (a) Shale Sample; (b) Pore with Water and Air Pressure; (c) System of Forces in a Pore with No Confinement Pressure; and (d) Pore with Air Pressure and External Confinement Pressure

Air pressure in the macropore under no confinement

The system of forces acting at the interface between the air and the water in a macropore under no confinement pressure are modeled in Fig. 4c. At equilibrium condition, the following equation applies (Means and Parcher 1963):

$$\pi d T_s - \frac{\pi d^2 p}{4} + \frac{\pi d^2 u}{4} = 0 \tag{1}$$

where d is the diameter of the macropore; T_s is the surface tension of water acting on the meniscus; p is the air pressure; and u is the pore water pressure.

From Eq. (1), the following relationship can be obtained:

$$p = u + \frac{4T_s}{d} \tag{2}$$

An analysis of Eq. (2) indicates that the air pressure, p, induced by water suction in the shale pore, increases as the diameter, d, of the pore decreases. Thus, the smaller the pore diameter, the larger the air pressure will be.

Since positive pore air pressure is enhanced by small pore radii, slaking of shales by air compression will be more pronounced in those shales containing small diameter pores. This conclusion is validated from an examination of the results of Table 2. The fourteen samples that slaked in the jar slake test had the smallest pore diameter.

Air pressure in the macropore under confinement

The air pressure, p, that develops in the pores of shales as a result of capillary suction will induce tensile stresses on the solid walls surrounding the pore (Figs. 4b and 4c). These tensile stresses will tend to break the shale. The tensile stresses will be somewhat counteracted within the fill mass by the confining compressive stress, p_o, at the boundary of the pore. This confining pressure is the result of the point loads from the fragments that surround the shale (Fig. 4d). The tensile strength, T_o, of the solid shale walls that surround the macropore will also counteract the air-induced tensile stresses. Taking into consideration all these stresses acting on the wall surrounding a cylindrical pore in the shale sample, "aerofracturing," or slaking, will occur when the pore air pressure, p, is equal to (Jaworski et al. 1981):

$$p = p_o + \frac{T_o}{2} \tag{3}$$

Thus, inside mine waste or other shale fills, the disintegration of shale samples decreases as the depth of burial inside the waste fill increases (more confinement pressure). Conversely, air pressure-induced slaking will be enhanced at shallow depths.

CONCLUSIONS

Settlement of shale embankments is directly related to degradation caused by point load crushing and slaking when the shales absorb water. An experimental and theoretical study of the mechanisms of shale degradation was carried out. Sixty-eight shale samples from surface coal mines in Kentucky, Tennessee, Virginia and West Virginia were used for the testing program. From the results of point load and jar slake tests, and thin section examinations, the following conclusions were made:

(1) The crushing by point loading of shale fragments in embankments was somewhat related to the diameter of the pore system in the shales. Some point load test results indicated that the larger the diameter of the shale pore system, the smaller their point load crushing strength.

(2) Slaking by pore air compression was the predominant mechanism for the

degradation of the shales. Pore air compression was the result of capillary suction. Slaking by pore air compression was directly related to the average diameter of the pore system in the shales. It was found that the smaller the pore diameter in a shale fragment, the more pronounced was its slaking by pore air compression (aerofracturing).

(3) Confinement of the shale fragments reduced slaking by pore air compression. Shale fragments at large depths (large confinement pressures) in fills will not slake as a result of pore air compression. Thus, aerofracturing is favored by shallow depths in waste fills.

(4) Settlement of shale fills is the result of point load crushing and slaking. Because point load strength and particularly slake durability are influenced by the diameter of the pore system of shales, it can be concluded that the settlement of shale fills is directly influenced by the diameter of their pore system.

Appendix - References

Broch, E., and Franklin, J. A. (1972). "The point load strength." *Int'l. J. Rock Mech. and Mining Science*, 9(6), 669-697.

Jaworski, G. W., Duncan, J. M., and Seed, H. B. (1981). "Laboratory study of hydraulic fracturing." *J. Geotech. Eng. Div.*, Proc. ASCE, 107(GT6), 713-732.

Lutton, R.J. (1977). "Design and construction of compacted shale embankments: slaking indices for design," *Report No. FHWA-RD-77-1*, Federal Highway Administration, Washington, D. C.

Means, R. E., and Parcher, J. V, (1963). *Physical Properties of Soils,* Charles E. Merril Books Inc., Columbus, Ohio.

Moriwaki, Y., 1974. "Causes of slaking in argillaceous materials," Ph.D. dissertation, Univ. of California, Berkeley.

Oakland, M. W., and Lovell, C. W.(1982). "Classification and other standard tests of shale embankments." *Joint Highway Research Project Report 82-4*, Purdue University, West Lafayette, Indiana.

Office of Surface Mining (1992). "Overburden strength-durability classification system for surface coal mining," *Open File Report 1992-1*, Office of Surface Mining, U.S. Department of the Interior, Pittsburgh, Pennsylvania.

Sowers, G. F., Williams, R. C., and Wallace, T. C. (1965). "Compressibility of broken rock and the settlement of rock fills." *Proc. 6th ICSMFE*, Montreal, II, 561-565.

Stanton, T. (1987). "Tablets for precision graphics." *PC Magazine*, 6(14), 159-181.

Strohm, W.E.(1978). "Design and Construction of Compacted Shale Embankments - Field and Laboratory Investigations." *Report No. FHWA-RD-78-140*, Federal Highway Administration, Washington D. C.

Vallejo, L. E., Welsh, R. A., Jr., Lovell, C. W., and Robinson, M. K. (1993). "The influence of fabric and composition on the durability of Appalachian shales." *Rock for Erosion Control*, ASTM STP 1177, ASTM, Philadelphia, Pennsylvania, 15-28.

Collapse Mechanism of Compacted Clayey and Silty Sands

T. A. Alwail,[1] C. L. Ho,[2] Member ASCE,
and R. J. Fragaszy,[3] Member ASCE

Abstract: The mechanism of wetting-induced collapse (hydrocompression) and the influence of fines ratios on compacted sandy soils are investigated. Various ratios of clay and silt are mixed with Ottawa sand to produce twenty-five soil combinations. Each soil is examined qualitatively under the Scanning Electron Microscope before and after collapse. Observed fabric alterations are supported by data obtained from Double-Oedometer Tests. Clay structures before collapse transform to a continuous or discontinuous clay blanket after collapse, depending on the initial clay quantity present in the soil. Magnitude of collapse increases with increasing clay/silt ratios. Silty sands with sharp silt grains and/or a higher clay content demonstrate greater collapse than soils with rounded silt grains and/or less clay content. This study suggests that collapse is profoundly influenced by the quantity of clay and the interaction of percolating water and the clay fractions of soils. A model showing fabric alterations of compacted clayey sands, silty sands, and clayey silty sands is presented.

Introduction

Compacted fill is widely used in the construction industry. It is the main constituent of earth structures such as highway embankments and earth dams.

---

[1] Civil Engineer (on leave), Ministry of Petroleum, P. O. Box 247, Riyadh 11191, Saudi Arabia.

[2] Associate Professor, Department of Civil and Environmental Engineering, Washington State University, Pullman, WA 99164-2910.

[3] Consulting Geotechnical Engineer, 1091 Summit Circle, Watkinsville, GA 30677.

Wetting-induced collapse (also called hydrocompression) and the resulting fill settlement may cause severe damage to these structures. Although most of the research on damage from collapse investigated collapse of natural soil, there is also documentation of damage from collapse of compacted soils. For example, damage was reported to have occurred in dam embankments (Peterson and Iverson 1953; Leonards and Narain 1963), road embankments (Knight and Dehlen 1963; Booth 1977), airfield runways (Novais-Ferreira and Meireles 1967), and residential houses (Lawton et al. 1989; Brandon and Duncan,1990).

The effects of various soil parameters on the magnitude of collapse were reported by several researchers. The investigated parameters include the following: placement water content (Holtz 1948; Booth 1975; Cox 1978; Lawton et al 1989; Huang 1989), overburden pressure (Booth 1975; Cox 1978; Lawton et al. 1989; Huang 1989), placement dry density (Booth 1975; Cox 1978; Lawton et al. 1989; Huang 1989), degree of saturation (Cox 1978; Houston and Houston 1988; Lawton et al. 1989; El-Ehwany and Houston 1990), compactive prestress (Witsman and Lovell 1979), principal stress ratio (Lawton et al. 1991), percentages of fines (Steadman 1987), and sample disturbance (Houston and El-Ehwany 1991). A recent review by Lawton et al. (1992) provides a summary of the current state-of-the-art.

The changes in fabric (arrangement of soil particles) of collapsed compacted soils is not well understood. Much of the research on collapse reported in the literature has been performed on naturally deposited clay and loess soils. There has been some research on the change in the fabric of compacted soils containing various amounts of silt and clay (Barden and Sides 1969; Barden et al. 1969). Explanations regarding the behavior of collapsible compacted soils were based on assumed probable structural arrangements of the fines and sand grains as deduced from behavior. The objective of the research program presented in this paper was to directly examine the alterations of fabric resulting from wetting-induced collapse for twenty-five compacted clay-silt-sand mixtures at a final applied vertical stress of 1600 kPa.

Description of Materials

The soils for this research program were fabricated in the laboratory by mixing different quantities of sand, silt, and clay. The major component of the mixtures is sand. Twenty-five different combinations were formed by using five percentages of fines: 15%, 25%, 35%, 45%, and 55% and five different clay/silt ratios (0:1; 1:3; 1:1; 3:1; 1:0). The Modified Proctor compaction test results (ASTM D-1557) were used to determine the compaction characteristic of these soils.

Equal masses of four types of Ottawa sand (Flint Shot, F75 series, F58 series, and Sawing) were mixed thoroughly. After mixing, the sand is classified as poorly graded SP according to the Unified Soil Classification System (USCS), (C_c = 1.0; C_u = 2.2).

A good source of silt in the Pacific Northwest is the sand bars in the southern edges of Wallowa Lake, Oregon. The freshly deposited silt-rich layer was mined, then washed and sieved. The specific gravity of the silt is 2.74. Hydrometer test results indicate that the silt contains approximately 3% clay-size particles.

X-ray diffraction and a binocular polarizing microscope show that the silt is composed of feldspars, calcite, iron minerals, and a very small percentage of kaolinite.

The clay used in this study is kaolinite (type hydrite-px) provided by Georgia Kaolin Company (GKC). The liquid limit is 60%, plastic limit is 26%, and plasticity index is 34%. The specific gravity is 2.58 (data provided by GKC).

Test Equipment and Procedures

Collapse potential of each soil mixture was evaluated using the Double-Oedometer Test. Initial soil fabric and changes in fabric resulting from wetting under load were evaluated qualitatively using Scanning Electron Microscope (SEM). Details of test equipment, specimen preparation, and test procedures are presented in Alwail (1990).

Test Matrix

Each soil was compacted into the consolidation rings by a static method of compaction at 3% dry of optimum and 90% relative compaction. Double-Oedometer Tests were conducted to provide specimens for the SEM investigation, the air conductivity measurements, and data for determination of maximum collapse. Special precautions were required to ensure the repeatability of soil properties (i.e., homogeneity, placement dry density and water content).

Results and Discussion

Clay-Sand Mixtures

The SEM examination of clay-sand mixtures (clay ranges from 15% to 55%) indicates that the clay fraction before collapse occurs in one or a combination of the following forms: bridging sand grains; coating sand grains; "clay balls" i.e. aggregates of clay platelet with face-to-face arrangement having preferred and/or random orientation; and flocculated structure (Fig. 1). Flocculated structure is detected only in soil containing 55% clay.

For soils with a small clay fraction (15% and 25%), the clay before collapse mainly bridges and coats sand grains. After collapse the clay forms a "discontinuous blanket" (Fig. 2a). As the quantity of the clay increases (25%), the pre-collapse structure exhibits more clay balls between sand grains, thus creating higher void ratios. The softening of the clay by the percolating water leads to the formation of a "continuous blanket" after collapse (Fig. 2b). The disintegration of clay balls leads to the formation of a denser and more stable structure after collapse. The findings of the SEM are also supported by the results of the oedometer tests. Maximum collapse of these soils increases with the clay quantity as shown in Fig. 3.

Silt-Sand Mixtures

For silt-sand mixtures (silt ranges from 25% to 55%), the silt is rounded to subrounded. There is little clay structure present and no conspicuous alternations

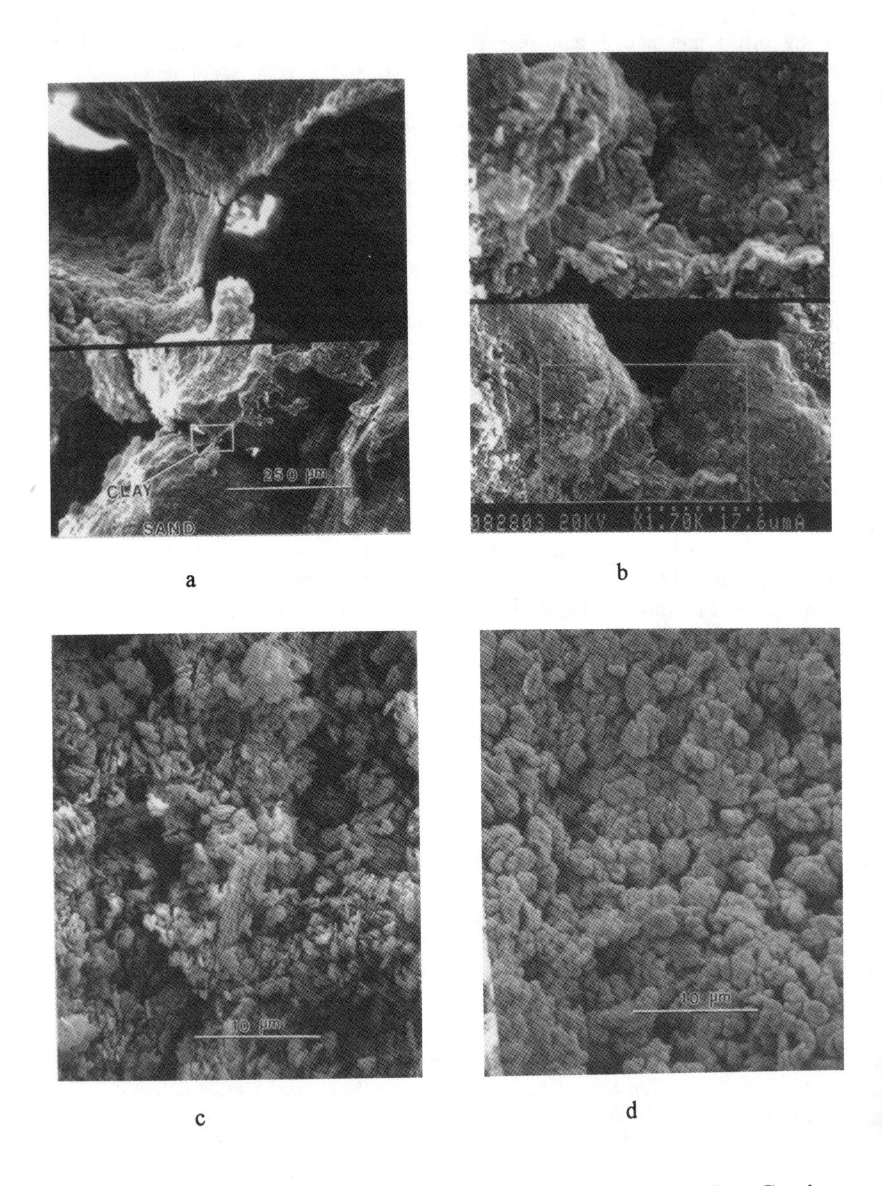

FIG. 1. Structure of Soils Before Collapse: (a) Clay Bridge; (b) Clay Coating Silt and Sand Grains; (c) Aggregation of Clay Balls; (d) Flocculated Structure

in the soil fabric after collapse. The only notable change that took place is a small reduction in the voids among silt grains. Fig. 4 shows examples of a sandy soil with only silt fraction before and after collapse. The maximum collapse for all of these soils is negligible (less than 1%).

a b

FIG. 2. Structure of Soils after Collapse: (a) Discontinuous Blanket; (b) Continuous Blanket

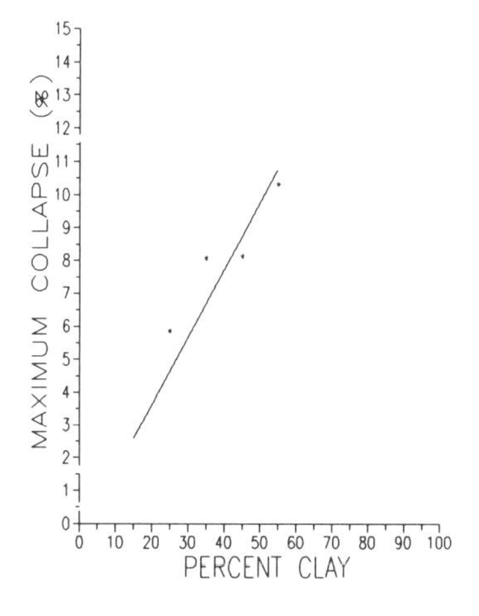

FIG. 3. Maximum Collapse (%) and Percent Clay for Clay-Sand Mixtures

FIG. 4. SEM Micrograph for Silt-Sand Mixture Group 1: (a) before Collapse; (b) after Collapse

A previous study of the collapse potential of silt-sand mixtures by Steadman (1987) showed greater magnitudes of collapse than obtained in this study, despite very similar test parameters. An investigation was made to determine the reason for the differences in results. Steadman's soils (designated Group 2 specimens) were composed of a very similar sand, also classified SP with $C_c = 1.19$ and $C_u = 2.67$, mixed with 10, 20, and 30% silt. The resulting mixtures were prepared for testing by a static compaction method in the oedometer rings to 90% Modified Proctor dry density at a water content 3% below optimum. The silt in Group 2 soils differs considerably from the silt used in this study (Group 1 specimens). The Group 2 silt contains 8.8% clay-size particles; has Atterberg limits of PL = 22; LL = 27, PI = 5; and has a specific gravity of 2.7. The predominant mineral of the silt is quartz with some pyrite. The clay fraction is smectite with appreciable amounts of illite.

The SEM investigation conducted by the authors on Group 2 specimens reveal that the silt grains are angular and flaky. The clay bridges most silt grains before collapse. The sharp edge of the silts coupled with the clay bridging lead to more open structure resulting in a greater collapse upon wetting. Fig. 5 shows examples of Group 2 specimens before and after collapse. The maximum collapse of these soils is remarkably higher than for Group 1 for all percentages of silt (Fig. 6).

a b

FIG. 5. SEM Micrograph for Silt-Sand Mixture Group 2: (a) before Collapse; (b) after Collapse

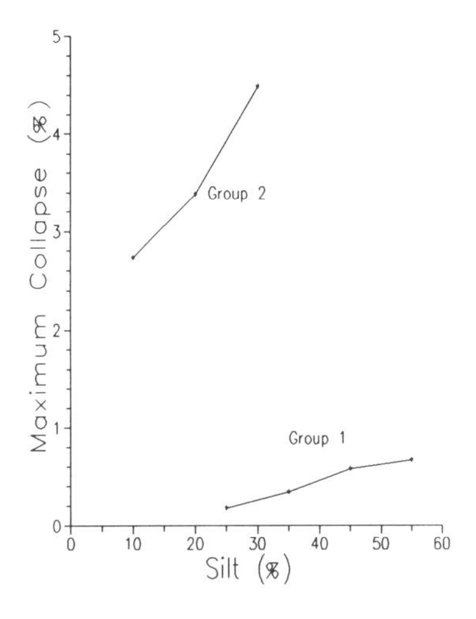

FIG. 6. Comparison of Maximum Collapse (%) versus Silt (%) for Group 1 and 2

Clay-Silt-Sand Mixtures

Clay structures observed in clay-sand mixtures before collapse were also noted in clay-silt-sand mixtures (clay/silt ratios: 1:3, 1:1, and 3:1). At low percentages of fines (15% and 25%), the destruction of clay bridges and the formation of a discontinuous blanket is the dominant mechanism of collapse. A more definite clay ball structure is evident when the fines reach 35%. Fig. 7 shows SEM micrographs for a soil with 35% fines before and after collapse.

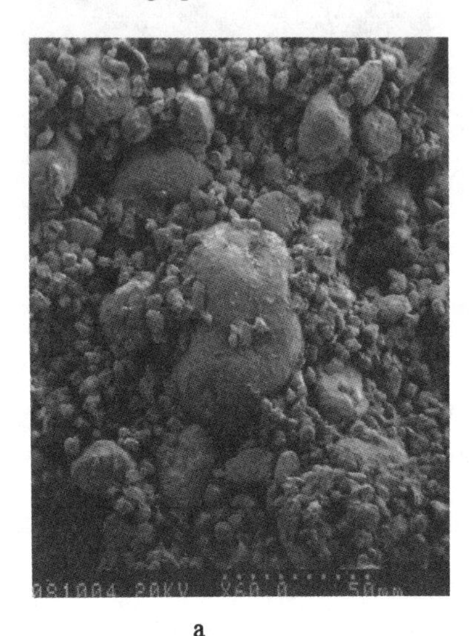

a

b

FIG. 7. SEM Micrograph for Clay-Silt-Sand Mixture Group 2: (a) before Collapse; (b) after Collapse

The results of oedometer tests for clay-silt-sand mixtures are shown in Fig. 8. At low percentages of fines (15% and 25%), there is no explicit relationship between the clay/silt ratio and maximum collapse. When the fines comprise at least 35% of the soil, the magnitude of collapse increases as the clay/silt ratio increases for a given percentage of fines.

DISCUSSION

Low quantities of clay (15% and 25%) before collapse mostly bridge or coat sand and silt grains. As the amount of clay increase, the clay begins to occur as clay balls between sand and silt grains. After collapse the clay form a continuous or discontinuous blanket depending on the initial quantity of clay. Soils with higher clay content exhibit greater collapse upon wetting (Figs. 3, 6 and 8).

The data of this study suggest that the quantity of clay in the soil plays a key role in determining the magnitude of collapse. The r^2 values from linear regression of maximum collapse vs. the percentages of clay in the fines and the percentages of the clay in total mass of the soils are 0.55 and 0.82, respectively. The latter r^2 value demonstrates that the variation in the magnitude of collapse is better explained by the variation of clay quantity in the total mass soil. Fig. 9 shows the maximum collapse vs. the percentages of clay in the total mass of the soils.

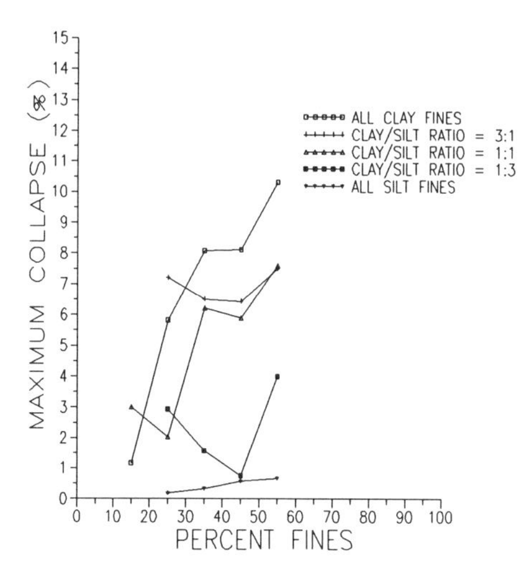

FIG. 8. Maximum Collapse (%) versus Percent of Clay of the Fines

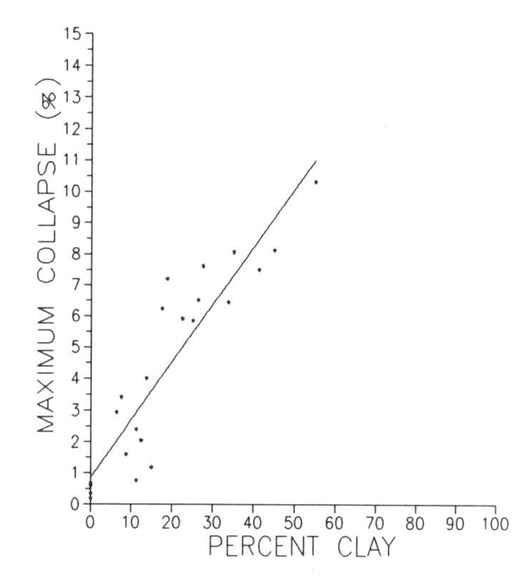

FIG. 9. Maximum Collapse (%) versus Percent Clay of the Total Mass of the Soil

The silt portion of the soils does not appear to contribute significantly to collapse, particularly if the silt grains are rounded or subrounded (Fig. 6). It was mentioned previously that Groups 1 and 2 contain an SP sand; however, the shape of silt grains, and quantity and type of clays are different. The SEM analysis suggest that the presence of clay as a binding agent between silt grains is the major contributor to collapse. Clay develops more open structures in Group 2 than Group 1 soils. Consequently greater collapse occurs upon wetting.

Based on the results of the SEM analysis and oedometer tests, a diagrammatic representation can be constructed showing the collapse mechanism for clay-silt-sand mixtures (Fig. 10). The alterations in the soil fabric are shown as the following: for silt-sand mixture (a), clay-sand mixtures (b), and clay-silt-sand mixtures (c). This model supports the contention that the principle of effective stress is maintained on the microscale level. No change in clay microstructure (flocculated to dispersed) was observed except for soils with 55% fines. Further study is needed to investigate the influence of clay type and water chemistry parameters such as pH, salinity, ionic strength on collapse.

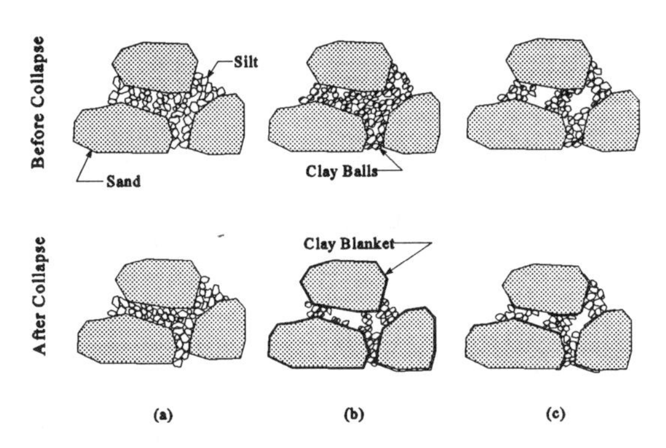

FIG. 10. Change in Fabric Caused by Wetting Induced Collapse for Compacted Sandy Soils Containing Various Amounts of Fines: (a) Only Silt Fines; (b) Only Clay Fines; and (c) Silt and Clay Fines

CONCLUSIONS

1. The dominant mechanism of the collapse of compacted sandy soils containing fines is the disintegration of the clay structure into either a discontinuous blanket or a continuous blanket, depending on the initial clay quantity present in the soil.
2. The transformation of flocculated clay structure before collapse to dispersed structure after collapse was only observed in soils with clay fraction of at least 55%.
3. The silt portion of the soils does not contribute significantly to collapse. Sandy soils with angular silt grains and higher clay content exhibit greater collapse upon wetting than soils with rounded silt and lower clay content.

4. The magnitude of collapse of compacted sandy soils containing clay but no silt increases with the percentage of clay.
5. Sandy soils with only silt fines exhibit little or no collapse upon wetting.
6. Soils with a higher clay/silt ratio experience greater magnitude of collapse than soils with a lower clay/silt ratio.

ACKNOWLEDGMENT

The authors would like to thank the Aerial Survey Department of the Ministry of Petroleum and Mineral Resources of Saudi Arabia and Washington State University for the financial support of this research project. Dr. David Mulla is acknowledged for his valuable suggestions. Special gratitude to Dr. Peter Larson and the staff of Electron Microscopy Center at WSU.

APPENDIX. REFERENCES

Alwail, T. A. (1990). "Mechanism and effect of fines on the collapse of compacted sandy soils," Ph.D. diss., Wash. St. Univ., Pullman, Wash.

Barden, L. and Sides, G. F. (1969). "The influence of structure on the collapse of compacted clay." *Proc. 2nd Int. Research and Engrg. Conf. on Expansive Clay Soils*, College Station, Texas, 317-326.

Barden, L., Madedor, A. O., and Sides, G. F. (1969). "Volume change characteristics of unsaturated clay." *J. Soil Mech. Found. Div.*, Proc. ASCE, 95(SM1), 33-51.

Booth, A. R. (1975). "The factors influencing collapse settlement in compacted soils." *Proc. 6th Regional Conf. for Africa on Soil Mech. and Found. Eng.*, 2, 57-63.

Booth, A. R. (1977). "Collapse settlement in compacted soils," *CSIR Research Report 324*, NITRR Bull. 13, Pretoria, South Africa.

Brandon, T. L. and Duncan, J. M. (1990). "Hydrocompression settlement of deep fills." *J. Geotech. Eng.*, ASCE, 116(10), 1536-1549.

Clemence, S. P. and Finbarr, A. O. (1981). "Design considerations for collapsible soils." *J. Geotech. Eng.*, ASCE, 107(3), 305-317.

Cox, D. W. (1978). "Volume change of compacted clay fill." *Proc. Conf. on Clay Fills*, Inst. of Civ. Engrs., London, 79-87.

Dudley, J. H. (1970). "Review of collapsing soils." *J. Soil Mech. Found. Div.*, ASCE, 96(3), 925-947.

El-Ehwany, M. and Houston, S. L. (1990). "Settlement and moisture movement in collapsible soils." *J. Geotech. Eng.*, ASCE, 116(10), 1521-1535.

Hilf, J. W. (1956). "An investigation of pore-water pressure in compacted cohesive soil." *Tech. Memo. 654*, US Dept. of Interior, Bur. of Rec., Denver, Colorado.

Holtz, W. G. (1948). "The determination of limits for the control of placement moisture in high rolled earth dams." *Proc.*, ASTM, Philadelphia, Pennsylvania, 1240-1248.

Houston, S. L. and Houston, W. N. (1988). "Prediction of field collapse of soils due to wetting." *J. Geotech. Eng..*, ASCE, 114(1), 40-58.

Houston, S. L. and El-Ehwany, M. (1991). "Sample disturbance of cemented collapsible soils." *J. Geotech. Eng.*, ASCE, 117(5), 731-752.

Huang, D. (1989). "A laboratory investigation on the behavior of collapsible soils," M.S. thesis, Colo. St. Univ., Fort Collins, Colo.

Knight, K. and Dehlen, G. (1963) "Failure of a road constructed on a collapsing soil." *Proc. 3rd Regional conf. for Africa on Soil Mech. and Found. Eng.*, 1, 31-34.

Lawton, E. C., Fragaszy, R. J., and Hardcastle, J. H. (1989). "Collapse of compacted clay sand." *J. Geotech. Eng.*, ASCE, 115(9), 1252-1268.

Lawton, E. C., Fragaszy, R. J., and Hardcastle, J. H. (1991). "Stress ratio effects on collapse of compacted clayey sand." *J. Geotech. Eng.*, ASCE, 117(5), 714-730.

Lawton, E. C., Fragaszy, R. J., and Hetherington, M. D. (1992). "Review of wetting-induced collapse in compacted soil." *J. Geotech. Eng.*, ASCE, 118(9), 1376-1394.

Leonards, G. A. and Narain, J. (1963). "Flexibility of clay and cracking of earth dams." *J. Soil Mech. Found. Div.*, Proc. ASCE 89(SM2), 47-98.

Novais-Ferreira, H. and Meireles, M. F. (1967). "On the drainage of muceque-a collapsing soil." *Proc. 4th Regional Conf for Africa on Soil Mech and Found Eng.*, 151-155.

Peterson, R. and Iverson, N. L. (1953). "Study of several low earth dam failures." *Proc. 3rd ICSMFE*, Zürich, 2, 273-276.

Steadman, L. (1987). "Collapse settlement in compacted soils of variable fines content," M.S. thesis, Wash. St. Univ., Pullman, Wash.

Witsman, G. R. and Lovell, C. W. (1979) "The effect of compacted prestress on compacted shale compressibility," *Report No. FHWA/IN/SHRP-79016*, Dept. of Civ. Eng., Purdue Univ., West Lafayette, Indiana.

DEFORMATIONS ABOUT EXCAVATIONS IN HIGHLY STRESSED ROCK

Paul Nash[1] and Garry Mostyn[2]

ABSTRACT: The central business district of Sydney, New South Wales (NSW), is dominated by the Hawkesbury Sandstone, a Triassic sedimentary formation. Measurements have confirmed the existence of a high in-situ horizontal (compressive) stress field. Average city basement excavations have deepened from 10 m to up to 30 m over the last two decades and rock stress problems have started to be encountered. This paper reports an elaborate and realistic investigation using UDEC (Universal Distinct Element Code) to consider the effects of rock mass defects which are such a dominant characteristic of Hawkesbury Sandstone. UDEC is an ideal modelling tool for this situation.

The investigation showed, as expected, that the deformations due to stress relief were influenced by the properties of the rock substance and the properties of the defects. The deformation was found to be a two stage process: firstly, the initial elastic dilation of the rock due to stress relief, which appear to be linear with depth; and, secondly, opening of vertical defects by rafting on horizontal defects near the excavation face, which is negligible at shallow depths (less than 10 m) and increases rapidly with depth. Other conclusions are presented. The work has shed considerable light on behavior of basement excavations in highly stressed rock masses.

---

[1] Engineer, DJ Douglas & Partners, 96 Hermitage Rd, West Ryde, NSW 2114, Australia.

[2] Senior Lecturer, School of Civil Engineering, The University of New South Wales, Kensington, NSW 2133, Australia.

INTRODUCTION

The central business district (CBD) of Sydney, NSW, is dominated by the Hawkesbury Sandstone, a Triassic sedimentary formation. This is characterized by regular near horizontal bedding planes and vertical joints, with spacings for both usually in the range of one to five meters. Measurements (Enever et al. 1990) have confirmed the existence of a high in-situ horizontal (compressive) stress field.

Basement excavations have progressively deepened over the last twenty years from typically 10 m to up to 30 m. In these deeper excavations problems of rock movements associated with stress relief have occurred (Pells 1990; Baxter and Nye 1990). The horizontal in-situ stresses have been concentrated by and around basement excavations which have resulted in horizontal movements of up to 45 mm and have caused damage to partially constructed footings, existing buildings and underground railway tunnels.

It is thought that rock mass defects, particularly horizontal weaknesses, can exacerbate deformations associated with the lateral stresses. It is known (Enever et al. 1990) that the stress field increases in magnitude with depth and experience suggests that problems increase with the depth of excavation. A schematic cross section showing the usual situation is given in Fig. 1. The stress fields in Fig. 1 represent the major principal stress which is assumed here to be normal to the long side of the excavation.

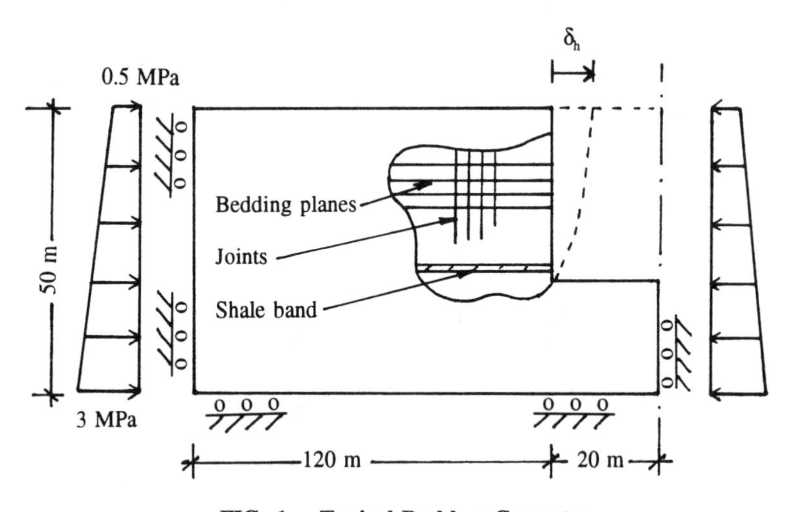

FIG. 1. Typical Problem Geometry

PHYSICAL BACKGROUND

The Sydney Basin is a major sedimentary unit which comprises rocks and sediments dating forward from the Permian. The dominant formation in the Sydney CBD is the Hawkesbury Sandstone, a Triassic age formation which is characterized by regular near horizontal bedding planes and vertical joints. Spacings for these rock mass defects are usually in the range of one to five meters, but can be up to fifteen meters. The strength of the rock is usually related to weathering, with moderately weathered to fresh rock providing excellent foundation conditions. Excavation sides are usually cut vertically and require no support.

Joints fall into two major sets, which are subvertical, approximately orthogonal and strike in north-south and east-west directions. A characteristic of Hawkesbury Sandstone is the presence of occasional layers or lenses of shale. These lenses are usually less than half a meter thick, have low strength and can be continuous over several hundreds of meters.

There is a growing body of evidence for the existence of a high in-situ horizontal (compressive) stress field throughout the Sydney Basin. In-situ rock stress measurements have been carried out at several sites in the basin and are summarized by Enever et al. (1990). Measurements have been carried out at surface and underground excavations using both overcoring and hydrofracture methods.

While there is some scatter in the stress measurement results, significant trends are evident. The major and intermediate principal stresses are both horizontal and are approximately north-south and east-west respectively. Several models of the stress field have been suggested (Pells 1990; Carter et al. 1990) in the form of:

$$\sigma_1 = \sigma_0 + K\sigma_v \tag{1}$$

where σ_1 is the major horizontal stress
σ_0 is the horizontal stress at the surface
K is the ratio of horizontal to vertical stress
σ_v is the overburden pressure.

Typically σ_v is 25 kPa/m depth with σ_0 ranging from 0.4 to 0.6 MPa and suggested values for K ranging from 1.5 to 2.5. Thus σ_1 at the base of a deep basement excavation is of the order of 2 MPa.

PREVIOUS WORK

While there has been a considerable amount of study into the effects of lateral stresses on underground excavations and construction, there has been much less study concerning surface excavations. Existing research has been in response to problems that have occurred in actual excavations and there has been relatively little theoretical study of stressed excavations.

Trow and Lo (1988) presented a case study of a 24 m deep excavation for a 69 storey building in Toronto, Canada. The local rocks have lateral stresses 10 to 30 times greater than the overburden stress. The excavation was bordered on three sides by existing high rise buildings which required excavation induced movements to be minimized. Movements were controlled by the use of careful rock bolting, and damage to the adjacent buildings was avoided.

Baxter and Nye (1990) detailed a major excavation in the Sydney CBD, which was adjacent to an underground railway tunnel. The railway authority imposed strict limits on movement of the unreinforced concrete tunnel to limit cracking and damage. In-situ stresses were measured and finite element modelling carried out to assess methods of controlling damage. The modelling results showed that if the rock behaved as a continuum then the movements would be within acceptable limits. A program of rock bolting above and around the tunnel was completed to bind the rock across discontinuities such as joints and bedding planes. Excessive deformation of the tunnel was avoided.

Pells (1990) presented several case studies of other deep excavations in the Sydney CBD. One involved a 30 m deep excavation, the deepest and largest in Sydney, adjacent to an underground railway tunnel. Field observations showed that the top edge of the excavation had moved inwards by 45 mm with significant movements occurring in the nearby rail tunnel and at the surface up to 40 m away. At another site, major fractures occurred in the sandstone floor of an excavation and were believed to be a result of stress concentrations in the sandstone base which was underlain by a weak shale seam. Pells presented results of finite element modelling of these cases and found that significant deformations will occur in nearby tunnels or services which are shallower than the excavation. The low shear strength clay/shale layer below the excavation produced significant stress increases which could approach the compressive strength of the rock. Pells also found that continuous seams caused much greater problems than seams which terminate beneath the excavation.

Carter et al. (1990) carried out a parametric study of excavations in Hawkesbury Sandstone close to underground railway tunnels using a boundary element model of a homogeneous, isotropic rock mass.

Carter and Alehossein (1990) extended the above work to include the effects of rock discontinuities by using an equivalence equation for a layered material whereby the modulus of the rock substance was reduced to allow for the effect of rock joints. Interestingly, the results of this study showed that the rock deformations were virtually the same with or without the correction.

METHOD OF STUDY

The above studies of excavations in the Sydney area have not specifically modelled the effects of rock defects and discontinuities. The finite and boundary element models adopted represented the rock as a continuum material, with allowances being made for defects by reducing the modulus of the material.

In the present study, a model was required that could accurately simulate the behavior of discrete rock blocks and joints as the excavation proceeded. Such a model is the Distinct Element Method. The program used was UDEC (Universal Distinct Element Code) V1.7, from the Itasca Consulting Group, a two dimensional geomechanics analysis program. UDEC has been specifically designed to model the behavior of discretely jointed rock masses and is thus an ideal simulation tool for Hawkesbury Sandstone.

Modelling consisted of examining a rectangular basement cross section in Hawkesbury Sandstone as shown in Fig. 1. Some rock mass parameters which are difficult to measure were modelled parametrically. Modelling of discrete and homogeneous rock masses as defined by Carter & Alehossein (1990) was completed to allow a direct comparison of the two. The following rock defect characteristics were studied:

- o Vertical joint spacing
- o Bedding plane properties
- o Combined joints and bedding
- o Joint stiffness
- o Joint friction
- o Low strength shale band.

The behavior of the excavation was assessed by observing displacement and stresses for the whole of the domain as well as monitoring certain points, most importantly the top edge of the excavation. Two methods were used to review the results: Observing changes in movement or stress at a specific location as the excavation deepens; and observing the deformed shape of the excavation face at a particular (30 m) depth of excavation. This is very useful for analyzing the mechanisms of movement, specifically when defects are involved.

INPUT PARAMETERS

UDEC allows the user to specify certain parameters relating to geometry, material properties and in-situ stresses. The "standard excavation" adopted was 40 m wide and up to 40 m deep. Since the program is two dimensional, the excavation is modelled as infinitely long. Only half the problem was modelled due to symmetry.

In assigning material properties for the rock, the values used have been based on field results for Hawkesbury Sandstone wherever possible. Two excellent sources of information about this rock are Pells et al. (1978) and Pells (1985), which provide a comprehensive summary of knowledge on Hawkesbury Sandstone.

For homogeneous rock, the following values for material properties were adopted: elastic modulus (*E*) of 2500 MPa; Poisson's ratio (μ) of 0.25; and density (γ) of 2500 kg/m^3. For jointed rock, additional properties must be specified. The properties used for modelling are given in Table 1.

TABLE 1. Properties Adopted for Materials

Property / Model	2.5 m Rock	4 m Rock	7.5 m Rock	Class I S'stone	Class II S'stone	Shale Band
Elastic modulus (*E*), MPa	5000	3000	3000	3333	2860	50
Defect spacing (s), m	2.5	4	7.5	5	2	0.1
Defect stiffness (*k*), MPa/mm	2000	3750	2000	6000	3333	1000
Defect angle of friction (ϕ), degrees	30	30	30	30	30	10

Properties for the 2.5, 4 and 7.5 m rock models were selected so that the models had an equivalent mass modulus of 2500 MPa. Poisson's ratio (μ) for all materials was taken to be 0.25 and the defects to have equal shear and normal stiffnesses (k_s & k_n) and zero cohesion. The bedding planes were assigned the same spacings and properties as the joints in this study, even though this is not generally the case in Hawkesbury Sandstone. Thus the *4 m Rock* model has both horizontal bedding and vertical joints at 4 m spacings with defect properties to give an equivalent mass modulus of 2500 MPa. The shale was given a very low shear strength to accentuate its effects. The in-situ stress field described in Eq 1 was adopted in the model. Both Pells (1990) and Carter et al. (1990) used a stress field with σ_0 of 0.5 MPa and K of 2, and these parameters have been adopted here for consistency. The excavation face was modelled as perpendicular to the major principal stress.

RESULTS

Unjointed massive rock

The first models that were investigated were homogeneous massive rock and a series of "equivalent" jointed rocks using an equivalent modulus from Carter and Alehossien (1990) denoted as E^*. The top corner movement for each of the five cases as excavation progresses is shown on Fig. 2 from which it can be seen that the movement of closely jointed rocks, say 2.5 m joint spacing, is quite different from that of the equivalent modulus (i.e. homogenous) case. Only in the unlikely case of rock with a 15 m joint and bedding plane spacing does the equivalent modulus model seem to work. Comparing the properties of these five materials from Table 1, the *2.5 m Rock* model has the highest elastic modulus and joint stiffness but has shown far greater deformation than the weaker rock materials.

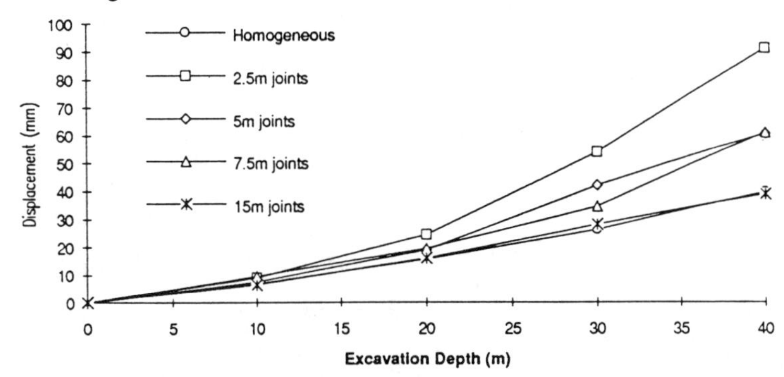

FIG. 2. Top Corner Displacement for Various "Equivalent" Models

Clearly, joint spacing is a major influence on the amount of deformation that occurs. The reason for the above can be seen on Fig. 3 which shows joint openings for the *2.5 m Rock* model. It should be noted that on the figures showing cross sections that the thickness of the joints represents the joint opening or shear displacement. Most of the joints within 30 m of the face have opened to some extent and this considerably increases the movement of the top corner. Fig. 2 also shows that as the joint spacing decreases the movement of the top corner becomes non-linear.

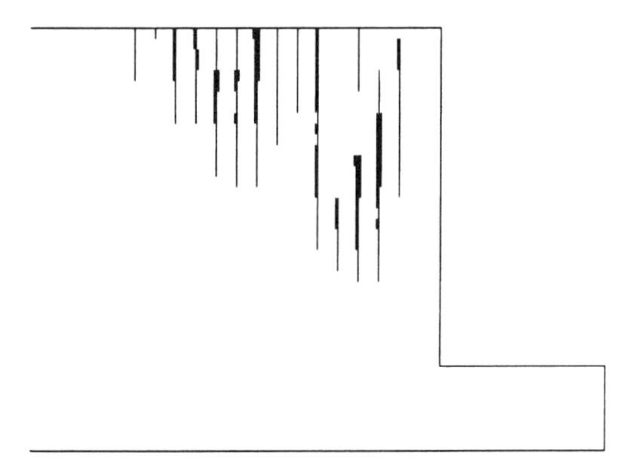

FIG. 3. Joint Opening for *2.5 m Rock* Model (Excavation 40 m Deep, Maximum Joint Opening 6 mm)

BEDDING PLANE SPACING

This was further investigated modelling two cases representing typical Hawkesbury Sandstone based on the classification system given in Pells et al. (1978) using properties derived for *Class I* (i.e. highest quality) and *Class II* (good) sandstone. Again the deformation of the top corner has been used to assess the two cases and is shown on Fig. 4. The most interesting feature of this graph is the rapid increase in deformation once the excavation becomes deeper than 20 m. This reinforces the idea that stress has become a problem since excavations passed the 20 m mark. A *rule of thumb* has been "1 mm of movement per meter of excavation", which is quite reasonable until 20 m is reached. The plots of joint openings for both cases are presented in Figs 5 and 6. There is considerable movement occurring due to joint opening and not just elastic dilation as is the case for a homogeneous material. The large movements near the surface create tensile strains of up to 4%, which could cause severe damage to above and below ground structures.

Braybrooke (1993) gives a good explanation of the mechanism and calls it "rafting". In this process the stress (or strain) is relieved progressively in the rock as the excavation proceeds downwards past each bedding plane. Thus the higher layers expand by elastic dilation but are also then carried, or rafted, by the underlying layers as they in turn dilate. Thus the movement of the top corner will be greater than the expected elastic movement and the difference will increase as more bedding partings daylight in the excavation. The greater frequency of bedding partings in the *Class II Sandstone* model compared with that for the *Class I Sandstone* model accounts for the divergence of the displacement curves in Figs 2 and 4 as the closely jointed rock deforms more than the widely jointed or homogeneous rock. Note the difference between the *Class I* and *II Sandstone* models is due to the fact that the bedding partings were taken to have the same spacing as the joints in these models; it should be noted that it is entirely possible to have closer spaced bedding partings in Class I sandstone than in Class II sandstone. For rafting to occur the rock must also have vertical defects which can open or the upper rock will be held in place by tension.

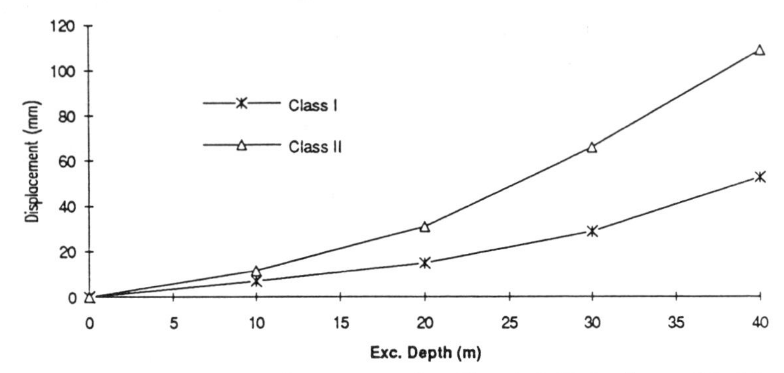

FIG. 4. Top Corner Displacement for *Class I & II Sandstone* Models

Shale bands

Fig. 7 shows the effect of a major horizontal weakness, such as a clay or weak shale band, placed at a depth of 32 m and it can be seen that it has little effect on the overall movement of the top of the excavation. This was against the authors' initial expectations and suggests that the shale band is acting in a similar manner to an ordinary bedding plane.

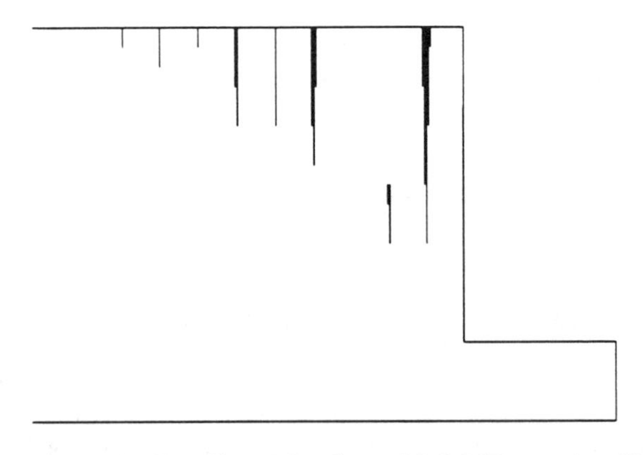

FIG. 5. Joint Opening for *Class I Sandstone* Model (Excavation 40 m deep, Maximum Joint Opening 5.4 mm)

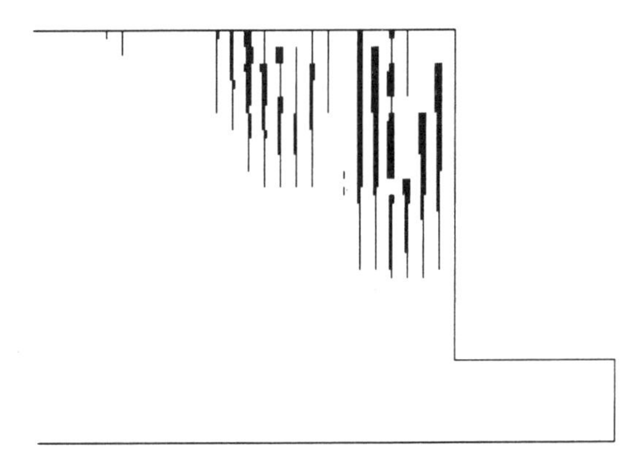

FIG. 6. Joint Opening for *Class II Sandstone* Model (Excavation 40 m Deep, Maximum Joint Opening 5.4 mm)

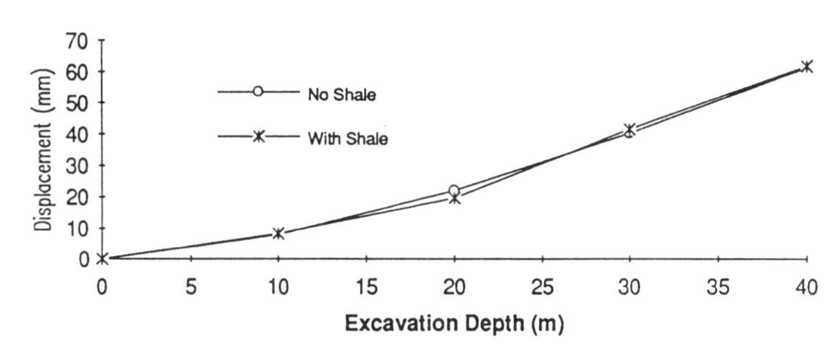

FIG. 7. Effect of Shale Band on Top Corner Movement

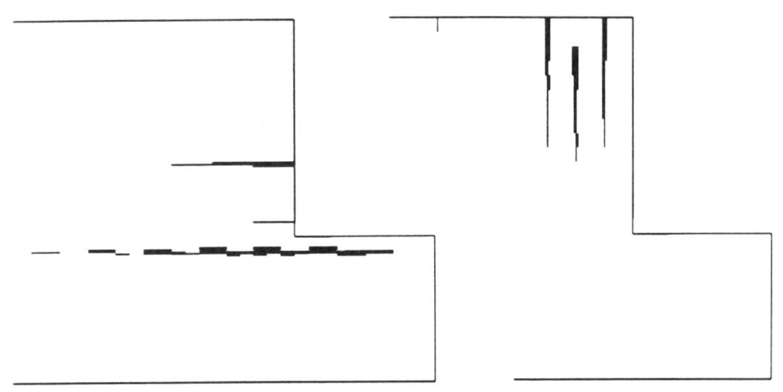

FIG. 8. Excavation 30 m Deep in *4 m Rock* Model With Shale Band at 32 m Depth: (a) Shear Displacement on Joints (maximum 12.5 mm); (b) Joint Opening (Maximum 5.2 mm)

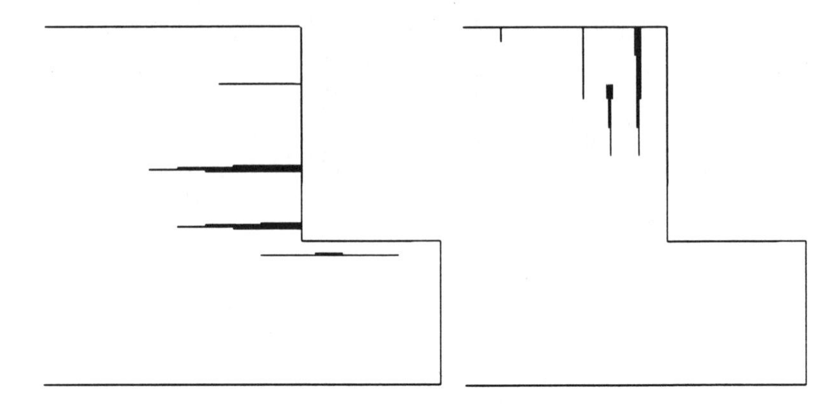

FIG. 9. Excavation 30 m Deep in *4 m Rock* Model Without Shale Band: (a) Shear Displacement on Joints (maximum 6.9 mm); (b) Joint Opening (Maximum 5.0 mm)

Fig. 8 shows both the shear displacements and joint openings on joint faces in the rock before a shale band at 32 m is reached and Fig. 9 the equivalent case with no shale band. While there has been some 12 mm of movement in the shale band itself it is only concentrating the movement which has been spread over several bedding planes in the other case. Figs 10 and 11 show the shear displacements once the shale band has been exposed, for each case the critical location is at the base of the excavation and not at the shale band.

Figs 12 and 13 show the deformed shape of the excavation face for both cases at excavation depths of 30 and 40 m. The minimal effect of the shale band can be seen clearly, as it is only a small change from the relatively smooth displacement profile of the no shale band case. For the deeper excavation the shale band just concentrates movement that would have happened on adjacent bedding planes.

DISCUSSION

The results of the work, and the shale band case in particular, suggest that the model of externally applied stresses of Fig. 1 is not strictly correct. If the model shown were correct then, in the shale band case discussed above, the applied stresses above the shale band must be carried by a diminishing thickness of overlying rock as the excavation progresses, and thus the stresses above the band should increase dramatically. The modelling shows that the stresses in the excavation floor when it is at 30 m depth (i.e. 2 m above the shale band) are 4.5 MPa with the shale band and 3.5 MPa without it. These can be compared with the virgin stress of 2 MPa and show that the stresses do not increase to carry the entire virgin in-situ stresses.

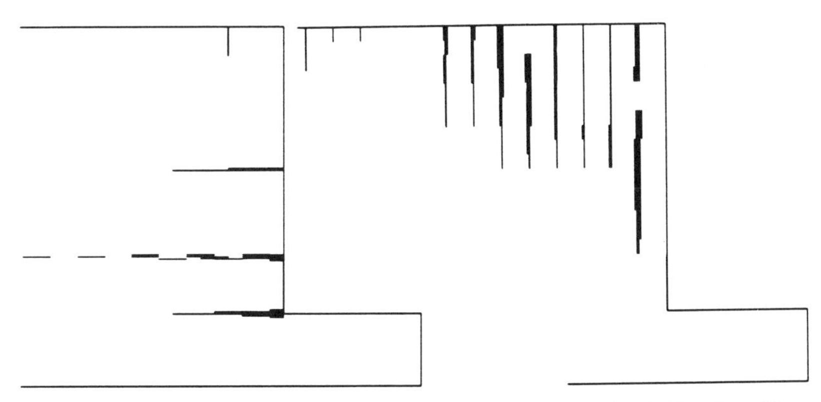

FIG. 10. Excavation 40 m Deep in *4 m Rock* Model With Shale Band at 32 m Depth: (a) Shear Displacement on Joints (Maximum 17.9 mm); (b) Joint Opening (Maximum 5.3 mm)

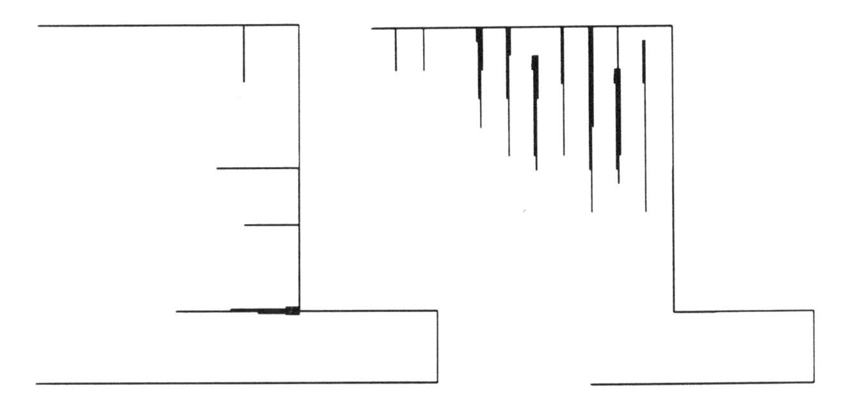

FIG. 11. Excavation 40 m Deep in *4 m Rock* Model Without Shale Band: (a) Shear Displacement on Joints (Maximum 20.0 mm); (b) Joint Opening (Maximum 5.2 mm)

A more appropriate model of the rock behavior is that of relieving internal strains. The in-situ internal strains are equivalent to that which would be produced by the applied stress field of Fig. 1. When the excavation is made the strain is relieved in the adjacent rock and no further elastic dilation of the rock takes place. In the shale case most of the rock strain has been relieved by the time the shale band is reached which is why there is only the very modest stress increase of 2.5 MPa over the virgin stress. Thus the in-situ stresses do not represent an externally applied stress field and once the strain has been released, the lateral stresses have dissipated.

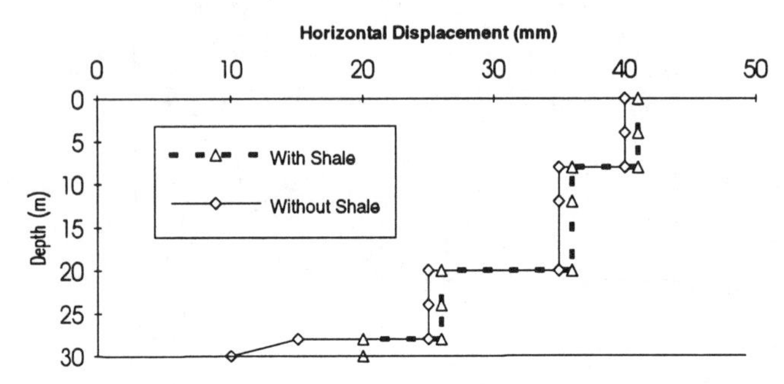

FIG. 12. Deformed Slope of Face of Excavation 30 m Deep With and Without Shale Band at 32 m Depth

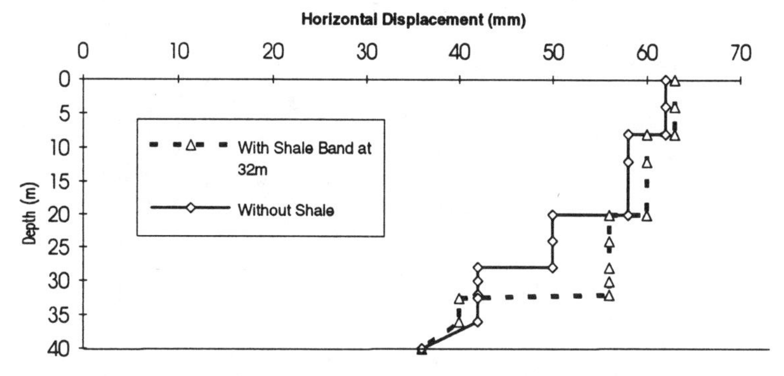

FIG. 13. Deformed Slope of Face of Excavation 40 m Deep with and without Shale Band at 32 m Depth

CONCLUSIONS

The following conclusions were drawn from the study:

- o The deformation of blocky rock masses cannot be accurately modelled using continuum methods as defects govern deformation in a non-linear manner which cannot be accounted for by a simple transform of the modulus.
- o Initial in-situ stresses are not external loads but represent internal strains which are relieved by excavation.
- o Deformation of excavations was found to be a two stage process. Firstly, initial elastic dilation of the rock occurs due to stress relief. This dilation appears to vary linearly with depth. Secondly, rafting of layers isolated by vertical defects occurs, this is negligible at shallow depths but increases rapidly with depth.

o Surprisingly, the presence of a weak shale band does not have a large influence on deformation of the excavation face.

ACKNOWLEDGEMENT

Much of the analysis presented in this paper was undertaken for an undergraduate thesis (Nash 1990) and the authors gratefully acknowledge the insights given by John Braybrooke.

APPENDIX - REFERENCES

Baxter, D. A., and Nye, E. J. (1990). "ANA Hotel, Sydney, excavation adjacent to a major railway tunnel." *Proc. 7th Australian Tunnelling Conf.*, Sydney, 250-257.

Braybrooke, J. C. (1993). *Personal communication.*

Carter, J. P., and Alehossein, H. (1990). "Basement excavations near tunnels in jointed rock." *Proc. 7th Australian Tunnelling Conference*, Sydney, 239-240.

Carter, J. P., Alehossein, H., Booker, J. R., and Balaam, N. P. (1990). "Elastic solutions for tunnels near excavations." *Civil Engineering Transactions*, Institution of Engineers, Australia, 32(8), 75-85.

Enever, J. R., Windsor, C. R., and Walton, R. J. (1990). "Stress regime in the Sydney Basin and its implications for excavation design and construction." *Proc. 7th Australian Tunnelling Conference*, Sydney, 45-50.

Nash, P., (1991). "The Influence of Lateral Stresses on and about Excavations in the Sydney Area," B.E. thesis, School of Civil Eng., University of New South Wales, Kensington, NSW, Australia.

Pells, P. J. N. (1985). "Engineering properties of the Hawkesbury Sandstone." *Engineering Geology of the Sydney Region*, Balkema, Rotterdam, The Netherlands, 179-198.

Pells, P. J. N. (1990). "Stresses and displacements around deep basements in the Sydney area." *Proc. 7th Australian Tunnelling Conference*, Sydney, 241-249.

Pells, P. J. N., Douglas, D. J., Rodway, B., Thorne, C., and McMahon, B. K. (1978). "Design loadings for foundations on shale and sandstone in the Sydney region." *Australian Geomechanics Journal*, G8, 31-39.

Trow, W. A., and Lo, K. Y. (1989). "Horizontal displacements induced by rock excavation: Scotia Plaza, Toronto, Ontario." *Can. Geotech. J.*, 26, 114-121.

RATE OF SETTLEMENT OF SILT ALLUVIUM AT TROJAN

Derek H. Cornforth,[1] Fellow, ASCE

ABSTRACT: Settlements of 1.83 m have been observed at the switchyard of the Trojan Nuclear Power Plant over a 22-year period since 1970. Rockfill had been placed on the site, a flood plain of the Columbia River, to a relatively uniform depth of 7 m. The fill bears directly on 33 m of soft, compressible silts and clays which are underlain by relatively incompressible sand deposits. Because the fill covers a very wide area (380 m times 750 m), the loading conditions offer a rare opportunity to observe the rate of settlement behavior of a deep stratum of layered alluvium under one-dimensional (vertical) settlement.

The results show that the alluvium essentially follows the Terzaghi theory of one-dimensional consolidation, i.e. the layered alluvium behaves as if one 'composite' material is being loaded. The practical significance of this result is that it demonstrates that the Terzaghi theory can be used to monitor settlements of alluvium in the field as, for example, in preloading construction sites. The settlement analysis also showed that the total settlement observed to date, at the end of the primary consolidation, is in good agreement with the calculated settlement from laboratory consolidation tests.

INTRODUCTION

Terzaghi's theory of one-dimensional consolidation considers a uniform applied stress acting on a horizontal, semi-infinite mass of soil in which the thickness of the compressible stratum is small relative to the width of the loading (Terzaghi 1943). These conditions are rarely found in practice, especially when the compressible soil stratum is thick.

---

[1] President, Cornforth Consultants, Inc., 10250 SW Greenburg Road, Suite 111, Portland, OR 97223.

The theory of one-dimensional consolidation (Terzaghi and Fröhlich 1936, Case 1) predicts a parabolic relationship between ground settlement δ and the elapsed time $(t-t_o)$ in the early phase of consolidation. Time t_o is the time of initial loading, assumed to occur instantaneously. Plots of δ versus $\sqrt{(t-t_o)}$ are linear, and this relationship is commonly used in laboratory tests of fine-grained soils to calculate the parameters of coefficient of consolidation, c_v, and coefficient of permeability, k_v.

Thick deposits of alluvium on the flood plains of rivers comprise soil layers of varying composition. Near a slow-moving river, the soils are predominantly clayey silts and very silty clays with partings, pockets and lenses of silt and sand. Organic content is variable. Given the inherently layered deposition of alluvium, and the different consolidation properties of these individual layers, how does the total stratum behave under vertical settlement? Also, how does the observed settlement compare with the calculated settlement based on laboratory consolidation tests? The main emphasis of the paper is on the *rate of settlement*, based on large numbers of settlement readings over a 22-year period.

Background Information

The Trojan nuclear power plant, near Rainier, Oregon, has a unique geological setting alongside the Columbia River. The main plant is built on a small island of basalt rock which protrudes through the flood plain of the river. During the late Pleistocene, the Columbia River was the principal conduit for immense quantities of ice melt which came from the ice masses to the north. At that time, the sea level is estimated to have been about 90 m (300 ft) below the present level, and periodic torrential floods (including ice dam collapses) scoured a deep river channel into the hard rocks of the region.

The area between the rock island and the Oregon river bank is an old channel of the river, about 800 m (2,600 ft) wide, which has infilled with river alluvium to depths of more than 60 m (200 ft). Prior to construction of the power plant, this area was a low-lying slough and flood plain, mostly a meter or so below normal river level.

In September and October 1970, angular basalt rockfill was taken from the power plant excavation and placed over an area 460 m times 230 m (1,500 ft times 750 ft) of the flood plain to create buildable land, part of which later became the switchyard (Fig. 1). The filled area was roughly graded to elev. +7.6 m (+25.0 ft) during the initial fill construction. The fill thickness was typically 6.1 m (20 ft), except at the access road and southwest corner, where it was thinner (Fig. 1).

The switchyard settled significantly over the next two years. To re-establish the grade to elevation +7.5 m (+24.67 ft), about 0.9 m (3 ft) of additional fill was placed over the switchyard, September 15 to 29, 1972.

Subsurface Conditions

Four borings were put down at the switchyard site, and their positions are shown on Fig. 1. The two earliest holes, B-50 and B-51, 20 m (66 ft) and 47 m (153 ft) deep, respectively, were put down by rotary drilling on slightly elevated ground in the flood plain prior to placing fill. Borings B-77 and B-78 were put down by a hollow stem auger drill rig approximately one year after the initial grading of the site and one year before the regrading work.

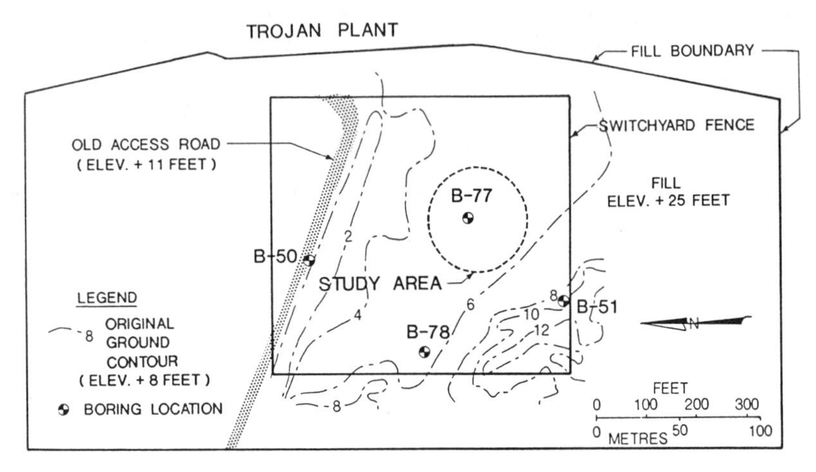

FIG. 1. Site Plan of Fill Construction

The subsurface conditions in B-77 (Fig. 2) were very similar to those in B-78 (not reproduced) and others at the Trojan site within the deep alluvium. Below the rockfill, there is a highly compressible stratum of thinly bedded sediments, 15.9 m (52 ft) thick. Most samples can be described as soft, grey, sandy, clayey SILT with numerous scattered black organics. The alluvium includes soils ranging from clays to silty fine sand. Organic content is highly variable, occasionally peaty.

Below this layer is a 2.1 m (7-foot) thick layer of very compact, light grey SILT (volcanic ash). It is a well-known marker bed in the area and probably represents a major volcanic eruption of the past.

A second stratum of alluvial silts and clays underlies the volcanic ash layer. It is similar in composition to the upper stratum but is noticeably stiffer, and the soils have lower water contents (Fig. 2). There is a greater frequency of sand lenses within the lower stratum. The thickness of the lower compressible stratum is 14.9 m (49 ft) in B-77.

The total thickness of fine-grained sediments is 32.8 m (107.5 ft). Below the silt and clay alluvium, there is very dense, fine-to-coarse SAND, which continued to the bottom of the hole at 48.9 m (160½ ft) in B-78. The sand stratum provides the lower drainage boundary.

The rockfill at the site is well-graded, angular basalt fragments (grading: silty, sandy gravel). The fines hold moisture above the water table. There are a few minor pockets of brown silt, estimated to be around 3% of the total fill.

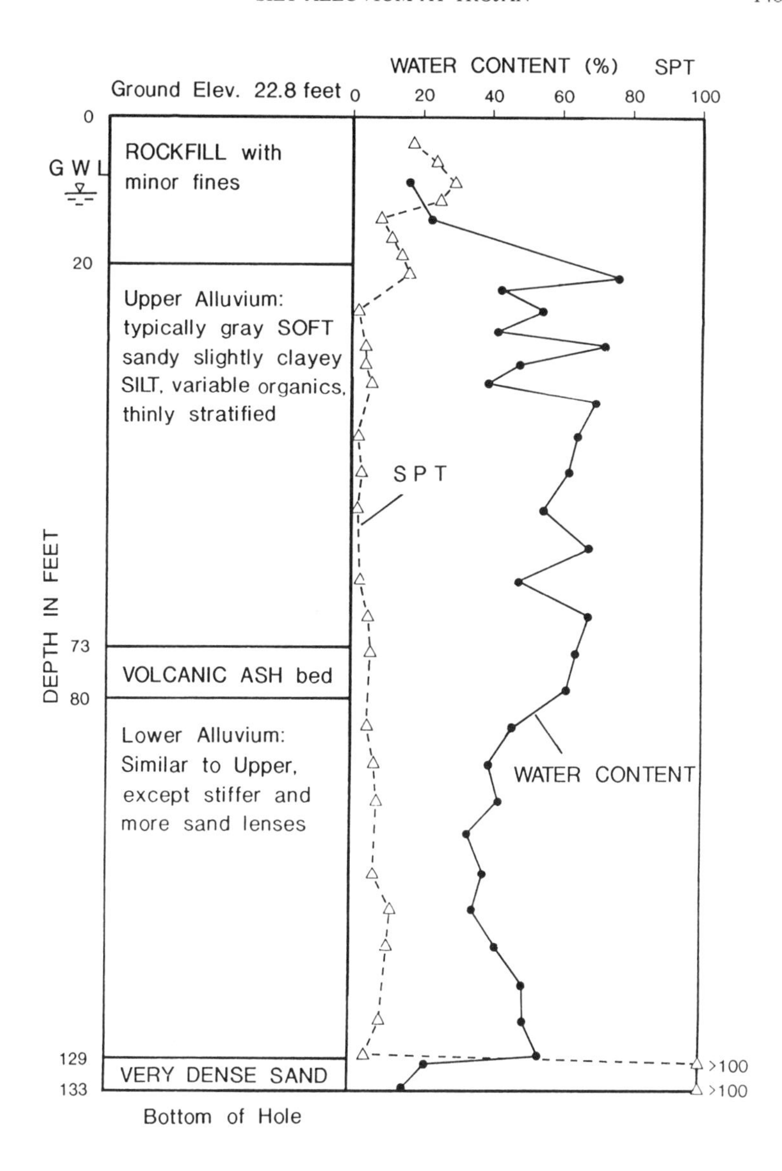

FIG. 2. Summary Boring Log B-77

Groundwater levels (GWL) are controlled by the Columbia River and the Reflecting Pond, a landscaping feature adjoining the switchyard (Fig. 1). All borings recorded groundwater at elev. +3.6 m (+12 ft).

Summary of Soil Properties

A comprehensive laboratory test program was performed on undisturbed samples taken by jacking thin-wall tubes into the soils. Particular emphasis was placed on the upper compressible stratum, which accounts for about 73% of the total calculated settlement. Some data is given below:

TABLE 1. Classification Properties for Upper Compressible Stratum

Property	Range of Values	Median Value (10 tests)
Liquid Limit (%)	40-148	58
Plastic Limit (%)	32-86	38
Plasticity Index (%)	8-62	20
Natural Water Content (%)	37-121	54
Liquidity Index	0.35-1.00	0.69

The *average* natural water content declined from 63.4% before construction to 57.9% after one year of settlement (when the results given in Table 1 were taken). This difference of 5.5%, if added to the median water content of 54% in Table 1, would bring the median natural water content *before construction* up to the median liquid limit, i.e. a liquidity index of around 1.0.

The undrained shear strength (Q-tests) in the upper compressible stratum increased from 20.6 kN/m^2 (430 lb/ft^2) before fill construction to 34.0 kN/m^2 (710 lb/ft^2) after one year of settlement. Eight consolidation (oedometer) tests were performed on undisturbed samples.

Ground Settlement Observations

The exploratory borings of October 1971, one year after the initial rough grading of the site, showed that there had been large ground settlements. Subsequently, spot levels were taken at nine points around the center and perimeter of the site in February, April, and July of 1972. Final (topping up) grading was performed September 15-29, 1972, and more frequent readings of settlement began in November 1972.

Piles were driven to support the 230 kv towers (Fig. 3) in October 1972. The piles are end-bearing in dense sand below the compressible soils. The switchyard structures on spread footings are very light and can be ignored for their contribution to settlement.

Settlement readings were taken at 19 locations between November 1972 and April 1975. The reading interval was initially two weeks but was changed to monthly readings in June 1973. In April 1975, the number of observation points was increased to 148. Reading intervals progressively increased from monthly to annually from 1975 to 1992. Some individual observation points have 84 settlement readings over the period 1972-1992.

There have been more than 6,000 settlement readings taken at the site. To simplify the analysis, an area 60 m (200 ft) in diameter and incorporating B-77 has been selected. It is shown by a broken circle on Figures 1 and 3 and is referred to as the *study area* hereafter.

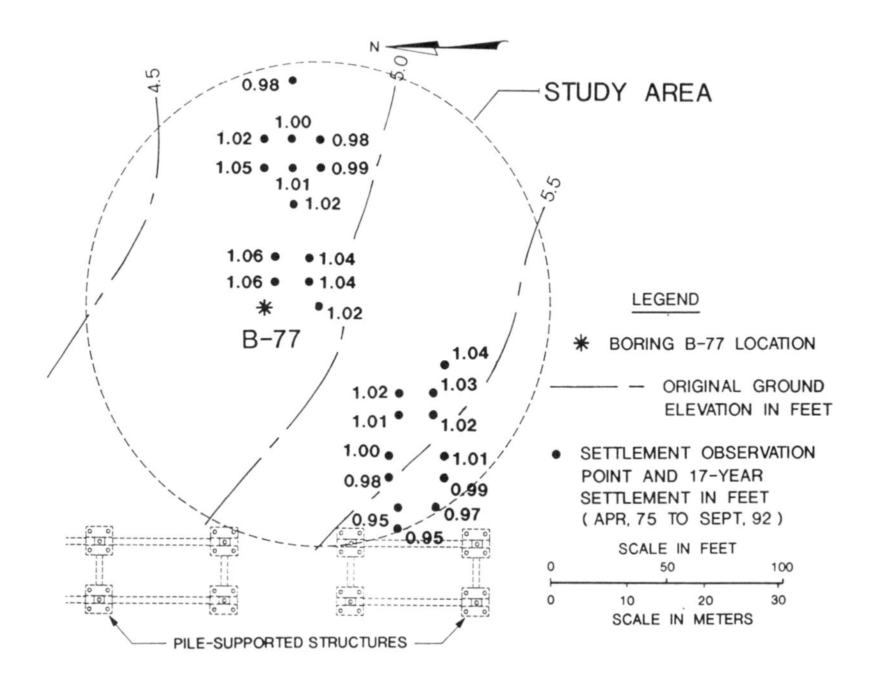

FIG. 3. Partial Plan of Switchyard Showing Details of the Study Area

RATE OF SETTLEMENT ANALYSIS: FIRST PERIOD OF LOADING

All the rough grading was placed during the two-month period of September and October 1970. It is assumed that the fill was built at a uniform rate and the equivalent time for placing an instantaneous load (time t_o for the Terzaghi one-dimensional consolidation theory) is midway through this fill period, i.e. October 1, 1970.

The spot elevations subsequently taken from February to July 1972 are very consistent with each other where the fill is around 6.1 m (20 ft) thick. Including the ground elevations taken during the drilling of B-77 and B-78, the results show a good linear relationship for settlement δ versus root elapsed time $\sqrt{(t-t_o)}$ on Fig. 4.

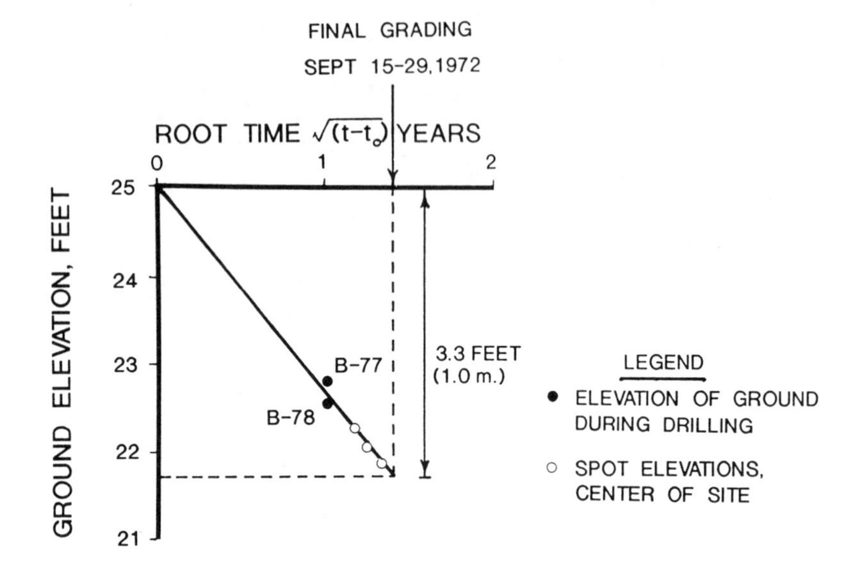

FIG. 4. Settlement Graph: First Period of Loading

Rate of Settlement Analysis: Second Period of Loading

The switchyard was topped up with additional fill to compensate for the settlement which had occurred. According to Fig. 4, the average amount of settlement at the time of the regrading work of September 15-29, 1972 was 1.00 m (3.3 ft). Since the regrading established a new grade of elev. +7.52 m (+24.67 ft) at the site, 0.9 m (3 ft) of additional rockfill was placed above the original fill thickness of 6.1 m (20 ft).

The original ground elevations of the flood plain within the study area are shown on Fig. 3 and average +1.52 m (+5.0 ft). The settlements over the 17-year period from April 1975 to September 1992 for 25 observation points within the circle of the study area are also shown on Fig. 3. The uniformity of the fill thickness and consistency of the observed settlement values within the study area are readily apparent and show that essentially one-dimensional settlement has occurred.

The analysis of the rate of settlement becomes more complicated for the period after the regrading because the additional fill speeds up the rate of settlement. There is also a small change in the applied vertical stress with time due to continuing settlement; this topic will be discussed first.

In the study area, the original roughly graded fill of 6.1 m (20 ft) had 2.14 m (7 ft) of fill below groundwater level (GWL) and 3.96 m (13 ft) above GWL. As settlement occurs, the height of fill passing below GWL increases, causing a slow reduction of the applied vertical effective stress, σ_v'. The average densities of the well-graded rockfill were computed from the average water contents above and below GWL to give the following results: damp density (above GWL) 18.86 kN/m^3 (120.0 pcf); buoyant density (below GWL) 10.38 kN/m^3 (66.0 pcf).

During the first period of loading, σ'_v declines parabolically from 96.7 to 88.1 kN/m^2 (1.01 to 0.92 tsf). After adding 0.91 m (3.0 ft) in regrading, σ'_v increases to 105.3 kN/m^2 (1.10 tsf), an increase of 19.5%.

To recap the train of thought, it has been shown on Fig. 4 that settlements over the first two years, when the original fill was on the site, approximately follow the classical one-dimensional consolidation relationship, i.e. settlement δ is proportional to $\sqrt{(t-t_o)}$ where $(t-t_o)$ is the elapsed time since loading. However, when the owner added more fill to the site to re-establish the grade, the additional fill speeded up the rate of settlement. It can be shown, using the Terzaghi theoretical model, that a subsequent increase in stress causes the settlements to eventually approach the curve that would have been obtained if the two stresses had been applied simultaneously at time t_o. This concept is illustrated on Fig. 5, and it can be used to determine whether, and for how long, the linear relationship between settlement δ and $\sqrt{(t-t_o)}$ persists at the Trojan site.

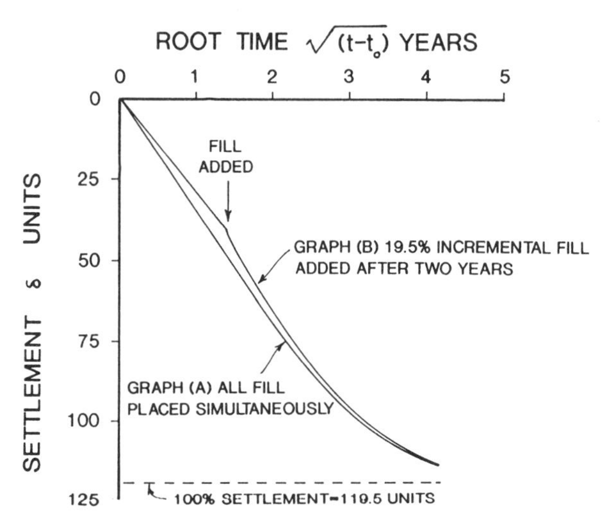

FIG. 5. Effect of an Incremental Applied Vertical Stress on the Rate of Settlement (illustration using Terzaghi theory of one-dimensional consolidation, with 50% average consolidation after three years and 19½% additional stress applied after two years)

The theoretical parameters used to produce Fig. 5 were selected to closely model the conditions at Trojan. It is based on soil reaching 50% of consolidation after three years (i.e. $\sqrt{(t-t_o)}$ = 1.73) and shows the effect of adding a 19.5% incremental stress increase after two years. Curve (B) represents the resulting theoretical settlement graph. Curve (A) is the theoretical graph obtained by placing both stresses simultaneously at time t_o. The two graphs assume that settlements are directly proportional to stress increases for incremental loading, i.e. a 19½% stress increase raises the final settlement from 100 units to 119½ units. By dividing the results of (A) by the results of (B) for the same elapsed times $(t-t_o)$, ratios are obtained by which *actual* settlements must be multiplied to give the *equivalent* settlement curve for the two fills placed simultaneously at time t_o.

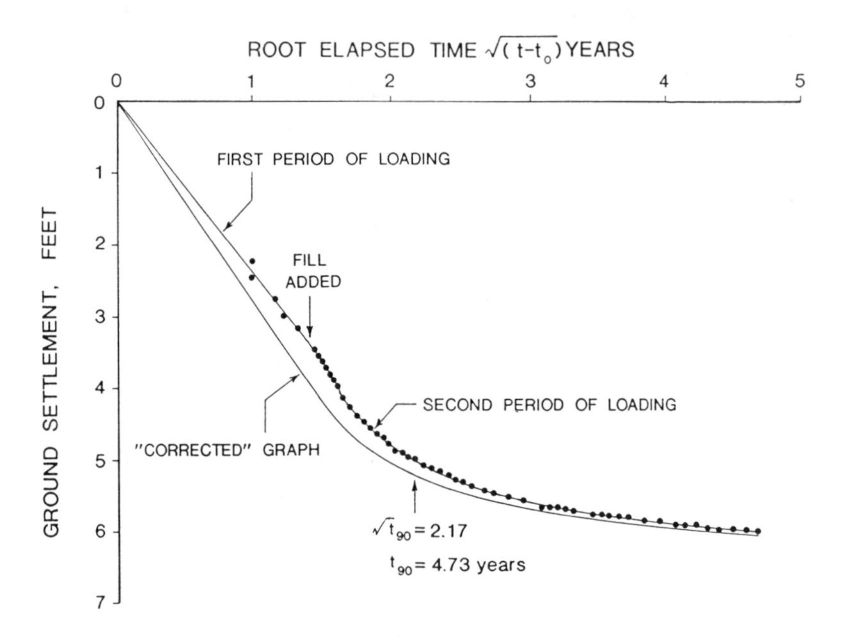

FIG. 6. Settlement Graphs: Actual and Corrected (Root Time Plots)

Over the first two years, the ratio is constant at 1.195 until the top up fill is added; after that, the ratio drops rapidly. The calculated difference is only 2% after seven years (five years after applying the extra fill) and drops below 1% difference after 11 years.

The actual and corrected settlements are plotted on Fig. 6. Many individual readings have been omitted to provide a clearer graph.

The corrected graph of Fig. 6 indicates that the straight-line portion continues to approximately $\sqrt{(t-t_o)}$ = 1.7, i.e. two years, eleven months after t_o (September 1973). The theoretical Terzaghi one-dimensional consolidation theory graph departs from linearity at an average degree of consolidation (U) of 60%. Using the offset method (departure of 15% from linearity at U = 90%), the calculated total settlement at the end of primary consolidation is 1.74 m (5.67 ft). Since the actual settlement in September 1992 was 1.83 m (6.0 ft), it can be concluded that primary consolidation is now complete, and the site is undergoing secondary compression due to the dispersed organics within the alluvium. The current (1992) rate of settlement is 0.006 m/year (0.02 ft/year).

Average vertical consolidation parameters are calculated as follows:

Coefficient of consolidation c_v = 0.0153 cm^2/s (520 $ft^2/year$)
Coefficient of permeability k_v = 0.85 times 10^{-6} cm/s

The average stratum properties obtained from this analysis are those which might be expected in a clayey to slightly clayey silt, which is the median soil type in the range of sample descriptions.

A common misconception in Oregon is that the Columbia River silt alluvium undergoes primary consolidation rapidly (90% within three months) when loaded by fills. As shown by this paper, primary consolidation is significantly slower than has been commonly assumed by geotechnical practitioners in the region.

This misconception seems to have been fostered by measurements of consolidation in the oedometer test using thin test specimens. For a c_v of 0.0153 cm^2/s, a 20 mm thick oedometer test specimen in double drainage reaches 50% of primary consolidation in 13 seconds, which is too fast for accurate measurement. A better laboratory test is to use the triaxial apparatus in single drainage to measure the coefficient of consolidation for silts and clayey silts (Cornforth 1961).

A more important result of this analysis is that the alluvial stratum, comprising many layers of varying soil and organic composition, essentially behaves as a composite single layer obeying the Terzaghi theory of one-dimensional consolidation. As a practical matter, this means that Terzaghi's theory can be used to determine the approximate degree of consolidation reached during field observations of settlement on silt alluvium.

Lateral Dissipation of Pore Pressures

The earlier calculations of composite soil properties assume that one-dimensional (vertical) *consolidation* is occurring during vertical *settlement*. Alluvial sediments are layered and it is of interest to analyze the individual sample descriptions to determine the extent of sand lenses within the alluvium. The average values were:

Silty clay, clayey silt, silt: Upper alluvium 92%; Lower 83%
Silty fine sand: Upper alluvium 8%; Lower 17%

The presence of silty fine sand lenses and pockets may cause *lateral* dissipation of excess pore water pressures during consolidation. Silty fine sand has a permeability of around 10^{-4} cm/s, i.e. about 100 times more permeable than the composite stratum. However, it seems unlikely that lateral pore pressure dissipation is significant at this site, because of the following factors: (i) a very wide loading area of 380 to 750 m compared to the 16 m deep drainage path, (ii) uniform loading of the large area, and (iii) the inherently poor lateral continuity of flood plain deposits.

Circumstantial evidence from the settlement data supports this conclusion. If significant lateral dissipation of pore water pressures were occurring, a pore pressure gradient would develop from the center to the edges of the loaded area. Thus, the edges of the fill area would consolidate more rapidly than the center. However, the 148 settlement observations across the site exhibit very uniform settlement during the period from 1975 to 1992. The data suggests that all parts of the site where the fill depth is the same are consolidating at near-identical rates. This evidence suggests that the water movement during consolidation is occurring vertically, not laterally.

TOTAL SETTLEMENT ANALYSES

The total settlement calculations followed established procedures of grouping the eight consolidation test results by initial water content and dividing the compressible strata into thin layers of similar water content. No corrections of the Schmertmann (1955) techniques were applied due to the difficulty of accurately determining the preconsolidation stresses from the consolidation tests. The *calculated* settlement was 1.80 m (5.90 ft) which agrees closely with the *observed* primary settlement of around 1.74 m (5.67 ft).

In the laboratory consolidation tests, each settlement graph was obtained from the 24-hour readings for each vertical stress increment of loading. These results include some secondary compression due to the dispersed organics in the soils. The relationship between secondary compression in the field and that measured in the laboratory is not well understood. It is beyond the scope of this study to make this interpretation.

CONCLUSIONS

1. Using the Terzaghi theoretical model for one-dimensional consolidation, it can be shown that the settlement curve for two separate fill placements eventually approaches the curve which would have resulted from the two fills being placed simultaneously. Therefore, by closely modeling the theoretical to the actual conditions at Trojan, the factors needed to convert the two-stage fill process to a single stage fill can be obtained. These factors were applied to the observed settlements to determine the extent of the linear relationship between settlement δ and root elapsed time. From this curve, the average properties of the thick compressible alluvium were evaluated as follows:

   Coefficient of consolidation, c_v = 0.0153 cm^2/s (520 $ft^2/year$)
   Coefficient of permeability, k_v = 0.85×10^{-6} cm/s

2. It is of considerable practical significance that a thick stratum of compressible alluvium, comprising many layers of varying soil compositions, produces a settlement graph which closely resembles the Terzaghi model for primary consolidation. This result means that primary consolidation can be monitored on-site (by settlement plates or stakes) using the root elapsed time analysis.

3. The total settlement calculations used the eight consolidation test results obtained from borings at the switchyard site. The *calculated* total settlement for the study area was 1.80 m which can be compared with the *observed* primary settlement of 1.74 m. Although agreement between the calculated and observed settlements is excellent, the role of secondary compression introduces an element into the comparison which is difficult to quantify.

APPENDIX - REFERENCES

Cornforth, D. H. (1961). Contribution to: "Soil properties and their measurement." *Proc. 5th ICSMFE*, Paris, 3, 121-122.

Schmertmann, J. H. (1955). "The undisturbed consolidation behavior of clay." *Trans. ASCE*, 120, 1201-1227.

Terzaghi, K. (1943). *Theoretical Soil Mechanics*, John Wiley & Sons, New York, New York.

Terzaghi, K., and Fröhlich, O. K. (1936). *Theorie der Setzung von Tonschichten*, F. Deuticke, Vienna, Austria.

Tree Root Induced Settlement of a Large Industrial Building Founded on Expansive Clay - A Case History

Aziz Aboaziza,[1] Member, ASCE,
and Hisham H. H. Mahmoud,[2] Associate Member, ASCE

Abstract: Trees planted adjacent to buildings have been known to cause significant structural distress. In this paper, a case study is presented where a large tilt-up concrete industrial building approximately 73 m (240 ft) × 122 m (400 ft) in plan dimension, founded on expansive clay, has settled differentially on the order of 64 mm (2.5 in.). The observed settlement is believed to be attributed to the differential shrinkage of the foundation clay soils due to a row of 45 trees planted along the distressed edge of the building. Geotechnical investigations were performed for the subject building. The results of these investigations as well as the mechanism of the observed distress are discussed.

Introduction

Although trees planted adjacent to structures have been known to be a source of severe structural distress and damage, this issue is often overlooked by engineers and planners. The main reason for the frequent reoccurrence of this problem is probably due to the information gap that exists between design engineers and landscape architects. The design engineer produces plans and specifications for building foundations and other structural elements that can sustain imposed stresses of the anticipated building loads for underlying subsurface soil conditions. The geotechnical project specifications typically deal with earthwork provisions,

---

[1] Associate, Woodward-Clyde Consultants, 1615 Murray Canyon Road, Suite 1000, San Diego, CA 92108.

[2] Project Engineer, Dames & Moore, 7500 N. Dreamy Draw Drive, Suite 145, Phoenix, AZ 85020.

foundation placement, and rarely discuss post construction landscape issues. The landscape architect's mission, on the other hand, is to provide aesthetically pleasing landscape growth around buildings, which may include large trees. The long-term interaction between trees, foundations, and clay soils is frequently ignored by the project design team.

Structural distress induced by tree roots typically occurs over several years after construction and depends on various factors including: tree type, soil type, availability of water, climate, etc. This makes the tree roots problem of significant consequence since by the time the tree damage is realized, the structures are occupied and in full operation. Repair and retrofitting of such structures are sometimes prohibitively expensive. Numerous litigation cases involving problems of this nature are filed every year. Geotechnical engineers are often implicated in these litigations and, to some extent, are held responsible. Landscape architects are typically held to a lesser degree of responsibility, if any.

Distress caused by tree roots can generally be classified in two categories: (1) distress due to uplift forces, and (2) distress due to induced differential settlement. In addition, tree roots have been known to block drains and damage sewer lines. Distress due to uplift forces is more frequently observed on sidewalks, driveways, pavements, and lightly loaded buildings. The uplift forces are generated due to near-surface tree root growth beneath structures resulting in cracking, differential lifting, and buckling.

The distress caused by induced differential settlement is mainly observed for structures founded on clayey soils. Differential settlement is due to tree roots extracting moisture from the underlying clay causing desiccation-shrinkage and, hence, settlement. This distress is typically more pronounced when expansive soils are involved. Tucker and Poor (1978) presented a case study where a residential development consisting of 69 homes founded on a high-plasticity clay experienced distress in the form of floor slab and wall settlements on the order of 51 to 127 mm (2 to 5 in.). The distress was attributed to the presence of about 90 trees within the residential development in close proximity to the homes. Cheney and Buford (1975) reported damage to a three-story building resulting from uplift on the order of 25 mm (1 in.) due to the removal of trees from the building site prior to construction.

The effect of trees and other vegetation on the desiccation of clayey soils has been addressed by many investigators including Hammer and Thompson (1966), Biddle (1979), Driscoll (1983), and Ravina (1984). Published case studies on the effect of trees on buildings share the following common scenario:

(1) A geotechnical investigation is performed for the proposed structures which identifies the presence of clay soils. Recommendations are developed for potential expansion of these soils, and settlement calculations are made using conventional procedures which only consider the structural loads.
(2) The structure is then constructed and landscaped. The landscaping would typically include trees in close proximity to the structures.
(3) Several years after construction, the trees mature. During dry seasons when the rate of evaporation exceeds rainfall, or for the case of a prolonged

drought, trees will advance their root system to reach beneath floor slabs and foundations in order to extract more moisture from an already dry soil, to satisfy their large water demand, causing differential shrinkage of the clayey soils. Distress to the overlying structure is then observed in the form of differential settlement, distortion, and cracks along walls and floor slabs.

Post-investigations of the cause of observed structural distress typically overlook the mechanism discussed above and frequently attribute distress to other soil mechanics factors. In this study, the writers examined the various possible causes of the building distress and, through a process of elimination, concluded that the effect of adjacent tree roots to be the most logical mechanism for the observed distress.

BACKGROUND

The subject building is a high ceiling, single story concrete tilt-up structure constructed in the summer of 1983 and is currently used as a warehouse. The building is about 73 m (240 ft) × 122 m (400 ft) in plan dimension founded on conventional shallow footings and slab-on-grade floors. The exterior and interior columns are spaced approximately 6 m and 12 m (20 and 40 ft) on centers. The exterior column loads range from approximately 240 to 374 kN (54 to 84 kips) with a foundation contact pressure on the order of 72 to 192 kPa (1.5 to 4.0 ksf). The exterior wall loads are on the order of 37 kN/m (2.5 kips/ft) with an average contact pressure of 72 kPa (1.5 ksf).

The area along the south edge of the building is landscaped with grass, low bushes, and 45 Eucalyptus and Bottle Brush trees. The trees are about 0.6 to 0.9 m (2 to 3 ft) in diameter and at a distance of about 1.5 to 2.1 m (5 to 7 ft) from the south edge of the building. The height of the trees ranges from approximately 7.6 to 9.1 m (25 to 30 ft). The areas along the north, east, and west edges of the building are paved.

Distress was observed along the south edge of the building in the form of cracks along the walls and floor slab, separation of the pre-cast concrete wall panels, spalling and buckling of concrete columns, and separation of the roof deck. The distress reportedly become more noticeable between 1989 and 1991. A floor level manometer survey was performed in 1991 which indicated a maximum differential settlement on the order of 64 mm (2.5 in.) along the southeast corner of the building (Fig. 1).

SUBSURFACE CONDITIONS

The initial geotechnical investigation for the site was performed in early 1982 and included drilling 7 borings to depths ranging from 9.1 to 15.2 m (30 to 50 ft). Several post-distress investigations were performed at the site between 1990 and 1992. A total of 18 hollow-stem auger borings, 3 Cone Penetration Tests (CPTs), 2 test pits, and 2 floor level surveys were performed in the area of the building. The borings and CPTs were advanced to depths ranging from about 3.0 to 15.2 m (10 to 50 ft) below existing grade. The approximate location of the borings, CPTs, and test pits are shown on Fig. 2.

The subsurface soil conditions in the building area, as revealed by the above investigations, generally consist of a shallow near surface fill approximately 0.6 to 1.2 m (2 to 4 ft) thick underlain by old alluvial fan deposits. The fill soils consist predominantly of silty to sandy clay and silty sands placed at water contents of 1 to 3 percent above optimum (average 2 percent) at a minimum of 90 percent relative compaction per ASTM Test Method D1557. The overall averages of the fill water contents and dry densities are 15 percent and 17.3 kN/m^3 (110 pcf), respectively.

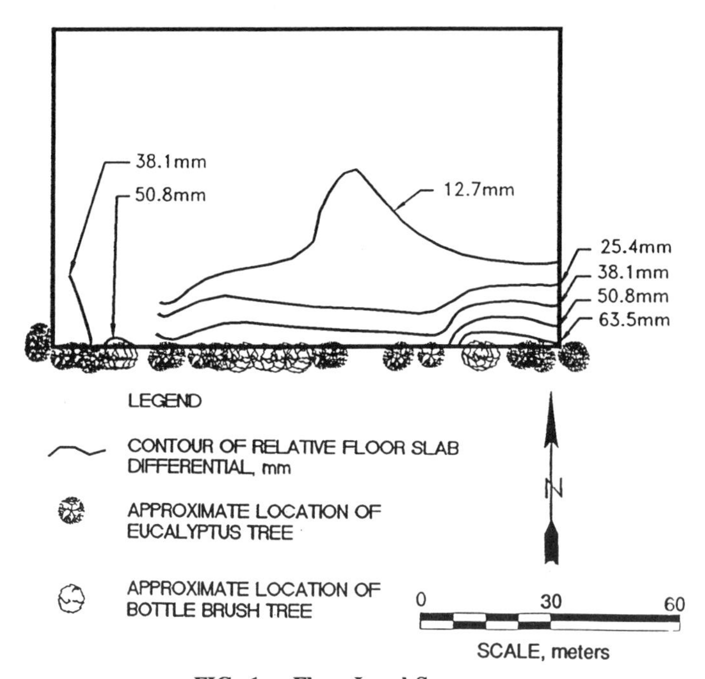

FIG. 1. Floor Level Survey

The old alluvial fan deposits are the basic geologic unit underlying the site at depth. The alluvial deposits generally consist of interbedded layers of stiff to hard silty to sandy clay and medium dense to dense silty sands (Fig. 3). The clay exhibits medium to high plasticity characteristics with liquid limits ranging from 45 to 56 percent and plasticity indices on the order of 20 to 30 percent. The Standard Penetration Test (SPT) blow counts in the alluvial deposits, throughout the maximum explored depth of 15.2 m (50 ft), ranged from 20 to 100 blows per 0.3 m (1 ft). The near surface clays indicate high swell potential of about 12 percent under a 6.9 kPa (144 psf) load.

POSSIBLE CAUSE AND MECHANISM OF OBSERVED DISTRESS

The building has experienced a differential settlement on the order of 64 mm (2.5 in.) over a six year period after construction. The observed settlement and structural damages were mostly along the south edge of the building. Several possible causes were examined including: (a) settlement of fill and native soils under imposed loads, (b) collapse of underlying soils due to wetting, (c) leaking sewer or water lines contributing to the distress, (d) seismically induced settlement, and (e) differential shrinkage of underlying clay causing settlement.

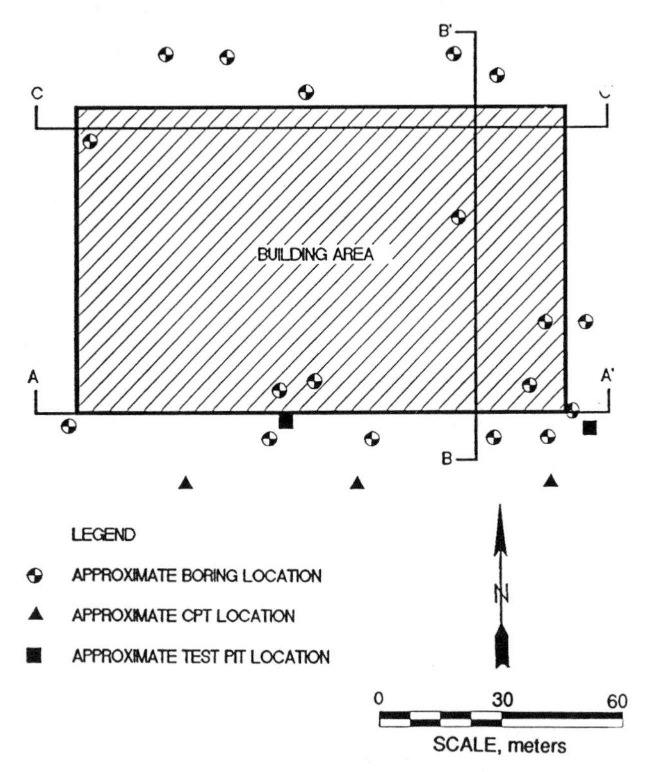

FIG. 2. Location of Test Borings and Geologic Cross-Sections

The compressibility of the near-surface fill and underlying alluvial deposits was evaluated using the results of laboratory one-dimensional consolidation tests. The natural water contents of the alluvial deposits ranged from about 10 to 24 percent and were close to the plastic limits. The results of the consolidation tests indicated an overconsolidated behavior. Upon compression, the laboratory samples generally indicated a low compressibility of 0.5 to 1 percent from existing water content to full saturation under a load of 192 kPa (4 ksf). At lighter loads of 48 to

96 kPa (1 to 2 ksf), the samples indicated a slight tendency to swell upon saturation. The near-surface fill records indicate that the fill is shallow and placed at a high water content and density. Based on this information, the contribution of the fill and underlying alluvial deposits to the observed settlement was considered minimal, particularly when considering the history and pattern of the settlement as well as the remainder of the building sections that exhibit no distress under similar loading and subsurface conditions. The conventional compressibility theory alone did not explain the observed settlement.

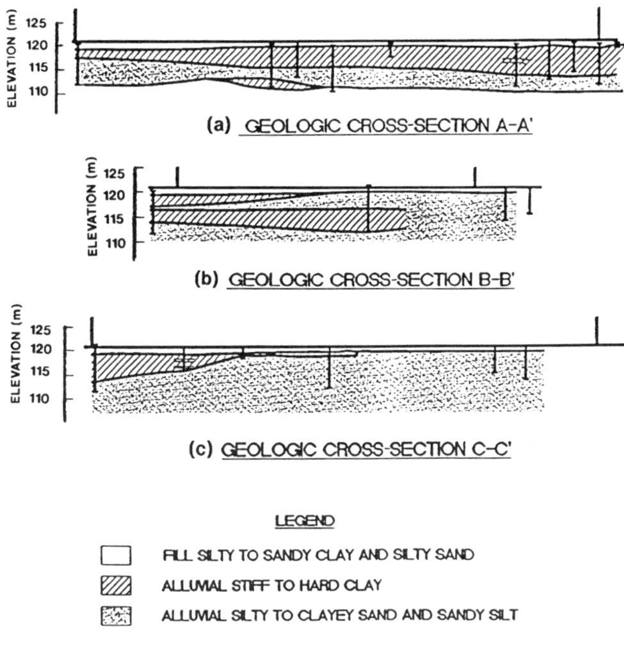

FIG. 3. Geologic Cross-Sections

The collapse potential of the underlying alluvial deposits upon wetting was evaluated as a possible mechanism of the observed settlement. Laboratory response-to-wetting tests were performed on relatively undisturbed samples of the alluvial silty to clayey sand and sandy silt. The tests did not indicate a potential for collapse. Considering these findings and the high density of the underlying natural deposits, the possibility of structural collapse of the older alluvial deposits as the source of the observed settlement was ruled out.

The possibility that the observed distress was caused by a local anomaly such as broken or leaking sewer or water lines was also evaluated. The building plans indicate a 152-mm (6-in.) diameter sewer pipe and a 102-mm (4-in.) water line along the south edge of the building. Review of the building's water consumption records and water content of soils near the utility lines did not indicate any unusual water loss or high water content of nearby soils.

The building is located in a seismically active region. The possibility that the observed settlement and distress was caused by a seismic event was discounted, however, due to the pattern of the distress and settlement.

The possibility of differential shrinkage due to desiccation of the underlying clay soils caused by the tree roots along the distressed south edge of the building was examined. A comparison was made of the soil water content profiles inside and outside the south edge of the building (Fig. 4). Fig. 4 indicates a distinct consistent difference in water contents with depth between the exterior and interior of the building along the south edge. The difference in water contents is about 5 to 7 percent, on average, and appears to occur over a depth of 6 m (20 ft) below existing grade, which closely corresponds to the bottom of the underlying clay layer. Below the depth of 6 m (20 ft), the interior and exterior water content profiles merge, indicating a uniform (equilibrium) water content pattern.

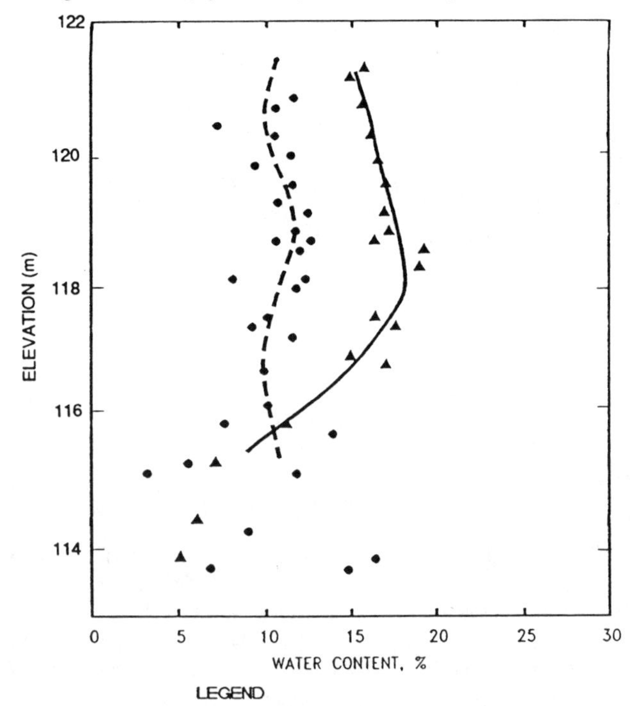

FIG. 4. Water Contents versus Depth along South Edge

The water content profile along the southern exterior edge of the building indicates an average water content of about 12 percent for the full thickness of the underlying clay layer, which is close to the shrinkage limit of the clay, while the water content profile along the interior of the building indicates an average water content of about 18 percent. Driscoll (1983) suggested that the onset of desiccation of a clayey soil occurs when the water content of the clay is about 50 percent of its liquid limit, and that desiccation becomes significant when the water content of the clay is about 40 percent of its liquid limit. This observation explains the gradual pattern of the observed settlement along the south edge of the building from the building's interior (plastic limit range) to the building's exterior (shrinkage limit range).

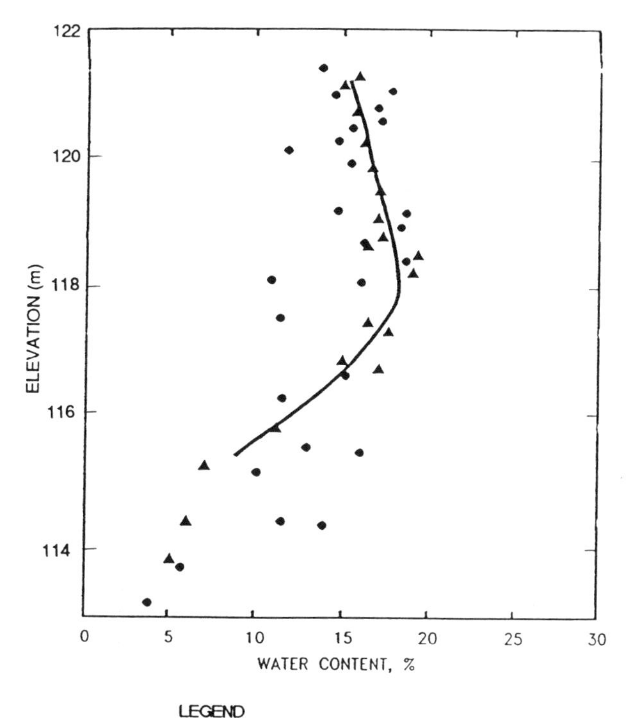

FIG. 5. Water Content versus Depth along North Edge

A similar comparison was made of the soil water content profiles inside and outside the north edge of the building (Fig. 5) where no settlement or structural distress was reported. The results did not indicate a significant change in the water content profile between the interior and exterior of the building. This moisture equilibrium may be due in part to the lack of tree roots and a thinner clay layer along the north edge of the building.

Based on the moisture content profile observations presented above, differential shrinkage of the underlying clay layer due to a significant reduction in water contents is believed to be the main cause of the observed settlement and distress at the southern portion of the building. The reduction in water contents of the clay layer is attributed to the root system of the trees along the south edge of the building.

CONCLUSION

A row of 45 trees along the south edge of an industrial building has caused structural distress and differential settlement on the order of 64 mm (2.5 in.) due to differential shrinkage of the underlying clay soils. A typical remedial action for this type of building distress may involve the removal of the trees. This would result in an increase in the water content of the desiccated clay causing it to eventually swell to its initial volume, provided that proper irrigation and drainage around the building is maintained. Repair of the structural damage would then be required. Caution should be taken to ensure moisture equilibrium of the foundation clay prior to building repair.

A possible approach to minimizing the potential for these types of problems is to narrow the information gap (lack of communications) between the design engineers and other members of the project team, including the landscape architects, during the design phase as well as throughout the execution of the project plans and specifications. Recommendations presented in geotechnical reports for buildings and other structures should address the potential effect of landscaping and irrigation, particularly when moisture-sensitive soils are involved.

APPENDIX - REFERENCES

Biddle, P. G. (1979). "Tree root damage to buildings - An arboriculturist's experience." *Arboricultural J.*, 3(6), 397-412.

Cheney, J. E., and Buford, D. (1975). "Damaging uplift to a three-story office block constructed on a clay soil following removal of trees." *Proc. Conf. Settlement of Structures*, Cambridge, 337-343.

Driscoll, R. (1983). "The influence of vegetation on the swelling and shrinkage caused by large trees." *Géotechnique*, 33(2), 93-105.

Hammer, M. J., and Thompson, O. B. (1966). "Foundation clay shrinkage caused by large trees." *J. Soil Mech. Found. Div.*, Proc. ASCE, 92(SM6), 1-17.

Ravina, I. (1984). "The influence of vegetation on moisture and volume changes." *The influence of vegetation on clays*, Thomas Telford Ltd., London, 62-68.

Richards, B. G., Peter, P., and Emerson W. W. (1983). "The effect of vegetation on the swelling and shrinking of soils in Australia." *Géotechnique*, 33(2), 127-139.

Tucker, R. L., and Poor, A. R. (1978). "Field study of moisture and volume changes." *J. Geotech. Eng. Div.*, Proc. ASCE, 104(GT4), 402-414.

DAMAGE EVALUATION OF BUILDING ON PIERS IN EXPANSIVE CLAY

Richard J. Finno,[1] Gary J. Klein,[2] Members, ASCE, and Paul J. Sabatini, Student Member,[3] ASCE

ABSTRACT: A number of columns in the crawl space of a pier-supported structure founded in expansive clays were damaged as a result of ground movements since the building was constructed in 1975. Results of investigations indicate that water contents within the swelling clay have not changed significantly in 17 years, implying that swelling had not caused the movements. The observed pattern of ground movements and damage are shown to be consistent with the movements expected from stress relief due to excavation of as much as 8.2 m of soil. Flexibility of the columns between the floor slab and excavated grade also influenced the observed patterns of damage.

INTRODUCTION

Thirty-eight of 115 interior columns of a pier-supported office building have sustained damage, apparently as a result of ground movements since the structure was constructed in 1975. A survey was made which indicated the locations of the damaged columns and estimates of the amount of damage to each column, the amount of distortion of the columns, and the interpreted directions of the column movements. The damage occurred to the reinforced concrete columns in the crawl space between the excavated grade and the bottom of the floor slab. Most of the damage occurred at columns closest to the exterior walls of the building. The soil in the crawl space typically heaved 50 to 75 mm, but heaved as much as 200 mm in the deepest part of the excavation.

A subsurface investigation was conducted to help identify the causes of the ground movements. A number of shallow boreholes were drilled in the crawl space

---

[1] Assoc. Prof., Dept. of Civ. Engrg., Northwestern Univ., Evanston, IL 60208.
[2] Sr. Consultant, Wiss, Janney, Elstner Associates, Inc., Northbrook, IL 60062.
[3] Res. Asst., Dept. of Civ. Engrg., Northwestern Univ., Evanston, IL 60208.

to develop shallow profiles through the building. Numerous water content determinations were made on samples obtained from these boreholes to evaluate the change in water content in these soils compared to those found during the original site investigation in 1975. Four deeper boreholes were drilled and sampled at locations outside the building to obtain samples for index property and consolidation testing. To help evaluate the cause of the apparent ground movements and related column damage, a finite element simulation of the excavation for the building was conducted. This paper summarizes the subsurface conditions at the site and the extent of the damage to the columns, describes the finite element analyses and discusses the apparent causes of the ground movements.

SUBSURFACE CONDITIONS

The subsurface conditions at the site relative to the foundations are idealized in Fig.1. The topography at the site is gently sloping; the original ground surface elevation varied from 114 to 110 m. The soils at the site are typically stiff to hard clays of the Yazoo formation. There is evidence of a shallow perched water table at approximately elev. 105 m. The piers are typically founded at elev. 100 m.

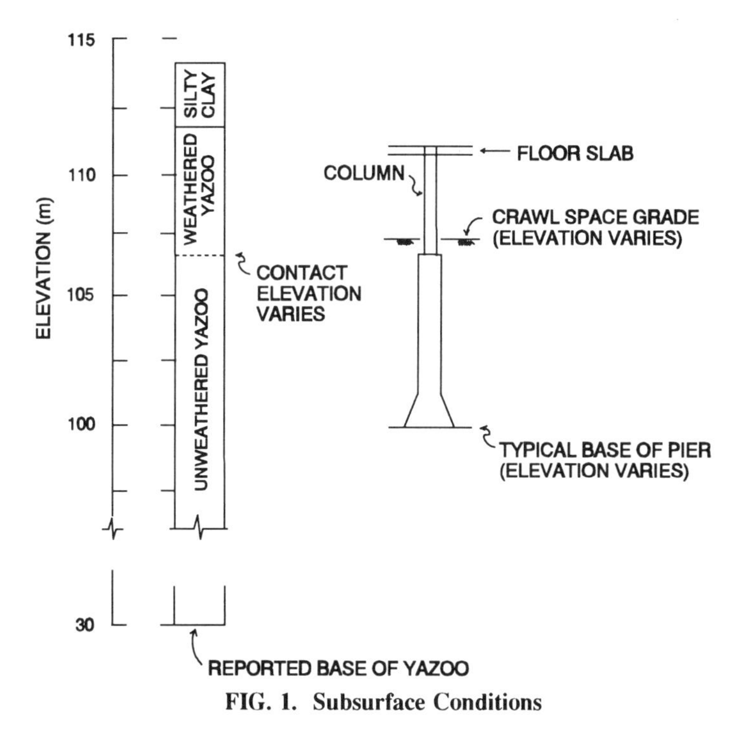

FIG. 1. Subsurface Conditions

The original surficial stratum of silty clay at the site was completely excavated at the building location. Beneath this stratum lies weathered and unweathered phases of the Yazoo formation. The former phase extends to approximately elev. 106.7 m and consists of very stiff to hard overconsolidated, active clays which contain fissures and slickensides. The liquid limits typically range from 100 to 150% and plastic limits varied from 30 to 40%. The natural water contents generally vary from 45 to 55%. The excavation for the building bottoms out in this stratum with approximately 1.5 to 4.9 m of weathered clay between the excavated grade and the unweathered clay. The unweathered Yazoo clays were encountered in the bottom of all boreholes and are reported to extend to elev. 40 m. These clays consist of very stiff to hard, overconsolidated clays. The liquid limits typically range from 100 to 115% and plastic limits varied from 30 to 35%. The natural water contents generally vary from 40 to 50%.

Summary of Damage

A survey was made of the conditions of the columns and within the crawl space. Results of the survey for the west portion of the building are shown in Fig. 2. These trends are typical and indicate that damage was concentrated near the exterior walls of the buildings, as evidenced by: (1) column cracking and distortions: damage was concentrated near the exterior walls and the most damage to a column was located on the face nearest an exterior wall; (2) gaps between the soil and base of columns: open gaps as wide as 50 mm were noted at the base of some columns, mostly adjacent to exterior walls and located next to the face of the column furthest from the nearest exterior wall; (3) relative vertical movements between the soil and the columns which indicated the soil at grade in the crawl space had displaced up relative to a column as much as 200 mm.

Damage to the columns was classified based on the widths of the widest cracks. The classifications on Fig. 2 are quantified as follows: "Severe damage" indicated cracks wider than 60 mils (1 mil = 0.025 mm), "significant" indicated cracks between 20 and 60 mils, "moderate" indicated cracks between 5 and 20 mils, and "limited" indicated cracks less than 5 mils. One of the columns (D7) failed in shear. Possible causes of these observed relative ground movements and related distress in the columns between the crawl space and the bottom of the floor slab include increases of in-situ water content causing swelling of the active clays and changes in effective stress in the foundation clays caused by the excavation for the building.

Swelling Potential

To assess the possible magnitude of soil deformations associated with swelling in the Yazoo clay, water contents determined from clay specimens obtained during 1973, 1975 and 1992 soil investigations were compared and swell tests were conducted on specimens obtained in 1992. Heave tests were conducted after specimens were subjected to a loading and unloading cycle in an oedometer and on specimens loaded directly to their approximate in situ vertical effective stress. After a specimen was consolidated to its final applied stress, water was added to the top and bottom of a specimen and vertical movements were measured until these movements ceased. No significant differences were observed in the results of the 8

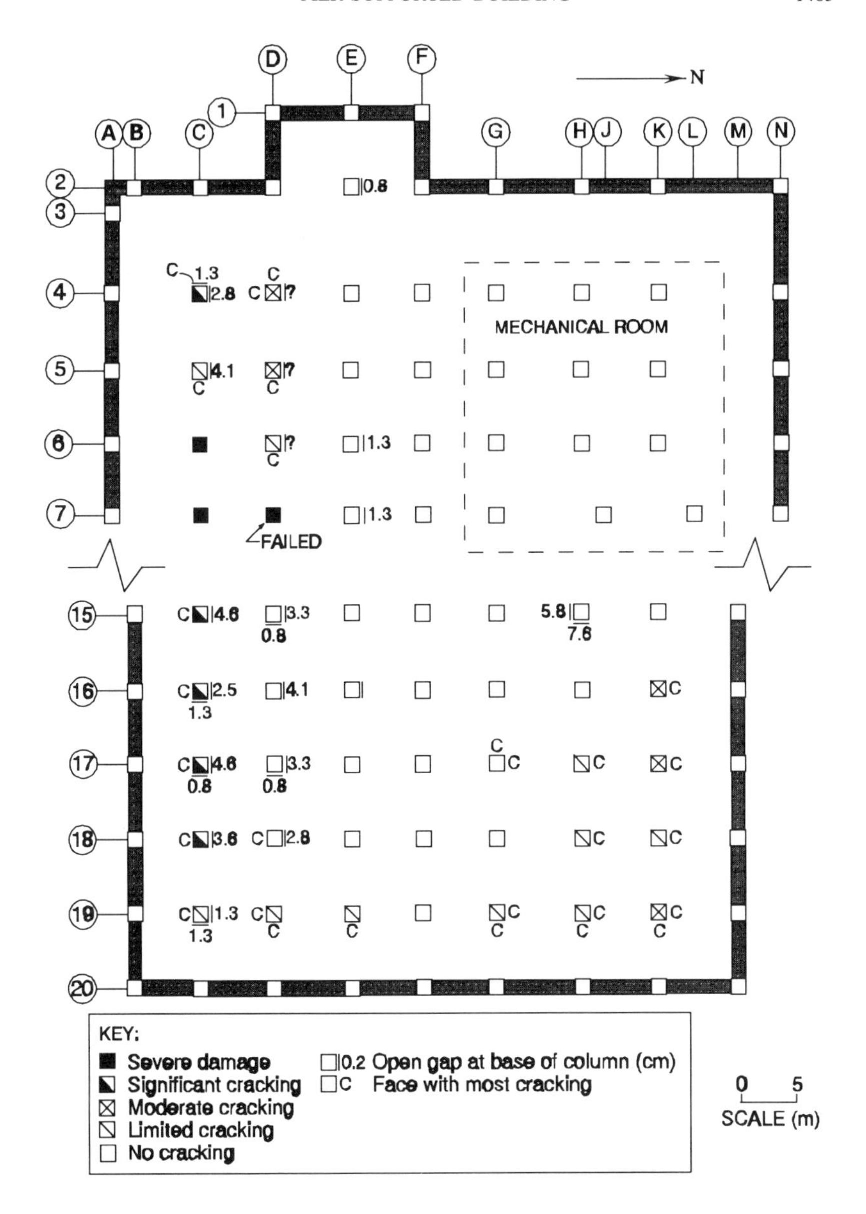

FIG. 2. Summary of Damage in West Part of Structure

tests. Results indicated that the percent swell varied from 0 to 0.41% and averaged 0.22% for the weathered phase and was about four times smaller for the unweathered phase. Using these results, if the weathered clay had access to water and assuming that 3 m of this clay existed below the foundation, approximately 12 mm of swelling could be expected; this would correspond to an increase of approximately 5 to 7 percentage points in water content.

Comparison of the water content data from the various field investigations indicates that in general there was very little change in water contents over the years. As indicated in Fig. 3, there is no difference in the water contents, except for those in a concentrated area. The increase of about 8 percentage points was consistent with the water content changes found in the swelling tests for the weathered Yazoo where water contents increased as much as 7 percentage points, with very little swelling. Damage to the columns in this concentrated area was not any different than in other areas, implying no cause and effect relation between damage and swelling at this site. In general, these data indicate that there has not been any appreciable change in water content over the years and thus not any significant amount of soil deformation, less than 12 mm, as a result of swelling movements. This small amount of swelling probably reflects the good performance of the vapor barrier installed over the soil in the crawl space.

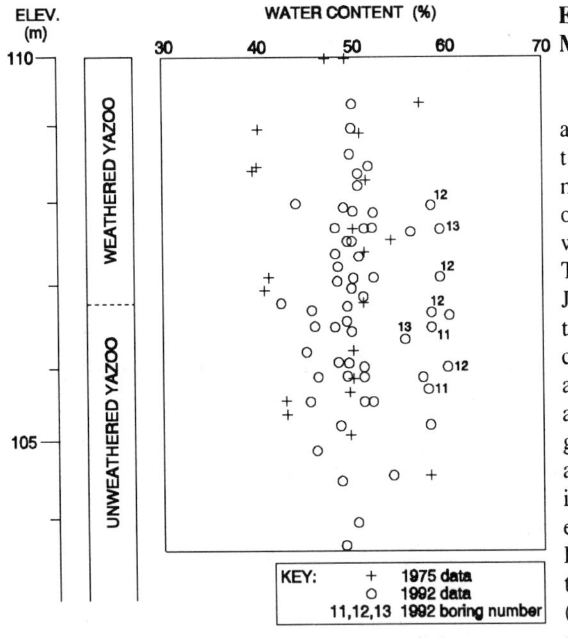

FIG 3. Water Content Versus Depth

EXCAVATION-INDUCED MOVEMENTS

Finite element analyses were conducted to evaluate the magnitudes and patterns of movements associated with the excavation. The finite element code JFEST was used to make the computations. The code has been used to analyze deformations associated with soft ground tunneling (Finno and Clough 1985), internally braced excavations (Finno and Harahap 1991) and tieback excavations (Finno et al. 1991). One of the main problems with simulating excavations in thick deposits of stiff clay is that the changes in stress due to excavation take a long time to equilibrate. For example, Skempton (1977) presented data from hydraulic piezometers with high air entry ceramic filters installed in stiff London Clay which indicated that 40 to 50 years were needed to equilibrate excavation-induced pore pressures. While it is difficult to conduct an analysis which explicitly accounts for these time-dependent effects under the best of circumstances, the soil at the site was not investigated in sufficient detail to depths greater than 15 m, making it impossible to identify flow boundary conditions needed for this type of analysis. Thus it was not possible to accurately make any computations regarding how long the excavation-induced movements would take to develop. To be consistent with the type of data available, several linear elastic, plane strain finite element simulations of the excavation were made. Parameters were selected to represent both short and long-term soil responses so that the response could be bounded.

Finite Element Mesh and Procedures

The finite element mesh encompassed the entire width of the excavation along column line 6 (Fig. 2) and extended from ground surface at elev. 114 m to elev. 40 m, the reported depth of the Yazoo clay. The induced stresses from the wide excavation (60 m by 90 m) have a significant effect on the clay even at the lower elevations of the Yazoo formation. The mesh represented a section through the deepest part of the excavation at the Mechanical Room location and included the most damaged columns. It represented the worst case scenario because of the combination of highest original grade and deepest excavation. Eight-noded, biquadratic isoparametric elements with two displacement degrees of freedom at each node were used to model the soil. Smooth boundaries, which prevent lateral but permit vertical deformations, were placed at the side of the mesh. The nodes at the bottom of the mesh were fixed so that no displacements occur at these locations.

The finite element procedure used to simulate excavation consisted of removing excavated elements and nodes from the solution routine and computing equivalent nodal forces to represent loads that are applied to the excavated surface to create a stress free boundary (Finno et al. 1991). To simulate the stiffening effect of the exterior walls and basement floor slab, the boundary conditions at the sides of the excavation were different for the short-term and long-term analyses. The short-term analyses represent conditions at the end of excavation prior to dissipation of excavation-induced pore pressures. At this time, the walls would not have been constructed, the piers would not have been drilled, and the slabs would not have been poured; in essence the stiffening effect of the structure would not impact these movements. The excavation was unsupported with slopes set back from the edge of the building. To simplify this analysis, the excavation side slopes were assumed vertical and unrestrained. In contrast, the long-term analysis represents the conditions when all excavation-induced pore pressures have dissipated. This condition is met some (undefined) time after construction. Thus the stiffening effect of the structure will have an effect on displacements, particularly the wall and basement floor slab. To simplify this analysis, the vertical excavation boundaries were idealized as rollers which prevented any lateral displacements while permitting vertical displacements.

Note that the piers were not explicitly included in the model. Displacements thus represent "free field" movements which would occur without the reinforcing effect of the piers and the weight of the structure. These displacements represent an upper bound model of the real situation, assuming soil parameters are adequately selected. The piers were not included in the simulations because the complexity of the mesh needed to conduct such analyses would not be consistent with the available soil data.

Soil Parameters

Because the clays at the site are overconsolidated and the excavation reduces the in situ stresses in the ground, Young's Modulus, E, is based on the recompression index, C_r, of the clay which is found from the unloading portions of the consolidation tests. The constrained modulus, D, found from drained compression tests with no lateral deformations is related to C_r. The drained Young's modulus, E_d, is related to D by:

$$E_d = \frac{(1+\nu)\,(1-2\nu)}{(1-\nu)} D = \frac{(1+\nu)\,(1-2\nu)}{(1-\nu)} \frac{(1+e)\,\Delta\sigma_V'}{C_r \log[\frac{\sigma_V' + \Delta\sigma}{\sigma_V'}]} \qquad (1)$$

where σ_v' is the vertical effective stress, e is the void ratio and $\Delta\sigma$ is the change in stress from excavation unloading. Eq. (1) is used to find the moduli used in the long term (drained) analyses. To find the short term (undrained) Young's moduli, E_u, the shear modulus, G, corresponding to a particular combination of E_d and ν can be calculated. Because G is the same for drained and undrained conditions, then E_u is simply equal to 3G and can be expressed as:

$$E_u = 3G = 3\frac{E_d}{2\,(1+\nu_d)} \qquad (2)$$

The consolidation data and the computed drained elastic parameters used in the finite element analyses are given in Table 1. The constrained modulus, D, and thus E, depends on σ_v' which increases with depth. Because the Yazoo formation is so deep, the variation of E with depth was approximated by assigning increasingly larger E values at greater depths in the soil profile. In addition, E varies with ν. For drained analyses, and thus for the long term conditions, ν was taken as 0.3. While it is expected that its effect is relatively minor, ν was also taken as 0.2 and 0.4, and the corresponding E values were used as input for a parametric study.

For undrained analyses, and thus for the short term conditions, ν is equal to one-half. To avoid numerical problems when solving undrained problems, ν was chosen as 0.495 for all undrained analyses. In summary, three finite element runs were made to evaluate the long term conditions (ν = 0.2, 0.3 and 0.4) and three runs were made to model the short-term case (ν = 0.495 and E based on Eq. (2)).

The displacements presented herein are a combination of the short and long term results for a given set of drained parameters, as indicated schematically in Fig.

TABLE 1. Soil Parameters for Finite Element Analysis

Stratum	Elevation (m)	C_r	Long-term Young's Modulus, E_d, (kPa) $\nu = 0.2$	$\nu = 0.3$	$\nu = 0.4$
Silty clay	114-111.6	0.16	1,450	1,200	760
Weathered Yazoc	111.6-104.6	0.16	4,960	4,100	2,570
Unweathered	104.6-95.7	0.095	15,120	10,350	7,850
Yazoo	95.7-81.4	0.095	21,900	18,100	11,400
	81.4-40	0.095	33,100	27,300	17,200

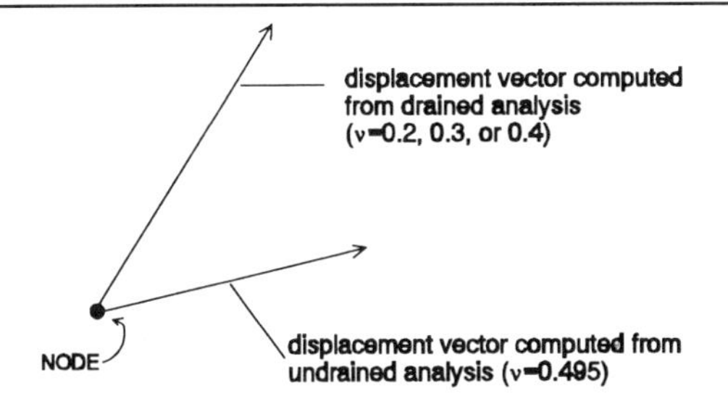

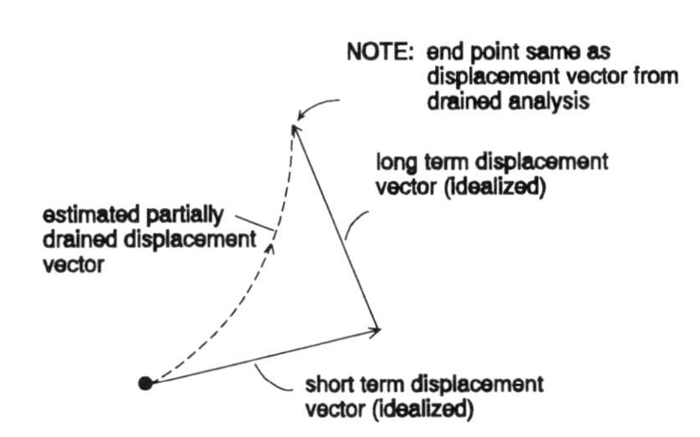

FIG 4. Computed Displacement Vectors

4. The displacement vectors shown in this figure consist of two parts. The short-term displacements are those directly from the undrained analysis with ν = 0.495. The long-term displacements represent those which occurred after the short term movements had developed. They were found by connecting the end points of the displacements from the undrained analysis to those found from the drained analysis (with ν = 0.2, 0.3 or 0.4). Thus the long term displacement vectors represent the incremental movements which arise from the dissipation of excess pore pressures. Note that these two displacement vectors represent an ideal case where it is assumed that excavation occurs instantaneously. Because an excavation of the size considered herein takes several months to complete, the conditions in the foundation soils are really partially drained and the displacements would not develop as indicate by the solid lines in Fig. 4b. A schematic representation of more realistic partially drained displacements is indicated by the dashed line in Fig. 4b. To compute these partially

drained displacements would require a finite element analysis with coupled displacements and pore pressures. This type of analysis is not warranted because of the lack of permeability and deep subsurface stratigraphic data. Therefore the limiting displacement vectors are indicated herein by the solid lines in Fig. 4b.

Note that the movements which impact the structure and the drilled pier foundations are those which occurred *after* they had been constructed. An unknown portion of the movements occurred as the excavation was made prior to drilling the foundations and constructing the superstructure. No data is available concerning these movements and thus absolute magnitudes of movements as they impact the structure are uncertain.

RESULTS OF FINITE ELEMENT ANALYSES

Results of the finite element analyses assuming ν of 0.3 for the drained conditions are shown in Fig. 5, a plot of computed displacements throughout the foundation soils above elev. 91.5 m. The pattern of movements shown here are the same as that for the analyses with the other values of ν. The movements associated with this excavation are quite deep seated. Movements greater than 25 mm are computed for depths more than 30 m below the excavated surface. Beneath the center of the excavation, the movements are essentially vertical. However, near the edges of the excavation the movements exhibit a marked horizontal component. The maximum lateral movements occurred at the south edge of the excavation where the most damaged columns were located. Also note that the ground surface will have heaved when all movements are complete. This latter trend has been observed when deep braced excavations have been made through stiff, overconsolidated clays like those present at this site (Ulrich 1989). The results of the computation indicates that the heave of the excavated grade could ultimately reach as much as 300 mm. While free field movements at the base of the pier reach as much as 230 mm for $\nu = 0.3$, the actual value will be less because of the restraint provided by the piers and the weight of the structure which rests upon them. Note that approximately 200 mm of soil movement was observed to have occurred relative to the column at the excavated grade.

Examination of Figs. 2 and 5 shows that most of the damaged columns are located within approximately 15 m of the edges of the excavation where most lateral movements have occurred. This is true for any location along the south side of the building. Note that very little damage has occurred in the columns located in the center of the building where very little lateral movements have been computed. No columns were damaged at the north side in this area. Here, the closest line of columns is farther from the edge of the wall than at the south side. Therefore the first row of columns at this side of the excavation would have been subjected to smaller maximum lateral movements ($\approx$ 50 mm computed) than at the south side ($\approx$70 mm computed), even though there was more soil excavated at the north side. Because the columns were longer at the north side, when the base of the column moved laterally, less distortion was induced in the columns because of the longer spans between excavated grade and the floor slab. Therefore the flexibility of the columns impacted the observed damage patterns.

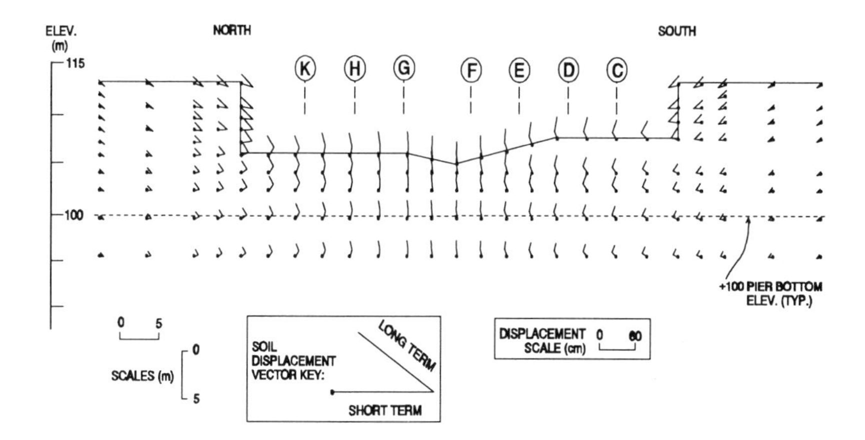

FIG 5. Computed Displacements at West Portion of Structure

The effects of the assumed values of ν on the computed heaves are indicated on Fig. 6 which shows a plot of maximum heaves at the ground surfaces, excavated grade in the crawl space and maximum short-term lateral movements near the excavated grade in the crawl space. Depending on the assumed value of ν, the finite element results indicate that the excavated grade in the crawl space will heave 300 to 400 mm, the ground surface will heave between 100 and 150 mm, and the soil near excavated grade will in the short term displace towards the center of the excavation between 100 and 200 mm. These ranges are relatively narrow, with a maximum deviation from the average value of ± 33%. Past experience indicates that 12 to 20 mm of heave can be expected for each 300 mm of excavated soil. For an average excavated depth of 6.1 m, experience would thus indicate that as much 240 to 380 mm of heave is ultimately expected. The finite element computations agree with this estimate.

DISCUSSION

Because the lateral movements directed towards the center of the excavation occurred primarily as a result of the short term response to excavation, there is a question of just how much lateral movement occurred before the piers and superstructure were constructed. It is also important to realize that the displacements computed by the finite element analysis are free field displacements that would occur if the drilled piers were not in place. Experience has shown that typical deep foundations will deform laterally with the soil except in the top 3 to 6 m of the foundation, depending on the relative stiffnesses of the pier and the soil. While the lower portions of the pier should move laterally with the soil, near the excavated surface, the piers are stiff relative to the soil and the piers will not deform with the

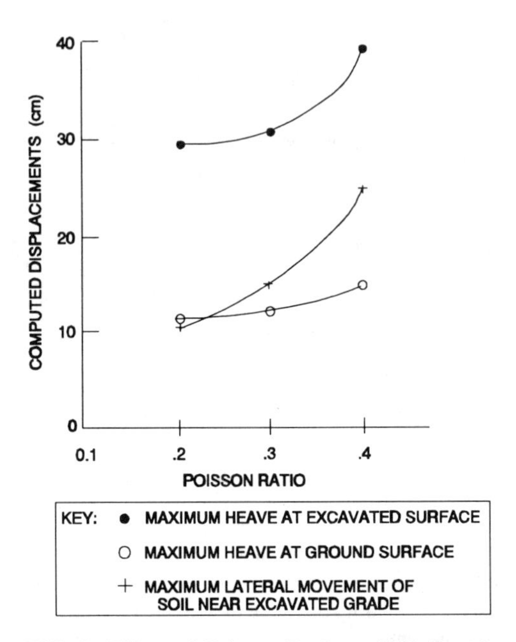

FIG 6. Effect of Poisson Ratio on F.E. Results

soil. Near the excavated surface the piers will resist the lateral ground movements and will move less than the soil.

With this in mind, the damage that occurred in the columns is consistent with the pattern of computed displacements. Considering the observations in the same order as before:

(1) Column cracking and distortions: Damage was concentrated near the exterior walls as was the largest lateral displacements. Assuming that some lateral displacements occurred after the columnswere connected to the piers and restrained by the basement floor slab, the pier and the bottom of the column would have displaced towards the center of the excavation. If the column connection at the slab was fixed (at least partially), then this pattern of displacements would have produced tensile strains on the face nearest to the exterior walls, consistent with field observations.

(2) Gaps between the soil and the base of the column: These gaps primarily reflect the inward movements of the soil and thus are apparent at the column locations close to the exterior walls and not in the center of the building. The gaps were generally observed at the face of the column furthest from the nearest exterior wall. While the lower portions of the pier would move with the soil, the upper portions would tend to resist the soil movement. Because between 100 and 200 mm of (free field) lateral movements might be expected at the excavated grade in the crawl space, it would be reasonable to expect that gaps as much as 50 mm could develop. These movements were large enough at column D7 to cause that column to fail in shear.

(3) Relative vertical movements between the soil and the base of the columns: Computed soil heaves at excavated grade vary between 300 and 400 mm. Because these are free-field, they represent upper bound estimates of heave. As much as 200 mm of relative movement were observed at some locations.

The computed excavation-induced displacements follow patterns which tend to explain the major damage observations. Because movements that may be attributed to swelling are so small relative to the these excavation-induced movements, it is likely that the damage to the columns at the building were caused by these latter movements.

CONCLUSIONS

Damage to columns of a pier-supported structure were observed in the crawl space between the excavated grade and the floor slab. Damage was concentrated in the columns near the exterior walls. Results of finite element computations indicate the damage is consistent with movements associated with excavation of its basement. Because the excavation was made in a deep deposit of stiff overconsolidated clay, the movements developed over a period of time. While the near surface weathered clay had a high swelling potential, water contents had not significantly changed in the 17 years between construction and the remedial work.

ACKNOWLEDGEMENTS

The authors thank Messrs. Robert E. Anderson and Vince Coleman of Allstate Insurance Company for their encouragement and permission to publish this study.

APPENDIX - REFERENCES

Finno, R. J., and Clough, G. W. (1985). "Evaluation of soil response to EPB shield tunneling." *J. Geotech. Eng.*, ASCE, 111(2), 155-173.

Finno, R. J., and Harahap, I. S. (1991). "Finite element analyses of the HDR-4 excavation." *J. Geotech. Eng.*, ASCE, 117(10), 1590-1609.

Finno, R. J., Lawrence, S. A., Allawh, N. F., and Harahap, I. S. (1991). "Analysis of performance of pile groups adjacent to a deep excavation." *J. Geotech. Eng.*, ASCE, 117(6), 934-955.

Skempton, A. W. (1977). "Slope stability of cuttings in brown London Clay." *Proc. 9th ICSMFE*, Tokyo, 3, 261-270.

Ulrich, E. J. (1989). "Tieback supported cuts in overconsolidated soils." *J. Geotech. Eng.*, ASCE, 115(4), 521-545.

Settlement of Building Foundations on Clay Soil Caused by Evapotranspiration

Vincent Silvestri,[1] Member, ASCE, and Claudette Tabib[2]

Abstract: This paper reports on the results of a field investigation and settlement monitoring program which involves five buildings founded on shallow foundations in the sensitive clay deposits of Montreal Island (Quebec, Canada). The purpose of this paper is to provide a better understanding of the shrinking behavior of these soils. Each site has been instrumented with foundation elevation pins, ground movement plates, deep settlement points, piezometers, and aluminum tubes for the measurement of volumetric weights and water contents in the clay deposits.

Introduction

Engineering problems associated with volume changes in clays usually arise from uneven shrinkage and swelling. Seasonal variation giving rise to changes in water content is not the only cause of ground movements. The water requirements of trees, for example, can be of such a magnitude that moisture depletion in the subsoil causes clay shrinkage and settlement of nearby structures (Ward 1953; Bozozuk and Burn 1960; Perpich et al. 1965; Hammer and Thompson 1966; Driscoll 1983).

In the summer of 1990, Ecole Polytechnique began a joint research program with the City of Montreal to study the effects of droughts on water content changes in typical clay deposits under an urban environment. This paper analyzes experimental data of settlements, pore water pressures and water contents gathered

---

[1] Professor, Civil Engineering Department, Ecole Polytechnique, P. O. Box 6079, Station Centre-Ville, Montreal, Quebec, Canada H3C 3A7.

[2] Professor, Department of Mathematics, College Edouard-Montpetit, 945 Chemin Chambly, Longueuil, Quebec, Canada J4H 3M6.

in 1991 on five sites located in the Mercier district, where a large number of residential buildings have experienced excessive settlements in recent years.

SITES, MATERIALS, AND METHODS

Five representative sites were selected in the Mercier district. The sites are referred to in this paper by the names of the streets on which they are located, namely, Site 1 - Taillon, Site 2 - Rameau, Site 3 - Honoré-Beaugrand, Site 4 - Beaurivage, and Site 5 - Souligny. On the basis of geotechnical studies, the five sites have been divided into two groups of similar properties: (1) Honoré-Beaugrand and Beaurivage; (2) Rameau, Souligny and Taillon. Considering the first group of sites, that is, Honoré-Beaugrand and Beaurivage, Fig. 1 indicates that the clay deposit has a thickness of about 3.5 m. In addition, there exists an upper 1.5 m thick oxidized clay crust in which the water content is relatively low and ranges between 42 and 44.5%. Between the depths of 2 and 4 m, the clay is plastic and its natural water content which varies from 46.5 to 60.5% is very close to the liquid limit. In this depth interval the undrained shear strength S_u has an average value of 78 kPa at Beaurivage and 85 kPa at Honoré-Beaugrand, and the preconsolidation pressure σ_p' ranges between 240 and 305 kPa for Honoré-Beaugrand, and 285 and 320 kPa for Beaurivage. The clay deposit is overconsolidated, with an overconsolidation ratio (OCR) varying between 3.4 and 6 for Honoré-Beaugrand, and 5.6 and 11 for Beaurivage. Thus, the ratio between the undrained shear strength S_u and the preconsolidation pressure σ_p', that is , S_u/σ_p', is equal to an average value of 0.27. The sensitivity of the clay deposit varies between 4 and 9.

For the second group of sites, that is, Rameau, Souligny, and Taillon, Fig. 2 indicates that the clay deposit has a thickness of 8 m and there is also an upper 1.5 m thick oxidized crust in which the water content is low, ranging between 36.6 and 51.1%. For depths exceeding 2.5 m, the water content increases steadily and reaches a maximum value of 71.9% at Rameau. Generally, the water content of the intact clay exceeds the liquid limit in the depth interval from 2.5 to 8 m. On the basis of in-situ vane tests carried out from 2.5 to 8 m, the undrained shear strength has an average value of 70 kPa Rameau, 74 kPa at Souligny, and 76 kPa at Taillon. The average preconsolidation pressures are also markedly similar for these three sites: 304 kPa for Rameau, 295 kPa for Souligny, and 318 kPa for Taillon. As a consequence, the S_u/σ_p' ratio is of the order of 0.24 in this case. In addition, the clay deposits are overconsolidated with OCR values ranging from 3.8 to 4.8 for Rameau, 2.4 to 6.9 for Souligny, and 3 to 10 for Taillon. The sensitivity of the clay deposits varies between 4 and 15 for Souligny and Taillon, and from 6 to 42 for Rameau.

For the measurements of settlements at each site, the following techniques were used: (a) reference points or pins were bolted on 1 to 1.5 m intervals to the structures along the first course of bricks above the concrete foundations; (b) surface settlement plates were anchored at various locations on the ground surface of the lots adjacent to the buildings; and (c) two Borros points were used on each site for the determination of deep settlements in the clay deposits: the first at a depth of 1.2 to 1.4 m, the second at a depth of 3 to 3.5 m. Each Borros point comprises a rod fixed into the ground by flexible metal anchors, an outer pipe being used to eliminate friction. In addition, permanent benchmarks were used on all five sites.

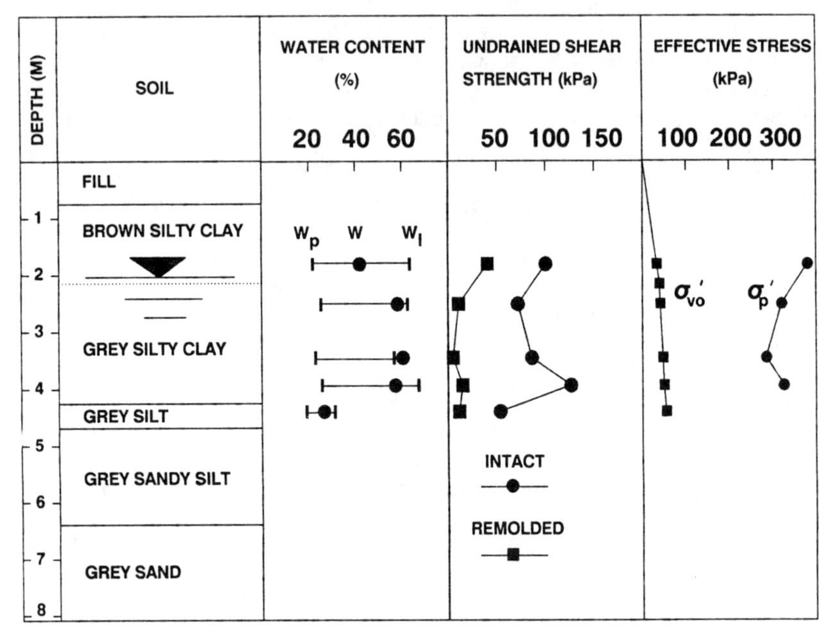

FIG. 1. Geotechnical Profile at Beaurivage and Honoré-Beaugrand

Each site was also instrumented with three electric piezometers for the monitoring of the pore water pressures, and two plastic tubes for the study of shallow water level fluctuations. Furthermore, three 6 m deep, 1.5 mm thick and 50 mm diameter aluminum tubes were permanently installed on each site for the measurement of both the volumetric weights and the volumetric water contents of the soil at different depths. A commercial nuclear depth probe was lowered in the aluminum tubes whenever density and moisture content readings had to be taken.

Finally, to assess the possibility of using irrigation to compensate for the water lost through evapotranspiration, three of the five sites were equipped with different automatic irrigation systems in 1991: (a) trickle or drip irrigation on Taillon; (b) 1.4 m deep recharging wells on Beaurivage; and (c) surface sprinklers on Honoré-Beaugrand. The remaining sites of Rameau and Souligny provided reference conditions.

Analysis of Experimental Results

On the basis of past studies (Bozozuk and Burn 1960; Silvestri et al. 1990), it has been found useful to relate seasonal variations of settlements and moisture contents of clay soils to cumulative rainfall deficit. Rainfall deficit accumulated up to and including a certain month "n" is determined by using the following equation

$$RD = \sum_{j=i}^{n} (PET_j - P_j) \tag{1}$$

where PET_j and P_j refer to potential evapotranspiration and precipitation, respectively, for each month "j". The units of RD, PET_j, and P_j are mm of water. Table 1 summarizes climatic data for Montreal, for both 1991 and long-term, 1930-1991, conditions. For the determination of RD, PET_j was calculated using an empirical equation proposed by Thornthwaite in 1948. From a consideration of the last two columns in Table 1, it appears that the summer of 1991 was somewhat drier than usual with a maximum rainfall deficit of 148.9 mm of water, reached at the end of August, as compared to a long-term average of 130.6 mm. According to an earlier study by Bozozuk and Burn (1960), such rainfall deficit for 1991 would imply moisture changes to a maximum depth of approximately 1.4 m in a typical clay deposit of Eastern Canada.

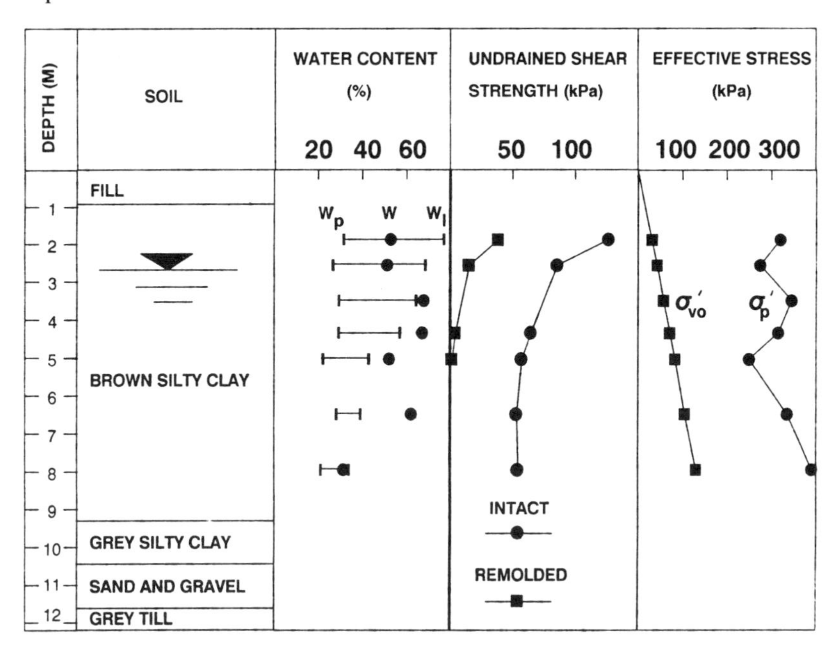

FIG. 2. Geotechnical Profile at Rameau, Souligny, and Taillon

In order to compare the behaviors of the structures founded on irrigated sites with those resting on non-irrigated sites, emphasis is in the following laid on the response of Taillon (irrigated) and Rameau (non-irrigated) during 1991. Please note that for the irrigated sites, the required amount of water was determined at the end of each week of observation by calculating the difference between potential evapotranspiration and precipitation fallen in Montreal. This quantity of water was provided by means of several irrigation cycles during the first two days of the subsequent week.

TABLE 1. Climate Characteristics for 1991 and Long-Term Averages

Month	Temperature (°C)		Precipitation (mm)		Rainfall deficit (mm)	
	1991	1930-1991	1991	1930-1991	1991	1930-1991
J	-10.5	-9.8	75.6	74.9	–	–
F	-5.5	-8.6	50.6	65.6	–	–
M	-0.7	-2.5	94.7	74.7	–	–
A	7.8	5.8	122.4	77.2	–	0.5
M	15.0	13.2	91.2	72.7	0.1	20.1
J	19.0	18.4	34.8	85.2	85.4	54.2
J	21.2	21.2	85.4	91.6	137.9	101.6
A	20.5	19.9	111.6	90.6	148.9	130.6
S	13.5	14.7	69.8	86.4	145.0	122.5
O	9.5	8.8	70.7	77.9	114.4	90.5
N	2.5	1.8	31.5	88.5	90.4	37.9
D	-7.3	-6.6	75.7	86.0	14.7	8.8

(a) Taillon (Fig. 3)

The building on this site is a three-storey apartment house with basement, built in 1967, rectangular in plan, 11.4 m by 14.2 m, of wood-frame construction and brick veneer, resting on cast-in-place foundation walls. The foundation walls are supported by a 0.6 m wide, 0.3 m thick strip footing at a bearing depth of 1.4 m below existing grade. The adjacent lot is covered with lawn and there is a row of trees growing on municipal property, along a line parallel to the street, at a distance of about 7 m from the front of the house. The two closest trees shown in Fig. 3 are red ashes (Fraxinus pennsylvanica), 13 m in height and 0.33 m in trunk diameter. Several 5 mm wide cracks are apparent along both the front (east side) and the transverse walls.

During the summer of 1991, in spite of irrigation, most of the reference points located along the foundation walls in the front portion of the house showed a maximum settlement of 5 mm which was followed also by a 5 mm rebound in early fall. However, reference points 4 to 6 and 34 to 36 settled between 10 and 15 mm, as shown in Fig. 4. Concerning ground movements, maximum settlements ranged between 20 and 40 mm with subsequent fall rebounds of 5 to 20 mm. For the two Borros points, only the one situated at a depth of 1.2 m showed some movement: first it settled 20 mm in the summer, then it rebounded about 15 mm in the fall. Concerning the pore water pressures, the piezometer P1 situated at a depth of 3.66 m showed the most variation with a decrease in water head of 1.3 m, compared to changes smaller than 0.8 m for the piezometers P2 at 7.31 m and P3 at 10.67 m. For the shallow water level tubes, the groundwater level dropped 1.7 m in the tube TO1-1 located in the front-yard and 0.5 m in TO1-2 located in the backyard.

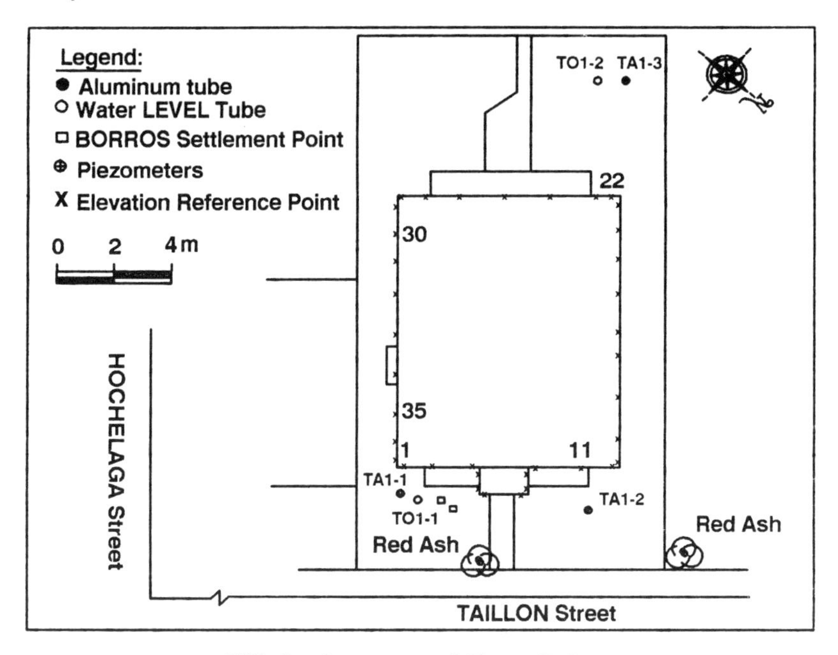

FIG. 3. Instrumented Site at Taillon

As for volumetric water content changes in the soil deposit, aluminum tube TA1-3 situated in the back-yard registered the least variation while aluminum tube TA1-1 in the front-yard showed the most variation. Fig. 5 which presents both the wettest and driest volumetric water content profiles registered in this latter tube, indicates that the maximum penetration of moisture change in the soil is about 2.2 m, in spite of sustained irrigation throughout the summer of 1991. The area between the two profiles in Fig. 5 corresponds to a moisture depletion of 101 mm in the depth interval from 0.2 to 2.2 m. Please note that because the top portions of the aluminum tubes were covered with heavy vandal-proof caps and 0.8 m long protecting casings, no volumetric water content measurements were taken in the top 0.8 m surface layer. On the basis of measurements carried out by Ward (1953), such water loss would give rise to a settlement of 25 to 34 mm, which compares well with the observed maximum settlement of the ground surface. As for the house foundations which rest at a depth of 1.4 m, the observed settlement is due to the water loss incurred between 1.4 and 2.2 m, and is thus less than that of the ground surface.

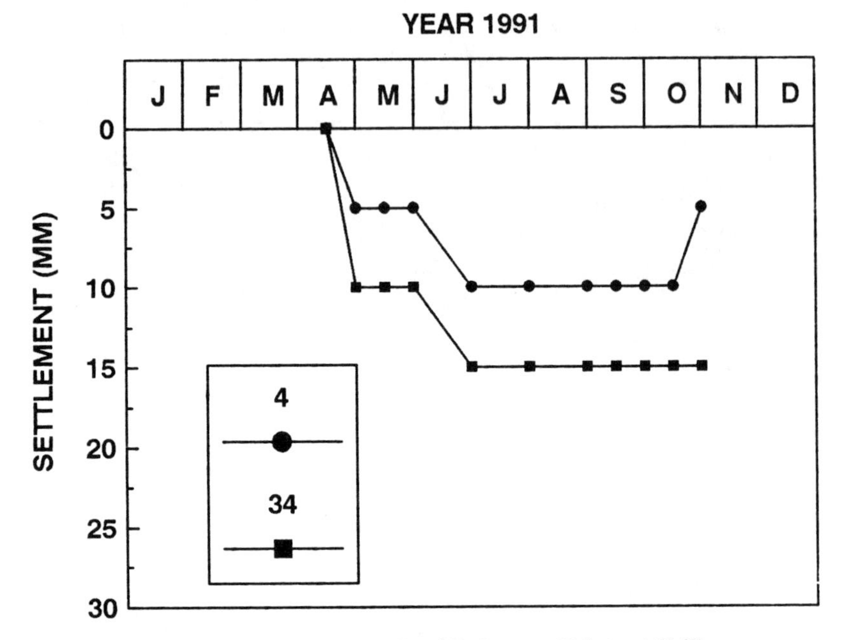

FIG. 4. Levelling Records of Reference Points at Taillon

(b) Rameau (Fig. 6)

The house on this site is a bungalow with basement, built in 1956, of wood-frame construction and brick veneer resting on cast-in-place foundation walls. It is L - shaped in plan; the wings measure 6.6 m by 12 m and 4.3 m by 6 m, respectively. The walls are supported on 0.6 m wide, 0.3 m thick strip footings at a depth of 1.5 to 1.8 m below existing grade. On the lot, vegetation consists of lawn, well-developed hedges along property lines, and a 4 m high white cedar at a distance of 2 m from the front wall. There is also a row of mature trees growing on municipal property, parallel to the street, at about 5.5 m from the front (east side). The two trees closest to the house are: (a) a red ash (Fraxinus pennsylvanica), 13.7 m in height and 0.36 m in trunk diameter; and (b) a silver maple (Acer saccharinum), 17 m in height and 0.57 m in trunk diameter. Numerous cracks are concentrated on the east side of the house, close to window openings, particularly along the north side of the front wing, near the north-east corner where a 30 mm wide vertical crack is found at the level of the foundation wall. The existence of several 5 to 20 mm wide cracks on either sides of the front wing indicates that the front portion of the house has rotated like a rigid block. The large width of the cracks at the level of the strip footings shows that the front portion of the house has also moved horizontally for about 20 to 30 mm towards the street.

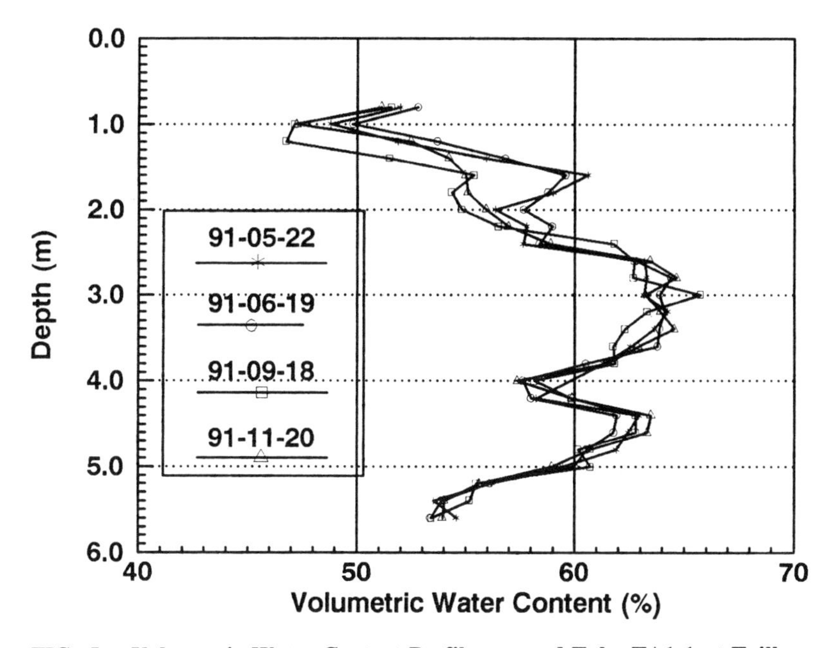

FIG. 5. Volumetric Water Content Profiles around Tube TA1-1 at Taillon

Levelling surveys in 1991 indicated that some portions of the walls moved during the period April to September. For example, along the front wall, reference points 1 to 5 showed a maximum settlement of 40 to 50 mm which was followed by a fall rebound of 10 to 25 mm, as shown in Fig. 7. Finally, along the south wall, the settlement increased steadily, starting from reference point 29, and reached 30 mm at reference point 32, near the south-east corner of the front wing. The maximum fall rebound along this same section did not exceed 10 mm. Along the remaining sections of the walls, there was practically no movement during the period of observation.

As for the ground surface in the front-yard, the settlement ranged between 25 and 85 mm and was followed by a 15 to 35 mm fall rebound. Concerning the Borros points, only the shallow one situated at a depth of 1.2 m moved: a maximum settlement of 60 mm followed by a rebound of 20 mm in the fall.

Among the three piezometers P1 (4.57 m), P2 (8.22 m), and P3 (12.19 m) installed on this site, only the latter was not affected by seasonal variation. However, P1 and P2 reacted swiftly, with a maximum decrease in head of water of 2.1 and 3.5 m, respectively. As for the shallow water levels, the groundwater level dropped 1.8 m in the front-yard and 0.5 m in the back-yard.

Concerning the volumetric water content measurements, only tube TA2-3 situated in the back-yard showed minor variations. The remaining tubes TA2-1 and

TA2-2, located in the front-yard, indicated considerable variations. For example, Fig. 8 which presents both the wettest and driest volumetric water content profiles recorded at the location of TA2-1 shows that shrinkage proceeded to a maximum depth of about 4.2 m in the soil deposit. The area between the two curves in Fig. 8 corresponds to a moisture depletion of 196 mm in the depth interval from 0.8 to 2.2 m. Again, according to the findings of Ward (1953), such water loss is equivalent to a settlement varying between 50 and 65 mm and which compares well with the observed movement of the ground surface.

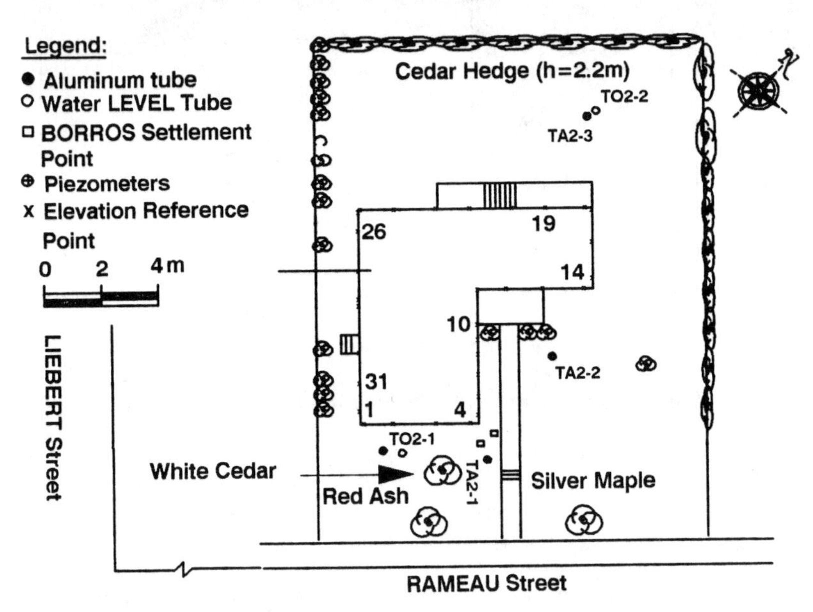

FIG. 6. Instrumented Site at Rameau

CONCLUSIONS

Detailed data is presented on two of the five experimental sites. The measurements indicate that, for the summer of 1991, ground settlements were much more important on the non-irrigated than on the irrigated sites, reaching 85 mm at Rameau. As for the settlements of foundation walls, the Rameau site showed settlements of 40 to 50 mm along sections closest to the row of mature trees, growing parallel to the street. Pore water pressures and shallow water levels also decreased during the summer of 1991, the most important changes always being registered on the front lawns of the two sites. Finally, the wettest and driest volumetric water content profiles allowed the moisture depletions suffered by the soil deposits at the two sites to be calculated and it was shown that the non-irrigated clay deposit of Rameau experienced an important water loss of 196 mm, even though the summer of 1991 was not extremely dry.

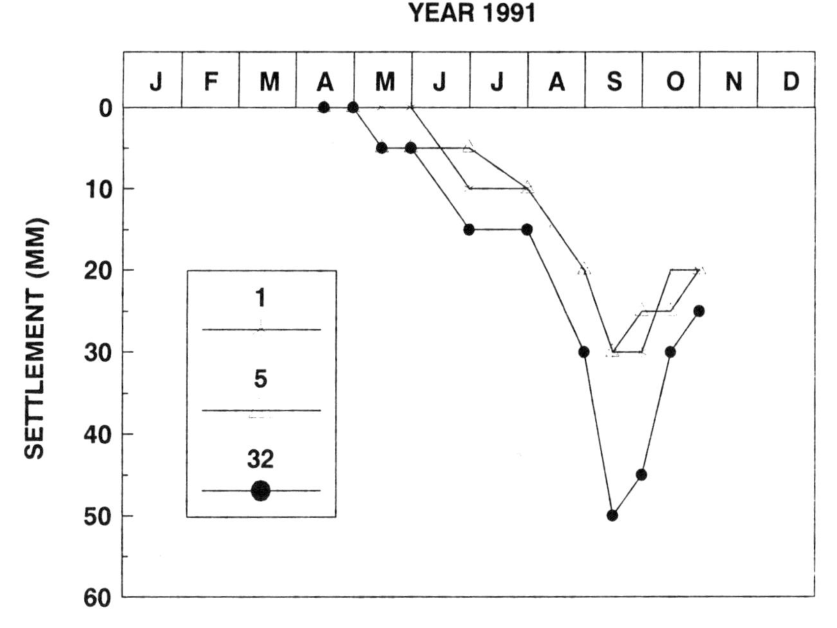

FIG. 7. Levelling Records of Reference Points at Rameau

ACKNOWLEDGEMENTS

The authors wish to express their deep gratitude to the City of Montreal for the financial support received in the course of this study. Mr. A. Campeau, P.Eng., was in charge of the project for the City of Montreal, and Mr. M. Tremblay, M.Sc.A., P. Eng., acted as the liaison officer. Levelling surveys were carried out by the technical staff of the City of Montreal, under the direction of Mrs. S. Gauthier, L. Surv. Financial assistance was also provided by FCAR of Quebec and NRC of Canada.

APPENDIX - REFERENCES

Bozozuk, M., and Burn, K. N. (1960). "Vertical ground movements near elm trees." *Géotechnique*, 10(1), 19-32.

Driscoll, R. (1983). "The influence of vegetation on the swelling and shrinking of clay soils in Britain." *Géotechnique*, 33(2), 93-105.

Hammer, M. J., and Thompson, O. B. (1966). "Foundation clay shrinkage caused by large trees." *J. Soil Mech. Found. Div.*, Proc. ASCE, 92(SM6), 1-17.

Perpich, W. M., Lukas, R. G., and Baker, C. N., Jr. (1965). "Dessication of soil by trees related to foundation settlement." *Can. Geotech. J.*, 2(1), 23-39.

Silvestri, V., Soulié, M., Lafleur, J., Sarkis, G., and Bekkouche, N. (1990). "Foundation problems in Champlain clays during droughts. I: Rainfall deficits in Montreal (1930-1988)." *Can. Geotech. J.*, 27(3), 285-293.

Thornthwaite, C. W. (1948). "An approach toward a rational classification of climate." *Geographical Review*, 38(1), 55-94.

Ward, W. H. (1953). "Soil movements and weather." *Proc. 3rd ICSMFE*, Zurich, 4, 477-482.

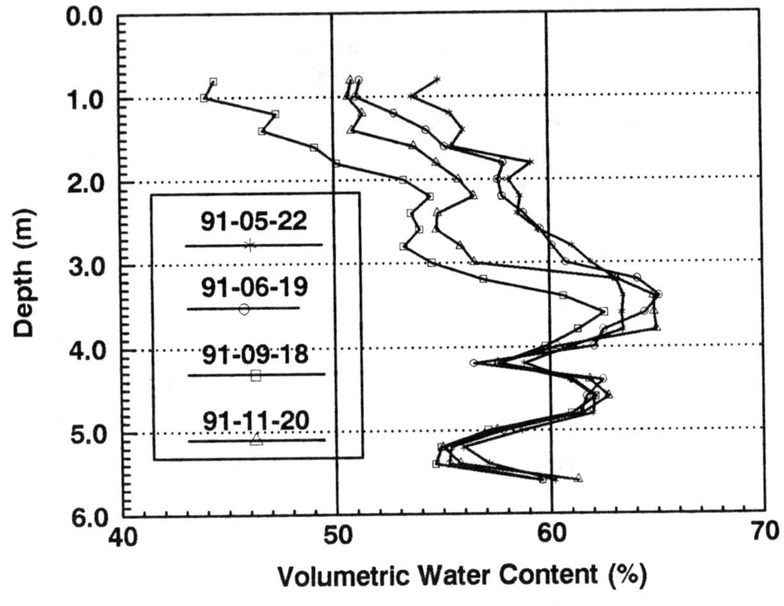

FIG. 8. Volumetric Water Content Profiles around Tube TA3-1 at Rameau

The Behavior of a Building with Shallow Foundations on a Stiff Lateritic Clay

Luciano Décourt,[1] Fellow, ASCE

ABSTRACT: The use of shallow foundations for buildings in São Paulo, either on soils of the Tertiary Sedimentary Basin of São Paulo (TSBSP) or on those of saprolitic origin is not as frequent as it could be.

The Rimini building would have been another one among those designed by the author on shallow foundations if fortunate decisions to perform unusual tests at this site were not taken. Not only the author but also many other colleagues elected this a research site. The objectives of the research were reached. But, the highlight was the extraordinary behavior, in terms of stiffness, of these lateritic soils, which are frequently found in tropical regions. This, to the authors knowledge has been detected for the first time.

The probable confirmation of this finding will in the near future allow substantial reductions in the cost of foundations on these soils.

In Situ and Laboratory Tests

An unusual amount and variety of tests were performed in situ and on samples, both undisturbed and remolded as further described, opposite to what is usually done. In routine jobs only Standard Penetration Tests (SPT) are performed. As far as the in situ tests were concerned, in addition to the SPTs the following tests were performed; SPT-Ts that are the traditional SPTs complemented by torque measurements (Décourt and Quaresma Filho 1991, 1994), Marchetti Dilatometer Tests (DMT), and Cross Hole Tests (CHT). As far as laboratory tests were concerned the following were carried out: Grain Size Analysis, Atterberg Limits,

---

[1] Director of Luciano Décourt Engenheiros Consultores Ltda., Av. Brig. Faria Lima, 1857 - 2º São Paulo, Brazil, CEP 01451-001.

Specific Gravity, Compaction, Unconsolidated Undrained Triaxial Compression (UUC), Oedometer and Resonant Column.

IN SITU TESTS

The Standard Penetration Test – SPT

The in situ investigation initially performed consisted, as it is usual in Brazil, of borings with Standard Penetration Tests performed at one meter depth intervals. The average system efficiency of the Brazilian SPT, e(avg.) is typically 72% (Décourt et al. 1989).

SPT-T

The SPT-T is the traditional SPT complemented by torque measurements (T). It is a very simple and cost-effective test, which was introduced in engineering practice, three years ago (Décourt and Quaresma Filho 1991). On the basis of the 10 borings carried out at the site, a typical soil profile was worked out and is given in Table 1.

TABLE 1. Soil Profile at the Rimini Building Site

Layer	Thickness (m)	N-SPT	Torque-T ($Nm \times 10^{-1}$)
Porous red clay	2.0	2	–
Stiff red clay	20.0	20	23

It is interesting to observe that the main layer consists of a relatively homogeneous stiff clay down to depths of 20.0 m. One of the first experiments carried out in this area was to determine the variation of the energy efficiency of the SPT (e) with depth.

According to the work of many researchers, (Schmertmann and Palacios 1979; Kovacs et al. 1981), this efficiency increases with depth up to the critical depth, where the ratio of the mass of the standard hammer M_h to the unit mass of used rod is one. But the torque ratio (TR) i.e., the ratio of the torque (T) required to overcome the friction between the sampler and the soil, measured in ($Nm \times 10^{-1}$) units, and the N-SPT value was found to be independent of the depth (Décourt and Quaresma Filho 1994). Since (e) and T/R varies similarly with depth, these measurements clearly suggest that the "measured" low efficiency of SPT for short rod lengths is not real.

The Marchetti Dilatometer Test (DMT)

In three of the borings, SPT-Ts were performed at one meter intervals, and flat dilatometer tests were performed at intermediate depths. For practical reasons the blade was driven into the soil instead of pushed as is considered preferable. Little is known about the possible damaging effects of dynamic penetrations as compared with quasi static ones. However, Schmertmann (1989) quoting Basnett suggested that for stiff clays they are probably small.

An apparently anomalous information derived from the measurements was that the values of the Material Index (I_d) were in the range $2.0 \leq I_d \leq 4.0$ which are typical of sandy soils and not of clayey soils as is the case at this site. This subject will be discussed further and a possible explanation for this discrepancy will be proposed.

Cross Hole Test (CHT)

Cross Hole tests were performed at the site. Two displays were considered, both consisting of one source and two receivers. Consistent measurements were obtained at three depths; 4.0, 5.0 and 6.0 m (the foundations levels). The average value of the maximum shear modulus, G_o computed using the equation $G_o = \rho V_s^2$ was 418.7 MPa.

Since for settlement computations the elastic modulus "E" must be known, their maximum values were computed for Poisson's ratios values of 0.2 and 0.3.

$$E_o = 2G_o(1+\mu) \tag{1}$$

$$E_o = 1{,}005 \text{ MPa} \qquad (\mu = 0.2) \tag{2}$$

$$E_o = 1{,}089 \text{ MPa} \qquad (\mu = 0.3) \tag{3}$$

Laboratory Tests

Standard Tests

At the beginning of this research, standard laboratory tests were performed on an undisturbed block sample from the foundation soil, recovered in a test pit. Grain size analysis indicated that 75% of the material was smaller than 0.074 mm (ASTM sieve #200) and that 100% passed through sieve #10. The values of Atterberg limits and specify gravity are given in Table 2.

TABLE 2. Soil Characteristics

ρ_s (kg/m^3)	LL* (%)	LP* (%)	LL** (%)	LP** (%)
2,745	65	46	80	47

* time of manipulation of 3 min. (NBR-6459)
** time of manipulation > 30 min. (Ignatius 1991).

Oedometer tests indicated preconsolidation pressures of 1,220 kPa. Unconsolidated undrained triaxial compression tests indicated C_u = 309 kPa. The measured degree of saturation of the samples was on average 95%.

Resonant column tests

Four Resonant Column tests were performed. For each test, three consolidation pressures, were used; 150 kPa, 300 kPa and 600 kPa. The measured and the computed average values of Go are presented in Table 3.

TABLE 3. G_o Values from Resonant Column Tests

Test No \ σ'_3 (kPa)	150	300	600
1	425	520	567
2	433	505	555
3	–	–	567
4	413	495	557
Average	424	507	561

In Fig. 1 the results of G_o from test No 4 are presented and in Fig. 2, the variation of G/G_o with the shear strain γ is given for σ'_3 = 600 kPa. As compared with G_o values determined from Cross Hole tests, G_o values from Resonant Column tests are somehow more troublesome to be used in engineering practice because a consolidation pressure has to be considered. For a long time the author has been using correlations with N-SPT to determine the pre-consolidation pressure (σ'_p) of soils (Décourt 1989).

$$\sigma_p' = 33.3 \ N_{72} \ (kPa) \tag{4}$$

For N_{72}-SPT = 20, results:

$$\sigma_p' = 666 \ kPa \tag{5}$$

On basis of this value and assuming $(K_o)_{nc}$ = 0.5, it was considered that the present effective octahedral effective stress (σ'_{oct}) would probably be:

$$\sigma'_{oct} = \frac{\sigma_v' + 2\sigma_h'}{3} \approx \frac{0 + 2 \times \frac{666}{2}}{3} \approx 222 \ kPa \tag{6}$$

Interpolating the data of Table 3, results:

$$G_o = 464.0 \ MPa \tag{7}$$

It is interesting to observe that this value of G_o is about 10% higher than the average value obtained from Cross Hole tests. For this value of G_o, the maximum values of the elastic modulus would be:

E_o = 1,206.4 MPa (μ = 0.3) and E_o = 1,113.6 MPa (μ = 0.2)

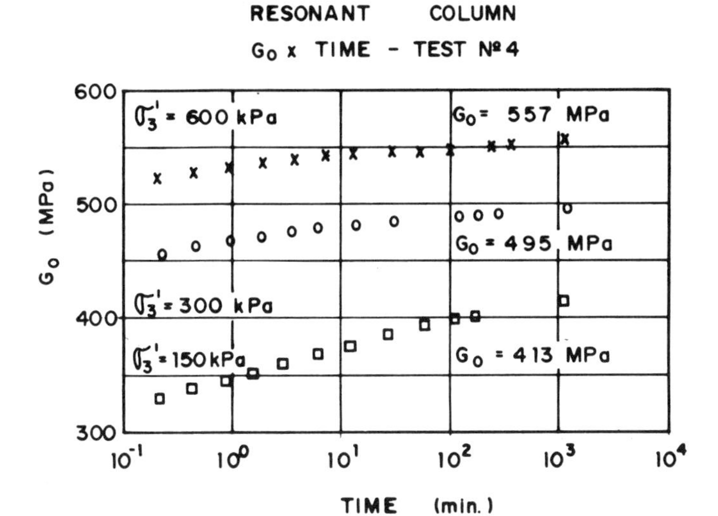

FIG. 1. G_o from Resonant Column Test

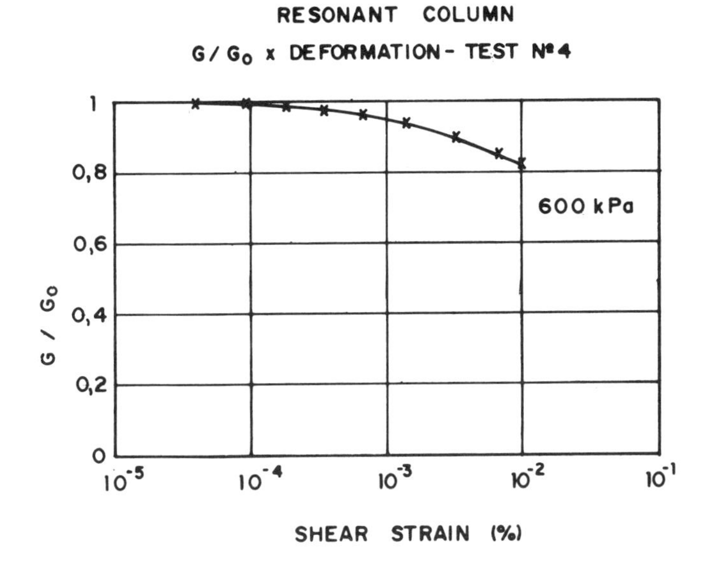

FIG. 2. G/G_o versus Shear Strain (%)

Assessment of the Elastic Modulus (E)

E from G_o

The present tendency in geotechnical engineering is to assess E using G_o taking into account its degradation with the shear strain level. After Burland (1989) it has been usual to consider that the shear strains under buildings on shallow foundations are of the order of 10^{-1} (%). The problem is estimating G_o, that corresponds typically to strains 10^{-5} (%) to 10^{-4} (%) and its degradation up to 10^{-1}(%). G_o may be obtained from CHT or Resonant Column tests and also through correlations with SPT (Barros 1992). Décourt (1991) suggested that for soils, with Plasticity Index PI ≈ 50%.

$$\frac{G}{G_o} = \frac{PI + 20}{100} \tag{8}$$

Once G is known, E is obtained through the theory of elasticity.

E from N-SPT

Décourt (1989) proposed direct correlations between E derived from plate load tests and N-SPT values. For the soils of the TSBSP (silty clays and clayey sands) and for saprolites of granite and gneiss.

$$E = 2.5\, N_{60}\ (\text{MPa}) \tag{9}$$

These E values correspond approximately to deformations (δ) of 6.0 mm of a steel plate with D = 0.80 m in diameter (δ/D ≈ 0.75%). Applying this correlation to the case of the Rimini building, N_{60} ≈ 24 the E value was expected to be equal 60 MPa.

Settlement Computations

In Brazil, the assessment of settlement of shallow foundations is not usually performed for routine jobs. It is assumed that when a footing is designed for a contact pressure considered allowable the differential settlements would certainly be within the allowable range. But the author recognizes that this is too simplistic an approach that might give good results only in the cases were the column loads do not differ very much from each other. Therefore, in many design jobs the author computes the settlement of the building using the theory of elasticity with E = 2.5 N_{60} (MPa) and taking into account the mutual influence or all footings. On the basis of some allowable differential settlement criteria the dimensions of adjacent footings are eventually changed. This approach is considered conservative since the beneficial influence of the structure rigidity in redistributing the loads in the columns when differential settlements occur is not taken into account. A computer program was recently developed by Gusmão (1990) for taking into account the rigidity of the structure on the settlements. It is expected that in the future such improvements will be much more frequently used in routine design jobs.

Settlement computations for the Rimini building were not performed during design. On the basis of the theory of elasticity, for a given footing, the settlement was computed disregarding the contribution of the loads of the adjacent footings. A maximum value of 40.0 mm was foreseen.

With the later decision of considering this building an experimental one, much more sophisticated tools were introduced in the computations. The FLEA program developed by Small (1991) at the University of Sidney was used. Also the graphs and tables found in the recent book of Milovic (1992) were considered.

Milovic tables and graphs are unique in the way that they are valid for foundations of any rigidity. The most important limitation of this later method is that only circular foundation could be considered. Therefore, for computation purposes, the rectangular footings of the Rimini building were transformed into circular footings of equivalent area. The errors introduced in such adjustment were supposed to be small.

The settlements of three columns, P-07, P-09 and P-10 were computed assuming that the values of the columns loads were 68% of their nominal values (structural designer recommendation) and that the elastic modulus of the soil was 1,000 MPa. The values of the computed settlements are given in Table 4. Comparisons between the computed settlements and the measured ones should allow assessments of the actual elastic modulus of the soil.

TABLE 4. Computed Settlements (mm)

Column	Milovic Tables		FLEA Program	
	$\mu = 0.2$	$\mu = 0.3$	$\mu = 0.2$	$\mu = 0.3$
P-07	2.222	2.099	1.894	1.758
P-09	2.164	2.019	1.786	1.660
P-10	1.330	1.251	1.004	0.935

SETTLEMENT MEASUREMENTS

Nineteen settlement pins were installed in the columns of the Rimini building. Three bench marks were provided in order to obtain the maximum possible precision. However, the pins were installed only on 06/08/1992 when the 3rd floor slab was being concreted and therefore these initial settlements could not be recorded. A hypothesis was assumed that during the 1st cycle of measurements the actual settlements were double the measured ones. With this assumption, the settlements have been corrected and the corresponding elastic modulus was derived by a back analysis. With this modulus value and an assessment of the column loads the column settlements were computed. The differences between computed and measured (06/08/92) settlements were assumed to be reasonable assessments of the settlements that occurred previously to the installation of the pins. An iterative procedure was used to optimize this procedure. Since it is well known that the elastic modulus is strain dependent it might be argued that the assumption of a constant modulus for both the initial and the final load is not correct. The author is aware of this limitation but since the values are very small, the error was assumed not only to be irrelevant but also conservative in the sense that the adjusted settlements are equal to or greater than the actual ones.

The measured settlements of columns P-07, P-09 and P-10, in 06/15/1993 were, respectively: 3.18 mm, 3.33 mm and 2.00 mm. At that time the column loads were typically 68% of their maximum design values.

Elastic Modulus (E) From Settlement Measurements

Comparing the computed settlements of columns P-07, P-09 and P-10 with the measured ones, the values of E were obtained for values of 0.2 and 0.3, Table 5.

TABLE 5. E (MPa) Values from Back Analyses

Column	Milovic Tables			FLEA Program		
	μ=0.2	μ=0.3	Av.	μ=0.2	μ=0.3	Av.
P-07	699	660	679	596	553	574
P-09	650	606	628	536	498	517
P-10	665	625	645	502	467	484
Overall average	651			525		

Analysis of all "E" Value Determinations

From the analysis of the Resonant Column test results and taking into account the proposed correlations (Décourt 1991) for assessing the variation of G/G_o with the shear strain, it might be assumed that for $=10^{-1}$ (%), $0.5 \leq G/G_o \leq 0.6$. Considering for comparison purposes a value of 0.25 and G/G_o values of 0.5 and 0.6 the following E values in MPa were derived, Table 6

TABLE 6. Range of Values of E as a Function of the Test

Test	E (MPa)
Cross Hole	523 to 628
Resonant Column	580 to 696
SPT	513 to 616
B. Analysis P-7,P-9, P-10 (Milovic)	628 to 679
B. Analysis P-7,P-9, P-10 (FLEA)	485 to 574

It is interesting to observe that all three tests provided estimates of E of the same order of magnitude of those obtained through back analyses of settlement measurements.

THE STIFFNESS OF LATERITIC SOILS

The stiffness of the clay on which the Rimini building was founded was much higher than any forecast. The reason for this extraordinary behavior of the soil was not immediately recognized. After some reasoning, the author decided to search the literature about the behavior of lateritic soils as bearing strata for foundations. It was found that practically nothing had been published on this subject. Even in the proceedings of the First International Conference on Lateritic and Saprolitic soils held in Brasilia, in 1985 very little information was found. An important exception was the paper by Velloso et al. (1978). Results of plate load tests on lateritic soils shown very high E values. But no one seemed to associate these high stiffnesses with the lateritic behavior of these soils.

Barros (1992) found that the Japanese correlations relating G_o with N (Osaki et al. 1973) were applicable to many Brazilian soils, either sedimentary or saprolitic. But for a group of soils classified as lateritic, these correlations were inapplicable, the G_o values measured in Resonant Column tests being much higher than those predicted on the basis of correlations. A correlation between G_o and N-SPT specific for lateritic soils was worked out.

$$G_o = 47.5\ N^{0.72}\ (\mathrm{MPa}) \tag{10}$$

Applying this correlation to the soil discussed herein gives: $G_o = 410.6$ MPa. This value is of the same order of magnitude of the values obtained from Cross Hole tests ($G_o = 418.7$ MPa) and Resonant Column tests ($G_o = 464.0$ MPa).

IDENTIFICATION OF LATERITIC SOILS

To define what is a lateritic soil is by no means an easy task. Chemically speaking a soil is considered lateritic if the ratio (Kr) between silicon dioxide and iron and aluminum oxide molecules is lower than two.

$$K_r = \frac{SiO_2}{Al_2O_3 + Fe_2O_3} < 2 \tag{11}$$

But under the point of view of the practicing engineer such classification is not satisfactory.

Nogami and Villibor (1981) presented a soil classification, originally devised for highway purposes that up to now has been considered in Brazil the best option for the identification of lateritic soils. The soil under study was submitted to these classification tests and the conclusion was that it was a lateritic clay.

Ignatius (1992) proposed a very simple method for this same purpose, which is based in the Standard Proctor test parameters, the maximum unit weight of dry soil, γ_{dmax} and the optimum moisture content h_{ot}, and also $\Delta\gamma_d$ and Δh. Ignatius defined a parameter L:

$$L = \frac{\Delta\gamma_d / \gamma_{dmax}}{\Delta h / h_{ot}} \quad (12)$$

For L > 0.3, the soil is lateritic. For the case of the soil under consideration the L value was 0.60, and therefore a lateritic soil. It is postulated that it is the lateritic behavior of such soil that is responsible for its stiffnesses 10 times higher than of other non lateritic soils, with the same penetration resistance.

The highly disrupting nature of the SPT and of other penetration tests preclude them from the detection of the high stiffness that are probably caused by micro cements that might subsist only under low stress/strain levels.

This very high stiffness might also be an explanation for the very high material index (I_d) of the Marchetti Dilatometer tests, that classified this soil as a sand and not as a clay. The difference between the corrected B and A readings ΔP, also designated P1 - P0 defines the dilatometer modulus and also the I_d. Since for this lateritic clay the stiffness is much higher than for other apparently similar clays the classification made according to criteria valid for non lateritic soils would certainly conduct to misleading results. Therefore a possibility might exist of utilizing the Marchetti Dilatometer for identifying lateritic clays. If the soil is surely a clay and the computed I_d is typical of a sand, it is reasonable to suppose that this clay is a lateritic clay.

Conclusions

The research on the Rimini building emphasized the advantages of using in situ and laboratory tests not usually carried out in routine jobs. However, the most important finding of this investigation was by far the discovery of a type of soil, probably not scarce in tropical regions, which presents stiffness properties much higher than common soils of the same penetration resistance.

For the reasons already discussed in this paper but evidently still lacking more clear explanations the penetration tests, SPT and possibly also the CPT do not recognize the extraordinary stiffness behavior of them.

Once these findings are confirmed through the analyses of more case records of buildings in soils presenting lateritic behavior the cost of foundations in such soils could safely be reduced to less than one half.

Acknowledgments

All in situ and laboratory tests and measurements were performed free of charge. To the owners and constructors of the Rimini building Mr. Giovanni Parasmo and Mr. Ricardo Parasmo, to Mr. A. R. Quaresma Filho, director of Engesolos that performed all the SPT/SPT-T borings, to Mr. L. C. Bottura, president director of Cota Territorial that performed the settlement measurements, to Prof. F. Bogossian, president director of Geomecânica S.A. that performed the DMT, to Mr. J. M. C. de Barros of the Institute for Technological Research, IPT, that performed most of the laboratory tests, to Mr. J. C. Dourado, also from IPT that performed the Cross Hole tests, to Mr. L. P. de Moraes Filho, the structural designer of the building and to Prof. H. G. Poulos, who suggested the use of the FLEA Program and of the Milovic tables, the gratitude of the author.

APPENDIX - REFERENCES

Barros, J. M. C. (1992). "Propriedades dinâmicas dos solos." *Feira da Dinâmica na Construção Civil*, São Paulo, Brazil.

Burland, J. B. (1989). "Small is beautiful, the stiffness of soils at small strains." *Can. Geotech. J.*, 26, 499-516.

Décourt, L. (1991). "Special problems on foundations." *Proc. 9th PAMCSMFE*, Viña del Mar, General Report, Vol. 4.

Décourt, L., and Quaresma Filho, A. R. (1991). "The SPT-CF, an improved SPT test." *Proc. of SEFE II*, São Paulo, 1, 106-110.

Décourt, L., Belincanta, A., and Quaresma Filho, A. R. (1989). "Brazilian experience on SPT." *Proc. 12th ICSMFE*, Supplementary contributions by the Brazilian Society for Soil Mechanics, Rio de Janeiro, 49-54.

Décourt, L., and Quaresma Filho, A. R. (1994). "Practical applications of the standard penetration test complemented by torque measurements, Present stage and future trends." *Proc. 13th ICSMFE*, New Delhi, 1, 143-147.

Gusmão, A. D. (1990). "A study of soil-structure interaction and its effect on the settlements of buildings," M.S. thesis. COPPE, Fed. Univ. Rio de Janeiro, Brazil.

Ignatius, S. G. (1991). "Solos tropicais, proposta de índice classificatório." *Solos e Rochas*, 14(2), 89-93.

Ignatius, S. G. (1992). "Limites de Atterberg, granulometria e classificação MCT de solos tropicais." *Proc. 9th COBRAMSEF, Brazilian Conf. on Soil Mech. and Found. Eng.*, Salvador, 2, 271-279.

Kovacs, W. D., Salomone, L. A., and Yokel, F. Y. (1981). "Energy measurements in the Standard Penetration Test." *Bldg. Sci. Ser. 135*, National Bureau of Standards, ashington, D. C.

Milovic, D. (1992). *Stress and Displacements for Shallow Foundations*, Elsevier, Amsterdam, The Netherlands.

Nogami, J. S., and Villibor, D. F. (1981). "Uma nova classificação de solos para finalidades rodoviárias." *Proc. Simpósio Brasileiro de Solos Tropicais de Engenharia do Rio de Janeiro*, 1, 30-41.

Schmertmann, J. H. (1989). *DMT Digest 11*, GPE Inc. Geotechnical Equipment, Gainesville, Florida.

Schmertmann, J. H., and Palacios, A. (1979). "Energy dynamics of SPT." *J. Geotech. Eng. Div.*, 105(GT8), 909-926.

Small, J. C. (1991). *The FLEA Program (Finite layer elastic analysis)*, the University of Sidney, Australia.

Velloso, P. P. C., Grillo, S. O., and Penedo, E. J. (1978). "Observações sobre a capacidade de carga e o módulo de deformação de solo por meio de provas de carga e ensaios de laboratório." *Proc. 6th COBRAMSEF*, Rio de Janeiro, 1, 305-328.

Settlement of Compacted Fills Caused by Wetting

Iraj Noorany,[1] Member, ASCE
and Jeffrey V. Stanley,[2] Assoc. Member, ASCE

Abstract: Post-construction wetting of fills can cause swell or compression, depending on soil composition and stress conditions just before wetting. The various factors affecting swell/compression of compacted cohesive soils are studied, and the steps in analyzing one-dimensional heave/settlement of compacted fills are described. Limitations of the analysis, and recommendations for reducing fill settlement are presented.

Introduction

The long-term performance of compacted fills and embankments is greatly influenced by post-construction wetting. Many structural fills in California, some deeper than 100 ft (30 m), have experienced damage several years after construction. This paper first explores the underlying reasons that so many structural fills have performed unsatisfactorily; second, it presents typical data regarding the swell/compression behavior of compacted expansive soils and shows how such data can be used for fill settlement analysis; and third, it offers recommendations for design of fills for reduced settlement.

Past Practice Related to Structural Fills

Geotechnical engineering practice related to structural fills in California prior to the 1990s was remarkably uniform. Actually, these types of fills were not

---

[1] Prof. of Civ Engrg., San Diego State Univ., San Diego, CA 92182; and Visiting Prof., Univ. of California, San Diego, La Jolla CA 92093.

[2] Project Engineer, Dames & Moore, 5600 B St., Suite 100, Anchorage, AK 99518; formerly Research Asst., Dept. of Civ. Engrg., San Diego State Univ., San Diego CA 92182.

designed to suit individual sites; rather, they were engineered according to a standard practice summarized below:

a. Generally, 90% relative compaction, based on the ASTM D1557 test, was specified, regardless of the type of material and fill thickness. Compaction water content was typically specified at optimum, or slightly higher to reduce soil expansion. The specified relative compaction was usually for *total material*; relative compaction of the *matrix* was not mentioned. This resulted in variable degree of compaction in the matrix (depending on the percentage of the oversize), hence, erratic and variable swell/compression behavior.

b. Swell characteristics of fill materials were evaluated qualitatively in terms of *swell potential* rather than quantitatively. Furthermore, it was assumed that expansive soils placed at depths greater than 2 or 3 feet (0.6 to 1 m) below the ground surface would not cause damage; use of 2 or 3 feet of non-expansive *select material* on top was considered to mitigate expansion problems. This was based on the assumption that with good surface drainage, the depth of wetting from rainfall and landscape irrigation would be limited to the upper 3 feet (1 m) of the fill.

c. Fill settlement during construction was considered to have no long-term consequences, except for very deep fills, which were monitored before a building permit was issued. However, this practice did not help with post-construction settlement caused by deep wetting from landscape irrigation.

d. Alluvial, colluvial, and topsoil deposits were either partially or totally removed before fill construction. Confined compression tests were run on *undisturbed* samples to find out whether wetting could cause *collapse* of these materials. Ironically, although wetting effects on the soils to be buried under the fill were considered, deep wetting of the fill itself was not considered.

e. Stability of fill slopes was analyzed for deep-seated and surficial failures. However, no analysis was made for lateral deformation of the fill slopes; the common practice in this regard was to recommend a minimum setback (horizontal distance from edge of base of foundation to face of slope) of 5 to 8 feet (1.5 to 2.5 m).

Although some fills constructed according to the practice discussed above performed satisfactorily, serious problems developed in many constructed in the late 1970s and the 1980s. Two important factors contributed to this turn of events: (1) the rapid pace of fill construction with bigger and heavier equipment, in larger and deeper canyons, made proper fill observation and testing difficult; and (2) the shift from single-family housing to high-density condominium developments with lush landscaped areas required heavy irrigation, equivalent to 60-plus inches of water per year (Sorben and Sherrod 1977; Brandon et al. 1990). Consequently, within a period of 5 to 10 years, many deep fills became fully wetted.

It is difficult to ascertain the impact of the first factor on the performance of deep fills, except that there is ample evidence of inadequate compaction, thick lifts, an abundance of nested oversize cobbles, highly expansive soils and not enough non-expansive select material, an inadequate number of field density tests, mismatch and other fill testing problems (Noorany 1987, 1990b). The extent of variability in compaction control of structural fills can be surmised from data obtained at three sites in San Diego, California (Noorany 1990a), as summarized in Table 1.

TABLE 1. Survey of Relative Compaction at Three Structural Fill Sites

Site	*Number of tests*	*Average compaction (%)*	*Range (%)*	*Range as percentage of mean*	*Standard deviation*	*Percent of data less than 90%*
1	24	91	15	17	3.9	17
2	16	88	23	26	6.4	62
3	22	94	14	15	3.4	14

As for the second factor, deep wetting of fills was undoubtedly responsible for serious damages to many fills, and costly litigation of unprecedented scale in California. The pattern of distress was different for confined fills than for fill slopes. In confined canyon fills with variable thickness, differential ground heave/settlement occurred as a result of cumulative swell (of shallow parts) and compression (of deep parts) of the fill. In fill slopes, deep wetting produced not only ground-surface differential heave/settlement, but also lateral deformation of the slope. This was typically manifested by ground-surface stretching in the direction of the slope, opening of joints of hardscape and separation of hardscape from building foundations, and cracking of slabs in the direction parallel to the slope. These distress features were observed at distances much greater than 5 or 8 feet (1.5 to 2.5 m) from the slope face.

Swell/Compression Mechanism

The compressibility of compacted sands, gravels, and rockfill has been studied by many researchers, notably Marachi et al. (1969), and Nobari and Duncan (1972). These studies show that cohesionless soils and rockfills exhibit low compressibility at low pressures, but there can be significant compression at high pressures due to grain crushing. Nobari and Duncan (1972) developed a finite-element method for computing vertical and horizontal deformation of granular embankments caused by wetting.

For cohesive fills, volume change caused by wetting can be swell or compression, depending on the stress level. Past studies of this phenomenon include: Jennings and Knight (1957), Leonard and Narain (1963), Booth (1975), Cox (1978), Maswoswe (1985), Nwabuokei and Lovell (1986), Alonso et al. (1988), Schreiner (1988), Lawton et al. (1989, 1991, 1992), Brandon et al. (1990), and Noorany and Stanley (1990). As pointed out by Burland (1965) and Mitchell (1993), partly saturated cohesive soils have *packets* of clay particles between contact points of silt and other granular particles. These packets have negative pore pressures and can expand if they absorb water under low total stresses, but will slip and distort if wetted under high total stresses. This slippage and distortion of clay packets can lead to an overall volume decrease (*collapse* or *hydrocompression*) of a soil's granular particles, even though individual clay packets take up water and expand. Thus, at microscale, the reduction in effective stress caused by wetting is compatible with the expansion of individual clay packets, but at macroscale, a net overall volume reduction occurs as granular particles slip at contact points. The transition from net swell to net compression behavior is smooth, and the *swell pressure* (stress required for preventing swell) depends on the soil type and as-compacted moisture, density and fabric.

The swell/compression mechanism described above can be further accentuated in the case of fills derived from hard claystones and siltstones. In fills of this type of soil, hard clumps do not break down adequately during field compaction, but they soften and disintegrate to soft clays and silts as a result of post-construction wetting (Sherard et al. 1963; Rogers 1992). This alteration from siltstone to silt or claystone to clay creates undesirable volume change characteristics for the fill.

Factors Affecting Swell/Compression Behavior

The amount of swell/compression of a compacted soil depends on a number of factors: soil type; as-compacted moisture, density and soil fabric; stress path from compaction to wetting, stress condition under which wetting occurs, and the extent of wetting. Because of the complexity of these factors, it is not easy to predict the amount of swell/compression without extensive laboratory tests and analysis. The example described in the following sections illustrates the trends of test results, and method of analysis.

Villa Trinidad is a residential area within the Tierra Santa development in San Diego, California, with fills deeper than 100 feet (30 m) in some areas. The fill soils consist primarily of expansive clayey sands (SC) and sandy clays (CL) with gravel and cobble, compacted to at least 90% of the maximum density in the ASTM D1557 test, and water contents ranging from slightly below to slightly above the optimum water content. By 1986, 10 years after construction, some fill areas had settled as much as 18 inches (46 cm), and the fill's moisture content had increased substantially throughout. The subsurface condition at a location where the fill depth was 70 feet (21.4 m) was investigated in December 1985. This particular location was selected because of its proximity to a monument that had measured a total ground settlement of 11.5 inches (29.2 cm). The fill's moisture profile at the time of placement and in 1985 are shown in Fig. 1 (Brandon et al. 1990).

A 30-inch (76-cm) diameter bucket-auger boring was made, and the entire soil column to a depth of 31 feet (9.5 m) was removed. The soil was uniform and consisted of wet clayey sand (SC) with 15% gravel and cobble coarser than the No. 4 sieve size. The percent finer than sieve No. 200 was 26, and the clay fraction finer than 0.002 mm was 17%. The soil had a liquid limit of 35, plasticity index of 21, and specific gravity of 2.66. A composite sample of this boring was used in this study.

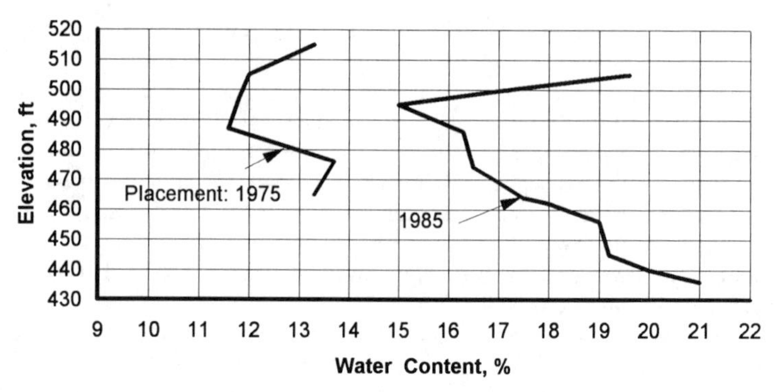

FIG. 1. Water Content Profile: Villa Trinidad (Data from Brandon et al. 1990)

Fig. 2 shows results of the ASTM D1557 compaction test on the fraction finer than the No. 4 sieve; also shown in this figure are the initial and final moisture-density values for eight series (four samples per series) of one-dimensional swell/compression tests. For these tests, samples were compacted using a moist-tamping procedure with a tamper having a foot size identical to the Harvard miniature kneading compactor. Specimens were assembled with dry porous stones and without filter paper to avoid errors caused by swell/compression of the paper. Load was applied to each sample in small increments up to a target stress, and left until equilibrium was reached. Normally, this did not take more than 30 to 60 minutes. The sample was then inundated with tap water, and its swell or compression was measured over 24 to 72 hrs until equilibrium was reached again. Sample thickness at the end of dry compression was used for computing the swell/compression strain after wetting. All samples were tested by wetting-after-loading, which correctly simulates field conditions; it was found that for expansive soils, loading-after-wetting gives different results than wetting-after-loading (Justo et al. 1984; Noorany 1992b), and is not appropriate for swell/compression tests.

Results of the swell/compression tests are shown in Fig. 3, and illustrate the pronounced influence of the as-compacted moisture and density on swell/compression. Fig. 4 shows that for a given degree of compaction, both swell and hydrocompression can be reduced by increasing the placement water content. For different vertical stress levels, corresponding to different fill depths, Fig. 5 shows plots of swell/compression strains as a function of various placement moisture-density combinations. These plots show that at shallow fill depths, swelling dominates the fill behavior, and low densities and high placement water contents are preferable; in deeper parts of the fill (higher vertical stresses), high densities and high water contents can reduce post-construction settlement caused by wetting.

The final moisture-density conditions of the samples after soaking by inundation are shown in Fig. 2. It can be seen that soaking increases a soil's degree of saturation, but does not cause 100% saturation. Furthermore, wetting by rainfall or irrigation in structural fills may result in a lower degree of saturation than that produced by inundation in the laboratory samples. This in turn may result in somewhat lower swell/compression than that measured in laboratory samples by full inundation.

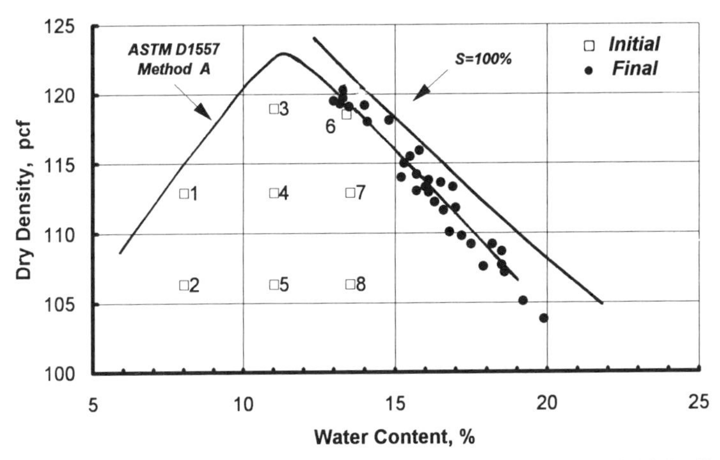

FIG. 2. Initial and Final Water Contents and Dry Densities of Eight Series of Samples Used in Swell/Compression Tests

The effect of partial wetting on swell/compression was studied by Cox (1978) by controlling the amount of water the sample absorbed during the swell/compression process. The trends of changes in moisture content and dry density as the soil approaches saturation under a constant overburden pressure are shown in Fig. 6. As can be seen in this figure, the change in dry density - or strain - is not necessarily a linear function of change in water content or degree of saturation, but over a broad segment of the curve, it is close to linear (Chen 1988).

Results of a series of partial wetting tests on samples of Villa Trinidad fill are shown in Fig. 7. Four specimens were compacted to the same water content and dry density, placed under identical loads, but were wetted by different amounts of water fed, drop per drop, from a tube through a hole in the top cap of the specimen. Over 24 hours was allowed for each specimen to absorb this limited amount of water uniformly, while evaporation was prevented by placing wet balls of filter paper in the annular space outside of the consolidometer mold and sealing it with plastic cover. The actual water content and degree of saturation of each specimen was measured at the end of the test, and specimens had uniform moisture throughout. Data from partial wetting tests can be used in heave/settlement analyses when the extent of wetting in the field is less than full inundation. For the worst-case scenario, however, full wetting should be considered in the design of fills and, of course, dam embankments.

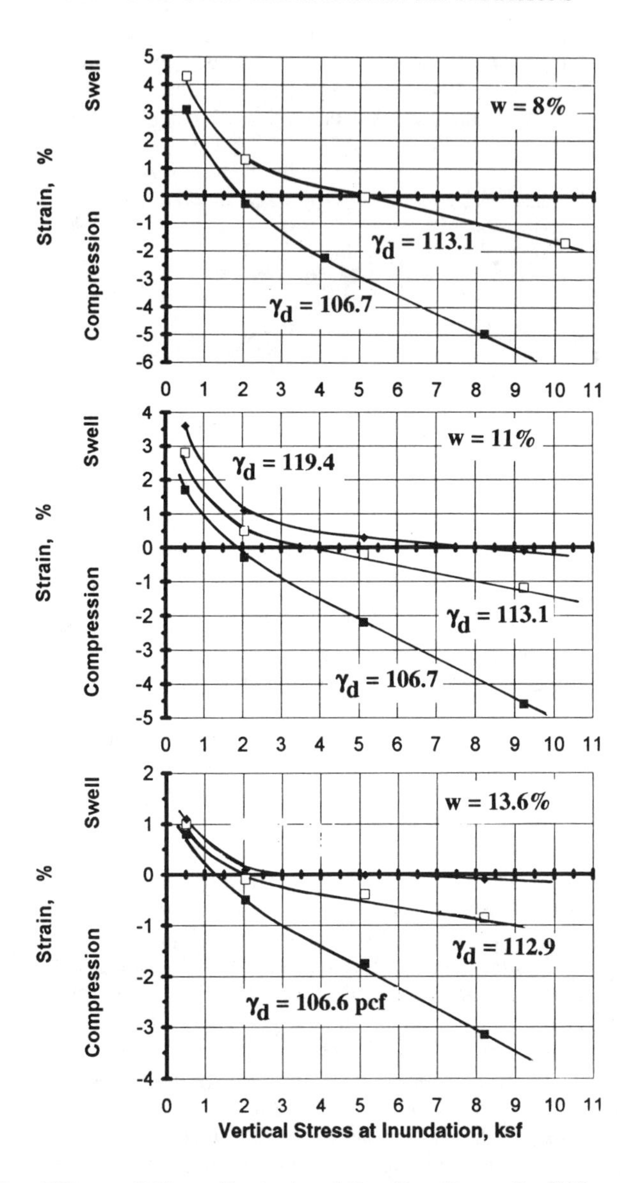

FIG. 3. Effects of Water Content and Dry Density on Swell/Compression

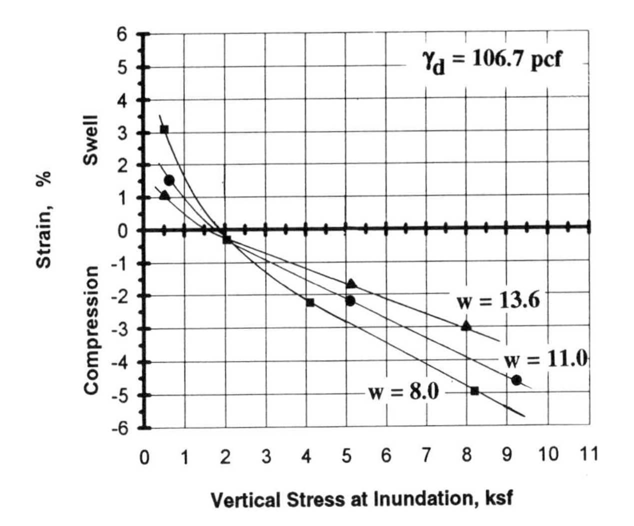

FIG. 4. Effect of Water Content on Swell/Compression

HEAVE/SETTLEMENT ANALYSIS

In order to use the swell/compression test results in Fig. 3 for analysis of Villa Trinidad fill, it was first necessary to evaluate the placement moisture and density of the fill's matrix (the minus No. 4 fraction). According to Brandon, et al. (1990), the fill's average placement water content was 11.7%, and the average placement density was 125.2 pcf (2 Mg/m^3), which corresponds to a dry density of 112.1 pcf (1.8 Mg/m^3). Considering that the fill in our large-diameter boring had 15% plus No. 4 gravel, and assuming a water content of 1% for gravel, we computed a water content of 13.6% for the fill's matrix. Also, the dry density of the fill matrix was computed from the following equation:

$$\gamma_{dm} = \frac{(1 - P_c)G_s\gamma_w\gamma_d}{G_s\gamma_w - P_c\gamma_d} \tag{1}$$

wherein γ_{dm} is the matrix dry density, γ_d total dry density, γ_w density of water, P_c is percent by dry weight of the oversize fraction, and G_s is the bulk specific gravity of the oversize material.

Substituting P_c=15% for percent of +4 oversize, G_s=2.66, γ_w = 62.4 pcf (1 Mg/m^3), and γ_d = 112.1 pcf (1.8 Mg/m^3), we compute γ_{dm} = 106 pcf (1.7 Mg/m^3) for the fill matrix. By extrapolation from the data in Fig. 3, the swell/compression curve in Fig. 8 was constructed for γ_{dm} = 106 pcf and w = 13.6% for the fill's matrix. Using this curve, and a fill unit weight of 125.2 pcf (2 Mg/m^3), wetting induced strains at various fill depths were computed and tabulated in Table 2.

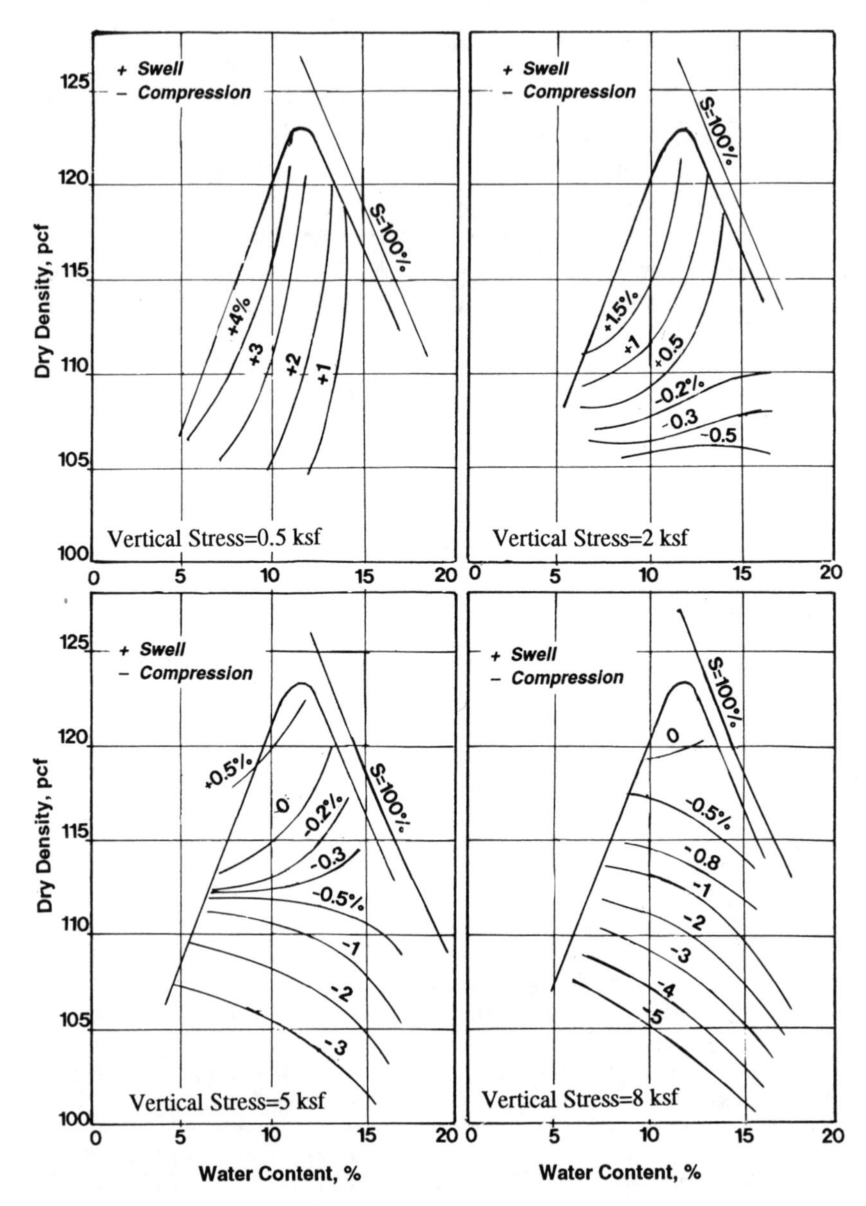

FIG. 5. Contours of Equal Swell and Equal Compression for Villa Trinidad Fill

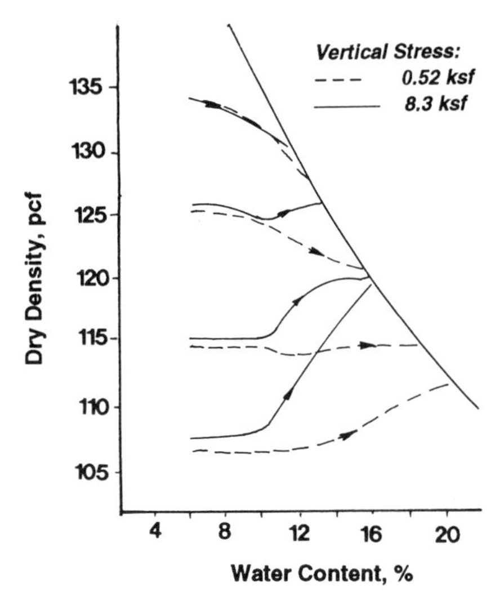

FIG. 6. Paths of Changes in Water Content and Dry Density in Swell/Compression (*Data from Cox 1978*)

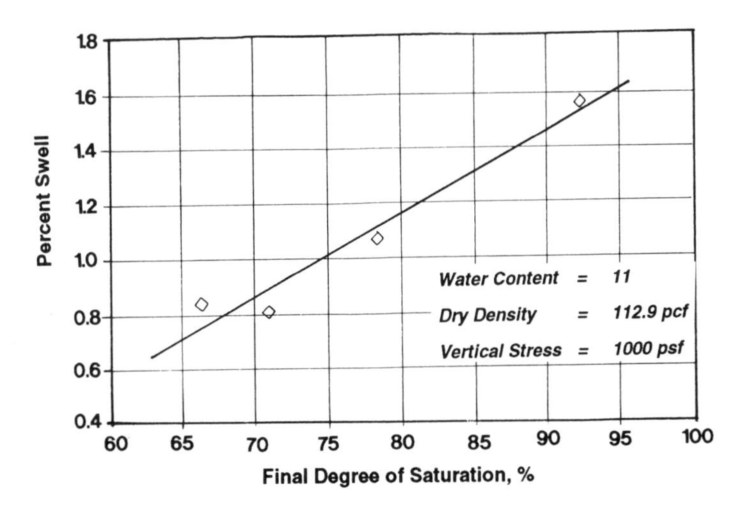

FIG. 7. Effect of Partial Wetting on Swell

TABLE 2. Estimated Swell/Compression Strains

Fill Depth *ft*	*Vertical Stress* *ksf*	*Strain* *%*	*Adjusted Strain* *%*
2	0.25	1.0	0.85
4	0.50	0.8	0.68
8	1.0	0.3	0.26
16	2.0	-0.5	-0.43
32	4.0	-1.4	-1.2
48	6.0	-2.3	-2.0
70	8.8	-3.6	-3.1

Because of the presence of 15% inert gravel (plus No. 4) in the fill, the swell/compression strains in the third column of Table 2 were multiplied by 0.85 to compute the adjusted strains in column four; adjusted strains are plotted against depth in Fig. 9. Total settlement for the 70-ft (21-m) fill can be computed as the net area under the curve in Fig. 9, excluding the upper 2 feet (0.6 m) which consisted of non-expansive select material. The calculated settlement is 10 inches (25.4 cm), which is close to the measured value of 11.5 inches (29 cm) when the fill settlement was essentially completed. The additional effect of increase in the fill's unit weight caused by wetting was not included in calculations, although it could easily be included in the manner suggested by Nwabuokei and Lovell (1986).

The preceding analysis involved some simplifying assumptions: First, we assumed strains in the confined fill to be one-dimensional as in the laboratory test. Second, we applied test results measured on a composite sample of the upper 31 ft to the entire 70-ft depth; there is justification for this, in view of the uniformity of the fill profile as reported by Brandon et al. (1990). Third, the actual degree of wetting in the field might have been less than in our laboratory tests, although data in Fig. 1 indicates an average water content of 18% for the fill profile in 1985, and that is very close to the inundation water contents in Fig. 2 for samples with initial densities in the range of 106 pcf (1.7 Mg/m^3). Despite these and other possible limitations, the example used here can serve as a guide for laboratory testing and analysis of settlement of confined fills modeled as a one-dimensional strain situation.

SUMMARY AND RECOMMENDATIONS

Performance of many deep structural fills in recent years have left no doubt that past practice in this regard needs major revisions. It is clear that every fill needs to be designed based on adequate test data and sound engineering principles. It seems that since all fills inevitably get wet sooner or later, one may as well design for that eventuality. The trends of swell/compression behavior presented in this paper are typical of all expansive fills, and similar curves for twelve other fills in California can be seen in Noorany and Stanley (1990). Some soils are so expansive that they should not be used, without treatment, in structural fills intended for shallow foundations and slab-on-grade construction.

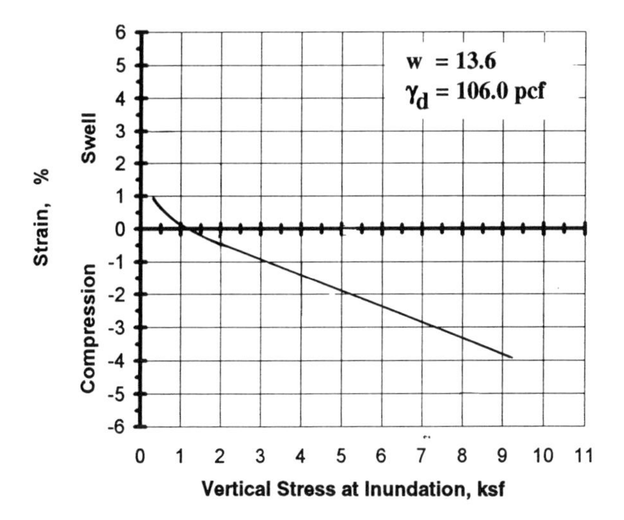

FIG. 8. Swell/Compression Curve for Villa Trinidad Fill Matrix

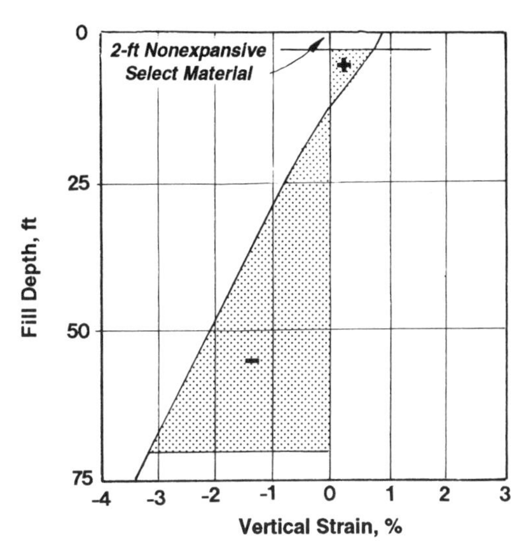

FIG. 9. Variation of Swell/Compression Strain with Fill Depth for Villa Trinidad Fill

By means of four or more series of tests and data analysis similar to those depicted in Figs. 2 through 5, it is possible to select the best placement moisture-density combinations for various fill depths so that detrimental effects of post-construction wetting can be minimized. Evidently, a single compaction standard, such as optimum water content and 90% relative compaction, cannot produce good results for all fills, and varying placement moisture-density with depth seems to be a good rational alternative.

Because swell/compression is caused by the soil's fine fraction, tests on small-diameter samples, using the soil fraction finer than the No. 4 or the No. 10 sieve size, seem to provide satisfactory results, except possibly for hard claystones and siltstones that do not break under compaction equipment in the field. For these kinds of soils, as well as those with abundant oversize, it would be preferable to use a large mold, such as the 4-ft (122-cm) in diameter mold described by Noorany (1987). Regardless of the size of the mold, it is important that kneading compaction be used. In our experience, moist-tamping by a foot identical to the Harvard miniature compactor gives results between kneading and static, but static compaction overestimates expansion and underestimates hydrocompression. Also, the double oedometer tests (loading-after-wetting) do not provide correct results for analysis of heave/settlement of expansive soils, and we recommend wetting-after-loading tests. It is also imperative that the test apparatus be calibrated for system compression.

In applying the test results to settlement analysis, it is necessary to use data from tests with moisture-density values corresponding to the fill's matrix moisture and density, so long as oversize fraction is not so much that they are interlocking. Also, from the standpoint of settlement evaluation, it is preferable for compaction control specifications to require a specific degree of compaction for a fill's matrix rather than for total material.

Aside from test-related factors, there are complex issues in the field, including the sequence and extent of wetting. If wetting begins at the bottom and rises to the top, the fill will experience settlement first and heave later; with wetting from the top, the reverse will be true. We recommend that the designer calculate ground-surface deformation for various plausible modes of wetting and extents of wetting, in order to see whether the maximum differential heave/settlement predicted would be tolerable for the structure.

The one-dimensional settlement analysis and related tests described in this paper are not directly applicable to the analysis of lateral deformation of compacted slopes. To evaluate heave/settlement and lateral deformation in slopes, triaxial swell/compression tests coupled with a finite-element analysis can be used (Noorany et al., 1992a). Both for confined fills and slopes, we should seize every opportunity to monitor fill deformation, so that the validity of these methods can be better evaluated.

ACKNOWLEDGMENTS

The laboratory tests were carried out at San Diego State University when Jeff Stanley was a graduate student. Partial wetting tests were performed by Canan Emrem. Field work was done by Woodward-Clyde Consultants, and financial support for the research was provided by a number of geotechnical firms. These contributions are gratefully acknowledged.

APPENDIX I. REFERENCES

Alonso, E. E., Batle, F., Gens, A., and Lloret, A. (1988). "Consolidation analysis of partially saturated soils-application to earth dam construction." *Numerical Methods in Geomechanics,* Balkema, Rotterdam, The Netherlands, 1303-1308.

Booth, A. R. (1975). "The factors influencing collapse settlement in compacted soils." *Proc. 6th Regional Conf. for Africa on Soil Mech. and Found. Engrg.*, South African Inst. of Civil Engineers, 2, 57-63.

Brandon, T. L., Duncan, J. M., and Gardner, W. S. (1990). "Hydrocompression settlement of deep fills." *J. Geotech. Eng.,* ASCE, 116(10), 1536-1548.

Burland, J. B. (1965). "Some aspects of the mechanical behavior of partly saturated soils." *Moisture Equilibria and Moisture Changes in the Soils beneath Covered Areas,* Butterworth, Sydney, Australia, 270-278.

Chen, F. H. (1988). *Foundations on Expansive Soils*, Elsevier Scientific Publishers, New York, New York.

Cox, D. W. (1978). "Volume change of compacted clay fill." *Proc. Conf. on Clay Fill,* Inst. of Civ. Engrs., London, England, 79-87.

Jennings, J. E., and Knight, K. (1957). "The prediction of total heave from the double oedometer test." *Transactions Symp. on Expansive Clays,* South African Inst. of Civ. Engrg., 13-19.

Justo, J. L., Delgado, A., and Luiz, J. (1984). "The Influence of Stress-path in the Collapse-Swelling of Soils at the Laboratory." *Proc. 5th Int'l. Conf. on Expansive Soils,* Adelaide, 67-71.

Lawton, E. C., Fragaszy, R. J., and Hardcastle, J. H. (1989). "Collapse of compacted clayey sand." *J. Geotech. Eng.*, ASCE, 115(9), 1252-1267.

Lawton, E. C., Fragaszy, R. J., and Hardcastle, J. H. (1991). "Stress ratio effects on collapse of compacted clayey sand." *J. Geotech. Eng.,* ASCE, 117(5), 714-730.

Lawton, E. C., Fragaszy, R. J., and Hetherington, M. D. (1992). "Review of wetting-induced collapse in compacted soil." *J. Geotech. Eng.*, ASCE, 118(9), 1376-1394.

Leonards, G. A., and Narain, J. (1963). "Flexibility of clay and cracking of earth dams." *J. Soil Mech. Found. Div.,* Proc. ASCE, 89(SM2), 47-98.

Mitchell, J.K. (1993). *Fundamentals of Soil Behavior*, 2nd Edition, John Wiley & Sons, New York, New York.

Marachi, N. D., Chan, C. K., Seed, H. B., and Duncan, J. M. (1969). "Strength and deformation characteristics of rockfill materials," *Report No. TE-69-5*, Dept. of Civ. Engrg., Univ. of Calif. at Berkeley, Berkeley, California.

Maswoswe, J. (1985). "Stress path for a compacted soil during collapse due to wetting," Ph.D. thesis, Imperial College, London University, England.

Nobari, E. S., and Duncan, J.M. (1972). "Effect of reservoir filling on stresses and movements in earth and rockfill dams," *Report No. TE-72-1,* Dept. of Civ. Engrg., Univ. of Calif. at Berkeley, Berkeley, California.

Noorany, I. (1987). "Precision and accuracy of field density tests in compacted soils," *Research Report*, Dept. of Civ. Engrg., San Diego State Univ., San Diego, California.

Noorany, I. (1990a). "Variability in compaction control." *J. Geotech. Eng.,* ASCE, 116(7), 1132-1136.

Noorany, I. (1990b). Discussion of "Compaction control and the index unit weight" by Steve J. Poulos, *Geotech. Testing J.,* ASTM 13(2). 146-147.

Noorany, I., and Stanley, J. V. (1990). "Swell and hydrocompression behavior of compacted soils: Test data," *Research Report,* Dept. of Civ. Engrg., San Diego State Univ., San Diego, California.

Noorany, I., Sweet, J. A., and Smith, I. M. (1992a). "Deformation of fill slopes caused by wetting." *Stability and Performance of Slopes and Embankments II,* Geotechnical Special Publication No. 31, ASCE, New York, New York, II, 1244-1257.

Noorany, I. (1992b). "Stress ratio effects on collapse of compacted clayey sand." *J. Geotech. Eng.,* ASCE, 118(9), 1472-1474.

Nwabuokei, S. O., and Lovell, C. W. (1986). "Compressibility and settlement of compacted fills." *Consolidation of Soils: Testing and Evaluation*, ASTM STP 892, Philadelphia, Pennsylvania, 184-202.

Rogers, J. D. (1992). "Long-term behavior of urban fill embankments." *Stability and Performance of Slopes and Embankments II,* Geotechnical Special Publication No. 31, ASCE, New York, New York, II, 1258-1273.

Schreiner, H. D. (1988). "Volume change of compacted highly plastic african clays," Ph.D. thesis, Imperial College, London University, England.

Sherard, J. L., Woodward, R. J., Gizienski, S. F., and Clevenger, A. C. (1963). *Earth and Earth Rock Dams*, John Wiley & Sons, New York, New York.

Sorben, D. R., and Sherrod, K. L. (1977). "Groundwater occurrence in the urban environment: San Diego, California." *Geology of Southwestern San Diego County, California and Northwestern Baja California,* San Diego Assoc. of Geologists, San Diego, California, 67-74.

APPENDIX II. CONVERSION TO SI UNITS

TO CONVERT	TO	MULTIPLY BY
ft	m	0.3048
in.	cm	2.54
lb	kg	0.4536
psf	kPa	0.04788
pcf	Mg/m^3	0.016

CASE HISTORY OF A COLLAPSIBLE SOIL FILL

**Alan L. Kropp,[1] David J. McMahon,[2]
and Sandra L. Houston,[3] Members, ASCE**

ABSTRACT: Collapse settlement of a deep compacted fill led to damage of a group of condominium units. The fill consisted of a highly heterogeneous mixture of coarse- to fine-grained soil containing angular fragments of gravel- to boulder-sized rock. Compaction specifications requiring 90 percent of the maximum dry density by ASTM method D-1557 were met or exceeded during fill placement, although the fill was generally placed dry of optimum water content, and corrections for rock content were not made in the field. Building distress ranged from mild to severe, depending upon the differential fill thickness beneath the structure. Extensive vertical and horizontal movement monitoring and two full-scale field wetting tests were conducted at the site following the onset of building distress. The results of the field investigations and laboratory response-to-wetting tests provided strong evidence that the building distress patterns were caused primarily by wetting-induced collapse settlement of the deep compacted fill. Based on the results of the field wetting tests, controlled wetting of the site with mud-jacking of the condominium units appears feasible as a mitigation alternative.

INTRODUCTION

A naturally-occurring collapsible soil is normally thought of as a loose, granular material containing small amounts of clay, silt, or other cementing agent. Such soil deposits are common in arid regions of the world where evaporation

---

[1] Principal and [2] Project Engineer, Alan Kropp and Associates, 2140 Shattuck Avenue, Berkeley, CA 94704.

[3] Associate Professor, Department of Civil Engineering, Arizona State University, Tempe, AZ 85287.

exceeds rainfall. Stable in their natural dry state, collapsible soils densify significantly and quickly upon wetting. Embankments of compacted soils may also experience wetting-induced collapse settlement, particularly when constructed as a deep fill. The same mechanism, primarily loss of soil suction upon wetting, causes collapse whether the soil is naturally-occurring or compacted. Though collapsible soils are normally thought of as granular, even a clayey soil can collapse if the stress-level at the time of wetting is sufficiently high. Thus, embankments of any compacted soil may expand or collapse upon wetting depending on the soil-type, the dry density and water content, and the confining pressure. In this paper, a case history of wetting-induced collapse of a deep compacted fill is presented.

Site Description

An old rock quarry site was filled with surplus material from a nearby housing subdivision grading operation. A typical cross section through the site showing the original quarry grade and the current fill grade is given in Figure 1. Subdrains were installed along the walls and base of the old quarry to collect water from groundwater seepage and surface infiltration. A subsurface investigation consisting of a series of borings and trenches revealed that the fill was a highly heterogeneous mixture of bedrock fragments and surficial soils. The fill consisted primarily of interbedded layers of silty gravel (GM), well-graded gravel (GW), poorly graded gravel (GP), silty and gravelly sand (SM), poorly graded sand (SP), and silt (ML). The finer-grained portion of the soils consisted of sand with silt and clay. Cobbles and boulders up to 1 m in diameter were encountered in trenches, and groups of nested boulders were observed. Photographs taken during grading show rocks and boulders pushed to the sides of the bulldozer blade, forming rows of very rocky material.

The fill was engineered and field-inspected, with the compaction specifications requiring at least 90 percent of maximum dry density by ASTM method D-1557, although a minimum water content was not specified. Moisture-density testing was performed during fill placement. Soils were typically placed at 1 to 3 percent below optimum, with some layers at 5 to 7 percent below optimum. The materials were compacted to about 92 to 96 percent of the maximum dry density, based on nuclear gauge readings. The percent rock was not obtained at the locations where the field densities were determined by the nuclear gauge. The fill met or exceeded the specifications described by the site geotechnical reports and construction plans and drawings.

The fill was originally intended to support a park area, and was completed in 1985 to an elevation approximately 2 m above the existing grade shown in Fig. 1. However, highly desirable views of the downtown San Francisco skyline increased the value of the property for condominium development. In 1987 the height of the fill was lowered to create building pads, while preserving the views of previously-constructed condominium units located uphill of the site. The final depth of fill at the site ranges from about 0 to 24 m. Condominiums were constructed and became occupied during 1988; signs of distress appeared within just a few months.

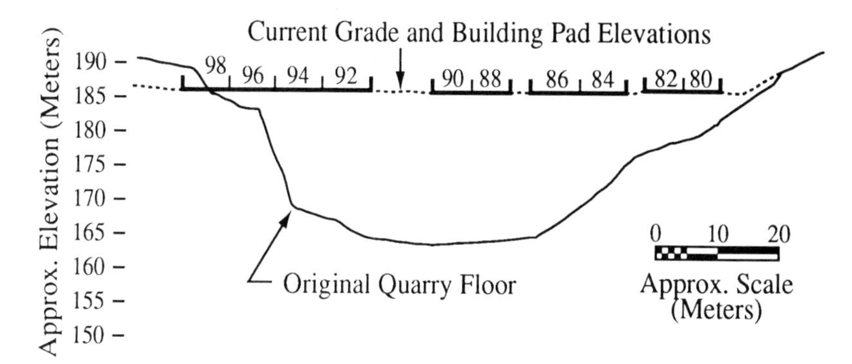

FIG. 1. Typical Cross Section of the Site

Building Distress And Movements

Distress at the site included arc-shaped cracks up to 3 cm wide and with up to 6 cm of vertical offset around the upslope limits of fill. Additional cracks ranging from hairline to 3 cm were found in exterior concrete and asphalt structures including roadways, sidewalks, storm drains, and curbs and gutters. Small sink holes with diameters less than 0.3 m were observed in poorly drained regions. The wood-framed structures were founded on post-tensioned slabs with perimeter and interior footings. Differential movement, cracking and structural failures of the post-tensioned slabs were noted, as well as cracks in the drywall-covered ceilings and walls.

The condominiums at the site were constructed on variable fill thickness, as depicted in Fig. 1. The degree of damage correlated closely with the variation of fill thickness beneath a given structural unit. For example, unit 96 (Fig. 1) was founded on fill ranging in thickness from about 2 m near the center of the structure to about 12 m at the southwest edge. This condominium suffered about 300 mm of differential movement, resulting in structural failure of the post-tensioned slab and severe cracking in walls and ceilings. In contrast, units 88 and 90 were placed on essentially constant 20 m fill thickness, resulting in larger total settlements, but only minor distress occurred because differential settlements were relatively small (less than 50 mm).

Distress began to be noticed at the site during 1988, and movement monitoring began in April, 1989. Horizontal and vertical monitoring points were positioned on the garage and front door sills of each unit. While the arc-shaped crack suggested landsliding, surface movement was generally laterally toward areas of the deepest fill, including some surface movement in an upslope direction. Stability calculations indicated that the site was stable, so landsliding was discounted as a possible cause of distress. Average settlements from April, 1989, through March, 1993, are shown in Fig. 2 for units 82, 86, 90, and 96. The survey data clearly indicate that settlements are largest in regions of greatest fill thickness, although some variation occurred due to nonhomogeneity and differential wetting of the fill. Horizontal movements (not shown) are toward the deepest fill, and are less than 50 mm. Although settlements have essentially ceased in regions of shallow fill (e.g. unit 82), settlement continues to occur in areas of thick fill.

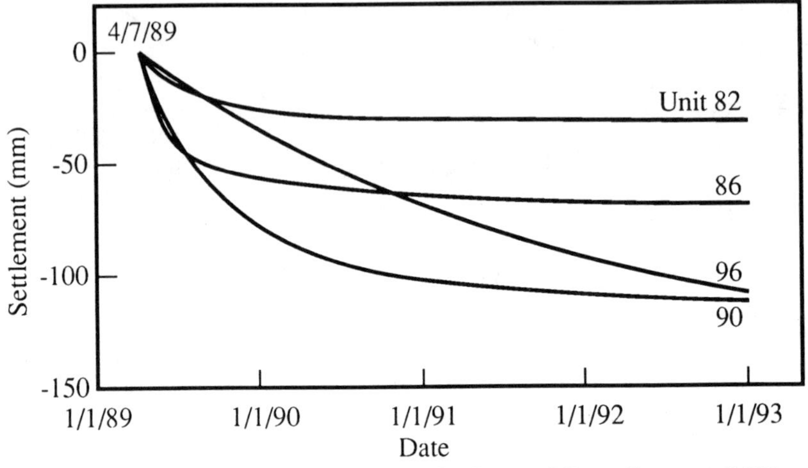

FIG. 2. Settlements for Buildings on Shallow and Deep Compacted Fill

Cause Of Distress

Both granular and clayey deep compacted fills are known to be susceptible to collapse upon wetting, particularly when compacted dry of optimum (Lawton et al. 1989; Brandon et al. 1990). A soil compacted to a low dry density is also susceptible to collapse (Noorany 1990). Low dry density and compaction dry of optimum lead to collapse vulnerability, particularly in the deep fill regions where overburden stress is relatively high.

Inadequate corrections to dry densities to account for the presence of large aggregate can also lead to lower overall density, and therefore collapse vulnerability. Typically the effect of increasing the rock content of a given soil is to increase the maximum dry density. This occurs because the specific gravity of the rock is usually higher than the bulk material between the rock fragments. Therefore, it is important that the reference dry density used in computing relative compaction be corrected for the appropriate percent rock, consistent with that of the field density test specimen. If for example, the reference dry density were obtained for 15 percent rock, and the field specimen actually contained 35 percent rock, then the computed relative compaction would be artificially high. This would result in a greater collapse-susceptibility of the compacted fill. In addition, several methods for determining maximum dry density of soils containing large aggregate are available. In many cases, the method for accounting for the rock fraction has a very significant impact on the reference dry density (Houston and Walsh 1993). Therefore, the engineer and field inspector must be clear on the technique that is to be used to make rock corrections. If, however, the rock percentage in the field is not known, then a correction (by any method) cannot be made.

The granular fill at this particular site contained silty and clayey fines, and was compacted 1 to 3 percent below optimum water content. In addition, because rock corrections were not used during fill placement, the fill probably had relatively low density. Therefore, wetting-induced collapse was considered the most likely source of distress.

Because of the presence of localized sinkholes, internal erosion was examined as a contributing settlement mechanism. It was considered likely that layers of relatively fine-grained soils might rest above layers of gravel, cobbles, and boulders. It was considered to be possible, although unlikely, that enough fine-grained soils might migrate into coarser soils to induce some additional settlement.

Wetting of the fill was considered to be the most likely settlement-triggering mechanism. Sources of water included groundwater seepage, rainfall, and landscape watering subsequent to owner occupation of the units. Color and false-color infrared aerial photographs showed zones of concentrated vegetation in landscaped portions of the condominium development and on natural slope faces in areas of seepage. Poor drainage was found in the landscaped areas and terrace drains, where depressions and differential settlements resulted in modifications to the original site drainage scheme. Localized ponding, sink holes, settlement around storm drains, and cracks in the soil and pavement surfaces all diverted additional water into the fill.

Two full-scale wetting tests and a series of laboratory response-to-wetting tests were conducted to evaluate the mechanisms of fill settlement and resulting distress. The field tests were also used to assess the feasibility of full-scale site wetting as a mitigation alternative.

LABORATORY TESTS

Laboratory specimens containing up to 13 mm gravel-sized rock fragments were compacted in 10 cm-diameter molds at 2 to 3 percent dry of optimum to a dry unit weight of approximately 20 kN/m^3. This corresponds to about 90 percent of the maximum dry density determined by ASTM method D-1557. Material larger than the 13 mm size was discarded. Response-to-wetting tests were conducted on the compacted specimens in 10 cm-diameter, 6.5 cm-high rings. Confining pressures of 25, 100, and 400 kPa were used for the response-to-wetting tests.

Because of the large rock fragments it was impossible to test specimens representative of the in-situ gradation. However, the laboratory-compacted specimens provided qualitative information on collapse potential of the site soils. The compacted specimens typically exhibited 3 to 5 percent collapse strain at 400 kPa (about 20 m of fill). No collapse occurred at 100 kPa, and a slight amount of swell (0.3 to 0.8 percent) occurred at 25 kPa, indicating the presence of at least some clay. A composite summary of the response-to-wetting test data is shown in Fig. 3.

To test the possible effects of fines migration, water was allowed to flow at gradients ranging from 4 to 320 through specimens previously subjected to response-to-wetting tests at 400 kPa confinement. These tests resulted in 0.5 to 1.5 percent of additional settlement in the laboratory specimens. However, for these soils, self-filtering occurred and settlement induced by fines-migration stopped, even at the highest gradients. In-situ gradients are much lower than the highest gradients used in the laboratory tests for fines migration, therefore the effects of fines migration in the field would likely be insignificant compared to the collapse settlement.

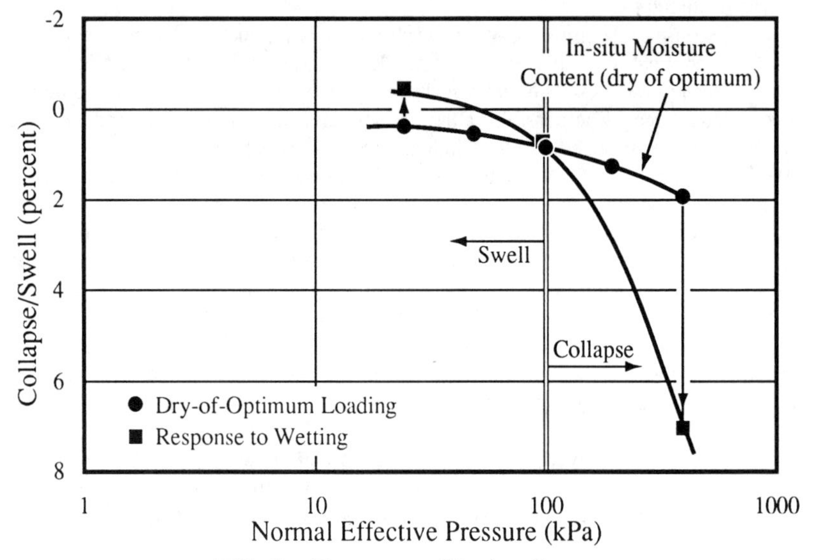

FIG. 3. Response to Wetting Curve

FIELD WETTING TESTS

Two large-scale field wetting tests were conducted at the site in a cul-de-sac area located several meters from any buildings, and in a region of relatively deep fill. Water was introduced to the soil at depths of 3, 9, and 15 meters through well points and piezometers shown in Fig. 4. Settlement was monitored with subsurface settlement indicators (S-1, S-2, and S-3, Fig. 4) and surface monitoring points spaced at 6 m on center, each way. The subsurface settlement indicators were located at depths of 3 m (S-1), 9 m (S-2) and 15 m (S-3). Piezometers were used to monitor the groundwater during the field wetting tests.

The wetting operation for the first field test was conducted in four phases, and was controlled using a valving system. Water was added at 15 m below the ground surface during Phase I from August 8 to August 23, 1989. Water was added at 9 and 15 m depths during Phase II (August 23 to September 7), and at 3, 9, and 15 m depths during Phase III (September 7 to September 21). Water was added at the central wetting points during Phases I, II, and III. During Phase IV water was added to the deepest piezometers around the perimeter of the test region to wet the soils over a wider area.

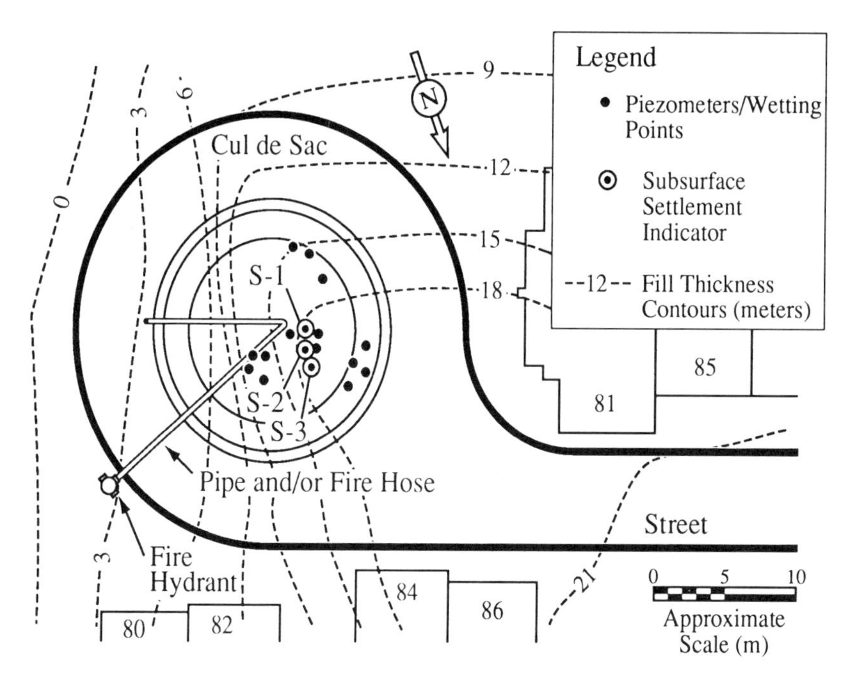

FIG. 4. Field Wetting Test Plan

The subdrain system was monitored during the field wetting test using a V-notch weir to evaluate the flow rate, and samples were periodically tested to determine the amount of fines migration. Phase IV of the field wetting test was terminated on September 28, 1989, due to homeowner concerns that the field wetting operation might result in further damage to surrounding structures.

The contours of surface settlements occurring after the first wetting test, but before the second wetting test, are shown in Fig. 5. Settlements of about 100 mm at the center of the wetted region occurred, with settlement decreasing radially outward from the points of wetting. The subsurface settlement indicator movements are given in Fig. 6, and a comparison of wetted versus near-by unwetted fill is shown in Fig. 7. Continued settlement after the end of Phase IV of the first wetting test is believed to be a result of wetting-induced settlements resulting from redistribution of moisture and the relief of arching. Arching probably resulted from wetting of a limited plan area, and probably induced horizontal movements toward the middle of the test area as described by Shmuelyam (1993).

Settlement markers S-1 (3 m) and S-2 (9 m) show essentially the same amount of vertical movement. Therefore, it is clear that the upper 9 meters of fill did not contribute to the overall surface settlement. The fill thickness in the vicinity of the subsurface settlement markers is approximately 18 m. Therefore, the lower

9 m of fill resulted in about 98 mm of settlement, with the deepest 3 m contributing approximately one-half of the total surface settlement. The lower 3 m of fill (average overburden stress of about 330 kPa) exhibited an average collapse strain of about 1.5 percent. These field collapse data are reasonably consistent with laboratory data.

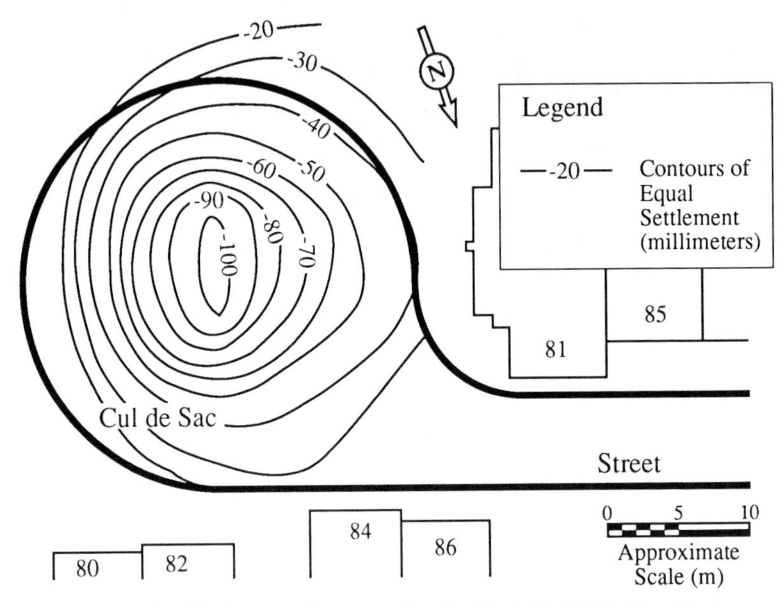

FIG. 5. Settlement Contours for the First Field Wetting Test

The second field wetting test was conducted beginning September 9, 1992 and ended on October 15, 1992. The main goal of the second wetting test was to determine whether re-wetting the fill would result in additional surface settlement. Water contents and degree of saturation were monitored during the second wetting test using neutron and gamma logging along with some direct soil sampling for correlation. The water was introduced using the same well point system and procedures as for the first wetting test. During the first three phases of wetting, flows were reduced to approximately 80 percent of those used in the first wetting test to minimize wetting areas not previously wetted. Flow rates during the final stage of wetting were increased to well above that achieved during the first wetting test.

Very little additional settlement resulted from the second field wetting test, as shown in Fig. 6. The surface and subsurface settlements at the end of the second wetting test were almost indistinguishable from those for the first field wetting test (Fig. 6), indicating that degrees of saturation during the first wetting test were approximately as great as those achieved during the second field wetting test. The

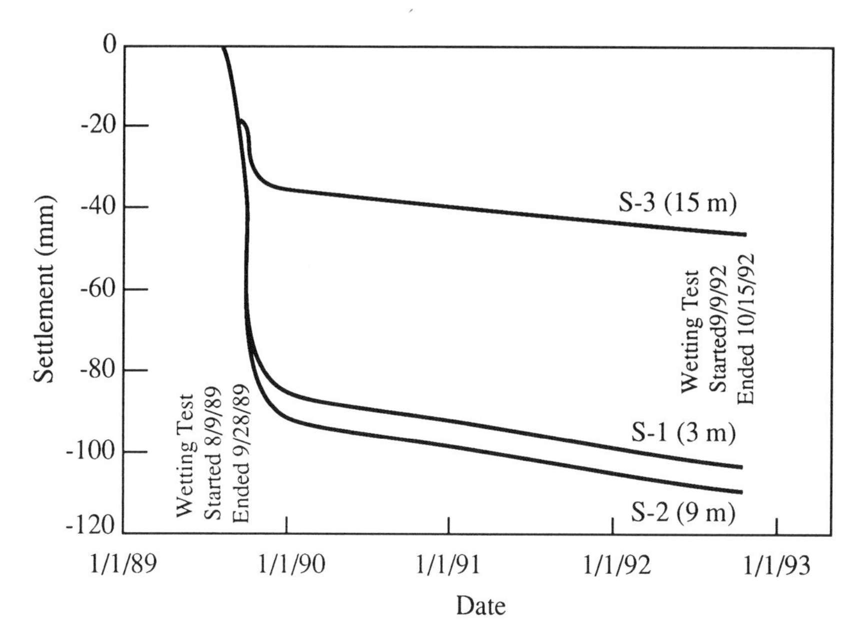

FIG. 6. Subsurface Settlement for Field Wetting Tests

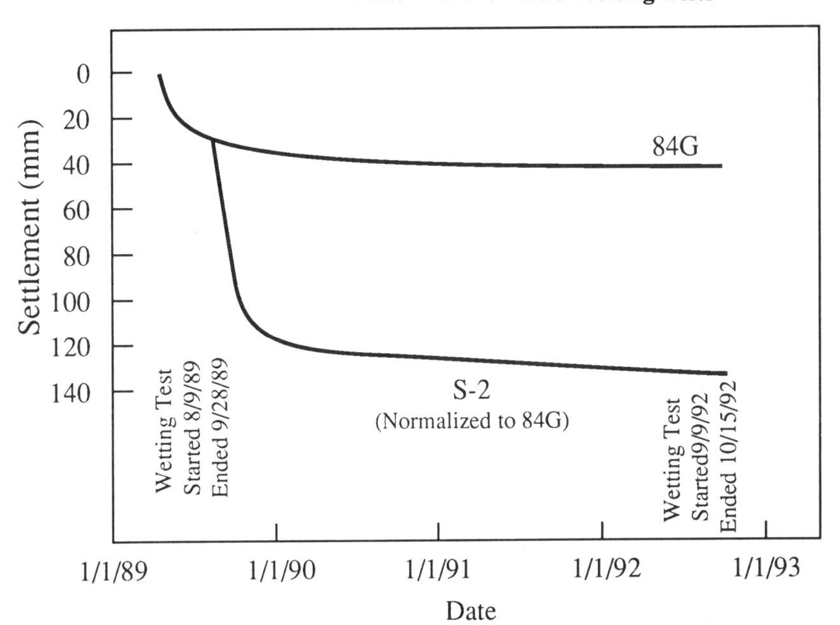

FIG. 7. Comparison of Wetted and Unwetted Fill Settlement

degree of saturation achieved during the second wetting test was essentially 100 percent in all but the very gravelly materials up to approximately 4.5 meters radially outward from the wetting points.

The flow measured by the V-notched weir in the subdrain correlated closely with the applied flow measured by a water meter. The fines testing performed on the subdrain effluent indicated that there was very little fines migration out of the fill, even during the highest flows.

Mitigation Considerations

Numerous mitigation measures have been applied to collapsible soil sites (Houston and Houston 1989). Two of these mitigation alternatives have been seriously considered for this site. One alternative is full-scale site wetting coupled with mud-jacking beneath the condominium units, and the second alternative is deep grouting. Of these two options, controlled wetting is almost certainly the most cost-effective long-term solution. Deep foundations extending through the fill are not considered a viable alternative because of great fill depths and anticipated drilling difficulties due to the rock fragments and boulders. The final decision on mitigation for this site has not yet been made. The controlled wetting operation being considered involves the installation of well points spaced at 6 m on center throughout the entire site. Based on the degree and extent of wetting observed during the second field wetting test, this configuration of well points should result in minimization of arching and corresponding time-delayed settlement. Wetting of the fill would be coupled with mud-jacking of the floor slabs to correct differential settlement. The mud-jacking during wetting is considered necessary due to the variation in fill thickness beneath the structures and because of the highly heterogeneous characteristics of the fill. Wetting would proceed gradually, to minimize differential movement and distress to the structures, particularly beneath structures founded on variable fill thickness.

Recommendations For Deep Fills

Several modifications to normal engineering and construction practices can greatly reduce the potential for collapse in deep compacted fills. The engineer should consider performing response-to-wetting tests on representative samples of the fill soils loaded to the expected overburden pressures. These should be performed at water contents that are representative of the driest soil conditions and lowest dry densities that would be accepted under the project specifications being considered. If the fill soil contains gravel or rock fragments the compaction specification should clearly indicate the technique to be used for correcting for the rock, and compaction of laboratory response-to-wetting specimens must be consistent with the specification (Houston and Walsh 1993). If the laboratory response-to-wetting data for the compacted specimens indicate that unacceptably high collapse settlements are likely, modifications to project compaction specifications should be made.

Possible modifications to compaction specifications include increasing the minimum allowable dry density and the minimum allowable water content, and restricting soil gradation (Noorany 1990). The effects of changes in compaction specifications should be evaluated with additional response-to-wetting tests, again for the case of minimal conformance to the proposed compaction specifications. This process should be repeated until acceptable performance criteria have been achieved with the proposed compaction specifications.

Increasing the minimum acceptable dry density and water contents will probably increase the costs of placing the fill. Larger equipment, additional passes, and thinner lifts may be required. In addition to construction costs, there will likely be increased construction monitoring costs. The cost of water may be significant, particularly in permeable soils or in arid areas. If response-to-wetting tests indicate that shallower portions of the fill will not collapse, then less rigorous specifications might be considered for shallower fill. Another alternative is to design the fill and/or structure to accommodate any expected collapse.

The engineer must balance the increased costs of construction, construction monitoring, and maintenance with improved performance. The increased design and construction costs are likely to be small compared to the costs of mitigation, particularly considering the possible additional costs of litigation. In any case, having properly characterized the fill for collapse potential by conducting an appropriate response-to-wetting testing program, the engineer and owner are in a position to make a rational decision with regard to design and construction alternatives by comparing estimated life-cycle costs of the various alternatives.

CONCLUSIONS

The settlement and resulting structural distress of the compacted fill at this site are primarily caused by wetting-induced collapse of the deep fill material. The fill material was susceptible to collapse in spite of compaction to at least 90 percent of ASTM method D-1557 maximum dry density, although rock corrections were not made in the field. The collapse is most likely due to the lack of moisture conditioning during fill placement which likely resulted from compaction specifications failing to address minimum water content requirements. In addition, if rock corrections had been made during field compaction inspection, it is possible that many of the relative compaction values would have been less than 90 percent.

The field wetting test provided a full-scale, representative response-to-wetting test for a highly heterogeneous, rocky fill material. This data was extremely valuable for evaluating the collapse potential of the existing fill. Laboratory tests indicated that fill within 5 m of the surface would not collapse when wetted, and would swell very slightly at the lightest overburden pressures. Field testing confirmed that fill above approximately 9 m did not contribute significantly to surface settlement. The fill was collapsible at higher overburden stress corresponding to fill depths greater than 9 m. The greatest damage resulted from differential settlement of structures founded on variable fill thickness, and very little damage resulted from settlement in deep fill of uniform thickness.

Due to variable fill thickness below the existing structures at the site, a controlled-wetting scheme, if employed for mitigation, should be used in conjunction with mud-jacking to minimize differential settlement of structures during the wetting operation. The two large-scale field wetting tests conducted at this site provided strong evidence that controlled site wetting represents a viable mitigation alternative.

In order to avoid problems caused by collapse settlement of deep fills, engineers should be familiar with a range of practical design and construction alternatives to minimize or accommodate the collapse. In almost all cases the increased costs of designing and constructing a deep fill to avoid problems with collapse are likely to be much less than the cost of remedial repair and litigation.

APPENDIX - REFERENCES

Brandon, T., Duncan, J. M., and Gardner, W. (1990). "Hydrocompression settlement of deep fills."*J. Geotech. Eng.*, ASCE, 116(10), 1536-1548.

Houston, W. N. and Houston, S. L. (1989). "State-of-the-practice mitigation measures for collapsible soil sites." *Proc. Foundation Engineering Congress*, Evanston, 161-175.

Houston, S. L. and Walsh, K. D. (1993). "Comparison of rock correction methods for compaction of clayey soils." *J. Geotech. Eng.*, ASCE, 199(4), 763-778.

Lawton, E., Fragaszy, R. J., and Hardcastle, J. H. (1989). "Collapse of compacted clayey sand." *J. Geotech. Eng.*, ASCE, 115(9), 1252-1268.

Nobari, E. S., (1968). "Effect of reservoir filling on stresses and movements in earth and rockfill dams," M.S. Thesis, University of California, Berkeley.

Noorany, I. (1990). "Swell and hydrocompression behavior of compacted soils - Test data," *Research Report,* Dept. of Civil Engrg., the University of California, San Diego.

Shmuelyam, A. Y. (1993). "In-situ investigation of horizontal subsidence deformation." *Proc. 1st Int'l. Symp. on Engineering Characteristics of Arid Soils*, London, 425-433.

Settlement of Peats and Organic Soils

Tuncer B. Edil,[1] Member, ASCE and Evert J. den Haan[2]

Abstract: New advances in investigation of settlement of peats and organic soils and issues related to construction over such soft ground are presented. Peats and organic soils are well known for their high compressibility and long term settlement. In many cases, the majority of settlement results from creep at constant vertical effective stress. Extensive studies of this creep behavior have been performed in recent years, resulting in important advances in our understanding and formulation of it. Laboratory and field evidence with regard to the one-dimensional compression behavior of peat deposits and the recent developments in formulation of peat compression are presented. The issues concerning the classification of peats and the experience in dealing with practical problems encountered in construction over peat are discussed.

Introduction

With the recent advances in soil reinforcement, construction of embankments over soft ground, e.g. peats and organic soils, has become primarily a problem of controlling excessive settlements. The interest in peats and organic soils is further reinforced by the problems of expanding existing embankments founded on such deposits many years ago and the diminishing availability of sites underlain by competent foundation soils in developing urban areas. Peats and organic soils are well known for their high compressibility and long-term settlement. In many cases, the majority of settlement results from creep at constant vertical effective stress.

---

[1] Professor, Department of Civil and Environmental Engineering, University of Wisconsin-Madison, 1415 Johnson Drive, Madison, WI 53706.

[2] Delft Geotechnics, P. O. Box 69, AB 2600, Delft, The Netherlands.

Extensive studies of this creep behavior have been performed in recent years, resulting in important advances in our understanding and formulation of it.

For field applications, peat soils must generally be improved for construction. Preloading techniques, in which a surcharge load is placed on a site prior to construction, have been used successfully in this regard. The basis of a successful preloading design for the improvement of soft ground is a method to predict both the magnitude and rate of settlement.

The objective of this paper is to present laboratory and field evidence with regard to the one-dimensional compression behavior of peat deposits and the recent developments in formulation of peat compression. The issues concerning the classification of peats and the experience in dealing with practical problems encountered in construction over peat are also discussed.

CLASSIFICATION, EMPIRICAL CORRELATIONS AND VARIABILITY

One of the issues still facing peat engineering is lack of a satisfactory and internationally accepted definition of terms and classification of peats. Most common definitions of peat are based on ash (or organic) content. The 1988 meeting of the International Peat Committee TC-15 of ISSMFE in Tallin revealed that the cutoff organic content for "peat" varied from 25 to 75% among the member countries. The term peat as used today includes a vast range of peats, peaty organic soils, organic soils and soils with organic content (Landva et al. 1983a). The geotechnical properties between these groups span a very wide range. For geotechnical purposes, the peat classification should delineate a certain class of mechanical behavior. Perhaps, any attempt for definition of peat and its classification should start by defining the type of characteristic behavior which sets "peats" apart from mineral soils and also from organic soils (which conform to general traits of clay behavior). The next step would be to identify those index properties that collectively identify materials with such behavior. Such properties may include ash content, fiber content, degree of humification similar to von Post's system but perhaps with fewer categories (Magnan 1994), water content, and origin (sedimentary versus sedentary peats).

The discussions on the subject of peat classification in a recent workshop on the mechanical behavior of peats held in Delft, the Netherlands indicated that the term *peat* should be reserved for those materials which do not conform to the general inorganic soil behavior both qualitatively and quantitatively. In the case of one-dimensional compression, this would mean the presence of *tertiary* compression (discussed below) and high degree of fiber-induced composite-material behavior (Venmans 1994). Organic materials that show such behavior are typically derived from plants and are therefore, in their non-humified or slightly humified state, very fibrous materials. As the degree of humification increases, the peat becomes less and less fibrous until it is transformed into an amorphous mass without any discernible structure. The mineral content also appears to increase with increasing humification. A *peat* definition based on a combination of organic content and fiber content, it seems, would delineate a material exhibiting a true *peat behavior* as distinct from soil behavior. The authors suggest for peat a minimum organic content of 75 to 80% and a fiber content of 45 to 50% (both on gravimetric basis). Water content, density, and degree of humification (a modified von Post system with a reduced number of classes) are simple indices that can aid classification. Adoption of such a definition is expected to end the tremendous confusion in the geotechnical community in dealing with peats and organic soils.

Empirical Correlations

Preliminary design of structures in and on peats can profit from empirical correlations sometimes even more so than when dealing with clays due to the difficulty of obtaining high quality peat samples and the extreme variability of peat deposits. There are not well established and widely known correlations between simpler index properties and more difficult mechanical parameters for peats. Numerous correlations based on local experience have been presented for many properties of peats. Useful overviews are given in the Muskeg Engineering Handbook (1969) and by Hobbs (1986) and Carlsten (1988). A general dependency of mechanical properties (compressibility and strength) on natural water content or corresponding void ratio appears to be well established (Magnan 1994; Farrell et al. 1994), while den Haan and El Amir (1994) normalize compression curves of varying peats by their organic content. Wide spread data make practical use of such correlations limited. Lack of understanding and consensus on both the definition of appropriate mechanical parameters for peat behavior and the basic index properties that delineate the character of peats, limits development of general correlations; however, local correlations for a deposit are often developed and successfully used by engineers.

Spatial Variability and Sampling

In general, peat deposits have significant spatial variability so that it is difficult or impossible to obtain representative samples for testing, unless a very large number of samples are taken (Magnan 1994). For this reason the laboratory behavior cannot be expected to yield more than a crude picture of the field behavior. The same concerns apply to test fills, even though a test fill is much more representative of the peat behavior than a small laboratory sample.

Peat is a difficult material to sample because of the presence of fibers and woody remnants. It generally exhibits anisotropy in its properties. Conventional sampling tubes (typically 75-mm in diameter) used for taking "undisturbed" samples of soft clays may not be suitable for sampling peats. Sample size is important with respect to both sampling disturbance and representative sample size. Peat samplers 100 to 250 mm in diameter have been developed, including a block sampler (Landva et al. 1983b).

Oedometer tests performed by Fox (1992) on the specimens (63.5 mm in diameter) trimmed from a block sample (400 mm square) indicated intrinsic differences in the rate of creep compression of these specimens that could not be accounted for by stress and temperature changes or differences in initial void ratio. These specimens could be considered "undisturbed" since they were trimmed from a large block sample and the variability could not be attributed to sample disturbance. A "fabric coefficient of creep", C_f, was defined to account for the microstructural effects on the rate of creep compression. C_f could be related to the position of the test specimens in the block sample and four zones of C_f values could be identified in the block sample, as shown in Fig. 1. There is clearly a need to determine a minimum representative volume for test specimens to account for spatial variations in laboratory testing.

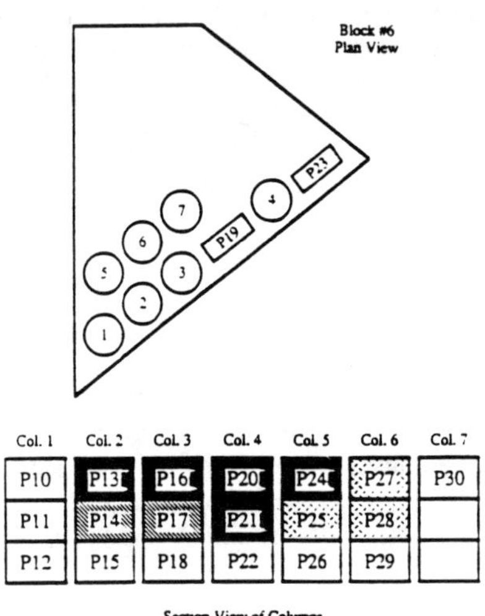

FIG. 1. Zones of Microstructure within a Peat Block (Fox 1992)

PEAT COMPRESSION BEHAVIOR

A large group of organic and peaty soils exhibit a one-dimensional compression behavior which is in general conformity with the behavior of clays typically encountered in practice. There may be differences in the magnitudes of various quantities measured but the general shapes of the consolidation curves appear reasonably similar and the formulations developed for clay compression can be used to predict the magnitude and rate of settlement. In general, these formulations treat primary (hydrodynamic) compression and secondary (creep) compression separately and decouple the stress and time effects. Typically, laboratory multiple-stage-load (MSL) oedometer tests are performed for a load-increment ratio (LIR) of unity and a load-increment duration (LID) of 24 hours. Alternatively, constant-rate-of-strain (CRS) tests are conducted to obtain equivalent information as generated by the MSL tests.

There are however a certain class of peats, typically high organic and fiber content materials with low degree of humification, that do not conform to the basic tenets of the conventional clay compression behavior because of their highly different solid phase properties and microstructure. The analysis of compression of such materials presents certain difficulties when the conventional methods are applied because the curves obtained from the conventional oedometer tests and the behavior exhibited by them show little resemblance to the clay behavior. The behavior of such peats and the recent advances in formulating their behavior is emphasized herein.

Fig. 2 shows a typical void ratio-logarithm of time (e-log t) curve along with the associated excess pore pressure dissipation curve for normally consolidated fibrous Middleton peat (a dark brown, poorly-humified, fibrous peat which has the following average index properties: gravimetric fiber content = 50%, water content = 550%, organic content = 93% and initial void ratio = 10.5). A significant amount of creep compression takes place after the dissipation of the excess pore pressure at time t_p. Primary and secondary stages are indistinguishable and the tangential slope ($C_\alpha = -\partial e/\partial \log t$) gradually increases with time, giving rise to a steeper *tertiary compression* segment on the logarithmic plot beginning at t_k (Edil and Dhowian 1979; Dhowian and Edil 1980). The load-increment-ratio (LIR) has an important influence on the overall shape of the resulting e-log t curve. Fig. 3 shows four curves from tests in which different specimens were loaded starting at 50 kPa with a single stress increment. The stress increment applied to each specimen varied from 25 to 250 kPa giving LIRs of 0.5 to 5. To allow comparison of their shapes, the curves are plotted using change of void ratio, Δe, from that at the beginning of the load increment (initial void ratios of the specimens varied considerably even though the specimens were trimmed from the same block sample). The end of primary consolidation, based on pore pressure measurement, is reached at time t_p as shown for each curve. Fig. 3 indicates that, in each case, the contribution of creep to total compression is large. For small LIR, the behavior shown in Fig. 2 is prevalent. On the other hand, for large LIR, end of primary consolidation (EOP) is indicated by a break in slope corresponding nearly to time t_p and tertiary compression is less pronounced.

TERTIARY COMPRESSION

Creep, as defined for one-dimensional compression of soils, is continuing volumetric compression under constant vertical effective stress. This time-dependent component of total settlement is referred to as secondary compression. Although it is thought to occur throughout the whole consolidation process, it is typically taken to occur after essentially all of the excess pore pressure has dissipated. Secondary compression carries with it the wide-spread connotation of constant rate with the logarithm of time which means decreasing true settlement rate with time. There is some evidence that C_α may change with time even for clays (Leonards and Girault 1961; Mesri and Godlewski 1977). However, there is a significant deviation from constant C_α for fibrous peats in most cases of stress application (Thompson and Palmer 1951; Dhowian and Edil 1980; Krieg and Goldscheider 1994). This deviation, which makes prediction of long-term settlements particularly difficult in such peats, must be clearly recognized. Field evidence of tertiary compression has been documented by Candler and Chartres (1988) in the U. K. The field settlement

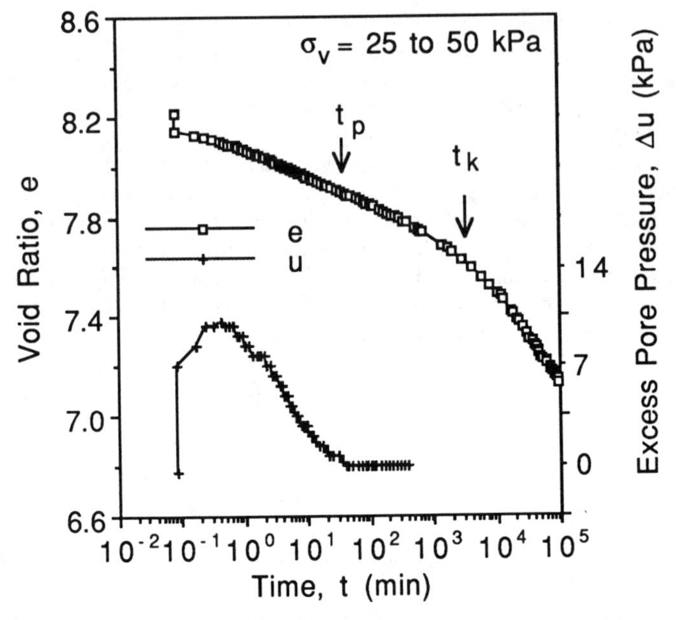

FIG. 2. Compression-Time Curve for Fibrous Middleton Peat

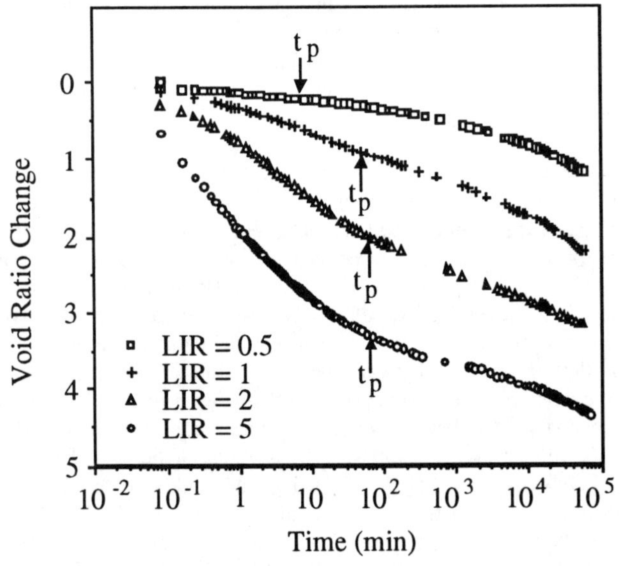

FIG. 3. Compression-Time Curves for Various Load Increment Ratios

data from a test embankment (13×13 m by 1.25 m high) founded on Middleton peats (Edil and Fox 1994) also exhibited tertiary compression as shown in Fig. 4.

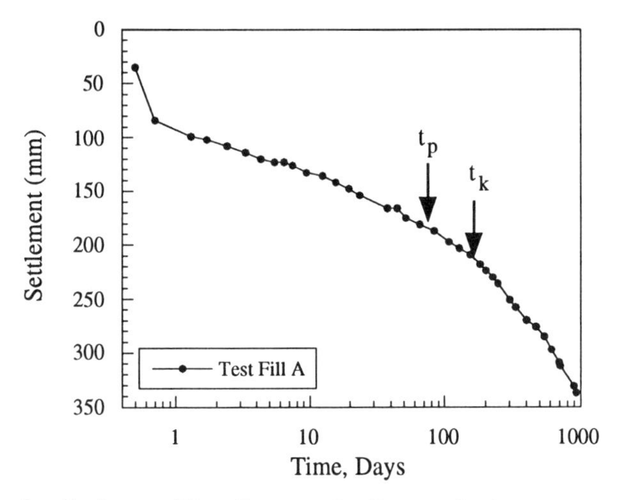

FIG. 4. Settlement-Time Curve at the Center of Middleton Test Fill

The term "tertiary" compression, as explained herein, was introduced (Edil and Dhowian 1979; Dhowian and Edil 1980) as a convenient device to designate the increasing C_α with time. Some confusion exists regarding primary, secondary and tertiary compression in the consolidation test on the one hand, and primary, secondary and tertiary creep in the undrained triaxial compression test on the other hand. These terms refer to successive stages observable in vertical strain versus time in either test and may or may not be correlated with a specific mechanism. In triaxial creep, these stages are identified in terms of the slope of log strain rate versus log time plots, m (Mitchell 1993). Using a similar approach, tertiary compression in one-dimensional laterally confined compression can be defined as a decrease of slope m in a plot of log strain rate versus log time, after a relatively constant stretch at $m \approx 1$ (den Haan 1994). Therefore, tertiary compression refers to a decreasing strain rate, which is changing at an increasing rate.

Compression-Stress and Compression-Time Relationships

The one-dimensional compression of many organic soils and some peaty soils can be analyzed following the conventional formulations involving decoupled stress and time effects if certain improvements are introduced. Such formulations are presented first and then are followed by the formulations proposed to couple the stress and time-dependent aspects of peat compression behavior.

Virgin Compression-Stress Relationship

When strains obtained in oedometer tests on peats are plotted against logarithm of stress, the resulting relationship in the virgin range past the preconsolidation pressure is often highly non-linear. However, the conventional compression index C_c presumes linearity and is thus not suitable for use with peats and highly compressible organic soils.

By simply replacing common engineering strain with *natural strain*, the virgin relationship is again rendered linear in many cases (den Haan 1992). Thus the *natural compression index* b which is the slope of the virgin curve in *natural* strain vs. log σ'_v, describes the complete virgin range adequately. Natural strain is obtained by the integration of increments of deformation relative to the current dimension. Thus,

$$\epsilon^H = -\int_{h_o}^{h} \frac{dh}{h} = -\ln\frac{h}{h_o} \tag{1}$$

The superscript H commemorates Hencky as the first to apply this measure of strain. Common engineering strain differs from natural strain in that deformation is related to the initial dimension h_o. Common (or Cauchy) strain is denoted by ϵ^C and is related to natural strain through

$$\epsilon^H = -\ln(1 - \epsilon^C) \tag{2}$$

In Fig. 5a, it is shown how ϵ^H increases to infinity as ϵ^C increases to 100%. By transforming sample height h to -ln (h/h_o), ϵ^H becomes linear, but ϵ^C curve bends off. It is this curvature which is observed at large compressions in peats and other soils. At low compression, the difference is negligible. Above approximately 10% compression, the difference becomes noticeable, and therefore the use of natural strain is of advantage in peats.

Lefebvre et al. (1984) made use of natural strain to obtain linear relationships between tangent modulus and effective stress for James Bay peats. They showed that this leads to a linear relationship between natural strain and logarithm of effective stress. Earlier, natural strain had been introduced in soil mechanics by Juárez-Badillo (1965) and Butterfield (1979). Den Haan (1992) shows that a very wide range of soil types are adequately formulated by natural strain rather than common strain. Therefore, there is sufficient backing at present to abandon the conventional ϵ^C-log σ'_v formulation in favor of the ϵ^H-log σ'_v formulation, replacing compression index C_c by *natural compression* index b. Natural compression index is shown to be related to organic content and vary in the range of 0.25 to 0.40 for peats (den Haan 1992).

Another virgin compression-stress relationship, originally proposed for clays by Hardin (1989), has been applied to peats with some degree of success (Liu and Znidarcic 1991; Lan 1992). Hardin's equation relates inverse of void ratio to effective stress through use of three model parameters and presents certain advantages over the conventional e-log σ'_v formulation. The main advantages of this formulation is that it behaves properly over a large range of effective stresses and it has proper values at the limiting stresses of zero and infinity.

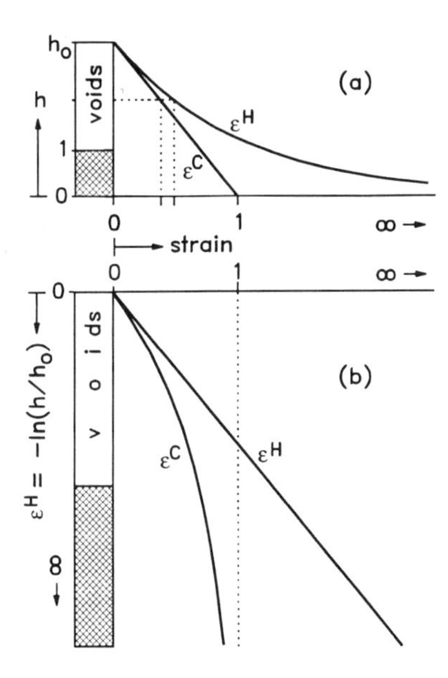

FIG. 5. Linear and Natural Strain: (a) as Function of Sample Height h; (b) as Function of -ln (h/h_o)

Compression-Time Relationship

Secondary compression is severe in peats and cannot be as easily ignored as is often done when dealing with more firm inorganic soils. Secondary compression occurring after the hydrodynamic primary period, is conventionally described by a linear void ratio - log time relationship with slope C_α where the time scale is taken to begin at the time of the load increment application. Due to the incompatibility of this choice of time scale with the inherently time-dependent behavior of soil, it has been shown that C_α cannot reflect a fundamental property of the soil (den Haan and Edil 1994). The need for a consistent formulation of secondary compression is therefore greater in peats. Additionally, there is a strong tertiary compression tendency in fibrous peats at certain stresses and small LIRs as described earlier.

The other problem associated in the conventional approach of separate treatment of hydrodynamic primary from secondary (creep) compression lies in the difficulty of establishing the time for end of primary (EOP) compression for peats. Often there are no inflection points on the laboratory void ratio - log time curve that relate to EOP as in the Casagrande construction. The Taylor construction method for EOP underestimates the time to EOP and overestimates the EOP void ratio (Edil et al. 1991a). The only reliable method of establishing time for EOP is by measurement of base pore pressures in the laboratory specimens. The exponent that relates the maximum drainage distance to the time to EOP is also found to be different than 2 due to the difficulties of applying the Terzaghi's one-dimensional consolidation theory to the hydrodynamic compression of peats (Hanrahan 1954; Lea and Brawner 1963; Samson and La Rochelle 1972; Edil et al. 1991a).

Early attempts to deal with these problems involved use of the linear structural viscosity model of Gibson and Lo (1961) to characterize the compression-time behavior of peat (Edil and Dhowian 1979; Edil and Mochtar 1984). The model consists of a linear spring in parallel with a linear dashpot, both of which are connected in series with another linear spring and involves three model parameters. While it was possible to represent the compression-time curves for a given increment of stress for a variety of peats, this approach does not offer any insights into the nature of the problem and lacks generality. Furthermore, due to the inherent non-linearity of model parameters and vast discrepancy of field and laboratory compression rates, Edil and Mochtar (1984) had to present graphs for correction of laboratory determined parameters to be used for field settlement prediction (see Fig. 14 as an example).

An improved formulation of the compression - time relationship of peats (and of many other soft soils) is obtained by replacing time by intrinsic time (den Haan and Edil 1994). Intrinsic time is denoted by τ and is defined by

$$\tau = t - t_r \tag{3}$$

where t is time since loading and t_r is a constant time shift. Fig. 6 shows how nonlinear ϵ - log t curves become linear ϵ - log τ curves due to the application of the constant time shift t_r. Note that the increments of intrinsic time and time since loading are equal on the secondary tail:

$$dt = d\tau \tag{4}$$

so that increments of intrinsic time and time since loading are equal.

Combining natural strain with intrinsic time by assuming a linear ϵ^H - ln τ relationship with a slope c (as in Fig. 6), we obtain

$$\epsilon^H - \epsilon_o^H = c \ln \frac{\tau}{\tau_o} = c \ln \frac{t - t_r}{t_o - t_r} \tag{5}$$

where the subscript o designates the initial condition on the intrinsic time line at the beginning of secondary compression. Strain is now linearly related to log $(t-t_r)$, not to log t. The time shift, t_r, is obtained by fitting Eq. (5) to the secondary tail with regression analysis, varying t_r until an optimal fit is obtained. It can be either positive or negative.

The rate of secondary compression $d\epsilon^H/dt$ or $\dot{\epsilon}^H$ is obtained by

$$\dot{\epsilon}^H = \frac{c}{\tau} \tag{6}$$

This equation explains the term intrinsic; it is the reciprocal of the rate of strain. Substituting this in Eq. (5) yields

$$\epsilon^H - \epsilon_o^H = -c \ln \frac{\dot{\epsilon}^H}{\dot{\epsilon}_o^H} \tag{7}$$

This is a linear plot of strain versus logarithm of strain rate. The slope is c here as well as in Eq. (5). This slope c is called the *natural secondary compression index*. Secondary compression is now related to rate or intrinsic time rather than to time since loading as with the conventional C_α, removing the difficulty of associating time since loading with the inherent time-dependence of the deformation of the soil.

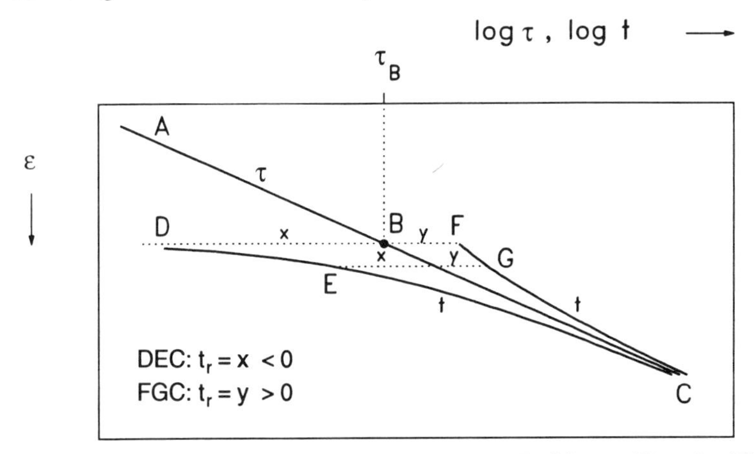

FIG. 6. Non-linearity of ϵ - log t Curves due to Positive or Negative Time Shifts

Due to diminishing influence of t_r as t increases in Eq. (5), this slope will eventually be linear with logarithm of time as well, but it is expected to be linear earlier when plotted against logarithm of intrinsic time. The value of the time shift t_r on the secondary tail is that it is not a soil constant. From Eqs. (3) and (5), it follows that t_r depends on the time necessary to arrive on the secondary tail, t_p, and the rate of deformation at the end of primary. In standard LIR=1 tests, the influence of t_r is usually quite small, and this confirms the success of the standard test in obtaining reasonably linear secondary tails. However, use of t_r leads to a consistent formulation irrespective of variables such as LIR, LID and distance to drainage boundaries.

Fig. 7 shows time - settlement curves obtained in a multiple-stage oedometer test on fibrous Middleton peat. Various scales are used to illustrate the advantage of natural strain and natural strain rate. In the conventional Fig. 7a, the final slopes (C_α) are clearly not parallel, however, they are parallel in the virgin range (above approximately 50 kPa) in Fig. 7b when natural strain is used. They remain parallel and are straight over longer sections when time is replaced by natural strain rate in

Fig. 7c. The slope of these lines, c, therefore formulates the complete virgin secondary behavior of this peat. The final slope at 400 kPa during prolonged loading steepens however, and this is a phenomenon which needs separate formulation (see the section on tertiary compression).

Not only are the lines in Figs. 7b and 7c parallel, they are also more or less equidistant. As the consecutive loads are doubled in each stage, this means that the slope b is independent of both stress and strain rate. This is revealed more clearly by replotting Fig. 7c in Fig. 8. Here we see that the constant strain rate lines are tolerably parallel in the virgin and secondary ranges, their slope being the natural compression index b. Fig. 8 however shows that in the area of precompression (below about 50 kPa) the constant strain rate lines are distorted. Also, after unloading at 400 kPa, subsequent reloading results in distortion of the constant strain rate lines. Therefore, reloading behavior requires separate formulation.

We conclude that the parameters b and c are powerful descriptors of peat compression. They are defined as

$$b = \frac{d\epsilon^H}{d\ln \sigma'_v} \tag{8}$$

and

$$c = \frac{d\epsilon^H}{d\ln \tau} \tag{9}$$

Parameter b successfully linearizes stress-strain behavior in the virgin compression range; c successfully linearizes strain-time behavior in the secondary compression range. Just how successful and powerful these parameters are is illustrated forcefully in Fig. 9 where Fig. 8 has been replotted on the conventional scales. Clearly the conventional description is highly non-linear during the last four loading stages.

FORMULATION OF STRESS-STRAIN-STRAIN RATE RELATIONSHIP

The behavior of the soil skeleton in most one-dimensional consolidation theories is expressed as a rheological function of two or more of the following variables: effective vertical stress, vertical strain (or void ratio), time, time rate of strain or void ratio, and time rate of vertical effective stress (Leroueil et al. 1985). Discussion continues in the literature over which formulation is appropriate for different types of soils. The classical Terzaghi formulation is the simplest relationship in which void ratio (or strain) is considered a unique function of effective stress. Recognizing the creep compression with time under constant effective stress (secondary compression), some researchers included elapsed time in the formulation (Bjerrum 1967; Garlanger 1972). It has been shown for natural clays that compression is influenced by strain rate (Suklje 1957; Crawford 1964; Vaid et al. 1979; Leroueil et al. 1985). Models utilizing strain rate show a unique relationship between effective stress, strain, and strain rate. This relationship on a diagram of strain versus effective stress can be represented by lines of constant strain rate called isotaches (Suklje 1957) and it broadens the conventional

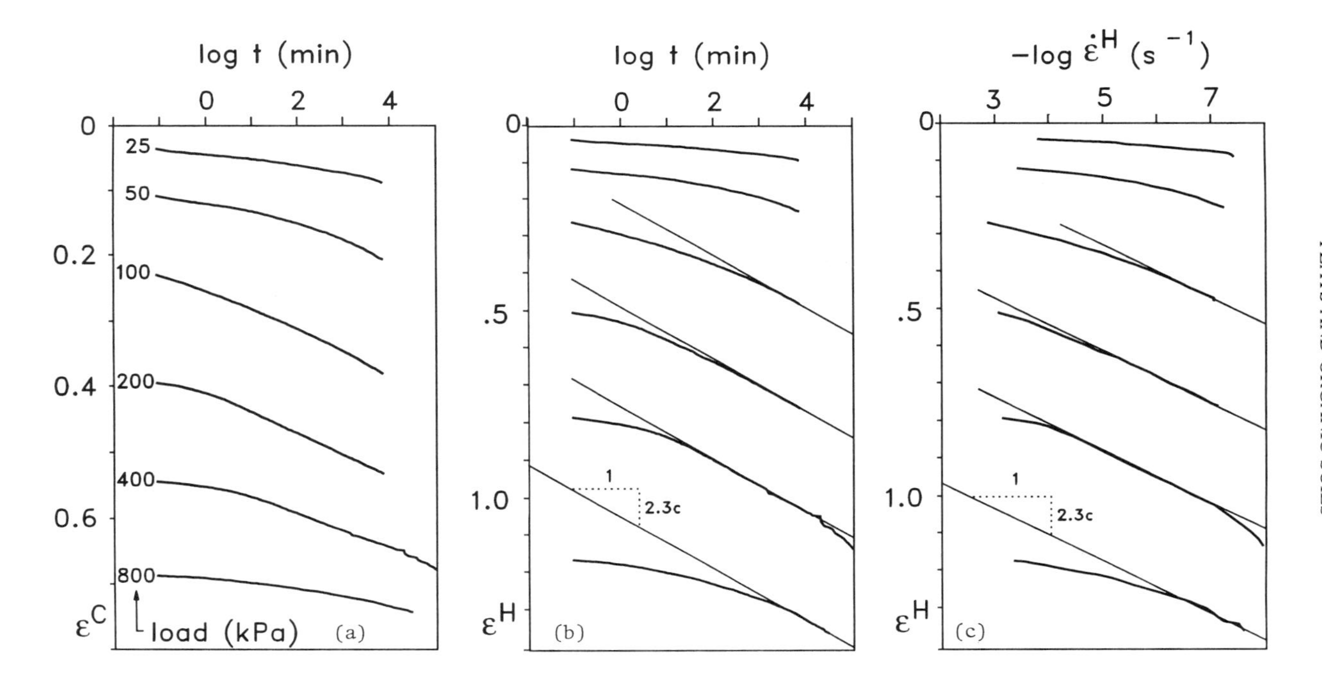

FIG. 7. Compression-Time Curves for Fibrous Middleton Peat for Various Strain and Time Scales

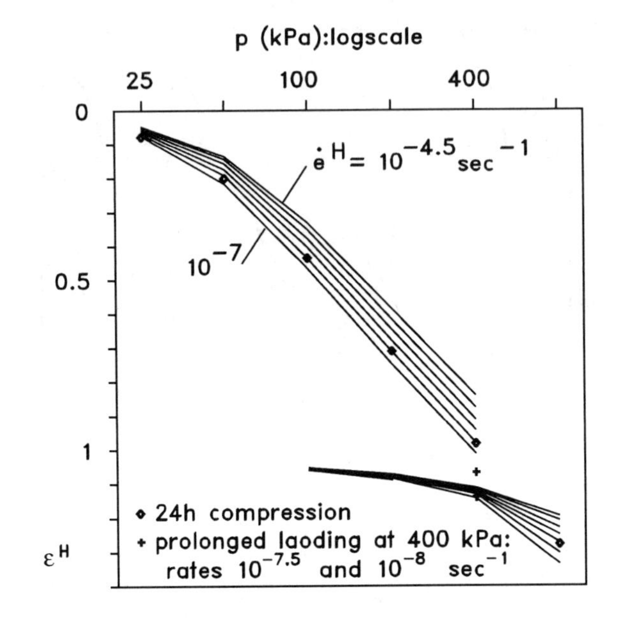

FIG. 8. Isotaches for Loading and Reloading for Fibrous Middleton Peat Using Natural Strain

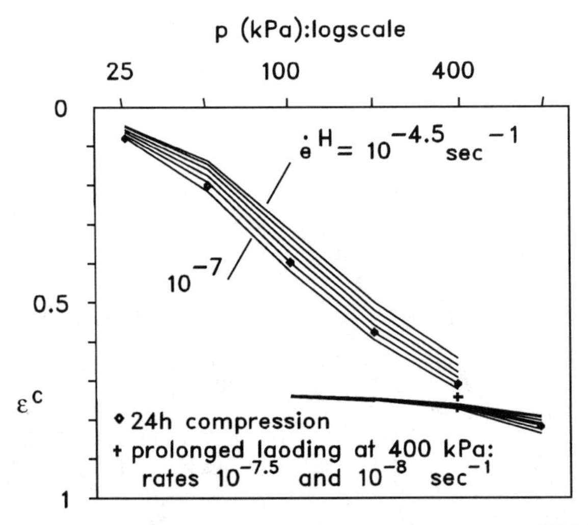

FIG. 9. Isotaches for Loading and Reloading for Fibrous Middleton Peat Using Common Strain

strain-stress relationship to include the time-dependent response. Some evidence reported in the literature clearly supports a unique stress-strain-strain rate relationship for normally consolidated soils while other results suggest that stress rate (loading rate) effects and stress history are important (Fox 1992). Three different formulations based on the concept of a unique stress-strain-strain rate relationship are presented along with the experimental evidence supporting this concept for peats.

The a-b-c Model

The system of parallel lines in Fig. 8 determine creep rate $\dot{\epsilon}^H$ as a function of natural strain and effective stress. Taking only the strain rates in the virgin stress and secondary compression ranges, results in a stress-strain-creep strain rate relationship as follows:

$$\epsilon^H - \epsilon_o^H = b \ln \sigma'_v + c \ln \left[\frac{\tau}{\tau_o} \right] \tag{10}$$

Here, τ_o is a reference value of intrinsic time. $\epsilon_o{}^H$ lies on its associated intrinsic time line, and follows from back-extrapolation of the natural strain scale to σ'_v = 1 kPa on this line. τ_o is taken at a natural strain rate of 10^{-7} s^{-1} because this rate often corresponds roughly to the rate at the end of 24 h. Much compression data have been reported in terms of 24h, and because 24h curves are essentially parallel to the constant creep rate lines (see Fig. 8), b can be determined fairly accurately from standard 24-h curves even in most peats.

The system of parallel lines forms a background pattern to the development of stresses and strains which is assumed to be valid in the primary (or hydrodynamic) period as well as the secondary period. In the primary period, strain rate then is assumed to consist of the sum of creep rate and rate due to compression induced by an increase in effective stress. For this, it is assumed

$$d\epsilon^H = a \, d\ln \sigma'_v \tag{11}$$

and therefore combining with Eq. (6), the total strain rate becomes

$$\dot{\epsilon}^H = \frac{a}{\sigma'_v} \frac{d\sigma'_v}{dt} + \frac{c}{\tau} \tag{12}$$

This equation is called the "a-b-c model" (den Haan and Edil 1994). It is similar in concept to Garlanger's (1972) formulation. However, Garlanger did not make use of the concepts of natural strain and intrinsic time.

The B Model

In an investigation of the uniqueness of the effective stress-strain-strain rate relationship, Edil et al. (1994a) reported results of multiple-stage-load (MSL), constant-rate-of-strain (CRS), and single-load-constant-stress (CS) oedometer tests on fibrous Middleton peat. The results are presented in terms of effective stress, void ratio, and void ratio rate (i.e., time rate of void ratio change). For the constant stress tests, a unique σ'_v-e-$\dot{e}$ relationship implies that regardless of the loading path, at the same effective stress and void ratio rate, the void ratio is uniquely defined. Fig. 10 shows void ratio versus effective stress curves corresponding to three void ratio rates. The data were obtained from a variety of tests. Tests #5 and #6 were MSL tests with load increment ratio, LIR = 1 but with different load increment durations (LID = 10 weeks for test #5 and 5 days for test #6). Test #7 was also an MSL test in which LID = 5 days and stress was applied in equal 25 kPa increments from 0 to 400 kPa, i.e., LIR $\neq$ 1. The remaining tests were CS tests conducted at final stress values ranging from 100 to 300 kPa. Fig. 10 shows reasonable groupings of the data at each of the three void ratio rates irrespective of loading path or duration.

Further evidence of a unique σ'_v-e-$\dot{e}$ relationship is provided by Fig. 11. The results of MSL and CS tests at a void ratio rate of 0.0005/min are superimposed over the e-log σ'_v relationship obtained from a CRS test at the same rate. The agreement of the tests suggests that a unique σ'_v-e-$\dot{e}$ relationship, independent of stress path, exists for Middleton fibrous peat in the normally consolidated (virgin loading) range. However, these test results may not be sufficient to entirely rule out the importance of time rate of effective stress change on compression behavior.

Based on this experimental evidence, a semi-logarithmic relationship was assumed between $\dot{e}$ and σ'_v through parameters A and B and the following σ'_v-e-$\dot{e}$ relationship was derived (Edil et al. 1994b):

$$\ln(-\dot{e}) = \frac{B(\sigma'_v/\sigma_a)}{D\left(\dfrac{e_o}{e} - \dfrac{1}{E}\right)} + A \tag{13}$$

The relationship between e and σ'_v in Eq. (13) is based on a normalized form of Hardin's relationship through parameters D and E (Lan 1992; Edil et al. 1994b) and σ_a is the atmospheric pressure. This model is referred as the "B model" and provides a useful relationship with potential for practical applications.

The C_σ Model

A more focused study of creep compression rate was conducted by means of step-stress (SS) and step-temperature (ST) oedometer tests on fibrous Middleton peat (Fox 1992). In a SS or ST test, the applied stress or temperature is increased by a small amount at some stage during creep process and the change of void ratio rate (i.e. $\dot{e}$) at constant temperature or stress, respectively, is evaluated. Its usefulness for creep studies is complicated by the unavoidable initial elastic compression and primary consolidation that accompany a change in total stress or temperature. Thus, the creep rate cannot be measured immediately after the stress or temperature increase. However, after the excess pore pressure has dissipated and creep again continues under conditions of constant effective stress, the creep rate is extrapolated back to the void ratio at which the change in stress or temperature occurred. Using this procedure, two void ratio rates are determined for two corresponding vertical

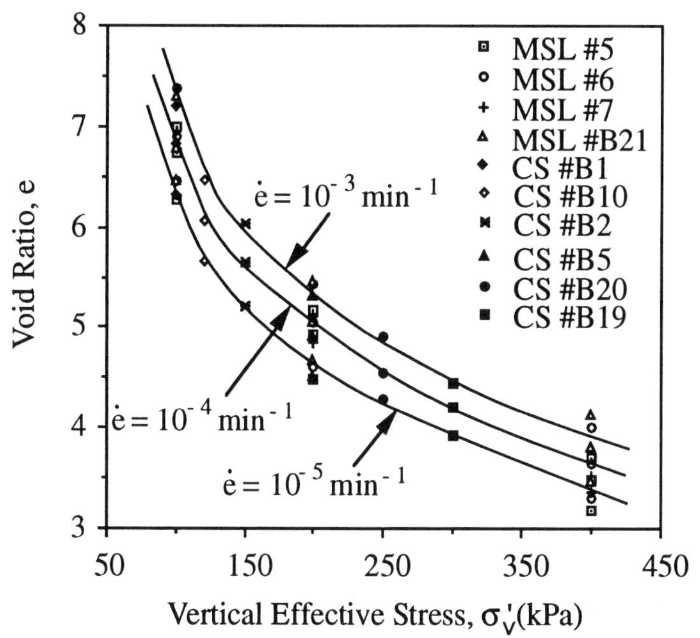

FIG. 10. Compression Curves for CS and MSL Tests on Fibrous Middleton Peat

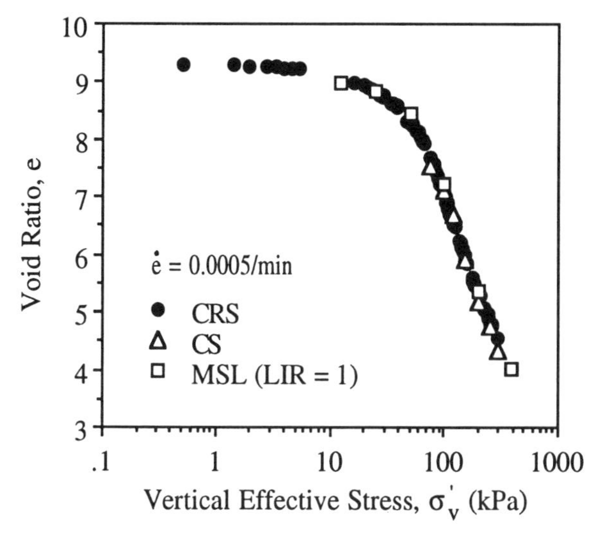

FIG. 11. Comparison of σ'_v-e-ė Relationship for CRS, CS and MSL Tests on Fibrous Middleton Peat

stress or temperature values, both at a constant void ratio and respectively constant temperature in the SS test and constant stress in the ST test (Fox 1992, Edil et al. 1994b; Fox and Edil 1994). These tests allow a detailed study of the creep rate response to changes in stress and temperature. Laboratory and field evidence indicates that one-dimensional creep compression rate of peat is a function of soil temperature as well as effective stress (Edil and Fox 1994).

Based on an isothermal step-stress test, a plot of log $\dot{e}$ versus σ'_v is obtained as shown in Fig. 12 by determining three $\dot{e}$ values corresponding to three stresses at given values of void ratio. Fig. 12 suggests a linear relationship between log$\dot{e}$ versus σ'_v at a given void ratio. Based on these results, a "stress coefficient of creep," C_σ is defined as (Fox 1992):

$$C_\sigma = \frac{d\ln(\dot{e})}{d\sigma'_v} \tag{14}$$

C_σ is the slope of the lines shown in Fig. 12 and varies with void ratio.

To correlate the temperature and void ratio rate increases, a "thermal coefficient of creep", C_T, is defined as (Fox 1992):

$$C_T = \frac{\ln\dot{e}_2 - \ln\dot{e}_1}{T_2 - T_1} \tag{15}$$

Fig. 13 shows that, for all step-temperature tests performed on fibrous Middleton peat, C_T is independent of void ratio. Similar plots indicate that C_T is also independent of vertical stress level and magnitude of temperature change (Fox 1992; Fox and Edil 1994). In all cases, $C_T = 0.26 \pm 0.02$ for Middleton peat.

Based on the experimental evidence shown in Figs. 12 and 13, the following equation is proposed for void ratio rate due to creep (secondary compression) (Fox, 1992):

$$\dot{e} = -C_f \exp(C_T T) \sinh(C_\sigma \sigma'_v) \tag{16}$$

where C_f is termed the "fabric coefficient of creep". The hyperbolic sine formulation, as opposed to simpler exponential form, is used because it provides zero void ratio rate under zero effective stress. This formulation is also suggested by the rate process theory based on triaxial creep tests (Mitchell et al. 1968). Settlement predictions using Eq. (16) require the acceptance of a unique σ'_v-e-$\dot{e}$ relationship under isothermal conditions. The fabric coefficient of creep, C_f, is a nonlinear function of void ratio and is partially dependent on fiber orientation. Evidence suggests that C_f is independent of stress history. The behavior of this coefficient is not simple to characterize as it seems to be sensitive to variations in sample fabric (see Fig. 1).

In order to determine creep settlement under isothermal conditions, Eq. (16) must be numerically integrated to determine void ratio as a function of time. Both C_f, and C_σ are functions of void ratio and must be updated using the void ratio from the previous time step. For total settlement, the primary compression during pore

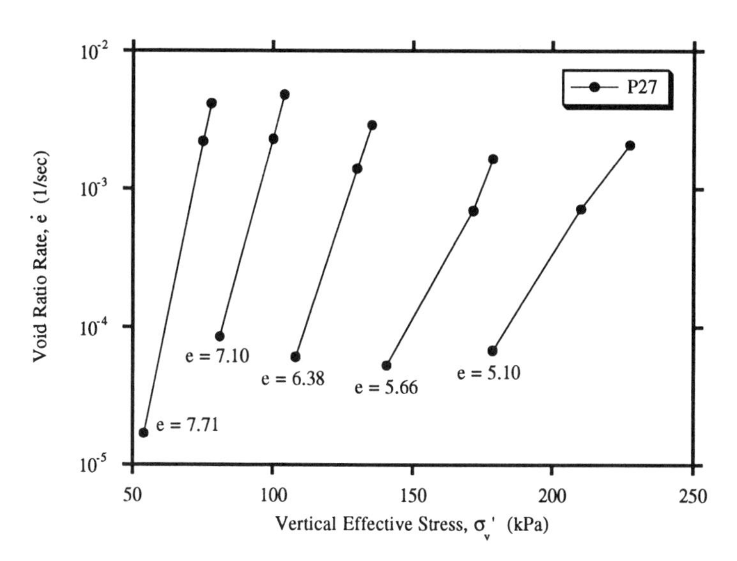

FIG. 12. Void Ratio versus Effective Stress at Five Void Ratios (after Fox 1992)

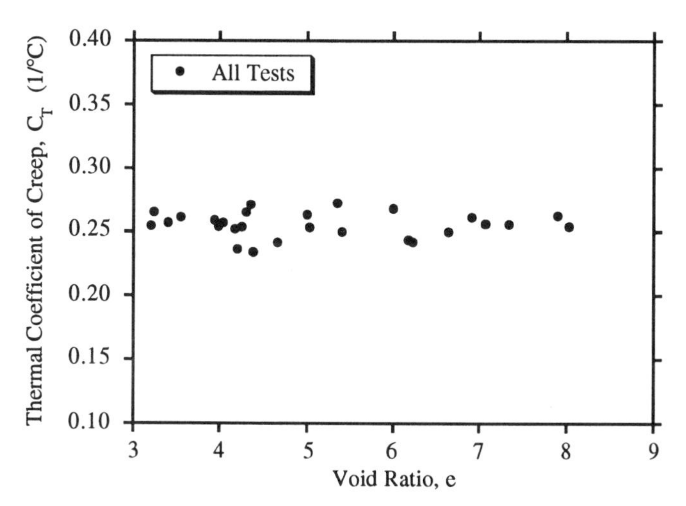

FIG. 13. Thermal Coefficient of Creep versus Void Ratio (after Fox 1992)

pressure dissipation must be added to the creep settlement. The total void ratio rate can be expressed for isothermal loading as

$$\dot{e} = -a_v \dot{\sigma}'_v - C_f \exp(C_T T) \sinh(C_\sigma \sigma'_v) \quad (17)$$

where a_v is the coefficient of compressibility and $\dot{\sigma}'_v$ is the time rate of effective stress change. a_v is an exponential function of void ratio over the entire compression range.

Synthesis

It should be noted that formulations given by Eqs. (12) and (17) obtain total strain or void ratio rate as the sum of creep and primary rates whereas Eq. (13) combines both rates in one expression. Combining the chosen stress - strain - strain rate equations with Darcy flow and the continuity equation according to now classical theory of large strain consolidation (Gibson et al. 1967), it is possible to calculate combined primary and secondary deformations following surface loading of peat deposits. The permeability of the soil must be known to calculate Darcy flow, and is typically taken to be a function of void ratio.

All three stress-strain-strain rate models expressed in Eqs. (12), (13), and (17) take creep to occur throughout the primary process. This identifies these models belonging to the Hypothesis B category (Ladd et al. 1977). In fact, the primary terms in Eqs. (12) and (17) involve only a very minor part of the total compression, most compression being creep in nature. Comparison of EOP void ratios from oedometer tests on thin and thick specimens of fibrous Middleton peat give support to Hypothesis B for peats (Edil et al. 1991a).

The parameters b, c, B and C_σ are connected. The three formulations that utilize these parameters basically represent a three-dimensional effective stress-strain-strain rate (or void ratio) surface. They differ only in the chosen formulation of the surface. The a-b-c model e.g. assumes the strain rate contours in Fig. 8 are parallel. For very fibrous peat this appears to be too much of a simplification, as the lines have a tendency to diverge in such material. Such divergence may be detected on careful examination of Fig. 8 and may be formulated by modifying the basic relationship. The final choice of formulation at present remains a matter of taste. Good fits to laboratory data can be obtained by all three models and their differences are therefore of minor importance. It should be noted that the a-b-c and C_σ models depart from a unique relationship between stress, strain and creep strain rate, and obtain total strain rate as the sum of creep and primary strain rates. The B and C_σ models have been shown to give good predictions for constant stress, multiple-stage loading and constant-rate-of-strain compression tests in the normally consolidated range using the same parameters (Fox 1992; Lan 1992).

Fig. 8 suggests that on reloading, the stress - strain - creep strain rate relationship is different than during first loading. The laws governing the change of the relationship are not yet clear. It is necessary to obtain formulations for the change in the relationship at any arbitrary stress or strain, to enable a general model to be constructed in which loading stages can be alternated by unloading stages.

The preloading technique often used over soft soils and peats is an obvious example calling for extension of the theory. Work is in progress to improve the models in this respect.

Construction Issues

This section's focus is on immediate issues faced in the engineering practice of construction of highways, dikes, buildings and structures over peat. Peat, having low bearing capacity and high compressibility, is considered to be among the worst foundation materials. It exhibits a high degree of spatial variability, generally much higher than exhibited by inorganic soils, and its properties can change drastically in response to stress application. However, earthen structures of great longitudinal extent (embankments, dikes, levees, etc.) often have to be placed directly on peat because of high cost and impracticality of using piling or replacing deep peat deposits.

Because of the known high degree of non-linearity of peat behavior as described above and large degree of peat heterogeneity and rather different microstructure, there is an on-going discussion as to whether the theories and procedures developed primarily for mineral soils can be directly applied to peat and, if not, what modifications of such theories and procedures can be made or if entirely new approaches are needed. The preceding sections addressed recent developments on such discussions. While discussions continue, engineers in many parts of the world have to make decisions to provide economical and tolerable solutions to the construction problems in hand. These decisions are constantly made and a body of field experience has emerged and constantly is expanding. In this section, practical issues in construction over peat are discussed and the recent experiences with respect to some of these issues are assessed.

The choice of construction method in areas underlain by peat deposits is a matter of finding optimal solutions between the economic and technical factors, available construction time, and the target performance standards. Avoidance of construction of fills over peat layers and replacement of surface peat layers by granular fill material have been the first choice of designers. Replacement is feasible typically for layers up to 5 to 6 m depth (Magnan 1994).

Stage-construction is used to overcome problems of instability in fills constructed over weak peat deposits (Edil 1988). Design of stage-construction requires assessment of the operating undrained strength (c_u), rate of undrained strength gain ($\Delta c_u/\Delta\sigma$), and time rate of excess pore pressure dissipation. Sand or wick (paper or polyester) drains are used to accelerate the dissipation of excess pore pressures and, therefore, the rate of construction of fills (Koda and Wolski 1994; Kurihara et al. 1994). There is some controversy regarding the effectiveness of vertical drains in peats. Surficial peats often exhibit very high permeability until they are compressed and a large portion of their total compression takes place under constant effective stress. Vertical drains may be effective in accelerating strength gain (Kurihara et al. 1994) but not total settlement. There has been some concern that the effectiveness of strip vertical drains may be additionally limited by deterioration and buckling of the drain due to typically large settlements and the consequent declined discharge capacity. However, recent investigations show that the performance of both paper and polyester drains remains satisfactory in spite of these problems (Koda and Wolski 1994) albeit smear effect remains to be an important factor in controlling pore pressure dissipation rates. General consensus is that vertical drains are effective tools for construction over peat (Venmans 1994). Typical drain spacing is 3 m for sand drains and smaller than this for synthetic strip drains.

Use of light-weight fill such as shredded waste tires (Edil and Bosscher 1992), wood chips (Bosscher and Edil 1988), geofoams (Flaate 1987; Monahan 1993; Horvath 1992), bales of peat (Hanrahan 1964), sawdust and expanded shale has been reported as a means of controlling stability and settlement problems in construction over soft ground. The lightweight fill concept, which is also referred to as "weight-credit construction," is a promising approach especially with the waste materials and geofoams since these materials provide economic benefits associated with conservation and recycling.

Stone and sand columns are also used to strengthen peat foundations of fills (Kurihara et al. 1994). This procedure may also aid in reducing settlements. Very weak surficial zones of peat may not have adequate strength to provide lateral support to sand columns. Use of geotextiles and synthetic fibers has been suggested to remedy this problem (Al-Refeai 1992). Use of stone or sand columns in peat needs to be explored more; however, the economics may limit their wide spread use to foundations of structures rather than earthen fills.

A very novel concept of improving peat soils, termed "thermal precompression", is being investigated at the University of Wisconsin-Madison. The laboratory tests performed by Fox (1992) indicated that moderate heating (15 to 40°C) accelerates settlement of peat soils significantly and further long-term settlement is greatly reduced upon cooling. This method is currently being evaluated in the field and the results to date are reported (Edil and Fox 1994). It could provide time savings over conventional stress-induced preloading, reduce stability issues associated with surcharging, and offer a means of perhaps completely arresting the long-term compression tendency of peats. Vacuum preloading is another precompression method that has been applied in some situations in Europe. In this procedure the site is covered with a geomembrane and subjected to a suction resulting in increase of effective stresses.

Piles provide another means of traversing peatlands. Friction piles have been used to cross a raised bog (Orr and McEnaney 1994; Crouse et al. 1993). Skin friction of peat-pile interface and creep settlement need to be assessed.

Rehabilitation and upgrading of existing fills over peat may pose special problems not normally encountered in new construction. For instance, differential settlement between the old and new sections and difficulty of removal of peat under the new section due to deeper slopes created by having an adjacent fill are cited among such problems. Solutions may be found in use of light-weight fill, stage construction, and thermal improvement.

Chemical and Biological Changes

An important characteristic of peats is their potential chemical and biological changes with time (Magnan 1994). Further humification of the organic constituents would alter the mechanical properties such as compressibility, strength and hydraulic conductivity. Lowering of groundwater may cause shrinking and oxidation of peat leading to humification with consequent increase in permeability (Vonk 1994) and compressibility. Oxidation also leads to gas formation which may contribute to excess pore pressures (Vonk 1994). The significance of these effects on the long term performance of the structures placed on peat are not taken into account rigorously. The opinion appears to be split regarding the long-term degradation of peats especially in the absence of any significant change in the submergence of these deposits (Venmans 1994). Magnan (1994) reports that the experience in France with respect to degradation of peats and organic soils is inconclusive. An embankment built of air-dried peat in the 18th century in Ireland has shown no signs of deterioration to date. However, dike breaches caused by humification of the dike body by air-rich water are reported in the Netherlands.

Field Observations

Because of the difficulties associated with the sampling and testing of peats and the problems of correlating the laboratory behavior to the field behavior, some engineers have been promoting the idea of using test fills or the data generated in the early stages of construction instead of laboratory testing in making the final adjustments to the preloading schemes and predicting settlements. Direct application of the mechanical parameters determined in the laboratory to the field, even if the questions of sample variability and disturbance are resolved and the stress histories are matched, appears to be questionable due to the nonlinear nature of peat behavior and the associated scale effects. Fig. 14 shows the difference in a parameter (λ/b) that reflects the rate of creep compression (secondary compression) as determined from the laboratory and the field data for a variety of peat soils from numerous sites as a function of the average strain rate (Edil and Mochtar 1984). There is a significant difference in the strain rates typically encountered in the laboratory and the field and the corresponding compression rate factors. This effect is basically ignored in the conventional analysis of creep. Furthermore, because of the often uncertain initial and boundary conditions of the consolidating stratum, the direct application of one-dimensional consolidation theory is difficult to justify.

Because of these difficulties associated with correlating the laboratory behavior to the field, it is desirable to use field observations in determining the mechanical parameters governing the settlement-time behavior. Asaoka's (1978) observational procedure uses early field settlement data to predict end-of-primary settlement and the in situ coefficient of consolidation. This method has been enjoying increasing popularity, especially for the consolidation of clays. Cartier et al. (1989) used Asaoka's method for the analysis of a test embankment on peat and reported a reasonable prediction of the settlement and time at 98% primary consolidation. Edil et al. (199lb), on the other hand, applied the procedure to a variety of clay and peat cases and questioned its applicability to peat settlement.

There is a need for a working observational method for peats to interpret the field settlement data correctly and to derive the compression parameters.

Additional difficulties in the field observations center around certain practical difficulties (Venmans 1994). For instance, if an embankment is built on a layered compressible foundation in which some layers are overconsolidated, the settlements measured at ground surface by settlement plates can be misinterpreted due to the superposition of component time-settlement curves of each layer. It is important in such cases to measure the settlement of each layer by in situ instrumentation. Another difficulty involves the determination of end of primary period from pore pressure measurements made by piezometers. Often excess pore pressures remain undissipated obscuring such determinations. Excess pore pressures may be induced by settling piezometers as they move into another soil.

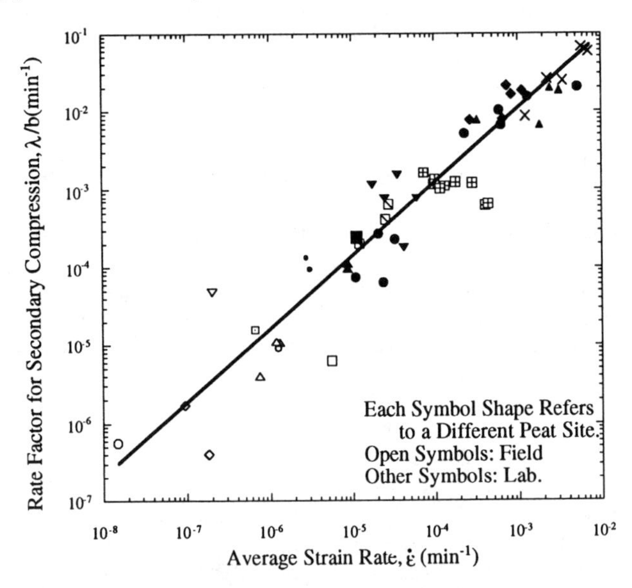

FIG. 14. Compression Rate Factor versus Average Strain Rate for Peats and Organic Soils (after Edil and Mochtar 1984)

SUMMARY AND CONCLUSIONS

Current issues regarding the compressibility of peats and organic soils are presented. These matters were discussed recently in considerable detail by peat investigators in the International Workshop on Advances in Understanding and Modeling the Mechanical Behavior of Peat held in Delft in June 1993. It may be necessary to consult the proceedings of this workshop as well as some of the references for a fuller understanding of the various formulations of one-dimensional compression of peats presented herein and the issues in engineering practice.

The issues covered included the problems encountered in definition and classification of peats based on mechanical behavior of peat. Certain classes of high organic and fiber content (greater than 75% and 45%, respectively) materials with a low degree of humification present a high degree of non-linearity in mechanical response coupled with a large degree of heterogeneity and rather different microstructure and consequently can not be dealt by the theories and procedures developed primarily for mineral soils. Modifications of existing approaches and some entirely new approaches are being developed to deal with such materials referred as *peat* herein.

Both the compression-stress and the compression-time behaviors show strong non-linearity such that the usual semi-logarithmic plots do not linearize the response. In particular, a change of the rate of decrease of strain rate with time termed "tertiary compression" is encountered in most cases of stress application in fibrous peats. Determining end of primary consolidation from compression-time curves is unsuitable for peats. Pore pressure measurement both in laboratory and field is essential for this purpose. A range of poorly understood mechanisms may be responsible for this significant and long-term highly non-linear creep compression of peats such as higher microstresses between fibers and a compressible and decomposible solid phase. Use of natural strain instead of common engineering strain and intrinsic time instead of real time with the origin at the beginning of the load increment application are proposed to linearize the behavior. This approach provides significant improvement in a majority of cases including some inorganic soils. However, for very fibrous peats, it appears to be too much of a simplification and require further modification of the formulation.

Several formulations are presented in which the uniqueness of effective stress-strain and strain rate (or void ratio and void ratio rate) is assumed. Detailed laboratory investigations support this concept under isothermal, normally consolidated conditions and indicate that the creep compression rate of peat increases exponentially with increasing vertical effective stress and void ratio. Furthermore, the creep behavior is thermally activated and the rate of creep compression increases exponentially with increasing temperature. The proposed formulations give good predictions for constant stress, multiple-stage loading and constant-rate-of-strain compression tests in the normally consolidated range using the same parameters. The generality of the models suggest that they should be considered even for mineral soils.

The choice of construction method in areas underlain by peat deposits is a matter of finding optimal solutions between the economic and technical factors, available construction time, and the target performance standards. Avoidance of construction of fills over peat layers, replacement of surface peat layers by granular fill materials, stage-construction, use of light-weight fill (weight-credit construction), in situ improvements (stone and sand columns and preloading) and piles are the various methods used in different countries in dealing with construction over peats. Geotextiles and strip drains (paper or polyester) are often incorporated into construction to improve stability, improve construct ability, and accelerate construction. A novel concept of in situ improvement by means of thermal compression is being currently investigated to significantly diminish the creep potential.

There appears to be a split of opinion with regards to long-term chemical and biological degradation of peats in the absence of any significant change in the submergence of these deposits. It is clear that lowering of groundwater may cause deterioration.

Finally, there appears to be a great need for working field observational methods in determining the mechanical parameters governing the settlement-time and settlement-load behavior based on test fill data or early-stage construction data on settlement. With a clearer understanding of the hydrodynamic and creep behaviors, it may be possible to interpret the field data quantitatively. Additionally, empirical correlations between simple index properties and the mechanical parameters that describe peat behavior correctly are needed in view of the significant variability of peats.

ACKNOWLEDGMENT

Tuncer B. Edil's investigation of peats is based upon work supported by the U. S. National Science Foundation Grants No. MSM-8617238 and MSS-9115315. Evert J. den Haan's investigation of peats is largely based on work supported by the Netherlands Directorate-General for Public Works and Water Management. The authors acknowledge Dr. Patrick J. Fox for his review of the manuscript and valuable comments and Mr. Xiaodong Wang for his assistance in manuscript preparation.

APPENDIX - REFERENCES

Al-Refeai, T. O. (1992). "Strengthening of soft soil by fiber-reinforced sand column." *Proc. Int'l. Symp. Earth Reinforcement Practice*, Fukuoka.

Asaoka, A. (1978). "Observational procedure of settlement prediction." *Soils and Foundations*, 18(4), 87-101.

Bjerrum, L. (1967). "Engineering geology of Norwegian normally consolidated clays as related to settlements of buildings." *Géotechnique*, 17(2), 214-235.

Bosscher, P. J. and Edil, T. B. (1988). "Performance of lightweight waste impoundment dikes." *Proc. 2nd Int'l. Conf. Case Histories in Geotechnical Engineering*, St. Louis, 1, 63-69.

Butterfield, R. (1979) "A natural compression law for soils (an advance on e-log p')." *Géotechnique*, 29(4), 469-480.

Candler, C. J. and Chartres, F. R. D. (1988). "Settlement and analysis of three trial embankments on soft peaty ground." *Proc. 2nd Baltic Conf. on Soil Mech. and Found. Engrg.*, Tallin, 1, 268-272.

Carlsten, P. (1988). "Geotechnical properties and up-to-date methods of design and construction." State-of-the-art report, Presented in the 2nd Baltic Conf. Soil Mech. Found. Engrg., Tallin.

Cartier, G., A., Allaeys, M. Londez and Ropers, F. (1989). "Secondary settlement of peat during a load test." *Proc. 12th ICSMFE*, Rio Janeiro, 3, 1721-1722.

Candler, C. J. and Chartres, F. R. D. (1988). "Settlement measurement and analysis of three trial embankments on soft peat ground." *Proc. 2nd Baltic Conf. Soil Mech. and Found Engrg on Peats and Deformations of Structures on Highly Compressible Soils*, Tallin, 1, 268-272.

Crawford, C. B. (1964). "Interpretation of the consolidation test." *J. Soil Mech. Found. Div.*, Proc. ASCE, 90(SM5), 87-102.

Crouse, C. B., Kramer, S. L., Mitchell, R., and Hushmand, B. (1993). "Dynamic test of pipe pile in a saturated peat." *J. Geotech. Eng.*, ASCE, 119(10), 1550-1567.

den Haan, E. J. (1992). "The formulation of virgin compression of soils." *Géotechnique*, 42(3), 465-483.

den Haan, E. J. (1994). "Summary of session 1: one-dimensional behavior." *Advances in Understanding and Modeling the Mechanical Behavior of Peat*, Balkema, Rotterdam, The Netherlands, 131-140.

den Haan, E. J. and Edil, T. B. (1994). "Secondary and tertiary compression of peat." *Advances in Understanding and Modeling the Mechanical Behavior of Peat*, Balkema, Rotterdam, The Netherlands, 49-60.

den Haan, E. J. and El Amir, L. S. F. (1994). "A simple formula for final settlement of surcharge loads on peat." *Advances in Understanding and Modeling the Mechanical Behavior of Peat*, Balkema, Rotterdam, The Netherlands, 35-48.

Dhowian, A. W. and Edil, T. B. (1980). "Consolidation behavior of peats." *Geotech. Testing J.*, ASTM, 3(3), 105-114.

Edil, T. B. (1988). "Construction of embankments on peat." *Proc. 2nd Baltic Conf. Soil Mech. Found. Eng.*, Tallin, 1, 158-169.

Edil, T. B. and Bosscher, P. J. (1992). "Development of engineering criteria for shredded waste tires in highway applications," *Final Report No. GT-92-9*, Wisconsin Department of Transportation.

Edil, T. B. and Dhowian, A. W. (1979). "Analysis of long-term compression of peats." *Geotechnical Engineering*, Southeast Asian Soc. of Soil Engineering, 10, 159-178.

Edil, T. B. and Fox, P. J. (1994). "Field testing of thermal precompression." *Vertical and Horizontal Deformations of Foundations and Embankments*, Geotechnical Special Publication No. 40, ASCE, New York, New York.

Edil, T. B. and Mochtar, N. E. (1984). "Prediction of peat settlement." *Proc., Symp. Sedimentation Consolidation Models*, San Francisco, ASCE, 411-424.

Edil, T. B., Fox, P.J. and Lan, L. T. (199la). "End-of-primary consolidation of peat." *Proc. 10th European Conf. on Soil Mech and Found. Eng.*, Florence, 1, 65-68.

Edil, T. B., Fox, P. J. and Lan, L. T. (199lb). "Observational procedure for settlement of peat." *Proc. Geo-Coast Conf.*, Yokohama, 2, 165-170.

Edil, T. B., Fox, P. J., and Lan, L. T. (1994a). "An assessment of one-dimensional peat compression." *Proc. 13th ICSMFE*, New Delhi, 1, 229-232.

Edil, T. B., Fox, P. J., and Lan, L. T. (1994b). "Stress-induced one-dimensional creep of peat." *Advances in Understanding and Modeling the Mechanical Behavior of Peat*, Balkema, Rotterdam, The Netherlands, 3-18.

Farrell, E. R., O'Neill, C., and Morris, A. (1994). "Changes in mechanical properties of soils with variations in organic content." *Advances in Understanding and Modeling the Mechanical Behavior of Peat*, Balkema, Rotterdam, The Netherlands, 19-26.

Flaate, K. (1987). "Superlight material in heavy construction." *Geotech. News*, 5(3), 22-23.

Fox, P. J. (1992). "An analysis of one-dimensional creep behavior of peat," Ph.D. thesis, Univ. of Wisconsin-Madison.

Fox, P. J. and Edil, T. B. (1994). "Temperature-induced one-dimensional creep of peat." *Advances in Understanding and Modeling the Mechanical Behavior of Peat*, Balkema, Rotterdam, The Netherlands, 27-34.

Fox, P. J., Edil, T. B., and Lan, L. T. (1992). "C_a/C_c concept applied to compression of peat." *J. Geotech. Eng.*, ASCE, 118(8), 1256-1263.

Garlanger, J. E. (1972). "The consolidation of soil exhibiting creep under constant effective stress." *Géotechnique*, 22(1), 71-78.

Gibson, R. E., England, G. L., and Hussey, M. J. L. (1967). "The theory of one-dimensional consolidation of saturated clays - 1. Finite non-linear consolidation of thin homogeneous layers." *Géotechnique*, 17, 261-273.

Gibson, R. E. and Lo, K. Y. (1961). "A theory of consolidation of soils exhibiting secondary compression." *Acta Polytechnica Scandinavica*, Ci. 10296, 1-16.

Hanrahan, E. T. (1954). "An investigation of some physical properties of peat." *Géotechnique*, 4, 108-123.

Hanrahan, E. T. (1964). "A road failure on peat." *Géotechnique*, 14(3).

Hardin, B. O. (1989). "1-D strain in normally consolidated cohesive soils." *J. Geotech. Eng.*, ASCE, 115(5), 689-710.

Hobbs, N. B. (1986). "Mire morphology and the properties and behavior of some British and foreign peats." *Quart. Jnl. Eng. Geol.*, 19(1), 17-80.

Horvath, J. S. (1992). "New developments in geosynthetics: 'lite' products come of age." *Standardization News*, ASTM, 50-53.

Juárez-Badillo, E. (1965). "Compressibility of soils." *Proc. 5th Symp. Behaviour of Soil under Stress*, Bangalore, Indian Institute of Science, A2/1-35.

Kabbaj, M., F. Oka, F., S. Leroueil and Tavenas, F. (1986). "Consolidation of natural clays and laboratory testing." *Consolidation of Soils*, ASTM STP 892, Philadelphia, Pennsylvania, 378-404.

Koda, E. and Wolski, W. (1994). "The influence of strip drains on the consolidation performance of organic soils." *Advances in Understanding and Modeling the Mechanical Behavior of Peat*, Balkema, Rotterdam, The Netherlands, 347-360.

Kurihara, N., Isoda, T., Ohda, H. and Sekiguchi, H. (1994). "Settlement performance of the central Hokkaido expressway built on peat." *Advances in Understanding and Modeling the Mechanical Behavior of Peat*, Balkema, Rotterdam, The Netherlands, 361-368.

Krieg, S. and Goldscheider, M. (1994). "Some results about the development of K_o during one-dimensional creep of peat." *Advances in Understanding and Modeling the Mechanical Behavior of Peat*, Balkema, Rotterdam, The Netherlands, 71-76.

Ladd, C. C., Foott, R., Ishihara, K., Schlosser, F. and Poulos, H. A. (1977). "Stress-deformation and strength characteristics." *Proc. 9th ICSMFE*, Tokyo, 2, 421-494.

Lan, L. T. (1992). "A model for one-dimensional compression of peat," Ph.D. thesis, Univ. of Wisconsin-Madison.

Landva, A. O., Korpijaakko, E.O. and Pheeney, P. E.. (1983a). "Geotechnical classification of peats and organic soils." *ASTM STP 820*, Philadelphia, Pennsylvania, 37-51.

Landva, A. O., Pheeney, P.E. and Merserau, D. E. (1983b). "Undisturbed sampling in peat." *ASTM STP 820*, Philadelphia, Pennsylvania, 141-156.

Lefebvre, G., Langlois, P., Lupien, C. and Lavalle, J. G. (1984). "Laboratory testing and in situ behavior of peat as embankment foundation." *Can. Geotech. J.*, 21(2), 322-337.

Lea, N. D. and Brawner, C. O. (1963). "Highway design and construction over peat deposits in lower British Columbia." *Hwy. Res. Record, No. 7*, Washington, D. C., 1-32.

Leonards, G. A. and Girault, P. (1961). "A study of the one-dimensional consolidation test." *Proc. 5th ICSMFE*, Paris, 1, 213-218.

Leroueil, S., Kabbaj, M., Tavenas, F., and Bouchard, R. (1985). "Stress-strain-strain rate relation for the compressibility of sensitive natural clays." *Géotechnique*, 35, 159-180.

Liu, J-C. and Znidarcic, D. (1991). "Modeling one-dimensional compression characteristics of soil." *J. Geotech. Eng.*, ASCE, 117(1), 162-169.

Magnan, J. P. (1994). "Construction on peat: state of the art in France." *Advances in Understanding and Modeling the Mechanical Behavior of Peat*, Balkema, Rotterdam, The Netherlands, 369-380.

Mesri, G. and Godlewski, P. M. (1977). "Time- and stress- compressibility interrelationship." *J. Geotech. Eng. Div.*, Proc. ASCE, 103(GT5), 417-430.

Mitchell, J. K (1993). *Fundamentals of Soil Behavior*, 2nd Edition. John Wiley & Sons, New York, New York.

Mitchell, J. K., Campanella, R. G., and Singh, A. (1968). "Soil creep as a rate process." J. *Soil Mech. Found. Div.*, Proc. ASCE, 94(SM1), 231-253.

Monahan, E. J. (1993). "Weight-credit foundation construction using artificial fills." *Paper No. 93-0157*, 72nd TRB Annual Meeting, Washington, D. C.

Orr, T. and McEnaney, W. (1994). "Friction piles for a walkway on a peat bog." *Advances in Understanding and Modeling the Mechanical Behavior of Peat*, Balkema, Rotterdam, The Netherlands, 381-388.

Samson, L. and La Rochelle, P. (1972). "Design and performance of an expressway constructed over peat by preloading." *Can. Geotech. J.*, 9, 447-466.

Suklje, L. (1957). "The analysis of the consolidation process by the isotaches method." *Proc. 4th ICSMFE*, London, 1, 200-206.

Thompson, J. B. and Palmer, L. A. (1951). "Report of consolidation tests with peat." *Consolidation Testing of Soils*, ASTM STP 126, Philadelphia, Pennsylvania, 4-8.

Vaid, Y. P., Robertson, P. K. and Campanella, R. G. (1979). "Strain rate behavior of Saint-Jean-Vianney clay." *Can. Geotech. J.*, 16, 34-42.

Venmans, A. A. M. (1994). "Summary of session 3: immediate issues in engineering practice." *Advances in Understanding and Modeling the Mechanical Behavior of Peat*, Balkema, Rotterdam, The Netherlands, 415-424.

Vonk, B.F. (1994). "Some aspects of the engineering practice regarding peat in small polder dikes." *Advances in Understanding and Modeling the Mechanical Behavior of Peat*, Balkema, Rotterdam, The Netherlands, 389-402.

PROBABILISTIC OBSERVATION METHOD FOR SETTLEMENT-BASED DESIGN OF A LANDFILL COVER

Wilson H. Tang,[1] Fellow, ASCE, Robert B. Gilbert,[2] Mauricio Angulo,[3] Associate Members, ASCE, and Richard S. Williams,[4] Member, ASCE

ABSTRACT: Uncertainties in subsurface conditions make predicting total and differential settlement a difficult task. The Observation Method, which is commonly implemented to acquire additional knowledge and reduce these uncertainties, can be enhanced by using probabilistic tools. A case study project involving differential settlement in a landfill cover is presented to demonstrate the application of a Probabilistic Observation Method and illustrate its benefits.

INTRODUCTION

Accurate prediction of the settlement of fills and structures is often difficult because of insufficient knowledge concerning subsurface conditions. Uncertainties arise from the use of a limited number of borings and small scale tests to represent the in situ conditions of the entire stratum. In order to proceed with a project in the face of these uncertainties, an observational approach is commonly adopted. By using the Observation Method (OM), one accepts that the site uncertainties can never be eliminated *completely*. Design and construction decisions are made based on an initial model while allowing for possible deviations from the predicted

---

[1] Prof. of Civ. Engrg., Univ. of Illinois at Urbana-Champaign, 3110 Newmark Civil Engrg. Lab., 205 N. Mathews Ave., Urbana, IL 61801.

[2] Asst. Prof. of Civ. Engrg., Univ. of Texas at Austin, 9.227E Cockrell Hall, Austin, TX 78712.

[3] Res. Asst., Univ. of Illinois at Urbana-Champaign, 2212 Newmark Civil Engrg. Lab., 205 N. Mathews Ave., Urbana, IL 61801.

[4] Principal, Golder Assoc. Inc., 1809 N. Mill St., Ste. C, Naperville, IL 60563.

performance. Contingency plans are established beforehand to cope with deviations. As construction proceeds, settlement is monitored and initial predictions are updated. Modifications are made in design or construction as required to account for the observed performance. Therefore, OM facilitates design and construction while avoiding large losses resulting from inappropriate designs, and it provides for safe construction without being overly conservative. These benefits can be enhanced significantly by incorporating probabilistic techniques into the OM. For example, Harchih and Vanmarcke (1983) demonstrate how probabilistic techniques can be used to update the hydraulic properties of an earth dam based on pore pressure measurements. Several researchers, such as Wu (1974), have also presented examples of decision analyses applied to geotechnical problems.

In this paper, a Probabilistic Observation Method (POM) is used to enhance the traditional OM in settlement related design problems. Probabilistic site characterization models, Bayesian updating methodologies and decision analysis techniques are applied within the OM to develop a general POM for tackling settlement problems. The POM approach is presented within the framework of a case study project involving the construction of a cover on an uncontrolled hazardous waste landfill. This case study project serves to demonstrate how the POM is applied and illustrate its benefits.

PROBLEM DESCRIPTION

The case study project site is a closed landfill located in Belvidere, Illinois. A detailed description of the site is contained in Golder Associates (1990). The landfill was operated from the early 1930's to the late 1970's when it was closed, and it is approximately 8 hectares in size and 10 to 15 m in thickness. The landfill contains both municipal and industrial wastes, some of which are currently classified as hazardous materials. Results from a series of site investigations performed during the early 1980's indicated that groundwater at the site was contaminated by leakage from the landfill. The selected remediation plan included constructing a cover system over the landfill to reduce infiltration into the waste and minimize leakage. A plan view of the landfill is shown in Fig. 1a, while the cover system configuration is shown in Fig. 1b.

A key issue in the cover system design was the potential for differential settlement. Differential settlement could potentially threaten the integrity of the components within the cover system (e.g. the compacted clay layer). In addition, differential settlement could also result in water ponding on the cover surface and in the drainage layer, increasing the rate of infiltration. In the preliminary design stage, cover grades ranging from 2 to 5 percent were identified as potential design alternatives. The estimated cost required to construct each alternative is summarized as follows: $2,180,000 for a 2-percent grade, $2,300,000 for a 3-percent grade, $2,420,000 for a 4-percent grade, and $2,540,000 for a 5-percent grade. Substantial amount of money could be saved by minimizing the design grade; however, the potential for differential settlement increases as the design grade decreases. Hence, the preliminary design process involved consideration of the construction cost, the likelihood that the design would not perform adequately, and the consequences

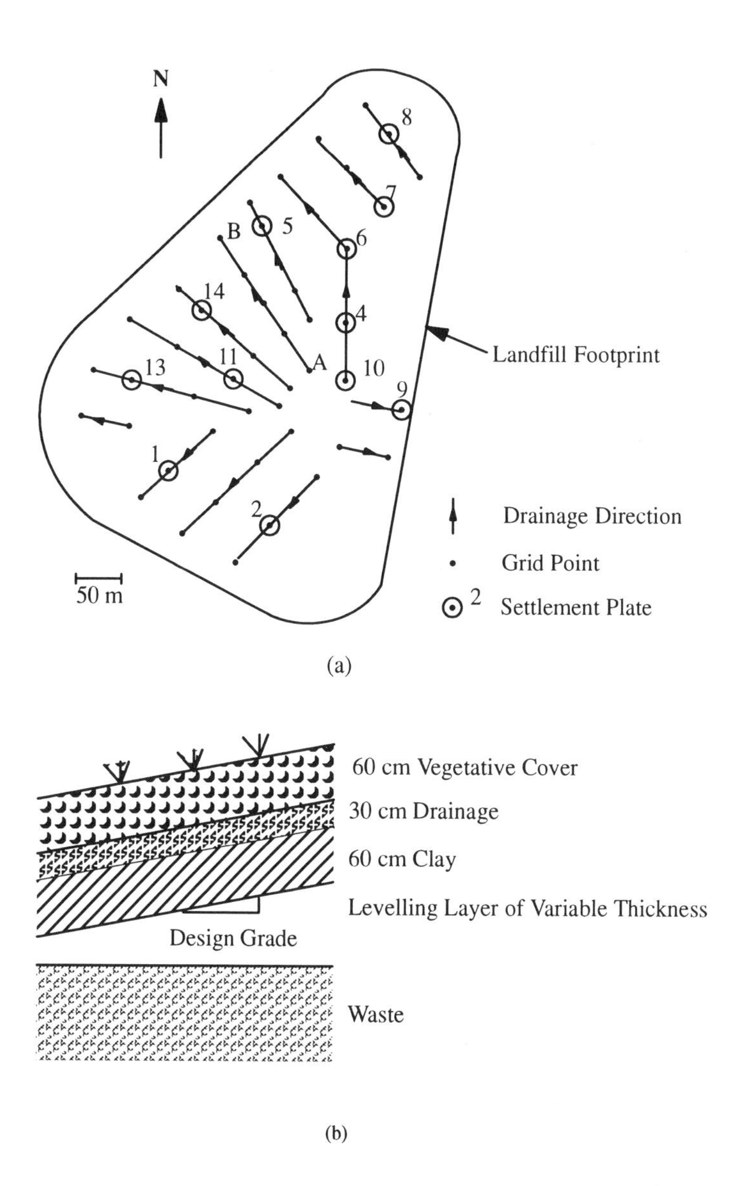

FIG. 1. (a) Site Layout; and (b) Cover System Configuration

associated with failure of the cover due to differential settlement.

TRADITIONAL OBSERVATION METHOD

The cover design was developed using a preliminary estimate of cover settlement. Settlement, s, at a given point on the cover surface was estimated according to the following model

$$s = C_c H \left\{ \log_{10}\left[\frac{p_0 + \Delta p}{p_0}\right] + \frac{C_\alpha}{C_c}\log_{10}\left[\frac{t}{t_p}\right] \right\} \tag{1}$$

where C_c is the coefficient of compression, H is the initial waste thickness, p_o is the initial vertical stress at the mid-point of the waste, Δp is the added stress due to the weight of the cover, C_α is the coefficient of secondary compression, t is the design life (i.e. 30 years), and t_p is the time required for primary compression. This settlement model is an empirical model that has proven to be valid for most natural soil types; however, extension of the model to waste materials is not necessarily appropriate. Several investigators, such as Oweis and Khera (1990), Morris and Woods (1990), and Landva and Clark (1990), have successfully applied this model to waste and verified its validity with field or laboratory data. Other empirical models of waste settlement have also been proposed. Edil et al. (1990) propose a power creep model to predict waste compression with time, while Edgers and Noble (1992) propose a similar model with an added expression to account for waste decomposition. The common aspect of all these potential settlement models is their empirical nature. As demonstrated in the literature, each model can generally be fit through field settlement data by adjusting the empirical constants, such as C_α and t_p in Eq. (1). The key to predicting settlement is in selecting appropriate values for the empirical constants in a given model, not in selecting the model itself. Therefore, the model given in Eq. (1) was selected because: (i) case studies in the literature indicate that this model is suitable; (ii) the waste at this site is 15 to 50 years old so that decomposition effects should be minimal; and (iii) we had successfully used this model to design covers for similar sites.

In order to select an appropriate design grade, the magnitudes of settlement were evaluated at the two ends of the maximum drainage path in the cover (i.e. points A and B in Fig. 1a). Since laboratory consolidation tests on waste materials are expensive and do not necessarily yield representative results, appropriate design values for C_c, C_α and t_p were selected based on published literature and experience for similar sites: C_c=0.25, C_α=0.1 and t_p=1 month. In addition, the waste density was assumed to be 12.6 kN/m^3. These parameters were taken to be the same at all points on the cover; therefore, differential settlement would result only from variations in the cover thickness (i.e. Δp) between points A and B. A 3-percent design grade was selected by using a design criterion of a 2-percent minimum grade after settlement.

To overcome the lack of knowledge concerning the appropriate site specific values for the empirical constants in Eq. (1), an Observation Method was implemented to obtain site specific information. Construction would proceed in two stages: Stage 1 consisted of the levelling layer while Stage 2 consisted of the remaining cover components (Fig. 1b). The levelling layer would be constructed at a 3-percent grade, and settlement would be measured at a number of points using settlement plates. Based on these results, the levelling layer grade would be increased if necessary before constructing the remainder of the cover system. The levelling layer was constructed in 1990 and 12 settlement plates were installed as shown in Fig. 1a. The magnitude of settlement under the weight of the levelling layer was measured at each location, and C_c was back-calculated knowing the applied weight and the waste thickness. The results are presented in Table 1. Since the average value of C_c from these results, 0.22, was close to and smaller than the assumed value, 0.25, the remainder of the cover was constructed at the 3-percent grade. Construction was completed in the fall of 1991, and no significant problems have been identified to date.

Table 1. Results from Settlement Plates

Settlement Plate	Applied Stress (kN/m^2)	Waste Thickness (m)	Settlement (m)	Estimated C_C
1	8.6	10.7	0.11	0.197
2	6.7	11.5	0.14	0.325
4	27.9	11.5	0.37	0.226
5	7.5	10.5	0.14	0.280
6	37.1	10.6	0.44	0.214
7	9.8	9.4	0.13	0.205
8	13.0	10.0	0.13	0.161
9	21.1	7.7	0.17	0.138
10	26.7	7.9	0.30	0.205
11	11.4	7.8	0.21	0.296
13	21.5	11.5	0.22	0.169
14	36.2	10.5	0.48	0.238

While this traditional design approach produced satisfactory results, some questions remain unanswered. First, the properties of the waste can vary from point to point in the landfill, as displayed by the range of measured C_c values in Table 1. This variability could cause differential settlement besides known variations in the cover geometry. Can this variability be accounted for in the design process? Second, only the parameter C_c could be evaluated from the field observation data. How do the uncertainties in C_α and t_p affect the design? Third, the level of conservatism in the design is not easily quantified. Would a less costly 2-percent design grade have been adequate? Would the additional capital cost associated with a 4-percent grade be justified in the long term? Finally, the value of using settlement plates in the Observation Method was not quantified. Was the OM worth the cost required to implement it? How could the measurement program be designed to most effectively implement the OM? Probabilistic tools can be used to address these questions, thus enhancing the Observation Method.

PROBABILISTIC OBSERVATION METHOD

Differential settlement between any two points will result from both known variations in the cover geometry *and* unknown or seemingly random variability in the waste properties. The variability in waste compressibility parameters (i.e. C_c, C_α and t_p) can be modeled probabilistically to evaluate the potential for differential settlement. A random field model is adopted to represent the spatial variability of C_c. The value of C_c at a point is described by a normal distribution with a mean, μ_{C_c}, and a standard deviation, σ_{C_c}, both of which are assumed to be constant over the site. The correlation coefficient between C_c at points i and j, $\rho_{i,j}$, is estimated by the following exponential model

$$\rho_{i,j} = e^{-(\tau_{ij}/b)^2} \tag{2}$$

where τ_{ij} is the distance between points i and j and b is equal to the scale of fluctuation, θ, divided by $\pi^{1/2}$ (Vanmarcke 1983). In physical terms, the scale of fluctuation represents the distance over which C_c is highly correlated. The exponential correlation model was chosen because of its mathematical simplicity and its general suitability in modelling spatial variabilities in soil materials. The secondary consolidation of the waste materials will be primarily due to mechanical creep. Since a material that exhibits high compressibility during the primary stage generally shows large creep deformation, the ratio of C_α/C_c is assumed to be a constant. This assumption is consistent with the observed behavior of many natural soil materials (Mesri and Castro 1987). It is also conservative because if C_α is assumed to vary independently of C_c, it will reduce the likelihood of having high (or low) primary and secondary compression at the same location, thus reducing also the potential for large differential settlement. Therefore, the spatial variability in both primary and secondary compressibility is described by the spatial variability in C_c. The above probabilistic model is deliberately assumed as the true model for describing the spatial variability of waste compressibility in this case to simplify the design process. However the general POM can accommodate the likelihood of other models as will be discussed in a later section.

This probabilistic model can now be used to evaluate the likelihood that the magnitude of differential settlement between any two points in the cover exceeds an allowable value. Since long term performance of the cover is related to maintaining a positive drainage grade, failure will be defined in terms of a drainage grade reversal. A grade reversal over a small distance (e.g. 1 m) will not cause significant damage and can be easily repaired. However, as the distance increases, the impact of the reversal also increases. Based on these considerations, a grade reversal over a minimum distance of 30 m is selected to define failure. The cover surface has been divided into a series of 50 grid points spaced at 30 m along the directions of drainage for each cover slope (Fig. 1a). Failure will occur if the grade between any pair of adjacent grid points along a drainage path becomes negative over the 30-year design life of the cover.

A Monte Carlo simulation procedure was used to determine the probability that grade reversal occurs. For each simulated field of C_c values, the magnitude of settlement is calculated at every point in the field to evaluate the potential for a grade reversal. For example, assuming $\mu_{C_c}=0.25$, $\sigma_{C_c}=0.1$, $C_\alpha/C_c=0.1$, $t_p=1$ month and $\theta=30$ m, simulation results indicate that grade reversal occurs in 260 of 1,000 simulations for the 2-percent design grade; hence the failure probability for this alternative is 0.260. The failure probabilities for the other design grades are: 0.165 for a 3-percent grade, 0.108 for a 4-percent grade and 0.073 for a 5-percent grade.

Preliminary Design

The optimal design grade is selected during the preliminary design stage considering both the cost of construction and the cost of failure. The cost of failure is however not readily defined. The potential consequences of failure could be: (1) increased maintenance effort due to erosion and ponding; (2) reconstruction of the cover system in the area of failure; and (3) increased environmental impact due to the ineffectiveness of the cover in minimizing infiltration into the waste. As an example, an assumed failure cost of \$2,500,000 (approximately the cost of construction) yields an expected cost, E(C), for the 2-percent grade of

$$E(C) = \$2{,}180{,}000 + 0.260(\$2{,}500{,}000) \tag{3}$$

or \$2,830,000. In a similar fashion, E(C) can be calculated for each design grade: \$2,713,000 for a 3-percent grade, \$2,690,000 for a 4-percent grade and \$2,723,000 for a 5-percent grade. The 4-percent design grade, which has the minimum expected cost, would be selected as the optimal grade based on this analysis.

Unfortunately, appropriate values for μ_{C_c}, σ_{C_c}, C_α/C_c and θ cannot be determined in the preliminary design stage with any certainty. A review of existing literature (Oweis and Khera 1990; Landva and Clark 1990) provides insight into the range of typical values for these parameters. Reported values of C_c for municipal waste range from 0.1 to 0.5, and values of C_α/C_c range from 0.05 to 0.15 while t_p is on the order of 1 month. Very little information is available concerning the spatial variability of C_c; however, experience indicates that the spatial pattern of differential settlement in covers is typically on the scale of tens of meters.

The uncertainty associated with selecting values for μ_{C_c}, σ_{C_c}, and C_α/C_c may affect the optimal design grade. One way to account for this source of uncertainty is to assume that μ_{C_c}, σ_{C_c}, and C_α/C_c are described by distributions of possible values. Based on the literature and relying heavily on engineering judgement and experience, μ_{C_c} is assumed to be equal to 0.15, 0.25 and 0.35 with equal likelihood; while, σ_{C_c} is equal to 0.05, 0.1 and 0.15 with equal likelihood. While discrete distributions for μ_{C_c} and σ_{C_c} are used here for simplicity, the approach is easily extended to continuous distributions. The distributions of μ_{C_c} and σ_{C_c} are taken to be independent; hence, there are nine possible combinations of μ_{C_c} and σ_{C_c} each having a probability of 1/9. Finally, C_α/C_c is described by a normal distribution with a mean value of 0.1 and a standard deviation of 0.02, representing a 95% chance that the actual value is between 0.06 and 0.14. It is important to recognize that the uncertainties associated with μ_{C_c}, σ_{C_c} and C_α/C_c are treated differently than the random variability in C_c. These uncertainties represent a lack of knowledge about the overall site characteristics, unlike the variability in C_c which accounts for variability from point to point in the site. The same values of μ_{C_c}, σ_{C_c} and C_α/C_c are used to generate the settlement at all points in the cover for each Monte Carlo simulation, while the value of C_c varies randomly between points in a given simulation. The values for t_p and θ are assumed to be known values of 1 month and 30 m, respectively. The sensitivity of these assumptions will be explored subsequently.

Selection of the preferred design grade can now be formulated within a decision analysis framework as shown in Fig. 2. For each design grade alternative, the expected cost including both construction and failure costs is calculated, and the alternative with the minimum expected cost is selected as the optimal alternative. The failure probability and expected cost are evaluated for each possible pair of μ_{C_c} and σ_{C_c}, and the costs are then combined in a weighted average where the cost for each pair is weighted by the respective likelihood. Using a failure cost of $2,500,000, the expected cost of the 3-percent design grade is $2,684,000 (Fig. 2). Similarly, the expected costs of the other alternatives are calculated, and the preferred design alternative is the 3-percent grade (Fig. 3).

The optimal design grade will be related to the cost of failure; as the failure cost increases, more conservative designs will be preferred. As an example, the expected cost for each design grade is plotted as a function of the failure cost in Fig. 3. Observe how the optimal design grade increases as the cost of failure increases. If the cost of failure is less than about $1,600,000 the optimal design grade is 2 percent; the optimal design grade increases to 5-percent if the cost of failure is greater than about $3,700,000. In this case study, a 3-percent design grade was selected as the preferred alternative; hence, it can be inferred from Fig. 3 that the perceived failure cost was between $1,600,000 and $2,700,000, assuming that all input values to the decision analysis represent our judgment realistically.

While the cost of failure may be difficult to establish with certainty at times, we should be able to narrow it down to a range together with an estimate of the likelihood over that range of values. This additional uncertainty can easily be included in the previous decision analysis. It requires the addition of a chance node for the potential failure costs and associated probabilities to the right of each branch containing a combination of (μ_{C_c}, σ_{C_c}) in Fig. 2. The expected total cost for each design grade is computed by weighing each total cost with these probabilities, and the alternative with the lowest expected total cost is selected. Suppose the failure cost is uniformly likely between $1,000,000 and $4,000,000, the resulting expected total cost is $2,746,000 for a 2-percent grade, $2,684,000 for a 3-percent grade, $2,694,000 for a 4-percent grade and $2,733,000 for a 5-percent grade. Hence, the 3-percent grade is the preferred design grade. It is interesting to observe that these expected total costs are exactly the same as those calculated earlier for a known failure cost of $2,500,000. In fact for the minimum expected cost decision criterion used here, the expected total cost calculation is only affected by the expected failure cost. Even if we are not certain about the exact failure cost, Fig. 3 is still useful

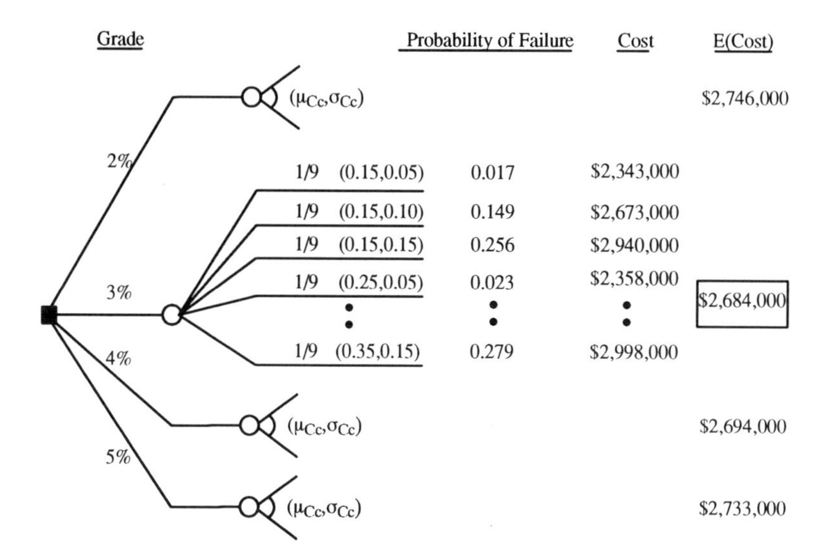

FIG. 2. Decision Tree for Preliminary Design (θ = 30.5 m, Failure Cost = $2,500,000)

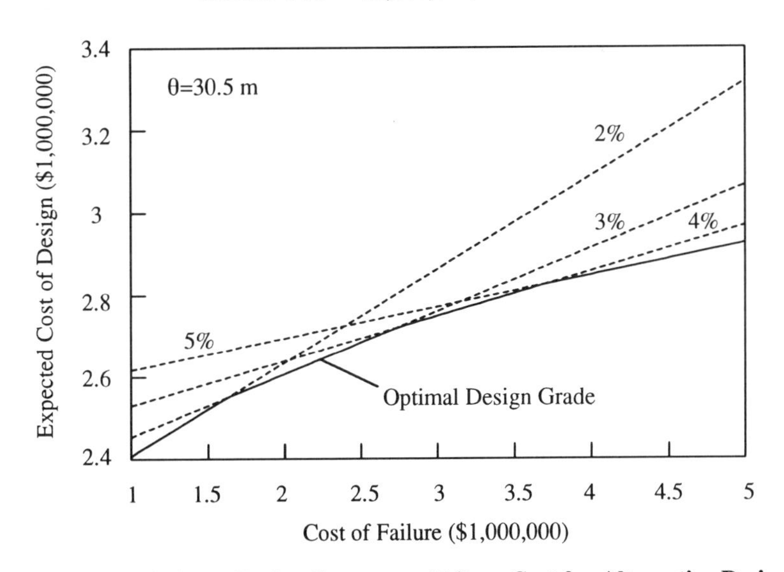

FIG. 3. Preliminary Design Cost versus Failure Cost for Alternative Design Grades

in determining the optimal grade by substituting the failure cost with its expected value. Since the expected failure cost for the above uniformly distributed failure cost is $2,500,000, the optimal design is the 3-percent grade.

A sensitivity study has been performed to assess the sensitivity of the design decision to assumptions in the analysis. Several important assumed parameters are θ, t_p and σ_{C_α/C_c}. Increasing θ increases the correlation in C_c between adjacent points in the grid; hence, the failure probability decreases because there is less chance for wide variability in C_c between points to cause differential settlement. The expected cost decreases from $2,684,000 to $2,579,000 if θ is doubled from 30 to 60 m, however the optimal design alternative remains the 3-percent grade. The magnitude of t_p controls the magnitude of secondary compression over the 30 year design life; as t_p decreases, the magnitude of settlement increases and the chance for differential settlement increases. The expected cost increases from $2,684,000 to $2,713,000 when t_p is reduced from 1 month to 1/2 month. In this case, the optimal design alternative switches to the 4-percent grade; however, E(C) for the 3-percent grade is only $3,000 greater than that for the 4-percent grade. Since C_α/C_c is assumed to be a constant value from point to point, this ratio affects the magnitude of secondary consolidation equally at every point in a manner similar to t_p. The net effect of σ_{C_α/C_c} on the failure probability and expected cost is negligible. In short, the 3-percent grade is considered satisfactory even if more precise estimates than these ranges are not available.

Value of Observed Information

The value of the Observation Method is to provide site specific performance information to confirm and/or modify the design during construction as required. Within a probabilistic framework, it is possible to quantify the value of the OM. Specifically, the observed information from the settlement plate measurements can be used to reduce the uncertainty in μ_{C_c} and σ_{C_c}. As shown in Fig. 2, this uncertainty can have a significant effect on the decision process. For example, the expected cost of the 3-percent grade ranges from $2,343,000 to $2,998,000 depending on the values of μ_{C_c} and σ_{C_c}. By reducing the uncertainty in μ_{C_c} and σ_{C_c}, it may be possible to reduce the expected cost of a given design alternative. Hence, the value of the observed information is measured by the reduction in the expected cost resulting from this information.

First, consider an extreme case where we are able to identify the appropriate combination of μ_{C_c} and σ_{C_c} with certainty (perfect information). The probability for this combination will be equal to 1, while that for all other combinations will be equal to 0. Without this information, the preferred design alternative is the 3-percent grade. Given the observed information, it may be more cost effective to increase the design grade. For example, if μ_{C_c} and σ_{C_c} are found to be 0.15 and 0.1, respectively, then the preferred alternative would be to raise the design grade from 3 to 4 percent. The expected cost for the 4-percent grade is $15,000 less than for the 3-percent grade using a $2,500,000 failure cost, or the value of this information is equal to $15,000. The value of perfect information on μ_{C_c} and σ_{C_c} is summarized for each possible combination of μ_{C_c} and σ_{C_c} in Table 2, and it ranges from $0 (the preferred design grade does not change) to \$65,000. Since each combination of μ_{C_c} and σ_{C_c} is equally likely based on the preliminary design analysis, the value of perfect information for each combination in Table 2 has a 1/9 probability of occurring and the expected value of perfect information is equal to \$23,610.

Table 2. Value of Information for Given Statistics of C_C

C_C Statistics		Updated Design Grade	Value of Information (VI)
Mean	Std. dev.		
0.15	0.05	3	\$0
0.15	0.10	4	\$15,000
0.15	0.15	5	\$65,000
0.25	0.05	3	\$0
0.25	0.10	4	\$22,500
0.25	0.15	5	\$40,000
0.35	0.05	3	\$0
0.35	0.10	4	\$57,500
0.35	0.15	5	\$12,500

While the value of perfect information establishes the maximum benefit associated with field observations, it will not typically be possible or feasible to obtain perfect information. A more appropriate question is: What value is associated with twelve sparsely spaced settlement plate measurements? The uncertainty in appropriate values for μ_{C_c} and σ_{C_c} will be reduced; and the probability mass functions (pmf's) for μ_{C_c} and σ_{C_c} can be updated by applying a Bayesian methodology (Ang and Tang 1975). Based on the mean value of the observed C_c values, m_{C_c}, and the associated standard deviation, s_{C_c}, the pmf is updated as follows

$$P''\left(\mu_{C_c}, \sigma_{C_c}\right) = kP\left(m_{C_c}, s_{C_c} \,|\, \mu_{C_c}, \sigma_{C_c}\right) P'\left(\mu_{C_c}, \sigma_{C_c}\right) \quad (4)$$

where $P''(\mu_{C_c},\sigma_{C_c})$ is the updated probability, $P(m_{C_c},s_{C_c}|\mu_{C_c},\sigma_{C_c})$ is the probability of observing sample mean value of m_{C_c} and standard deviation of s_{C_c} in the measured results *given* the statistics for C_c, $P'(\mu_{C_c},\sigma_{C_c})$ is the prior pmf of μ_{C_c} and σ_{C_c}, and k is a normalizing constant so that the probabilities of each combination sum to 1.0. The updated probability of a given (μ_{C_c} and σ_{C_c}) combination is proportional to the likelihood of obtaining the measured statistics given that combination multiplied by the prior probability of the combination.

To evaluate the conditional probability term in Eq. (4), the distributions of m_{C_c} and s_{C_c} for given values of μ_{C_c} and σ_{C_c} are required. A convenient assumption is that the population of field measurements follows a normal distribution. This assumption is generally reasonable for many soil properties. By assuming further that the measurements are statistically independent (the settlement plates are spaced further apart than the correlation distance for C_c, θ), it can be shown (Ang and Tang 1975) that m_{C_c} is normally distributed while s_{C_c} follows a chi-square distribution.

As the number of independent measurements increases, the variance in each of these distributions decreases and we can identify the appropriate combination of μ_{C_c} and σ_{C_c} with more certainty. As an example, consider the case where the sample mean of the measured test results is equal to 0.25 while the sample standard deviation is equal to 0.1. The updated marginal pmf's for μ_{C_c} and σ_{C_c} are shown in Fig. 4 as a function of the number of tests, n. For n=0, the updated pmf's are the same as the prior pmf's (i.e. equal likelihood for each combination). As n increases, the likelihood of $\mu_{C_c}=0.25$ and $\sigma_{C_c}=0.1$ generally increases relative to the other combinations.

Using these results, the value of information can be calculated as a function of the number of tests, n. The results are shown in Fig. 5. With more tests, the value of information increases and eventually approaches the value of perfect information. There is a cost associated with obtaining the information. Each settlement plate is estimated to cost approximately $500, and the cost of information is also plotted on Fig. 5. For n less than about 4, the value of information is smaller than the cost and it is not worthwhile to obtain the field measurements. Similarly, the cost of information exceeds the value of information for n greater than about 40. The value of information relative to the cost of information is maximized at n equal to about 20. Hence, this analysis indicates that the most cost effective measurement program would consist of approximately 20 settlement plates (plates spaced on a 60 m grid). It is important to note that the measurements are assumed to be independent. As n exceeds 40, the settlement plate spacing becomes smaller than the scale of fluctuation and individual measurements are no longer independent. In this case, the value of information will be less than that shown in Fig. 5.

Once the Observation Method is implemented, the final step is to incorporate the actual field observations into the decision process. The distribution of μ_{C_c} and σ_{C_c} is updated according to a modified form of Eq. (4), in which $P(m_{C_c}, s_{C_c} | \mu_{C_c}, \sigma_{C_c})$ is substituted by

$$P\left(x_1, \ldots, x_n \mid \mu_{C_c}, \sigma_{C_c}\right) = \prod_{i=1}^{n} f_X\left(x_i \mid \mu_{C_c}, \sigma_{C_c}\right) \tag{5}$$

where each x_i represents a measured value of C_c and $f_X(x_i | \mu_{C_c}, \sigma_{C_c})$ is a normal density function with mean value μ_{C_c} and standard deviation σ_{C_c} evaluated at x_i. The updated pmf's for μ_{C_c} and σ_{C_c} are first evaluated by using the observed results summarized in Table 1. Comparing adjacent pairs of C_c measurements reveals that these values are essentially independent, indicating that θ is less than 60 m as assumed in our analysis. Next, the expected costs for each design alternative are calculated assuming a $2,500,000 failure cost. The results are: $2,487,000 for a 2-percent grade, $2,423,000 for a 3-percent grade, $2,482,000 for a 4-percent grade, and $2,574,000 for a 5-percent grade. Hence, the 3-percent design grade should not be increased based on this analysis according to the minimum expected cost decision criterion. However for a conservative decision maker who is somewhat hesitant in taking risk, the 5%-grade may be the preferred alternative on the basis of a small premium in the expected cost. This alternative decision criterion could easily be

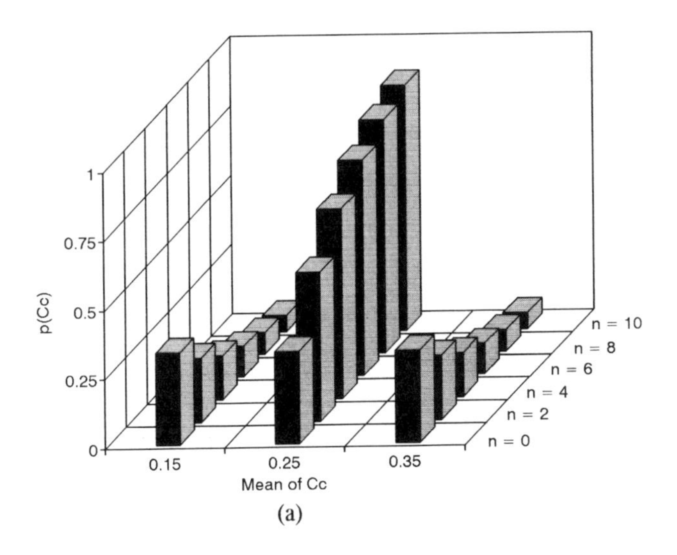

(a)

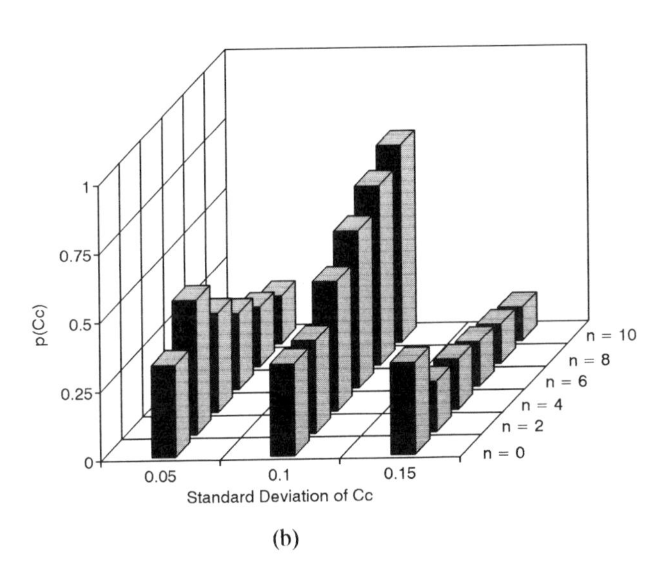

(b)

FIG. 4. Updated Probabilities of: (a) Mean Values of C_c; and (b) Standard Deviation of C_c (Based on $m_{C_c} = 0.25$, $s_{C_c} = 0.1$)

incorporated into the POM decision process using an expected utility analysis to reflect risk aversive behavior (Ang and Tang 1984).

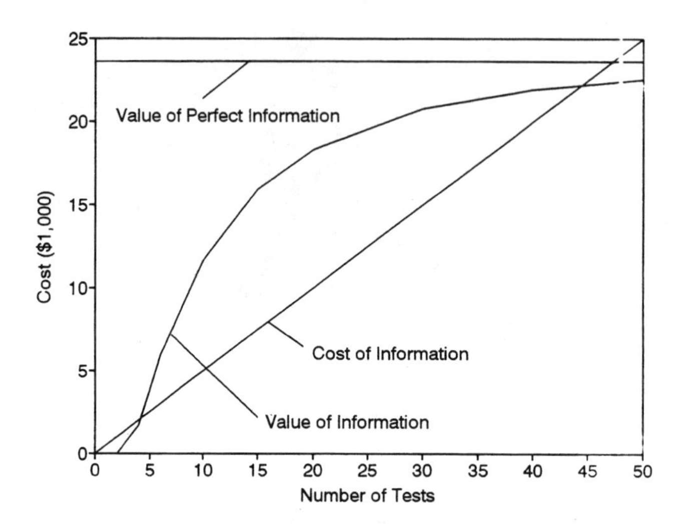

FIG. 5. Value of Information versus Testing Effort

In evaluating the value of the observed information on the parameters μ_{C_c} and σ_{C_c}, we have assumed that the cost of failure is known with certainty. What then is the effect of an uncertain cost of failure on the value of observed information? It can be shown again that for the minimum expected cost decision criterion used here, the value of observed information is influenced only by the expected failure cost. When the failure cost cannot be pinpointed, we can use the expected failure cost to evaluate the value of observed information. For instance, when the cost of failure is uniformly distributed between $1,000,000 and $4,000,000, the expected failure cost is $2,500,000; hence the value of perfect information on μ_{C_c} and σ_{C_c} is the $23,610 value obtained previously by assuming a known failure cost of $2,500,000.

The discussion in this section has focused on the value of obtaining more information on the parameters μ_{C_c} and σ_{C_c}. However the value of reducing the uncertainties of other model parameters can also be studied. For instance, consider the case where we cannot pinpoint the failure cost but instead estimate it to be between $1,000,000 and $4,000,000. Although it was shown that the 3-percent grade is the optimal design choice for an expected failure cost of $2,500,000, the optimal design would be other grades if the actual failure cost is different from $2,500,000. For example, if the cost of failure is $1,000,000 exactly, the optimal alternative is the 2-percent grade at an expected cost of $2,410,000 from Fig. 3. This is $50,000 less than the expected cost of $2,460,000 associated with using the 3-percent grade at this failure cost. Therefore a savings of $50,000 is realized if the failure cost is exactly $1,000,000. Similarly, it can be shown that a savings of only $10,000 is achieved if the failure cost is known to be $3,000,000 with certainty. Now the question is: Since the actual failure cost can be any value from $1,000,000 and $4,000,000, what is the expected worth of attempting to predict the failure cost accurately? The value of identifying the failure cost with certainty is computed by weighing the savings according to the probabilities of the respective failure cost values. For the uniformly distributed failure cost between $1,000,000 and $4,000,000, the expected value of pinpointing the failure cost is $16,700.

DISCUSSION

The advantage of the Probabilistic Observation Method (POM) is its flexibility to incorporate engineering judgment. It requires the active involvement of all parties including designers and possibly owners to assess the range and likelihood of likely outcomes and to assess the potential consequences. Sometimes we are not certain of which physical model to use; for instance, we may have other viable models for predicting the settlement of the landfill cover besides that given in Eq. (1). In this case we can perform a sensitivity analysis within the previous POM analysis to study whether there will be any change in the optimal design if any one of these alternatives is indeed correct. However a better approach is for the engineer to assess the relative likelihood between these hypotheses based on judgment and experience, and to include the alternative hypotheses directly into the POM analysis. In this way we can determine the optimal alternative reflecting the relative confidence in each model. To go one step further, it is possible within the POM framework to use the observed performance during construction to revise the relative likelihood of the correct hypothesis. This information can be then used to analyze if any change is necessary in the preliminary design. Conceptually this process is similar to the updating of the probability of μ_{C_c} and σ_{C_c} values as shown in Fig. 4. In fact, the POM can even help in determining which monitoring scheme is the most cost effective during construction. The above approach can be also applied to account for uncertainty in the relationship between C_α and C_c, the stationarity of the random field model for C_c, and the parameters describing the spatial correlation of C_c. For instance, besides the assumption that C_α/C_c is a constant, one may believe that it is 10% likely that the secondary compression parameter C_α is independent of the primary compression parameter C_c. That assumption can be incorporated into the POM analysis by first calculating the probability of failure assuming a new model with a separate random field for C_α and then weighing this failure probability by the likelihood of this model, 10 percent. This result would be combined with 90 percent of the previous failure probability to obtain the new probability of failure for each path in Fig. 2. In any event, the POM provides a systematic way to process the uncertainties and to study their effect on the optimal design or decision.

CONCLUSIONS

In summary, a probabilistic approach is proposed to complement (not to replace) the traditional Observation Method in settlement related design applications.

The Probabilistic Observation Method consists of three main components: site characterization models, Bayesian updating methods, and decision analysis tools. The implementation of the POM and its benefits are illustrated within the framework of a case study project. These benefits include the following:

(1) Uncertainties in the subsurface can be efficiently accounted for and quantified in developing the initial model and identifying possible deviations;
(2) The worth of various observation and monitoring programs in reducing uncertainties can be evaluated prior to implementation in order to design an optimal observation scheme;
(3) The validity of the initial model and possible deviations, including associated uncertainties, can be systematically updated based on the observed data; and
(4) Engineering judgement, an essential component to successfully implementing OM, can be rationally incorporated into all stages of the process in order to make the most effective use of this subjective information.

ACKNOWLEDGEMENTS

The research reported herein is supported by the National Science Foundation (Project # MSS 92-04433).

APPENDIX. REFERENCES

Ang, A. H-S., and Tang, W. H. (1975). *Probability Concepts in Engineering Planning and Design, Vol. 1 - Basic Principles*, John Wiley & Sons, New York, New York.

Ang, A. H-S., and Tang, W. H. (1984). *Probability Concepts in Engineering Panning and Design, Vol. 2 - Decision, Risk and Reliability*, John Wiley & Sons, New York, New York.

Edgers, L., Noble, J. J., and Williams, E. (1992). "A biologic model for long term settlement in landfills." *Environmental Geotechnology*, Balkema, Rotterdam, The Netherlands, 177-183.

Edil, T. B., Ranguette, V. J., and Wuellner, W. W. (1990). "Settlement of municipal refuse." *Geotechnics of Waste Fills - Theory and Practice*, ASTM STP 1070, American Society for Testing and Materials, Philadelphia, Pennsylvania, 240-258.

Golder Associates Inc. (1990). "Remedial design report, Belvidere Municipal No. 1 Landfill, Belvidere, IL," report submitted to U. S. EPA, Chicago, IL.

Hachich, W., and Vanmarcke, E. H. (1983). "Probabilistic updating of pore pressure fields." *J. Geotech. Eng.*, ASCE, 109(3), 373-387.

Landva, A. O., and Clark, J. I. (1990) "Geotechnics of Waste Fill." *Geotechnics of Waste Fills - Theory and Practice*, ASTM STP 1070, American Society for Testing and Materials, Philadelphia, Pennsylvania, 86-103.

Mesri, G., and Castro, A. (1987), "C_α/C_c concept and K_0 during secondary compression." *J. Geotech. Eng.*, ASCE, 113(3), 230-247.

Morris, D. V., and Woods, C. E. (1990). "Settlement and engineering considerations in landfill and final cover design." *Geotechnics of Waste Fills - Theory and Practice*, ASTM STP 1070, American Society for Testing and Materials,

Philadelphia, Pennsylvania, 9-21.

Oweis, I. S., and Khera, R. P. (1990). *Geotechnology of Waste Management*, Butterworths, London, England.

Vanmarcke, E. H. (1983). *Random Fields: Analysis and Synthesis*, MIT Press, Cambridge, Massachusetts.

Wu, T. H. (1974). "Uncertainty, safety, and decision in soil engineering," *J. Geotech. Eng. Div.*, Proc. ASCE, 100(GT3), 329-348.

SETTLEMENT OF DYNAMICALLY COMPACTED DEPOSITS

Robert G. Lukas,[1] Member, ASCE, Norman H. Seiler[2]

ABSTRACT: This paper compares predicted and measured settlements of deposits that were densified by means of dynamic compaction. Settlement records are presented for sites consisting of old, midage, and new landfill deposits, building rubble, and mine spoil. For sites that are comprised of inert materials, settlement after loading was rapid and directly related to the increase in pressure. Settlement predictions based upon standard penetration resistance or pressuremeter testing after site improvement are in reasonable agreement with measured values. At newer landfill sites, immediate compression as a result of new loading occurs relatively quickly, but long term compression from decomposition is large. Prediction of settlement from decomposition is possible but only approximately correct because of the variable ingredients of landfills and the difficulty of accurately measuring index parameters such as void ratio, unit weight or specific gravity from which estimates of secondary compression indices can be made.

INTRODUCTION

Dynamic compaction has been undertaken for a number of purposes including:

- Reduction of settlement under new loading.
- Reduction in liquefaction potential.
- Densification of collapsible soils.
- Collapse of voids in karst deposits or in formations overlying sinkholes.

---

[1] Consultant, Ground Engineering Consultants, Inc., Deerfield, IL 60015.
[2] Senior Engineering Technician, STS Consultants, Ltd., Northbrook, IL 60062.

This paper is limited to a discussion of the first item listed which is the reduction in settlement of compressible deposits under new loading. Case histories are presented for seven projects where both short and long term settlement observations have been obtained.

Settlement predictions are also presented where measured settlement data was available. Immediate compression predictions are based upon standard penetration resistance (SPT) and pressuremeter tests (PMT) taken before and after dynamic compaction. At the active landfill sites, settlement prediction as a result of decomposition was also undertaken. The purpose of including the settlement predictions is to establish the validity of conventional procedures for estimating settlements under new loading, as well as to compare the predicted settlements with the induced ground compression during the dynamic compaction.

PROJECT DESCRIPTIONS

Sites A and B - Older Landfills

Both of these sites are similar in that they were former clay pits ranging in depth from 9 to 18 m (30 to 60 ft) into which refuse was deposited without a significant amount of control. Daily cover was not applied and the only cover that was placed over the landfill was after they were completed. Each landfill was approximately 25 to 30 years old at the time the sites were developed. At both sites, when soil borings were made, methane gas was not detected. There were no particles of refuse, paper, or other highly perishable goods in the samples and the deposits looked relatively inert.

Site A was developed for a shopping center with one to two story buildings. Columns loads were typically in the range of 1 to 1.33 MN (225 to 300 kips). Dynamic compaction was undertaken at this site with a 6.1 Mg (6 ton) tamper with a 12 m (40 ft) drop height. This is a low level of energy for the relatively great depth of this landfill, but deep densification was not required. The original ground surface at this landfill site ranged from a high of 12 m (70 ft) above street level to a low of 18 m (60 ft) below street level, and land balancing was undertaken before dynamic compaction. In the area where the top of the hill was lowered, this portion of the landfill had been preloaded by the weight of debris removed. The lower portion of the site was densified by compaction of the landfill debris in layers with conventional compaction equipment. Dynamic compaction was used to densify the surface deposits after land balancing and in transition areas that were not preloaded or compacted. A detailed description of this project is presented by Lukas (1985a).

Site B was developed for a high one story commercial building with column loads on the order of 355 kN (80 kips), wall loads approximately 8 kN/m (6 kips/ft), and a slab loading of 19 kPa (400 psf). This site was densified with a 15 Mg (15 ton) tamper with a 23 m (75 ft) drop. A detailed description of this project is presented by Steinberg and Lukas (1984).

At both Sites A and B, the SPT and PMT results in borings undertaken before site improvement were extremely variable. The SPT values ranged from 2 to over 30 and the PMT modulus values ranged from 0.5 to 2.9 MPa (5 to 30 tsf).

Variable support conditions were encountered during dynamic compaction. Crater depths of 2.1 to 3 m (7 to 10 ft) were measured following a single drop at some locations within Site B. Placement of crushed stone and additional energy was required to improve these loose areas.

Site C - New Landfill

A highway embankment was constructed on top of a former landfill with a thickness ranging from 3 to 10.7 m (10 to 35 ft). This project is described in detail by Arkansas State Highway and Transportation Department (1982) and Welsh (1983). Dynamic compaction was undertaken when the landfill had only been closed for approximately 3 years. Decomposition was still active and there was a heavy odor from the gases generated by decomposition at the time of the site improvement. The unit weight of the landfill was determined to be on the order of 0.48 Mg/m^3 (30 pcf).

The compressibility and effectiveness of dynamic compaction was determined from field loading tests rather than conventional borings and in-situ tests. Before dynamic compaction, a static load test was performed over a 9 to 10.7 m (30 to 35 ft) deep portion of the landfill with a 10.7 m (35 ft) high conical fill pile. One day after loading, 26.9 cm (10.6 in) of settlement was recorded. After a period of 7 days, the settlement was 29 cm (11.5 in). This indicates that compression under new loading is rapid. After dynamic compaction, the load test was run again and at this time, the settlement was reduced to 1.5 cm (0.6 in). Since the new roadway embankment was to be only 4.6 m (15 ft) in height, the load test indicated a favorable ground improvement had been accomplished.

Sites D and E - Midage Landfills

Dynamic compaction was undertaken at Site D for a 6.1 m (20 ft) highway embankment. The landfill at this site was 7.6 to 9.1 m (25 to 30 ft) in thickness and site improvement was undertaken approximately 11 to 13 years after closure. Gases were still being generated from the landfill and vented with riser pipes at the time the initial soil boring investigation was made. Details of this project are given by Lukas (1992).

Before dynamic compaction, the SPT values generally ranged from 10 to 25. The PMT modulus values ranged from 1.0 to 4.8 MPa (10 to 50 tsf). After dynamic compaction, the PMT modulus values were on the order of 1.7 to 2.7 MPa (70 to 110 tsf), but the SPT values remained about the same, ranging from 18 to 28.

To further evaluate the compressibility of the landfill, three static load tests were performed before dynamic compaction using a 10.7 m (35 ft) high conical fill pile with a settlement plate installed at the base of the piles. The measured settlement ranged from 0.4 to 0.6 m (1.22 feet to 1.89 ft). Most of this settlement occurred one day after loading, although the readings were taken for 6 to 8 days.

After dynamic compaction, the three load tests were repeated at slight offset locations. Settlements ranged from 0.07 to 0.14 m (0.23 to 0.46 ft).

Site E, involved construction of an embankment to a height of 7.6 m (25 ft) above grade over a former landfill that ranged in depth from 7.6 to 15.2 m (25 to

50 ft). This landfill was closed for 11 years when dynamic compaction was started. Methane gas was still being generated when dynamic compaction was started. SPT test values before dynamic compaction were on the order of 5 to 20. No SPT tests were obtained after improvement.

An inexperienced contractor did the dynamic compaction and did not apply the required energy. In spite of this, induced ground compressions were on the order of 0.76 to 0.9 m (2.5 to 3 ft). The embankment was constructed approximately 6 months prior to construction of the roadway as to allow settlement to occur before final paving. Settlement observation points were established after the roadway was completed.

Site F - Building Rubble

At Site F dynamic compaction was used to densify a 4.6 m (15 ft) deep rubble filled basement for an eight story nursing home building. Wall loads were on the order of 28.5 kN/m (21 kips/ft). Consideration was originally given to deep foundations, but the cost for extending them through the rubble and the old footings which were left in place below basement level were sufficiently high so as to rule out deep foundations for this project. Excavation and replacement would have required bracing to minimize movement of adjacent structures and the streets, so dynamic compaction to permit the use of spread footing and a slab-on-grade was the only reasonable alternative.

Unfortunately, settlement sensitive structures were located immediately adjacent to the site. The conventional procedure of energy application from existing grade could not be undertaken because this would have required a heavy tamper and large drop height, which would induce damaging off site vibrations. For this project, the solution consisted of removal of the upper 2.4 m (8 ft) of rubble, followed by dynamic compaction of the lower 2.1 m (7 ft) with a 4.1 Mg (4 ton) steel tamper and a 9.1 m (30 ft) drop. The upper 2.4 m (8 ft) of rubble was replaced into the excavation in approximately 0.76 m (2.5 ft) lifts and densified with the a 6.1 Mg (6 ton) tamper with a drop height of 6.1 to 7.6 m (20 to 25 ft). This reduced energy minimized off-site ground vibrations and no damage or even complaints were registered.

After dynamic compaction, PMT modulus values in the rubble fill were found to range from 12.4 to 25.3 MPa (130 to 265 tsf). The SPT values ranged from 35 to 75.

Site G - Mine Spoil

A highway embankment, Site G, was constructed over a former mine spoil which was placed somewhere during the interval of 1930 to 1940 when coal mining occurred in this area. The mine spoil can be classified as a mixture of reworked clays, broken shales, occasional pieces of sandstone and inclusions of trash or waste. The typical liquid limit ranged from 30 to 40%, and the typical plastic limit ranged from 20 to 25%. The water content of the clayey portions was on the order of 15 to 20%.

Standard penetration resistance tests taken prior to dynamic compaction ranged from 1 to 20 with a mean of 14. Some voids were detected within the mass of mine spoil. Pressuremeter tests performed prior to dynamic compaction at a test section indicated limit pressures on the order of 0.67 to 1.05 MPa (7 to 11 tsf) and a pressuremeter modulus on the order of 9.6 to 14.4 MPa (100 to 150 tsf).

The mine spoil thickness at the test section was on the order of 12.2 m (40 ft) with the water table located at a depth of 9.8 m (32 ft). The final embankment in this area extended to a height of 9.1 m (30 ft) above present grade. Because of the relatively low standard penetration resistance values in the mine spoil and the presence of intermittent voids, dynamic compaction was undertaken to reduce the potential for significant differential settlement. The work was undertaken with a 179 kN (18 metric ton) tamper with a drop height of 22.9 m (75 ft). Details of the test sections are presented by Snethen and Homan (1986) and Lukas (1985-b).

SETTLEMENT PREDICTIONS AND MEASUREMENTS

Ground settlements obtained at various intervals of time after start of loading at each of the sites is shown in Fig. 1. The settlements have been normalized by dividing by the original thickness of the deposit. Where multiple settlement observation points were established at a site, the average value was plotted on Fig. 1.

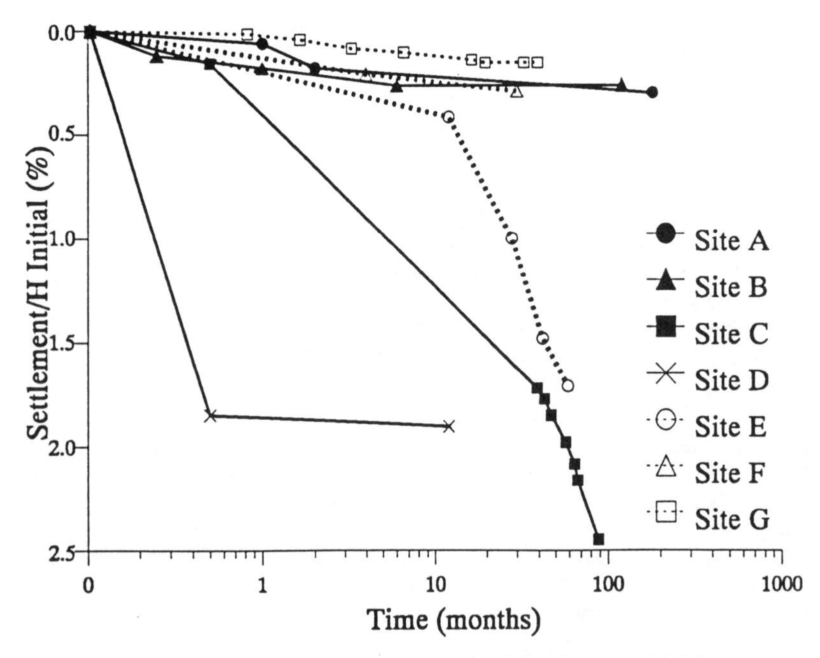

FIG. 1. Variation in Average Normalized Settlement with Time

Two different settlement patterns can be distinguished on Fig. 1. At Sites A, B, F, and G, most of the settlement occurs during the normal construction loading period of two to four months and there is very little additional long term settlement. Settlements at these sites are called immediate settlements but could also be described as a distortion or pseudo-elastic settlement as a result of loading. Dynamic compaction previously consolidated these deposits. Excess pore water pressures generated by the tamping had dissipated well in advance of construction. In addition, the upper portion of many of these deposits were generally partially saturated being elevated above the water table. Thus there should be very little, if any, primary settlement that occurs from dissipation of excess pore water pressures generated from the new loading.

At Sites C and E, there is some immediate or distortion settlement during the first few months of loading but this is minor in comparison with the long term settlement. These sites are active landfills of young to mid age where decomposition is still occurring. Settlement in landfill deposits, Sowers (1973), is due to:

- Mechanical compression due to distortion, bending, crushing, and reorientation of the materials under self weight
- Biological decomposition of organic wastes
- Physico-chemical change such as oxidation, corrosion and combustion
- Ravelling of fines into larger voids

Settlement from the above listed sources usually occurs over a period of decades and has been compared to secondary compression of organic soils, Sowers (1973). In this paper, the settlement that occurs after the immediate settlement is also called secondary compression even though there likely was no primary compression as a result of the new loading. A more appropriate label for this settlement might be creep settlement.

Prediction of settlements based upon tests performed before and after dynamic compaction are summarized in Fig. 2. Wherever SPT data was available, the predictions were based upon a procedure outlined by Meyerhof (1965). Where PMT data was available, the settlement predictions were based upon conventional pressuremeter prediction formulas presented by Baguelin et al. (1978). Both the SPT and PMT predictions estimate settlement due to immediate compression under loading. Estimates of secondary compression were made for the landfill deposits using values of secondary compression indices that were estimated from the initial void ratio of the landfills, Sowers (1973). The predicted settlements are for the time corresponding to the latest settlement observation.

During dynamic compaction, ground compression occurs and the average ground subsidence following dynamic compaction is called the induced settlement. The amount of induced settlement for each of the sites is also shown on Fig. 2. At some locations, the induced settlement was somewhat difficult to determine, since crushed stone fill was placed on ground surface and added as the project was underway. In these cases, the best estimate of induced ground settlement is shown in Fig. 2.

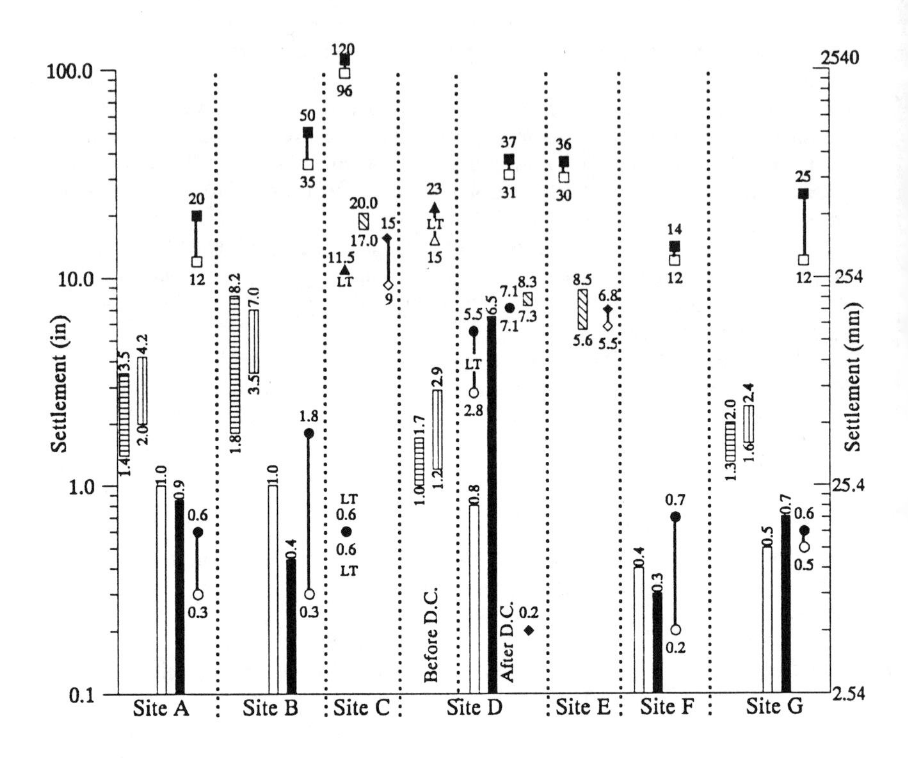

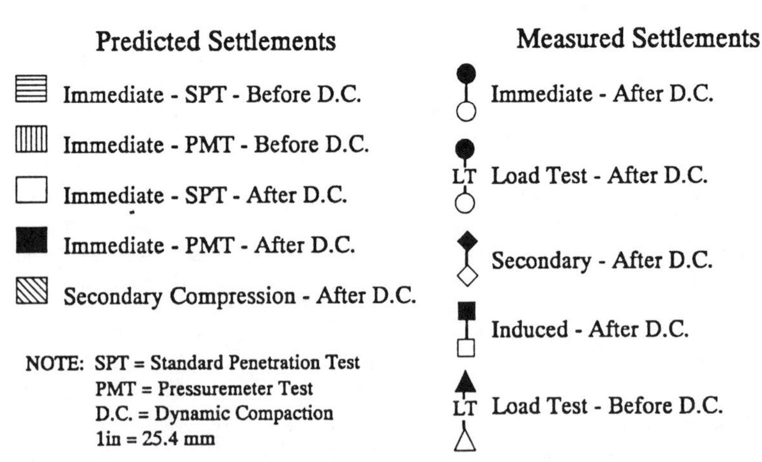

FIG. 2. Settlement Predictions and Measurements

The range in the measured immediate and secondary settlements are also shown in Fig. 2 to allow for a comparison with the predicted and measured settlements. The predicted settlements are shown in the form of bar graphs. Where there is a considerable variation in soil properties resulting in a range of predicted settlements, the bar graphs show maximum and minimum predictions. The measured settlements are shown by symbols with the solid symbol showing the maximum observed settlement and the open circle the minimum.

The data for Site D is separated into two portions. On the left side of the plot, predictions of settlement based on initial SPT and PMT tests are compared with the load test completed before dynamic compaction. On the right side of the plot, predictions of settlement based upon SPT and PMT data after dynamic compaction are compared to the measured embankment settlement and the load test conducted after site improvement.

INTERPRETATION OF SETTLEMENT PREDICTIONS AND MEASUREMENTS

Immediate Settlement

Dynamic compaction was instrumental in reducing the immediate settlement at all seven sites to tolerable values. At Sites A, B, and F, which are building sites, the settlements were reduced to values on the order of 3.8 cm (1.5 in) or less. At the remaining sites which are roadway embankments, immediate settlement was reduced to values ranging from 2.5 to 17.9 cm (1 to 7 in). Wherever SPT and PMT data were available, prediction of immediate settlement based upon tests after dynamic compaction was in reasonable agreement with the measured values. An exception occurred for the SPT prediction of Site D. The low settlement prediction is attributed to high SPT values obtained in miscellaneous fill deposits that contain large size particles of materials.

At Sites A and B which are old landfills, settlement predictions using the SPT and PMT tests before site improvement indicate a wide range in predicted settlements under the proposed loading. At Site A, the predicted range in settlement was from 3.6 to 10.7 cm (1.4 to 4.2 in) and at Site B, the range was from 4.6 to 20.8 cm (1.8 to 8.2 in). No load tests were performed at these two sites to confirm these predictions, but the predictions do show the wide variation in predicted settlement that could occur without site improvement. During dynamic compaction at these two sites, extremely loose pockets of fill were encountered in localized areas requiring extra energy for densification, as well as the addition of crushed stone into the depressed area. Thus, the actual settlements in these localized areas could have been significantly more than predicted by the SPT and PMT tests.

At Site D, both SPT and PMT data obtained prior to dynamic compaction was used to estimate immediate compression to compare with the field load tests performed prior to dynamic compaction. These estimates of immediate compression were only on the order of 10% of the measured values. In the case of the pressuremeter, this poor prediction is attributed to the inappropriate use of the pressuremeter test for deposits that are still decomposing and consolidating under their own weight. In order for the PMT settlement prediction to be valid, the deposits should be pre-consolidated, since the pressuremeter modulus is only valid up to the creep pressure which is considered to be the same as the pre-consolidation pressure (Lukas, 1976). In the case of the SPT prediction, the values obtained before site improvement can be influenced by large size particles within the landfill. In addition, it has been found that in loose consolidating deposits, there are zones of extremely loose pockets that may not be identified by the boring program. Thus, prediction of immediate compression before site improvement based on conventional tests for under consolidated deposits such as Sites C, D, and E, does not appear appropriate.

Induced Settlement

At all of the sites, the induced settlements were significantly more than settlements predicted by either the SPT and PMT tests, or the load tests performed prior to site improvement. As shown in Table 1, the ratio of induced settlement to predicted settlement after dynamic compaction ranged from 5.8 to 15.5. This large amount of induced settlement indicates that site improvement was definitely necessary at these project sites, and also indicates that a significant amount of pre-consolidation of the deposit has taken place during dynamic compaction.

The ratio of induced settlement to thickness of the densified deposit is also shown in Table 1. This ratio was typically on the order of 8 to 10 percent. At the young landfill, Site C, the ratio is 25 percent which is indicative of the loose condition of these deposits. At the minespoil, Site G, the ratio was only 4 percent confirming the initial SPT and PMT test results that this deposit was in a medium dense condition prior to site improvement.

TABLE 1. Settlement Ratios

Site	Average Induced Settlement Divided by:		
	Measured Immediate after D.C.	Predicted Immediate before D.C.	Densified Depth
A	36	5.8	0.10
B	40	8.5	0.10
C	180	9.4	0.25
D	4.8	15.5	0.10
E	–	–	0.11
F	29	–	0.08
G	34	10.1	0.04

Note: D.C. = Dynamic Compaction

Secondary Compression in Landfills

At Sites C and E, the settlement readings indicate that secondary compression is occurring. Based upon the rate of settlement with time, the calculated modified secondary compression index, ranges from 0.016 to 0.025 for these two sites, respectively, which is approximately the same as one would predict for landfills without site improvement (Sowers 1973). It is therefore concluded that dynamic compaction did not have a significant effect upon settlement due to secondary compression. Dynamic compaction did result in a reduction of immediate settlement as evidenced by the large induced settlement and the low settlement observed during the load test, but secondary settlement from decomposition and chemical changes are still occurring.

Site D is similar in age to Site E and some secondary compression was anticipated at this site. However, the long term settlement readings indicate very little secondary compression. It is possible that the portion of the landfill where the settlement plates were installed was more completely decomposed than originally thought, so secondary compression would not be a factor. There is also the possibility that the settlement readings were discontinued too soon after embankment construction and that if additional readings had been taken, more movement may have been measured.

Sites A and B are also landfill deposits, but there is no apparent secondary compression occurring at these locations. The initial investigation at these sites confirmed that the landfills were in an inactive stage. Determination of whether a landfill is active or inactive is difficult, but the following guidelines can be used.

The active stage of a landfill is defined as the period of time where decomposition and compression of the landfill under its own self weight is still taking place. During this time, methane gas is generally produced, leachate is generated, and the temperature within the landfill can rise to 22 to 33°C (40 to 60°F) above ambient. For these sites, secondary compression will be large.

The landfill becomes inactive when the biological and chemical decomposition processes are essentially completed. The remaining materials in the landfill can be considered to be relatively inert, although some very slow decaying substances such as wood or leather may still be present. Lukas (1985a) performed gradation tests upon samples from an old landfill and found the particle sizes to consist of approximately 40% in the sand size range, 20% each of silt and gravel size, and 15% in the clay size. There were also large chunks of concrete, occasional timbers, numerous bottles, rubber tires and pieces of metal. The time required for a landfill to become inactive is a function of the climate and in particular the amount of water which is necessary for decomposition. In the Chicago area where landfills have been placed in former clay pits and the water table is only 1 to 3 m (3 to 10 ft) below grade, the biological decomposition process is essentially complete after a minimum of 20 to 25 years following completion of the landfill. Yen and Scanlon (1975) report on settlement rates of landfills. Based upon extrapolation of 9 year settlement readings, they conclude that secondary compression of landfills will last about 21 years. However, in some parts of the United States where rainfall is very low, much longer periods of time would be required for decomposition.

CONCLUSIONS

Based upon the long term settlement readings obtained at seven sites, it is concluded that:

1. Dynamic compaction significantly reduces the settlement of inactive soil deposits. The amount of ground compression induced by dynamic compaction is generally much greater than the settlement predicted for the new loading indicating that the deposits are generally preconsolidated as a result of dynamic compaction.
2. Settlement predictions of dynamically compacted inactive deposits can be made based upon soil parameters obtained after dynamic compaction. Long term settlement readings agree reasonably well with these predictions. The measured settlement divided by the thickness of the densified deposit was found to be on the order of 0.25% for both embankment loads and building column loads. This settlement occurs rapidly and additional long term settlement is minimal.
3. Settlement predictions by SPT or PMT tests on deposits that are still consolidating under their own weight are misleading. The predictions are much less than settlement measured by load tests.
4. Dynamic compaction of active landfills results in large induced compressions in these deposits, thereby reducing immediate compression. However, secondary compression can still be quite large. It is possible to estimate the secondary compression using conventional procedures. Unfortunately, the amount of published data regarding properties of landfills such as void ratio and secondary compression rate of landfills is meager.

Appendix - References

Arkansas State Highway and Transportation Department. (1982). "Dynamic compaction of a sanitary landfill," *Report No. DTFH61-81-C-00117*, Federal Highway Administration, Washington, D. C.

Baguelin, F., Jezequel, J. F., and Shield, D. H. (1978). *The Pressuremeter and Foundation Engineering*, Trans Tech Publications, Clausthal, Germany.

Lukas, R. G., deBussy, B. L. (1976). "Pressuremeter and laboratory test correlations for clays." *J. Geotech. Eng. Div.*, Proc. ASCE, 102(GT9), 945-962.

Lukas, R. G. (1985a). "Densification of a decomposed landfill deposit." *Proc. 11th ICSMFE*, San Francisco, 3, 1725-1728.

Lukas, R. G. (1985b). "Working draft of the report on the dynamic compaction measurements undertaken in Tulsa, Oklahoma," Federal Highway Administration Contract FHWA/RD-86/133.

Lukas, R. G. (1992). "Dynamic compaction for transportation projects." *Geotechnical Lecture Series*, Illinois Section ASCE, Chicago, Illinois.

Meyerhof, G. G. (1965). "Shallow foundations." *J. Soil Mech. Found. Div.*, Proc. ASCE, 91(SM2), 21-31.

Snethen, D. R., and Homan, M. H. (1986) "Evaluation of dynamic compaction in Tulsa, Oklahoma," *Report FHWA/OK 86-(3)*, Federal Highway Administration, Washington, D. C.

Sowers, G. F. (1973). "Settlement of waste disposal fills." *Proc. 8th ICSMFE*, Moscow, 1, 207-210.

Steinberg, S. B., and Lukas, R. G. (1984). "Densifying a landfill for commercial development." *Proc. 2nd Int'l. Conf. on Case Histories in Geotechnical Engineering*, Rolla, 3, 1195-1200.

Welsh, J. P. (1983). "Dynamic deep compaction of sanitary landfill to support superhighway." *Proc. 8th European Conf. on Soil Mech. and Found. Eng.*, Helsinki, 319-321.

Yen, B. C., and Scanlon, B. (1975). "Sanitary landfill settlement rates," *J. Geotech. Eng. Div.*, Proc. ASCE, 101(GT5), 475-487.

Experimental Study of the Settlement of Shallow Foundations

Samuel Amar,[1] François Baguelin,[2] Yves Canépa,[3] and Roger Frank[4]

Abstract: This article describes the main results obtained experimentally by the French Bridge and Road Research Laboratories on the settlement of shallow foundations under centered vertical loads. Some 80 loading tests performed at 6 sites, representing 4 types of soil, were analyzed. The authors propose a method for estimating the long-duration settlements of shallow foundations on silts and sands.

Introduction

Between 1980 and 1990, the French Bridge and Road Research Laboratories performed about a hundred loading tests of shallow foundations on natural soils. These tests were performed at 6 different sites representing 4 types of soil.

The results of these tests have been interpreted previously primarily in terms of bearing pressure at failure (Amar et al. 1984; Amar et al. 1987).

The object of this article is to describe in more detail and to summarize the results obtained concerning the settlements observed under centered vertical loads. Altogether, 80 short-duration and 6 long-duration tests under centered vertical load were performed. A review of the main characteristics of the soils tested and a description of the experimental conditions are followed by a presentation of the results typically observed. The time-settlement curves obtained under centered vertical load are analyzed.

---

[1] Research Engineer, LCPC, 58 Boulevard Lefèbvre, 75015 Paris, France.

[2] Scientific Director, TERRASOL, Immeuble Hélios, 72 Avenue Pasteur, 93100 Montreuil, France.

[3] Research Engineer, LREP, 319 Avenue Georges Clémenceau, 77530 Vaulx-le-Pénil, France.

[4] Director, CERMES (ENPC/LCPC), La Courtine, 93167 Noisy-le-Grand, France.

An engineering method using the classical formulae of linear elasticity is proposed; in it, the Young's modulus selected is adjusted according to the pressuremeter modulus E_M, for a loading level applied to the foundation q equal to half of the loading at failure q_r ($q = q_r/2$)

The practical results derived from this study are given in conclusion.

SITES AND EXPERIMENTAL CONDITIONS

There was a detailed geotechnical survey of each of the sites tested. Tables 1 and 2 summarize the main characteristics measured at each of the sites.

TABLE 1. Identification of Soils

Site	Soil type	w (%)	W_L (%)	I_p (%)	γ_d (kN/m^3)
PLANCOET (PLA)	Soft clay	37	35	16	13.0
JOSSIGNY (JOS)	Silt	22	38	14	16.0
LOGNES (LOG)	Stiff clay	25	85	52	16.0
LABENNE (LAB)	Dune sand	5	-	-	16.0
PROVINS (PRO)	Stiff clay	19	94	60	16.0
CHATENAY (CHA)	Chalk	27	-	-	15.5

TABLE 2. Mechanical Characteristics of Soils

SITE	TRIAXIAL TESTS UU and CD			PMT (1)		CPT	DPA (2)	SPT
	c_u kPa	c' kPa	ϕ' °	P_l kPa	E_M kPa	q_c kPa	q_d kPa	N
PLA	46	0	34	280	2040	500	550	6
JOS	38	12	32	510	6500	2100	1900	7
LOG	80	0	22	740	12000	2500	3500	16
LAB	-	0	32	910	8400	5600	3600	15
PRO	145	-	-	1290	24500	-	5700	-
CHA	-	-	-	1350	19500	-	3500	-

(1) PMT: Ménard preborehole pressuremeter tests.
(2) DPA: dynamic penetration tests.

All of the loading tests were performed on square or rectangular foundations having dimensions of the order of: 0.7 m < B < 1.6 m and 1 < L/B < 3. Embedment varied between 0 and 1.7 m depth. The foundations were made of steel (6 cm thick) or reinforced concrete (20 cm thick). They can all be considered as rigid plates with regard to the soils involved.

Two types of tests were performed: short-duration tests (SD) conducted to failure of the soil; and "creep" tests, in other words long-duration tests under constant load (LD).

To each of these two types of test there corresponds a particular experimental arrangement. These two experimental arrangements have been described by Amar et al. (1987)

In the first case (SD), the foundation is loaded gradually by a classical hydraulic jack, in steps up to failure of the soil. Failure is defined for settlements of the foundation equal to 10 % of its width B (S = 0.1 B). Each loading step was kept constant for a duration of 30 min.

In the case of the long-duration tests (LD), the load is applied to the foundation by means of a transmission rod linked to a lever. One of the ends of the lever is loaded with a dead weight (like in a oedometer test arrangement), and the other end is anchored into the soil. The long duration loads were monitored by means of an electric load cell ($\Delta Q \pm 2.5$ %).

Fig. 1 gives, by type of soil, the number of short-duration tests under centered vertical load.

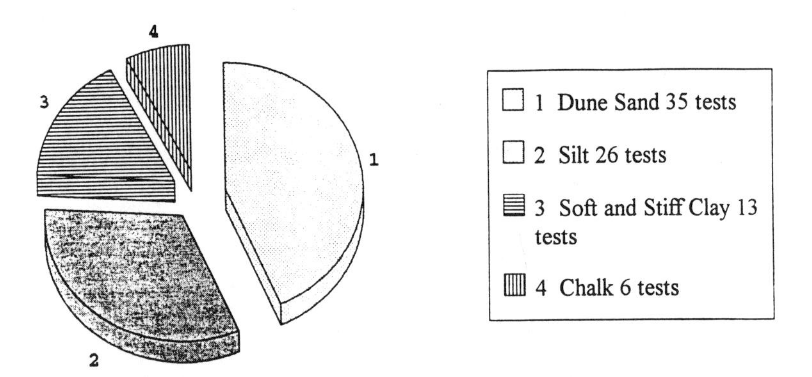

FIG. 1. Breakdown of Short-Duration Tests Under Centered Vertical Load

EXPERIMENTAL RESULTS

Behaviour during Short-Duration Tests (SD)

Fig. 2a shows a typical loading curve, obtained by increasing the load in steps held for 30 min; the settlements are given at the end of the steps (S_o). Fig. 2b gives, for the same test, curves showing the settlement of the foundation versus time for each loading step.

A look at the loading curve of Fig. 2a shows that its first part is roughly linear; subsequently, the settlements increase for a given load increment.

This curve does not, however, exhibit a clear break that could be used to determine a failure load unambiguously. However, the failure load q_r will be defined conventionally as the load corresponding to a settlement of 0.1 B.

Examination of Fig. 2b shows that the settlements of the foundation, under a constant load, increase linearly with the logarithm of time. A straight line of "stabilization" is thus obtained for each loading step.

The slope of these straight lines is called A_o and the reference settlement (that obtained at the end of the step) is called S_o (30 min):

$$S = A_o \times \log t + A_1 \quad (1)$$

$$S_o = A_o \times \log 30 + A_1 \quad (2)$$

Table 3 sums up, for all sites, the characteristic ratios A_o/S_o measured. Note that this ratio is very near to be constant for Jossigny, Lognes and Labenne, whatever the loading level, and for a large number of tests.

As for Provins clay and Chatenay chalk, this ratio increases with q/q_r from 5 (low values of q/q_r) to 20 ($q/q_r \approx 1$).

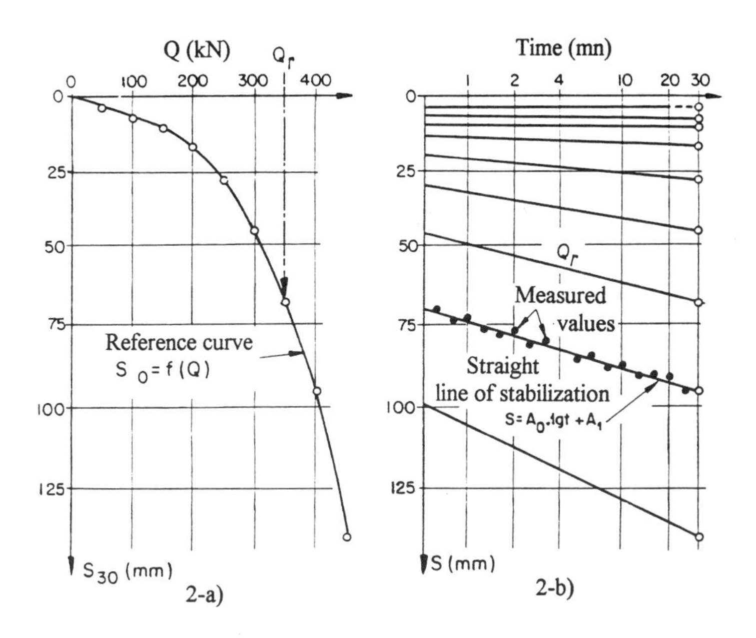

Labenne Dune Sand Site: B = 0.7 m and L/B = 1

FIG. 2. Typical Loading Curve of a SD Test and Corresponding Stabilization Lines

TABLE 3. Values of Ratio A_0/S_0

Site	Soil type	A_0/S_0
JOSSIGNY	Silt	15
LOGNES	Stiff Clay	15
LABENNE	Sand	5
PROVINS	Stiff Clay	5 to 20
CHATENAY	Chalk	5 to 20

Results of Long-Duration Tests under Constant Load (LD tests)

Fig. 3 shows the evolution of settlements S versus the logarithm of time in the 6 long-duration tests. It is found that, for the clay and silt sites (Fig. 3a), for the time domains tested (t ≈ 500 to 800 days), the settlements are at first linear with the logarithm of time; and then follows an "acceleration" of the settlement with respect to this function.

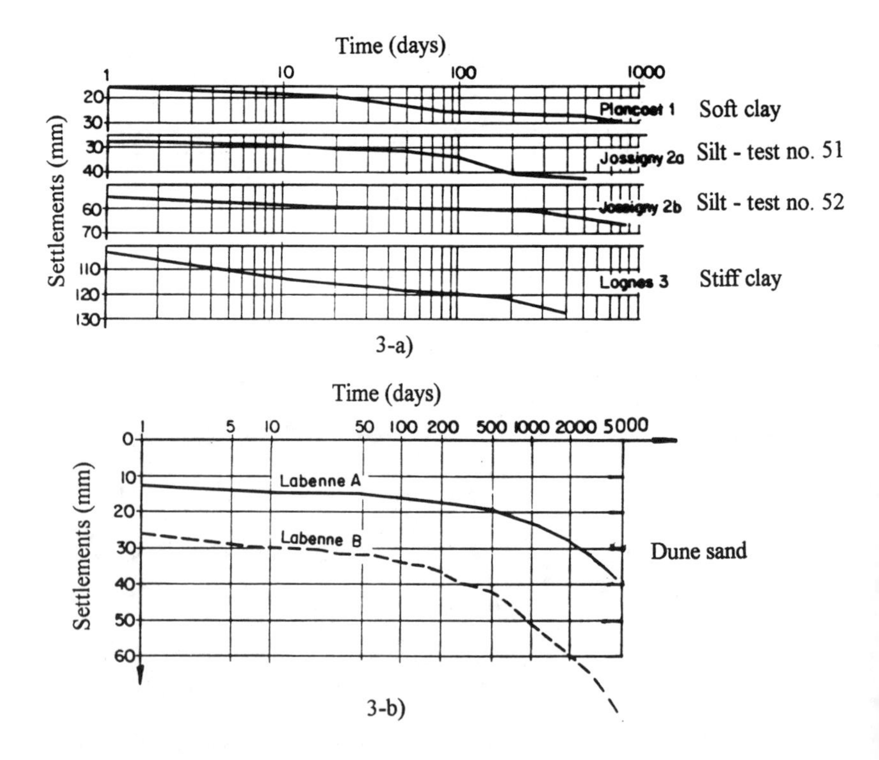

FIG. 3. Settlement versus the Logarithm of Time from LD Tests

In the tests performed on sand, at Labenne (Fig. 3b), a clearly marked break appears between 1 and 2 years after the application of the load. This phenomenon could be ascribed to various factors (variation of water content over the years, logarithmic fitting law poorly chosen, etc.), and at present it is not possible to decide among them. It is noted, however, that the measurements are not disturbed by vibration, traffic, etc.

ANALYSIS OF RESULTS

Figs. 4 and 5, respectively, group the results obtained at the sites of Jossigny (silt) and Labenne (sand). These two figures show various loading curves, with the loading level ($x = q/q_r$, in %, q_r being the pressure at failure) on the x-axis and the settlement as a percentage of the settlement at failure S_r defined as $S_r = 0.10$ B ($y = S/S_r$, in %) on the y-axis. The curves plotted are, on the one hand, the loading curves of the short-duration tests (30 min steps) in the form of a range and, on the other, the curves extrapolated to 10 years from the straight line of stabilization of the SD tests at 30 min.

These same figures also show the results obtained in the 2 long-duration tests (LD) performed at each of the sites. In the case of Jossigny they were extrapolated from data at 558 days (test no. 51, see Fig. 3a) and at 1911 days (test no. 52). In the case of Labenne they are directly the measurements at 9 years approximately.

A look at the Jossigny tests (Fig. 4) shows good agreement between the results extrapolated from the short-duration tests (SD) and the values measured in the long duration tests (LD). Analysis of these results shows that, for the range of loads tested in the long-duration tests (LD) and for loading levels q/q_r between 50 and 75%, the ratios between the settlements at 10 years and those at 30 min are of the order of 2:

$$S_{10\,\mathrm{years}} = 2 \times S_{30'} \tag{3}$$

On the other hand, for the tests performed at Labenne on sand, Fig. 5 shows that the values measured in the long-duration tests (LD) are nearly twice as high as those obtained by extrapolation from the short-duration tests (SD).

The ratios between the "final" measured settlement (at nearly 10 years, Fig. 3b and Table 4) and the "initial" ones (at 30 min) are, for both foundations A and B (respectively $q/q_r = 45\%$ and 60%):

$$S_{10\,\mathrm{years}} = 3.3 \times S_{30'} \tag{4}$$

TABLE 4. Settlements versus Time, Measured at Labenne

Time	1 min	30 min	360 days	1090 days	3565 days
Labenne A	10.8	11.8	18.9	23.4	39.5
Labenne B	19.8	23.3	39.7	51.3	77.3

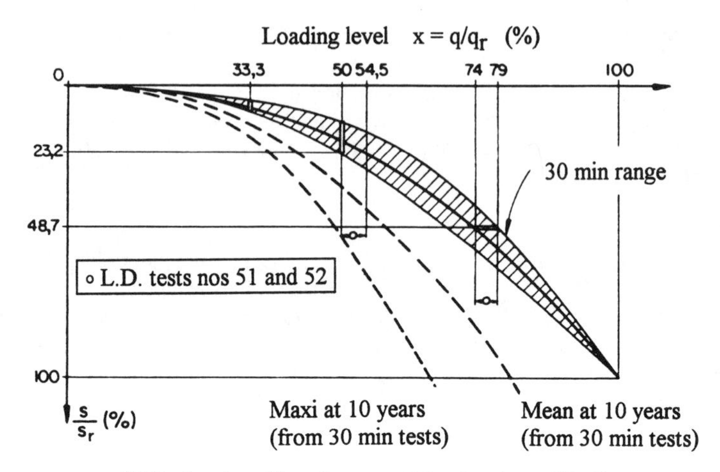

FIG. 4. Loading Curves of the Jossigny Silt Site

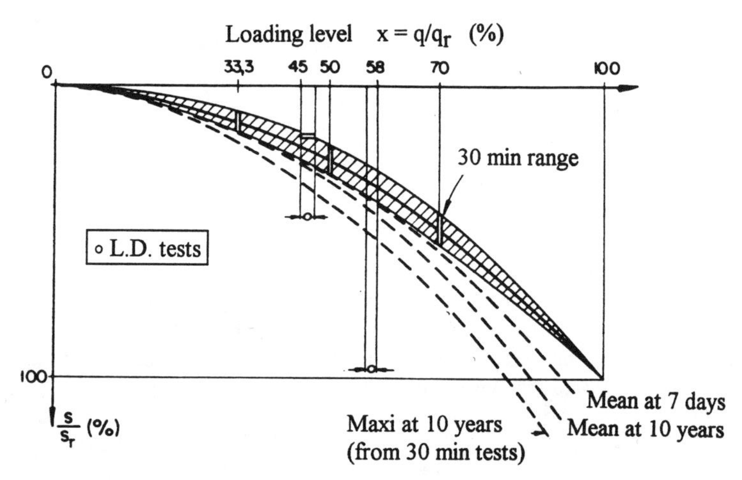

FIG. 5. Loading Curves of the Labenne Dune Sand Site

To make it possible to compare these results with those reported by other authors (Burland and Burbidge 1985), the evolution of the settlements is represented, between times t_0 and t_1, by a function of the type

$$S_1 = S_0 \left[1 + m \log \frac{t_1}{t_0} \right] \tag{5}$$

The following values of m are obtained (Frank 1994):

- Foundation A m = 0.50 between 1 and 3 years
  m = 1.32 between 3 and 10 years

- Foundation B m = 0.61 between 1 and 3 years
  m = 0.97 between 3 and 10 years

These values are notably larger than the value proposed by Burland and Burbidge (1985), i.e. m = 0.15 under static loading for t_0 = 3 years.

This difference can be explained by the fact that the loading levels are quite different, if one looks at the settlement ratios (S/B) obtained in both studies.

PROPOSED PREDICTION METHOD

The results measured in the short-duration tests, with steps of 30 min, together with the long-duration measurements, have been used to propose a prediction method. This method uses the classical isotropic linear elastic approach in which the modulus of elasticity is replaced by the pressuremeter modulus E_M (weighted versus depth according to Boussinesq's stress attenuation laws, for a flexible foundation and Poisson's ratio ν = 0.33). The pressuremeter modulus E_M is to be determined according to French standard (Afnor 1991). For the present soils and sites preboreholes were always used.

The measured values can be approximated by multiplying the moduli E_M by a coefficient α.

Table 5 gives the values of α obtained on the Jossigny silt and the Labenne sand for a loading level q/q_r equal to 0.5.

TABLE 5. Values of α in $E = \alpha E_M$

Type of soil	Value of α for 30 min	Value of α for 10 years
Silt	3	1.5
Sand	3.5 to 3	1.3 to 1

In practice, the same coefficients a can be used for silts and sands, namely α equal 3 for short-duration settlements and 1.5 for the long-duration settlements. This last value, proposed for the long-duration settlements for sands, amounts to neglecting a part of the immediate settlements. It should in fact be noted that the fitting was done on settlement measurements that include not only delayed settlements (creep) over several years but also immediate settlements (at 0 minute) and short-duration settlements (between 0 and 30 minutes). But, in most cases, it is the delayed settlements, after the end of construction, that it is important to determine correctly for the life of most civil engineering structures.

Furthermore, it is likely that the delayed settlement is relatively large in the case of the foundations laid on the Labenne sand.

CONCLUSIONS

Analysis of the experimental data presented here yields information about the settlement of shallow foundations under centered vertical load:

(1) whatever the type of soil (sand, silt, or clay), delayed settlements can not be neglected, and must be taken into account when designing a shallow foundation; and

(2) for foundations of which diameter B is less than 2 m, it seems possible to obtain a good estimate of the long-duration settlements on silts and sands using the classical elastic formulae and taking 1.5 times the pressuremeter modulus E_M as the value of the Young's modulus ($E = 1.5\ E_M$).

It must be noted that the proposed method uses the usual pressuremeter modulus, in other words the one determined on the curve of first loading between the "initial" and "creep" pressures of the standardized pressuremeter curve. This modulus is very sensitive to remoulding, which depends not only on how the pressuremeter probe is inserted into the ground, but also on the type of soil.

The measurement of a modulus on an unloading-reloading cycle included in the test would doubtless be preferable (Frank 1991). In this case the correlations with the Young's modulus proposed here for estimating the long-duration settlement of shallow foundations would obviously have to be reconsidered.

Finally, it would be desirable to conduct a similar analysis on large instrumented structures, such as nuclear power plants, before generalizing the method proposed or refining it by type of soil.

APPENDIX - REFERENCES

Afnor (1991). "Essai pressiométrique Ménard." *Norme Française NF P 94110*, Paris La Défense, 1-32 (in French).

Amar, S., Baguelin, F., and Canépa, Y. (1984). "Etudes expérimentales du comportement des fondations superficielles." Serie Sols et Fondations 189, *Annales de l'I.T.B.T.P.*, Paris, No. 427, 82-109 (in French).

Amar, S., Baguelin F., and Canépa, Y. (1987). "Comportement des fondations superficielles sous differents cas de chargement." *Actes du Colloque Interactions Sols Structures*, Presses de l'ENPC, Paris, 15-22 (in French).

Burland, J. B., and Burbidge, M. C. (1985). "Settlement of foundations on sand and gravel." *Proc. Instn. Civ. Engrs*, 78, Part 1, 1325-1381.

Frank, R. (1991). "Some recent developments on the behaviour of shallow foundations." P*roc. 10th European Conference on Soil Mechanics and Foundation Engineering*, Florence, 3, 1003-1030 (in French; English translation to appear in Vol. 4).

Frank, R. (1994). "Reflexions sur le tassement des fondations superficielles." Panelist Contribution, *Proc. 13th ICSMFE*, New Delhi, 5, 83-84 (in French).

TOLERABLE DEFORMATIONS

Harvey E. Wahls,[1] Fellow, ASCE

ABSTRACT: Current criteria for tolerable movements of buildings, bridges and other structures are reviewed. Tolerable values of total and differential settlement, relative rotation, relative deflection and tilt are reported and discussed. The tolerable movement criteria for buildings are traced from their origins in the 1950s to the current development of Eurocode 7, which presents these criteria in the context of limit state design. Tolerable movement criteria for bridges are based on a 1985 FHWA study by Moulton. The criteria for both buildings and bridges are shown to be based primarily on studies of the settlement records of damaged and undamaged structures. The relations of tolerable movements to levels of damage and other factors are discussed.

INTRODUCTION

The tolerable movements of buildings and other structures have been a continuing concern to geotechnical and structural engineers, as well as to architects. For example, Peck (1948) described the attempts of engineers to deal with the effects of settlement on buildings in Chicago in the late 19th century. These engineers expected buildings to settle and developed designs that attempted to minimize the differential settlement. Their experience demonstrated that significant settlement can be accommodated without impairing the safety and function of a structure.

Between 1955 and 1975, there were several significant investigations of the tolerable movements of buildings, e.g., Skempton and MacDonald (1956), Polshin and Tokar (1957), Burland and Wroth (1974) and Grant et al. (1974). Periodically,

---

[1] Professor, Civil Engineering Dept., Box 7908, North Carolina State University, Raleigh, NC 27695.

the results of these studies have been reviewed critically (Feld 1964; Golder 1971; Burland, Broms and DeMello 1977; Wahls 1981). These reviews generally note the complexities of determining the tolerable deformations of a structure, which are affected by many factors, including the type and size of the structure, the properties of the structural materials and the subsurface soils, and the rate and uniformity of settlement. In general, both vertical movements, including settlement and heave, and horizontal movements are of concern, and in some instances, vibrations also may be important. Furthermore, it must be determined whether the tolerable movements should be based on consideration of the safety, function and serviceability, or appearance of the structure. Clearly, movements that affect safety are unacceptable for every structure, and also the function of the structure must be maintained. For many types of buildings, significant cracking of architectural elements or finishes, which may affect appearance but not safety or function, may be the governing concern. However, for some buildings, minor cracking or impairment of function, which can be economically repaired, may be tolerable. Thus, it is important to distinguish between structures for which aesthetic appearance is important and those for which it is not.

This paper reviews the current status of guidelines for tolerable deformations of structures. The primary emphasis is on vertical movements of buildings and both vertical and horizontal movements of bridges. Also, there is a brief discussion of the tolerable movements of other structures, such as steel tanks, chimneys, and silos.

The tolerable movements of embankments, retaining structures and pavement systems have been excluded from the scope. Also, excavation-induced movements of buildings are omitted because the effects of excavations have been discussed in excellent recent papers by Clough and O'Rourke (1990) and Boscardin and Cording (1989).

DEFINITIONS

Settlement is the downward vertical displacement of the foundation, and heave is upward vertical displacement. Conventionally, settlement and heave are considered to be positive and negative displacements, respectively. Because damage commonly is related to variations in the displacement with location in the building, the differential settlement (or heave) between two locations usually is more important than total displacement at one location. Common terminology for describing total and differential settlement (or heave) is defined for cases without and with a component of rigid body rotation or tilting in Fig. 1a and 1b, respectively, where ρ_i = vertical displacement at "i"; δ_{ij} = differential displacement between "i" and "j"; Δ = relative deflection = maximum displacement from a straight line connecting two reference points that are a distance, L, apart; ω = tilt = rigid body rotation; $\beta_{ij} = \delta_{ij}/l_{ij} - \omega$ = relative rotation (angular distortion); l_{ij} = the distance between points "i" and "j"; and Δ/L = deflection ratio, which is an approximate measure of the curvature of the settlement curve.

The definition of tilt, ω, as shown in Fig. 1b, is appropriate for mat foundations but will be more complex for frames supported on individual footings. Also, Burland et al. (1977) suggest that the term "relative rotation" is preferable to "angular distortion" because the latter should be limited to structures, such as framed buildings, where shear distortion clearly is present.

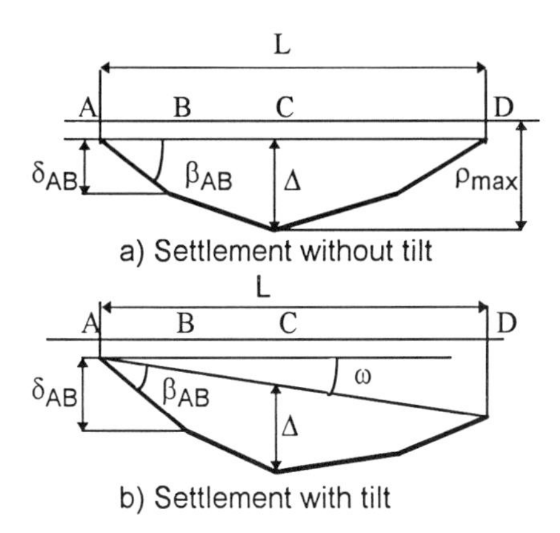

FIG. 1. Definitions of Settlement Terminology

Limit States

The current procedures for the design of structures are based on the consideration of acceptable performance for two limit states. The ultimate strength limit state is based on full mobilization of the ultimate strength and considers the safety against instability and total collapse of the structure. The ultimate strength limit state can be produced by large deformations that overstress one or more structural elements, and thus tolerable deformations can be established for the ultimate limit state.

The serviceability limit state considers deformations that affect the function and/or appearance of the structure and, in some instances, the comfort of the people who occupy the structure. Maximum tolerable deformations for serviceability limits may depend on the deformations, displacements and damage, including cracking, that affect the use or appearance of the structure, or cause damage to finishes or non-structural elements. They are more restrictive than limiting deformations based on ultimate strength criteria and thus usually are the more critical deformation conditions. The establishment of tolerable deformations for the serviceability limit states also is more subjective than for ultimate strength limit state. While deformations that endanger the stability of the structure are universally unacceptable, the tolerable level of damage to appearance and function varies with the type and function of a structure and, for a given structure, may vary among individuals.

Relation of Tolerable Movement to Damage

The establishment of tolerable deformations for serviceability limits requires consideration of tolerable levels of damage, which must be related to the consequences of the damage and the ease and economics of repair. This concept was adopted by a Transportation Research Board (TRB) committee for the evaluation of movements of highway bridges. Each highway agency was asked to judge qualitatively the tolerability of the movements of its bridges in accord with the following definition:

> "Movement is *not* tolerable if damage requires costly maintenance and/or repairs *and* a more expensive construction to avoid this would have been preferable."

Using this definition, relatively large movements often were judged to be tolerable. Applications of this concept to investigations of bridges are presented in a subsequent section of this paper.

If this concept is used to define tolerable movements for the serviceability limit of buildings, the costs of maintenance and repair should be weighed against the increased initial construction cost required to reduce the level of damage. For example, if slight cracking is caused by the settlement of a spread footing, the cracking and settlement might be regarded as intolerable if it can be reduced to a tolerable level at little cost by slightly enlarging the footing. However, if a very costly deep foundation system is required, the original settlement and the slight cracking might be regarded as tolerable.

This approach requires a classification system for the level of damage in buildings. Table 1 is a modified version of a classification of damage to masonry and plaster walls, which was presented by Burland et al. (1977). Tolerable deformations for such walls could be defined on the basis of "very slight"to "moderate" damage, depending on the nature of the structure. For example, moderate damage might be tolerable in a storage warehouse, but intolerable in an office, commercial or residential building.

TOLERABLE DEFORMATIONS OF BUILDINGS

Relative Rotation (Angular Distortion)

Skempton and MacDonald (1956) reported observed settlement and damage of 98 buildings, including both steel and reinforced concrete frame structures and a few structures with load-bearing walls. Damages to panel walls, interior partitions, floors or primary structural members were reported for 40 structures. Because most of the damage appeared to be related to distortional deformations, "angular distortion", β, (hereafter referred to as "relative rotation") was selected as the critical index of settlement. They concluded that cracking of load-bearing walls or panel walls in frame structures is likely when β exceeds 1/300 and that structural damage is probable when β exceeds 1/150. Finally, Skempton and MacDonald recommended $\beta = 1/500$ as a design criterion that provides some factor of safety against any cracking. These criteria subsequently were incorporated into recommended limiting values of relative rotation proposed by Bjerrum (1963). Grant et al. (1974) reviewed settlement and damage data for an additional 95 buildings, of which 56 reportedly had suffered some damage. This study supported the Skempton-MacDonald conclusion that cracking should be anticipated when β exceeds 1/300. However, Burland et al. (1977) discussed limitations of the data bases used

TABLE 1. Classification of Visible Damage to Plaster and Masonry Walls[a]

Degree of Damage	Description of typical damage[b]	Approx. Crack width[c]	Ease of repair
	Negligible	Hairline (0.1 mm)	
1. Very Slight	Cracks in external brickwork visible on close inspection. Possible isolated slight fracture in building.	> 1 mm	Fine cracks can be easily treated during normal decoration.
2. Slight	Cracks are visible externally. Several slight fractures show within building. Doors and windows may stick slightly.	> 5 mm	Cracks easily filled. Re-decoration probably required.
3. Moderate	Doors and windows sticking. Weathertightness often impaired. Service pipes may fracture.	5-15 mm or a number of cracks > 3 mm	Cracks require some opening up and can be patched by a mason. Recurrent cracks can be masked by suitable lining. Repointing of external brickwork likely and possibly a small amount of brickwork to be replaced.
4. Severe	Windows and door frames distorted, floors sloping noticeably. Walls leaning or bulging noticeably, some loss of bearing in beams. Service pipes disrupted.	15-25 mm, also depends on number of cracks	Requires extensive repair work involving breaking-out and replacing sections of walls, especially over doors and windows.
5. Very Severe	Beams lose bearing, walls lean badly and require shoring. Windows broken with distortion. Danger of instability.	usually > 25 mm, also depends on number of cracks	Requires major repair job involving partial or complete rebuilding.

[a] Modified from Burland et al. (1977). [b] Location of damage in the structure affects the assessment of the degree of damage. [c] Level of damage should not be assessed only on the basis of crack width, which is only one attribute of damage.

by Skempton and MacDonald (1956) and Grant et al. (1974) and suggested that their conclusions were more applicable to traditional framed structures than to load-bearing walls.

Both Skempton and MacDonald (1956) and Grant et al. (1974) removed the differential settlement due to tilting from the computed values of relative rotation. Leonards (1975) noted that tilting contributes to the stress and strain in the frame unless each individual footing tilts or rotates through the same angle as the overall structure. Because this is unlikely to occur, the effects of tilt should be included in the differential settlement criteria for framed structures on isolated footings.

Polshin and Tokar (1957) presented tolerable settlement criteria based on 25 years of Soviet experience. Framed structures and load-bearing walls were treated separately. For frames, the tolerable settlement was expressed in terms of the relative rotation between adjacent columns without correction for tilt. The limiting values ranged from 1/500 for infilled steel and concrete frames to 1/200 for frames where there is no infill and were similar to the criteria of Skempton and MacDonald (1956).

For load-bearing walls, Polshin and Tokar (1957) defined the tolerable settlement in terms of the deflection ratio, Δ/L, at which cracking occurred. Cracking was assumed to occur when a limiting level of tensile strain was reached in the wall. Using a limiting tensile strain of 0.05% for unreinforced brick walls, the tolerable deflection ratio was related theoretically to the length to height ratio, L/H, of the wall. This relation was shown to agree well with observations of damaged and undamaged buildings.

Finally, for multistory brick buildings, a larger deflection ratio was allowed for structures supported on plastic clay than for structures founded on sand or hard clay. Presumably, the slower rate of settlement for plastic clays allowed time for creep of the structure and increased the limiting level of tensile strain and hence the deflection ratio at which cracking begins. Subsequently, very little field evidence has been reported in support of this concept.

Based on the Polshin-Tokar concepts, the tolerable deflection ratios for unreinforced load-bearing brick walls ranged from approximately 1/1400 to 1/3300, depending on (L/H) and the foundation soil type. These limits are significantly more stringent than the Skempton-MacDonald relative rotation criteria.

Burland and Wroth (1974) used the concept that maximum tolerable deformations can be related to the onset of visible cracking, which occurs when a limiting value of tensile strain, ϵ_{lim} is reached at some point in the structure. A building was represented by a simple rectangular beam to illustrate the factors that affect the settlement at which cracking begins. In this idealization, the maximum tolerable deflection ratio, (Δ/L), occurs when the limiting tensile strain, ϵ_{lim}, develops either by direct bending tension in the extreme fiber or by diagonal tension along the neutral axis of the beam, depending on the geometry (L/H) and the relative stiffness of the beam in bending and shear. Using elastic beam deflection theory, the deflection ratio, (Δ/L), and the slope, or relative rotation, β, of the deflected beam can be expressed in terms of the elastic properties of the beam and the limiting tensile strain in bending or diagonal tension. Typical values of ϵ_{lim} were suggested as 0.05-0.10% for brick and concrete block and 0.03-0.05% for reinforced concrete, and an average value of ϵ_{lim} = 0.075% was adopted. For a given value of L/H, the limiting Δ/L is directly proportional to ϵ_{lim} for the material used.

Parametric studies can be used to assess the role of various factors in the establishment of the tolerable (Δ/L). For low values of L/H, diagonal strain is critical, and for large values of L/H, direct bending strain is critical. The allowable Δ/L increases as L/H increases and the beam becomes more flexible. Structures that are more flexible in shear than in direct tension, such as typical frames, were modeled by varying the ratio E/G of the rectangular beam. The diagonal strain becomes more critical as the beam is assumed more flexible in shear. The allowable Δ/L increases as the value of E/G increases and also increases rapidly with increasing L/H.

For a mat foundation, which is very stiff in the lateral direction, Burland and Wroth assumed the neutral axis is at the bottom of the beam. For the usual sagging (concave upward) pattern of settlement, the entire beam is in compression, and the diagonal strain condition will be critical for all values of L/H. However, if the settlement curve is concave downward (hogging), i.e., the settlement of the ends is greater than the settlement of the center, direct bending strains become critical except for very small values of L/H. For this condition, the tolerable deformations are significantly smaller than for the case with the neutral axis at the middle, which suggests that the tolerable differential settlement will be smaller for cases in which the settlement curve is concave downward, e.g., subsidence of ends due to adjacent excavations or differential heave due to swelling soils.

On the basis of this simple beam analogy, Burland and Wroth developed limiting deflection ratio criteria for three different cases:

1. *Diagonal strain* will be critical for framed structures, which typically are relatively flexible in shear, and reinforced load-bearing walls, which are relatively stiff in direct tension. This case may be approximated by the behavior of the rectangular beam with a high equivalent E/G, as shown by curve 1 in Fig. 2.
2. *Bending strain* will be critical for unreinforced masonry walls and structures, which have relatively low tensile resistance.
    a. For the *sagging mode* (settlement curve concave upward), the behavior may be modeled by the isotropic rectangular beam with low equivalent E/G and the neutral axis at the mid-depth, as shown by curve 2 in Fig. 2.
    b. For the *hogging mode* (settlement curve concave downward), the behavior may be approximated by the rectangular beam with low equivalent E/G and the neutral axis at the bottom, as illustrated by curve 3 in Fig. 2.

In Fig. 2, Curves 1-3 are based on the assumption that ϵ_{lim} = 0.075%. Fig. 2 also includes Polshin and Tokar's criterion for cracking of unreinforced load-bearing walls, which assumes ϵ_{lim} = 0.05%, and the Skempton-MacDonald criteria of 1/500 and 1/300 for relative rotation. The maximum relative rotation of the beam is assumed equal to the slope at the end supports for the condition of maximum deflection ratio.

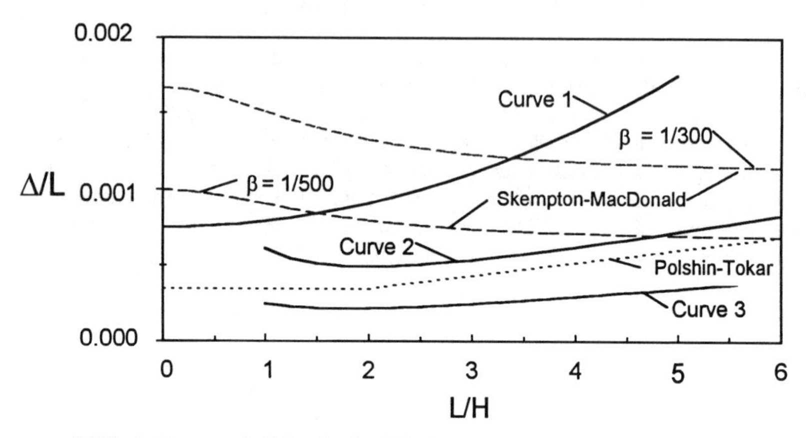

FIG. 2 Proposed Criteria for Maximum Tolerable Values of Δ/L

Burland and Wroth compared the curves of Fig. 2 with observed settlements and damage of real structures. For frame buildings, the Burland-Wroth curve 1 and the Skempton-MacDonald criterion both provided reasonable and satisfactory limits for the deflection ratio when L/H is less than 3. For L/H greater than 3, the Skempton-MacDonald criterion was the more conservative of the two, but no field observations were available for comparison. For load-bearing walls in the sagging mode, the Skempton-MacDonald criterion appeared to be unconservative while Polshin and Tokar's limit and curve 2 both provided reasonable limits. Finally, load-bearing walls in the hogging mode cracked at deflection ratios that are much smaller than the Polshin-Tokar limits. Curve 3 appeared to provide a reasonable limit for this case, but relatively few data were available to verify this case.

In summary, the Burland-Wroth beam analogy provides a conceptual mechanism for understanding the factors that influence tolerable settlements. The model can be used to demonstrate the effects of the limiting tensile strain of the structural materials, the L/H ratio of the structure, the relative stiffness of the structure in shear and direct tension and the mode of the settlement curve. It also provides some insight into the type of the information that should be included in meaningful settlement case histories.

Tolerable Differential and Total Settlement

There have been attempts to relate the tolerable relative rotation to limiting levels of total or differential settlement. While there is agreement that no reliable general relationships have been established, some guidelines have been suggested (Skempton and MacDonald 1956; Polshin and Tokar 1957; Terzaghi and Peck 1967;

Grant et al. 1974). Burland et al. (1977) presented a critical discussion of these guidelines.

For foundations on sand, Burland et al. (1977) suggested that a relative rotation of 1/500 develops when the differential settlement reaches about 20 to 25 mm and the maximum total settlement reaches 25 to 40 mm for isolated footings and 50 to 65 mm for raft foundations. However, they noted that these levels of settlement rarely are reached for foundations on sand. Also, because settlement occurs rapidly, much of the total settlement occurs before installation of the cladding and architectural finishes that are most sensitive to distortion. Thus, there are few, if any, reported cases of significant damage to conventional buildings founded on deep layers of sand.

For foundations on clay, a limiting differential settlement of 40 mm and maximum total settlements of 65 mm and 65 to 100 mm are suggested for isolated footings and rafts, respectively. Burland et al. (1977) noted that the settlement is more uniform for buildings founded on a stiff layer overlying the clay than for buildings founded directly on clay. Also, they reported cases of undamaged buildings that have experienced settlement in excess of the suggested limits. Specifically, they stated that no cases of damaged buildings on raft foundations have been reported for differential and total settlements less than 125 mm and 250 mm, respectively.

ISSMFE Technical Committee 1 (1982-1985)

In 1982 the International Society for Soil Mechanics and Foundation Engineering (ISSMFE) established Technical Committee 1 (TC 1) on Allowable Deformations of Buildings and Damages under the sponsorship of the Mexican member society and the chairmanship of P. Girault. During the period 1982-1985, the committee restricted its deliberations to the relation of damage to differential deformation. The committee summarized its findings in an unpublished draft report (ISSMFE 1985). The report noted that most current tolerable movement criteria are based on the premise that no visible cracking is tolerable and recommended that these criteria should be restricted to buildings for which aesthetic appearance is important. For some buildings, e.g., industrial plants, warehouses, silos and temporary structures, the report suggested that significant savings could be realized by using more liberal tolerable deformation criteria that are based on consideration of structural integrity and function rather than the prevention of minor cracking. Some other recommendations of the TC1 report are discussed in the following comments.

Load-bearing walls

The report suggested that cracking of reinforced concrete walls is of minor concern because it can be controlled, or prevented, by the size and spacing of reinforcement. However, it recommended that unreinforced brick walls supported on compressible soils should be limited to heights of approximately three meters. Brick walls higher than three meters should be designed with sufficient reinforcement to minimize cracking, as in reinforced concrete, and then deformation criteria will not govern the design.

Frames

The report recommended a tolerable relative rotation of 1/100 for open ductile frames, which are designed using ultimate strength concepts and can withstand large deformations without affecting structural integrity, and for simply supported beams and two-hinged rigid frames with space to tilt. However, this value must be reduced if aesthetics govern. For infilled frames, the tolerable relative rotation will be controlled by the cracking of non-load bearing and panel walls between the frames.

Panel walls

The relative rotation that will cause cracking of panel walls is a function of the material used in the wall. A report by Bozozuk (1965), which is referenced but not discussed in the TC1 report, presented the results of laboratory tests of 2.4 × 2.4 m. wall sections which indicated that the relative rotation at cracking was 1/60 to 1/170 for plywood or fiberboard on wood frames and 1/500 to 1/1000 for brick with cement-lime mortar. However, Bozozuk also reported field observations of masonry walls with relative rotations of 1/180 without cracking and attributed the difference to the slow rate of field deformation (60 years) as compared to laboratory loading of several hours.

Effect of soil type

Because most construction materials creep, larger strains can be accommodated without cracking when the deformations occur very slowly. Polshin and Tokar (1957) used this concept to recommend higher tolerable deformation criteria for foundations on plastic clay than for those on sand. Grant et al. (1974) questioned the importance of this factor, and Burland et al. (1977) suggested that there was little field evidence in support of this approach. Bozozuk (1965), as reported in the preceding section, provides one example of the benefits of a slow rate of deformation.

The amount of settlement that occurs during construction is a more important consideration. For buildings supported on sand, most of the dead load of the structure and, hence, most of the deformation is developed before the partition walls and non-structural finishes are installed. Thus, the relative rotation based on the total deformation is significantly larger than value to which these elements are subjected. However, for buildings on clay, most of the deformation will occur after construction of the non-structural elements, and these elements are likely to be subjected to the relative rotation based on total settlement of the foundation. Load-bearing walls on sand or clay obviously are subjected to the deflection ratio, or relative rotation, corresponding to the total deformations of the wall.

Analytical approaches

Based on the results of the very simple Burland-Wroth beam analogy, the TC1 report recommended the development of more complex models to consider such factors of geometry, diagonal bracing, openings, and secondary and non-structural components. Recent developments of modern numerical methods and the expanding computational capacity of computers have made analyses of very sophisticated models more feasible. However, it was recognized that applications of such models to real structures still require the stress-strain characteristics and limiting tensile strains for all elements of the model. Also, the nature and relative rigidity of connections, which may be affected by construction procedures, will influence the interaction between individual elements. Nevertheless, parametric studies of these factors with more complex models were regarded as useful research.

ISSMFE Technical Committee 1 (1985-1989)

Technical Committee 1 was continued for the period 1985-89, but the sponsorship was transferred to the USSR member society with S. N. Klepikov as chairman. There was very little continuity in the committee membership, and a new agenda was adopted. The committee activities for this period have been reported by Klepikov (1989). In this report, a distinction was made between allowable deformations, which do not cause damage, and ultimate deformations, which cause significant damage to individual structural elements or the entire structure. Tables of allowable (tolerable) and ultimate values of average settlement, relative deflection, relative rotation, and tilt were compiled for various types of buildings and other structures, e.g., towers, smokestacks, tanks, etc. The recommended values were obtained from Russian, Canadian, Australian, Finnish and U. S. sources. The similarities among the recommended values from different sources and their general similarity to the Skempton-MacDonald and Polshin-Tokar criteria were noted.

Eurocode Criteria

In recent years the European Committee for Standardization (CEN) has attempted to formulate standards for design and construction of structures. Eurocode 1: Part 1: Basis of Design (1993) addresses design requirements, including the definition of limit states, from the perspective of the structural engineer, and Eurocode 7: Geotechnical Design (1993) provides geotechnical design requirements. Both documents provide limiting values for foundation movements. The referenced versions of these documents are drafts that still may be revised. The final approval of Eurocode 7 is anticipated early in 1994.

The "fourth and final draft" of Eurocode 7 (1993) defines both Principles and Application Rules. Principles are "general statements and definitions for which there is no alternative" and "requirements and analytical models for which no alternative is permitted unless specifically stated." Application Rules are "examples of generally recognized rules which follow the Principles and satisfy their requirements." Alternative rules from those given in Eurocode 7 may be used, provided that they are in accord with the relevant Principles. With respect to limiting values for movements, some of the relevant Principles are:

- The selection of design values for limiting movements should consider the confidence with which the acceptable value can be specified, the type and proposed use of the structure, the types of construction materials, the types of foundation and ground and the mode of deformation.
- The differential settlements and relative rotation shall be estimated to ensure that these do not lead to the occurrence of an ultimate or serviceability limit state.
- Calculations shall take account of random or systematic variations in ground properties, the loading distribution, the construction method and the stiffness of the structure.

The related Applications Rules include:

- The maximum acceptable relative rotations for open frames, infilled frames and load bearing or continuous brick walls are unlikely to be the same but are likely to range from about 1/2000 to about 1/300 to prevent the occurrence of a serviceability limit state in the structure. A maximum relative rotation of 1/500 is acceptable for many structures. The relative rotation likely to cause an ultimate limit state is about 1/150.
- For normal structures with isolated foundations, total settlements up to 50 mm and differential settlements between adjacent columns of 20 mm are often acceptable. Larger total and differential settlements may be acceptable provided the relative rotations remain within acceptable limits and provided the total settlements do not cause problems with the services entering the structure, tilting, etc.
- The above guidelines concerning settlement apply to normal routine structures. They should not be applied to buildings or structures which are out of the ordinary or for which the loading-intensity is markedly non-uniform.

Similar limiting values for serviceability limit states are given in the sixth draft of Eurocode 1 (1993). These values, which are summarized in Table 2, appear to be defined more specifically than in Eurocode 7; however, they are described as "guidance for project specifications" and are regarded as "simplifications which may be overruled by more appropriate model and definitions." Furthermore, the recommended values are assumed to apply to foundations on sand. For foundations on clay, higher settlements may be allowed provided the limits for relative rotation and tilt are satisfied and the design indicates that service entries to the building will not be adversely affected.

The tolerable movement criteria in the Eurocode documents are very similar to those recommended more than 35 years ago by Skempton and MacDonald (1956) and Polshin and Tokar (1957). However, the Eurocode statements are worded very carefully to emphasize that the criteria are only guidelines, which may be appropriate for many conventional structures but should not be used for unusual structures.

TABLE 2. Guidelines for Tolerable Foundation Settlement (Eurocode 1 1993)

Total settlement	
Isolated foundation	25 mm
Raft foundation	50 mm
Differential settlement between adjacent columns	
Open frames	20 mm
Frames with flexible cladding or finishes	10 mm
Frames with rigid cladding or finishes	5 mm
Relative rotation (Angular distortion) β	1/500
Tilt ω	determined in design

TOLERABLE MOVEMENTS OF BRIDGES

All bridge abutments and their foundations are likely to move, and tolerance limits must be established to maintain the safety and function of the bridge. Significant horizontal and vertical abutment movements may seriously affect the safety and the rideability of the bridge. Also, it is necessary to consider the relative settlement of the abutment and the approach fill.

A 1975 survey conducted by the Transportation Research Board (TRB) Committee A2K03 on Foundations of Bridges and Other Structures was the first significant attempt to establish rational tolerable movement criteria for bridges. The committee proposed that the tolerability of the movement should be judged qualitatively by the agency responsible for each bridge using the following definition (Walkinshaw 1978):

"Movement is *not* tolerable if damage requires costly maintenance and/or repairs *and* a more expensive construction to avoid this would have been preferable."

Bozozuk (1978) summarized the survey and concluded that the results could be separated into three levels of damage, as shown in Table 3.

TABLE 3. Level of Damage versus Movement (Bozozuk 1978)

Level of Damage	Vertical Movement, mm	Horizontal Movement, mm
Tolerable	< 50 mm	< 25 mm
Harmful but Tolerable	50–100 mm	25–50 mm
Not tolerable	> 100 mm	> 50 mm

A very comprehensive study of bridge movements was conducted for the Federal Highway Administration (FHWA) by Moulton and his colleagues at West Virginia University (Moulton et al. 1985; Moulton 1986). Measurable movements were reported for 439, or approximately 75%, of 580 abutments but for only 269, or 25%, of 1068 piers included in the study. The majority (357) of the observed movements were for perched abutments, but full-height and spill-through abutments also were included. Vertical movements were reported for 379 abutments and 234 piers, while horizontal movements were observed at 138 abutments and 52 piers. Significant movements were reported for abutments and piers supported on piles, as well as for those on spread footings.

The tolerability of these movements also was judged qualitatively by the agency responsible for each bridge in accord with the TRB definition. Using this criterion, the movement was regarded as tolerable for 90% of the cases for which the vertical movement was less than 100 mm and the horizontal movement was less than 50 mm. The movement was considered intolerable in approximately 80% of the cases in which these limits were exceeded. These findings are consistent with Bozozuk's recommendations, as shown in Table 3. However, Moulton recommended that the tolerable horizontal movements should be reduced to 40 mm when both vertical and horizontal movements are likely to occur in the same span. More distress was observed for concrete bridges than for steel bridges. Continuous-span bridges were affected more often and more severely than single span structures.

The tolerable differential settlement increased with span length. Moulton (1986) suggested a tolerable relative rotation (differential settlement/span length) of 1/250 for continuous-span bridges and 1/200 for simply supported spans. These values are slightly larger but relatively consistent with the criteria commonly used for tolerable settlement of buildings. However, Duncan and Tan (1991) concluded that the recommendation for simple spans was unduly influenced by one case and recommended a tolerable relative rotation of 1/125 for simple spans.

Most bridges are less complex structures than buildings and thus would appear more amenable to analytical studies of the effects of movements. However, attempts, e.g., Moulton (1986), to establish tolerable movements from analyses of the effects of differential settlement on the stresses in bridges have significantly underestimated the criteria established from field observations (Duncan and Tan 1990).

The preceding criteria do not address the differential settlement between the bridge deck and the approach pavement. A differential settlement of 12 mm between these components is likely to produce a "bump at the end of the bridge" that will require maintenance (Wahls 1990). However, in accord with the TRB definition, such movement would not be considered intolerable unless the cost of maintenance is greater than the cost of the design modifications necessary to reduce the differential settlement.

Tolerable Movements of Other Structures

For steel tanks, ultimate limit states rather than serviceability limits usually govern. Meyerhof (1979) recommended a danger (ultimate) limit for relative rotation of 1/150 and a safe limit of 1/250, which were identical to his recommendations for building frames. Klepikov (1989) includes limiting values of δ_{max}/D from Konovalov and Ivanov (1985) that range from 1/100 to 1/250, depending on the volume of the tank and apparently the shape of deformation curve. However, no justification for the values is provided.

Marr et al. (1982) reviewed the tolerable deformation criteria for steel tanks and studied performance data for 90 large steel tanks. Tensile rupture of the bottom plate due to non-planar settlement was identified as the most critical deformation problem. Limiting settlement criteria, based on the factor of safety and tensile strength of the plate, were presented for the cases of classical dish-shaped settlement, where the maximum settlement is at the center, and localized settlement where the maximum settlement occurs at some point between the edge and the center of the tank. For a factor of safety of one, i.e., the ultimate limit state, the recommended maximum tolerable differential settlement is D/44 for the dish-shaped case and d/51 for the local settlement case, where D = diameter of the tank and d = the diameter of a local depression. If these criteria are restated in terms of relative rotation (angular distortion), the limiting values are 1/22 and 1/25 for the dish-shaped and local settlement cases, respectively.

D'Orazio and Duncan (1987) examined performance records for 31 tanks and identified three typical deformation profiles. Profile A is the classical dish-shaped pattern with the maximum settlement at the center. For Profiles B and C, the maximum settlement is away from the center; B is relatively flat on the interior with the settlement decreasing rapidly near the edge, while C has its maximum settlement about 2/3 of the distance from the center to the edge of the tank. The occurrence of these profiles was related to the minimum factor of safety, F_{min}, based on undrained strength, and (D_e/H_c) where D_e = the effective diameter of the tank, which is the diameter of the tank, D, plus the thickness, H_p, of any granular or compacted clay pad between the tank and the compressible clay, and H_c = the thickness of the compressible clay. Tolerable settlement criteria then were developed for each profile and expressed in terms of the ratio δ_{cen}/D_e, where δ_{cen} is the differential settlement between the center and the edge. The recommendations are summarized in Table 4. The tolerable value of δ_{cen}/D_e decreases as the profile shifts from A to B to C because, for a given settlement at the center, the severity of the relative rotation increases as the point of maximum settlement shifts away from the center. The equivalent relative rotation is 1/20 for Profile A and is estimated at about 1/30 for Profile C. The recommendations in Table 4 are comparable to those of Marr et al. (1982).

TABLE 4. Tolerable Deformation Criteria For Steel Tanks (after D'Orazio and Duncan 1987)

D_e/H_c	F_{min}	Anticipated Profile	δ_{cen}/D_e
< 4	> 1.1	A	1/40
> 4	> 1.1	B	1/67
> 4	< 1.1	C	1/200

For tall rigid structures, e.g., brick or reinforced concrete chimneys, silos, and towers, tilt is the critical criterion, and several authors recommend the maximum tolerable tilt as 1/250 (Polshin and Tokar 1956; Sowers 1962; Bjerrum 1963; Meyerhof 1979). Klepikov (1989) reported that Russian codes specify maximum tolerable tilts ranging from 1/100 for H < 20 m. to 1/250 for H = 100 m, where H is the height of the structure. For H > 100 m, the recommended maximum tolerable tilt is 1/2H, which limits the lateral movement of the top to 0.5 m.

Various authors, e.g., Sowers (1962) and Klepikov (1989), have included maximum tolerable movements for steel TV/radio towers or for specialized industrial equipment, e.g., crane rails, compressors, turbo-generators, etc. However, the requirements for various types of industrial equipment are highly dependent on the individual specifications of the equipment. In such cases, it is essential that the geotechnical and structural engineers discuss the specific requirements with the suppliers of the equipment. Often the manufacturers specifications are very restrictive and will require uneconomical foundation designs. Sometimes the equipment support system can be modified to accommodate more foundation movement than was initially specified.

CONCLUSIONS

1. Despite significant critical reviews and additional studies over the past 30 years, most current guidelines for the tolerable movements of buildings continue to be based on the work of Skempton and MacDonald (1956) and Polshin and Tokar (1956). Since 1980, these criteria have been reviewed extensively by ISSMFE Committee TC1 and the European standards committees. Eurocode 7 provides good statements of these criteria in the context of modern limit state design practices and clearly identifies the published criteria as "guidelines" rather than rigid standards.
2. Reasonable guidelines for the tolerable movements of bridges are provided by the FHWA study by Moulton (1986). The maximum tolerable deformations generally are larger for bridges than for buildings.
3. D'Orazio and Duncan (1987) provide reasonable and easy-to-use criteria for the tolerable deformation of steel tanks.
4. Current criteria for tolerable deformations are based primarily on field observations of movements of damaged and undamaged structures. Analytical studies, e.g., Burland and Wroth (1974), have been helpful in identifying factors that affect the relation of damage to deformations, but to date analytical attempts to establish tolerable movement criteria have significantly underestimated the values observed in the case histories.
5. There is a need for additional complete case histories of both damaged and undamaged structures. These case histories should include reliable settlement records, detailed assessments of any damage, and complete structural and subsurface information.
6. The tolerable movements of buildings should be defined in terms of tolerable damage, which should be based on the relative costs of repair of the damage and the initial construction required to eliminate the damage. This definition should parallel the TRB definition for tolerable movements of bridges.

7. There is a need to develop classifications for levels of damage in buildings, considering the effects of the damage on the use of the structure and the ease and cost of repairs.

APPENDIX - REFERENCES

Bjerrum, L. (1963). "Allowable settlement of structures." *Proc. European Conf. on Soil Mech. and Found Engrg.*, Weisbaden, 3, 135-137.

Boscardin, M. D. and Cording, E. J. (1989). "Building response to excavation-induced settlement." *J. Geotech. Engrg.*, ASCE, 115(1), 1-21.

Bozozuk, M. (1962). "Soil shrinkage damages shallow foundations at Ottawa, Canada." *The Engineering Journal*, 45(7), 33-37.

Bozozuk, M. (1978). "Bridge foundations move." *TRR 678*, Transportation Research Board, 17-21.

Burland, J. B. and Wroth, C. P. (1974). "Allowable and differential settlement of structures, including damage and soil-structure interaction." *Proc. Conf on Settlement of Structures*, Cambridge University, 611-654.

Burland, J. B., Broms, B. B., and DeMello, V. F. B. (1977). "Behavior of foundations and structures." State-of-the-Art Report, *Proc. 9th ICSMFE*, Tokyo, 2, 495-546.

Clough, G. W. and O'Rourke, T. D. (1990). "Construction induced movements of in situ walls." *Design and Performance of Earth Retaining Structures*, Geotechnical Special Publication No. 25, ASCE, New York, New York, 439-470.

D'Orazio, T. B. and Duncan, J. M. (1987). "Differential settlements in steel tanks." *J. Geotech. Engrg.*, ASCE, 113(9), 967-983.

Duncan, J. M. and Tan, C. K. (1991). "Engineering manual for estimating tolerable movements of bridges." *Manuals for the Design of Bridge Foundations*, NCHRP Report No. 343, Transportation Research Board, National Research Council, Washington, D. C., 219-228.

Eurocode 1 (1993). "Basis of design and actions on structures." Sixth Draft, European Committee for Standardization, TC 250.

Eurocode 7 (1993). "Geotechnical design." 4th and Final Draft, European Committee for Standardization, TC 250/SC 7.

Feld, J. (1964). "Tolerance of structures to settlement." *Proc. Specialty Conf on Design of Foundations for Control of Settlement*, ASCE, Northwestern University, 555-569.

Golder, H. Q. (1971). "The allowable settlement of structures." *Proc. 4th Pan-American Conf. on Soil Mech. and Found. Engrg.*, Puerto Rico, 1, 171-187.

Grant, R., Christian, J. T., and Vanmarcke, E. H. (1974). "Differential settlement of buildings." *J. Geotech. Engrg. Div.*, Proc. ASCE, 100(GT9), 973-991.

ISSMFE Technical Committee 1 on Allowable Deformations of Buildings and Damages. (1985). "General report," unpublished draft committee report.

Klepikov, S. N. (1989). "Performance criteria-allowable deformations of buildings and damages." General Report, ISSMFE Technical Committee 1, *Proc. 12th ICSMFE*, Rio de Janeiro, 5, 2735-2744.

Konovalov, P. A. and Ivanov, Y. K. (1985). 'Ultimate values of average and nonuniform settlements of steel tanks." *Bases, Found. and Soil Mech.*, Moscow, 5, 27.

Leonards, G. A. (1975). Discussion on "Differential settlement of buildings." *J. Geotech. Engrg. Div.*, Proc. ASCE, 101(GT7), 700-702.

Marr, W. A., Ramos, J. A., and Lambe, T. W. (1982). "Criteria for settlement of tanks." *J. Geotech. Engrg. Div.*, Proc. ASCE, 108(GT8), 1017-1039.

Meyerhof, G. G. (1979). "Soil-structure interaction and foundations." *Proc. 6th Pan-American Conf. on Soil Mech. and Found. Engrg.*, Lima, 1, 109-140.

Moulton, L. K. (1986). "Tolerable movement criteria for highway bridges," *Final Report No. FHWA-TS-85-228*, Federal Highway Administration, Washington, D. C.

Moulton, L. K., GangaRao, H. V. S., and Halvorsen, G. T. (1985). "Tolerable movement criteria for highway bridges," *Report No. FHWA/RD-85/107*, Federal Highway Administration, Washington, D. C.

Peck, R. B. (1948). "History of building foundations in Chicago," *Engineering Exp. Station Bulletin No. 373*, University of Illinois, Urbana-Champaign.

Polshin, D. E. and Tokar, R. A. (1957). "Maximum allowable non-uniform settlement of structures." *Proc. 4th ICSMFE*, London, 1, 402-406.

Skempton, A. W. and MacDonald, D. H. (1956). "Allowable settlement of buildings." *Proc. Institution of Civil Engineers*, Part III, 5, 727-768.

Sowers, G. F. (1962). "Shallow foundations." *Foundation Engineering*, McGraw-Hill Book Co., New York, New York, 525-632.

Terzaghi, K. and Peck, R. B. (1967). *Soil Mechanics in Engineering Practice*, 2nd Edition, John Wiley and Sons, New York, New York.

Wahls, H. E. (1981). "Tolerable settlement of buildings." *J. Geotech. Engrg. Div.*, Proc. ASCE, 107(GT11), 1489-1504.

Wahls, H. E. (1990). "Design and construction of bridge approaches," *NCHRP Synthesis of Highway Practice 159*, Transportation Research Board, National Research Council, Washington, D. C.

Walkinshaw, J. L. (1978). "Survey of bridge movements in the western United States." *TRR 678*, Transportation Research Board, 6-12.

SETTLEMENT PREDICTION FOR DRIVEN PILES AND PILE GROUPS

Harry G. Poulos,[1] Fellow, ASCE

ABSTRACT: Various methods of settlement prediction are reviewed for single piles and pile groups, and it is demonstrated that several of the methods give similar results. It is also found that the representation of a pile group by an equivalent pier provides a useful practical means of estimating the settlement behavior of pile groups, and can be used for either linear or nonlinear analyses.

Simple-to-use dimensionless design charts are presented for piles and piers in typical clay and sand profiles. The assessment of soil parameters is then discussed, and it is emphasized that distinctions must be made between the values of soil modulus along the shaft, immediately below the tip, well below the tip, and between piles in a group.

The practical applicability of some of the theoretical approaches is examined for a number of published case histories. It is found that the predicted settlements are far more sensitive to the soil stiffness or modulus than to the method of analysis. Nonlinear analysis appears to be desirable for both single piles and pile groups in sand, although linear analyses provide adequate settlement predictions for piles in clay.

INTRODUCTION

There exist many procedures for estimating pile foundation settlements, ranging from relatively simple hand calculation methods to sophisticated nonlinear finite element analyses, and it seems appropriate to review and assess some of these

---

[1] Senior Principal, Coffey Partners International, 12 Waterloo Road, North Ryde, Australia, 2113; and Professor of Civil Engineering, University of Sydney, Australia, 2006.

methods. The primary purpose of this paper is to carry out such an assessment and to compare the predicted settlements from some of the available methods when applied to both idealized hypothetical situations and real problems. Consideration is given to both linear and nonlinear methods, and the circumstances under which nonlinear methods may be desirable are examined. Single piles and pile groups are treated in turn. The critical question of estimation of soil parameters is also addressed, and attention is focussed on correlations between these parameters and commonly available in-situ test data.

Because of the limitations of space, no attempt is made to explore in detail the various numerical analyses mentioned, nor the theoretical and experimental characteristics of single pile and pile group behavior. For detailed discussion of some of the above issues, reference may be made to comprehensive reviews by O'Neill (1983), Van Impe (1991), Poulos (1989), and Fleming et al. (1985).

SETTLEMENT OF SINGLE PILES

This section categorizes and discusses some of the available methods of settlement analysis of single piles, including both linear and nonlinear methods. Attention is confined to methods which have a proper theoretical basis (Category 2 and 3 methods in the terminology of Poulos 1989).

There are four main means of modelling pile-soil interaction:

(a) via load-transfer ("t-z") curves, which relate local shear stress to local displacement along the pile; these are, in effect, nonlinear independent spring support characteristics describing the soil-pile response at a particular location on the pile (e.g. Kraft et al. 1981; Randolph 1986).

(b) via elastic theory, which idealizes the soil as an elastic continuum, and allows consideration of interaction between different portions of the pile through the soil (e.g. Butterfield and Banerjee 1971; Banerjee and Davies 1977; Poulos and Davis 1980)

(c) via simplified analysis methods which consider localized shear around the pile and can lead to convenient closed-form solutions (e.g. Randolph and Wroth 1978; Fleming 1992).

(d) via numerical analyses which utilize advanced constitutive models of soil behavior (e.g. Jardine et al. 1986; Trochanis et al. 1991).

Different methods which utilize the theory of elasticity in general give very similar solutions for the pile settlement. For example, for a typical pile in a soil whose modulus increases linearly with depth, four independent linear analysis (Banerjee and Davis 1977; Randolph and Wroth 1978; Poulos 1979, 1989) give solutions for settlement which agree to within $\pm 7\%$, which is adequate for practical purposes.

Jardine et al. (1986) have employed a finite element analysis using approach (d) above to analyze a pile in a homogeneous clay layer. This problem has been analyzed by various alternative methods in categories (a) and (b) above by Poulos (1989), and for a realistic pile stiffness, the resulting load-settlement curves have been found to agree well with the finite element solutions of Jardine et al. (1986). There is therefore a strong indication from these comparisons that the method of analysis is not a critical factor in single pile settlement prediction, provided that the method is soundly-based.

Requirements for Successful Prediction of Single Pile Settlements

Based on the extensive research into single pile behavior carried out by many people over the past three decades, it is possible to identify those factors which are critical for a good settlement prediction, and those which are less important. Some of the factors influencing single pile settlement behavior are summarized in Table 1, together with an assessment of the relative importance of each factor. Table 1 indicates that the factors which are of primary importance are the stiffness of the soil mass around the pile shaft, the soil stiffness below the pile tip, and nonlinear soil-pile response. The latter two factors are generally more significant for piles in sand or piles which derive the majority of their resistance from end-bearing.

Table 1 therefore suggests that, provided suitable assessments can be made of soil stiffness adjacent to and below a single pile, it should be possible to make a reasonable prediction of settlement at working loads using a relatively simple method of calculation. It is worthy of note that Frank (1985) suggests the settlement of an isolated driven pile under its design load can be estimated to between 0.8 and 1.2% (average 0.9%) of its diameter. This value is consistent with the order of settlement predicted by most of the available methods when appropriate soil stiffness parameters are used as input.

Design Charts Derived from Theoretical Solutions

For rapid practical estimation of settlement, it is convenient to develop design charts from the theoretical analyses. Such charts, in dimensionless form, are shown in Figs. 1 and 2 for typical values of pile stiffness relative to soil stiffness. Fig. 1 is for a pile in a homogeneous clay deposit, while Fig. 2 is for a pile in sand whose modulus increases linearly with depth. The solutions have been derived via spreadsheet evaluation of the Randolph and Wroth (1978) solutions. In each case, the pile is assumed to bear on a stratum of equal or greater stiffness to that of the soil above. These charts are necessarily simplified and use idealized representations of the soil profile, but nevertheless provide a useful means of at least preliminary estimation of settlement when used with appropriate engineering judgement.

SETTLEMENT OF PILE GROUPS

At least seven broad categories of analysis procedures have been employed for group settlement calculation:

(a) simplified procedures, which reduce the group to an equivalent raft (e.g. Terzaghi and Peck 1967; Tomlinson 1986).

(b) simplified methods which reduce the group to an equivalent pier (e.g. Poulos and Davis 1980; Poulos 1993; Van Impe 1991).

(c) the settlement ratio method, in which the settlement of a single pile (at the average load level) is multiplied by a settlement ratio which is usually derived from one of the methods outlined below (e.g. Poulos 1979; Fleming et al. 1985).

TABLE 1. Relative Importance of Various Factors on Pile Group Settlement Prediction

Factor	Relative Importance – Single Piles		Relative Importance – Pile Groups		Remarks
	Driven Piles in Clay	Driven Piles in Sand	Driven Piles in Clay	Driven Piles in Sand	
Average soil modulus along shaft	1	1	1	1	
Distribution of soil modulus along shaft	2	2	1	1	
Soil Modulus just below pile tip	2–3	1	1	1	
Stiffness of soils well below pile tip	3	3	1	1	Depends on group geometry
Stiffness of soils between piles	3	3	1–2	2	Influences interaction between piles
Installation effects on soil stiffness around piles	2	1	2	1	
Nonlinear pile-soil behavior	2	1	2	1–2	Depends on load level
Young's modulus of pile	2	2	3	3	More important as pile length increases
Method of analysis	2	2	2	1–2	Method becomes more important if nonlinear load-settlement response is required
Consolidation and time effects	3	3	2	3	Depends on group size and geometry

Notes: 1 = very important
2 = moderately important
3 = relatively unimportant

(d) methods which compute the response of a single pile via elastic theory, and which consider pile-soil-pile interaction via interaction factors which are also computed from some form of elastic theory (e.g. Poulos 1968; Randolph and Wroth 1979; Hirayama 1991; Polo and Clemente 1988).

(e) *hybrid* methods which compute the single pile response from a load-transfer analysis, while pile-soil-pile interaction is allowed for via the use of elastic theory (e.g. O'Neill et al. 1977; Chow 1986; Lee 1993).

(f) *complete* boundary element methods, in which each pile is divided into discrete elements and pile-soil-pile interaction is considered between each of these elements via the use of elastic theory (e.g. Butterfield and Douglas 1981; Poulos and Hewitt 1986).

(g) finite element methods, which often simplify the group to an equivalent plane-strain or axisymmetric system (e.g. Ottaviani 1975; Pressley and Poulos 1986).

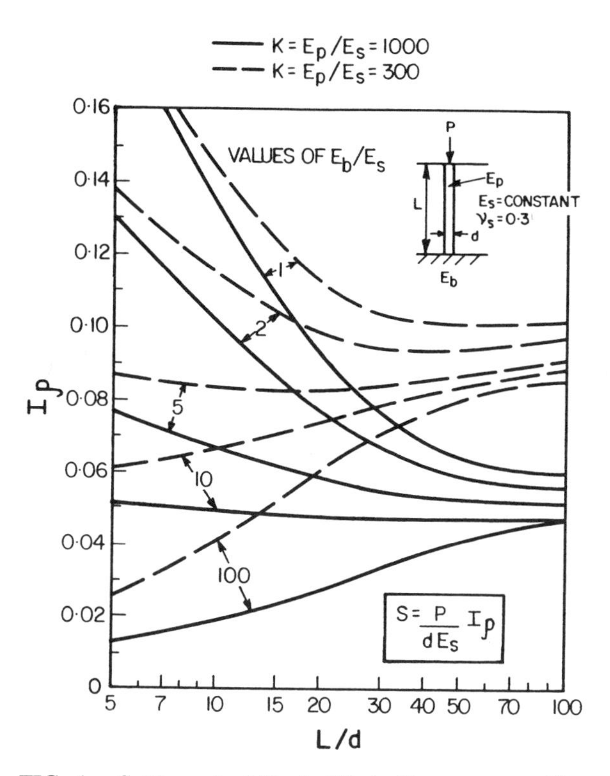

FIG. 1. Settlement of Single Pile in Homogeneous Clay

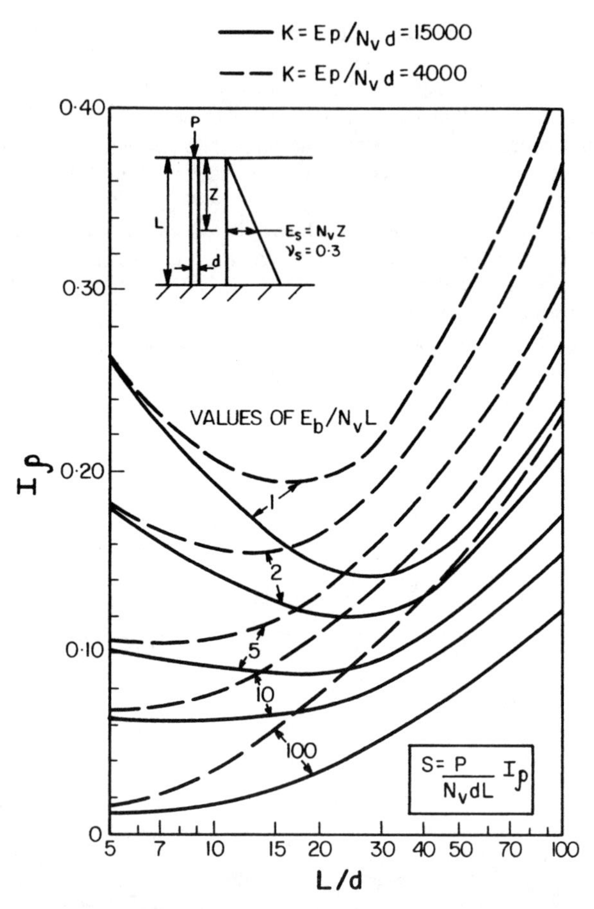

FIG. 2. Settlement of Single Pile in Soil with Linearly Increasing Modulus

A number of comparisons have been published between various methods, and it has been found that a similar settlement is predicted by the various methods based on elastic theory, e.g., Poulos and Randolph (1983), Chow (1986), and Lee (1993). Poulos (1993) has found that, for end bearing groups, the equivalent raft method tends to overestimate the settlement of groups containing a small number of piles, although it gives reasonable values for groups containing 16 or more piles. The equivalent raft method is found to be better suited to analyzing a group of friction piles, and can give adequate solutions for groups as small as four piles. O'Neill and Ha (1982) have found that hybrid and elastic methods can give similar results, provided that the input soil parameters are chosen appropriately, while Pressley and Poulos (1986) have demonstrated that the load-settlement response of a pile group, using a linear finite element analysis, is similar to that predicted via an interaction factor analysis.

For a 9-pile group in clay at two different spacings, Fig. 3 compares the nonlinear finite element solutions of Pressley and Poulos (1986) with two simplified equivalent pier analyses:

(i) the hand calculation procedure suggested by Poulos (1972);

(ii) a nonlinear boundary element analysis using the computer program PIES which assumes a hyperbolic interface between the soil and the equivalent pier.

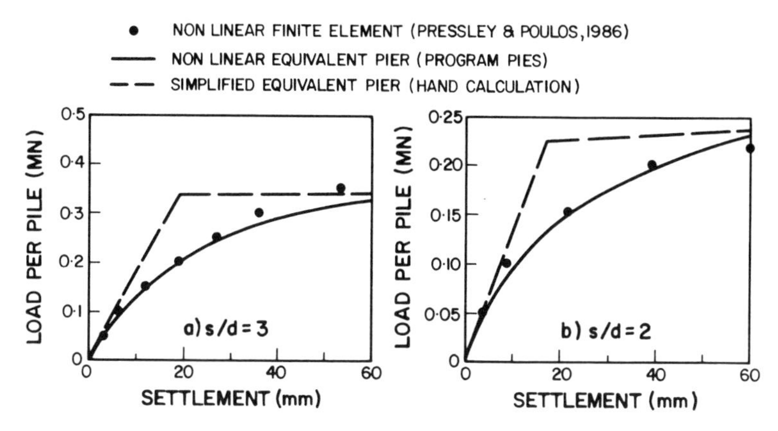

FIG. 3. Theoretical Solutions for Load - Settlement of Group in Clay

In each case, the shaft and base resistances of the equivalent pier have been computed by multiplying the single pile values by 9 (the number of piles) and by an efficiency factor (0.9 for 3 diameter spacing, 0.6 for 2 diameter spacing). The hand calculation method gives a good estimate of settlement at low loads, but underestimates the settlement at higher loads (it gives an almost linear load-settlement curve to failure). The nonlinear boundary element analysis give remarkably good agreement with the finite element solutions for both spacings. Fig. 3 therefore suggests that, if only the settlement of a group in clay is required, the equivalent pier method can provide a reasonable estimate, especially if a nonlinear analysis is carried out. Of course, if the detailed load distribution within the group is required, a more complete analysis, incorporating pile-soil-pile interaction, is necessary.

Factors Influencing Accuracy of Group Settlement Predictions

Table 1 identifies the relative importance of various factors which may influence the load-settlement behavior of a group. It is immediately obvious that many more factors have an important influence on the settlement of pile groups than on single piles. In particular, it will be noted that the soil modulus around and beneath the pile,and also the soil moduli between the piles and well below the pile tip, are important.

The effects of installation on the soil characteristics within the pile group can be very important, particularly for piles in sand. It is now well-understood that the effects of driving a group of piles in sand is to densify, stiffen and strengthen the soil around the interior piles in the group. Thus, the load-settlement behavior of a pile within a group may be quite different to that of a single isolated pile, and the group settlement ratio may be less than unity in some cases. O'Neill (1983) summarizes published data that demonstrates that the group settlement ratio decreases as the average group breadth decreases, and may be as low as 0.2 for closely-spaced piles driven into loose sand. It is therefore clear that injudicious application of theoretical analyses, without a clear understanding of the significance of some of the factors listed in Table 1, may lead to highly inaccurate estimates of group settlement, particularly in sands (e.g. Poulos 1968; Leonards 1972). Fortunately, such inaccuracies tend to be on the conservative side, i.e., the predicted settlements are larger than the true settlements.

Rapid Practical Estimation of Group Settlements

For rapid practical estimation of group settlements without recourse to a computer, there are at least three convenient methods which may be employed: the equivalent raft method, the equivalent pier method, and the settlement ratio method. For estimating the settlement of driven pile groups, it is suggested that the equivalent pier method has the following advantages:

(a) it does not require detailed consideration of the response of single piles within the group, which may have been influenced by installation effects;
(b) it reflects the increasing proportion of the load which is carried by the pile tips because of pile-soil-pile interaction effects;
(c) it can be extended to include nonlinear load-settlement response;
(d) it allows an assessment to be made of the rate of settlement of pile groups in clay.

Fig. 4 presents dimensionless solutions for a pier in a homogeneous soil, bearing on a stratum of equal or greater stiffness. The compressibility of the pier has been chosen to be representative of the average value of a pile and soil *block* with piles at a spacing of about 3 diameters. For short piers, the relative compressibility is unimportant unless the pier is very compressible, or unless it is founded on a very stiff stratum. Fig. 4 may be used with sufficient accuracy for a pier in a non-homogeneous soil, by using an average soil modulus along the shaft of the pier. Solutions for the proportion of base load are given in Fig. 5, and are useful if the approximate approach suggested by Poulos (1972) is used to estimate the load-settlement curve to failure.

In utilizing the equivalent pier approach, the following points should be noted:

1. the diameter d_e of the equivalent pier should be such that it has an equal total surface area (shaft and base) to the enclosed "block" of piles and soil. For

a block of square plan area B × B, d_e will lie between 1.13B and 1.27B, depending on the length of the block.

2. the Young's modulus of the pier is taken as the area-weighted average value for the pile-soil block;
3. in selecting the Young's modulus E_b of the bearing stratum, consideration needs to be given to the effects of installation. Also, an average value (weighted with respect to the relative depth below the base of the pier) should be used as discussed in more detail in the following section; and
4. for a nonlinear analysis, the ultimate shaft and base resistances of the group (which are not necessarily the same as those for the equivalent pier) are computed. If a computer analysis is to be performed, the average ultimate skin friction and end bearing values can be obtained by dividing the computed shaft and base resistances by the shaft and base areas of the pier respectively.

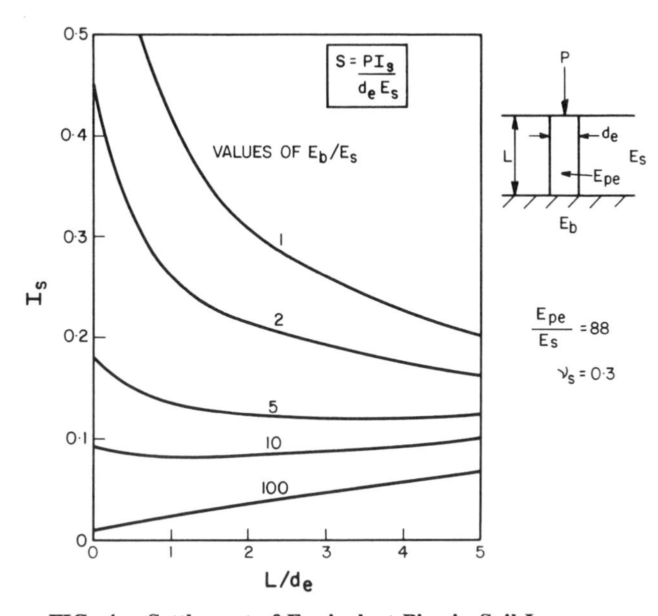

FIG. 4. Settlement of Equivalent Pier in Soil Layer

Assessment of Parameters

For predictions of pile settlement, the key geotechnical parameter required is the stiffness of the soil. If an analysis based on elastic continuum theory is used, then the soil stiffness can be expressed in terms of a Young's modulus E_s or shear modulus G_s. Both the magnitude and distribution of these moduli are important. It cannot be emphasized too strongly that E_s (or G_s) are not constants, but depend on many factors, including soil type, initial stress state, stress history, the method of installation of the pile, the stress system and stress level imposed by the pile or pile group, and whether short-term or long-term conditions are being considered. The most satisfactory procedure for assessing the soil modulus is to carry out pile load tests on prototype piles and backfigure the modulus from the observed load-settlement response, using the same theory that will be used for the actual settlement prediction. Because this is not always possible, especially in the preliminary stages of design, it is usual practice to correlate the soil modulus with the results of simple in-situ tests such as the standard penetration test (SPT) and cone penetration test (CPT), or with the results of field or laboratory strength tests.

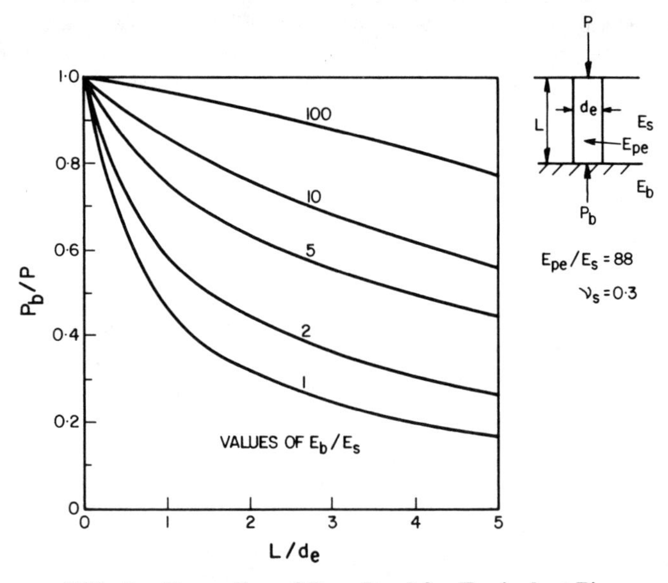

FIG. 5. Proportion of Base Load for Equivalent Pier

Four different values of Young's modulus can be distinguished for pile settlement analysis:

1. the value E_s for the soil in the vicinity of the pile shaft. It will tend to influence strongly the settlement of a single pile and pile groups;
2. the value E_{sb} immediately below the pile tip; this will also tend to influence the settlement of single piles and pile groups;
3. the initial tangent value E_i for the soil between the piles; this will reflect the small strains in this region and will affect the settlement interaction between the piles; and
4. the value E_{sl} for the soil well below the tips. This value will influence the settlement of a group increasingly as the group size increases.

E_s and E_{sb} will both be influenced by the installation process, and would be expected to be different for bored piles and for driven piles. On the other hand, E_{si} and E_{sl}

are unlikely to be influenced by the installation process, but rather by the initial stress state and stress history of the soil. An interesting corollary is that the method of installation is likely to have a much more significant effect on the settlement of a single pile, (which depends largely on E_s or E_{sb}) than on the settlement of a pile group, which may depend to a large extent on E_{sl} and E_{si}.

Table 2 summarizes some suggested correlations for E_s, E_{si} and E_{sl}. In all cases, the correlations relate to the drained Young's modulus, and therefore to the calculation of final settlements. There appears to be little information available on the modulus E_{sb} below the tip of a driven pile. It is suggested that, for clays, the same correlation be used for E_{sb} as for E_s, while for sands, E_{sb} should be 3 to 5 times that given by the correlation for E_s.

The values of E_s and E_{sb} in Table 2 are meant to be used in elastic or elastic-plastic analyses of settlement, and represent a **secant modulus** at typical working load levels of one-third to one-half of ultimate load. However, if a nonlinear analysis is employed, the initial tangent values of E_s and E_{sb} should be greater than the value in Table 2. For example, if a hyperbolic model is used, the initial tangent modulus values along and beneath the pile tip should be increased by a factor of between about 1.4 and 1.6 over the values in Table 2.

If the simple theoretical solutions such as those in Figs. 1, 2 and 4 are used, it is necessary to estimate average values of Young's modulus along the shaft and below the pile tip. Unless the piles are very slender or compressible, it is usually adequate to adopt an average modulus along the shaft of the pile (or the equivalent pier). Below the pile tip, a weighted average modulus $E_{sb(e)}$ can be estimated as follows:

$$E_{sb(e)} = \frac{\sum_{i=1}^{n} W_i h_i}{\sum_{i=1}^{n} \frac{W_i h_i}{E_{sli}}} \tag{1}$$

where h_i = thickness of layer i, E_{sli} = Young's modulus of layer i, W_i = weighting factor for layer i, n = total number of layers within zone of influence of pile tip or pier base.

W_i can be evaluated approximately from the vertical strain distribution derived from elastic theory for an elastic half-space, and is plotted in Fig. 6 as a function of the relative depth of the center of a layer beneath the pile tip.

For driven piles or pile groups in sand, account should be taken of the likely depth of influence of the driving on the modulus of the sand below the pile tips in assessing $E_{sb(e)}$ from Eq. (1) Typically, the depth of influence extends to between 2.5 and 5 diameters below the pile tip, the larger values being associated with greater initial relative density (Poulos and Davis 1980).

Poisson's ratio of the soil is not a very significant parameter for pile settlement prediction if the soil stiffness is expressed in terms of Young's modulus, rather than shear modulus. For saturated clays under undrained conditions, a value of 0.5 is relevant while, for most clays and sands, the drained value is usually in the range 0.3-0.4.

TABLE 2. Summary of Some Correlations for Drained Young's Modulus for Pile Settlement Analysis

(a) Clays

Near-shaft Modulus E_s	Small strain Modulus E_{si}	Modulus well below pile tips E_{sl}
$(2.5 \pm 0.5)N$ MPa (Decourt et al. 1989)	14N MPa (Hirayama 1991)	$(0.5 \pm 0.2)N$ MPa (Stroud 1974)
$(500 \pm 300)\ c_u$ (Callanan and Kulhawy 1985)	$1500\ c_u$ (Hirayama 1991)	$(150 \pm 50)\ c_u$
$(15 \pm 5)\ q_c$ (Poulos 1989)	$49.4 q_c^{0.695} e_o^{-1.13}$ MPa (Mayne and Rix 1993) q_c in MPa e_o = initial void ratio	$(7.5 \pm 2.5)\ q_c$
–	–	(0.5 - 0.7) M (M = constrained modulus)

(b) Silica Sands

Near-shaft Modulus E_s	Small strain Modulus E_{si}	Modulus well below pile tips E_{sl}
$(2.5 \pm 0.5)N$ MPa (Decourt et al. 1989)	$16.9N^{0.9}$ MPa (Ohsaki and Iwasaki 1973)	$7\sqrt{N}$ MPa (Denver 1982)
$(7.5 \pm 2.5)q_c$ MPa (Poulos 1989)	$53 q_c^{0.61}$ MPa (Imai and Tonouchi 1982)	$(7 \pm 4)q_c$ (Jamiolkowski et al. 1988)

Notes:
1. Values of E_s and E_{si} for sands are for single isolated pile. In a group, the values may be increased, depending on pile spacing and initial density.
2. Below pile tip, E_{sb} can be taken as equal to E_s for clays, and 3 to 5 times E_s for sands.
3. Above values of E_s and E_{sb} are for use in an elastic analysis. Higher values are appropriate for nonlinear analyses e.g. the initial tangent values for a hyperbolic model should be 1.4 - 1.6 times the values in this Table.
4. N is the SPT value (blows/300 mm), and should be corrected to a rod energy of 60%.

For a nonlinear analysis of load-settlement behavior, it is also necessary to estimate the pile shaft resistance and the end bearing resistance. There is a vast body of literature on this subject, and some of the available information is summarized by Fleming et al. (1985), Poulos (1989), Decourt (1982), and Bustamante and Gianeselli (1982), among others.

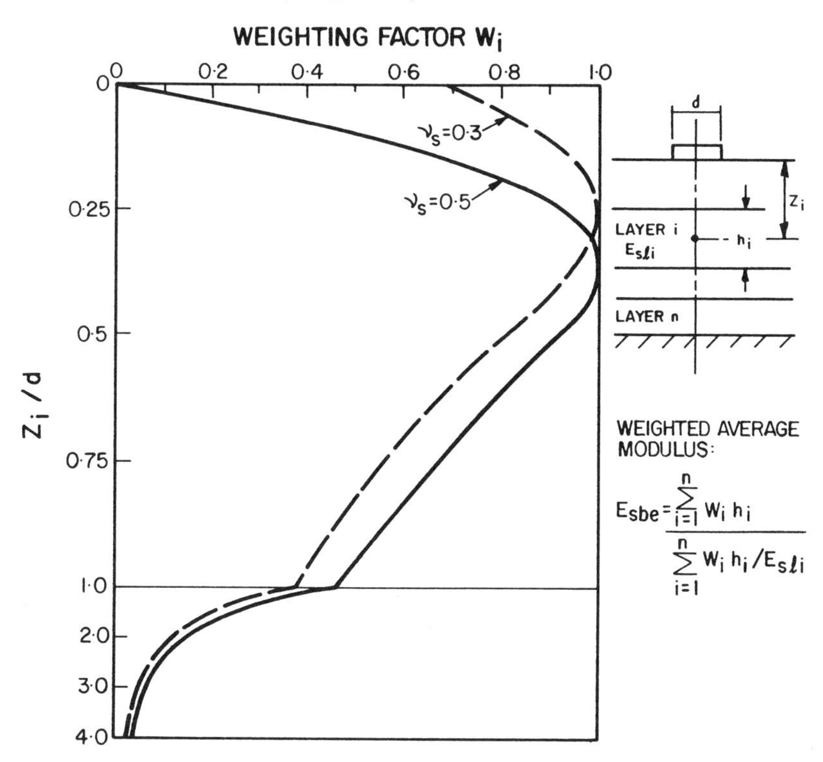

FIG. 6. Weighting Factor for Estimation of Equivalent Modulus below Circular Area

Applications to Some Case Histories

This Section will explore the capability of various methods of settlement prediction when applied to published field case histories. Four such cases will be considered, involving single piles and pile groups in both clay and sand deposits. In each case, the rationale for the assessment of the soil parameters will be described briefly, and the settlement from selected methods of analysis will be compared with the field measurements.

Single Pile Test in Clay

Trochanis et al. (1991) have described the case of a concrete test pile which was loaded to failure in Mexico City. The pile had a 300 mm wide square section and was 15 m long. Fig. 7 shows the available data on the soil profile,which consisted of layers relatively soft Mexico City clay, with an average undrained shear strength of about 40 kPa.

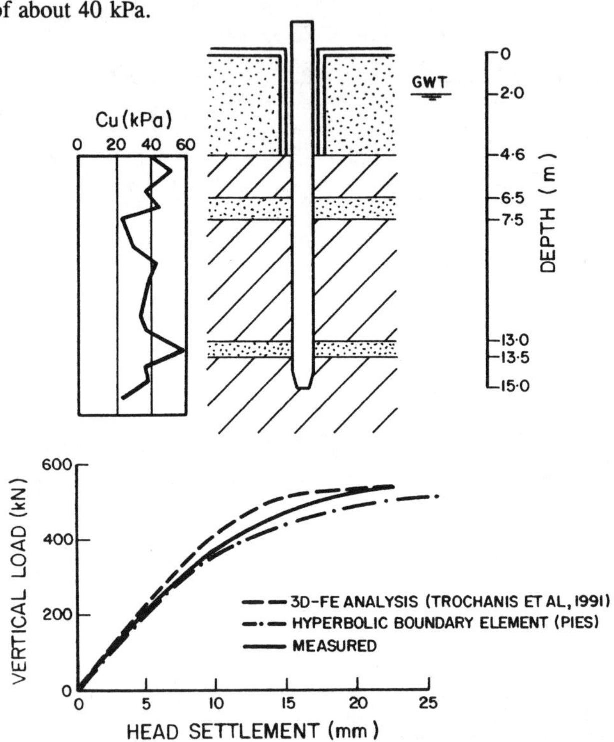

FIG. 7. Measured and Predicted Load-Settlement Curves for Mexico City Load Test

Analyses were carried out via the following methods:

(i) a linear analysis using Fig. 1;

(ii) a linear boundary element analysis using the computer program PIGLET (Randolph and Wroth 1979); and

(iii) a nonlinear boundary element analysis, using the program PIES (Poulos 1989) in which the soil is considered as an elastic continuum, but slip is allowed at the pile-soil interface, and hyperbolic relationships are used to express the nonlinearity of the soil modulus along the shaft and below the tip.

For the linear analyses, on the basis of a lower-bound correlation of Callanan and Kulhawy (1985) in Table 2, the near-pile Young's modulus E_s of the soil was taken to be 220 times the average undrained shear strength c_u, and the modulus at the pile tip E_{sb} was assumed to be equal to E_s. For the nonlinear analysis using PIES, an initial tangent of modulus of 300 c_u was chosen (from Table 2). The ultimate skin friction was assumed equal to c_u, and the ultimate end bearing resistance was taken as $9c_u$. A hyperbolic interface model was used, as described by Poulos (1989), with a hyperbolic factor of 0.9 for the pile tip and 0.5 for the shaft.

Fig. 7 shows the load-settlement curve predicted by PIES and the measured load-settlement curve. These two curves are in good agreement. Also shown is the solution from a nonlinear finite element analysis obtained by Trochanis et al. (1991) using an elastic-plastic soil model. This analysis is in good agreement with both the field measurements and the PIES analysis, but it is interesting to note that it required about 2.5 hours to obtain the solution on a CRAY supercomputer. The solution using PIES required only about 1.5 minutes on a 386 microcomputer.

At a typical working load level of 250 kN, the following settlements were calculated by the linear analysis methods:

Fig. 1:	6.5 mm
PIGLET:	5.8 mm

These values also compare reasonably well with the measured value of about 5.5 mm.

This case demonstrates clearly that it is not always essential to employ a sophisticated numerical analysis to obtain satisfactory settlement predictions for a single pile in the clay. A nonlinear analysis based on the boundary element method gives a good representation of the overall load-settlement behavior, while simple linear methods give adequate predictions of settlement at the working load.

Single Pile Test in Sand

Vesic (1969) described the results of a series of tests on piles in sand, and here one of those has been analyzed by various methods. The pile (H13) was a steel tube 8.9 m long with 0.457 m outer diameter and a 12.7 mm wall. Calculations of settlement were carried out using three methods:

1. the design curves in Fig. 2;
2. the approach of Randolph and Wroth (1978); and
3. a nonlinear boundary element analysis via the program PIES.

In all cases, the parameters were selected via the same rationale using Table 2 as a basis. The near pile Young's modulus of the soil was taken as 2.5 times the SPT value while the modulus below the pile tip was assumed to be 10 times the average SPT value in the vicinity of the pile tip. For the first two methods, it was necessary to approximate the near-pile soil modulus distribution as linear, commencing at zero at the surface and finishing at 80 MPa at the bottom of the shaft. The value E_{sb} below the pile tip was taken to be 300 MPa. For the PIES analysis, using the recommendations in Table 2, the initial tangent moduli were taken as 1.5 times the values used for the elastic analysis, and the hyperbolic parameters were taken as 0.5 for the shaft and 0.9 for the base. It was also necessary to assess values of ultimate skin friction and end bearing pressure, and based on the test results, values of 4N kPa and 0.4N MPa respectively were selected.

At a typical working load of 1 MN, the computed settlements were as follows:

Fig. 3:	3.3 mm
Randolph and Wroth:	3.4 mm
PIES:	4.5 mm

These values compare reasonably well with the recorded settlement of about 5 mm.

Fig. 8 compares the measured load-settlement curve to failure with that computed from PIES. The agreement is quite good over the whole range of load. Also shown in the computed load-settlement curve for the case in which the soil Young's modulus below the pile tip is the same as that above the tip i.e. if no account is taken of installation effects. Clearly, this assumption leads to unsatisfactorily large settlements at loads above about 1 MN.

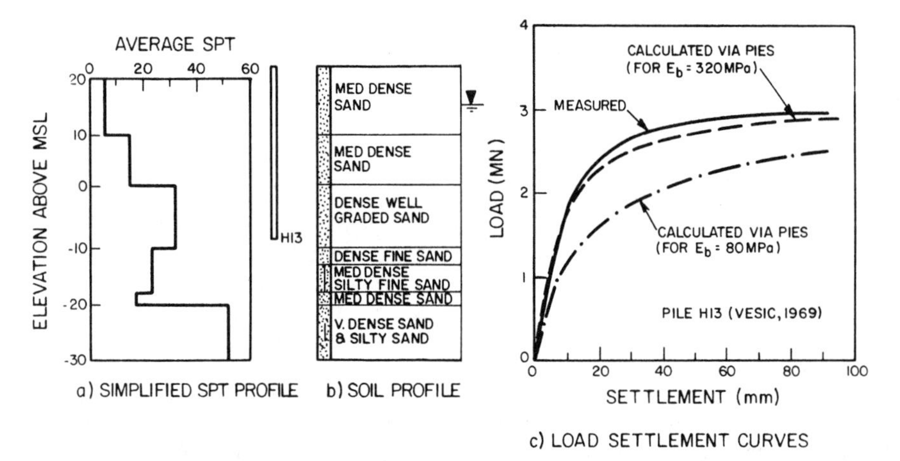

FIG. 8. Measured and Computed Load-Settlement Curves for Single Pile in Sand

Pile Group in Clay

Borsetto et al. (1991) have presented the results of settlement measurements on a number of buildings in the Ostiglia power plant in Italy and have compared these with finite element analyses. The 200 m high chimney of units 3 and 4 has been chosen for analysis here. The chimney was supported on 281 Frankipiles 0.52 m in diameter and 25 m in length, below a circular concrete mat foundation 30.4 m in diameter and 4.25 m thick. The total design load was about 235 MN.

The soil profile consisted of a layer of clayey silt about 10 m thick underlain by a silt layer 14 m thick, overlying a sand layer about 14 m thick, which was in turn underlain by stiff clayey silt. Cone penetration test data were presented in the paper, and, on the basis of the correlations shown in Table 2, the near-pile soil

Young's modulus in the silty soils was taken to be 7.5 times the cone resistance q_c, the modulus below the pile tips was assumed to be $15q_c$ (to a distance of 2 m below the tips), and the modulus well below the pile tips was taken as $5q_c$. The Young's modulus values were considered to be long-term (drained) values, and the consequent settlements were total (immediate plus consolidation) values.

Settlement calculations have been carried out using a linear equivalent raft analysis, the equivalent pier method (Fig. 4), the settlement ratio method, and a nonlinear equivalent pier analysis using the program PIES. For the equivalent raft method, the raft was assumed to act at a depth of 2/3 of the pile length, and corrections were made for raft rigidity and embedment. In the settlement ratio analysis, the single pile settlement was computed using the chart in Fig. 2, and the settlement ratio was estimated as the square root of the number of piles. In the PIES analysis, using correlations collected by Poulos (1989) and also Table 2, the ultimate skin friction and end bearing were estimated from the CPT data, the initial tangent modulus values were taken as 1.4 times the values used in the linear analysis, and the hyperbolic parameters for the shaft and base were taken as 0.5 and 0.9 respectively.

Table 3 shows the computed settlements of the group for the total load of 235 MN. These range between about 35 and 52 mm, and compare reasonably with the measured settlement, 6.5 years after construction, of about 40 mm; this settlement appeared to still be increasing with time. The settlements in Table 3 agree reasonably well with those computed from Borsetto et al. (1991) from three different finite element analyses, which ranged between about 39 and 52 mm. It should be noted that the latter calculations were based on a completely independent assessment of the soil Young's moduli by those authors.

TABLE 3. Computed and Measured Settlements for Large Pile Group in Clay

Method	Settlement	Remarks
Equivalent Pier	47.2	Linear analysis, using Fig. 5
Equivalent Raft	34.7	Linear analysis for subsurface rigid raft
Equivalent Pier	42.8	Nonlinear analysis via program PIES
Settlement Ratio	51.8	$R_s = n^w$; w = 0.5 assumed
Measured (Borsetto et al. 1991)	40	Approx. 6.5 years after construction

Pile Group in Sand

Leonards (1972) has described briefly the case of a 10-pile group of piles driven through a 2.4 m thick layer of stiff silty clay into a dense silty sand layer underlain by very dense gravel. The piles were 305 mm diameter shell-type piles,with lengths ranging between about 3.3 and 4.0 m. Average SPT values were about 6 in the stiff silty clay, 30 in the dense silty sand and about 90 in the very dense gravel.

The group settlement for this case has been computed by three approaches: the equivalent pier (linear) method, using Fig. 4, the settlement ratio method, and a nonlinear boundary element analysis of the equivalent pier using the program PIES. In the linear analyses, using the correlations in Table 2, the soil Young's modulus along the pile shaft has been estimated as 2.5N MPa (where N = average SPT) while the value below the pile tips has been assumed to be 7.5 N, to a distance of 4 diameters below the tip, and then 2.5 N below this depth. In the PIES analysis, the initial tangent moduli along the shaft and immediately beneath the tips were taken as 1.5 times the value used for the linear analysis. For the chart solution for the equivalent pier, an average value of E_b was computed using Fig. 6, and a value of 168 MPa was obtained. For the settlement ratio method, the single pile settlement was computed from Fig. 2, and the settlement ratio was taken to be equal to the cube root of the number of piles (Poulos 1989). An average pile length of 3.35 m has been assumed.

Fig. 9 summarizes the computed load-settlement curves, together with the measured relationships. The agreement between the two linear methods is very close, and at the maximum load of 3.8 MN, the computed settlements agree very well with the measured value. This agreement is however fortuitous, as the actual load-settlement curve is strongly nonlinear. The nonlinear equivalent pier analysis gives larger settlements, but gives a closer representation of the actual load-settlement behavior. If the values of initial tangent Young's modulus had been further increased by about 50-100%, the agreement with the measured load-settlement response would have been very close.

This and the preceding case history demonstrate that accurate prediction of settlements is more difficult to achieve for pile groups than for single piles, a characteristic which was noted earlier in discussing Table 1.

CONCLUSIONS

This paper has reviewed some of the available methods of load-settlement prediction for piles and pile groups and has demonstrated that similar answers can be obtained from a variety of approaches. Design charts enabling rapid estimation of pile and settlements have been presented, and it has been demonstrated that the representation of a group by an equivalent pier is a convenient means of carrying out both linear and nonlinear analyses of group load-settlement response.

Procedures for estimating the required soil parameters have been reviewed, and it has been emphasized that different values of soil Young's modulus may be required along the pile shaft, immediately below the pile tips, well below the pile tips, and between the piles.

The ability of several methods to predict settlements has been examined by applying them to a number of published case histories. For a single pile in clay, acceptable estimates of settlement can be obtained from relatively simple linear methods of analysis. Nonlinear analysis methods however appear to be desirable for piles and pile groups in sand, as the load settlement behavior can become strongly nonlinear at conventional design load levels. In such cases, a nonlinear boundary element analysis appears to be capable of providing an adequate load-settlement prediction, although careful selection of the soil Young's moduli is necessary. Indeed, the selection of soil moduli appears to be far more critical to the success of pile settlement prediction than is the method of analysis, provided that a soundly based method is employed.

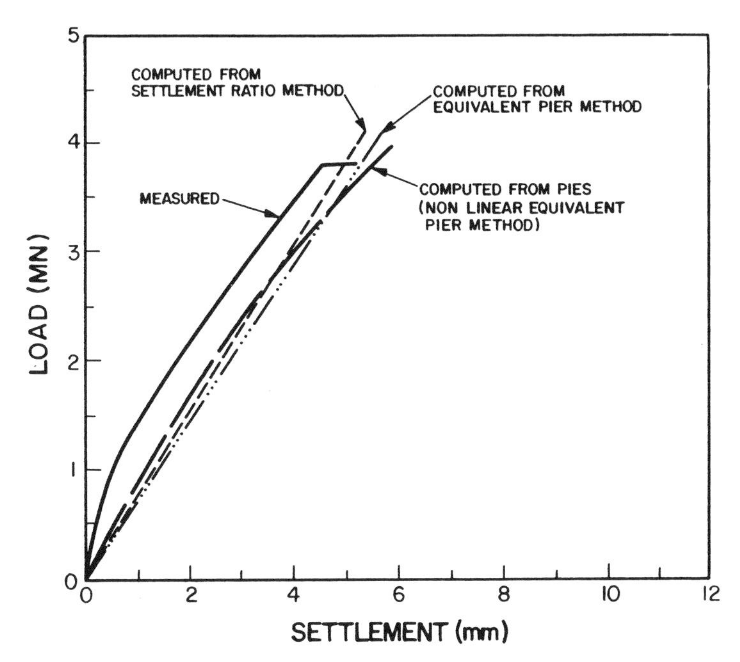

FIG. 9. Measurement and Computed Load-Settlement Curves for 10-Pile Group in Sand (Leonards 1972)

APPENDIX - REFERENCES

Banerjee, P. K., and Davies, T. G. (1977). "Analysis of pile groups embedded in Gibson soil." *Proc. 9th ICSMFE*, Tokyo, 1, 381-386.

Borsetto, M., et al. (1991). "Settlement analysis of main buildings in power plants by means of 2-D and 3-D modes." *Proc. 10th Eur. Conf. Soil Mech. Found. Eng.* Florence, 1, 323-328.

Bustamante, M., and Gianeselli, L. (1982). "Pile bearing capacity prediction by means of static penetrometer CPT." *Proc. ESOPT II*, Amsterdam, 2, 493-500.

Butterfield, R., and Banerjee, P. K. (1971). "The elastic analysis of compressible piles and pile groups." *Géotechnique*, 21(1), 43-60.

Butterfield, R., and Douglas, R. A. (1981). "Flexibility coefficients for the design of piles and pile groups." *Tech. Note 108*, CIRIA, London, England.

Callanan, J. F., and Kulhawy, F. H. (1985). "Evaluation of procedures for predicting foundation uplift movements." *Report to EPRI, EL-4107*, Cornell Univ., Ithaca, New York.

Chow, Y. K. (1986). "Analysis of vertically loaded pile groups." *Int. J. Num. Anal. Meth. Geomechs.*, 10(1), 59-72.

Decourt, L. (1982). "Prediction of the bearing capacity of piles based exclusively on N values of the SPT". *Proc. ESOPT II*, Amsterdam, 1, 29-34.

Decourt, L., Belicanta, A., and Quaresma Filho, A. R. (1989). "Brazilian experience on SPT." *Proc 12th ICSMFE*, Rio De Janeiro, Supp. Contributions by Braz. Soc. for Soil Mechs. 49-54.

Denver, H. (1982). "Modulus of elasticity determined by SPT and CPT." *Proc. ESOPT II*, Amsterdam, 1, 35-40.

Fleming, W. G. K. (1992). "A new method for single pile settlement prediction and analysis." *Géotechnique*, 42(3), 411-425.

Fleming, W. G. K., Weltman, A. J., Randolph, M. F., and Elson, W. K. (1985). *Piling Engineering*, Halsted Press, New York, New York.

Frank, R. (1985). "Recent developments in the prediction of pile behavior from pressuremeter results." *Proc. Symp. from Theory to Practice on Deep Foundns*, Porto Alegre, 1, 69-99.

Hirayama, H. (1991). "Pile group settlement interaction considering soil non-linearity." *Comp. Meth. Adv. Geomechs.*, A. A. Balkema, Rotterdam, The Netherlands, 1, 139-144.

Imai, T., and Tonouchi, K. (1982). "Correlation of N value with S-wave velocity and shear modulus." *Proc. ESOPT II*, Amsterdam, 1, 67-72.

Jamiolkowski, M., Ghionna, V., Lancellotta, R., and Pasqualini, E. (1988). "New correlations of penetration tests for design practice." *Proc. ISOPT-1*, Orlando, 1, 263-296.

Jardine, R. J., Potts, D. M., Fourie, A. B., and Burland, J. B. (1986). "Studies of the influence on non-linear stress-strain characteristics in soil-structure interaction." *Géotechnique*, 36(3), 377-396.

Kraft, L. M., Ray, R. P., and Kagaura, T. (1981). "Theoretical t-z curves." *J. Geotech. Eng. Div.*, Proc. ASCE, 107(GT12), 1543-1561.

Lee, C. Y. (1993). "Pile group settlement by hybrid layer approach." *J. Geotech. Eng.*, ASCE, 119(6), 984-997.

Leonards, G. A. (1972). "Settlement of pile foundations in granular soil." *Proc. Conf. Perf. Earth and Earth-Supp. Structs*, ASCE, 2, 1169,1184.

Mayne, P. W., and Rix, G. J. (1993). "G_{max} -q_c relationships for clays." *Geotech. Test. J.*, ASTM, 16(1), 54-60.

Ohsaki, Y., and Iwasaki, R. (1973). "On dynamic shear moduli and Poisson's ratio of soil deposits." *Soils and Found.*, 13(4), 61-73.

O'Neill, M.W. (1983). "Group action in offshore piles." *Proc. ASCE. Conf. Geot. Prac. in Offshore Eng.*, Austin, 25-64.

O'Neill, M. W., Ghazzaly, O. I., and Ha, H. B. (1977). "Analysis of three-dimensional pile groups with nonlinear soil response and pile-soil-pile interaction." *Proc. 9th OTC,* Houston, 245-256.

O'Neill, M.W., and Ha, H.B. (1982). "Comparative modelling of vertical pile groups." *Proc. 2nd Int. Conf. Num. Meth. Offshore Piling*, Austin, 399-418.

Ottaviani, M. (1975). "Three-dimensional finite element analysis of vertically loaded pile groups." *Géotechnique*, 25(2), 159-174.

Polo, J. M., and Clemente, J. L. M. (1988). "Pile group settlement using independent shaft and point loads." *J. Geotech. Eng.*, ASCE, 114(4), 469-487.

Poulos, H. G. (1968). "Analysis of the settlement of pile groups." *Géotechnique*, 18(4), 449-471.

Poulos, H. G. (1972). "Load-settlement prediction for piles and piers." *J. Soil Mech. Found. Div.*, Proc. ASCE, 98(SM9), 879-897.

Poulos, H. G. (1979). "Settlement of single piles in non-homogeneous soil." *J. Geotech. Eng. Div.*, Proc. ASCE, 105(GT5), 627-641.

Poulos, H. G. (1989). "Pile behavior-theory and application." *Géotechnique*, 39(3), 365-415.

Poulos, H. G. (1993). "Settlement prediction for bored pile groups." *Deep Found. on Bored and Auger Piles,*, A.A. Balkema, Rotterdam, The Netherlands, 103-117.

Poulos, H. G., and Davis, E. H. (1980). *Pile Foundation Analysis and Design*, John Wiley & Sons, New York, New York.

Poulos, H. G., and Hewitt, C. M. (1986). "Axial interaction between dissimilar piles in a group." *Proc. 3rd Int. Conf. Num. Method in Offshore Piling,* Nantes, 253-270.

Poulos, H. G., and Randolph, M. F. (1983). "A study of two methods for pile group analysis." *J. Geotech. Eng.*, ASCE, 109(3), 355-372.

Pressley, J.S., and Poulos, H.G. (1986). "Finite element analysis of mechanisms of pile behavior." *Int. Jnl. Num. Anal. Meth. Geomechs.* 10, 213-221.

Randolph, M. F. (1986). "RATZ - load transfer analysis of axially loaded piles," *Rep. Geo. 86033*, Dept. Civ. Eng., Univ. West. Australia, Perth.

Randolph, M. F., and Wroth, C. P. (1978). "Analyses of deformation of vertically loaded piles." *J. Geotech. Eng. Div.* Proc. ASCE, 104(GT12), 1465-1488.

Randolph, M. F., and Wroth, C. P. (1979). "An analysis of the vertical deformation of pile groups." *Géotechnique*, 29(4), 423-439.

Stroud, M. A. (1974). "The standard penetration test in insensitive clays and soft rocks." *Proc. ESOPT*, Stockholm, 2.2, 367-375.

Terzaghi, K., and Peck, R. B. (1967). *Soil Mechanics in Engineering Practice*, 2nd Ed. John Wiley & Sons, New York, New York.

Tomlinson, M. J. (1986). *Foundation Design and Construction*, 5th Ed., Longman Scientific and Technical, Harlow, England.

Trochanis, A. M., Bielak, J., and Christiano, P. (1991). "Three-dimensional nonlinear study of piles." *J. Geotech. Eng.*, ASCE, 117(3), 429-447.

Van Impe, W.F. (1991). "Deformations of deep foundations." *Proc. 10th Eur. Conf. Soil Mechs. Foundn. Eng.*, Florence, 2, 1031-1062.

Vesic, A.S. (1969). "Experiments with instrumented pile groups in sand." *ASTM STP 444*, Philadelphia, Pennsylvania, 177-222.

Settlement Analysis for 450 Meter Tall KLCC Towers

Clyde N. Baker, Jr.,[1] Fellow ASCE, Ir. Tarique Azam,[2]
and Len S. Joseph,[3] Member ASCE

Abstract: The site evaluation and settlement analysis performed for the two tallest buildings in the world currently under construction, are described.

The site evaluation resulted in shifting building locations to facilitate minimizing differential settlement. Two full scale 30,000 kN instrumented bored pile load tests are evaluated to determine foundation design parameters.

An elastic soil modulus back-calculated from the pile load test results compared reasonably to the average rebound reload modulus from an extensive in-situ pressuremeter testing program performed at representative borehole locations.

The rationale for the foundation solution of a raft on variable length post-grouted bored piles (or barrettes) is outlined and the results from the SAP90 computer program used to model the foundation subsoil system is presented, as well as a discussion of the selection of the appropriate modulus values used in the computer program. The treatment of underlying limestone cavities and overlying slump zones to help assure substrata deformation properties consistent with foundation design assumptions is briefly described.

Introduction

The twin Petronas Towers as currently under construction by developer Kuala Lumpur City Centre Berhad will be the tallest buildings in the world when

---

[1] Senior Principal Engineer, STS Consultants, Ltd., Northbrook, Illinois 60062.

[2] Chief Geotechnical Engineer, Ranhill Bersekudah, Sdn, Bhd, Kuala Lumpur, Malaysia.

[3] Vice President, Thornton-Tomasetti, New York, New York 10011.

completed with a height of 450 meters (7 meters taller than the current record holder, Sears Tower). The 4 million square feet towers are the first phase of a 17 million square foot complex planned for a 97 acre site in Kuala Lumpur, Malaysia. The towers will have 88 occupied stories above grade and 5 levels of below grade parking. Each tower has perimeter columns on a 46 meter diameter base with an adjacent 21 meter diameter, 45 story bustle. The towers stand 55 meters apart and are connected by a bridge at the 41st and 42nd floors.

Due to the high slenderness ratio and the structural interconnection of the towers, the developer and the designer were very interested in minimizing differential settlement. The goal was to have the calculated design differential settlement be as close to zero as reasonably practical (less than 1/2 inch (12.7 mm) across the base of the towers). Meeting this desirable goal was made technically very challenging due to the known geologic site conditions. Ten to 20 meters of water bearing alluvium is underlain by variable thickness residual soils of Metasedimentary formations; namely siltstone, sandstone, shale and occasionally phyllite (known as the Kenny Hill formation), followed by the Kuala Lumpur limestone formation which can vary dramatically with regard to surface elevation and solution activity (rock elevations varying 140 meters over a distance of less than 50 meters). The interface is often overlain by erratic slump zones where Kenny Hill material has softened and eroded into limestone cavities.

With this anticipated geology and the goal of minimizing differential settlement, the foundation design concepts studied included a "floating" raft, a system of bored piles socketed into limestone past any significant cavities, and a raft on friction piles located in the Kenny Hill well above the limestone (but with cavities and slump zones grout-filled, if necessary), with pile lengths varied to minimize differential settlement. During the preliminary design and soil exploration phase, conditions were found to be so variable at the initially planned tower locations as to make achieving design objectives at reasonable cost impractical. The tower locations were then shifted approximately 60 meters to where the thickness of Kenny Hill formation was sufficient to support a raft on bored friction piles. Architectural and space planning requirements limited the amount of relocation feasible. Settlement analyses then depended primarily on the properties of the supporting Kenny Hill formation.

Site Geology and Soil Profile

The site geology and soil profile were determined in some detail by means of several hundred borings and probes throughout the site area. The work was performed in three basic phases. Phase I consisted of 28 split-barrel sample borings with coring when hard rock was encountered.

Fig. 1 shows a location diagram of both the final and initial planned tower locations. Based on this preliminary investigation, a definite difference in the Kenny Hill formation was observed with weaker Kenny Hill noted in the area underlain by shallower limestone noted as Zone A and stronger Kenny Hill indicated where the limestone was deep (more than 80-100 meters), noted as Zone B. Boring locations

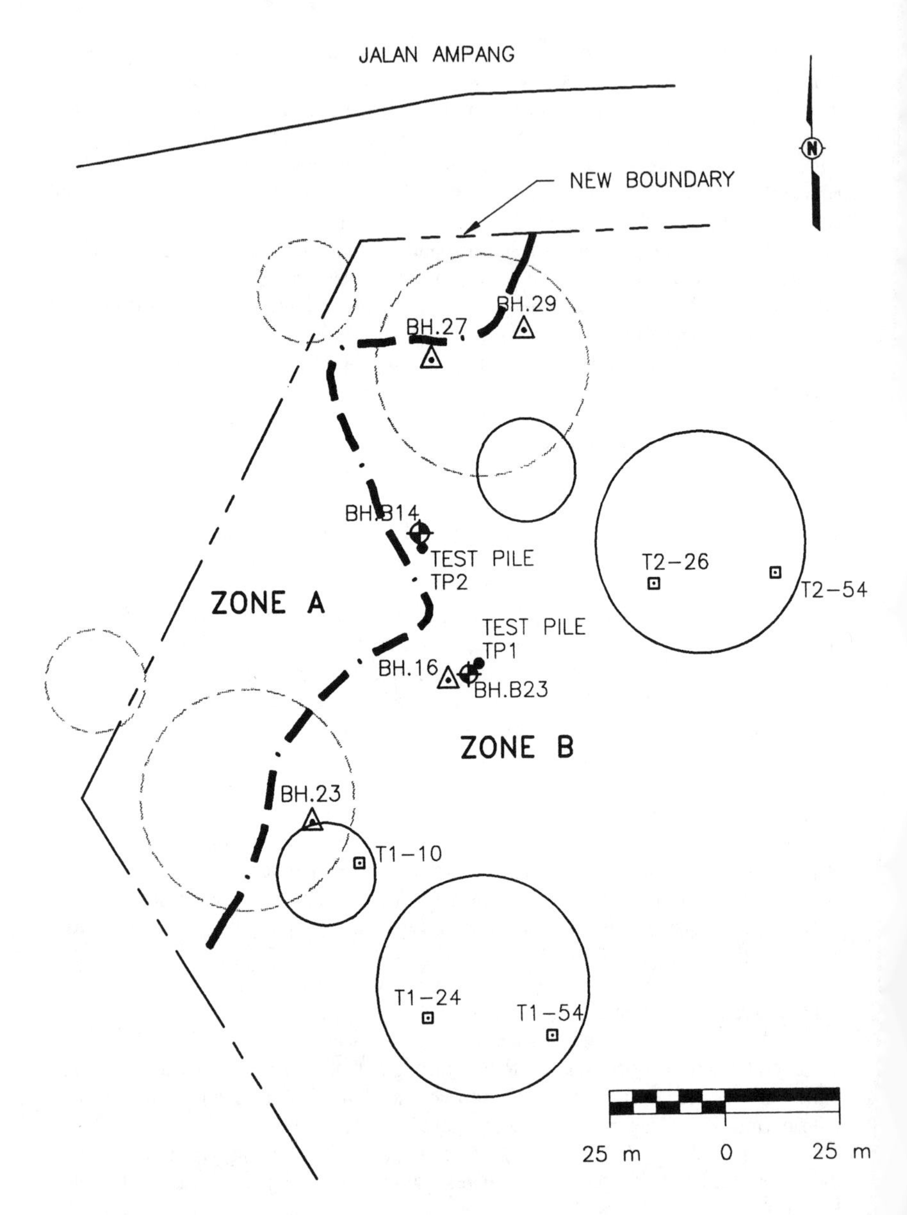

FIG. 1. Location Diagram

where data has been mentioned in this paper are the only ones noted for easy reference.

The Phase II exploration program consisted of split-barrel sample borings on approximate 16 meter centers at the planned tower locations with additional rock probes at actual column locations coring 20 meters into limestone to check for major cavities.

The Phase III exploration program was required at the revised tower locations and consisted of split-barrel sample borings on 8 meter centers extended to a maximum depth of 180 meters or 20 feet of limestone coring. A cross section profile across the towers in a northeast-southwest direction is shown in Fig. 2. Typical standard penetration resistance profiles are shown in Fig. 3.

Based on the initial exploration and laboratory testing program, the soil properties selected for initial design are shown in Table 1. The final elastic soil deformation modulus selected for design was subsequently doubled (from 125 MPa to 250 MPa) based on the results of the test pile program and in-situ pressuremeter testing program described in the next sections.

TABLE 1. Preliminary Geotechnical Design Parameters

Kenny Hill Formation	Zone A	Zone B
Residual Soils		
Elevation (depth below G.S.)	12±2 to 25±5	8±2 to 25±5
Bulk Density	19 kN/m^3	19 kN/m^3
Drained Conditions	$c' = 0$, $\phi = 30°$	$c' = 0$, $\phi = 32°$
Undrained Conditions	$c_u = 100$ kPa, $\phi = 0$	$c_u = 125$ kPa, $\phi = 0$
Deformation Modulus	75 MPa	85 MPa
Weathered Rock		
Elevation	25±5 to 30-90	25±5 to 100+
Bulk Density	20 kN/m^3	20 kN/m^3
Drained Conditions	$c' = 0$, $\phi' = 33°$	$c' = 0$, $\phi' = 34°$
Undrained Conditions	$c_u = 200$ kPa, $\phi = 0$	$c_u = 250$ kPa, $\phi = 0$
Deformation Modulus	100 MPa	125 MPa*

* Raised to 250 MPa after test pile program and pressuremeter test program.

A summary of laboratory tests for moisture content, plasticity and grain size gradation from two representative borings is shown in Table 2.

In-Situ Pressuremeter Testing Program

During the phased exploration program, more than 260 in-situ pressuremeter tests were performed at nine representative boring locations; two at the original tower locations, two near the test piles, and five at the final tower locations. The

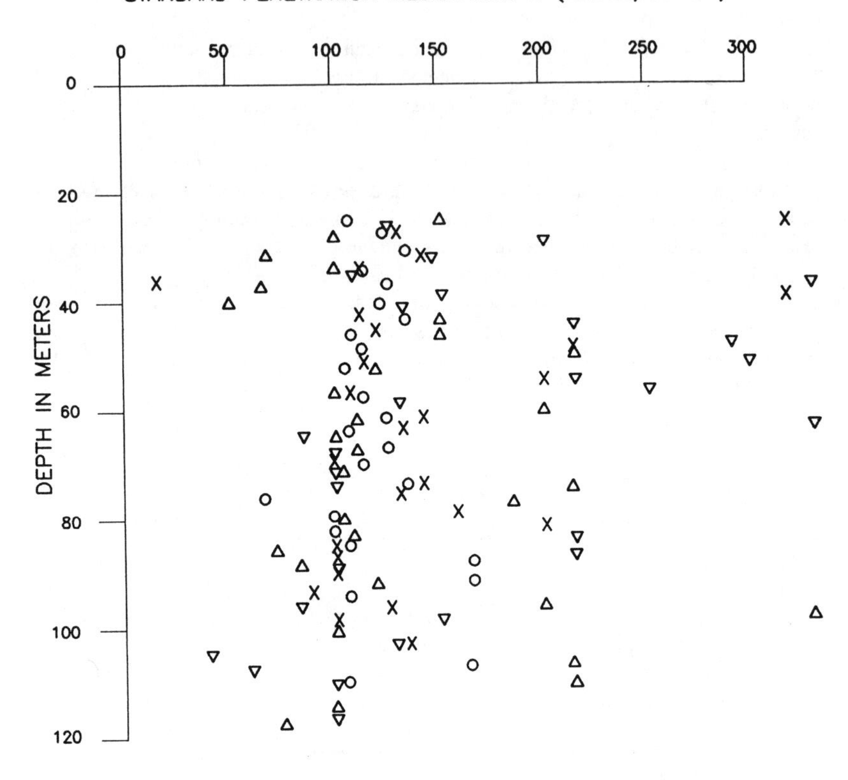

FIG. 3. Standard Penetration Resistance Profile

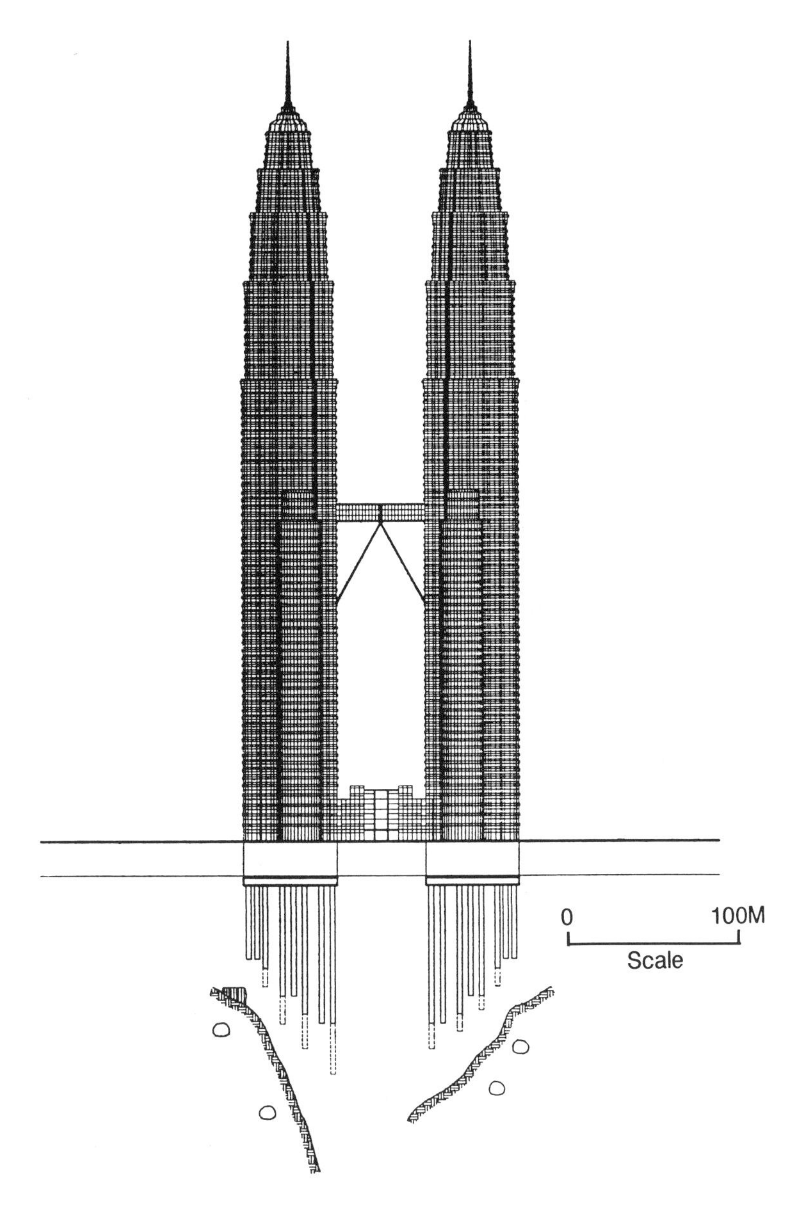

FIG. 2. Tower Foundation Profile

TABLE 2. Master Testing Services SDN.BHD

Laboratory Tests

Borehole No	Sample No	Depth (m)	M/C %	LL %	PL %	PI %	Clay %	Silt %	Sand %	Gravel %
BH-23	D1	33.40	22		LS		27	48	23	2
	D3	39.40	27	39	25	14	24	45	19	12
	D5	45.40	29	39	26	13	28	46	22	4
	D7	51.40	22	39	25	14	26	48	17	9
	D9	57.00	24		LS		19	63	17	1
	D10	60.00	-		-		20	60	14	6
	D11	64.90	24	40	26	14	15	67	14	4
	D12	75.90	-		-		27	51	17	5
	D13	78.40	-		-		30	58	10	2
BH-29	D1	1.50	24	43	26	17		58	37	5
	D3	4.50	14	43	25	18		27	54	19
	UD1	9.00	29	46	29	17	58	17	1	
	D8	12.00	30	46	27	19	30	56	12	2
	D11	16.50	25	46	30	16		80	16	4
	D14	21.00	27	40	26	14		68	24	8
	D17	25.50	11		LS			36	30	34
	D19	30.00	–		–			45	30	25
	D21	36.00	17	43	25	18		62	17	21
	D23	42.00	44	56	35	21	18	54	19	9
	D25	48.00	–		–			70	15	15
	D27	54.00	26	56	35	21		77	18	5
	D29	60.00	25		LS		23	48	21	8
	D31	64.50	29	57	37	20	18	52	23	7
	D34	72.00	30		LS		14	32	30	24
	D36	75.00	40	56	34	22		68	17	15
	D38	78.00	48	57	37	20	16	40	24	20
	D40	81.00	–		–			59	21	20
	D43	85.50	50		LS		26	34	21	19
	UD2	88.50	34	44	27	17	29	41	21	9

Remarks: L.S. = Limited Sample

pressuremeter tests were performed with a Menard type pressuremeter in accordance with ASTM designation D4719-87. The Phase I pressuremeter tests were performed without any unload/reload cycles. The remaining pressuremeter tests were typically performed with an unload/reload cycle in the pseudo elastic range below the creep pressure value. The results of the pressuremeter tests at four representative locations at the final tower locations are plotted versus depth in Figs. 4 and 5 and all test results are summarized in Table 3. For comparison, the pressuremeter test results at the original tower locations are shown in Fig. 6.

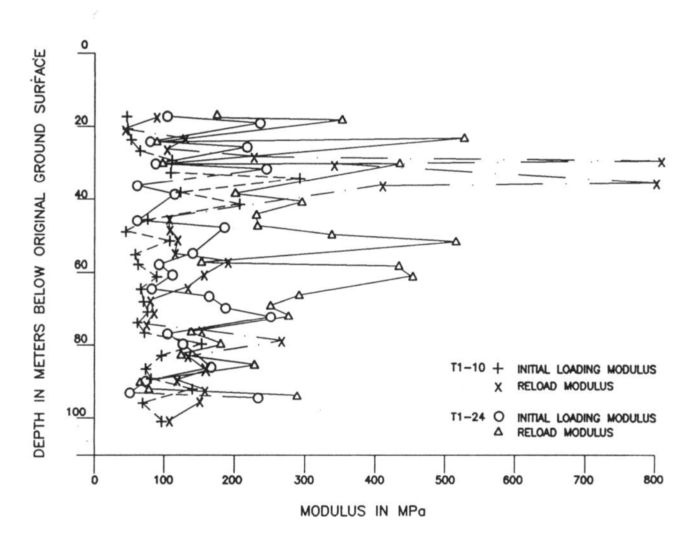

FIG. 4. Pressuremeter Test Profile at Boring Locations T1-10 and T1-24

The pressuremeter tests confirm a wide variation in the modulus properties of the Kenny Hill formation, resulting from the weathering process, varying parent material and distance above any slump zone activity. This confirmed the division of the Kenny Hill into an A zone at the extreme northwestern portion of site, encompassing significant portions of the initial tower locations, and a B zone for the remainder of the site encompassing the final tower locations. The locations closest to the A zone (Borings B14 and 27), gave the lowest average pressuremeter modulus, less than 40 MPa. However, the average pressuremeter modulus at the final tower locations is 97 MPa, and the average unload/reload modulus is 269 MPa.

At the boreholes located close to the test pile locations, the average pressuremeter modulus was 86 MPa and the average unload/reload modulus was 289 MPa. However, there are many significantly softer and harder spots throughout the Kenny Hill formation, with pressuremeter modulus values from 9 to 693 MPa and unload/reload modulus values from 22 to 3,800 MPa.

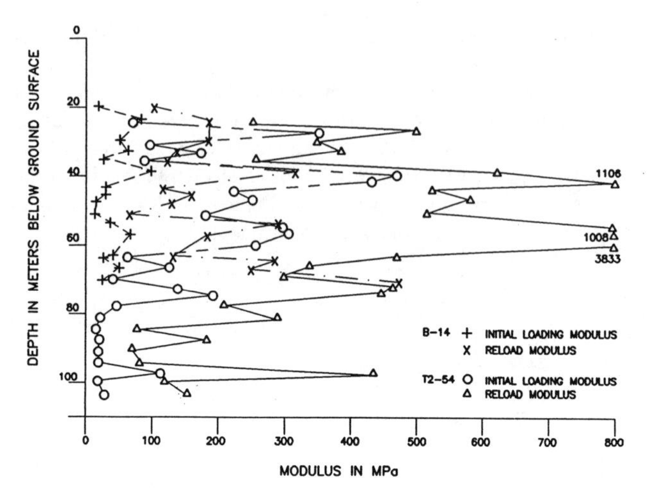

FIG. 5. Pressuremeter Test Profile at Boring Locations B-14 and T2-54

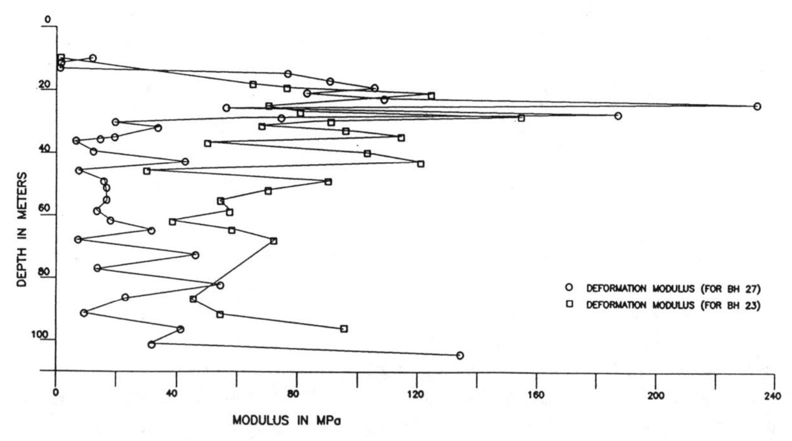

FIG. 6. Plot of Pressuremeter Modulus with Depth

Test Pile Program

Two full scale bored pile load tests were conducted to determine design friction values and Kenny Hill modulus values. The test piles were 1.2 m in diameter extending to a depth of 75 m, with the surface 20 meters cased off and isolated by double casing. The test pile locations were near the center of the tower site, between the towers and in Zone B close to the boundary with Zone A. The test piles were installed under bentonite slurry following production piling specifications. Both piles were constructed with full length rebar cages, ten levels

TABLE 3. Pressuremeter Test Results

Boring	B14	B23	T1-10	T1-24	T1-54	T2-26	T2-54
E_d Min.	9.3 MPa	10 MPa	32 MPa	17.8 MPa	38.5 MPa	18.3 MPa	11.7 MPa
Max.	99	309	683	222	199.4	157	470
# of Tests	18	15	27	26	26	31	27
Avg.	37.6 MPa	133.9 MPa	67.9 MPa	109.8 MPa	101.8 MPa	64.1 MPa	149 MPa
E_R Min.	27.5	22.3	55	32	57.7	47.8	68.3
Max.	479	931	851	496	590.3	495	383.3
# of Tests	17	15	27	25	25	31	27
Avg.	186.9 MPa	391.8 MPa	176 MPa	226 MPa	223 MPa	190 MPa	535 MPa

Overall weighted
E_d Avg. = 94.3
E_R Avg. = 267

of vibrating wire gage instrumentation, and soft toes with built in load cells. The perimeter of Test Pile 1 was post-grouted to determine its effect on friction capacity and load transfer. A reaction in excess of 30,000 kN was provided for the test piles by a Kentledge system using stacked concrete blocks. Test Pile 2 was constructed first, taking six days with bentonite in the shaft and likely thick filter cake development. Test Pile 1 was constructed in three days with undoubtedly less filter cake development. Test Pile 1 was loaded to 30,000 kN in three load cycles, whereas Test Pile 2 took five load cycles to reach 30,000 kN. The observed load versus top deflection curves are shown in Fig. 7 and the calculated load transfer curves based on the observed strain gage data are shown in Figs. 8 through 11.

The effect of the thick filter cake on Test Pile 2 is clearly evident in the observed results and the benefit of post-grouting to break through any filter cake is evident in Test Pile 1.

Comparing Test Pile 2 with Test Pile 1, it appears that Test Pile 2 initially measured primarily friction in the filter cake, but as cycles increased and deflection increased, more of the friction capacity in the Kenny Hill was developed. Because of the filter cake, more load was transferred initially over the full length of the pile, but with increasing deflection, relatively greater load was carried by the upper Kenny Hill and the load actually decreased over the bottom 20 meters of pile.

It is interesting to note that the initial slopes of the load deflection curves on the later cycles are quite similar between the grouted and ungrouted pile.

The maximum skin friction recorded at Test Pile 1 was 311 kN per square meter, compared to 215 kN per square meter at Test Pile 2, at maximum applied load of about 30,000 kN. However, since the maximum friction capacity measured was still increasing, it is likely that if additional reaction had been available, somewhat higher friction values could have been developed. This is particularly true with Test Pile 1 where the bottom 15 meters of pile carried negligible load.

At maximum applied load of approximately 30,000 kN, the bottom of Test Pile 2 recorded a settlement of 118 mm compared to only 0.4 mm at Test Pile 1, with measured toe load cell readings of 1,042 kN and 168 kN, respectively. The significant bottom settlement at Test Pile 2 could indicate excessive sediment accumulation prior to concreting.

The toe settlement was measured with a tell tale and linear displacement Transducer to an accuracy of 0.002 mm.

Using the results from Test Pile 1 as more fairly representing the capacity of the Kenny Hill, an elastic soil modulus can be back-calculated by several different techniques. Using the load and deflection measured at the pile tip, and the formula

$$E = \mu_o \mu_1 \frac{qB}{S} \tag{1}$$

where q = bearing pressure; B = width; $\mu_o\mu_1$ = influence factors and S = settlement. As described by Christian and Carrier (1978), an elastic modulus of 267 MPa is calculated.

Using the measured top deflection and the Poulos and Davis procedures (1980) for the case of a floating pile corrected for a partially isolated top 20 meters,

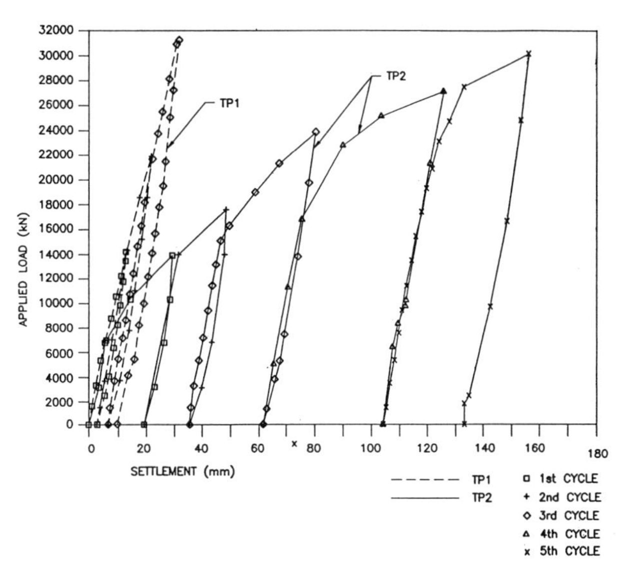

FIG. 7. Relationship between Applied Load with Settlement (Transducer) – TP1 and TP2

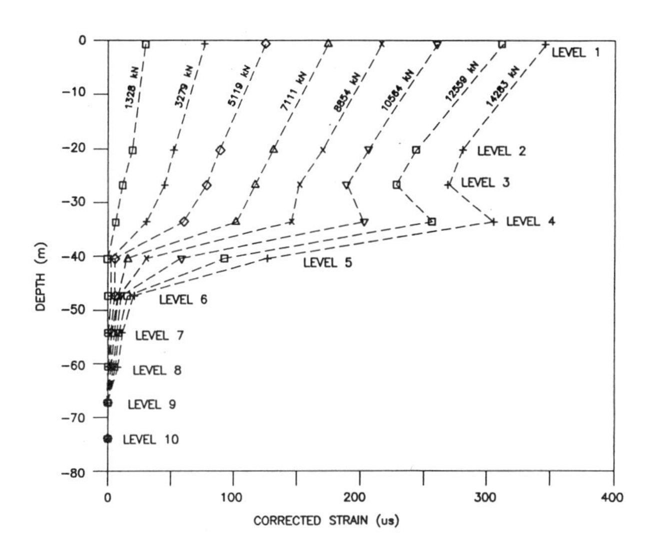

FIG. 8. Relationship of Depth with Strain and Applied Load – TP1 (1st Cycle)

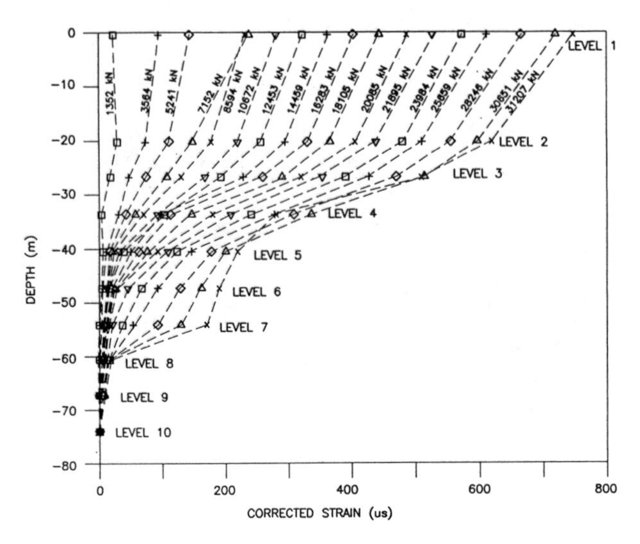

FIG. 9. Relationship of Depth with Strain and Applied Load − TP1 (3rd Cycle)

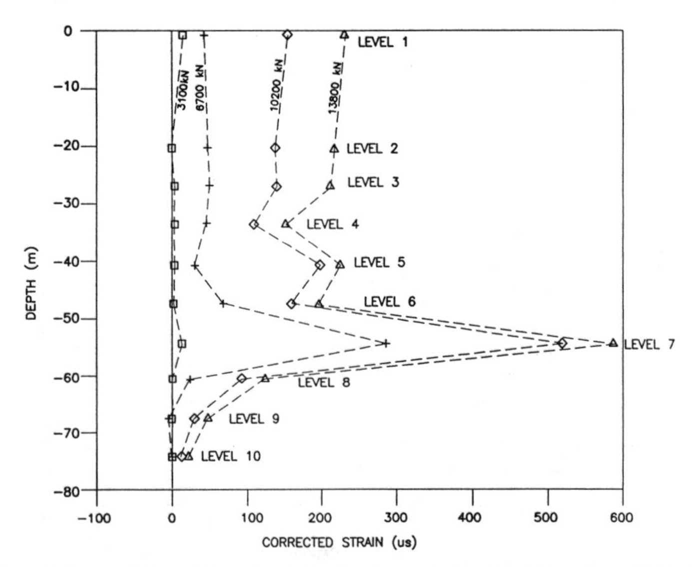

FIG. 10. Relationship of Depth with Strain and Applied Load − TP2 (1st Cycle)

an elastic modulus of approximately 200 MPa is calculated. Since the upper portions of the Kenny Hill are typically more weathered and compressible than the deeper portions and since the test piles are near the boundary with the weaker A zone Kenny Hill and since the revised tower locations are further into the better B zone Kenny Hill, an elastic soil modulus of 250 MPa was selected as representative for final settlement analyses. This value is slightly less than the overall average pressuremeter unload reload modulus of 269 MPa. Baker (1993) reported that settlement predicted based on elastic theory of a 65 story building supported on very dense silt in Chicago compared favorably (on the conservative side) with measured performance utilizing the average reload modulus from pressuremeter tests as the modulus of elasticity E.(Any anisotropic effects are considered not significant (in very dense silt or highly weathered silty shale) or conveniently compensated for or balanced by other factors in some way unknown to the writers since the method seems to give reasonable results.) A SAP90 finite element analysis using different E's above and below 20 meters to simulate casing friction, also back calculated an E of just below 200 MPa.

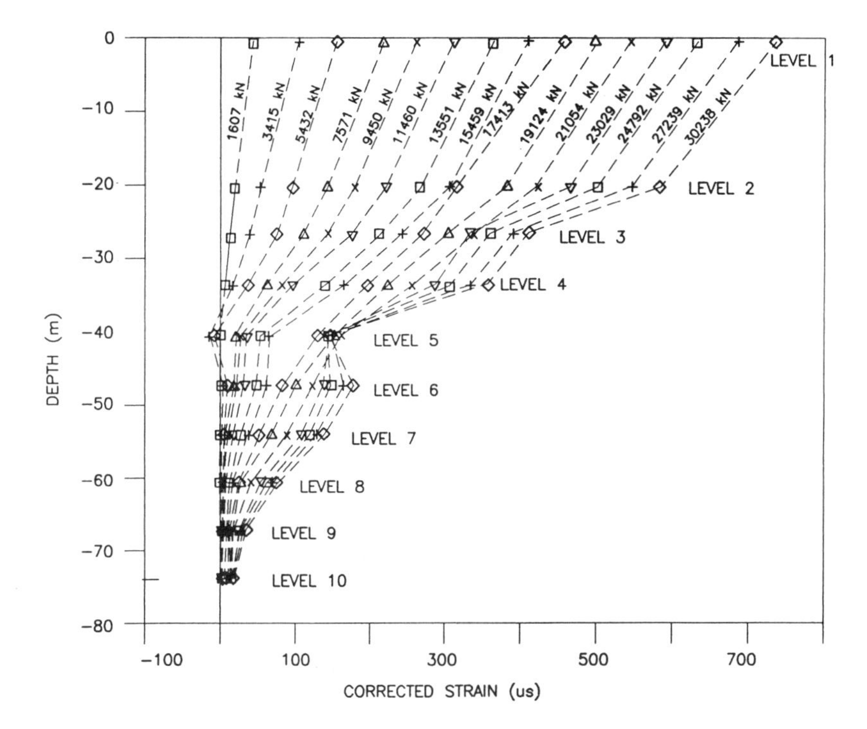

FIG. 11. Relationship of Depth with Strain and Applied Load - TP2 (5th Cycle)

A design friction value of 110 kPa was also selected, providing a factor of safety of almost 3 based on the maximum developed friction by the skin grouted test pile.

A modulus can also be estimated from the average standard penetration values. The average blow count in two borings relatively close to the test pile in the 20 to 50 meter depth range where most of the pile load was carried was 80. This yields an E/N ratio of about 2.5 MPa which is midway between values recommended by other local practitioners. Toh et al. (1991) back-calculated from several pile load tests in Kenny Hill on other sites and observed an average value of E/N = 5.7. However, Chang and Broms (1990) observed on pile load tests performed in Singapore in dense residual soils that E/N = 1 all measured in MPa. Thus, a ratio of 2.5 for this site appears reasonable.

No specific reduction in elastic modulus was made to allow for stress relief effects caused by the drilling under slurry of the bored piles. It was assumed that any such relief would be partly compensated for by the soil and slurry replacement by the fluid concrete and that the final value selected was sufficiently conservative.

Foundation Analysis and Settlement Analysis

At the final selected tower locations, the depth to limestone varied from approximately 80 meters to over 180 meters, making practical a raft on piles solution for support of the towers. Based upon a tower load of 2,680,000 kN and a raft diameter of 53.7 meters, the average loading under the raft is approximately 1,180 kN per square meter. Accepted local practice limits the allowable bearing for footings and mats near the surface of the Kenny Hill formation to no more than 500 kN per square meter. Enlarging the raft size to reduce bearing pressure was investigated, but the larger mat would not be stiff enough to redistribute bearing loads and control differential settlement, making a mat-only solution impractical. Thus, the mat on bored piles was the selected foundation system, with the primary question being the pile penetration required. Based on friction capacity utilizing the design friction value of 110 kPa, 1.3 meter diameter piles on 4.7 meter centers extending 33 meters below the mat would be adequate based on bearing capacity considerations. Fig. 12 illustrates this case.

With the above case in mind, a simple calculation can be made for predicted settlement based on an average elastic soil modulus of 250 Mpa. Using Eq. (1), but with a reduced combined influence factor of 0.47 resulting from the different geometry, a predicted settlement of 98 mm is calculated assuming an average load spread of 1H:4V from the mat down and acting at the 2/3 depth point (22 meters below the mat).

The 1H:4V stress spread from mat level to the pile equivalent mat level 22 meters below the mat is a conservative first approximation of the effect of the friction load carried around the perimeter of the block of piles and is only useful as a first quick analysis.

This magnitude of settlement was considered excessive, particularly when recognizing the lighter, rigidly attached bustle would tend to cause tilt in the tower. For this reason, extension of piles was considered as a method for reducing total

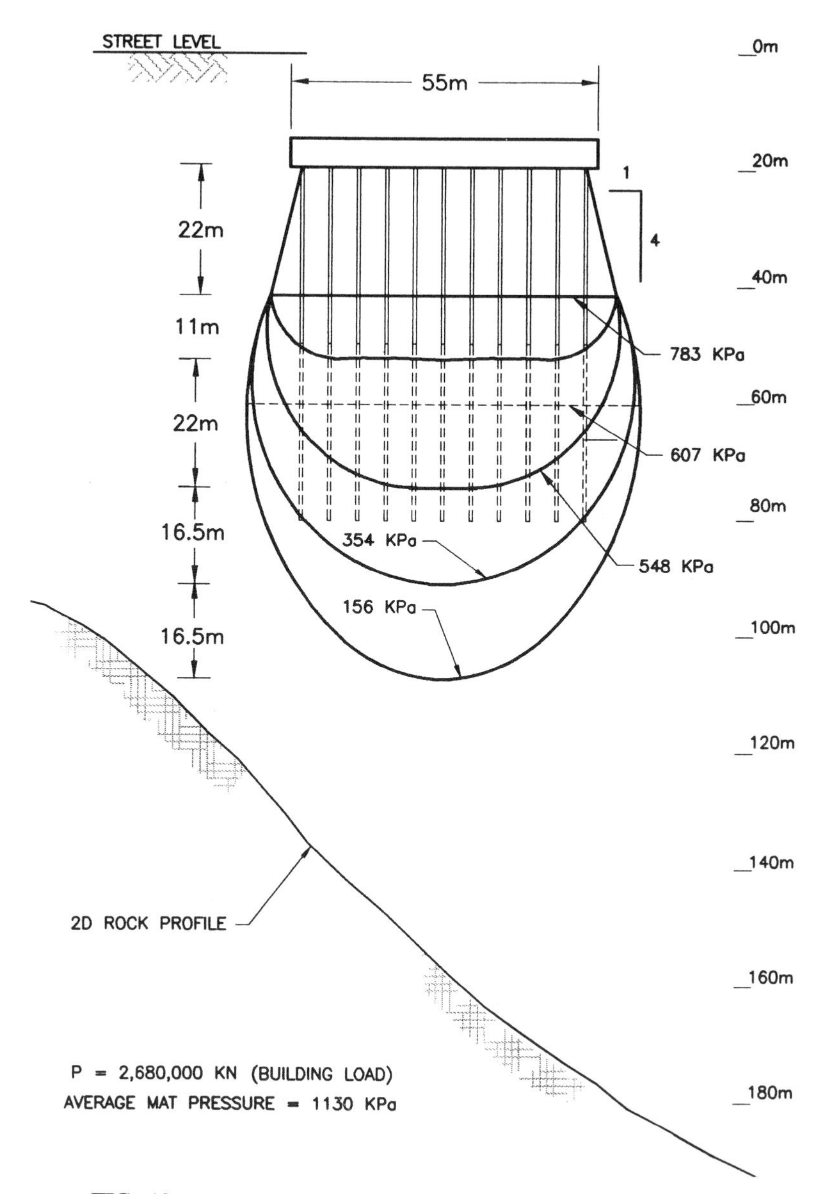

FIG. 12. Foundation Schematic for Simple Settlement Analysis

settlement and thereby reducing differential settlement. Elastic calculations made for a 55 meter diameter raft on 60 meter piles indicated a reduction in total settlement from 98 mm to 65 mm. A separate independent similar elastic calculation for a 25 meter diameter raft on 20 meter piles for the bustle indicated a settlement of 34 mm. These simple calculations assume uniform depth to rock. In actuality, the rock slopes so uniform pile lengths would result in a tendency to tilt.

Settlement analyses performed independently for tower and bustle with the PLAXIS 3D finite element computer program from Delft Laboratories using an axisymmetrical geometric model because of radial geometry and a nonlinear (elastic-plastic mohr-coulomb)soil model gave slightly higher values for similar cases under the tower and lower values under the lighter loaded bustle as summarized in Table 4. Included is a case where the modulus of a 20 m zone directly under the piles has been effectively doubled by jet grouting. This concept for reducing settlement eventually lost out to the concept of lengthening piles which can accomplish the same effect.

TABLE 4. Summary of Settlement Analysis (Plaxis)

Scheme No.	Description	Main Tower Design Load = 1150 kN/m² Rock Depth 133 m Center Settlement	Bustle Design Load = 645 kN/m² Rock Depth 90 m Center Settlement
1	Mat only		35 mm
2	Mat +15 m piles		22 mm
3	Mat + 30 m piles		16 mm
4	Mat + 45 m piles	115 mm	
5	Mat + 60 m piles	82 mm	
6*	Mat + 45 m piles + 20 m deep jet grout zone beneath pile toe	84 mm	
7*	Mat + 60 m piles + 20 m jet grout zone beneath pile toe	50 mm	

* Assumed composite modulus of the jet grout zone is twice the Kenny Hill modulus, or 500 MPa.

In order to further refine the settlement prediction based on varying pile lengths, a three dimensional finite element computer program called SAP90 was used. To develop confidence in the computer model and check for sensitivity to changes in parameters, nearly 100 analyses were performed with variations in modulus (125,250 MPa), Poisson's ratio (0.5 short term, 0.2 long term), Kenny Hill depth, finite element size (4 to 11 m), overall soil block size, boundary conditions, mat stiffness and pile properties. The final models for Towers 1 and 2 reflected their individual rock profiles, and included a laterally restrained plane of symmetry at the side of the soil mass facing the other tower, a 4 meter thick mat, bustle piles 20 meters long, tower piles 40 to 105 meters long, and rock 20 times stiffer than Kenny Hill. With 11,000 nodes and 30,000 degrees of freedom, each final model analysis required 8 hours on a 486DX50 PC with 16 MB RAM and 500 MB hard drive space. The results of the computer runs in the form of deformed shape and deflection diagrams are shown in Figs. 13 through 16.

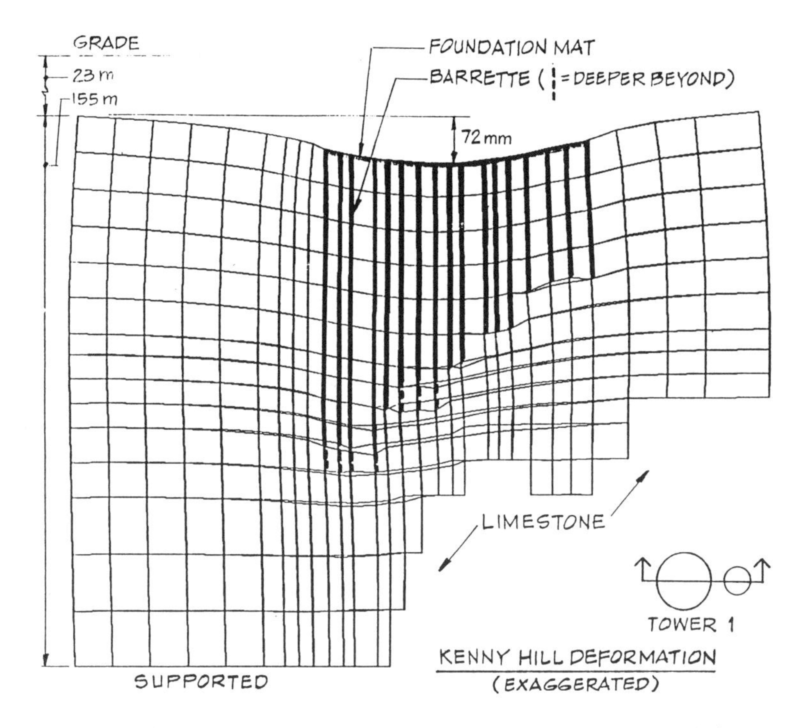

FIG. 13. Tower 1 Foundation Deformation Profile - Long Axis

The results indicate a maximum combined settlement under the tower of approximately 72 mm and minimum settlement under the bustle of approximately 44 mm for a maximum differential settlement along the axis of the tower and bustle of approximately 28 mm, or slightly over 1 inch over a distance of approximately 40 meters. However, under the tower proper, the maximum edge to edge differential settlement was only 11 mm, or less than 1/2 inch. This satisfied the design goal.

Consideration was given to the possible destructuring effect on the Kenny Hill (with resultant decreased modulus and increased settlement) of excavating 20 meters off the site for the basement and raft. To minimize this effect, the piles were required to be installed prior to significant basement excavation (a 3.5 meter advance excavation was permitted as part of the contractor's construction sequencing plan) in order to reinforce the ground and restrain it from heaving. Provided destructuring did not occur, small pseudo elastic heave (based on the effective composite modulus of the Kenny Hill and concrete piles) would be compressed back on reloading without effecting predicted elastic settlement of the towers.

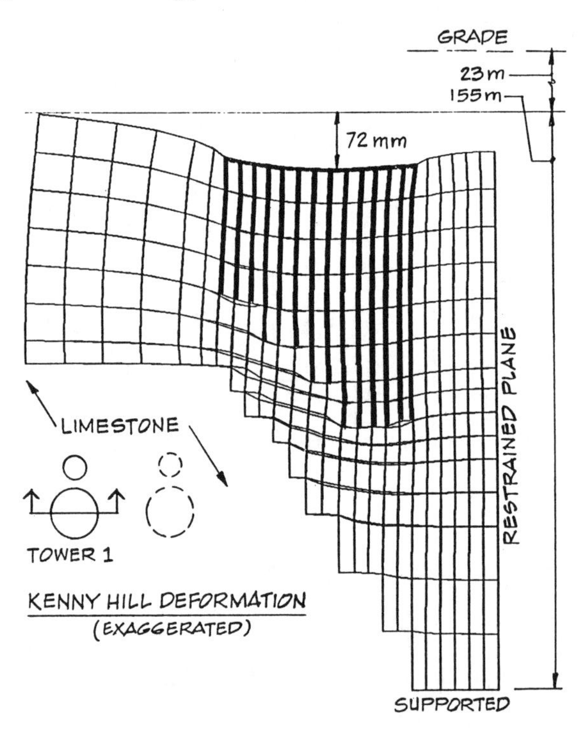

FIG. 14. Tower 1 Foundation Deformation Profile - Short Axis

Vibrating wire strain gage instrumentation has been installed in the foundations to permit monitoring actual load transfer into the ground and a settlement observation program is planned to compare actual settlement with predicted.

Ground Improvement Program

In order to have confidence in the assumed modulus values taken for the Kenny Hill formation and underlying rock, a grouting program was required to fill

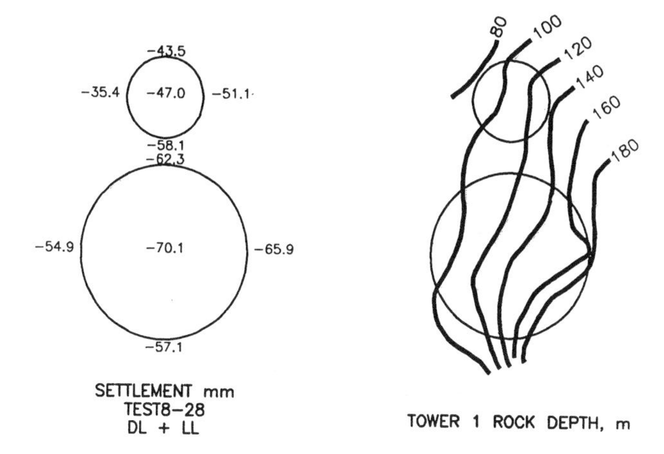

FIG. 15. Tower 1 Settlement Map and Rock Contour Plan

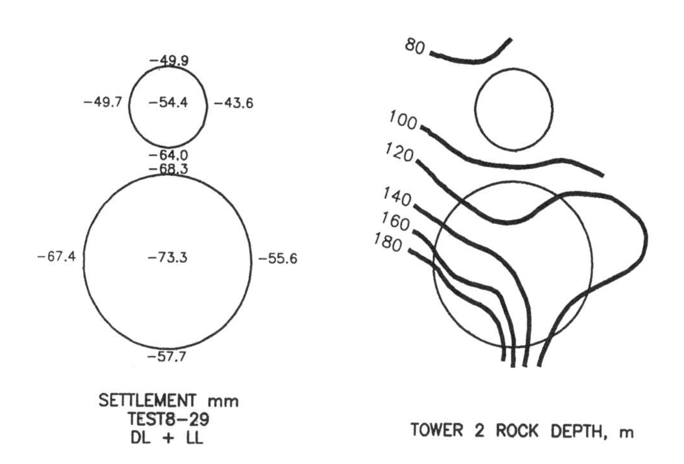

FIG. 16. Tower 2 Settlement Map and Rock Contour Plan

cavities in the rock within the tower zone of influence and to improve the slump zone areas occasionally found immediately above the limestone, where Kenny Hill has eroded into cavities and solution channels in the limestone. The goal of the grouting program was to improve the slump zone areas to the equivalent of adjacent Kenny Hill and to make certain that large cavities in the limestone within the structure influence zone could not suddenly collapse causing unanticipated settlement. Details of the grouting program are beyond the scope of this paper. However, in summary, 2,300 cubic meters of grout was placed in 15 limestone cavities in the Tower 1 area, with no slump zone areas encountered. In the Tower 2 area, 1,100 cubic meters of grout was placed in limestone cavities and 900 cubic meters of compaction grout placed in 6 slump zone areas (5 under the bustle and 1 under the main tower).

In the limestone cavities, either fluid cement grout with a water cement ratio of 0.6:1 or a moderate slump compaction grout (mortar grout) was utilized, typically in a multi-stage program to minimize off-site grout loss.

In the slump zone areas, only low slump compaction grout was utilized in an effort to both densify slump zone material and to create a column of grout to improve the composite effective modulus of the slump zone grout material.

A check hole program was utilized to confirm success in both the cavity grouting and slump zone grouting.

CONCLUSIONS

1. In-situ pressuremeter tests utilizing unload-reload cycles and back-calculation from full scale pile load tests provide reasonably comparable modulus of soil elasticity values for predicting foundation settlement, for a raft on bored piles supported in a weathered shale and sandstone formation, similar to dense to very dense soil consisting of a mixture of silt, sand and clay.
2. Simple hand calculations using elastic theory and simplifying assumptions yield settlement prediction values consistent with refined three dimensional finite element computer programs, modeling the raft foundation, bored piles, supporting soil, and underlying rock. However, with the three dimensional program, it is possible to take into account varying loads in different areas and varying pile lengths, as well as varying rock depths and thus refine both total and differential settlement predictions.

ACKNOWLEDGEMENTS

The authors wish to acknowledge the support of Kuala Lumpur City Centre Berhad for the geotechnical investigations and analyses, and their kind permission for the publication of this paper.

APPENDIX - REFERENCES

Baker, C. N., Jr. (1993). "Use of pressuremeter in mixed high rise foundation design." *Design and Performance of Deep Foundations: Piles and Piers in Soil and Soft Rock*, Geotechnical Special Publication No. 38, ASCE, New York, New York, 1-13.

Chang, M. F., and Broms, B. B. (1991). "Design of bored piles in residual soils based on field-performance data." *Can. Geotech. J.*, 28(2), 200-209.

Christian, J. T., and Carrier, W. D. (1978). "Janbu, Bjerrum and Kjaernki's chart reinterpreted." *Canadian Geotechnical Journal*, 15(1), 124-128.

Poulos, H. G., and Davis, E. H. (1980). *Pile Foundation Analysis and Design*, John Wiley & Sons, New York, New York.

Toh, C. J., Ooi, T.A., Chiu, H.K., Chee, S.K., Ting, W.H. (1989). "Design parameters for bored piles in a weathered sedimentary formation." *Proc. 12th ICSMFE*, Rio de Janeiro, 2, 1073-1078.

Vibration Induced Settlement from Blast Densification and Pile Driving

Charles H Dowding,[1] Member, ASCE

Abstract: This paper describes settlement and vibration environments produced by two construction activities, blast densification and pile driving. Blasting is of interest because of the significant energy released during the process, and piling effects are of great interest because of their universal use in construction. The paper begins with a consideration of the distribution of surface settlement from other construction induced losses of ground that are more familiar. Details of several case studies form the basis for the following general discussion of the settlements resulting from blasting and piling. These two techniques are related in that they have been found to be most effective only with liquefiable sands. Finally, the paper ends with a comparison of piling induced settlements with those resulting from expected loss of lateral support during excavation in sands.

Introduction

Familiar settlement distribution curves from tunneling and mining subsidence provide a useful model for discussion of dynamically induced construction settlements. As Peck (1969) demonstrated in his "state-of-the-art" paper on soft ground tunneling, the surface manifestation of a subsurface volume loss takes the form of a normal distribution curve centered over the lost volume. As shown in Fig. 1, mining of a tabular body (Brauner 1973) produces settlements, which are maximum over the center and extend outward a distance nearly equal to the depth of the lost volume. The critical distance, B, is approximately the radius of the volume lost at which the settlement is 1/2 that of the maximum. Furthermore these past studies show that the maximum settlement, s_{max}, is less than the height of the

[1] Professor, Dept. of Civil Engineering, Northwestern Univ., Evanston, IL 60208.

of the volume lost. The reduction in apparent settlement increases as the depth increases at which the volume is lost and as the width of the volume lost decreases relative to the depth. In mining and rock mechanics, this reduction is largely the result of bulking of the material above the volume loss. In soil mechanics, it is often conservatively assumed that there is no volume reduction.

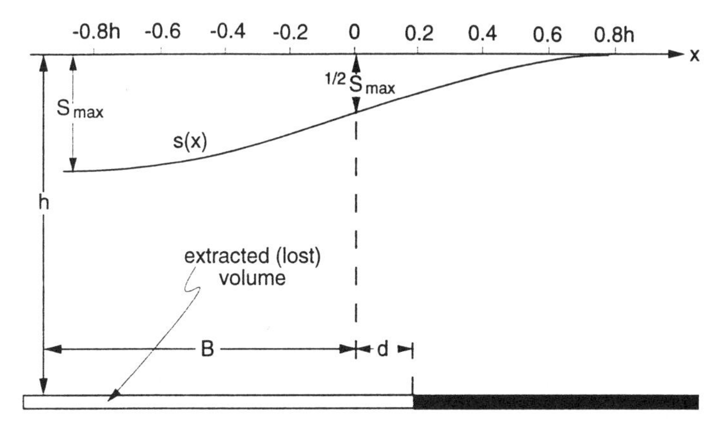

FIG. 1. Subsidence Profile Function from Longwall Mining Showing the Relationship of the Lateral Extent of Settlement Relative to the Extent of the Volume Extracted (Lost or Densified) (from Brauner 1973)

These empirical observations indicate that important parameters are the volume lost as well as the depth and width over which the volume is lost. In the following discussion, volume lost will be described as densification. Unfortunately, the volume densified is not known precisely because most measurements are made at the surface and therefore strain cannot be calculated directly. Because of the lack of data on strains, only the extent of surface settlement will be traced throughout this paper. Even this restriction poses some problems as the extent of surface settlements will be a function of the extent of the zone of subsurface densification.

BLAST DENSIFICATION (from Hryciw 1986)

"In a typical blast-densification project, 1 to 4 kg (2.2 to 8.8 lb) charges are placed in the sand at a depth equal to two-thirds of the layer to be densified. Holes are drilled or jetted for the charges on 5 to 10 m (16.4 to 32.8 ft) centers. Solymar (1984) documented the successful densification of loose river bed alluvium at unprecedented depths of 25 to 40 m (82 to 131 ft) at the Jebba Dam Site in Nigeria. By comparison,the deepest charges at previously successfully densified dam sites, Franklin Falls and Karnafuli, were 4.5 and 15.2 m (15 and 50 ft) (Lyman 1940; Hall 1962). As shown below, it's more likely that only some 10 m of the Nigerian 40 m profile was densified.

If [densification] is achieved in a particular zone, increasing the size of the charge will not improve densification. The zone of [densification] however will be increased. Prugh (1963), Kummeneje and Eide (1961), Ivanov (1967) and others have shown that further [densification] will be achieved by detonating a second and third series of charges after the soil is reconditioned and residual pore pressures have dissipated. The first set of charges will produce 50 to 60 % of the surface settlement, the second 20 to 30% and subsequent series will have progressively reduced effects. Laboratory studies of shock induced [densification] have been conducted by Charlie et al. (1981). Studer and Kok (1980) have documented increases in residual pore water pressure as a function of charge weight and distance from field experience on the Netherlands.

Effectiveness of blast-densification has commonly been judged by surface settlements and in-situ tests such as the standard penetration test (SPT) and cone penetration test (CPT). A typical settlement profile from multiple, single delay blasts is shown in Fig. 2. However, despite obvious densification as verified by surface settlements, cone penetration tests conducted after blasting often register significantly lower penetration resistances than before the blasts. Mitchell and Solymar (1984) [and Dowding and Hryciw (1986)] have reported such decreases followed by steady increases in penetration resistance over a period of months after blasting. These increases were attributed to gradual reformation of cementatious bonds at particle contacts which had been disrupted by blasting and/or [the solution of the gas bubbles produced by the large volume of gas released by the explosives]."

COMPARISON OF BLAST DENSIFICATION STUDIES

Table 1 presents studies in which sufficient information on blasting procedures, induced ground motions, soil type, and settlement were presented to allow comparison. As can be seen by the vacancies in the table, even these carefully chosen studies are not complete. These voids underscore the difficulty of completely instrumenting a blast densification site to sufficiently document both the immediate site and blast information as well as the external effects of the procedure. Components of the table are arranged in a consequential order. The blast generates the ground motions, which interact with the soil to produce surface settlement and increased density of sensitive sand strata.

As shown in the blast section, the charge weight, W, per delay and delay interval is as important in soil as it is in rock. It is the subtle interaction of the stress waves in the near field that produce the densification and it is their addition on the surface that control their external effects. Unlike rock, short delay intervals, while separating the body wave pressure pulses in the near field, do not separate the far field surface waves.

While both near field subsurface body waves and far field surface waves are important, this table summarizes only the relatively far field motions. At scaled distances greater than 25, strains are too small to densify even loose clean sands, however, they are large enough to be perceptible and are therefore important to monitor. Only motions at far field scaled distances are tabulated to highlight the large distances out to which significant motions occur with this process. Distances are scaled by division of either the square or cube root (Dowding 1985) of the explosive energy (proportional to the charge weight per delay, W) and computed in terms of $ft/lb^{1/2}$ in the table; ($10\ ft/lb^{1/2} = 4.9\ m/kg^{1/2}$, with kg weight).

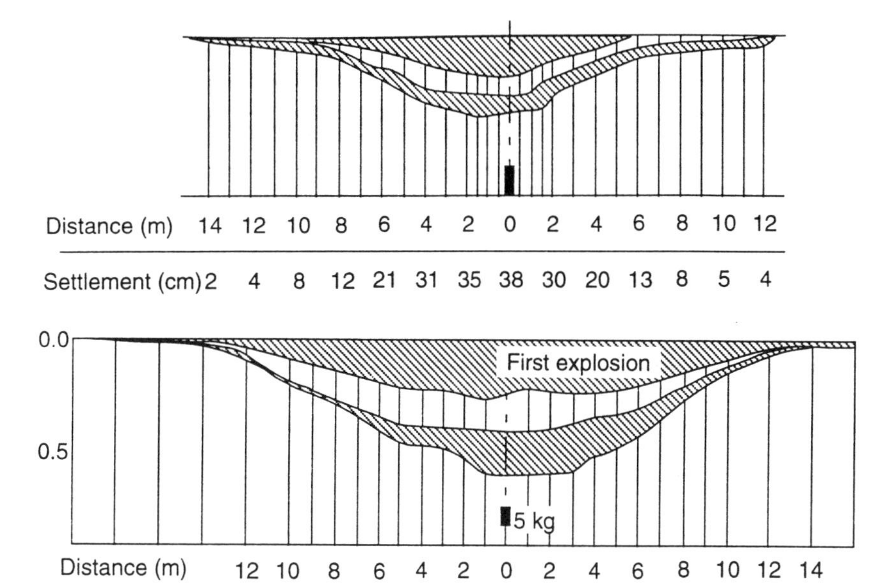

FIG. 2. Surface Settlement of Submerged Sand from 5-kg, Single Delay Blasts 4 to 5 m below the Surface (▮ = Explosives): (Upper) D_r = 0.5, Loose Sand; (lower) D_r = 0.2, Very Loose Sand (from Ivanov 1967)

There are a number of soil conditions that apparently need to be satisfied for blast densification to be effective. As shown in the table, the soil must be a clean sand (D_{10}, size of fines, greater than 0.1 mm),loose (D_r, relative density, less than 75%), and below the water table. If any of these soil property factors is not satisfied, then results are discouraging.

Finally, the extent of the densification is important. Densification is measured in the field by increases in penetration resistance or direct manifestation in surface settlements. Surface settlements achieved during these studies are summarized at the limit of significant strains ($R/W^{1/2}=20$). As will be discussed later, this scaled distance was found by Ivanov to correlate with the distance out to which 10 mm of settlement could be expected from relatively shallow placement of explosives. Lyman's work produced a maximum of 381 mm, in loose, uniform, alluvial sand, whereas Charlie's experience in a fairly dense, gravelly sand showed

TABLE 1. Comparison of Example Blast Densification Studies

Case	Blast Information			Ground Motion surface particle velocity		Soil Information		
	W/delay (kg)	delay interval (ms)	W total (kg)	R/W = 50 (mm/s)	R/W = 100 (mm/s)	D_{10} (mm)	cone tip penetration resistance kg/cm^2	relative density D_r (%)
Lyman (1940)	3.6		75.6			0.15		27
Solymar (1984)	?			?	?	0.11	50@ 33 m	
Dowding (1986)	0.0005	25 & 50				0.15		50
Hryciw (1986)	1.7	0 & 0 0 & 25	3.4	119(1) 66	33(2) 23	0.07-0.1	25@ 2 m	
La Fosse (1991)	3.4	8 125	10.2	64 46	46 33		35@10 m	
Charlie (1992)	5	0	5	41@20		0.5	76@4 m	75

Notes: (1) @15 m (50 ft) (2) @ 30 m (100 ft)

no significant settlement. La Fosse was able to achieve a maximum of 250 mm of settlement. Neither Solymar nor Hryciw measured settlement, and Dowding's lab scale results will not be relevant. More effort is needed to collect information on the extent of the zone of densification (as indicated by the loss of volume models) to assist in the prediction of settlement.

GROUND MOTIONS FROM BLAST DENSIFICATION

The type and depth of burial of transducers to measure ground motions is important. Both subsurface and surface particle velocities measured with specialized, high frequency, transducers (Hryciw and Dowding 1986) at Harriet's Bluff are compared with others in Fig. 3. These other data include that measured by Long et al. (1981) and Solymar (1984) with standard, low frequency, geophones (which measure particle velocity), as well as that measured by Drake and Little (1983) with high frequency, sensitive accelerometers for defense testing. Also presented is the DuPont Blaster's Handbook (1980) standard attenuation curve, used by some for prediction.

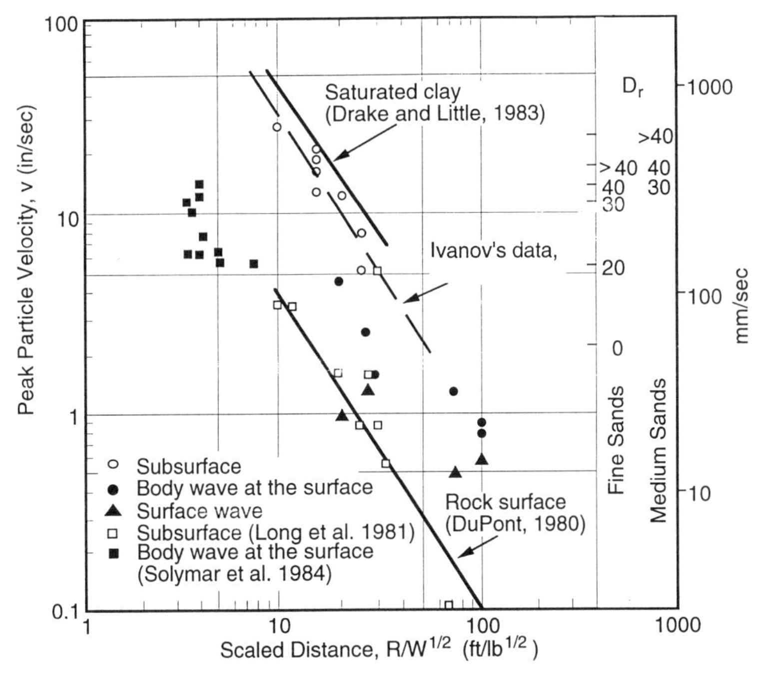

FIG. 3. Attenuation of Peak Particle Velocity Measured with High Frequency Accelerometers (Labeled "Drake and Little" and "○ Subsurface") and Typical Geophones with Limited High Frequency Response (from Hryciw 1986)

In Fig. 3 Harriet's Bluff, subsurface, body-wave, data agree well only with Drake and Little's observations for subsurface motions in saturated clay. This type of agreement is to be expected for the near-field ($R/W^{1/2} < 25$) subsurface data, as both studies employed high frequency transducers (Hryciw 1986). Standard particle velocity geophones cannot record the high frequency pulses found at small scaled distances. Secondly, measurement of body waves at the surface for cases with significant travel distance through unsaturated material (as the case with the Long and Solymar data), there will be greater attenuation in the more compressible, unsaturated materials (Hryciw 1986). Thus these data fall below even further below those measured by Drake.

A comparison in Fig. 4 of surface waves measured at Harriet's Bluff at some distance from the blast with those measured by La Fosse and Gelormino (1992) shows the effect of delay interval. It also confirms the tendency to produce relatively large amplitude, surface wave ground motions when blasting below the water table in loose saturated sands. La Fosse was forced to greatly increase the inter-hole delay beyond 100 ms to reduce off-site, surface wave ground motion. La Fosse data were apparently collected at distances out to 250 m. The standard rock blasting attenuation curve (DuPont 1980) is repeated for reference to show the significant surface ground motions. Charlie's much lower motions seem to represent a lower bound of expectation (Dowding 1995).

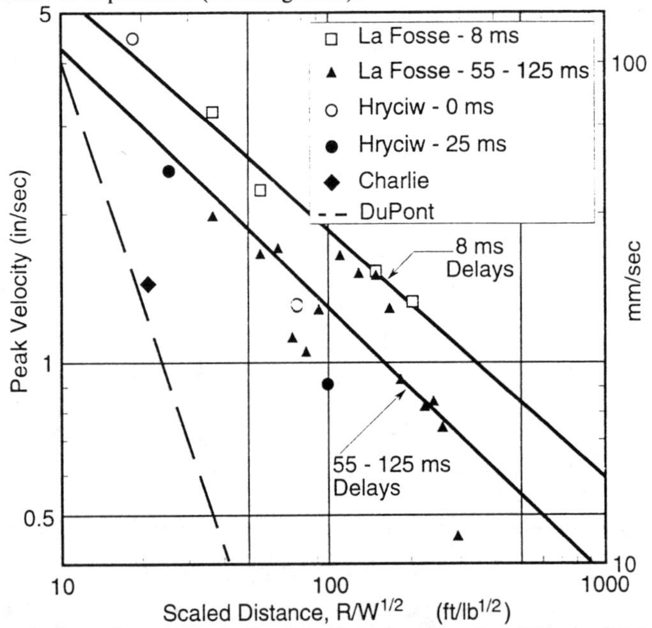

FIG. 4. Attenuation of Peak Particle Velocity Showing Effect of Multiple Holes and Short Delay Intervals (after La Fosse and Gelormino 1991)

Blast Induced Settlement

Settlement profiles produced by 5 kg (11 lb) single delay blasts in loose and medium dense sands are shown in Fig. 2. Measurable settlement extended the furthest (18 m) in the loosest sand and the least in the sand with the greatest density. The cube root scaled distance at 18 m is 26 ft/lb$^{1/3}$. According to the attenuation plots of surface motions in Figs. 3 and 4, particle velocity at the surface, as measured with standard transducers, would range between 25 and 75 mm/s (1 and 3 ips). These sands in Fig. 2 were very loose to loose and would have uncorrected standard penetration resistance (Peck et al. 1974) blow count numbers, N, of less than 4 and 10 respectively. Medium dense sands with estimated, uncorrected N's between 10 and 30 were also found to be densifiable by blasting. These geologically young sands are generally found in river valleys and associated estuaries.

In addition to being loose, geologically recent, and below the water table, these sands are also relatively clean as shown be the grain size distributions in Fig. 5. These grain size distributions are those of the sands for the cases compared in Table 1. This plot of the percent of the soil passing a certain sieve opening (or grain size) is similar to that in Fig. 8, for soils sensitive to vibratory densification from pile driving. In both cases the fines (those smaller than 0.075 mm) do not constitute more than 10% of the grain sizes. The only documented case for fine grained soils displaying tendencies for vibratory densification involved glacial silt deposit that was interfingered with fine grained sands (Lacey and Gould 1985).

Data obtained by Hryciw (1986) can be employed to determine the scaled distance out to which densification can occur by employing Ivanov's (1967) charge weight, W, and radius of influence, R_c, relationship for settlements greater than 1 cm (0.4 in.). The resulting scaled distance relationship is plotted as the dashed line (labeled "Ivanov") for sands of varying relative density on the right hand side of Fig. 3. It shows that increasing particle velocities are necessary to densify sands of increasing relative density.

Fig. 3 can now be employed to calculate a maximum radius of influence, if 1 cm is considered the limit of significance. This is not B, introduced in Fig. 1, but is the distance out to which the surface settles. The radius out to which the sand is actually densified, B, at depth is probably a good deal smaller as indicated in Fig. 1. Since sands with relative densities of less than 20% are relatively rare, a relative density of 25% and scaled distance (SD) of 22 was chosen for this calculation. Therefore a charge of 5 kg (11 lb) will produce densification out to

$$R = SD \times W^{1/2} = 22 \times 11^{1/2} = 22 \times 3.3 = 73 \text{ ft } (22 \text{ m})$$

for a fine sand with relative density of 20% (for metric conversion of $R/W^{1/2}$: 1 ft/lb$^{1/2}$ = 0.49 m/kg$^{1/2}$). This distance should represent the MAXIMUM distance to which densification affects are noticeable at the surface since most sands will have greater absolute and relative densities than implied by this choice. Applicable particle velocities should be measured in the subsurface with accelerometers that can faithfully measure high frequency excitation.

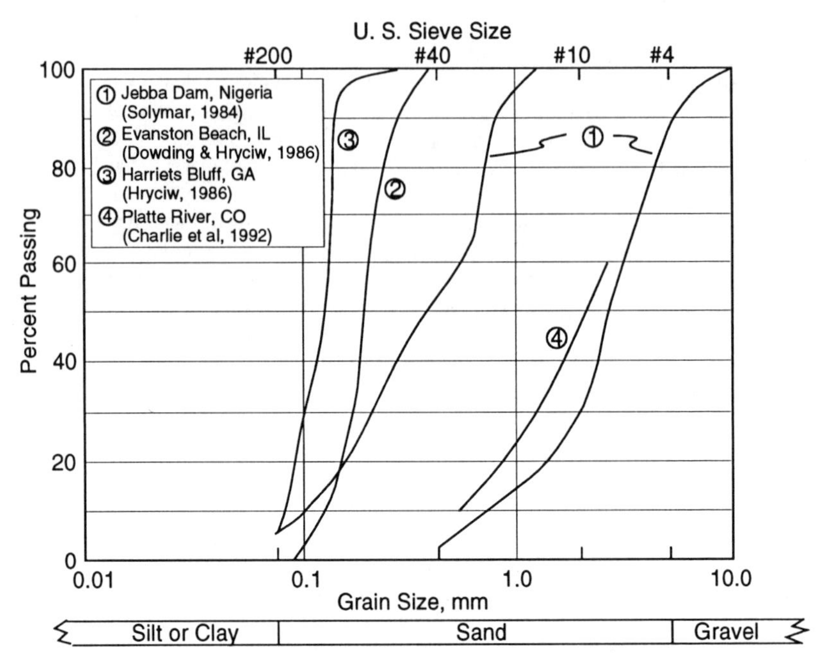

FIG. 5. Grain Size Distribution Curves Showing Soils Susceptible to Blast Densification When Submerged

Settlement Induced during Pile Driving

Adjacent construction, which often involves pile restrained excavation, produces permanent deformation outside of the temporary support following excavation. The literature is replete with case after case of the lateral deformation and settlement produced by the reduction of lateral support during construction of temporary retaining walls. The proper question is not,"Will it occur?, but rather "How much will occur?" One of the most practical summaries of this inevitable consequence of adjacent excavation is contained in a Federal Highway Administration report on Lateral Support Systems and Underpinning by Goldberg et al. (1976).

To compound the confusion between permanent deformation and transient vibratory deformation of adjacent facilities, the pile driving/vibrating induced motions can also produce settlement of certain, liquefiable soils. One of the most complete summaries of cases involving such "Settlement from Pile Driving in Sands" was published by Lacy and Gould (1985).

Distinction between Vibratory and Excavation Induced Settlement

In any discussion of pile driving induced settlement, three types of deformation are intimately intertwined. These three are: (1) transient, vibratory, ground motion; (2) permanent, vibratory induced settlement; and (3) permanent, settlement from loss of lateral support. "Wall insertion" may or may not produce "vibratory ground motion" depending upon the method of pile placement. This " transient, vibratory, ground motion" may or may not produce "permanent, vibratory settlement" depending upon the soil type and intensity of ground motion. Finally the "adjacent excavation" produces the inevitable "permanent settlement from loss of lateral support. This loss of support may or may not be large depending upon construction technique and type of soil. This loss of lateral support may have no relation to the transient vibratory ground motion.

Displacement that results from the loss of lateral support can occur with all adjacent construction, methods of wall insertion, and soil types. While this paper focuses upon displacement associated with cohesionless soils (sands) that are potentially liquefiable, Peck (1969) and Goldberg et al. (1976) should be consulted for excavation induced displacements of cohesive or clay soils.

It is important to consider which parties in the design-construction team have responsibility to ensure that the possible negative consequences of pile restrained excavation do not occur. In most cases the contractor is responsible for design and placement of temporary lateral support for most excavation because of the inseparable association of deformation and the timing and details for erection of temporary support. Thus, most specifications require that the contractor provide a support system that limits permanent deformation of adjacent facilities. However, the implied specification of a construction method when a pile foundation system is required complicates the issue. For instance if piling is required by the engineer/designer, then they assume responsibility for a typical level of vibration that is associated with insertion. Specification of an arbitrarily low allowable vibration level may preclude the use of piling altogether, and thus contradict the requirement of a bid based upon pile driving. Fortunately, specification of prebored piling can avoid this implied allowance of vibration.

Case History Illustrating the Importance of Permanent Deformation

A concern for pile driving vibration arose because new abutments for a railroad bridge involved driving within a meter of a 1 m diameter, pressurized natural gas pipeline. A vibration limit for safety of the pipeline to control the driving of the adjacent sheeting and "H" piles was needed. A study of the crack susceptibility of piping from blasting vibrations indicated that pipelines could withstand vibration levels as high as 150 to 210 mm/s (6 to 8.5 ips) (Dowding 1995), which allowed specification of a conservative control limit of 50 mm/s (2 ips). The report presenting the vibration controls warned the owner that there was a significant probability that driving into the underlying gravely sands with N's between 10 and 20 would produce vibratory densification, that this possibility should be investigated, and that deformations be monitored. The following describes the ensuing permanent deformation and its measurement (Linehan et al. 1992).

Driving of the east abutment cofferdam sheeting and center pier piles produced 12 mm (0.5") and 25 mm(1") of settlement, respectively. More importantly, a combination of center pier "H" and coffer dam sheet pile driving and center pier cofferdam deformation lead to the lateral and vertical deformation of the pipe as shown in Fig. 6. While the settlements were distributed so as to produce a relatively large radius of curvature, the lateral deformations were concentrated so as to produce a kink or small radius of curvature and thus high calculated bending stresses in the pipe. Conservative calculation of pipe bending stresses described in the Linehan et al. (1992) article, indicated that the elastic limit of the steel was being exceed,and the pipe was excavated to relieve the stresses. Subsequent driving of the east abutment piles resulted in total settlement that exceed 50 mm (2 in.).

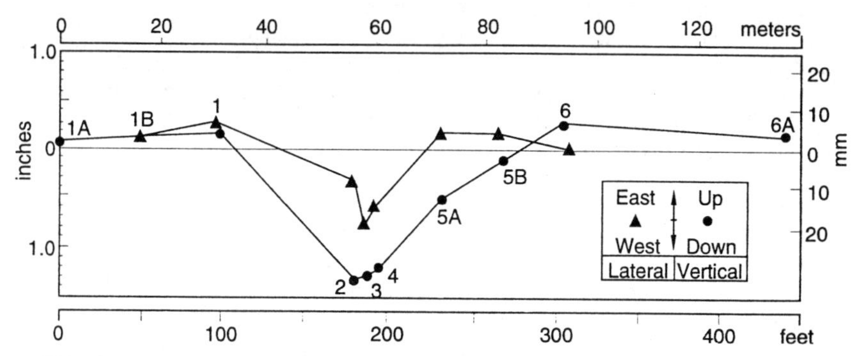

FIG. 6. Comparison of Lateral and Vertical Deformation of the Pipe Showing Localization of Lateral Deformation that Led to the Uncovering of the Pipe Line to Relieve Construction Induced Stresses (from Linehan et al. 1992)

Despite the high initial concern for a vibration control limit, the main risk resulted form the permanent deformation, which resulted from vibratory densification and, more importantly in this case, loss of lateral support during excavation necessary to place the piles for the center pier. Meanwhile, vibrations on the pipe and at the ground surface above the pipe compared to the distance from driving in Fig. 7 showed that pipe motion was considerably below the control limit of 50 mm/s (2 ips) at the pipe. Further discussion of the attenuation characteristics of pile driving motions can be found elsewhere (Dowding 1995 and others).

WHAT SOILS ARE SUSCEPTIBLE TO VIBRATORY DENSIFICATION ?

Obviously, the gravely sand underlying the pipeline described above was susceptible to vibratory densification. The question remains, what general characteristics can be employed as indices of such a response?

Synthesis of the cases summarized by Lacy and Gould (1985) reveals that in general these densifiable soils are "narrowly-graded, single sized clean sands with relative densities (corrected for depth) of 50 to 55% [or less]." All of these cases involve late glacial outwash sands and uniform grain size silts below the water table, as did the case above. Vibratory densification has been reported also in recent alluvial sand deposits of the Mississippi River (Holloway et al. 1980) and San Francisco Bay sands and muds (Clough and Chameau 1980).

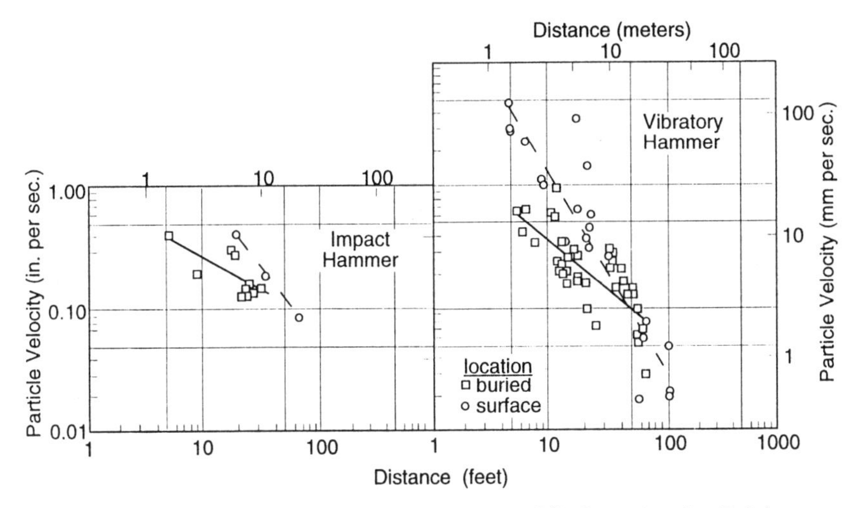

FIG. 7. Peak Particle Velocities Produced by Pile Insertion by Driving or Vibrating that Shows Attenuation with Distance and Compares Magnitudes of Motion Measured at the Surface and on the Pipe Line (after Linehan et al. 1992)

Uncorrected standard penetration test (SPT) resistance, N, should not be employed alone to gage susceptibility through relative density correlations, especially in non uniform gravelly deposits. Many glacial outwash deposits contain gravel and cobble sized particles that raise N values by blocking the penetration of the spilt spoon sampler. Thus the high N values do not accurately reveal the true and lower density of the sand matrix within which the gravel lies. In uniform sands and silts, the N values should be corrected for depth effects (Bazara 1967) as done by Lacy to determine relative density, D_r.

Grain size distribution should be determined. Fig. 8 shows the grain size distribution for several of the cases studied by Lacy and Gould, which is comparable with those in Fig. 5 that are susceptible to blast densification. The sands fall within the bounds which define soils that are liquefiable (Bhandari 1981). The two curves for case A are those for the sand and uniform grain size silt. In general sands must be clean, contain less than 10 % fines or that passing the #200 sieve. If the fines are silt rather than clay, densification may still be possible, provided the silt is of a uniform grain size.

ATTENUATION OF PILE DRIVING VIBRATIONS

As shown in Fig. 7, vibratory ground motion (reported herein in terms of particle velocity) can be as high as 100 mm/s (4 ips) within 1.5 m (5 ft) of the pile but decreases rapidly to 25 mm/s (1 ips) at 3 m (10 ft). Data are presented for both impact and vibratory driving, and a more extensive explanation of is available in

Dowding (1995). At 3 m from pile driving, vibrations themselves are unlikely to cause even cosmetic cracking, since they are below 25 mm/s (1 ips) (Dowding 1985). Cosmetic cracking is hair sized and almost indistinguishable from cracking caused by natural expansion and contraction of structures. Detailed discussion of vibration induced cracking can be found elsewhere.

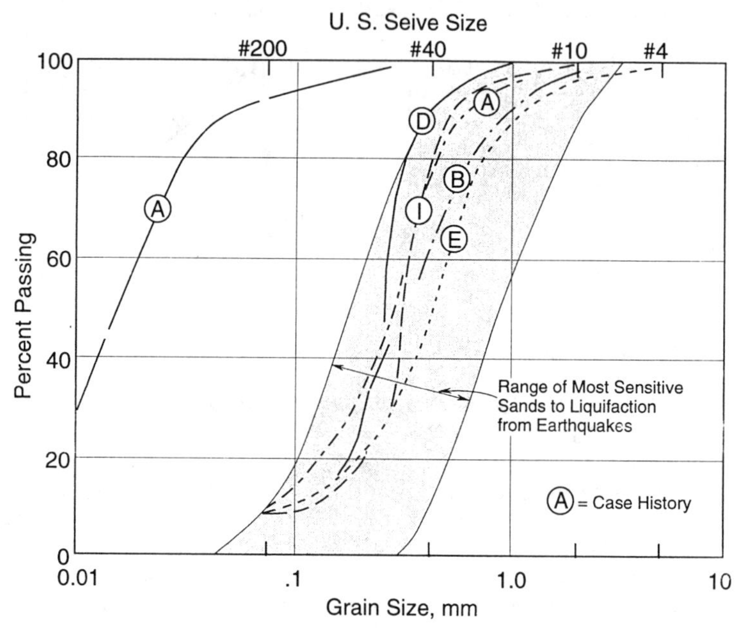

FIG. 8. Grain Size Distribution Curves for Cases in Which Sands were Sensitive to Vibratory Densification (Lacy and Gould 1985) that Show the Similarity of Liquefiable and Vibratorily Densifiable Sands

TABLE 2. Comparison of Distance to Zero Settlement with Pile Length (m)

Case	Pile Length	Distance of Effect
Lacy and Gould (1985)		
A	30	45
D	40	37
Clough and Chameau (1985)	12	11-15
Linehan et al. (1992)	< 23	18

Unfortunately vibratory densification occurs at distances where vibration amplitudes are lower than that necessary to induce cosmetic cracking directly. Densification can extend approximately as far as the piling is long. The cases shown in Table 2 that include some measure of the attenuation of settlement with distance form the basis of this "rule of thumb." This extension of settlement out to one pile length follows from the initial discussion of the extension of settlement from lost ground out a distance equal to the depth of the lost volume.

As shown by the case above and Lacy and Gould study, densification occurs with both impact and vibratory driving. There is some anecdotal evidence that impact driving may result in less settlement than vibratory driving. Densification and thus settlement results from a complex combination of vibration amplitude, number of repetitions, soil properties, and position of the water table. The number of repetitions or pulses depends upon the number of piles, their length, and the number of blows or vibratory cycles required per unit penetration. Thus it appears that the number of piles or the size of the project can affect the extent of the deformation.

COMPARISON OF VIBRATION SETTLEMENT WITH DEFORMATION RESULTING FROM LOSS OF LATERAL SUPPORT DURING ADJACENT EXCAVATION

Fig. 9 compares vibratory settlements with those resulting from loss of lateral support during adjacent excavation as a function of distance from the excavation. Typical settlements associated with excavation are shown by the thick solid lines, while settlements produced by pile driving alone referenced in this study are shown by the solid symbols connected by the thin solid line. Settlements are plotted as increasing downward and are related by the left vertical axis. Vibration produced by pile driving is shown by the open symbols connected with the thin dashed lines with particle velocity increasing upward as described by the left vertical axis. Typical settlements for 6 and 12 meter excavations in sand were taken from construction case histories by Goldberg et al. (1976). Vibratory settlements from the Clough and Chameau (1980) study shown by the thin solid lines are similar to the excavation settlements beyond 4 m and exceed them within 2 m of the wall. Settlements resulting mainly from pile driving from other studies, Linehan et al. (1992) and Holloway et al. (1980), are also shown with the filled circles and squares respectively. The filled circles with the superscript "l's" illustrate increasing vibratory settlement for the case presented herein as the construction progressed from the initial driving of the east cofferdam sheet piling, 10 mm, to the final insertion of the east abutment piling, 100 mm. The squares represent particle velocities and settlements associated with vibratory driving of an H pile (Holloway et al. 1980).

Also shown in Fig. 9 by the open symbols, connected by the thin dashed line, is the vibration environment associated with the settlements. Settlements as large as 25 to 30 mm were measured at distances, 4 m, where the particle velocities at the ground surface, 13 to 19 mm/s, were less than that necessary for cosmetic cracking of structures.

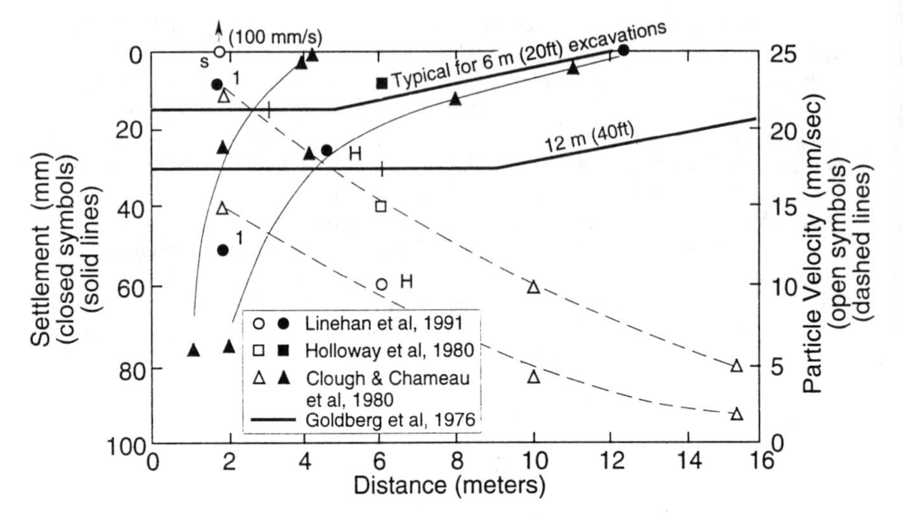

FIG. 9. Comparison of Settlement and Particle Velocities Produced by Pile Driving Vibrations (Thin, Solid and Dashed Lines Respectively) with Typical Settlements Produced by Braced Excavation in Sand (Thick, Solid Lines)

CONCLUSIONS

Settlement and ground motions from two vibratory construction activities, blast densification and pile driving, were compared to determine common trends, which form the basis for the following conclusions. Geologically young, loose, clean, sands below the water table are most susceptible to blast and vibratory densification. Grain size distribution curves for sands densifiable by either technique were found to be similar. At small distances particle velocities, which are proportional to dynamic strains, should be measured with high frequency transducers. These small distances at which densification is expected can be described in terms of scaled distance for blasting and pile length for piling. Surface settlement and transient ground motion extend beyond the zone of densification. Permanent settlement resulting from loss of lateral support and vibratory densification can be more significant than building vibration from piling and should not be overlooked. At relatively large distances from blast densification activities, surface waves involve sufficient low frequency amplitudes to produce significant building vibration, and must be considered in feasibility studies.

APPENDIX - REFERENCES

Bazara, A. R. S. S. (1967). "Use of the standard penetration test for estimating settlements of shallow foundations on sand," PhD Thesis, Department of Civil Engineering, University of Illinois, Urbana, Illinois.

Bhandari, R .K. M. (1981). "Dynamic consolidation of liquefiable sands." *Proc. International Conference on Recent Advances in Geotechnical Earthquake Engineering and Soil Dynamics*, St. Louis.

Brauner, G. (1973). *Subsidence due to Underground Mining*, Information Circular 8571, U. S. Bureau of Mines, Washington, D. C.

Charlie, W. A., Mansoori, T. A., and Ries, E. R. (1981). "Predicting liquefaction induced by buried charges." *Proc. 10th ICSMFE*, Stockholm, 1, 77-80.

Charlie, W. A., Mutabihirwa, F. J. R., and Doehring, D. O. (1992). "Time dependent cone penetration resistance due to blasting." *J. Geotech. Eng.* ASCE, 118(8), 1200-1215.

Clough, G. W. and Chameau, J. L. (1980). "Effects of vibratory sheet pile driving - a case study." *Minimizing Detrimental Construction Vibrations*, ASCE Special Technical Publication, New York, Nw York, Preprint 80-175.

Dowding, C. H. (1985). *Blast Vibration Monitoring and Control*, Prentice Hall, Englewood Cliffs, New Jersey.

Dowding, C. H. (1995). *Construction Vibrations*, Prentice Hall, Englewood Cliffs, New Jersey.

Dowding, C. H. and Hryciw, R. W. (1986). "A laboratory study of blast densification of saturated sand." *J. Geotech. Eng.*, ASCE, 112(2), 187-199.

Drake, J. L. and Little, C. D. (1983). "Ground shock from penetrating conventional weapons." *Proc. Symposium on the Interaction of Non-Nuclear Munitions with Structures*, U. S. Air Force Academy, Colorado Springs.

DuPont (1977). *Blasters Handbook*, Technical Services Div., E.I. DuPont, Wilmington, Delaware.

Goldberg, D. T., Jaworski, W. E., and Gordon, M. D. (1976). "Lateral support systems and underpinning. Vol. II. Design fundamentals," *Report No.FHWA-RD-75-129*, Federal Highway Administration, Washington, D. C.

Hall, C. E. (1962). "Compacting a dam foundation by blasting." *J. Soil Mech. Found. Div.*, Proc. ASCE, 88(SM3), 33-51.

Holloway, D. M., Moriwaki, Y., Demsly, E., Moore, B. H., and Perez, J. Y. (1980). "Field study of pile driving effects on near by structures." *Minimizing Detrimental Construction Vibrations*, ASCE Special Technical Publication, New York, New York, Preprint 80-175.

Hryciw, R. D. (1986). "Deep blast-densification of sand with millisecond delays," Ph.D. thesis, Dept. of Civil Engineering, Northwestern Univ., Evanston, Illinois.

Hryciw, R. D. and Dowding, C.H. (1986). "Dynamic pore pressure and ground motion instrumentation during blast densification of sand." *Proc. 1st International Symposium on Environmental Geotechnology*, Bethlehem, 620-629.

Ivanov, P. L. (1967). "Compaction of noncohesive soils by explosions." Translated from Russian, Indian National Scientific Documentation Center (available from United States Water and Power Resources Service, Denver, Colorado).

Kummeneje, O. and Eide, O. (1961). "Investigation of loose sand deposits by blasting." *Proc. 5th ICSMFE*, Paris, 1, 491-497.

Lacy, H. S. and Gould, J. P. (1985). "Settlement from pile driving in sands." *Vibration Problems in Geotechnical Engineering*, ASCE Special Technical Publication.

Linehan, P. W., Longinow, A., and Dowding, C. H. (1992). "Pipeline response to pile driving and adjacent excavation." *J. Geotech. Eng.*, ASCE, 118(9), 300-316.

Long, J. H., Ries, E. R., and Michalopoulos, A. P. (1981). "Potential for liquefaction due to construction blasting." *Proc. International Conference of Recent Advances in Earthquake Engineering and Soil Dynamics*, St. Louis.

LaFosse, U. and Gelormino, T. A. (1991). "Soil improvement by deep blasting - a case study." *Proc. 17th Annual Symposium on Explosives and Blasting Technique*, International Society of Explosive Engineers, Montville.

Lyman, A. K. B. (1941). "Compaction of cohesionless foundation soils by explosives." *Proc. ASCE*, 67, 769-780.

Mitchell, J. K. and Solymar, Z.V. (1984). "Time-dependent strength gain in freshly deposited or densified sand." *J. Geotech. Eng.*, ASCE, 110(11), 1559-1576.

Peck, R. B. (1969). "Deep excavations and tunneling in soft ground." State of the Art Report, *Proc. 7th ICSMFE*, Mexico City, 1, 225-281.

Peck, R. B., Hanson, W. R., Thornburn, T. H. (1974). *Foundation Engineering*, 2nd Edition, John Wiley & Sons, New York, New York.

Prugh, B. J. (1963). "Densification of soils by explosive vibrations." *J. Const. Div.*, Proc ASCE, 89(CO1), 79-100.

Solymar, Z. V. (1984). "Compaction of alluvial sands by deep blasting." *Can. Geotech. J.*, 21(2), 305-321.

Studer, J. and Kok, L. (1980). "Blast induced excess porewater pressure and liquefaction experience and application." *Proc. International Symposium on Soils under Cyclic and Transient Loading*, Swansea, 581-593.

Residual Soil Settlement Related to the Weathering Profile

George F. Sowers,[1] Honorary Member, ASCE

Abstract: Residual soils are the in-place residue from rock weathering. They differ greatly from the better understood deposited soils in that their particle arrangements and mineralogy reflect both the degree of weathering and the structure of the parent rock. They are neither homogeneous nor isotropic. Like deposited soils they consolidate under load, but the soil compressibility is related empirically to the void ratio of the undisturbed material instead of to the usual index properties. Because most of the more compressible residual soils are partially saturated, their settlements develop far more rapidly than those of clays, usually within the first year after loading. Conventional laboratory tests and analyses based on in-situ pressuremeter tests can be used for predicting settlements. The conventional analyses tend to over predict settlement by about 25 percent; the pressuremeter results appear to be more scattered, but on the average the predictions are closer to the measured settlements.

Introduction

Residual soils are the products of the in-place weathering of rock that have not moved significantly from their point of origin. Their texture and mineralogy depend on the degree of breakdown as well as the minerals present in the rock. The physical properties vary with the position in the weathering profile as well as the variations in the rock mineralogy.

---

[1] Senior Geotechnical Consultant, Law Companies Group, Inc., 114 TownPark Drive, Kennesaw, GA 30144.

WEATHERING

Weathering takes place from the ground surface down, and inward from the cracks that penetrate most rocks. Air, water and humic acids from vegetation decay seep downward to the rock surface to attack its minerals. New minerals are created from the old. Many of the minerals in igneous and metamorphic rocks are complex aluminum silicates with varying amounts of sodium, potassium, magnesium and iron. The weathering products are complex hydrous aluminum silicates, collectively known as the clay minerals, plus soluble carbonates and bicarbonates of calcium, sodium and potassium. Some of the clay minerals produced by the weathering may be further altered by cations such as magnesium that are released into solution by the chemical changes at another rock horizon. This often alters the mineralogy of the original clays (toward the smectite or montmorillonite group).

The chemical decomposition is augmented by mechanical forces from temperature changes, water pressure, freezing and from the expansion of the rock minerals as they decompose. Some of the decomposition products are soluble. They are leached from their site of origin and deposited elsewhere. The weathering environment may be either oxidation or reduction depending on the oxygen and organic acids present and on the rate of percolation through the rock cracks. Iron-bearing minerals oxidize and create the red colors present in many residual soils. More important, the precipitation of the oxidized iron between the unweathered particles remaining from the original rock can cement them together in a more coherent mass.

MECHANICAL EFFECTS OF ROCK WEATHERING

The weathering process locally disrupts the continuity of the unweathered remainder of the rock mass, breaking it into smaller particles of the unweathered rock and reasonably intact particles of the minerals that resist the weathering such as quartz and some of the micas. These are surrounded by a matrix of the decomposed minerals, predominantly new clay minerals whose particles are orders of magnitude smaller than the mineral grains from which they were weathered. Most of the mineral bonds in the original rock are broken, causing the new mass to exhibit little or no tensile strength or cohesion. This makes undisturbed sampling and laboratory testing difficult.

Decomposition of the feldspars and other more complex aluminum silicates is often accompanied by expansion despite the loss of soluble weathering products by solution. Theoretical computations based on the chemical reactions in weathering suggest that void ratios in the weathered materials that exceed about 1.0 involve a volume increase in the mass. The amount of the rock volume increase varies with the relative volume of the decomposition-resistant minerals such as quartz, the volume of the clay minerals and the volume of the soluble weathering products that leach away. If the rock is free to expand during weathering, the shear strains from non-uniform expansion causes new fractures in the residual soils. Many residual soils from granite, gneiss and schist contain slickensided shear surfaces which are not inherited from the rock from which they were derived nor from subsequent landsliding. Differential expansion is one reasonable explanation.

When the weathering material is restrained, such as by overburden stress, in situ stresses develop that are greater than would be expected from the overburden stress and Poisson's ratio. Such stresses are probably partially responsible for the erratic "preconsolidation" stresses observed in residual soils. Some weathering is accompanied by non-uniform contraction accompanied by erratic tension cracking.

SECONDARY OR PEDOLOGIC WEATHERING

Near the ground surface the local environment and topography dominate the weathering. The minerals formed by the deeper, primary weathering are altered and additional volume changes take place. The environment is modified by any organic debris that accumulates at the ground surface. This is termed secondary or pedologic weathering. It produces a near-surface soil profile that is the focus of "Soil Science". In flat, humid regions the horizon may be 1.5 to 3 m (5 to 10 ft) thick. The pedologic horizon seldom play a part in the foundation performance of major structures.

SIGNIFICANCE OF WEATHERING PROFILES

The weathering profile is the key to interpreting site exploration data and for characterizing the three dimensional geometry of the soil and rock for engineering analyses. The weathering generally proceeds from the ground surface downward. Therefore, the degree of alteration of the rock properties decreases with the depth below the ground surface. The weathering also progresses outward from cracks, complicating the pattern of rock alteration.

The weathering profile can be divided into horizons for description and characterization. These are illustrated in Figs. 1 and 2. The changes in the degree of weathering and the associated soil properties with increasing depth are generally continuous but often irregular.

The deepest horizon is relatively unweathered rock in which the engineering properties have not been significantly altered. Rock core recoveries are high, approaching 100 percent.

The RQD (rock quality designation) is usually but not necessarily high because it reflects the crack spacing in unweathered rock (although rock cracking does influence weathering).

Above lies the partially weathered horizon where the mineral alteration is not complete and which varies from level to level and point to point. The rock core recovery is often less than 75 percent. In most rocks, the degree of weathering decreases with increasing depth. This is reflected in an increasing percentage of rock core recovery. The variations in weathering are usually three dimensional. The weathering becomes less progressing outward from the cracks, often with unweathered rock "core-stones" midway between the cracks.

The materials above the partially weathered horizons have been weathered to the degree that the texture and strength of the materials are those of soil. This can be defined by a Standard Penetration Resistance, N, less than about 300 mm (1A), an unconfined compressive strength less than 1.5 MPa (200 lb per sq. in.) and a maximum particle size of about 300 mm (12 in.).

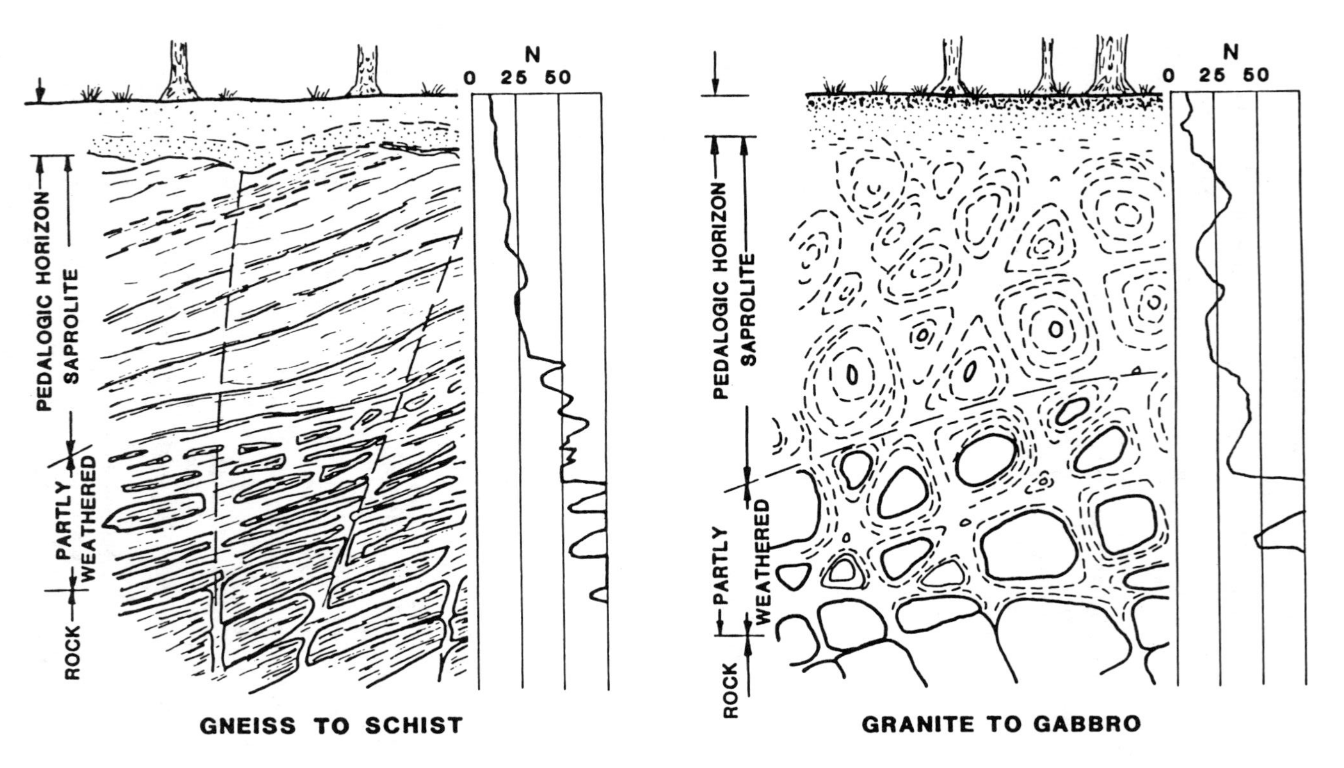

FIG. 1. Weathering Profiles in Crystalline Rock: Gneiss to Schist, Granite to Gabbro

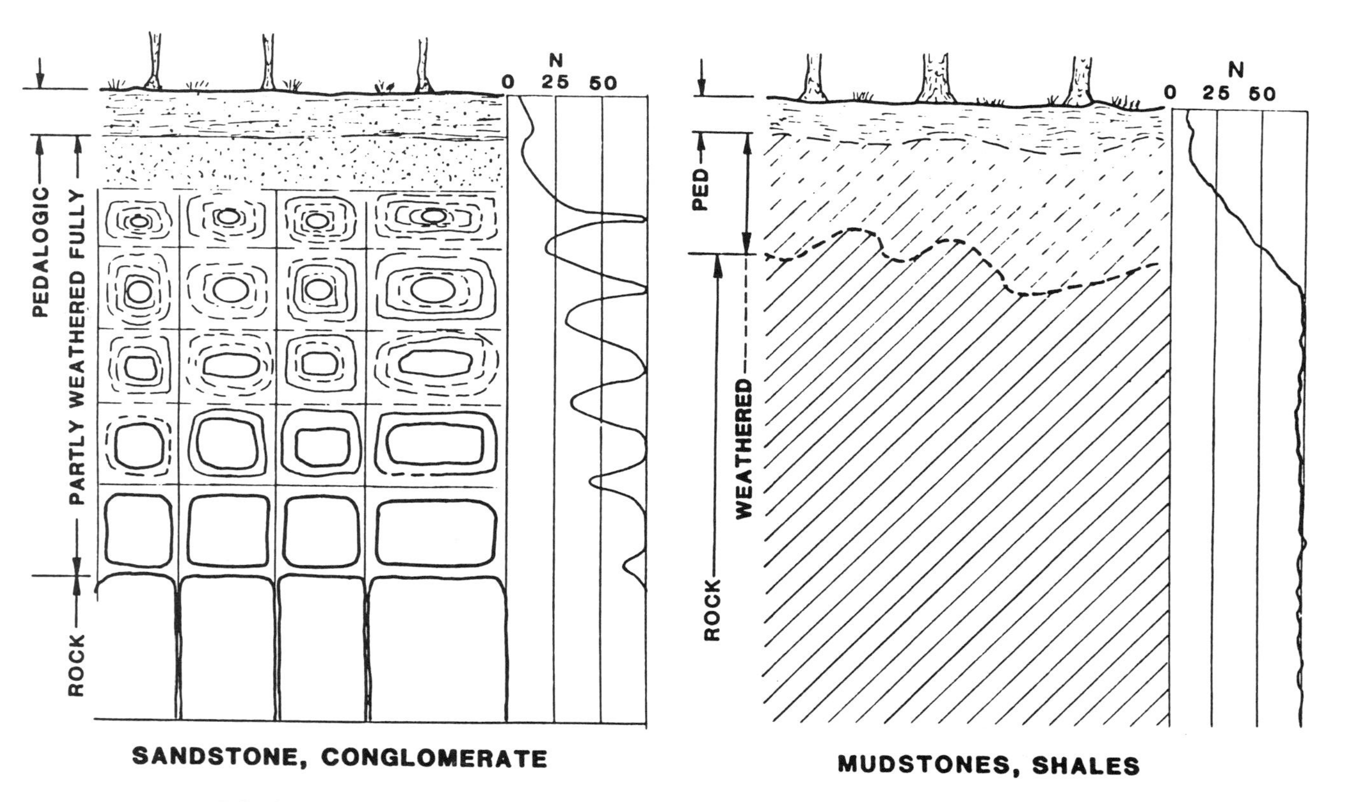

FIG. 2. Weathering Profiles in Clastic Sedimentary Rock: Sandstones and Shales

When the original rock is prominently stratified or layered, the completely weathered rock will reflect those strata or layers in its variations in new mineral content, texture and strength. Such soils are termed "Saprolites". Viewed at a distance they appear to be rock but with bands of different hues that are more typical of soils: white, yellow, tan, brown, purple and red-brown. Texturally they are soils that retain the relict structure of the original rock.

The pedologic horizons have lost any resemblance to the original rock. Moreover, the surficial secondary weathering has developed new layering that reflects the downward leaching of soluble and fine grained particles during wet weather and local accumulations of finds and precipitated soluble materials near the base of the horizon during dry weather. These horizons are typically 1 to 3 m (3 to 10 ft) thick in temperate regions, but sometimes thicker in the tropics.

Categories of Horizons

There are three broad categories of horizons depending on the nature of the source rock. The igneous and metamorphic rocks, sometimes termed the crystalline rocks, produce similar profiles. Their minerals have not been subject to decomposition since their formation under heat, pressure and shear. They undergo substantial changes in their mineral nature during weathering.

Most sandstones and mudstones (including shales) are derived from sediments that were previously weathered materials. Therefore, their chemical decomposition is less significant than with the crystalline rocks. Instead the interparticle bonding and cementing is destroyed. In the process there is usually some fracture of the intact particles, and a corresponding decrease in particle sizes, compared to the original sediment.

The carbonate rocks or limestones produce a distinctly different profile because the dominant weathering mechanism is solution. The residual soil is the insoluble part of the rock, negligible in some limestones and a major part of others.

Gneiss and Schist: Metamorphic Rock Profiles

Gneiss, schist, phyllites (and some slates) are similar in that the weatherable and resistant minerals are segregated in narrow often undulating layers resembling strata. The unweathered rock is non-homogeneous and strongly anisotropic, parallel to the foliation. All have been metamorphosed from their original igneous or sedimentary formations. These materials have been folded and distorted tectonically, creating numerous cracks in both regular and random patterns.

Weathering invades the formations along three fronts: from the surface down, from the cracks outward and along the foliation surfaces. The result is a complex three dimensional pattern as illustrated in Fig. 1. The partially weathered zone above the rock consists of layers of reasonably sound rock in elongated or lenticular core-stones separated by layers of more weathered rock. The proportion of the more weathered materials and their degree of weathering increases with decreasing depth until only fragments of partially weathered rock remain in the profile. Thus the depth to "rock" and the thickness of the partially weathered rock are not only arbitrary but also vary greatly from one location to another.

Decomposition in the partially weathered zone is accompanied by differential volume changes that locally enlarge existing cracks and generate new thin lenticular openings or voids along the foliation surfaces. These may be 300 to 600 mm (1 to 2 ft) wide and up to 25 mm (1 in.) thick.

The saprolite horizon consists largely of soil, texturally and strength-wise. The values of N range between 5 and 50 blows per 300 mm (1 ft), often with no particular pattern to the variations. The saprolites are very non-homogeneous and anisotropic. Typical void ratios are 0.6 to 1.2, and occasionally as great as 2. The thickness of the saprolite zone is highly variable, even within one building site. In downtown Atlanta depths from 3 m (10 ft) to more than 30 m (100 ft) occur within a horizontal distance of 60 m (200 ft), probably reflecting the crack patterns in the underlying rock.

The saprolites are compressible; their void ratios decrease with confining stress, with stress-strain or stress void ratio curves that appear identical to those of clays. The settlement mechanism appears to be distortion of the micas and clay minerals, probably accompanied by small reorientation of the larger, more equidimensional grains. The log stress-void ratio curve shapes include a sharp transition into a steep straight line that mimics the preconsolidation of clays. However, the amount of the "preconsolidation" (more properly termed pseudo-preconsolidation) varies erratically from level to level with no apparent relation to past or present overburden stress. In the author's opinion it reflects stresses generated by weathering volume changes, residual bonds between particles and possibly residual lateral tectonic stress associated with formation uplift and folding.

The compression index, representing the slope of the log stress-void ratio curve, has been studied by Sowers (1963) and Dib (1985). Both Dib and Sowers found a reasonably consistent relation between void ratio and the compression index in soils derived from a variety of crystalline and metamorphic rocks, as shown in Fig. 3. Their correlations are remarkably similar. Both show significant scatter, but increased compressibility with increasing void ratio. Dib also found a marked but more scattered relation between liquid limit and void ratio, similar but somewhat different to that for clays published by Terzaghi and Peck (1947). This might be expected because the saprolites tested by Dib include the basalts and andesites that produce very clayey soils, whereas those tested by Sowers are largely micaceous sandy silts and silty sands.

The time rate of consolidation is usually much faster than for clays, because the hydraulic conductivity is several orders of magnitude greater than for clays. Moreover, upper horizons are often only partially saturated. Both the percentage of initial or immediate consolidation range from 25 to 75 percent of the total, depending on the stress level and the degree of saturation. Consolidation tests, therefore, should be programmed to reflect the loading changes that are anticipated in the proposed project as well as the likely range in soil saturation. Most of the settlement of saprolites in the Southeastern USA is likely to occur within a year of the completion of construction. There is some continuing long term settlement, resembling secondary compression of clays in that the log time-settlement relation is a reasonably straight line. This rate can be approximated by conventional laboratory testing.

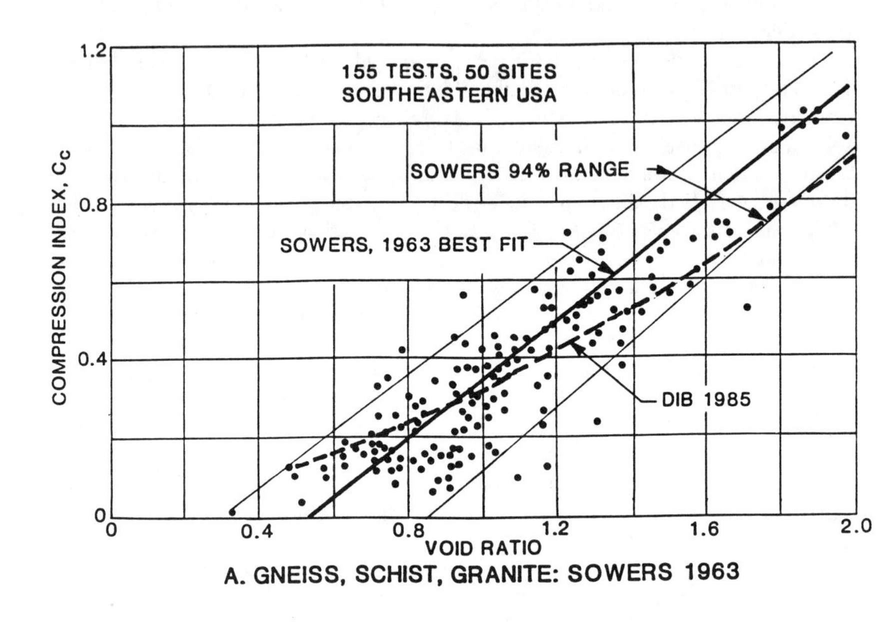

A. GNEISS, SCHIST, GRANITE: SOWERS 1963

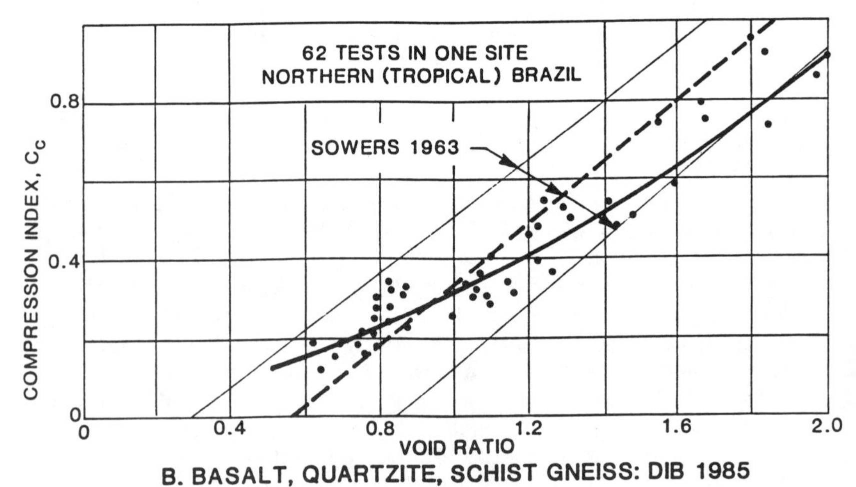

B. BASALT, QUARTZITE, SCHIST GNEISS: DIB 1985

FIG. 3. Relation Between the Void Ratio and Compression Index in Residual Soils Derived from Crystalline (Igneous and Metamorphic) Rocks

Settlement of Saprolites from Metamorphic and Igneous Rocks

Structures up to 10 stories built before 1950 were supported on spread foundations with very conservative bearing pressures 200 to 300 kPa (4-6 ksf) and settlements not observed unless severe cracking occurred. Modern multi-story buildings are supported by piers or piles to rock. Therefore, few settlement data are available.

The settlement of large fills is seldom noticed because it occurs largely during construction and is reflected only in a modest shortfall in the fill borrow compared to the volume of the completed fill. However, Mello et al. (1985) presented settlement data for 8 locations beneath surcharge embankments from 13 to 20 m (43 to 66 ft) high on saprolites 6 to 24 m (20 to 80 ft) thick in northern Brazil. Half of the areas involved excavation before filling of from 0.5 to 0.7 of the embankment height. In effect, the excavation represents preconsolidation. The settlements of the fills on unexcavated saprolite were 400 to 800 mm (1.3 to 2.6 ft). Those settlements where the upper part of the saprolite had been removed were 27 to 44 mm (1.1 to 1.7 in.). These settlements where the upper saprolite has been removed are about 6 percent of those for the fills where there had been no excavation. Part of the difference was that the excavation both reduced the thickness of the saprolite and removed the softest and most compressible upper horizons. If the settlements are normalized to equal saprolite thickness, surcharging with excavation of from 0.5 to 0.8 of the future fill weight reduces the settlement to about 10 percent of that without surcharge.

The larger measured saprolite settlements in the Southeastern USA have occurred from substantial water table drawdown. The footings for an 8-story department store in Atlanta were supported on 16 to 18 m (53 to 70 ft) of micaceous sandy silt saprolite derived from gneiss. The footings were about 1.5 m (5 ft) below the ground surface and 1 m (3 ft) below the ground water level. Settlement analyses based on conventional consolidation tests showed that lowering the ground water table to the rock surface in order to permit excavation of the adjoining subway would cause 100 to 125 mm (4 to 5 in.) of settlement. In order to minimize settlement, the contract for excavation bracing required a slurry-emplaced concrete bracing wall reinforced with H-piles. Unfortunately, the slurry wall seal on the rock was ineffective because the contractor did not have the tools to cut through the hard seams in the partially weathered horizons. During construction the water level dropped about 8 m (25 ft) below the spread footings nearest the slurry wall. This was accompanied by 65 to 75 mm (2 1/2 to 3 in.) of settlement, about 2/3 of that predicted, but enough to damage the building.

The settlements of four tall chimneys at a power plant were supported on octagonal mat foundations on 10 to 12 m (33 to 40 ft) of micaceous sandy silt saprolite with N of 8 to 35 blows, 300 mm, increasing uniformly with depth. Standard ASTM consolidation tests were made of 12 representative undisturbed samples. The first two chimneys had a dead load of 1700 metric tons (3700 kips) and the second pair, about 15 percent greater. The foundation design pressure was 120 kPa (2500 lb per sq ft). The computed settlements for immediate and primary consolidation were 15 to 25 mm (0.6 to 1.0 in.), respectively reflecting the small differences in soil profile and chimney load. The measured settlements averaged 72 percent of the computed values. From 60 to 80 percent of the consolidation took place during construction and the primary consolidation was complete within two months of the completion of loading, about the same as predicted from the test data. The additional settlement from secondary compression was between 1 and 3 mm for the first three years after construction was complete, the same as predicted from the consolidation tests.

Martin (1977) utilized pressuremeter data to estimate settlements of two multi-story office buildings; one a mat foundation, the other with footings. The mat settlement was 36 mm (1.4 in.). Two different pressuremeter analyses predicted 1.6 times the actual and close to the actual settlement. The building on footings settled 8 mm (0.3 in.). The two different pressuremeter analyses found from 0.75 times the actual to the actual settlement.

Because the measured settlements on saprolites from gneiss and schist based on laboratory test of undisturbed samples are consistently from 2/3 to 3/4 of those computed from conventional laboratory tests of representative undisturbed samples in the various horizons with significant void ratios, a correction factor of about 0.8 is applied to the computed values by many geotechnical engineers in the Southeastern USA. The limited pressuremeter results suggest that they are more scattered in the results, but on the average, in good agreement.

GRANITE TO GABBRO

These igneous rocks usually occur in massive bodies that are mineralogically reasonably homogeneous and isotropic. Their weathering and soil profile is largely governed by the patterns of joint cracks, many of which occur in somewhat rectilinear patterns that divide the rock into blocks with dimensions of less than a meter to tens of meters. The fracture surfaces are the focus of weathering. This is illustrated in Fig. 1. The relative sound rock may have core recoveries of 100 percent and seismic compression wave velocities exceeding 4000 m/s (12,000 ft/s). The weathering proceeds inward from the joints with each block developing an outer skin of decomposition and increase of void ratio from loss of weathered material by leaching of soluble weathering products. As weathering proceeds inward from the crack surfaces, the innermost unweathered block or core-stone becomes smaller and more rounded and boulder like. The successive layers of weathering become onion-like. The saprolite zone retains the remnants of the core-stones as small boulders or coarse gravel that float in a matrix of soil-like weathered rock. The residual soil texture is that of a silty to clayey sand with kaolinitic low plasticity clays for granites. The soils of the gabbros range from clayey sands to sandy high plasticity smectite clays.

In the continental United States the residual soil derived from granite is seldom very thick. The pedologic horizon is typically 1 m (3 ft) thick at the most and consists of silty sand. The saprolite horizon may be as much as 6 m (20 ft). In tropical areas, such as Malaysia, the saprolite thickness may be tens of meters thick. Because of the lower hydraulic conductivity of the more plastic residual clays, the saprolites over gabbro and andesite are thinner than those above granite.

There are few data on the settlement of the residual soils derived from granite. The compression index can be estimated from Fig. 2a. When the void ratios are less than about 0.7, they consolidate similar to dense sands. Most of the settlement occurs during the load application from grain repositioning and to a smaller degree from grain edge fracture. Unless the soil contains at least 30 percent fines, there is no delay from pore water pressure dissipation. There is some small continuing settlement whose total is typically a few percent of the primary and initial settlement.

In 50 years of experience with residual soils, the author has not encountered a building settlement problem from foundations supported on the residual soils derived from granite. However, in tropical climates of alternately very dry and very wet seasons, some of the silty kaolin helps bond the sandy particles. During wet weather, portions of the silty kaolin clays are slowly leached away leaving the residual soil mass as a lightly cemented but loose sand. Collapse settlement can occur in unusually wet weather or upon inundation. Such settlement has been reported in South Africa and Malaysia.

The residual soils from gabbro are highly plastic clays. These can be sampled and tested for consolidation in the same way as deposited soils. The author knows of no case histories of consolidation settlement of structures supported on residual clay from gabbro and the similar and more widespread andesite in the continental United States. Because these soils are highly plastic, they are prone to expansion when dry and given access to moisture and to shrinkage when wet and allowed to desiccate. Swelling and shrinking appear to be a more serious concern than settlement.

Lava

Flow lava (basalt and some andesites) have closely spaced, deep joint cracks perpendicular to their upper surface. Their weathering profiles, where developed, resemble those in closely jointed granite or gabbro, with elongated core-stones.

Sandstones and Shales

The sandstones and mudstones including shales are derived from sediments that had been previously weathered. Therefore, changes during reweathering of sediments are usually not so profound as those in crystalline rock. Instead the interparticle bonding is destroyed accompanied by an increase in the proportion of fine particles. In addition, some of the clay minerals may be altered. Arkosic sandstones (containing feldspar grains) weather into clayey sands or sandy clays. Their weathering profiles are more similar to granite.

A typical sandstone weathering profile is shown in Fig. 2. The weathering focuses on cracks, particularly normal to the bedding and along bedding, that break the rock into large rectangular blocks. The rounded weathering core-stones that often develop are similar to, but more regular than, those in granites. The settlements are similar to those that are experienced in sand. They occur immediately upon loading and are usually inconsequential; the amount depends on the density of the residual sand. If the sandstone is bonded with soluble materials, repeated saturation and drying can cause collapse and catastrophic structural damage.

Settlements in residual soils over shale are typically small because of the depth of weathering is limited by the low hydraulic conductivity of the rock. However, weathering can be as deep as 10 m (33 ft), when the shale is badly cracked or contains numerous thin seams of sandstone. The more pervious sandstone brings air and water to the shale seams, allowing them to revert to clay soils. In such cases clay settlement can be evaluated by laboratory consolidation tests.

THE CARBONATE ROCKS: LIMESTONES, DOLOMITES AND MARBLE

The carbonate rocks, collectively termed "limestone", weather differently from the other rocks because their solution takes place far more rapidly than either the mechanical or chemical weathering previously described. Solutioning occurs at the rock surface and along cracks where the predominant carbonate minerals dissolve, leaving any insoluble materials in the original rock behind to blanket the rock surface and fill the cracks as silty clay and clayey sand residual soils, Fig. 4. The thickness of the residual soil blanket depends of the proportion on non-soluble minerals in the rock, the rate of solution, and the rate that residual soils are eroded away at the soil surface. In many limestone regions in the USA, residual soil may be as deep as 30 m (100 ft).

Unlike most other residual soils, those derived from limestones retain little or no evidence of the original rock structure. This is because the insoluble part of most limestones is a very small part of the rock volume. However, if the insoluble minerals are a significant part of the rock, they may be concentrated in layers or lenses of the coarser insoluble minerals, such as quartz sand, chert and some fossils. Alternating strata of limestone and mudstone (including shale) will retain remnants of the mudstone, usually distorted by the varying rates of carbonate solution. The residual soil blanket has an inverted strength profile, as seen in Fig. 4. The oldest soil is at the ground surface and is usually stiff. The youngest is at the soil-rock interface, and is soft.

Consolidation settlement in residual soils from limestone can be estimated from the results of laboratory tests on undisturbed samples. However, such settlements are seldom of concern because of the inverted soil profile. The most important problem is collapse of domes within the residual soils and the development of a sinkhole accompanied by catastrophic foundation subsidence, often with little warning. Such a dome is shown in Figure 4. Dome formation commences with downward seepage through the residual soil that erodes the soil into solution enlarged cracks and cavities in the limestone. The erosion precipitates soil ravelling and formation of a dome-shaped cavity in the soil that propagates upward until the dome roof cannot support the weight of the soil (and a structure) above. Any increase in surface infiltration or large fluctuations in groundwater levels aggravate the process and precipitate dome collapse.

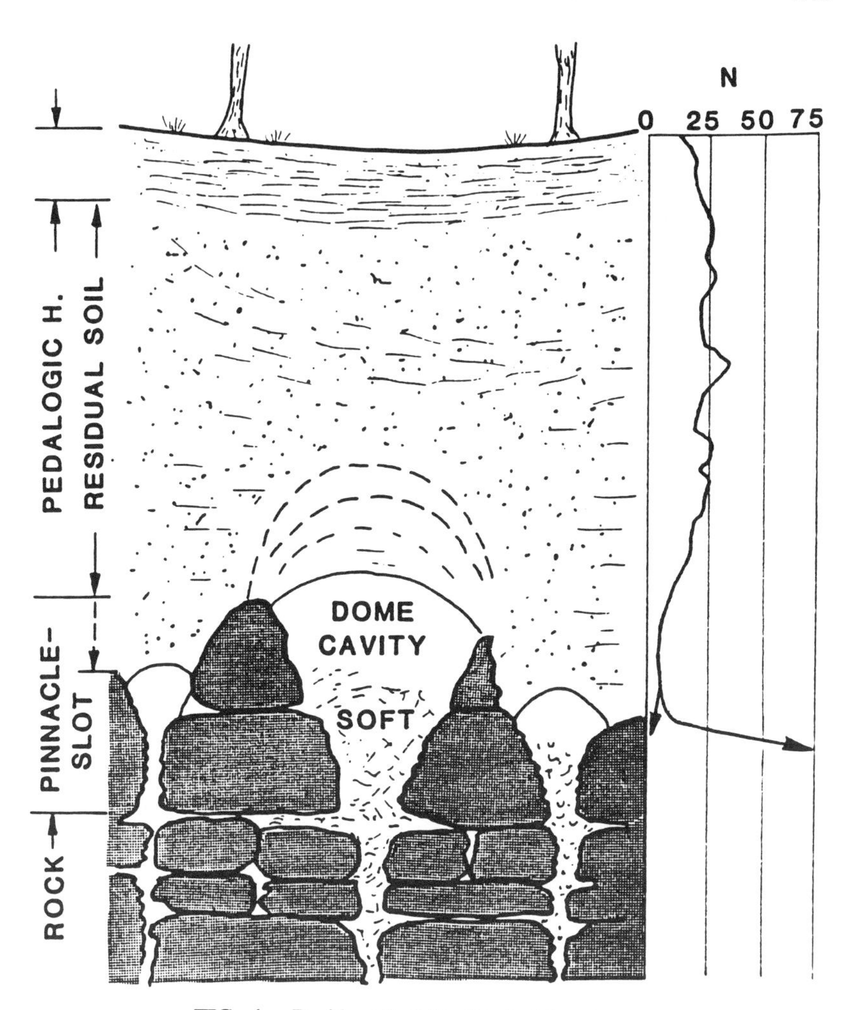

FIG. 4. Residual Soil Profile in Limestones

SUMMARY

Settlement of residual soils is greatly dependent on the nature of the parent rock, its weathering profile, and the degree of weathering. Unlike the deposited clays whose compressibilities can be related to their index properties, the compressibility of most residual soils is related to their mineral composition and particularly void ratios. The stress compressibility relationships are similar to those of other soils, including what appears to be preconsolidation. However, this pseudo preconsolidation load is not related to loading and usually varies erratically with depth.

The time rate of consolidation is usually far greater than for clays because the hydraulic conductivity is usually orders of magnitude greater, and most residual soil profiles exhibit considerable depths where the soils are not saturated.

APPENDIX - REFERENCES

Dib, P. S. (1985). "Compressibility characteristics of tropical soils making up the foundation the Tucouri Dam in Azamonas, Brazil." *Proc. 1st International Conference on Geomechanics*, Brasilia, 2, 131-141.

Martin, R. E. (1977). "Estimating foundation settlements in soils." *J. Geotech. Eng. Div.*, Proc. ASCE, 103(GT3), 197-212.

Mello, G. F. S., Cepollina, M., and Oliveira, F. J. P. (1985). "Use of surcharges as treatment of residual soils foundations, a case history." *Proc. 1st International Conference on Geomechanics*, Brasilia, 2, 87-100.

Sowers, G. F. (1954). "Soil problems in the Southern Piedmont region." *Proc. ASCE*, 80(416), 4161-41619.

Sowers, G. F. (1963). "Engineering properties of residual soils derived from igneous and metamorphic rocks." *Proc. 2nd Pan American Conference on Soil Mechanics and Foundation Engineering*, Rio de Janiero, 183-191.

Sowers, G. F. (1965). "Settlement of tall chimneys." *Proc. 5th Symposium of Civil Engineering*, Indian Institute of Technology, Bangalore.

Sowers, G. F. (1985). "Correction and protection in limestone terrain." *Proc. 1st Multidisciplinary Conference on Sinkholes*, Florida Sinkhole Institute, Orlando, 373-379.

Sowers, G. F. (1985). "Residual soils in the United States (except Alaska)." *Sampling and Testing of Residual Soils*, Report of the Committee on Sampling and Testing of Residual Soils, ISSMFE, Scorpion Press, Hong Kong.

Terzaghi, K. and Peck, R. B. (1948). *Soil Mechanics in Engineering Practice,* John Wiley and Sons, New York, New York.

Movement of Foundations on Rock

Ian W. Johnston[1]

Abstract: This paper reviews methods for estimating the settlement of vertically loaded foundations on rock, with particular emphasis on shallow and piled foundations in and on soft rock.

Introduction

For as long as man has been building structures, it has been recognised that foundations placed on rock are generally superior to foundations placed on soils. Even the Bible has recognised this geotechnical rule of thumb (Matthew 7:24-27) where it was noted that despite a range of adverse conditions, the house built on rock by the wise man "fell not", whereas the house built on sand by the foolish man "fell and great was the fall of it".

Although houses on rock may not pose many more problems than in biblical times, structures in general have become very much bigger, heavier and subject to many more performance requirements. Furthermore, it has been recognised that rocks are not always hard, intact and incompressible but can be soft, weak and extremely weathered and intersected by numerous and highly variable discontinuities. All of these factors have meant that foundations on rock can experience major problems, or in the words of Heyward and Gershwin concerning some matters considered in the Good Book, "... it ain't necessarily so".

The two principal criteria for foundation design require that there is an adequate factor of safety against bearing failure and that settlements do not exceed tolerable limits. While both criteria must be satisfied, the general experience for rocks is that settlement usually governs. Therefore, although mention will be made of bearing capacity, for this reason and in keeping with the theme of the conference, the main thrust of this contribution will be a consideration of deformation or movement of vertically loaded foundations on rock.

---

[1] Professor and Dean of Engineering, Victoria University of Technology, P. O. Box 14428, Melbourne, Victoria 3000, Australia.

SHALLOW FOUNDATIONS

Use of Design Codes

The simplest means of designing foundations on rock is by making use of bearing values provided in codes. It is usual to find that these are expressed as allowable pressures implying that both bearing and settlement criteria should be satisfied. The codes do not usually permit the calculation of settlement.

An example of a general guide which makes use of a maximum settlement of 50 mm, is provided in Table 1. This is a relatively large settlement and the effect of a smaller value on the "presumed" bearing values is not known. Despite this anomaly, it is clear that the values given are very conservative. In defence of such a table, it must be stated that such a simplistic approach must be conservative so that it reflects the lowest common denominator of the wide range of rock types that could be represented in each simple classification.

TABLE 1. Presumed Bearing Values for Foundations on Rock (after Tomlinson 1975)

Type of rock	Presumed bearing value in MPa
Massive hard igneous and metamorphic rocks, massive hard limestones, hard strongly cemented sandstone.	Exceeds safe working stress on concrete or masonry sub-structure
Schists or slates free from crumbly or weathered material.	4
Interbedded hard mudstones, hard shales and sand-stones of coal measures (free from coal or fireclay), friable weakly cemented sandstones (free from uncemented sand layers).	3
Hard unweathered Keuper marl, hard unweathered chalk (free from wide fissures)	2
Clayey shales and soft mudstones	1

Empirical Approach

Another approach that is often employed makes use of empirical correlations between a relatively easily obtained geotechnical parameter and some more complex aspect of performance. Such a correlation has been provided between RQD and allowable contact pressure (Peck et al. 1974) as shown in Fig. 1. Although as before, this relationship gives an allowable bearing pressure, it is stated that settlements should not exceed about 13 mm.

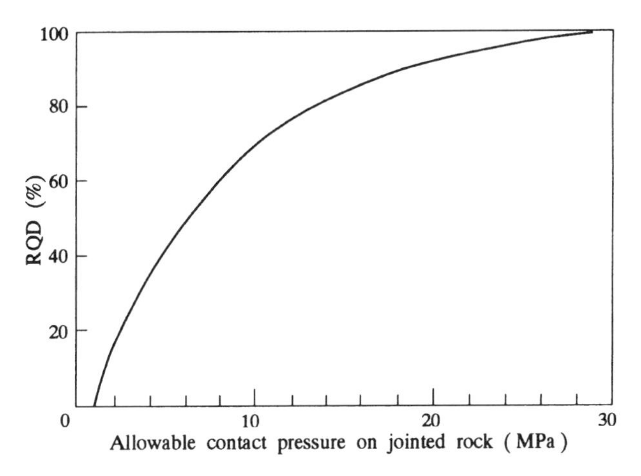

FIG. 1. Allowable Contact Pressure on Jointed Rock as a Function of RQD (after Peck et al. 1974)

It should be noted that this relationship applies to rocks containing joints which are "tight or not wider than a fraction of an inch". It would also appear that this correlation is principally for rocks that are relatively hard with a uniaxial compressive strength of at least 25 MPa. It would be quite feasible to obtain an RQD of 70% for a soft mudstone of uniaxial compressive strength of, say, 1 MPa. However, while such a material may support an applied stress of 10 MPa without failing, the resultant settlements could be well in excess of 13 mm. It follows then that if the allowable bearing values given in Fig. 1 are in excess of the uniaxial compressive strength of the intact rock, in order to limit settlements, the allowable bearing pressure should be reduced. Although it has been suggested that this limit should be the uniaxial compressive strength (Peck et al. 1974), there are arguments to suggest that such a value could be excessively restrictive (Williams et al. 1980).

The RQD parameter is a very simple rock mass characteristic which is primarily controlled by joint spacing. While this must have some influence, settlement is more likely to be affected by the condition of the joints. It follows that more complex classifications are needed to reflect this situation.

The RMR system (Bieniawski 1973) which involves the evaluation of five rock parameters (intact strength, RQD, discontinuity spacing, discontinuity condition and ground water flow) has been related to the deformation of foundations, as shown in Figs. 2 and 3. However, these correlations are based on the results of field tests which were relatively small and therefore not representative. The tests were also made at a number of dam and tunnel sites where the intact rock involved was relatively hard with uniaxial compressive strengths generally greater than 25 MPa.

The results of various in-situ loading tests in moderately weathered Melbourne mudstone of UCS in the range 2 to 3 MPa (Johnston et al. 1980) have yielded in-situ moduli of about 0.5 GPa for estimated RMRs of about 70. This seems well below the values presented in Fig. 2, probably reflecting the influence of the greater compressibility of intact soft rock. Therefore, at this stage it would seem prudent to be a little careful with the application of these figures, particularly when dealing with softer rocks.

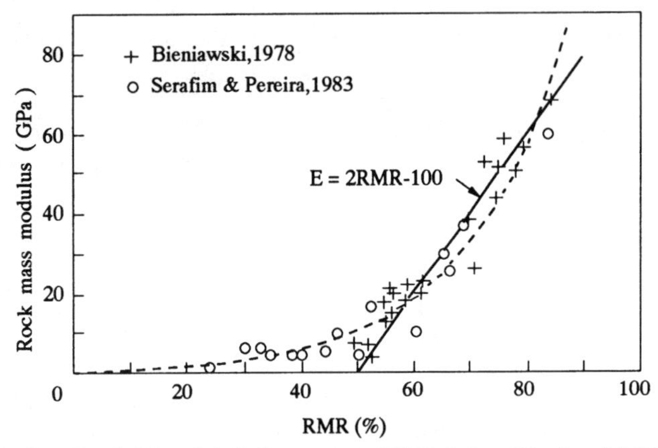

FIG. 2. Rock Mass Modulus against RMR (after Bieniawski 1984)

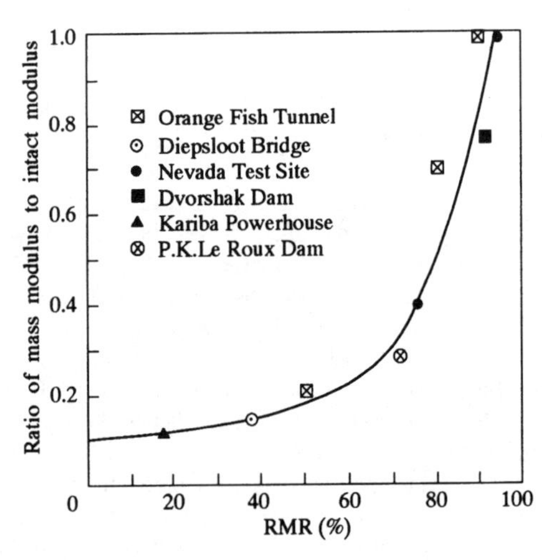

FIG. 3. Influence of RMR on Ratio of Mass Modulus to Intact Modulus (after Bieniawski 1975)

Analytical Solutions

Solutions Using the Rock Mass Modulus

As has been argued, the settlement of foundations bearing on rocks can be estimated by the application of elastic analysis. Equations for settlement take the form

$$\rho = \frac{qB}{E_m}(1 - \nu^2) I_\rho F_R \qquad (1)$$

where q is the net applied stress, B is the width of the foundation, E_m is the elastic modulus of the rock mass, ν is the Poisson's ratio of the rock mass, I_ρ is an influence factor reflecting the geometry of the foundation system and F_R is a factor which allows for the rigidity of the foundation.

There are two general approaches by which E_m can be determined. The first is through the use of empirical relationships as has already been discussed in connection with Figs. 2 and 3. The second method, involving some form of direct measurement of the rock mass modulus, must represent a superior alternative. As the sampling and laboratory testing approach is generally considered as unsatisfactory, in-situ tests, such as the plate bearing test, the pressuremeter test and the borehole jack test must be conducted.

Although plate load tests are very useful, particularly at relatively large scales, the cost of conducting them can be very high. There are many different ways in which plate load tests can be carried out as well a number of other advantages and disadvantages. More details of some of these many be found in Pells (1983). The pressuremeter is also a useful test technique and because it provides its own reaction within a borehole, it is relatively cheap and easy to perform. However, as has been discussed by Haberfield and Johnston (1993), there are a number of factors which should be taken into account when the results of pressuremeter tests are interpreted. The borehole jack test is similar to the pressuremeter test and therefore must be subject to the same general comments as above.

Solutions Using the Rock Intact Modulus with a Reduction Factor

Kulhawy (1978) has suggested an analytical model which derives the rock mass modulus from the elastic properties of the intact rock and the discontinuities contained in the rock mass. By making a number of simplifying assumptions, the modulus reduction factor, α_E, which is the ratio of mass modulus, E_m in the i direction to the intact modulus, E_r, is related to the discontinuity spacing in the i direction, S_i, and the normal stiffness of the discontinuities perpendicular to the i direction, K_{ni}, by the expression

$$\alpha_E = \frac{E_{mi}}{E_r} = \left[1 + \frac{E_r}{S_i K_{ni}}\right]^{-1} \qquad (2)$$

Kulhawy (1978) argued that discontinuity spacing, although preferred, is not particularly easy to obtain. Instead it is more common to find the RQD specified for a site. On the basis of statistical models, it was argued that general relationships existed between RQD and number of discontinuities. By combining these relationships, Fig. 4 was derived to give a relationship between the modulus reduction factor, α_E, and RQD for various values of core loss and E_r/K_{ni}. Kulhawy (1978) gives a solution for a transversely isotropic rock mass.

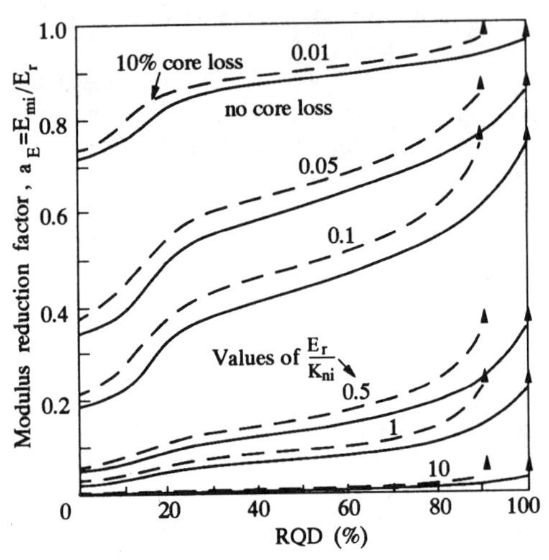

FIG. 4. Modulus Reduction Factor as A Function of RQD (after Kulhawy 1978)

While the above solutions could be described as neat and precise, there are some major shortcomings with their practical application. For the isotropic case, although the RQD and the percentage core loss may be easily obtained for a given site, the value of normal stiffness, K_{ni}, for each discontinuity is far from easily evaluated. Recognising this, Kulhawy (1978) suggested a range of typical values for these properties. However, since the more important values seem to vary by at least an order of magnitude or more, it becomes very difficult to make a meaningful selection for a particular site. There are also some broad generalisations made in the derivation of the method, such as the relationships between RQD and discontinuity spacing, which must considerably reduce its relevance to specific sites. Despite these criticisms, the method does give an insight into the mechanisms of rock deformation as well as a potentially very useful technique for future development, especially in combination with complimentary loading tests.

Numerical Solutions

With the rapidly increasing power of computers and the availability of many specialist programs, the solution of problems in geotechnical engineering by numerical methods has become very important. However, despite the potential of these methods, there are still some major problems relating to parameter definition and determination, particularly for rocks containing significant defects.

In view of these comments, it seems inappropriate that this contribution should become involved in this very extensive topic. The author would, however, like to make some general comments relating to rock foundations.

In many of the numerical models which are applied to rock foundations, the material is assumed to be an elasto-plastic continuum with ductile failure occurring. While this might be considered reasonable for soils, rocks are not continuous and often perform in a brittle, dilatant manner. These factors lead to mechanisms of deformation dominated by brittle crack formation and propagation (Haberfield and Johnston 1991). It follows that for numerical models to be relevant, the development of cracks should also be included.

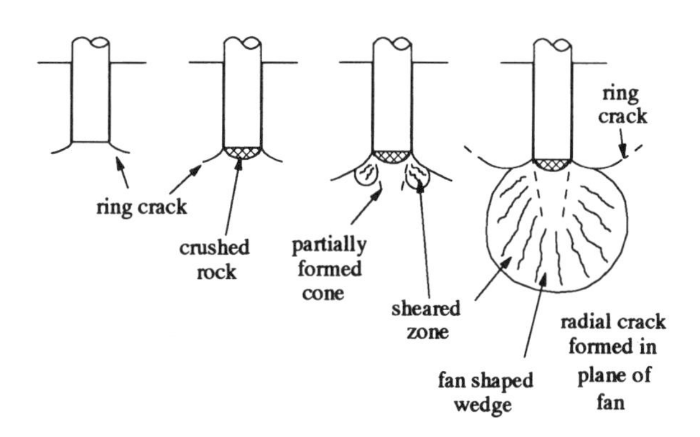

FIG. 5. Failure Mechanisms under a Loaded Footing

An example of the importance of this principle has been provided by the author and his colleagues through their work on engineering foundations on the soft rock, Melbourne mudstone. Following a large number of field tests on footings and end bearing piles (Williams 1980; Williams et al. 1980), attempts were made to model performance using finite element techniques. Initially, the methods used did not consider the development of tensile cracks nor their propagation, and failure was assumed to be by shearing. The load-displacement predictions made by these models were far from satisfactory. Subsequently, it was observed that tensile cracking had a major influence on the bearing response (Johnston and Choi 1985) as is shown in Fig. 5 for a short end bearing pile, and this had to be included in the models. Significant improvements in prediction were made, but because of problems in modelling crack propagation, these were still not as good as was hoped. It then became necessary to apply fracture mechanics theory to the problem and introduce further refinements (e.g. Lim et al. 1992) which allowed cracks to grow. The situation at the time of writing is that the numerical model is now producing load-displacement predictions that are much closer to the field test results with mechanisms that appear much more realistic (Lim 1992).

Load Testing

Although load testing, particularly in the form of the plate load test, was discussed earlier, this was aimed at providing data from which relevant rock mass moduli could be derived. Load testing can be used in a much more direct way to determine foundation settlements. The technique ideally requires a prototype foundation to be subjected to at least working load and monitoring the resultant settlements. Unfortunately, it is not always possible to accomplish this ideal with the result that smaller scale tests must be used in areas that may not be representative of final construction conditions. When this happens, interpretation of results becomes a problem once again.

PILED FOUNDATIONS

General

Where a pile simply bears against the upper surface of rock, settlements can be estimated by the same principles and methods that have been discussed above. Where a pile is formed by drilling a socket or shaft into the rock, the additional, and probably more important, side resistance component must be considered. In order to develop methods capable of estimating the movements involved in the development of side resistance in a socketed pile, it is necessary to appreciate the mechanisms that occur.

Before a rock socketed pile is loaded, the side of the concrete pile will be in full contact with the rough as drilled wall of the socket as is illustrated in Fig. 6a. When a structural load is applied to the pile, depending on the roughness of the interface and the characteristics of the materials involved, the socket will dilate as the concrete of the pile slides over the rock asperities as illustrated in Fig 6b. As the dilation proceeds, the area of rock in contact with the concrete decreases until a point is reached at which the resistance to asperity shearing becomes less than the resistance to continued sliding. It follows that the initially dominant sliding mechanism gives way to a shearing mechanism.

It has been demonstrated on many occasions (e.g. Johnston 1977; Johnston and Lam 1989) that this development of shear resistance is under conditions of constant normal stiffness. This requires that the change in normal stress, $\Delta\sigma_n$, with dilation, Δr, is given by

$$\frac{\Delta\sigma_n}{\Delta r} = K = \frac{E_m}{1+\nu}\,\frac{1}{r} \qquad (3)$$

where K is the normal stiffness, E_m is the rock mass modulus, ν is the Poisson's ratio of the rock and r is the original radius of the socket.

The influence that the above dilatant mechanism has on the performance of a pile is shown in Fig. 7. The initial part of the curve is represented by an elastic portion OA which corresponds to the pile/rock interface showing no relative movement. At A, the bonding between the rock and the pile is broken allowing the dominant sliding mechanism to begin. This sliding causes dilation against the rock stiffness to occur as discussed above, and this results in an increase in stresses acting normally to the pile surface. Consequently, there is an increase in pile load for the portion AB. At B, a peak load is reached when the shearing mechanism dominates, and the load drops back towards a residual at C.

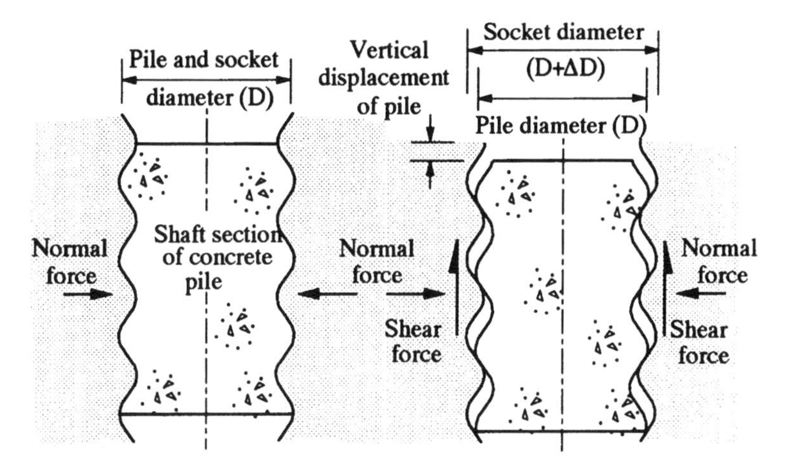

FIG. 6. Idealised Displacement Behaviour of the Side of a Socketed Pile

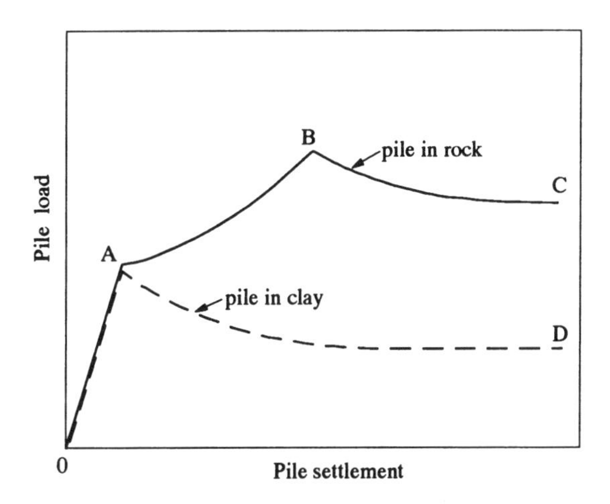

Fig. 7. Idealised Load-Settlement Curves for Piles in Rock and Clay

For a pile in clay, OA must also occur. However, at A, piles in clay would develop a shearing mechanism immediately with a shear plane forming through the clay surrounding the pile and no socket dilation occurring. At this point, the pile load will drop from A towards a residual at D. While this argument has been oversimplified, it is clear that piles in rock show marked work strengthening which can lead to significantly greater peak side resistances at greater shear displacements than might be expected. It is clear that the above mechanisms must have a major impact on the load-deflection characteristics of piles socketed into rock.

Use of Design Codes

Similar to allowable bearing pressures on shallow foundations, the side resistances of socketed piles have been limited by design codes and practices to maximum allowable values. Where piles have been considered to have only a small end bearing component, the allowable side resistances have tended to be fairly small and of the order of 100 or 200 kPa. Where there has been a large end bearing component, these resistances have been noticeably higher. However, there appears to be no means of estimating settlement apart from the suggestion that elastic methods could be used. On the basis that the above allowable stresses seem to be very low, it is perhaps quite reasonable that elastic methods are appropriate for settlement calculations. However, the sole use of these methods will almost certainly lead to very conservative designs.

Empirical Approach

There would appear to be one relatively comprehensive empirical method for the design of piles socketed into rock. The method devised by Williams et al. (1980) acknowledges that settlement is normally the controlling criterion for design, and is the main input parameter for the design process. The design method also accounts for the base resistance should it be present.

The principle of the method involves the calculation of the load required to cause a factored amount of settlement for the selected trial dimensions of a pile acting under purely elastic conditions. Once this elastic load has been divided between the base and the side resistance components as stresses, they are relaxed according to a series of design curves which have been derived from a large number of field tests. The resultant relaxed stresses are then converted into loads and summed to find the pile capacity. If this is outside the range required for design, the dimensions are adjusted and the process repeated. An estimate of the factor of safety against bearing failure is checked at the end.

The major advantages of this method are that it is simple to use, it makes use of only a few, easily derived engineering properties and it has been based on a large number of actual pile tests. The major problems with the method are that it still uses some empirical input data, it requires detailed calibration against pile test results in a wider range of rock types and there are unusual pile geometries, such as long slender piles when the solution becomes unstable.

Analytical Solutions

Elastic Analyses

As noted above, for lightly loaded socketed piles where the sides of the piles are still in perfect contact with the socket walls, elastic conditions are still likely to apply. Therefore, it seems reasonable that elastic methods of analysis are used in much the same way as was discussed with shallow foundations.

For piles socketed in rock, non-elastic conditions are developed along the side of a pile soon after loading commences (Williams et al. 1980). This is particularly true for rough sockets in fine grained rocks such as mudstone which does not form a significant bond with the concrete of the pile. For smoother sockets in stronger and more porous rock, because of wet cement bleeding into the rock, a significant bond can develop making elastic analyses much more relevant at working loads.

Solutions Allowing for Slip

On the basis of a number of simplifying assumptions which allow slip at the socket wall, Kulhawy and Carter (1992) have proposed a series of analytical solutions which permit an estimation of vertical pile movement. While these appear useful for assessing some of the observed trends which have been noted with pile tests, considerably more validation is required before the method could be considered a serious contender for practical pile design.

Solutions Based on τ - z Curves

The application of τ - z curves for the iterative determination of the load-settlement response of piles in soils (Coyle and Reese 1966) has been known for many years. While there are a number of sources of inaccuracy with the technique, it has, nonetheless, been used with some reasonable success for piles in rock (Johnston et al. 1987).

As the technique for compressible piles is iterative, discussion would perhaps be better undertaken under the heading of numerical solutions. However in addition to the numerical approach, a series of analytical solutions have been developed by Kodikara and Johnston (1994). These solutions are based on an idealised strain hardening τ - z curve of the form shown in Fig. 8 and derived from tests on a concrete rock interface. Clearly the parameters given in Fig. 8 can have a range of values. However, in order to provide some indication of magnitudes, for a large diameter bored pile in weathered mudstone of uniaxial compressive strength of about 1 MPa, $f_i \approx 200$ kPa, $f_p \approx 400$ kPa, $z_i \approx 2$ mm, $z_p \approx 20$ mm and $\zeta \approx 0.6$.

Numerical Solutions

Over the years there have been many numerical solutions suggested for the prediction of the development of the side resistance of rock socketed piles. Some of these have been based on models of shear behaviour of rough rock joints (e.g. Chiu and Dight 1983), some have made use of established numerical modelling techniques (e.g. Rowe and Armitage 1987) and others have been based on wear theory (e.g. Leong and Randolph 1991).

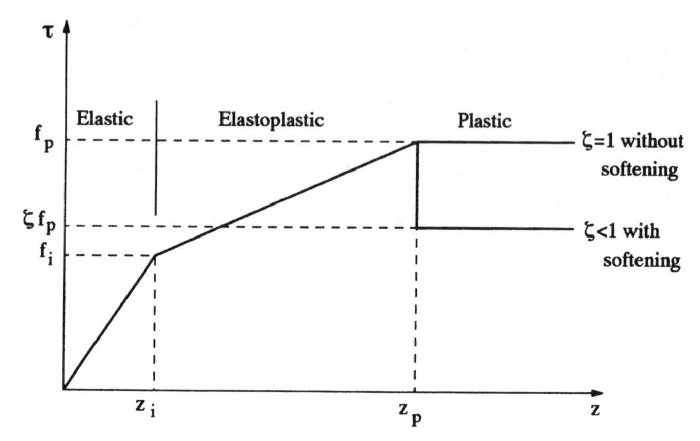

FIG. 8. Idealised τ - z Curve for Use with Kodikara and Johnston (1994) Analysis of Compressible Piles

Unfortunately, to this point in time, none of the above solutions seems to have provided a workable means of predicting the load-settlement response of socketed piles. Objections to these methods include an inability to reproduce mechanisms, a need for full scale tests with which to calibrate predictions, and complex and often subjective input data.

A numerical solution which has shown considerable promise has been progressively developed by the author and his colleagues over many years. The basis of the approach was a realisation that any numerical model would require a large number of assumptions if it were to satisfactorily describe the very complex behaviour of a rough rock interface. Clearly the success of any model would depend on the accuracy of these assumptions and a few incorrect choices could lead to major errors.

Therefore, the model has been developed in a series of progressively more complex stages which can now handle a randomly irregular compressible rock interface along the side of the pile as well as including the effects of any radial cracking that may be induced due to socket dilation. A more detailed account of the development of this model is provided by Johnston and Haberfield (1992).

Load Testing

As with shallow foundations, load testing may also form an important part of the design process for deep foundations in rock. Unfortunately, because rock socketed piles are often high capacity piles, the costs involved with dead load testing, even to working loads, can become very significant. Therefore, reliable alternatives can be very attractive. One such technique involves a retrievable test rig which can be relatively rapidly employed for testing in a contract socket (Johnston et al. 1980).

Another technique that has become an acceptable alternative for large diameter socketed piles is dynamic pile testing. This is a process that once again removes the need for expensive dead loading systems, although there can be some problems associated with generating sufficient dynamic energy at the head of the pile to mobilise all of its resistance. More details of the processes involved may be found in Balfe (1984).

Although relatively large scale load testing is normally regarded as being an in-situ technique, the application of large scale laboratory tests can also be very useful for design studies. As has been discussed above, there are a number of situations for which detailed performance characteristics (e.g. the derivation of τ -z curves and the identification of mechanisms) are required for the rough concrete-rock interface between the pile and the surrounding rock. Fig. 9 shows such a laboratory device. This is a computer controlled 250 kN cyclic constant normal stiffness direct shear device shear device which was recently designed and constructed for testing soft rock (Johnston et al. 1993).

FIG. 9. Cyclic Constant Normal Stiffness Direct Shear Device

CONCLUSIONS

Rock has generally been regarded as providing excellent support for engineering foundations. While this might be true for moderate loadings, when considering some of the very large structures being built, the consequences of unsatisfactory behaviour and the characteristics of weak rock, these materials can provide significant engineering challenges. This contribution has outlined some of the methods for estimating the movement of vertically loaded foundations on rock.

However, it must be observed that the highly variable nature of a rock, especially with regard to the much weaker discontinuities, makes such estimations much less reliable than would be expected when dealing with a soil. It is important that design procedures give recognition to this with perhaps a number of approaches being used for any one application so that performance bounds may be established and risks of unsatisfactory performance evaluated. It would also be wise in critical situations to consider the use of load testing at as large a scale as is possible. Where this is not feasible, a liberal factor of safety should always be employed.

APPENDIX - REFERENCES

Balfe, P. J. (1984). "Dynamic testing of piles socketed into weak rock." *Proc. 4th Australia-New Zealand Conf. on Geomech.*, Perth, 2, 361-365.

Bieniawski, Z. T. (1973). "Engineering classification of jointed rock masses." *Trans. South African Inst. Civ. Engrs.*, 15(12), 335-344.

Bieniawski, Z. T. (1975). "Prediction of rock mass behaviour by geomechanics classification." *Proc. 2nd Australia-New Zealand Conf. on Geomech.*, Brisbane, 1, 36-41.

Bieniawski, Z. T. (1984). *Rock Mechanics Design in Mining and Tunnels*, A. A. Balkema, Rotterdam, The Netherlands.

Chiu, H. K., and Dight, P. M. (1983). "Prediction of the performance of rock-socketed side-resistance-only piles using profiles." *Int. Journ. Rock Mech. and Min. Sci.*, 20(1), 21-32.

Coyle, H. M., and Reese, L. C. (1966). "Load transfer for axially loaded piles in clay." *J. Soil Mech. Found. Div.*, Proc. ASCE, 92(SM1), 1-26.

Haberfield, C. M., and Johnston, I. W. (1991). "Numerical modelling for weak rock." *Proc. 7th Int. Conf. Comp. Meths and Adv. in Geomech. Res.*, Cairns, 1159-1164.

Haberfield, C. M., and Johnston, I. W. (1993). "Factors influencing the interpretation of pressuremeter tests in soft rock." *Proc. Int. Symp. Hard Soil -Soft Rock*, Athens, 1, 525-531.

Johnston, I. W. (1977). "Rock socketing down-under." *Contract J.*, 279, 50-53.

Johnston, I. W., and Choi, S. K. (1985). "Failure mechanisms of foundations in soft rock." *Proc. 11th ICSMFE*, San Francisco, 3, 1397-1400.

Johnston, I. W., Donald, I. B., Bennett, A. G., and Edwards, J. (1980). "The testing of large diameter pile rock sockets with a retrievable test rig." *Proc. 3rd Australia-New Zealand Conf. on Geomech.*, Wellington, 1, 105-108.

Johnston, I. W., and Haberfield, C. M. (1992). "Side resistance of piles in weak rock." *Proc. Conf. on Piling: European Practice and Worldwide Trends, London*, 52-58.

Johnston, I. W., Lam, T. S. K., and Williams, A. F. (1987). "Constant normal stiffness direct shear testing for socketed pile design in weak rock." *Géotechnique*, 37(1), 83-89.

Johnston, I. W., and Lam, T. S. K. (1989). "Shear behaviour of regular triangular concrete-rock joints - analysis." *J. Geotech. Eng.*, ASCE, 115(5), 711-727.

Johnston, I. W., Seidel, J. P., and Haberfield, C. M. (1993). "Cyclic constant normal stiffness direct shear testing of soft rock." *Proc. Int. Symp. Hard Soil - Soft Rocks*, Athens, 2, 977-984..

Johnston, I. W., Williams, A. F. and Chiu, H. K. (1980). "Properties of soft rock relevant to socketed pile design." *Proc. Int. Conf. on Struct. Foundns on Rock*, Sydney, 1, 55-64.

Kodikara, J. K., and Johnston, I. W., (1994). "Analysis of compressible axially loaded piles in rock." *Int. Journ. for Numerical and Analytical Methods in Geomechanics*, in press.

Kulhawy, F. H., (1978). "Geomechanical model for rock foundation settlement." *J. Geotech. Eng. Div.*, Proc. ASCE, 104(GT2), 211-227.

Kulhawy, F. H., and Carter, J. P. (1992). "Socketed foundations in rock masses." *Engineering in Rock Masses*, Butterworth-Heinemann, Oxford, U.K., 509-529.

Leong, E. C., and Randolph, M. F. (1991). "Modelling sliding behaviour of rock interfaces." *Proc. 7th Int. Conf. Comp. Meths and Adv. in Geom. Res.*, Cairns, 1, 365-369.

Lim, I. L. (1992). "Fracture propagation in soft rock," Ph.D. thesis, Monash University, Melbourne, Victoria, Australia.

Lim, I. L., Johnston, I. W., and Choi, S. K. (1992). "Comparison between various displacement based stress intensity factor computational techniques." *Int. Journ. of Fracture*, 58(1), 193-210.

Peck, R. B., Hanson, W. E., and Thornburn, T. H. (1974). *Foundation Engineering*, 2nd Ed., John Wiley and Sons, New York, New York.

Pells, P. J. N. (1983). "Plate load tests on soil and rock." *In-situ Testing for Geotechnical Investigations*, A. A. Balkema, Rotterdam, The Netherlands, 73-86.

Rowe, R. K., and Armitage, H. H. (1987). "Theoretical solutions for axial deformation of drilled shafts in rock." *Can. Geotech. J.*, 24(2), 114-125.

Seraphim, J. L., and Pereira, J. P. (1983). "Consideration of the geomechanics classification of Bieniawski." *Proc. Int. Symp. on Engng. Geol. and Underground Const.*, LNEC, Lisbon.

Williams, A. F. (1980). "The design and performance of piles socketed into weak rock," Ph.D. thesis, Monash University, Melbourne, Victoria, Australia.

Williams, A. F., Johnston, I. W., and Donald, I. B. (1980). "The design of socketed piles in weak rock." *Proc. Int. Conf. on Struct. Foundns on Rock*, Sydney, 1, 327-347.

Software for Settlement Analysis

John T. Christian,[1] Fellow ASCE

Abstract: The software most widely used in engineering practice as well as for general purposes consists of word processors, spreadsheets, and databases, and these are useful for settlement analysis as well. There are several systems for processing data from borings and entering the results in databases. These, when combined with Geographical Information Systems, provide engineers with powerful tools for performing their first task — understanding the soil profile. Recently some software has also become commercially available for performing conventional settlement analysis; two packages in particular are described briefly.

Introduction

This conference meets thirty years after the first ASCE specialty conference devoted to the design of foundations for the control of settlement. In that time the profession has seen many developments in both practice and research. Laboratory testing equipment has become much more sophisticated, and the profession's understanding of soil behavior has become deeper and more elaborate. It is now conventional practice to use, in both the laboratory and the field, instrumentation that would have been beyond the capabilities of the most advanced researchers at the time of that earlier conference. Subspecialties that were not even thought of in 1964 have become major areas of practice.

Of all the technical developments in the last generation, surely none has been more dramatic and has had a more profound effect on research, practice, and our daily lives than those involving computers. Engineers can now have on their desks computational power that greatly exceeds what was available in the largest and most expensive machines then available, and the capabilities of the highly portable

---

[1] Vice President, Stone & Webster Engineering Corporation, P. O. Box 2325, Boston, MA 02107.

notebook computers border on the miraculous. It is appropriate to ask at this conference devoted to the study of settlements how this revolutionary development has affected the practice of geotechnical engineering.

One obvious use of computers in foundation engineering has been to perform sophisticated analyses of stresses and deformation, using advanced constitutive models, finite elements, boundary elements, discrete elements, or any of the other mathematical tools that have proliferated as computers became more widely available, more powerful, and cheaper. Several papers at this conference describe such developments. However, these applications require the attention of trained, experienced persons, who understand how the analytical techniques work and how they can go wrong. This paper is not concerned with that type of software. It deals instead with software developed to assist the engineer in performing conventional tasks — in this case, settlement analysis.

GENERAL ENGINEERING USE OF COMPUTERS

Engineers spend surprisingly little of their time doing calculations. Most of the working day is consumed in writing reports, conferring with clients and colleagues, investigating field conditions, and so on. Even when calculations are performed, they are often simple and repetitive. In a provocative book, Ferguson (1992) points out that, at its best, engineering is a tactile, graphical, intuitive, and artistic endeavor, rather than merely an application of scientific principles. This is not a surprising conclusion to those who have practiced engineering, but it is not often seen in print. Others have observed that the actual work of engineering has much more to do with the mobilization, organization, and direction of people than with engineering science. The way engineers use their time is a reflection of this fact.

It is generally recognized that the two most widely used types of software for microcomputers are word processors and spreadsheets. Of course, this is to be expected of the general population of computer users, but it is also true of engineers, and the reasons are not hard to find. Engineers write a lot of material — reports, letters, memoranda, specifications, and so on. A word processor is a natural tool for this work, and engineers increasingly do it directly on the word processor instead of passing hand-written, and often barely legible, drafts to a secretarial staff for typing and iterative correction.

Many calculations lend themselves naturally to the tabular format of the electronic spreadsheet; indeed, before computers ever appeared on the scene, this was the preferred way to set up many engineering calculations. The format is also ideal for managerial tasks. Since modern spreadsheet programs include excellent graphical capabilities and some sophisticated programming tools, it is now possible to perform very advanced engineering calculations and plot the results using nothing other than an electronic spreadsheet. This is true of settlement calculations, as it is for other areas of engineering.

Since one of the main products of an engineering office consists of drawings, it is to be expected that two-dimensional computer aided drafting (CAD) systems are also widely used in practice. These systems have now become the standard way to produce drawings for major engineering projects.

The fourth most widely used type of software is the simple database manager. The theory and practice of database management can be very complicated, but simple databases are relatively easy to implement. Modern database software implemented on a microcomputer makes possible some very powerful and simple manipulation of engineering data. This is particularly true when the data consist of a mixture of numbers, qualitative information, notations, and other information. Boring logs and the related soil profiles are classic examples.

It is worth noting in passing that the first electronic spreadsheet was invented in the late 1970s and that, although primitive word processors have existed for a long time, most people did not know what the term meant until about ten years ago. The popularity of both is a creation of the microcomputer world.

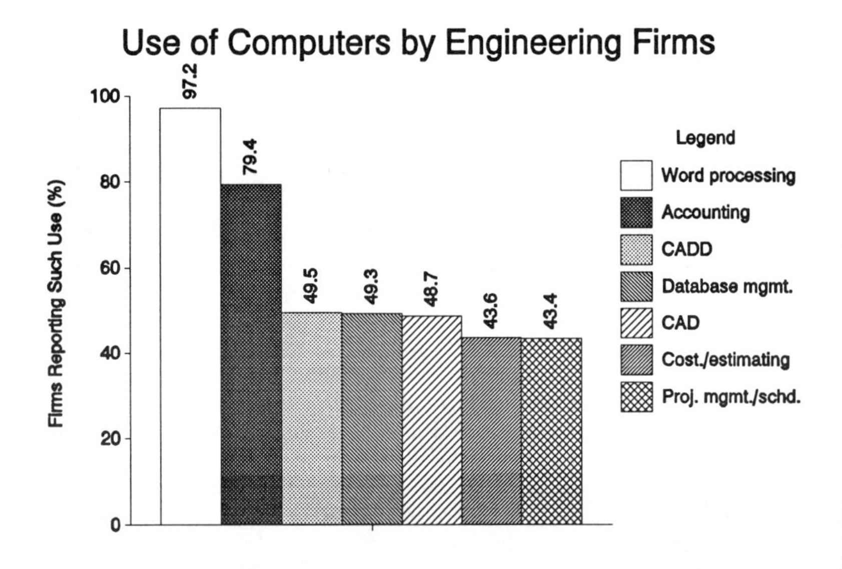

FIG. 1. Percentage of Firms Reporting Various Types of Computer Usage (after Backman 1993)

Backman (1993) summarizes the results of a study conducted for ASCE that confirms the above observations. The study covered a far wider range of firms than those involved in geotechnical engineering, but the results are probably typical of most specialties within civil engineering. Fig. 1 shows the percentage of responding firms that indicated they used each of several categories of software. Spreadsheets are not identified *per se*, so their use is included in several of the categories. According to Backman,

Three-dimensional and solid modeling applications comprised only 13.9% of applications owned, and only 9% of engineering firms used geographical information systems (GIS). The study found that the number of firms using GIS would double next year.

BORING LOGS AND GEOGRAPHICAL INFORMATION SYSTEMS

Before the engineer can analyze or predict settlements, he or she must establish the soil profile and the properties of the soils. In many cases this involves processing data from a boring program to prepare boring logs and idealized soil profiles. In recent years several programs have become available that reduce some of the tedium in these activities.

Although some software for preparing boring logs is available commercially, the writer's firm has found the most useful program to be one that it wrote internally called BORLOG. With this program, the user performs data entry and editing of boring log information on a portable DOS-based computer and generates report-quality logs using an HP LaserJet printer. Data are input through menus and screens that provide the necessary user prompts. Data include the usual heading and sample information, including depths, blow counts, group symbols, sample descriptions, and strata descriptions. The program was developed using the Clipper® compiler (Clipper is a registered trademark of the Nantucket Corporation, 1255 W. Jefferson Blvd., Los Angeles, CA 90066) and produces database files that are compatible with dBASE III+® (dBASE is a registered trademark of Borland International, Inc., 1800 Green Hills Rd., P. O. box 660001, Scotts Valley, CA 95067-0001). Therefore, dBASE III+ can be used to reformat the data for use in other applications, such as spreadsheets, word processors, or text editors, or to export the data for use in other widely used CAD systems. The number of borings is limited only by the disk space available. It has been used with up to 100 borings in one database, but the operation is more efficient if the number is limited to 10 or 20 per database. Unlike some commercially available programs, it will operate in 640 K of RAM on systems without hard disk drives, so field personnel can use it on relatively inexpensive portable machines.

Geographic Information Systems (GIS) represent a very exciting development that has many possible applications to geotechnical engineering. A recent workshop at Georgia Institute of Technology (Frost and Chameau 1993) examined their use in geotechnical earthquake engineering. These are systems in which the computer manipulates, analyzes, and synthesizes surficial and underground geographic data to provide a variety of representations that can be grasped by the engineer. In particular, they provide graphical representations in situations where the sheer mass of data makes other forms of interpretation very difficult.

Software to capture the results of boring programs combined with GIS technology promises to provide the geotechnical engineer with powerful tools to understand the subsurface conditions. Graphical representation of the data is compatible with the way engineers have traditionally treated such data (see, for example, Ferguson 1992). For large, complicated projects this capability is probably more significant than the ability to perform conventional settlement analyses rapidly and accurately by computer.

Despite the great promise of GIS, there are inevitably some limitations with which the user has to deal. First, GIS software and hardware is expensive to acquire, and most systems require skilled and trained operators. The cost may be justified only for large or complicated projects. Second, the technology of data representation and manipulation in GIS is still evolving. For example, Adams (1993) presents a very interesting discussion comparing the relative merits of relational representation of the data with those of an object-oriented feature-based system. She concludes,

> Furthermore, because the relational data model was designed for nonspatial business applications, semantic relationships between spatial features or cartographic data are not easily defined. Semantic relationship can enhance the modeling of geographical data, so that in addition to topological relations, map features can be effectively prescribed. Feature-based systems attempt to add more meaning to data bases that contain natural resource or geographic data.

Software Designed Specifically for Settlement Analysis

While conventional software can deal effectively with many engineering problems, there is still a need for software designed to handle the specific calculations involved in estimating settlement. The limited market has restricted development of software specifically aimed at geotechnical practice, but several systems for geotechnical engineering applications have appeared in recent years. For the present purpose the writer examined two programs for conventional analysis of settlement under foundation and embankment loads. Prototype Engineering, of Winchester, Massachusetts, has released the SAF-I (Urzua 1991) and SAF-TR (Urzua 1992) programs for settlement analysis and the STRESS program for calculation of elastic stress distribution. Unisoft, of Ottawa, Ontario, Canada, has released UNISETTLE (Goudreault and Fellenius 1992) for stress and settlement analysis and UNIPLOT (Goudreault and Fellenius 1993) for plotting the results. Both systems run on DOS-based microcomputers. There are other similar systems available, but these illustrate the current state of the art.

SAF-I allows the user to specify the layering of the soil profile, the water table, soil properties, and foundation loads in a manner similar to what the engineer would do by hand. The options for the soil properties of the compressible layers include most of the conventional descriptions of clay behavior, including maximum past pressures, virgin compression parameters, rebound parameters, recompression parameters, and a choice of representation in terms of e-log p curves or incremental strain-log p curves. Incompressible layers are allowed. The program does not include exponential stress-strain relations such as Janbu's (1965) equations.

Increments of vertical stress can be computed for a wide range of footing shapes and load distributions. The STRESS program incorporates a larger suite of elastic solutions, whose results can be introduced into SAF-I as specified vertical stress increments for each layer. Both Boussinesq and Westergaard solutions are available.

Settlements are computed by one-dimensional compression theory. That is, the increments of vertical stress are combined with the soil properties to obtain increments of vertical strain for each compressible layer. The sum over all the layers represents the estimated settlement. This is, of course, the conventional method of estimating settlement, which is taught in all introductory courses in soil mechanics and used every day in hundreds of design offices.

SAF-TR computes the time-dependent settlement and pore pressure distribution from Terzaghi's one-dimensional consolidation theory. The program can handle a variety of initial distributions of excess pore pressure. The conventional solution, in which the initial excess pore pressure distribution is expanded in a Fourier series and each term decays exponentially, converges rapidly for moderate to late times because very few terms in the infinite series need to be used. At early times this solution requires many more terms to converge, and the computer time needed to achieve convergence becomes noticeably longer. To avoid this problem, the program uses for early times an equally correct solution in terms of a series of complementary error functions, which converges much faster early in the consolidation process.

This writer has found the user interface to the SAF-I, SAF-TR, and STRESS programs fairly easy to use. It does not use the currently fashionable graphical user interfaces familiar to users of Windows and similar operating systems, but the combination of menus, input screens, and help features is well designed. Some learning and experimentation are always needed when a new system is being used, but navigating among the various menus and screens was generally straightforward and intuitive. The suite of programs is based on conventions used in the United States. The programs give quite clear references for the solutions and procedures they employ. The manuals and help screens cite references commonly used in American education and practice. The manuals also present examples showing in detail how each problem is described to the computer.

For what they do, the programs have few limitations. The most important one is that SAF-I will not handle mixed loadings. That is, the program will accept a set of various rectangular loads or a set of various circular loads, but it will not accept a combination of rectangular loads, circular loads, and so on. This is not a very severe limitation, for the user can usually replace one type of loading with an equivalent area of another type. Nevertheless, the user's interaction with the program would be improved if it would accept mixed loadings. The suite of stress-strain relations is also somewhat limited and, as is stated above, it does not include relations involving exponential powers of the confining stress. A third restriction is that there is no plotting capability within the programs themselves. This is a problem primarily for SAF-TR, for it would be helpful to be able to obtain plots of the isochrones. However, all the programs do provide output in ASCII format, which can be edited and loaded into the user's stand-alone plotting program.

UNISETTLE covers much of the same territory. Inevitably, it does some things SAF-I does not do, and vice versa. It does include constitutive relations based on exponential powers of the confining stress, such as Janbu's equation. The constitutive relations and algorithms for computing settlement in UNISETTLE are keyed to the provisions of the Canadian Foundation Engineering Manual (1985). It has a companion program, UNIPLOT, that can generate more elaborate or specialized plots than UNISETTLE alone.

The stress distribution portion of UNISETTLE is more limited than the corresponding features of SAF-I. For example, it is much easier to describe multiple rectangular surface loads in SAF-I. Also, the stress computations in SAF-I appear to be instantaneous on a computer using an Intel 486 chip running at 33 MHz, but there is a noticeable delay for the same calculations with UNISETTLE. On the other hand, the plotting features of UNISETTLE permit the user to see graphically what the loading pattern looks like, once it has been described. UNISETTLE does include a one-on-two stress distribution algorithm in addition to the Boussinesq and Westergaard solutions. The program can plot the distribution of stress versus depth. UNISETTLE does not deal with consolidation or other time-dependent effects.

This writer's evaluation of UNISETTLE was somewhat limited because he was using a demonstration copy of the program, in which the printer functions were disabled. Demonstration copies of programs are often awkward to use. They seldom have the most recent modifications and corrections, and sometimes the disabling process affects more than the distributor intended so that the sample problems do not run as they should. A full, uncrippled version of the program is likely to be much more responsive than the demonstration copy, and readers should bear this in mind when reading the evaluation of the program's interface.

The user interface for UNISETTLE looks a bit more modern than that for SAF-I; for example, it allows the use of a mouse. Nevertheless, this writer found it difficult to use. The program comes with sample problems in separate files, but the actual interactive use of the program is not well explained or demonstrated in the manual. In some cases the input is counter-intuitive. Part of the difficulty is that some input parameters are called on to perform double duty. This can increase the efficiency of the program, and it can accelerate the input for an experienced user, but it confuses and frustrates the occasional or novice user. Two examples indicate the nature of the problem.

When describing a loading condition the user does not select from a menu a point, line, strip, triangular, or quadrilateral load and then describe the necessary parameters in a separate menu. Instead, the program determines the geometry from the coordinates that are input. A single point (x1, y1) means a point load. A single point (x1, y1) and an additional value (x2) mean a circular load with radius x2 centered at the point. Two points, (x1, y1) and (x2, y2), mean a line load. Three points mean a triangular loaded area, and four points mean a quadrilateral loaded area. A zero for coordinate x1 means to ignore the load. The input does not explicitly include loading distributed over the entire site, such as due to a layer of fill, and it does not include embankment loads.

The default option for the stress-strain properties is the Janbu relation expressed in terms of j, m, and m_r, the exponent, one-dimensional compression modulus coefficient, and rebound modulus coefficient, respectively. When $j = 0$, the program assumes that the relation between stress and strain is logarithmic. To input an e-log p relation in terms of C_c, the user must input 0 for j and h, which is normally a call for help, for the value of m; then a window appears requesting the values of the appropriate parameters. To express the relation in terms of the E modulus alone, the user must input 1 for j and then use the same help approach to get the input window. To express the Janbu relation in terms of E, the user must input 0.5 for j and call for help. The program recalculates the Janbu parameters for the system of units being employed. When the writer tried to input an e-log p relation in terms of C_c so as to compare results with those from one of SAF-I's sample problems, the program accepted the input, recalculated the parameters, and then defaulted back to other values of the stress-strain parameters.

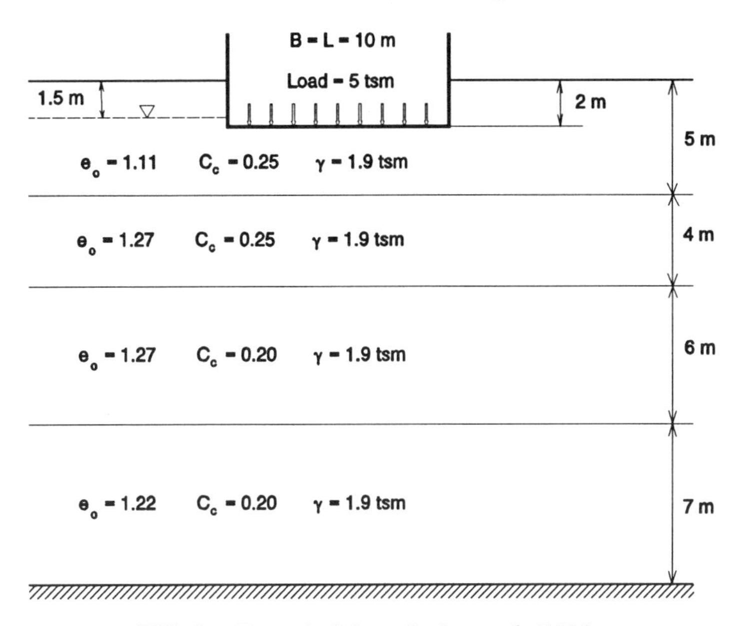

FIG. 2. Example 7 from Janbu *et al.* (1956)

Both suites of programs are useful tools for the geotechnical engineer who is performing conventional, one-dimensional settlement analysis. A user familiar with one set of programs will probably find the other awkward to use at first. The writer ran several problems on both systems and found they gave comparable results, except when the demonstration version of UNISETTLE would not accept the necessary input.

The first problem is illustrated in Fig. 2, which is based on Example 7 from NGI Publication 16 (Janbu *et al.* 1956). The results of analyzing this problem with UNISETTLE are included in the UNISETTLE User's Manual. The results from

SAF-I are reproduced in Fig. 3, which shows both the description of the problem's parameters and the results of the calculations. The settlement reported in the original publication and in the UNISETTLE manual is 8.9 cm; this is identical to the SAF-I result to two significant figures. Both programs permit the user to change the number of sublayers within each major deposit and to do other parametric studies. It turns out that using a finer division of the soil layers leads to a small increase in the estimated settlement, and the two programs give nearly identical results in all cases. The output can also be edited in a word processor to fit it in a report; that is in fact what was done to create Fig. 3.

```
ONE DIMENSIONAL SETTLEMENT ANALYSIS/PROTOTYPE ENGINEERING INC.
                    RECTANGULAR LOADS

Project Name   : NGI Pub. 16        Client          : Testing
Project Number : Ex. 7              Project Manager : J. T. Christian
Date           : 9/16/93            Computed by     : JTC

        Increment of stresses obtained using : Boussinesq

        Settlement for X =   5.00 (m)        Y =   5.00 (m)

    Footing #      Corner Point P1     Corner Point P2       Load
                   X1(m)   Y1(m)       X2(m)   Y2(m)        (Tsm)
        1           0.00    0.00       10.00   10.00         5.00

Foundation Elev.    =   -2.00 (m)   Ground Surface Elev.=   0.00 (m)
Water table Elev.   =   -1.50 (m)   Unit weight of Wat. =   1.00 (Tcm)

     LAYER              COEFFICIENT          UNIT   SPECIFIC  VOID
Nº.  TYPE   THICK.  COMP. RECOMP. SWELL.   WEIGHT  GRAVITY   RATIO  Settlement
             (m)                            (Tcm)                      (cm)

1  INCOMP.  1.5   -----  -----  -----     1.90    ----      ----      ----
2  INCOMP.  0.5   -----  -----  -----     1.90    ----      ----      ----
3   COMP.   3.0   0.250  0.000  0.000     1.90    2.65      1.11      4.73
4   COMP.   4.0   0.250  0.000  0.000     1.90    2.65      1.27      2.72
5   COMP.   6.0   0.200  0.000  0.000     1.90    2.65      1.27      1.04
6   COMP.   7.0   0.200  0.000  0.000     1.90    2.65      1.22      0.39

                                       Total Settlement =            8.88

       SUBLAYER                    SOIL STRESSES
  Nº.  THICK.  ELEV.     INITIAL  INCREMENT  MAX.PAST PRESS.    SETTLEMENT
        (m)     (m)       (Tsm)     (Tsm)        (Tsm)             (cm)

   1   INCOMP.
   2   INCOMP.
   3   3.00    -3.50      4.65      1.67         4.65              4.73
   4   4.00    -7.00      7.80      1.19         7.80              2.72
   5   6.00   -12.00     12.30      0.57        12.30              1.04
   6   7.00   -18.50     18.15      0.26        18.15              0.39

                                     Total Settlement =          8.88 (cm)

   Hit arrow keys to display next screen. <F8> Print. <F10> Main Menu
```

Fig. 3. Settlement at Centerline for Example in Fig. 2 - SAF-I Output

Fig. 4 shows another example from the UNISETTLE User's Manual. The vertical stress distribution is to be found beneath the specified point due to a uniform load of 250 kPa on the irregular area. Fig. 5 presents the results from the two programs. While the results are not identical, they are comparable. Other comparisons between the results from the two programs showed similar agreement.

CONCLUSIONS AND FINAL COMMENTS

In the thirty years since the last ASCE geotechnical specialty conference devoted to settlements many technical developments have taken place, but none has affected engineering and society at large as much as the computer. Computer technology is widely used today to automate the preparation of boring logs and to move the resulting data into databases or CAD systems. Recent advances in GIS technology promises to make it possible for the engineer to manipulate and assimilate a large quantity of data from the field and display the results graphically. This is a developing technology in which several technical issues still need to be resolved.

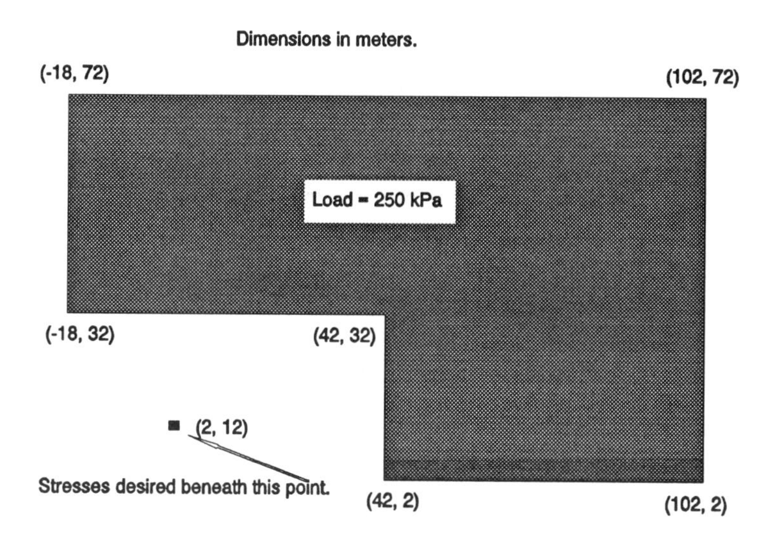

FIG. 4. Surface Loading and Point beneath which Vertical Stresses are to be Computed

Although software for geotechnical applications has been relatively slow to appear in the market, there are now several packages available commercially for analysis of settlements by conventional methods. Two systems in particular are the programs SAF-I, SAF-TR, and STRESS from Prototype Engineering and the programs UNISETTLE and UNIPLOT from Unisoft. The programs differ in the range of problems they can handle, their interfaces to the user, their ease of use, and their speed of execution. Nevertheless, they do deal with much the same material for the most part. Where comparisons can be made, they give similar results.

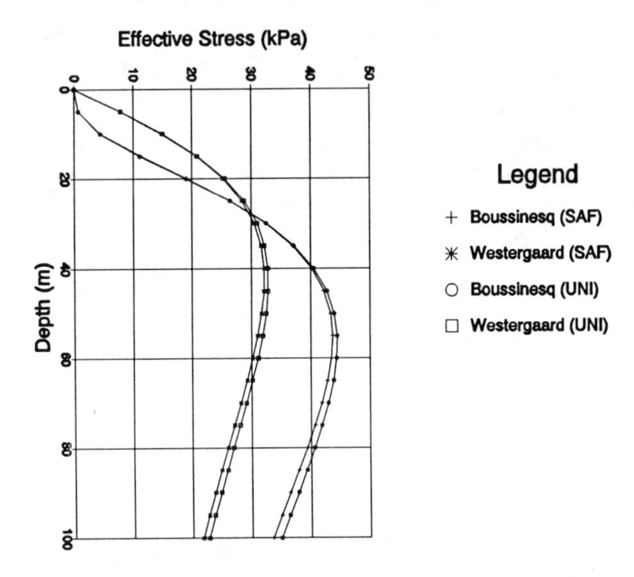

FIG. 5. Vertical Stress Calculated with SAF-I and UNISETTLE for Boussinesq and Westergaard Solutions

ACKNOWLEDGEMENTS

Dr. Alfredo Urzua of Prototype Engineering and Dr. Bengt Fellenius of Unisoft generously provided the author with copies of their software. Mr. Paul J. Trudeau of Stone & Webster Engineering Corporation provided several very helpful suggestions.

APPENDIX - REFERENCES

Adams, T. M. (1993). "Spatial data models for managing subsurface data." *Journal of Computing in Civil Engineering*, ASCE, 7(3), 260-277.

Backman, L. (1993). "How engineers use computers." *Civil Engineering Computing Review*, ASCE, 5(9), 7.

Canadian Geotechnical Society (1985). *Canadian Foundation Engineering Manual*, Second edition, Bi-Tech Publishers, Vancouver, B. C., Canada.

Christian, J. T. (1991a). "Geotechnical engineering analysis in the age of the modern computer." *Geotechnical Engineering Congress*, Geotechnical Special Publication No. 27, ASCE, New York, New York, I, 468-478.

Christian, J. T. (1991b). "Software for geotechnical problems." *Geotechnical News*, 9(4), 53-54.

Ferguson, E. S. (1992). *Engineering and the Mind's Eye*, MIT. Press, Cambridge, Massachusetts.

Frost, J. D., and Chameau, J.-L. (1993). Editors. *Geographic Information Systems and their Application in Geotechnical Earthquake Engineering*, ASCE, New York, New York.

Goudreault, P. A., and Fellenius, B. H. (1992). *UNIPLOT, Version 1.0 - User Manual*, Unisoft, Ottawa, Ontario, Canada.

Goudreault, P. A., and Fellenius, B. H. (1993). *UNISETTLE, Version 1.0 - Background and User Manual*, Unisoft, Ottawa, Ontario, Canada.

Janbu, N. (1965). "Consolidation of clay layers based on non-linear stress-strain," *Proc. 6th ICSMFE*, Montreal, 2, 83-87.

Janbu, N, Bjerrum, L. and Kjaernsli, B. (1956). "Veiledning ved losning av fundamenteringsoppgaver." *Publication No. 16*, Norwegian Geotechnical Institute, Oslo, Norway, (in Norwegian).

Urzua, A. (1991). *SAF-I - User's Manual*, Prototype Engineering, Winchester, Massachusetts.

Urzua, A. (1992). *SAF-TR - User's Manual*, Prototype Engineering, Winchester, Massachusetts.

Stress-Deformation Behavior of an Embankment on Boston Blue Clay

Charles C. Ladd,[1] Fellow, ASCE, Andrew J. Whittle,[2] Associate Member, ASCE, and Dante E. Legaspi, Jr.,[3] Student Member, ASCE

Abstract: An 11 m high I-95 embankment was thoroughly instrumented with piezometers, settlement rods and inclinometers to measure the behavior of the underlying 40 m thick deposit of Boston Blue Clay (BBC) during staged loading and four years of subsequent consolidation. This paper uses updated estimates of the engineering properties of BBC at the site, based on several major in situ and laboratory test programs, in order to perform finite element analyses that incorporate coupled consolidation with two generalized effective stress soil models, Modified Cam Clay (MCC) and MIT-E3. Comparison of predicted versus measured performance during 1967-1973 led to four principal conclusions: (1) The analyses give reasonable predictions of pore pressures and the overall magnitudes of deformations, except in the upper 15 m of the clay crust where uncertainties in the stress history were evaluated using two preconsolidation pressure profiles. (2) Differences between the two soil models lie primarily in the predicted horizontal displacements, for which MIT-E3 gives better agreement with measured behavior. (3) Measurements of outward horizontal displacements (lateral spreading) after the end of construction are predicted as a consequence of anisotropic yield in the MIT-E3 model and are most probably not related mainly to creep of the clay as postulated by others. (4) During consolidation, the analyses predict a significant

---

[1] Professor, [2] Associate Professor, [3] Graduate Research Assistant, Dept. of Civil and Environmental Engineering, Massachusetts Institute of Technology, Cambridge, MA 02139.

reduction of centerline vertical total stress due to arching which contradicts the conventional assumption of constant total stress.

INTRODUCTION

Construction of a 3.9 km extension of Interstate Highway I-95 north of Boston started in 1965 across a tidal swamp area underlain by a thick deposit of Boston Blue Clay (BBC). A portion of the embankment, called the MIT-MDPW Test Section, was thoroughly instrumented with piezometers, settlement rods and inclinometers (Fig. 1). Placement of the embankment fill occurred in three stages over a period of 1.5 years, followed by monitoring of consolidation behavior for an additional 4 years. The engineering properties of the BBC deposit at the Test Section also have been thoroughly studied by several in situ and laboratory test programs during 1966 to 1980. This case history therefore offers an excellent opportunity to evaluate the profession's ability to predict the performance of a major embankment during and after loading using generalized effective stress soil models that can simulate real time behavior.

Prior analyses of this Test Section are quite limited. D'Appolonia et al. (1971) evaluated the pore pressure behavior during loading using an undrained total stress soil model, with a focus on the occurrence of local yielding within the clay. Whittle (1974) used a hyperbolic total stress-strain model (FEECON, as described by Simon et al. 1974) to predict undrained pore pressures and deformations during construction and also attempted to back-calculate field values of compressibility and coefficients of consolidation using one-dimensional and uncoupled, elastic two-dimensional consolidation solutions. On the other hand, ten analyses were made in conjunction with a Foundation Deformation Prediction Symposium (MIT 1975) wherein a portion of the I-95 embankment located 0.5 km from the Test Section was loaded to failure in 1974. Much less field monitoring data were available at that site and none of these original analyses used an effective stress soil model with coupled consolidation. However, researchers at Stanford University have conducted finite element analyses of the foundation response during the staged construction and subsequent consolidation period prior to the 1974 failure using effective stress soil models and also including creep (e.g., Kavazanjian and Poepsel 1984; Kavazanjian et al. 1985; Borja et al. 1990).

This paper presents comparisons of measured pore pressures, settlements and horizontal displacements at the Test Section versus those predicted from finite element analyses using two soil models: the Modified Cam Clay (MCC) that has been widely used in practice for over 20 years; and a recently developed model called MIT-E3 that better simulates anisotropic and non-linear clay behavior. The paper has four main objectives:

1. To document the measured behavior from an unusually comprehensive set of field measurements.
2. To compare, in detail, all major aspects of embankment performance, i.e., pore pressures and deformations measured both during and after loading at many locations within the clay.

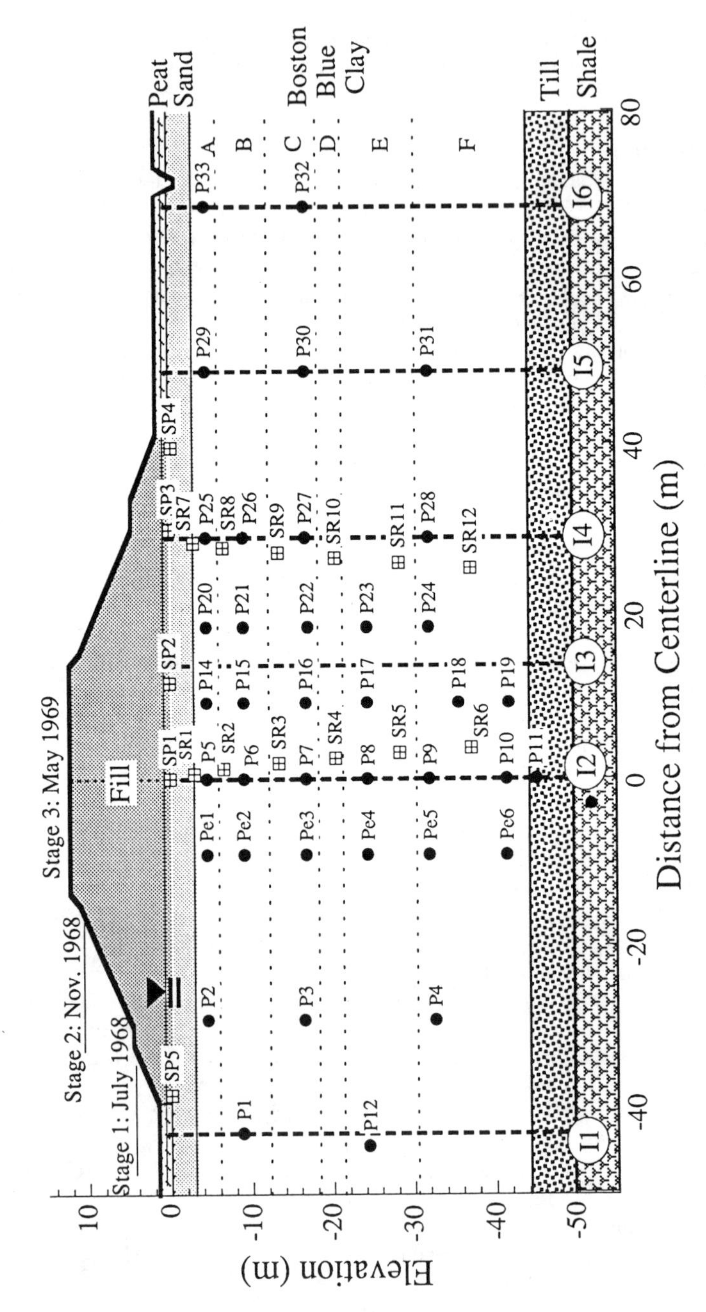

FIG. 1. MIT-MDPW Test Embankment (Sta. 246): Cross Section and Instrumentation

3. To compare the relative merits of using a simple versus complex soil model for predicting field behavior; and
4. To help determine if the measured increases in horizontal displacements with time after loading are primarily due to creep (viscous) effects (e.g., as concluded by Borja et al. 1990) or if they should be expected from a two-dimensional, non-linear consolidation analysis which couples the effective stress-strain properties of the soil skeleton to the flow of pore water.

SITE CONDITIONS AND TEST PROGRAMS

The I-95 alignment started from the south (Sta. 190) at Cutler Circle in Revere and ended northward at the Saugus River in Saugus, MA, under supervision of the Massachusetts Department of Public Works (MDPW). The MIT-MDPW Test Section is located at Sta. 246 about 400 m north of Pine River. It was constructed during 1967 to 1969 to provide data for evaluating techniques for predicting deformations, pore pressures and stability (e.g., MIT 1969; D'Appolonia et al. 1971). After a moratorium on highway construction into Boston, fill was removed from the Test Section in 1974 in order to load Sta. 263 to failure as part of a prediction symposium (MIT 1975) sponsored by the MDPW and the US FHWA.

The subsurface conditions at the Test Section determined from borings and instrumentation installation are shown in Fig. 2 and consist of ground surface El. = +1.5 m (mean sea level); soft peat above El. 0 m; 3 m of a poorly graded marine sand having an average blow count of N $\approx$ 20; 41 m of Boston Blue Clay (BBC), a post-glacial illitic marine clay; dense glacial till composed of clayey sand and gravel; and gray shale bedrock (Cambridge Argillite). The mean water table is located at El. +0.76 m and the underlying till has an artesian head of 1.52 m. Subsequent analyses assume that this artesian pressure decreases linearly through the BBC deposit.

Three consolidation test programs were run on 65-mm diameter by 20-mm thick specimens cut from undisturbed tube samples taken in 1966, 1977 and 1980, as follows:

- 1966 – incremental oedometer tests on 125-mm diameter samples taken at the centerline of the Test Section (Guertin 1967);
- 1977 – oedometer and constant rate of strain (CRS) consolidation tests on 75-mm diameter fixed piston samples taken in a wash boring with heavy drilling mud located 60-m to the right of the centerline (Ladd et al. 1980). Note that the embankment fill had been reduced to El. +5.5 m in 1974; and
- 1980 – oedometer and CRS consolidation tests on both vertical and horizontal specimens of samples obtained 25 m north of the 1977 location and taken in a similar fashion (Ghantous 1982). This program also included constant head permeability tests on vertical and horizontal specimens.

Guertin (1967) reports results from isotropically consolidated-undrained triaxial compression (CIUC) tests at varying overconsolidation ratios (OCR = 1 to 12), and K_0-consolidated undrained triaxial compression (CK_0UC) tests at OCR = 1, run on the 1966 undisturbed samples. However, the undrained shear properties required as input parameters for the two soil models were obtained from CK_0U triaxial compression and extension tests run on resedimented BBC. Whittle (1993) and Whittle et al. (1994) summarize these data and compare MIT-E3 predictions with CK_0U tests having other modes of failure and varying OCR.

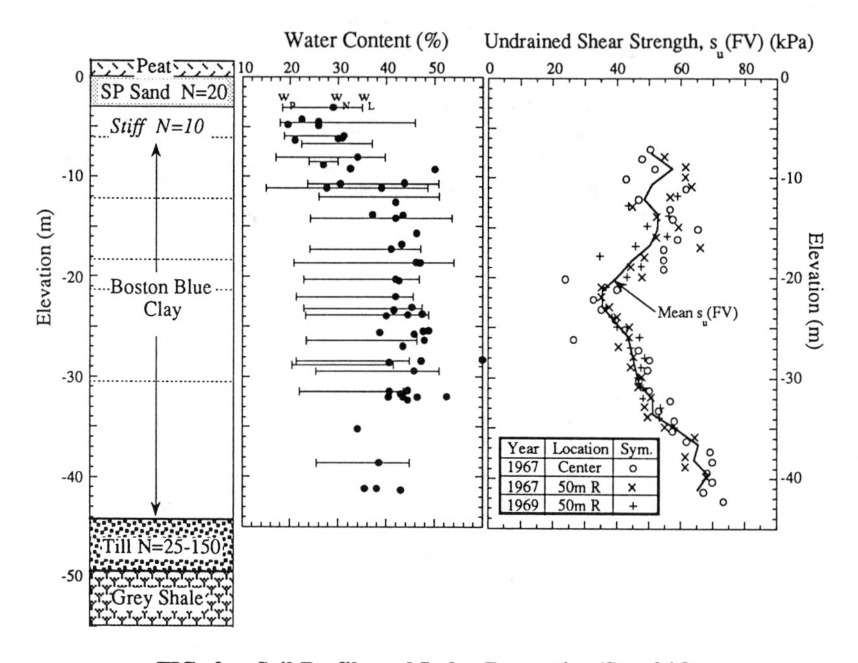

FIG. 2. Soil Profile and Index Properties (Sta. 246)

Geonor field vane tests were run at the centerline and at 50 m right of the centerline before and after construction of the embankment. A 10-m wide strip adjacent to the ditch shown in Fig. 1 was used for research with a variety of other in situ testing devices during 1975 to 1982. Morrison (1984) summarizes typical data from the following devices: earth pressure cell, self-boring pressuremeter, Dutch and piezocone, and piezo-lateral stress cell (model pile).

Properties of Boston Blue Clay

Fig. 2 plots elevation versus water content and Geonor field vane strengths. The natural water content (w_N) increases rapidly within the top 10 m of the deposit and then typically averages between 40% and 45%. The deposit is a CL clay with mean $\pm$ standard deviation values of liquid limit, $w_L = 45\% \pm 7\%$ and plasticity index, $I_p = w_L - w_P = 23\% \pm 6\%$. The mean field vane strength decreases with depth within the upper crust, reaches a minimum of $s_u(FV) = 35$ kPa near El. -22 m, and then increases approximately linearly with depth.

Stress History

Fig. 3 shows the calculated effective overburden stress (σ'_{v0}) and this stress plus the increment of total vertical stress ($\Delta\sigma_v$) at the centerline at the end of construction (EOC) obtained from the four finite element analyses described later. Note that these analyses give almost identical values of $\Delta\sigma_v$. The figure also plots values of preconsolidation pressure (σ'_p) estimated from compression curves measured in the three consolidation test programs. These values were obtained via the conventional Casagrande (1936) technique, except for highly rounded curves in the upper crust where the "strain energy" technique of Becker et al. (1987) was considered more reliable. The large scatter in σ'_p within the upper crust is typical of BBC deposits and probably reflects varying degrees of desiccation. Values of σ'_p that plot below σ'_{v0} are attributed to excessive disturbance (either due to sampling or extrusion from the tubes) even though compression curves having obvious signs of disturbance were excluded.

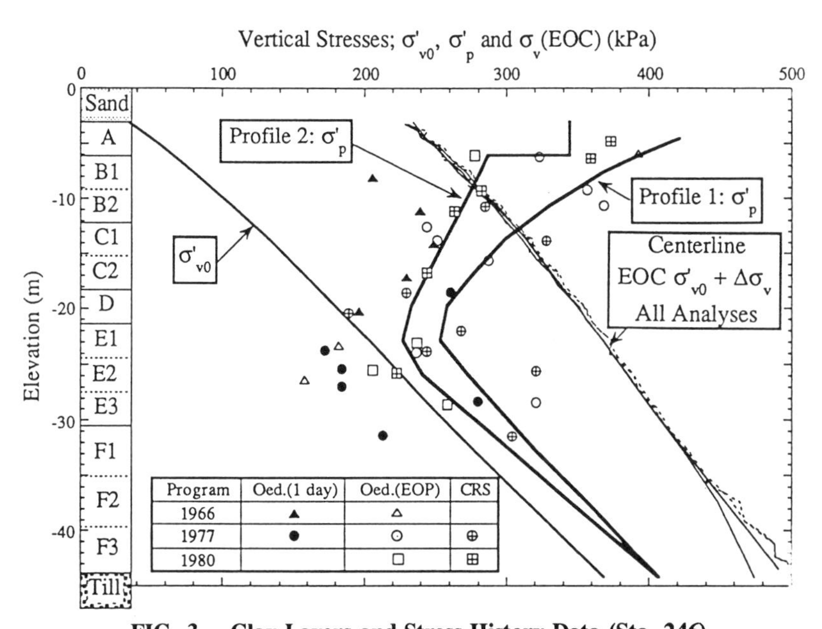

FIG. 3. Clay Layers and Stress History Data (Sta. 246)

Fig. 3 shows two σ'_p profiles selected for the finite element analyses. The upper bound Profile 1 emphasized the σ'_p data from the 1966 and 1977 programs above El. -30 m and assumed a constant amount of precompression ($\sigma'_p - \sigma'_{v0}$) at greater depths. The lower bound Profile 2 emphasized the 1980 σ'_p data and considered the fact that $s_u(FV)/\sigma'_{v0}$ increased with depth from 0.18 to 0.21 below El. -23 m. The soil profile was divided into 3.05 or 4.6 m (10 or 15 ft) intervals for the finite element analyses and are designated as layers A, B1, etc., through F3 in Fig. 3.

Consolidation Properties

The left side of Fig. 4 plots elevation versus the maximum slope of strain vs. log vertical consolidation stress (σ'_{vc}) curves during virgin compression, denoted as $CR_{max} = C_c/(1 + e_0)$. Tests in the upper crust usually produced a linear virgin compression line and CR_{max} increases with depth from about 0.1 to 0.3. But tests in the lower clay below El. -20 m often had S-shaped curves, with much higher values of CR_{max} just beyond σ'_p. The maximum stress at the end of consolidation within the lower clay exceeds σ'_p by only 25 to 50% (see Fig. 3) and the maximum compressibility occurs within this range of stresses. Proper definition of these S-shaped curves requires oedometer tests with small increments or preferably continuous loading as in CRS tests. Due to the lack of data below El. -30 m, Fig. 4 also includes results from CRS and automated K_o-consolidated triaxial tests (CK_0-TX) performed as part of a Special Test Program for Boston's Central Artery/Third Harbor Tunnel (CA/T) project (Haley & Aldrich, Inc. 1993). These tests on high-quality samples from East and South Boston near Logan Airport give values of CR_{max} in good agreement with the I-95 data above El. -30 m and hence were used to define compressibility below that depth. The solid line in Fig. 4 designates the selected values of CR_{max} and corresponds to soil layers A through F shown along the left axis of Fig. 3.

The right side of Fig. 4 plots values of the coefficient of consolidation for loading into the normally consolidated range [c_v(NC)]. As was done for compressibility, Fig. 4 also includes data from the CA/T Special Test Program. The selected values of c_v(NC) represent approximate averages of the measured data, but with considerable judgment within the top 10 m of the deposit, where the selected values are probably too low.

Consolidation Properties Selected for Finite Element Analyses

Table 1 lists the selected properties for layers A through F3 shown in column 1. Columns 2, 3, 4 and 6 come directly from Fig. 3. Values of the initial void ratio (e_0) were obtained from the w_N data in Fig. 2 using a specific gravity of 2.75 and the values of $C_c = CR_{max}$ $(1 + e_0)$ come from Fig. 4. A constant magnitude was selected for the recompression index (C_r = 0.078) based on Whittle's (1987) evaluation of unload-reload cycles on the 1980 samples after loading into the virgin compression range. The values of K_0 listed in columns 5 and 7 come from one-dimensional swelling behavior described by the MIT-E3 model (and controlled by the input parameters K_{0NC} and 2G/K listed in Table 2).

The analyses require separate input of the coefficient of permeability for vertical and horizontal flow (k_v and k_h, respectively). A linear void ratio vs log k relationship was obtained for clay layers A through F using Eq. (1):

$$(e_0 - e_p) = C_k \log_{10}\left[\frac{k_{v0}}{k_{vp}}\right] \quad (1)$$

where k_{v0} = initial k_v at initial void ratio e_0; k_{vp} = k_v at void ratio e_p corresponding to the preconsolidation pressure; and C_k = slope of the e vs. log k_v relationship. The following steps were used in applying Eq. (1):

1. Calculate $m_v = -de/d\sigma'_{vc} = [0.434\ C_c/(1+e_0)\ \sigma'_p]$ = coefficient of volume change for normally consolidated clay at a consolidation stress just exceeding σ'_p (based on the upper bound values in Fig. 3);
2. Calculate $k_{vp} = c_v\ m_v\ \gamma_w$, using $c_v = c_v$ (NC) from Fig. 4;
3. Calculate $e_0 - e_p = C_r \log \sigma'_p/\sigma'_{v0}$ = decrease in void ratio for recompression from overburden stress to preconsolidation pressure (Note: The actual calculations used $C_r = (1 + e_0)$ (RR = 0.015) that was selected for preliminary analyses. Using C_r = 0.078 would have increased k_{v0} by 80%, 25% and 7% for layers A, B and C, respectively); and
4. Calculate k_{v0} from Eq. (1) using $C_k = C_c$, which gives a constant c_v for normally consolidated clay as typically measured in the CRS tests.

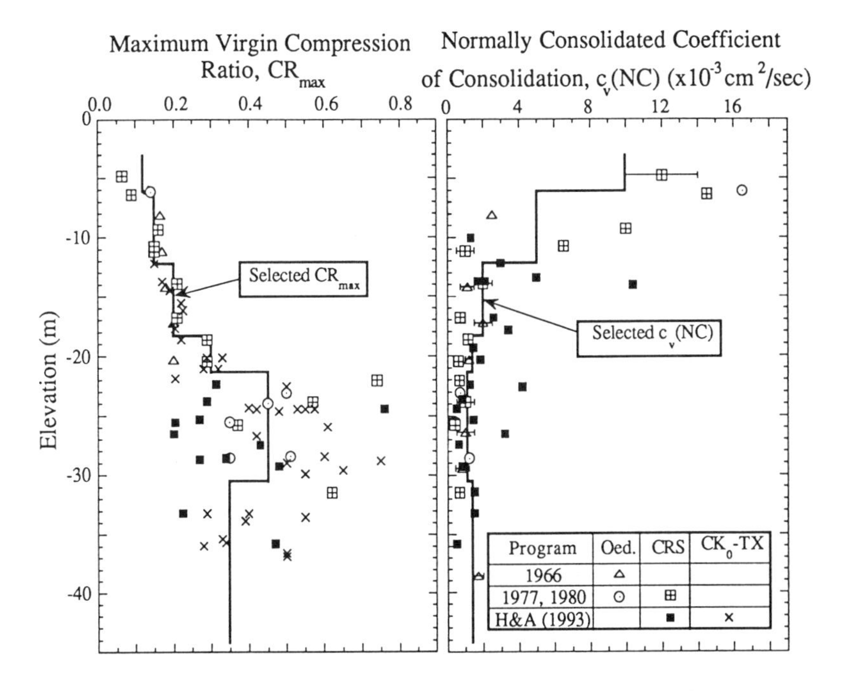

FIG. 4. Compressibility and Coefficient of Consolidation

The computed values of k_{v0} listed in column 10 of Table 1 agree with measurements from constant head permeability and CRS consolidation tests summarized by Baligh and Levadoux (1986). The k_h/k_v ratios listed in column 11 were obtained from similar tests run on horizontal specimens summarized in the same reference.

TABLE 1. Stress History and Consolidation Properties of BBC

Layer	Center El. (m)	σ'_{v0} (kPa)	Profile 1		Profile 2		e_0	C_c $(=C_k)$	k_{v0}[(b)] (10^{-8} cm/s)	k_h/k_v
			OCR	K_0[(a)]	OCR	K_0[(a)]				
(1)	(2)	(3)	(4)	(5)	(6)	(7)	(8)	(9)	(10)	(11)
A	-4.6	50	8.34	1.35	6.82	1.24	0.615	0.195	15.8	2.0
B1 B2	-7.6 -10.7	80 108	4.6 3.07	1.08 0.85	3.52 2.50	0.92 0.76	0.88	0.28	10.4	2.0
C1 C2	-13.7 -16.8	133 157	2.25 1.77	0.80 0.65	1.95 1.57	0.68 0.62	1.17	0.435	6.3	1.5
D	-19.8	180	1.44	0.60	1.30	0.58	1.17	0.65	7.1	1.5
E1 E2 E3	-22.9 -25.9 -29.0	203 226 249	1.25 1.21 1.18	0.57 0.57 0.56	1.12 1.12 1.08	0.55 0.55 0.55	1.255	1.015	7.8	1.5
F1 F2 F3	-32.8 -37.3 -41.9	278 314 350	1.16 1.14 1.11	0.56 0.56 0.55	1.09 1.10 1.10	0.55 0.55 0.55	1.115	0.74	5.9	1.5

Notes:
(a). K_0 values are computed by the MIT-E3 model.
(b). Calculated from upper OCR and values of $C_r = (1+e_0)RR$ (where RR = 0.015)

CONSTRUCTION HISTORY AND INSTRUMENTATION

As shown in Fig. 5, the Test Section embankment was constructed in three stages. Stage 1 consisted of replacing the peat layer with SW sand fill (probably end-dumped) and continued filling to El. +2.75 m during Construction Day (CD) 92 to 123. The Stage 2 filling to El. +11.0 m started about 6 months later, during CD 298 to 461 (hence over a period of 5.3 months). After waiting for 4.5 months, the final Stage 3 loading to El. +12.2 m occurred during CD 598 to 620. The sand fill above El. +1.5 m was compacted and had an average total unit weight of γ_t = 18.7 kN/m^3.

Spencer (1967) circular arc stability analyses using the field vane strength profile in Fig. 2 give a factor of safety at the end of construction of 1.25. With likely three-dimensional end-effects (Azzouz et al. 1983) and the significant consolidation that occurred during construction, the minimum factor of safety probably exceeded 1.35.

Wolfskill and Soydemir (1971) describe the instrumentation devices and installation procedures. Fig. 1 shows the instrument locations, while Fig. 5 shows when the initial readings were taken. The following summarizes the main features.

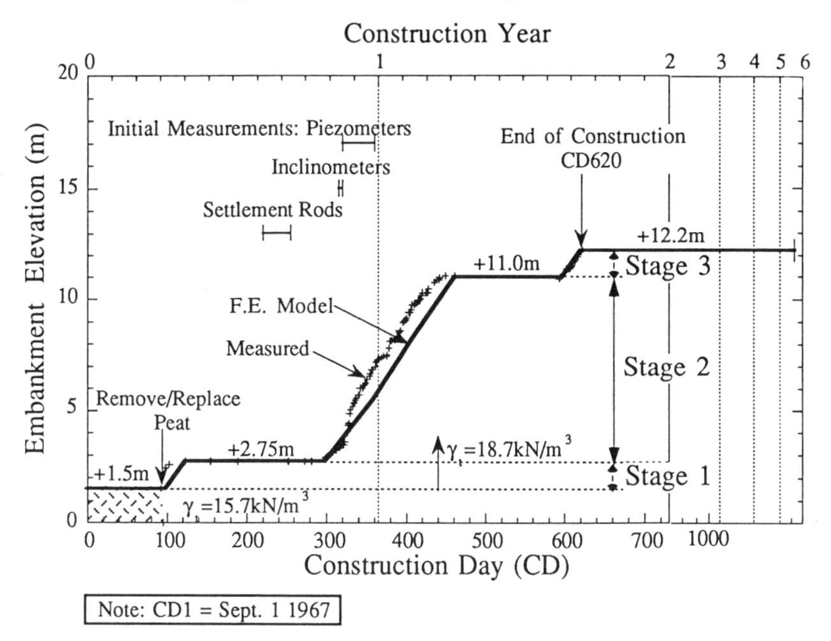

FIG. 5. Construction History (after Whittle 1974)

- 5 surface settlement platforms (SP) on top of the natural sand.
- 12 settlement rods (SR) at the centerline and 29 m Right. These consisted of 25 mm O.D. pipe with a 65 mm steel disk welded 0.9 m above the tip within a 76 mm O.D. casing pipe. Due to inadequate clearance between the bottom of the casing and the steel disk, SR4, 6, 10 and 11 recorded settlements that were too large (Whittle 1974). The reported settlement data for these four devices are shown as dashed lines after CD 1200 $\pm$ 200 assuming a linear settlement vs. log time relationship as recorded by the other devices. A settlement rod installed in the till provided the benchmark.
- 39 piezometers of two types: 6 Geonor vibrating wire (P_e) and 33 Casagrande hydraulic (P) with double leads to enable flushing read via pressure gauges. About half of these were installed at the centerline and offsets of $\pm$9 m, with the others located up to nearly 70 m Right of the centerline. In addition, 5 well points recorded pore pressures in the upper sand (not shown in Fig. 1, but

located between the centerline and 50 m Right). Of the 30 piezometers within 30 m of the centerline, about 85% provided useful data during construction; about 70% did likewise during subsequent consolidation (including three replacements at the centerline).

- 6 inclinometers (I) manufactured by Slope Indicator Company located up to almost 70 m from the centerline. They were anchored 1.5 m into the till and consisted of a 76 mm diameter aluminum casing surrounded by pea gravel in a 150 mm diameter hole. Readings during construction generally had a precision of about 1 cm, but more than doubled thereafter, due to the use of several different Wilson "torpedoes" (Whittle 1974). Reported horizontal displacements at CD 2000 are based on linear regression analyses.

Dowding et al. (1974) describe the 1.5 m diameter by 33 m long corrugated metal tunnel that provided access to most of the instrumentation. The tunnel protected the instrumentation from vandalism and eliminated the need to add extensions during placement of the fill and to add antifreeze in the hydraulic piezometers.

Finite Element Model

Finite element analyses of the I-95 embankment at Sta. 246 have been carried out using the ABAQUS program (HKS 1989) together with subroutines ('user materials') for the effective stress models being used at MIT (see Hashash 1992; Hashash and Whittle 1993). The principal features of the finite element model can be summarized as follows:

1. The cross-section in Fig. 1 shows that the soil strata at Sta. 246 are approximately horizontal (this is in marked contrast to conditions nearby at Sta. 263, the site of the 1974 prediction symposium) and the embankment cross-section is nearly symmetric. Hence, the finite element model assumes symmetry and extends from the crest centerline laterally to a distance of 165 m where the effects of embankment construction on in situ pore pressures and stresses are negligible. The bottom boundary of the model corresponds to a rough interface between the till and the underlying rock. In the farfield, the distribution of (non-hydrostatic) initial pore pressures are maintained throughout the analysis.
2. Coupled analysis of fluid flow and deformation in the clay is simulated using mixed isoparametric elements with 8-displacement nodes and 4-corner pore pressure nodes. These elements provide quadratic interpolation of displacements, strains and effective stresses; and bi-linear interpolation of pore pressures. The calculations are performed with full integration of element stiffness matrices. Solid, 8-noded, isoparametric elements are used to represent the embankment fill, sand and underlying till layers which are all effectively free draining materials. The overall mesh comprises 336 elements (36 of which model the embankment itself) with a total of 2,721 nodal degrees of freedom.
3. Fig. 5 shows the history of embankment construction used in the finite element model which compares closely with the actual record of fill elevation

versus time. Embankment construction is simulated by activation of new elements in an initial strain free state (using multi-point constraints as described by Legaspi 1994).

The effective stress-strain-strength properties of the Boston Blue Clay are represented by two generalized soil models:

1. Modified Cam Clay (MCC; Roscoe and Burland 1968) is the most widely used effective stress model in geotechnical analysis. The model formulation uses the incremental theory of rate independent elasto-plasticity and is characterized by an isotropic yield function, associated plastic flow and density hardening. The particular version of the model implemented in this study uses a von Mises generalization of the yield surface.
2. MIT-E3 (Whittle 1993; Whittle and Kavvadas 1994) is a significantly more complex elasto-plastic model which describes many aspects of the rate-independent behavior of K_0-consolidated clays, which exhibit normalized behavior (Ladd and Foott 1974), including: (a) small-strain non-linearity; (b) anisotropic stress-strain-strength; (c) hysteretic and inelastic behavior due to cyclic loading. The model uses 15 input parameters which can be evaluated from standard types of laboratory tests comprising: (a) 1-D compression tests with load reversals and lateral stress measurements; (b) resonant column (or similar tests) to estimate the small strain elastic shear modulus; and c) undrained triaxial shear tests on K_0-consolidated clay in compression (at OCR's = 1, 2) and extension (at OCR = 1) modes of shearing.

Whittle et al. (1994) have given full details of the selection of model input parameters for representing the normalized behavior of resedimented Boston Blue Clay. Table 2 summarizes the model parameters used for the non-homogeneous deposit at the MIT-MDPW Test Section. The initial stress state (σ'_{v0}, K_0) and compressibility of normally consolidated clay (e_0, $\lambda = 0.434\ C_c$) are specified for each of the sub-layers in Table 1, while the remaining stiffness and strength parameters (Table 2) are assumed constant throughout the profile (in the absence of more reliable laboratory test data). The permeability of Boston Blue Clay is represented by the non-linear (e - log k) relation in Eq. (1), together with selected input parameters summarized in Table 1.

Strength and deformation properties are especially difficult to estimate for the fill, marine sand, peat and underlying till due to the lack of laboratory test data on intact/undisturbed material. For these materials much simpler soil models have been used:

1. The compacted fill and end-dumped, replacement fill (below El. +1.5 m) are modeled as isotropic elastic materials with shear modulus, $G/\gamma_t z = 170$, where z is the depth below final crest elevation (El. +12.2 m), and Poisson's ratio, $\nu = 0.4$.
2. The marine sand is modeled as a poro-elastic, perfectly plastic material (EP-DP, after Whittle et al. 1993) with $G/\sigma' = 170$, $\nu = 0.3$, and plane strain friction angle, $\phi'_{PS} = 37°$, with initial $K_0 = 0.45$.
3. The peat contributes a surcharge load with unit weight, $\gamma_t = 11.8\ kN/m^3$, with nominal elastic shear modulus, G = 8 kPa, and $\nu = 0.3$.

4. The underlying till is modelled as an EP-DP material with parameters proposed by Whittle et al. (1993): $G/\sigma' = 110$, $\nu = 0.3$, $\phi'_{PS} = 43°$ and $K_0 = 1.0$.

TABLE 2. Selection of MCC and MIT-E3 Input Parameters for BBC

Laboratory Test	MIT-E3		MCC	
(1)	Parameter (2)	BBC (3)	Parameter (4)	BBC (5)
1. One-dimensional Compression	e_0	see Table 1	e_0	see Table 1
	λ	see Table 1	λ	see Table 1
- (oedometer, CRS)	C	22.0	κ	0.034
- with swelling and reloading	n	1.6		-
	h	0.2		-
2. K_0-Triaxial with Swelling	K_{0NC}	0.53		-
	2G/K	1.05	2G/K	1.05
3. Undrained Triaxial Shear Tests	ϕ'_{TC}	33.4°	ϕ'_{TC}	33.4°
	ϕ'_{TE} (b)	45.9°		-
CK_0UC (OCR=1.0)	c	0.86		-
CK_0UC (OCR≈2.0)	S_t	4.5		-
	ω	0.07		-
CK_0UE (OCR=1.0)	γ	0.5		-
4. Small Strain Modulus (a)	κ_0	0.001		-
5. Drained Triaxial	ψ_0	100.0		-

Notes:
(a) Resonant column, local strain measurement in laboratory or field shear wave velocity measurement in field.
(b) Recent, more reliable data show $\phi'_{TE} \approx \phi'_{TC}$

Additional analyses (Legaspi 1994) have shown that the selected stiffness and strength parameters of the fill, sand and peat have minimal effect on the predictions described in this paper. In practice, permeability properties of the underlying till are a major source of uncertainty. However, the data at Sta. 246 support the assumption of free drainage conditions used in the analyses.

COMPARISON OF PREDICTIONS AND MEASUREMENTS

Pore Pressures

Figs. 6a and 6b summarize measurements of the excess piezometric head ($h_e = (u-u_0)/\gamma_w$, where u_0 is the *hydrostatic* pore water pressure) at the end of construction (CD 620) and at CD 2000, corresponding to the end of the field monitoring. The data are presented at three vertical sections corresponding to conditions along the centerline (which includes data from piezometers offset at ±9 m; i.e., Pe1-Pe6, and P14-P19, Fig. 1), at ±19 m, and ±29 m. Note that h_i represents the artesian pressure at the site. The following points are observed from the measured data.

1. At the end of construction, the largest excess piezometric heads ($h_e \approx 14$ m) are measured at the center of the clay layer at El. -25 m (layer E) under the crest of the embankment. Very small excess heads ($h_e \leq 4$ m) remain in the upper, highly overconsolidated layers (A, B), while the overall distribution of pore pressures below El. -10 m can be well described by a parabolic-shaped isochrone. These general observations are consistent with classical one-dimensional consolidation theories with partial dissipation of pore pressure as the pore water moves towards the upper and lower drainage horizons (sand and till layers).
2. At CD 2000, there is negligible excess head remaining in layers A and B. However, measurements between El. -15 m and El. -35 m show relatively small reductions in the piezometric head compared to conditions at the end of construction ($\Delta h_e = h_e(2000) - h_e(620) \approx$ -2 to -4 m). Hence, the field measurements during Stage 3 consolidation represent only a small fraction of the time required for complete dissipation of the construction pore pressures.

Figs. 6a and 6b also show predictions of the excess piezometric head from the finite element analyses using the MCC and MIT-E3 soil models, respectively. Compared to the scatter in the measured data, there is remarkable agreement between the results from the two models. The coupled analyses describe the characteristic parabolic isochrone of excess pore pressure at the end of construction, which could not be interpreted from previous undrained analyses reported by D'Appolonia et al. (1971). The analyses also give reasonable estimates of the overall reduction in pore pressure during Stage 3 consolidation to CD 2000. Results for Profile 2 (lower estimate of σ'_p) give consistently higher magnitudes of excess pore pressure (at both CD 620 and CD 2000) than Profile 1, but have little impact on the overall pattern of behavior. More detailed comparisons between the predictions and measurements show the following:

1. Excess piezometric heads measured between El. -15 m and El. -35 m are well described by the analyses. Profile 2 predictions match the measured

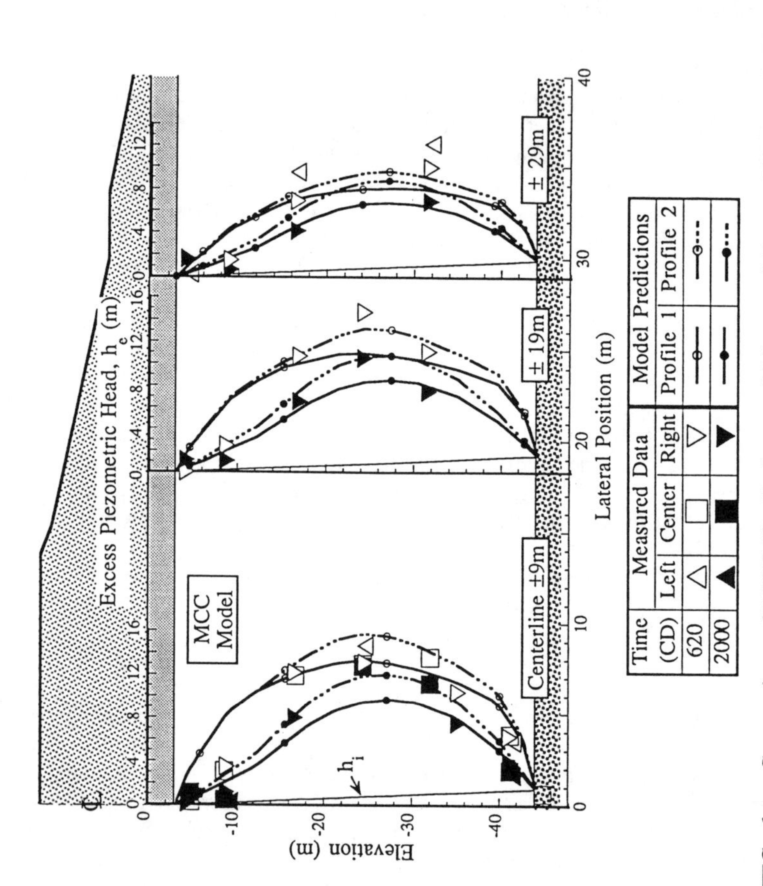

FIG. 6a. Comparison of MCC Predictions and Measured Excess Piezometric Heads

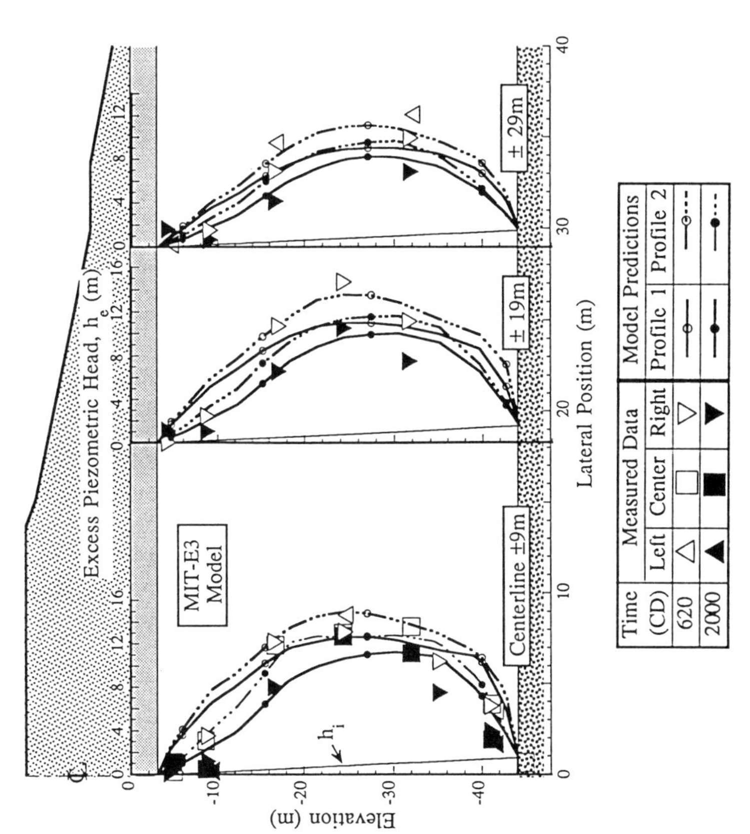

FIG. 6b. Comparison of MIT-E3 Predictions and Measured Excess Piezometric Heads

data more closely than Profile 1 at the end of construction, but tend to overestimate the excess pressures at CD 2000. However, collectively the results from the two stress histories bracket most of the measured data.

2. The results from all four analyses greatly overestimate the pore pressures measured in the upper soil layers (A and B) at the end of construction. For example, at El. -10 m, the predicted centerline excess heads range from $h_e = 7 \pm 1$ m (MIT-E3) to 9 m (MCC) compared to the measured data, $h_e = 3.5 \pm 0.5$ m.

   Fig. 7 illustrates this behavior more clearly through comparison of MIT-E3 predictions and measurements from 6 individual piezometers from CD 300 to CD 700. For piezometers P7, P9, P27, and P28 located within the middle of the clay, MIT-E3 gives very good predictions of the excess pore pressures generated during Stage 2 loading (up to CD 460), the small reductions in h_e during Stage 2 consolidation (CD 460 - CD 600) and Stage 3 loading to the end of construction at CD 620. In contrast, the measurements for piezometer P6 (located along the centerline at El. -8.7 m) show a much lower rate of pore pressure generation during Stage 2 loading, compared to model predictions, and almost no change in excess head during Stage 2 consolidation and Stage 3 loading. The analyses also overestimate the pore pressure generation for piezometer P10, located close to the bottom of the clay layer (El. -41.2 m), for embankment construction in Stages 2 and 3, but show less discrepancy during the subsequent consolidation.

3. There are relatively small differences in the magnitude of excess pore pressures predicted by the two models at the end of construction, however, MCC shows significantly larger reductions in excess piezometric head during Stage 3 consolidation than MIT-E3.

Settlements

Figs. 8a and 8b compare model predictions (MCC and MIT-E3, respectively) of the settlement versus time with measurements at three clay elevations and two lateral locations (centerline and 28 m R, which roughly corresponds to the toe of the Stage 2 fill). One consistent observation from these results is that the analyses underestimate the overall settlements measured at the top of the clay layer (SR1, SR7) during Stage 3 consolidation (CD 620 - CD 2000). There is generally much better agreement between predictions and measurements at the middle (El. -21 m; SR4, SR10) and bottom of the clay layer (El.-37 m; SR6, SR12). These results, and those in Fig. 6, suggest that the primary cause for underprediction of surface settlements is related to the compressibility and/or permeability of the upper clay layers (A and B).

In contrast to the pore pressure data, there are significant differences in predictions of settlements from the two soil models which can only be appreciated through detailed observations:

1. Both models predict very similar behavior in the lower part of the clay (layer F). During construction (up to CD 620) the analyses tend to overestimate the measured settlement at SR6, but give excellent agreement with the data

at SR12 throughout construction and Stage 3 consolidation. Tne MIT-E3 model predicts slightly larger settlements than MCC during Stage 3 consolidation, although there is greater dissipation of pore pressures in the latter (cf. Fig. 6).

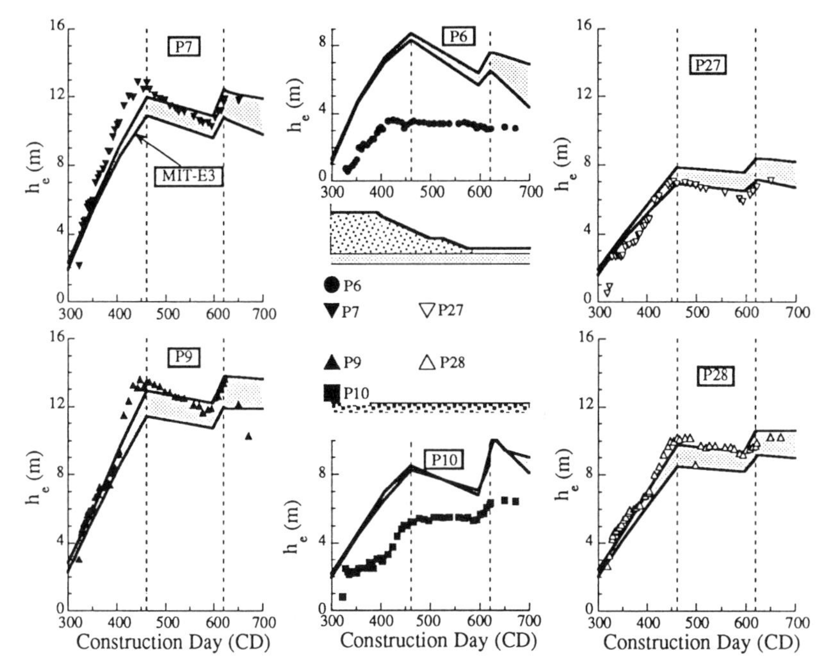

FIG. 7. Comparison of MIT-E3 Predictions and Measured Excess Heads during Construction

2. In the middle of the clay layer, MIT-E3 gives very good settlement predictions (i.e., agree with the data) at the centerline and offset locations (SR4 and SR10, respectively). The MCC model also gives reasonable estimates of behavior, but tends to underestimate the measured behavior during Stage 3 consolidation.
3. At the top of the clay, the MCC model gives a deceptively close prediction of the centerline settlement during construction (SR1), but underestimates the subsequent Stage 3 consolidation movement. In contrast, MIT-E3 underestimates settlements from CD 450 onwards (following Stage 2 loading), except eventually at the centerline for Profile 2.
4. The predictions for stress history Profiles 1 and 2 provide some insight into the importance of soil compressibility on the settlement behavior. When lower pre-consolidation pressures are used in the analysis (Profile 2), much larger Stage 3 consolidation settlements are predicted close to the centerline

in the upper part of the clay layer (SR1). However, the stress history has very little effect on the settlements predicted during construction (up to CD 620) or on the behavior measured under the toe of the Stage 2 fill (SR7, etc.). These results, combined with the previous observations regarding pore pressures, suggest the need for further refinement of permeability as well as stiffness properties in layers A and B.

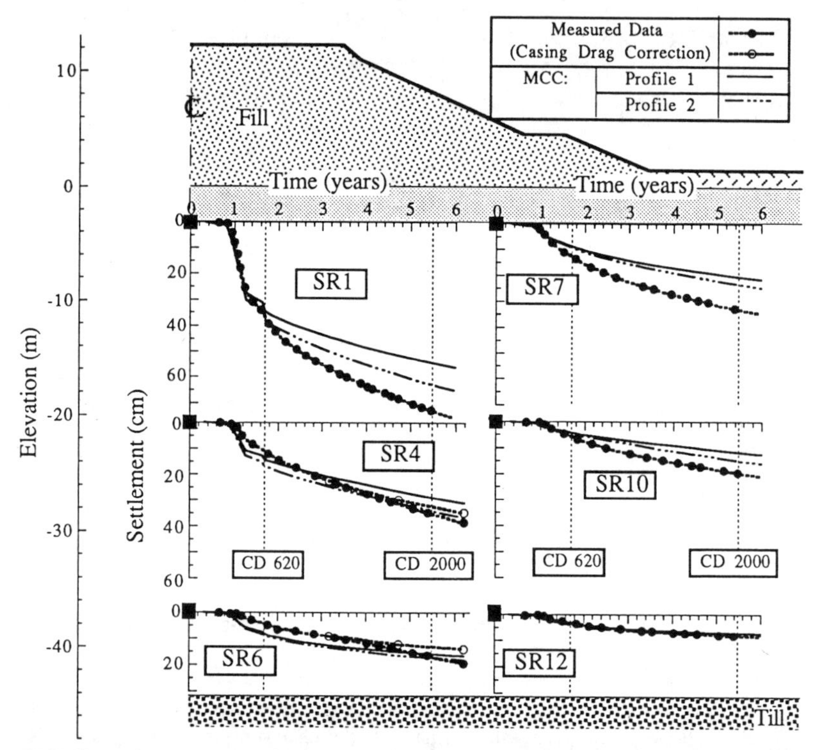

FIG. 8a. Comparison of MCC Predictions and Measured Settlement vs. Time

Horizontal Displacements

It is well established in the literature (e.g., Poulos 1972) that it is difficult to achieve reliable predictions of lateral deflections under an embankment, especially after the end of construction. As a result, several authors have proposed empirical and semi-empirical methods for estimating lateral deflections (e.g., Tavenas et al. 1979; Sekiguchi et al. 1988). Previous studies of the nearby I-95 embankment at Sta. 263 (see MIT 1975), including coupled finite element analyses with the MCC model (Borja et al. 1990), have attributed the measured lateral spreading during consolidation to creep (viscous) properties of BBC.

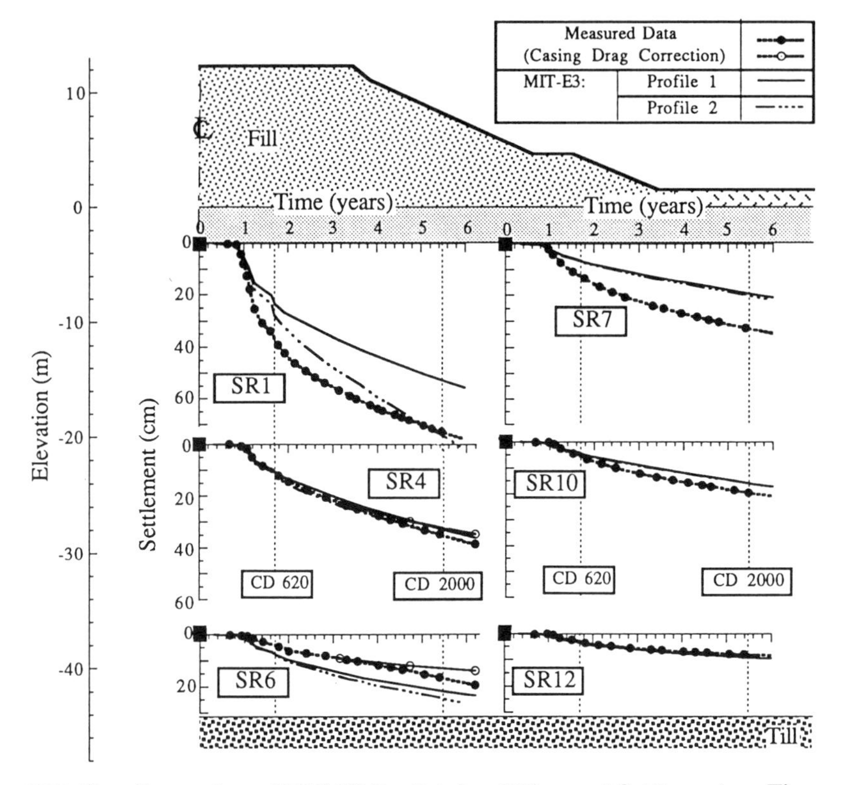

FIG. 8b. Comparison of MIT-E3 Predicted and Measured Settlement vs. Time

Figs. 9a and 9b compare predictions of horizontal displacements with measurements from inclinometers I3, I4, I5 and I6 (Fig. 1) at the end of construction (CD 620) and at CD2000. The largest lateral movements ($h_m \approx 17$ cm) are measured in the upper clay (layer B) by inclinometer I4 located close to the toe of the embankment. All of the inclinometers measure progressive outward displacements at all elevations in the clay throughout the entire monitoring period (up to CD 2000). There are large differences in predictions of lateral deflections from the two soil models.

1. The MCC model greatly overestimates the measured outward deflections at the end of construction (Fig. 9a), especially in the bottom half of the deposit (below El. -20m). During Stage 3 consolidation (CD 620 - CD 2000), the model predicts small amounts (< 1-2 cm) of lateral spreading in the lower soil layers (E and F), while predicting inward movements in layers A, B and C. These two serious discrepancies partially compensate each other leading to relatively reasonable estimates of maximum lateral movements at CD

2000. The pattern of lateral deflections reflects the elastic behavior described by MCC for overconsolidated clay.

2. The MIT-E3 model gives much more realistic predictions of horizontal displacements at the end of construction at all four inclinometer locations. The model also gives reasonable predictions of the general magnitude of lateral spreading during Stage 3 consolidation. The results for Profiles 1 and 2 show the importance of the stress history on the magnitude of the lateral deflections, but has little effect on the distribution of deformations. In general, the model overestimates the lateral spreading in layer F, with maximum outward displacements occurring in layer D under the embankment. In contrast, the mode shape of the measured data show maximum movements in layer B.

The results in Fig. 9 highlight the need for reliable modeling of soil behavior in order to make realistic predictions of the lateral deformations, and show the severe limitations of the MCC model for this purpose. The results from MIT-E3 are encouraging and show that lateral spreading during consolidation can be described by a model which includes anisotropic yielding and is not necessarily caused by creep (viscous) effects. However, underprediction of both the vertical settlements and lateral deflections in layers A and B can only be addressed through refinement in the selection of material properties.

Horizontal Displacement versus Settlement

Tavenas et al. (1979) analyzed the relationship between the maximum horizontal displacement (h_m) measured by inclinometers at the toe of 21 embankments and the maximum (circa centerline) settlement (s_m). In particular they evaluated changes in the slope of h_m versus s_m, which Ladd (1991) termed the Deformation Ratio ($DR = dh_m/ds_m$), observed during and after placement of the fill. For about a dozen embankments constructed on overconsolidated deposits, they reported mean and standard deviation values of DR:

- During filling, DR = 0.18 ± 0.09 for initial loading that was accompanied by significant drainage, increasing to DR = 0.9 ± 0.2 when undrained shear deformations predominated, and overall (average) DR = 0.4 ± 0.3 at the end of filling.
- During subsequent consolidation over several years, h_m increases linearly with s_m and DR = 0.16 ± 0.07.

Fig. 10 plots h_m for inclinometer I4 versus s_m for SR1 during (i) Stage 2 loading (El. 4.5 m to 11.0 m, Fig. 5); (ii) Stage 2 consolidation (five month construction hiatus); (iii) Stage 3 loading to El. 12.2 m; and (iv) Stage 3 consolidation (to CD 2000). The data for the last period were obtained by linear regression on rather scattered measurements of h_m versus log t. For the Stage 2 filling and consolidation, DR = 0.3 ± 0.05. As most of the BBC deposit remains overconsolidated, this value is larger than expected from the correlations of Tavenas et al. (1979). But the measured DR = 0.19 during Stage 3 consolidation agrees with prior experience.

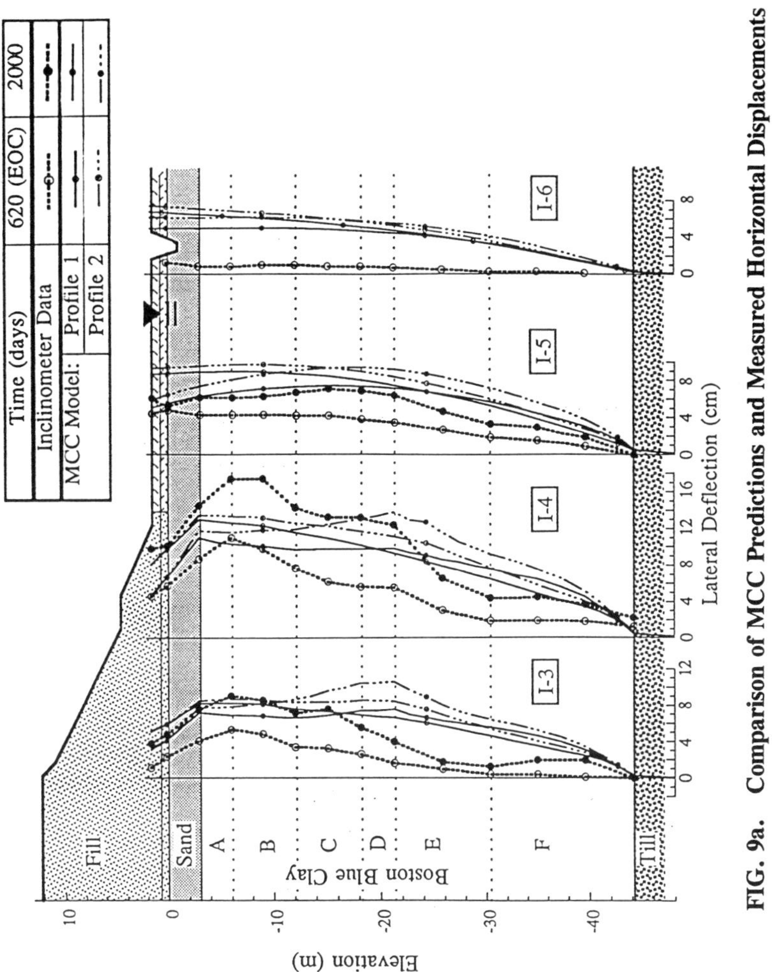

FIG. 9a. Comparison of MCC Predictions and Measured Horizontal Displacements

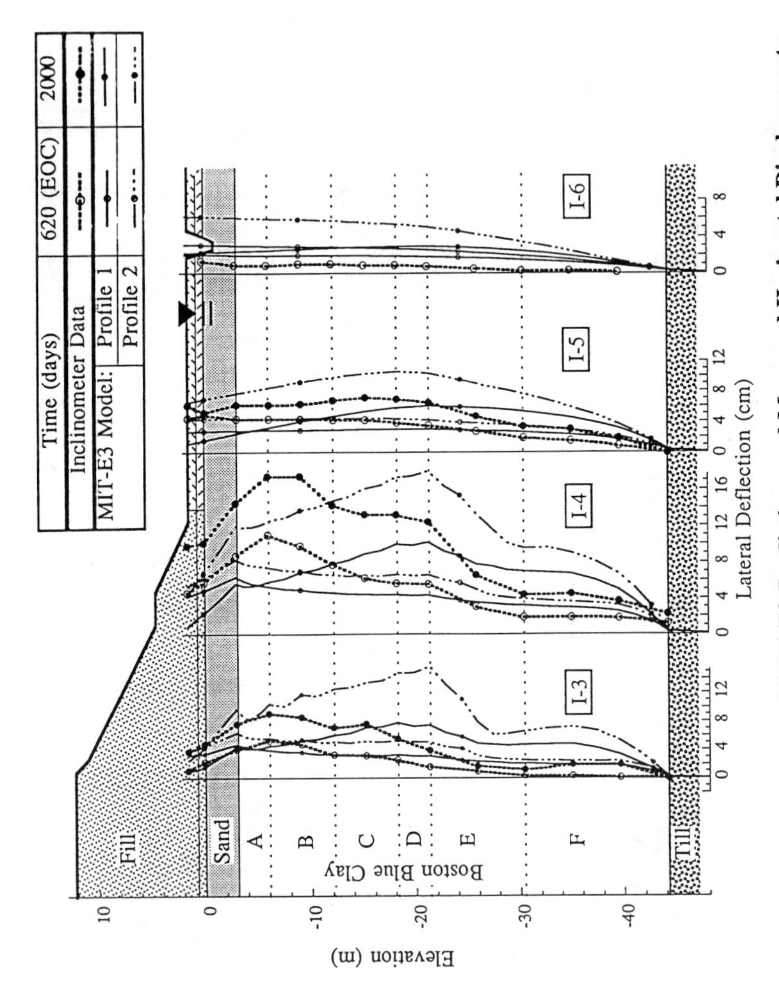

FIG. 9b. Comparison of MIT-E3 Predictions and Measured Horizontal Displacements

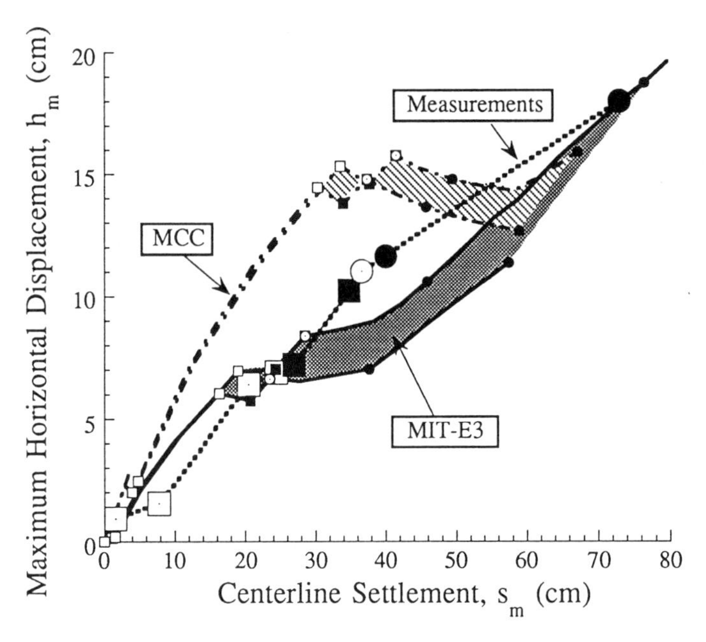

Stage:	2		3	
	Loading	Consol.	Loading	Consol.
Measured Data				
MCC Predictions				
MIT-E3 Predictions				

FIG. 10. Comparison of Predicted and Measured Maximum Horizontal Displacement versus Centerline Settlement

Fig. 10 also shows predictions from the finite element analyses for both stress history profiles. The MCC model greatly overestimates h_m during Stage 2 construction, and predicts unrealistic changes in lateral deflection during the Stage 3 consolidation (up to CD 2000). In contrast, the general trends of the measured behavior are well described by MIT-E3 throughout the staged construction and monitoring period, but tends to underestimate the magnitude of both h_m and s_m during Stage 3 as noted above.

Interpretation of Predictions

The underlying mechanisms which give rise to the finite element predictions described above can be examined through detailed interpretation of the stress, strain and pore pressure changes occurring at individual locations within the clay. Studies of this type have been performed by Magnan et al. (1982) who plot total and effective stress paths for their analysis of the test embankment at Cubzac-les-Ponts and by Almeida et al. (1986) in the interpretation of centrifuge model embankments on kaolin.

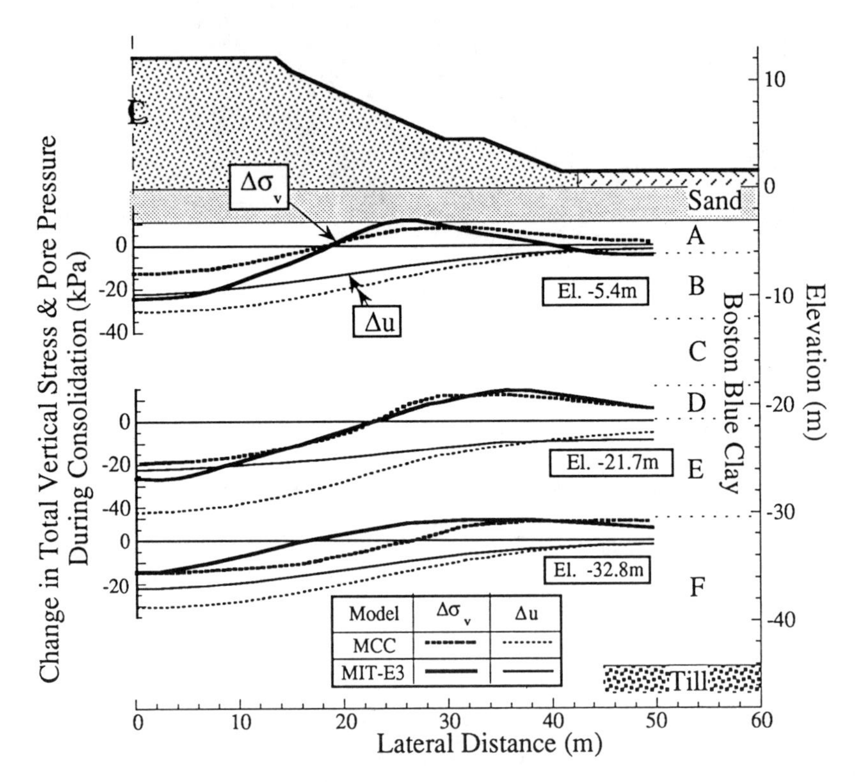

FIG. 11. Predictions of Total Vertical Stress and Pore Pressures during Stage 3 Consolidation

Fig. 11 investigates the changes in vertical effective stress, $\Delta\sigma'_v$, occurring during Stage 3 consolidation (from CD 620 to CD 2000). The figure reports changes in the vertical total stress, $\Delta\sigma_v$ (= σ_v[CD 2000] − σ_v[CD 620]) and pore pressure, Δu, predicted by the MCC and MIT-E3 soil models at three elevations in the clay (with Profile 1 stress history). The results show a net reduction in pore pressures at all locations in the clay, while both models predict significant redistribution ('arching') of total vertical stresses. A large zone of soil extending from the centerline laterally a distance of 20 - 30 m experiences a reduction in total vertical stress, while there is a corresponding increase in stress in the soil beneath the slope and toe of the embankment. Although both soil models show qualitatively similar behavior there are important differences, most notably in the net change in effective stress ($\Delta\sigma'_v = \Delta\sigma_v - \Delta u$). For the MCC model, $\Delta\sigma_v/\Delta u \leq 50\%$ at locations close to the centerline and there is a large (16-24 kPa) increase in the effective vertical stress which is approximately uniform over the full width of the embankment. In contrast, MIT-E3 predicts $\Delta\sigma_v/\Delta u \geq 100\%$ close to the centerline in the upper part of the clay, which implies a net reduction in the effective vertical stress between CD 620 and CD 2000 at these locations! However, there are large concomitant increases in vertical effective stress occurring under the slope and toe of the embankment. The non-uniform distribution of $\Delta\sigma'_v$ may be an important factor in model predictions of lateral displacements in the clay.

Overall, the results in Fig. 11 suggest serious limitations of conventional analyses which assume $\Delta\sigma_v = 0$ during consolidation. Further studies are currently in progress to investigate the factors controlling total stress redistribution within a non-homogeneous clay layer.

Summary and Conclusions

This paper analyzes the performance of the I-95 MIT-MDPW test embankment using finite element methods which incorporate coupled consolidation, and generalized effective stress models to describe the behavior of the underlying, deep deposit of Boston Blue Clay (BBC). Fig. 1 shows the cross-section of the completed 11 m high embankment and the location of the 17 settlement points, 37 piezometers and 6 inclinometers. Placement of the fill occurred over a period of 1.5 years (Fig. 5) and monitoring continued for an additional four years.

The soil profile under the embankment consists of 3 m of a marine sand and 41 m of BBC, a post-glacial marine (CL) clay, overlying till having a small artesian pressure (Fig. 2). Results from several in situ and laboratory test programs conducted during 1966 to 1980 provided unusually detailed engineering properties of the clay. Figs. 3 and 4 and Table 1 show how the deposit was subdivided into layers to represent the very large variation in the stress history and compressibility-permeability properties of the deposit. It should be noted that the test section is located at Sta. 246 of the I-95 alignment, which is 0.5 km south of Sta. 263 that was loaded to failure as part of a prediction symposium (MIT 1975).

The finite element analyses modeled the fill as elastic and the marine sand as an elasto-plastic material. The analyses use two effective stress soil models for the clay: Modified Cam Clay (MCC); and MIT-E3, which extends the basic concepts of MCC by incorporating anisotropic stress-strain-strength behavior, small strain non-linearity and strain softening. In both cases, analyses were made using upper and lower bound estimates of the preconsolidation pressure, Profiles 1 and 2 in Fig. 3. The analyses closely simulated three stages of the actual construction (Fig. 5), which ended on CD 620, with subsequent consolidation measured until CD 2000.

The paper presents detailed comparison of predictions with field measurements of pore pressures, settlements and horizontal displacements. These comparisons and other results from the analyses lead to the following observations and conclusions.

1. The piezometer measurements (Figs. 6 and 7) show very small excess pore pressures in the clay close to the upper and lower drainage boundaries (i.e., very rapid consolidation during construction of the embankment). In contrast, piezometers near the center of the clay typically show less than 20% dissipation of excess pore pressures during almost four years of subsequent consolidation.
2. The analyses give very reasonable predictions of pore pressures and overall magnitudes of the vertical and horizontal movements, but generally underpredict deformations within the upper 15 m of the clay crust, even with the lower bound, Profile 2 stress history (Figs. 8 and 9). The results emphasize the need for further measurements of the stress-strain and permeability properties of the upper desiccated crust.
3. Predictions of changes in horizontal displacements during the Stage 3 consolidation period (CD 620 to CD 2000) show significant differences between the two soil models (Fig. 9 and 10). Whereas the inclinometers recorded large increases in lateral spreading within the clay, the MCC model predicts much smaller displacements, including *inward* movements associated with elastic modeling of the overconsolidated clay. In contrast, the MIT-E3 model, which accounts for anisotropic yielding of the clay, predicts continued lateral spreading during consolidation that agrees reasonably well with the measured data. These results imply that creep (viscous) effects may not be a major factor during consolidation of BBC as previously proposed based on MCC analyses of the I-95 embankment at Sta. 263 (e.g., Borja et al. 1990).
4. Analyses with both models predict a significant redistribution in total vertical stress during Stage 3 consolidation (Fig. 11). The resultant decrease in vertical stress beneath the crest of the embankment represents a large component of the predicted pore pressure "dissipation". This form of arching is associated with coupled two dimensional consolidation and suggests serious limitations in conventional analyses which assume constant vertical stress during consolidation.

All of the soil properties and field data reported in this paper are stored in digital form and can be made available to others wishing to evaluate this well documented case history.

ACKNOWLEDGMENTS

The installation and readings of the instrumentation for the MIT-MDPW Test Section were performed by the MIT Department of Civil Engineering under a research contract funded by the Massachusetts Department of Public Works (now the Massachusetts Highway Department) and the U. S. Federal Highway Administration. T. W. Lambe supervised the research and L. A. Wolfskill directed the field work with the assistance of the late W. R. Beckett. The monitoring data and construction history were obtained from the 1974 SM Thesis by J. F. Whittle, Jr. MIT staff and students deserving special recognition for the in situ and laboratory testing programs include M. M. Baligh, J. T. Germaine, I. M. Ghantous, J. D. Guertin, S. M. Lacasse and M. J. Morrison.

The finite element analyses were conducted as part of MIT's Consortium Cooperative Research on Arctic Offshore Engineering and Construction supported by Amoco, Arco, BP, Chevron, Conoco, Exxon, Mobil and Texaco. M. P. Whelan performed the stability analyses.

The opinions and conclusions expressed in this paper are those of the authors and not necessarily those of any of the sponsors.

APPENDIX - REFERENCES

Almeida, M. S. S., Britto, A. M. and Parry, R. H. G. (1986). "Numerical modelling of a centrifuged embankment on soft clay." *Can. Geotech. J.*, 23(1), 103-114.

Azzouz, A. S., Baligh, M. M. and Ladd, C. C. (1983) "Corrected field vane strength for embankment design." *J. Geotech. Eng.*, ASCE, 109(5), 730-734.

Baligh, M. M. and Levadoux, J.-N. (1986). "Consolidation after undrained piezocone penetration, II. Interpretation." *J. Geotech. Eng.*, ASCE, 112(7), 727-745.

Becker, D. E., Crooks, J. H. A., Been, K. and Jefferies, M. G. (1987). "Work as a criterion for determining in situ and yield stresses in clays." *Canadian Geotech. J.*, 24(4), 549-564.

Borja, R. I., Hsieh, H. S. and Kavazanjian, E., Jr. (1990). "Double-yield-surface model. II: Implementation and verification." *J. Geotech. Eng.*, ASCE, 116(9), 1402-1421.

Casagrande, A. (1936). "The determination of the pre-consolidation load and its practical significance." *Proc. 1st ICSMFE*, Cambridge, 3, 60-64.

D'Appolonia, D. J., Lambe, T. W. and Poulos, H.G. (1971). "Evaluation of pore pressures beneath an embankment." *J. Soil Mech. Found. Div.*, Proc. ASCE, 97(SM6), 881-987.

Dowding, C. H., Lambe, T. W. and Wolfskill, L. A. (1974). "Embankment instrumentation tunnel performance." *J. Soil Mech. Found. Div.*, Proc. ASCE, 100(SM5), 567-570.

Ghantous, I. M. (1982). "Prediction of in situ consolidation parameters of Boston Blue Clay," C.E. Thesis, MIT, Cambridge, Massachusetts.

Guertin, J. D. (1967). "Stability and settlement analyses of an embankment on clay," S.M. Thesis, MIT, Cambridge, Massachusetts.

Haley & Aldrich Inc. (1993). "Special laboratory and in situ testing program for Central Artery (I-93)/Tunnel (I-90) Project, Boston," Final Report submitted to Massachusetts Highway Department.

Hashash, Y. M. A. (1992). "Analysis of deep excavations in clay," Ph.D. Thesis, MIT, Cambridge, Massachusetts.

Hashash, Y. M. A. and Whittle, A. J. (1993). "Integration of the Modified Cam-Clay model in non-linear finite element analysis." *Computers and Geotechnics*, 14(1), 59-83.

HKS (1989). *ABAQUS Version 4.9 User's Manual*, Hibbitt, Karlsson and Sorensen, Inc., Providence, Rhode Island.

Kavazanjian, E. Jr. and Poepsel, P. H. (1984). "Numerical analysis of two embankment foundations." *Proc. Symp. on Sedimentation and Consolidation Models, Prediction and Validation*, ASCE, San Francisco, 84-106.

Kavazanjian, E., Jr., Borja, R. I. and Jong, H.-L. (1985). "Time-dependent deformations in clay soils." *Proc. 11th ICSMFE*, San Francisco, 2, 535-538.

Ladd, C. C. (1991). "22nd Terzaghi Lecture: Stability evaluation during staged construction." *J. Geotech. Eng.*, ASCE, 117(4), 540-615.

Ladd, C. C. and Foott, R. (1974). "New design procedure for stability of soft clays." *J. Geotech. Eng. Div.*, Proc. ASCE, 100(GT7), 763-786.

Ladd, C. C., Germaine, J. T., Baligh, M. M. and Lacasse, S. M. (1980). "Evaluation of self-boring pressuremeter tests in Boston Blue Clay," *Research Report R79-4*, Dept. of Civil Engrg., MIT, Cambridge, Massachusetts, and *FHWA/RD-80/052*, Federal Highway Administration, Washington, D. C..

Legaspi, D. E. (1994). *Ph.D. Thesis* in progress, MIT, Cambridge, Massachusetts.

Magnan, J.-P., Humbert, P., Belkeziz, A. and Mouratidis, A. (1982). "Finite element analysis of soil consolidation with special reference to the case of strain hardening elastoplastic stress-strain models." *Proc. 4th Intl. Conf. on Numerical Meth. in Geomechs.*, Edmonton, 327-336.

MIT (1969). "Performance of an embankment on clay, Interstate 95," *Research Report R69-67*, Dept. of Civil Engrg., MIT, Cambridge, Massachusetts.

MIT (1975). "Proceedings of the foundation deformation prediction symposium," *Report No. FHWA-RD-75-515*, FHWA, Washington, D. C., 2 Volumes.

Morrison, M. J. (1984). "In situ measurements on a model pile in clay." Ph.D. Thesis, MIT, Cambridge, Massachusetts.

Poulos, H. G. (1972). "Difficulties in prediction of horizontal deformations of foundations." *J. Soil Mech. Found. Div.*, Proc. ASCE, 98(SM8), 843-848.

Roscoe, K.H., and Burland, J.B. (1968). "On the generalized stress-strain behaviour of 'Wet' clay." *Engineering Plasticity*, Cambridge University Press, Cambridge, England, 535-609.

Sekiguchi, H., Shibata, T. and Mimura, M. (1988). "Effects of partial drainage on the lateral deformation of clay foundations." *Proc. Intl. Conf. on Rheology and Soil Mechs.*, Coventry, 164-181.

Simon, R. M., Christian, J. T. and Ladd, C. C. (1974). "Analysis of undrained behavior of loads on clay." *Proc. Conf. on Analysis and Design in Geotech. Engrg.*, ASCE, 1, 51-84.

Spencer, E. (1967). "A method of analysis of the stability of embankments assuming parallel inter-slice forces." *Géotechnique*, 17(1), 11-26.

Tavenas, F., Mieussens, C. and Bourges, F.(1979). "Lateral displacements in clay foundations under embankments." *Canadian Geotech J.*, 16, 532-550.

Whittle, A. J. (1987). "A constitutive model for overconsolidated clays with application to the cyclic loading of tension piles in clay," *Sc.D. Thesis*, MIT, Cambridge, Massachusetts.

Whittle, A. J. (1993). "Evaluation of a constitutive model for overconsolidated clays." *Géotechnique*, 43(2), 289-315.

Whittle, A. J., DeGroot, D. J., Ladd, C. C. and Seah, T.-H. (1994). "Model prediction of the anisotropic behavior of Boston Blue Clay." *J. Geotech. Eng.*, ASCE, 120(1), 199-224.

Whittle, A. J., Hashash. Y. M. A. and Whitman, R. V. (1993). "Analysis of deep excavation in Boston." *J. Geotech. Eng.*, ASCE, 119(1), 69-90.

Whittle, A. J. and Kavvadas, M. (1994). "Formulation of the MIT-E3 constitutive model for overconsolidated clays." *J. Geotech. Eng.*, ASCE, 120(1), 173-198.

Whittle, J. F. Jr. (1974). "Consolidation behavior of an embankment on Boston Blue Clay." M.S. Thesis, MIT, Cambridge, Massachusetts.

Wolfskill, L. A. and Soydemir, C. (1971). "Soil instrumentation for the I-95 MIT-MDPW test embankment." *J. Boston Soc. Civil Engrs.*, 58(4), 193-229.

Stress and Settlement of Footings in Sand

Bengt H. Fellenius,[1] Member, ASCE and Ameir Altaee[2]

Abstract: In current engineering practice, the magnitude of the settlement of a footing in sand, as compared to the settlement of a different size footing in the same sand, is considered to be a non-linear function of the footing width. Further, the settlement is considered to be proportional to the density of the sand. Results of finite element analysis of settlement for footings of three sizes placed in two different sand types show that the settlement in sand is a direct function of neither footing size nor soil density. Instead, the settlement should be related to the steady state line of the sand and to the upsilon distance of the sand, that is, the initial void ratio distance to the steady state line at equal mean stress and at homologous points. This requirement imposes scaling-rules for model tests and it limits the range of application of a small scale test to a prototype behavior. Moreover, it imposes boundaries on the geometric scale, because a model test can not realistically be carried out in a sand that is looser than the maximum void ratio, and it is meaningless if performed in a sand close to the minimum void ratio, because it would then not be representative for any prototype.

Introduction

In current foundation engineering practice, when assessing settlement of footings in sand, conditions are normally so favorable that it is obvious that the settlement will not exceed the usual one-inch limit. However, sometimes, existence

---

[1] Professor, Department of Civil Engineering, University of Ottawa, 161 Louis Pasteur St., Ottawa, Ontario, Canada K1N 6N5; and Anna Geodynamics Inc., 5350 Canotek Road, Unit 22, Ottawa, Ontario, Canada K1E 9E2.

[2] Anna Geodynamics Inc., 5350 Canotek Road, Unit 22, Ottawa, Ontario, Canada K1E 9E2.

of a favorable situation is not that obvious and a closer look is required. The closer look, invariably, involves calculation and analysis.

Most calculation of settlement of footings in sand involves empirical methods whereby the settlement is determined in relation to an average N-value, cone penetrometer data, or other indirect method based on in-situ testing. Sometimes, a modulus of elasticity of the sand is estimated and combined with the Boussinesq stress distribution below the center of the footing (or below some point between the center and a side, such as the so-called characteristic point). In both cases, the settlement of the footing is the accumulated compression calculated for a series of sub-layers.

Generally, agreement between calculated settlement and reality has little correlation to whether an empirical method or a sophisticated method is used, or to the degree of complexity of the approach. The governing aspect is the experience base of the person making the settlement estimate.

For qualitative assessment of settlement, it is generally considered that the denser the sand, the smaller the settlement for a given applied stress and footing size. The density of the sand is usually expressed in terms of its density index, I_D (formerly called 'relative density'). Unfortunately, many reports omit to mention the values of actual density, void ratio, or porosity, forgetting that the I_D-value alone has little meaning. Sometimes, of course, the reason for omitting the actual soil density may be because reliable values of the in-situ conditions of a sand are difficult to determine directly. However, there is no good reason for not including the maximum and minimum boundaries used in determining the density index values. (When also, inexplicably, the vital pore pressure information is missing, such as whether the test is performed in dry, wet, or saturated sand, and where the groundwater table is located, the value of the test data is seriously impaired).

It is also generally recognized that sands of different degree of uniformity, angularity of the grains, etc., will behave differently under otherwise equal conditions. Of particular importance are the geographical/geological/mineralogical aspects of the sand; a footing in a calcareous sand can hardly be expected to settle similarly to one placed in a silica sand, be density indices, density values, coefficients of uniformity, etc. ever so equal. Fortunately, foundations on dense sands only rarely entail major concerns about settlement and loose sands can be densified, lessening the severity of the consequence of an inaccurate prediction of settlement. However, ground improvement treatment costs money and before recommending this, or some other solution, to a settlement problem, the need for it must be shown. Often, a small-scale test is performed and the results are extrapolated to the prototype (full-scale) condition.

The most relevant experience base consists of results from observations of the behavior of existing footings under known loads and/or of results from loading-tests on footings. Because one rarely has the means to perform a full-scale test, a footing test is usually performed at some ratio of scale to the actual footing considered. Moreover, few engineers are fortunate enough to be able to support a settlement assessment by means of a project-specific footing test at any scale.

Instead, they have to rely on experience of tests from other projects or on more or less applicable information found in the literature.

LITERATURE

As to extrapolation of results from a small-scale footing test to the behavior of a prototype footing, the Terzaghi and Peck (1967) relation between the settlement of a footing in terms of the settlement for a one-foot reference footing is probably the best known such relation. This relation has been used to state the conclusion that however large a footing, its settlement will never be larger than four times that of a one-foot diameter footing. This conclusion is incorrect, of course, the relation is only intended to apply to relatively small footings. Another well-known relation is that proposed by Bond (1961), who indicated that for footings in dense sands, the ratio of settlement is equal to the square root of the ratio of the footing width and that for loose sand the width ratio exponent is smaller than 0.5.

Of course, extrapolating results from a small-scale test to a full-scale (prototype) test must be with the small-scale footing test performed in the same type of sand as the prototype footing. It has also been taken as self evident that the test should be performed in a sand of the same density as the sand at the prototype footing. The latter postulation is a fallacy, however, which will be addressed in this paper.

The technical literature abounds with reports on footing tests. Most available references reporting results from tests on footings do not isolate one parameter at a time, making the results difficult to use as support for generalized conclusions. One exception to this is Vesic (1967; 1975) who presented results from a comprehensive series of tests on model footings tested in sands of different densities. Fig. 1 compiles load-settlement curves from Vesic's tests on 150 mm diameter circular model footings tested on the surface of a sand. The influence of the sand density is clearly evident. (The dry density of the sand ranged from 1,360 kg/m^3 through 1,540 kg/m^3, corresponding to void ratios ranging from 0.96 through 0.73. The maximum and minimum void ratios were 1.10 and 0.62, respectively).

Another exception is Ismael (1985), who published results from a very useful and conclusive series of field tests in Kuwait on rigid footings on "compact fine to medium non plastic cohesionless windblown sand with little silt". The silt content ranged from 5 percent through 12 percent and the tests were performed above the groundwater table. The tests consisted of measuring the settlement induced by incremental loading of four square footings placed at a depth of 1.0 m. The footing diameters were 0.25 m, 0.50 m, 0.75 m, and 1.00 m. The results of the tests are shown in Fig. 2 as contact stress versus settlement and indicate that the larger the footing diameter, the larger the settlement for a certain contact stress. It is of interest to note that despite the large relative deformation, 16 percent for B = 0.25 m, the applied load is well below the capacity of the footing. Additional insight into the results can be obtained if the results are normalized to show the contact stress versus the settlement divided by the footing diameter, as is shown in Fig. 3. The normalization appears to suggest that, for a given contact stress, the settlement is proportional to the footing width.

Ismael (1985) also tested 0.5 m and 1.0 m footings at depths of 0.50 m, 1.00 m, 1.50 m, and 2.00 m below the ground surface at the site, that is, at depth ratios ranging from 0.5 through 4.0. The results are shown in Fig. 4 and indicate that the influence of footing depth is very small. Note, that if the settlement values would be normalized to footing diameter, the two groups of curves would plot within a common band.

Steady State Soil Mechanics Approach

When extrapolating results from small-scale model tests to the behavior of prototype footings, the scale requirement is not limited to the geometric scale, there is also a stress-scale to consider. This requirement can be addressed by performing the small-scale test in the centrifuge keeping the stress-scale equal to unity (equal stress at homologous points between model and prototype). The centrifuge test is performed using the same sand density and stress field for test and prototype. Small-scale tests outside the centrifuge, however, are performed at normal gravity and the stress-scale is not unity. For test result to represent the prototype behavior requires recognition that the density (void ratio), geometric scale, and stress-scale are related and must be considered together. A detail explanation and discussion of this statement is presented by Altaee and Fellenius (1994) and only major points are presented in this paper.

A first step toward understanding the behavior of sands was taken by Casagrande (1936), who showed that the behavior of a sand can be either contractant, dilatant, or neither. Casagrande established the term "critical void ratio" or "critical density", which denotes the void ratio or density of a soil subjected to continuous shear under neither dilatant nor contractant behavior. A next step was by Roscoe et al. (1958) who established the Critical State Soil Mechanics, which explains the fundamental behavior of a clay as a function of the void ratio and the mean stress. Later, Roscoe and Poorooshasb (1963) suggested that this principle could be extended to the behavior of non-cohesive soil.

Been and Jefferies (1985) indicated that, in a void ratio versus mean stress plot, the distance between the actual void ratio and the void ratio at the critical (or steady) state is an important parameter. They demonstrated that a similarity of behavior would occur between samples of the same sand tested at different void ratio and mean stress as long as the states are at equal void ratio difference. Building on Altaee (1991), Altaee and Fellenius (1994) developed scaling relations for small-scale model testing and analysis and presented a soil model based on steady state behavior as developed by Bardet (1986) and modified and implemented by Altaee (1991) in the AGAC93 finite element program. The principle of the critical state, or steady state, for sands is illustrated in Fig. 5, showing that the critical void ratio at the critical or steady state of the sand (when it shears with no further volume change) is a linear function of the logarithm of mean stress. The line is defined by its critical void ratio value, Γ (at the reference mean stress of 100 kPa) and its slope, λ. The compression of the sand follows a line with slope κ. When stress is introduced to a sand, the behavior of the sand is a function of its state location in this void-ratio versus mean-stress diagram and the void ratio distance (the upsilon

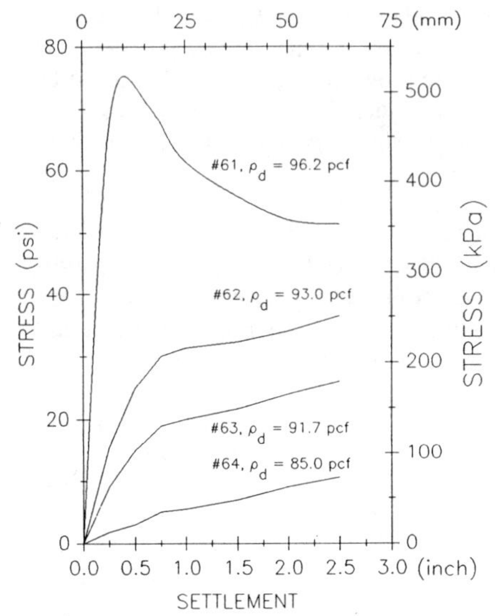

FIG. 1. Contact Stress versus Settlement of 150 mm Footings (Vesic 1967)

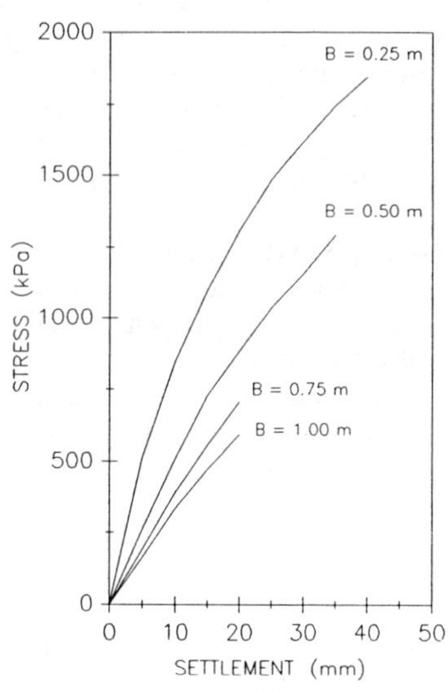

FIG. 2. Contact Stress versus Settlement of 0.25 - 1.00 m footings (Ismael 1985)

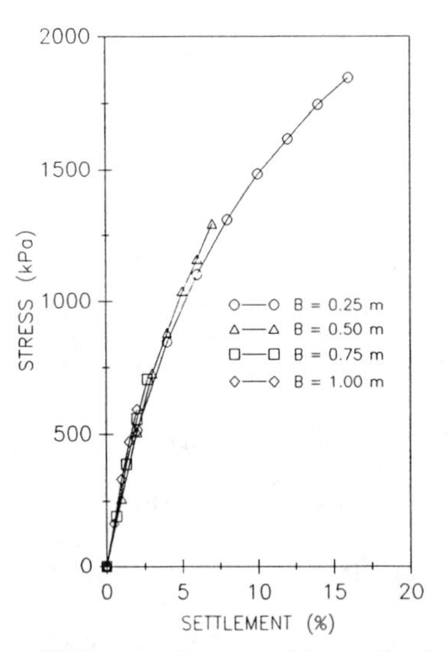

FIG. 3. Stress vs. Normalized Settlement of the Data Shown in Fig. 2

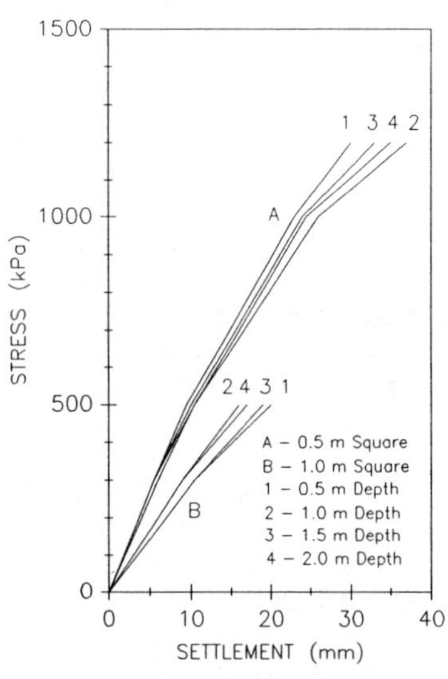

FIG. 4. Influence of Footing Depth on Stress-Settlement (Ismael 1985)

value, Y) from the initial void ratio to the void ratio along the steady state line at the same mean initial pressure. Positive upsilon values indicate contractant behavior and negative values dilatant behavior.

Fig. 6 shows several steady-state lines compiled from various sources by the authors (Altaee and Fellenius 1994). The compilation indicates a vast variety of slope and critical void ratio values, which demonstrate the very variable behavior exhibited by different sands. No two sands can a *priori* be assumed similar in behavior.

To demonstrate the importance of the steady-state approach in analyzing the behavior of a sand, we will discuss the load-settlement behavior of footings in two of the sands whose steady state lines are shown in Fig. 6: Fuji River and Kogyuk sands.

A summary of the soil parameters pertaining to the two sand types is presented in Table 1. The two sands are very different. (They are described in more detail by Altaee 1991, and Altaee and Fellenius 1993, respectively). The steeply sloping line of the Fuji River Sand, as opposed to the flat slope of that of the Kogyuk sand, indicates that the former sand is much more compressible than the latter. Also the peak strengths, as indicated by the peak friction angle, differ considerably. Detailed data on these sands are presented by Tatsuoka and Ishihara (1974), Ishihara et al. (1991), and Been and Jefferies (1985), respectively.

Agreement (not here documented) has been established between computed behavior (simulated tests) and the reported behavior in laboratory tests, which confirms the adequacy and relevance of the soil model and analysis method (Altaee 1991; Altaee and Fellenius 1993).

TABLE 1. Comparison of Sand Parameters

Parameter	Soil Type	
	Fuji River	Kogyuk
Mean particle size (mm)	0.22	0.35
Uniformity coefficient, C_u	2.21	1.80
Maximum void ratio, e_{max}	1.08	0.83
Minimum void ratio, e_{min}	0 53	0 47
Effective angle of friction (°) from triaxial testing		
Ultimate Compression	36.9	30.5
Ultimate Extension	32.0	30.5
Peak	38.0	35.0
Critical void ratio at 100 kPa, Γ	0.920	0.713
Slope of critical line in e - ln(p) plane, λ	0.120	0.029
Slope of unloading-reloading line in e-ln(p) plane, κ	0.010	0.006
Poisson's ratio, ν	0.30	0.10
Aspect ratio of bounding surface, ρ	2.2	2.0
Hardening parameter, h_o	1.0	1.0

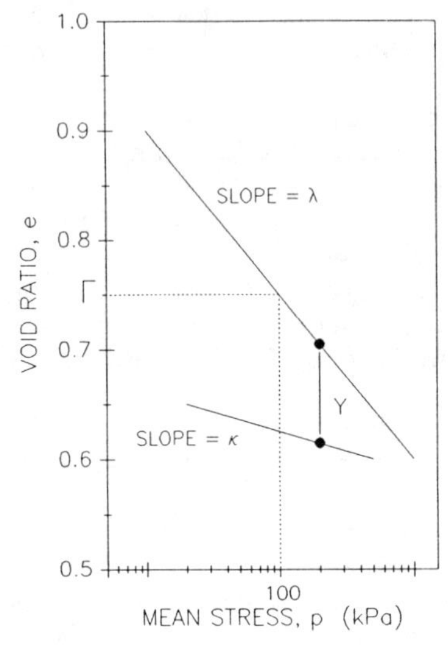

FIG. 5. Definition of Steady-state Lines in the e-ln(p) Plane

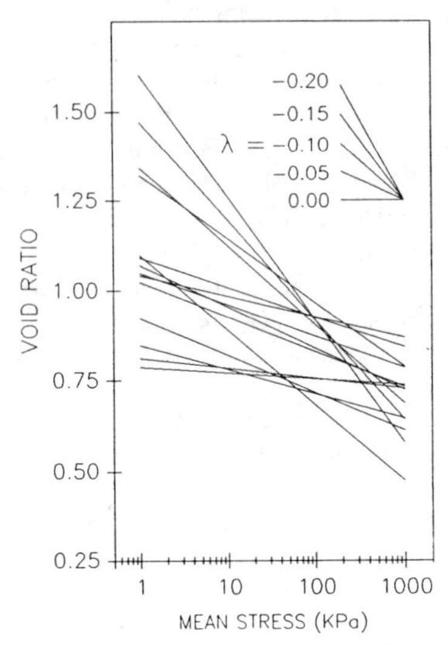

FIG. 6. Compilation of Steady State Lines (Altaee and Fellenius 1994)

NUMERICAL ANALYSIS

The settlement behavior of footings placed on the two sand types is investigated numerically in a three-part analysis series. The analysis is a plane strain (two-dimensional) finite element analysis incorporating the bounding surface plasticity model for sand (Bardet 1986; Altaee 1991) employing the AGAC93 computer program. The finite element mesh consists of 300 nodal points selected by means of a parametric study to determine the size of the soil mass to include in the analysis as well as the geometric boundaries. The footings are rigid and continuous with rough contact surface.

Fig. 7 presents a diagram of the data for the Fuji River Sand plotted as void ratio versus mean stress with the initial void ratio and initial mean stress below the footing base for each analysis. The similar plot for the Kogyuk data is not shown. (Note, for reasons of achieving clarity, Fig. 7 shows the conditions at a depth of 3B below the footing; the initial void ratio distance – Upsilon value – to the steady state line is different at different depths).

In a first series on each sand type, footings of three sizes (B = 0.5 m, B = 1.0 m, and B = 2.0 m) are placed at the ground surface, at a depth equal to the footing size, and a depth equal to twice the footing size. For all these nine footings (numbered from 1 through 9), the initial void ratio of the soil is essentially the same (note, as the mean stress increases, the void ratio decreases slightly due to compression of the soil). The density indices are also essentially equal. However, the upsilon values (void ratio distance to the steady state line) vary somewhat and more so for the Fuji River Sand than the Kogyuk Sand, because of the different slopes of the steady state lines.

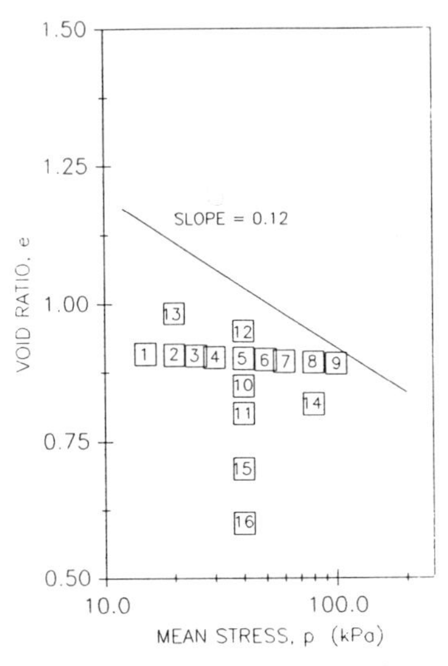

FIG. 7 e-ln(p) Diagram for the Initial State of Homologous Points for the Analyses

In a second series, to demonstrate clearly the effect of varying upsilon value, six footings of equal size (1.0 m) are placed at equal depth (1.0 m), but in sands of different upsilon values, initial void ratios, and initial mean stresses. These analysis cases are numbered 10, 11, 12, 15, and 16. Case 5 of Series 1 fits into this series too.

The third analysis series includes the footing of different sizes placed at a one-diameter depth in the sand having the same upsilon value, but different void ratios. These analyses are numbered 13 and 14. Again, analysis No. 5 of Series 1 fits into this series.

The void ratios, density indices, and upsilon values pertaining to the initial states of the three analysis series at a depth of 1B below the footing base are listed in Table 2. Note that the void ratio values vary with the depth and mean stress.

Analysis Results

The computations proceeded by calculating the contact stress for imposed values of settlement. Figs. 8 and 9 show the results of the calculations of for Series 1 (different size footings placed at different depths, i.e., different stress at homologous points) for footings in the Fuji River and Kogyuk sands, respectively. The shape of the curves is very similar to that of the field tests presented in Fig. 2 taken from the work of Ismael (1985). Note, that no sign of impeding failure can be seen despite the settlement ratio reaching a value of 10 percent of the footing width.

TABLE 2. Void Ratios, Density Indices, and Upsilon Values

Analysis		B	z	Fuji River Sand			Kogyuk Sand		
#	Series	(m)	(m)	e	Y	I_D (%)	e	Y	I_D (%)
#1	1	0.5	0	0.909	-0.239	21	0.680	-0.088	42
#2	1	0.5	B	0.907	-0.206	21	0.678	-0.079	42
#3	1	0.5	2B	0.905	-0.181	32	0.677	-0.076	42
#4	1	1.0	0	0.902	-0.162	32	0.676	-0.072	43
#5	1,2,3	1.0	B	0.900	-0.130	33	0.676	-0.063	43
#6	1	1.0	2B	0.898	-0.104	33	0.675	-0.058	43
#7	1	2.0	0	0.895	-0.086	34	0.671	-0.057	44
#8	1	2.0	B	0.893	-0.054	34	0.671	-0.048	44
#9	1	2.0	2B	0.891	-0.029	34	0.670	-0.043	44
#10	2	1.0	B	0.850	-0.180	42	0.650	-0.089	50
#11	2	1.0	B	0.800	-0.230	51	0.625	-0.114	60
#12	2	1.0	B	0.950	-0.080	24	0.725	-0.014	29
#13	3	0.5	B	0.983	-0.130	18			
#14	3	2.0	B	0.817	-0.130	48			
#15	2	1.0	B	0.700	-0.330	69			
#16	2	1.0	B	0.600	-0.430	97			

To study the effect of the footing size, the Series 1 stress-settlement data are normalized to the footing widths in Figs. 10 and 11, respectively, for the two sand types (three pairs of data in each diagram). The rather small differences in the normalized behavior appear to suggest that, in conformity with the field tests (Fig. 3), the stress-settlement behavior of footings in sands of equal density might be independent of the footing width and difference in initial stress at homologous points. Note, however, that the behavior of the footing at the ground surface of the Kogyuk sand deviates from its counterpart on the Fuji River sand, which shows that the footing width is not always even approximately useful as a normalizing parameter.

Fig. 12 shows the contact stress versus settlement for footings of equal width (B = 1.0 m) placed at ground surface and at depths of 1.0 m and 2.0 m. The void ratios and the density indices are essentially equal for the sands of each of the two series. The diagrams show that while the influence of depth, that is, of initial mean stress and upsilon values, is small, it is not insignificant. Note that although the difference in density index between the two sands is not large, the settlement (at equal contact stress) in the Fuji River Sand is several times larger than that in the

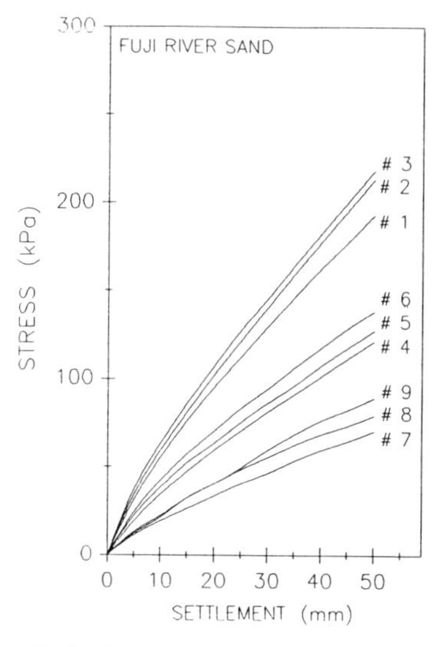

FIG. 8. Contact Stress vs. Settlement Series 1, Fuji River Sand

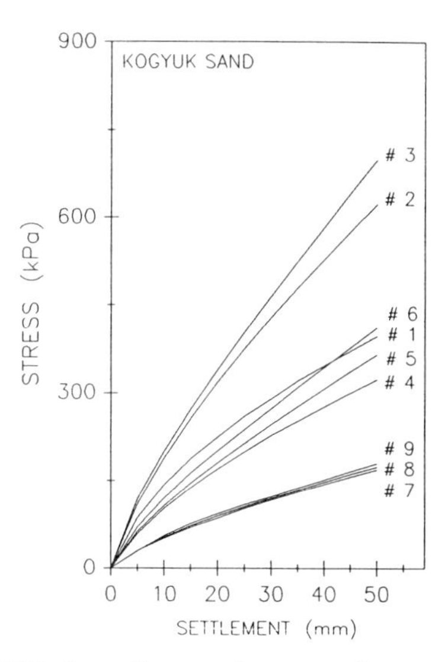

FIG. 9. Contact Stress vs. Settlement Series 1, Kogyuk Sand

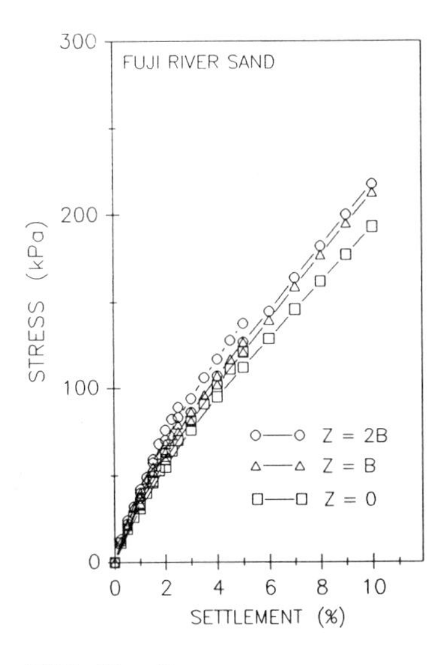

FIG. 10. Stress vs. Normalized Settlement Series, Fuji River Sand

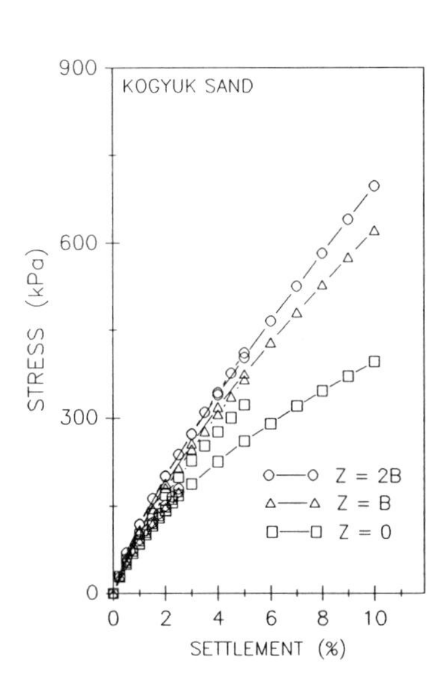

FIG. 11. Stress vs. Normalized Settlement Series 1, Kogyuk Sand

Kogyuk Sand. Note also that the higher friction angle of the Fuji River Sand appears to be of no consequence.

The results of Series 2 (same size footing placed at equal depth, but in sands of varying density – varying upsilon value) are shown in Figs. 13 and 14. The results indicate that the density of the sand has a significant influence on the stress-settlement behavior – not particularly novel a discovery, of course. Notice also that the upsilon value at homologous points of the sand differs between the different footings.

Fig. 15 shows the results from Series 3 (different footing width, different initial density, different initial stress, but same upsilon value). The stress-settlement behavior of these three analyses are equal, which demonstrates that the important parameter governing the settlement behavior is the upsilon value. In fact, for the behavior of a model test to agree with the behavior of its prototype requires that the test is performed at a density (void ratio) that has an equal distance to the steady state line, that is, the test has to be made at an upsilon distance that for homologous points is equal to that of the prototype.

Furthermore, the requirement of equal upsilon value means that a model test in a soil of some density emulates the behavior of a prototype (if larger than the model, of course) in a denser soil. Therefore, small-scale model tests in a dense soil have very limited application, because when applied to a prototype of some size, the density of the relevant prototype soil very quickly exceeds the maximum density of the soil. At the same time, when investigating settlement, as well as capacity, in small-scale tests, the emulation of large prototypes must be performed in soils very much looser than that of the prototype soil. The ratio of geometrical scale is determined by the practical limit of how loose the sand can be. Very small footing tests, for example, have limited application to the behavior of full size foundations.

Note that the stress-settlement behavior for the analyses results shown in Fig. 15 is the same for all curves, in contrast to the results shown in Figs. 13 and 14, although the initial void ratios are different. This means that, when comparing footings of different size in the same sand, the settlement is not a function of the density per se, nor is it a function of the density index.

CONCLUSIONS

The settlement of footings of different widths placed in sand of different void ratios and at different depths below the ground surface can be directly related if the conditions of steady state are considered and if the tests are performed at equal void ratio distance to the steady state line – equal upsilon value. Small-scale models will only be representative for prototype behavior if this requirement is fulfilled.

When comparing the behavior of footings of different size in the same sand, the settlement is not a function of the density per se, nor is it a function of the density index (the relative density).

The requirement of equal upsilon value means that the small-scale model test must always be performed in a soil that is looser than the prototype soil. This imposes boundaries on the geometric scale, because, first, a model test can not be carried out in a sand that is looser than the maximum void ratio. Second, a model

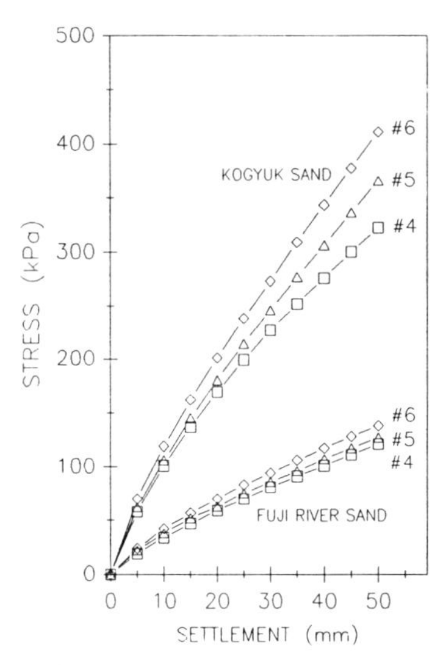

FIG. 12. Stress versus Settlement for Footings at Different Depth (z = 0; z = B; z = 2B) Series 1 Fuji River and Kogyuk Sands

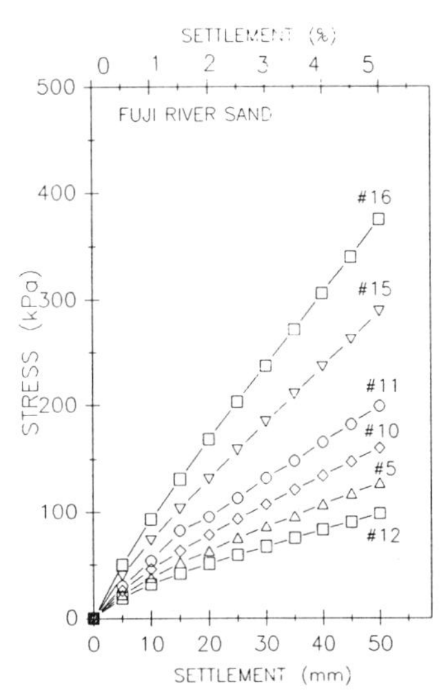

FIG. 13. Stress vs. Settlement Series 2, Fuji River Sand

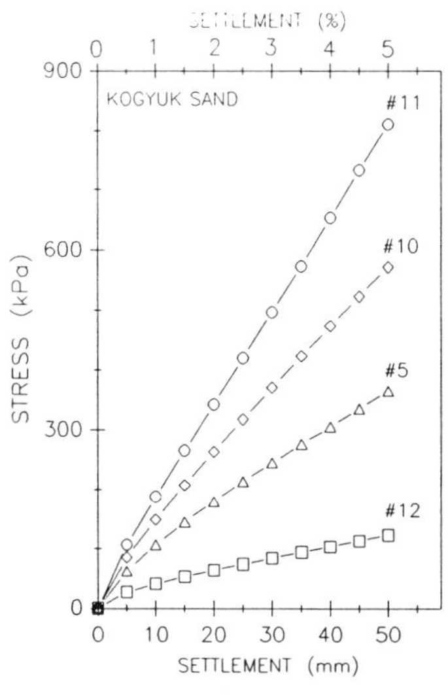

FIG. 14. Stress vs. Settlement Series 2, Kogyuk Sand

test must not be performed in a soil that is denser than what corresponds to realistic density of its prototype soil. Therefore, a model test is meaningless – has no corresponding prototype – if it is performed in a sand close to the minimum void ratio, because, if so, the density of the prototype sand would have to approach zero void ratio, that is, cease to be a soil.

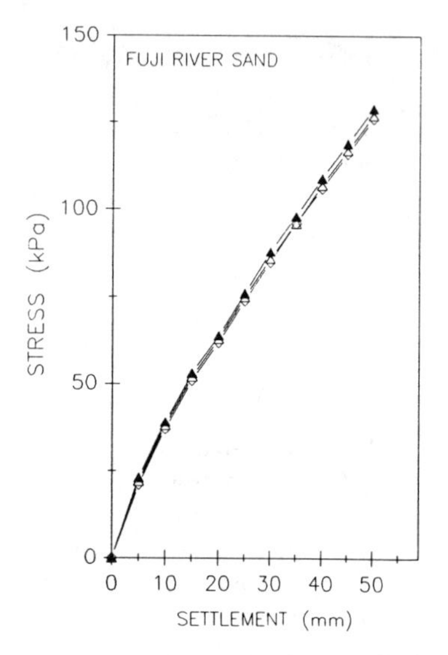

FIG. 15. Stress-Settlement Series 3, Cases 5, 13, and 14 Fuji River Sand

Furthermore, although the stress-settlement relation is approximately similar for footings of varying size in sand of uniform density, normalization to footing size does not strictly provide a similitude of results.

Appendix - References

Altaee, A. (1991). "Finite element implementation, validation, and deep foundation application of a bounding surface plasticity model," Ph.D. thesis, University of Ottawa, Ottawa, Ontario.

Altaee, A. and Fellenius B. H. (1993). "Cyclic performance of an earth fill retention arctic offshore structure." *Proc. 4th Canadian Marine Geotechnical Conf.*, St.John's, 20 pp.

Altaee, A. and Fellenius B. H. (1994). "Physical modeling in sand." *Can. Geotech. J.*, 31(3), in press.

Bardet, J. P. (1986). "Bounding surface plasticity model for sands." *J. Eng. Mech.*, ASME, 112(EM11), 1198-1217.

Been, K. and Jefferies, M. G. (1985). "A state parameter for sands." *Géotechnique*, 35(1), 99-112.

Bond, D. W. (1961). "Influence of foundation size on settlement." *Géotechnique*, 11(2), 121-143.

Casagrande, A. (1936). "Characteristics of cohesionless soils affecting the stability of slopes and earth fills." *Contribution to Soil Mechanics 1925-1940*, Boston Society of Civil Engineers, Boston, Massachusetts, 257 - 276.

Ishihara, K., Verdugo, K., and Acacio, A. A. (1991). "Characterization of cyclic behavior of sand and post-seismic stability analyses." *Proc. 9th Asian Regional Conf. on Soil Mech. and Found. Eng.*, 2, 45-67.

Ismael, N. F. (1985). "Allowable bearing pressure from loading tests on Kuwaiti soils." *Can. Geotech. J.*, 22(2), 151-157.

Roscoe, K. H., Schofield, A. N., and Wroth, C. P. (1958). "On the yielding of soils." *Géotechnique*, 8(1), 22-53.

Roscoe, K. H. and Poorooshasb, H. (1963). "A fundamental principle of similarity in model test for earth pressure problems." *Proc. 2nd Asian Regional Conference on Soil Mechanics*, Bangkok, 1, 134-140.

Tatsuoka, F. and Ishihara, K. (1974). "Drained deformation of sand under cyclic stress reversing direction." *Soils and Foundations*, 14(1), 51- 65.

Vesic, A. S. (1967). "A study of bearing capacity of deep foundations," *Final Report Project B-189*, Georgia Institute of Technology, Engineering Experiment Station, Atlanta.

Vesic, A. S. (1975). "Bearing capacity of shallow foundations." *Foundation Engineering Handbook*, Van Nostrand Reinhold Co., New York, New York, 121-147.

Load Settlement Curve Method for Spread Footings on Sand

Jean-Louis Briaud,[1] Fellow, ASCE, and Philippe Jeanjean[2]

Abstract: A newly developed method is presented to predict the complete load settlement curve for square spread footings on sand loaded at their center with a vertical load and with an embedment from 0.25B to 0.75B where B is the footing width. This method also allows one to obtain the settlement of the footing as a function of time for a given load. The basis for this new prediction technique is a complete preboring pressuremeter curve. The method is developed after performing finite element simulations and after performing 5 full-scale footing load tests pushed to 0.15 m of vertical displacements. Footing widths vary from 1 to 3 m. The new method satisfies observations from the experimental results including the fact that there is no scale effect on the load settlement curves for these footings if those curves are plotted as pressure versus settlement over width.

The Idea

This paper deals with square spread footings on sand loaded at their center with an axial load and having embedments varying from 0.25B to 0.75B where B is the footing width. The idea is to develop a method to predict the complete load-settlement curve for such footings. For piles subjected to axial monotonic loading this problem was solved by Seed and Reese (1957) and led to the t-z curve concept for axially loaded piles. The idea is to do for spread footings what was done for piles a long time ago.

---

[1] Buchanan Prof. of Civ. Engrg., Texas A&M Univ., College Station, TX 77843-3136.
[2] Res. Engr., AMOCO Production Co., 4502 East 41st St., Tulsa, OK, 74102-3385.

In a typical design for a spread footing on sand both the bearing capacity p_u and the settlement s are calculated. A number of methods exists to calculate these quantities (e.g. Table 1) and these methods have been evaluated for their accuracy (Tan and Duncan 1991; Amar et al. 1984; Jeyapalan and Boehm 1986). The bearing capacity methods often do not indicate what displacement is necessary to reach the calculated p_u. Fig. 1 represents a load settlement curve for a load test performed on a 12.5 mm diameter minicone point. The CPT point was loaded to plunging failure in a pressure chamber filled with a medium dense uniform fine dry sand confined at 55 kPa. While there is a difference between a pointed cone and a flat bottom spread footing, the cone test results may shed some light on the footing behavior at least at large strains. As can be seen from Fig. 1, the displacement necessary to reach plunging failure can be larger than two times the foundation diameter. For a 3 m × 3 m footing this pressure would become meaningless since the associated displacement would be very large.

TABLE 1. Some Methods for Predicting the Settlement of Spread Footings on Sand

Method	Test	Method	Test
Alpan (1964)	SPT	Oweis (1979)	SPT
Briaud (1992)	PMT	Parry (1971)	SPT
Burland and Burbidge (1985)	SPT	Peck and Bazaraa (1969)	SPT
D'Appolonia, D'Appolonia, and Brissette (1968)	SPT	Peck, Hanson and Thornburn (1974)	SPT
De Beer and Martens (1957)	CPT	Sanglerat (1972)	CPT
Marchetti (1980)	DMT	Schmertmann (1978)	CPT
Menard and Rousseau (1962)	PMT	Schultze and Sherif (1973)	SPT
Meyerhof (1965)	SPT	Terzaghi and Peck (1967)	SPT
NAVFAC (1982)	SPT		

In order to place the purpose of this article in perspective, one must note that the rest of the discussion focuses on the early part of the curve (0 to B in Fig. 1); indeed all measurements and predictions contained herein deal with displacements which range from 0 to 0.15B. One must also note that very often bearing capacity is defined as the pressure reached for a displacement of 0.1B. On Fig. 1 this would represent only one third of the true plunging load.

Settlement methods are often based on elasticity or elasticity modified formulas. Most of them predict a linear load settlement curve which gives the correct answer at the point of intersection between that straight line and the measured load settlement curve. While most methods seem to aim at predicting the

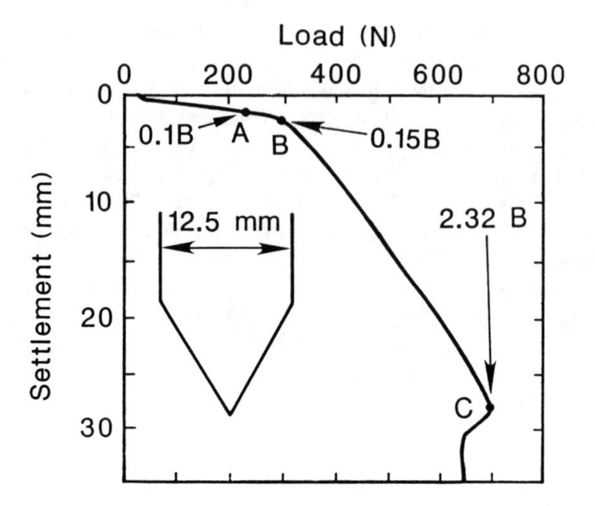

FIG. 1. Load Settlement Curve for a Mini Penetrometer

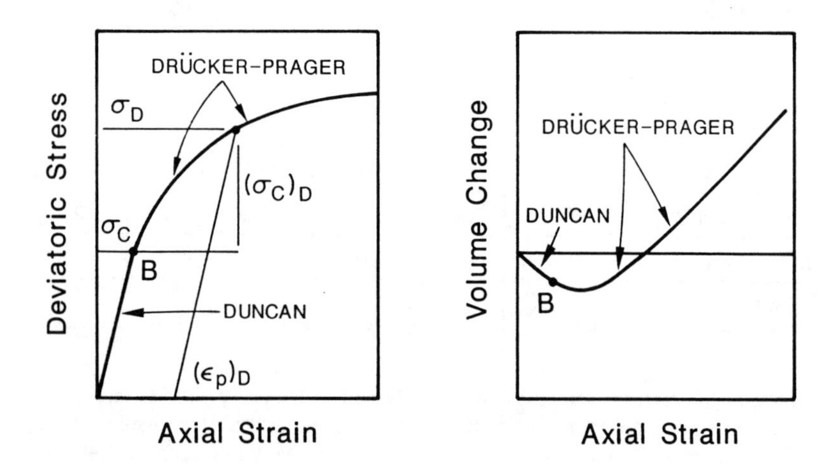

FIG. 2. Elements of the Duncan/Drücker Prager Model

load for a settlement equal to 25 mm, few methods actually specify which point on the load-settlement curve is being predicted. Moreover, few methods make recommendations for predicting the time dependency of the settlement. In short, most current methods predict only two points on the load-settlement curve: the pressure to cause 25 mm of settlement and the pressure to cause failure of the soil. Considering these shortcomings, the idea was to develop a method to predict the complete load-settlement curve for a footing on sand including time effects.

There are several approaches to solving a geotechnical problem. One is the fundamental approach where the soil is modelled by discrete elements and the continuum behavior is theoretically assembled: this is the case of the finite element method. One is the empirical approach where a correlation is used between a test result and the parameter to be predicted: this is the case of Peck's method (1974) for predicting the pressure corresponding to 25 mm of settlement for footings on sand on the basis of SPT blow counts. Another method which fits possibly between those two approaches is the approach by analogy where the soil test which is performed closely resembles the loading imposed on the soil by the foundation and where the remaining difference is bridged by using theoretically and experimentally based corrections: this is the case of the cone penetrometer method used to predict the axial capacity of a pile or the pressuremeter method used to predict the behavior of a horizontally loaded pile.

The fundamental approach and the approach by analogy were used in this study.

Fundamental Approach: Finite Element Model

The concept in this approach is to identify or develop a continuum mechanics model which will closely represent the soil behavior, take soil samples, perform laboratory tests such as the triaxial test under proper stress conditions, match the nonlinear stress-strain curves obtained from the laboratory tests and from the theoretical model in order to back calculate all the soil-model parameters, and finally perform a finite element simulation of the foundation problem using those site specific soil-model parameters.

Several models exist for sands. One commonly used model is the hyperbolic 7 parameter model which will be called here the Duncan model (Duncan and Chang 1970; Wong and Duncan 1974). Some of the advantages of this model are its relative simplicity, its ability to account for a large number of important influencing factors, and the extensive experience accumulated in using it. One drawback is that this model does not give great flexibility in accounting for dilatancy due to shear because it uses an incremental form of Hooke's law; indeed the value of Poisson's ratio ν_t loses significance as soon as $\nu_t > 0.5$. Plasticity offers an alternative to describe dilatancy in sands. One common plastic model which can handle dilatancy is the Drücker-Prager model (Drücker and Prager 1952). This 5-parameter model, however, is usually combined with linear elasticity until the yield point. This is not consistent with the fact that sands have stress-dependent moduli as acknowledged by the Duncan model. Therefore it was felt that a better model would combine the advantages of the Duncan model at small strains and the Drücker-Prager model at large strains.

A combined 7-parameter model called the Duncan/Drücker-Prager model was created and used in this study. This model is described in detail in Jeanjean (1993). In the pre-yield behavior (small strain) an incremental form of Hooke's law is used with stress dependent modulus E_t and Poisson's ratio ν_t.

$$E_t = K p_a \left(\frac{\sigma_3}{p_a} \right)^n \tag{1}$$

$$\nu_t = G - F \log_{10} \left(\frac{\sigma_3}{p_a} \right) \tag{2}$$

where K is the modulus number, p_a the atmospheric pressure, σ_3 the minor principal stress, n the modulus exponent, G the Poisson's ratio when $\sigma_3 = p_a$ and F the decrease in Poisson's ratio for a ten-fold increase in σ_3. These parameters are obtained by performing consolidated drained triaxial tests at different confining pressures while measuring volume change. The early part of the deviatoric stress versus axial strain curve for different σ_3 gives K and n while the early part of the volume change versus axial strain curve for different σ_3 gives G and F. For simplicity, no strain dependency is accounted for since for small strains this may not be a severe limitation as shown by Cosentino (1987). On the stress strain curve and the volume change curve, this pre-yield model accounts for the early part of the curve (until point B on Fig. 2).

Initial yield occurs at point B. This point corresponds to the end of the seemingly linear part of the stress-strain curve obtained in a triaxial test. Judgement is involved in determining this point. The stresses at point B are used to determine the initial yield effective stress friction angle δ. Note that because of this definition, δ will be smaller than the commonly used friction angle ϕ which corresponds to the maximum deviatoric stress. The extended Drücker-Prager yield surface is defined as

$$t - p\tan\beta - \left(1 - \frac{1}{3}\tan\beta\right)\sigma_c = 0 \tag{3}$$

where t is a measure of the deviatoric stress (Jeanjean 1993), p the first stress invariant, β the friction angle in the p-t plane, σ_c the yield stress in uniaxial compression. The parameters β and σ_c are obtained by matching Eq. (3) with the Mohr-Coulomb criterion and with the experimental data.

The hardening rule for a given σ_3 allows to describe the post yield stress vs strain curve and is defined by 6 points read on the corresponding triaxial test curve beyond point B. For each point the plastic strain ($(\epsilon_p)_D$ on Fig. 2) is calculated with an unload modulus equal to the first loading modulus, and the hardening stress ($(\sigma_c)_D$ on Fig. 2) is the current stress minus the yield stress.

The flow rule is a non-associated flow rule and allows to describe the post yield volume change vs strain curve. The plastic strain increment $d\epsilon_{ij}^{pl}$ is:

$$d\epsilon_{ij}^{pl} = \frac{d\bar{e}^{pl}}{(1 - \frac{1}{3}\tan\psi)} \frac{\partial g}{\partial \sigma_{ij}} \quad (4)$$

where $d\bar{e}^{pl} = \| d\epsilon_{ij}^{pl} \|$ in uniaxial compression, g is the flow potential $= t - p \tan\psi$, ψ is the dilation angle in the p-t plane. The dilation angle ψ is obtained by curve fitting the volume change vs strain curves.

FUNDAMENTAL APPROACH: PREDICTING THE LOAD SETTLEMENT CURVE

In order to evaluate the accuracy of this method, a class A prediction was made for one of the large spread footings tests performed in the fall of 1993 at the National Geotechnical Experimentation Site at Texas A&M University (Briaud 1993). The soil is a medium dense fine silty silica sand. This middle Eocene sand was formed in a coastal plain environment. The grain size distribution curve is relatively uniform with most of the grain sizes between 0.5 mm and 0.05 mm. The sand is probably lightly overconsolidated by desiccation of the fines and removal of about 1 m of over-burden at the location of the spread footing tests. A series of soil tests were performed in the spring of 1993 at the site (Gibbens and Briaud 1993). The test layout is shown in Fig. 3, the soil cross section in Fig. 4, SPT results in Fig. 5 and CPT results in Fig. 6. The range of SPT and CPT soundings shows the subsurface variability across the site. Disturbed samples were retrieved with a hand auger down to 4 m and with the SPT below that. The in situ moisture content above the ground water table was measured to average 5%. The water table is 4.9 m deep.

In order to reconstruct the samples for triaxial tests, the relative density was estimated from SPT data (Peck et al. 1974) and from CPT data (Briaud and Miran 1992), to be approximately 55%. The minimum and maximum void ratios were measured according to ASTM D4254 and D4253: $e_{max} = 0.925$, $e_{min} = 0.635$, $G_s = 2.65$. The samples were reconstructed in the triaxial mold at the estimated natural unit weight and water content. This natural unit weight averaged 15.5 kN/m^3. The oven-dried sand was mixed with the proper amount of water to reach the desired water content of 5%. Then the moist sand was tamped into the triaxial test sample mold in small layers. By trial and error the right amount of tamping was found to bring the overall sample to the correct unit weight inside the mold which had a known volume. The complete set of results for 2 sample depths, 2 relative densities, and 3 confining pressures are included in Jeanjean (1993). Figs. 7 and 8 show the results of 6 of the 12 tests performed.

By curve fitting the experimental triaxial test results as explained previously, the Duncan/Drücker-Prager parameters were back calculated:

$K = 600$, $n = 0.64$, $G = 0.36$, $F = 0.08$, $\delta = 23.7°$,
$\psi = 0.5°$, $\sigma_c = f(\sigma_3, \epsilon_p)$ (see Table 2)

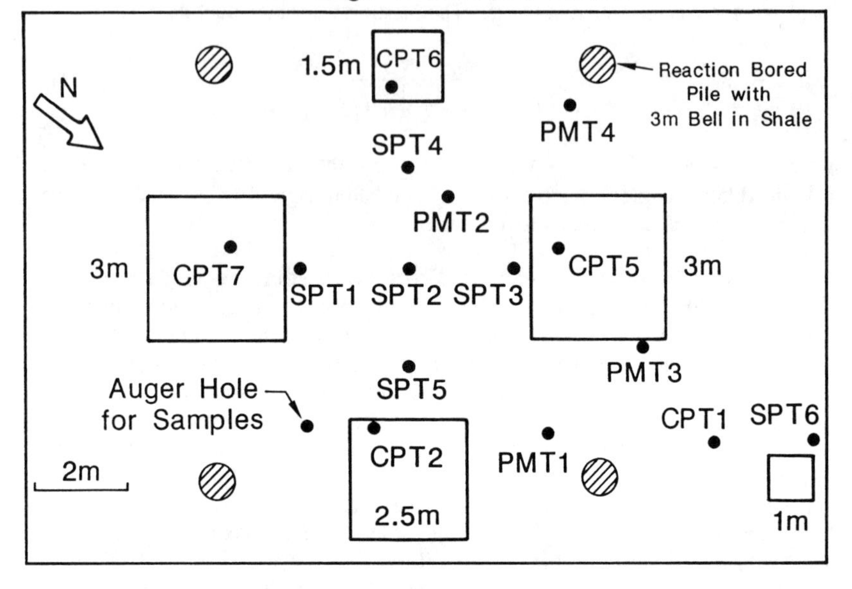

FIG. 3. Spread Footing Tests: Plan View

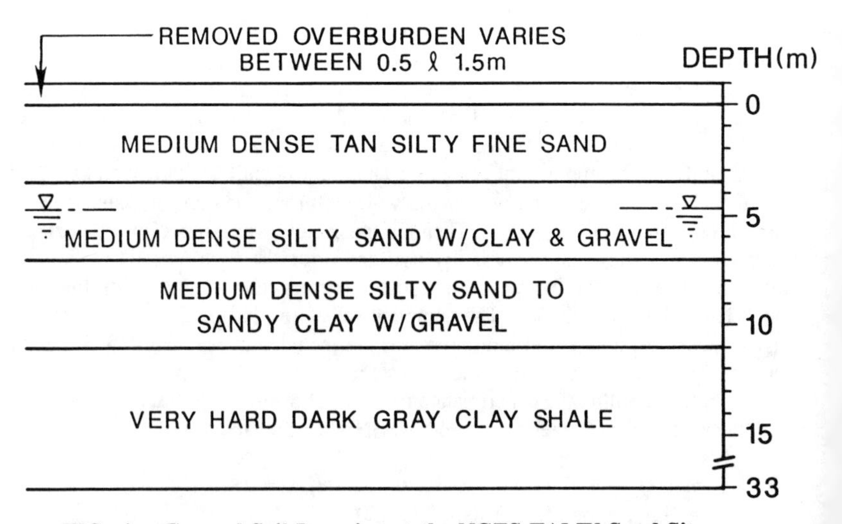

FIG. 4. General Soil Layering at the NGES-TAMU Sand Site

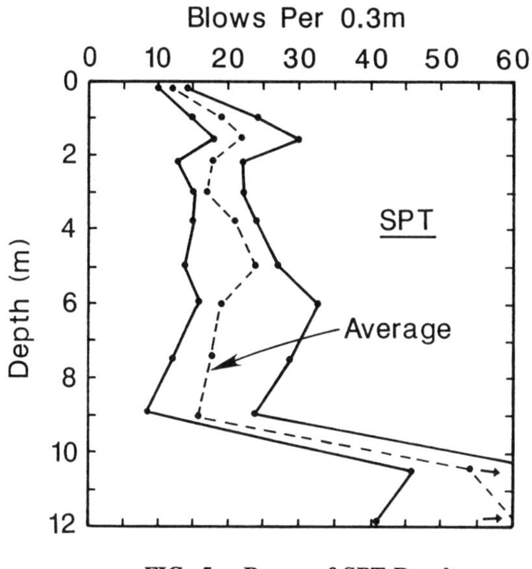

FIG. 5. Range of SPT Results

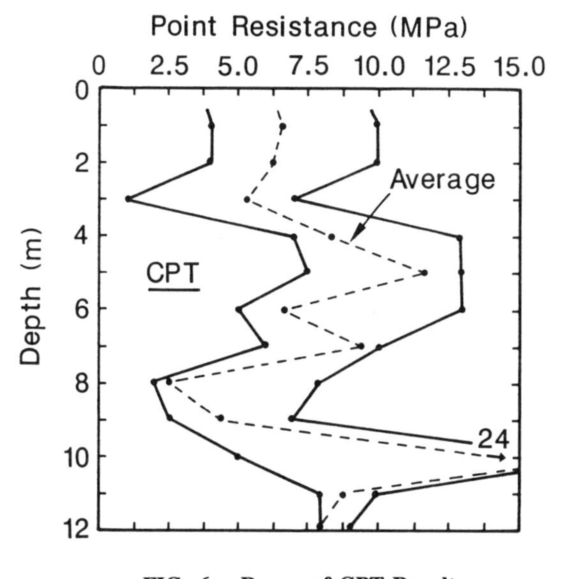

FIG. 6. Range of CPT Results

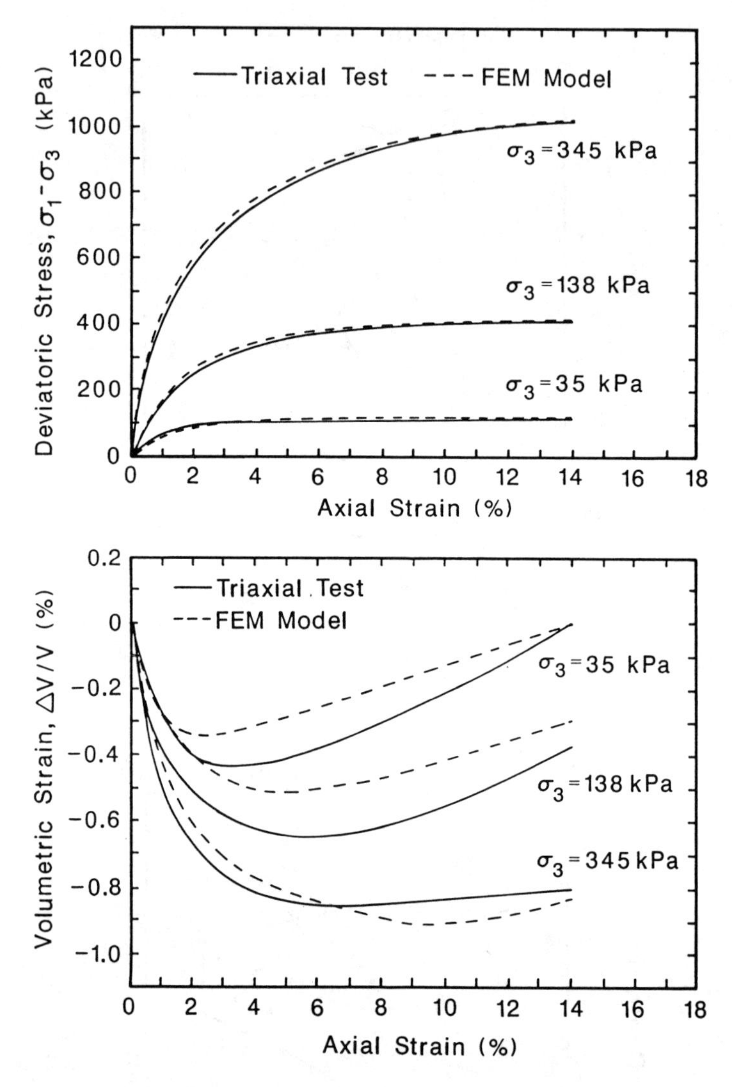

FIG. 7. Triaxial Test Results for the Medium Dense Sand and FEM Model Matching (D_r = 55%, w = 5%, z = 3 m)

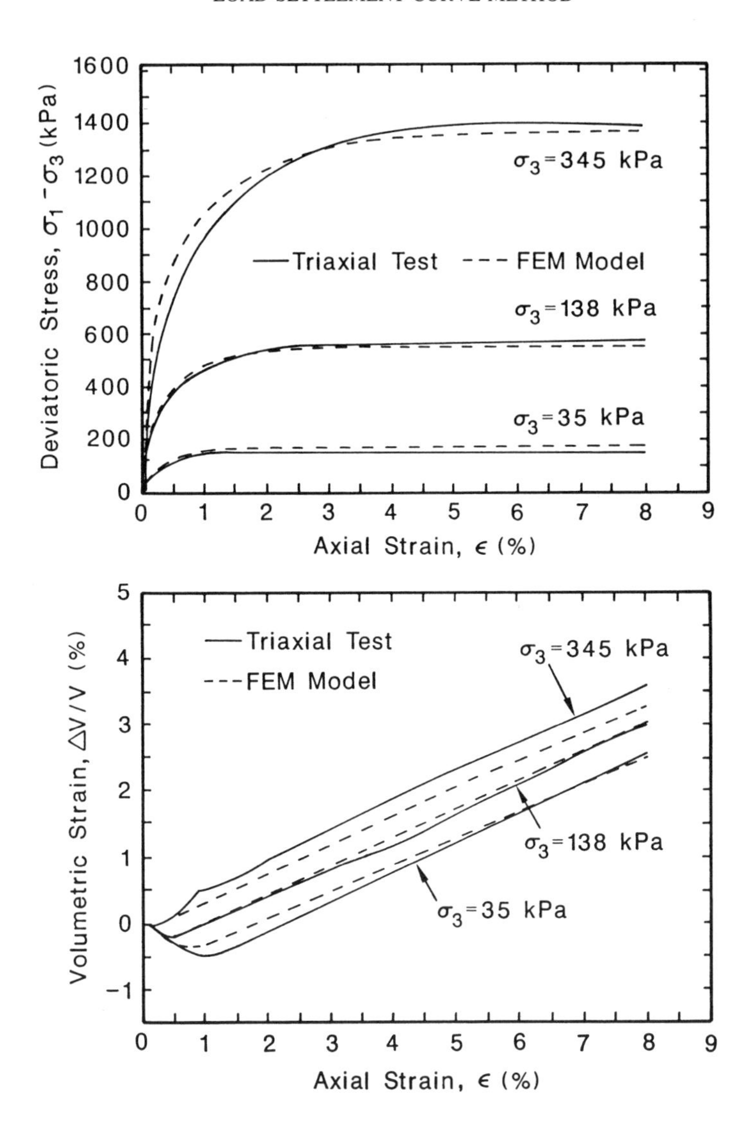

FIG. 8. Triaxial Test Results for the Dense Sand and FEM Model Matching (D_r = 95%, w = 5%, z = 3 m)

TABLE 2. Hardening Stress of Medium Dense Sand as a Function of the Plastic Strain

	Hardening Stress σ_c, (kPa)		
Plastic Strain, ϵ_p (%)	σ_3 = 34 kPa	σ_3 = 138 kPa	σ_3 = 345 kPa
0.8	25	100	260
1.7	36	150	370
2.7	46	185	440
3.8	46	205	500
5.1	46	220	540
6.6	46	235	580
8.7	47	240	620
13.0	49	250	660

Using those parameters, a finite element analysis for a 3 m × 3 m spread footing, embedded 0.75 m into the sand, was performed using the program ABAQUS (1991) with the specially prepared subroutine for the Duncan/Drücker-Prager model. The size of the mesh was increased until the boundaries were far enough to bring the relative error between the close form elastic solution and the FEM elastic solution down to 1%. This was achieved when the mesh was 40 times wider than the footing and had a depth equal to 25 times the footing width. The three dimensional nonlinear analysis was performed on a CRAY supercomputer which produced the load settlement curve shown on Fig. 9. This process was repeated for a relative density of 95% for parametric purposes and led to the predicted curve on Fig. 10. It took about one year to reach this point in the project.

Approach by Analogy: Developing the Method

The penetration of a footing into a sand deposit can be considered as a cavity expansion phenomenon. Therefore a cavity expansion test would make sense for the approach by analogy. Since the complete load settlement curve is desired, this cavity expansion test should give a complete pressure-deformation curve. This led to the choice of the pressuremeter test (PMT) for the method by analogy. The concept is then to perform pressuremeter tests within the zone of influence of the spread footing, generate an average PMT curve by using an influence factor distribution for a weighted average of the individual curves, and transform the average PMT curve into a spread footing curve by using a correction factor obtained from theoretical and experimental considerations.

It is recommended that pressuremeter tests be performed at 0.5B, 1B and 2B below the footing level where B is the footing width. Each test leads to a pressure p versus relative increase in probe radius $\Delta R_p/R_p$ as shown on Fig. 11 where R_p is

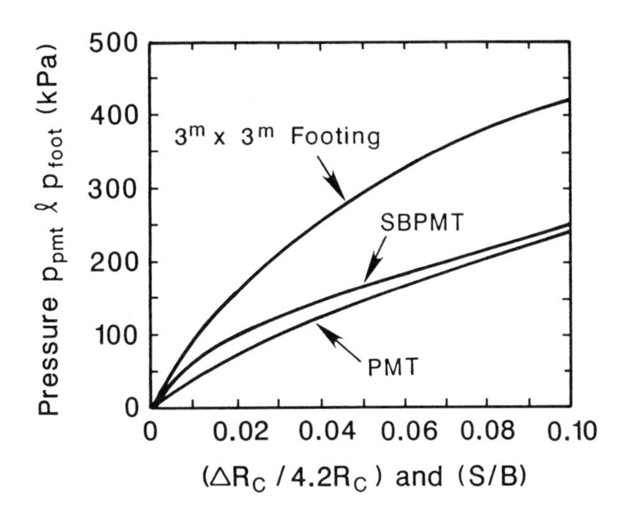

FIG. 9. FEM Results for the Medium Dense Sand

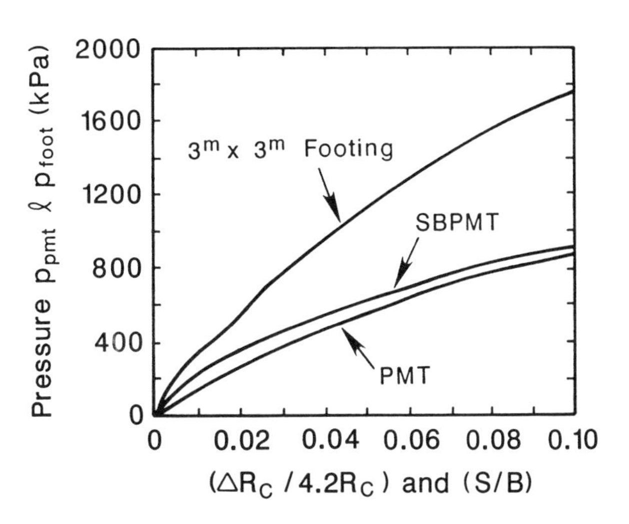

FIG. 10. FEM Results for the Dense Sand

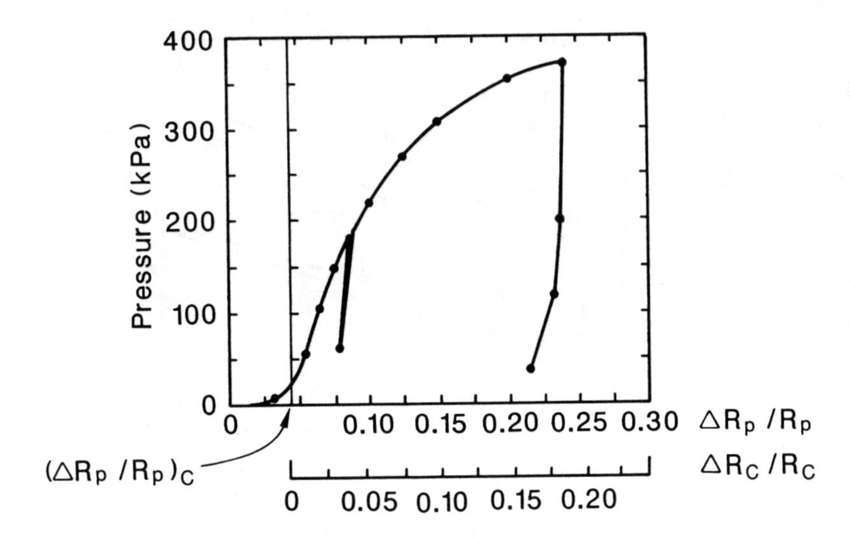

FIG. 11. Pressuremeter Curve (PMT 1, z = 0.6 m)

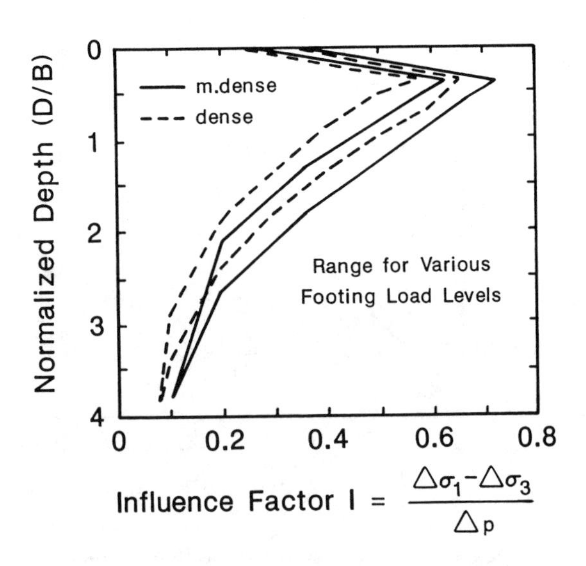

FIG. 12. Influence Factor Distribution with Depth

the radius of the deflated probe and ΔR_p the increase in probe radius. Each curve is then adjusted for the size of the borehole (Fig. 11) by extending the straight line part of the PMT curve to p = 0, shifting the vertical axis to that new origin and calculating the relative increase in cavity radius ($\Delta R_c/R_c$) as:

$$\frac{\Delta R_c}{R_c} = \frac{\dfrac{\Delta R_p}{R_p} - \left[\dfrac{\Delta R_p}{R_p}\right]_c}{1 + \left[\dfrac{\Delta R_p}{R_p}\right]_c} \quad \textbf{(5)}$$

where R_c is the initial radius of the borehole (cavity), ΔR_c the increase in cavity radius, and $(\Delta R_p/R_p)_c$ the value of the relative increase in probe radius corresponding to the initial radius of the cavity.

The corrected curves (p vs $\Delta R_c/R_c$) obtained at the testing depths below the footing are then averaged into a single curve called the average PMT curve on the basis of a stress influence factor I similar to the one proposed by Schmertmann (1970). This factor is defined as:

$$I = \frac{\Delta\sigma_1 - \Delta\sigma_3}{\Delta p} \quad (6)$$

where $\Delta\sigma_1$ and $\Delta\sigma_3$ are the increase in major and minor principal stress respectively at a depth z below the footing when the footing is loaded with a pressure Δp. A three dimensional non-linear finite element simulation was performed using the model and the parameters described previously. The results are shown on Fig. 12. These profiles appear to be relatively independent of the pressure level on the footing and of the sand density. Therefore, a unique simplified diagram is used (Fig. 13).

For any value of $\Delta R_c/R_c$, the pressures p_i for each of the pressuremeter curves within the depth of influence are averaged as follows to give the pressure p on the average PMT curve corresponding to this $\Delta R_c/R_c$:

$$p = \frac{1}{A} \Sigma\, p_i a_i \quad (7)$$

where a_i is the tributary area under the influence diagram for the PMT test at depth z_i; and A is the total area under the diagram (A = 1.125). In order to obtain the a_i values, the depths at which the PMT tests were performed are identified, then boundaries are established at mid-height between two consecutive PMT test depths, and a_i values are calculated as the areas between the boundaries (Fig. 13). This process leads to the average PMT curve for that footing (Figs. 9 and 10).

In order to transform the average PMT curve into the footing curve, the following was considered. First, the PMT limit pressure which is found at $\Delta R_c/R_c$ = 42% is generally associated with the bearing capacity of the footing p_u defined here as the footing pressure reached at a settlement over width ratio (s/B) equal to 10%. In order for those two points to match, the $\Delta R_c/R_c$ axis is transformed into

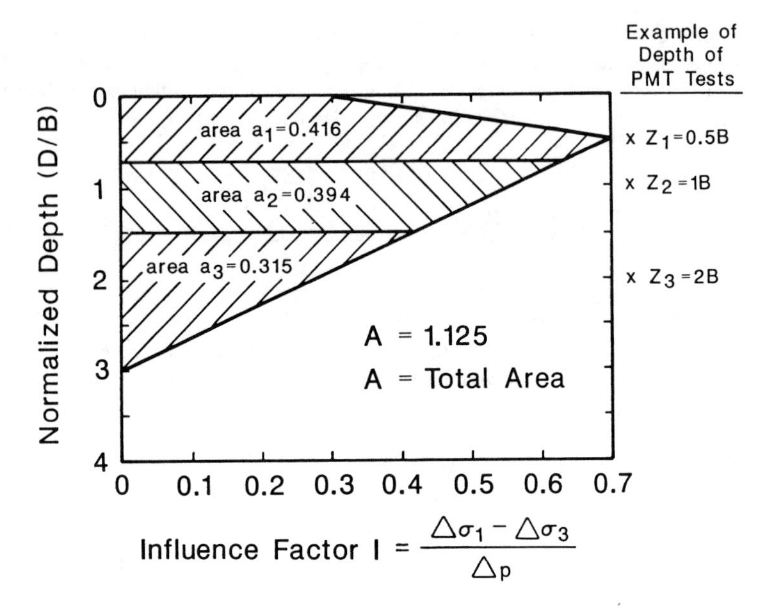

FIG. 13. Recommended Influence Factor Distribution, and Example of the Determination of the a_i Factors.

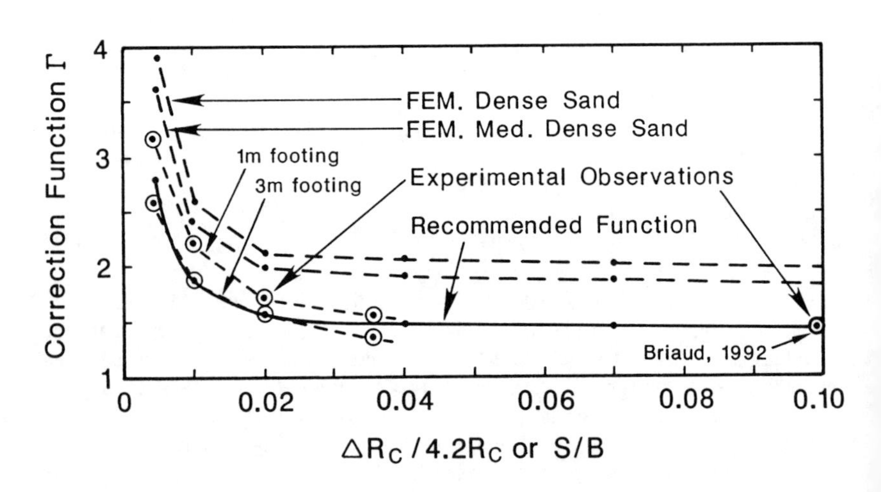

FIG. 14. The Correction Function Γ(s/B)

a ($\Delta R_c/4.2R_c$) axis. Second, a three dimensional nonlinear finite element simulation of the footing and of PMT tests at 0.5B, 1B and 2B was performed for both sand densities (D_r = 55% and 95%). For the pressuremeter tests, both the preboring and selfboring tests were simulated. The mesh was first validated against the elastic theory, then the geostatic stresses were turned on in the whole mass. Then for the selfboring test the geostatic stresses which would have existed against the borehole wall had the borehole not been drilled were applied against the borehole wall (this step was bypassed for the simulation of the preboring test). Then the pressure exerted by the pressuremeter probe was applied in increments on the borehole wall. This allowed to generate the PMT curves as p versus ($\Delta R_c/4.2R_c$) mentioned above at the depths 0.5B, 1B, and 2B below the 3 m × 3 m footing. The weighted average of these 3 curves was determined using the influence factor process described earlier. The weighted average PMT curves for both pressure meter types are shown on Figs. 9 and 10 together with the 3 m × 3 m footing pressure versus relative settlement curve. These curves allowed to develop the transformation function Γ to go from the weighted average PMT curve to the footing pressure vs relative settlement curve.

The transformation is performed by using the function Γ as follows for each point on the average PMT curve:

$$\frac{s}{B} = \frac{\Delta R_c}{4.2R_c} \tag{8}$$

$$p_{footing} = \Gamma p_{pmt} \tag{9}$$

where s is the footing settlement, B the footing width, ($\Delta R_c/R_c$) and p_{pmt} correspond to a point on the average PMT curve, $p_{footing}$ is the pressure on the footing corresponding to s/B, and Γ the transformation function which depends on s/B. The function Γ was obtained from the finite element simulation by taking the ratio of $p_{footing}/p_{pmt}$ for a given s/B (Figs. 9 and 10). The results are shown on Fig. 14 for the preboring pressuremeter and for the medium dense and dense sand.

At the limit pressure where ($\Delta R_c/4.2R_c$) and s/B are equal to 0.1, it is known from footing load tests (Briaud 1992) that for sand, the Γ factor is about 1.4 as shown on Fig. 14 for relative embedments (D/B) between 0.25 and 0.75. Indeed in this case the Γ factor is the bearing capacity factor k used in the design of spread footings based on pressuremeter data. This is lower than the FEM Γ value which is about 2. Furthermore, the results of the full scale load tests which are described later allowed to generate the Γ values shown on Fig. 14 for the 3 m footing and the 1 m footing. The FEM results, the full scale tests performed in this study, and the previous experimental data at s/B = 0.1 led to the choice of a recommended conservative function Γ (solid line on Fig. 14). Because the load tests results were known at the time of the prediction and because they were used to develop this method, the following section is not a class A prediction. The true value of this new method can only be evaluated on future class A predictions. There is a significant difference between the theoretical FEM values of Γ and the experimental values of Γ (although the general shape is the same). This difference is yet unexplained: research is continuing. The recommended function Γ (solid line on Fig. 14) was used for the calculations in the next section.

In order to predict the settlement as a function of time for a given pressure, the model proposed by Briaud (1992) is used.

$$s(t) = s(1\ \text{min}) \left(\frac{t}{1\ \text{min}} \right)^{n_{footing}} \tag{10}$$

where s(t) is the settlement after a time t in minutes, s(1 min) is the settlement predicted by the proposed method, and $n_{footing}$ is the time exponent. The value of $n_{footing}$ is related to the value of n_{pmt} obtained from a creep step towards the end of the linear phase in the pressuremeter test; the pressure is held constant and the relative increase in radius is recorded as a function of time (Briaud 1992). The modulus is then plotted as shown on Fig. 15 and the values of the time exponent n_{pmt} are given by the slopes of those lines. The relationship between $n_{footing}$ and n_{pmt} is discussed later.

APPROACH BY ANALOGY: PREDICTING THE LOAD-SETTLEMENT CURVE

As pointed out in the previous section, the load test results were known at the time of the predictions. Therefore, the predictions are not class A predictions. Furthermore, the load test results were used to develop the method, therefore the predictions do not allow to evaluate the method. They do, however, allow to illustrate the process.

The load settlement curves were generated using this method for two of the 5 large spread footings tested in the fall of 1993 at the National Geotechnical Experimentation Site at Texas A&M University (Briaud 1993; Gibbens and Briaud 1993). A series of preboring pressuremeter tests were performed in 4 borings. The hole was prepared with a hand auger in the dry down to 4 m. The deeper tests were done in holes prepared by the wet rotary method with a drilling rig. The results are summarized in Fig. 16, with profiles of first load modulus (E_o), reload modulus (E_r) and limit pressure (p_L).

The step-by-step procedure for the PMT load-settlement curve method was followed:

- Correct the PMT curves to pressure on the cavity wall p versus relative increase of the cavity radius ($\Delta R_c/R_c$) by making use of Eq. (5).
- Select the PMT curves relevant to the problem. For this prediction, since several PMT soundings were performed in a small area, the PMT curve for any depth was the average of all the curves measured at that depth. A single profile of curves was generated in this manner for the site.
- For a given footing the PMT curves within the depth of influence (3B) were averaged according to the influence factor distribution (Fig. 13, Eq. (7) and Fig. 17). This led to the weighted average PMT curve for each footing giving p_{pmt} vs ($\Delta R_c/R_c$).

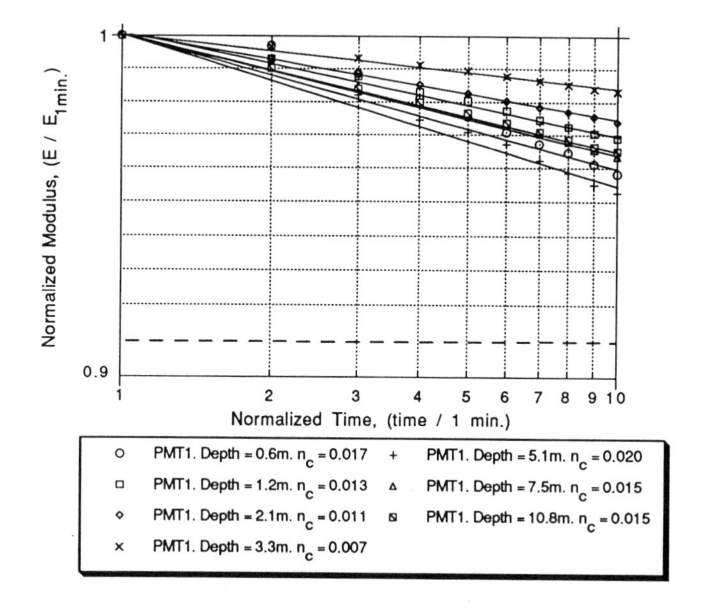

FIG. 15. Pressuremeter Modulus as a Function of Time

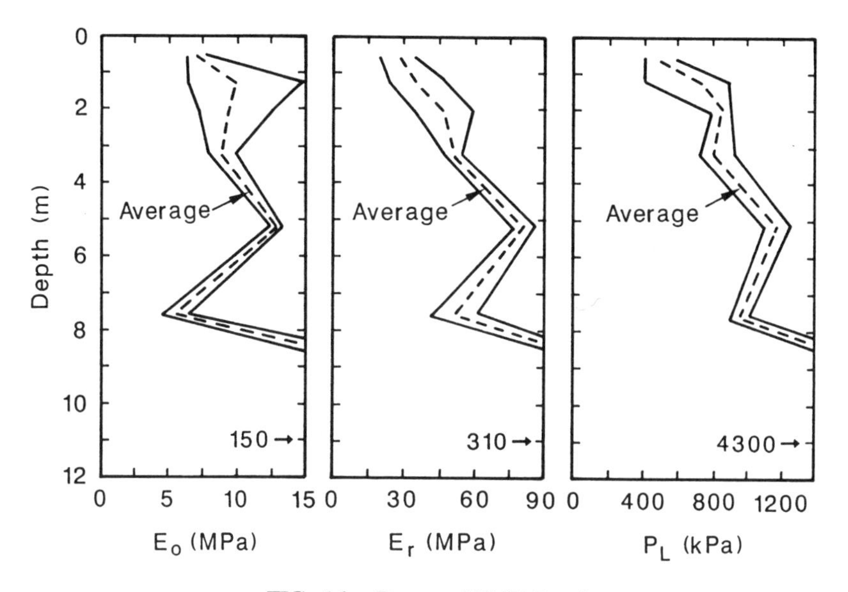

FIG. 16. Range of PMT Results

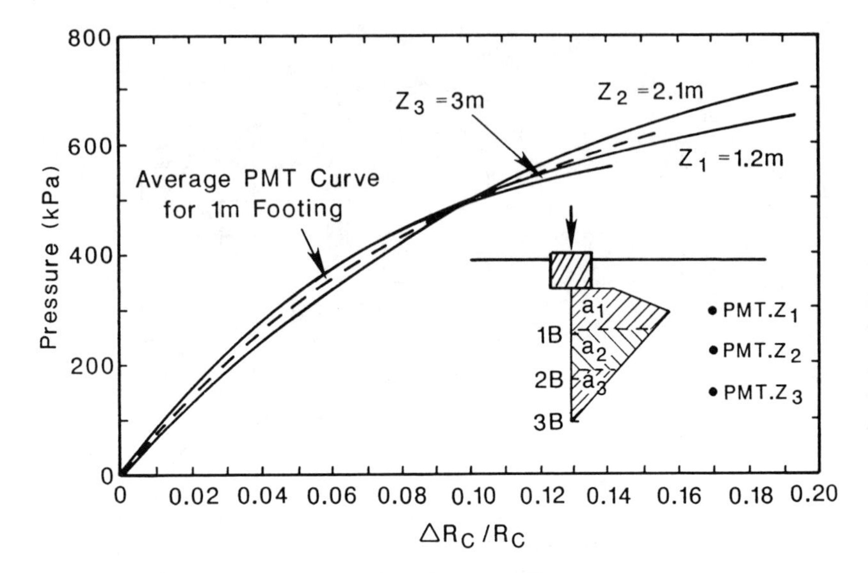

FIG. 17. Generating the Average PMT Curve for a Footing

- The weighted average PMT curve was then transformed into the footing curve p_{foot} vs s/B by using Eqs. (8) and (9). The function Γ recommended on Fig. 14 was the one used for the predictions.
- The predicted curves are shown on Fig. 18 for the 1 m and 3 m footings which were load tested. Note that the predicted curves stop at s/B = 0.035 because the pressuremeter tests were performed only to $\Delta R_c/R_c$ of 0.15. For future use the PMT test should be carried out to the maximum $\Delta R_c/R_c$ possible.
- The time rate of settlement is then estimated by using the power law model proposed in Eq. (10).

The TAMU Large Scale Footing Load Tests and the Scale Effect

Five spread footings were load tested at one of the two National Geotechnical Experimentation Sites at Texas A&M University. The footings were 1×1 m, 1.5×1.5 m, 2.5×2.5 m, 3×3 m (south) and 3×3 m (north) as shown on Fig. 3. They were all embedded 0.75 m. The load was applied vertically at the center of each footing and the settlement recorded after 30 minutes at each corner of the footing and averaged. The pressure vs settlement curves are shown on Fig. 19. Note that the curve for the 3 m South footing is much weaker than the 3 m North footing because the soil was much weaker under the 3 m South footing than under the 3 m North footing as given by the CPT profiles.

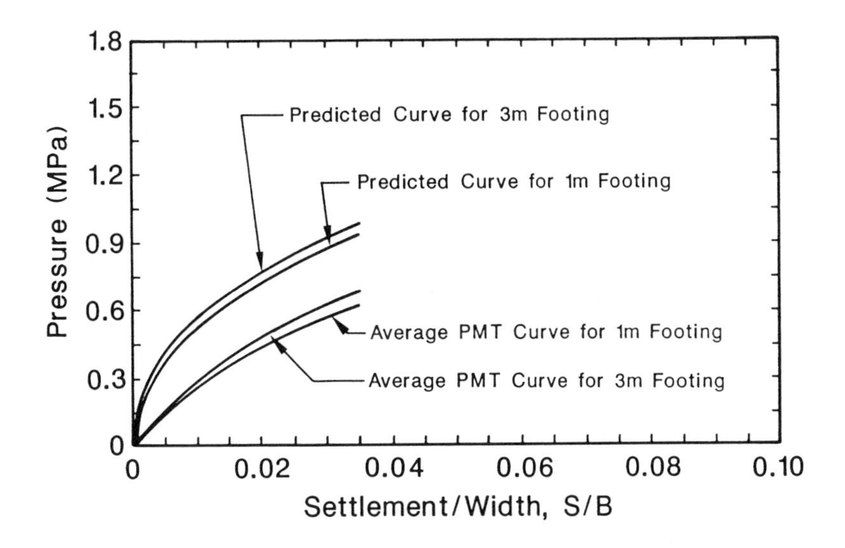

FIG. 18. Predicted Footing Response by Proposed PMT Method

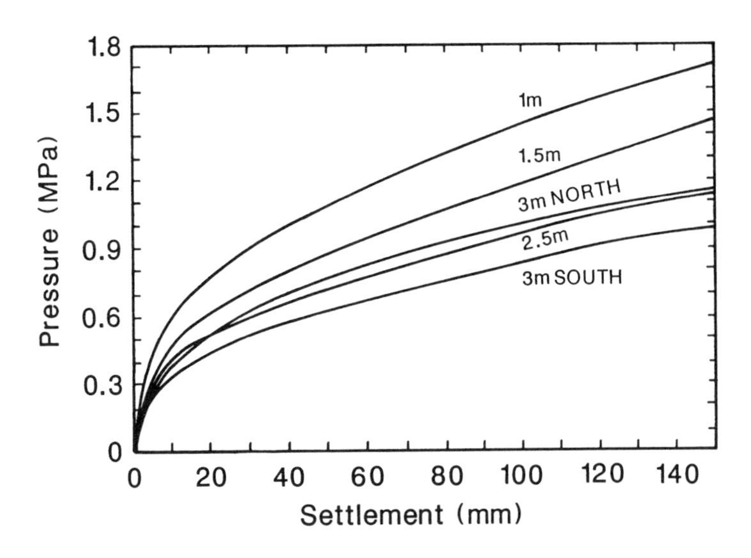

FIG. 19. Pressure vs Settlement Curves for the NGES-TAMU Load Tests

The scale effect on bearing capacity has been acknowledged for spread footings on sand. If bearing capacity is defined as the pressure reached for a displacement of 150 mm, the scale effect is obvious (Fig. 19); indeed the larger the footing, the smaller the pressure that can be applied before reaching a given settlement. In Fig. 20, this is illustrated by the curve with solid dots, which shows the classical shape (Meyerhof 1983).

Now let us imagine that one is performing a 1 m diameter and a 3 m diameter triaxial test where the top platen is a spread footing (Fig. 21), while the confining pressure and the sand used are the same. One would expect to obtain the same stress-strain curve for both tests regardless of the scale. Therefore one would expect in either case the same stress for the same strain. However, if one was to plot the stress-displacement curves, different curves would be obtained and the stress read at the same displacement would depend on the size of the footing and would be lower for the larger footing.

This was the clue which led to comparing the bearing capacity at the same relative s/B, a measure of the average strain below the footing. If the bearing capacity is defined at a settlement over width (s/B) ratio equal to 0.05 the scale effect disappears (Fig. 20). In fact, if the load settlement curves are plotted as pressure vs settlement/width curves, the curves almost converge to a unique curve (Fig. 22). This proves that there is no scale effect if the load tests results are plotted as "stress-strain" curves. This experimental finding was verified by FEM runs.

A literature search conducted after this finding led to a discussion by Osterberg (1947) who presented load-settlement curves for plate tests of varying diameters (0.25, 0.5, 0.75, 1 m) on clay. Osterberg shows that if the data is plotted as pressure versus settlement over diameter, the load-settlement curves for the 3 largest plates collapse into one curve while the load-settlement curve for the 0.25 m plate remains somewhat stiffer. Palmer (1947) also presents the load-settlement curves for 3 plates with varying diameters (0.38, 0.61, 0.76 m) on a pavement. If those curves are plotted as pressure versus settlement over diameter, they collapse into one unique curve. Burmister (1947) goes on to propose a method to obtain the load-settlement curve for a footing from a triaxial stress-strain curve. Skempton (1951) dealing with spread footings on clay also proposed a method to obtain the load settlement curve from triaxial test results and showed, using linear elasticity, that the average vertical strain under the footing is equal to S/2B.

If the triaxial test analogy of Fig. 21 points in the direction of the uniqueness of the p vs s/B curve, one factor points in the other direction; the depth of influence of a small footing is much shallower than the depth of influence of a large footing. Since the stiffness of a sand is dependent on the mean stress level the large footing should exhibit a somewhat stiffer p vs s/B curve than the small footing. For the footings tested, this does not seem to be a major factor.

Note that if bearing capacity is defined as the pressure for an s/B ratio of 0.1 (1.5 MPa on Fig. 22) and if a factor of safety of 3 is applied to obtain a safe pressure (0.5 MPa), then the safe pressure corresponds to an s/B ratio equal to 0.007. For all the footings in this study, this leads to a settlement smaller than 25

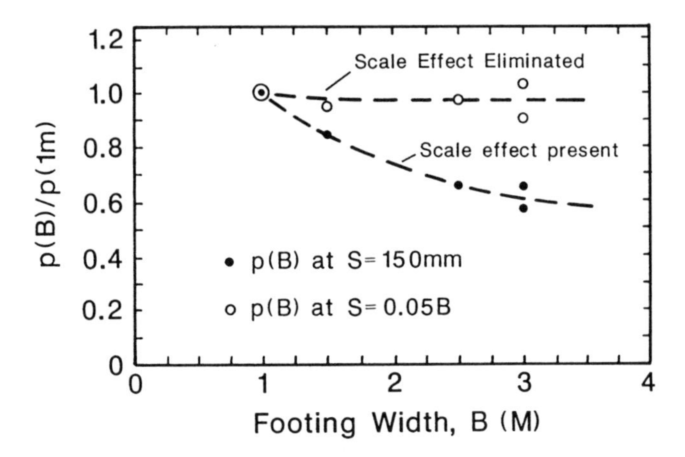

FIG. 20. Scale Effect at Large Displacement: NGES-TAMU Load Tests

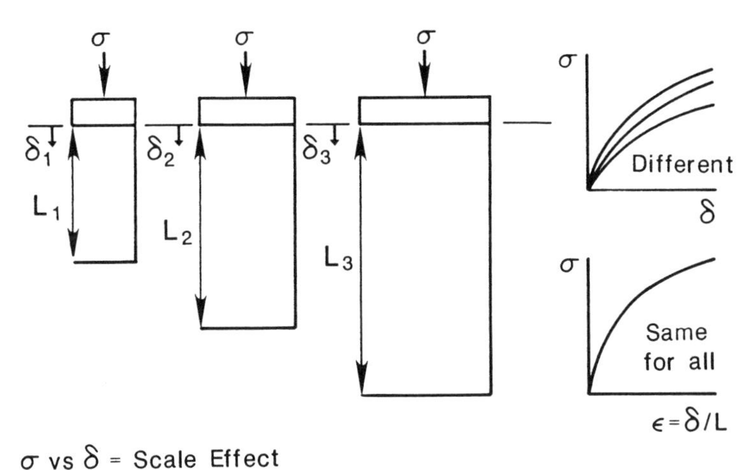

FIG. 21. Triaxial Test/Spread Footing Analogy

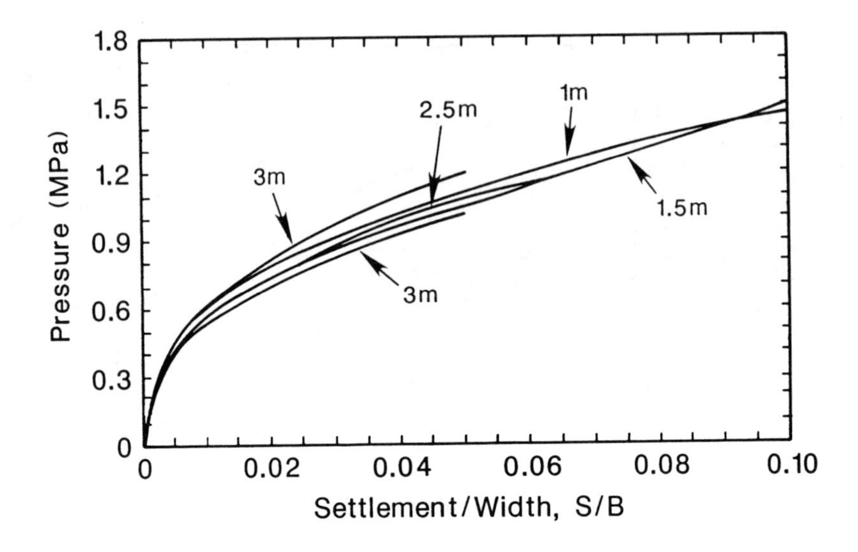

FIG. 22. Pressure vs Settlement/Width Curves for Footing Tests

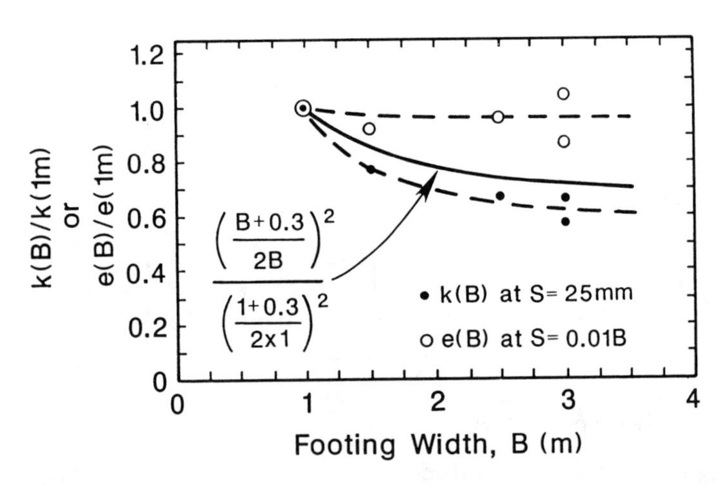

FIG. 23. Scale Effect on Modulus of Subgrade Reaction (NGES-TAMU Load Tests)

mm. Therefore, for those 5 footings the bearing capacity criterion would control the design, not the settlement criterion. This is contrary to common belief for footings on sand but is due to the definition used for the bearing capacity (s/B = 0.1) rather than plunging failure (point A instead of point C on Fig. 1).

If the p vs s/B curve is unique for a given deposit then the bearing capacity p_u is independent of the footing width. For footings at the surface of a sand, the general bearing capacity equation gives:

$$p_u = \frac{1}{2}\gamma B N_\gamma \quad (11)$$

where γ is the unit weight and N_γ a parameter depending only on the sand strength. If $\frac{1}{2}\gamma BN\gamma$ is a constant independent of B, then N_γ cannot be a constant and must carry a scale effect in 1/B similar to the classic trend on Fig. 20. This major shortcoming plus the difficulty in obtaining an accurate value of the needed soil parameter ϕ and the documented poor accuracy of this method (Amar et al. 1984) lead to the recommendation that the use of this equation should be discontinued. A doubt can be cast on the above recommendation if it is argued that equation 11 gives a pressure which corresponds to plunging failure and not to s/B = 0.1 (point C instead of point A on Fig. 1), and that the load-settlement curves could diverge past s/B = 0.1 (end of the experimental data) to reestablish the validity of Eq. (11).

By comparing the bearing capacity value p_u read at an s/B ratio equal to 0.1 with the SPT blow count N and CPT point resistance q_c taken as averages within the zone of influence of the footings, it appears that the following rules of thumb are reasonable for the medium dense sand at this site:

$$p_u = \frac{N}{12} \text{ with N in blows/0.3m and } p_u \text{ in MPa} \quad (12)$$

$$p_u = \frac{q_c}{4} \quad (13)$$

One can also give the following approximations for the pressure p_a which leads to an s/B ratio equal to 0.01:

$$p_a = \frac{N}{36} \text{ with N in blows/0.3 m and } p_a \text{ in MPa} \quad (14)$$

$$p_a = \frac{q_c}{12} \quad (15)$$

Another acknowledged scale effect phenomenon exists at the settlement level. Terzaghi and Peck (1967) and NAVFAC (1982) give the following relationship:

$$k_{(B)} = k_{(0.3m)} \left[\frac{B+0.3}{2B}\right]^2 \quad (16)$$

where $k_{(B)}$ and $k_{(0.3m)}$ are the moduli of subgrade reaction for footings of diameter B (in meters) and 0.3 m respectively. The parameter $k_{(B)}$ is defined as the pressure on the footing divided by the settlement of the footing under that pressure; the unit is kN/m^3. The values of $k_{(B)}$ can be calculated for a given settlement (25 mm) for the five footings tested. Fig. 23 shows that in this case there is a scale effect and that this scale effect is well represented by the function $[(B+0.3)/(2B)]^2$. If however, the values of $k_{(B)}$ are compared at the same s/B or better if a new parameter e is defined as

$$e = \frac{p}{s/B} \tag{17}$$

and if $e_{(B)}$ values are compared for the same s/B, then the scale effect disappears. In fact, e is closely related to the soil modulus E. Indeed in elasticity:

$$e = \frac{p}{s/B} = \frac{E}{I(1-\nu^2)} \tag{18}$$

Therefore, for square footings, e is independent of B while k is dependent on B. The use of k should be discontinued because it induces an unnecessary difficulty and e, a true modulus of subgrade reaction should be used in its place because it essentially represents a soil property.

These observations and Fig. 22 point in the direction of the uniqueness of the p vs s/B curve for a given sand deposit. If this finding is corroborated by future research it will greatly simplify the prediction of spread footing behavior. Note that the proposed method by analogy satisfies this basic observation and induces scale effect only if the deeper layers involved by the larger footings are stronger or weaker.

If one recalls the heterogeneity of the sand deposit as shown by the CPT soundings (Figs. 3 and 6), it may be surprising to see how well behaved the pressure versus relative settlement curves are for the 5 footings (Fig. 22). In fact the heterogeneity decreases when going from the CPT (Fig. 6) to the SPT (Fig. 5), to the PMT (Fig. 16), and to the footings (Fig. 22). This is attributed to the gradual increase in scale and in the volume of soil tested from one type of test to the next. This tends to show that a sand deposit which is apparently very heterogeneous at the scale of the CPT point (36 mm) may be quite homogeneous at the scale of a spread footing (3000 mm). The implication is that reasonably accurate predictions should be possible even for apparently heterogeneous deposits, that differential settlement between adjacent footings may not be as large as calculated on the basis of separate borings, and that the test which involves the largest soil volume has an advantage.

COMPARISON BETWEEN MEASURED AND PREDICTED BEHAVIOR

The comparison between the FEM class A predictions and the load test on the 3 m footings shows a drastic underprediction (Fig. 24) for the medium dense sand, $D_r = 55\%$, which was determined as the in situ relative density based on SPT and CPT correlations. Interestingly, the prediction for the dense sand ($D_r = 95\%$) comes much closer to the measurements although the shape of the curve is still not

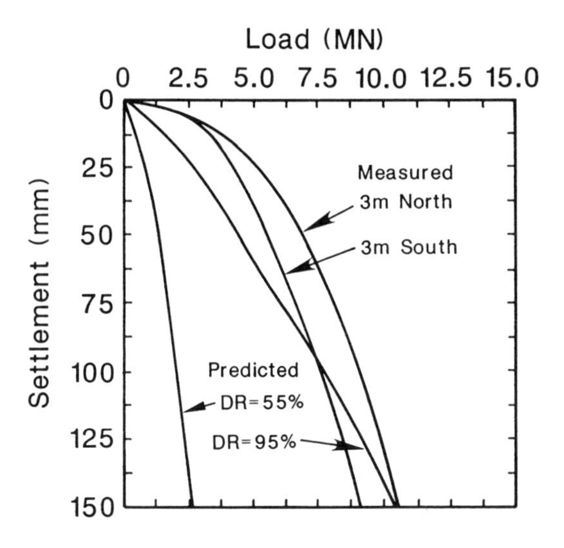

FIG. 24. Measured versus FEM Predicted Load Settlement Curves

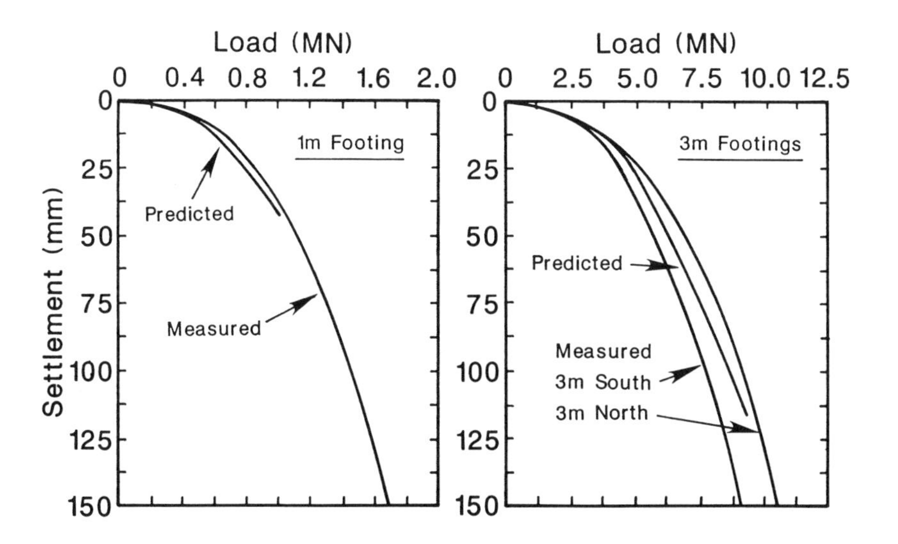

FIG. 25. Measured versus PMT Predicted Load Settlement Curves

the right one. Improvements in the approach taken would be to include the influence of the strain level in the pre-yield behavior or more generally to choose a better constitutive model. This may help in obtaining the proper shape of the curve.

Another improvement would be to obtain the model parameters not only by matching the triaxial test results but also and maybe more importantly by matching in situ tests, like the pressuremeter test curve, which are more likely to represent the in situ conditions. Indeed, while better models can always be used, it appears that the major weakness of this technique, aside from the difficulties encountered in using the procedure, is the uncertainty in reproducing a representative sample. The major advantage, however, is its tremendous potential in solving more complex problems such as inclined eccentric loadings near a slope. Therefore the FEM should be the method of choice for the future. However, the single most important step in drastically improving the FEM approach is not the improvement of the models but finding a routine way to take and test undisturbed samples of sand.

The comparison between the method by analogy and the load tests is very good (Fig. 25) as can be expected since the function Γ was chosen after analysis of the full-scale tests. This new method needs to be checked against other data and may represent an improvement in the short term while the FEM problems are being solved. The creep settlement of the 3 m footing during load holding tests can be compared with the creep pressuremeter test results by comparing the time exponents obtained from the footing tests $n_{footing}$ and the pressuremeter tests n_{pmt}. This was done and led to an average n_{pmt} value of 0.014 and an average $n_{footing}$ value of 0.027. Therefore while waiting for further experimental evidence and for an analytical explanation of this difference, it is suggested that

$$n_{footing} = 2n_{pmt} \tag{19}$$

Conclusions

It must be kept in mind that the following conclusions are based on 5 load tests on 1 m to 3 m wide footings which were pushed to a maximum settlement of 0.15 m and on associated theoretical analyses:

(1) The load tests have shown that there is no scale effect if comparisons are made for the same average strain in the soil mass below the footing. This average strain is conveniently represented by the ratio s/B. In the future all spread footing load settlement curves should be presented as pressure versus settlement over footing width. This p vs s/B curve appears to be unique and only related to the sand characteristics within the depth of influence of the footing.

(2) A corollary of conclusion (1) is that the use of the general bearing capacity equation should be discontinued; indeed this equation does not describe properly the findings in conclusion 1. Also the equation which involves the term $[(B+0.3)/(2B)]^2$, while reasonably accurate, induces an unnecessary complexity by comparing settlement values instead of settlement over width ratios. A new modulus of subgrade reaction concept is proposed.

(3) The load test results indicate that if the bearing capacity p_u is defined as the pressure reached for a settlement equal to 10% of the width (s/B = 0.1) then the bearing capacity criterion using a factor of safety of 3 controls the design, not the 25 mm settlement criterion.

(4) The bearing capacity p_u defined as the pressure at s/B = 0.1 satisfies the following approximate relationships for all footing sizes:

$$p_u = \frac{N}{12}$$ with N in blows per 0.3m and p_u in MPa

$$p_u = \frac{q_c}{4}$$

the pressure p_a defined at s/B = 0.01 leads to:

$$p_a = \frac{N}{36}$$ with N in blows per 0.3m and q_c in MPa

$$p_a = \frac{q_c}{12}$$

(5) The heterogeneity of a deposit is a concept relative to the scale of the volume of soil involved in the test. The CPT soundings showed significant heterogeneity. This heterogeneity was less significant for the SPT, even less for the PMT and nearly disappeared for the footing tests which showed very little scatter in the p versus s/B curves. As a result the differential settlement between adjacent footings will be less than calculated using a CPT boring at each footing location.

(6) The finite element method leads to a drastic under-prediction when using the estimated in situ relative density (55%). This is attributed to the uncertainty in preparing a representative sand sample. The comparisons improved when using a much higher relative density (95%). It is felt that the FEM approach is the method of choice but that one cannot take full advantage of it until a routine way of taking and testing undisturbed samples of sand is found. An alternative is to back figure the model parameters by matching in situ tests rather than laboratory tests.

(7) A method by analogy is proposed to obtain the complete load-settlement-time curve for a square spread footing loaded vertically in its center, on flat ground and with relative embedments from 0.25 to 0.75. This method makes use of the complete pressuremeter curve and of a transformation function Γ based on a combination of theory and load test data. This method obeys the finding in conclusion 1 and benefits from the analogy between a cavity expansion test (the pressuremeter) and a cavity expansion problem (the spread footing penetration). It performed well in predicting the NGES-TAMU spread footing load test curves since that data was used to develop the method.

(8) The method by analogy is limited to soil conditions where a suitable borehole can be drilled for the pressuremeter. This rules out soft clays and loose sands (especially under the water table). However, these are soil conditions where spread footings are not usually used. Note that this method is relatively inexpensive since a drill rig may not be necessary. Only shallow tests are required and a hand auger may be all that is needed.

ACKNOWLEDGMENTS

This project was completely sponsored by the Federal Highway Administration (Mr. Albert DiMillio) through Geotest Engineers Inc. (Dr. Vijay Vijayvergiya). Mr. Robert Gibbens was in charge of the spread footing tests performed at the National Geotechnical Experimentation Site. This site is jointly sponsored by the National Science Foundation (Dr. Mehmet Tumay) and the Federal Highway Administration (Mr. Albert DiMillio). Mr. George Nasr contributed significantly in the pressuremeter prediction calculations and Mr. Kabir Hossain made additional FEM runs.

APPENDIX - REFERENCES

ABAQUS (1991). "User's and Theory Manuals: Version 4.9," Hibbitt Karlsson and Sorensen Inc., Rhode Island.

Alpan, I. (1964). "Estimating the settlements of foundations on sands." *Civil Engineering and Public Works Review,* November, London, U. K.

Amar, S., Baguelin, F., and Canepa, Y. (1984). "Etude experimentale du comportement des fondations superficielles." *Annales de l'ITBTP*, Paris.

Briaud, J. L. (1992). *The Pressuremeter*, Balkema, Rotterdam, The Netherlands.

Briaud, J. L. (1993). "The National Geotechnical Experimentation Sites at Texas A&M University: Data Collected until 1992," *Research Report*, Civil Engineering, Texas A&M University.

Briaud, J. L., and Miran, J. (1992). "The cone penetrometer test," Report No. FHWA-SA-91-043, Federal Highway Administration, U. S. Dept. of Transportation, Washington, D. C.

Burland, J. B., and Burbidge, M.C. (1985). "Settlement of foundations on sand and gravel." *Proc. Inst. of Civil Engrs.*, (Part 1), 78, London, 1325-1381.

Burmister, D. M. (1947), Discussion in Symposium on Load Tests of Bearing Capacity of Soils, *ASTM STP No. 79*, ASTM, Philadelphia, Pennsylvania, 139-146.

Cosentino, P. J. (1987). "Pressuremeter moduli for airport pavement design," Ph.D. dissertation, Civil Engineering, Texas A&M University.

D'Appolonia, D. J., D'Appolonia, E., and Brissette, R. F. (1968). "Settlement of spread footings on sand." *J. Soil Mech. Found. Div.*, Proc. ASCE, 94(SM3), 735-760.

DeBeer, E. E. and Martens, A. (1957). "Method of computation of an upper limit for the influence of the heterogeneity of sand layers on the settlement of bridges." *Proc. 4th ICSMFE*, London, 1, 276-282.

Drücker, D. C. and Prager, W. (1952). "Soil mechanics and plastic analysis or limit design." *Quarterly of Applied Mathematics*, 10(2), 157-175.

Duncan, J. M. and Chang, C. Y. (1970). "Nonlinear analysis of stress and strain in soils." *J. Soil Mech. Found. Div.*, Proc. ASCE, 96(SM5).

Gibbens, R., and Briaud, J. L. (1993). "Data and prediction request for the Spread Footing Prediction Event sponsored by FHWA at the occasion of the ASCE Specialty Conference: Settlement 94," Civil Engineering, Texas A&M University.

Jeanjean, Ph. (1993). "Load settlement curves for spread footings on sand from the pressuremeter test," Ph.D. dissertation, Civil Engineering, Texas A&M University.

Jeyapalan, J. K., and Boehm, R. (1986). "Procedures for predicting settlement in sands." *Settlement of Shallow Foundations on Cohesionless Soils: Design and Performance*, Geotechnical Special Publication No. 5, ASCE, New York, New York.

Marchetti, S. (1980). "In situ tests by flat dilatometer." *J. Geotech. Eng. Div.*, Proc. ASCE, 106(GT3), 299-321.

Menard, L., and Rosseau, J. L. (1962). "L'evaluation des tassements-tendances nouvvelles." *Sols-Soils*, 1(1).

Meyerhoff, G. G. (1965). "Shallow foundations." *J. Soil Mech. Found. Div.*, Proc. ASCE, 91(SM2).

Meyerhoff, G. G. (1983). "Scale effects of ultimate pile capacity." *J. Geotech. Eng.*, ASCE, 109(6), 797-806.

NAVFAC (1982). *Soil Mechanics, Design Manual 7.1*, U. S. Dept. of Navy, U. S. Govt. Printing Office, Washington D. C.

Osterberg, J. O. (1947), Discussion in Symposium on Load Tests of Bearing Capacity of Soils, ASTM STP 79, ASTM, Philadelphia, Pennsylvania, 128-139.

Oweis, J. S. (1979). "Equivalent linear model for predicting settlement of sand bases." *J. Geotech. Eng. Div.*, Proc. ASCE, 105(GT12), 1525-1544.

Palmer, L. A. (1947), "Field loading tests for the evaluation of the wheel load capacities of airport pavements." ASTM STP 79, ASTM, Philadelphia, Pennsylvania, 9-30.

Parry, R. H. G. (1971). "A direct method of estimating settlements in sand from SPT values." *Proc. Symp. Interaction of Structures Foundations*, 29-37.

Peck, R. B., and Bazaraa, A. S. (1969). "Discussion: Settlement of spread footings on sand." *J. Soil Mech. Found. Div.*, Proc. ASCE, 95(SM3), 905-909.

Peck, R. B., Hanson, W. E., and Thornburn, T. H. (1974). *Foundation Engineering*, 2nd Edition, John Wiley & Sons, New York, New York.

Sanglerat, G. (1972). *The Penetrometer and Soil Exploration*, Elsevier, Amsterdam, The Netherlands.

Schmertmann, J. H. (1970), "Static cone to compute static settlement over sand." *J. Soil Mech. Found. Div.*, Proc. ASCE, 96(SM3), 1011-1043.

Schmertmann, J. H. (1978). "Guidelines for cone penetrometer test: Performance and design," *Report No. FHWA-TS-78-209*, Federal Highway Administration, U. S. Dept. of Transportation.

Schultze, E., and Sherif, G. (1973). "Prediction of settlements from evaluated settlement observation for sand." *Proc 8th ICSMFE*, Moscow, 1.3, 225-230.

Seed, H. B., and Reese, L. C. (1957). "The action of soft clay along friction piles." *Trans. ASCE*, Paper 2882, 731-764.

Skempton, A. W., (1951), "The bearing capacity of clays." *Proc. Building Research Congress*, Institution of Civil Engineers, London, Division I, 180-189.

Tan, C. K., and Duncan, J. M. (1991). "Settlement of footings on sand, accuracy and reliability." *Geotechnical Engineering Congress*, Geotechnical Special Publication No. 27, ASCE, New York, New York, 2, 446-455.

Terzaghi, K., and Peck, R. B. (1967). *Soil Mechanics in Engineering Practice*, 2nd Edition, John Wiley & Sons, New York, New York.

Wong, K. S., and Duncan, J. M. (1974). "Hyperbolic stress-strain parameters for non-linear finite element analyses of stresses and movement in soil masses," *Report No. TE-74-3*, submitted to National Science Foundation, Civil Engineering, University of California at Berkeley.

DOWNDRAG ON PILES: REVIEW AND RECENT EXPERIMENTATION

John Anthony Little[1]

ABSTRACT: A review of the literature on pile downdrag is provided, within the context of a recent full-scale study involving two pile groups driven in soft clay, subsequently surcharged by a 2.5 m high embankment.

INTRODUCTION

Whilst the phenomenon of downdrag on piles has long been recognised, it is only during the last three decades that systematic studies on negative skin friction and its effects on the behaviour of piles have been carried out.

In this paper, a brief review is provided of the studies carried out by investigators on both full-scale and model piles. In addition, some of the existing theories for determining negative skin friction on single piles and pile groups are outlined. A description is given of recent full-scale experimentation at the U.K. Science and Engineering Research Council soft-clay site at Bothkennar, Scotland, and conclusions are drawn regarding the observed field measurements.

REVIEW OF LITERATURE ON NEGATIVE SKIN FRICTION: THEORY

In the effective stress method the maximum negative skin friction F_{ns} is calculated according to the following equation:

$$F_{ns} = K\sigma_v \tan\phi' \qquad (\beta = K\tan\phi') \tag{1}$$

Full mobilisation of the negative skin friction along the entire length of the pile is assumed. It is applicable when significant relative movement or slippage occurs between pile and soil. An overestimation of the negative skin friction will occur when the surface settlement is not sufficient to provide full slippage.

[1] Prof. and Head of Civil Engrg., University of Paisley, Paisley, PA1 2BE, Scotland.

In the total stress method maximum negative skin friction is evaluated from the initial undrained shear strength of the soil (c_u) according to the equation

$$F_{ns} = \alpha c_u \tag{2}$$

where α is an adhesion factor which expresses the mobilised pile/soil adhesion as a ratio of the undrained shearing resistance of the soil. The value of α depends on the properties of the pile material and of the surrounding soil, the stress history of the soil, and the method of pile installation.

Other methods of analysis have been proposed on the basis of elastic theory. The pile-soil system is analysed as a continuous elastic medium using either Mindlin's (1936) equation or a finite element formulation. Solutions developed from the theory of elasticity can only be used at small ground settlements before the maximum skin friction resistance has been mobilised (i.e. when partial slip occurs), close to the ground surface. In addition, difficulties are experienced in applying these solutions to non-homogeneous soil profiles.

Terzaghi and Peck (1948, 1967) proposed one of the earliest analytical approaches. In essence, this is a conservative method since full mobilisation of the soil shear strength at the pile-soil interface along the full length of the individual piles in a pile group or along the perimeter of the pile group is assumed. Hence, the neutral point (where relative movement of pile/soil is zero) is assumed to be located at the bearing stratum.

Zeevaert (1959) suggested a theoretical approach for the evaluation of downdrag forces on end bearing piles which takes into account the 'hang-up' tendency of the settling soil.

Buisson et al. (1960) noted that the shear stress along the pile must be related to the shear strain and that at some point along the pile the relative movement would be zero, and no load would be transferred between soil and pile. This method was essentially a refinement of Terzaghi's method (1948), except for the recognition of the existence of the neutral point.

Brinch Hansen (1968) postulated a theoretical method for calculating the skin friction for single piles and pile groups and applied the results to negative skin friction. Brinch Hansen applied the above method to the data obtained from the pile tests reported by Johannessen and Bjerrum (1965). Close agreement between the measured and calculated values was reported.

Fellenius (1971,1972) presented a general discussion on the determination of the allowable working load on piles and suggested that many factors, including negative skin friction, should be considered in the design. It was shown that negative skin friction was a settlement problem and not a failure problem.

Bozozuk (1972) reported that downdrag could be related to the horizontal effective stress acting on the pile surface. The position of the neutral point, where the downdrag was a maximum, was observed to be a function of the friction angle between the soil and the pile, the angle of internal friction of the soil, the submerged unit weight of the soil, and the coefficient of at-rest earth pressure.

Bjerrum (1973) also recognised that negative skin friction was a function of the effective horizontal stresses acting on the pile. No recognition, however, was made of the reduction in the vertical stresses by virtue of soil resting on the pile.

Garlanger and Lambe (1973) and Lambe et al. (1974) proposed a method for the prediction of pile downdrag based on an effective stress analysis.

Burland (1973) adopted the effective stress approach for calculating the shaft resistance of piles in clay. Based on the analysis of a large number of pile tests carried out on a variety of soft clays, he demonstrated that the ratio between the average shaft friction and the mean effective overburden pressure τ_s/p (= ß) lies between about 0.25 and 0.40, irrespective of the clay type. This represented a very much smaller spread than the equivalent α values, which typically would lie between 0.5 and 1.6. The author suggested that the approach could also be used to estimate negative skin friction, essentially being the same method as that adopted by Johannessen and Bjerrum (1965). An upper limit of ß=0.25 was therefore proposed by Burland for negative skin friction on piles in soft clay.

Tomlinson (1975) suggested a design method for calculating the magnitude of the downdrag forces of single piles. Two cases were considered: a) when a pile is driven into a relatively incompressible layer or b) when the pile is driven in a compressible layer.

Broms (1969, 1976, and 1979) also proposed a method to calculate the increase of the axial load caused by negative skin friction. Three cases were recognised, depending on the spacing of the piles and the magnitude of the load causing the settlement.

Vesic (1977) reported that negative skin friction can be expected above the point of the pile where the relative downdrag displacement of the soil exceeds 15 mm. The intensity of the downdrag was considered to be proportional to the effective vertical ground stress. The skin resistance factor, N_s (i.e. ß), for ordinary, uncoated piles in soft compressible strata of clay and silts was suggested to fall in the 0.15- to 0.30 range, generally increasing with ϕ^e_u

Bowles (1988) suggested an approach for the estimation of downdrag force on piles based on the effective stress analysis. Two cases of common interest were investigated and the negative skin resistance force was calculated for cohesive fill overlying cohesionless soils, and for cohesive soil underlying cohesionless fill.

Briaud et al. (1991) proposed a method for predicting downdrag loads on individual piles within a pile group. The authors stated that at spacings larger than 5 diameters there is very little group effect and consequently, the group should be designed as a cluster of single piles. However, at spacings smaller than 2.5 diameters there is a definite group effect. It was suggested that the group should be designed for the following downdrag loads

$$
\begin{aligned}
F_{n(corner)} &= 0.75 \times F_{n(single)} \\
F_{n(edge)} &= 0.5 \times F_{n(single)} \qquad (3) \\
F_{n(interior)} &= q_o s^2
\end{aligned}
$$

where q_o = the surcharge applied to the ground surface; and s^2 = the centre to centre spacing.

In addition, the authors recommended that the neutral point for the group could be taken at the point for groups bearing on a perfectly rigid base and at a depth equal to 66% of the pile group length for friction pile groups with no point bearing resistance. It is interesting to note that the recommended pile spacing at which a group effect is believed to occur (i.e. 2.5 diameters), is the same as that reported by Koerner (1972) for his model tests on groups of piles in clay.

Model and Field Tests

Measurements of negative skin friction on instrumented piles have been carried out by many investigators both in the laboratory and the field.

Gant et al. (1958) reported the first field measurements of pile downdrag due to settlement of clayey silt and overlying fill around three instrumented Monotube steel piles, two vertical piles and one batter pile. The results clearly demonstrated the existence of large downdrag forces and indicated that the drag caused by the soil above the clayey silt was accumulated in the first three weeks, whereas the drag from the compressible clayey silt took place at a slower rate.

Tschebotarioff (1958) in a discussion following the article by Gant et al. (1958), suggested that only a part of the measured axial pile forces could have been caused by drag forces while the rest may have been due to high bending stresses in the piles.

El Masry (1963) conducted the first reported model pile tests for the determination of negative skin friction. Numerous tests were performed to evaluate effects such as varying the thicknesses of the compressible layer, its unit weight, and the water content, among others.

Johannessen and Bjerrum (1965) investigated negative skin friction on two steel piles, 0.47 m in diameter, with a length of about 55 m, which were driven to bedrock through a layer of soft marine clay of about 50 m thickness. The stress distribution along the piles was measured using a mechanical system of steel rods. Thirteen months after the placement of a fill, the total settlement of the clay was about 1.7 m, which had caused a corresponding total compression of 14.3 mm in the pile. The drag forces under the effect of the settling clay reached about 250 tons. The observed distribution of stresses in the pile indicated that adhesion between the pile and the soil was governed by the effective vertical stress.

Negative skin friction on two 82 m long, 1 m diameter composite piles was investigated by Bozozuk and Labrecque (1969). Piles were driven through a deep layer of soft clay to bedrock. The site was preloaded for one year with 4.6 m of fill, during which time about 1.5 m of settlement occurred. Subsequently another 6.4 m of fill was placed. After 6 months the steel piles were driven and filled with concrete.

The measurements obtained in this field study revealed some unexpected results. Initially in the six-week period following placement of the concrete, the piles elongated (about 1.5 mm) due to concrete hydration. Five months later this had reduced to about 1.0 mm as negative skin friction loads overcame the expansive forces of hydration. During this period the subsoils consolidated by more than 150 mm.

Endo et al. (1969) carried out in-situ measurements of negative skin friction using four kinds of uncoated steel pipe piles at a site in Tokyo. The piles were 43 m long, 600 mm diameter, driven at 10 m spacing to avoid any interference from adjacent piles. The piles were both friction and end-bearing. Although the data showed a considerable scatter, the authors suggested that average values for ß were between 0.2-0.35. The observed distribution of the axial force in each pile was approximately symmetrical about the neutral point.

Bjerrum et al. (1969) reported a major field testing program for the measurement of negative skin friction on steel pipe piles in Norway. The piles were driven to rock through deposits of soft clay at five different sites where settlements had occurred due to recent filling. A method of reducing the negative friction on piles by coating the piles with bitumen was reported. In all cases the drag loads were very large and caused yielding in the lower part of the piles. The settlements of the piles (before any load was applied) varied from 10 to 100 mm, depending on the conditions. Negative friction developed very quickly and only small relative movements were required to fully develop its maximum value which corresponded to 10-15% of the effective overburden pressure, corresponding to, ß=0.20 to 0.25.

Negative skin friction was investigated in two instrumented precast concrete piles by Fellenius and Broms (1969), Fellenius (1971) and Fellenius (1972). The hexagonal piles were about 55 m long with a cross sectional area of 80000 mm^2 and a circumference of 1.05 m. The soil at the test site consisted of 40 m of uniform normally consolidated clay through which the test piles were driven into underlying silt and sand.

During the driving, the clay close to the piles was remoulded and displaced. The soil reconsolidated due to pile driving, and the resulting settlements transferred load to the pile by negative skin friction. After 495 days the total drag load was about 55 tons, 30 tons of which corresponded only to reconsolidation of the clay due to pile driving. The corresponding negative friction was about 30% of the undrained shear strength or to about 10% of the effective overburden pressure.

Bozozuk (1972) measured the downdrag on a 305 mm diameter and 49 m long steel pipe pile "floating" in marine clay. The size and distribution of skin friction generated in the floating pile over a period of 5 years was reported and compared with predicted values based on in-situ horizontal effective stresses and the influence of embankment loading. The analysis of axial compressions of the pile after 5 years indicated that a peak compressive load of 140 tons developed at the neutral point 22 m below the top of the pile. Analysis indicated that there was little or no relation between the skin friction exerted on the pile and the in-situ shear strength of the soil. Where relative movement between pile and soil was small, and excess pore pressures had dissipated, skin friction approached, but did not exceed, the drained strength. On the other hand, where relative movements and excess pore pressures were large, mobilised skin friction corresponded to the remoulded strength of the soil.

Koerner (1972) studied the development of negative skin friction on model piles in organic clayey silt. The influence of pile batter, pile group spacing, soil-water content, and pile material on average negative skin friction was investigated.

Walker and Darvall (1973) investigated negative skin friction on two driven steel pipe piles of 0.76 m diameter. The soil profile consisted of about 6 m of sand then 15 m of soft to firm silty clay overlying dense gravels and sands. The maximum drag load reported for an uncoated pile at the lower third point of the pile was 180 tons. The maximum load in a coated pile was 3 tons, with the maximum load occurring just below mid-depth. Thus the negative skin friction for the coated pile was only 2% of that for the uncoated pile.

Garlander and Lambe (1973), and Lambe et al. (1974) investigated negative skin friction on a steel H-pile supporting a bridge abutment (Cutler Circle Bridge). Negative skin friction was evaluated by measuring the elastic rebound of the top of the pile when an electric current was passed through the pile and the pile was freed from the surrounding soil. The measurements indicated that the negative skin friction in the fill and the sand was about 32-45 tons and that the negative skin friction in the clay was about 60 to 72 tons. The corresponding values of ß were 0.20 to 0.25 for the clay and 0.30 to 0.40 for the sand and fill. Subsequently, a symposium on negative skin friction was held at the Massachusetts Institute of Technology to compare different methods of prediction for the downdrag loading on piles at the Cutler Circle Bridge. The results of this symposium emphasised the uncertainties in the design of pile groups subjected to negative skin friction and suggested that the fundamentals concerning downdrag were not well understood.

Field measurements of negative skin friction on a bored cast-in-situ pile were undertaken by Mohan et al. (1981) in the marine clay deposits of Visakhapatnam dockyard in India. The pile was 43 cm diameter and 5 m long, cast in a bentonite filled borehole in ground reclaimed by depositing dredged material from sea over very soft marine silty clay. Based on the results, the authors concluded that the coefficients α and ß increased with increase in preload. The values of the coefficient α were found to be more or less in agreement with the values suggested in the literature, but ß was about half of those suggested for clays of a similar nature (about 10 to 15 percent of the effective stress for the preload of 10 to 30 kPa).

Bozozuk (1981) carried out a comprehensive load-testing program to investigate the load-carrying capability of an instrumented steel pipe pile first reported in 1972. The steel pile was 49 m long, driven through a highway embankment into a deep deposit of compressible marine clay. Downdrag loads generated in the pile reached a peak 1.52 MN ten years after installation with a corresponding pile settlement of 694 mm.

Significant downdrag loads, varying according to the seasons of fluctuations of the water table were measured by Auvinet and Hanell (1981) during a four year period on instrumented precast concrete piles driven in an area of the former Texcoco Lake in Mexico. The area was affected by an intense pumping-induced consolidation. Two piles with triangular cross-sections were used under no external load, a 30.5 m friction pile and 32 m point-bearing pile. It was concluded that an adhesion factor value $\alpha=0.8$ would be in good agreement with the measured values.

In addition, the authors concluded that a relative displacement of about 2 cm between the point-bearing pile and the clay was required to fully mobilise skin friction.

Bhandari et al. (1984) reported the results of a full-scale field study undertaken to monitor negative skin friction on a large diameter, bored cast-in-place reinforced concrete pile installed to rock in a deep deposit of soft marine clay with undrained shear strength in the range of 14-20 kPa. The test pile was 0.66 m diameter and 28.4 m deep with 0.4 m of pile toe embedded in weathered rock. The value of the adhesion factor α estimated in terms of total stress from the pullout tests and from the instrumented pile test were 0.46 and 0.53 respectively. The corresponding negative drag on the pile for c_u = 20 kPa was 350 kN. On the other hand, the negative drag on the test pile estimated in terms of the effective stress, was between 400-480 kN. The value of the coefficient ß was found to vary between 0.2 and 0.24.

Clemente (1984) carried out a field testing programme in Honolulu to measure downdrag on prestressed concrete piles in under-consolidated, highly plastic silty clay. The rate of ground settlement due to self-weight of the soil was about 12-18 mm/year. Five piles were employed to monitor the development of downdrag and to examine the application, behaviour and effectiveness of bitumen coatings to reduce downdrag. The piles, about 52 m long and 420 mm diameter each, were driven through a compressible layer, about 43 m deep. Values for ß of 0.47 and 0.31 were deduced from the load distributions. Based on the findings of this study, the author concluded that ß values may vary significantly from one site to another, and even within the same site. The variation was suggested to be due to many factors including pile lengths, degree of driving, location of neutral points, and the effective stresses acting on the piles.

In Hong Kong, Lee and Lumb (1988) investigated negative skin friction on two steel tube piles with of 610 mm diameter. One of the piles was coated with a bituminous slip layer. Immediately after driving, an earth embankment of 2 m height was formed around the test piles so as to induce consolidation of the compressible layer and hence induce settlements of the ground surface. The settlement at and below the ground surface was monitored using settlement gauges. The total settlement of the ground surface 341 days after filling was 463 mm, which was mainly due to the consolidation of the soft clay marine deposit layer. The consolidation of the fill and the underlying residual soil were negligible. The maximum axial compressive force recorded on the plain pile was 1960 kN and that on the coated pile was 380 kN. The coefficient ß for the fill material and marine clay were 0.61 and 0.21 respectively. The combined coefficient for the full length of the test pile was 0.30.

Little and Toma (1989) carried out a laboratory-scale programme to investigate the phenomenon of negative skin friction on an instrumented end bearing pile embedded in soft clay undergoing surcharge-induced consolidation. The soil was made up from a remoulded soft to firm intact silty clay, sampled from a site located at the Forth Estuary in Scotland adjacent to the SERC soft clay site reported in the following section. The pile was instrumented internally with electrical resistance strain gauges. Two testing programmes were conducted. Each testing programme consisted of applying loading increments on the soil up to 90 kPa. The consolidation pressure was applied by means of a cylindrical rubber bag.

Using an effective stress analysis, an average value of $\beta = 0.20$ was determined. For a total stress analysis, α values were obtained by two different methods depending on the undrained shear strength of the clay layer: from the initial surrounding shear strength of the soil at the beginning of each loading stage, and from the calculated value of the undrained shear strength appropriate to the changing degree of consolidation. Assuming a constant initial undrained shear strength α ranged between 0.27 and 2.13. Taking into account the changing shear strength due to consolidation resulted in calculated adhesion values in the range 0.29 to 0.58.

Adhesion values by both methods were expressed as a function of the degree of consolidation of the clay layer and the settlement ratio (defined as the settlement of the clay surface divided by the depth of clay layer). These indicated that the rate of increase of α based on the measured initial undrained shear strength was approximately 25% for every 2% increase of the settlement ratio; the rate of increase of α determined from calculated values of undrained shear strength reached a limiting value of approximately 0.6 at a settlement ratio of 10%.

The literature indicates that a considerable amount of research has been carried out in the last 20 years describing the effects of downdrag on piles generated by the consolidation of soft clay deposits. The importance of downdrag as a factor to be taken into account in deep foundation design has been confirmed by both model and full-scale tests. Field observations for piles bearing on incompressible strata indicate that negative skin friction along the pile surface can be large, and may exceed the design load for the superstructure. It has been suggested that single piles and pile groups subjected to negative skin friction should be designed for maximum skin friction above the neutral point, and that the resulting downdrag may be estimated by either a total stress (α) or an effective stess (β) approach. Values for α and for β have been reported by a number of authors. There is a consensus view that the β method is the more appropriate, a view perhaps substantiated by the relatively narrow range of β values reported from the various field studies. Theoretical methods based on elastic theory show reasonable agreement with field measurements, but only when representative elasticity parameters have been obtained.

There is still a need for basic data describing the effects of downdrag, particularly from field studies on pile groups. The following sections describe the results of such a full-scale study, on pile groups, currently in progress in Scotland. The opportunity presented by this conference has enabled the presentation of some of the previously unpublished data from this study. The ongoing nature of the project means that further analysis of the results will be reported at a later date.

FULL-SCALE STUDY OF PILE DOWNDRAG AT BOTHKENNAR, SCOTLAND

In 1988 an opportunity arose to carry out a full-scale study of the effects of negative skin friction on pile group behaviour on a green field site underlain by

Postglacial clays on the south side of the Forth Estuary, Scotland. The site was purchased by the Science and Engineering Research Council of the United Kingdom in 1987 for the purpose of providing a facility to enable researchers in geotechnical engineering to undertake studies, at large- and full-scale, in soft clay. The 11 ha. site is underlain by a variable thickness (av. 15 m) of relatively uniform, soft to firm, normally consolidated estuarine clay deposited within the last 9000 years, (Hawkins et al. 1989). The top 1 m of soil comprises a weathered and desiccated stiff crust, immediately below which is the water table.

Between 35-50% of the material is clay sized; natural water contents vary between 40-60%. The Atterberg limits are sensitive to the method of sample preparation but on average the liquid limit is 65%, and the plastic limit is 32%. The organic content of the material is approximately 5%. Undrained shear strengths determined in the soft clay underneath the stiff crust by the in situ vane method were less than 40 kPa. The results of cone penetration testing in the clay indicated an almost constant friction ratio of 2%. Oedometer tests on high quality piston samples of the clay indicated a coefficient of volume compressibility in the range 1.0 - 2.0 $\times 10^{-3}$ kPa^{-1}, thus indicating its highly compressible nature and, therefore, suitability for a study into the development of negative skin friction on piles.

The study was planned to investigate the interaction between pile shaft and surrounding soil, for two groups (end-bearing, friction) of driven cast-in-place (Westpile) piles, as the ground surface around the piles was surcharged via a 2.5 m high embankment. Further details concerning the site and the ground and pile instruments were reported by Little et al. (1991); Little and Ibrahim (1993) reported the results of a prediction exercise, associated with this study, held at the Wroth Memorial Symposium in Oxford, during July 1992.

A schematic plan of the piling and embankment layout is shown in Fig. 1. The twenty-three piles in the two groups, including 5 single piles, were driven during July 1990. A detailed record was made of the excess pore pressures generated in the soil at different depths close to the piles, and the total horizontal stresses on the pile walls measured using radiused Glotzl cells, during driving and subsequently, Fig. 2. So as to separate the effect of the soil reconsolidation from that of the loading from the embankment on downdrag, the embankment was not constructed until full dissipation of excess pore pressures, due to driving, had taken place. (This took approximately 1 year to achieve, Fig. 2).

In Fig. 1, X, Y, Z refer to positions adjacent to a single end-bearing pile, and the centre piles in each of the friction and end-bearing groups, respectively. Figs. 3 and 4 show the observed axial loads in the centre, edge and corner piles in the friction group, and end-bearing group respectively. The force profiles are with respect to both depth and time.

These profiles show the expected distribution with depth for friction piles, that is an increasing axial force (downdrag) to a depth corresponding to the neutral point ($\simeq 10$ m), reducing to zero at the pile point. The separate profiles for each pile clearly indicated the time-dependency of downdrag. For pile G5 (centre of friction group) for example, the downdrag force increased from 100 kN to 187 kN over the two year period after surcharging. Note, however, the relatively large downdrag (about 50% of the maximum) evident immediately following the construction of the 2.5 m embankment. This downdrag component was mobilised in a (largely) undrained state; embankment construction took merely 5 days to complete.

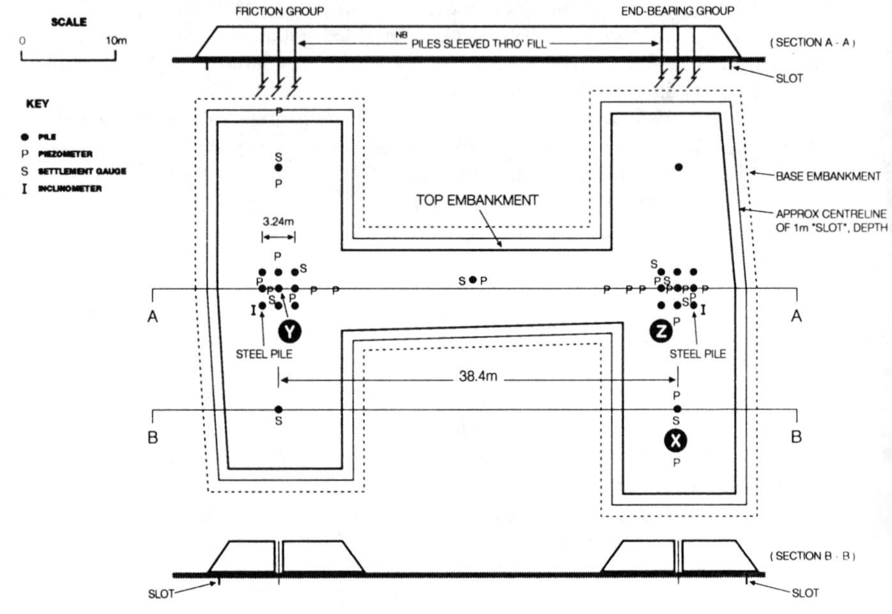

FIG. 1. Bothkennar Downdrag Project: Experimental Layout

Fig. 4 shows the corresponding data for the centre pile (G15) in the end-bearing group. Unlike its corresponding friction pile (G5) the expected force depth profile is not what might have been expected. That the neutral point was not nearer the pile toe could be a reflection of the relatively small surface loading applied (approximately 40 kPa), or the degree of penetration of the pile point into the gravel layer at that depth.

The soil surface settlement, pile head settlement, excess pore pressure, and mobilised shear stresses are also shown plotted against the same time base, for piles G5, G15, Figs. 5 and 6. For pile G5 (friction) it would appear that full mobilisation of downdrag had occurred when a total differential settlement corresponding to approximately 30% of the pile diameter had been achieved. (The pile heads settled about 5 mm throughout). However, 1/3 of this settlement had been achieved immediately after surcharging, during undrained loading.

Table 1 shows a comparison between the observed maximum downdrag forces for the centre edge and corner piles with those predicted according to Briaud et al. (1991).

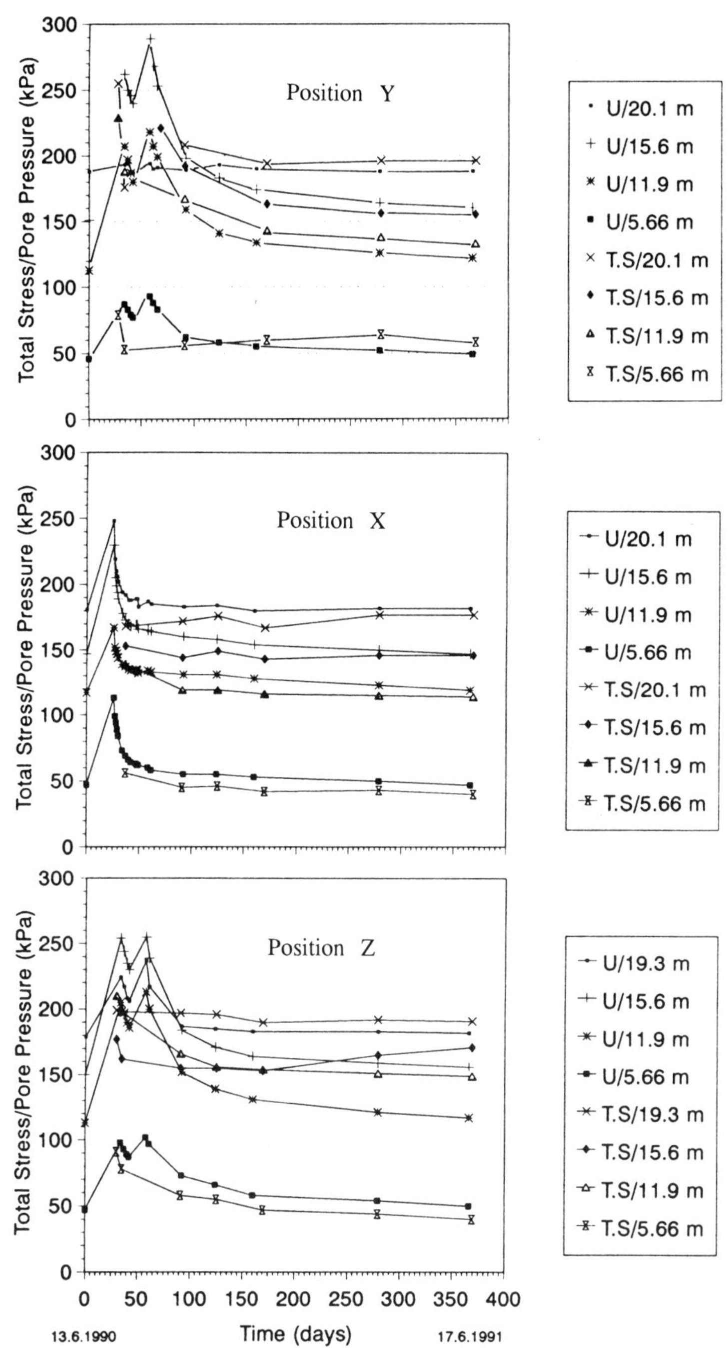

FIG. 2. Dissipation of Excess Pore Pressures Post-Driving

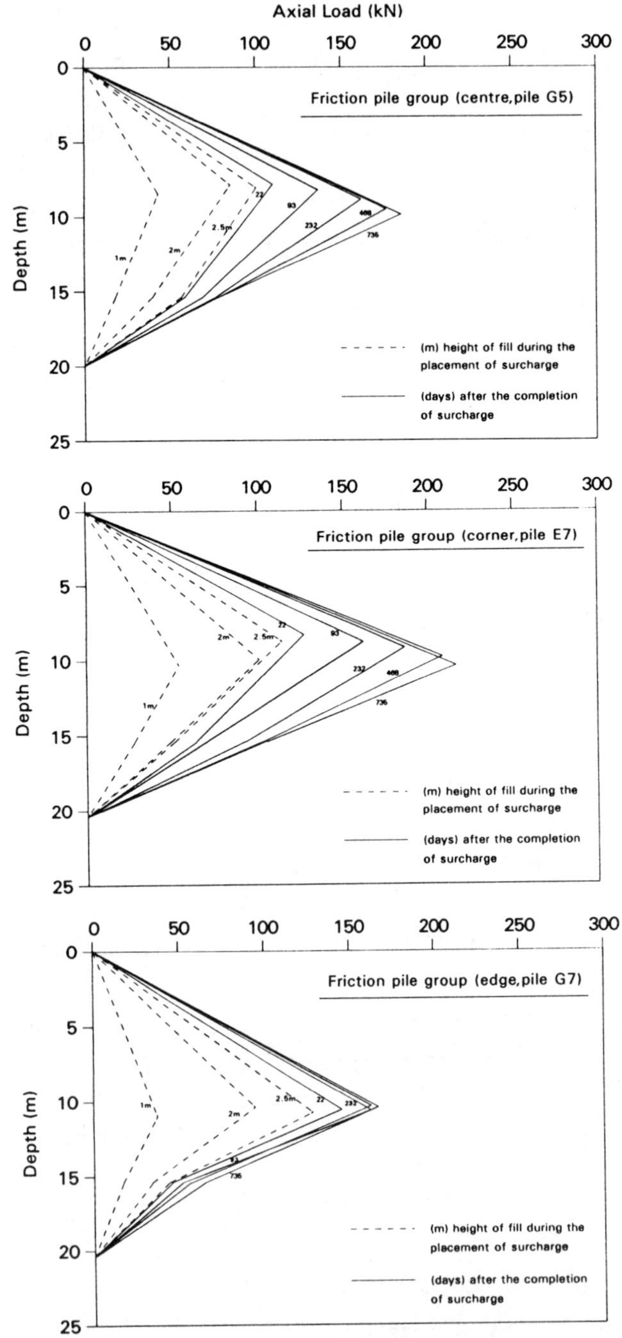

FIG. 3. Axial Force versus Depth, Friction Piles

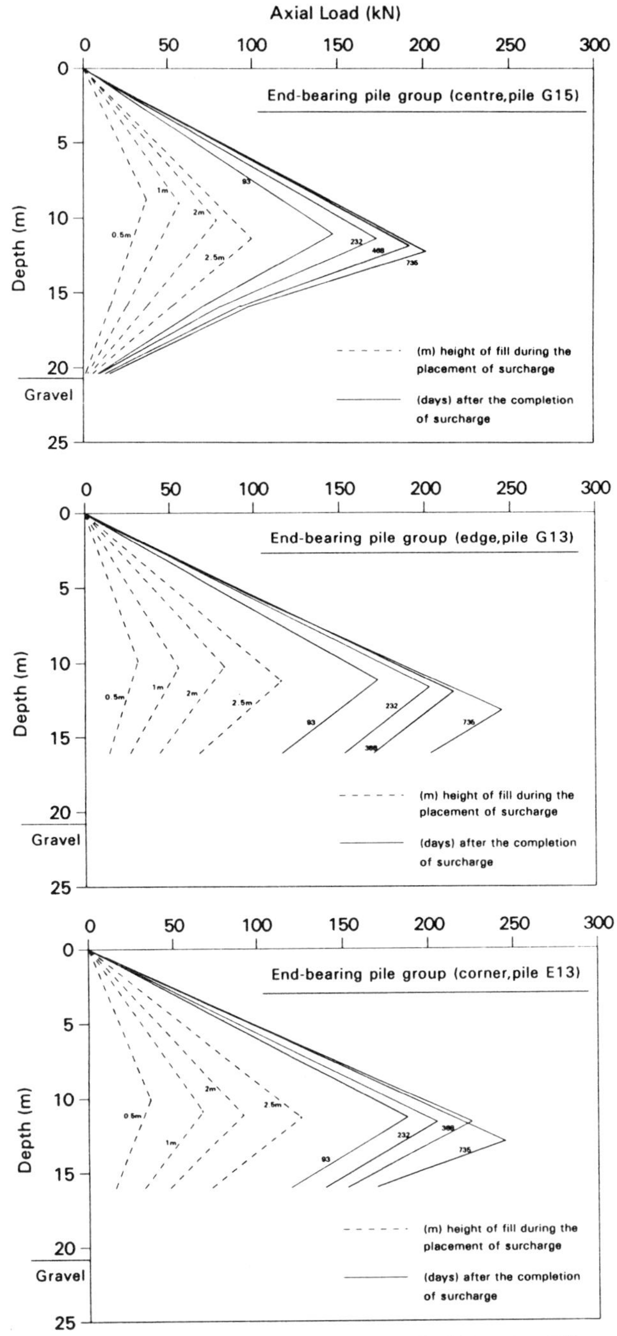

FIG. 4. Axial Force versus Depth, End-Bearing Piles

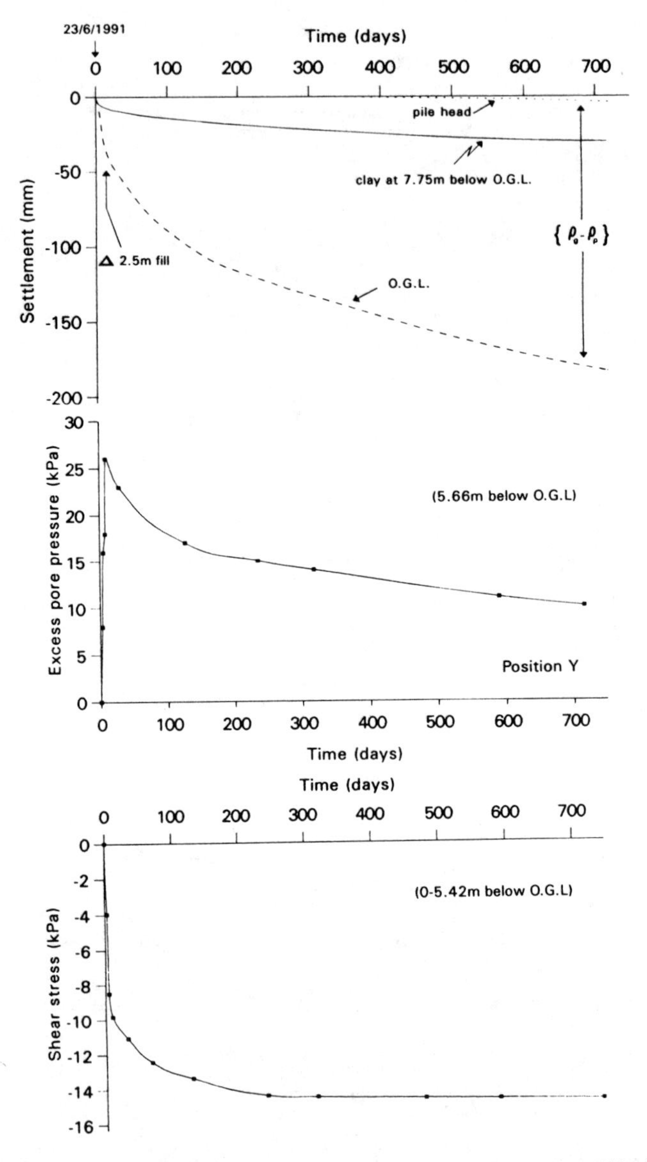

FIG. 5. Settlement, Pore Pressure, Shear Stress, Friction Pile G5

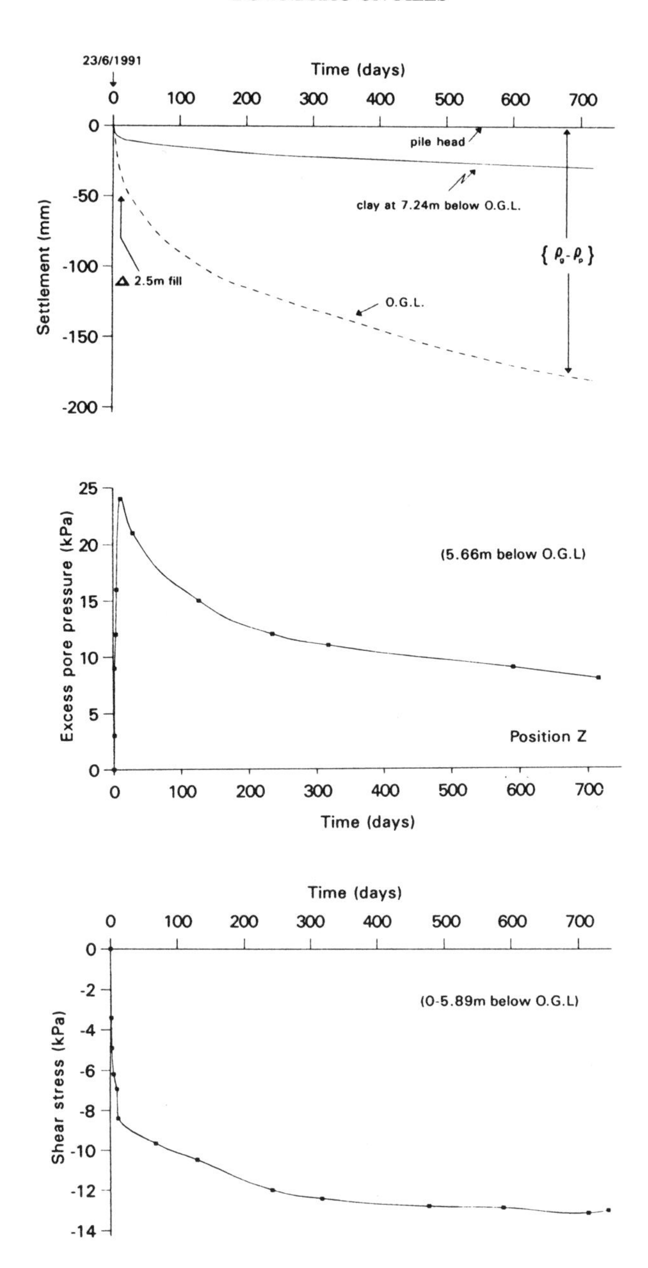

FIG. 6. Settlement, Pore Pressure, Shear Stress, End-Bearing Pile G15

TABLE 1. Comparison between the Observed Maximum Downdrag Forces with Those Predicted According to Briaud et al. (1991)

	$F_{n(corner)}/F_{n(edge)}$	Briaud et al. (1991)	$F_{n(centre)}$ (kN)	$q_o s^2$ (kN)
Friction	1.31	1.50	187	106
End-bearing	1.00		202	

Whilst the maximum downdrag force on corner and edge piles could be overestimated by as much as 50% using these guidelines, it would appear that the downdrag on the centre piles may be underestimated by 75-90% at 4 pile spacings.

Figs. 7 and 8 show the distribution of negative and positive shear stress with depth, and time, for piles G5 and G15. On these figures are shown the vane undrained shear strength profiles for the soil, both peak and remoulded.

The initial response to the undrained embankment loading indicated a mobilised negative shear stress of 7-10 kPa down to about 10 m depth. Towards the tops of the piles, where relative pile/soil displacements were largest, this corresponded to an undrained adhesion factor $\alpha \simeq$ 0.4-0.5, and an average β value = 0.2.

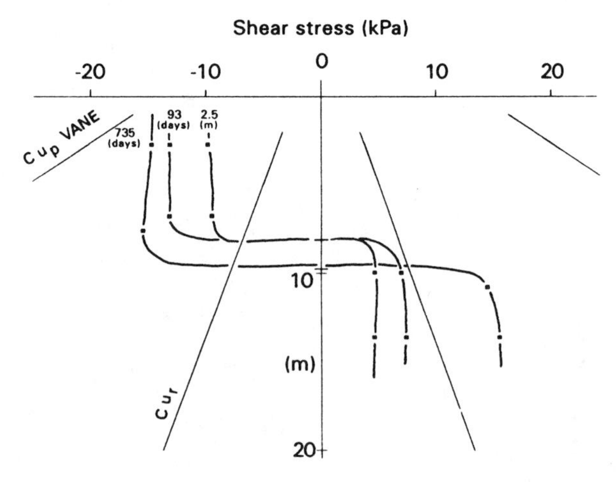

FIG. 7. Shear Stress versus Depth, Pile G5

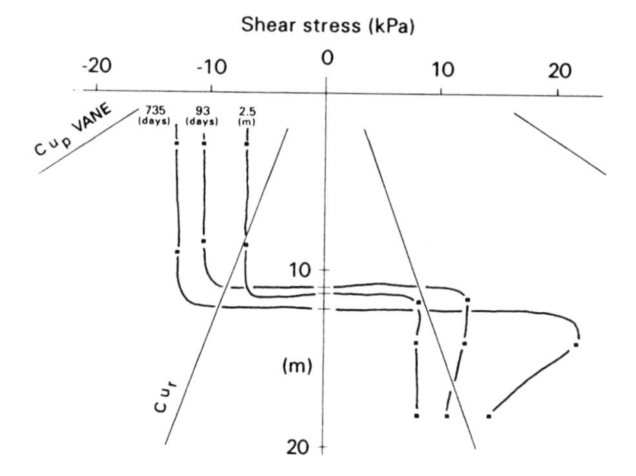

FIG. 8. Shear Stress versus Depth, Pile G15

It has been shown (Toma 1989) that the relationship between the negative shear stress τ_f, the initial undrained shear strength c_{uo}, the average isotropic consolidation pressure p', and the current undrained shear strength c_u, for the Postglacial clays of the Forth estuary is of the form

$$\log_{10}\left(\frac{\tau_f}{c_{uo}}\right) = -\left[1.289 + \log_{10}\left(\frac{c_u}{p'}\right)\right] \tag{4}$$

letting $m = \tau_f/c_u$,

$$\log_{10}\left(\frac{m\,c_u}{c_{uo}}\right) = -\left[1.289 + \log_{10}\left(\frac{c_u}{p'}\right)\right] \tag{5}$$

whence $(m/c_{uo})(c_u^4/p'^3) = 0.0514$ or

$$\frac{c_u}{p'} = 0.476\,m^{-1/4}\left(\frac{c_u}{p'}\right)^{1/4} \tag{6}$$

Substituting Eq. (6) into Eq. (4) yields

$$\log_{10}\left(\frac{\tau_f}{c_{uo}}\right) = -\left[1.289 + \log_{10}\left[0.476^3\, m^{-3/4}\left(\frac{c_{uo}}{p'}\right)^{3/4}\right]\right] \tag{7}$$

or

$$\frac{\tau_f}{c_{uo}}\left(\frac{c_{uo}}{p'}\right)^{3/4} = 0.476\, m^{3/4} \tag{8}$$

The value m = 0.495 can be determined experimentally,

$$\therefore \quad \tau_f = \frac{0.281\, c_{uo}}{(c_{uo}/p')^{3/4}}\ \text{kPa} \tag{9}$$

assuming $p' = 1/3(\sigma_v' + 2\sigma_h')$ and inserting known quantities, yields an average τ_f = 5.16 kPa between 0-6 m depth. This represents the expected (additional) negative skin friction achieved during the consolidation stage, following the undrained (embankment) loading. The observed values were 4.90 kPa for G5 and 6.0 kPa for G15. The total negative shear stress for an initial undrained loading, followed by a drained embankment-type loading for this clay may therefore be approximated by:

$$\tau_{f\ total} = \alpha c_{uo} + c_{uo}\left(\frac{0.281}{(c_{uo}/p')^{3/4}}\right)\ \text{kPa} \tag{10}$$

CONCLUSIONS

There is an extensive literature describing negative skin friction and pile downdrag. Estimates of downdrag magnitude are still based largely on either a total stress (undrained) approach utilising a clay/pile adhesion factor or alternatively, on an effective stress approach, incorporating appropriate drained soil and pile/soil shearing parameters.

The use of the effective stress method is usually preferred as values for ß lie within a relatively narrow range, for most soil types.

Whilst this approach has the benefit of simplicity, nevertheless there remain difficulties in making accurate predictions of downdrag, as has been shown at 2 prediction symposia on this subject, held in the USA (1973) and in the UK (1992).

It is important to separate the pore pressure effects of downdrag resulting only from soil reconsolidation after pile driving, from subsequent downdrag resulting from, say, an embankment loading. A separate study, on adjacent site, involving longer piles (70 m) is currently being planned for this purpose.

The Bothkennar downdrag project permitted a period of time post-driving to enable full dissipation of excess pore pressures, prior to embankment loading.

Under such conditions it has been possible to identify both an undrained phase of downdrag (immediately post-surcharge) and a drained (strictly, "draining") progression of negative skin friction with time thereafter.

At Bothkennar approximately 2 years was required to fully mobilise negative skin friction under an embankment of 2.5 m height; about half of this occurred during the 5 days of surcharging.

The use of *either* the α method *or* the β method in the prediction of negative skin friction in this context therefore is considered inappropriate. *Both* methods should be employed.

ACKNOWLEDGEMENTS

The experimental work reported in this paper was supported by SERC under grants GR/E25337, GR/F97850, GR/J45810. The support and assistance of Westpile Ltd. is gratefully acknowledged. The author thanks Mr. K. Ibrahim for help in the preparation of this paper, and Mrs. E. Fletcher for typing the manuscript.

APPENDIX - REFERENCES

Aldrick, H. P., Jr. (1970). "Back Bay, Boston-Part 1." *Journal of the Boston Society of Civil Engineers*, 57(1), 1-33.

Auvinet, G. and Hanell, J. J. (1982). "Negative skin friction on piles in Mexico City clay." *Proc. 10th ICSMFE*, Stockholm, 2, 599-604.

Bhandari, R. K., Soneja, M. R., Sharma, D. (1984). "Downdrag on an instrumented bored pile in soft clay." *Proc. Int'l. Conf. on Case Histories in Geotechnical Eng.*, St. Louis, 3, 1019-1026.

Bjerrum, L. (1969). "Discussions." *Can. Geotech. J.*, 6(3), 366-367.

Bjerrum, L., Johannessen, I. J., and Edie, O. (1969). "Reduction of negative skin friction on steel piles to rock." *Proc. 7th ICSMFE*, Mexico City, 2, 27-33.

Bowles, J. E. (1988). *Foundation Design and Analysis*, 4th Edition, McGraw-Hill, New York, New York.

Bozozuk, M. and Labreque, A. (1969). "Downdrag measurements on 270 composite piles performance of deep foundations." *Performance of Deep Foundations*, ASTM STP 444, ASTM, Philadelphia, Pennsylvania, 15-40.

Bozozuk, M. (1970). "Field observations of negative skin friction loads on long piles in marine clay." *Proc. Conf. on Design and Installation of Pile Foundations and Cellular Structures*, Bethlehem, 273-280.

Bozozuk, M. (1972). "Downdrag measurements on a 160 ft. floating pipe test pile in marine clay." *Can. Geotech. J.*, 9(2), 127-136.

Bozozuk, M. (1981). "Bearing capacity of pile preload by downdrag." *Proc. 10th ICSMFE*, Stockholm, 2, 631-636.

Briaud, J. L., Jeong, S., and Bush, R. (1991). "Group effect in the case of downdrag." *Geotechnical Engineering Congress*, Geotechnical Special Publication No. 27, ASCE, New York, New York, 1, 505-518.

Broms, B. B. (1969). "Design of pile groups with respect to negative skin friction." Specialty Session on Negative Skin Friction and Settlement of Pile Foundations, Contribution to Discussion, Proc. 7th ICSMFE, Mexico City, 2, 1-2.

Broms, B. B. (1976). "Pile foundations - pile groups." *Proc. 6th European Conf. Soil Mech. Found. Eng.*, Vienna, 2.1, 103-132.

Broms, B. B. (1979). "Negative skin friction." State-of-the-art Report, *Proc. 6th Asian Regional Conf. Soil Mech. Found. Eng.*, Singapore, 2, 41-75.

Buisson, M., Ahu, J., and Habib, P. (1960). "Le Frottement Negatif." *Annales de l'Instit ut Francis du Batiment et des Travaux Publics*, Sols et Foundations, No. 31, 29-46.

Bukholdin, B. F. and Berman, V.L. (1975). "Investigation of negative skin friction on piles and suggestions on its calculation." (Translated from Russian), *Soil Mech. Found. Eng.*, Plenum Publishing Corporation, New York, New York, 11(4), 238-244.

Burland, J. B. (1973). "Shaft friction of piles in clay." *Ground Engineering*, 6(3), 30-42.

Chellis, R. D. (1961). *Pile Foundations*, 2nd Edition, McGraw-Hill, New York, New York.

Clemente, F. M., Jr. (1984). "Downdrag, negative skin friction and bitumen coatings on prestressed concrete piles," Ph.D. thesis, University of Tulane, New Orleans, Louisiana.

Elmasry, M. A. (1963). "The negative skin friction of bearing piles," Ph.D. Thesis, Swiss Federal Institute of Technology, Zurich, Switzerland.

Endo, M., Minov, A., Kawasaki, T. and Shibata, T. (1969). "Negative skin friction acting on steel pipe-piles in clay." *Proc. 7th ICSMFE*, Mexico City, 2, 85-92.

Fellenius, B. H. (1969). "Negative skin friction on piles in clay - a literature survey." Specialty Session, Negative Skin Friction and Settlement of Pile Foundations, *Proc. 7th ICSMFE*, Mexico City, 2, 1-8.

Fellenius, B. H. and Broms, B. B. (1969). "Negative skin friction on long piles driven in clay." *Proc. 7th ICSMFE*, Mexico City, 2, 93-98.

Fellenius, B. H. (1971). "Negative skin friction on long piles driven in clay: I. Results of a full scale investigation, II. General views and design recommendations." *Proc. Swedish Geotechnical Institute*, 25, 38.

Fellenius, B. H. (1972). "Downdrag on piles in clay due to negative skin friction." *Can. Geotech. J.*, 9(4), 325-337.

Fellenius, B. H. (1986). "Unified design of piles and pile groups considering capacity, settlement, and negative skin friction." *Proc. Seminar and Workshop on Foundation Design and the 1985 Canadian Foundation Engineering Manual*, Nova Scotia, 1-17.

Gant, E. V., Stephens, J. E., and Moulton, L. K. (1958). "Measurement of forces produced in piles by settlement of adjacent soil." *Analysis of Soil Foundation Studies*, Bull. 173, Highway Research Board, Washington, D. C., 20-37.

Garlanger, J. E. and Lambe, T. W. (1973). "Proceedings of a symposium on downdrag of piles," *Research Report No. 73-56*, MIT, Cambridge, Massachusetts.

Hansen, J. B. (1968). "A theory for skin friction on piles." *Bulletin No. 25*, Danish Geotechnical Institute, 5-12.

Hawkins, A. B., et al. (1989). "Selecting the location, and the initial investigation of the SERC soft clay test bed site." *Quart. J. Eng. Geol.*, 22(4), 281-316.

Johannessen, I. J. and Bjerrum, L. (1965). "Measurement of the compression of a steel pile rock due to settlement of the surrounding clay." *Proc. 6th ICSMFE*, Montreal, 2, 261-264.

Kerisel, J. (1965). "Vertical and horizontal capacity of deep foundations in clay." *Proc. Symposium on Bearing Capacity and Settlement of Foundations*, Duke University, Durham, 45-61.

Kerisel, J. (1976). "Deep foundations - basic experimental facts." *Proc. North American Conf. on Deep Foundations*, Mexico City, 121-137.

Koerner, R. M. and Mukhopadhay, C. (1972). "Behaviour of negative skin friction on model piles in medium plasticity silt." *Highway Research Record No. 405*, 34-44.

Lambe, T. W., Garlanger, J. E., and Leifer, S. A. (1974). "Prediction and field evaluation of downdrag forces on a single pile," *Research Report R74-27, Soils Publication 339*, Dept. of Civil Engrg., MIT, Cambridge, Massachusetts.

Lee, P. K. K. and Lumb, P. (1982). "Field measurements of negative skin friction on steel piles in Hong Kong." *Proc. 7th South East Asian Geotechnical Conf.*, Hong Kong, 363-374.

Little, J. A. and Toma, T. M. (1989). "The development of shaft adhesion with onset of negative skin friction for a fixed base model pile." *Proc. 2nd Int'l. Conf. Found. Tunnels, Eng.*, Technics Press, 1, 111-117.

Little, J. A., Price, G., Ibrahim, K. (1991). "Geotechnical instrumentation for a full-scale study of negative skin friction in clay." *Int'l. Conf. Piling and Deep Foundations*, Stresa, DFI, 1, 491-496.

Little, J. A. and Ibrahim, K. (1993). "Predictions associated with the pile downdrag study at the SERC soft clay site at Bothkennar, Scotland." *Proc. Wroth Mem. Sym.*, Oxford, 796-818.

Locher, H. G. (1965). "Combined cast-in-place and precast piles for reduction of negative skin friction caused by embankment fill." *Proc. 6th ICSMFE*, Montreal, 2, 290-294.

Mindlin, R. D. (1936). "Force at a point in the interior of a semi-infinite solid." *Physics*, 7(5), 195-202.

Mohan, D., Bhandari, R. K., Sharma, D., and Soneja, M. R. (1981). "Negative drag on an instrumented pile - a field study." *Proc. 10th ICSMFE*, Stockholm, 2, 787-790.

Narasimha, R. S. and Krishnamurthy, N. R. (1982). "Studies of negative skin friction in model piles." *Indian Geotech. J.*, 13(Part 1), 83-91.

Parry, R. H. G. and Swain, G. W. (1977a). "Effective stress methods of calculating skin friction on driven piles in soft clays." *Ground Engrg.*, 10(3), 23-26.

Parry, R. H. G. and Swain, G. W. (1977b). "A study of skin friction on piles in stiff clay." *Ground Engineering*, 10(5), 33-37.

Puri, V. K., Das, B. M., and Karna, U. (1991). "Negative skin friction on coated and uncoated model piles." *Proc. 4th International DFI Conf.*, Balkema, Rotterdam, The Netherlands, 627-632.

Silva, A. J. (1966). "Downdrag on piles," Ph.D. Thesis, University of Connecticut, Storrs.

Terzaghi, K. and Peck, R. B. (1967). *Soil Mechanics in Engineering Practice*, 2nd. Edition, John Wiley and Sons, New York, New York.

Toma, T. M. (1989). "A model study of negative skin friction on a fixed base pile in soft clay," Ph.D. Thesis, Heriot-Watt University.

Tomlinson, M. J. (1957). "The adhesion of piles driven in clay soil." *Proc. 4th ICSMFE*, London, 2, 66-71.

Tomlinson, M. J. (1971). "Some effects of pile driving on skin friction." *Proc. Conf. on Behaviour of Piles*, London, Inst. Civil Engrs., 107-114.

Tomlinson, M. J. (1975). *Foundation Design and Construction*, 3rd Edition, Pitman Publishing, London, England.

Vesic, A. S. (1977). "Design of pile foundations," *Synthesis of Highway Practice No. 42*, National Cooperative Highway Research Program, Transportation Research Board, Washington, D. C.

Walker, L. K. and Dravall, P. L. P. (1973). "Dragdown on coated and uncoated piles." *Proc. 8th ICSMFE*, Moscow, 2.1, 257-262.

Zeevaert, L. (1957a). "Foundation design and behaviour of Tower Latino Americana in Mexico City." *Géotechnique*, 7(1), 115-133.

Zeevaert, L. (1957b). "Discussion on negative friction and reduction of point bearing capacity." *Proc. 4th ICSMFE*, London, 3, 188.

Zeevaert, L. (1959). "Reduction of point bearing capacity because of negative friction." *Proc. 1st Pan-Amer. Conf. Soil Mech. Found. Eng.*, Mexico City, 3, 1145-1152.

Zeevaert, L. (1973). *Foundation Engineering for Difficult Subsoil Conditions*, Van Nostrand Reinhold, New York, New York.

PREDICTION OF MOVEMENT IN EXPANSIVE CLAYS

Robert L. Lytton,[1] Fellow, ASCE

ABSTRACT: The movement of expansive soils is usually due to a change of suction near the soil surface. The properties of the soil that govern the amount and rate of movement are the suction compression index, and the unsaturated permeability and diffusivity. Methods of using these to determine suction and heave (or shrinkage) profiles with depth are outlined. Methods of estimating these properties using simple laboratory tests, namely Atterberg limits, water content, dry density, porosity, sieve analysis, and hydrometer analysis are presented. Differential movement governs the design of slabs-on-ground, highway and airport pavements and canal linings, which are themselves controlled by the edge moisture variation distance. Graphs of the edge moisture variation distance as it changes with the unsaturated diffusivity and the Thornthwaite Moisture Index are presented for both the center lift and edge lift distortion modes. The values were computed using a coupled unsaturated moisture flow and elasticity finite element program which had been calibrated to match reasonably well the measured suctions in an extensive field study involving several pavement sites in a number of different climatic zones in Texas.

INTRODUCTION

The prediction of movement in expansive soils is important principally for the purpose of designing foundations or other ground supported structural elements. In design, the principal interest is in making an accurate estimate of the range of movement that must be sustained by the foundation. It is for that reason that envelopes of maximum heave and shrinkage are important for design purposes. For

---

[1] A. P. and Florence Wiley Professor of Civil Engineering, Texas A&M University, College Station, Texas 77843-3136.

slab-on-ground design, differential movements are important. For highway and airport pavements, canals, and pipelines, the wave spectrum of differential movements versus wave lengths are the desirable design characteristic. Structural floors suspended above expansive clays must be provided with a gap that exceed the total expected heave. Drilled piers (or shafts) must be designed to resist simultaneously a vertical movement profile and a horizontal pressure profile, both of which change with wetting and drying conditions. Retaining structures, basement walls, rip rap, and canal linings must be designed to withstand lateral movements. Finally, all foundations must be designed against the time-dependent vertical and horizontal curvature that is generated by down hill creep.

Each of these types of movement is of sufficient importance and complexity to warrant a separate paper of its own. Differential movement is selected as the topic of this paper principally because it involves the prediction of the total movement at two different locations which are separated by a characteristic distance. This distance depends upon how pervious the soil is. Understanding differential movement and how to predict heave and shrinkage envelopes of it provides much of the information needed for most types of foundation design.

This paper provides results of a multiple year study of differential movements of pavements on expansive soils as they are affected by vertical moisture barriers, and of a computer study of the horizontal zone of influence that is affected by changes of moisture. The first section presents a summary of the theoretical relationships between volume change, suction change, and total stress changes. The second section summarizes material property relationships that were developed during the vertical barrier study. The material properties that can be predicted are the volume change coefficients, unsaturated permeability and diffusivity, and characteristics of the suction-versus-water content relation. The third section presents the results of the computer study of the size of the moisture influence zone for edge lift and center lift conditions. The concluding section comments upon the significance of these results for the prediction of differential movements.

Expansive Clay Volume Change

Movements in expansive soils are generated by changes of suction which are brought about by the entry or loss of moisture. The volume change that accompanies the change of suction (and water content) depends upon the total stress states that surround the soil. Within a soil mass, a decrease of the magnitude of suction results in an increase of water content. The volume of the soil also increases unless the surrounding pressure is sufficient to restrain the swelling.

Suction is defined by the Kelvin equation:

$$h = \frac{RT}{mg} \ln \frac{H}{100} \qquad (1)$$

where h = the total suction in gm-cm/gm, a negative number;
R = the universal gas constant, 8.314×10^7 ergs-K/mole;
T = absolute temperature, degrees K;

m = gram-molecular weight of water, 18.02 gm/mole;
g = 981, conversion from grams mass to grams force; and
H = relative humidity, in percent.

"Suction" is a term used principally by engineers for the thermodynamic quantity, Gibbs free energy which is inherently negative, as seen in Eq. (1), and generates tension in the pore water stretching between soil particles.

Total suction may have two components: matrix suction, which is due to the attraction of water to the soil particle surfaces and osmotic suction, which is due to dissolved salts or other solutes in the pore water. A complete discussion of suction and its measurement is found in the book by Fredlund and Rahardjo (1993), and will not be explained in more detail here.

A common measure of suction is the pF-scale, in which pF is defined as:

$$pF = \log_{10}|h| \tag{2}$$

where $|h|$ = the magnitude of suction in cm of water, a positive value.

Fig. 1 illustrates the suction-vs-water content curve for a natural soil under wetting and drying conditions. Hysteresis is commonly observed between these two conditions with the water content upon wetting being lower than that upon drying at the same level of suction. The relation between the soil volume and water content rises from the dry volume to its maximum value around field capacity as long as it is not constrained from doing so by external pressure. When the water content is above the shrinkage limit, the volume change-vs-water content line is roughly parallel to the zero air voids line, gaining one cubic centimeter of volume for each cubic centimeter of water increased. Various suction levels corresponding to the field capacity (pF = 2.0); plastic limit (pF = 3.5 for clays); wilting point for plants (pF = 4.5); tensile strength of confined water (pF = 5.3); air dry at 50% relative humidity (pF - 6.0); and oven dry (pF - 7.0) are marked on the suction-vs-water content curve.

A conceptual graph of suction-versus-volume can be drawn using the relations of each to water content. This is illustrated in Fig. 2 on the plane corresponding to zero pressure. A similar graph can be drawn relating pressure (total stress) - versus-volume on the plane corresponding to zero suction. The simultaneous change of the magnitude of suction (decrease) and pressure (increase) results in a small change of volume, following the path from Point A to Point C on the pressure-suction-volume surface. The magnitude of suction decreases from Point A′ to Point B′ while the pressure increases from Point B′ to Point C′. The volume change process can be viewed as the net result of two processes:

a. Increase of volume from A to B at constant mechanical pressure or total stress.
b. Decrease of volume from B to C at constant suction.

For small increments of volume change on this surface, the volume strain, $\Delta V/V$, is linearly related to the logarithms of both pressure and | suction |. The general relation between these, and a change of osmotic suction, π, is:

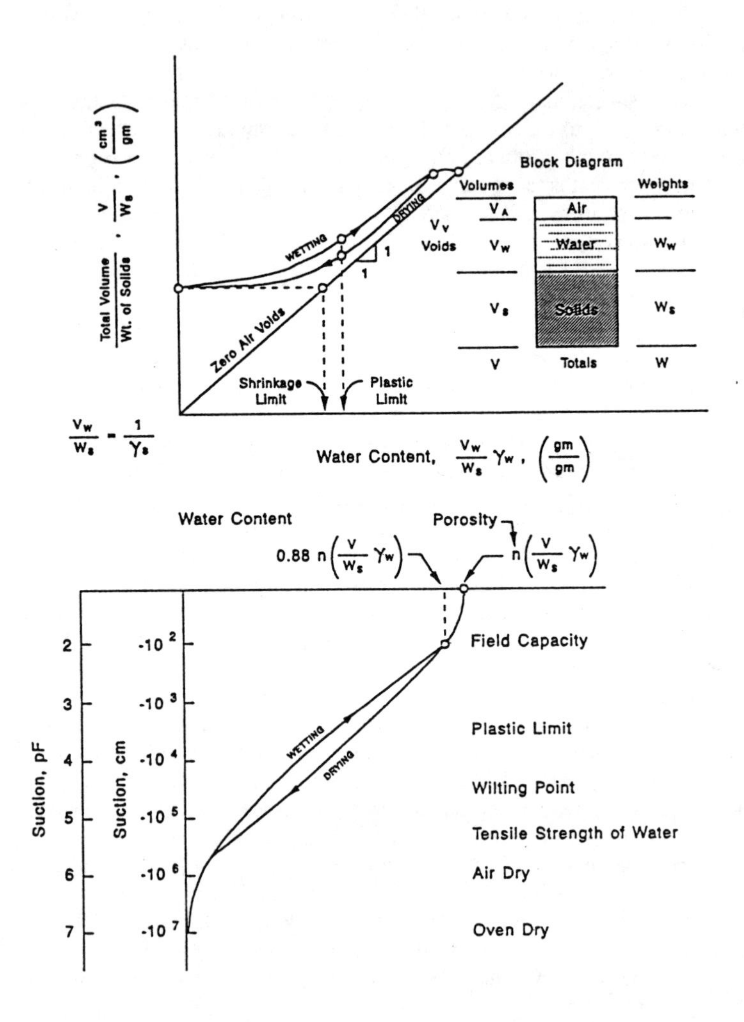

FIG. 1. Suction-Water-Content-Volume Relationships in the Absence of Total Stress

$$\frac{\Delta V}{V} = -\gamma_h \log_{10}\left[\frac{h_f}{h_i}\right] - \gamma_\sigma \log_{10}\left[\frac{\sigma_f}{\sigma_i}\right] - \gamma_\pi \log_{10}\left[\frac{\pi_f}{\pi_i}\right] \qquad (3)$$

in which:

$\Delta V/V$ = the volume strain;
h_i, h_f = the initial and final matrix suction;
σ_i, σ_f = the initial and final values of mean principal stress;
π_i, π_f = the initial and final values of osmotic suction;
γ_h = the matrix suction compression index;
γ_σ = the mean principal stress compression index; and
γ_π = the osmotic suction compression index.

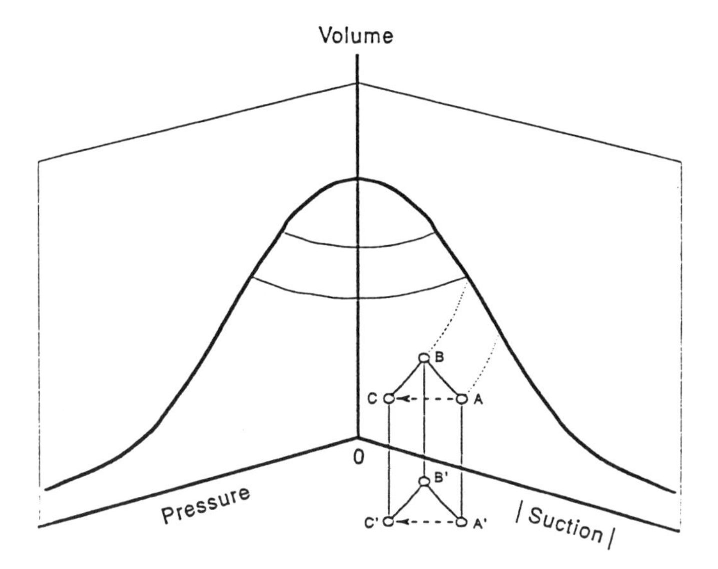

FIG. 2. Pressure-Suction-Volume Surface for Expansive Soil

The mean principal stress compression index is related to the commonly used compression index, C_c, by:

$$\gamma_\sigma = \frac{C_c}{1 + e_o} \qquad (4)$$

where e_o = the void ratio.

In order to predict the total movement in a soil mass, initial and final values of matrix suction, osmotic suction, and mean principle stress profiles with depth must be known. It is the change of matrix suction that generates the heave and shrinkage while osmotic suction rarely changes appreciably, and the mean principal stress increases only slightly in the shallow zones where most of the volume change takes place. It is commonly sufficient to compute the final mean principal stress, σ_f, from the overburden, surcharge, and foundation pressure and treat the initial mean principal stress, σ_i, as a constant corresponding to the stress-free suction-vs-volume strain line represented by Eq. (3). Because there is no zero on a logarithmic scale, σ_i may be regarded as a material property, i.e., a stress level below which no correction for overburden pressure must be made in order to estimate the volume strain. It has been found to correspond to the mean principal stress at a depth of 40 cm. This is illustrated in Fig. 3.

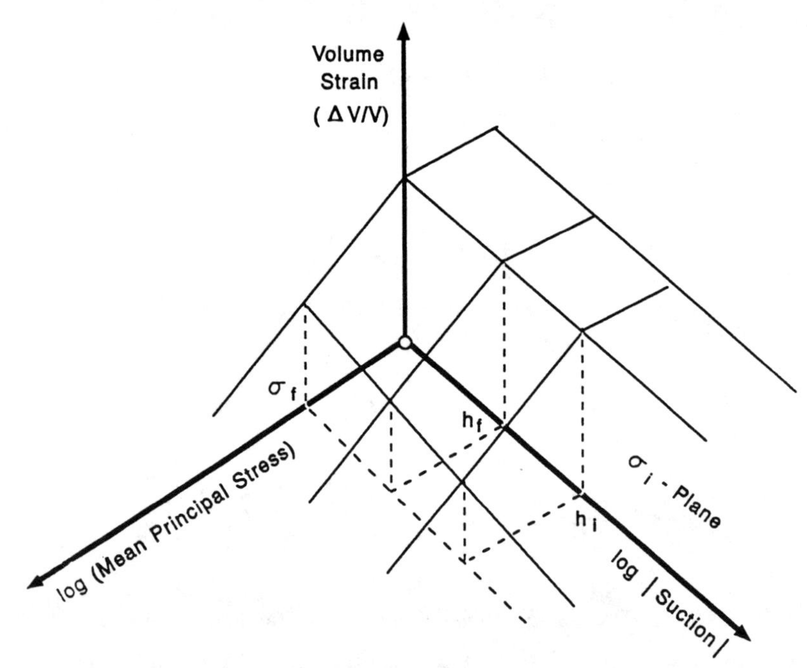

FIG. 3. Graph of Volume Strain as a Function of Log | Suction | and Log (Mean Principal Stress)

The mean principal stress is estimated by:

$$\sigma = \left[\frac{1+2K_o}{3}\right]\sigma_z$$

where σ_z = the vertical stress at a point below the surface in a soil mass; and
K_o = the lateral earth pressure coefficient.

With an active soil which can crack itself in shrinking and generate large confining pressures in swelling, the lateral earth pressure coefficient, K_o, can vary between 0.0 and passive earth pressure levels. Typical values that have been back-calculated from field observations of heave and shrinkage are as follows:

K_o = 0.00 when the soil is badly cracked.
K_o = 0.33 when the soil is drying.
K_o = 0.67 when the soil is wetting.
K_o = 1.00 when the cracked are closed and the soil is swelling.

The vertical strain is estimated from the volume strain by using a crack fabric factor, f.

$$\frac{\Delta H}{H} = f\left[\frac{\Delta V}{V}\right] \tag{6}$$

Back-calculated values of **f** are 0.5 when the soil is drying and 0.8 when the soil is wetting. The level to which the lateral pressure rises is limited by the Gibbs free energy (suction) released by the water; the level to which it drops on shrinking is limited by the ability of the water phase to store the released strain energy. The total heave or shrinkage in a soil mass is the sum of the products of the vertical strains and the increment of depth to which they apply, Δz_i.

$$\Delta = \sum_{i=i}^{n} f_i \left[\frac{\Delta V}{V}\right]_i \Delta z_i \tag{7}$$

where n = the number of depth increments;
Δz_i = the ith depth increment; and
$(\Delta V/V)_i$ = the volume strain in the ith depth increment.

The principal material property needed to compute the vertical movement is the suction compression index, γ_h. This may be estimated with the chart developed by McKeen (1981), shown in Fig. 4. The two axes are given by the activity ratio, Ac, and the Cation Exchange Activity ratio, CEAc, which are defined as follows:

$$Ac = \frac{PI\,\%}{\dfrac{(\%\ -2\ \text{micron})}{(\%\ -\text{No.}\,200\ \text{sieve})} \times 100} \tag{8}$$

$$CEAc = \frac{CEC\ \dfrac{\text{milliequivalents}}{100\ \text{gm of dry soil}}}{\dfrac{(\%\ -2\ \text{micron})}{(\%\ -\ \text{No. }200\ \text{sieve})} \times 100} \tag{9}$$

where PI = the plasticity index in percent.

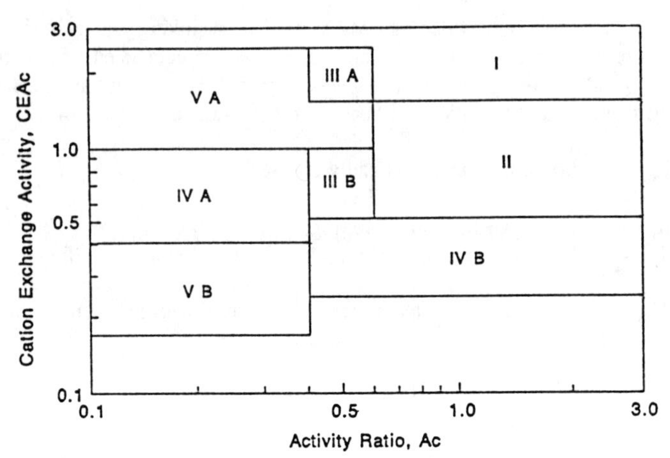

FIG. 4. Chart for the Prediction of Suction Compression Index Guide Number

The denominator of both activity ratios is known as the "percent fine clay" and represents that percent of the portion of the soil which passes the No. 200 sieve which is finer than 2 microns.

The Cation Exchange Capacity (CEC), may be measured with a spectrophotometer (Mojeckwu 1979) or it may be estimated with sufficient accuracy by Eq. (10) which was developed by Mojeckwu (1979):

$$CEC \cong (PL\%)^{1.17} \tag{10}$$

The regions on the chart each have a volume change guide number corresponding to the suction compression index of a soil with 100 percent fine clay. Values of the guide numbers are given in Table 1. The actual suction compression index is proportional to the actual percent of fine clay in the soil. Thus the actual γ_h is:

$$\gamma_h = \gamma_o \times \left[\frac{\% - 2\ \text{micron}}{\% - \text{No. } 200}\right] \tag{11}$$

for the soil portion finer than the No. 200 sieve. A method for estimating γ_h for soils containing coarse-grained particles was developed by Holmgreen (1968).

The mean principal stress compression index, γ_σ, is related to γ_h by the following equation:

$$\gamma_\sigma = \gamma_h \frac{1}{1 + \dfrac{h}{\Theta\left[\dfrac{\partial h}{\partial \Theta}\right]}} \tag{12}$$

where Θ = the volumetric water content; and

$\partial h/\partial \Theta$ = the slope of the suction-versus-volumetric water content curve.

TABLE 1. Values for a Soil with 100% Fine Clay Content

Region	Volume Change γ_o Guide Number
I	0.220
II	0.163
IIIA	0.096
IIIB	0.096
IVA	0.061
IVB	0.061
VA	0.033
VB	0.033

Suction Profiles

For design purposes, it is desirable to compute the total heave that occurs between two steady state suction profiles, one given by a constant velocity of water entering the profile (low suction levels due to wetting) and the other given by a constant velocity of water leaving the profile (high suction levels due to drying). Steady state conditions are given by Darcy's law:

$$v = -k \left[\frac{\partial H}{\partial Z} \right] \tag{13}$$

The total head, H, is made up of the total suction, h, and the elevation head, Z:

$$H = h + Z \tag{14}$$

The gradient of total head is:

$$\frac{\partial H}{\partial Z} = \frac{\partial h}{\partial Z} + 1 \tag{15}$$

Solving for the change of suction as a function of the change of elevation gives:

$$\partial h = -\partial Z \left[1 + \frac{v}{k} \right] \tag{16}$$

Use of Gardner's equation for the unsaturated permeability (Gardner, 1958) gives:

$$\Delta h = -\Delta Z \left[1 + \frac{v}{k_o} (1 + a|h|^n) \right] \tag{17}$$

where a, n = 10^{-9}, 3.0 typically; and
k_o = saturated permeability, cm/s.

The sign of the velocity, v, is positive for water leaving the soil (drying) and negative for water entering the soil. Using Mitchell's equation for the unsaturated permeability (Mitchell 1980) gives:

$$\Delta h = -\Delta Z \left[1 + \frac{v}{k_o} \left[\frac{h}{h_o} \right] \right] \tag{18}$$

where h_o = about -100 cm. in clays.

Mitchell's expression takes into account, to some extent, the increased permeability of the soil mass due to the cracks that become open at high suction levels. This is illustrated in Fig. 5 which contrasts the permeability of intact soil with the Mitchell unsaturated permeability formulation. The increased permeability due to cracks begins to develop at approximately a pF of 3.5. It is speculated that in general, the pF-level where cracks begin to form is the equilibrium pF-value which corresponds to the local value of the Thornthwaite Moisture Index (Thornthwaite 1948). The velocity of water entering or leaving the soil may be estimated from Thornthwaite Moisture Index moisture balance computations.

The suction profiles for two transient states can be predicted approximately using Mitchell (1980):

$$U(Z,t) = U_e + U_o \exp\left[-Z\sqrt{\frac{n\pi}{\alpha}} \right] \cos\left[2\pi nt - Z\sqrt{\frac{n\pi}{\alpha}} \right] \tag{19}$$

where U_e = the equilibrium value of suction expressed as pF;
U_o = the amplitude of pF (suction) change at the ground surface;
n = the number of suction cycles per second (1 year = 31.5×10^6 seconds);
α = the soil diffusion coefficient using Mitchell's unsaturated permeability (ranges between 10^{-5} and 10^{-3} cm^2/s); and
t = time in seconds.

Tables of values of U_e and U_o for clay soils with different levels of Mitchell's unsaturated permeability have been found using a trial and error procedure. The dry suction profile has a U_e-value of 4.5 and a U_o-value of 0.0. The wet suction profile has U_e and U_o-values that vary with the soil type and Thornthwaite Moisture Index. Typical values are shown in Table 2.

Values of n are 1 cycle per year for all Thornthwaite Moisture Indexes (TMI) less than -30.0 and 2 cycles per year for all TMI greater than -30.0.

Eq. (16) shows that the equilibrium suction profile corresponds to a vertical velocity of zero and that it has a slope of 1 cm more negative suction for every 1 cm higher in elevation.

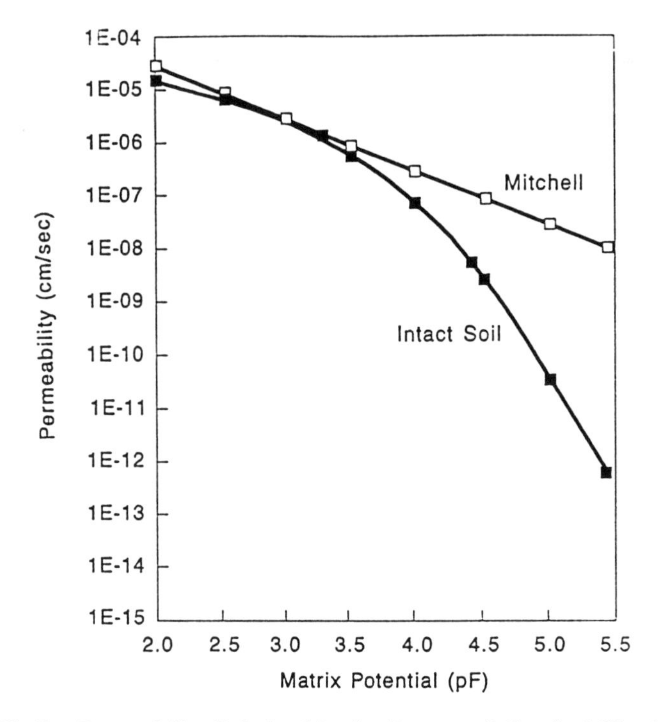

FIG. 5. Permeability Relationships for Intact and Cracked Clay Soil

Use of Mitchell's unsaturated permeability formulation in a finite element simulation of suction changes on each side of a vertical moisture barrier produced reasonable predictions of the measured values except in the vicinity of cracks that were open to the air. The pattern of measured versus predicted suctions are as shown in Fig. 6. Actual data for a monitoring site near Seguin, Texas are shown in Fig. 7, (Jayatilaka et al. 1993). A crack that is open to the atmosphere gets much wetter and drier with fluctuations of the weather than does the cracked soil in which the cracks are not open to the air. The close correspondence between the predicted and measured values of suction in all other instances lends support to the practical use of Mitchell's relationship for unsaturated permeability.

The values of the equilibrium suction U_e that may be used to estimate suction profiles vary with the Mitchell unsaturated permeability, p(cm^2/sec), and the Thornthwaite Moisture Index. Typical values are given in Table 3.

Heave (or shrinkage) from a present condition in the soil uses as the initial value of suction, h_i, the value measured from samples taken. The suction can be measured by any of a number of acceptable means. The filter paper method is the simplest.

TABLE 2. Wet Suction Profile Values

Thornthwaite Moisture Index	Mitchell Unsaturated Permeability (cm^2/s)	U_e (pF)	U_o (pF)
-46.5	5×10^{-5} 1×10^{-3}	4.43 4.27	0.25 0.09
-11.3	5×10^{-5} 1×10^{-3}	3.84 2.83	1.84 0.83
26.8	5×10^{-5} 1×10^{-3}	3.47 2.79	1.47 0.79

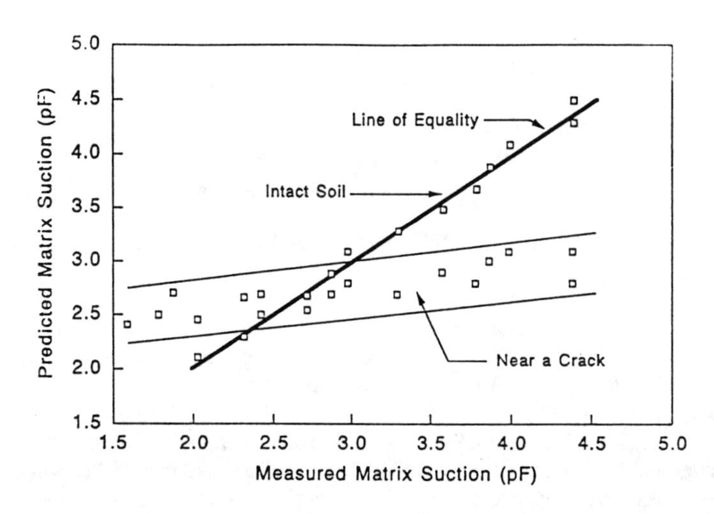

FIG. 6. A Typical Pattern of Measured Soil Suction vs. Predicted Soil Suction

If the suction profile is not controlled by the evapotranspiration at the soil surface but by a high water table, this fact can be discovered by measuring the suction on a Shelby tube sample. If the magnitude of the suction is lower than that expected when the suction profile is governed by surface evapotranspiration, then it is controlled by a high water table. This will usually be within about 10 m (30 feet) below the surface.

If the suction is higher than expected then there is osmotic suction present. Osmotic suction levels may be measured with vacuum desiccators.

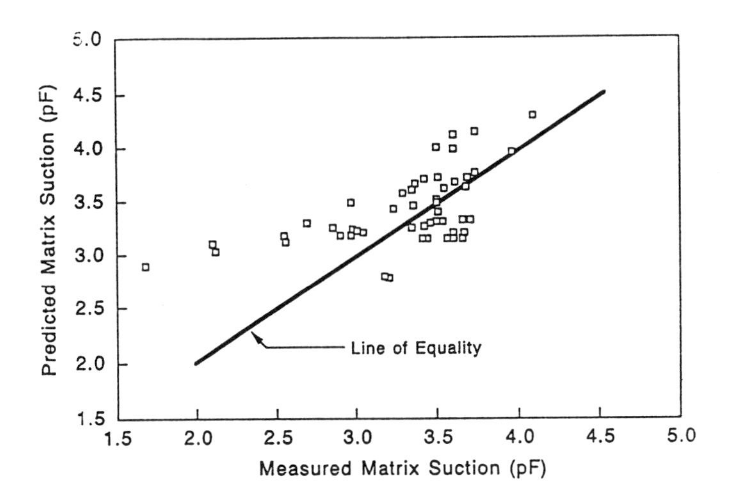

FIG. 7. Measured Soil Suction vs. Predicted Soil Suction at Seguin (Jayatilaka et al. 1993)

TABLE 3. Equilibrium Suction Values, U_e

TMI	Mitchell Unsaturated Permeability, cm^2/s		
	5×10^{-5}	2.5×10^{-4}	1.0×10^{-3}
-46.5	4.27	4.32	4.43
-30.0	3.80	3.95	4.29
-21.3	3.42	3.64	4.20
-11.3	2.83	3.10	3.84
26.8	2.79	3.05	3.47

Estimates of Unsaturated Soil Properties

The fundamental definition of p is :

$$p = \frac{k_o |h_o|}{0.4343} \tag{20}$$

where $|h_o| = 100$ cm for clays.

The units of k_o, the saturated permeability, (cm/s), and $|h_o|$, the suction at which the soil desaturates (cm) produce units of (cm^2/s) for the Mitchell unsaturated permeability.

The Mitchell unsaturated permeability, p, is estimated by:

$$p = \frac{\alpha \gamma_d}{|S| \gamma_w} \quad (\text{cm}^2/\text{s})$$

where γ_w = the unit weight of water;
α = the Mitchell diffusion coefficient, cm^2/s, which is used in Eq. (19);
$|S|$ = the absolute value of the slope of the pF-vs-gravimetric water content, w line; and
γ_d = the dry unit weight of the soil.

The value of α can be estimated from:

$$\alpha = 0.0029 - 0.000162(S) - 0.0122(\gamma_h) \tag{21}$$

The value of S is negative and can be estimated from:

$$S = -20.29 + 0.1555(LL\%) - 0.117(PI\%) + 0.0684(\% -\#200) \tag{22}$$

where LL = the liquid limit in percent;
PI = the plasticity index in percent; and
-#200 = the percent of the soil passing the #200 sieve.

The slope of the suction-versus-volumetric water content curve is given by:

$$\left[\frac{\partial h}{\partial \Theta}\right] = \frac{1}{0.4343} \frac{S \gamma_w}{\gamma_d} h \tag{23}$$

Because both S and h are negative, the slope is inherently positive as illustrated in Fig. 1. The correction term in the relation between γ_h and γ_σ given in Eq. (12) is found by:

$$\frac{h}{\Theta \left[\frac{\partial h}{\partial \Theta}\right]} = \frac{0.4343}{S w} \tag{24}$$

where w = the gravimetric water content.

Because S is negative, so is the correction term.

An approximate suction (pF)-versus-volumetric water content curve can be constructed with the empirical relationships given above and the saturated volumetric water contents given in Table 4. The construction is illustrated in Fig. 8. First, point A is located at the intersection of the field capacity volumetric water content ($= 0.88\ \Theta_{sat}$) and a pF of 2.0. Second, a line with a slope of $S\gamma_w/\gamma_d$ is drawn from point A to its intersection with the vertical axis. Third, point C is located at a volumetric water content of $0.10\ \Theta_{sat}$ and the tensile strength of water (pF = 5.3 or 200 atmospheres). Fourth, point D is located at zero water content and a pF of 7.0, corresponding to oven dry. Fifth, a straight line is drawn between points C and D to its intersection with the first line.

This construction makes it possible to estimate water contents once the computed suction profiles are known. This allows measured water contents to be compared with the predicted values.

TABLE 4. Ranges of Saturated Volumetric Water Content by Unified Soil Class (Mason et al. 1986)

Unified Class	Ranges of Θ_{sat}*
GW	0.31 - 0.42
GP	0.20
GM	0.21 - 0.38
GM-GC	0.30
SW	0.28 - 0.40
SP	0.37 - 0.45
SM	0.28 - 0.68
SW-SP	0.30
SP-SM	0.37
SM-SC	0.40
ML	0.38 - 0.68
CL	0.29 - 0.54
ML-CL	0.39 - 0.41
ML-OL	0.47 - 0.63
CH	0.50

* Θ_{sat} = n (porosity)

Differential Movement

Differential movement which affects the performance of a ground-supported slab may take numerous shapes but the most important shapes for design purposes are those which generate the maximum values of moment, shear, and differential deflection of the slab. The two shapes that can be generated by water entering or leaving the soil beneath a slab are the edge lift and center lift conditions.

If a slab is cast on dry ground, the entire slab may move upward until an equilibrium suction profile is established, after which the edges will move up and down in response to the seasonal changes. If the same slab were cast on wet ground, the entire slab will move downward until an equilibrium profile is established. Once more, the edges will move up and down in response to the seasonal moisture changes. Thus, a major concern for design is whether these seasonal movements will cause moment, shear, and differential deflections that exceed the capacity of the designed slab cross-section. The distance within which these changes take place has been named the "edge moisture variation distance".

An empirical relation between this distance and the Thornthwaite Moisture Index has been used in the Post-Tensioning Institute Manual for the Design and Construction of Post-Tensioned Slabs-on-Ground (1980). Because it is known that the "edge moisture variation distance" depends upon the permeability of the soil as well, it is important to determine that relation.

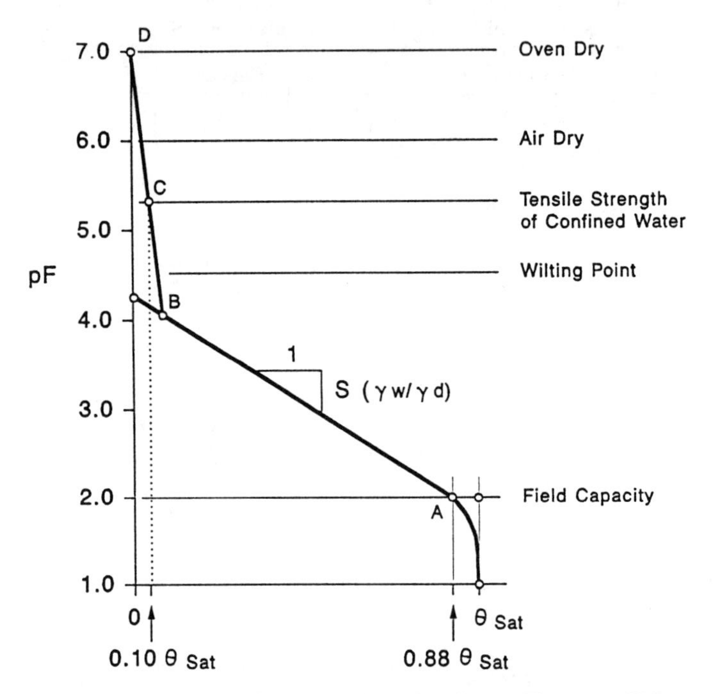

FIG. 8. Approximate Construction of a Suction (pF)-versus-Volumetric Water Content Curve

The calibrated finite element program with coupled transient moisture flow and elasticity that had been used in the study of vertical moisture barriers provided an ideal means to study the edge moisture variation distance. A full range of α and p values were used to determine the relation of the moisture distance and the Thornthwaite Moisture Index and unsaturated soil properties. Both edge lift and center lift conditions were explored using several hundred runs with the program. Center lift conditions were simulated by a one year dry spell following a wet suction profile condition. Edge lift conditions were simulated by a one year wet spell following a dry suction profile condition. The edge moisture variation distance was considered to be that distance between the edge of the foundation and the point beneath the covered area where the suction changed no more than 0.2 pF during the entire period of simulation.

The dry and wet conditions used annual suction variation patterns that were appropriate for each of nine different climatic zones ranging from a Thornthwaite Moisture Index of −46.5 to +26.8, spanning the range found in Texas. The resulting edge moisture variation distances are shown in Figs. 9 and 10. Seven different soils were used in the study. No distance less than 2.0 feet (0.6 m) was considered to be adequate for design purposes.

In Fig. 9 for the center lift condition, Soils No. 1, 2, and 3 are highly pervious and Soils No. 5, 6, and 7 are practically impervious. Only soils with properties between No. 3 and No. 4 have edge moisture variation distances in the range presently used in the PTI manual (1980).

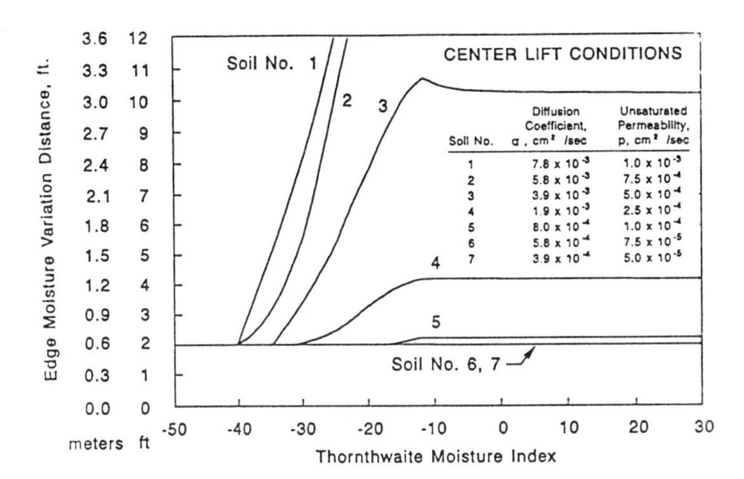

Soil No.	Diffusion Coefficient, α, cm²/sec	Unsaturated Permeability, p, cm²/sec
1	7.8×10^{-3}	1.0×10^{-3}
2	5.8×10^{-3}	7.5×10^{-4}
3	3.9×10^{-3}	5.0×10^{-4}
4	1.9×10^{-3}	2.5×10^{-4}
5	8.0×10^{-4}	1.0×10^{-4}
6	5.8×10^{-4}	7.5×10^{-5}
7	3.9×10^{-4}	5.0×10^{-5}

FIG. 9. Edge Moisture Variation Distances for the Center Lift Moisture Condition

In Fig. 10 for the edge lift condition, Soils No. 5, 6, and 7 are practically impervious while Soils No. 2, 3, and 4 have edge moisture variation distances in the range presently used in the PTI manual. Soil No. 1 is more pervious and outside the range presently used in the PTI manual.

The edge moisture variation distances of soils with unsaturated permeabilities different than these seven soil types can be found by interpolation on these two figures. The edge moisture variation distance in center lift mode, in which the soil around the edge of the slab is drier than the soil supporting it, is more sensitive to changes in the unsaturated permeability than with the edge lift mode.

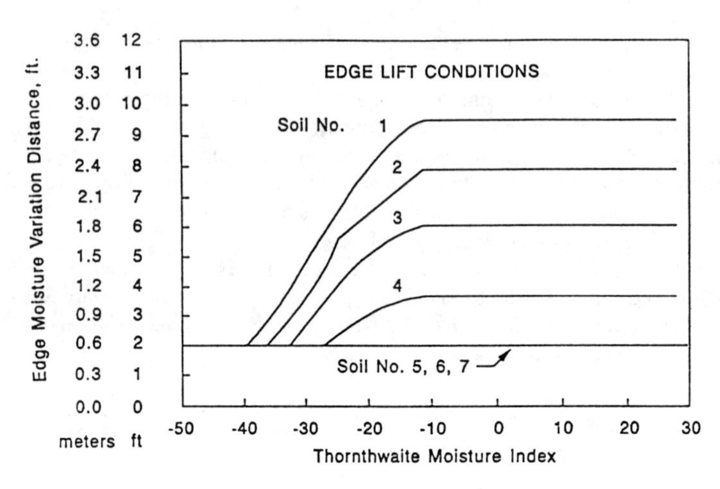

FIG. 10. Edge Moisture Variation Distances for the Edge Lift Moisture Condition

CONCLUSIONS

Simple laboratory tests can be used to determine important properties of expansive soils including the compression indices due to matrix suction and mean principal stress, the slope of the suction-versus-water content curve, and the unsaturated permeability and diffusivity. The tests are the Atterberg limits, hydrometer test, water content, dry density, and sieve analysis.

Prediction of differential movement depends strongly upon the edge moisture variation distance which, in turn, depends upon the Thornthwaite Moisture Index and the unsaturated permeability of the soil. Tree roots penetrating beneath the edge of a building will have a zone of moisture influence beyond the edge of the root zone equal to the edge moisture variation distances shown in Figs. 9 and 10. This explains the unusually destructive effect that trees have when they grow near enough to the edge of a foundation to have their roots intrude beneath the edge. It also explains the effectiveness of vertical root and moisture barriers around the perimeter of the foundation in reducing the moisture variation distance and the differential movement. A vertical barrier carried to a depth of a 4 feet (1.2 meters) excludes many roots, makes the edge moisture variation distance predictable, and reduces the differential movement that a foundation must be designed to withstand.

ACKNOWLEDGEMENTS

The author gratefully acknowledges D. A. Gay for developing the coupled unsaturated moisture flow and elasticity program used in the analysis reported in this paper and R. Jayatilaka for making the multiple runs of the program which resulted in the family of curves in Figs. 9 and 10. The Texas Department of Transportation supported the research project that produced the data shown in Fig. 7.

APPENDIX - REFERENCES

Post-Tensioning Institute (1980). *Design and Construction of Post-Tensioned Slabs-on-Ground*, Phoenix, Arizona.

Fredlund, D. G. and Rahardjo, H. (1993). *Soil Mechanics for Unsaturated Soils*, John Wiley & Sons, New York, New York.

Gardner, W. R. (1958). "Some steady state solutions of the unsaturated moisture flow equation with application to evaporation from a water table." *Soil Science*, 85(4), 223-232.

Holmgreen, G. G. S. (1968). "Nomographic calculation of linear extensibility in soils containing coarse fragments." *Soil Science Society of America Proceedings*, 32(4), 568-570.

Jayatilaka, R., Gay, D. A., Lytton, R. L., and Wray, W. K. (1992). "Effectiveness of controlling pavement roughness due to expansive clays with vertical moisture barriers," *Research Report No. 1165-2F*, Texas Transportation Institute, College Station, Texas.

Mason, J. G., Ollayos, C. W., Guymon, G. L., and Berg, R. L. (1986). *User's Guide for the Mathematical Model of Frost Heave and Thaw Settlement in Pavements*, U. S. Army Cold Region Research and Engineering Laboratory, Hanover, New Hampshire.

McKeen, R. G. (1981). "Design of airport pavements on expansive soils," *Report No. DOT/FAA-RD-81-25*, Federal Aviation Administration, Washington, D.C.

Mitchell, P. W. (1980). "The structural analysis of footings on expansive soil," *Research Report No. 1*, 2nd Edition, K. W. G. Smith and Associates.

Mojeckwu, E. C. (1979). "A simplified method for identifying the predominant clay mineral in soil," M.S. Thesis, Texas Tech University, Lubbock.

Thornthwaite, C. W. (1948). "Rational classification of climate." *Geographical Review*, 38(1), 55-94.

FOUNDATION DEFORMATION DUE TO EARTHQUAKES

Ricardo Dobry,[1] Member, ASCE

ABSTRACT: Permanent foundation deformation and associated damage to buildings due to earthquakes is discussed. Ground straining is a main cause of distress, especially that associated with liquefaction-induced lateral spreading. Descriptions and examples are presented for the various mechanisms of foundation deformation, followed by a discussion of the correlation between ground displacement and foundation/building damage. A review is included of available methods for the engineering evaluation of liquefaction effects.

INTRODUCTION

Deformation of foundations of buildings, bridges and other structures is a common occurrence during earthquakes and is responsible for much of the damage and economic losses caused by seismic events. Mizuno (1987) reviewed damage to pile foundations for earthquakes in Japan between 1923 and 1983. Youd (1989) discussed a number of case histories of building distress caused by ground displacement during strong earthquakes in several countries. Two volumes edited by Hamada and O'Rourke (1992) and O'Rourke and Hamada (1992) have summarized extensive evidence of ground and foundation deformation due to liquefaction for ten strong earthquakes in the United States, Japan and Philippines between 1906 and 1990. Most of the evidence in the U. S. corresponds to California and Alaska. However, similar liquefaction phenomena including ground and/or foundation movement have been reported for earthquakes in New Madrid, Missouri in 1811-1812, in Charleston, South Carolina in 1886, and in Quebec, Canada in 1988 (Fuller 1912; Peters and Herrmann 1986; Tuttle et al. 1990). As in most cases the cause of damage is the permanent movement of the foundation

---

[1] Prof. of Civ. Engrg., Rensselaer Polytechnic Institute, Troy, NY 12180-3590.

rather than its cyclic component during shaking, this paper focuses on permanent deformation.

Earthquake-induced deformation has affected both shallow and deep foundations in different soils. Fig. 1 shows the failure of the Marine Sciences Laboratory at Moss Landing, in the epicentral area of the 1989 Loma Prieta earthquake in California. This group of low, modern 1- and 2-story structures founded on concrete slabs was destroyed beyond repair by liquefaction of the cohesionless foundation soil, which experienced a permanent lateral displacement of more than 1 m; this displacement literally pulled the foundations apart (Seed et al. 1991; Mejia et al. 1992). Figs. 2 and 3 illustrate the settlement and tilting of a multistory office building founded on a concrete slab and friction piles in soft clay in Mexico City, several hundred kilometers from the epicenter of the 1985 earthquake. This foundation had a low factor of safety against static bearing capacity failure. The building tilted 3.3%, with a maximum settlement of 0.78 m measured in the SW corner, when the static plus inertial loads (especially the dynamic loads created by the large overturning moment) exceeded the capacity of the soil under the slab and around the piles. The situation was aggravated by the smaller number and diameter of piles along the periphery of the slab (Mendoza and Auvinet 1988).

FIG. 1. Lateral Spread Failure due to Liquefaction, Marine Sciences Laboratory at Moss Landing, CA, 1989 Loma Prieta Earthquake (Youd, Personal Communication; Photo Taken by G. Castro)

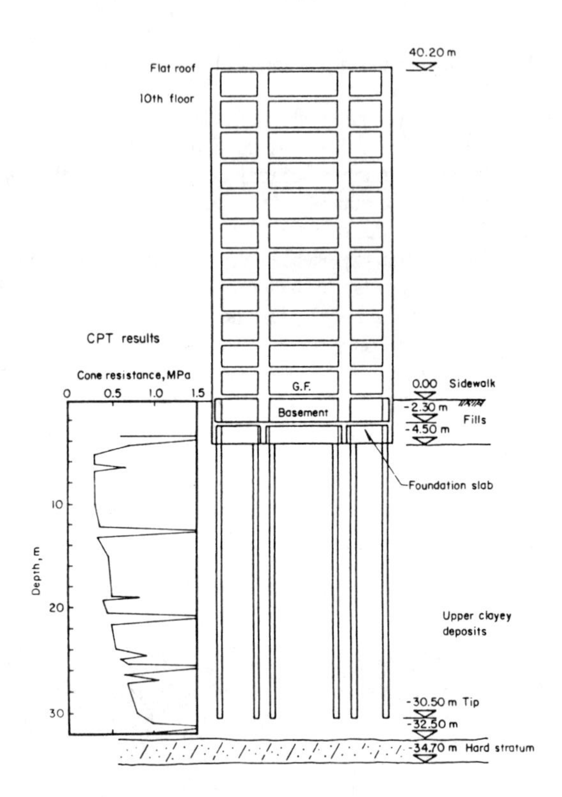

FIG. 2. Multistory Building on Slab and Friction Piles in Soft Clay that Settled and Tilted in 1985 Mexico City Earthquake (Mendoza and Auvinet 1988)

Causes of Foundation Deformation

Table 1 lists six main mechanisms of foundation deformation and failure during earthquakes. In the first five of them, the foundation deforms because the ground moves and carries the foundation along with it; this is by and large the most general and pervasive contributor to foundation damage (Fig. 1). The last mechanism in Table 1 corresponds to dynamic or static soil failure, associated respectively with the structure's inertia forces during shaking (as in Figs. 2 and 3), and with bearing capacity failure due to the reduced stiffness and strength of a soil liquefied by the earthquake shaking. In the rest of this paper, brief descriptions are first provided for each one of these mechanisms and its manifestations including appropriate references. This is followed by comments on the correlation between ground displacement and damage, and by a discussion on engineering evaluation of liquefaction effects and use of centrifuge model testing.

TABLE 1. Main Causes of Foundation Deformation and Failure during Earthquakes

*	Surface faulting
*	Compaction settlement of loose cohesionless soil
*	Lateral spreading of loose saturated soil due to liquefaction
*	Landsliding
*	Movement or failure of nearby retaining structure
*	Dynamic or static soil failure associated with structural forces

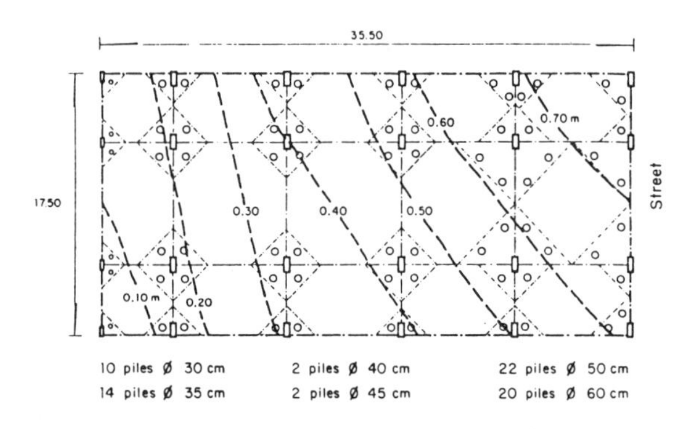

FIG. 3. Settlement and Tilting of Foundation Slab of Building in Fig. 2, 1985 Mexico City Earthquake (Mendoza and Auvinet 1988)

Surface Faulting

Vertical and/or horizontal ground displacements can occur at or near active faults, affecting structures either during earthquakes or as a result of fault creep. The phenomenon is important in seismic regions where surface faulting has occurred repeatedly, such as California, but it is restricted to the fault rupture zone, with the ground displacement attenuating rapidly with distance from the fault. The value of the maximum displacement is strongly correlated with the size of the earthquake, varying from a few centimeters to more than 10 m for very large magnitude events. To avoid this problem, the State of California generally requires a minimum setback distance of 15 m (50 ft) between a building and a well-defined zone containing the traces of an active fault. The relative importance of horizontal versus vertical ground displacement and the type of predominant surface ground straining (shear, compression or extension), exhibit a general correlation to the style of faulting: strike-slip, normal, thrust, etc. (Bonilla 1970; Youd 1980; Bonilla et al. 1984; BSSC 1992).

Compaction Settlement

Vertical ground displacement due to the shaking and compaction of loose (dry or saturated) sand and other cohesionless soils. The settlement is typically larger when associated with liquefaction, compared with the same soil in dry condition. Settlements as much as 5% or more of the thickness of the loose sand layer have been reported. Differential settlements and associated vertical shear straining of the ground and of foundations placed on it can occur in areas where the thickness or density of the compacting soil changes rapidly over short distances (Tokimatsu and Seed 1987; Ishihara and Yoshimine 1990; O'Rourke et al. 1992a).

Lateral Spreading

Lateral spreading of loose saturated sand and other cohesionless soils due to liquefaction is the single most important cause of large ground displacement and associated damage to foundations and structures. Most of the effects summarized in the books by Hamada and O'Rourke (1992) and O'Rourke and Hamada (1992) are associated with lateral spreading. A sketch of the phenomenon is presented in Fig. 4, while its consequences are illustrated by Figs. 1 and 5. It induces mostly horizontal displacements from a few centimeters to more than 10 m, and it can affect a large area which moves, either downslope along a slope as small as 0.5%, or toward a free face. The amount of lateral displacement typically increases with slope and height of the free face and decreases with distance from the free face. Extensional ground straining including fissures, as well as vertical settlements, tend to occur at the head of the spread while compression and ground uplifting appears at the toe. Ground shear develops especially at the spread margins. Fig. 6 shows the pattern of lateral ground displacements for the 1971 San Fernando, California earthquake, obtained mainly by comparison of air photos before and after the earthquake in a large area of more than 1 km^2. Fig. 7 presents a map of the corresponding surficial ground cracks. Although most of the lateral displacements were due to liquefaction and lateral spreading of a loose alluvium layer, they also included a tectonic (faulting) component. The average ground surface in the area was 1.5°, with a maximum slope through the Juvenile Hall of about 3° (Youd 1973; Youd and Perkins 1987; Bartlett and Youd 1992; O'Rourke et al. 1992).

Landsliding

Landslides (sometimes massive) can occur during earthquakes in slopes of soils ranging from clays and sands to rock. Several examples are cited by Seed (1970) from the 1964 Alaskan and the 1960 Southern Chile events; many small and some large landslides also occurred during the 1989 Loma Prieta earthquake (Seed et al. 1990). In some of these cases liquefaction may have contributed to the failure. Slides generate vertical and horizontal ground displacements. Although typically vertical settlement, shear and extension predominate at the head of the slide and behind the slope, with compression appearing at the toe, the detailed pattern of ground straining can be quite complicated. In the Turnagain Heights area of Anchorage, Alaska, several dozen houses were destroyed by the 1964 slide, while the 1989 Loma Prieta slides also damaged or destroyed a number of single family homes.

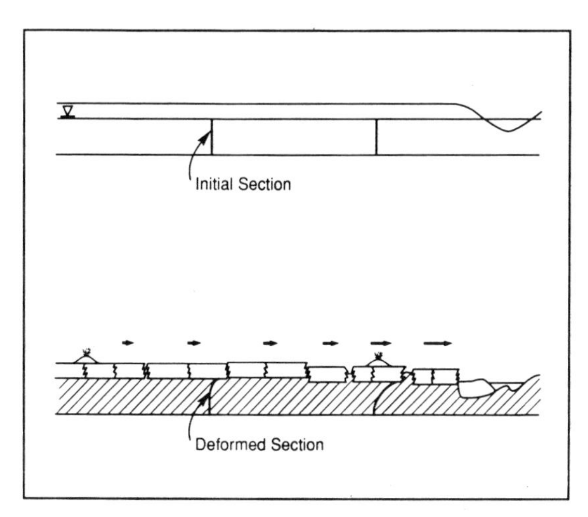

FIG. 4. Sketch of Lateral Spread before and after Failure; Liquefaction Occurs in the Cross-Hatched Zone (Youd 1984)

Movement or Failure of Nearby Retaining Structure

The movement of retaining structures and associated effects (illustrated by Fig. 8), is often caused by liquefaction of the backfill. It typically occurs near bodies of water with damage to waterfront facilities (Seed 1970; Hamada 1992).

Dynamic or Static Soil Failure due to Structural Forces

Movement has been observed of shallow and deep foundations, as well as of bridge abutments, due to soil yielding under the static and/or inertia forces of the structure during shaking. Failure during shaking due to structural inertia forces was the main mechanism of damage to both shallow foundations and friction piles in soft sensitive clay in Mexico City in 1985 (Figs. 2 and 3). A common feature of the damaged foundations for that event is that they had a low static factor of safety before the earthquake. For tall buildings founded on piles in clay or sand, the peripheral piles may be especially sensitive to the effect of inertia forces caused by a high overturning moment. At the other extreme is the static bearing failure of shallow foundations due to liquefaction of the soil, such as observed for many buildings that settled and tilted in the 1964 Niigata, Japan earthquake. The decrease in lateral stiffness of piles due to sand liquefaction may increase the cyclic lateral movements of the foundation during shaking, while liquefaction of the soil under the pile tips may induce penetration during or after the shaking (Seed 1970; Youd 1980; Mizuno 1987; Mendoza and Auvinet 1988; Hamada 1992a).

FIG. 5. Reinforced-Block Building Pulled Apart by about 0.6 m Ground Extension due to Lateral Spreading, 1971 San Fernando, CA Earthquake (Youd 1980)

Ground Displacement and Foundation Damage

Similar to the case of static settlements, the cause of earthquake damage to foundations and buildings is not so much the ground displacement itself, but the ground straining. For example, the destruction of shallow foundations by lateral spreading in Figs. 1 and 5 was caused by horizontal extension of the ground. Differential lateral displacements – such as associated with the variation with distance of the magnitudes of the vectors in Fig. 6 – can produce horizontal extension, compression or shear, while differential vertical displacements cause vertical shearing of the ground. As noticed by Youd (1989), in general, shallow foundations are most sensitive to ground extension and vertical shear, and somewhat less sensitive to horizontal shear and compression. A main cause of damage to pile foundations is the variation of lateral ground displacement with depth.

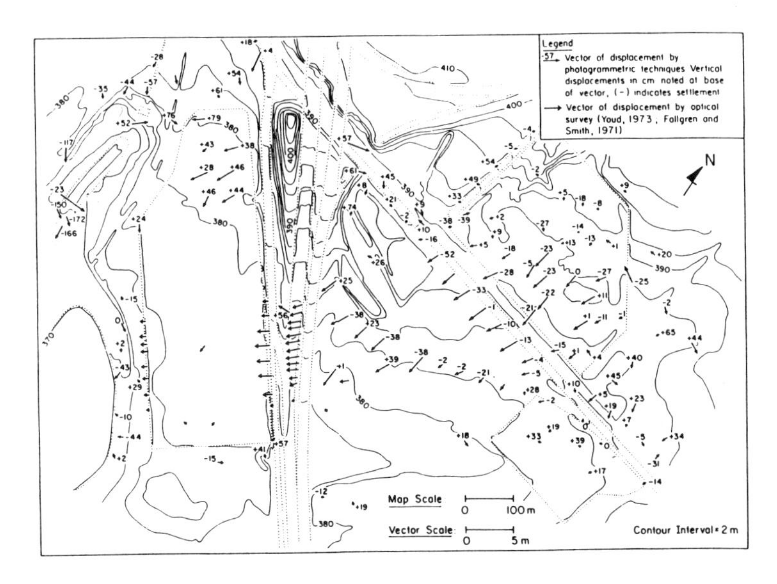

FIG. 6. Lateral Displacement Vectors Obtained from Air Photo Analyses and Optical Surveys, Juvenile Hall and Nearby Areas, 1971 San Fernando, CA Earthquake (O'Rourke et al. 1992)

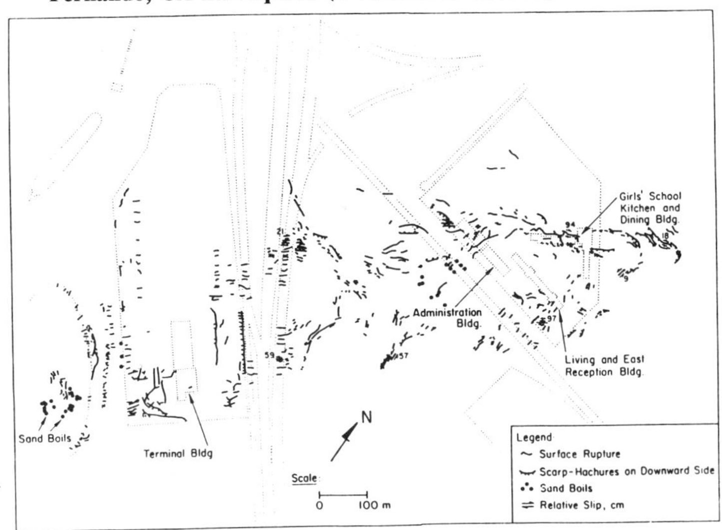

FIG. 7. Map of Surficial Ground Cracks, Sand Boils, and Pressure Ridges for the Same Area of Fig. 6, 1971 San Fernando, CA Earthquake (O'Rourke et al. 1992)

FIG. 8. Damage to Sheet Pile Quay Wall (Nakajima Pier), 1983 Nihonkai-Chubu, Japan Earthquake (Hamada 1992)

Therefore, any indication of the type of ground surface straining expected due to the design earthquake is useful to the engineer and should help his/her judgment when making design or retrofitting decisions for shallow foundations. The next logical steps should be to have methods to predict both the amount of ground strain of that type in the free field, and the degree of foundation/building damage associated with such free field strain. Susuki and Masuda (1991) have studied the measured surface ground movements due to lateral spreads at two Japanese cities after earthquakes, and have attempted to model analytically the corresponding patterns of ground straining. Unfortunately, this is a difficult problem, and with a couple of exceptions discussed in the next section, ground surface strain is generally difficult to measure and even more difficult to predict. As a result, foundation and building damage have been correlated with ground displacement rather than with strain (Table 2 and Fig. 9). Again, the use of ground displacement as in Table 2 is similar to the standard static design procedure for shallow footings on sand, where an acceptable settlement of 2.5 cm (1 inch) is taken to imply that the differential settlements/vertical shear straining of ground and foundation will also be small and acceptable. The above discussion applies especially to shallow foundations; the case of piles is somewhat different as discussed in the next section.

Engineering Evaluation of Liquefaction Effects

For structures on shallow or deep foundations in loose saturated sand or other cohesionless soil, the first step in evaluating possible liquefaction effects is to decide if liquefaction will be triggered at the site by the design earthquake. For horizontal or mildly sloping sites such as those typically affected by lateral spreading, this determination can usually be accomplished by the use of empirical charts such as those proposed by Seed et al. (1983), based on standard penetration or static cone measurements at the site. If liquefaction is not triggered, no significant ground displacement associated with lateral spreading is predicted. Also, even if liquefaction is triggered, Bartlett and Youd (1992) have found after reviewing several hundred case histories, that no significant horizontal displacements have occurred for earthquakes of moment magnitude M_w up to about 8 if the sand is dense to very dense with $(N_1)_{60}$ > 15 blows/ft, where $(N_1)_{60}$ is the normalized standard penetration index corresponding to 60% energy defined by Seed et al. (1984). Therefore, even if liquefaction triggering is anticipated, no significant lateral ground displacements due to liquefaction are likely in those cases where $(N_1)_{60} > 15$ and $M_w < 8$.

TABLE 2. Approximate Amounts of Ground-Failure Displacement Required to Cause Repairable and Irreparable damage (Youd 1989)

Type of Deformation	Foundation	Displacement Required to Cause: Repairable Damage (m)	Displacement Required to Cause: Irreparable Damage (m)
Shear	Poorly-Reinforced[1] Well-Reinforced[2]	0.1 > 0.3	> 0.3 ?
Extension	Poorly-Reinforced Well-Reinforced	< 0.05 > 0.1	> 0.3 ?
Compression	Poorly-Reinforced Well-Reinforced	< 0.3 > 0.5	> 0.5 ?
Compression with Vertical	Poorly-Reinforced Well-Reinforced	< 0.2 < 0.3	> 0.2 > 0.3
Vertical	Poorly-Reinforced Well-Reinforced	< 0.05 < 0.1	> 0.2 > 0.3

[1] Foundations with minimal or no temperature reinforcing steel.
[2] Foundations with adequate reinforcing steel to provide considerable structural strength.

The next step is to determine the expected free-field vertical settlement due to compaction of cohesionless layers by the shaking. Although compaction settlement can occur in dry sand, and also in saturated sand even if liquefaction is not triggered, the largest settlements are induced in saturated sands with $(N_1)_{60}$ < 15 after liquefaction has occurred. This is illustrated by Fig. 10, which shows the chart proposed by Tokimatsu and Seed (1987) to predict compaction vertical strain ϵ_v for earthquakes of magnitude $M = 7.5$. The ground surface settlement is obtained by adding up the contributions $\epsilon_v H$ of all cohesionless layers, where H = layer thickness. The paper by Tokimatsu and Seed gives a complete engineering procedure to evaluate compaction settlement for a range of earthquake magnitudes, and it also includes the cases of saturated sand which does not liquefy, as well as of dry sand. Once the ground surface settlement due to compaction is determined (and if no other sources of vertical movement are anticipated), it can be compared with charts such as Table 2. Differential settlements due to the thinning out or disappearance of the loose sand layer(s) over a short distance under the structure can be especially dangerous.

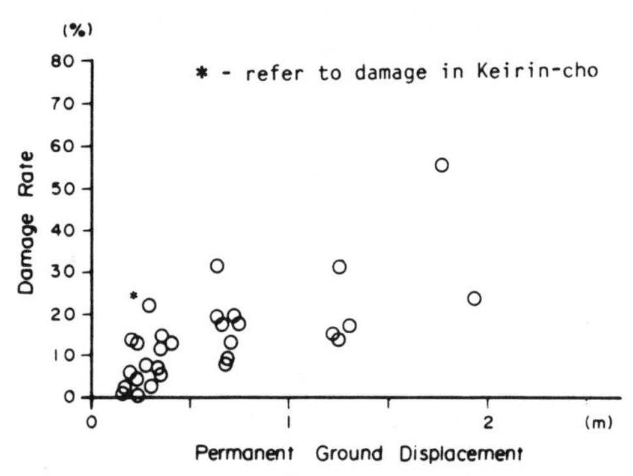

FIG. 9. Relation between Damage Rate to Houses and Permanent Ground Displacements, 1983 Nihonkai-Chubu, Japan Earthquake (Hamada 1992)

Vertical settlements – often accompanied by tilting – of structures on liquefied soil are often observed due to the static bearing capacity effect previously discussed. This possibility should be considered only if liquefaction triggering is anticipated. The vertical settlement under the structure can be significantly higher than the free-field compaction settlement mentioned in the previous paragraph. This additional vertical settlement of the building is difficult to predict, and evaluation techniques based on centrifuge modelling are being developed, see Fig. 11 (Liu and Dobry 1994).

Finally, the evaluation of the free-field horizontal ground displacement D_H associated with lateral spreading is most critical. As previously noticed, lateral ground displacements are not likely unless two conditions are satisfied: (i) liquefaction is triggered at the site, and (ii) $(N_1)_{60}$ < 15 blows/ft (for earthquake moment magnitudes, M_w up to about 8). If these conditions are fulfilled, then the probable value of D_H must be evaluated. This is not a trivial task, as the state-of-the-art on the subject is still under development, and the evaluation may involve considering the topography, geology, and soil conditions/deposition history over a relatively large area extending much beyond the site and neighboring buildings (see Fig. 6). A study of case histories, and especially of Japanese and United States earthquakes where the air photo technique developed by Hamada et al. (1986) has been applied to measure the actual pattern of horizontal ground displacement over large areas, can be very helpful in guiding the engineer's judgment (Hamada and O'Rourke 1992; O'Rourke and Hamada 1992). Specific available methods to evaluate D_H at a site include: (i) an empirical correlation with ground slope and thickness of liquefied layer proposed by Hamada (1992, 1992a); (ii) an upper bound value for very loose, geologically recent alluvial sand deposits with shallow water table, labelled Liquefaction Severity Index (LSI) by Youd and Perkins (1987) and correlated by these authors with magnitude and distance for the Western United States (Fig. 12); (iii) empirical correlations between D_H and factors such as earthquake magnitude and distance, ground slope, thickness of liquefied layer, etc. published by Bartlett and Youd (1992); and (iv) the sliding block, Newmark-type analysis procedure developed by Dobry and Baziar (1992), see Fig. 13. All these methods are strongly influenced by the available case histories, and in some cases have been calibrated versus each other. For example, the line in Fig. 12 is the general Youd and Perkins' LSI prediction for $M_w = 6.5$ earthquakes, while the data points were calculated by Dobry and Baziar using sliding block analysis and earthquake records from the 1979, $M_w = 6.5$ Imperial Valley earthquake. Table 2 can be used as a guide for evaluation of possible effect of D_H on shallow foundations.

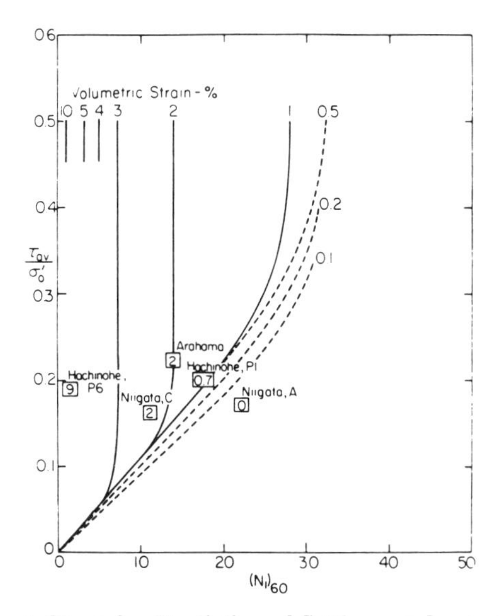

FIG. 10. Proposed Chart for Prediction of Settlement due to Compaction of Saturated Sands due to Earthquakes of Magnitude M = 7.5 (Tokimatsu and Seed 1987)

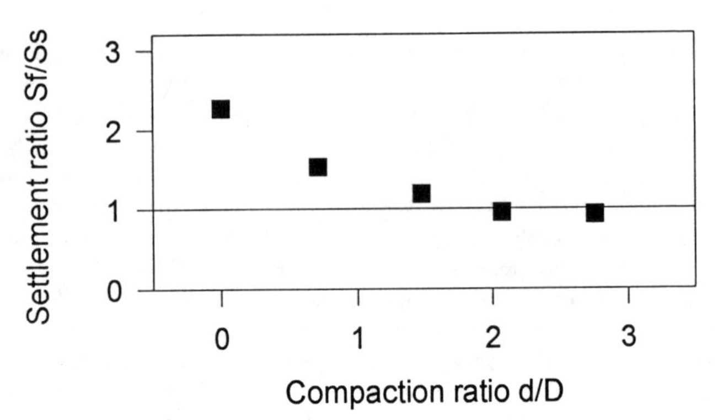

FIG. 11. Effect on Foundation Settlement, Sf, of Compacting Sand Column to Depth d under Circular Surface Foundation of Diameter D, where Ss = Free Field Settlement of Loose Saturated Sand. Obtained from Centrifuge Modelling Tests Which Included In-Flight Earthquake Shaking (Liu and Dobry 1994)

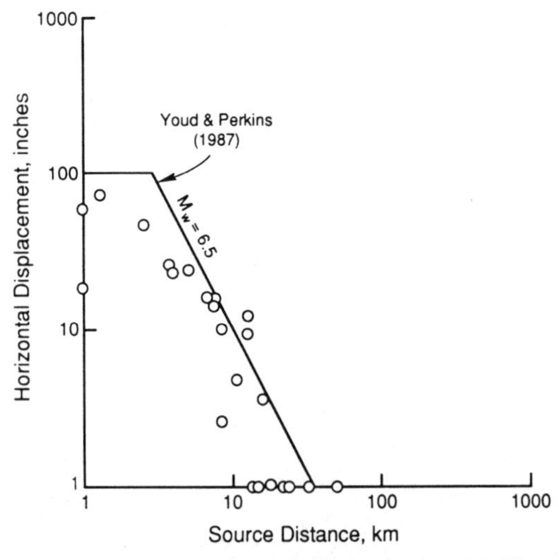

FIG. 12. Comparison between Predicted LSI for M_w = 6.5 (line – Youd and Perkins 1987) and Results of Sliding Block Analyses Using 1979 Imperial Valley, CA Earthquake Records (Data Points – Dobry and Baziar 1992)

Once D_H for the ground surface due to lateral spreading has been predicted, its effect on pile foundations can also be evaluated. As reasonable assumptions are possible for the distribution of lateral displacement with depth – based on the location and thickness of the liquefiable layer – the analysis of piles is more straightforward than that of shallow foundations. Fig. 14 shows the observed damage to reinforced concrete point bearing piles 350 mm diameter produced by $D_H \approx 1.2$ m in the 1964 Niigata earthquake. Fig. 15 presents the results of the predicted pile bending moments by using a numerical modal developed at Cornell University by Miura and O'Rourke (1991) and Meyersohn et al. (1993). This model – implemented in computer program B-STRUCT – accounts for geometrical and material nonlinearities of both piles and soils. The flexural characteristic of the reinforced concrete piles are modelled by moment-curvature relationships, which are obtained by appropriate selection of stress-strain curves of concrete under compressive and tensile stress. The model has been used to develop a design methodology for piles subject to lateral spread which accounts for excessive bending, buckling, and soil flow. Dimensionless charts have been proposed which help define limit states of pile performance as a function of pile stiffness, axial load, subsurface conditions, soil properties, and magnitude of lateral displacement (Meyersohn et al. 1992; Meyersohn 1994). Simplified models of pile group performance have also been proposed. This analytical procedure for piles and pile groups subjected to lateral spreading has been calibrated by field case histories such as that of Fig. 14, and is being further refined based on centrifuge modelling performed at Rensselaer Polytechnic Institute.

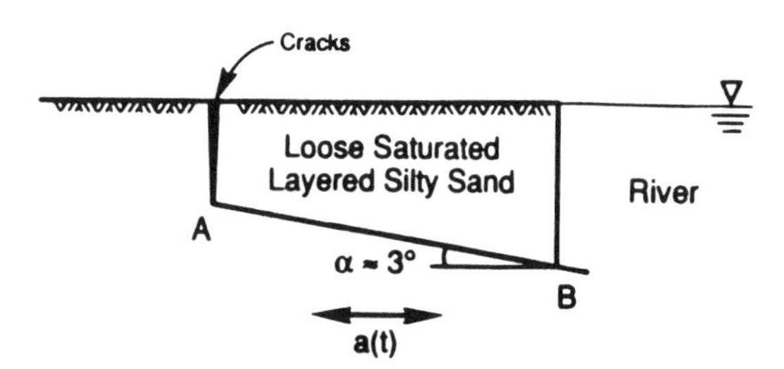

FIG. 13. Sliding Block Model Used for Analysis of Lateral Spreading and LSI Prediction (Dobry and Baziar 1992)

Summary and Conclusions

Deformation of foundations of buildings and other structures occurs often during strong earthquakes and is responsible for much of the damage and economic loss caused by seismic events. Ground deformation and straining in the free field due to liquefaction induced lateral spreading, compaction settlement, faulting, landsliding, and movement of nearby retaining structures, is a main mechanism responsible for structural distress, followed by static or dynamic soil failure under the structural forces. A preliminary correlation available from Youd (1989) between ground displacement and the threshold of damage to foundations and buildings is

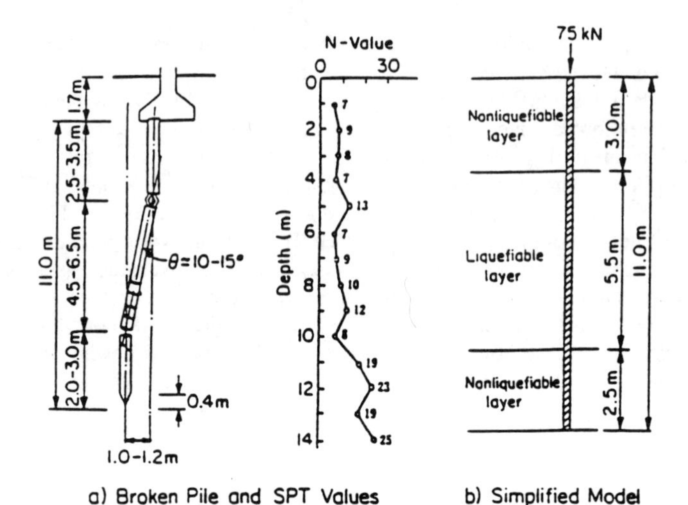

FIG. 14. Observed Damage to Reinforced-Concrete Pile Foundation at NHK Building due to Liquefaction-Induced Lateral Spreading, 1964 Niigata, Japan Earthquake (Hamada et al. 1986; Meyersohn 1994)

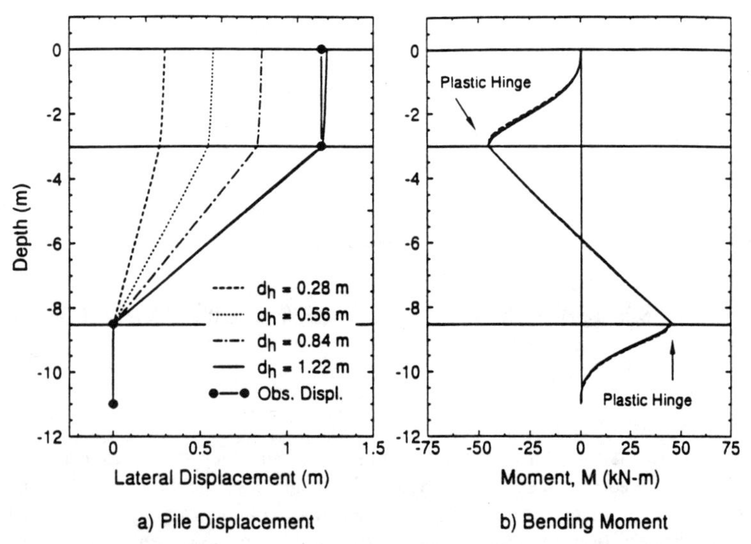

FIG. 15. Analytical Results for the NHK Building Pile Foundation (Meyersohn et al. 1992; Meyersohn 1994)

reproduced in Table 2. This table can be especially useful as a guide when evaluating shallow foundations. The engineering evaluation of a site for liquefaction effects caused by a given design earthquake involves first determining if liquefaction will trigger, followed by estimates of permanent vertical and lateral movements using various recommended procedures. Even if liquefaction triggers, the existing case histories show that for earthquake moment magnitudes, $M_w < 8$, no significant lateral displacements are likely unless $(N_1)_{60} < 15$ blows/ft. Analytical techniques for piles subjected to lateral spreading are available and are being rapidly improved.

ACKNOWLEDGMENTS

The author is grateful to M. J. O'Rourke, T. D. O'Rourke, A. Papageorgiou, and T. L. Youd for their suggestions and help during the preparation of this paper, and to G. Castro, M. Hamada and T. L. Youd for the photos in Figs. 1, 5, and 8.

APPENDIX - REFERENCES

Bartlett, S. F., and Youd, T. L. (1992). "Empirical analysis of horizontal ground displacement generated by liquefaction induced lateral spreads," *Technical Report NCEER 92-0021*, National Center for Earthquake Engineering Research, State University of New York, Buffalo, New York.

Bonilla, M. G. (1970). "Surface faulting and related effects." *Earthquake Engineering*, Prentice-Hall, Englewood Cliffs, New Jersey, 47-74.

Bonilla, M. G., Mark, R. K., and Lienkaemper, J. J. (1984). "Statistical relations among earthquake magnitude, surface rupture length, and surface fault displacement." *Bull. Seis. Soc. of America*, 74(6), 2379-2411.

BSSC (1992). *1991 Edition of NEHRP Recommended Provisions for the Development of Seismic Regulations for New Buildings*, Part 2 (Commentary), Ch. 7, Rept. No. FEMA 223, Building Seismic Safety Council, Washington, D.C.

Dobry, R., and Baziar, M. H. (1992). "Modeling of lateral spreads in silty sands by sliding soil blocks." *Stability and Performance of Slopes and Embankments - II*, Geotechnical Special Publication No. 31, ASCE, New York, New York, 1, 625-652.

Fuller, M. L. (1912). "The New Madrid Earthquake." *U. S. Geological Survey Bulletin 494*, U. S. Dept. of the Interior, Washington, D. C.

Hamada, M. (1992). "Large ground deformations and their effects on lifelines: 1983 Nihonkai-Chubu earthquake." Ch. 4 of Hamada and O'Rourke (1992), 4-1 to 4-85.

Hamada, M. (1992a). "Large ground deformations and their effects on lifelines: 1964 Niigata earthquake." Ch. 3 of Hamada and O'Rourke (1992), 3-1 to 3-123.

Hamada, M., and O'Rourke, T. D. (1992). Eds. "Case studies of liquefaction and lifeline performance during past earthquakes, 1: Japanese case studies," *Report No. NCEER-92-0001*, National Center for Earthquake Engineering Research, State University of Buffalo, Buffalo, New York.

Hamada, M., Yasuda, S., Isoyama, R., and Emoto, K. (1986). "Study on liquefaction induced permanent ground displacements," *Research Report*, Assoc. for the Development of Earthquake Prediction, Tokyo, Japan.

Ishihara, K., and Yoshimine, P. (1991). "Evaluation of settlements in sand deposits following liquefaction during earthquakes." *Soils and Foundations*, 32(1), 173-188.

Liu, L., and Dobry, R. (1994). "Seismic settlements and pore pressures of shallow foundations." *Proc. Int'l. Conf. Centrifuge '94*, Singapore, in press.

Mejia, L. H., Hughes, D. K., and Sun, J. I. (1992). "Liquefaction at Moss Landing during the 1989 Loma Prieta earthquake." *Proc. 10th World Conf. on Earthquake Engineering*, Madrid, 3, 1435-1440.

Mendoza, M. J., and Auvinet, G. (1988). "The Mexico earthquake of September 19, 1985 – behavior of foundations in Mexico City." *Earthquake Spectra*, 4, 835-853.

Meyersohn, W. D. (1994). "Pile response to liquefaction induced lateral spread," Ph.D. thesis, School of Civil and Environmental Engineering, Cornell University, Ithaca, New York.

Meyersohn, W. D., O'Rourke, T. D., and Miura, F. (1992). "Lateral spread effects on reinforced concrete pile foundations." *Proc. 5th U. S. – Japan Workshop on Earthquake Disaster Prevention for Lifeline Systems*, Tsukuba, 173-196.

Miura, F., and O'Rourke, T. D. (1991). "Nonlinear analysis of piles subjected to liquefaction-induced large ground deformation." *Proc. 3rd Japan – U. S. Workshop on Earthquake–Resistant Design of Lifeline Facilities and Countermeasures for Soil Liquefaction*, Tech. Rept. NCEER 91-0001, National Center for Earthquake Engineering Research, State University of New York, Buffalo, New York, 497-512.

Mizuno, H. (1987). "Pile damage during earthquakes in Japan (1923-1983)." *Dynamic Response of Pile Foundations*, Geotechnical Special Publication No. 11, ASCE, New York, New York, 53-77.

O'Rourke, T. D., and Hamada, M. (1992). Eds. "Case studies of liquefaction and lifeline performance during past earthquakes, 2: U. S. Case studies," *Report No. NCEER-92-0002*, National Center for Earthquake Engineering Research, State University of Buffalo, Buffalo, New York.

O'Rourke, T. D., Roth, B. L., and Hamada, M. (1992). "Large ground deformations and their effects in lifeline facilities: 1971 San Fernando earthquake." Ch. 3 of O'Rourke and Hamada (1992), 3-1 to 3-85.

O'Rourke, T. D., Pease, J. W., and Stewart, H. E. (1992a). "Lifeline performance and ground deformation during the earthquake." *The Loma Prieta, California Earthquake of October 17, 1989 – Marina District*, USGS Professional Paper 1551-F, U. S. Dept. of the Interior, Washington, D. C., 155-179.

Peters, K. E., and Herrmann, R. B. (1986). (compilers and eds.) *First Hand Observations of the Charleston Earthquake of August 31, 1886, and other Earthquake Materials*, South Carolina Geological Survey, Bull. 41.

Seed, H. B. (1970). "Soil problems and soil behavior." *Earthquake Engineering*, Prentice-Hall, Englewood Cliffs, New Jersey, 227-251.

Seed, H. B., Idriss, I. M., and Arango, I. (1983). "Evaluation of liquefaction potential using field performance data." *J. Geotech. Eng.*, ASCE, 109(3), 458-482.

Seed, H. B., Tokimatsu, K., and Harder, L. (1984). "The influence of SPT Procedures in evaluating soil liquefaction resistance," *Report No. UCB/EERC 84-15*, Earthquake Engineering Research Center, University of California, Berkeley, California.

Seed, R. B., Dickenson, S. E., and Riemer, M. F. (1991). "Liquefaction of soils in the 1989 Loma Prieta earthquake." *Proc. 2nd Int'l. Conf. on Recent Advances in Geotechnical Earthquake Engineering and Soil Dynamics*, Rolla, II, 1575-1586.

Seed, R. B., Dickenson, S. E., Riemer, M. F., Bray, J. D., Sitar, N., Mitchell, J. K., Idriss, I. M., Kayen, R. E., Kropp, A., Harder, L. F., Jr., and Power, M. S. (1990). "Preliminary report on the principal geotechnical aspects of the October 17, 1989 Loma Prieta earthquake," *Report No. UCB/EERC-90/05*, University of California, Berkeley, California.

Susuki, N., and Masuda, N. (1991). "Idealization of permanent ground movement and strain estimation of buried pipes." *Proc. 3rd Japan – U. S. Workshop on Earthquake-Resistant Design of Lifeline Facilities and Countermeasures for Soil Liquefaction*, Tech. Rept. NCEER 91-0001, National Center for Earthquake Engineering Research, State University of Buffalo, Buffalo, New York, 455-469.

Tokimatsu, K., and Seed, H. B. (1987). "Evaluation of settlements in sands due to earthquake shaking." *J. Geotech. Eng.*, ASCE, 113(8), 861-878.

Tuttle, M., Law, T., Seeber, L., and Jacob, K. (1990). "Liquefaction and ground failure induced by the 1988 Saguenay, Quebec, earthquake." *Can. Geotech. J.*, 27, 580-589.

Youd, T. L. (1973). "Ground movements in Van Norman Lake vicinity during San Fernando earthquake," San Fernando, *California Earthquake of February 9, 1971*, U. S. Dept. of Commerce, NOAA, Washington, D. C., 3, 197-206.

Youd, T. L. (1980). "Ground failure displacement and earthquake damage to buildings." *Proc. 2nd ASCE Conf. on Civil Engineering and Nuclear Power*, Knoxville, 7-6-1 to 7-6-26.

Youd, T. L. (1984). "Geologic effects – liquefaction and associated ground failure." *Proc. Geologic and Hydrologic Hazards Training Program*, Open-File Report 84-760, U. S. Geological Survey, Menlo Park, California, 210-232.

Youd, T. L. (1989). "Ground failure damage to buildings during earthquakes." *Foundation Engineering: Current Principles and Practices*, Geotechnical Special Publication No. 22, ASCE, New York, New York, 758-770.

Youd, T. L., and Perkins, D. M. (1987). "Mapping of liquefaction severity Index." *J. Geotech. Eng.*, ASCE, 113(11), 1374-1392.

Deformations in Granular Soils due to Cyclic Loading

P. M. Byrne[1] and J. McIntyre[2]

Abstract: An incremental stress-strain model for granular soils based on fundamental soil mechanics principles is presented. The model captures the drained skeleton behaviour observed in laboratory tests under cyclic loading. The undrained behaviour is captured using the same skeleton stress-strain relation together with the volumetric constraint imposed by the porewater fluid. The model predicts cyclic simple shear response in close agreement with observed cyclic test data in terms of porewater pressure rise, cycles to trigger liquefaction, as well as the characteristic post-liquefaction response. Finally, the model is incorporated in a dynamic analyses procedure and applied to the field case history recorded at the Wildlife site. The recorded downhole time history was used as input and the predicted response compared with the field observation. In general, the agreement is good except for the porewater pressure response, which showed a more rapid rise than was observed.

Introduction

A major concern for soil-structures comprised of granular materials is the movements induced by cyclic loading. These may be the transient movements that occur during the loading sequence or the permanent movements after the loading. They comprise of both horizontal displacements that are mainly associated with shear strains and vertical displacements or **settlements** that often arise from volumetric strains, but which on occasion can be largely due to shear strain.

---

[1] Professor, Department of Civil Engineering, University of British Columbia, 2324 Main Mall, Vancouver, B.C., Canada V6T 1Z4.

[2] Graduate Student, Department of Civil Engineering, University of British Columbia, 2324 Main Mall, Vancouver, B.C., Canada V6T 1Z4.

Cyclic loading may arise from seismic, water wave, ice loading (ice quake), or any other source of oscillating load. In this presentation only the **shear** component of the cyclic loading is considered. It is realized that the cyclic normal or volumetric component of the loading could also induce displacements, particularly in unsaturated soils, but this is not considered herein.

Cyclic shear loading causes a tendency for volumetric compaction or contraction of granular material, whether it be loose or dense. If the pores of the material are filled with a fluid that can either compress or escape during the loading, then an actual volumetric contraction will occur. If, on the other hand, the pores are filled with an essentially incompressible fluid, such as water, and if this fluid cannot escape during the period of shaking, then the tendency for volume change will transfer the normal load from the soil skeleton to the water, causing a rise in porewater pressure and a reduction in effective stress.

As the effective stress reduces, both the modulus and strength reduce leading to increased shear strains. If the effective stress drops to zero, the shear modulus will also be essentially zero and the soil will behave as a liquid - a state of transient liquefaction. This state only exists at the instant when the shear stress is zero. As the soil undergoes large shear strain at low confining stress, it will dilate causing the porewater pressure to drop and the effective stress to rise. This in turn causes the element to strain harden and develop some stiffness and strength depending on its density. Loose sands may develop only a small amount of stiffness and strength, whereas dense sands will quickly develop a high strength and stiffness.

This paper presents a fundamental examination of the response of a granular medium to cyclic load. The shear and volumetric strain responses of the dry or drained skeleton are first examined and a relatively simple incremental elastic-plastic model that captures the laboratory data proposed.

The saturated or partially saturated undrained response to cyclic loading is captured using the same skeleton model as for the drained condition, by applying the volumetric constraint that arises from the presence of the porewater fluid. The undrained model is validated by comparison with laboratory cyclic test data in terms of both porewater pressure rise and triggering of liquefaction as well as the post-triggering response.

Finally the model is used in the dynamic mode to predict the response at the Wildlife site in California where liquefaction occurred during an earthquake in 1987. Accelerations measured both at the surface and below the depth of the liquefied layer, as well as porewater pressure measurement in the liquefied layer, allow a comparison between predicted and measured field response.

Rational Analysis Procedure

The analysis of a soil-structure system subjected to earthquake loading is complex. The structure can be modelled as comprising a number of elements that prior to the earthquake loading, are generally under a range of static stresses (Fig. 1). Under earthquake loading each element will be subjected to a time history of normal and shear stresses (cyclic stresses) starting from a different static bias. In addition, these cyclic stresses themselves depend on the stress-strain response of the

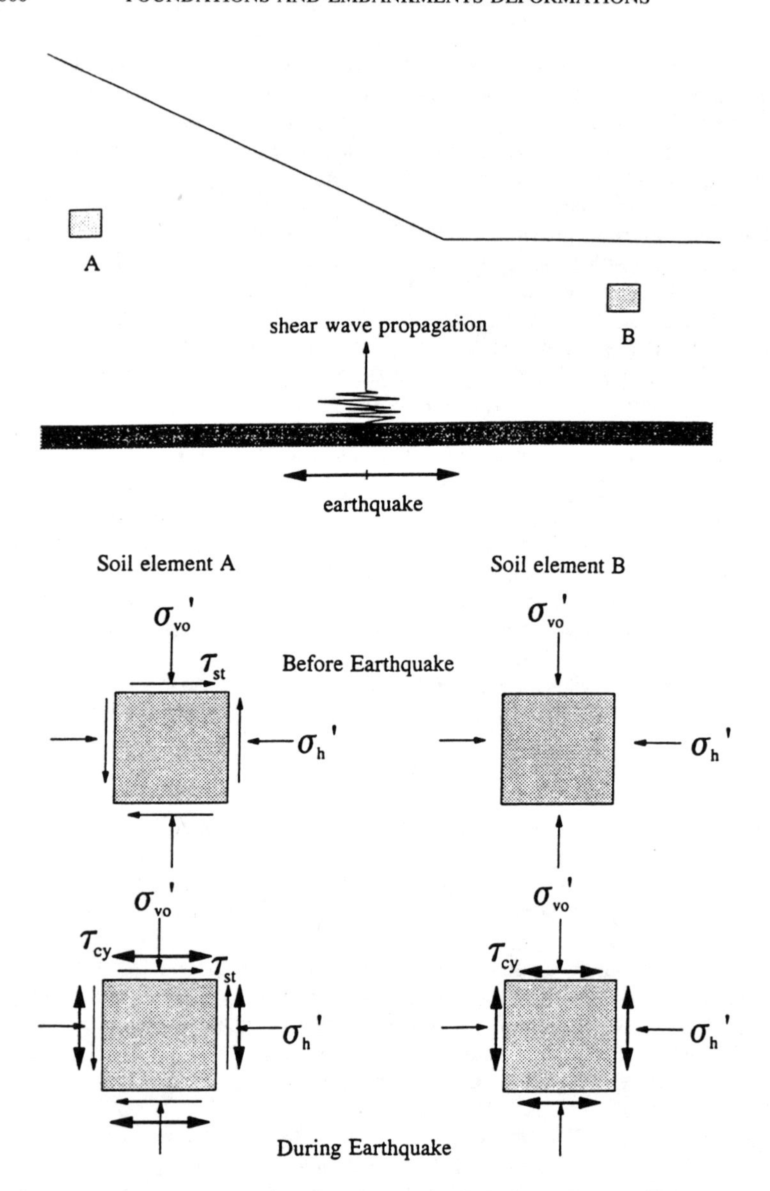

FIG. 1. Elements under Static and Seismic Loading Conditions

elements. As the stiffness of the element drops due to rise in porewater pressure, the overall period of the structure will increase. This in turn may increase or decrease the structure response and element dynamic stresses, depending on the predominant period of the input motion.

A rational response analysis requires a solution in the time domain taking into account the stress-strain-porewater pressure response of each element. Therefore, the essence of the problem is the formulation of an element stress-strain and porewater pressure model that captures the observed laboratory element response up to and including triggering of liquefaction, as well as the post-triggering phase. Once the element behaviour is captured, it can be incorporated in a finite element or finite difference code to predict the displacement response of the soil-structure system to the specified time history of loading.

The purpose here is to present an incremental stress-strain model that captures the element or laboratory test response for both drained and undrained conditions and validate it by comparison with observed field behaviour. A key factor in the response of granular material to both monotonic and cyclic loading is the coupling that occurs between shear and volumetric strains, i.e., shear strains induce volumetric strains and this is called shear-volume coupling. It is this coupling that induces the rise in porewater pressure when the porewater is prevented from leaving the skeleton by a drainage constraint.

SHEAR-VOLUME COUPLING

Empirical Approach

Shear strains induce volumetric strains in unbonded granular soil, and cyclic shear strains cause an accumulation of volumetric compaction strain with number of cycles (Fig. 2). The first shear-volume coupling model was presented by Martin-Finn-Seed (1975). This model was based on simple shear test data and would simulate earthquake loading under level ground conditions. The increment of volumetric strain per cycle of shear strain, γ, was expressed as follows:

$$(\Delta\epsilon_v)_{cycle} = C_1(\gamma - C_2\epsilon_v) + \frac{C_3\epsilon_v^2}{\gamma + C_4\epsilon_v} \tag{1}$$

in which

$(\Delta\epsilon_v)_{cycle}$ = the increment of volumetric strain in percent per cycle of shear strain;

ϵ_v = the accumulated volumetric strain from previous cycles in percent;

γ = the amplitude of shear strain in percent for the cycle in question; and

C_1, C_2, C_3, C_4 = constants for the sand in question at the relative density under consideration.

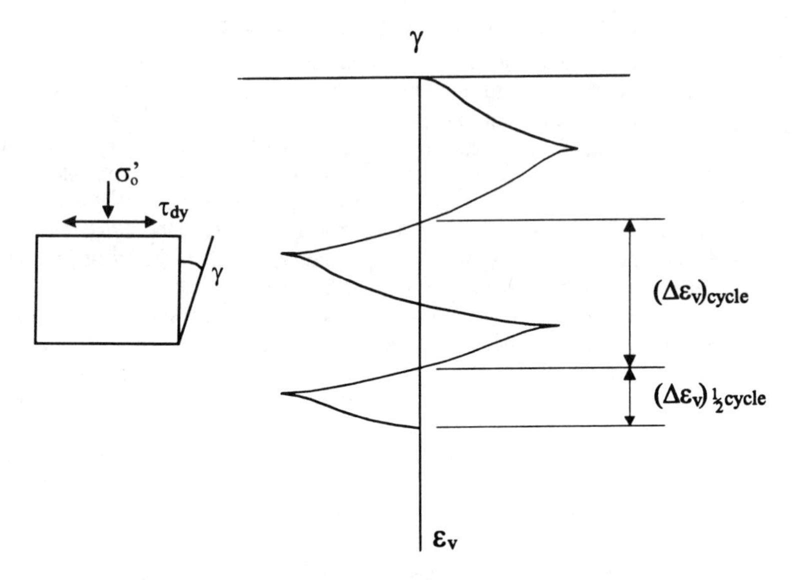

FIG. 2. Accumulation of Volumetric Strain due to Cyclic Shear Strains

This formulation as discussed by Byrne (1991) is unnecessarily complex and is not generally stable. He proposed that the data base presented by Martin et al. (1975) can be better modelled by

$$(\Delta\epsilon_v)_{1/2\ \mathrm{cycle}} = 0.5\gamma C_1 \exp\left(\frac{-C_2\epsilon_v}{\gamma}\right) \tag{2}$$

in which $(\Delta\epsilon_v)_{1/2\ \mathrm{cycle}}$ is the additional volumetric strain per half cycle of strain, γ, and ϵ_v is the accumulated strain. C_1 and C_2 are constants that depend on the relative density of the sand. This formulation was shown by Byrne to be in good agreement with available data (Figs. 3 and 4).

Best fit values of C_1 and C_2 as a function of relative density, D_r, or normalized standard penetration value, $(N_1)_{60}$ are shown in Table 1.

While this formulation can be used in a loose coupled analysis, it is not appropriate for a coupled incremental analysis. For such a formulation, the increment of volumetric strain, $d\epsilon_v$, as a function of the increment of shear strain, $d\gamma$, is required for each time step rather than every 1/2 cycle. The simplest incremental formulation based on Eq. (2) is obtained by assuming that the volumetric strain develops linearly with shear strain during any half-cycle, as shown in Fig. 5, from which:

$$d\epsilon_v \quad = \quad 0.25\, d\gamma\, C_1 \exp\left[\frac{-C_2\,\epsilon_v}{\gamma}\right] \quad = \quad d\gamma \cdot D_t \tag{3}$$

in which γ is the largest strain in the current 1/2 cycle. The terms associated with $d\gamma$ can be lumped into a single shear-volume coupling term, D_t.

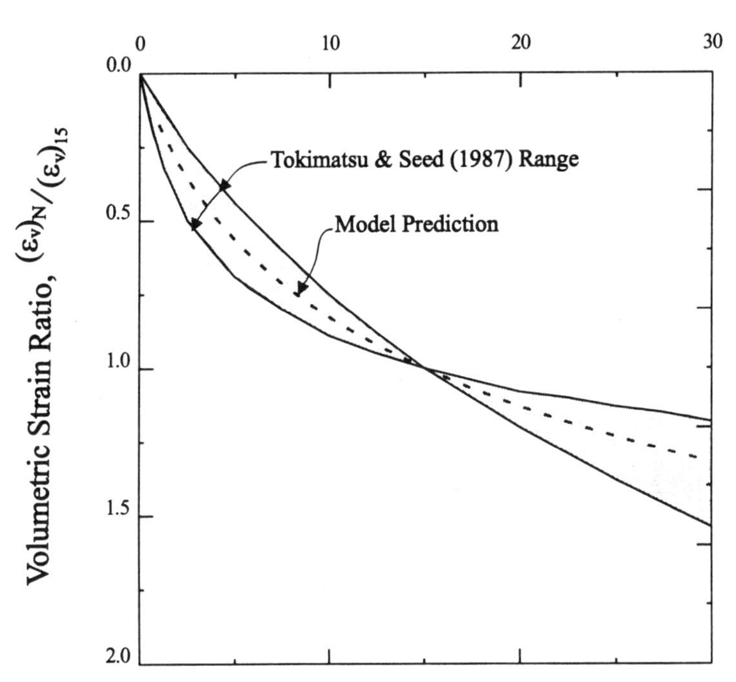

FIG. 3. Relationship between Volumetric Strain Ratio and Number of Cycles for Dry Sands (Test Data from Tokimatsu and Seed 1987)

This is a satisfactory approach when the shear strain sequence is known a priori and gives essentially the same result as Eq. (2). However, for the earthquake problem, the strain sequence is not known ahead of time. One solution is to assume γ to be the largest strain in the current or previous cycle, whichever is larger.

This empirical approach is satisfactory for simple shear conditions in the absence of a static bias. For a more general initial stress state, and for conditions after triggering of liquefaction, a more fundamental approach is desirable.

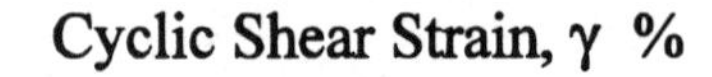

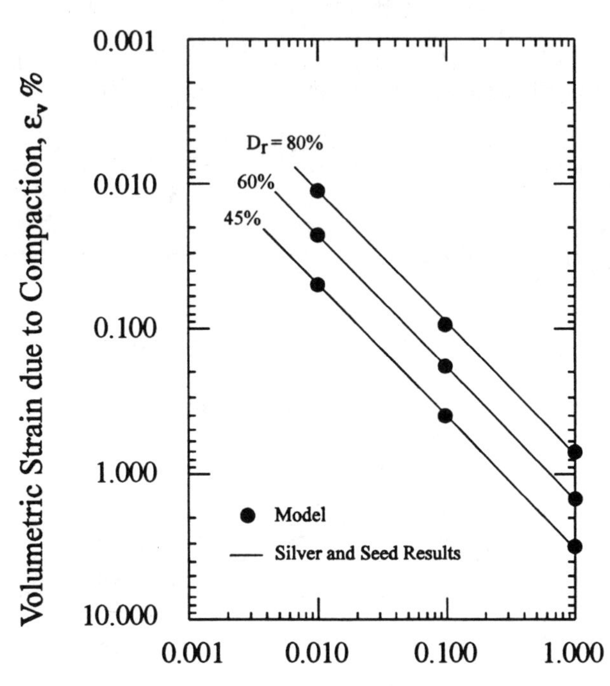

FIG. 4. Relationship between Volumetric Strain and Shear Strain for Dry Sands (Test Results from Silver and Seed 1971)

TABLE 1. C_1 and C_2 in Terms of D_r and $(N_1)_{60}$

$(N_1)_{60}$	D_r	C_1	C_2
5	34	1.00	0.40
10	47	0.50	0.80
20	67	0.20	2.00
30	82	0.12	3.33
40	95	0.06	6.66

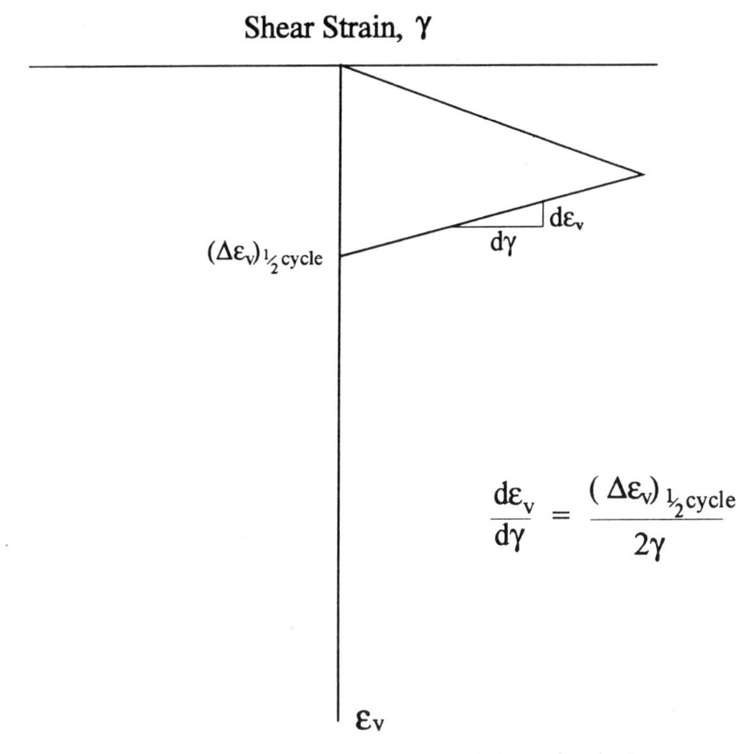

FIG. 5. Volumetric Strain as a Function of Shear Strain Increment

Fundamental Approach

(a) Monotonic

The characteristic stress-strain and volume change response of the granular skeleton to monotonic simple shear loading is shown in Fig. 6. The effective stress ratio τ/σ' and the density state basically control the shear strains as shown in Fig. 6a. These shear strains in turn induce volumetric strains as shown in Fig. 6b. Basically, loose material is contractive when sheared to any effective stress ratio state, while dense material is contractive for stress ratio states below the phase transformation or constant volume friction angle, and dilative above (Fig. 6d). The dilation rate as expressed in terms of the strain increment ratio $d\epsilon_v/d\gamma$ is linearly related to the effective stress ratio τ/σ' as shown in Fig. 6c. This relationship appears to be true for all densities and all confining stresses for a given sand and can be approximated by:

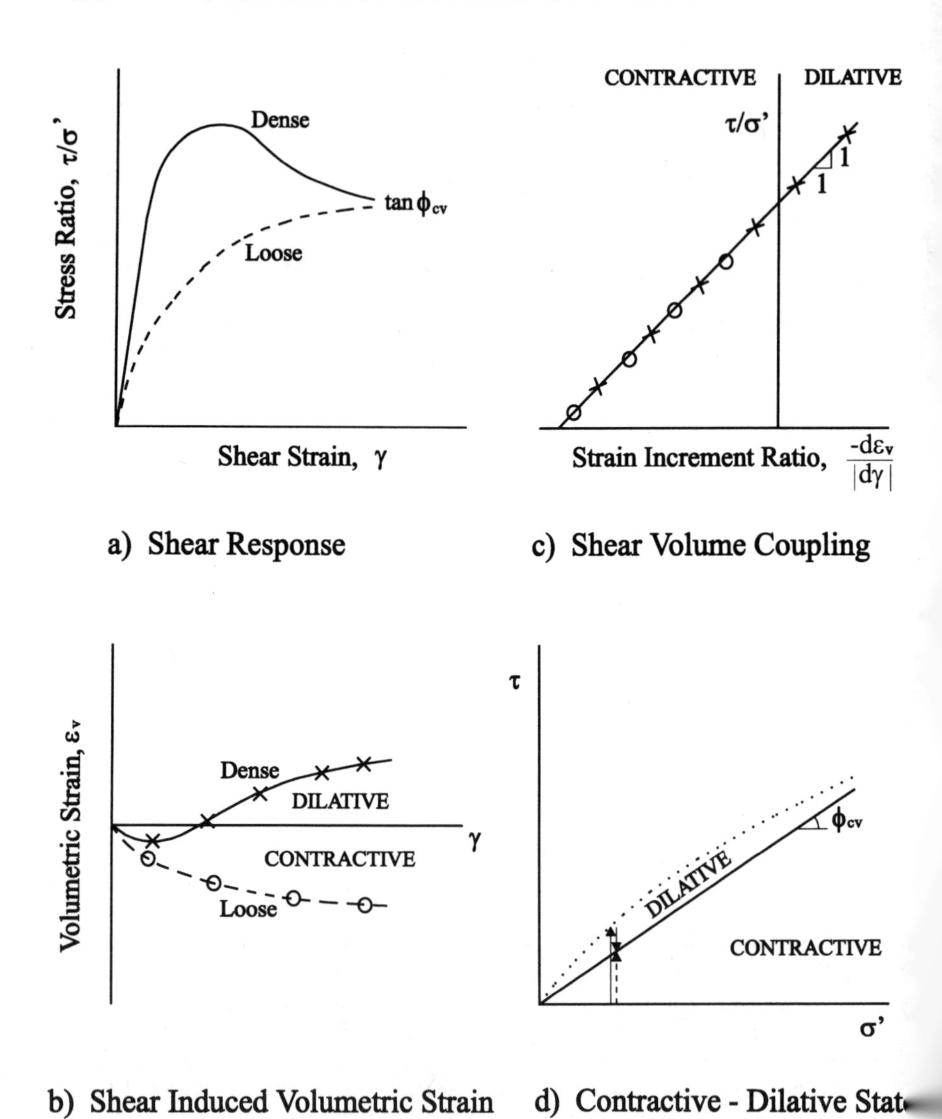

FIG. 6. Stress-Strain and Volume Change Response of the Granular Skeleton to Monotonic Simple Shear Loading

$$\frac{\tau}{\sigma'} \approx -\frac{d\epsilon_v}{d\gamma} + \tan\phi_{cv} \tag{4}$$

This is similar to the model proposed by Matsuoka and Nakai (1977).

Based on energy considerations as outlined in Appendix II, it is shown that the relationship from mechanics principles should be related to the plastic rather than the total strains, thus

$$\frac{\tau}{\sigma'} = -\frac{d\epsilon_v^p}{d\gamma^p} + \tan\phi_{cv} \tag{5}$$

where $d\epsilon_v{}^p$ and $d\gamma^p$ are the plastic volumetric and shear strain increments. The plastic strain increment is, therefore:

$$d\epsilon_v^p = d\gamma^p \left[\tan\phi_{cv} - \frac{\tau}{\sigma'}\right] = d\gamma^p \cdot D_t \tag{6}$$

(b) Cyclic

Test data presented by Lee (1991) indicate that Eq. (6) is also valid for cyclic loading with a slight modification. Upon reversal of the shear stress or strain (Fig. 7a), the element becomes highly contractive as depicted in Fig. 7b, and the τ/σ' vs. $-d\epsilon_v{}^p/d\gamma^p$ follows an hourglass pattern that is true for all densities and stress states. The increment of plastic volumetric strain, $d\epsilon_v{}^p$, for unloading is given by:

$$d\epsilon_v^p = d\gamma^p \left[\tan\phi_{cv} - \frac{\tau}{\sigma'}\right] = d\gamma^p \cdot D_t \tag{6a}$$

where $d\gamma^p$ is always the absolute value.

Neither Matsuoka and Nakai (1977) or Lee (1991) actually decomposed the strains into their elastic and plastic components, but for the strain levels they were considering, the elastic strains would have been small and the relationship similar whether total or plastic strains are considered.

Eq. (6), therefore, gives a simple expression for the incremental plastic volumetric strain, $d\epsilon_v{}^p$, due to an increment of plastic shear strain $d\gamma^p$ under either monotonic or cyclic loading. The expression is independent of stress level and density. This indicates that a given increment of plastic shear will cause the same plastic volumetric strain in a loose or a dense sample. This appears to contradict the empirical Eq. (3), which in turn is based on laboratory test measurements. However, Eq. (3) involves **total** shear strains rather than **plastic** shear strains and when this is considered, Eqs. (3) and (6) do give similar accumulations of volumetric strain under cyclic loading.

Shear volume coupling effects under strain controlled conditions can be computed from the empirical Eq. (3) which captures the laboratory data. They can also be computed from the more fundamental Eq. (6). For load controlled conditions which arise in earthquake and other cyclic load situations it is first necessary to compute the increment of shear strains from the shear stress-strain law and this is next addressed.

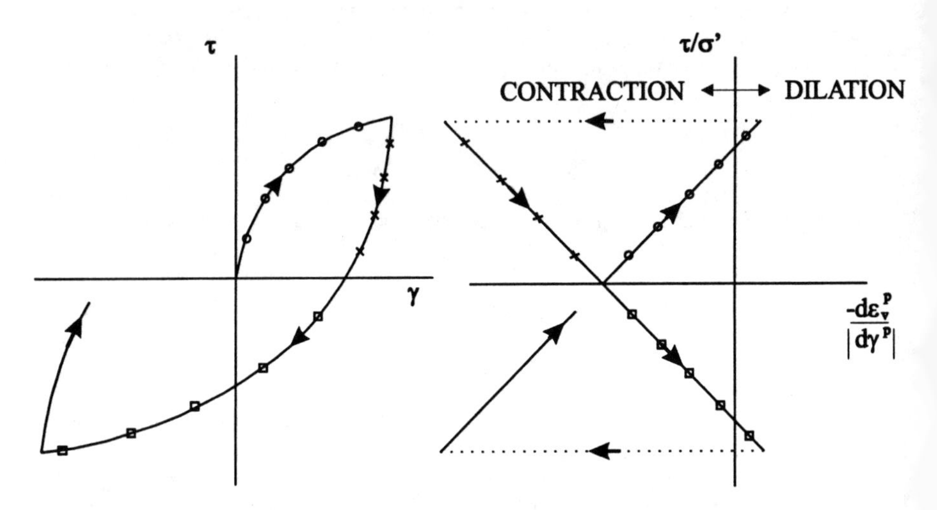

a) Shear Stress vs. Shear Strain b) Shear Volume Coupling

FIG. 7. Cyclic Shear-Volume Coupling

Shear Stress-Strain Law

The simplest shear stress-strain law that is in reasonable accord with laboratory data is the Hyperbolic Formulation. The tangent stiffness G_t at any stress state is defined as:

$$G_t = \frac{d\tau}{d\gamma} = G_{max}\left[1 - \frac{\tau}{\tau_f}R_f\right]^2 \qquad (7)$$

in which

G_{max} = the maximum shear modulus that occurs at zero shear strain (Appendix III);

τ = the shear stress;

τ_f = the failure shear stress;

R_f = the failure ratio τ_f/τ_{ult} in which τ_{ult} is the ultimate strength from the best fit hyperbola (Fig. 8a). R_f may also be considered a factor that defines the strain, γ_f, at which the failure stress occurs; and

γ_f = $\tau_f/G_{max}\,(1 - R_f)$.

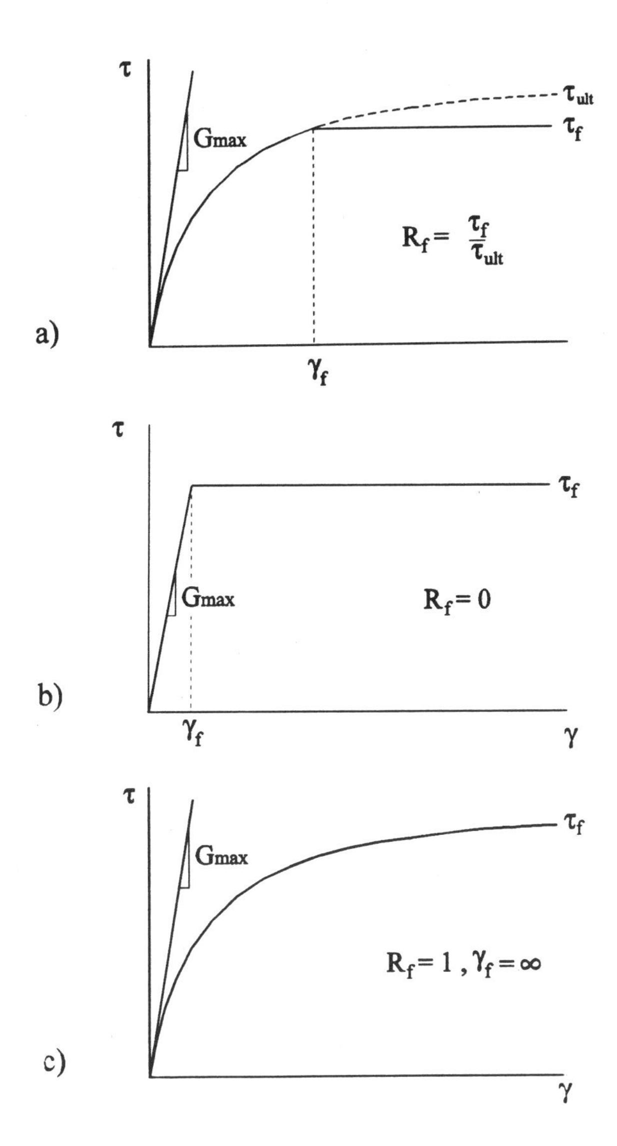

FIG. 8. Modified Hyperbolic Stress-Strain Formulation

The parameter R_f is used to modify the hyperbola to fit the laboratory data. R_f = 0 specifies a linear elastic plastic material with $\gamma_f = \tau_f/G_{max}$ as shown in Fig. 8b. R_f = 1 specifies a strain hardening material with γ_f equal to infinity. For most sands, R_f lies between 0.5 and 0.9.

The shear stress-strain relation under cyclic loading can also be modelled as a hyperbola and the tangent stiffness at any stress level expressed by

$$G_t = G_{max}\left[1 - \frac{\tau^*}{\tau_f^*}R_f\right] \tag{8}$$

where

G_{max}	=	the shear modulus immediately upon unloading;
τ^*	=	$(\tau_A + \tau)$;
τ_f^*	=	$(\tau_A + \tau_f)$; and
τ_A	=	the shear stress at the reversal point as shown in Fig. 9.

The increment of shear strain, $d\gamma$, for an applied increment of shear stress, $d\tau$, is, therefore,

$$d\gamma = \frac{1}{G_t}d\tau \tag{9}$$

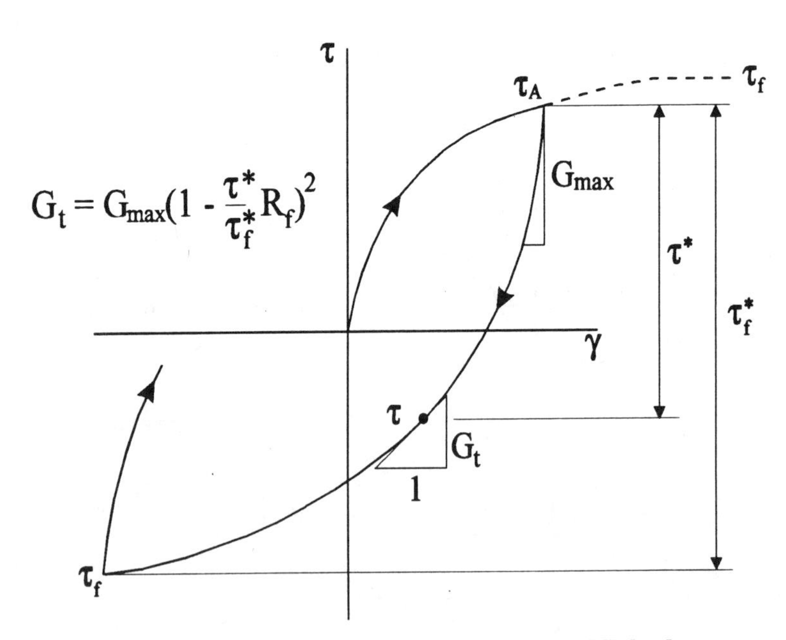

FIG. 9. Shear Stress-Strain Model for Unload-Reload

A more general formulation for sand can be obtained by expressing the shear-strain response in dimensionless form, namely: η vs γ, as shown in Fig. 10 where $\eta = \tau/\sigma'$, the effective stress ratio.

Test data indicate that the shape is approximately the same whether the sand is tested drained or undrained, but will differ for loose and dense states as shown on the figure. A more fundamental approach would be to consider the η vs γ curves to vary with the state parameter, the distance from the steady state line, as this would account for both stress and density states (Been and Jefferies 1985).

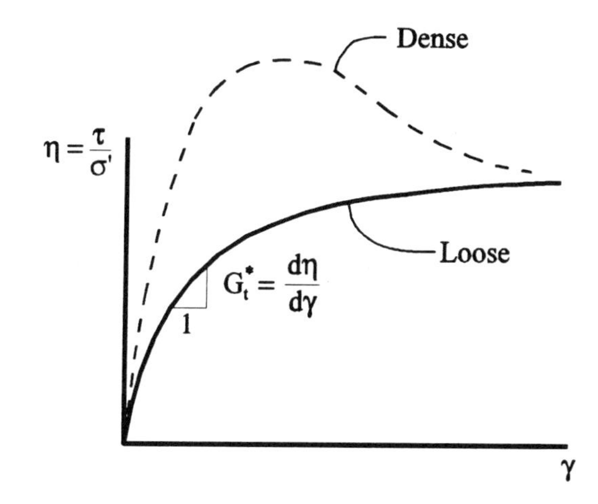

FIG. 10. Shear Strain Response in Dimensionless Form

The characteristic η vs γ relationship can also be captured by a hyperbolic formulation, and the tangent stiffness expressed as

$$G_t^* = \frac{d\eta}{d\gamma} = \frac{G_{max}}{\sigma'}\left[1 - \frac{\eta}{\eta_f}R_f\right]^2 = \frac{G_{max}}{\sigma'}\left[1 - \frac{\tau}{\tau_f}R_f\right]^2 = \frac{G_t}{\sigma'} \quad (10)$$

where

η = the current τ/σ' effective stress ratio state; and

η_f = the effective stress ratio at the failure state, τ_f/σ'.

Since $\eta = \tau/\sigma'$, a change in stress ratio $d\eta$ can be brought about by a change in $d\tau$ or a change in $d\sigma'$, i.e.,

$$d\eta = \frac{d\tau}{\sigma'} - \frac{\tau}{(\sigma')^2}d\sigma \quad (11)$$

hence from Eqs. (11) and (12),

$$d\gamma = \frac{1}{G_t}\left[d\tau - \frac{\tau}{\sigma'}d\sigma'\right] \tag{12}$$

if $d\sigma' = 0$, as it would be for a drained simple shear test, then the usual definition of G_t is obtained. If $d\tau = 0$, then a reduction in effective stress ($-d\sigma'$) will still cause a shear strain $d\gamma$ (Fig. 11). For undrained tests, in which a significant porewater pressure rise occurs, $d\sigma'$ will be negative and a significant reduction in G_t will result. Positive values of $d\sigma'$ are neglected in Eq. (12).

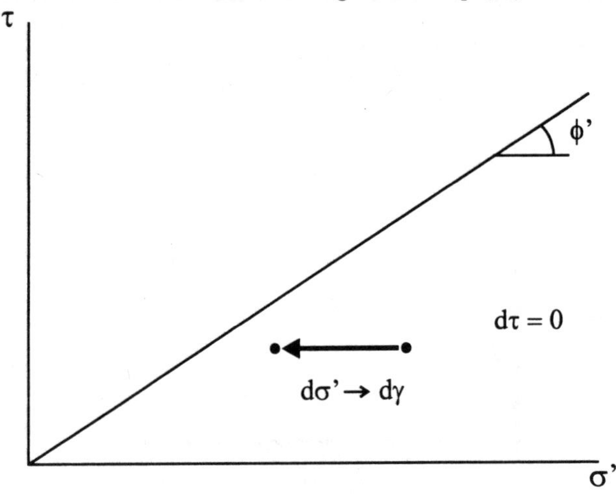

FIG. 11. Strain Increment due to a Reduction in Normal Effective Stress

Hence, an increment of strain is caused by either an increment of shear stress, $d\tau$, or reduction of normal effective stress, $d\sigma'$. The effect of the increment of normal stress is fundamental to liquefaction analysis as will be discussed later.

Skeleton Stress-Strain Relations

The skeleton stress-strain model described thus far involves an incremental elastic-plastic stress-strain law. In plasticity nomenclature the yield loci are lines of constant stress ratio, the flow rule is non-associated, and the hardener is the plastic shear strain.

The response of the skeleton to simple shear loading is obtained as follows:

- The increment of shear strain $d\gamma$ is computed from Eqs. (9) or (12), and the plastic shear strain increment $d\gamma^p$ from

$$d\gamma^p = d\gamma - d\gamma^e = d\gamma - \frac{d\tau}{G_{max}} \tag{13}$$

- The increment of plastic volumetric strain from Eq. (6), namely,

$$d\epsilon_v^p = |d\gamma^p| \left[\tan\phi_{cv} \pm \frac{\tau}{\sigma'}\right] = |d\gamma^p| \cdot D_t \tag{14}$$

The increment of elastic volumetric strain, $d\epsilon_v^p$ from

$$d\epsilon_v^e = \frac{d\sigma'}{M} \tag{15}$$

where M is the constrained modulus described in Appendix IV.

For a drained simple shear test, the stress increment $d\tau$ can be selected ($d\sigma'=0$), and $d\gamma$ and $d\epsilon_v^p$ computed from Eqs. (12), (13) and (14). The stress-strain and volumetric strain response is then obtained by summation of the increments.

UNDRAINED RESPONSE

If drainage of porewater fluid is prevented from occurring during the application of a load increment, a volumetric constraint is imposed on the skeleton. The response of the skeleton is predicted here using the same skeleton model described in the previous section but taking into account the volumetric constraint of the porewater fluid.

If the porewater fluid and solids are assumed incompressible, the overall volumetric strain will be zero. However, grain slip will still occur within the skeleton causing both plastic shear and plastic volumetric strains such that:

$$d\epsilon_v = d\epsilon_v^e + d\epsilon_v^p = 0 \tag{16}$$

Now $d\epsilon_v^e = d\sigma'/M$, hence from Eq. (16),

$$d\sigma' = -M d\epsilon_v^p = -M D_t d\gamma^p \tag{17}$$

Since $d\sigma' = d\sigma - du$, and for conventional simple shear tests or 1-D field conditions $d\sigma = 0$, the rise in porewater pressure $du = -d\sigma'$, therefore

$$du = M d\epsilon_v^p = M D_t d\gamma^p \tag{18}$$

Because the application of a shear stress increment also causes a rise in porewater pressure and a change in effective stress $d\sigma'$, this should be considered when computing $d\gamma$ in Eq. (12). By substituting for $d\sigma'$ from Eq. (17) and for $d\epsilon_v^e$ from Eq. (15) and re-arranging:

$$d\gamma^p = d\tau \cdot \frac{\left[\dfrac{1}{G_t} - \dfrac{1}{G_{max}}\right]}{1 - \dfrac{\tau}{\sigma'} \cdot D_t \dfrac{M}{G_t}} \tag{19}$$

$$d\gamma = d\gamma^e + d\gamma^p = d\tau \left[\frac{1}{G_{max}} + \frac{\left[\frac{1}{G_t} - \frac{1}{G_{max}} \right]}{1 - \frac{\tau}{\sigma'} \cdot D_t \frac{M}{G_t}} \right] \tag{20}$$

Eqs. (17) to (20) define the complete stress-strain behaviour of the soil under conditions of no volume change in terms of fundamental stress state and soil parameters. If the shear-volume coupling term, D_t, is zero, then the original $d\tau = G_t\, d\gamma$ results and there is no change in effective stress or porewater pressure.

With the shear volume coupling term defined as $D_t = (\tan \phi_{cv} \pm \tau/\sigma')$, it is the ratio M/G_t that controls the shape of the stress-strain curve. Laboratory data suggest that M is essentially independent of density while G_t increases with increased density. Thus the ratio M/G_t increases with reduced density and can cause the tangent stiffness to go to zero, and in fact have negative values as shown in Fig. 12.

The characteristic undrained stress-strain and effective stress paths predicted from Eqs. (17) to (20) for monotonic loading, are shown in Fig. 12, and capture the form of the observed test data.

An alternative simpler approach is to define the undrained shear-stress strain relation in terms of G_{max}, an undrained strength s_u, and R_f, i.e.

$$d\tau = G_t d\gamma = G_{max} \left[1 - \frac{\tau}{s_u} R_f \right]^2 d\gamma \tag{21}$$

The porewater pressure rise can then be computed from Eq. (18). The empirical shear-volume coupling term from Eq. (3) instead of Eq. (6) can also be used in Eq. 18. This simpler approach may be satisfactory for denser materials that do not strain-soften, but cannot directly capture the behaviour of strain softening material (Fig. 12a) unless the portion of the curve beyond the peak is separately defined.

The model described in Eqs. (17) to (20) or (21) are applicable to cyclic loading conditions when G_t and D_t are appropriately defined for unloading and reloading as discussed earlier (Eqs. (8) and (14)).

MODEL CALIBRATION

The key input parameters to the model are:

(1) The maximum shear modulus, G_{max}.
(2) The peak friction angle of skeleton, ϕ'_p, or the undrained strength, s_u.
(3) The constant volume friction angle ϕ'_{cv}.
(4) The failure ratio, R_f, that controls the shape of the shear stress-strain curve.
(5) The rebound constrained modulus, M.

These have been related to the normalized penetration resistance value $(N_1)_{60}$ as described in detail by Byrne (1991).

The model has been calibrated against the Seed et al. (1984) liquefaction chart, so that, for example, a simple shear element under a vertical effective stress of 1 T/ft^2, and subject to a cyclic stress ratio of 0.11 will liquefy in about 15 cycles in agreement with the chart.

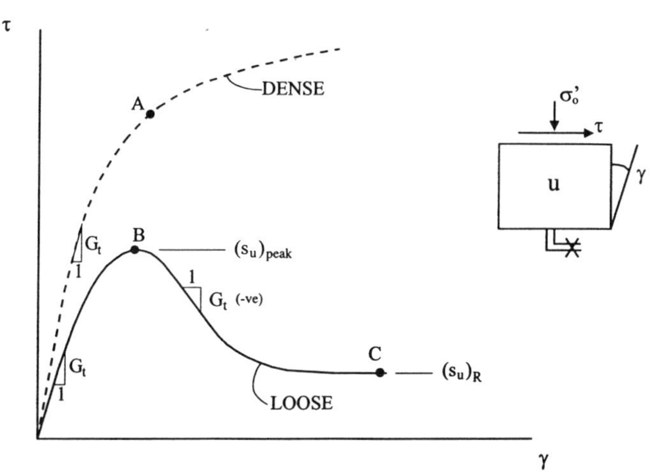

a) Shear Stress versus Shear Strain

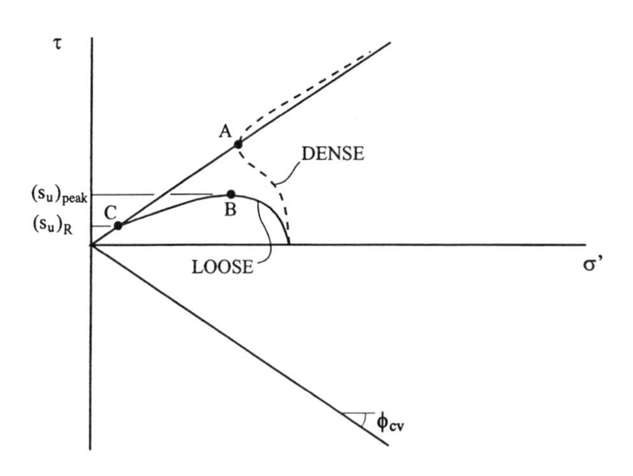

b) Shear Stress versus Effective Stress

FIG. 12. Undrained Response

The predicted cyclic shear stress-strain and effective stress path response for a cyclic load controlled test are shown in Fig. 13. It may be seen (Fig. 13a) that the stress-strain response remains stiff for a number of cycles until the porewater pressure rise causes the effective stress state to reach the ϕ'_{cv} line (point A, Fig.

13b). The material then deforms with a low modulus and large strain as it moves up the ϕ'_{cv} line to point B. Upon unloading from B, the soil is initially very stiff but is also very contractive so that it generates a large porewater pressure rise that may drive the effective stress point to zero at point C, when $\tau = 0$. The material is essentially a liquid at this point ($G_t \approx 0$) and large strains occur. With further increase in strain, the material eventually dilates causing a drop in porewater pressure and the stress point moves up the ϕ'_{cv} line, stiffening as it goes. Upon unloading from D, the process is repeated leading to large cyclic strains commonly referred to as cyclic mobility or cyclic liquefaction.

This predicted behaviour is in agreement with observed laboratory cyclic simple shear data. In fact, the model has been calibrated to capture the laboratory data.

Having captured the element behaviour, the next check in the validation procedure is to incorporate the element response in a dynamic analysis and predict and compare with a field event, and this is described in the section which follows.

FIELD VERIFICATION

Background

The model was used to predict the dynamic response of the Wildlife site - an instrumented site where liquefaction occurred during an 1987 earthquake. The Wildlife site is located in southern California in the seismically active Imperial Valley. The site was instrumented in 1982 by the U.S. Geological Survey using accelerometers and piezometers in an effort to record ground motions and porewater pressures during earthquakes.

Site Description

The Wildlife site is located in the floodplain of the Alamo River approximately 36 km north of El Centro. Although the site is on level ground, it is located in close proximity (about 20 m) to the river's western bank. As the river is incised to a depth of about 3.7 m, there exists the opportunity for lateral spreading towards this free field (Holzer et al. 1989). In-situ and laboratory investigations (Bennett et al. 1984) have shown that the site stratigraphy consists of a surficial silt layer approximately 2.5 m thick underlain by a 4.3 m thick layer of loose silty-sand. Underneath the sand layer from a depth of 6.8 m to 11.5 m is stiff to very stiff clay. The groundwater table fluctuates within the surficial silt layer and is located at an approximate depth of 2.0 m.

Instrumentation

The liquefaction array at the Wildlife site consists of two 3-component accelerometers and six electric piezometers. One accelerometer was mounted at the surface on a concrete slab supporting an instrument shed. The second accelerometer was installed in a cased hole beneath the liquefiable layer at a depth of 7.5 m. Five of the six piezometers were installed within the liquefiable sand layer. In addition to the above, an inclinometer casing was installed approximately 12 m to the west of the array to allow measurement of permanent horizontal displacements in the subsurface. Details about the instrumentation and the installation procedure are given by Youd and Wieczorek (1984).

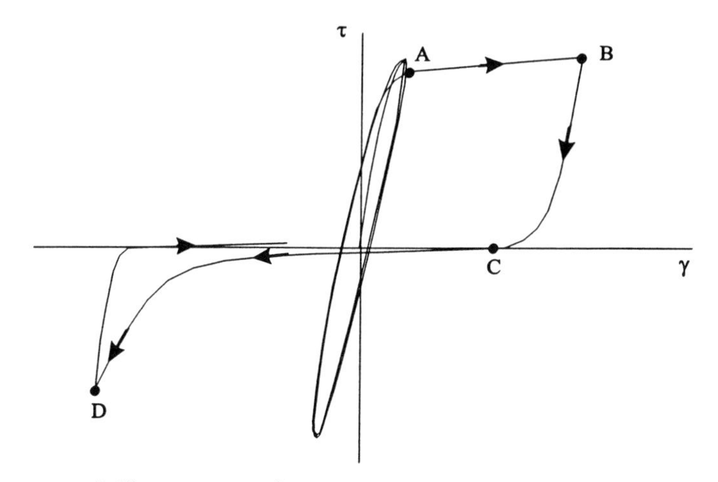

a) Shear stress-strain response

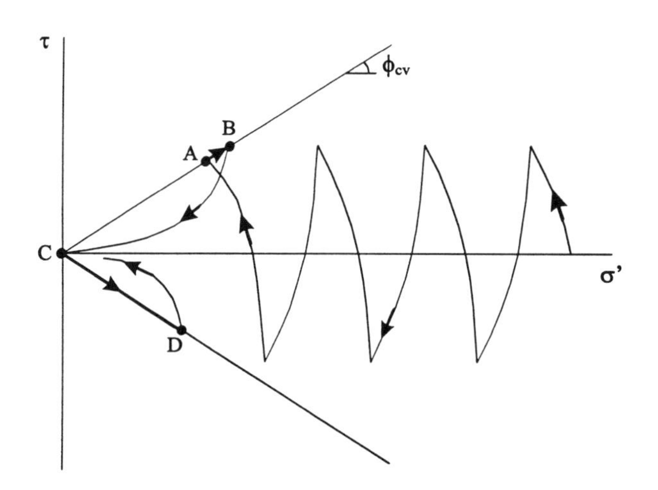

b) Effective stress path response

FIG. 13. Predicted Cyclic Shear Stress-Strain and Effective Stress Path Response for a Cyclic Load Controlled Test

Recorded Site Response

In November, 1987 the Wildlife site was shaken by two earthquakes - the Elmore Ranch earthquake and the Superstition Hills earthquake. Both events triggered the instrumentation at the site; however, only the Superstition Hills earthquake (M = 6.6) generated dynamic porewater pressures. Subsequent site investigations showed evidence of liquefaction in the form of sand boils and small ground fissures (Zeghal and Elgamal, in review). Relative displacements of 180 mm were measured by means of the inclinometer casing; however, within the array, surface displacements were considerably smaller. Holzer et al. (1989) report that the cumulative opening across cracks at the array was 126 mm.

Fig. 14 shows the measured acceleration time histories for the North-South component of the Superstition Hills quake. Fig. 14a shows the surface time history while the downhole time history is shown in Fig. 14b. Ground motions in the East-West direction were smaller, and were not analyzed using the model. Surface and downhole displacement time histories were obtained by integration of the acceleration time histories, and are shown in Fig. 15a and Fig. 15b respectively.

Relative displacements between the surface and the stiff base are of prime interest and these were obtained by subtracting the surface and downhole displacement time histories at each time increment. The resulting relative displacement time history is shown in Fig. 15c. Note that the relative displacements were essentially zero for about the first fourteen seconds of shaking despite the fact that significant displacements were measured both at the surface and downhole. This indicates that up until fourteen seconds, the soil units above and below the liquefiable sand layer essentially moved together. After fourteen seconds, significant relative displacements occurred, indicating the uncoupling of the soil units above and below the sand layer.

The recorded time history of surface acceleration versus relative displacement is shown in Fig. 16. This plot is similar to a shear stress versus shear strain plot, as shear stress would simply be the surface acceleration multiplied by the soil mass, and the strains would be the relative displacements divided by the thickness of the liquefied layer. Since neither the soil mass nor the thickness of the liquefied layer are known with certainty, presenting the data in this form introduces less error.

By isolating brief segments of the data from Fig. 16, it is possible to see how the soil modulus changes with cycles. Fig. 17 shows four discrete cycles at different times during the earthquake. For about the first 14 seconds of shaking, the soil is stiff as shown in Fig. 17a and there has been little degradation of modulus. At about 16 seconds (Fig. 17b) significant degradation of modulus has occurred. At 35 seconds (Fig. 17c), further degradation of modulus has occurred with a flat zero modulus zone followed by strain-hardening and an abrupt increase in modulus upon unloading.

The behaviour shown in Fig. 17c is typical of a cyclic laboratory simple shear response after liquefaction has been triggered, and is caused by repeated dilatant and contractant response as the stress point cycles through the zero effective stress state as discussed previously. This same behaviour is seen in Fig. 17d, except the base accelerations are considerably smaller at this stage of the earthquake.

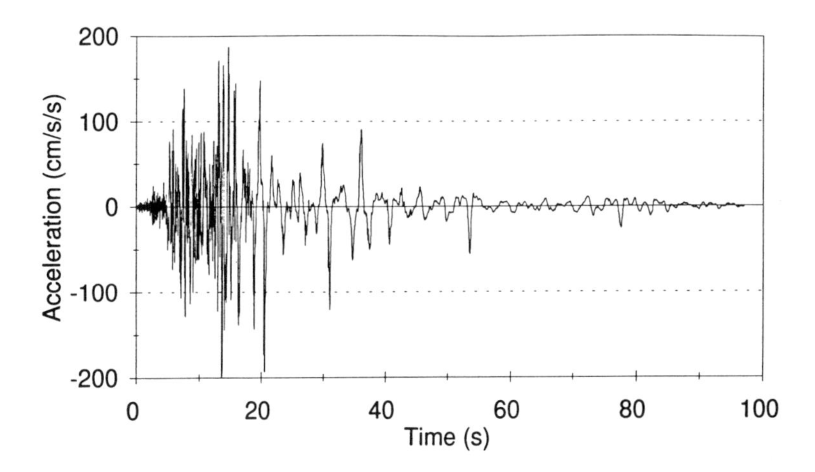

a) Surface acceleration time history, N-S component

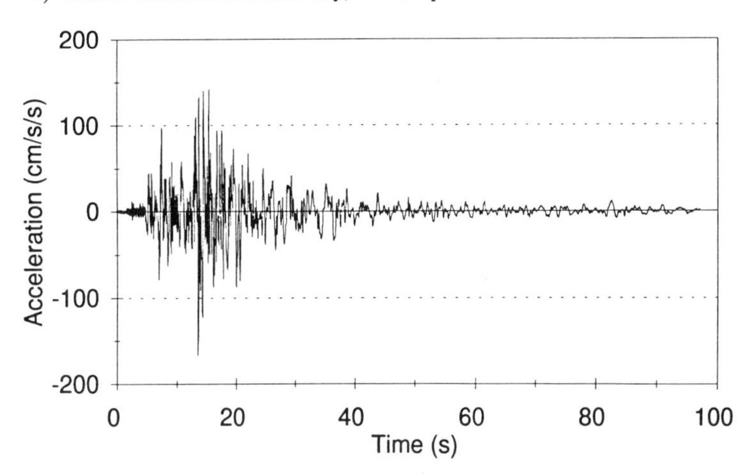

b) Downhole acceleration time history, N-S component

FIG. 14. Acceleration Time Histories - Wildlife Site, 1987 Superstition Hills Earthquake

The approximate 500 fold reduction in soil stiffness that occurs after roughly 18 seconds of shaking is a clear indication to the authors that effective stresses have reduced to near zero and liquefaction has been triggered, at least in some zones of the soil profile.

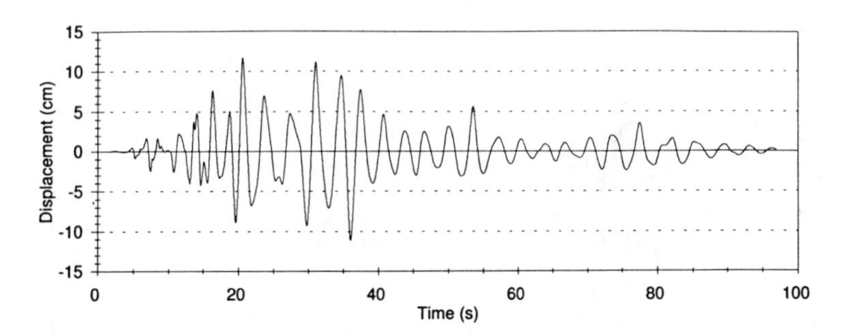

a) Surface displacement time history

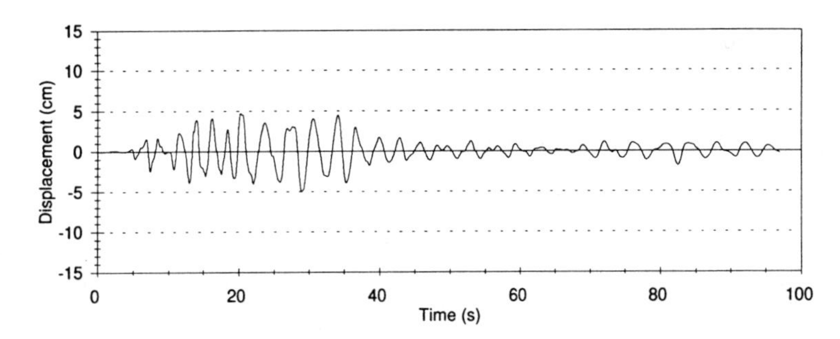

b) Downhole displacement time history

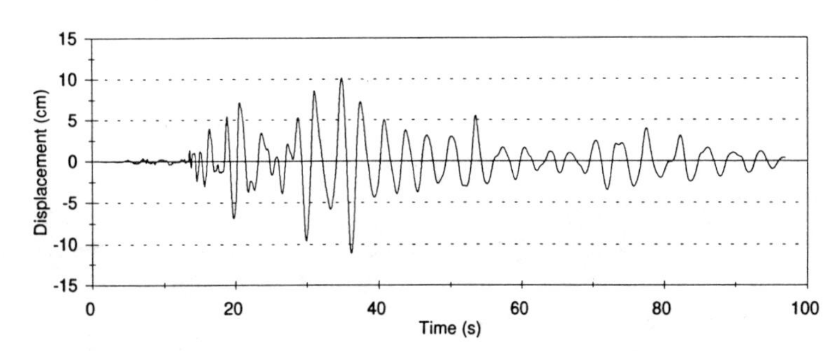

c) Relative displacement time history

FIG. 15. Displacement Time Histories - Wildlife Site, 1987 Superstition Hills Earthquake

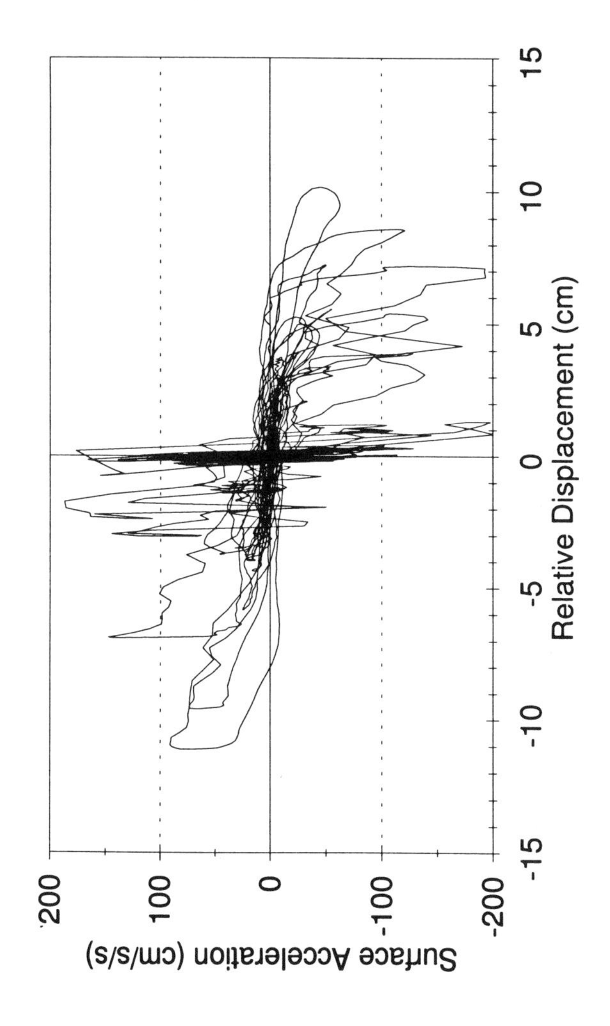

FIG. 16. Surface Acceleration versus Relative Displacement; N-S Component, Wildlife Site, 1987 Superstition Hills Earthquake

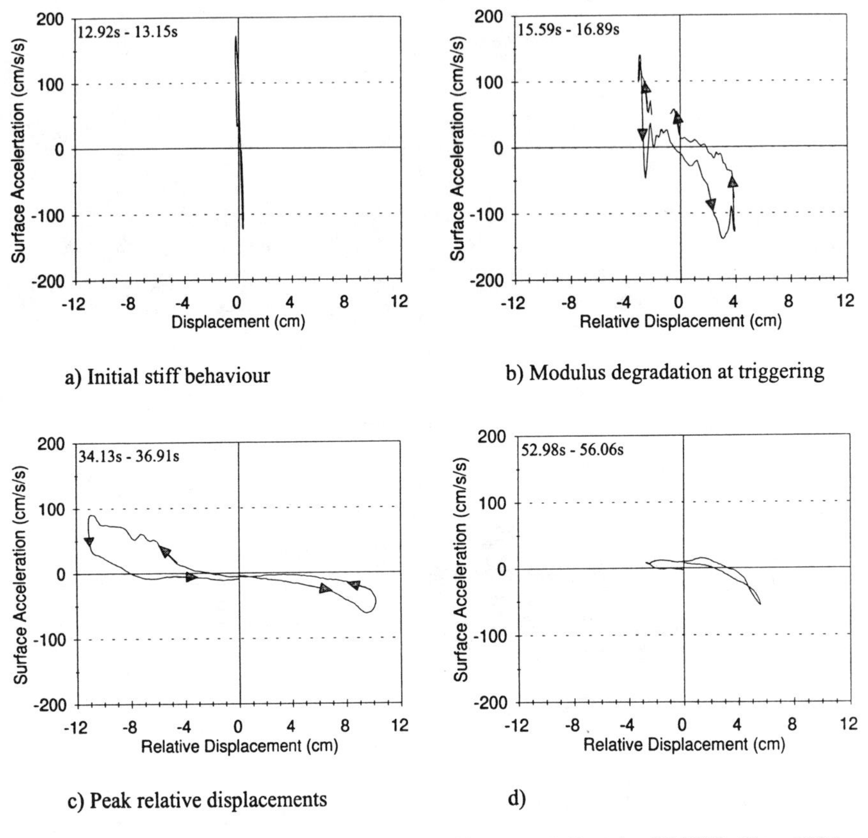

a) Initial stiff behaviour

b) Modulus degradation at triggering

c) Peak relative displacements

d)

FIG. 17. Change in Soil Stiffness during Selected Cycles - Wildlife Site, 1987 Superstition Hills Earthquake

Analysis Procedure

The dynamic analysis of the site was carried using a single-degree-of-freedom lumped mass and spring model. The lumped mass involved both the mass of the 2.5 m surficial crust and the 1/2 thickness of the 4.3 m liquefiable layer. The spring was nonlinear and represented the stiffness of the liquefiable layer by incorporating the stress-strain model discussed earlier. The downhole time history of acceleration was applied as base input motion and the response of the system obtained by step-by-step integration in the time domain. The computed response in terms of surface accelerations, relative displacements, and porewater pressures are compared in the next section.

Results

The predicted and observed surface accelerations are shown in Fig. 18a where it may be seen that the general form of the predicted response is in reasonable accord with the observation. The predicted and observed relative displacements are shown in Fig. 18b, where it may be seen that up to about 17 seconds both computed and measured relative displacements are very small. After 17 seconds relatively large displacement oscillations are predicted. It may be seen that both the pattern and magnitude of predicted and observed displacements are in reasonable accord.

The surface acceleration versus relative displacement pattern is shown in Fig. 19a. This corresponds with a shear-stress versus shear-strain plot. Prior to about 17 seconds the loops are very steep. At this point, liquefaction is triggered causing very flat loops that are in general accord with the observed pattern shown in Fig. 16. However, Fig. 16 shows a less abrupt degradation of modulus than the model prediction. This may be the result of a gradual spreading of the zone of liquefaction with time as compared to the assumption made in the analysis that the whole zone liquefied at one time.

The predicted effective stress path is shown in Fig. 19b. It may be seen that the effective stress point gradually worked its way back from an initial state of $\sigma'_{vo} = 66$ kPa and $\tau_{st} = 0$. This occurred as the shaking caused cyclic shear stress pulses and associated porewater pressure rise. It may be seen that the stress point reached the phase transformation or ϕ'_{cv} line a few times before the developed strain was sufficient to trigger a large porewater pressure rise and drive the stress point to the zero effective stress state upon unloading. Once this state was reached, subsequent butterfly loops up the ϕ'_{cv} line and down within the ϕ'_{cv} line, are predicted to occur with accompanying porewater pressure oscillations.

The predicted and observed porewater pressure ratios are shown in Fig. 20. It may be seen that the predicted porewater pressure rise is much faster than the measured rise and shows significant oscillations due to dilation after liquefaction has been triggered. The measured porewater pressure response does show significant pulses after about 30 seconds when about 80% porewater pressure rise occurs. These would appear to correspond with dilation pulses after liquefaction has been triggered. Numerous explanations have been proposed to explain the apparent lag in porewater pressure rise (Thilakaratne and Vucetic 1989; Zeghal and Elgamal, in review).

The measured relative displacements shown in Fig. 18b indicate that liquefaction and 100% porewater pressure rise was triggered at some depth within the liquefiable layer at about 17 seconds. If piezometer #5 is reading correctly, it would suggest that liquefaction did not occur at this depth or location at this time, but that migration of porewater from adjacent liquefied zones occurred starting at about 17 seconds and caused liquefaction to occur here at a later time.

SUMMARY

The characteristic shear stress-strain and volumetric response of the granular skeleton is captured using an incremental elastic-plastic stress-strain law. The normalized shear behaviour in both loading and unloading is modelled by modified

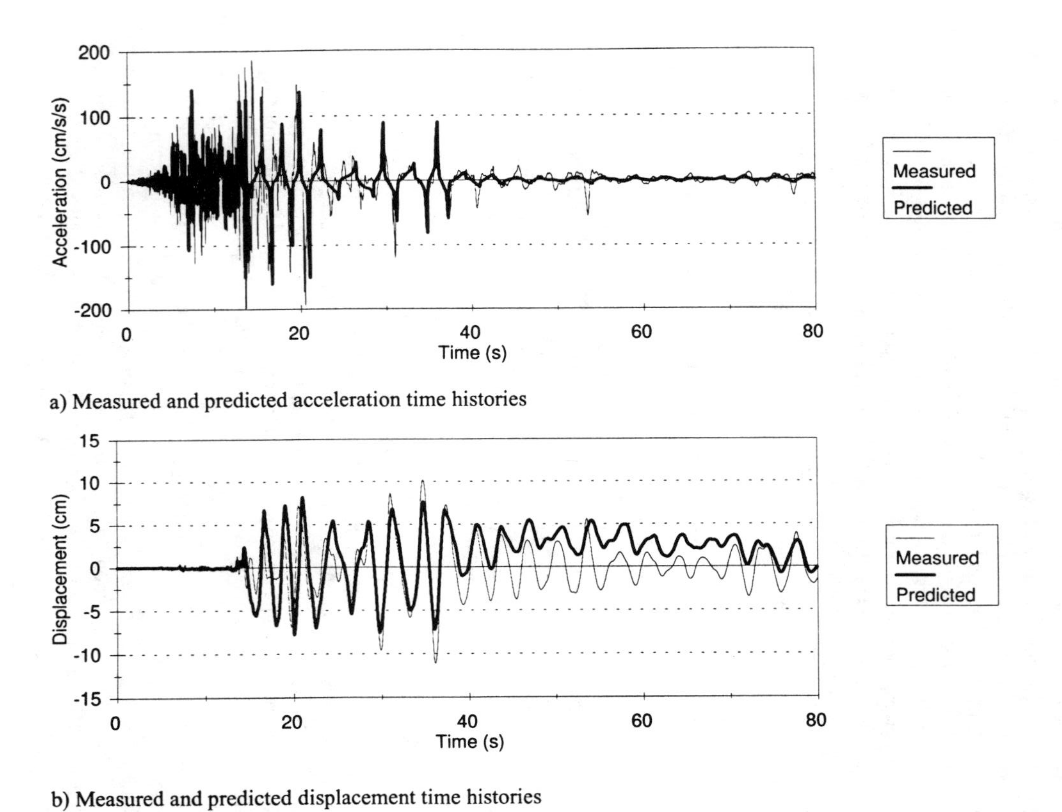

FIG. 18. Comparison of Measured and Predicted Time Histories - Wildlife Site, 1987 Superstition Hills Earthquake

hyperbolas. The shear-volume coupling is derived from energy dissipation considerations and captures the concept of contraction below the phase transformation or ϕ_{cv} line, and dilation above.

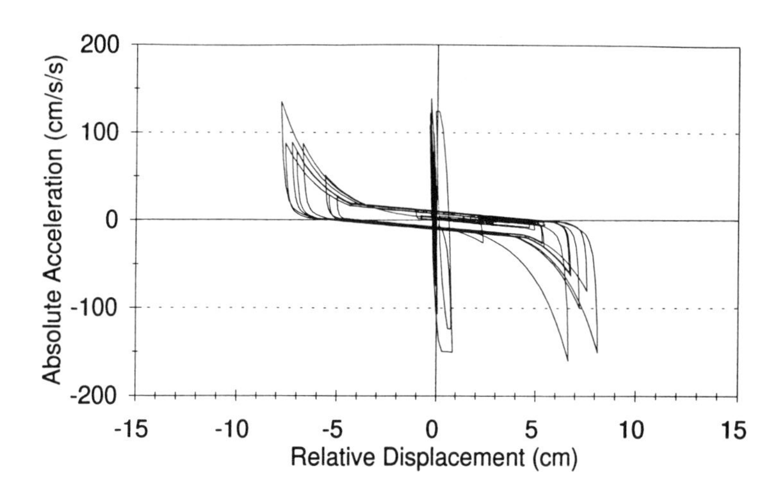

a) Acceleration versus relative displacement

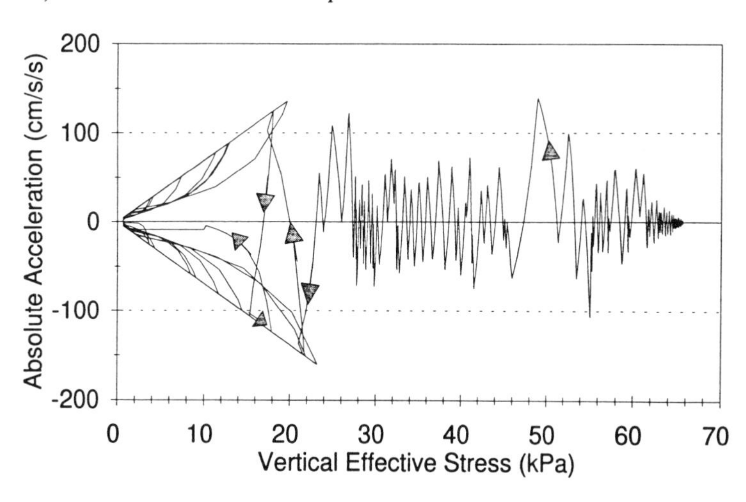

b) Acceleration versus vertical effective stress

FIG. 19. Predicted Dynamic Response of Wildlife Site for 1987 Superstition Hills Earthquake

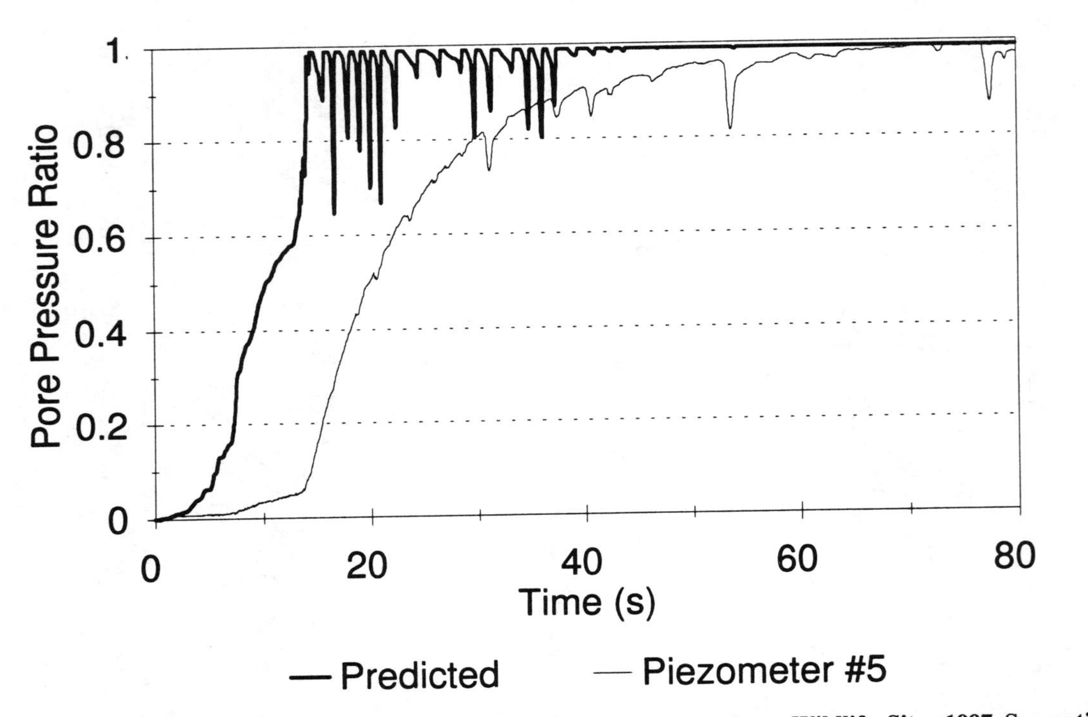

FIG. 20. Comparison between Measured and Predicted Pore Pressure Ratios - Wildlife Site, 1987 Superstition Hills Earthquake

In plasticity nomenclature, the yield loci are lines of constant stress ratio, the flow rule is non associated, and the hardener is the plastic shear strain.

The model is first calibrated in the drained mode by comparison with available laboratory data on volumetric accumulation with cycles of shear strain. Undrained response is predicted by imposing a volumetric constraint on the granular skeleton The model captures the undrained laboratory test data in terms of the degradation of the shear stress-strain response with porewater pressure rise during cyclic loading. It also captures the effective stress path followed as the stress state moves to the phase transformation or ϕ_{cv} line to trigger liquefaction, as well as the complex butterfly loops observed after triggering.

Finally, the model is incorporated in a dynamic analysis procedure and applied to the field case history recorded at the Wildlife site in California in 1987. The recorded downhole time history of acceleration was used as input to the dynamic model and the predicted response, in terms of surface acceleration, relative displacement, and porewater pressure compared with the measurements.

The predicted and observed surface acceleration are in reasonable agreement in terms of both the amplitude and characteristic frequency of response. The relative displacements are also in reasonable agreement with observations. In particular, the relative displacement pattern after 17 seconds, at which time we believe liquefaction was triggered, is in good agreement.

The predicted acceleration versus relative displacement (stress versus strain) curves are in very good agreement and indicate that prior to t = 17 seconds the stress-strain response is very stiff, whereas after this time a major reduction in stiffness by a factor of about 500 occurs. This indicates that liquefaction and essentially 100% porewater pressure rise was triggered at least in some zones at about t = 17 seconds.

The predicted porewater pressures are not in good agreement with the measurements. The predicted porewater pressure rise is much faster than the measured values. The slower measured response is thought to be due to either compliance in the measuring system or to the possibility that liquefaction did not occur simultaneously at all points in the liquefied layer. Thus, while the surface response characteristics would be largely controlled by the first continuous layer to liquefy, it is quite possible that the porewater pressure probe was not located in this layer and would consequently show a "delayed" porewater pressure response.

ACKNOWLEDGEMENT

The authors acknowledge the financial support of NSERC Canada in this work. They are grateful to Mrs. K. Lamb for her typing and presentation of the paper.

Appendix I. REFERENCES

Been, K. and Jefferies, M. G. (1985). "A state parameter for sands." *Géotechnique*, 35(2), 99-112.

Bennett, M. J., McLaughlin, P. V., Sarmiento, J. and Youd, T. L. (1984). "Geotechnical investigation of liquefaction sites, Imperial Valley, California," *Open File Report 84-242*, U. S. Geological Survey, 1-103.

Byrne, P. M. (1991). "A cyclic shear volume - coupling and porewater pressure model for sand." *Proc. 2nd Int'l. Conf. on Recent Advances in Geotechnical Earthquake Engineering and Soil Dynamics*, St. Louis, Report 1.24, 1, 47-56.

Holzer, T. L., Bennett, M. J. and Youd, T. L. (1989). "Lateral spreading field experiments by the US Geological Survey." *Proc. 2nd US-Japan Workshop on Liquefaction, Large Ground Deformation and Their Effects on Lifelines*, Buffalo, 82-101.

Lee, C.-J. (1991). "Deformation of sand under cyclic simple shear loading." *Proc. 2nd Int'l. Conf. on Recent Advances in Geotechnical Earthquake Engineering and Soil Dynamics*, St. Louis, Report 1.12, 1, 33-36.

Martin, G. R., Finn, W. D. L., and Seed, H. B. (1975). "Fundamentals of liquefaction under cyclic loading." *J. Geotech. Eng. Div.*, Proc. ASCE, 101(GT5), 423-438.

Matsuoka, H. and Nakai, T. (1977). "Stress-strain relationship of soil based on SMP." *Proc. Specialty Session 9*, 9th ICSMFE, Tokyo, 153-162.

Schofield, A. and Wroth, P. (1968). *Critical State Soil Mechanics*, McGraw-Hill, London, England, 124-127.

Seed, H. B., Tokimatsu, K., Harder, L. F. and Chung, R. M. (1985). "Influence of SPT procedures in soil liquefaction resistance evaluations." *J. Geotech. Eng.*, ASCE, 111(12), 1425-1445.

Taylor, D. W. (1948). *Fundamentals of Soil Mechanics*, John Wiley & Sons, New York, New York, 346-347.

Thilakaratne, V. and Vucetic, M. (1989). "Liquefaction at the wildlife site - effect on soil stiffness on seismic response." *Proc. 4th Int'l. Conf. on Soil Dynamics and Earthquake Engineering*, Mexico City, 37-52.

Tokimatsu, K. and Seed, H. B. (1987). "Evaluation of settlements in sands due to earthquake shaking." *J. Geotech. Eng.*, ASCE, 113(8), 861-878.

Youd, T. L., Holzer, T. L. and Bennett, M. J. (1989). "Liquefaction lessons learned from Imperial Valley, California." Earthquake Geotechnical Engineering, *Proc. of Discussion Session*, 12th ICSMFE, Rio de Janeiro, Brazil, 47-54.

Youd, T. L. and Wieczorek, G. F. (1984). "Liquefaction during the 1981 and previous earthquakes near Westmorland, California," *Open-File Report 84-680*, U. S. Geological Survey.

Zeghal, M. and Elgamal, A. W. (in review). "Analysis of Wildlife Site using earthquake records," submitted to *J. Geotech. Eng.*, ASCE for review and possible publication.

APPENDIX II.

The simple linear relationship between stress ratio τ/σ' and strain increment ratio ($d\epsilon_v/d\gamma$) as shown in Fig. 6c, is similar to a flow rule in plasticity theory.

Such a relationship can be established from energy considerations following the approach by Taylor (1948), and Schofield and Wroth (1968), using critical state soil mechanics concepts. For a simple shear element under a normal stress, σ, and shear stress τ, and subjected to strain increments $d\epsilon_v$ and $d\gamma$, the work done, dE, by the external forces per unit volume is,

$$dE = \sigma d\epsilon_v + \tau d\gamma \tag{A1}$$

The energy dissipated dW is

$$dW = \sigma d\epsilon_v^p + \tau d\gamma^p \tag{A2}$$

where $d\epsilon_v{}^p$ and $d\gamma^p$ are the plastic strain increments defined as

$$d\epsilon_v^p = d\epsilon_v - d\epsilon_v^e \tag{A3}$$

$$d\gamma^p = d\gamma - d\gamma^e \tag{A4}$$

where $d\epsilon_v{}^e$ and $d\gamma^e$ are the elastic volumetric and shear strain increments.

If we assume that energy is dissipated as if the element were at a critical state, $\tau_{cs} = \sigma \tan \phi_{cv}$, then

$$dW = \tau_{cs} d\gamma^p = \sigma \tan\phi_{cv} d\gamma^p \tag{A5}$$

From Eqs. (A2) and (A5),

$$\frac{\tau}{\sigma'} = -\frac{d\epsilon_v^p}{d\gamma^p} + \tan\phi_{cv} \tag{A6}$$

i.e., a straight line relationship between stress ratio and strain increment ratio similar to Fig. 6c except that the strains are now the plastic strains rather than the total strain. For significant strain levels, the plastic strain and the total strain are very similar. Thus, this simple theory based on energy concepts gives predicted response in good agreement with the observed data.

The plastic volumetric strain increment in turn is given by:

$$d\epsilon_v^p = d\gamma^p \left[\tan\phi_{cv} - \frac{\tau}{\sigma'}\right] = d\gamma^p \cdot D_t \tag{A7}$$

This is similar to Matsuoka and Nakai (1977) equation for sand. Matsuoka and Nakai worked with stresses and strains on the mobilized plane, and for plane strain conditions this was assumed to be the plane of maximum principle stress ratio or plane of maximum obliquity. Herein, it is assumed that the plane of maximum shear strain increment, rather than the plane of maximum obliquity is the plane of concern.

It is more versatile to assume that,

$$d\epsilon_v^p = \alpha d\gamma^p \left[\beta \tan\phi_{cv} - \frac{\tau}{\sigma'}\right] = d\gamma^p \cdot D_t \tag{A8}$$

where α and β are additional parameters to best fit the laboratory data.

APPENDIX III.

G_{max} depends on density and current effective stress as follows:

$$G_{max} = 21.7(k_2)_{max} P_A \sqrt{\frac{\sigma'_m}{P_A}} \tag{B1}$$

and

$(k_2)_{max} = 3.5(D_r)^{2/3}$ in which D_r = relative density in %; or
$(k_2)_{max} = 20(N_1)_{60}^{1/3}$;
P_A = atmospheric pressure in selected units; and
σ'_m = the current mean normal effective stress.

APPENDIX IV.

The constrained modulus, M, is modelled by

$$M = K_M P_A \sqrt{\frac{\sigma'_v}{P_A}} \tag{C1}$$

where

K_M = the constrained modulus number
P_A = atmospheric pressure, and
σ'_v = the current vertical effective stress.

K_M should be obtained from unload-reload tests. Some evidence suggests that K_M is independent of density. However, our modelling indicates that the observed laboratory liquefaction behaviour can be captured by assuming that $M = 1.4\ G_{max}$ for all stress and density conditions.

TEST AND PREDICTION RESULTS FOR FIVE SPREAD FOOTINGS ON SAND

Results of a Spread Footing Prediction Symposium
Held at Texas A&M University on June 16, 17, and 18, 1994
Sponsored by the
Federal Highway Administration

ASCE Geotechnical Special Publication No. 41

Jean-Louis Briaud,[1] Fellow, ASCE, and
Robert M. Gibbens,[2] Student Member, ASCE, Editors

ABSTRACT

A separate volume summarizes this prediction symposium. It is available from ASCE in New York and is entitled as listed above. This prediction event took place during the conference and the following is an abstract of the content of the volume which includes the prediction package as sent to the predictors, the measured behavior of the footings, a comparison of the predicted and measured responses, and all the prediction papers. The prediction volume is approximately 250 pages in length.

Five square spread footings were load tested to 0.15 m of penetration. The footings ranged in size from 1 m to 3 m, were embedded 0.75 m, and were loaded vertically at their center. A detailed soil investigation program was undertaken and led to numerous in situ and laboratory soil test results. The soil test results and the spread footing details were sent to engineers around the world.

Participants were asked to predict the loads (Q_{25} and Q_{150}) necessary to create 25 mm and 150 mm of settlement after 30 minutes of load application as well as the

---

[1] Buchanan Prof. of Civil Engrg., Texas A&M Univ., College Station, TX 77843-3136.

[2] Staff Engineer, Geotest Engineering Inc., 5600 Bintliff Drive, Houston, TX 77036.

creep settlement over 30 minutes for the 25 mm load Q_{25} and the settlement by the year 2014 under the 25 mm load Q_{25}. A total of 31 predictions were received from 8 countries, half from consultants and half from academics. The conclusions reached from comparing the predictions and the measurements are as follows:

1. Nobody gave a complete set of answers which consistently fell within ±20% of the measured values. Two participants had 80% of their answers falling within the ±20% margin of error.
2. The load creating 25 mm of settlement, Q_{25}, was underestimated by 27% on the average. The predictions were 80% of the time on the safe side. The scale effect was properly predicted since this number (80%) was consistent for all sizes.
3. The load creating 150 mm of settlement, Q_{150}, was underestimated by 6% on the average. The predictions were 63% of the time on the safe side. The scale effect was not properly predicted and there was a trend towards overpredicting Q_{150} for the larger footings.
4. A large variety of methods were used and it was not possible to identify the most accurate method because most people used published methods modified by their own experience or used a combination of methods. The most popular method was Schmertmann's method using CPT data. Of all the soil tests performed, the most used one was the CPT; then came the SPT, the PMT and the DMT.
5. The creep settlement over the 30 minute load step for Q_{25} was predicted reasonably well considering the limited data available for this prediction. The average prediction for the settlement by the year 2014 under Q_{25} is 35 mm or an additional 10 mm over the next 20 years.
6. The design load Q_d for each footing and each participant was defined as $MIN(Q_{25(predicted)}, Q_{150(predicted)}/3)$. The factor of safety F was defined as the ratio of the measured Q_{150} over Q_d. Since 31 participants predicted the behavior of 5 footings, there was a total 155 values of the factor of safety F. Only once out of 155 was F less than 1, the next worse case was 1.6; the average was 5.4. Therefore it appears that our profession knows how to design spread footings very safely.
7. The settlement S_d under the design load Q_d was read on the measured curves at the value of the predicted design load for each footing and for each participant. The overall average was 10.3 mm which is much smaller than 25 mm. Considering the high factors of safety and the low settlement values, the design load could have been significantly higher. Therefore it appears that our profession could design spread footings more economically.

Subject Index

Page number refers to first page of paper

Author Index

Page number refers to the first page of paper